Handbook of
Biosensors and Biochips

We would like to thank the following institutions for assistance with cover images:

American Institute of Physics for allowing us to reproduce a figure from H. Taha, R.S. Marks, L.A. Gheber, I. Rousso, J. Newman, C. Sukenik, and A. Lewis, Protein printing with an atomic force sensing nanofountainpen. *Applied Physics Letters*, 2003, **83**(5), 1041–1043.

American Chemical Society for allowing us to reproduce a figure from B. Polyak, S. Geresh, and R.S. Marks, Synthesis and characterization of a biotin-alginate conjugate and its application in a biosensor construction. *Biomacromolecules*, 2004, **5**(2), 389–396.

Handbook of Biosensors and Biochips

Editors

Robert S. Marks

Department of Biotechnology Engineering and National Institute for Biotechnology in the Negev, Ben-Gurion University of the Negev, Israel

David C. Cullen

Cranfield Health, Cranfield University, UK

Isao Karube

School of Bionics, Tokyo University of Technology, Japan

Christopher R. Lowe

Institute of Biotechnology, University of Cambridge, UK

Howard H. Weetall

National Association for Hispanic Elderly Senior Environmental Program Assisting the US Environmental Protection Agency, USA

VOLUME 2

This publication is designed to provide accurate and authoritative information in regard to the subject matter covered. It is sold on the understanding that the Publisher is not engaged in rendering professional services. If professional advice or other expert assistance is required, the services of a competent professional should be sought.

Other Wiley Editorial Offices

John Wiley & Sons Inc., 111 River Street,
Hoboken, NJ 07030, USA

Jossey-Bass, 989 Market Street,
San Francisco, CA 94103-1741, USA

Wiley-VCH Verlag GmbH, Boschstr. 12,
D-69469 Weinheim, Germany

John Wiley & Sons Australia Ltd, 42 McDougall Street,
Milton, Queensland 4064, Australia

John Wiley & Sons (Asia) Pte Ltd, 2 Clementi Loop #02-01,
Jin Xing Distripark, Singapore 129809

John Wiley & Sons Canada Ltd, 6045 Freemont Boulevard,
Mississauga, Ontario, Canada L5R 4J3

Wiley also publishes its books in a variety of electronic formats. Some content that appears in print may not be available in electronic books.

Wiley Bicentennial logo: Richard J. Pacifico.

Library of Congress Cataloging-in-Publication Data

Handbook of biosensors and biochips / editors, Robert S. Marks ... [et al.].
    p. ; cm.
  Includes bibliographical references and index.
  ISBN-13: 978-0-470-01905-4 (hardcover : alk. paper)
  ISBN-10: 0-470-01905-0 (hardcover : alk. paper)
1. Biosensors–Handbooks, manuals, etc. 2. Biochips–Handbooks, manuals,
etc. I. Marks, Robert (Robert S.)
  [DNLM: 1. Biosensing Techniques–methods. QT 36 2355 2007]
  R857.B54H357 2007
  610.28–dc22   2007025204

British Library Cataloguing in Publication Data

A catalogue record for this book is available from the British Library

ISBN-13 978-0-470-01905-4 (HB)

Typeset in 10 / 12 pt Times by Laserwords Private Limited, Chennai, India.
Printed and bound by Markono Group, Singapore.
This book is printed on acid-free paper responsibly manufactured from sustainable forestry in which at least two trees are planted for each one used for paper production.

Contents

Preface

The role of analytical technologies in collecting chemical intelligence in a variety of biomedical, veterinary, agri-food, industrial, health and safety, security, environmental, sports, and leisure sectors has gained so much importance of late that a dedicated handbook devoted to one of the key platform technologies, biosensors and biochips, has become long overdue. The technology for measuring and monitoring chemical and biological agents is not new and to a large extent can be visualized like the Olympics, where there have been impressive bursts intermingled with evolutionary progress and more than a few U-turns, full circles, and disappointing performances. Perhaps the evolution of this field can best be illustrated by recanting the history of glucose measurement for the diagnosis of diabetes mellitus. This can easily be traced back several millennia to early Chinese and Egyptian physicians who observed that ants accumulated near the urine of sufferers of diabetes mellitus. Much later, human tasters were used in medieval Britain to detect the presence of sugar in the urine of diabetics.

The introduction of the first chemical test for glucose by von Fehling in the mid-nineteenth century revolutionized chemical analysis and largely removed the biological element from the analytical process. However, selectivity was difficult to achieve by purely chemical procedures and enzymes were introduced for the determination of complex biological molecules such as starch. The enzymatic analysis of glucose, specifically β-D-glucopyranose, was introduced in a spectrophotometric linked assay using glucose oxidase and peroxidase by Keston in 1956, but it was the pioneering work of Clark and Lyons in 1962 that stimulated the use of immobilized enzymes for analysis and created the field we now know as biosensors. Since then, the research field has blossomed into a truly multidisciplinary endeavor involving biologists, chemists, materials scientists, physicists, engineers, computer scientists, and mathematicians to create a range of novel configurations exploiting biological recognition systems allied with physicochemical transducers. It is particularly noticeable in this field, as in imaging and several others, that the physicists and engineers are riding to the rescue of the biologists, with the introduction of a range of optical, acoustic, magnetic, thermal, and electrical technologies, coupled with microelectronic and micromachined devices.

However, despite this progress, dark clouds loom on the horizon and two key and related questions need to be asked. Firstly, does the market require such devices, and secondly is this technology up to the job? The overall diagnostics market approached $100 billion in 2005, although only a very small proportion was accounted for by biosensors, with the majority of the latter being glucose biosensors. Why is this? It is primarily because few, if any, of the current technologies live up to the much-vaunted expectation that such systems are inexpensive, rapid, sensitive, facile to operate, and power frugal, and these technologies are therefore not competitive with other more established technologies. It is also because such devices are exceptionally "dim", that is, they have an IQ approaching zero, and cannot differentiate easily between specific and nonspecific interactions. This lack of intelligence has prompted the development of array technologies to improve differentiation of the required response from the background noise and a collection of sophisticated signal processing and chemometric techniques to further enhance the signal-to-noise ratio. However, these approaches carry a significant price penalty and newer materials and production techniques are required to make these biosensor and biochip technologies economically viable while still retaining the ability to measure analytes sensitively and selectively in a multiplexed fashion.

The author of this preface felt that the time was right to review the whole field of biosensors and biochip technologies in light of this long and distinguished history and to see which of the many techniques proposed had actually made it to the marketplace. The concept for this handbook came from a discussion with an old (not in age, but in years of acquaintance!) colleague from John Wiley & Sons, Martin Röthlisberger, and was in a first step elaborated by myself together with Dave Cullen. This first concept was then refined by the other members on the editorial team, Robert Marks, Howard Weetall, and Isao Karube. All of us were chivied along very firmly but professionally by Layla Harden, without whom this handbook would probably not have appeared. Each contribution has been written by renowned scientists, who, in the majority of cases, were the key authors of the seminal works in their particular subfields. The contributions have been arranged in subsections with a few choice words from the editors as an introduction to each of these. Finally, the handbook finishes with a forward look into what the future may or may not hold.

Christopher R. Lowe
Cambridge

October 2007

Contributors

Ibrahim Abdulhalim *Department of Electrooptics Engineering, Ben-Gurion University of the Negev, Beer-Sheva, Israel*

Stanley Abramowitz *Advanced Technology Group, Silver Spring, MD, USA*

Tamar Amir *Department of Virology, Ben-Gurion University of the Negev, Beer-Sheva, Israel*

George P. Anderson *Center for Bio/Molecular Science and Engineering, US Naval Research Laboratory, Washington, DC, USA*

Bernardita Araya-Kleinsteuber *Institute of Biotechnology, University of Cambridge, Cambridge, UK*

Valentyna N. Arkhypova *Institute of Molecular Biology and Genetics, National Academy of Sciences of Ukraine, Kiev, Ukraine*

Sunil K. Arya *Biomolecular Electronics and Conducting Polymer Research Group, National Physical Laboratory, New Delhi, India*

Danit Atias *Department of Virology, Ben-Gurion University of the Negev, Beer-Sheva, Israel and Department of Biotechnology Engineering, Ben-Gurion University of the Negev, Beer-Sheva, Israel*

Keith L. Aubin *School of Applied and Engineering Physics, Cornell University, Ithaca, NY, USA*

Leonidas G. Bachas *Department of Chemistry, University of Kentucky, Lexington, KY, USA*

Ron L. Bardell *MicroPlumbers Microsciences LLC, Seattle, WA and Minneapolis, MN, USA*

Laurence Baril *Unit of Infectious Disease Epidemiology, Institut Pasteur, Dakar, Sénégal*

Paul W. Barone *Department of Chemical Engineering, Massachusetts Institute of Technology, Cambridge, MA, USA*

Christa Baumstark-Khan *Radiation Biology Department, Institute of Aerospace Medicine, Köln, Germany*

Sina Bavari *United States Army Medical Research Institute of Infectious Diseases, Frederick, MD, USA*

Shimshon Belkin *Institute of Life Sciences, Hebrew University of Jerusalem, Jerusalem, Israel*

Kathryn M. Bell *Center for Strategic and Innovative Technologies, University of Texas at Austin, Austin, TX, USA*

Anne-Sophie Belmont *Compiègne University of Technology, Compiègne, France*

Conrad Bessant *Cranfield Health, Cranfield University, Silsoe, UK*

Loïc J. Blum *Laboratoire de Génie Enzymatique et Biomoléculaire, Université Claude Bernard Lyon 1, Villeurbanne Cedex, France*

Anja Boisen *Department of Micro- and Nanotechnology, Technical University of Denmark, Kongens Lyngby, Denmark*

Catherine Cabrera *Biosensor and Molecular Technologies, MIT Lincoln Laboratory, Lexington, MA, USA*

S. J. Carlisle *Unipath Ltd., Bedford, UK*

Tony Cass *Institute of Biomedical Engineering, Imperial College London, London, UK*

Sonia Centi *Department of Chemistry, University of Florence, Florence, Italy*

Irwin Chaiken *Department of Biochemistry and Molecular Biology, Drexel University College of Medicine, Philadelphia, PA, USA*

Trevor Chapman *Drug Discovery (Neurodegeneration Research), GlaxoSmithKline, Harlow, UK*

Wilfred Chen *Department of Chemical and Environmental Engineering and Center for Nanoscale Science and Engineering, University of California, Riverside, CA, USA*

Stuart D. Collyer *Institute of Bioscience and Analytical Technology, Cranfield University, Silsoe, UK*

Pascal Colpo *Institute for Health and Consumer Protection, Joint Research Centre, Ispra, Italy*

Alan Cooper *WestChem Department of Chemistry, University of Glasgow, Glasgow, Scotland*

Bruce A. Cornell *Surgical Diagnostics Pty Ltd., St Leonards, New South Wales, Australia*

Serge Cosnier *Institut de Chimie Moléculaire, Université Joseph Fourier, Grenoble, France*

Harold G. Craighead *School of Applied and Engineering Physics, Cornell University, Ithaca, NY, USA*

Graham H. Cross *Department of Physics, Durham University, Durham, UK*

David C. Cullen *Cranfield Health, Cranfield University, Silsoe, UK*

Bengt Danielsson *Department of Pure and Applied Biochemistry, Lund University, Lund, Sweden*

Sylvia Daunert *Department of Chemistry, University of Kentucky, Lexington, KY, USA*

R. J. Davies *Unipath Ltd., Bedford, UK*

Frank Davis *Institute of Bioscience and Analytical Technology, Cranfield University, Silsoe, UK*

Sapna Deo *Department of Chemistry, Indiana University Purdue-University Indianapolis, Indianapolis, IN, USA*

Emre Dikici *Department of Chemistry, University of Kentucky, Lexington, KY, USA*

Sergei V. Dzyadevych *Institute of Molecular Biology and Genetics, National Academy of Sciences of Ukraine, Kiev, Ukraine and CEGELY, Ecole Centrale de Lyon, Ecully Cedex, France*

S. S. Eapen *Unipath Ltd., Bedford, UK*

Anna V. El'skaya *Institute of Molecular Biology and Genetics, National Academy of Sciences of Ukraine, Kiev, Ukraine*

Tatsuro Endo *Department of Mechano-Micro Engineering, Tokyo Institute of Technology, Yokohama, Japan*

Mark Ensor *Department of Chemistry, University of Kentucky, Lexington, KY, USA*

Randall M. Erb *Department of Mechanical Engineering and Materials Science, Duke University, Durham, NC, USA*

Katalin Erdélyi *MicroVacuum Ltd., Budapest, Hungary*

Stuart A. G. Evans *Oxford Biosensors Ltd., Yarnton, UK*

John B. C. Findlay *Institute of Membrane and Systems Biology, University of Leeds, Leeds, UK*

William J. Fitzgerald *Department of Engineering, University of Cambridge, Cambridge, UK*

Neville J. Freeman *Farfield Scientific Ltd., Crewe, UK*

Anthony G. Frutos *Corning Life Sciences, Corning Incorporated, Corning, NY, USA*

Rahela Gašparac *Department of Chemistry, Tufts University, Medford, MA, USA*

Güenter Gauglitz *Institute of Physical and Theoretical Chemistry, Eberhard Karls University of Tuebingen, Tuebingen, Germany*

Natalie Gavrielov *Department of Virology, Ben-Gurion University of the Negev, Beer-Sheva, Israel*

Levi A. Gheber *Department of Biotechnology Engineering, Ben-Gurion University of the Negev, Beer-Sheva, Israel*

Electra Gizeli *Institute of Molecular Biology and Biotechnology, Foundation for Research and Technology, Crete, Greece and Department of Biology, University of Crete, Crete, Greece*

Rick Gussio *Target Structure Based Drug Discovery Group, National Cancer Institute–SAIC, Frederick, MD, USA*

Daniel Haefliger *Department of Micro- and Nanotechnology, Technical University of Denmark, Kongens Lyngby, Denmark*

Steven Hagens *Institute of Food Science and Nutrition, ETH Zurich, Zurich, Switzerland*

Elizabeth A. H. Hall *Institute of Biotechnology, University of Cambridge, Cambridge, UK*

Kelly M. Halverson *United States Army Medical Research Institute of Infectious Diseases, Frederick, MD, USA*

Erol C. Harvey *MiniFab (Aust) Pty. Ltd., Scoresby, Victoria, Australia and Faculty of Engineering and Industrial Science, Swinburne University of Technology, Hawthorn, Victoria, Australia*

Karsten Haupt *Compiègne University of Technology, Compiègne, France*

Daniel A. Heller *Department of Chemistry, Massachusetts Institute of Technology, Cambridge, MA, USA*

Christine E. Hellweg *Radiation Biology Department, Institute of Aerospace Medicine, Köln, Germany*

Sarah E. Henrickson *Electronics and Electrical Engineering Laboratory, National Institute of Standards and Technology, Gaithersburg, MD, USA*

Keith E. Herold *Department of Bioengineering, University of Maryland, College Park, MD, USA*

Sebastien Herrmann *Department of Biotechnology Engineering, Ben-Gurion University of the Negev, Beer-Sheva, Israel*

Séamus P. J. Higson *Institute of Bioscience and Analytical Technology, Cranfield University, Silsoe, UK*

Atsunori Hiratsuka *Research Center of Advanced Bionics, National Institute of Advanced Industrial Science and Technology (AIST), Tsukuba, Japan*

Wah O. Ho *Research and Development, Hypoguard Ltd., Woodbridge, UK*

Bertold Hock *Center of Life Sciences, Technical University of Muenchen, Freising, Germany*

Bojan Ilic *School of Applied and Engineering Physics, Cornell University, Ithaca, NY, USA and Cornell Nanoscale Science and Technology Facility, Cornell University, Ithaca, NY, USA*

Rodica E. Ionescu *Centre de Genie Electrique de Lyon, Ecole Centrale de Lyon, Lyon, France*

Yasunori Iribe *Departmemt of Electric and Electronic Engineering, University of Toyama, Toyama, Japan*

Tetsuya Ishino *Department of Biochemistry and Molecular Biology, Drexel University College of Medicine, Philadelphia, PA, USA*

Angela Ivask *National Institute of Chemical Physics and Biophysics, Tallinn, Estonia*

Edmund S. Jackson *Department of Engineering, University of Cambridge, Cambridge, UK*

Paul Jacobs *Innogenetics NV, Zwijnaarde, Belgium*

Nicole Jaffrezic-Renault *CEGELY, Ecole Centrale de Lyon, Ecully Cedex, France*

Esther S. Jeng *Department of Chemical Engineering, Massachusetts Institute of Technology, Cambridge, MA, USA*

Anne Kahru *National Institute of Chemical Physics and Biophysics, Tallinn, Estonia*

Isao Karube *School of Bionics, Tokyo University of Technology, Tokyo, Japan*

John J. Kasianowicz *Electronics and Electrical Engineering Laboratory, National Institute of Standards and Technology, Gaithersburg, MD, USA*

Kagan Kerman *School of Materials Science, Japan Advanced Institute of Science and Technology (JAIST), Ishikawa, Japan*

Hideki Kinoshita *Research Center of Advanced Bionics, National Institute of Advanced Industrial Science and Technology (AIST), Tsukuba, Japan*

Steven E. Kornguth *Center for Strategic and Innovative Technologies, University of Texas at Austin, Austin, TX, USA*

Karl Kramer *Center of Life Sciences, Technical University of Muenchen, Freising, Germany*

Akhlesh Lakhtakia *Department of Engineering Science and Mechanics, Pennsylvania State University, University Park, PA, USA*

Wim Laureyn *NEXT-Nano-Enabled Systems, Leuven, Belgium*

Coulton Legge *Pharmaceutical Development, GlaxoSmithKline, Ware, UK*

Jeffery C. Lerman *Electronics and Electrical Engineering Laboratory, National Institute of Standards and Technology, Gaithersburg, MD, USA*

Patricia Lisboa *Institute for Health and Consumer Protection, Joint Research Centre, Ispra, Italy*

Leslie Lobel *Department of Virology, Ben-Gurion University of the Negev, Beer-Sheva, Israel*

Martin J. Loessner *Institute of Food Science and Nutrition, ETH Zurich, Zurich, Switzerland*

Stefan Löfås *Biacore AB, Uppsala, Sweden*

Lee O. Lomas *Ciphergen Biosystems Inc., Fremont, CA, USA*

Christopher R. Lowe *Institute of Biotechnology, University of Cambridge, Cambridge, UK*

Michael E. G. Lyons *School of Chemistry, University of Dublin, Dublin, Ireland*

Moni Magrisso *National Institute for Biotechnology in the Negev, Ben-Gurion University of the Negev, Beer-Sheva, Israel*

Georg Mahlknecht *Center of Life Sciences, Technical University of Muenchen, Freising, Germany*

Bansi D. Malhotra *Biomolecular Electronics and Conducting Polymer Research Group, National Physical Laboratory, New Delhi, India*

Robert S. Marks *Department of Biotechnology Engineering and National Institute for Biotechnology in the Negev, Ben-Gurion University of the Negev, Beer-Sheva, Israel*

Christophe A. Marquette *Laboratoire de Génie Enzymatique et Biomoléculaire, Université Claude Bernard Lyon 1, Villeurbanne Cedex, France*

Christopher H. Marrows *School of Physics and Astronomy, University of Leeds, Leeds, UK*

Claude Martelet *CEGELY, Ecole Centrale de Lyon, Ecully Cedex, France*

Marco Mascini *Department of Chemistry, University of Florence, Florence, Italy*

Paula McCourt *Department of Biochemistry and Molecular Biology, Drexel University College of Medicine, Philadelphia, PA, USA*

David A. McCrae *Research International, Monroe, WA, USA*

Anat Meir *Department of Biomedical Engineering, Ben-Gurion University of the Negev, Beer-Sheva, Israel*

Kathryn A. Melzak *Institute of Molecular Biology and Biotechnology, Foundation for Research and Technology, Crete, Greece*

Maria Minunni *Department of Chemistry, University of Florence, Florence, Italy*

Martin Misakian *Electronics and Electrical Engineering Laboratory, National Institute of Standards and Technology, Gaithersburg, MD, USA*

Kohji Mitsubayashi *Institute of Biomaterials and Bioengineering, Tokyo Medical and Dental University, Tokyo, Japan*

Hirotaka Miyachi *School of Bionics, Tokyo University of Technology, Tokyo, Japan*

Elizabeth A. Moschou *Department of Chemistry, University of Kentucky, Lexington, KY, USA*

Ashok Mulchandani *Department of Chemical and Environmental Engineering and Center for Nanoscale Science and Engineering, University of California, Riverside, CA, USA*

Lindy J. Murphy *Oxford Biosensors Ltd., Yarnton, UK*

Nosang V. Myung *Department of Chemical and Environmental Engineering and Center for Nanoscale Science and Engineering, University of California, Riverside, CA, USA*

Jeffrey D. Newman *Cranfield Health, Cranfield University, Silsoe, UK*

Tam Nguyen *Target Structure Based Drug Discovery Group, National Cancer Institute–SAIC, Frederick, MD, USA*

Joseph Nickels   *Department of Biochemistry and Molecular Biology, Drexel University College of Medicine, Philadelphia, PA, USA*

Petr I. Nikitin   *Natural Science Center of General Physics Institute, Russian Academy of Sciences, Moscow, Russia*

Yoko Nomura   *Department of Biomedical Engineering, University of California, Davis, CA, USA*

Rekha G. Panchal   *Target Structure Based Drug Discovery Group, National Cancer Institute–SAIC, Frederick, MD, USA*

Ash Patel   *Drug Discovery (Biopharmaceuticals), GlaxoSmithKline, Beckenham, UK*

John C. Pickup   *Metabolic Unit, King's College London School of Medicine, London, UK*

Arshak Poghossian   *Institute of Nano- and Biotechnologies and Research Center Jülich, Aachen University of Applied Sciences, Jülich, Germany*

Boris Polyak   *Department of Cardiology Research, The Children's Hospital of Philadelphia, Philadelphia, PA, USA*

Rachela Popovtzer   *Department of Physical Electronics, Tel Aviv University, Ramat-Aviv, Israel*

Güenther Proll   *Institute of Physical and Theoretical Chemistry, Eberhard Karls University of Tuebingen, Tuebingen, Germany*

Xiaoge Qu   *Department of Chemistry, University of Kentucky, Lexington, KY, USA*

Jeremy J. Ramsden   *Department of Materials, Cranfield University, Cranfield, UK*

Avraham Rasooly   *Center for Devices and Radiological Health, FDA, Silver Spring, MD, USA and Diagnostic Biomarkers and Technology Branch, NIH-National Cancer Institute, Rockville, MD, USA*

John J. Rippeth   *Research and Development, Hypoguard Ltd., Woodbridge, UK*

Judith Rishpon   *Department of Molecular Microbiology and Biotechnology, Tel Aviv University, Ramat-Aviv, Israel*

Kim R. Rogers   *National Research Exposure Laboratory, Environmental Protection Agency, Las Vegas, NV, USA*

François Rossi   *Institute for Health and Consumer Protection, Joint Research Centre, Ispra, Italy*

Laura Rowe   *Department of Chemistry, University of Kentucky, Lexington, KY, USA*

Israel Rubinstein   *Department of Materials and Interfaces, Weizmann Institute of Sciences, Rehovot, Israel*

Helge R. Schnerr   *Department of Food Safety and Food Quality, Leatherhead Food International Ltd., Leatherhead, UK*

Michael J. Schöning   *Institute of Nano- and Biotechnologies and Research Center Jülich, Aachen University of Applied Sciences, Jülich, Germany*

Matthias Schuenemann   *MiniFab (Aust) Pty. Ltd., Scoresby, Victoria, Australia*

Nikolay V. Sergeev   *Center for Devices and Radiological Health, FDA, Silver Spring, MD, USA*

Yosi Shacham-Diamand   *Department of Physical Electronics, Tel Aviv University, Ramat-Aviv, Israel*

Devanand K. Shenoy   *Naval Research Laboratory, Center for Bio/Molecular Science and Engineering, Washington, DC, USA*

Mifumi Shimomura-Shimizu   *School of Bionics, Tokyo University of Technology, Tokyo, Japan*

Mark R. Sims *Department of Physics and Astronomy, University of Leicester, Leicester, UK*

Surinder P. Singh *Biomolecular Electronics and Conducting Polymer Research Group, National Physical Laboratory, New Delhi, India*

Alexey P. Soldatkin *Institute of Molecular Biology and Genetics, National Academy of Sciences of Ukraine, Kiev, Ukraine*

Ursula E. Spichiger-Keller *Centre for Chemical Sensors and Chemical Information Technology, Swiss Federal Institute of Technology (ETH), Zurich, Switzerland*

Tim Stakenborg *NEXT-Nano-Enabled Systems, Leuven, Belgium and Veterinary Research Institute, Brussels, Belgium*

Vincent M. Stanford *Information Technology Laboratory, National Institute of Standards and Technology, Gaithersburg, MD, USA*

Adrian C. Stevenson *Institute of Biotechnology, University of Cambridge, Cambridge, UK*

Milan N. Stojanovic *Department of Medicine, Columbia University, New York, NY, USA*

Michael S. Strano *Department of Chemical Engineering, Massachusetts Institute of Technology, Cambridge, MA, USA*

Masayasu Suzuki *Departmemt of Electric and Electronic Engineering, University of Toyama, Toyama, Japan*

Yoshio Suzuki *Research Center of Advanced Bionics, National Institute of Advanced Industrial Science and Technology (AIST), Tsukuba, Japan*

Marcus J. Swann *Farfield Scientific Ltd., Crewe, UK*

István Szendrő *MicroVacuum Ltd., Budapest, Hungary*

Sabine Szunerits *Institut National Polytechnique de Grenoble, Domaine Universitaire, Saint Martin d'Hères, France*

Eiichi Tamiya *Department of Applied Physics, Osaka University, Osaka, Japan*

Lisa Tang *Institute of Membrane and Systems Biology, University of Leeds, Leeds, UK*

Krystal Teasley *Department of Chemistry, University of Kentucky, Lexington, KY, USA*

Leon A. Terry *Plant Science Laboratory, Cranfield University, Silsoe, UK*

Tatsuya Tobita *NTT Advanced Technology Corp., Atsugi, Japan*

Sara Tombelli *Department of Chemistry, University of Florence, Florence, Italy*

Jens Tschmelak *Bierlachweg, Erlangen, Germany*

Anthony P. F. Turner *Cranfield Health, Cranfield University, Silsoe, UK*

Shimon Ulitzur *Department of Biotechnology and Food Engineering, Technion Institute, Haifa, Israel*

Keisuke Usui *Research Center of Advanced Bionics, National Institute of Advanced Industrial Science and Technology (AIST), Tsukuba, Japan*

Andrea Valsesia *Institute for Health and Consumer Protection, Joint Research Centre, Ispra, Italy*

Alexander Vaskevich *Department of Materials and Interfaces, Weizmann Institute of Sciences, Rehovot, Israel*

Marko Virta *Department of Applied Chemistry and Microbiology, University of Helsinki, Helsinki, Finland*

Guy Voirin *CSEM Centre Suisse d'Electronique et de Microtechnique SA, Neuchâtel, Switzerland*

David R. Walt *Department of Chemistry, Tufts University, Medford, MA, USA*

Adam K. Wanekaya *Chemistry Department, Missouri State University, Springfield, MO, USA*

Howard H. Weetall *Under Cooperative Agreement with the National Association for Hispanic Elderly Senior Environmental Program Assisting the US Environmental Protection Agency, Las Vegas, NV, USA*

Bernhard H. Weigl *Department of Bioengineering, University of Washington, Seattle, WA, USA*

Scot R. Weinberger *GenNext Technologies Inc., Montara, CA, USA*

Graham Whyteside *Institute of Membrane and Systems Biology, University of Leeds, Leeds, UK*

Bin Xie *Department of Pure and Applied Biochemistry, Lund University, Lund, Sweden*

Victoria Yavelsky *Department of Virology, Ben-Gurion University of the Negev, Beer-Sheva, Israel*

Benjamin B. Yellen *Department of Mechanical Engineering and Materials Science, Duke University, Durham, NC, USA*

Kenji Yokoyama *Research Center of Advanced Bionics, National Institute of Advanced Industrial Science and Technology (AIST), Tsukuba, Japan*

Mohammad Zourob *Biosensors Division, Biophage Pharma, Montreal, Quebec, Canada*

Abbreviations and Acronyms

AA	Arachidonic Acid
AAO	Anodic Aluminum Oxide
Ab	Antibody
ABEI	N-(4-Aminobutyl)-N-Ethylisoluminol
ac	Alternating Current
AChE	Acetylcholinesterase
ACMG	American College of Medical Genetics
ACOG	American College of Obstetricians and Gynecologists
AD	Alzheimer's Disease
ADC	Analogue-to-Digital Converter
ADDL	Amyloid-β-Derived Diffusible Ligand
ADH	Alcohol Dehydrogenase
ADHD	Attention-Deficit/Hyperactivity Disorder
ADME	Absorption, Distribution, Metabolism, Excretion
ADP	Adenosine Diphosphate
AFL	Astrobiology Field Laboratory
AFM	Atomic Force Micrograph
AFM	Atomic Force Microscopy
Ag	Antigen
AHL	N-Acyl-ʟ-homoserine Lactone
αHL	Alpha-Hemolysin
AIBN	Azo-Bis-Isobutyronitrile
ALA	δ-Aminolevulinic Acid
ALDH	Aldehyde Dehydrogenase
ALFMED	Apollo Light Flash Moving Emulsion Detector
AMASE	Arctic Mars Analog Svalbard Expedition
AMP	Adenosine Monophosphate
AMR	Anisotropic Magnetoresistance
ANNs	Artificial Neural Networks
ANOVA	Analysis of Variance
anti-LH	Anti-Luteinizing Hormone
AOAC	Association of Analytical Communities

AOC	Assimilable Organic Carbon
AOD	Alcohol Oxidase
AP	Alkaline Phosphatase
APD	Avalanche Photodiode
APM	Acoustic Plate Mode
ApoE	Apolipoprotein E
AR	Ankyrin Repeat
AR	autoregressive
ARX	autoregressive with external input
AsA	Ascorbic Acid
ASPE	Allele-Specific Primer Extension
ASTP	Apollo–Soyuz Test Project
ASV	Anodic Stripping Voltammetry
ATIR	Attenuated Total Internal Reflection
ATP	Adenosine 5′-Triphosphate
ATP	Applied Technology Program
ATV	Automated Transfer Vehicle
AW	Acoustic Wave
AWACSS	Automated Water Analyzer Computer Supported System
BaP	Benzo(a)Pyrene
BAR	Bin/Amphiphysin/Rvs
BARC	Bead Array Counter
BAW	Bulk Acoustic Wave
BB	Borate Buffer
BBP	Bilin-Binding Protein
BCG	Bacillus Calmette–Guerin
BeP	Benzo(E)Pyrene
β-gal	β-Galactosidase
bFGF	Basic Fibroblast Growth Factor
BGB	BioGlovebox
BGMs	Blood Glucose Meters
bHLH-ZIP	Basic Region Helix-Loop-Helix-Leucine Zipper
BIA	Biomolecular Interaction Analysis
BIBIC	Bioluminescent Bioreporter Integrated Circuit

BioMEMS	Bio Microelectromechanical Systems		CGMS	Continuous Glucose Monitoring Systems
BLM	Bilayer Lipid Membrane		CI	Confidence Interval
BM	Biological Molecules		CIP	Cleaning-in-Place
BMBF	Bundesministerium Für Bildung Und Forschung		CITAC	Cooperation on International Traceability in Analytical Chemistry
BOD	Biochemical Oxygen Demand			
BOD	Biological Oxygen Demand		CL	Chemiluminescence
bp	Base Pairs		CME	Chemically Modified Electrode
bPBP	Bacterial Periplasmic Binding Protein		CMOS	Complementary Metal Oxide Semiconductor
BRET	Bioluminescence Resonance Transfer		CN	Cellulose Nitrate
			CNC	Computer-Numerical-Control
BSA	Bovin Serum Albumin		CNS	Central Nervous System
BTEX	Benzene, Toluene, Ethylbenzene, And Xylenes		CNTs	Carbon Nanotubes
			Co-TPP	Co-Tetraphenylporphyrin
BTK	Bacillus Thuringiensis Kurstaki		COC	Cyclic Olefin Copolymer
BUN	Blood Urea Nitrogen		CODIS	Combined DNA Index System
BWA	Biological Warfare Agent		ConA	Concanavalin A
			COP	Cyclic Olefin Polymer
CA	Cellulose Acetate		COSPAR	Committee on Space Research
CaM	Calmodulin		CP NWs	Conducting Polymer Nanowires
CAPD	Continuous Ambulatory Peritoneal Dialysis		CPE	Carbon Paste Electrode
			CPE	Constant Phase Element
CARS	Coherent Anti-Stokes Raman Scattering		CPG	Controlled Pore Glass
			CPK	Corey, Pauling, and Koltun
CBB	Coomassie Brilliant Blue		CPMV	Cowpea Mosaic Virus
CBD	Cell-Wall-Binding Domain Proteins		CPs	Conducting Polymers
			CPWR	Coupled Plasmon Waveguide Resonance
CCA	Crystalline Colloidal Array			
CCD	Charge Coupled Device		CRM	Cytokine Recognition Motif
CD	Circular Dichroism		CS	Condroitin Sulfate
CDC	Centers for Disease Control and Prevention		CSF	Cerebrospinal Fluid
			CT	Charge Transfer
cDNA	Complementary Deoxyribonucleic Acid		CT	Cholera Toxin
			CTLA4	Cytotoxic T-Lymphocyte Associated Antigen
CDR	Complementarity-Determining Regions			
			CV	Coefficient of Variation
CE	Capillary Electrophoresis		CV	Cyclic Voltammetry
CE-EC	Capillary Electrophoresis with Electrochemical Detection		CVD	Cardiovascular Disease
			CVD	Chemical Vapor Deposition
CE-LIF	Capillary Electrophoresis with Laser-Induced Fluorescence		CVL	Copper Vapor Laser
CERASP	Cellular Responses to Radiation in Space		DAAO	D-Amino acid Oxidase
			DAD	Diode Array Detection
CFTR	Cystic Fibrosis Transmembrane Conductance Regulator		DARPA	Defense Advanced Research Projects Agency
CFU	Colony-Forming Unit		DAT	Direct Agglutination Test
CGC	Chirped Grating Couplers		DBP	DNA-Binding Protein
CGD	Chronic Granulomatous Disease		DCF	2,7-Dichlorofluorescein
CGI	Computer Generated Image			

DCFH — Nonfluorescent 2,7-dichlorofluorescein

DCFH-DA — 2,7-Dichlorofluorescein-diacetate

DCIP — 2,6-Dichlorophenolindophenol

DDT — Dichloro-Diphenyl-Trichloroethane

DE — dimensional electrophoresis

DEAE — DiEthylAminoEthyl

DEC — Disposable Enteric Card

DELFIA — Dissociation-Enhanced Lanthanide Fluoroimmunoassay

DEN — Dengue Virus

DEP — Dielectrophoresis

DFG — Deutsche Forschungsgesellschaft

DI — Deionized

DL — Detection Limit

DLP — Digital Light Processor

DLR — German Aerospace Center

DLS — Dynamic Light Scattering

DMAEM — Dimethylaminoethyl Methacrylate

DMD — Digital Micromirror Device

DME — Dropping Mercury Electrode

DMEWS — Dual Mode Evanescent Wave Spectroscopy

DMF — Dimethyl Formamide

DMSO — Dimethyl Sulfoxide

DNA — Deoxyribonucleic Acid

DNP-PEG4-C11thiol — Dinitrophenyl Poly(Ethylene Glycol) Undecanthiol

DOC — Dissolved Organic Carbon

DOPC — 1,2-Dioleoyl-Sn-Glycero-3-Phosphocholine

DOSTEL — Dosimetry Telescope

DPA — Differential Pulse Amperometry

DPI — Dual Polarization Interferometry

DPN — Dip-Pen Nanolithography

DPP — Differential Pulse Polarography

DPV — Differential Pulse Voltammetry

DSC — Differential Scanning Calorimetry

dsDNA — Double Stranded DNA

DSP — 3,3′-Dithiodipropionic Acid-Di(N-Succinimidyl Ester)-Bound Streptavidin

EC — Effective Concentration

EC-OWLS — Electrochemical-Optical Waveguide Lightmode Spectroscopy

ECL — Electrochemiluminescence

ECLSS — Environmental Control and Life Support Systems

ECs — Experiment Containers

EDC — Endocrine Disrupting Chemical

EDC/NHS — 1-Ethyl-3-(3-Dimethylamino-propil) Carbodiimide/N-Hydroxysuccinimide

EDMA — Ethylene Glycol Dimethacrylate

EDP — Electrodeposition Paints

E-DPN — Electrochemical AFM Dip-Pen Nanolithography

EET — Estimated Efficiency of the Test

EF — Edema Factor

EF — Exposed Facility

E3G — Estrone-3-glucuronide

EGFP — Enhanced Green Fluorescent Protein

EIS — Electrochemical Impedance Spectroscopy

EIS — Electrolyte–Insulator–Semiconductor

EL — Electroluminescent

ELISA — Enzyme-Linked Immunosorbent Assay

ELM — Experiment Logistic Module

EM — Expectation Maximization

EMCCD — Electron-Multiplying Charge-Coupled Device

EMDG — Enhanced Magnetic Direct Generation

EnFETs — Enzyme-Modified Field-Effect Transistors

E-NFP — Electrochemical Nano Fountain Pen

EOF — Electroosmotic Flow

EPA — Environmental Protection Agency

EPM — European Physiology Modules

EPU — Experiment Preparation Unit

ER — Endoplasmic Reticulum

ES — Endocervical Swabs

ESA — European Space Agency

ESEM — Environmental Scanning Electron Microscopy

ET — Electron Transfer

ET — Enzyme Thermistor

ETC — Electron Transport Chain

ETV — Environmental Technology Verification

EVA — Extravehicular Activities

FAD	Flavin Adenine Dinucleotide	gDNA	Genomic Deoxyribonucleic Acid
FAST	Fast Agglutination Screening Test	GDP	Guanosine-Diphosphate
FATS	Fish Acute Toxicity Syndrome	GE	General Electric
FAW	Fibre Acoustic Wave	GEO	Gene Expression Omnibus
FBAR	Film Bulk Acoustic Resonator	GFP	Green Fluorescent Protein
FDA	Food and Drug Administration	GFR	Glomerular Filtration Rate
FDH	Formaldehyde Dehydrogenase	GGA	Glucose-Glutamine
FED	Field-Effect Devices	GLOD	L-glutamate Oxidase
FEP	Fluorinated Ethylene Propylene	GM-CSF	Granulocyte Macrophage-Colony Stimulating Factor
FET	Field Effect Transistor	GMM	Gaussian Mixture Model
Ff	F-Specific Filamentous	GMO	Genetically Modified Organism
FFT	Fast Fourier Transform	GMP	Good Manufacturing Principals
FIA	Flow Injection Analysis	GMR	Giant Magnetoresistance
FIS	Faradaic Impedance Spectroscopy	GOx	Glucose Oxidase
FISH	Fluorescent in Situ Hybridization	GOX-B	Biotinylated Glucose Oxidase
FITC	Fluorescein Isothiocyanate	GPA	Glycine, Proline, And Alanine
FLD	Fluorescence Detection	GPCR	G-Protein Coupled Receptor
FLIM	Fluorescence-Lifetime Imaging	GPS	Global Positioning System
FMN	Flavin Mononucleotide	GTP	Guanosine 5′-Triphosphate
FMO	Flavin-Containing Monooxygenase	GWS	Grating Waveguide Structures
FNF	False Negative Fraction	H-PTFE	Hydrophilic Polytetrafluoroethylene
FOP	Fiber-Optical Picoscope	hAR	Human Androgen Receptor
FORRAY	Fluorescence Orbital Radiation Risk Assessment Using Yeast	HBDH	3-Hydroxybutyrate Dehydrogenase
FPF	False-Positive Fraction	HBM	Hybrid Bilayer Membranes
FPRL-1	Formyl Peptide Receptor like-1	HCA	Hierarchical Cluster Analysis
FPW	Flexural Plate Wave	HCF(III)	Hexacyanoferrate(III)
FRET	Fluorescence Resonance Energy Transfer	HCG	Human Chorionic Gonadotropin
FRET	Forster Resonance Energy Transfer	HCV	Hepatitis C Virus
FSD	Full Scale Deflection	HD	Huntington's Disease
FSL	Fluid Science Laboratory	HDL	High-Density Lipoprotein
FSV	Fast Scan Voltammetry	HDT	Hexadecanethiol
FTIR	Fourier Transform Infrared	HEG	Hexaethylene Glycol
FTIR	Frustrated Total Internal Reflection	HEMA	Hydroxyethyl Meth- acrylate
		$hER\alpha$	Human Estrogen Receptor α
GANIL	Grand Accelerateur National d' Ions Lourds	$hER\beta$	Human Estrogen Receptor β
		HGMS	High-Gradient Magnetic Separation
GAP	Gas Analysis Package	HIV	Human Immunodeficiency Virus
GB	Glycine Buffer	HIV-1	Human Immunodeficiency Virus 1
GC	Gas Chromatography	HLA	Human Leukocyte Antigen
GCE	Glassy Carbon Electrode	HMDS	Hexamethyldisiloxane
GCR	Galactic Cosmic Radiation	HMM	Hidden Markov Model
GDH	Glucose Dehydrogenase	HMW	High Molecular Weight
		HPLC	High Performance Liquid Chromatography

hPSTI	Human Pancreatic Secretory Trypsin Inhibitor	IPTG	Isopropyl β-D-1-Thiogalactopyranoside
HPVs	Human Papillomaviruses	IRMM	Institute for Reference Materials and Measurements
HQ	Hydroquinone		
HRF	Human Research Facility	ISE	Ion Selective Electrode
HRP	Horseradish Peroxidase	ISFET	Ion Selective Field Effect Transistor
HRSEM	High-Resolution Scanning Electron Microscope		
		ISO	International Standardization Organization
HS	Heparin Sulfate		
HSA	Human Serum Albumin	ISPRs	International Standard Payload Racks
HSV	Herpes Simplex Virus		
HTS	High-Temperature Superconductor	ITAR	International Traffic in Arms Regulations
HTS	High-Throughput Screening	ITC	Isothermal Titration Calorimetry
HZE	particles of high charge Z and high Energy		
		ITO	Indium Tin Oxide
		IUPAC	International Union of Pure and Applied Chemistry
ICS	Ion Channel Switch		
ICT	Immunochromatographic Test	IVD	In Vitro Diagnostic
IDA	Iminodiacetic Acid		
IDAs	Interdigitated Microarray	JEM	Japanese Experiment Module
IDE	Interdigitated Electrode	JIS	Japanese Industrial Standard
IDT	Interdigitated Transducers		
IEF	Isoelectric Focusing	KS	Keratin Sulfate
IFC	Integrated Fluidic Handling Cartridge	KS	Kolmogorov–Smirnov
IFCC	International Federation of Clinical Chemistry	LAL	Limulus Amebocyte Lysate
		LAPS	Light-Addressable Potentiometric Sensor
IFN-γ	Interferon-γ		
IgA	Immunoglobulin A	LAS	Linear Alkylbenzene Sulfonates
IgE	Immunoglobulin E	LB	Langmuir–Blodgett
IGFET	Insulated-Gate Field-Effect Transistor	LB	Luria Broth
		LbL	Layer-by-Layer
IgG	Immunoglobulin G	LC	Liquid Chromatography
IgM	Immunoglobulin M	LCL	Luminol-Dependent Chemiluminescence
IHR	International Health Regulations		
IL	Interleukin	LCP	Liquid Crystal Polymer
IL2	Interleukin 2	LDA	Linear Discriminant Analysis
IMAC	Immobilized Metal Affinity Capture	LDCL	Lucigenin-Dependent Chemiluminescence
IMPDH	Inosine Monophosphate Dehydrogenase	LDH	Lactate Dehydrogenase
		LDL	Low-Density Lipoprotein
IMS	Immunomagnetic Separation	LEDs	Light-Emitting Diodes
IOW	Integrated Optical Waveguide	LEO	Low Earth Orbit
IPA	Isopropyl Alcohol	LET	Linear Energy Transfer
IPCCA	Intelligent Polymerized Crystalline Colloidal Array	LF	Lethal Factor
		LFiA	Lateral-Flow Immunochromatographic Assay
IPG	Immobilized pH Gradient		
IPO	Initial Public Offering	LH	Leutenising Hormone
IPR	Isopropylamino Residue	LLoQ	Lower Limit of Quantitation

LLR	Limit of Linear Range	MET	Mediated Electron Transfer
LMC	Life Marker Chip	MFN	Meniscus Force Nanolithography
LMR	Lower End of the Measuring Range	MFP	Micromachined Fountain Pen
LMW	Low Molecular Weight	MGED	Microarray Gene Expression Data
LoB	Limit of Blank	MGIT	Mycobacteria Growth Indicator Tubes
LOD	Level of Detection		
LOD	Limit of Detection	MHD	16-Mercaptohexadodecanoic Acid
LOQ	Limit of Quantification		
LOX	Lactate Oxidase	MIAME	Minimum Information about a Microarray Experiment
LPCVD	Low-Pressure Chemical Vapor Deposition		
		MIMIC	Micromolding in Capillaries
LPS	Lipopolysaccharide	MIPs	Molecularly Imprinted Polymers
LRP	Luciferase Reporter Phage	MIS	Metal–Insulator–Semiconductor
LSPR	Localized Surface Plasmon Resonance	ML	Mycobacterium Leprae
		mLab	Micro-Lab
LWs	Leaky Waveguides	MLST	Multilocus Sequence Typing
		MM	Methyl Mercaptan
mAb	Monoclonal Antibody	MMC	Mitomycin C
MAbs	Monoclonal Antibodies	MNNG	N-Methyl-N'-Nitro-N-Nitrosoguanidine
MAGE-ML	Microarray and Gene Expression Markup Language		
		MOSFET	Metal-Oxide-Semiconductor Field-Effect Transistor
MALDI	Matrix-Assisted Laser Desorption/ionization		
		MPA	Mercaptopropionic Acid
MAO	Monoamine Oxidase	MPLM	Multi-Purpose Logistics Module
MAO-A	Monoamine Oxidase Type A	MPO	Myeloperoxidase
MAP	Mitogen-Activated Protein	MRAM	Magnetic Random Access Memory
MAPK	Mitogen-Activated Protein Kinase		
		mRNA	Messenger Ribonucleic Acid
MARS	Magnetic Acoustic Resonator Sensor	MRSA	Methicillin Resistant Staphylococcus Aureus
MAS	Marker-Assisted Selection	MS	Mass Spectrometry
MAS	Maskless Array Synthesizer	MS	Multiple Sclerosis
MASSE	Modular Assays for Solar System Exploration	MSFC	Marshall Space Flight Center
		MSL	Membrane Spanning Lipid
MAT	Microscopic Agglutination Test	MSL-EML	Material Science Laboratory Electromagnetic Levitator
MB	Magnetic Bead		
MB	Methylene Blue	MST	Microsystem Technology
MB	Molecular Beacon	MTB	Mycobacterium Tuberculosis
MC	Measurement Cantilever	MTJ	Magnetic Tunnel Junction
MCLW	Metal-Clad Leaky Waveguides	μCP	Microcontact Printing
MDC	Minimum Detectable Concentration	μEDM	Microelectrodischarge Machining
MDCK	Madin-Darby Canine Kidney	MUF	Methylumbelliferone
MDG	Magnetic Direct Generation	MUP	Methylumbelliferoyl Phosphate
MDR1	Multidrug Resistant Enzyme	μTAS	Miniaturized Total Analytical Systems
MEA	Microelectrode Array		
MELFI	Minus Eighty Laboratory Freezer for ISS	μTM	Microtransfer Molding
		MW	Molecular Weight
MEMS	Microelectromechanical Systems	MWCNTs	Multiwalled Carbon Nanotubes

MWCO	Molecular Weight Cutoff		NP	Nanoparticle
MZI	Mach–Zehnder Interferometer		NPC	Niemann Pick Type C
			NPP	Normal Pulse Polarography
NA	Neutroavidin		NPRW	Nanopen Reader and Writer
NA	Nucleic Acid		NRL	Naval Research Laboratory
NA	Numerical Aperture		$nS\ s^{-1}$	Response Slope in Nanosiemens/s
NADH	Nicotinamide Adenine Dinucleotide		NSB	Nonspecific Binding
NAP	p-Nitro-α-Acetylamino-β-Hydroxy Propiophenone		NSF	National Science Foundation
			NSL	Nanosphere Lithography
NAR	Nonlinear Autoregressive		NSOM	Near-Field Scanning Optical Microscopy
NASA	National Aeronautics and Space Administration		NTA	Nitrilotriacetic Acid
NASBA	Nucleic Acid Sequence–Based Amplification		NW	Nanowire
NBT	Nitroblue Tetrazolium		OCP	o-Chlorophenol
NC	Nitrocellulose		ODN	Oligodeoxynucleotide
NC	Noncomplementary		ODN	Oligonucleotide
NCI	National Cancer Institute		OECD	Organization for Economic Co-operation and Development
ND	Neurological Disease			
nDNA	Noncomplementary Deoxyribo-nucleic Acid		OEM	Original Equipment Manufacturer
NDNAD	National DNA Database		OHHL	N-3-(Oxohexanoyl)-L-Homoserine
NEDO	New Energy and Industrial Technology Development Organization		OPEEs	Organic-Phase Enzyme Electrodes
NEMS	Nanoelectromechanical Systems		OR	Odds Ratio
NF-κB	Nuclear Factor Kappa B		OVA	Ovalbumin
nFMLP	n-Formyl-Methionyl-Leucyl-Phenylalanine		OWLS	Optical Waveguide Light Mode Spectroscopy
NFP	Nanofountain Pen		PA	Polyamide
NGF	Nerve Growth Factors		PA_{63}	Protective Antigen 63
NHGRI	National Human Genome Research Institute		PAA	Polyacrylic Acid
NHLBI	National Heart Lung Blood Institute		PAB	Potentiometric Alternating Biosensing
NHS	N-Hydroxysuccinimide		pAEMA	Poly (2-Hydroxyethyl Methacry-late-Co-2-aminoethyl Methacrylate-Co-Ethylene Dimethacrylate)
NIAID	National Institute of Allergy and Infectious Diseases			
NIEHS	National Institute of Environmental Health Sciences		PAHs	Polycyclic Aromatic Hydrocarbons
NIH	National Institutes of Health		PAN	Poly(Acrylamide-Co-N-acryloxysuccinimide)
NIMH	National Institute of Mental Health		PAP	p-Aminophenol
NIPALS	Nonlinear Iterative Partial Least Squares		PAPG	p-Aminophenyl-β-D-Galactopyranoside
NIR	Near-Infrared		PAT	Process Analytical Technology
NIST	National Institute of Standards and Technology		PBS	Phosphate Buffer Saline
NL	Normalized Luminescence		PBT	Polybutylene Terephthalate
NMI	National Metrology Institute		PC	Polycarbonate

PCA	Principal Components Analysis	POC	Point-of-Care
PCB	Printed Circuit Board	POCT	Point-of-Care Testing
PCBs	Polychlorinated Biphenyls	POIC	Payload Operations and Integration Center
PCC	Premature Chromosome Condensation	Poly-Si TFT	Polycrystalline Silicon Thin-Film Transistor
PCR	Polymerase Chain Reaction	POM	Polyoxymethylene
PCs	Principal Components	PP	Plasma-Polymerization
PDA	Personal Digital Assistants	PP	Polypropylene
PDAs	Pin-Diode Arrays	PPC	Pressure Perturbation Calorimetry
PDB	Protein Data Bank		
PDGF	Platelet-Derived Growth Factor	PPE	Polyphenylene Ether
PDH	Pyruvate Dehydrogenase	PPF	Plasma-Polymerized Film
PDMS	Polydimethyl Siloxane Elastomer	PPH	Primary Pulmonary Hypertension
PDMS	Polydimethylsiloxane	PPO	Polyphenol Oxidase
PDT	Photodynamic Therapy	PPO-B	Biotinylated Polyphenol Oxidase
PE	Polyethylene		
PE-CVD	Plasma-Enhanced Chemical Vapor Deposition	PPY	Polypyrrole
		PQQ	Pyrroloquinoline Quinone
PEEK	Polyetheretherketone	(PQQ)GDH	Pyrroquinoline Quinone Glucose Dehydrogenase
PEG	Polyethylene Glycol		
PEI	Polyetherimide	PR	Piezoresistor
PEO	Polyethylene Oxide	PS	Packaging Signal
PET	Polyethylene Terephthalate	PS	Polystyrene
PETN	Pentaerythritol Tetranitrate	PSA	Prostate Specific Antigen
Pf HRP-2	P. Falciparum Histidine-Rich Protein-2	PSD	Power Spectral Density
		PSI	Photosystem I
PGL-I	Phenolic Glycolipid-I	PSU	Polysulfone
PhACs	Pharmaceutical Active Compounds	PTF	Polymer Thick-Film
		PTFE	Polytetrafluoroethylene
PHE	Planar Hall Effect	PV	Pressure-Volume
PI	Polyimide	PVA	Poly(Vinyl Alcohol)
PIC	Plasmon Intensity Change	PVA-SbQ	Poly(Vinyl Alcohol)-Quaternized Stilbazol
PIWAS	Parallisiertes Immunreaktions-basiertes Wasseranalysator-system		
		PVDC	Polyvinylidene Chloride
		PVDF	Polyvinylidene Fluoride
PLC	Programmable Logic Controller	PVF	Poly(Vinylferrocene)
PLD	Pulsed Laser Deposition	PVN	Negative Predictive Value of the Test
PM	Pressurized Module		
PMA	Phorbol 12-Myristate 13-Acetate	PVP	Polyvinylpyrolydine
PMB	Polymyxin B	PVP	Predictive Value of a Positive Test Result
pMBA	*Para*-mercaptobenzoic Acid		
PML	Promyelocytic Leukemia	PWs	Planar Waveguides
PMMA	Polymethyl Methacrylate		
PMN	Polymorphonuclear	QC	Quality Control
PMT	Photomultiplier Tube	QCM	Quartz Crystal Microbalance
PNA	Peptide Nucleic Acid	Qdots	Quantum Dots
PNIPAAm	Poly(*N*-Isopropylacrylamide)	QSAR	Quantitative Structural and Functional Relationship
PNNL	Pacific Northwest National Laboratory		

Rabbit IgG	Rabbit Immunoglobulin G
RC	Reference Cantilever
RDL	Reliable Detection Limit
REM	Replica Molding
RF	Radio Frequency
RF	Replicative Form
RFLP	Restriction Fragment Length Polymorphism
RFP	Red Fluorescent Protein
RH	relative humidity
rhs	Right-Hand Side
RI	Refractive Index
RI	Research International
RIANA	River Analyzer
RIAs	Radioimmunoassays
RIU	Refractive Index Unit
RM	Reference Materials
RM	Resonant Mirror
ROC	Receiver Operating Characteristic
RR	Relative Risk
RRE	Resonant Raman Effect
RSD	Relative Standard Deviation
RT-CES	Real-Time Cell Electronic Sensing
RT-PCR	Reverse Transcriptase-Polymerase Chain Reaction
RU	Resonance Unit
RVC	Reticulated Vitreous Carbon
RWG	Resonant Waveguide Grating
S1P	Site-1 Protease
S2P	Site-2 Protease
SA	Streptavidin
SAA	South Atlantic Anomaly
SAM	Self-Assembled Molecular Monolayers
SAMIM	Solvent-Assisted Micromolding
SAMs	Self-Assembled Monolayers
SANS	Shallow Angle Neutron Scattering
SARS	Severe Acute Respiratory Syndrome
SASS	Smart Air Sampler System
SAW	Surface Acoustic Wave
SAXS	Shallow Angle X-Ray Scattering
SBE	Single-Base Extension
SBIR	Small Business Innovation Research
SC	Spectral Correlation

SCAP	Srebp Cleavage Activating Protein
SCE	Saturated Calomel Reference Electrode
SCE	Silver Chloride Electrode
scFv	Single-Chain Variable Antibody Fragment
SCLM	Scanning Chemiluminescence Microscopy
SCR	Solar Cosmic Radiation
SD	Standard Deviation
sdAb	Single Domain Ab
SDS	Sodium Dodecyl Sulfate
SDS-PAGE	Sodium Dodecyl Sulfate-Polyacrylamide Gel Electrophoresis
SDV	Standard Deviation Value
Se	Sensitivity
SE	Spectroscopic Ellipsometry
SEAC	Surface Enhanced Affinity Capture
SEB	Staphylococcal Enterotoxin B
SECM	Scanning Electrochemical Microscopy
SEF	Surface-Enhanced Fluorescence
SELDI	Surface-Enhanced Laser Desorption/Ionization
SELDI-TOF	Surface-Enhanced Laser Desorption/Ionization Time-of-Flight
SELEX	Systematic Evolution of Ligands by Exponential Enrichment
SEM	Scanning Electron Microscopy
SEND	Surface Enhanced Neat Desorption
SEPS1	Selenoprotein S
SERRS	Surface-Enhanced Resonance Raman Spectroscopy
SERS	Surface-Enhanced Raman Scattering
SERS	Surface-Enhanced Raman Spectroscopy
SH	Shear-Horizontal
SH-APM	Shear-Horizontal Acoustic Plate Mode
SH-SAW	Shear-Horizontal Surface Acoustic Wave
SH3	Src Homology 3
Si NWs	Silicon Nanowires
SIMCA	Soft Independent Modeling of Class Analogies
SIP	Selectively Infective Phage
SIP	Surface-Initiated Polymerization

SL	Sensing Layer
SLB	Supported Lipid Bilayer
SLD	Superluminescent Light Emitting Diode
SMILE	*S*pecific *M*olecular *I*dentification of *L*ife *E*xperiment
SNP	Scanning Near-Field Photolithography
SNP	Single Nuclear Polymorphism
SOLID	Signs-of-Life Detector
SP	Screen-Printed
Sp	Specificity
SP	Surface Plasmon
SPADs	Single-Photon Avalanche Diodes
SPE	Screen-Printed Electrodes
SPE	Solid Phase Extraction
SPEs	Solar Particles Events
SPI	Spectral-Phase Interference
SPM	Scanning Probe Microscopy
SPM	Surface Probe Microscopy
SPR	Surface Plasmon Resonance
SPW	Surface Plasma Wave
SQUID	Superconducting Quantum Interference Device
SRCR	Scavenger Receptor Cysteine Rich
SRE-1	Sterol Response Element-1
SREBP	Sterol Regulatory Element-Binding Protein
SRMs	Standard Reference Materials
SSBW	Surface Skimming Bulk Waves
ssDNA	Single-Stranded Deoxyribonucleic Acid
ssODN	Single-Stranded Oligodeoxynucleotide
STAT	Signal Transducer and Activator of Transcription
STDs	Sexually Transmitted Diseases
STEL	Short-Term Exposure Limit
STM	Scanning Tunneling Microscope
STR	Short Tandem Repeat
STW	Shear Transverse Wave
STW	Surface Transverse Wave
SUMO-1	Small Ubiquitin Modifier-1
SVD	Singular Value Decomposition
SWNT	Single-Walled Carbon Nanotube
TAM	Thermal Activity Monitor
Taq	Thermus Aquaticus
TB	Tuberculosis

TBR	TrisBipyidyl Ruthenium
TCA	Tricarboxylic Acid Cycle
TCD	Temperature Coefficient Delay
TCUs	Temperature-Controlled Units
TDM	Therapeutic Drug Monitoring
TE	Transverse Electric
TELISA	Thermometric Enzyme-Linked Immunosorbent Assay
TFB	Trifluorobenzene
TIR	Total Internal Reflection
TIRF	Total Internal Reflection Fluorescence
TLD	ThermoLuminescence Dosimeters
T-LSPR	Transmission Localized Surface Plasmon Resonance
TM	Transverse Magnetic
TMA	Trimethylamine
TMAO	Trimethylamine N-Oxide
TMOS	Tetramethoxysilane
TMR	Tunneling Magnetoresistance
TMV	Tobacco Mosaic Virus
TNF	Tumor Necrosis Factor
TNF-α	Tumor Necrosis Factor Alpha
TNT	2,4,6-trinitrotoluene
TNT	Trinitrotoluene
TO	Thiazole Orange
tPA	Tissue Plasminogen Activator
TPA	Tripropylamine
TRF	Time-Resolved Fluorescence
TRIM	Trimethylolpropane Trimethacrylate
TSM	Thickness Shear Mode
TSMR	Transverse Shear Mode Resonator
TSS	Total Soluble Solids
TT	Topical Teams
TTA	Total Titratable Acidity
TTF-TCNQ	Tetrathiafulvalinium-Tetracyanoquinomethane
TWA	Time-Weighted Average
UAV	Unmanned Aerial Vehicle
UHMW	Ultrahigh Molecular Weight
ULQ	Upper Limit of Quantitation
ULR	Upper Limit of the Linear Range
UME	Ultramicroelectrode
UMIST	University of Manchester Institute of Science and Technology
USB	Universal Serial Bus

UTFA	Uniform Thin Film Approximation	WHO	World Health Organization
		WIOS	Wavelength-Interrogated Optical Sensing System
UV	Ultra Violet	WIP	Wasp-Interacting Protein
V	Variable	WNV	West Nile Virus
V_H	Variable Heavy	WPI	World Precision Instruments
V_L	Variable Light		
VCSEL	Vertical Cavity Surface-Emitting Laser	XPS	X-ray Photoelectron Spectroscopy
VL	Visceral Leishmaniasis		
VLDL	Very Low Density Lipoprotein	YF	Yellow Fever
VOCs	Volatile Organic Compounds	YFP	Yellow Fluorescent Protein
		YSI	Yellow Springs Instruments
WASP	Wiskott-Aldrich Syndrome Protein	2,4-D	2,4-Dichlorophenoxyacetic Acid
WD	Working Distance	2-AA	2-Aminoanthracene
WGG	Whole-Genome Genotyping	4-NQO	4-Nitroquinoline-N-Oxide
WGMs	Whispering Gallery Modes	6-keto-PGF1α	6-Keto-Prostaglandin-F1-alpha

Part Five

Miniaturized, Microengineered, and Particle Systems

Miniaturized, Microengineered, and Particle Systems

David C. Cullen

Cranfield Health, Cranfield University, Silsoe, UK

One of the current themes within the biosensor sector is to reduce critical size scales and thus there is a resultant emphasis on miniaturization and production of microengineering based systems. In part, this theme is greatly benefiting from developments in many other sectors such as electronics, optics, and mass-manufacture of micro-scale devices and the science of miniaturization and micro-scale phenomena. These sectors provide both fabrication and manufacturing techniques at the micro- and smaller size scales as well as specific components such as miniaturized transducers. For the biosensor and biochip community, obvious advantages of these approaches include the need for reduced sample volumes, new approaches to the deployment of sensors, i.e. their small size enabling novel applications including implantable sensors, distributed sensor systems, instrumentation of real world processes, and integration of sensors into array-based systems. Additionally as size scales are reduced, new basic phenomena appear that can be exploited. Examples range from how the behavior of liquids changes through to new optical and electronic properties resulting from the spatial containment of electrons and photons. Therefore the following collection of chapters highlights a number of the key sub-themes within the application of miniaturized and microengineered systems within a biosensor and biochip context.

Microfluidics is a major topic within modern biosensor and biochip activities. Underpinning this topic, consideration of the theoretical aspects of microfluidics is crucial to optimize the resources deployed to implement the use of microfluidic components and devices in the biosensor and biochip community (*see* **Introduction to Microfluidic Techniques** by Bernhard Weigl, Ron Bardell, and Catherine Cabrera and **Practical Aspects of Microfluidic Devices: Moving Fluids and Building Devices** by the same authors). This situation arises from the behavior of fluids changing drastically at small size scales with one unable to simply scale our everyday observations of fluid flow to the micro-scale.

To allow the development, and real-world impact, of miniaturized and microengineered systems, fabrication and manufacturing approaches are required that are compatible with commercial realities. The historically used silicon photolithographic based fabrication approaches from the traditional microelectronics sector are unlikely to be appropriate for widespread application in the biosensor sector. Thus, low cost mass-produced polymeric microfluidic devices are one approach that is gaining widespread acceptance (*see* **Polymer-Based Microsystem Techniques** by Matthias Schuenemann and Erol Harvey). Furthermore, as one moves to the sub-micrometer size scale, further issues arise that require additional developments for the fabrication of biosensor

Handbook of Biosensors and Biochips. Edited by Robert S. Marks, David C. Cullen, Isao Karube, Christopher R. Lowe and Howard H. Weetall.
© 2007 John Wiley & Sons, Ltd. ISBN 978-0-470-01905-4.

and biochip components (*see* **Nanobiolithography of Biochips** Levi Gheber), (*see* **Nanosphere Lithography-Based Chemical Nanopatterns for Biosensor Design** by Pascal Colpo, Andrea Valsesia, Patricia Lisboa and François Rossi). Further examples of the fabrication, integration and usage issues of miniaturized systems can be seen in the areas of electrochemical systems (*see* **Microelectrochemical Systems** by Stuart Evans and Lindy Murphy) and electro-mechanical systems (*see* **Micro- and Nanoelectromechanical Sensors**, Keith Aubin, Bojan Ilic, Harold Craighead).

Bead- or particle-based technology is another growth area. Some particulate materials have benefits due to the emergence of new properties such as the optical properties of quantum dots and their exploitation as optical labels (*see* **Quantum Dots: Their Use in Biomedical Research and Clinical Diagnostics** by Stanley Abramowitz) whilst others enable new physical manipulations as well as new labeling possibilities (*see* **Manipulation and Detection of Magnetic Nanoparticles for Diagnostic Applications**, by Benjamin Yellen and Randall Erb).

The changing electronic properties of materials at the nanometer scale are also being utilized for biosensor development. The novel electronic properties of polymeric nanowires (*see* **Conducting Polymer Nanowire-Based Biosensors** by Adam Wanekaya, Wilfred Chen, Nosang Myung and Ashok Mulchandani) and carbon nanotubes (*see* **Biosensors Based on Single-Walled Carbon Nanotube Near-Infrared Fluorescence** by Paul Barone, Esther Jeng, Daniel Heller and Michael Strano) are two examples that highlight the exploitation of new nano-scale phenomena for biosensing. Additionally, more traditional biological nano-scale phenomena are being used (*see* **The Detection and Characterization of Ions, DNA, and Proteins Using Nanometer-Scale Pores** by John Kasianowicz, Sarah Henrickson, Jeffery Lerman, Martin Misakian, Rekha Panchal, Tam Nguyen, Rick Gussio, Kelly Halverson, Sina Bavari, Devanand Shenoy and Vincent Stanford).

Introduction to Microfluidic Techniques

Bernhard H. Weigl,[1] Ron L. Bardell[2] and Catherine Cabrera[3]

[1] *Department of Bioengineering, University of Washington, Seattle, WA, USA,* [2] *MicroPlumbers Microsciences LLC, Seattle, WA and Minneapolis, MN, USA and* [3] *Biosensor and Molecular Technologies, MIT Lincoln Laboratory, Lexington, MA, USA*

1 INTRODUCTION

Microfluidics is the science of fluid flow in structures that have at least one dimension in the microscale (between $1\,\mu m$ and $1\,mm$).

According to this very broad definition, capillary electrophoresis (CE), flow injection analysis (FIA), small-diameter versions of high-pressure liquid chromatography (HPLC), and many other techniques would fall under the microfluidics umbrella.

Therefore, a device is more commonly considered to be microfluidic if it has two or more of the characteristics listed below:

- It comprises a channel network wherein the channels have microdimensions.
- It is microfabricated into or from a solid substrate.
- It integrates two or more discrete laboratory functions on a single chip.
- Fluid flow in a microstructure is a required element of the analytical or preparative function of the device. This excludes microarrays, microtiter plates, and so on, from this more narrow definition.

Other definitions related to microfluidics are also in common use. For example, a "lab on a chip" integrates *several* laboratory processes on a single chip. A "micro–total analysis systems (μTAS)" integrates *all* laboratory processes required for an analysis on a single chip.[1]

However, in reality, many microfluidics researchers use the terms *microfluidics, microfluidic devices, lab on a chip,* and *µTAS* interchangeably. Furthermore, conferences under the microfluidics heading will usually feature talks on microarrays, and vice versa.

In this chapter, we will therefore use a more practical approach to delineate microfluidics: *microfluidics is what the literature and the researchers in the field call microfluidics.*

Many of today's microfluidics researchers started their careers in optical or electrochemical microsensors. This is no coincidence. Microsensors hold much of the same promise that is now expected from microfluidic circuits—the possibility of designing a small, easy-to-use, fully integrated, chemical–analytical device.

Microsensors do not only, to some extent, compete with microfluidic systems as tools for chemical and biological analysis, but they also frequently play an important part either during the development of microfluidics devices (e.g., pressure, flow, and temperature sensors) or as detector components in microfluidic systems. And conversely, microfluidic systems can provide an environment that enhances the performance characteristics of microsensors (e.g., by providing a constant and laminar flow past the sensor head).

Handbook of Biosensors and Biochips. Edited by Robert S. Marks, David C. Cullen, Isao Karube, Christopher R. Lowe and Howard H. Weetall.
© 2007 John Wiley & Sons, Ltd. ISBN 978-0-470-01905-4.

Microsensors can be physically placed in contact with microfluidic flow in a number of different ways. For example, both optical and electrochemical sensors can form one wall of a channel. True microdiameter sensors can also be placed in the center of a channel, although this has rarely proved beneficial, frequently causing the sensor to be less stable and prone to mechanical damage from dispersed particles and cells.

Pressure and flow sensors are used to verify actual flow rates in microfluidic systems, and to calibrate micropumps. Honeywell, Inc. has developed a microflow sensor based on a differential heat-loss measurement[2,3] that has been used in the development of microfluidic flow-cytometer circuits.[4] Temperature probes are typically used as feedback systems for temperature-controlled microfluidic chambers, for example for polymerase chain reaction (PCR) heat-cycling applications.[5]

Many microsensors have also been used as detector elements in microfluidic circuits. Electrochemical and optical sensors are frequently used for detection of chemical species in small-diameter and microfluidic channels, such as in glucose monitors[6] as well as in biotechnical process monitoring applications.[7]

Microfluidic devices are often described as miniature versions of their macroscale counterparts. While this analogy is true for some aspects of microfluidic devices, many phenomena do not simply scale linearly from large to small implementations. Examples include the following: increased surface area-to-volume ratio (actually this does scale linearly, but what it affects may not) and the omnipresence of laminar flow.

The first commercial microfluidic lab-on-a-chip-based systems were introduced for life science applications less than 3 years ago. Since that time the field has also seen the formation of a number of diverse microfluidics companies, the publication of more than 3000 scientific papers on microfluidics, and development of several additional microfluidics-based products.[8,9]

The *MIT Technology Review* named microfluidics 1 of 10 technologies that will change the world and one of the many areas in which this will be seen is in the life sciences sector.[10] In 2001, Larry Kricka surveyed the range of microanalytical devices, from microchips and gene chips to bioelectronics chips, and their impact on diagnostic testing.[11] He predicted a move of clinical testing from central laboratory to nonlaboratory settings with a positive impact on healthcare costs.

During the bubble years of the late 1990s and early 2000s, the market growth for microfluidics-based products was forecast to be several billion US dollars by 2004. Now, as with other markets, current growth forecasts are a much more modest, yet healthy, increase from US $127.8 million (2002) to $709.9 million by 2008.[12]

To date, several lab-on-a-chip companies, including Aclara (Mountain View, CA), Caliper (Mountain View, CA), Gyros AB (Uppsala, Sweden), and Orchid Biosciences (Princeton, NJ), have developed microfluidic technologies that work for highly predictable and homogeneous samples that are common in the drug discovery process, whether in compound screening, genomic analysis, or proteomics.[13–19] The first generations of many of these systems address the non-FDA-regulated life sciences research market. One of the primary challenges for homogenous sample-based lab chip providers, however, is their inability to perform analysis on-chip directly from normal, complex, and heterogeneous clinical samples, such as whole blood. Other companies active in the microfluidics area (e.g., Micronics, Redmond, WA; Cepheid, Sunnyvale, CA; MicroPlumbers Microsciences, Seattle, WA; and Fluidigm, San Francisco, CA) have tried to address this issue by their choice of chip materials, structure, and dimensions, by selecting a fluid transport method that is compatible with biological fluids, by employing various novel methods for sample preparation upstream of analysis, and by constructively allowing for the handling of blood and cell-laden streams on microchips.[5,20–26]

2 CHARACTERISTICS OF FLUID FLOW IN MICROCHANNELS

The behavior of fluids in the microscale is quite different from the macroscale behavior we are familiar with in our everyday lives, for example, water flowing from the tap into the kitchen sink or a spoon stirring cream and coffee in a cup. It is the interplay between fluid properties, flow properties, and the scale of the fluid passage that determines the way fluids behave. In this section we discuss how fluid passage dimensions affect both fluid properties and flow properties.

The term *fluids* includes both liquids and gases. We can utilize the kinetic theory of gases (there is no corresponding theory for liquids) to conceptualize important fluid properties. By this theory gas consists of molecules in constant motion, frequently colliding with walls and with each other.

Pressure is the force per unit area imparted by the number of collisions between molecules and a unit area of surface. *Temperature* relates to the speed with which the molecules are traveling; higher temperature gives higher speed. *Density* is the product of molecular mass and the number of molecules per unit volume. The ideal gas law states that pressure is linearly proportional to the product of temperature and density. From the kinetic theory perspective it is easy to see how an increase in temperature or density would increase pressure.

Unlike the molecules in gases, which have a nonzero mean free path (the average distance a molecule can travel between collisions), the molecules in liquids are so close together they are always exchanging momentum. However, liquids do not have the rigid structure of a solid and thus are able to deform when the physical shape of their container changes or a physical object moves through them (e.g., a spoon in a cup of coffee). *Viscosity* is the measure of the effort required to deform a fluid. Compare the effort required in walking on land (in air) with walking in neck-deep water. Viscosity is often introduced by discussing Couette flow, in which fluid is contained between a moving wall and a parallel stationary wall that are

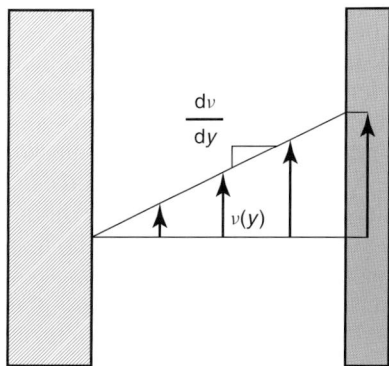

Figure 1. Couette flow is the motion of fluid between two parallel plates, one stationary plate (on the left) and one moving. A Newtonian fluid exhibits a constant spatial gradient of its velocity.

a distance y apart (see Figure 1). In a macroscale Couette flow device, the fluid velocity immediately next to the wall will equal the wall velocity. This is referred to as *zero slip*.

If the fluid is a *Newtonian* fluid, such as water, the fluid velocity will change smoothly from zero at the stationary wall to the velocity v_{wall} at the moving wall. We can say the spatial gradient of the fluid velocity dv/dy is a constant. The viscosity μ is simply the proportionality constant between the shear stress τ applied to the fluid and the resulting velocity gradient, and thus $\tau = \mu \, dv/dy$. Shear stress is what you apply to lotion when you rub it between your hands (i.e., the lotion is experiencing Couette flow). But many fluids, particularly biological fluids, are not Newtonian. They may be shear thinning, shear thickening, dependent on shear history, or may require an initial shear stress that must be applied before they begin flowing, for example, blood.

It is often helpful to use a consistent set of units that is tailored for the physical scale in which the processes of interest occur. We prefer the system shown in Table 1 that is based on grams, millimeters, and seconds.

The flow behavior in a microdevice can be understood through dimensionless parameters such as the Knudsen, Peclet, Reynolds, and Bond numbers. Their definition includes a length scale, a characteristic dimension L, that varies with the geometrical shape of the fluid. In a droplet or round pipe, the diameter is appropriate. In other shapes, the "hydraulic diameter" concept of $D_H = 4 \times$ area/wetted perimeter is useful. In a rectangular cross section, $D_H = 2/(1/\text{height} + 1/\text{width})$.

2.1   The Knudsen Number – Fluid Continuum or Discrete Fluid Particles?

Though fluids are collections of discrete molecules, each with individual properties, it is mathematically simpler to define velocity, temperature, density, and pressure as statistically based properties in a *continuum approximation*. For the continuum concept to be valid, the smallest region of the flow field to which we assign averaged property values must contain a statistically significant number of molecules. At standard temperature and pressure, a 1-μm-sided cube contains 25 million air molecules or 34 billion water molecules. The continuum approximation can become invalid for

Table 1. Microfluidic units

Property	Unit	Name	Definition
Mass	g	Gram	Base unit
Length	mm	Millimeter	Base unit
Time	s	Second	Base unit
Volume	μl	Microliter	mm^3
Force	μN	Micronewton	$g\ mm\ s^{-2}$
Pressure	Pa	Pascal	$\mu N\ mm^{-2}$
Energy	nJ	Nanojoule	$\mu N\ mm$
Power	nW	Nanowatt	$nJ\ s^{-1}$

a gas flow but still be acceptable in a liquid-filled passage that is 100 times smaller. When it becomes invalid, a kinetic theory–based analysis is needed.

When the physical dimensions of a microdevice become so small that there are few molecules near the wall, the assumption of zero slip between fluid and wall can no longer be made. The *Knudsen number*, $Kn = L_{\mathrm{mfp}}/L$, compares intermolecular spacing to the characteristic dimension. For gases, the mean free path is $L_{\mathrm{mfp}} = kT/(\sqrt{2}\pi P d^2)$ in which k is the Boltzmann constant, T is absolute temperature, P is absolute pressure, and d is molecular diameter. For $Kn < 0.001$, the continuum approximation holds and zero slip boundary conditions are appropriate; for $0.001 < Kn < 0.1$, the continuum approximation holds, but there is finite slip; for $0.1 < Kn < 10$, the continuum approximation is invalid and a particle-based method (e.g., direct simulation Monte Carlo) should be used for flow characterization; for $Kn > 10$, free molecular flow requires Boltzmann equation modeling or another method also based on kinetic theory.[27] The situation is less clear cut for liquid flows. Since the molecules are always in collision state, the mean free path is roughly equivalent to the molecular diameter. From experimental results, it appears that 10–20 molecular layers is the minimum height without invalidating the continuum approach or the zero slip boundary condition. For most liquids, minimum flow passage dimensions as small as 1–2 μm will meet the molecular layer requirement. For even smaller channels, molecular dynamics simulations may be the appropriate analysis method.

2.2 The Peclet Number – Which Transport Mechanism?

The constant motion of molecules in fluids ensures that they will intermingle. When we employ the continuum concept, it is convenient to separate the actual mixing process into two conceptual transport mechanisms: *diffusion*, a molecular process modeled as a statistical random walk that is proportional to the degree of kinetic energy in the system, and *advection* (*convection* if heat is being transferred), in which molecules are carried along by the local velocity of the fluid. The relative importance of these two conceptual transport mechanisms is the Peclet number, the ratio of advection and diffusion, $Pe = vL/D$, in which v is the fluid velocity and D is the diffusion coefficient of the solute in the solvent. When $Pe < 1000$, molecular diffusion becomes more effective than stirring for mixing.

2.3 The Reynolds Number – Laminar or Turbulent?

The relative importance of the inertial versus viscous forces in the flow, (i.e., the ratio of the momentum of the fluid to the friction force imparted on the fluid by the walls), is described by the Reynolds number $Re = \rho v L/\mu$, originally proposed by Osborne Reynolds in 1883,[28] in which v is bulk velocity of the flow, ρ is fluid density, and μ is fluid viscosity. A low Reynolds number flow is a laminar, or layered, flow in which fluid streams flow parallel to each other and mix only through advective and molecular diffusion (see Figure 2). Laminar flow is dominated by viscous forces and has fluid velocity at all locations invariant with time when boundary conditions are constant. There is advective mass transport only in the direction of fluid flow. Laminar flow is rarely observed in everyday life unless the fluid is very viscous, for example, pouring honey or squeezing toothpaste from a tube.

In contrast, a high Reynolds number flow is a turbulent flow in which inertial forces dominate and various-size parcels of fluid exhibit motions that are simultaneously random in both space and time. Significant advective mass transport occurs in all directions. We are familiar with this regime from watching water fill a sink or stirring cream in our coffee.

The transition between laminar and turbulent flow typically occurs above $Re = 2000$, though some experiments suggest transition in gas flows in microchannels may occur at Re as low as

Figure 2. A glacier illustrates laminar flow. No mixing occurs between the two side-by-side streams of ice.

400.[29] A flow is identified as laminar or turbulent by either experimental or computational methods. Using experimental data, a laminar flow is identified by a linear proportionality between the log of the pressure loss in the channel and the log of the volume flow rate. If the flow transitions to turbulence, the proportionality constant would change at that flow rate. Transition to turbulence can also be identified using numerical techniques like the finite element or finite volume methods to simulate the flow, because as Hinze[30] states, turbulence is defined as irregular flow with random variation of flow properties (e.g., velocity, pressure, etc.) in both time and space coordinates simultaneously. A numerical simulation based on solving the appropriate conservation of mass and momentum equations will not converge to a steady solution if the flow is randomly varying. Time averaging of the flow properties or some other technique must be used to mathematically model a turbulent flow. Flow in microchannels is virtually always laminar, unless the fluid is driven at very high velocity.

2.4 The Bond Number – How Critical is Surface Tension?

Another flow characteristic that becomes important in microscale passages is the interfacial tension between gas and liquid phases or between immiscible fluids. In flow in porous media, the Capillary number, the ratio of viscous forces to interfacial tension forces, is important. For droplet breakup,

the Weber number, the ratio of inertial and interfacial tension forces is a useful parameter. For microfluidic circuits with changes in elevation, the Bond number, is given by $Bo = \rho g L^2 / \sigma$, the ratio of gravity to interfacial tension forces, in which g is the acceleration of gravity and σ is surface tension. A low Bond number flow responds more to change in surface energy than to change in elevation of the free surface between the phases. Thus, the liquid rises in a capillary tube in spite of gravitational force.

2.5 Example of Flow Characterization

As an example of using nondimensional parameters to predict flow behavior, we investigate a flow rate of $0.05 \,\mu l\, s^{-1}$ in a microchannel that is 1 mm in width and 0.050 mm in height. First, we imagine the channel is filled with water vapor and compute the Knudsen number. The Boltzmann constant in microfluidic units is $k = 1.3806 \times 10^{-14} \, nJ\, K^{-1}$ and the diameter of a water molecule is $d = 0.25 \times 10^{-6}$ mm. If absolute pressure is $P = 101325$ Pa and temperature is $T = 293$ K, the mean free path is $L_{mfp} = 0.000144$ mm. Basing the Knudsen number on the smallest dimension, the channel height, gives $Kn = 0.0029$. According to the Knudsen number ranges discussed above, we can utilize the continuum approximation in this case, but would need to calculate the finite slip between the fluid and the channel walls.

Now imagine our fluid passage is filled instead with a dilute saline solution at $20\,°C$. As a liquid, we replace L_{mfp} with d giving $Kn = 0.00025$, which allows both the continuum approximation and zero slip boundary conditions. The diffusion coefficient for NaCl in water is $D = 1.74 \times 10^{-3}\,\mathrm{mm^2\,s^{-1}}$. Since the width is an order-of-magnitude larger than the height, the hydraulic diameter concept suggests a characteristic length that is approximately twice the smaller dimension, or $L = 0.1\,\mathrm{mm}$. From the ratio of the flow rate and the flow area, we calculate the bulk fluid velocity as $u = 1\,\mathrm{mm\,s^{-1}}$. Thus, the Peclet number is $Pe = 57$, suggesting that diffusion is an effective mass transport mechanism in this case. (Indeed, a diffusion front of NaCl would cross the channel height in less than 2 s.) Assuming essentially water properties, the fluid density is $\rho = 0.001\,\mathrm{g\,\mu l^{-1}}$ and dynamic viscosity is $\mu = 0.001\,\mathrm{Pa\,s}$. Thus, $Re = 0.1$, a very laminar flow. Two streams carrying different solutes would flow side by side in this channel with their components mixing only by diffusion. The Bond number depends on gravity ($g = 9810\,\mathrm{mm\,s^{-2}}$) and surface tension, which for water is $\sigma = 72\,\mathrm{\mu N\,mm^{-1}}$, giving $Bo = 0.0014$. Gravity will be a weak mechanism compared to the capillary effect.

Determining dimensionless parameters is a good way to start the initial design of a microfluidic circuit. This microchannel contains a highly laminar flow in which solutes mix only by diffusion and channel wetout will depend on surface energy, not elevation change. In addition, we also know how changing channel dimensions or fluid properties will impact this flow behavior.

3   USING MICROSCALE EFFECTS TO DESIGN FASTER, MORE ACCURATE, AND LESS EXPENSIVE DEVICES

Researchers developing microfluidic devices frequently face obstacles that are directly related to the fundamental physics of microscale flow. Here are a few practical examples:

- Fluids that are joined in a microfluidic circuit do not mix easily.
- Sample particles that are heavier than the surrounding fluid settle to the channel bottom very quickly.
- Microfluidic devices tend to have a very large surface-to-volume ratio, thus providing much wall space for particles to stick to for a given volume.
- A small drop of fluid placed in the inlet of a microfluidic device can evaporate very rapidly.
- In microdevices, capillary force and surface energy effects are large forces compared to gravity. Depending on their direction and nature, they may move fluids upward and sideways, or block fluid movement, even downward.
- Small fluid volumes will almost immediately take on the temperature of the environment, and cool down or heat up very quickly.

However, those superficially unfortunate effects can be turned into extremely powerful tools in microfluidic devices:

- Flow is usually laminar, allowing the parallel flow of several layers of fluid, thus enabling the design of separation and detection devices based on laminar fluid diffusion interfaces, which will be discussed later.
- At micrometer dimensions, diffusion becomes a viable approach to move particles, mix fluids, and control reaction rates. Typical small drug molecules (e.g., cephradine with a MW of 349) diffuse about $14.3\,\mathrm{mm\,s^{-1}}$ at $25\,^\circ\mathrm{C}$ in aqueous solutions. This allows the establishment of controlled concentration gradients in flowing systems, as well as complete equilibration of the molecule across a $100\,\mathrm{\mu m}$ channel in less than one minute.
- Unaided by centrifugation, sedimentation becomes a viable means to separate dispersed particles by density across small channel dimensions. For example, red blood cells sediment in a 100-$\mathrm{\mu m}$-deep channel in about 1 min and generate a 50-$\mathrm{\mu m}$ layer of plasma in the process.
- In microchannels, the reactor size (and thus the diffusion distance) can be made extremely small, particularly if fluid streams are hydrodynamically focused. Thus, diffusion-controlled chemical reactions occur more rapidly than in comparable macroscopic reaction vessels. For example, Hatch et al.[31] have shown a microfluidic immunoassay that was completed in less than 25 s, as opposed to more typical immunoassay reaction times of 10 min or more.
- Evaporation of small quantities of fluids can be extremely rapid because of a typically large surface-to-volume ratio. This effect can be use both for concentration of sample particles, as

well as for the movement of fluids through a disposable "evaporation pump".[32]

- Active particle transportation and separation methods, such as CE, show greatly enhanced separation performance in small channels.

Other positive characteristics of microfluidic devices that are derived from economics, convenience, and safety include the following:

- Plastic microfluidic structures can be mass-produced at very low unit cost, allowing them to be made disposable.
- Microfluidic devices are somewhat amenable to high throughput by processing multiple samples and assays in parallel.
- Microdevices require only small volumes of sample and reagents, and produce only small amounts of waste, which can often be contained within the disposable device.
- The small scale of the various components of microfluidic systems allows them to be integrated into total analysis systems (μTAS) capable of handling all the steps of the analysis on-chip, from sampling, sample processing, separation and detection to waste handling. This integration also makes complex analyses potentially simpler and safer to perform.
- It is possible to design passive fluidic devices that utilize inherent properties of the fluid and its microenvironment (capillary force, evaporation, wicking, heat transfer, diffusion, etc.) for fluid movement, mixing, heating, cooling, and catalyzing chemical reactions. Thus, disposable stand-alone devices can be designed that require no external power source or instrumentation, yet still perform many, if not all, of the functions typically associated with full-scale automated chemical analysis devices containing pumps, mixers, heat elements, readout electronics, and so on.
- Passive microfluidic devices with integrated detection have the potential to be compatible with very small amounts of fluid, thus allowing ultralow-pain (yield low-volume) blood extraction methods to be used in an integrated fashion. This potentially opens up the regular home use of medical diagnostic assays such as cholesterol, antibodies for sexually transmitted diseases (STDs), and other tests that are currently only performed in laboratories.

4 THE PREFERRED SCALE FOR BIOSCIENCE APPLICATIONS

There is quite a bit of confusion in the microfluidics world about the use and applicability of the various dimensional prefixes such as macro, meso, micro, nano, and pico. The problem stems in part from their use for both dimensional parameters (length, width, and diameter) as well as for volume parameters (i.e., the volume pumped by a micropump per second). Thus, what is considered a microvolume (e.g., a few microliters) can be contained in a device with dimensional parameters in the millimeters.

Further, many authors use a more colloquial definition of macro, micro, and mesoscale. For example, the macroscale sometimes is defined as "human dimension", the microscale as "atomic dimensions", and the mesoscale as bridging these two regimes.[33]

The authors prefer a definition that refers to the smallest dimensional parameter within a particular microfluidic flow structure since that dimensional parameter typically controls the physical behavior of the fluid. Therefore, in a *macroscale* fluidic device, *no part of the flow structure* (channel diameter, height, width, or, conceivably though unlikely, length) has a *dimension of less than 1 mm*. A *microfluidic device* has a smallest dimension of somewhere *between* 1 *and* 999 μm, and nano- and picofluidic devices have equivalent definitions. The authors also accept the use of the term *mesoscale structure* for a *subset of microscale devices that have a smallest dimension of between* 100 *and* 999 μm.

On the basis of these definitions, we now explore some dimensions relevant to biosciences and list appropriate device dimensions for fluidic applications.

Bioanalytical methods generally analyze and manipulate biological particles ranging from tissues (greater than 1 mm in diameter), cells (typically ranging from 0.1 to 30 μm in diameter), and molecules (ranging from a few nanometers for small molecules such as drugs and hormones to hundreds of micrometers for large proteins and nucleic acids).

A number of research groups[34] have used microfluidic devices to expose immobilized tissues to various concentrations of agents. While the channels leading up to the tissue, as well as any mixing or diluting structures present in such devices frequently are microfluidic, the actual chamber that holds the tissue typically is not strictly microfluidic.

Cells, on the other hand, have long been manipulated using microscale structures. For example, most conventional flow cytometers use a microscale nozzle that focuses cells in a single line inside a fluid jet. More recently, several researchers have developed microfluidic flow cytometers that either simulate such a jet inside a mesoscale microfluidic structure,[20] or align cells by squeezing them through smaller, cell-sized microstructures.[35–37]

Large to midsize proteins such as proteins and nucleic acids are generally, for the purposes of separating, mixing, reacting, or detecting them, handled in bulk solution in microfluidic circuits, and not manipulated individually. However, some researchers have designed devices with dimensions in the very low microscale range (1–10 μm,

thus still considered microfluidic) in order to manipulate those molecules individually.[38] For example, the folding of individual proteins can be studied in such devices,[39] and DNA can be stretched out,[40] for example for the purposes of analyzing its code using an atomic force microscope. Small molecules and ions are practically always manipulated in bulk within microfluidic devices.

Many papers have recently been published claiming that their device is "nanofluidic", or uses nanostructured materials. In many cases this does not mean that their devices have a manufactured dimension in the nanoscale, but simply refers to the use of, for example, molecular monolayers of a material inside the structure, or the immobilization of individual molecules (proteins, DNA) inside the channels.[41] In other cases, researchers use the term *nanofluidic* or even *picofluidic* for devices (e.g., ink-jet heads) that generate or manipulate nano- or picoliter volumes of a fluid. Again,

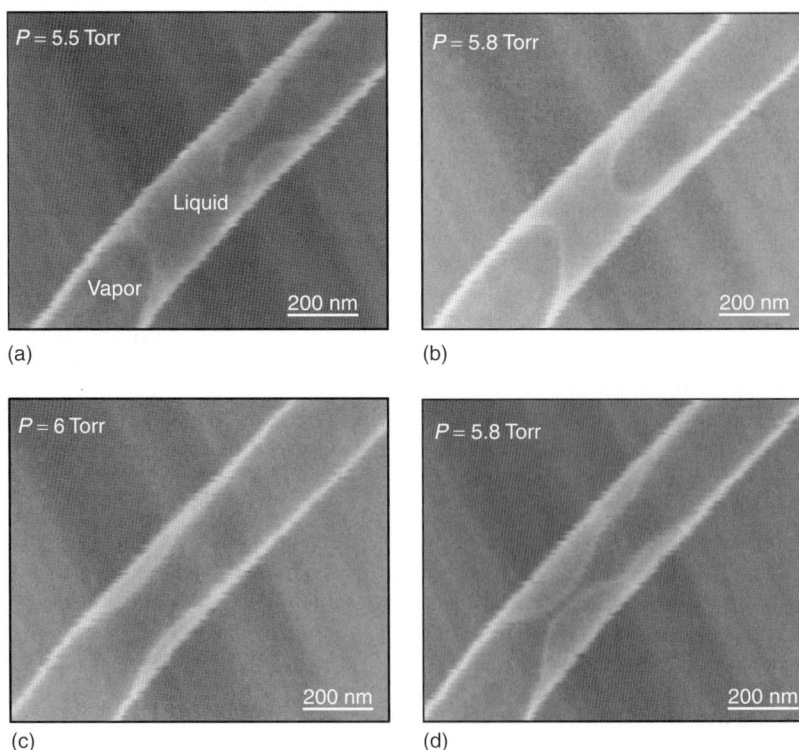

(a)

(b)

(c)

(d)

Figure 3. Sequence of environmental scanning electron microscopy (ESEM) images obtained when partial pressure of water in the ESEM chamber was gradually raised in a controlled manner, while observing a single open carbon nanotube filled with water (a–c). Note the liquid-volume recovery during subsequent pressure decrease (c–d).

this usually does not mean that any fabricated part of the device has nano- or picometer dimensions.[42]

Also, some researchers claim devices as being nanofluidic if they use, for example, nanoporous materials as filters or immobilization matrices. While true in the strictest sense of our definition, we are somewhat hesitant to accept this reading. Nanoporous materials are generally generated using a statistical process (γ ray irradiation, chemical etching), and not a linear design process.

However, there are indeed a few microfabricated fluidic devices that have dimensions of less than $1\,\mu\text{m}$, for example, the carbon nanotube shown in Figure 3.[43] These devices, however impressive in their complexity of manufacturing, have found only limited application in biosciences so far.[44]

In summary, the microfluidic (1–$999\,\mu\text{m}$) scale is uniquely suited for most, if not all, processes and devices needed for biosciences applications. Manufacturing at this scale is now routine, at least for research devices, and within the next 5 years will be routine for low-cost mass manufactured devices as well.

The authors believe that further advances on the nanoscale may bring great insights on specific problems such as single-molecule detection, but the vast majority of all devices used in biosciences will migrate from the macroscale to the microscale within the next decades, and remain there for the foreseeable future.

5 THEORIES OF MICROFLUIDIC PUMPING AND FLUID MOVEMENT

5.1 Pressure-driven Flow

Fluids, by definition, cannot sustain shear and will flow in response to a sufficient difference in pressure between the inlet and outlet of a fluid passage to accelerate the fluid mass and counter fluid friction with the walls. The pressure can be applied by an external source, by the weight of the fluid itself (hydrostatic head), or even by an expanding air bubble in the passage. To determine the fluid velocity, it might be tempting to use Bernoulli's equation,

$$(P + \frac{1}{2}\rho U^2 + \rho g h)_{\text{inlet}} = (P + \frac{1}{2}\rho U^2 + \rho g h)_{\text{outlet}} \quad (1)$$

in which ρ is fluid density, U is mean velocity of the cross section, g is the acceleration of gravity, and h is fluid height. Bernoulli's equation states that the sum of the three pressure terms: the thermodynamic pressure P, the velocity pressure $(1/2\,\rho U^2)$, and the hydrostatic pressure $(\rho g h)$ is the same at any cross section of the fluid passage. It is derived from a conservation of energy statement by assuming the fluid has no viscosity and energy dissipation due to shear stresses in the fluid is negligible, both inappropriate assumptions for low Reynolds number flows.

Instead of using energy conservation to determine a low Reynolds number flow response to applied pressure, it is more accurate to use momentum conservation, which leads to the Stokes or Navier–Stokes equations.[45]

$$\frac{\partial}{\partial t}\rho\boldsymbol{u} + \boldsymbol{u} \cdot \nabla\rho\boldsymbol{u} + \nabla P - \mu\nabla^2\boldsymbol{u} = 0 \quad (2)$$

This is a general equation for an incompressible Newtonian fluid and can be used for complex three-dimensional flow fields with vector velocity $\boldsymbol{u} = (u_1, u_2, u_3)$. The z-direction component, for example, is

$$\frac{\partial}{\partial t}\rho u_3 + u_1\frac{\partial}{\partial x}\rho u_3 + u_2\frac{\partial}{\partial y}\rho u_3 + u_3\frac{\partial}{\partial z}\rho u_3$$
$$= -\frac{\partial P}{\partial z} + \mu\left(\frac{\partial^2 u_3}{\partial x^2} + \frac{\partial^2 u_3}{\partial y^2} + \frac{\partial^2 u_3}{\partial z^2}\right) \quad (3)$$

The Navier–Stokes equation is typically simplified to suit the particular case of fluid flow. For example, when the Reynolds number $Re \ll 1$, the inertial force term $\boldsymbol{u} \cdot \nabla\rho\boldsymbol{u}$ is so small compared to the viscous force term $\mu\nabla^2\boldsymbol{u}$ that it can be neglected. This is termed *Stokes flow*.

The flow in a straight fluid passage of constant cross section is unidirectional. If we use Cartesian coordinates (x, y, z) and take z to be the direction of fluid motion, the fluid velocity components are $\boldsymbol{u} = (0, 0, u)$. If we choose an incompressible fluid, then, by conservation of mass in the constant cross-section passage, the fluid cannot be changing speed as it travels down the passage. Thus, $\partial u/\partial z = 0$ and $\boldsymbol{u}$ may be a function of x and y, but not z. The three components of the Navier–Stokes equation simplify to $\partial P/\partial x = 0$, $\partial P/\partial y = 0$, and

$$\rho\frac{\partial u}{\partial t} + \frac{\partial P}{\partial z} = \mu\left(\frac{\partial^2 u}{\partial x^2} + \frac{\partial^2 u}{\partial y^2}\right) \quad (4)$$

If the flow is steady, laminar, and the y dimension of the passage is at least an order-of-magnitude smaller than the x dimension, then the two-dimensional solution for the velocity in the fluid passage is a good approximation and is simply

$$u = \frac{\mathrm{d}P}{\mathrm{d}z}\left(\frac{d^2}{2\mu}\right)\left(\frac{y^2}{d^2} - \frac{y}{d}\right) + u_{\text{slip}} \qquad (5)$$

in which the passage height is $0 < y < d$. This solution reveals the parabolic velocity profile of steady laminar flow (recall that a parabola is $u = a\delta^2 + b\delta + c$) as a natural consequence of viscous shear and momentum conservation in a straight channel. Figure 4 illustrates that the velocity profile is indeed unchanging across the width of the channel, except near the endwalls.

If the x and y dimensions of a rectangular cross-section channel are comparable in size, the exact analytical solution for velocity can be defined as an infinite series.[46] A good approximation of the ratio between mean velocity U and the maximum velocity U_{max} at the center of the channel is

$$\frac{U}{U_{\text{max}}} = \frac{2}{3}\left(1 - \frac{1}{3}\frac{h}{w}\right) \qquad (6)$$

in which height h and width w are chosen so that $h < w$. This ratio is always $4/9 \leq U/U_{\text{max}} \leq 2/3$. Also, if the Knudsen number $Kn < 0.001$, the continuum approximation holds and there is no slip velocity at the walls, $u_{\text{slip}} = 0$. If $0.001 < Kn < 0.1$, the continuum approximation holds, but a nonzero value of u_{slip} would need to be calculated. For higher Kn, these equations would be inappropriate, as the flow could not be modeled as a continuum.

5.2 Calculating Volume Flow Rate or Pressure Difference in Straight Channels

An efficient method to determine the flow rates in a microfluidic device is to calculate the flow resistance of each passage and then combine them via a circuit diagram and the electric–hydraulic circuit analogies: *current ↔ volume flow rate* and *voltage ↔ pressure difference* between inlet and outlet. Thus, analogous to *electrical resistance = voltage/current*, we have *flow resistance = pressure difference/flowrate*.

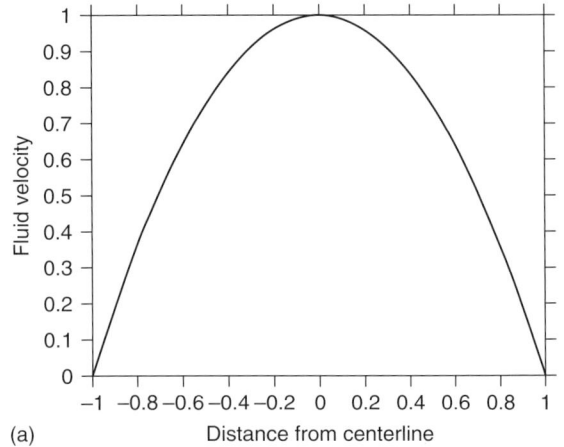

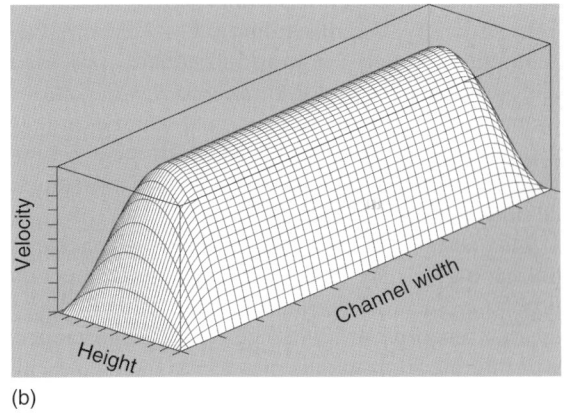

(a)

(b)

Figure 4. Parabolic velocity profiles. (a) Velocity varies parabolically between the two parallel walls. (b) Velocity map in a rectangular channel shows the flow is really two-dimensional anywhere along the width except within a distance of height/2 from the endwalls.

The flow resistance in a straight channel can be derived from the Darcy friction factor,

$$f = \frac{(D_{\text{H}}\Delta P/L)}{(1/2\rho U^2)} \qquad (7)$$

which is a ratio of the energy dissipated in shear to the kinetic energy of the fluid. The hydraulic diameter D_{H} was introduced previously, the pressure gradient is the ratio of the pressure difference ΔP over the channel length L, and U is the mean fluid velocity of the cross section. Experiments over many years have shown that for laminar flow in straight channels, $f = 64/(\varphi Re)$, in which the aspect ratio factor is $\varphi = 1$ for circular pipes or

$$\varphi \approx \frac{2}{3} + \frac{11}{24} AR(2 - AR) \qquad (8)$$

for passages with rectangular cross sections.[47] The aspect ratio AR is the ratio of the channel height and width, or its reciprocal, so that $AR \leq 1$. Combining these to eliminate f gives a general equation for flow resistance

$$R = \frac{128 \, \mu L}{(4 \, \text{Area} \, \varphi D_H^2)} \qquad (9)$$

When $AR \ll 1$, the hydraulic diameter of a rectangular channel depends mainly on the smaller dimension and the flow resistance is inversely proportional to the third power of the smaller dimension.

It is a simple matter to add up the resistances of each section of fluid path, analogous to an electric circuit, and calculate the necessary pressure difference ΔP between inlet and outlet to obtain a required volume flow rate Q. For example, for two different channels connected in series, we use

$$\Delta P = (R_1 + R_2)Q \qquad (10)$$

When the flow is unsteady, for example, starting, stopping, or oscillatory, additional fluid circuit elements modeling inertance and capacitance must be added to the fluid circuit model.

5.3 Gravity-driven Flow

The driving pressure ΔP can be produced by a variety of different sources: syringe pump, hand-held syringe, tank pressurized by gas, or simply gravity, also called *hydrostatic head*. In this last case the driving pressure is

$$\Delta P = \rho g h \qquad (11)$$

where ρ is the density of the fluid, g is acceleration of gravity, and h is the elevation difference between the inlet and outlet of the fluid passage. Other elevation changes completely within the passage are immaterial.

5.4 Examples of Pressure-driven Flow

Regardless of how the driving pressure is created, whether by pump, gravity, or compressed air cylinder, the amount of pressure needed to produce a required flow rate is calculated in the same way. Utilizing the previous equations in this section, we determine the driving pressure needed to generate a flow of $Q = 1 \, \mu l \, s^{-1}$ of water at 25 °C. in a $w = 1$-mm-wide by $h = 0.1$-mm-high by $L = 100$-mm-long rectangular channel. The channel aspect ratio is $AR = h/w = 0.1$ so the aspect ratio factor is calculated as $\varphi \approx 0.75$. Taking the viscosity of water as $\mu = 0.001 \, \text{Pa s}$ and calculating the hydraulic diameter (see Section 2) as $D_H = 2/(1/h + 1/w) = 0.182$, gives the resistance to flow in the channel as $R = 1290 \, \text{Pa s} \, \mu l^{-1}$. To achieve the required flow rate requires a driving pressure of $\Delta P = 1290 \, \text{Pa}$. If we supply this by the hydrostatic pressure of a column of water, it would need to be $h = \Delta P/(\rho g) = 131 \, \text{mm}$ high, (assuming density $\rho = 0.001 \, \text{g} \, \mu l^{-1}$ and $g = 9810 \, \text{mm s}^{-2}$). Clearly, not much pressure is needed to drive this flow rate in this microfluidic channel. Here we assumed that surface tension forces at the inlet and outlet are either equal or insignificant, but they should be included in modeling a low Bond number flow.

5.5 Forced Flow through Packed Bed

There are applications, such as affinity chromatography[48] or enzyme kinetics,[49] in which the microcircuit designer may wish to capture a component of the solution on solid support. Columns with packed media can be implemented in a microdevice by filling a channel with microbeads or silica. The pressure required to drive the solution at the desired flow rate Q through porous media is often the largest pressure drop in the system and should be calculated before the fluid circuit is built to avoid an unreasonably high back pressure. A packed bed can be characterized by the following parameters: A_{bed}, the cross-sectional flow area of the packed bed; L_{bed}, the flow-direction length of the bed; D_p, the diameter of its particles; ε, the void fraction (volume fraction of the empty space between particles through which fluid can flow); and ϕ, the sphericity of the particles. For spherical particles, $\phi = 1$. For nonspherical particles: D_p is the diameter of an equivalent sphere that has the same volume as the actual particle, and $\phi = 6v_p/(D_p S_p)$, the ratio of surface areas of the equivalent sphere and the actual particle S_p, where v_p is the volume of particle. A packed-media Reynolds number is defined as $Re_{PM} = \rho u_{bed} D_p/\mu$, where $u_{bed} = Q/(A_{bed}(1-\varepsilon))$. The pressure drop in the packed bed is determined from the Ergun equation, which is appropriate for both creeping (Stokes) flow ($Re_{PM} < 10$) and laminar flow ($Re_{PM} < 1,000$).

$$\Delta P_{PM} = \left(\frac{150}{\phi^2 Re_{PM}} + \frac{1.75}{\phi} \right)$$
$$\times \left(\frac{\rho(1-\varepsilon)/\varepsilon^3 L_{bed}}{D_p/A_{bed}^2} \right) Q^2 \quad (12)$$

A slightly different approach is useful when more than one size of microbeds are used.[50]

Channeling due to irregular packing and nonuniformly sized media can be problematic and greatly lower the effectiveness of a packed bed. Some consideration should be applied to the fact that microchannels can have a large surface-to-volume ratio. If the volume of fluid to be processed is very small and the height of the channel can be decreased sufficiently, the walls of the channel may perform as the solid support.

5.6 Example of Flow in Packed Media

Let us revisit the previous pressure-driven flow example to see how much additional pressure is required if the channel is filled with media. We assume spherical particles (so $\phi = 1$) of diameter $D_p = 0.003$ mm and a bed void fraction of $\varepsilon = 0.35$, for a tightly packed bed. The area of the bed is the cross-sectional area of the channel, $A_{bed} = 0.1$ mm^2. Using the same flow rate as the previous example, $Q = 1\,\mu l\,s^{-1}$, gives a bed velocity of $u_{bed} = 15.4$ mm s^{-1} and a packed-media Reynolds number of $Re_{PM} = 0.046$. If we take the same density and viscosity as the previous example, the required driving pressure is $\Delta P_{PM} = 5750$ Pa, or 4.5 times larger because of the addition of the media. To drive this flow hydrostatically would require a 586-mm-high column of water. Also note that the packed-media driving pressure is a

function of the square of the flow rate, unlike the linear relationship in a channel without media. Thus, increasing the flow rate by a factor of 10 would increase the packed-media driving pressure by a factor of 100.

5.7 Electroosmotic Flow (EOF)

Electroosmotic flow (EOF) occurs when the channel walls of a microfluidic device are charged at the local buffer pH. The fluid proximal to the channel walls will not be neutral but rather will contain a higher-than-bulk concentration of counterions (balancing the opposing charge of the channel wall).

When an electric field is applied parallel to the surface, the charged fluid will move in bulk toward the complimentary electrode, resulting in convective fluid flow (see Figure 5).[51] In the case of a negatively charged surface the fluid will have a net positive charge and will migrate toward the cathode (negative electrode). Initially only a thin layer of fluid begins to migrate, creating a velocity gradient that results in momentum transfer to the adjacent fluid through viscous shear forces, which then begins to move. In other words, the charged fluid layer "drags" the adjacent fluid layer along, until finally the entire channel moves at a uniform velocity.[51] Essentially, one creates the "one fixed wall, one moving wall" situation described during the discussion of viscosity. Note that this uniform velocity occurs if the channel characteristic diameter is at least seven times that of the electric double layer proximal to the channel walls[52] and if other sources of fluid acceleration, such as convective currents due to Joule heating, are negligible. This velocity profile is very different from that of pressure-driven flow, which has a nonuniform, parabolic profile.

The effective charge of a surface, called *zeta potential* (ζ), is defined as the potential difference at the interface between the tightly held layer of counterions immediately proximal to the surface and the bulk solution (this interface is called the *surface of shear*).[53] In addition to the surface charge of the material, three fluid parameters influence the ζ potential: pH, dielectric constant, and ionic strength.[54] The velocity of fluid undergoing EOF is a function of the ζ potential of the walls,

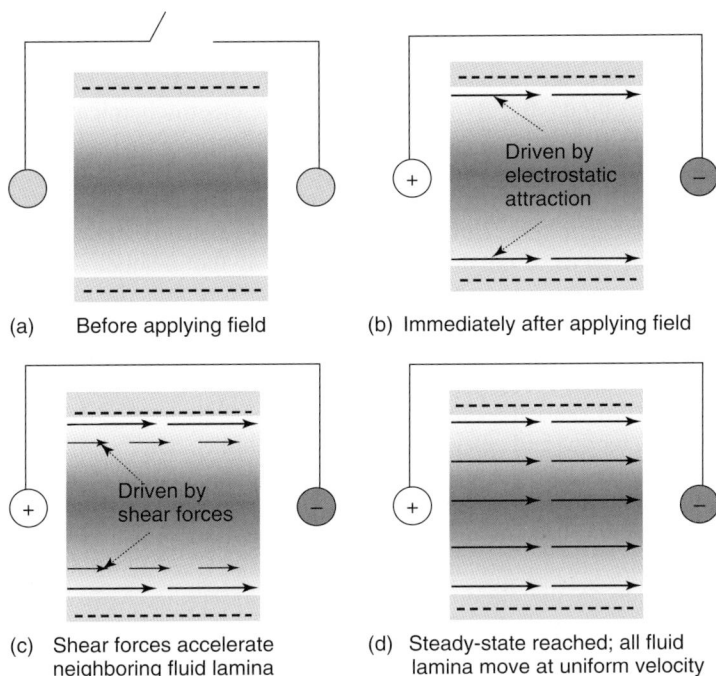

Figure 5. Schematic of electroosmotic flow in a microfluidic channel. White = positive, dark gray = negative, light gray = neutral. The channel walls are negatively charged; the bulk fluid is neutral. Arrow size corresponds to fluid velocity.

the electric field strength, and the fluid properties themselves, as

$$v_{EOF} = \frac{\zeta D F}{\mu} \qquad (13)$$

in which v_{EOF} is the velocity of the fluid due to EOF in a circular capillary, ζ is the zeta potential of the charged wall, D is the dielectric constant of the fluid, and μ is the viscosity of the fluid.

EOF velocity depends on characteristics of both the fluid undergoing transport and the channel wall material. The dielectric constant and the viscosity of the fluid undergoing transport directly affect EOF. In addition, other fluid characteristics (pH, ionic strength, composition) can have additive or opposing effects. For example, in a bare glass capillary, the surface charge increases with pH. But if the pH is increased through addition of high concentrations of a metallic salt, then the ionic strength will increase, which acts to decrease EOF.

Up until fairly recently, the vast majority of microfluidic devices were made from glass, which has a well-characterized surface charge that varies in a predictable way with changes in the fluid.

Under physiologic conditions, glass and silica have a negative ζ potential; as the local pH drops the silanol groups on the glass surface become protonated and the ζ potential drops in magnitude. Many different surface modification techniques have been developed for glass, which allow the user to change the surface charge and/or apply a nonfouling coating. EOF in glass capillaries is therefore a fairly straightforward technique, commonly used in many microfluidic devices, particularly those used for CE.

As newer devices begin to incorporate less expensive materials, particularly polymeric components such as Mylar and silicone, the associated EOF behavior becomes more difficult to predict. In fact, a major obstacle to the use of materials other than glass for microfluidic devices has been the lack of information on the surface properties of these materials.[55] Compared to glass and silica, there are significantly fewer established techniques for surface modifications of polymers. One should rely on empirical data specific to the material of interest when designing polymeric devices to work with EOF.

Owing to increasing interest in the use of polymeric materials for microfluidic devices, new surface modification techniques are continually being developed[56] as are methods for dynamic coatings,[57] for example, some researchers suggest that oxidation of polydimethylsiloxane (PDMS) increases its ζ potential and therefore the EOF velocity.[58] Certain fabrication techniques can alter the surface properties of some areas of the device, which creates a nonuniform surface and can lead to nonplug flow.[59]

The use and treatment history of the device also influence EOF velocity. Fluid components, particularly protein, can adsorb onto the channel walls, thus altering their charge and therefore changing the associated double layer. For EOF to operate consistently over time, this surface fouling must be avoided, either through treating the channel walls to minimize adsorption and/or through addition of materials to the fluid itself that prevent adsorption. Alternately, if the degree of fouling is well known, the device can be operated for a fixed period of time and then flushed with a cleaning solution that returns the walls to a pristine state. Biological fluids, in particular, create significant fouling concerns, primarily due to the relatively high concentration of proteins in solution. If the device is single use, fouling concerns are reduced but not eliminated, since EOF could be affected by fouling during a single assay.

5.8   Advantages and Disadvantages of EOF

There are three main advantages to using EOF to drive fluids: uniform flow profile, no-moving-part pumping, and a simplified fluidic interface. Uniform velocities result in uniform retention times for all particles in a given section of the device, which can greatly simplify calculations and analysis. Because fluids undergoing EOF move as a bolus, the leading and trailing edges of materials are minimized, which reduces the time and material required to change solutions in a device. Unlike pressure-driven flow, EOF does not require a leak-tight interface between the source of the hydraulic driving force (electrodes in this case) and the fluid being driven. Therefore, the interface between the source of pumping and the fluid being pumped can be as simple as two wires placed into holes in the device.

A disadvantage of EOF is its strong dependence on the electrochemical properties of both channel walls and the fluid. If a device is expected to process a variety of fluids or a fluid of unknown pH/ionic strength, the EOF velocity will be unpredictable. Since EOF depends on the material properties of the channel wall, any changes in device manufacture must be analyzed for potential effects on EOF. In addition, EOF often requires high voltages (typically in the kilovolt to megavolt range), which requires isolation of the electrodes from the sample fluid to avoid the products of electrolysis (bubbles, acid/base) from entering the sample, while also retaining electrical connectivity. Heat produced by the high electric field may also have to be dissipated.

Finally, if a microfluidic device will be incorporating an applied voltage, the possibility of EOF, intentional or not, must be considered during the design stage. This concern is particularly important when applying an electric field perpendicular to the direction of fluid flow, since the induced EOF will itself be perpendicular to the direction of fluid flow, at least initially. If one observes unusual flow patterns in a microfluidic device to which electricity is being applied, particularly recirculating flows, one should consider EOF as a likely cause of the phenomenon.

Because of the utility and ubiquity of EOF in the microfluidic community, a significant amount of research has been done on many aspects of EOF. A full review is beyond the scope of this text; for good starting points into the related scientific literature, see the excellent review by Manz and colleagues.[51]

5.9   Capillary Flow

Molecules at a liquid–gas interface experience molecular attraction only from the liquid side. In response to this imbalance, the liquid surface contracts like a stretched membrane in tension; this is surface tension σ, a force per unit length. An analogous argument can be made on the basis of energy instead of force.[60] Organic compounds with hydrophilic heads and hydrophobic hydrocarbon tails, such as detergents, are strongly adsorbed at the liquid–gas interface and tend to form a monolayer on the liquid surface. Even a very low concentration of these "surface-active" molecules, *surfactants*, (e.g., 0.004 M sodium dodecyl sulfate) interferes with hydrogen bonding and can lower the surface tension of water from $72.9\,\mu\text{N mm}^{-1}$ (at 20 °C) to 40–$50\,\mu\text{N mm}^{-1}$.

A drop of liquid placed on a solid surface may spread out and wet the surface or form a static contact angle at the three-phase boundary between the wetted solid and the liquid–gas interface. The pressure rise due to surface tension in the general case of an aspherical drop with principal radii of curvature, r_1 and r_2, (i.e., the radii of curvature along any two orthogonal tangents) is $\Delta P = \sigma = 1/r_1 + 1/r_2$), the Young–Laplace equation.

The contact angle is a sensitive measure of surface energy and depends, not only on the chemical composition of the fluid (e.g., its pH), but also on the surface condition of the material (Figure 6). The same material will result in different contact angle measurements depending on its cleanliness, its previous contact with hydrophobic materials that can alter the surface (e.g., Teflon and silicone oil), whether it is already wetted, and even the humidity of the surrounding gas. An advancing contact angle (e.g., an expanding drop during condensation) is typically significantly larger than a receding contact angle and both change with time as the precursor film[61,62] that extends out beyond the drop is spreading or retracting. Table 2 lists advancing contact angles for a water drop on various materials.

The capillary pressure in a rectangular channel with height h and width w (see Figure 7) can be determined from a force balance, $-\Delta Phw = 2(h + w)\sigma \cos\theta$, which can be simplified to

$$\Delta P = -2\sigma \cos\theta (1/h + 1/w) \qquad (14)$$

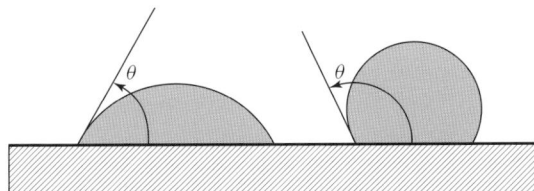

Figure 6. Contact angles, θ, for liquid water drops on solid surfaces. The solid on the left is hydrophilic (i.e., $\theta < 90°$), while that on the right is hydrophobic.

Table 2. Approximate advancing contact angles of a water drop. Glass, polyethylene terephthalate (PET), and polymethyl methacrylate (PMMA) are hydrophilic and should be naturally wetting, while Teflon and poly-dimethylsiloxane (PDMS) would require addition of surfactants to the aqueous solution or a surface-energy-increasing treatment to lower the contact angle

Solid	Contact angle (°)
Glass, borosilicate (Pyrex)	14
PET	50
Acrylic (PMMA)	75
PS	86
Paraffin	109
Poly(tetrafluoroethylene) (Teflon)	112
PDMS	115

PS: polystyrene.

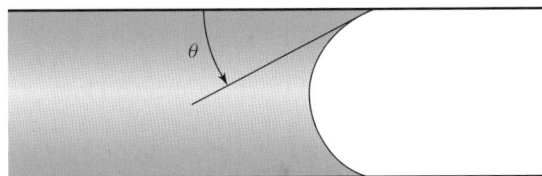

Figure 7. Liquid filling a channel. The small contact angle means the surface tension is pulling the liquid to accomplish the wetout.

Why is the minus sign needed here? When the contact angle is small, the fluid is being pulled into the channel by surface tension and the pressure in the liquid is actually lower than in the downstream gas. Conversely, when the contact angle is $\theta > 90°$, a positive pressure difference would be required to force liquid into the channel against surface tension.

Compare capillary behavior in the same rectangular channel in two different materials using the contact angle values from Table 2. The capillary pressure in water in a 1-mm-wide by 0.1-mm-high channel fabricated of PET is $\Delta P = -2(72.9)\cos 50°(1/0.1 + 1/1) = -1031\,\text{Pa}$ with respect to the gas pressure downstream of the meniscus. If the water at the upstream end of the channel is at ambient pressure (i.e., 0 Pa) and the channel elevation is constant, the pressure difference in the liquid between the upstream and downstream end at the meniscus will push the liquid into the channel. On the other hand, if the channel is fabricated of PDMS, the capillary pressure is 678 Pa, which would oppose liquid flow

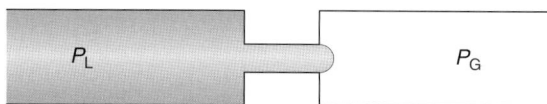

Figure 8. Area discontinuity can create an unstable "surface tension stop" that sustains a small pressure difference, $P_L > P_G$.

into the channel. This positive capillary pressure would not be likely to push the liquid out of the PDMS channel however, because the receding contact angle would likely be less than 90°.

Another interaction between surface tension and channel geometry is a *surface tension stop* as shown in Figure 8. A desirable (or unintentional) pressure difference can be sustained by an area discontinuity.

5.10 Absorptive and Wicking-driven Flow

Uptake of fluid by capillary pumping of a hydrophilic polymer is the mechanism for fluid movement in the lateral flow strip, a qualitative immunological assay in a passive handheld format, such as a home pregnancy test.

Most current lateral flow strip assays are based on membranes composed of cellulose nitrate or cellulose acetate depending on the degree of protein binding desired. A similar functionality is obtainable from hydrogels, cross-linked polymer networks surrounded by an aqueous solution. When wetted by a solvent, the chains in the network are solvated, but do not mix due to the cross-linking, which provides an elastic restoring force to counter swelling. Hydrophilic polymers can be divided into categories based on the relaxation time of the polymer and the diffusion time of the solvent. One parameter is the Deborah number, the ratio of the rates of solvent penetration and polymer relaxation, $De = \lambda D/\delta^2$, in which λ is the characteristic polymer relaxation time from swelling stresses, D is the diffusion coefficient of the solvent, and δ is the diffusional distance at time $= \lambda$.[63] Case I transport ($De \ll 1$) occurs when the diffusion time is much slower than the polymer relaxation time, leaving diffusion as the controlling mechanism. This is typical in nonswelling systems. In case II transport ($De \gg 1$), the rate limiting process is polymer relaxation. In other hydrogels,

De is on the order of 1 and the two processes occur on the same time scale, leading to anomalous transport behavior. When *De* is very large, this is sometimes referred to as *super case II behavior.*

A simple description of the time-dependent swelling of a polymer is $M_t/M_\infty = kt^n$, in which M_t/M_∞ is the fractional uptake (or release) of solvent normalized by the equilibrium conditions and k and n are constants dependent on solvent diffusion coefficient and type of transport process.[64–66] Figure 9 plots the fractional uptake of solvent for various transport types. For diffusion-controlled (type 1) transport, $n = 0.5$, while for polymer relaxation controlled (type II) transport, $n = 1$. Anomalous transport has $0.5 < n < 1$, and super type II has $n > 1$.

The uptake of solvent and solute in swelling polymeric systems can be numerically modeled by the species conservation equation[67,68] if the solvent velocity u is known from the solution of the momentum conservation equations, from Darcy's equation for fluid flow through a porous medium and the permeability and porosity of the medium, or from experimental data to determine the constants. For example, data from experiment with $5 \mu m$ pore size nitrocellulose polyether sulfone exhibits Type I transport behavior[69] in which wetting speed is proportional to the square root of wetting time. Wetting time data for a typical lateral flow strip based on Whatman nitrocellulose "Purabind" is shown in Figure 10.

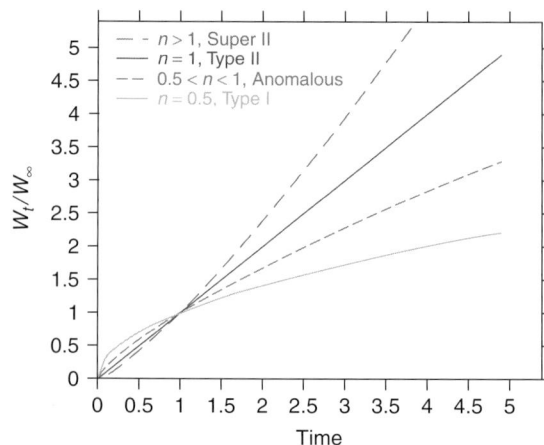

Figure 9. Fractional uptake of solvent versus time for various transport types.

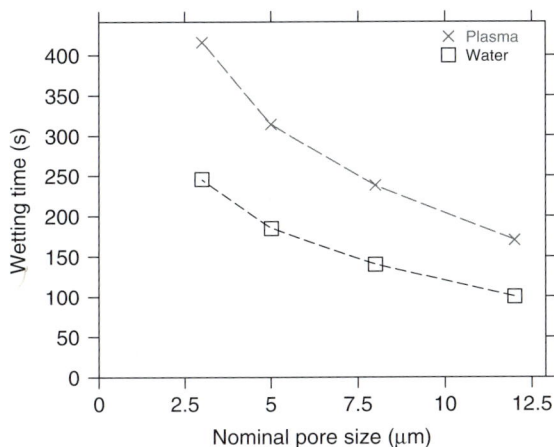

Figure 10. Wetting times for 45-mm-long lateral flow strip of Whatcom nitrocellulose "Purabind" for water at 25 °C. The data is also scaled by viscosity ratio to predict behavior with blood plasma as solvent.

6 USING APPROPRIATE STRATEGIES FOR DEVELOPMENT OF MICROFLUIDIC DEVICES

Microfluidics brings unique capabilities to assays and sensors in bioscience applications, such as: lower cost assays, more rapid results, smaller reagent volumes, and less hazardous waste. But fluids in the microscale do not display the familiar behavior of fluids in the human-scale world. Understanding the physics of microfluidics is one key to successful development of microdevices. It is a nearly impossible task to replicate macroscale fluid-management strategies in the microscale and far more fruitful instead to leverage its unique behaviors and inherent capabilities. In the microscale, inertial and gravitational forces are often much less important than surface tension and viscous forces. Alternatives to pressure-driven flows exist and are robust and reliable, such as electrokinetics and wicking. Dimensionless parameters can be utilized to show which physics will be dominant in your device. Then mathematical modeling becomes an effective way to estimate performance in the deterministic world of microfluidics. In this way, a new microfluidic device has the potential to become a novel solution in its application area.

REFERENCES

1. B. H. Weigl, R. L. Bardell, and C. R. Cabrera, Lab-on-a-chip for drug development. *Advanced Drug Delivery Reviews (invited review article)*, 2003, **55**(3), 349–377.
2. E. Cabuz, J. Schwichtenberg, B. DeMers, E. Satren, A. Padmanabhan, and C. Cabuz, *MEMS-Based Flow Controller for Flow Cytometry Honeywell International*, 2002 http://content.honeywell.com/sensing/solutions/markets/medical/88PAD_HH_2002_twopage_fullpaper.pdf.
3. U. Bonne et al., *Microsensor Housing*, US Patent 6,322,24, Issued November 27, 2001.
4. http://www.darpa.mil/MTO/bioflips/presentations/2001-1/index.html, (DARPA BioFlip Program), 2001.
5. P. Belgrader, M. Okuzumi, F. Pourahmadi, D. Borkholder, and M. A. Northrup, A microfluidic cartridge to prepare spores for PCR analysis. *Biosensors and Bioelectronics*, 2000, **14**, 849–852.
6. B. H. Weigl, R. L. Bardell, T. Schulte, D. C. Cullen, and J. Demas, Modeling of microscale processes speeds development and enhances performance of medical diagnostic product. *In Vitro Diagnostics Technology*, 10, June 2004, 41–49, invited.
7. B. H. Weigl, A. Holobar, W. Trettnak, and O. S. Wolfbeis, Optical triple sensor for measuring pH, oxygen and carbon dioxide. *Journal of Biotechnology*, 1994, **32**, 127–138.
8. B. H. Weigl, *New Assays and Separations Based on Laminar Fluid Diffusion Interfaces—Results From Field Trials for Cell Analysis, HTP Screening, and Medical Diagnostics*, in *Micro Total Analysis Systems 2002*, D. J. Harrison and A. Van den Berg (eds), Kluwer Academic Publishers, Dordrecht, 2002.
9. P. Mitchell, Microfluidics—downsizing large-scale biology. *Nature Biotechnology*, 2001, **19**, 717–721.
10. J. Benditt, Ten Emerging Technologies that will Change the World, MIT Technology Review: January/February, 2001.
11. L. J. Kricka, Microchips, microarrays, biochips and nanochips: personal laboratories for the 21st century. *Clinica Chimica Acta*, 2001, **307**(1–2), 219–223.
12. Frost and Sullivan, U.S. Point-Of-Care Retail Diagnostics Markets, 3/15/2003, available from hyperlink http://www.marketresearch.com (this is a market research report), 2003.
13. R. L. Chien and J. W. Parce, Multiport flow-control system for lab-on-a-chip microfluidic devices. *Fresenius Journal of Analytical Chemistry*, 2001, **371**(2), 106–111, (eng).
14. L. Bousse, C. Cohen, T. Nikiforov, A. Chow, A. R. Kopf-Sill, R. Dubrow, and J. W. Parce, Electrokinetically controlled microfluidic analysis systems. *Annual Review of Biophysics and Biomolecular Structure*, 2000, **29**, 155–181.
15. M. T. Cronin, T. Boone, A. P. Sassi, H. Tan, Q. Xue, S. J. Williams, A. J. Ricco, and H. H. Hooper, Plastic microfluidic systems for high throughput genomic analysis and drug screening. *Journal of the Association for Laboratory Automation*, 2001, **6**(1), 74–78.
16. M. T. Cronin, M. Pho, D. Dutta, F. Frueh, L. Schwarcz, and T. Brennan, Utilization of new technologies in drug trials and discovery. *Drug Metabolism and Disposition*, 2001, **29**(4), 586–590.

17. A. J. Ricco, T. D. Boone, Z. H. Fan, I. Gibbons, T. Matray, S. Singh, H. Tan, T. Tian, and S. J. Williams, Application of disposable plastic microfluidic device arrays with customized chemistries to multiplexed biochemical assays. *Biochemical Society Transactions*, 2002 **30**(2), 73–78.

18. W. W. Weber and M. T. Cronin, *Pharmacogenetic Testing, in Encyclopedia of Analytical Chemistry*, R. A. Meyers (ed), John Wiley & Sons, Sussex, 2000, pp. 1506–1531.

19. M. T. Boyce-Jacino, J. E. Reynolds, T. T. Nikiforov, Y. H. Rogers, C. Saville, T. C. McIntosh, P. Goelet, and M. R. Knapp, High volume molecular genetic identification of single nucleotide polymorphisms using genetic bit analysis: application to human genetic disease. *American Journal of Human Genetics*, 1994, **55**(3), 18–22.

20. B. H. Weigl, R. L. Bardell, N. Kesler, and C. J. Morris, Lab-on-a-chip sample preparation using laminar fluid diffusion interfaces—computational fluid dynamics model results and fluidic verification experiments. *Fresenius Journal of Analytical Chemistry*, 2001, **371**(2), 97–105.

21. A. Y. Fu, C. Spence, A. Scherer, F. H. Arnold, and S. R. Quake, A microfabricated fluorescence-activated cell sorter. *Nature Biotechnology*, 1999, **17**, 1109–1111.

22. J. W. Hong and S. R. Quake, Integrated nanoliter systems. *Nature Biotechnology*, 2003, **21**, 1179–1183.

23. M. T. Taylor, P. Belgrader, R. Joshi, G. A. Kintz, and M. A. Northrup, Fully automated sample preparation for pathogen detection performed in a microfluidic cassette. *Micro Total Analysis Systems*, 2001, 670–672.

24. M. A. Northrup, L. Christel, W. A. McMillan, K. Petersen, F. Pourahmadi, L. Western, and S. Young. *A New Generation of PCR Instruments and Nucleic Acid Concentration Systems*, *PCR Applications—Protocols for Functional Genomics*, 1999, 105–125.

25. T. Thorsen, S. J. Maerkl, and S. R. Quake, Microfluidic large scale integration. *Science*, 2002, **298**, 580–584.

26. P. Jandik, B. H. Weigl, N. Kesler, J. Cheng, C. J. Morris, T. Schulte, and N. Avdalovic, Initial study of using laminar fluid diffusion interface for sample preparation in HPLC. *Journal of Chromatography A*, 2002, **954**, 33–40.

27. M. Gad-el-Hak, The fluid mechanics of microdevices—the freeman scholar lecture. *Journal of Fluids Engineering-Transactions of the ASME*, 1999, **121**, 5–33.

28. O. Reynolds, An experimental investigation of the circumstances which determine whether the motion of water in parallel channels shall be direct or sinuous and of the law of resistance in parallel channels. *Philosophical Transactions of the Royal Society of London*, 1883, **174**, 935–982.

29. P. Wu and W. A. Little, Measurement of friction factors for the flow of gases in very fine channels used for microminiature Joule-Thomson refrigerators. *Cryogenics*, 1983, **23**, 273–277.

30. J. O. Hinze, *Turbulence*, 2nd Edn, McGraw-Hill, 1987, pp. 1–4.

31. A. Hatch, A. E. Kamholz, K. R. Hawkins, M. S. Munson, E. A. Schilling, B. H. Weigl, and P. Yager, A rapid diffusion immunoassay in a t-sensor. *Nature Biotechnology*, 2001, **19**(5), 461–465.

32. D. J. Beebe and G. M. Walker, An evaporation-based microfluidic sample concentration method. *Lab on a Chip*, 2002, **2**, 57–61.

33. National Academy of Science, *Summary, the Impact of Materials—from Research to Manufacturing*, http://www.national-academies.org. 2002.

34. A. Folch and M. Toner, Microengineering of cellular interactions. *Annual Review of Biomedical Engineering*, 2000, **2**, 227.

35. J. P. Brody, P. Yager, R. Goldstein, and R. H. Austin, Biotechnology at low Reynolds numbers. *Biophysical Journal*, 1996, **71**, 3430–3441.

36. R. H. Carlson, J. P. Brody, S. Chan, C. Gabel, J. Winkleman, and R. H. Austin, Self-sorting of white blood cells in a lattice. *Physical Review Letters*, 1997, **79**, 2149–2152.

37. E. Altendorf, E. Iverson, D. Schutte, B. H. Weigl, T. Osborn, R. Sabeti, and P. Yager, *Optical Flow Cytometry Utilizing Microfabricated Silicon Flow Channels*, in *Advanced Techniques in Analytic Cytology (BIOS 96) SPIE Proceedings*, SPIE (formerly the International Society for Optical Engineering), Vol. 2678.

38. T. A. J. Duke and R. H. Austin, Microfabricated sieve for the continuous sorting of macromolecules. *Physical Review Letters*, 1998, **80**, 1552–1555.

39. J. B. Knight, A. Vishwanath, J. P. Brody, and R. H. Austin, Hydrodynamic focusing on a silicon chip: mixing nanoliters in microseconds. *Physical Review Letters*, 1998, **80**, 3863–3866.

40. R. H. Austin, J. P. Brody, E. C. Cox, T. Duke, and W. Volkmuth, Stretch genes: aligning single molecules of DNA. *Physics Today*, 1997, 32–36.

41. J. S. Kuo, Interfacing Chip-Based Nanofluidic-Systems to Surface Desorption Mass Spectrometry, JIN Reports, Pacific Northwest National Laboratory, 2003, http://www.pnl.gov/nano/institute/2003reports/.

42. T.-C. Kuo, D. M. Cannon Jr, M. A. Shannon, J. V. Sweedler, and P. W. Bohn, Hybrid three-dimensional nanofluidic/microfluidic devices using molecular gates. *Sensors and Actuators A*, 2003, **102/3**, 223–233.

43. Y. Gogotsi, C. M. Megaridis, H. Bau, J.-C. Bradley and P. Koumoutsakos, *Carbon Nanopipes for Nanofluidic Devices and In-situ Fluid Studies*, In: *NSF Nanoscale Science and Engineering Grantees Conference*, National Science Foundation, Arlington, Virginia, 2003, Dec 16–18.

44. L. Wanli, J. O. Tegenfeldt, L. Chen, R. H. Austin, S. Y. Chou, P. A. Kohl, J. Krotine, and J. C. Sturm, Sacrificial polymers for nanofluidic channels in biological applications. *Nanotechnology*, 2003, **14**, 578–583.

45. R. L., Panton, *Incompressible Flow*, John Wiley & Sons, 1984, p. 154.

46. F. M. White, *Viscous Fluid Flow*, 2nd Edn, John Wiley & Sons, 1991, p. 120.

47. O. C. Jones Jr, An improvement in the calculation of turbulent friction in rectangular ducts. *Journal of Fluids Engineering-Transactions of the ASME*, 1976, **98**, 173–181.

48. N. Malmstadt, P. Yager, A. S. Hoffman, and P. S. Stayton, A smart microfluidic affinity chromatography matrix composed of poly(N-isopropylacrylamide)—coated beads. *Analytical Chemistry*, 2003, **75**, 2943–2949.

49. G. H. Seong, J. Heo, and R. M. Crooks, Measurement of enzyme kinetics using a continuous-flow microfluidic system. *Analytical Chemistry*, 2003, **75**, 3161–3167.

50. M. J. MacDonald, C. F. Chu, P. P. Guilloit, and K. M. Ng, A generalized Blake-Kozeny equation for multisized spherical particles. *AIChE Journal*, 1991, **37**, 1583–1588.

51. A. Manz, C. S. Effenhauser, N. Burggraf, D. J. Harrison, K. Seiler, and K. Fluri, Electroosmotic pumping and electrophoretic separations for miniaturized chemical analysis systems. *Journal of Micromechanics and Microengineering*, 1994, **4**, 257–265.

52. T. S. Stevens and H. J. Cortes, Electroosmotic propulsion of eluent through silica-based chromatographic media. *Analytical Chemistry*, 1983, **55**, 1365.

53. G. V. Sherbert, *The Biophysical Characterisation of the Cell Surface*, Academic Press, 1978.

54. R. C. Boltz and T. Y. Miller, *A Citrate Buffer System for Isoelectric Focusing and Electrophoresis of Living Mammalian Cells*, in *Electrophoresis '78*, N. Catsimpoolas (ed), Elsevier North Holland Biomedical Press (New York), 1978, 345–355.

55. L. E. Locascio, C. E. Perso, and C. S. Lee, Measurement of electroosmotic flow in plastic imprinted microfluid devices and the effect of protein adsorption on flow rate. *Journal of Chromatography A*, 1999, **857**, 275–284.

56. S. L. R. Barker, D. Ross, M. J. Tarlov, M. Gaitan, and L. E. Locascio, Control of flow direction in microfluidic devices with polyelectrolyte multilayers. *Analytical Chemistry*, 2000, **99A**, 5925–5929.

57. Y. Liu, J. C. Fanguy, J. M. Bledsoe, and C. S. Henry, Dynamic coating using polyelectrolyte multilayers for chemical control of electroosmotic flow in capillary electrophoresis microchips. *Analytical Chemistry*, 2000, **72**, 5939–5944.

58. S. Wang, C. E. Perso, and M. D. Morris, Effects of alkaline hydrolysis and dynamic coating on the electroosmotic flow in polymeric microfabricated channels. *Analytical Chemistry*, 2000, **72**, 1704–1706.

59. F. Bianchi, F. Wagner, P. Hoffmann, and H. H. Girault, Electroosmotic flow in composite microchannels and implications in microcapillary electrophoresis systems. *Analytical Chemistry*, 2001, **73**, 829–836.

60. R. F. Probstein, *Physicochemical Hydrodynamics*, 2nd Edn, Wiley-Interscience, 2003, 305–306.

61. P. M. Ball, Spreading it about. *Nature (London)*, 1989, **338**, 624.

62. K. Kaski, *Europhysics News*, 1995, **26**, 23.

63. C. S. Brazel and N. A. Peppas, Dimensionless analysis of swelling of hydrophilic glassy polymers with subsequent drug release from relaxing structures. *Biomaterials*, 1999, **20**, 721–732.

64. N. A. Peppas and R. W. Korsmeyer, *Dynamically Swelling Hydrogels in Controlled Release Applications*, in *Hydrogels in Medicine and Pharmacy*, N. A. Peppas (ed), CRC Press Boca Raton, 1987, Vol 3.

65. A. R. Berens and H. B. Hopfenberg, Diffusion and relaxation in glassy polymer powders: 2. Separation of diffusion and relaxation parameters. *Polymer*, 1978, **19**(5), 489–496.

66. S. J. Kim, K. J. Lee, I. Y. Kim, and S. I. Kim, Swelling kinetics of interpenetrating polymer hydrogels composed of poly(vinyl alcohol)/chitosan. *Macromolecular Science*, 2003, **A40**(5), 501–510.

67. D. Berger and D. C. T. Pei, Drying of hygroscopic capillary porous solids: a theoretical approach. *International Journal of Heat and Mass Transfer*, 1973, **16**, 293–302.

68. H. B. Hopfenberg and H. L. Frisch, Transport of organic micromolecules in amorphous polymers. *Polymer Letters*, 1969, **7**, 405–409.

69. S. Krishnamoorthy, V. B. Makhijani, M. Lei, M. G. Giridharan, and T. Tisone, *Computational Studies of Membrane-Based Test Formats*. In: *Technical Proceedings 2000 International Conference on Modeling and Simulation of Microsystems, MSM*, San Diego, CA, 2000, 590–593.

42

Practical Aspects of Microfluidic Devices: Moving Fluids and Building Devices

Bernhard H. Weigl,[1] Ron L. Bardell[2] and Catherine Cabrera[3]

[1] *Department of Bioengineering, University of Washington, Seattle, WA, USA,* [2] *MicroPlumbers Microsciences LLC, Seattle, WA and Minneapolis, MN, USA and* [3] *Biosensor and Molecular Technologies, MIT Lincoln Laboratory, Lexington, MA, USA*

1 MOVING FLUIDS IN A MICROFLUIDIC DEVICE

"Fluid" and "fluid movement" lie at the heart of microfluidics. Various types of fluid motion are required, including moving fluids in a continuous stream, dispensing a controlled bolus of liquid, and getting the fluid into the device from the macroscopic world in the first place. Each of these topics is discussed below.

Techniques for controlling the motion of fluids in microfluidic devices can be as simple as harnessing the hydrostatic pressure of a column of liquid in a tube placed above a port in the device or as complicated as a microfabricated multistage pump. A discussion of laminar flow in general and the theoretical basis for several means of generating laminar flow in a microfluidic device can be found in **Chapter 41, Introduction to Microfluidic Techniques**.

1.1 Dispensing Discrete Volumes

Many biomedical microfluidic applications require dispensing precise volumes of fluids, perhaps to another fluid, a dried reagent pellet, or to or from the outside world. Electroosmotic flow (EOF) can

be used to perform "valveless" switching to inject fixed amounts of a fluid into a flow stream.[1] By controlling the voltage and duration of the applied field, the fluid volume dispensed can be precisely and reproducibly controlled (see Figure 1).

Alternate approaches rely on pressure-driven flow to actuate fluid dispensation. Microscale fluid movers are small enough to be included on-chip in a microfluidic device, though the additional cost may not be justified in a single-use chip. They can be fabricated from a number of different materials, including silicon, brass, glass, and polymers, and usually function by changing volume, usually by flexing a wall. They typically have some type of valving to direct the fluid flow and are driven by electrical energy. They can be categorized in a number of different ways, by: purpose, output type, working fluid, and physical mechanism. We may call them micropumps, though not all of them conform to the usual concept of a machine, especially those in which the physical mechanism has no moving parts whatsoever.

1.2 Micropumps

Depending on their purpose, micropumps can be divided into *continuous-flow* pumps and *batch-flow* pumps. We can categorize micropumps by

Handbook of Biosensors and Biochips. Edited by Robert S. Marks, David C. Cullen, Isao Karube, Christopher R. Lowe and Howard H. Weetall.

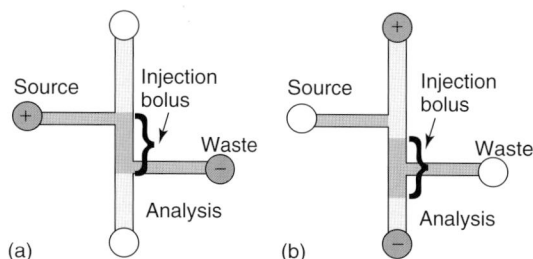

Figure 1. Electrokinetic injection of fluid sample. (a) Formation of injection bolus. (b) Switch voltage and direct bolus to analysis region.

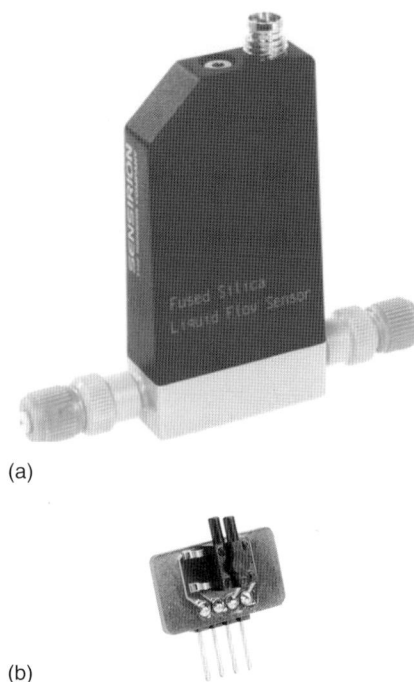

Figure 2. Liquid mass flow meter (a) for flow rates from $50 \, \text{nl s}^{-1}$ to $40 \, \mu\text{l s}^{-1}$ based on calorimetric sensing. Sensor body with control electronics is $35 \times 70 \, \text{mm}$ (Sensirion AG, Zürich, CH). (b) A differential pressure sensor for resistance-based flow metering. Sensor body is $6 \times 7 \times 8 \, \text{mm}$ (Honeywell Sensing and Control, Freeport, IL).

output type, either as a *displacement source* or as a *pressure source*. A displacement source changes the volume of the pump to "displace" a chosen amount of fluid; pressure is not controlled. For accurate delivery in an *open-loop* control system, the pump should be paused after fluid displacement until fluidic compliances (e.g., bubbles, flexible membranes) discharge and fluid motion ceases. A pressure source can provide continuous flow in an open-loop control system, but needs an integrated flow sensor in a *closed-loop* feedback system to deliver a precise volume of fluid. Flow-rate sensors (as shown in Figure 2) can be fluid-resistance based (viscosity dependent) or calorimetric (specific-heat dependent) and are calibrated for the particular fluid, but can self-adjust to temperature variation if temperature sensing is integrated.

If the sample is of unknown or variable composition (e.g., blood), one approach to achieve dispensed volume accuracy is to avoid pumping the actual fluid and instead pump a "pusher" fluid, such as water or Fluorinert™. Disadvantages include contamination and/or dilution of the sample as the sample/pusher interface is stretched by the parabolic velocity profile of pressure-based flow and interchange by diffusion occurs. This is ameliorated if the pusher fluid and sample are immiscible. Air can be used as a pusher fluid or as a small bubble separating pusher and sample fluids, but its compliance lengthens time constants for starting/stopping flow and, if dispensing pressure is low enough, surface tension effects (e.g., variation in capillary pressure at the liquid meniscus) may make control of fluid position difficult.

Micropumps can be separated into several categories by their physical mechanism: *electromechanical, pneumatic, capillary,* and *osmotic.* Most

of these are pressure sources, but electromechanical micropumps can be designed as displacement sources and pneumatic or capillary micropumps can function as displacement sources if the back-pressure opposing fluid motion is very small. Electromechanical micropumps form the most varied category, but in shear number of units the capillary type is most common due to the ubiquity of lateral flow strips.

1.3 Continuous-flow Micropumps

Electromechanical micropumps typically use electrical energy to expand and/or contract the pump body and use check valves to direct the resulting flow downstream. There are three types of check valves: *fixed, passive,* and *active. Fixed* valves have no moving parts. Their inner channel is shaped to dissipate more viscous energy in upstream than in downstream flow. Turning off a fixed-valve micropump does not stop the flow; a separate valve must be actuated. A major advantage is that they can pass particles nearly the size of their throat (typically> $50\,\mu$m) without wear.

Passive and *active* valves typically employ moving flaps that seal against the valve seat. The clearance between the seat and the flap is generally quite small and the seat and flap can be damaged by particles. The hydrodynamic forces moving the flap of a *passive* valve are small and can be overwhelmed by high fluid viscosity or surface tension, especially once the valve seat is wetted but the pump body contains air. An *active* valve can overcome surface tension at the expense of mechanical and control complexity.

A robust micropump is *self-priming* and *bubble tolerant*. Its compression ratio should be large enough to pump air against the upstream pressure drop. The most common actuator for electromechanical micropumps is a piezoelectric disk bonded to the flexible wall(s) of the pump body.[2–7] These are often referred to as membrane pumps. Figures 3 and 4 show examples of electromechanical micropumps using fixed valves and actuated by piezoelectric disks. Some designs indirectly couple the piezoelectric actuator to the flexible wall by means of a lever to produce proportionally larger motion[8,9] (Figure 5).

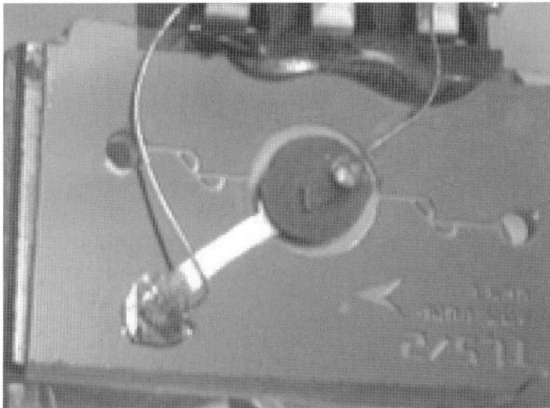

(a)

(b)

Figure 4. (a) Prototype piezoelectric-disk fixed-valve micropump in plastic with 12-mm-diameter chamber by Bartels Mikrotechnik GmbH, Dortmund. (b) Piezoelectrically activated diaphragm micropump for liquids and gases by thinXXS Microtechnology AG, Zweibruecken.

Figure 3. Piezoelectric-disk fixed-valve micropump with a 3-mm-diameter chamber and 0.114-mm-wide curly channels that operate as leaky check valves, providing a higher pressure drop in the reverse (left-to-right) than in the forward direction. Highest efficiency is achieved when the actuation frequency is set at the system resonance of the pump. [Reprinted with permission Morris and Forster[14] copyright 2003, IEEE.]

Another actuation method for electromechanical micropumps is electrostatic force obtained by applying opposite charges on parallel conducting plates, one of which is rigid and the other is a flexible wall of the pump body.[10,11] This electric field can cause electrokinesis of particles in the fluid being pumped, which may be an undesirable effect.

Electromechanical pumps that utilize electromagnetic actuators, similar to audio speaker technology, have also been developed.[12,13] Though bulkier than an electrostatic actuator or piezoelectric disk, this design offers low voltage–high current operation, which may be more appropriate for in vivo use.

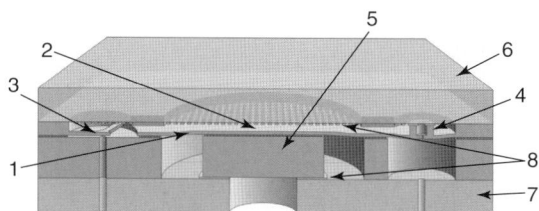

Figure 5. Piezoelectric-lever micropump with passive valves. Illustrated are pump membrane (1), chamber (2), inlet (3) and outlet (4) check valves, membrane stiffener block (5), top (6) and bottom (7) glass cover plates, and membrane-motion limiters (8). During operation, the membrane is pushed up against the top glass cover plate by a piezoelectric lever actuator (not shown) that pokes through the center hole and pushes against the membrane stiffener block, which is bonded to the membrane. [Reprinted with permission Maillefer et al.[9] copyright 1999, IEEE.]

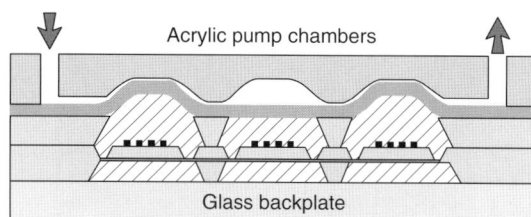

Figure 6. Illustration of actuation of electrothermal peristaltic pump. [After Grosjean.[15]] The working fluid (e.g., air) is contained in the crosshatched volumes below the membrane. Heaters in each volume operate in sequence to heat and expand the working fluid, pressing and sealing the membrane against the upper chamber surface and moving the pumped fluid from chamber to chamber. Arrows show pumped fluid direction.

Other novel electromechanical pumps have been developed, including an electrothermal peristaltic pump.[15] In this design, three sequential chambers are sequentially compressed by a flexible membrane as illustrated in Figure 6. Each chamber has its own heater. The movement of the membrane is accomplished by heating air to lower its density and increase its volume. The actuation frequency is low, but it can be operated as a self-priming pump.

1.4   Batch-flow Micropumps

All the electromechanical micropumps described thus far have been *continuous-flow* pumps. However, it is possible to develop *batch-flow* pumps, based on a chamber with one fluid connection and flexible walls that are actuated by any of the

electromechanical methods described above. They can be operated reversibly with fluid intake on volume expansion and fluid output during contraction. A ubiquitous form of batch pump is the printhead of ink-jet printers (e.g., Bio-Dot™), which are excellent at placing precise drops a sample solution on a solid or liquid surface.

Not all micropumps are electromechanical. Some examples include thermally induced bubbles.[16] For example, Liu et al. have developed batch-flow *pneumatic* micropumps,[17] which use electrothermal actuation. A second design is an electrochemical pump that produces gas by electrolysis. It is more efficient, but cannot be operated reversibly. A third design uses phase-change actuation, essentially boiling the working fluid.

A *capillary* micropump is a batch-flow, nonreversible device usually implemented as an absorbent pad of nitrocellulose as in a lateral flow strip. The fluid flow rate is controlled by the cross section of the pad, the duration of flow by its length.

When a single, very-low, flow rate (e.g., $1 \, ml \, h^{-1}$) is desired, an *osmotic* pump may be appropriate (e.g., ALZET® 1003D). These are essentially fluid-displacement pumps driven by chemical potential. They consist of an inner chamber containing the fluid to be pumped and an outer chamber with a rigid semipermeable outer wall and a flexible impermeable inner wall. On exposure to water, the concentrated salt solution in the outer chamber expands and the inner chamber is compressed, forcing out the pumped fluid.

1.5   Macro-to-micro Interface for Transferring Fluids

The interface between the macroscopic and microscopic worlds is a major design challenge of microfluidic devices.[18,19] The simplest solution is the open reservoir with pipette access to introduce or remove sample. Examples include microarrays and the Agilent LabChip® and 2100 Bioanalyzer system. Another approach is to epoxy plastic or glass tubing directly to the device, but machining holes in glass or silicon, positioning the tube over the hole, or selecting an epoxy that will bond to a silicone polymer without clogging the hole make this approach nonoptimal. A better approach is to insert the microfluidic chip into a void within a

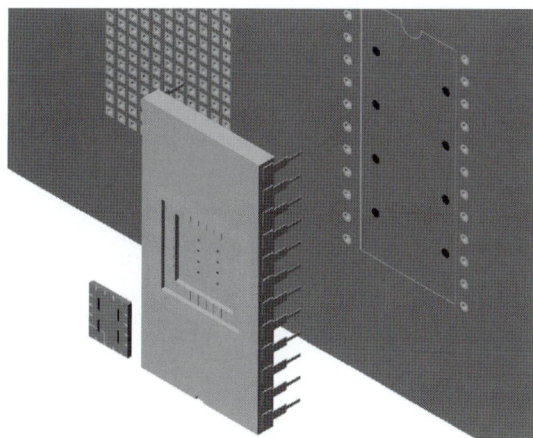

(a)

(b)

Figure 7. (a) An illustration of a combined electromicrofluidic packaging architecture. The fluidic printed wiring board (FPWB), the largest object, has fluidic connections (the black holes) as well as electrical connections. The electromicrofluidic dual in-line package plugs into the FPWB. The microfluidic integrated circuit (MIC) is the smallest object shown.[20] (b) A pneumatic–microfluidic manifold that connects air and liquid lines to inlet/outlet ports on the edge of a clear plastic laminate microfluidic card that is held along that edge by a spring-loaded clamp. The air lines operate pneumatically actuated valves on the card to direct the fluid flow (Micronics, Inc., Redmond, WA).

soft lithography or plastic laminate structure that provides a macro–micro interface, such as wells that align with access holes in the chip.

Once the fluid connection positions become fixed in a final microdevice design, a standardized layout fixture that supplies multiple interfaces: fluidic, electrical, vacuum, and hydraulic connections between microchip and reader instrument is advantageous. An example is the Caliper sipper chip that has a glass capillary attached to the single fluid inlet port. Figure 7 shows two more examples.

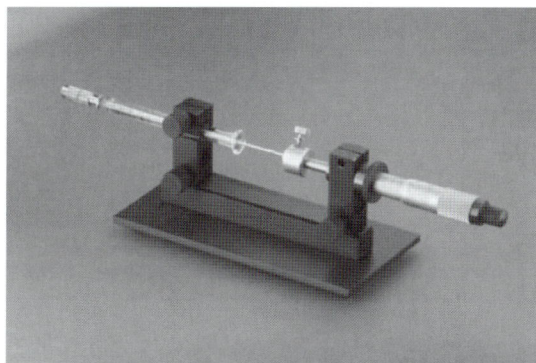

Figure 8. Manual microsyringe pump (Stoelting Co., Wood Dale, IL).

A more automated fluid delivery system would load only the sample fluid by pipette and load all reagents from off-card syringe pumps through small-bore rigid tubing. Figure 8 shows a syringe pump that, while manual, is inexpensive and accurate. Automatic computer-controlled dispensing from multiple syringe pumps is available with pump modules like those shown in Figures 9 and 10.

2 MICROCONSTRUCTION TECHNIQUES AND DEVICE EXAMPLES

Fabrication of prototype microdevices is more like watchmaking than like automobile engine repair. Features that are $10\,\mu m$ wide are barely visible to the unaided eye; they look like scratches. Electrical and fluid inlet and outlet connections are often susceptible to damage. Dust control is essential, usually by deploying ionizers at the assembly bench; it is generally impossible to remove dust from a completed structure, even with compressed air. Achieving uniform surface energy throughout the internal passages of the microdevice is often critical to proper filling of the channels with a polar liquid or aqueous solution. Care must be taken to avoid resting microdevice parts on any surface that has molecules eager to migrate to the higher-energy surfaces of your microdevice, thus lowering the surface energy and making wetting more difficult.

There is a wide range of technologies available for constructing microdevices: *traditional lithography, soft lithography*, or *machining* of plastics,

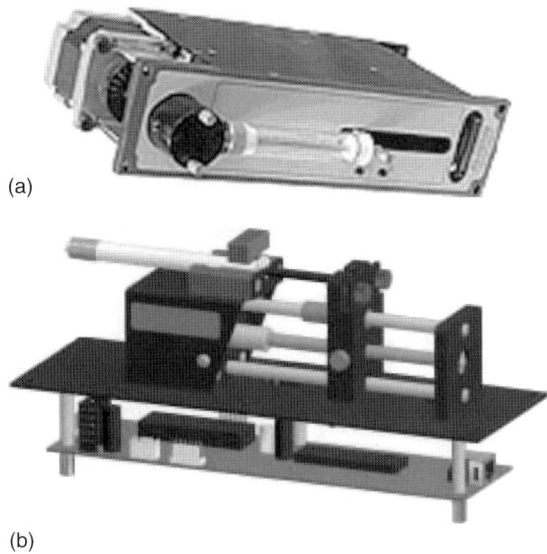

(a)

(b)

Figure 9. Syringe pump modules that can deliver microliter fluid volumes. (Kloehn, Ltd, Las Vegas, NV (a), Harvard Apparatus, Inc., Holliston, MA (b).)

metals, or glass by laser ablation or computer-numerical-control (CNC) mill to produce finished parts or primary molds. Polymer films can be laser-cut, knife-cut, or stamp-cut with patterns that when stacked, aligned, and bonded together create a three-dimensional *laminate* structure that can contain fluidic channels, pneumatically activated valves, flex circuits to support sensors, porous membranes, and so on. Almost any object can be positioned within a molded part or a laminate.

2.1 Traditional Lithographic Techniques

Microelectromechanical systems (MEMS), of which microfluidics is a subcategory, began as a offshoot of the computer chip processing industry, in which hard substrates, most commonly silicon, are the primary construction material. The basic paradigm of standard lithographic techniques involves the use of electromagnetic radiation, typically ultraviolet (UV) light, to transfer a pattern to a light-sensitive polymer (also known as *photoresist*), which undergoes a chemical response upon exposure to the radiation (see Figures 11–13). The exposed areas are defined by means of a photomask, usually consisting

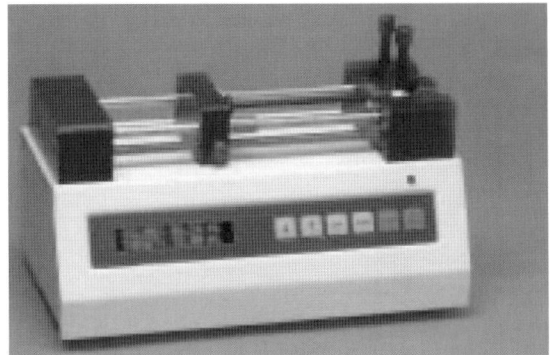

(a)

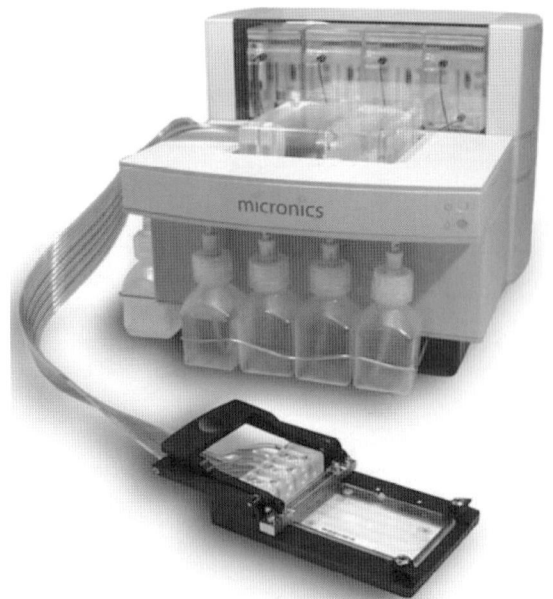

(b)

Figure 10. Complete microsyringe pump systems. If multiple fluids can be injected simultaneously at the same flow rate into your microfluidic device, the system in (a) offers complete control of flow rates and volumes (Harvard Apparatus, Inc., Holliston, MA). If up to four fluids need to be controlled independently, the device in (b) controls flow rates, volumes, and timing, as well as automatic reloading of the syringes from supply bottles (Micronics, Inc., Redmond, WA). It also offers multiple pneumatic lines that can be switched independently between positive, negative, and ambient pressure. These are used to control the pneumatic liquid-control valves on the disposable plastic laminate cards that fit in the manifold (foreground).

of a patterned thin chromium film on a quartz plate; areas to be exposed are defined by open areas in the mask plate. After light exposure, the polymer-coated disk must be developed, similar to developing a photograph, such that the areas exposed to the electromagnetic radiation behave in an opposite manner to unexposed areas; one set of areas polymerizes and remains on the surface while the other set is washed away. The item being machined then undergoes either material removal (e.g., etching of exposed substrate) or material addition (e.g., metal deposition) Finally, the remaining photoresist is removed, typically with a solvent, and the substrate is ready for the next round of machining.

Traditional lithography methods are amenable to parallel processing and can produce mechanically strong and chemically resistant devices, with feature sizes as small as 0.1 μm. However, these methods are also expensive, require significant chemical processing equipment outlays, and often require toxic chemicals. In addition, the processes are often time consuming and not amenable to rapid design iterations. New methods of MEMS fabrication, categorized as "soft lithography", address many of these concerns. For many BioMEMS applications, traditional lithographic techniques may be most applicable for creating a stamp or mold that is then used to produce large numbers of devices via soft lithography.

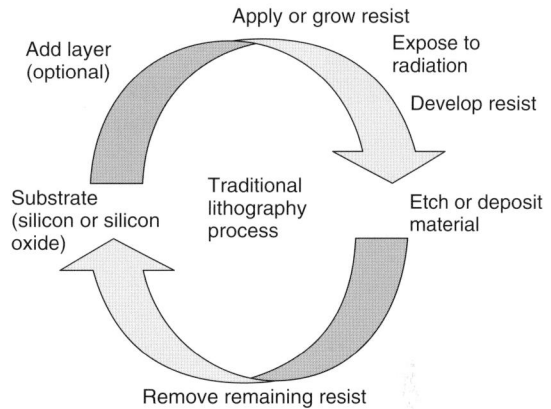

Figure 11. General process of traditional lithography.

In that context, there are several common modifications to traditional lithography when used for bioMEMS applications.

Given the relatively large feature size required for bioMEMS devices, the mask that contains the pattern can be as simple as an overhead transparency on which the design has been drawn by hand. The desired feature size is the single most important consideration in selecting a mask and illumination source, with budget and time

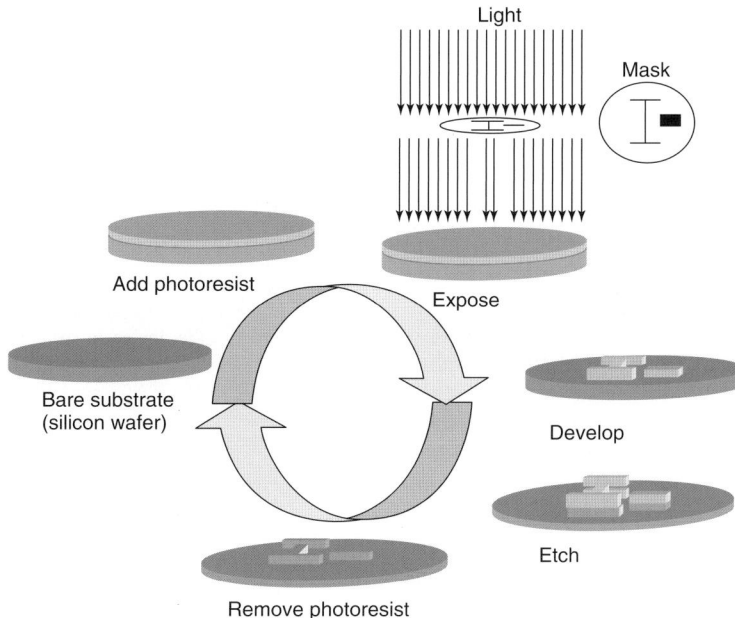

Figure 12. The photolithographic process, using positive photoresist followed by an etch step, is shown.

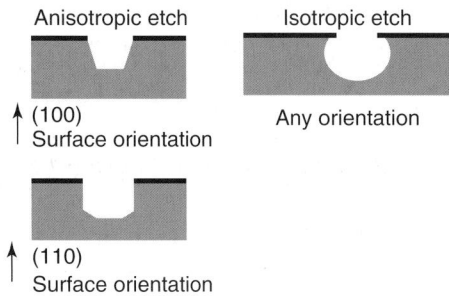

Figure 13. Etch patterns based on silicon type.

considerations coming a close second. Typically ultraviolet (UV) light is used, although electromagnetic waves with narrower wavelengths, such as X rays, have been used to achieve a finer resolution.

In traditional photolithography, the photoresist layer thickness is typically on the order of 1 μm and is completely removed by the end of a micromachining cycle. In contrast, by using special photoresists at a thickness on the order of hundreds of microns, the photoresist itself can be patterned and used as a mold. For example, SU-8 photoresist can be layered up to 450 μm thickness and can be used to achieve aspect ratios of 15:1.[21] SU-8 components can be used as molds for soft lithography or assembled together to form a microfluidic device. A complete integrated microfluidic system for mass spectrometry has been constructed using SU-8 to form multiple layers of channels sandwiched between silicon and Pyrex wafers.[22]

Two excellent sources of additional information on traditional lithography are *VLSI Technology*[23] and *Introduction to Microelectronic Fabrication*.[24]

2.2   Soft Lithography

Though well-developed, traditional lithographic techniques are expensive and require a relatively elaborate fabrication facility. Soft lithography attempts to address these concerns and offer additional flexibility, such as the ability to work with nonplanar surfaces and to use primary molds that are either positive (i.e., the mold is the same shape as the final part) or negative (i.e., the void in the mold is the same shape as the final part).

A foundation paper with many useful references is from Xia and Whitesides.[25] They define soft

lithography as "an elastomeric stamp with patterned relief structures on its surface that is used to generate patterns" and claim reproduction of feature sizes as small as 30 nm and up to 500 μm. A commonly used elastomer is a silicone rubber, polydimethylsiloxane (PDMS), but other elastomers are available, such as polyurethanes, polyimides, and phenol formaldehyde polymers. They describe a variety of soft lithography techniques: cast molding, replica molding (REM), microcontact printing (μCP), microtransfer molding (μTM), micromolding in capillaries (MIMIC), and solvent-assisted micromolding (SAMIM), but the most commonly used techniques are injection molding and embossing. In hot embossing (see Figure 14) a hard negative mold is pressed into a heat-softened thermoplastic polymer, which on cooling retains the pattern. In soft embossing a flexible rubber mold is filled with hot polymer under pressure (see Figure 15).

Unlike hard polymers, the deformability (elasticity) of soft polymers like PDMS can be problematic for accurate registration between parts and limits the aspect ratio of features to between 0.2 and 2.0 to ensure structural integrity. A useful reference for practical tips is *My Little Guide to Soft Lithography* (or *Soft Lithography for Dummies*) from Krogh and Åsberg.[27]

Commonly used materials for primary molds are SU-8, a negative photoresist, which can be patterned using the standard lithographic processes[21]

Figure 14. SEM image of pyramids with a 30-μm base width fabricated in PMMA by hot embossing with a silicon mask formed by standard lithography with wet-chemical etching.[26] [Reprinted with permission Lin et al.[26] copyright 1998, Springer Verlag.]

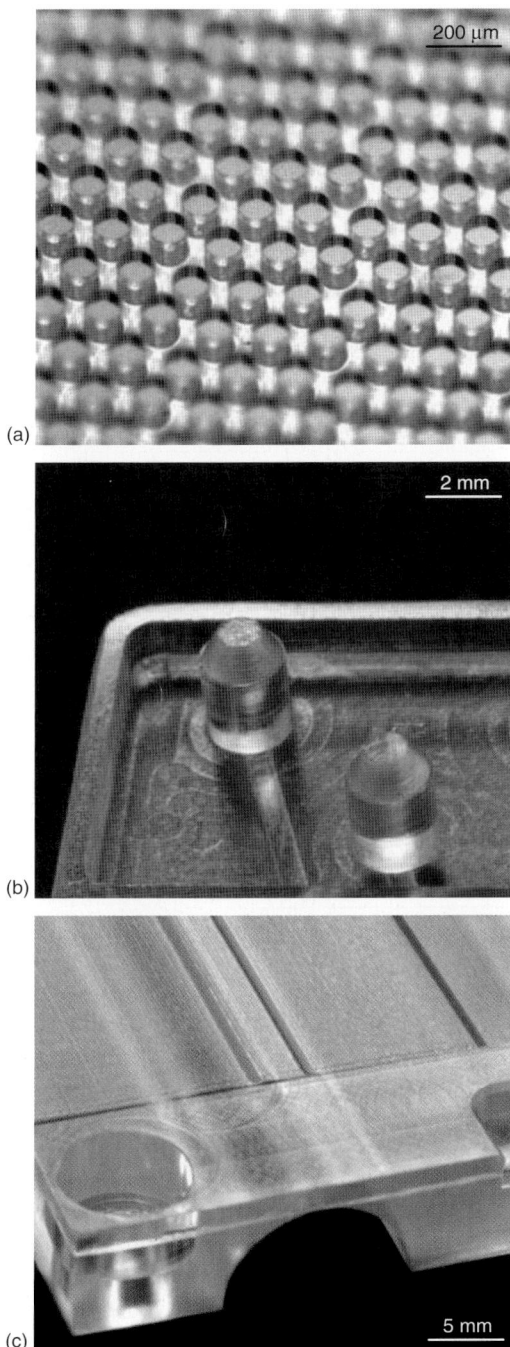

(a)

(b)

(c)

Figure 15. (a) Soft embossing of heated plastic resins in a hard rubber mold can produce high aspect ratio (height/width) features, such as these micropillars; (b) vertical features such as these posts do not need the draft angle that is required in injection molding since the mold can flex when pulling the part; (c) features can be placed on both sides of the part. [Photos courtesy of Edge Embossing LLC, Medford, MA.]

and used as the primary mold in any of the soft lithography processes; and high-temperature epoxy, which can be used as a stamp for hot embossing.[28] Mold wear can be reduced by making parts from negative secondary molds that were formed from a primary positive mold.

2.3 Machining, Cutting, and Laser Ablation

Tools for direct shaping of materials for microdevice constructions include lasers, knife plotters, and miniature CNC milling machines. All use a software "mask" that can be easily altered. Miniature CNC mills (e.g., Sherline 5400 CNC Mill with tolerance $\sim25\mu$m, Taig CNC Mill with tolerance $\sim15\mu$m) are low-cost (US $2000) means to shape metals, glass, silicon, polymers, and plastic sheets with small-diameter diamond-tipped bits (e.g., 50, 75, 100, 125 μm diameter). Figure 16 shows an example of machined acrylic sheet. Small knife plotters (e.g., SummaCut D60 vinyl cutter with tolerance $\sim25\mu$m) are available for a similar price to cut plastic films up to 0.030 in. (30 mils, 800 μm) thick.

There are several types of lasers capable of shaping parts for microdevices. Lasers operating at IR or visible light wavelengths provide lower resolution and remove material by melting, vaporization, and pyrolysis (thermal decomposition of chemical bonds). Lasers that operate in the UV range provide high-resolution photolytic (direct decomposition of chemical bonds due to absorption of single or multiple photons) micromachining. At high power density, both pyrolytic and photolytic decomposition create a rapid rise in pressure and temperature that ejects material in a process called *ablation*. The energy is absorbed into a depth of the material, depending on the first-order absorption cross section of the material. The absorbed energy is converted to heat with a temperature profile that decays exponentially into the surrounding material. The material directly exposed to the radiation heats and vaporizes rapidly, ejecting gas and particles while cooling the surrounding material. As the laser is pulsed, it moves in steps to adjacent uncut material, and the laser is pulsed again. The laser repetition rate, the pulse width, and the pulses per inch all affect the quality of the cut. (L. Levine, private communication.[30]) Figure 17 compares cut quality

Figure 16. Close-up of CNC-machined fluidic prototype device.[29] Machining is typically a more time-intensive process than laser ablation, but the edges of the machined features are sharper and the substrate material properties have not been altered by the high temperature characteristic of the ablation process.

Laser type	CO$_2$	Tripled YAG	KrF excimer
Cut quality	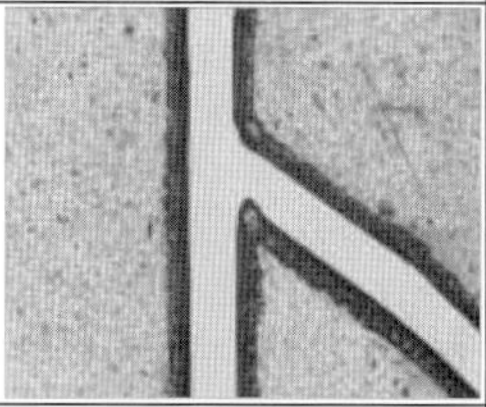	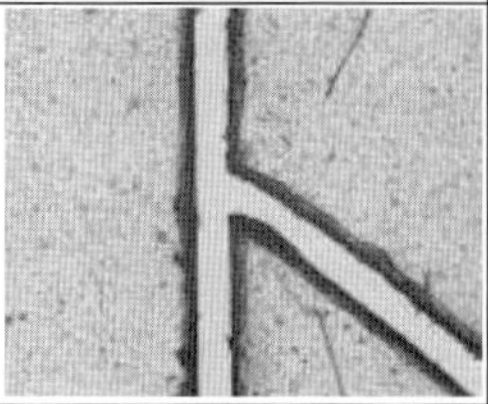	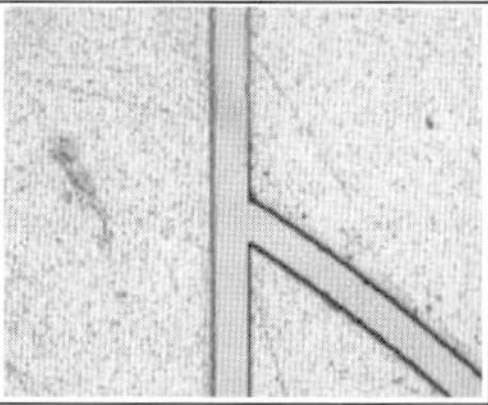
Cutting time (s)	0.7	18	>4000

Figure 17. Trade-off between cut quality (a function of laser type, power, and spot size) and cutting time with Mylar as the substrate. [From Photomachining Inc.]

Table 1. Comparison of wavelengths and spot sizes of various lasers. The smaller spot sizes give increased precision, but also require increased processing time

Laser type	CO$_2$	YAG	Tripled YAG	KrF excimer
Color	IR	Near IR	Near UV	UV
Wavelength (μm)	10.6	1.06	0.36	0.25
Approximate spot size (μm)	120	12	4	3

From Laserod Inc.

and cutting time of different laser types when set up for high speed cutting. Table 1 compares the wavelengths and spot sizes of different types of lasers.

CO$_2$ lasers (e.g., Universal Laser M-360) can cut or engrave any of a wide variety of materials, such as: aluminum, brass, titanium, stainless steel, glass, silicon, and polymer or plastic sheet. The Nd:YAG (neodymium:yttrium–aluminum–garnet) laser is typically used to cut or weld metals. The tripled YAG can also cut some polymers, such as: Kapton, polycarbonate, and polyimide. An excimer laser is compatible with many polymers, including: fluorinated ethylene propylene (FEP), Kapton, parylene, polyethylene terephthalate (PET, Mylar),

polymethyl methacrylate (PMMA, acrylic), polycarbonate, polyester, polyethylene, polyimide, polyurethane, polyvinyl alcohol (PVA), and Teflon. Table 2 lists the preferred sizes and unique properties of the most commonly used materials.

Cut edge quality depends on laser type. The ablation process of the UV lasers can produce a very clean edge, while the IR lasers tend to melt, instead of vaporize, the material.

2.4 Laminate Technologies

A very useful and adaptable construction technique for microfluidic devices is the concept of building

Table 2. Commonly available plastic sheet and film stock from suppliers/distributors: McMaster-Carr, CS Hyde Industries, and Sheffield Plastics

Material	Available thickness (μm)	Unique properties
Polyester (PET or Mylar)	12.5, 25, 50, 75, 100, 125, 250	Okay optical properties, autofluorescence, most widely available in a variety of film stocks
Polycarbonate	75, 125, 250	Good optical quality, low fluorescence
Acrylic, clear (cast)	500, 1000, 1250, 1500	Good optical quality, thickness highly variable, can come in UV grade transparent to 350 nm
Acrylic, clear (extruded)	1000, 1500	Good optical quality, grazes readily with exposure to even dilute alcohol
Acrylic, black (extruded)	250, 500, 1000, 1500	
Polyimide	25, 50, 75	Orange color, often thermally bonded at high temperature, not very suitable for CO_2 laser
COP	50, 125, 250	Thickness variability around 20%
Silicone	125, 250, 500, 1000	Makes dust when cutting, requires air—assist to avoid flaming
Polypropylene (clear)	45	
Polypropylene (hazy)	250, 500, 1500	Usually low-quality material
Polyethylene (LMW, HMW, or UHMW)		Readily available in food grade
PVDF (clear)	50, 75	
FEP (clear)	50, 75	
Urethane	50, 75+	
Acetal (white, black)	250, 500, 1000, 1500	

COP: cyclic olefin polymer.
LMW: low molecular weight.
HMW: high molecular weight.
UHMW: ultrahigh molecular weight.
PVDF: polyvinylidene fluoride.

up a device by stacking layers, each of which has its own planar pattern of channels and chambers to hold fluids or simply holes (vias) to allow fluid communication between neighboring layers. (This stacking technique was intensively developed by engineers in the 1950s and 1960s for fluidic amplifiers in the flight control circuits of jet aircraft.)

A basic microfluidic device can comprise a primary layer into which a pattern defining the fluid channels has been etched, machined, cast, or embossed and a second layer that serves as a cap to close off the channels. The layer materials can be any combination of silicon, glass, metals, polymers, or plastic films. If the pattern passes completely through the primary layer, then a third layer is needed as a capping layer. The choice of two-layer or three-layer construction depends mainly on the choice of pattern-making method. Adding additional layers allows overlapping channels and complex features like valves, mixing structures, and pumps, or the opportunity for easy parallelization.

2.5 Plastic Film and Sheet Stock (L. Levine, private communication)

Thinner materials are classified as films; those thicker than 0.010 in. are generally considered sheets. Other than custom stock, few materials are available in thicknesses between 0.010 and 0.015 in. Sheet materials up to 0.080 in. thick can be readily handled for laser cutting and lamination. Film thickness can be defined either by gauge (i.e., the thickness in inches multiplied by 100), or by "mil" thickness (i.e., 1 mil is equivalent to 0.001 in. = 25.4 μm and 40 mils (0.040 in.) = 1 mm). The thickness tolerance of film stock is generally 5–10%. Extruded sheet stock is similar, but the thickness tolerance of cast sheet stock can be as much as 25% of material thickness (e.g., cast acrylic).

Other material characteristics of potential importance are the addition of flame retardants, typically brominated compounds, that may become chemical interferents in an assay and the autofluorescence of the material that may interfere with high-sensitivity fluorescence detection. Materials specified as medical or food grade are often the better choice.

The most widely available material is polyester film (i.e., PET, Mylar®). Widely used in the graphics industry, it can be purchased with surface treatments that enhance the adhesion of aqueous-based inks (usually corona treated), and is available in heat-sealable grades with a thin layer of low molecular weight (MW) polymer, usually a polyolefin or a polyvinylidene dichloride/PVA, or oligomers of PET itself.

2.6 Layer Bonding

The appropriate bonding method depends on the type of bond (permanent or reversible), the tolerance of the material to high temperatures, the deformability of the material, and solubility of the bonding agent to the fluids of interest (e.g., water, solvents). Some fabrication technologies, such as injection molding, plastic-film laminates, and soft lithography, have their own native bonding methods.

Plastic-film laminates are typically fabricated by alternating layers of plastic film or rigid sheet with pressure-sensitive adhesive (PSA) film. For increased durability during assembly, the adhesive film is often a composite structure of a polyester carrier film coated with PSA on both sides, essentially double-sided sticky tape. One drawback is that, if fluid channels are cut in one of the layers to be bonded, or in the PSA film itself, the adhesive will be exposed to the fluid, which introduces a potential incompatibility to the use of organic solvents or the long-term storage of aqueous solution. Figure 18 shows a stack-up of layers to form a microfluidic card.

Injection-molded parts are usually assembled by solvent bonding, diffusion bonding, ultrasonic bonding, or laser bonding, though PSA can be used as well.

For devices implemented in PDMS by soft lithography, a low-temperature permanent bond is obtained by treating the bonding surfaces with an oxygen plasma to raise their surface energy. If the surfaces are placed in contact with each other within 3 min, a permanent bond is formed between the layers of PDMS (a rubbery transparent polymer) and glass, silicon, or itself. If a reversible bond to glass is desired, the glass surface should be cleaned with isopropyl alcohol (IPA) to improve adhesion. A PSA film or double-stick tape (e.g., Adhesives Research, ARcare 7841) can also be used to bond PDMS to many materials, such as metals or printed-circuit boards.

A reversible bond can be created between rigid materials that can tolerate a temperature of 100 °C, including metal, glass, and ceramic, by using a wafer-mounting wax, such as Crystalbond 509, which is transparent in thin sections, dissolves in acetone, and, though it has a flow point at 120 °C, is workable at 100 °C. No bonding agent at all is necessary if the layers have optically smooth surfaces; they can be pressed together mechanically in a jig to form a watertight seal if they are completely dry when assembled.

Permanent bonds can be formed with a low-viscosity epoxy (e.g., $\mu < 100$ cP), which will wick over the smooth surface between parts, but be prevented from filling wider gaps by surface tension. Silicon and glass can be permanently bonded by anodic bonding, in which a high temperature (~ 400 °C) and a DC electrical potential ($250 < V < 1000$) cause sodium ions to migrate across the boundary from the glass to create a permanent electrostatic bond. Silicon can be bonded to silicon at lower temperatures by using an intermediate layer of sputtered lithium borosilicate.[31]

2.7 Surface Modification

In materials with naturally hydrophilic surfaces, channels that are designed to carry liquids are easily wetted if their width-to-height aspect ratio is near unity. Variations in surface energy will affect the wetting speed as the liquid proceeds through the channel, but the channel will wet as long as the meniscus contact angle remains less than 90°. However, even slight variations in surface energy between different locations along the wall can make wetting of channels with aspect ratios greater than 5 quite difficult (see Figure 19).

There are several ways to achieve uniform surface energy. Sheet PET can be purchased with a hydrophilic surface coating, but manufacturing procedures are needed to ensure that the material does not contact any surface that will change its surface energy, such as a surface shedding hydrophobic particles or contaminated with skin

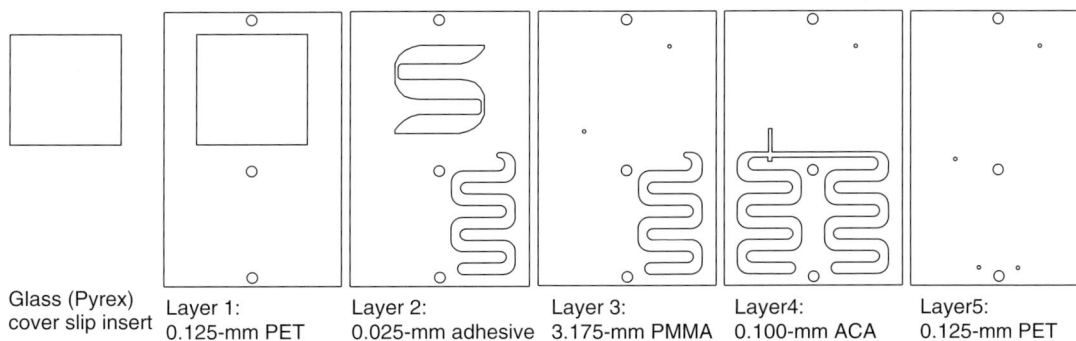

Glass (Pyrex) cover slip insert | Layer 1: 0.125-mm PET | Layer 2: 0.025-mm adhesive | Layer 3: 3.175-mm PMMA | Layer4: 0.100-mm ACA | Layer5: 0.125-mm PET

Figure 18. Layout of a microdevice built up from five plastic laminate layers and a glass insert. Layers 1 and 5 are polyethylene terephthalate (PET), layer 2 is a pressure-sensitive adhesive (PSA), layer 3 is polymethyl methacrylate (PMMA), and layer 4 is a sandwich of PSA with a PET center layer.

oils. These cautions should also be followed during assembly of parts that have undergone plasma treatment. Both oxygen and fluorine plasmas can create a temporary uniform hydrophilic surface. A direct-current corona torch forms hydroxyl groups on the surface of PET that, even after several days, can halve the contact angle of aqueous solutions. Another method to achieve uniform hydrophilic surfaces is to use stable coatings of hydrophilic polymers (e.g., AST HydroLast) of the channel walls after device assembly.

2.8 Integration of Heterogeneous Materials

It is possible to integrate a wide variety of materials as inserts into engineered voids in microdevices. Calculated tolerance stackups of inserts and part thicknesses should ensure that the insert is thinner than the void into which it fits. Any fluid inlet and outlet ports to the insert should be located on one side of the insert and pressed against or bonded to a mating surface of the surrounding part.

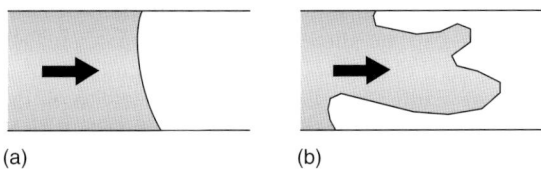

(a)   (b)

Figure 19. Wetting of two high width-to-height aspect ratio channels with hydrophilic walls. Channel walls in (a) have uniform surface energy; channel walls in (b) have variation in surface energy and, even though they are hydrophilic, air pockets are likely to form.

Thin-film membranes, proton-exchange membranes, and membranes for ultrafiltration, nanofiltration, and microfiltration can be integrated into a plastic-laminate or soft lithography system (see Figure 20). Applications such as ligand assays, enzymatic processes sensitive to trace metals, electrophoresis, degassing of aqueous solutions, and fluorometry are enabled in this way.

Microdevices, such as lab-on-a-chip type devices, typically employ optically clear materials, but the feature size is small. Spherical or cylindrical lenses can be integrated to enable visual inspection with the unaided eye.

Electronics can be integrated as polymer thick-film (PTF) flexible circuits (i.e., flex circuits), which are screen-printed thick-film conductive inks on a low-cost polyester dielectric substrate. Multilayer circuits are produced with dielectric materials as insulating layers, and double-sided circuits with printed through-hole technologies. Lead-free, silver-loaded isotropic conductive adhesive provides both electrical and mechanical connection of active and passive surface-mounted components for applications such as optoelectronics, electrokinetics, and liquid crystal displays.

Nanogen has been a leader in the field of integrating polymeric materials and IC chips for biological applications. In a collaboration with Genoptix and UC Irvine, researchers developed a microfluidic device that successfully isolated bacteria from a spiked blood sample and provides an excellent example of a heterogeneous microfluidic device, consisting of a variety of materials each machined in a different way (see Figure 21).[32] The bottom layer, made of polycarbonate, contains

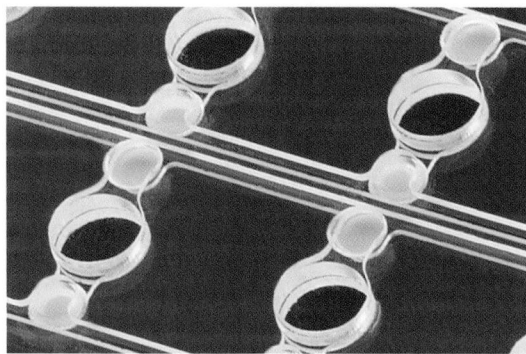

(a)

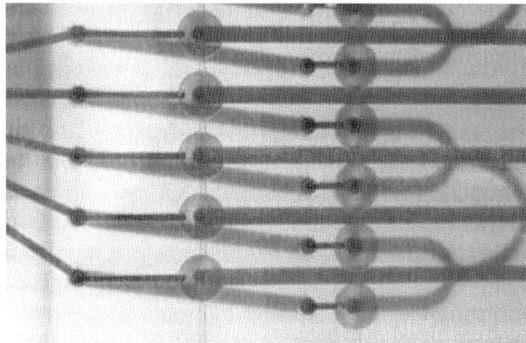

(b)

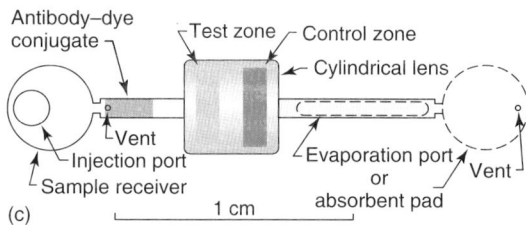

(c)

Figure 20. A wide variety of materials can be integrated into microdevices: (a) porous membranes (the white circles) can sequester live bacteria in neighboring wells (NASA GeneSat card); (b) elastomers (the gray disks) can form the flexible membranes of pneumatically actuated valves (ALine, Inc., Redondo Beach, CA); (c) a cylindrical lens inserted in a micro lateral flow strip device to magnify test and control zones for viewing with the unaided eye (MicroPlumbers Microsciences LLC, Seattle, WA).

machined fluidic channels. The next layer, a PSA, adheres the bottom layer to the heart of the device, a dielectrophoresis (DEP) chip. The DEP layer is itself heterogeneous, consisting of a flexible polyimide layer patterned with electrical circuits and a silicon chip with electrodes fabricated using repeated cycles of traditional UV photolithography

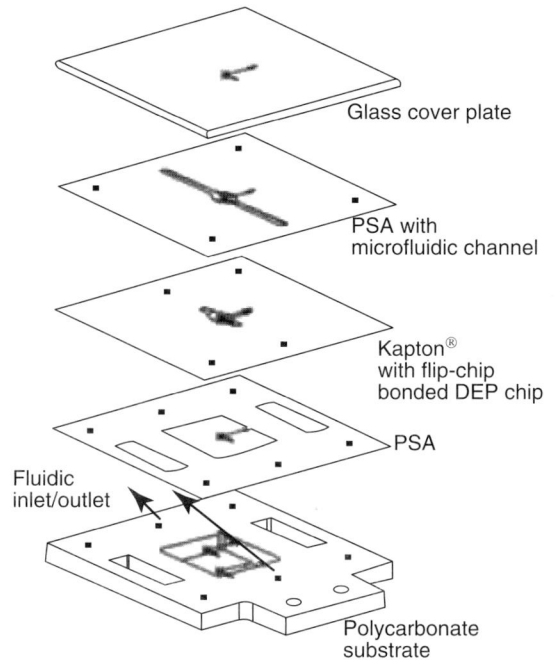

Figure 21. Schematic of a heterogeneous microfluidic device used to extract bacteria from a blood sample. [Reprinted with permission Huang et al.[32] copyright 2003, Springer Science and Business media.]

and sputtered deposition of a titanium–tungsten adhesion layer followed by deposition of a platinum layer. The electrodes are $50\,\mu m$ squares, spaced $50\,\mu m$ apart. The DEP chip is attached to the polyimide layer via "bump" bonding with silver epoxy. To complete the device, a glass cover plate is sealed to the top by a second PSA layer with cutouts for fluidic channels.

2.9 Immobilization of Biological Material

There are a number of ways to immobilize bioactive molecules on solid substrates in microdevices. Proteins can be directly adsorbed on a surface, but surface hydrophobicity, charge, and chemical makeup can affect both their stability and orientation. Protein–material interactions may result in decreased protein activity and nonspecific protein adsorption can change the intended biological activity.

Another approach, in addition to adsorption and covalent coupling, is tethering with an intermediate linker molecule such as polyethylene oxide (PEO) chains that reduce nonspecific protein adsorption and denaturing of the target protein that is caused by interactions with the substrate. Compared to direct adsorption, tethered proteins have greater mobility, which avoids steric hindrance of binding processes and allows clustering of ligand-bound receptors within the cell membrane, which is known to be a requirement for activation of some intracellular signaling pathways. Substrate materials include polystyrene, PMMA, polyurethane, polyamide, and hydrophobized glass.

Mesoporous silica can be integrated into a plastic-laminate or soft lithography structure. With negatively charged functional groups on its surface, a favorable chemical environment is created for proteins. The Pacific Northwest National Laboratory (PNNL) has demonstrated that high concentrations of an active enzyme can be immobilized in a mesoporous, functionalized silica structure while exhibiting higher activity than they would in aqueous solution.

DNA can also be immobilized onto a surface, most commonly in the form of a DNA chip used for hybridization microarray assays. DNA chips can be functionalized in two different ways. Various DNA probes can be synthesized off-chip and applied to the chip surface using a linker chemistry to immobilize the DNA. Alternatively, the DNA can be synthesized in situ, base pair by base pair, an approach commercialized by Affymetrix. Photolithographic techniques are used to selectively protect and expose different regions of the chip, typically 18- to 20-μm squares. A solution containing a single deoxynucleotide (A, C, T, or G) linked to a removable protection group is then washed over the entire chip. Exposed squares participate in linking reactions, thus extending one base pair, while the rest of the chip remains unmodified.

2.10 Separation Matrices

Hydrogels can be cast into wells in plastic-laminate and soft lithography structures. This includes agarose gel separation matrix designed for the separation of nucleic acid (NA) fragments. Applications include separation of base pairs ranging in length from 100 to 1200 and separation of polymerase chain reaction (PCR) products.

3 PUTTING IT ALL TOGETHER – AN EXAMPLE OF THE DEVELOPMENT OF A FULLY INTEGRATED MICROFLUIDIC DEVICE

One of us (Weigl), with collaborators (collaborators are Micronics, Inc, for microfluidic card development, Yager Group, University of Washington, for dry-down reagent development, and Dr. Tarr, Washington University, for clinical validation and support), is developing a lab-on-a-card platform to identify enteric bacterial pathogens in patients presenting with acute diarrhea, with special reference to infections that might be encountered in developing countries.[33,34] Component functions that are integrated on this platform include on-chip capture and lysing of pathogens, multiplexed NA amplification and on-chip detection, sample processing to support direct use of clinical specimens, and dry reagent storage and handling. All microfluidic functions are contained on the lab card. This new diagnostic test will be able to rapidly identify and differentiate *Shigella dysenteriae* serotype 1, Shigella toxin–producing *Escherichia coli, E. coli* 0157, *Campylobacter jejuni*, and *Salmonella* and *Shigella* species.

The multiplex disposable enteric card (DEC) test (see schematic in Figure 22) will be an automated, rapid, easy-to-use, point-of-care platform to simultaneously detect multiple enteric pathogens causing disease with similar symptoms. It will provide a mechanism for accurate, point-of-care diagnosis with a rapid turnaround time for results. The method is based on laminate microfluidic lab card technology developed at Micronics and the University of Washington. This technology has been used in many different applications ranging from diffusion-based separation and detection to projects involving flow cytometry on a chip and NA–based amplification and detection techniques.

The individual microfluidic subcircuits of the DEC card were initially designed and validated with both pathogen isolates as well as stool samples before integration of the subcircuits into a single disposable unit. The subcircuits are (i) capture and lysis of pathogens, (ii) NA extraction, (iii) NA amplification, and (iv) visual detection of amplified NA.

A subcircuit card that can purify NAs from lysed leukocytes or bacteria (Figure 23) is loaded with specimen in a lysis solution, which allows RNA

Feces extract

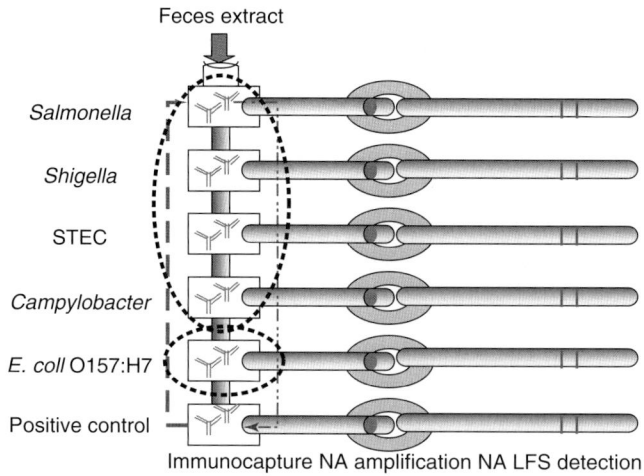

Salmonella

Shigella

STEC

Campylobacter

E. coli O157:H7

Positive control

Immunocapture NA amplification NA LFS detection

Figure 22. Schematic of DEC approach showing a combination of pathogen capture and lysis, nucleic acid extraction, PCR, and visual detection of amplicons.

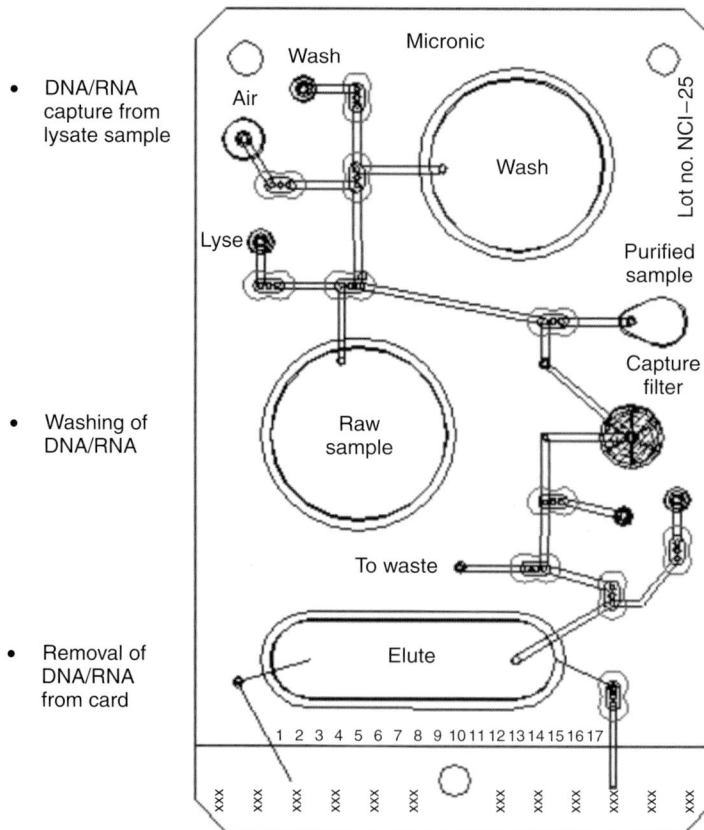

- DNA/RNA capture from lysate sample

- Washing of DNA/RNA

- Removal of DNA/RNA from card

Micronic

Wash

Air

Lot no. NCI−25

Wash

Lyse

Purified sample

Capture filter

Raw sample

To waste

Elute

1 2 3 4 5 6 7 8 9 10 11 12 13 14 15 16 17

XXX XXX XXX XXX XXX XXX   XXX XXX XXX XXX XXX XXX

Figure 23. Credit-card-sized microfluidic lab card that automates NA extraction.

Rapid PCR amplification
breadboard and lab card
8 min for 35 cycles (nonoptimized)

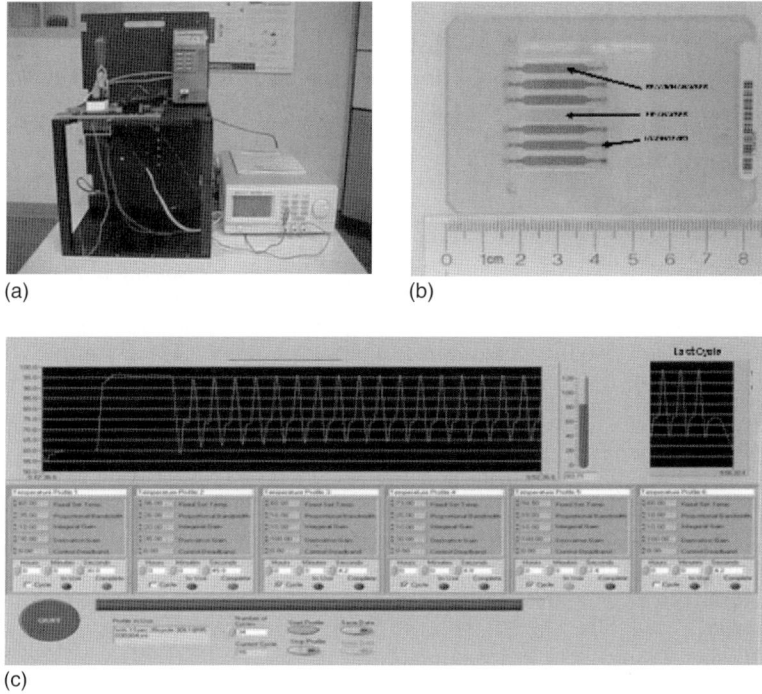

(a)

(b)

(c)

Figure 24. Microfluidics-enabled rapid PCR amplification lab card and breadboard designed by Micronics, Inc. Prototype Thermal Electric Cooler (a), lab card with PCR reaction chambers (b), and thermal couple trace showing 60-s reverse 60 °C reverse transcription followed by 16-s PCR cycles (c).

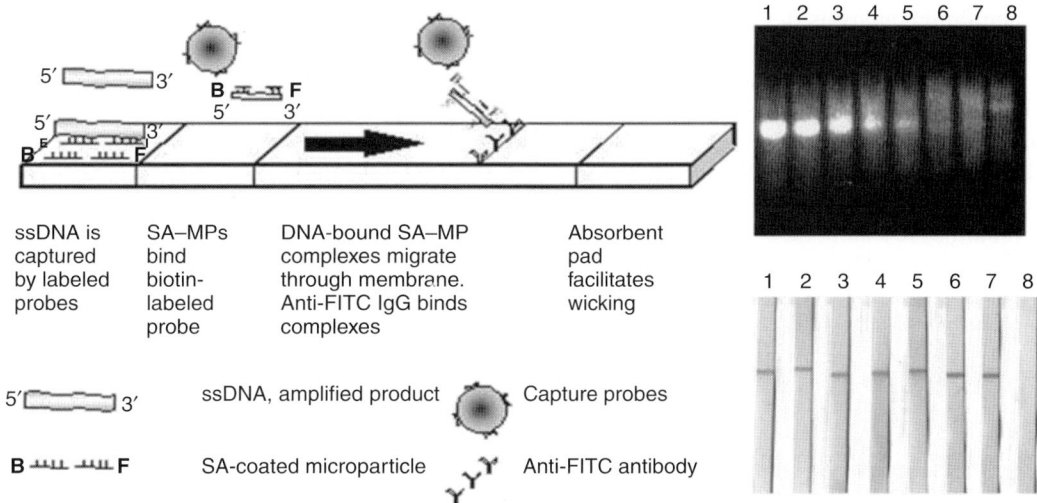

| ssDNA is captured by labeled probes | SA–MPs bind biotin-labeled probe | DNA-bound SA–MP complexes migrate through membrane. Anti-FITC IgG binds complexes | Absorbent pad facilitates wicking |

5′ □□□ 3′   ssDNA, amplified product   ● Capture probes

B ⊥⊥⊥ ⊥⊥⊥ F   SA-coated microparticle   Y Y Anti-FITC antibody

Figure 25. Visual lateral flow-based amplicon detection process developed by Micronics Inc.

from the sample to bind to silica. An on-card silica filter and microfluidic valves provide fluid control to automate the RNA binding, washing, drying, and elution steps. The card was initially validated in experiments in which 10^6 white blood cells were processed using commercial kits or suspended in lysis buffer prior to loading on the lab card. Real-time PCR assays determined that the microfluidic card solutions were detected a few cycles earlier than the control RNeasy solutions. This card has also been validated for lysis and detection of gram-negative bacteria isolated from feces. Similarly, the NA amplification subcircuit has been designed (Figure 24) and tested for each of the pathogens.

The PCR product generated by the subcircuit shown in Figure 24 is detected using a microfluidic visual amplicon detection method. Analogous to immunochromatographic strip tests, this method allows multiplexed detection of PCR-amplified products without instrumentation or software. As amplicons bound to colored microparticles accumulate on an antibody stripe immobilized in the microchannel, a colored line becomes visually apparent, indicating successful NA amplification and the presence of target (Figure 25). Procedural control lines have been included and results are available in a few minutes.

This example demonstrates typical subcomponents that have to be integrated to form a fully functional microfluidic device. Further, this example is quite representative of the status of integrated microfluidic analysis devices today—many are in development in both corporate and academic settings, but few, if any, are currently in production or use. Given the enormous progress that has been made in the microfluidics and lab-on-a-chip fields over the 15 years since its inception, and the enormous and broad efforts that currently go into microfluidics research, the authors believe that fully integrated microfluidic devices will indeed become mainstream analytical tools within the decade.

REFERENCES

1. A. Manz, C. S. Effenhauser, N. Burggraf, D. J. Harrison, K. Seiler, and K. Fluri, Electroosmotic pumping and electrophoretic separations for miniaturized chemical analysis systems. *Journal of Micromechanics and Microengineering*, 1994, **4**, 257–265.

2. F. Forster, R. Bardell, M. Afromowitz, and N. Sharma, *Design, Fabrication and Testing of Fixed-valve Micropumps*, In: *Proceedings of the ASME Fluids Engineering Division, 1995 IMECE*, San Francisco in November 1995, Vol. 234 pp. 39–44.

3. A. Olsson, P. Enoksson, G. Stemme, and E. Stemme, *A Valve-less Planar Pump in Silicon*, In: *The 8th International Conference on Solid-state Sensors and Actuators, and Eurosensors IX*, Stockholm, Sweden, 1995 June 25–29, pp. 291–294.

4. T. Gerlach and H. Wurmus, Working principle and performance of the dynamic micropump. *Sensors and Actuators A (Physical)*, 1995, **50**(1–2), 135–140.

5. Y. H. Mu, N. P. Hung, and K. A. Ngoi, *Simulation and Optimization of a Piezoelectric Micro-pump*, In: *International Conference of ASME*, Anaheim, California, 1998, November 15–20.

6. R. Linnemann, M. Richter, A. Leistner, and P. Woias, *A Full Wafer Mounted Self-priming and Bubble-tolerant Piezoelectric Silicon Micropump*, In: *Proceedings of the Actuator '98 Conference*, Bremen, Germany, 1998, June 17–19, pp. 78–81.

7. P. Woias, R. Linnemann, M. Richter, A. Leistner, and B. Hillerich, *A Silicon Micropump with a High Bubble Tolerance and Self-priming Capability*, in *Micro Total Analysis Systems*, J. Harrison and A. Van den Berg (eds), Kluwer Academic Publishers, Dordrecht, 1998, pp. 383–386.

8. C. R. Tamanaha, L. J. Whitman, and R. J. Colton, Hybrid macro-micro fluidics system for a chip-based biosensor. *Journal of Micromechanics and Microengineering*, 2002, **12**, N7–N17.

9. D. Maillefer, S. Gamper, B. Frehner, P. Balmer, H. van Lintel, and P. Renaud, *A High Performance Silicon Micropump for an Implantable Drug Delivery System Technical Digest MEMS'99*, 1999, pp. 541–546.

10. A. Richter and R. Zengerle, *Properties and Applications of a Micro Membrane Pump with Electrostatic Drive*, In: *3rd International Conference of New Actuators (ACTUATOR '92)*, Bremen, Germany, 1992, pp. 28–33.

11. M. T. A. Saif, E. Alaca, and H. Sehitoglu, Analytical modeling of electrostatic membrane actuator for micro pumps. *Journal of Microelectromechanical Systems*, 1999, **8**(3), 335–345.

12. C. Yamahata and G. Gijs, *Integrated Plastic Micropumps with Magnetic Actuation*, In: *NanoTech 2003, 7th Annual European Conference on Micro & Nanoscale Technologies for the Biosciences*, Montreux, Switzerland, 2003 November 25–27.

13. H. J. Yoon, J. M. Jung, J. S. Jeong, and S. S. Yang, Micro devices for a cerebrospinal fluid (CSF) shunt system. *Sensors and Actuators A*, 2004, **110**, 68–76.

14. C. J. Morris and F. K. Forster, Low-order modeling of resonance for fixed-valve micropumps based on first principles. *Journal of Microelectromechanical Systems*, 2003, **12**, 325–334.

15. C. Grosjean and Y. C. Tai, *A Thermopneumatic Peristaltic Micropump*, In: *1999 International Conference on Solid-state Sensors and Actuators (Transducers '99)*, Sendai, Japan, 1999 June, pp. 1776–1779.

16. J. H. Tsai and L. W. Lin, Micro-to-macro fluidic interconnectors with an integrated polymer sealant.

Journal of Micromechanics and Microengineering, 2001, **11**, 577–581.

17. R. H. Liu, J. Yang, R. Lenigk, J. Bonanno, and P. Grodzinski, Self-contained, fully integrated biochip for sample preparation, polymerase chain reaction amplification, and DNA microarray detection. *Analytical Chemistry*, 2004, **76**, 1824–1831.

18. D. Ross and L. E. Locasio, Rapid microfluidic mixing. *Analytical Chemistry*, 2002, **74**, 45–51.

19. J. Liu, C. Hansen, and S. R. Quake, Solving the "World-to-Chip" interface problem with a microfluidic matrix. *Analytical Chemistry*, 2003, **75**, 4718–4723.

20. P. Galambos and G. Benavides, *Electrical and Fluidic Packaging of Surface Micromachined Electro-microfluidic Devices*, In: *SPIE Micromachining and Microfabrication Conference*, San Jose, California, 2000 September.

21. H. Lorenz, M. Despont, N. Fahmi, N. LaBianca, P. Renaud, and P. Vettiger, SU-8: a low-cost negative resist for MEMS. *Journal of Micromechanics and Microengineering*, 1997, **7**, 121–124.

22. J. Carlier, S. Arscott, V. Thomy, J. C. Fourrier, F. Caron, J. C. Camart, C. Druon, and P. Tabourier, Integrated microfluidics based on multi-layered SU-8 for mass spectrometry analysis. *Journal of Micromechanics and Microengineering*, 2004, **14**, 619–624.

23. S. M. Sze, *VLSI Technology*, McGraw-Hill Science/Engineering/Math, 1988.

24. R. C. Jaeger, *Introduction to Microelectronic Fabrication*, 2nd Edn, Prentice Hall, 2001, p. 232.

25. Y. Xia and G. M. Whitesides, Soft lithography. *Angewandte Chemie International Edition*, 1998, **37**, 550–575.

26. L. Lin, Y. T. Cheng, and C. J. Chiu, Comparative study of hot embossed micro structures fabricated by laboratory and commercial environments. *Microsystem Technologies*, 1998, **4**, 113–116.

27. L. Krogh and P. Åsberg, *My Little Guide to Soft Lithography (or Soft Lithography for Dummies)*, Linköping University, website, http://www.ifm.liu.se/~petas/microsystem/Links/Material_files/Soft_Lithography_for_Dummies.pdf, 2003.

28. T. Koerner, L. Brown, and R. D. Oleschuk, *Prototyping of Polymeric Microfluidic Devices with Hot Embossing*, In: *Third Canadian Workshop on MEMS*, Ottawa, Canada, 2003 August 22.

29. *Photo Courtesy of MicroPlumbers Microsciences LLC*, http://www.microplumbers.com, 2007.

30. Private communication from L. Levine, ALine Inc, Redondo Beach, CA, Hyperlink http://www.alineinc.com/.

31. A. Gerlach, D. Maas, D. Seidel, H. Bartuch, S. Schundau, and K. Kaschlik, Low-temperature anodic bonding of silicon to silicon wafers by means of intermediate glass layers. *Microsystem Technologies*, 1999, **5**, 144–149.

32. Y. Huang, J. M. Yang, P.J. Hopkins, S. Kassegne, M. Tirado, A. H. Forster, and H. Reese, Separation of simulants of biological warfare agents from blood by a miniaturized dielectrophoresis device. *Biomedical Devices*, 2003, **5**(3), 217–225.

33. P. Yager, T. Edwards, E. Fu, K. Helton, K. Nelson, M. R. Tam, and B. H. Weigl, Microfluidic diagnostic technologies for global public health. *Nature*, 2006, **442**(7101), 412–418.

34. B. H. Weigl, J. Gerdes, P. Tarr, P. Yager, L. Dillman, R. Peck, S. Ramachandran, M. Lemba, M. Kokoris, M. Nabavi, F. Battrell, D. Hoekstra, E. J. Klein, and D. M. Denno, Fully integrated multiplexed Lab-on-a-chip assay for enteric pathogens. *Proceedings of the SPIE*, 2006, **6112**, 1–11.

43

Polymer-Based Microsystem Techniques

Matthias Schuenemann[1] and Erol C. Harvey[1,2]

[1] *MiniFab (Aust) Pty. Ltd., Scoresby, Victoria, Australia and* [2] *Faculty of Engineering and Industrial Science, Swinburne University of Technology, Hawthorn, Victoria, Australia*

1 INTRODUCTION

The need for point-of-care or point-of-use biosensors and bioanalytical devices in health care, in the food industry as well as for environmental testing has been a major factor in the development of low-cost microfluidic devices. The advantages of miniaturization (e.g., rapid analysis times, higher achievable analytical performance facilitated by changed fluid dynamics, low sample/reagent volumes, cost-effective reagent usage, reduced sample wastage, and reduced contamination and cross-contamination) have been attracting increasing attention from research groups as well as from commercial device manufacturers.[1–3]

Point-of-care devices enable diagnostic procedures at or close to the actual point of application (point-of-care, point-of-test). Extensive cleaning and test preparation procedures for setting up point-of-care tests are neither acceptable nor practical. For this reason, disposable devices are favored by the end user. On the other hand, it is very rarely economically feasible to integrate all technical components and subsystems required for complex sophisticated tests into disposable products. Therefore, biosensor systems tend to be divided into disposable chips or cartridges (or similar technical solutions) and a nondisposable host device (or instrument). The very essence of a successful design of a commercially viable point-of-care analysis system lies in the smart division between the disposable cartridge and the reusable instrument, and the careful design of the interface between the two. Figure 1 shows a microscope slide-sized disposable biosensor chip with integrated micromixers, delay lines, passive valves, and electrochemical sensors.

Economical restrictions dictate cost-efficient materials and technologies for the manufacturing of disposable cartridges. Many of the already published approaches to highly miniaturized bioanalytical systems are realized either in silicon[4,5] or in glass,[6,7] mostly relying on adapted silicon micromachining technologies and resulting in highly integrated miniaturized devices. Unfortunately, the respective manufacturing technologies as well as the utilized materials are rather expensive. For many potential applications of miniaturized bioanalytical devices, such high production costs cannot be justified. Polymer materials have been demonstrated to be a very versatile and cost-effective material choice, and the use of polymer-based microsystem techniques leads to bioanalytical devices with a very competitive cost of ownership per test.

2 MANUFACTURING OF LOW-COST POLYMER MICROFLUIDIC BIOCHIPS

One of the greatest challenges in developing low-cost disposable polymer microfluidic biosensors

Handbook of Biosensors and Biochips. Edited by Robert S. Marks, David C. Cullen, Isao Karube, Christopher R. Lowe and Howard H. Weetall.

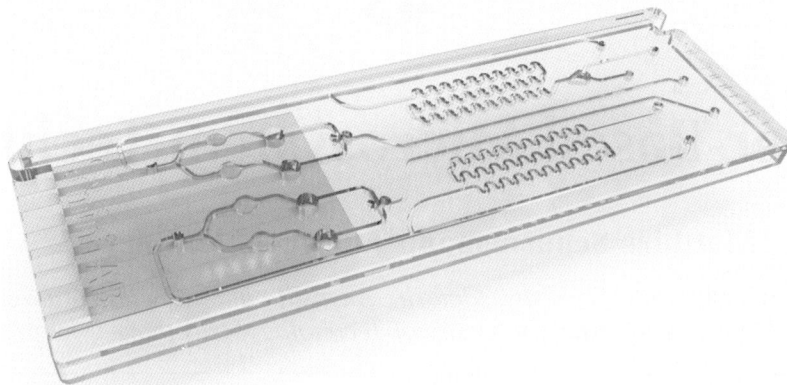

Figure 1. Disposable biosensor chip with integrated micromixers, delay lines, passive valves, and electrochemical sensors.

is to select a set of manufacturing processes that can be readily scaled in volume for each of the development stages, and that from the start, uses materials that will form the final device. This is essential in order to make cost effective the process of developing a biosensor that will pass regulatory compliance to become a commercial product.

Assuming that the bioassay has already been demonstrated at the laboratory scale, and that the appropriate reporting mechanism has been chosen (e.g., electrochemical, fluorescence, gravimetric), the first stage is to demonstrate the incorporation of these processes within a microfluidic system. The fabrication processes used for this stage must allow rapid iteration of design. Only a few working units, typically fewer than 10, are required to demonstrate the proof-of-concept. If the assay involves several discrete processing steps requiring specialist fluidic solutions, for example, passive valving, filtering, or metering, these steps may be demonstrated separately to show proof-of-principle. We generally refer to this as a Stage 0 development and use it to give confidence that the assay can be transferred to a disposable biosensor.

Although desirable that the materials used to fabricate the proof-of-concept devices are the same as those to be used in the final manufacture, this becomes an essential requirement for the next stage. In Stage 1 each of the individual components of the disposable biosensor are brought together to demonstrate a working assay and explore its performance in terms of sensitivity,

selectivity, specificity, and performance variation. This information is compared to the performance of the assay at the laboratory bench if the design requirement was set on the basis of the bench-scale performance. Often the microfluidic design must be further improved as sources of manufacturing variation are identified. These might include variation in channel dimensions, electrode surface areas, optical interference, or variation in component adhesion. The manufacturing process typically produces batches of 10–100 units for this stage and could be considered "prototyping" since the devices are manufactured either individually or, perhaps using thermal embossing techniques in small batches of 10–50 per sheet.[8]

For Stage 2 the manufacturing process must incorporate some element of batch or volume manufacturing as the biosensor testing process should now obtain statistically significant performance variation data on several thousand devices. Clearly there is less opportunity, and hopefully less need, for design change; and high-speed replication processes such as injection molding can be used. The tooling for this replication step must be robust enough to produce repeatable results and can be created in a number of ways including nickel electroforming from a master, electrodischarge machining, or diamond-tipped precision milling.[9] Stage 2 development is used to eliminate sources of variation in the manufacturing process such as polymer shrinkage, warpage, component misalignment during assembly, or irreproducible bonding.

After this stage the biosensors may be ready for early stage field trials so that Stage 3 must be able to produce runs of up to 10 000 devices in a few months. If the developer aims for a low-cost disposable polymer biosensor there should be a clear manufacturing path to achieve production volumes of 10^5–10^7 devices per year; clearly only achievable with high-speed replication, minimal assembly, and considerable process automation. This is best achieved using highly integrated polymer devices.

3 POLYMER MATERIALS

The selection of a suitable polymer material for a disposable biosensor cartridge is highly dependent on the intended application, the sample preparation, amplification, and detection process as well as the design complexity of the device. Almost every bioanalytical application will introduce specific technical demands for the chosen polymer material. Nonetheless, many applications share a common set of general requirements.[10]

Most polymer-based microfluidic chips and cartridges are manufactured from thermoplastic polymers. Thermoplastic polymers are characterized by first- and second-order thermal transitions. Whereas the first-order transition temperature usually corresponds to melting and allotropic transformation and is usually well outside the thermal working range for most materials, the second-order transition temperature characterizes the point above which the fixed molecular structure of the material is partially broken down by a combination of thermal expansion and thermal agitation.[11] Most thermoplastic polymers soften above this glass transition temperature, T_g, and exhibit a rubberlike behavior. Only when a material is highly crystalline are mechanical properties maintained above the glass transition temperature. Certain biological processes such as polymerase chain reaction (PCR) are performed at temperatures as high as 96 °C, while other devices potentially need to survive hostile storage temperatures. The polymer material has to be selected so that the fabricated device does not deform or disintegrate at the maximum specified temperatures.

Microfluidic structures and highly miniaturized fluidic devices are characterized by a surface-to-volume ratio that is at least 1 or 2 orders of magnitude higher than in conventional laboratory equipment. The large surface-to-volume ratio leads to a much stronger interaction of sample material and reagents with the substrate material, compared with standard laboratory equipment. The diagnostic integrity of a device is therefore heavily influenced by biocompatibility of the polymer material as well as by material properties such as water vapor permeability, gas permeability, and water absorption.

Biocompatibility of the substrate materials with the assay is an especially important requirement for a good analytical performance. Protein adsorption and cell adhesion are common phenomena interfering with bioanalytical processes. Protein adsorption is affected by factors such as surface energy/tension, surface charge, roughness, crystallinity, and entropy. Additionally, additives embedded in the polymer material such as plasticizers or UV stabilizers might lyse from the substrate material when subjected to elevated temperatures or contact with fluidic samples and reagents and inhibit biochemical reactions. The increased surface-to-volume ratios as well as the small amounts of target molecules characteristic of microfluidic systems make them particularly susceptible to these processes.[12]

A major drawback of polymers is their relatively high water vapor permeability rates, making polymers more difficult to use in bioanalytical devices compared to glass and metals. Any loss of water or water vapor from a sample or reagent may lead to a drift in pH or osmolarity and interfere with the analytical process on the device. Again, the large surface-to-volume ratio in miniaturized bioanalytical devices increases their susceptibility to this. The rate of potential fluid loss depends upon material properties such as water vapor permeability and water absorption rates as well as on design factors such as the diffusion path length from the microfluidic structure to the external environment.

Detection of reaction products forms a fundamental part of the bioanalytical device. Many detection methods rely on optical techniques, such as detecting fluorescent dyes attached to proteins or optical measurement of spots on microarrays.[1,13] An optically transparent material is essential for devices that depend on such detection techniques. However, many polymer materials are characterized by background fluorescence (or autofluorescence), resulting either from fluorescence intrinsic to the bulk polymer or from additives, impurities, or degradation products. The small sample and reagent volumes in miniaturized biosensors as well as the comparatively small amounts of fluorophores attached to the respective targets will result in small fluorescence signals. Background fluorescence decreases the signal-to-noise ratio significantly. It is therefore important to analyze, understand, and consider autofluorescence properties for successful system design as well as for suitable material selection.[14]

A well-informed selection of a suitable polymer material is critical not only to the performance but also to the manufacturability of the biosensor or biochip. During development and prototyping, significant research efforts have to be invested to overcome technical challenges resulting from imperfect material behavior. Having to change materials en route from Stage 1 to Stage 3 is likely to impose considerable cost as well as significant time delays in any development project. Manufacturing-related material requirements may include laser machinability, low shrinkage during injection molding, and sufficient resistance to cleaning agents. The material should be commercially available and reasonably well introduced into the market. A material with only a single supplier can potentially become temporarily or permanently unavailable, putting the commercial prospects of a device in a highly dangerous position. The selection of a rarely used material may also delay regulatory approval processes.

Polymeric materials like polycarbonate (PC),[15] polydimethylsiloxane (PDMS),[16] polyethylene terephthalate (PET),[17] polymethyl methacrylate (PMMA),[18] cyclic olefin copolymer (COC),[19] and polyimide (PI)[20] have commonly been used to fabricate prototypes of microfluidic bioanalytical devices. Less frequently, other standard organic polymeric materials such as polyethylene (PE), polyetheretherketone (PEEK), polystyrene (PS), polyamide (PA), polyetherimide (PEI), liquid crystal polymer (LCP), polypropylene (PP), polybutylene terephthalate (PBT), polyoxymethylene (POM), polyphenylene ether (PPE), and polysulfone (PSU) have successfully been used in microtechnology, mainly in a research environment outside the miniaturized biosensor domain.[8]

PMMA and PC meet the basic material requirements for miniaturized biosensors, are widely available and comparatively easy to machine and are therefore favored by the research community.[10,12] Another material receiving increasing attention is PET,[17,21,22] its main advantages being its widespread use in the packaging and printing industry and the extensive fabrication and processing knowledge base. PDMS is another transparent, elastomeric material that is used in soft lithography and molding processes for prototyping.[23] Although quite suitable for the fabrication of prototypes, its disadvantages include limited mechanical strength, limited ability to bond to other polymer materials, and very high water vapor permeability. A newer group of materials, COC, has great potential in microanalytical processes for its high chemical stability, low permeability rates, and very good optical properties,[24] but suffers from high costs and limited availability.

Table 1 shows a qualitative overview of the suitability of selected polymer materials for biochip manufacturing. The analysis shows that there is no one ideal polymer able to meet all requirements. Rather, the device or package designer has to find a technically and economically valid compromise for his specific application, taking into account such factors as the temperature regime of the analytical process, the storage conditions for the disposables prior to their use, the emission wavelength of the utilized fluorophores, the design complexity, and the fabrication technology most suitable for the expected production volumes.[12]

4 MICROSTRUCTURING OF POLYMERIC MATERIALS

One of the greatest challenges in developing low-cost disposable polymer microfluidic biosensors is to select a set of manufacturing processes that can be readily scaled in volume for each of the development stages. For this reason direct machining, laser cutting, and thermal embossing are attractive methods for development since each is able to structure a bulk polymer sheet that is available in high volume with excellent uniformity. Also, in the case of thermoplastics, the sheet form of the polymer can be considered to have properties similar to that achieved by injection molding (although in some specific details this may not be a valid assumption as UV laser cutting can leave

Table 1. Suitability of selected polymer materials for biochip manufacturing

Material	Properties					Machinability							
	Thermal stability	Water absorption	Water vapor permeability	Optical transparency	Autofluorescence	Laser CO₂ (λ = 10.6 µm)	Laser 3ω Nd:YAG (λ = 355 nm)	Laser excimer (λ = 248 nm)	Hot embossing	Injection molding	Micromilling	Cutting/blanking	Cost
COC	●	●	●	●	●	●	◐	○	●	●	●	●	●
PC	●	◐	◐	●	●	●	●	●	●	●	●	●	◑
PDMS	●	●	○	●	●	○	●	○	○	○	○	●	○
PEEK	●	◐	◐	○	n/a		○	●	○	●	●		●
PET	●	●	◐	◐	◐	●	●	●	●	●	◐	●	●
High-density PE	◐	●	●	◐	n/a	●	●	●	●	●	○	●	◐
Low-density PE	●	●	●	◐	n/a	●	●	◐	◐	●	●	●	●
PI	●	◐	○	◐	n/a	●	◐	◑	○	◐	◐	○	◐
PMMA	◐	◑	●	●	●	●	●	●	●	●	●	○	●
PS	◑	◐	●	◐	●	●	○	●	●	●	◑	●	●
Polytetrafluoroethylene	●	●	●	○	n/a	●	○	◐	◐	◐	●	●	●
Polyvinylchloride	◑	◐	●	◐	n/a	●	○	◐	●	●	◐	●	●
PVDC	◑	◐	●	●	n/a	○	●	●	●	●	●	●	●
Polyvinylidene fluoride	●	●	●	◐	n/a	○	○	○	●	◐	○		●

Material characteristics and machinability of polymer materials can vary significantly depending on manufacturer and material grade, due to the influence of additives such as plasticizers or UV stabilizers. Especially laser micromachining processes are highly sensitive to the presence of such additives. Figures in the table are therefore given for base polymers/most common polymer grades.
Most favorable material ●▸◑▸◐▸○ least favorable material.

a temporary surface activation not produced by molding techniques).[25]

4.1 Laser Fabrication

A wide range of lasers able to cut and pattern an even wider range of materials suitable for polymer biosensor packaging is now available.[26] Laser micromachining systems have the advantage of being noncontact tools that can be rapidly reprogrammed to produce varied patterns, making them particularly suitable for the design and development phase of the microfluidic biosensor.

Lasers are available that produce either pulsed or continuous radiation and are characterized by the wavelength of the light they produce. For micromachining applications pulsed lasers are essential since they allow greater control of the molten, or heat-affected zone in comparison to continuous systems. Generally the shorter the pulse length, the smaller the heat-affected zone. Table 2 shows a range of the popular pulsed lasers used for microfabrication and their operating characteristics. It should be obvious that for any material to be able to be laser machined it should absorb light at the wavelength of the laser to be used. For this reason, many infrared laser sources (e.g., carbon dioxide lasers ($10.6\,\mu m$) or fiber lasers ($1.09–1.55\,\mu m$ depending upon type)) will produce wildly varying results depending on the nature of the polymer used or the precise detail of additives in the polymer. Unfortunately the buyer of the bulk polymer often is unable to know what these additives are and in what concentration they are present, therefore much laser work tends to be by trial and error.

The most popular pulsed visible lasers are specially modified Nd:YAG (neodymium–yttrium–aluminum–garnet) lasers that can produce powerful pulses of green (533 nm) or ultraviolet (256 nm) light at high repetition rates. The alternate wavelengths are produced by placing wavelength doubling or tripling crystals in the beam, a technique that can be applied to some other lasers, for example the infrared fiber laser or the green copper vapor laser (CVL), enabling each to produce ultraviolet pulses. Ultraviolet pulses are the most useful wavelength for micromachining applications since they produce the least thermal damage and, since most polymers strongly absorb ultraviolet radiation, can provide control of the depth of laser machining.

Excimer lasers are pulsed gas lasers that produce ultraviolet light without the aid of doubling crystals. Another major difference from other laser sources is the large rectangular beam produced by excimer lasers. Hence rather than being used as a focused spot, excimer lasers are generally used as an illumination source for a stencil or mask that is imaged onto the polymer workpiece. If an image-reducing lens is used, the features produced at the workpiece can be of submicron size, and for most polymer materials the machined depth is typically less than a micron per laser pulse.

Table 2. Pulsed laser types and their operating characteristics

Pulsed laser source	Type	Frequency multiplied	Wavelength	Pulse width[a]	Repetition rate[a]
Carbon dioxide (CO_2)	Infrared	Fundamental	$9.24–10.64\,\mu m$	$25\,\mu s–1\,ms$	20 kHz
Ti:Sapphire	Infrared	Fundamental (tunable)	700–1080 nm	20–100 fs	75–120 MHz
Nd:YAG	Infrared	Fundamental	$1.064\,\mu m$	10–300 ns	2–100 kHz
Nd:YAG 2 ω	Visible	Doubled	532 nm		
Nd:YAG 3 ω	Ultraviolet	Tripled	355 nm		
Nd:YAG 4 ω	Ultraviolet	Quadrupled	266 nm		
Copper Vapor (CVL)	Visible	Fundamental	511 nm	20 ns	10–20 kHz
	Visible	Fundamental	578 nm		
Copper Vapor 2 ω (CVL 2 ω)	Ultraviolet	Doubled	255 nm		
	Ultraviolet	Doubled	271 nm		
Excimer (XeCl)	Ultraviolet	Fundamental	308 nm	20 ns	100 Hz–6 kHz
Excimer (KrF)	Ultraviolet	Fundamental	248 nm		
Excimer (ArF)	Ultraviolet	Fundamental	193 nm		
Excimer (F2)	Ultraviolet	Fundamental	157 nm		

[a] Typical specifications. Values will vary depending upon configuration and operating conditions.

Hence by computer-controlled manipulation of the number of pulses and the mask shape, complex three-dimensional shapes such as channels, ports, weirs, wells, mixers, and bifurcators can be rapidly fabricated in polymers.[27]

It is necessary to implement some form of polymer replication process once increasing manufacturing volumes are required. A range of microreplication processes for thermoplastic polymers are available that are generally smaller-scale implementations of their macroworld counterparts. These include hot embossing, injection molding, reaction injection molding, injection compression molding, and thermoforming.[28]

4.2　Hot Embossing

Hot embossing is a popular replication process since it is relatively easy to tool-up for and is a comparatively easy process to execute. It is able to achieve excellent replication of high-aspect-ratio microstructures, for example 8-μm-wide beams 150 μm tall in PMMA,[8] but has the disadvantage of a slow cycle time that can be up to 20–30 min.

In the hot-embossing process a mold tool is created that has the inverse features of the desired shape. This can be done by direct precision machining of metals or can be a nickel electroform grown from a previously microfabricated master. This master could be made in a number of ways including laser ablation of polymers, wet or dry lithographic etching in silicon,[29] UV lithography in thick SU-8 photoresist, or synchrotron exposure using the LIGA process (a German acronym for lithography, electroplating and replication). In some cases the polymer or silicon master can itself be used as the embossing tool.[29–31] The tool is mounted into a press and heated to a temperature slightly above the glass transition temperature, T_g of the polymer to be embossed (PMMA 106 °C, PC 150 °C). Polymer sheet is introduced into the press, which is closed, and a force of between 20–30 kN over a 4-in.-diameter area is applied under a vacuum of around 10^{-1} mbar. After a hold time of a few minutes and with the force still applied, the tool is then cooled to below T_g to stabilize the polymer before opening and demolding. Optimization of the process will reduce the thermal stresses in the part realizing improved replication, but usually at the cost of increased cycle time.

Nanosized features are readily reproduced by hot embossing. While this may be useful it also means that imperfections and roughness in the tool are also readily reproduced.

4.3　Injection Molding

Injection molding is a highly developed process for macroreplication and is now increasingly available for microscale thermoplastic replication.[28] The process has the advantage of extremely fast cycle times, of the order of a few seconds per cycle, but at the cost of a considerably more complex molding tools. In this process, a microstructured mold insert is placed within a specially formed mold cavity within the injection-molding machine. Polymer beads are heated above T_g and forced to flow into the mold cavity at high pressure where they rapidly cool to form a solid component that is ejected from the tool. This cyclic temperature control is called *variotherm* (*variothermal*). The resulting parts can have high degrees of internal stress and variable rates of shrinkage due to the rapid cooling of the polymer in the tool. Minimizing these effects as well as creating an effective ejection system for removing the part from the tool becomes part of the skill in designing good injection-molding systems. The ability to produce many millions of parts per year at relatively low cost makes this an important part of the industrial manufacturing process.

Table 3 compares selected manufacturing technologies for polymer biochips and rates them according to their suitability for tool making, prototyping, and volume production.

Having created the polymer components of the microfluidic system they must then be assembled and bonded together to form complete units.

5　SURFACE MODIFICATION OF POLYMER MATERIALS

Surface modification techniques change the surface characteristics of a material for a specific application without severely affecting the bulk properties of the polymer substrate. A range of biological, physical, and chemical methods are employed to modify surface properties such as wettability, permeability, biocompatibility,

Table 3. Selected manufacturing technologies for polymer biochips

Technology	Minimum structure width	Precision	3D capabilities	Process flexibility	Process time/throughput	Investment costs/operational costs	Application
LIGA	●	●	◕	○	◔	○/○	Toolmaking
Silicon bulk micromachining	●	●	◔	◔	◔	◑/◑	Toolmaking
Laser micromachining (CO_2, $\lambda = 10\,\mu m$)	◔	◑	○	●	◕	◑/●	Prototyping production
Laser machining (Nd:YAG, $\lambda = 355\,nm$)	◑	◑	○	●	◑	◑/◕	Prototyping toolmaking
Laser machining (Excimer, $\lambda = 248\,nm$)	◕	◕	◑	◕	◕	◔/○	Prototyping toolmaking
Micromilling	○	◑	◑	●	◔	◕/◕	Prototyping toolmaking
μEDM	◔	◕	◕	◕	◔	◔/◕	Toolmaking
Plasma etching	◑	◑	◔	◔	●	◑/◑	Prototyping production
Hot embossing	●	●	●	◔	◑	◔/◕	Toolmaking production
Injection molding	◑	◕	◕	◔	●	◑/●	Production
Roto-cutting blanking	○	◑	○	○	●	◑/●	Production
Stereolithography	◕	◕	●	●	◔	◔/◑	Prototyping toolmaking

μEDM: microelectrodischarge machining.
Most favorable process ●►◕►◑►◔►○ least favorable process.

chemical inertness, bondability, electrical characteristics, or optical properties.[32]

The majority of polymer packaging technologies use surface modification techniques in order to condition the polymer surfaces for the bonding process. Most polymer surfaces are hydrophobic, leading to poor wetting of the surface and therefore a poor spreading of adhesives or poor adhesion during bonding. One commonly used surface modification method is the use of a corona discharge to oxidize the polymer surface by ionized particles.[33] Although corona treatment is very cost effective compared with other surface modification methods, its short shelf-life and limitations in the treatable thickness of polymer sheets (up to $250\,\mu m$) restrict its applicability.

An alternative to corona discharge is gas plasma treatment. Typically, gas plasma treatment of polymers are utilized to ablate surface contamination, introduce chemically functional groups to the surface, and/or to introduce cross-linking.[34] Common gases utilized in this process include oxygen, nitrogen, and argon. One of the most widely used applications of gas plasma treatment is the oxygen plasma modification of PDMS utilized to convert hydrophobic Si–C siloxane groups to hydrophilic SiO_x groups.[35] Similar to the corona discharge process, the surface modification can be short lived due to polymer chain mobility.

Another major disadvantage of gas plasma treatment is the requirement for an evacuated environment. In an atmospheric plasma treatment process a polymer film can be passed through the plasma beam without the need for a vacuum, allowing for continuous in-line processing.[36]

Many surface modification methods combine a chemical surface treatment with physical changes to surface properties. Polymer surfaces can be grafted with chemicals that provide excellent adhesion to a large range of materials. This process starts with a surface activation step, followed by the deposition of chemicals dissolved in a water solution (e.g., silanes) which bond to the activated polymer.[37] In another surface modification approach, PET surfaces have been modified using a saponification reaction, in which polymer substrates were immersed in a bath of highly concentrated NaOH to etch the surface immediately before bonding.[21] The effect of photodegradation has been utilized to modify the surface of PMMA by exposing the polymer film to UV light to soften the top surface of the substrate.[38]

The nonspecific adsorption of proteins to polymer surfaces (and surfaces of other materials) is a significant problem encountered in a variety of biotechnological and bioanalytical applications. The behavior of hydrophilic polymer surfaces can vary considerably from nonfouling to selectively binding to high binding, depending on their respective physical and chemical surface characteristics. Proteins also adsorb to hydrophobic surfaces, but tend to denature, resulting in a thin, denatured, but persistently attached protein film preventing subsequent protein adsorption. This behavior may lead to a depletion of available target proteins and may significantly distort the measurements. For processes such as PCR, the use of different additives, for example, polyethylene glycol (PEG) or bovine serum albumin (BSA) in the PCR buffer has been shown to reduce nonspecific binding of key assay components to polymer surfaces and to improve process yield.[15,39–41] The grafting of polyethylene oxide (PEO) has also been demonstrated to reduce nonspecific binding.[42]

Permeability of the substrate material is another essential property to control. Microsized bioassays only handle small amount of fluids and reagents. It is therefore essential to maintain the volume of these fluids. A small fluid reduction caused by permeable materials affects the integrity of the process. Modifying the surface to become less permeable assists in avoiding this problem. Barrier layers on flexible polymer substrates are usually formed by depositing a thin layer of inorganic material like aluminum or silicon oxide on commodity polymers, such as PE or PET.[43] A major disadvantage of this method is the loss of optical transparency preventing the use of optical detection techniques. In another approach, high barrier polymers are formed by co-extruding commodity polymers with polymer barrier layers, such as polyvinylidene chloride (PVDC).

A very promising approach to surface modification is surface coating with parylene, a conformable, transparent coating based on polymerized para-xylylene. Parylene is deposited via chemical vapor deposition. In addition to preventing fluid losses through substrate materials, parylene reduces protein adsorption and cell adhesion.[44] Since the process requires vacuum

Table 4. Surface modification of polymer materials for biochip manufacturing

Surface treatment	Strength of effect	Sustainability of effect	Option for selective treatment	Process compatibility	Investment/operational costs
Surface modification techniques for adhesion promotion					
Saponification	◑	◑	◕	◑	●/◑
Photogradiation (UV)	◕	◑	●	◕	●/●
Corona treatment	◕	◕	◕	◕	●/●
Plasma treatment	●	◕	◕	◑	◕/◑
Polymer grafting	◕	◕	◑	◑	◕/◑
Primer deposition	◕	◕	◑	◑	●/◑
Surface modification techniques for biocompatibility					
PEO	◕	◑	◕	◕	◕/◑
PEG	◕	◕	◕	◑	●/◕
BSA	◕	◕	◕	◑	●/◕
Parylene deposition	●	●	◑	◕	◕/◑
Surface modification techniques for water vapor permeability reduction					
Barrier layer deposition	●	●	◑	◑	◕/◑
Parylene deposition	●	●	◑	◕	◕/◑
Barrier layer co-extrusion	◕	●	○	◕	◑/◕

Most favorable process ●▶◕▶◑▶◔▶○ least favorable process.

conditions, it is difficult to integrate into continuous productions systems.

Surface modification techniques are usually applied before bonding the prefabricated polymer layers or parts together. Unfortunately, surface requirements for bondability, biocompatibility, and permeability are usually contradictory and potentially conflicting. Surface modifications to condition the polymer surfaces for the bonding process often leave reactive functional chemical groups on the surface, which may inhibit bioanalytical processes. The same inhibition can occur when treating a surface to minimize loss of reagent. Again, the consequences of these contradictions are amplified by the large surface-to-volume ratio typical for miniaturized microfluidic devices. Each surface modification addressing one requirement therefore needs cross-checking to ensure compatibility with other requirements. Table 4 shows a qualitative assessment of selected surface modification techniques of polymer materials for biochip manufacturing.

6 ASSEMBLY AND PACKAGING OF POLYMER-BASED MICRODEVICES

Almost all microfluidic devices are based at least partly on fully enclosed and sealed microfluidic structures (i.e., channels, reservoirs, process chambers). Polymer microfabrication techniques, however, are usually only capable of generating open fluidic structures, and rely on bonding and sealing technology for the completion of the microfluidic device.

The simplest way to bond and seal a microfluidic structure is to cap a single planar microstructured polymer substrate on one or both sides with an unstructured cover layer. More sophisticated devices may be assembled from several stacked layers of microstructured polymer films or substrates, creating true three-dimensional microfluidic systems. The number of layers that can be bonded together is only limited by the applied bonding technique, the complexity of the design, and the feature size of the microfluidic structures. As an example, Figure 2 shows a PCR cartridge, driven by pneumatically actuated peristaltic on-chip pumps and controlled by pneumatically actuated on-chip valves. The device is fabricated from seven vertically assembled polymer layers.

The actual microfluidic reactor consists of three microstructured layers. The pneumatic control circuit is realized by another three microstructured layers, and an elastomeric membrane layer joins and separates the two three-layer prefabricates.

Primary functions for assembly, bonding, and sealing are to realize fully functional microfluidic devices by joining microstructured prefabricates, to prevent leakage from microfluidic features and to provide sufficient structural integrity within the assembled device.

Several bonding technologies can be used for assembly, bonding, and sealing of prestructured polymer layers (see Table 5). Most of these are adapted from standard polymer manufacturing technologies. For microfluidic circuits, bonding technologies that enable a selective bonding and sealing only at preselected areas (e.g., around the channel walls) are especially interesting. Although these technologies are usually more costly than bulk bonding techniques, many of them reduce the risk of involuntarily blocking channels and microfluidic structures and/or avoid accidental exposure of biological fluids to potentially non-biocompatible auxiliary materials (e.g., adhesives, solvents).

The joining of polymer substrates using adhesives is widely used during prototyping of polymer microfluidic devices. Adhesives are capable of bonding between dissimilar polymers as well as bonding polymers to metal layers or polymer prefabricates with large metallization areas. Additionally, adhesive bonding processes do not require extensive capital investment.

The direct application of an adhesive layer onto the surface of the polymer substrate carries a high risk of channel blocking. More commonly, adhesives (i.e., ultraviolet curable adhesives) are selectively applied to the bond surface using screen-printing techniques.[45] However, adhesives spread after application and clamping, potentially entering microsized features and clogging channels, mixers, or fluidic junctions. A voidless adhesive joint with liquid or thixotropic adhesives and screen printing is difficult to realize around complex or densely packed microstructures. The use of adhesive films instead of liquid adhesives prevents the undesired flow of adhesives into microfeatures, but necessitates prestructuring of adhesive film to create the required microfluidic vias between the polymer layers and careful alignment of the

(a)

(b)

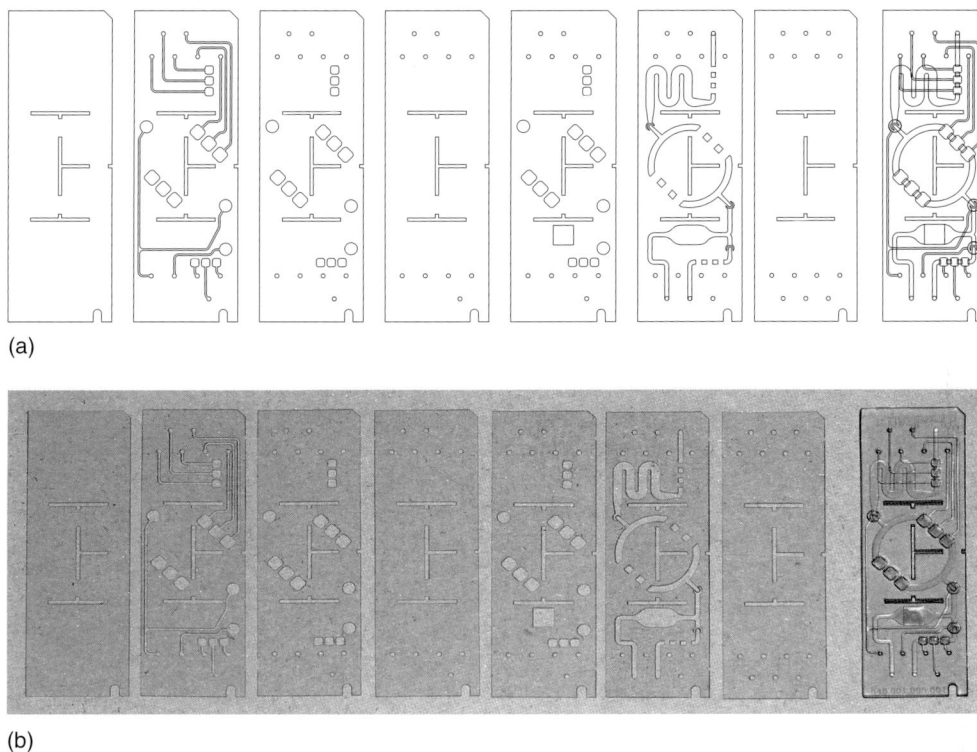

Figure 2. (a) Design for a polymer biochip fabricated from vertically assembled microstructured polymer layers. (b) Polymer biochip fabricated from vertically assembled microstructured polymer layers—manufactured layers and completed device.

Table 5. Systematization of polymer bonding techniques

	Direct bond between structural layers	Bond mediated by auxiliary materials
Bulk bonding	Lamination	Adhesive bonding
	Thermal diffusion bonding	Adhesive tape bonding
	UV-assisted thermal diffusion bonding	Solvent bonding
	Plasma-assisted thermal diffusion bonding	Chemical etching–assisted thermal diffusion bonding
	Ultrasonic welding	Screen-printed adhesives and adhesive bonding
Selective bonding		Prestructured adhesive tape and adhesive bonding
	Transmission laser welding	Light-absorbing dyes laser welding
	Reverse conductive laser welding	Microwave absorber
		Microwave welding

adhesive film relative to the structures on the polymer substrates, thus increasing manufacturing costs significantly. Additionally, adhesives have to be selected carefully as there is a high risk of undesired molecular interaction of the biological assay with the surface chemistry of the adhesive layer, which may influence or even inhibit the biochemical processes on the device.[12]

Thermal lamination is a simple, effective method to cap single planar layers that contain microfluidic structures. A very common laminate consists of a thin PE/PET film thermally bonded to PET substrates using a hot laminator.[22,46,47] Thermal lamination is especially suited to seal simple devices consisting of only one microstructured layer, but it has considerable limitations in sealing large, shallow features (reservoirs, reaction chambers) without any additional structural support, as the laminating film sags into these structures during lamination and potentially interferes with their intended function. Additionally, multilayer devices usually cannot as easily be sealed with thermal lamination since the laminating film isolates microfluidic layers from each other by blocking microfluidic vias between layers. One approach to realize vertically integrated microfluidic multilayer devices via lamination is to microstructure (relatively thick) PE/PET/PE films (i.e., by laser manufacturing or roto-cutting) and laminate them together (i.e., in a reel-to-reel system).

For prototype or small series production, thermal diffusion bonding, realized by applying heat and pressure over a given time to preassembled polymer slides, is a suitable method to bond prestructured polymer layers to each other. As no auxiliary materials such as adhesives are required, potential channel blocking is avoided, and biocompatibility is maintained. The bonding success depends heavily on the mobility of molecular chains in the polymer. Thus, only very similar materials with identical glass transition temperature can be bonded together. Successful thermal diffusion bonding has been reported for PC, PET, PMMA, and COC.[12,48,49] Bonding temperature and applied pressure are critical as an unsuitable parameter combination will deform the material and collapse the channels. For the device in Figure 2, two three-layer PC prefabricates were manufactured separately by thermal diffusion bonding. The laser-machined PC layers, which had alignment features incorporated into the design, were placed into an in-house developed hot embossing tool. A temperature of 135 °C and a pressure of 4.2–4.5 MPa were applied for 20 min.[12]

A major disadvantage of thermal diffusion bonding is the required process time of up to 30 min for a bond with sufficient strength. UV treatment prior to bonding allows for bonds to form at a significantly faster rate compared to thermal diffusion bonding of untreated surfaces.[38] The number of layers that can be bonded together by this technology is limited by the complexity of the design and the feature size of the cut-out structures. The pressure distribution to any bond area situated above or below a void in the structure (e.g., a channel or a process chamber) is very uneven and might locally prevent successful bonding. Strict observation of design rules is required.

Solvent bonding involves the exposure of a polymer surface to a suitable solvent. Upon joining two solvent-exposed surfaces, the interfaces of both substrates diffuse in one another and form a bond after the solvent evaporates from the assembly. Although solvent bonding is a common joining method for the assembly of polymer parts with many material/solvent combinations being available, it has rarely been used for bioanalytical devices. One approach is to deposit a thin layer of COC with a lower glass transition temperature on a thick layer of COC with a higher glass transition temperature by dissolving it in toluene and spin-coating it on the thicker substrate. The solvent-bonded parts were subsequently thermal diffusion bonded to each other.[50]

The optical transparency of many polymer materials has been used for a number of selective bonding techniques. A common bonding technique is through-transmission laser welding used to bond two polymer parts with different optical transmission characteristics. A laser transmits energy through the transparent layer. The laser energy is absorbed by the subjacent opaque polymer, causing the material to heat past its melting temperature. As a result, the two substrates will locally join. Scanning a focused laser beam around microsized features enables selective sealing and bonding. In reverse conductive welding, the energy-absorbing layer is not part of the device, but forms a workbench that heats up during energy absorption. From there, the thermal energy is conducted back to the interface between the polymer layers. This process creates a large heat-affected zone, which leads to distortion of microsized features in the vicinity of the bond. Another way to weld polymer layers together is based on an energy-absorbing dye. The dye is deposited onto at least one of the surfaces to be bonded and subsequently heated by a laser source with a wavelength corresponding to the absorption wavelength of the dye.[51]

In ultrasonic welding, high frequency mechanical energy is applied via an acoustic horn to

Table 6. Process performance of selected polymer bonding techniques

	Technical maturity	Geometrical resolution	Structural complexity	Channel clogging / distortion	Compatibility with metallization	Transparency	Biocompatibility	Technical flexibility	Process time and throughput	Cost
Adhesive bonding (bulk)	●	○	◔	○	●	◕	◑	◕	◕	●
Adhesive bonding (screen-printed adhesive)	●	◑	◕	◑	●	◕	◑	◑	◕	◕
Adhesive tape bonding (bulk)	●	◔	◑	◑	●	◕	◑	◕	●	●
Adhesive tape bonding (prestructured tape)	◕	◑	◔	◕	●	◕	◑	◔	◑	◑
Lamination	●	◔	◑	◑	●	◕	◑	◕	●	◕
Thermal diffusion bonding	◕	◕	◕	◕	◑	●	●	●	◔	◑
Thermal diffusion bonding (UV-assisted)	◕	◕	◕	◕	◑	◑	◑	●	◑	◑
Thermal diffusion bonding (O₂ plasma-assisted)	◕	◕	◕	◕	◑	◕	◑	●	◑	◔
Thermal diffusion bonding (chemical etch–assisted)	◔	◑	◑	◕	◑	◕	◔	◕	◑	◔
Solvent bonding	◑	○	◑	◑	◑	◕	◔	◑	◕	◕
Laser welding (transmission)	◕	●	◕	●	◑	◑	●	◑	◑	◑
Laser welding (reverse conductive)	◔	◔	◑	◔	◔	◑	●	◑	◑	◑
Laser welding (absorbent dye)	◔	◕	◑	◑	◑	◑	●	◕	◕	◔
Ultrasonic welding	◕	◑	◑	◑	◑	◕	●	◕	●	◑
Microwave welding (microwave absorber)	◔	◑	◑	◕	○	◑	◑	◑	◕	◔

Most favorable process ●►◕►◑►◔►○ least favorable process.

the polymer assembly creating frictional heat between molecules and causing the polymer to melt and join.[52] Another bonding technique utilizes microwave technology to join polymer layers in microfluidic devices. Most polymers are transparent to microwave radiation. If a microwave absorber such as a conductive polymer or a metal film is added to the joint interface between two polymer layers, it will selectively absorb the microwave energy. This interaction results in a local heat generation at the interface, leading to bulk polymer flow across the joint and formation of a weld.[12,53]

Table 6 qualitatively assesses the process performance of selected polymer bonding techniques. The analysis shows that again, owing to the strengths and weaknesses of each of the discussed bonding techniques, a single preferred bonding technique cannot be identified. The selection of a suitable bonding technique depends heavily on the complexity of the design, the utilized materials, and the fabrication technology most suitable for the respective production volumes. A more detailed discussion of polymer bonding techniques is available in the literature.[54–56]

7 SUMMARY

Polymer microfabrication offers a wide variety of techniques for fabricating microdevices. Some of the techniques are relatively new and are borrowed from other microfabrication processes while the vast majority are adaptations of techniques already established in the macromanufacturing environment. We can expect to see even more innovations arising from the combination of other traditional manufacturing processes, for example printing processes, that when combined with polymer fabrication will produce increasingly integrated microsystems. The development of new polymer materials will further accelerate this development. The low material cost and great structural resolution possible with polymers makes for a highly cost-effective approach to designing and fabricating complex devices. As with all product development, careful attention must be paid to the way the manufacturing process is scaled in volume. However, low cost, great design flexibility, and the ability to cost-effectively achieve high production volumes mean that we are seeing an increasing introduction of innovative and commercially successful disposable biosensor systems into the market.

REFERENCES

1. D. R. Reyes, D. Iossifidis, P.-A. Auroux, and A. Manz, Micro total analysis systems. 1. Introduction, theory, and technology. *Analytical Chemistry*, 2002, **74**, 2623–2636.

2. P.-A. Auroux, D. Iossifidis, D. R. Reyes, and A. Manz, Micro total analysis systems. 2. Analytical standard operations and applications. *Analytical Chemistry*, 2002, **74**, 2637–2652.

3. L. J. Kricka, Miniaturization of analytical systems. *Clinical Chemistry*, 1998, **44**, 2008–2014.

4. M. A. Northrup, M. T. Ching, R. M. White, and R. T. Watson, *DNA Amplification in a Microfabricated Reaction Chamber*, In: *Digest of Technical Papers: Transducers '93*, Yokohama, 1993, pp. 924–926.

5. M. A. Northrup, B. Bennett, D. Hadley, P. Landre, S. Lehew, J. Richards, and P. Stratton, A miniature analytical instrument for nucleic acid based on micromachined silicon reaction chambers. *Analytical Chemistry*, 1998, **70**, 918–922.

6. E. T. Lagally, P. C. Simpson, and R. A. Mathies, Monolithic integrated microfluidic DNA amplification and capillary electrophoresis analysis system. *Sensors and Actuators. B*, 2000, **63**, 138–146.

7. E. T. Lagally, C. A. Emrich, and R. A. Mathies, Fully integrated PCR-capillary electrophoresis microsystem for DNA analysis. *Lab on a Chip*, 2001, **1**, 102–107.

8. H. Becker and U. Heim, Hot embossing as a method for the fabrication of polymer high aspect ratio structures. *Sensors and Actuators. A*, 2000, **83**, 130–135.

9. W. Schomburg, Review of micro moulding of thermoplastic polymers. *Journal of Micromechanics and Microengineering*, 2004, **14**, R1–R14.

10. M. L. Hupert, M. A. Witek, Y. Wang, M. W. Mitchell, Y. Liu, Y. Bejat, D. E. Nikitopoulos, J. Goettert, M. C. Murphy, and S. A. Soper, *Polymer-Based Microfluidic Devices for Biomedical Applications*, in *Microfluidics, BioMEMS, and Medical Microsystems*, H. Becker and P. Woias (eds), SPIE, Bellingham, 2003, SPIE Vol. 4982, pp. 52–64.

11. M. Chanda and S. K. Roy, *Plastics Technology Handbook*, 4th Edn, CRC Press, Boca Raton, 2007.

12. M. Schuenemann, D. Thomson, M. Atkin, S. Garst, A. Yussuf, M. Solomon, J. Hayes, and E. Harvey, *Packaging of Disposable Chips for Bioanalytical Applications*, in *Proceedings of the IEEE 54th Electrical Components and Technology Conference*, P. Thompson (ed), IEEE Press, Piscataway, 2004, pp. 853–861.

13. M. A. Burns, B. N. Johnson, S. N. Brahmasandra, K. Handique, J. R. Webster, K. Madhavi, T. S. Sammarco, P. M. Man, D. Jones, D. Heldsinger, C. H. Mastrangelo, and D. T. Burke, An integrated nanoliter DNA analysis device. *Science*, 1998, **282**, 484–487.

14. K. R. Hawkins and P. Yager, Nonlinear decrease of background fluorescence in polymer thin-films—a survey of materials and how they can complicate fluorescence detection in μTAS. *Lab on a Chip*, 2003, **3**, 248–252.

15. J. Yang, Y. Liu, C. B. Rauch, R. L. Stevens, R. H. Liu, R. Lenigk, and P. Grodzinski, High sensitive PCR assay in plastic micro reactors. *Lab on a Chip*, 2002, **2**, 179–187.

16. X. Yu, D. Zhang, T. Li, L. Hao, and X. Li, 3D microarrays biochip for DNA amplification in polydimethylsiloxane (PDMS) elastomer. *Sensors and Actuators. A*, 2003, **108**, 103–107.

17. J. S. Rossier, M. A. Roberts, R. Ferrigno, and H. H. Girault, Electrochemical detection in polymer microchannels. *Analytical Chemistry*, 1999, **71**, 4294–4299.

18. S. Qi, X. Liu, S. Ford, J. Barrows, G. Thomas, K. Kelly, A. McCandless, K. Lian, J. Goettert, and S. A. Soper, Microfluidic devices fabricated in poly (methyl methacrylate) using hot-embossing with integrated sampling capillary and fiber optics for fluorescence detection. *Lab on a Chip*, 2002, **2**, 88–95.

19. J. Steigert, S. Haeberle, T. Brenner, C. Müller, C. P. Steinert, P. Koltay, N. Gottschlich, H. Reinecke, J. Rühe, R. Zengerle, and J. Ducrée, Rapid prototyping of microfluidic chips in COC. *Journal of Micromechanics and Microengineering*, 2007, **17**, 333–341.

20. S. Metz, R. Holzer, and P. Renaud, Polyimide-based microfluidic devices. *Lab on a Chip*, 2001, **1**, 29–34.

21. M. Atkin, J. P. Hayes, N. Brack, K. Poetter, R. Cattrall, and E. Harvey, *Disposable Microchip Fabrication for DNA Diagnostics*, in *Biomedical Applications of Micro- and Nanoengineering*, D. V. Nicolau (ed), SPIE, Bellingham, 2002, SPIE Vol. 4937, pp. 125–135.

22. J. Rossier, F. Reymond, and P. E. Michel, Polymer microfluidic chips for electrochemical and biochemical analyses. *Electrophoresis*, 2002, **23**, 858–867.

23. C. S. Effenhauser, G. J. M. Bruin, A. Paulus, and M. Ehrat, Integrated capillary electrophoresis on flexible silicone microdevices: analysis of DNA restriction fragments and detection of single DNA molecules on microchips. *Analytical Chemistry*, 1997, **69**, 3451–3457.

24. C. H. Ahn, J. W. Choi, G. Beaucage, J. H. Nevin, J. B. Lee, A. Puntambekar, and J. Y. Lee, Disposable smart lab on a chip for point-of-care clinical diagnostics. *Proceedings of the IEEE*, 2004, **92**, 154–173.

25. M. A. Roberts, J. S. Rossier, P. Bercier, and H. Girault, UV laser machined polymer substrates for the development of microdiagnostic systems. *Analytical Chemistry*, 1997, **69**, 2035–2042.

26. C. Khan Malek, Laser processing for bio-microfluidics applications (Part 1). *Analytical and Bioanalytical Chemistry*, 2006, **385**, 1351–1361.

27. E. C. Harvey, P. T. Rumsby, M. C. Gower, and J. L. Remnant, *Microstructuring by Excimer Laser*, in *Micromachining and Microfabrication Technology*, K. W. Markus (ed), SPIE, Bellingham, 1995, SPIE Vol. 2639, pp. 266–277.

28. M. Heckele and W. K. Schomburg, Review on micro moulding of thermoplastic polymers. *Journal of Micromechanics and Microengineering*, 2004, **14**, R1–R14.

29. J. Elders, H. V. Jansen, M. Elwenspoek, and W. Ehrfeld, *DEEMO: A New Technology for the Fabrication of Microstructures*, in *Proceedings of the IEEE Micro Electro Mechanical Systems*, IEEE Press, Piscataway, 1995, pp. 238–243.

30. J. Narashimhan and I. Papautsky, Polymer embossing tools for rapid prototyping of plastic microfluidic devices. *Journal of Micromechanics and Microengineering*, 2004, **14**, 96–103.

31. C. Khan Malek and R. Duffait, Packaging using hot-embossing with a polymeric intermediate mould. *International Journal of Advanced Manufacturing Technology*, 2006. DOI 10.1007/s00170-006-0595-2.

32. S. L. McArthur and K. M. McLean, *Surface Modification*, in *Encyclopedia of Biomaterials and Biomedical Engineering*, G. E. Wnek and G. L. Bowlin (eds), Marcel Dekker, New York, 2004.

33. C. M. Chan, T. M. Ko, and H. Hiraoka, Polymer surface modification by plasmas and photons. *Surface Science Reports*, 1996, **24**, 3–54.

34. P. K. Chu, J. Y. Chen, L. P. Wang, and N. Huang, Plasma-surface modification of biomaterials. *Materials Science and Engineering R*, 2002, **36**, 143–206.

35. D. C. Duffy, J. C. McDonald, O. J. A. Schueller, and G. M. Whitesides, Rapid prototyping of microfluidic systems in poly(dimethylsiloxane). *Analytical Chemistry*, 1998, **70**, 4974–4984.

36. M. J. Shenton and G. G. Stevens, Surface modification of polymer surfaces: atmospheric plasma versus vacuum plasma treatments. *Journal of Physics D: Applied Physics*, 2001, **34**, 2761–2768.

37. W. S. Gutowski, S. Li, C. Filippou, P. Hoobin, and S. Petinakis, Interface-interphase engineering of polymers for adhesion enhancement: Part II. Theoretical and technological aspects of surface-engineered interphase-interface systems for adhesion enhancement. *Journal of Adhesion*, 2003, **79**, 483–519.

38. R. Truckenmüller, P. Henzi, D. Herrmann, V. Saile, and W. K. Schomburg, Bonding of polymer microstructures by UV irradiation and subsequent welding at low temperatures. *Microsystem Technologies*, 2003, **10**, 372–374.

39. E. Kiss, J. Samu, A. Toth, and I. Bertoti, Novel ways of covalent attachment of poly(ethylene oxide) onto polyethylene: surface modification and characterization by XPS and contact angle measurements. *Langmuir*, 1996, **12**, 1651–1657.

40. M. Zhang, T. Desai, and M. Ferrari, Proteins and cells on PEG immobilized silicon surface. *Biomaterials*, 1998, **19**, 953–960.

41. B. C. Giordano, E. R. Copeland, and J. P. Landers, Towards dynamic coating of glass microchip chambers for amplifying DNA via the polymerase chain reaction. *Electrophoresis*, 2001, **22**, 334–340.

42. P. Kingshott and H. J. Griesser, Surfaces that resist bioadhesion. *Current Opinion in Solid State and Material Science*, 1999, **4**, 403–412.

43. M. Hanika, H.-C. Langowski, U. Moosheimer, and W. Peukert, Inorganic layers on polymeric films—influence of defects and morphology on barrier properties. *Chemical Engineering and Technology*, 2003, **26**, 605–614.

44. Y. S. Shin, K. Cho, S. H. Lim, S. Chung, S. Park, C. Chung, D. Han, and J. K. Chang, PDMS-based micro PCR chip with Parylene coating. *Journal of Micromechanics and Microengineering*, 2003, **13**, 768–774.

45. J. Han, S. Lee, A. Puntambekar, S. Murugesan, J.-W. Choi, G. Beaucage, and C. H. Ahn, *UV Adhesive Bonding Techniques in Room Temperature for Plastic Lab-on-a-Chips*, in *Proceedings of Micro Total Analysis Systems 2003*, M. A. Northrup, K. F. Jensen, and D. J. Harrison (eds), Transducers Research Foundation, San Diego, 2003, pp. 1113–1116.

46. M. A. Roberts, J. S. Rossier, P. Bercier, and H. H. Girault, UV laser machined polymer substrates for the development of microdiagnostic systems. *Analytical Chemical*, 1997, **69**, 2035–2042.

47. J. S. Rossier, G. Gokulrangan, S. Svojanovsky, G. S. Wilson, and H. H. Girault, Characterization of protein adsorption and immunosorption kinetics in photoablated polymer microchannels. *Langmuir*, 2000, **16**, 8489–8494.

48. J. Yang, Y. Liu, C. B. Rauch, R. L. Stevens, R. H. Liu, R. Lenigk, and P. Grodzinski, High sensitivity PCR assay in plastic micro reactors. *Lab on a Chip*, 2002, **2**, 179–187.

49. X. Zhu, G. Liu, Y. Guo, and Y. Tian, Study of PMMA thermal bonding. *Microsystem Technologies*, 2007, **13**, 403–407.

50. F. Bundgaard, T. Nielsen, D. Nilsson, P. Shi, and G. Perozziello, *Cyclic Olefin Copolymer (COC/Topas®)—an Exceptional Material for Exceptional Lab-on-a-chip Systems*, in *Proceedings of Micro Total Analysis Systems 2004*, T. Laurell, J. Nilsson, K. Jensen, D. J. Harrison, and J. P. Kutter (eds), Royal Society of Chemistry, Cambridge, 2004, Vol. 2, pp. 372–377.

51. L. Dosser, K. Hix, K. Hartke, R. Vaia, and M. Li, *Transmission Welding of Carbon Nanocomposites with Direct-diode and Nd:YAG Solid State Lasers*, in *Photon Processing in Microelectronics and Photonics III*, P. R. Herman, J. Fieret, A. Pique, T. Okada, F. G. Bachmann, W. Hoving, K. Washio, X. Xu, J. J. Dubowski, D. B. Geohegan, and F. Traege (eds), SPIE, Bellingham, 2004, SPIE Vol. 5339, pp. 465–474.

52. R. Truckenmüller, Y. Cheng, R. Ahrens, H. Bahrs, G. Fischer, and J. Lehmann, Micro ultrasonic welding: joining of chemically inert polymer microparts for single material fluidic components and systems. *Microsystem Technologies*, 2007, **12**, 1027–1029.

53. A. A. Yussuf, I. Sbarski, J. P. Hayes, M. Solomon, and N. Tran, Microwave welding of polymeric microfluidic devices. *Journal of Micromechanics and Microengineering*, 2005, **15**, 1692–1699.

54. C. A. Harper, *Plastics Joining*, in *Handbook of Plastics, Elastomers, and Composites*, 4th Edn, C. A. Harper (ed), McGraw-Hill, New York, 2002, pp. 507–560.

55. T. Velten, H. H. Ruf, D. Barrow, N. Aspragathos, P. Lazarou, E. Jung, C. Khan Malek, M. Richter,

J. Kruckow, and M. Wäckerle, Packaging of Bio-MEMS: strategies, technologies and applications. *IEEE Transactions on Advanced Packaging*, 2005, **28**, 533–546.

56. S. Garst, M. Schuenemann, M. Solomon, M. Atkin, and E. Harvey, *Fabrication of Multilayered Microfluidic Packages*, in *Proceedings of the IEEE 55th Electrical Components and Technology Conference*, P. Thompson (ed), IEEE Press, Piscataway, 2005, pp. 853–861.

FURTHER READING

H. Becker and C. Gartner, Polymer microfabrication methods for microfluidic analytical applications. *Electrophoresis*, 2000, **21**, 12–26.

D. Thomson, J. P. Hayes, and H. Thissen, *Protein Patterning in Polycarbonate Microfluidic Channels*, in *BioMEMS and Nanotechnology*, D. V. Nicolau (ed), SPIE, Bellingham, 2004, SPIE Vol. 5275, pp. 161–167.

44

Microelectrochemical Systems

Stuart A. G. Evans and Lindy J. Murphy

Oxford Biosensors Ltd., Yarnton, UK

1 INTRODUCTION

The use of microelectrodes in the field of biosensors has led to increasingly lower detection limits and sample volumes, due to their small dimensions and high sensitivity of measurement. Detection limits as low as femtomolar concentrations of DNA or zeptomolar concentrations of analytes, and sample volumes as low as picoliters have been reported. In addition, advances in microfabrication techniques have resulted in increasing numbers of lab-on-a-chip-type devices with inbuilt electrochemical detection being reported, some of which are commercially available. Microelectrodes have also been fundamental to the development of the technique of scanning electrochemical microscopy (SECM), which allows investigation of redox processes at electrode surfaces with high resolution. This article describes the electrochemical response and methods of fabrication of microelectrodes, and outlines some of the recent applications of microelectrodes in the field of bioelectrochemistry.

2 MICROELECTRODES: DEFINITION AND PROPERTIES

Microelectrodes, as their name suggests, differ from conventional electrodes (macroelectrodes) with respect to their size. Macroelectrodes typically have dimensions in the meters to millimeters scale, depending on their application,

whereas microelectrodes (which are also known as *ultramicroelectrodes* or *UMEs*) are regarded as having at least one dimension in the micrometer range. The question of how small an electrode must be in order to be defined as a microelectrode has been discussed in great detail, but with no clear resolution. Part of the reason being that the term *microelectrode* was initially used in the 1940s for electrodes with dimensions in the millimeter range, but in the late 1970s it was used for smaller electrodes with dimensions in the micrometer range. It is generally accepted that for an electrode to be considered a microelectrode it must have at least one dimension, the critical dimension, smaller than the diffusion layer thickness, under the experimental conditions employed.[1] For the purpose of this review, a microelectrode is defined as an electrode having at least one dimension smaller than 25 μm but greater than 10 nm. It will therefore not include the so-called nanodes,[2] with critical dimensions that reside in the nanometer range. For clarity, the critical dimension can be the thickness of the electrode for microring, microband, or tubular microband electrodes, or the radius of the electrode for microdisc, hemisphere, or spherical microelectrodes.

When one electrode dimension is below the critical size, the electrode response has been shown to deviate from the standard theory for macroelectrodes and to exhibit some unique properties. Under appropriate experimental conditions, for example during slow scan voltammetry, the voltammetric response of microelectrodes is very

Handbook of Biosensors and Biochips. Edited by Robert S. Marks, David C. Cullen, Isao Karube, Christopher R. Lowe and Howard H. Weetall.
© 2007 John Wiley & Sons, Ltd. ISBN 978-0-470-01905-4.

different to that observed at electrodes of conventional size because the diffusion layer thickness can greatly exceed the dimensions of the microelectrode (see Figure 1).[3] When this occurs, the microelectrode attains a time-independent steady-state response, characterized by a sigmoidal-shaped voltammogram. This is similar to the polarograms obtained with a dropping mercury electrode or the current-voltage curves obtained with a rotating disc electrode, but in this case it is due to high diffusion rates under quiescent conditions. To explain this phenomenon fully, a simple model will be described where a microelectrode is immersed in a solution of an oxidizable redox species, with the microelectrode poised at a potential sufficient to oxidize the redox species at a diffusion-controlled rate. Initially, after application of the potential, the electrode perturbs the solution and causes the formation of a diffusion layer that moves out from the electrode into solution. At short times, the diffusion layer thickness, δ, is very thin and so the electrode is much larger than the diffusion layer thickness. Consequently, the nonuniform current distribution resulting from high mass transport of redox species to the edge of the microelectrode (the edge effect) has little contribution to the measured current and the electrode response is described by that of an infinitely large planar electrode (see Figure 2a and b). As time progresses, the diffusion layer thickness increases and eventually exceeds the dimensions of the microelectrode. Under these conditions, the edge effect becomes dominant and results in the

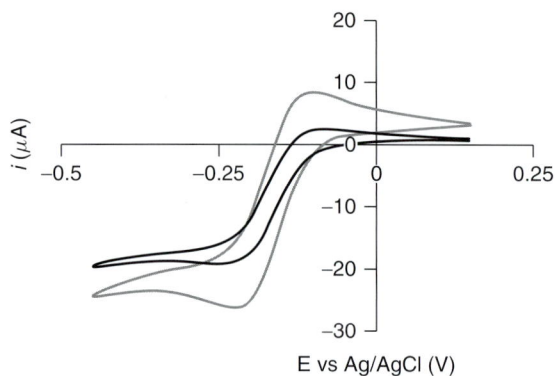

Figure 2. Illustration depicting the diffusion fields to (a) a macroelectrode (planar diffusion), (b) a microelectrode at short time after application of a potential step (planar diffusion), and (c) a microelectrode at long times after a potential step (radial diffusion).

formation of a spherical diffusion field and the attainment of a steady-state response with high current density (Figure 2c).

Conversely, for short timescale experiments, for example with cyclic voltammetry recorded at high scan rate, the diffusion layer thickness is smaller than the size of the microelectrode, semi-infinite planar diffusion is dominant and the voltammetry reverts to the peak-shaped behavior seen at electrodes of conventional size. Figure 3 compares cyclic voltammograms recorded using a 14-μm thick carbon microband electrode in a 10-mM $Ru(NH_3)_6Cl_3$ solution recorded with fast and slow scan rates. The slow scan rate voltammogram displays the characteristic sigmoidal shape consistent with radial diffusion to a microelectrode, while the fast scan voltammetry (FSV) has peaks consistent with planar diffusion.

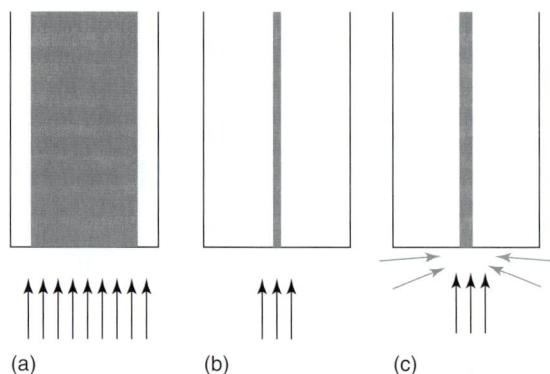

Figure 1. Cyclic voltammograms for a 14-μm-thick screen printed carbon microband electrode in 10 mM $Ru(NH_3)_6Cl_3$, recorded with scan rates of 10 mV s^{-1} (black line) and 100 mV s^{-1} (gray line).

3   ADVANTAGES OF MICROELECTRODES

Microelectrodes, because of their small size, have several advantages compared with macroelectrodes and, as a result they have been the focus of several comprehensive review articles.[4–6]

As the electrolysis currents recorded during microelectrode experiments are typically small, the impact of ohmic phenomena (iR drop) is greatly reduced making microelectrodes amenable to undertaking electrochemistry in a wide variety

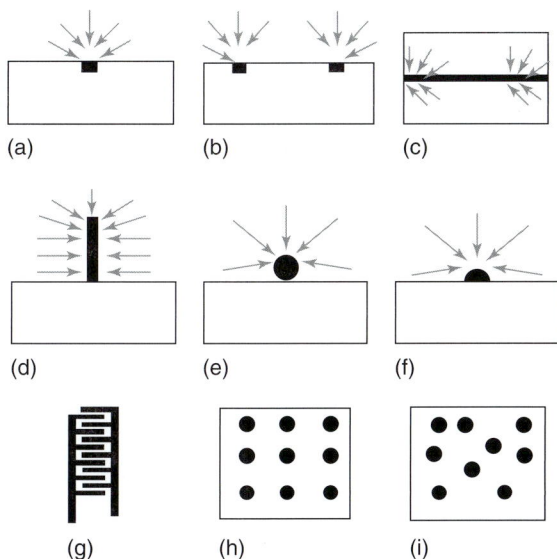

Figure 3. Showing the most commonly used geometries of microelectrodes and microelectrode arrays; (a) microdisc, (b) microring, (c) microband, (d) microcylinder, (e) microsphere, (f) microhemisphere, (g) interdigitated array, (h) microdisc array, (i) random array of microdiscs.

4 TYPES OF MICROELECTRODES AND THEIR FABRICATION

Microelectrodes fall into one of two main categories; single or array microelectrodes. Figure 3 shows the most common geometries of microelectrode, which have been fabricated and utilized in the literature. Of these, the most commonly used microelectrode geometry, accounting for approximately 50% of all microelectrode studies,[11] is the microdisc electrode (also known as *inlaid microdisc*). Of the other common geometries, the microcylinder accounts for 20%, microarray electrodes (both random and uniform) account for a further 20%, while the remaining 10% is split mainly between the microband and microring, with a small percentage attributable to the microsphere and microhemisphere electrodes.

Fabrication of the single microelectrodes is typically undertaken by sealing a microwire, thin foil, or fine fiber into an insulating material such as glass or epoxy resin. The reviews by Zoski and Murray and coworkers provide highly detailed and comprehensive discussion of the topic including the design, fabrication, and characterization of microelectrodes.[12,13] Briefly, microdisc electrodes are fabricated by either heat sealing a microwire into a glass capillary (under vacuum) and then polishing the end of the capillary to yield a microdisc electrode or by inserting the microwire into the capillary and then pulling the metal/glass assembly using a pipette puller. To prepare microcylinder electrodes, the microwire is again inserted into the glass capillary, but in this case a small length of microwire (<1 mm) is left to protrude from the end of the glass thereby forming a microcylinder of exposed wire. Microband electrodes are most commonly fabricated by sandwiching metal foils (or thin metal films) between glass or epoxy insulating layers. Spherical and hemispherical microelectrodes are typically fabricated by electrodeposition of mercury films onto platinum microdisc supports, the radius of the sphere or hemisphere being determined by the amount of mercury deposited. Finally, in order to fabricate microring microelectrodes (also known as *inlaid ring microelectrodes*), the interior or exterior walls of a pulled glass capillary (or rod) is coated, either by painting with an organometallic compound or by vapor deposition of a thin film of conducting material, for example Au, Pt,

of chemical media including nonaqueous solvents, gas, ice, polymer films, and in low-conductivity aqueous solutions with little or no supporting electrolyte.[7] The minimal distortion from iR drop also enables the use of microelectrodes for fast scan rate voltammetry with scan rates of over $1 \times 10^6 \, \text{V s}^{-1}$.[8] In addition to this, the response time is reduced since the capacitive (nonfaradaic) response of an electrode decreases with electrode radius, so more information can be gained in the early part of chronoamperometric transient responses and in fast scan rate voltammetry for the investigation of high-speed electron transfer reactions that were previously inaccessible with macroelectrodes.

The small physical size also makes them ideally suitable for experimental conditions where either space or sample volume is at a premium, for example during single cell studies in nanoliter volumes.[9] Finally, the steady-state response obtained with microelectrodes makes them ideally suitable for electroanalytical applications, since the limiting current is directly proportional to the analyte concentration giving an excellent signal-to-noise ratio and as low as zeptomole detection limits.[10]

or C. The coated capillary (or rod) is then sealed into a larger glass tube with epoxy resin. Polishing the end of the capillary exposes the microring electrode.

Arrays of microelectrodes fall into one of two groups; regular arrays, where identically sized electrodes are positioned periodically with uniform separation or random arrays, where either the electrodes are of uniform size but spaced randomly or alternatively where both the size and spacing of the electrodes are random. Regular arrays with well controlled electrode size and distribution are generally constructed using standard photolithographic techniques,[14] whereas random arrays are generally prepared by sealing small conducting particles (e.g., graphite powder) or a large number of microwires (or fibers) into a nonconducting support. Polishing the surface exposes the random array of disc electrodes.

5 CHARACTERIZATION OF MICROELECTRODES

A number of methods have been devised to characterize the size, shape, and quality of microelectrodes after fabrication. The electrodes are typically inspected using scanning electron microscopy (SEM) to determine the quality of the seal between the conducting and insulating materials and to approximate the dimensions of the microelectrode. In order to characterize the electrochemical response of the microelectrodes, steady-state voltammetry is undertaken in an aqueous solution containing a well-characterized reversible redox couple possessing a fast heterogeneous electron transfer rate, for example, 1 mM $Ru(NH_3)_6Cl_3$.[15] The voltammogram obtained with a slow scan rate (typically $<10\,mV\,s^{-1}$) should be sigmoidal in shape with the reverse scan retracing the forward sweep. A separation between the forward and reverse scans indicates that either the scan rate is too fast, or that the seal between the conducting and insulating materials is poor. The magnitude of the diffusion limiting current will depend on the geometry and size of the microelectrode used and the diffusion coefficient of the redox species.

6 EXPERIMENTAL SETUP FOR SCANNING ELECTROCHEMICAL MICROSCOPY (SECM)

SECM is a scanning probe microscopy (SPM) technique where a microelectrode probe (typically a microdisc electrode) is scanned in close proximity to a sample (the substrate) immersed in an electrolyte solution. The technique differs from other SPM techniques because it relies on the electrochemical response of the microelectrode probe. SECM can therefore be used to perform almost any type of electrochemical experiment with the electrochemical probe above a micrometer size area of the sample. AC voltammetry, amperometry, potentiometry, and microfabrication can be undertaken using SECM. The technique was originally devised by Engstrom (in 1986) to probe the diffusion layer of a large electrode,[16] although it was not until later that year that Bard and coworkers named the technique *scanning electrochemical microscopy* (*SECM*).[17] Since then, over 700 research articles, several comprehensive reviews, and one book have been written about the technique.[18,19]

6.1 The Feedback Mode of SECM Operation

The most popular operation of SECM is in the amperometric mode where the microelectrode tip acts as an active probe for oxidizing or reducing redox active species in solution. The interaction of the microelectrode diffusion layer and the sample surface forms the basis of the so-called feedback mode.

When a microdisc electrode is immersed in a solution containing an oxidizable species (R) and poised at a potential sufficient to oxidize the redox species at a diffusion-controlled rate, a quasihemispherical diffusion layer builds up around the tip of the microelectrode. After a short time (of the order of tens of a^2/D, where a is the radius of the microdisc and D is the diffusion coefficient of the species of interest), the size of this diffusion layer becomes constant (at approximately $7 \times a$) and the steady-state current, $i_{T\infty}$, is obtained as depicted by Figure 4(a). In this situation, the current attained at the tip is proportional to the concentration of R, C_R, and

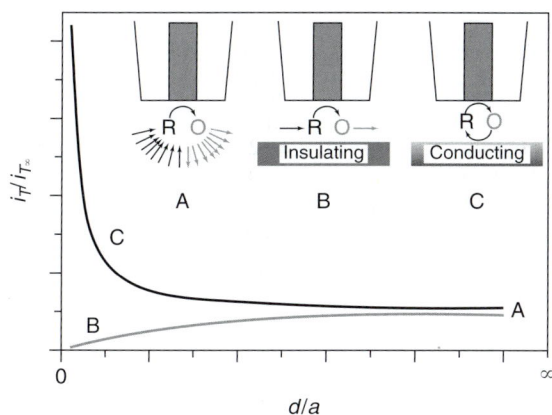

Figure 4. Basic principles of the feedback mode of SECM and the corresponding approach curves for each limiting case where (A) shows the response of the microelectrode in the bulk solution, where diffusion leads to a steady-state current, $i_{T\infty}$; (B) when the microelectrode is moved nearer to an insulating substrate, hindered diffusion leads to $i_T < i_{T\infty}$; (C) when the microelectrode is moved closer to a conductive substrate, positive feedback leads to $i_T > i_{T\infty}$.

diffusion coefficient, D_R, of the oxidizable species in the solution. The faradaic current recorded at the microelectrode can therefore be given by equation (1):

$$i_{T\infty} = 4nFD_R C_R^\infty a \qquad (1)$$

where n is the number of electrons transferred in the reaction and F the Faraday constant. This current, $i_{T\infty}$, assumes that the microelectrode is located in the bulk far away from the substrate and that the rate of mass transport is diffusion controlled.

If the tip of the microdisc electrode is then moved closer to the substrate surface, so that the diffusion layer of the microelectrode interacts with the substrate surface, the tip current becomes dependent on the conductivity or reactivity of the substrate. The resulting response forms the basis of the most commonly used mode of SECM operation—the feedback mode.

For example, when the tip approaches an insulating substrate, the steady-state current (i_T) flowing through the microelectrode becomes less than $i_{T\infty}$. The decrease in current is attributable to the insulating surface blocking the diffusion of R to the tip and causing the tip to see a decreased concentration of the reduced species; an effect

called *negative feedback* or *hindered diffusion* (Figure 4B). A typical hindered diffusion approach curve is depicted in Figure 4(B) and shows that as the tip/substrate distance decreases, $i_{T\infty}$ tends to zero. If the tip is approached toward a very reactive substrate, the steady-state current (i_T) flowing through the tip becomes greater than $i_{T\infty}$. The increased current is attributable to the substrate converting the oxidized species back into R, which then diffuses back to the tip. The flux of R from the substrate, combined with the flux of R from the solution around the tip causes the tip to see an increased concentration of reduced species and results in i_T being greater than $i_{T\infty}$, an effect called *positive feedback* (Figure 4C). A typical positive feedback approach curve is shown in Figure 4(C).

The actual feedback current can be more complicated than the two limiting cases described in the preceding text. For example, if electron transfer reaction from $O \rightarrow R$ at the substrate surface is kinetically controlled rather than diffusion controlled, then the tip current not only reflects the tip substrate distance but also the rate of regeneration of R at the substrate surface. This can therefore be used to determine the heterogeneous kinetics of reaction at the substrate.

When operated in the amperometric mode, the SECM can be used to study homogenous electrochemical reactions, electrochemical processes within films, as a tool for microfabrication (etching and deposition of metals, polymers, or biological systems), and to image maps of surface reactivity in both biological (single cell, enzyme, and antibody) and chemical systems. The versatility of SECM also makes it amenable to the study of a range of different samples including the liquid/liquid, gas/liquid and liquid/ice interfaces.

7 APPLICATIONS OF MICROELECTRODES TO BIOELECTROCHEMISTRY

One of the main applications of microelectrodes for biosensing arises from the small electrode size, permitting electrochemical detection in very small volumes such as single cells or in microanalytical devices. The small size also allows high spatial resolution of the response, which is directly of use in SECM and in vivo monitoring. Another feature of microelectrodes, the high sensitivity of measurement, allows detection of very low currents and hence very low detection limits. These applications are discussed in more detail in the subsequent text.

7.1 Low Detection Limit

The low detection limit of microelectrodes is one intrinsic feature of their response characteristics. Detection at the picomolar and femtomolar range in µl size droplets is possible. One example of this is the highly sensitive detection of 3000 copies of DNA in a 10-µl droplet at 0.5 fM concentration, using a 10-µm-diameter glassy carbon electrode.[20] The microelectrode was coated with redox polymer onto which was immobilized a DNA capture sequence, which after hybridization with the target DNA was exposed to a detection sequence labeled with horse radish peroxidase. Although this is a multistage measurement, the technique could lend itself to a miniaturized device. This methodology can also be extended to the detection of immunoreactions with high sensitivity and low detection limits.

Also of interest is the electrochemical monitoring of a single cell with a 3-µm radius carbon fiber electrode in a vial of 100–200 picoliters.[21] A layer of mineral oil was used to reduce evaporation and permitted measurements for several minutes. Although the small volume allowed complete electrochemical exhaustion of the redox species within 60 s, FSV allowed continuous measurement without depletion.

7.2 In Vivo Sensing

Microelectrodes have been used in individual cells, in tissue samples, and in vivo to directly measure a variety of target analytes, including pH change, O_2, CO_2, superoxides, NO, and chemical messengers such as catecholamines (dopamine, epinephrine and norepinephrine), serotonin and histamine. Peptides with a tryptophan or tyrosine reside are also electroactive and can be monitored. Microbiosensors modified with enzymes such as glucose oxidase have also been used. Monitoring the analytes themselves can be of interest, or they may be measured to determine the effect of an external stimulant such as application of a therapeutic or recreational drug or a physiological event such as a tail pinch.[22]

In vivo sensing of glucose for the monitoring and control of glucose levels in diabetic patients is an active area of research, because of the potential benefits of improved control of glucose levels and the possible incorporation in a feedback system with an artificial pancreas. However, this method typically uses sensors with dimensions of tens of micrometers in conjunction with immobilized enzyme and an outer protective layer, so that the potential advantages of microelectrodes of fast response time and high spatial resolution of the response are not fully exploited. In vivo glucose sensing has recently been covered in an excellent review by Wilson.[23]

The investigation of neurochemical processes is the most active area of in vivo microelectrode research. Neurochemistry has been investigated using enzyme modified microelectrodes to detect analytes, in particular ATP.[24] An ATP-sensing microelectrode developed by Dale's group is now commercially available from world precision instruments (WPI). Use of microelectrodes for the real-time direct detection of neurochemicals has been reviewed by Wightman.[25] Also of interest is the detection of exocytosis events at the single cell and single vesicle level which has been made possible by microelectrodes. This has been applied to the monitoring of the neurotoxic effects of environmental pollutants and drugs of abuse on vesicular catecholamine release.[26]

Wightman has pioneered the use of FSV to increase the temporal resolution of in vivo microelectrode responses, in addition to the high spatial resolution of the microelectrode signal. In vivo sensors for direct electrochemical detection often have a protective polymer coat to reduce the amount of interferent species reaching the electrode. However in FSV, higher temporal resolution is obtained for uncoated electrodes due to faster diffusion of electroactive species to the electrode surface.[27]

In vivo measurement of NO with microelectrodes also merits discussion. The biological role of NO as a physiological messenger molecule was first discovered in the 1980s, and subsequently several methods of preparation of microelectrochemical sensors with modified surfaces for in vivo determination of NO have been reported. Electrochemistry with microelectrodes is the only technique currently available for the quantitative detection of in vivo NO levels, which are in the nanomolar concentration range. NO sensors must be able to measure the low levels of NO in a background of other electroactive species such as ascorbate, and the small size and hydrophobic nature of NO naturally lends itself to direct electrochemical detection at microelectrodes modified by use of size exclusion and/or hydrophobic coatings. Alternatively, NO can be detected catalytically at chemically modified sensors. The preparation and use of in vivo NO sensors has been reviewed by Bedioui.[28]

8 MICROFABRICATED DEVICES

The technique of electrochemistry can be readily applied to miniaturized devices, since fabrication of microelectrodes and cheap and small-scale integrated instrumentation for detection of electrochemical responses is relatively facile. This compares favorably with other detection techniques such as optical or mass spectrometry. Although these techniques can use miniaturized chips to perform measurements with very low detection limits, sometimes at the single molecule level, the instrumentation required to perform the techniques themselves are not easily miniaturized.

The use of microelectrodes in microelectromechanical systems (MEMS) (also known as *miniaturized total analytical systems* (*μTAS*)) extends beyond electrochemical detection to methods of manipulating cellular material. The techniques of electroporation and dielectrophoresis have been extended to the individual cellular level by the use of microelectrodes, and can be used as part of a μTAS. In electroporation, a strong electric field (hundreds to thousands of volts) is applied between two electrodes, in between which is placed a population of cells. The electric field causes part of the cell membrane to break down so that exogenous chemicals such as fluorescent markers can be incorporated into the cells and the functioning of the cell compartment investigated. To investigate a single cell without a microelectrode, either a single cell needs to be isolated and placed between the macro electrodes, or focusing of the electric field is required. Microelectrodes have enabled individual cells or even parts of the cell membrane to be electroporated.[29]

Dielectrophoresis is the motion of particles caused by dielectric polarization in a nonuniform electric field. The degree of motion is determined by the magnitude and polarity of the charges induced in a particle in an electric field, where the particle can be a cell, microorganism or other bioparticle. The induced charges impart an electric dipole to the particle, equivalent to approximately 0.1% of the net surface charge usually carried by the particle. Dielectrophoresis can be used to manipulate cells and particles in microfluidic devices, and use of microelectrodes to apply the electric field can allow individual cells or particles to be manipulated.[30]

Microelectrodes have also been used to monitor extracellular species in the region of a single cell, using very low volume "petri dishes". By sampling the microelectrode response at shorter times, high temporal resolution of cellular processes can be achieved.[31] Microelectrode arrays (MEAs) have also been used to interrogate tissue slices and the intra and extracellular biochemistry of cells. MEAs are commercially available from Multichannel Systems (Reutligen, Germany) and Panasonic (Tokyo). The use of MEAs to interrogate neuronal cell activity and the effect of drugs and toxins on tissue slices has recently been reviewed.[32] MEAs can be used extensively for pharmaceutical applications, allowing screening of prospective therapeutic agents and monitoring of potentially adverse tissue reactions, for example, the potential effect of novel drugs on cardiac function.

DiagnoSwiss have developed disposable microtiter plates for electrochemical immunoassays, using a plasma etching process to make microchannels in which conventional immunoassay reagents are placed. Microelectrodes in the channel detect the immunoreactions within 15 min, because of the small volume of sample in the microchannels resulting in fast equilibration between the sample and the reagents.[33,34]

Microelectrodes also enable the technique of CE-EC (capillary electrophoresis with electrochemical detection) to be extended to miniaturized systems. Use of individually addressable MEAs combined with electrophoretic separation can allow detection of multiple products in a sample.[35]

Portable DNA detectors using microelectrodes in microsystems and biochips has been reviewed by Lee and Hsing.[36] Electrochemical detection of DNA can be achieved by direct or catalyzed oxidation of DNA bases, or by the electrochemical response generated by enzyme or other redox markers by a specific binding event with the target DNA. There have recently been several excellent articles reviewing the electrochemical detection of DNA, including the use of nanoparticles that can increase the sensitivity of detection.[37–39]

9   SECM

The main application of SECM to biosensors has been to probe the surface activity, and hence the surface reaction kinetics, of immobilized biological systems, in particular immobilized enzymes. Readers are referred to the large series of papers by Bard, one of the originators of SECM, which explore and expand the possibilities of the technique. Enzyme reactions can be measured by electrochemical detection of a substrate or product of an enzyme reaction, for example hydrogen peroxide, or alternatively a redox mediator can be used. Measurements with the probe can be made amperometrically or potentiometrically. Two modes of SECM can be used to image enzyme activity, the enzyme-mediated feedback mode first utilized by Pierce and coworkers,[40] or the generator/collector mode.[41] The generator/collector mode is used most frequently with enzyme samples as the enzyme reaction rates are often too slow for the feedback mode, though the spatial resolution of this technique somewhat worse.

Electrochemical reactions at the probe can also be used to alter the solution composition at precisely defined areas of the sample, such as by generation of hydroxide ions to alter the local pH. The probe can then be used to monitor any change in biochemical activity. Detection of immuno or protein-binding events such as DNA hybridization can also be made by the use of enzyme labeling of a relevant protein species.[42,43] Individual cells can also be investigated.

SECM is also used to electrochemically pattern the electrode surface at the micrometer scale and to investigate the biochemical activity of miniaturized sensor arrays, including cross talk between neighboring immobilized species.[44] Micropatterning of biomolecules is a necessary requirement for the assembly of biochips and micro sensing arrays. Electrochemical treatment of small, precise areas of an electrode surface by the probe can alter the surface properties to promote attachment of biomolecules. Electropolymerization of redox monomers such as pyrrole can also be patterned onto an electrode surface, and can then be used for attachment of biomolecules.

10   CONCLUSIONS

The field of biosensors and biochips continues to be an exciting and rapidly expanding area of research and the potential for miniaturization of multiparametric sensing systems is very attractive. The combination of microelectrodes with the technologies of microfluidics and MEMS will lead to a range of novel and multistep electrochemical assays such as the production of cheap and disposable DNA chips, immunoassays, and μTAS for an array of different analytes. These chips will have several advantages over current technologies because they will not only require small sample volumes, but also provide high sensitivity and low detection limits.

ACKNOWLEDGMENT

The authors would like to thank Professor Allen Hill for helpful discussions and suggestions.

REFERENCES

1. K. Stulik, C. Amatore, K. Holub, V. Marešek, and W. Kutner, Microelectrodes. Definitions, characterization, and applications. *Pure and Applied Chemistry*, 2000, **72**, 1483–1492.

2. R. M. Penner, M. J. Heben, T. L. Longin, and N. S. Lewis, Fabrication and use of nanometer-sized electrodes in electrochemistry. *Science*, 1990, **250**, 1118–1121.

3. M. A. Dayton, J. C. Brown, K. J. Stutts, and R. M. Wightman, Faradaic electrochemistry at microvoltammetric electrodes. *Analytical Chemistry*, 1980, **52**, 946–950.

4. J. Heinze, Ultramicroelectrodes in electrochemistry. *Angewandte Chemie International Edition in English*, 1993, **32**, 1268–1288.

5. C. Amatore, *Electrochemistry at Ultramicroelectrodes*, in *Physical Electrochemistry—Principles, Methods and Applications*, I. Rubinstein (ed), Marcel Dekker, New York, 1995, pp. 131–208.

6. J. Wang, *Analytical Electrochemistry*, 2nd Edn, Wiley-VCH, New York, 2000, pp. 128–134.

7. A. M. Bond, Past, present and future contributions of microelectrodes to analytical studies employing voltammetric detection, a review. *Analyst*, 1994, **119**, R1–R21.

8. C. P. Andrieux, D. Garreau, P. Hapiot, and J. M. Saveant, Ultramicroelectrodes: cyclic voltammetry above one million V s^{-1}. *Journal of Electroanalytical Chemistry*, 1988, **248**, 447–450.

9. N. Gao, M. Zhao, X. Zhang, and W. Jin, Measurement of enzyme activity in single cells by voltammetry using a microcell with a positionable dual electrode. *Analytical Chemistry*, 2006, **78**, 231–238.

10. S. E. Hochstetler, M. Puopolo, S. Gustincich, E. Raviola, and R. M. Wightman, Real-time amperometric measurements of zeptomole quantities of dopamine released from neurons. *Analytical Chemistry*, 2000, **72**, 489–496.

11. R. J. Forster, Microelectrodes: new dimensions in electrochemistry. *Chemical Society Reviews*, 1994, **23**, 289–297.

12. C. G. Zoski, Ultramicroelectrodes: design, fabrication, and characterization. *Electroanalysis*, 2002, **14**, 1041–1051.

13. R. L. McCarley, M. G. Sullivan, S. Ching, Y. Zhang, and R. W. Murray, *Lithographic and Related Microelectrode Fabrication Techniques*, in *Microelectrodes: Theory and Applications*, M. I. Montenegro, M. A. Queiros, and J. L. Daschbach (eds), Kluwer Academic Publishers, Dordrecht, 1991.

14. R. Feeney and P. Kounaves, Microfabricated ultramicroelectrode arrays: developments, advances, and applications in environmental analysis. *Electroanalysis*, 2000, **12**, 677–684.

15. R. M. Wightman and D. O. Wipf, in *Electroanalytical Chemistry*, A. J. Bard (ed), Marcel Dekker, New York, 1989, Vol. 15, p. 267.

16. R. C. Engstrom, M. Weber, D. J. Wunder, R. Burgess, and S. Winquist, Measurements within the diffusion layer using a microelectrode probe. *Analytical Chemistry*, 1986, **58**, 844–848.

17. H. Y. Liu, F. R. F. Fan, C. W. Lin, and A. J. Bard, Scanning electrochemical and tunneling ultramicroelectrode microscope for high-resolution examination of electrode surfaces in solution. *Journal of the American Chemical Society*, 1986, **108**, 3838–3839.

18. A. J. Bard, F. R. Fan, and M. V. Mirkin, *Scanning Electrochemical Microscopy*, in *Electroanalytical Chemistry*, A. J. Bard (ed), Marcel Dekker, New York, 1994, Vol. 18, pp. 242–373.

19. M. V. Mirkin and A. J. Bard (eds), *Scanning Electrochemical Microscopy*, Marcel Dekker, New York, 2001.

20. Y. Zhang, H.-H. Kim, and A. Heller, Enzyme-amplified amperometric detection of 3000 copies of DNA in a 10-μL droplet at 0.5 fM concentration. *Analytical Chemistry*, 2003, **75**, 3267–3269.

21. K. P. Troyer and R. M. Wightman, Dopamine transport into a single cell in a picoliter vial. *Analytical Chemistry*, 2002, **74**, 5370–5375.

22. R. M. Wightman, Probing cellular chemistry in biological systems with microelectrodes. *Science*, 2006, **311**, 570–574.

23. G. S. Wilson and R. Gifford, Biosensors for real-time in vivo measurements. *Biosensors and Bioelectronics*, 2005, **20**, 2388–2403.

24. N. Dale, S. Hatz, F. Tian, and E. Llaudet, Listening to the brain: microelectrode biosensors for neurochemicals. *Trends in Biotechnology*, 2005, **23**, 420–428.

25. K. P. Troyer, M. L. A. V. Heien, B. J. Venton, and R. M. Wightman, Neurochemistry and electroanalytical probes. *Current Opinion in Chemical Biology*, 2002, **6**, 696–703.

26. R. H. S. Westerink, Exocytosis: using amperometry to study presynaptic mechanism of neurotoxicity. *Neurotoxicology*, 2004, **25**, 461–470.

27. B. J. Venton and R. M. Wightman, Psychoanalytical electrochemistry: dopamine and behaviour. *Analytical Chemistry*, 2003, **75**, 414A–421A.

28. F. Bedioui and N. Villeneuve, Electrochemical nitric oxide sensors for biological samples—principle, selected examples and applications. *Electroanalysis*, 2003, **15**, 5–18.

29. J. Olofsson, K. Nolkrantz, F. Ryttsen, B. A. Lambie, S. G. Weber, and O. Orwar, Single-cell electroporation. *Current Opinion in Biotechnology*, 2003, **14**, 29–34.

30. R. Pethig and G. H. Markx, Applications of dielectrophoresis in biotechnology. *Trends in Biotechnology*, 1997, **15**, 426–432.

31. J. M. Cooper, Towards electronic petri dishes and picolitre-scale single-cell technologies. *Trends in Biotechnology*, 1999, **17**, 226–230.

32. A. Stett, U. Egert, E. Guenther, F. Hofmann, T. Meyer, W. Nisch, and H. Haemmerle, Biological application of microelectrode arrays in drug discovery and basic research. *Analytical and Bioanalytical Chemistry*, 2003, **377**, 486–495.

33. J. Rossier, F. Reymond, and P. E. Michel, Polymer microfluidic chips for electrochemical and biochemical analyses. *Electrophoresis*, 2002, **23**, 858–867.

34. J. S. Rossier, C. Vollet, A. Carnal, G. Lagger, V. Gobry, H. H. Girault, P. Michel, and F. Reymond, Plasma etched polymer microelectrochemical systems. *Lab on a Chip*, 2002, **2**, 145–150.

35. J. Wang, Electrochemical detection for capillary electrophoresis microchips: a review. *Electroanalysis*, 2005, **17**, 1133–1140.

36. T. M.-H. Lee and I.-M. Hsing, DNA-based bioanalytical microsystems for handheld device applications. *Analytica Chimica Acta*, 2006, **556**, 26–37.

37. T. G. Drummond, M. G. Hill, and J. K. Barton, Electrochemical DNA sensors. *Nature Biotechnology*, 2003, **21**, 1192–1199.

38. J. Wang, Nanoparticle-based electrochemical DNA detection. *Analytica Chimica Acta*, 2003, **500**, 247–257.

39. A. Markoci, M. Aldavert, S. Marin, and S. Alegret, New materials for electrochemical sensing V: nanoparticles for DNA labeling. *Trends in Analytical Chemistry*, 2005, **24**, 341–349.

40. D. T. Pierce, P. R. Unwin, and A. J. Bard, Scanning electrochemical microscopy. 17. Studies of enzyme-mediator kinetics for membrane- and surface-immobilized glucose oxidase. *Analytical Chemistry*, 1992, **64**, 1795–1804.

41. G. Wittstock, R. Hesse, and W. Schuhmann, Patterned self-assembled alkanethiolate monolayers on gold patterning

and imaging by means of scanning electrochemical microscopy. *Electroanalysis*, 1997, **9**, 746–750.

42. M. V. Mirkin and B. J. Horrocks, Electroanalytical measurements using the scanning electrochemical microscope. *Analytica Chimica Acta*, 2000, **406**, 119–146.

43. G. Wittstock, Modification and characterization of artificially patterned enzymatically active surfaces by scanning electrochemical microscopy. *Fresenius Journal of Analytical Chemistry*, 2001, **370**, 303–315.

44. R. E. Gyurcsanyi, G. Jagerszki, G. Kiss, and K. Toth, Chemical imaging of biological systems with the scanning electrochemical microscope. *Bioelectrochemistry*, 2004, **63**, 207–215.

45

Micro- and Nanoelectromechanical Sensors

Keith L. Aubin,[1] Bojan Ilic[1,2] and Harold G. Craighead[1]

[1] *School of Applied and Engineering Physics, Cornell University, Ithaca, NY, USA and* [2] *Cornell Nanoscale Science and Technology Facility, Cornell University, Ithaca, NY, USA*

1 INTRODUCTION

In general, biosensors consist of certain elements regardless of the interrogation method employed. Specifically, the recognition of specific analytes out of many that may exist in the medium of interest, be it blood, air, or water, or any other substance from which knowledge of its constituents is desired, is for the most part dependent upon nature herself, that is to say, the biochemistry of life. Indeed, one of the most heralded biosensors known is right in front of the reader. As she/he is undoubtedly well acquainted with this organ, it still does some justice to point out here the great sensitivity of the human olfactory gland. Researchers across the globe seek an artificial scheme to replicate such sensitivity to the myriad of delights and repugnances that we experience every day and with every whiff. However, for simpler schemes of detection, say for the sensing of a single analyte of interest, it is quite adequate to borrow from the diversity of recognition that arises from the immune system, whose abilities are equally impressive, if not more so.

Biosensors in general (artificial ones that is) require at the very least sensing and signal elements. The former usually relies upon this biochemistry existent in nature to specifically detect the analyte of interest. For example, antibodies, which are proteins used in the immune system to recognize unwanted entities, or antigens, can be manufactured and used to coat the surface of a biosensor so as to make that surface receptive to a very specific target. Once (and if) that target analyte is present on the surface of this sensor, there must be a scheme in place to sense its presence. There are many ways presently in practice to accomplish this. One popular method is to use fluorescently labeled secondary antibodies that will also specifically bind to the analyte of interest. At a later step, this so-called "tagged" antibody is introduced and its incandescent properties are interrogated to reveal the presence of the analyte being sought.

Although this method is quite useful, it does suffer from several difficulties. Arguably, the worst among these is the time required to incubate the sample so that a detectable amount of analyte adheres to the functionalized sensor surface. This time could be many hours, depending on the concentrations and analytes in question. If a method existed that could detect a smaller number or even single binding events, this would be a marked improvement over methods currently employed. Moreover, if such a method could be miniaturized and made

Handbook of Biosensors and Biochips. Edited by Robert S. Marks, David C. Cullen, Isao Karube, Christopher R. Lowe and Howard H. Weetall.

at such a cost so as to be relatively expendable, the frequency, and availability of these tests would benefit society in ways too numerous to state.

Among the sensors being developed, the emerging field of microelectromechanical systems (MEMS) and nanoelectromechanical systems (NEMS) has demonstrated a number of recent significant scientific advancements, translating into a wide range of potential chemical and biological sensing applications.[1–18] Two alternate detection methods of note are outlined below. Briefly stated, these work by either detecting added mass through a shift in a natural resonance or detecting surface stresses brought about through receptor–ligand interaction. These devices, made by lithographic techniques, can be formed in highly uniform arrays in a form that can be readily integrated with motion transduction and microfluidic systems.[19]

In the case of oscillators, the types of materials that can be structured in this way have low mechanical losses providing a high mechanical quality factor and therefore well-defined resonant frequencies. The very specific resonant frequencies coupled with the low mass of the oscillator enable the detection of small amounts of additional bound mass. Experimental investigations illustrate that the ability to engineer nanoscale features on the surface of NEMS devices, combined with localized chemical functionalization, allows for specificity and calibration of these devices as detectors.[20,21]

For deflection devices, materials can be selected and layered in such a way as to maximize the binding and measured surface stresses. Although the nature of this detection method makes measuring single molecules impossible, they have been shown to be quite effective in measuring low concentrations of relevant analytes.[22,23] They have the added advantage of having the ability to operate in liquid and thus measure analytes in real time.

2   RESONANCE DETECTION OF BIOLOGICAL ANALYTES

2.1   Introduction

The frequency of vibration of any resonating body obeys precisely known mathematics, which depend upon many factors, such as the material of the structure, its shape, its mass, and so on. This is clearly seen in the example of a violin, whose strings, although of the same material, of the same length, and under similar tension, emit very different tones on account of the differing densities between the strings. This analogy can be generalized by saying structures of different masses, with all other parameters similar, will resonate at different frequencies (with some exceptions, of course, most notably the simple pendulum).

It is this effect that is currently being investigated at the micro- and nanoscale. The motivation behind this is simple. All biosensors detect the presence of a mass, generally through secondary effects due to the presence of that mass. If one could detect this mass directly, it would simplify the system through the absence of this secondary "probe".

One possible way of doing this is to measure the frequency of a vibrating structure that is functionalized against an analyte of interest. As the analyte binds to the sensor, its frequency changes by virtue of the finite added mass of that analyte. The sensitivity to this mass change can be seen both mathematically and intuitively to be largely dependent upon the mass of the sensor itself, that is, the less massive the sensor, the smaller the detectable bound mass. Since the masses of biological analytes are exceedingly small, to detect them, one would need a very small vibrating device, indeed. This effect has therefore been investigated using micro- and nanoelectromechanical resonating systems. Most popular among these types of resonating sensors is a cantilever (or diving board–like) shape.

To perform their specialized functions, resonant sensors and actuators must reliably store and convert different forms of energy, transduce signals, and respond in a repeatable manner to external chemical and biological environments. For instance, biomolecular adsorption of target analytes to treated regions of a cantilever-based sensor can alter mechanical stress within the oscillator as well as its total mass and thus influence both the bending and the natural frequency of the cantilever, respectively. Since both the deflection and resonant frequency shift are highly dependent upon the position of the adsorbed material, it is difficult to determine the exact amount of additional mass present without microscopic inspection.[24–34]

To circumvent these limitations, one can construct arrays of surface micromachined oscillators with precisely positioned chemically functionalized anchors. In this scenario, binding events are confined to a particular portion of the device and do not occur anywhere else on the surface.

Although many other methods for signal transduction exist (such as piezoresistive, capacitive, and magnetomotive), for the cases described below, signal transduction was achieved by employing an optical-deflection or interferometric system to measure the mechanical bending or the frequency change in the out-of-plane translational vibrations resulting from additional loading by the specifically adsorbed mass.[11–13,20,21,35–40] Within such a configuration, a collimated laser beam is focused onto the free end of the cantilever and is reflected onto a split photodiode. The difference signal between the two cells of the position-sensitive detector determines the cantilever bending while the AC signal corresponds to vibrations of the cantilever (see Figure 1a). In the case of interferometric detection, reflectance variations from the incident He–Ne laser are measured using a single-cell photodetector (see Figure 1b). The measured vibrations are induced through electrostatic, magnetic, piezoelectric, or optical actuation.[18,40–49]

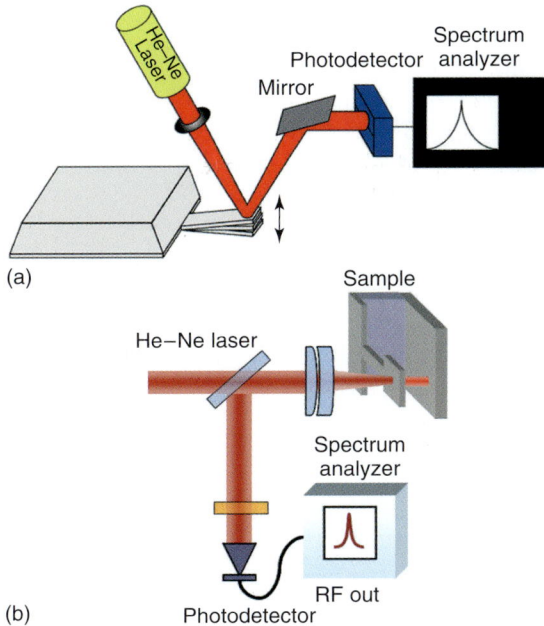

Figure 1. Schematic of (a) deflection and (b) interferometric optical measurement apparatus.

2.2 Mass Sensitivity of Resonant Detection

In order to estimate the lower detectable mass limit, surface-machined NEMS oscillators with integrated circular Au contacts and sub-attogram mass detection sensitivity were used in a study by researchers at Cornell University (see Figure 2).[20] In order to maximize the sensitivity, the Au dots were placed in close proximity to the free end of the oscillator, where the amplitude of the oscillation is maximized (see equation 1). Mass loading effects were illustrated through selective immobilization of dinitrophenyl poly(ethylene

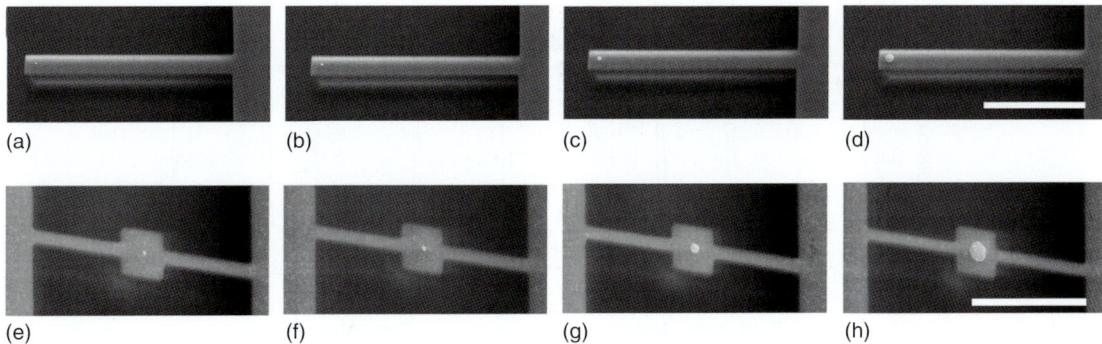

Figure 2. Scanning electron micrograph (SEM) micrographs of cantilever (a–d) and bridge (e–h) type oscillators where the scale bars correspond to 5 μm and 2 μm, respectively. The diameters of the Au pads were 50, 100, 200, and 400 nm, from left to right. [Reprinted from Ilic et al.,[20] with permission from American Institute of Physics.]

glycol) undecanthiol (DNP-PEG4-C11thiol)-based molecules to prefabricated Au contacts on the surface of the NEMS resonator.

Following measurement of baseline frequencies, the gold dot was then removed using a wet gold etch. Subsequent measurements provided calibration data of corresponding frequency shifts that depended on the size of the original gold dot (Figure 3). Similar devices were immersed in a thiol solution to facilitate the selective binding of

DNP-PEG4-C11thiol self-assembled monolayers to the gold nanodots. Frequency shifts due to this additional mass loading were measured. For the frequency shifts of 125 Hz and 1.10 kHz demonstrated in Figure 4(a) and (b), the corresponding additional mass loading, calculated using Equation (1), was 6.3 and 213.1 ag, respectively. Within the linear elastic limit, the resonant frequency shift due to additional mass loading, assuming the bound mass is much less than the mass of the

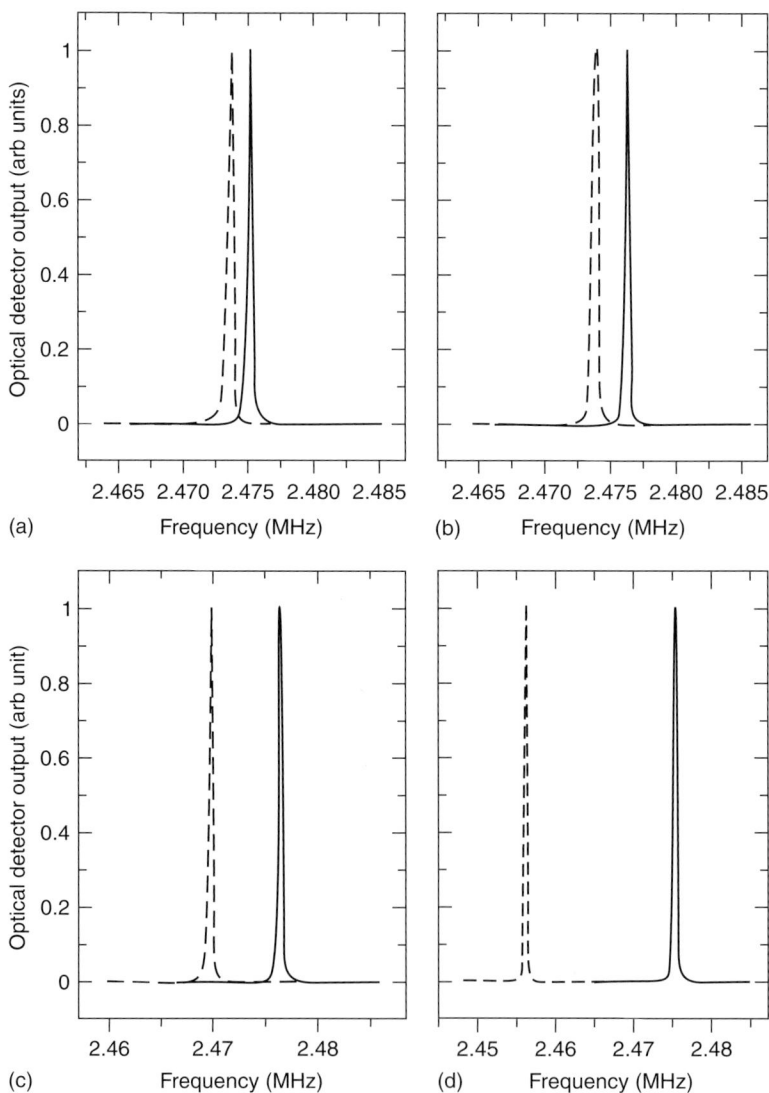

Figure 3. Calibration frequency spectra of 10-μm-long rectangular cantilevers before (dashed line) and after (solid line) the removal of the (a) 50-, (b) 100-, (c) 200-, and (d) 400-nm-diameter gold dots. [Reprinted from Ilic et al.,[20] with permission from American Institute of Physics.]

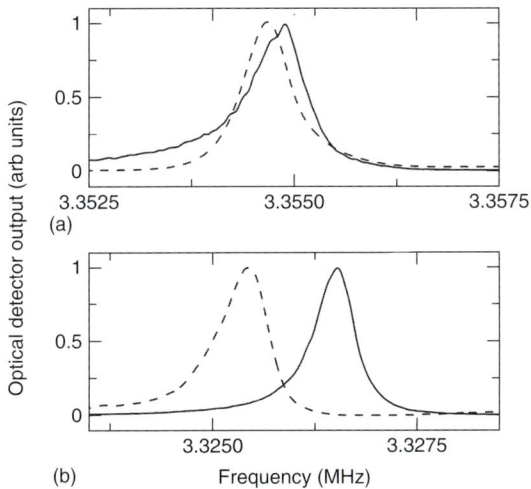

Figure 4. Experimentally measured frequency spectra before (solid line) and after (dashed line) the adsorption of the thiolate on (a) 50- and (b) 400-nm-diameter gold dot. [Reprinted from Ilic et al.,[20] with permission from American Institute of Physics.]

oscillator, is given by

$$\Delta f = 0.279 m_{\text{eff}} \sqrt{\frac{EI}{l^3 m_{\text{o}}^3}} \qquad (1)$$

where I is the moment of inertia of the cantilever, E is the Young's modulus of low-stress silicon nitride ($E_{\text{measured}} \sim 110\,\text{GPa}$ assuming a silicon nitride density of $3.4\,\text{g cm}^{-3}$), $m_{\text{eff}} = m_{\text{bound}}(x/l)$ is the effective mass of the mass bound to the cantilever, x is the position of the bound mass measured from the base of the cantilever, l is the total length of the cantilever, and m_{o} is the mass of the cantilever without bound mass.

The same researchers, in a different study, set about to show how this exquisite sensitivity could be used to detect a single bound biological analyte using similar devices.[21] Figure 5(a–c) shows cantilever devices fabricated from 90-nm-thick, low-stress silicon nitride with a 40-nm Au dot near the free end. Thiolate functionalized double-stranded 1587-bp DNA (dsDNA) molecules were used to illustrate the ability of single-molecule detection (see Figure 6). The resonant frequency of individual oscillators in an array of resonator devices was measured by thermo-optically driving the individual devices and detecting their motion by optical interference. The number of bound molecules was quantified from the measured frequency shift of the oscillator. Figure 7 illustrates the frequency shift due to a single dsDNA molecule bound to the catalyzing Au dot.

2.3 Measurement of Biological Analytes

As a proof of principle, different types of biological analytes were detected using resonant NEMS/MEMS structures. These analytes included cells and viruses. The importance of detecting these types of analytes is clear when applied to areas of public health. Although the specific species detected in the studies described subsequently were not necessarily harmful to humans

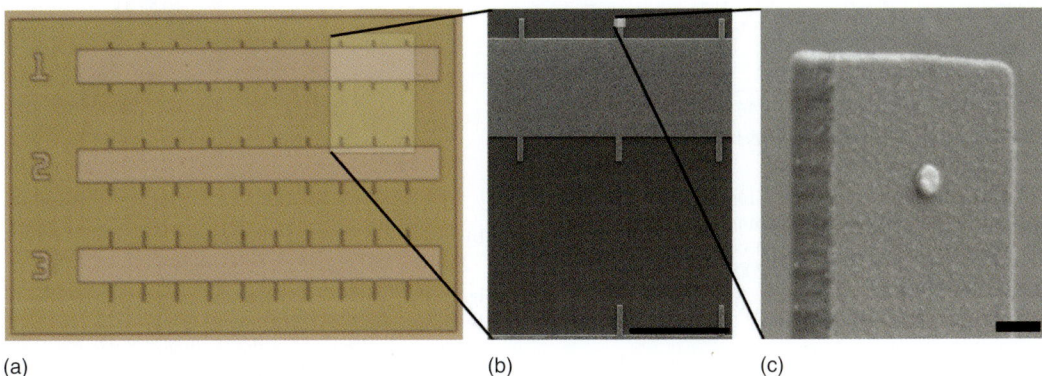

Figure 5. (a) Optical and (b) and (c) scanning electron micrographs highlighting cantilevers of various length with 40-nm Au dots centered 300 nm away from the free end of the cantilever. [Reprinted with permission Ilic et al.[21] copyright 2005, American Chemical Society.]

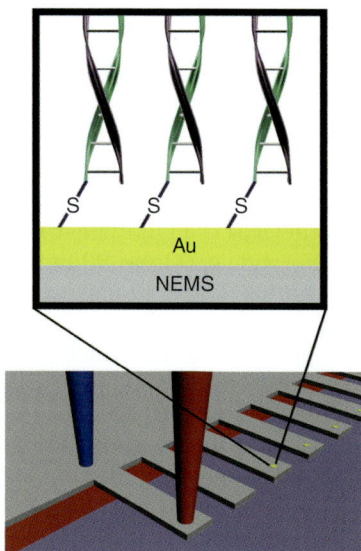

Figure 6. Schematic of the optically driven interferometric setup employing a red He–Ne laser and a modulated blue diode laser for motion detection and excitation, respectively. Zoomed-in schematic shows the binding dynamics of the thiolated dsDNA molecules to the Au dots. [Reprinted with permission Ilic et al.[21] copyright 2005, American Chemical Society.]

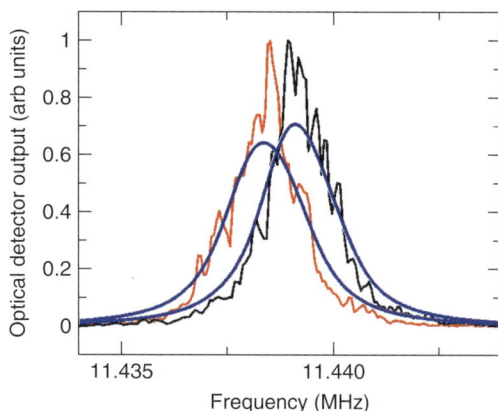

Figure 7. Measured frequency spectra for a NEMS resonator before (black) and after (red) the binding of a single DNA molecule. The blue lines are Lorentzian curve fits. [Reprinted with permission Ilic et al.[21] copyright 2005, American Chemical Society.]

bulk micromachined silicon nitride cantilevers, the presence of a single bacterium of *Escherichia coli* O157 : H7 was detected.[30] This type of bacteria is known to cause severe illness through the ingestion of undercooked meat. The cantilevers were coated with antibodies reactive against *E. coli* and then immersed into solutions containing the cells at concentrations varying from 10^5 to 10^9 colony-forming units/ml. Resonant frequency spectra were taken before and after antibody coating and after exposure to cells. The measured vibrational mode was actuated entirely because of thermal noise and ambient vibrations in air.

A single *E. coli* cell bound to a cantilever is shown in Figure 8(a). The measured frequency

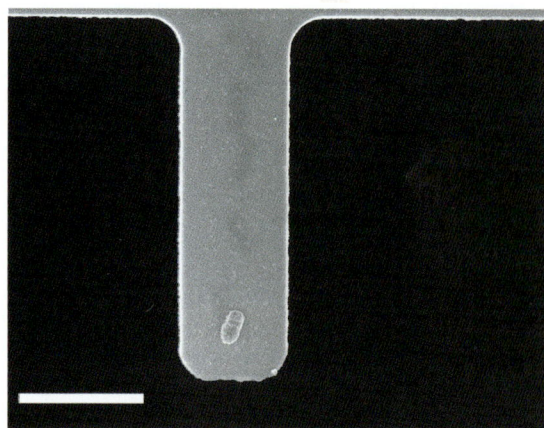

(a)

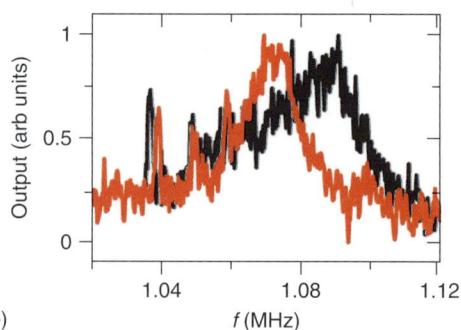

(b)

Figure 8. (a) Scanning electron micrograph (SEM) of a single *E. coli* O157:H7 cell bound to the immobilized antibody layer near the free end of the oscillator. Scale bar corresponds to 5μm. (b) The corresponding thermal and ambient noise spectra due to the transverse vibrations of the cantilever before (black) and after (red) antibody immobilization and single cell attachment. [Reprinted from Ilic et al.,[31] with permission from AVS The Science & Technology Society.]

in their measured state, the detection methods used could be adapted through different biochemistry to detect analytes of more relevance. Using an array of relatively large (15–500-μm long)

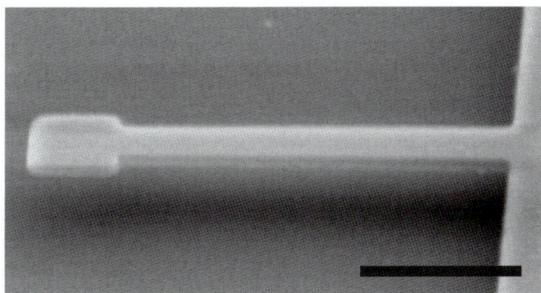

Figure 9. A cantilever beam with a $1 \times 1 \mu m^2$ paddle, defined using electron beam lithography. The scale bar represents $2 \mu m$. [Reprinted from Ilic et al.,[32] with permission from AVS The Science & Technology Society.]

shift of 4.6 kHz due to the immobilization of a single cell corresponds to a mass of 665 fg, which is consistent with other reports and the estimated volume of this cell. The measured resonant frequency spectra of the cantilever, in air, before and after antibody and cell attachment, are plotted in Figure 8(b).

A similar experiment requiring more sensitive devices was performed to detect the presence of a single virus. As a nonpathogenic model the group used the insect phage baculovirus.[32] In this case, sensitivity enhancement was accomplished by employing smaller, surface-machined polycrystalline silicon NEMS devices

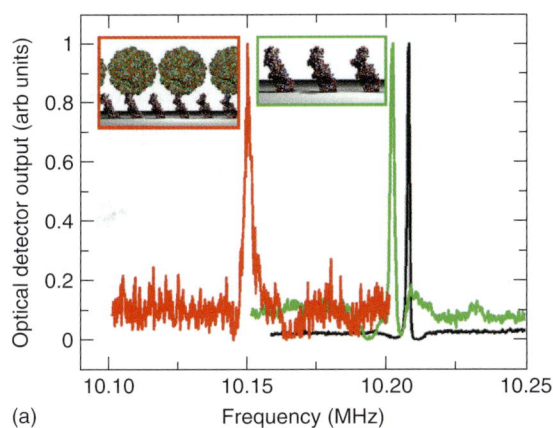

(a)

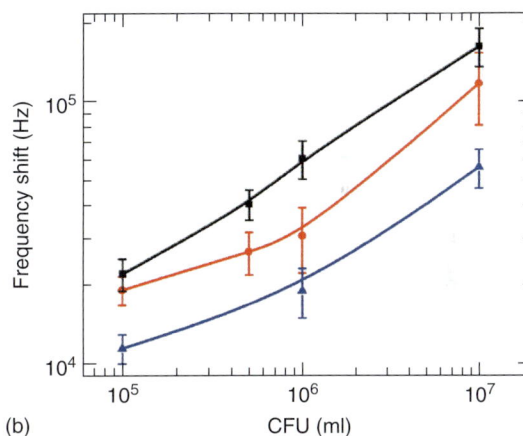

(b)

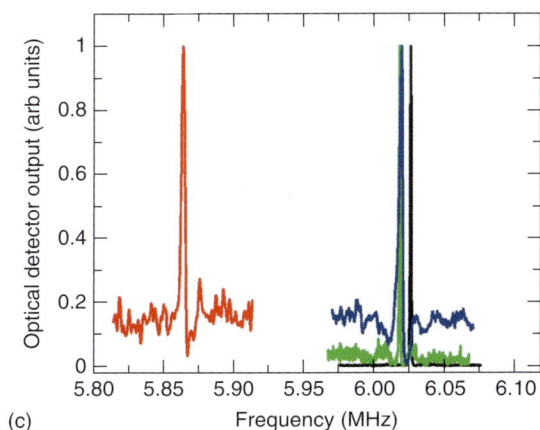

(c)

Figure 10. Data showing detection of baculovirus. (a) Frequency spectra. The baseline frequency measurement is shown as the black peak. Subsequent frequency measurements following antibody (green) and virus binding (red) show easily measurable frequency shifts. The insets are cartoon schematics of the antibody–surface and virus–antibody–surface interactions (from right to left). (b) Frequency shifts of 6 (black), 8 (red), and 10 (blue) μm long cantilevers as a function of virus concentration in solution. (c) Frequency shift due to the control experiment using a buffer solution without baculovirus (blue). [Reprinted from Ilic et al.,[32] with permission from AVS The Science & Technology Society.]

(see Figure 9). Following baseline frequency measurements in vacuum (10^{-6} Torr, in order to remove viscous damping effects), the devices were first functionalized by submersion in an antibody solution (AcV1 antibody against baculovirus gp64 envelope protein). Their frequency shift due to antibody binding was then measured. Similar steps using a baculovirus solution showed additional frequency shift, thereby sensing the specific binding of virus particles to the functionalized cantilever (see Figure 10a). Figure 10(b) shows the frequency shift variation with the baculovirus concentration for three different cantilever lengths. Nonspecificity of the binding was evaluated using a buffer solution without baculovirus. The frequency shifts from the control experiments were negligible compared to a shift from the binding of the baculovirus.

3 STATIC-DEFLECTION-BASED DETECTION

3.1 Introduction

Most surface-stress-based MEMS detection systems work by functionalizing a single side of a cantilevered beam. As the target ligand binds to an immobilized receptor on this surface, a stress develops across that face of the beam. Since only one side of the beam experiences this effect, the differential stress between the top and bottom faces of the beam cause a measurable bending of the cantilever. This slight bending is governed by Stoney's equation:

$$\Delta z = 3\frac{(1-v)}{E}\frac{L^2}{t^2}\Delta\sigma \qquad (2)$$

Here, Δz is the deflection of the cantilever tip, v is Poisson's ratio of the device material, t is its thickness, L its length, and $\Delta\sigma$ is the differential change in surface stress (in J m^{-2}) between the top and bottom surfaces of the cantilever.

Generally, this slight bending is interrogated through the use of optical methods, similar to those used in atomic force microscopy (AFM), where a laser, impinging upon the cantilever with oblique incidence, is deflected into a split photodiode. The translated signal is thereby the difference in potential between the two adjacent photodiodes.

A notable difficulty with this method is the tendency for an uncoated cantilever, otherwise resting at its equilibrium position, to succumb to thermal effects and deflect without the stress of bound species. The deflection method in general is especially prone to this problem mainly because coating one side of the cantilever, designed to secure only the receptor of interest, by definition means that the cantilever will be a bilayer structure. Unless by some fortunate chance that the two layers share in common a near exact coefficient of thermal expansion, any temperature change will incite bending of the cantilever and be a frustrating source of noise. Several solutions to this problem have been implemented in the literature. These involve either controlling the temperature of the system or incorporating a reference device into the sensor.[22,23] The latter method would also help minimize false signals due to nonspecific binding of nontarget material.

3.2 Detection of DNA Hybridization

By exploiting the effect outlined in the preceding text, the detection of hybridization events between probe and sample single-stranded DNA (ssDNA) was accomplished using 500-μm-long by 100-μm-wide by 1-μm-thick silicon cantilever devices (Figure 11).[23] DNA hybridization is normally detected using fluorescent tags for the measurement of gene expression. The cantilever beams used in this study were purposely made long and thin (see equation 2) and consequently, to avoid stiction (i.e., the permanent and catastrophic attachment of the cantilever to an underlying substrate) these devices were made using bulk micromachining methods, thereby removing the substrate beneath the device altogether.

Not only were hybridization events detected, but two separate types of DNA were detected in solution by functionalizing neighboring cantilevers with different probe capture strands of ssDNA. In serial measurements, these neighboring devices served as reference devices in order to minimize false signals due to thermal effects or nonspecific binding (Figure 12).

Figure 11. Scanning electron micrograph of an array of microfabricated silicon cantilevers (1-μm thick, 500-μm long, and 100-μm wide; Micro- and Nanomechanics Group, IBM Zurich Research Laboratory, Switzerland). [Reprinted with permission from J. Fritz, et al. *Science*, 2000, **288**, 316. Copyright 2000 AAAS.]

3.3 Protein Detection (Prostate Specific Antigen)

The importance of the detection of small quantities of proteins in biological fluids is showcased by the example of prostate specific antigen (PSA). Blood levels of PSA have been shown to be elevated for people with prostate cancer. Such levels have been used to help diagnose the presence of this disease before the onset of symptoms. Indeed it is also known that such marker proteins exist for other types of cancer. As a competitive technology to enzyme-linked immunosorbent assays (ELISA) for protein detection, the deflection method of detection holds several advantages. First, where the former requires many twofold binding events for a successful detection (antigen to capture antibody and labeled probe antibody to antigen), the latter is a label-free method. Furthermore, since the cantilever devices are fabricated using well-established methods from the microelectronics industry, an array of such devices would be quite smaller than the 96-well microtiter plates used in ELISA, thus allowing for a high density of tests in a small area. An impressive example of protein detection using microfabricated cantilevers was that of the prostate cancer marker PSA. This small protein was detected using a commercially available AFM cantilever (Figure 13) functionalized with a coating of monoclonal antibodies specific to PSA. This cantilever was placed in a temperature-stabilized flow cell and was shown to detect PSA concentrations down to $0.2 \, \text{ng ml}^{-1}$ in simulated

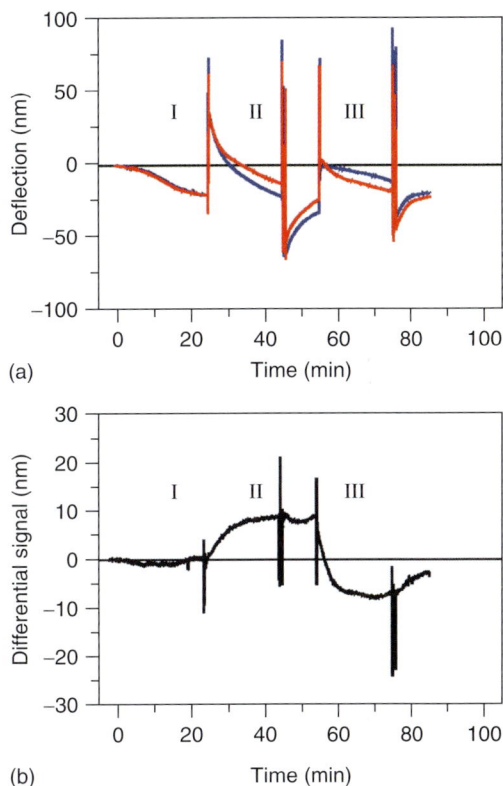

Figure 12. Hybridization experiment using two cantilevers functionalized with different sequences. Interval I is a baseline measurement followed by injection of a complementary sequence to one of the cantilevers (interval II). After purging, the second complementary sequence was injected (interval III), followed 20 min later by another purge. (a) Absolute deflection versus time of two individual cantilevers covered with two different oligonucleotides (red and blue). (b) Corresponding differential signal. [Reprinted with permission from J. Fritz, et al. *Science*, 2000, **288**, 316. Copyright 2000 AAAS.]

human serum (Figure 14).[22] This concentration is within the range of clinical interest and its successful detection makes this method competitive to those presently used.

4 FLUIDIC NEMS AND MEMS

As described earlier in the chapter, resonant NEMS systems have been demonstrated as sensitive mass detectors with sub-attogram and even single-molecule sensitivity. Sample delivery is generally difficult in such cases requiring the entire

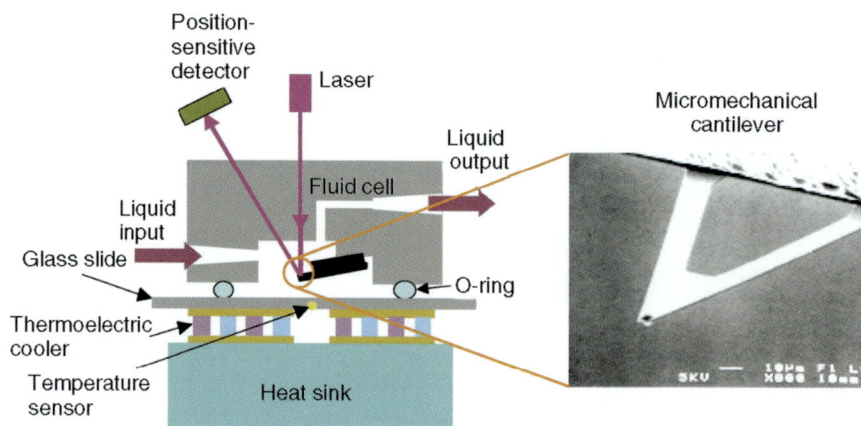

Figure 13. Schematic diagram of the experimental setup where a microcantilever is mounted in a temperature-controlled fluid cell. The scanning electron micrograph on the right shows the geometry of a gold-coated silicon nitride cantilever beam (200-μm long, 0.5-μm thick, and with each leg 40-μm wide). Deflection measurements were made using a laser beam that was reflected off the back of the cantilever and focused onto a position-sensitive detector. [Reprinted with permission Wu et al.[22] copyright 2001, Nature Publishing Group.]

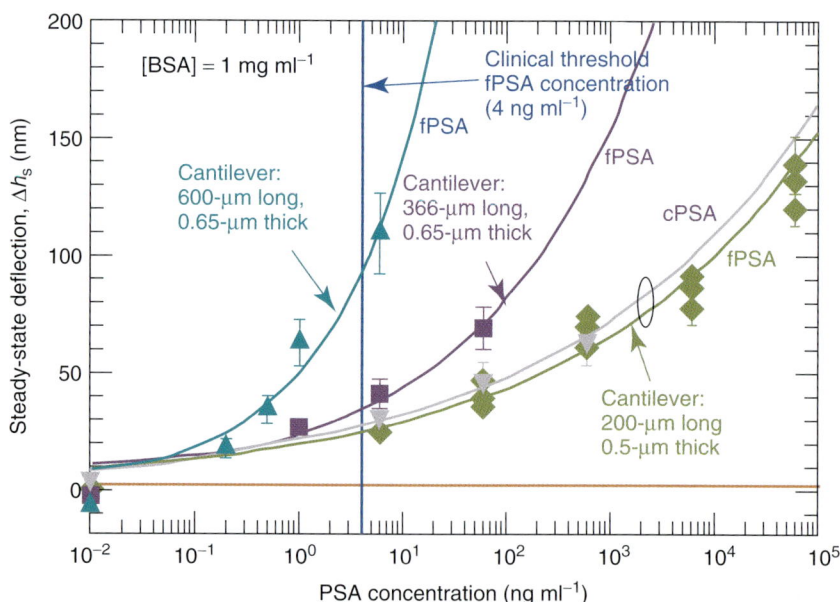

Figure 14. Steady-state cantilever deflections as a function of fPSA and cPSA concentrations for three different cantilever geometries. Note that longer cantilevers produce larger deflections for the same PSA concentration, thereby providing higher sensitivity. Each point represents an average of cantilever deflections obtained in multiple experiments done with different cantilevers. The error bars represent the range of deflections obtained from these experiments. The data (green diamonds) for fPSA detection, however, is from multiple experiments at a given concentration and is shown as a cluster plot. The error bar in each of these data points represents the fluctuation of the cantilever during the particular measurement. [Reprinted with permission Wu et al.[22] copyright 2001, Nature Publishing Group.]

device chip to be submersed into an analyte-containing mixture. Additionally, high vacuum is required to remove viscous damping to improve sensitivity.

Recently, researchers at Cornell University have made progress showing that these devices can be a useful part of an on-chip analysis system by demonstrating the ability to encapsulate resonant NEMS in part of a microfluidic network (Figure 15). On-chip implementation of these systems and other microscale systems is a highly sought after solution to the bulky laboratory apparatus that is presently required to perform bioassays. Since chips in the microelectronics industry are mass produced in a highly reproducible and low-cost fashion, the advantages of placing highly sensitive bioanalysis systems in this format are clear. For the NEMS system described here, microchannels were used for delivery of liquids and nitrogen (for drying) and the channels could be pumped down to pressures where viscous damping effects are negligible (Figure 16).

The devices were successfully operated under vacuum conditions while encapsulated within the microfluidic channels. Low operating pressures inside the channels eliminated viscous damping effects that would degrade the quality factor of resonance and thus reduce the mass sensitivity of the sensor if operated in air. This was confirmed by measuring the quality factor of a resonating structure while monitoring the pressure at the vacuum pump while the system was allowed to slowly leak

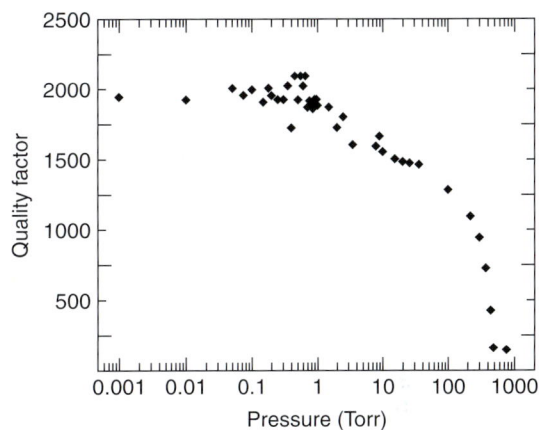

Figure 16. Quality factor as a function of pressure.

to atmospheric pressure (see Figure 16). These leaks were mostly from the external plumbing network, as the leak rate was not significantly different with or without the channels attached.

The encapsulated devices were shown to detect baculovirus using a method similar to those described earlier in the chapter with the exception that the fluids containing antibodies for device functionalization and virus particles for detection were delivered via the microfluidic network. Washing agents, nitrogen (for drying), and vacuum were also applied via microfluidics to achieve desirable measurement conditions.

Other advances in microfluidics have shown that it is possible to create on-chip sample preparation methods including preconcentration, filtration, solid phase extraction, and cell manipulation.[50–52] Microfluidic on-chip pumps and flow sensors have also been demonstrated, with the latter consisting of an encapsulated MEMS.[19] Such systems could be incorporated into a microfluidic network, part of which would encapsulate NEMS resonators for mass sensing applications.

Researchers at MIT were successful at doing the converse of what is described in the preceding text. By creating a microscale cantilevered resonator constructed out of a microfluidic channel, researchers under the direction of S. Manalis were able to perform real-time resonant detection of binding events.[53]

Keeping biological molecules in solution is crucial in preserving their maximum function so that processes such as receptor–ligand binding

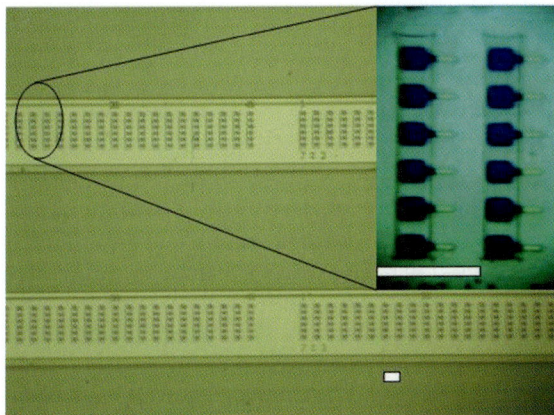

Figure 15. Optical micrograph of encapsulated devices. The scale bars are 50 μm. The inset shows 12 NEMS devices which are part of a larger array.

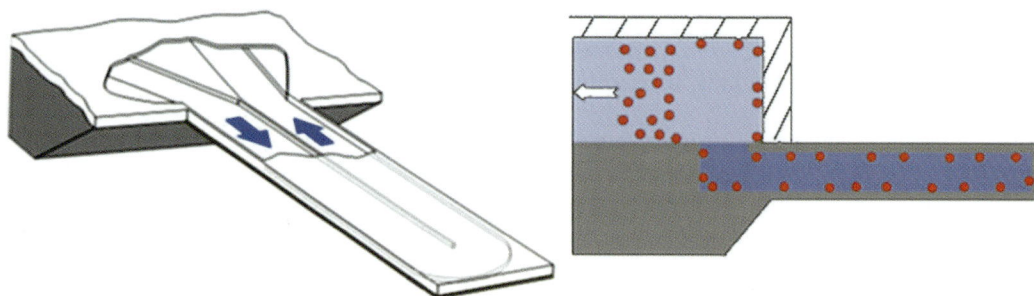

Figure 17. Cantilevered microfluidic channel. [Reprinted from Burg and Manalis[53], with permission from American Institute of Physics.]

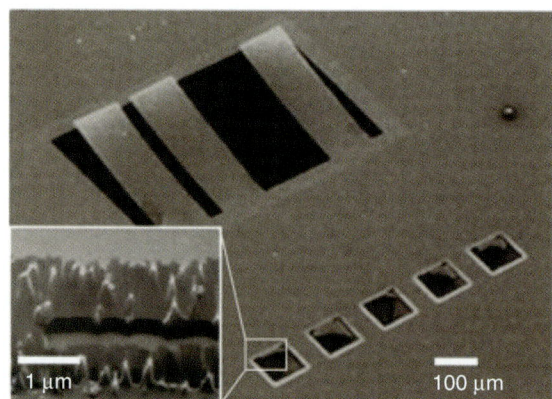

(a)

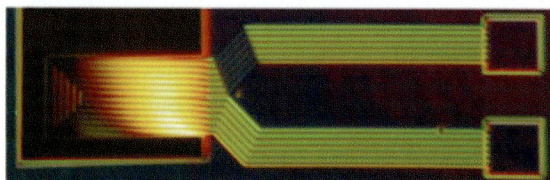

(b)

Figure 18. Cantilevered microfluidic channel. [Reprinted from Burg and Manalis[53], with permission from American Institute of Physics.]

can take place with an effectiveness close to that which occurs inside the body. This is problematic with the majority of resonant detection systems employing MEMS or NEMS oscillators since they require the removal of viscous damping forces to operate at peak sensitivity. Such can only be accomplished by the reduction of the scores of impinging molecules that would pelt the surface of the oscillator in a fluid environment. Thus,

the creation of reasonably high vacuum in the vicinity of the resonator is necessitated. Such an environment would be prohibitive to the retention of liquid on the sensor's surface, meaning that any biological molecules fixed there would be left dry, without the benefit of the ionic solution that would otherwise help it retain its very particular folded shape. Therefore, it is hypothesized that the drying of such molecules is detrimental to their function and may reduce the effectiveness of a sensor employing these techniques.

The advantage that the Manalis system holds is that since the liquid runs through the resonator, and not over it, resonant motion of the device can take place in vacuum, thereby allowing for the resonant detection of binding events as they occur inside the cantilever. With this system, such protein interactions as streptavidin–biotin binding were measured. Figures 17 and 18 show schematics and images of the device.

5   CONCLUSION

As the above examples illustrate, there exists a considerable effort underway in providing miniaturized, highly sensitive biological sensors. It should be noted that many other efforts exist, not listed here, and were omitted mainly for the sake of brevity. Progress has been made in the ability to detect low analyte concentrations or even single molecules. To achieve this ever-increasing sensitivity, more complicated design, fabrication, transduction methods, and/or environmental control was necessary. Such requirements may seem daunting when thinking of real-world consumer applications for the detection of such analytes as disease protein markers, pathogenic contaminates, or biological warfare agents. However these engineering issues will undoubtedly be overcome. Research and commercial tools based on this technology may be soon realized.

REFERENCES

1. W. E. Newell, Miniaturization of tuning forks. *Science*, 1964, **161**, 1320–1326.

2. H. C. Nathanson and R. A. Wickstrom, A resonant-gate silicon surface transistor with high-q band-pass properties. *Applied Physics Letters*, 1965, **7**, 84–86.

3. H. C. Nathanson, W. E. Newell, R. A. Wickstrom, and J. R. Davis, The resonant gate transistor. *IEEE Transactions on Electron Devices*, 1967, **14**, 117.

4. K. E. Peterson, Micromechanical light modulator array fabricated on silicon. *Applied Physics Letters*, 1977, **31**, 521–523.

5. K. E. Peterson, Silicon torsional scanning mirror. *IBM Journal of Research and Development*, 1980, **24**, 631–637.

6. K. E. Peterson, Silicon as a mechanical material. *Proceedings of the IEEE*, 1982, **70**, 420–457.

7. R. A. Buser and N. F. de Rooij, Resonant silicon structures. *Sensors and Actuators A*, 1989, **17**, 145–154.

8. N. V. Lavrik, M. J. Sepaniak, and P. G. Datskos, Cantilever transducers as a platform for chemical and biological sensors. *Review of Scientific Instruments*, 2004, **75**, 2229–2253.

9. H. G. Craighead, Nanoelectromechanical systems. *Science*, 2000, **290**, 1532–1535.

10. A. N. Cleland and M. L. Roukes, Fabrication of high frequency nanometer scale mechanical resonators from bulk Si crystals. *Applied Physics Letters*, 1996, **69**, 2653–2655.

11. D. W. Carr and H. G. Craighead, Fabrication of nanoelectromechanical systems in single crystal silicon using silicon on insulator substrates and electron beam lithography. *Journal of Vacuum Science and Technology B*, 1997, **15**, 2760–2763.

12. D. W. Carr, L. Sekaric, and H. G. Craighead, Measurement of nanomechanical resonant structures in single-crystal silicon. *Journal of Vacuum Science and Technology B*, 1998, **16**, 3821–3824.

13. D. W. Carr, S. Evoy, L. Sekaric, H. G. Craighead, and J. M. Parpia, Measurement of mechanical resonance and losses in nanometer scale silicon wires. *Applied Physics Letters*, 1999, **75**, 920–922.

14. X. M. H. Huang, C. A. Zorman, M. Mehregany, and M. L. Roukes, Nanoelectromechanical systems: nanodevice motion at microwave frequencies. *Nature*, 2003, **421**, 496.

15. P. A. Williams, S. J. Papadakis, A. M. Patel, M. R. Falvo, S. Washburn, and R. Superfine, Torsional response and stiffening of individual multiwalled carbon nanotubes. *Physical Review Letters*, 2002, **89**, 255502.

16. D. Qian, G. J. Wagner, W. K. Liu, M.-F. Yu, and R. S. Ruoff, Mechanics of carbon nanotubes. *Applied Mechanics Reviews*, 2002, **55**, 495.

17. S. J. Papadakis, A. R. Hall, P. A. Williams, L. Vicci, M. R. Falvo, R. Superfine, and S. Washburn, Resonant oscillators with carbon-nanotube torsion springs. *Physical Review Letters*, 2004, **93**, 146101.

18. V. Sazonova, Y. Yaish, H. Üstünel, D. Roundy, T. A. Arias, and P. L. McEuen, A tunable carbon nanotube electromechanical oscillator. *Nature*, 2004, **431**, 284–287.

19. D. Czaplewski, B. R. Ilic, M. Zalalutdinov, W. L. Olbricht, A. T. Zehnder, H. G. Craighead, and T. A. Michalske, A micromechanical flow sensor for microfluidic applications. *Journal of Microelectromechanical Systems*, 2004, **13**, 576.

20. B. Ilic, H. G. Craighead, S. Krylov, W. Senaratne, C. Ober, and P. Neuzil, Attogram detection using nanoelectromechanical oscillators. *Journal of Applied Physics*, 2004, **95**, 3694–3703.

21. B. Ilic, Y. Yang, K. Aubin, R. Reichenbach, S. Krylov, and H. G. Craighead, Enumeration of DNA molecules bound to a nanomechanical oscillator. *Nano Letters*, 2005, **5**, 925–929.

22. G. Wu, R. H. Datar, K. M. Hansen, T. Thundat, R. J. Cote, and A. Majumdar, Bioassay of Prostate-Specific Antigen (PSA) using microcantilevers. *Nature Biotechnology*, 2001, **19**, 856–860.

23. J. Fritz, M. K. Baller, H. P. Lang, H. Rothuizen, P. Vettiger, E. Meyer, H.-J. Guntherodt, Ch. Gerber, and J. K. Gimzewski, Translating biomolecular recognition into nanomechanics. *Science*, 2000, **288**, 316–318.

24. H. P. Lang, R. Berger, C. Andreoli, J. Brugger, M. Despont, P. Vettiger, Ch. Gerber, J. K. Gimzewski, J. P. Ramseyer, E. Meyer, and H.-J. Guntherodt, Sequential position readout from arrays of micromechanical cantilever sensors. *Applied Physics Letters*, 1998, **72**, 383–385.

25. Z. Hu, T. Thundat, and R. J. Warmack, Investigation of adsorption and absorption-induced stresses using microcantilever sensors. *Journal of Applied Physics*, 2001, **90**, 427–431.

26. L. A. Pinnaduwage, V. Boiadjiev, J. E. Hawk, and T. Thundat, Sensitive detection of plastic explosives with self-assembled monolayer-coated microcantilevers. *Applied Physics Letters*, 2003, **83**, 1471–1473.

27. D. R. Baselt, G. U. Lee, and R. J. Colton, Biosensor based on force microscope technology. *Journal of Vacuum Science and Technology B*, 1996, **14**, 789–793.

28. G. Wu, H. Ji, K. Hansen, T. Thundat, R. Datar, R. Cote, M. F. Hagan, A. K. Chakraborty, and A. Majumdar, Origin of nanomechanical cantilever motion generated from biomolecular interactions. *Proceedings of the National Academy of Sciences of the United States of America*, 2001, **98**, 1560.

29. C. A. Savran, S. M. Knudsen, A. D. Ellington, and S. R. Manalis, Micromechanical detection of proteins using aptamer-based receptor molecules. *Analytical Chemistry*, 2004, **96**, 3194–3198.

30. B. Ilic, D. Czaplewski, H. G. Craighead, P. Neuzil, C. Campagnolo, and C. Batt, Mechanical resonant immunospecific biological detector. *Applied Physics Letters*, 2000, **77**, 450–452.

31. B. Ilic, D. Czaplewski, M. Zalalutdinov, H. G. Craighead, P. Neuzil, C. Campagnolo, and C. Batt, Single cell detection with micromechanical oscillators. *Journal of Vacuum Science and Technology B*, 2001, **19**, 2825–2828.

32. B. Ilic, Y. Yang, and H. G. Craighead, Virus detection using nanoelectromechanical devices. *Applied Physics Letters*, 2004, **85**, 2404–2406.

33. N. V. Lavrik and P. G. Datskos, Femtogram mass detection using photothermally actuated nanomechanical resonators. *Applied Physics Letters*, 2003, **82**, 2697.

34. A. Gupta, D. Akin, and R. Bashir, Single virus particle mass detection using microresonators with nanoscale thickness. *Applied Physics Letters*, 2004, **84**, 1976–1978.

35. G. Meyer and N. M. Amer, Novel optical approach to atomic force microscopy. *Applied Physics Letters*, 1988, **53**, 1045–1047.

36. S. Alexander, L. Hellemans, O. Marti, J. Schneir, V. Ellings, P. K. Hansma, M. Longmire, and J. Gurley, An atomic-resolution atomic-force microscope implemented using an optical lever. *Journal of Applied Physics*, 1989, **65**, 164–167.

37. D. Rugar, H. J. Mamin, and P. Guethner, Improved fiber-optic interferometer for atomic force microscopy. *Applied Physics Letters*, 1989, **55**, 2588–2590.

38. P. K. Hansma, B. Drake, D. Grigg, C. B. Prater, F. Yashar, G. Gurley, V. Ellings, S. Feinstein, and R. Lal, A new optical-lever based atomic force microscope. *Journal of Applied Physics*, 1996, **76**, 796–799.

39. D. W. Burns, J. D. Zook, R. D. Horning, W. R. Herb, and H. Guckel, Sealed-cavity resonant microbeam pressure sensor. *Sensors and Actuators, A*, 1995, **48**, 179–186.

40. B. Ilic, S. Krylov, K. Aubin, R. Reichenbach, and H. G. Craighead, Optical excitation of nanoelectromechanical oscillators. *Applied Physics Letters*, 2005, **86**, 193114.

41. D. Rugar, R. Budakian, H. J. Mamin, and B. W. Chui, Single spin detection by magnetic resonance force microscopy. *Nature*, 2004, **430**, 329–332.

42. T. C. Nguyen, Micromechanical filters for miniaturized low-power communications. *Proceedings of SPIE*, 1999, **3673**, 55.

43. S. Evoy, D. W. Carr, L. Sekaric, A. Olkhovets, J. M. Parpia, and H. G. Craighead, Nanofabrication and electrostatic operation of single-crystal silicon paddle oscillators. *Journal of Applied Physics*, 1999, **86**, 6072–6077.

44. C. W. Yuan, E. Batalla, M. Zacher, A. L. de Lozanne, M. D. Kirk, and M. Tortonese, Low temperature magnetic force microscope utilizing a piezoresistive cantilever. *Applied Physics Letters*, 1994, **65**, 1308.

45. V. Kaajakari and A. Lal, Parametric excitation of circular micromachined polycrystalline silicon disks. *Applied Physics Letters*, 2004, **85**, 3923–3925.

46. M. Zalalutdinov, K. L. Aubin, R. B. Reichenbach, A. T. Zehnder, B. Houston, J. M. Parpia, and H. G. Craighead, Frequency entrainment for micromechanical oscillator. *Applied Physics Letters*, 2003, **83**, 3815.

47. R. B. Reichenbach, M. K. Zalaludinov, K. L. Aubin, D. A. Czaplewski, B. Ilic, B. H. Houston, H. G. Craighead, and J. M. Parpia, Resistively actuated micromechanical dome resonators. *Proceedings of SPIE*, 2004, **5344**, 51.

48. M. Zalalutdinov, A. Zehnder, A. Olkhovets, S. Turner, L. Sekaric, B. Ilic, D. Czaplewski, J. M. Parpia, and H. G. Craighead, Autoparametric optical drive for micromechanical oscillators. *Applied Physics Letters*, 2001, **79**, 695.

49. K. L. Aubin, M. Zalalutdinov, T. Alan, R. B. Reichenbach, R. Rand, A. Zehnder, J. Parpia, and H. G. Craighead, Limit cycle oscillations in CW laser-driven NEMS. *Journal of Microelectromechanical Systems*, 2004, **13**, 1018.

50. S. Song, A. K. Singh, and B. J. Kirby, Electrophoretic concentration of proteins at laser-patterned nanoporous membranes in microchips. *Analytical Chemistry*, 2004, **76**, 4589–4592.

51. Y. Yang, C. Li, K. H. Lee, and H. G. Craighead, Coupling on-chip solid-phase extraction to electrospray mass spectrometry through an integrated electrospray tip. *Electrophoresis*, 2005, **26**, 3622–3630.

52. B. J. Kirby, A. R. Wheeler, R. N. Zare, J. A. Fruetel, and T. J. Shepodd, Programmable modification of cell adhesion and zeta potential in silica microchips. *Lab on a Chip*, 2003, **3**, 5–10.

53. T. P. Burg and S. R. Manalis, Suspended microchannel resonators for biomolecular detection. *Applied Physics Letters*, 2003, **83**, 2698–2700.

46

Nanobiolithography of Biochips

Levi A. Gheber

Department of Biotechnology Engineering, Ben-Gurion University of the Negev, Beer-Sheva, Israel

1 WHY GO NANO?

The introduction of the arrayed biosensor, the so-called DNA chips and variations, has brought tremendous advances in the field of genomics, proteomics, diagnosis, and drug development during the last decade. Currently, more and more pharmaceutical companies are using this technology for developing new drugs, more and more hospitals are using them for diagnostic purposes. However, the present technology limits the use of these valuable tools to large research laboratories, major pharmaceutical companies, or advanced hospitals. In order for these instruments to provide maximum societal benefit, there is a pressing need to make them portable, point-of-care devices. In order to realize this transition, it is important to realize what precisely the limitations of the present technology are and to devise ways of overcoming them.

1.1 Limitations of Current Techniques

The array spots are *large*, $\sim100\,\mu$m in width, placed at a spacing of $\sim300\,\mu$m. In principle such separations can be resolved with the naked eye (which can easily resolve 0.3 mm). An objective is needed in a scanner because the emitted light levels are very low, so high light-collection power is needed. To achieve a high light-collection power, a high numerical aperture (NA) is required, which leads to a large magnification and a small view field. For example, a 20× objective on a typical microscope will provide a ~1-mm-diameter view-field, so it will image approximately nine spots at once, at best. In order to image the whole array, it is necessary to mechanically scan the sample across the objective. Scanning requires complex mechanics; in fact a chip scanner includes sophisticated robotics, which leads to large size and heavy weight of the instrument. This renders the chip nonportable, despite the relatively small size and light weight of the microscope slide on which the array is printed.

Miniaturization serves the purpose of portability in a number of ways, beyond the obvious advantages of smaller size and lighter weight. As explained in the preceding text, the bottleneck in achieving true portability of chips is not the size of the chips per se, but the implications for reading that are imposed by this size. Reducing the size of the arrays themselves offers several advantages, as described in the subsequent text.

1.2 Foreseen Advantages of Future Nanobiochips

1.2.1 Optical Advantage (No Scanning)

Smaller spots, ~100-nm diameter at a spacing of ~400 nm, are resolvable with conventional optical

Handbook of Biosensors and Biochips. Edited by Robert S. Marks, David C. Cullen, Isao Karube, Christopher R. Lowe and Howard H. Weetall.

microscopy (diffraction limit for high-NA objectives is ~250 nm). The same 20× objective mentioned in the preceding text, providing a viewfield with a diameter of ~1 mm can image 2500 spots at once, at this spacing. This means that there is no need to scan the sample; the whole array can be read at once for spots with these dimensions.

1.2.2 Light-collection Advantage (High NA Possible Due to No Scanning)

Oil objectives are not used with current scanners because their typical working distance is ~170 μm with a depth of field of ~1 μm, which makes it extremely difficult to avoid collisions of the sample with the objective while scanning, and movement of the array out of the focal plane of such objectives is practically inevitable. If scanning is not required, the use of high-NA, immersion oil objectives is possible. The great advantage in using high-NA objectives is their light-collection power. In epifluorescence, where excitation light is delivered through the objective that is also collecting the emitted fluorescence light, brightness of collected image is proportional to NA^4 and inversely proportional to M^2 (M: magnification). For example: a 20× objective, $NA = 0.6$ collects ~7.7 times less light than a 40× objective, $NA = 1.4$ (oil), despite the higher magnification of the 40×.

Reduction of the size of spots and the spacing between them makes scanning unnecessary, opening the way to portability. In addition, the decrease in fluorescence signal following the reduction of spot size can be compensated to some extent by the use of high-NA oil objectives, which is possible once scanning is not required. Obviously, if an objective is implemented in the reading device, the portability is still limited, although much improved in comparison with the current technology (a scanner weighs somewhere between 8 and 15 kg, measures some 0.5 m in length and costs ~€50 000, an oil objective is ~5 cm in length and diameter, weighs a few hundreds of grams and costs ~€1000).

1.2.3 Speed Advantage

Reactions on a smaller chip are faster, because mass transfer (governed by diffusion of target molecules) and heat exchange (dissipation of heat generated by exothermic reactions) are faster.

1.2.4 Weight Advantage

Reduction in the linear dimensions of spots by 3 orders of magnitude leads to reduction in surface area by 10^6. An additional reduction of ~10× in thickness (from a standard ~1-mm thick microscope slide to ~0.15 mm of a cover glass) leads to an expected reduction in the weight of the carrier glass of ~10^7. Although the weight of one microscope slide is not a problem in its own, these first-principle calculations make the point that within the same light weight, one could accommodate at least 1 million miniaturized nanoarrays.

1.2.5 Cost Advantage

The biological components of an array represent a disproportionately large fraction of the materials costs. Miniaturization would enable the consumption of as much as 10^6 times less biological probe molecules, yielding a substantial cost saving to offset the increased cost of more sophisticated processing methods.

1.3 The "Lab-on-a-Chip" Concept and Nanobiolithography

The current biochip technology requires, in fact, a fully equipped laboratory and trained personnel in order to use the microarrays. The operations required typically include separation and purification of the sample, amplification (PCR in the case of DNA), blocking, hybridization, and washing. True portability of biochips would not be achieved even if the chip-reading equipment is portable (as explained in the preceding text), owing to this fact. This realization constitutes the main reasoning for the need to include a whole laboratory on a chip, able to perform the various tasks that are nowadays performed in the laboratory surrounding the microarray scanner. Such a lab-on-a-chip should include fluidic systems, like channels, pumps, and valves, able to perform tasks of separation, transfer of liquids, purification, amplification, and so on, in addition to bioarrays (perhaps several, perhaps some DNA arrays and some protein arrays) and the array-reading systems (light sources, filters, light sensors, etc.).

Current microfluidic devices have typical dimensions of millimeters to centimeters, with microchannels of micrometer-scale diameters. Pumps, valves, and other mechanical devices are manufactured using the so-called microelectromechanical systems (MEMS) techniques, which are basically borrowed from microelectronics technology. Their dimensions, as the name indicates, are of tens of micrometers. Clearly, in order to tackle the concept of a portable lab-on-a-chip there is a pressing need to move from microfluidics to nanofluidics and from MEMS to NEMS (nanoelectromechanical systems), and this constitutes yet another significant reason to "go nano".

2 HOW TO GO NANO?

2.1 Nanobiolithography Techniques

Biolithography differs from the lithography employed in micro- and nanoelectronics in the fact that it aims at producing the same kind of features *with* biomolecules, *on* biomolecules, or *in* biomolecules. Since biomolecules are much more sensitive to their environment than inorganic materials (such as semiconductors and noble metals), most of the lithography technologies used in micro- and nanoelectronics are unsuitable for biolithography. High or ultrahigh vacuum is unacceptable, evaporation/sputtering is inapplicable, etching with strong acids would damage biomolecules, and irradiation with UV light would destroy DNA and proteins. Biolithography techniques, therefore, must operate in close to ambient atmosphere and/or in liquid, close to room temperature and moderate pH. *Nano*biolithography techniques have to comply with all the preceding conditions *and* provide nanometer precision of positioning and nanometer-size dimensions of features. This is the reason that the most natural candidate for nanobiolithography is the scanning probe microscope (SPM), which enjoys all these abilities. This chapter therefore concentrates on SPM-based techniques.

Several *SPM-based methods* have been demonstrated during the last decade. We shortly describe

them in the subsequent text, accompanied by representative examples, and discuss the advantages and disadvantages of each.

2.1.1 Nanografting

This approach is based on "nanoshaving" a self-assembled monolayer (SAM)[1] on the surface of the sample, thus dislocating molecules of the SAM and exposing the substrate for adsorption of other molecules from solution (Figure 1). It is not a "direct-write" technique, and the quality of the protein nanostructures depends on the spatial precision of SAM nanopatterns and on the selectivity of protein adsorption. The adsorption of a biomolecule is performed in a second step and is based on the interaction between the inserted molecule (Z in Figure 1) and the biomolecule.

The advantage of the technique is the simplicity of the SPM tools needed, basically just a simple Atomic Force Microscope (AFM) probe is needed in order to create the negative patterns in the SAM. The chemical details, however, render it less widely used.

2.1.2 Dip-pen Nanolithography (DPN)

Dip-pen nanolithography (DPN) has been historically the first demonstrated technique to use SPM in order to directly write nanoscale patterns of molecules.[2] It uses an AFM probe that is dipped into an "ink" and is then precisely positioned on a surface onto which it writes the molecules, much

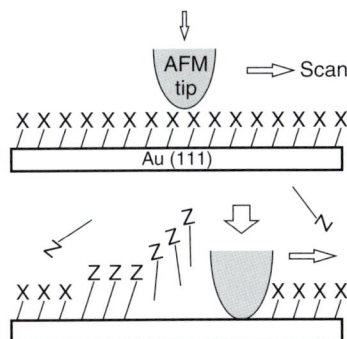

Figure 1. Basic principle of nanografting. [Reproduced with permission Wadu-Mesthrige et al.[1] copyright 1999, American Chemical Society.]

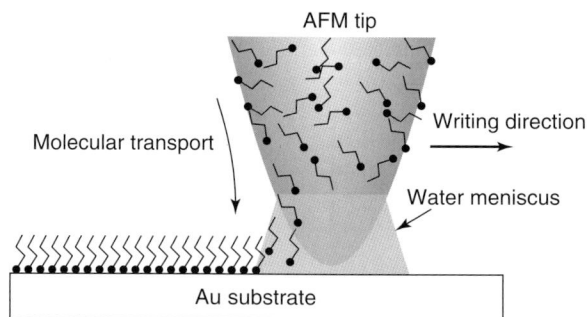

Figure 2. Schematic process of DPN writing. [Reprinted with permission from Piner et al.[2] Copyright 1999 AAAS.]

like a conventional dip pen, with a nanometer resolution, hence the name of the technique.

The writing process is mediated by the (spontaneous) formation of a narrow capillary "neck" of water that is formed between the AFM probe tip and the sample, when the experiment is conducted in ambient atmosphere (in air). This capillary bridge allows the transport of molecules from the tip to the sample (and vice versa) and in the case where the molecules attach themselves to the surface, stable surface structures are formed on the surface, with nanometric dimensions (Figure 2). The choice of an "ink" designed to react with the surface on which it is written is very important, since it provides the chemical driving force that favors the transport of molecules from the tip to the surface.[3]

Another important aspect of DPN is that the humidity and temperature control provide a means of controlling the basic writing process and size of printed features. On the other hand, the need for controlled-environment chambers imposes some limitations on the technique.

DPN can be extended in principle to a parallel lithography process,[4,5] by using arrays of AFM cantilevers. Owing to the fact that AFM cantilevers are manufactured using conventional lithographic methods borrowed from the microelectronics industry, it is relatively easy to manufacture such arrays.

2.1.3 Nano Fountain Pen Nanolithography (NFP)

The "fountain pen" technique typically use a glass or quartz capillary drawn into a sharp tip, with an aperture ranging between a few tens of nanometers up to a few hundreds of nanometers (Figure 3). This nanopipette is mounted as the probe of an SPM and precisely controlled, to deliver a liquid filled in the capillary to the surface. In contrast to DPN, the probe is not dipped into an ink vessel and can in principle write continuously, similar to the differences between the macro dip pen and fountain pen.

The use of micropipettes as probes for SPMs had been originally developed for near-field scanning optical microscopy (NSOM, or SNOM) as early as 1993,[6] however, originally their purpose was to serve as light waveguides with subwavelength apertures. It was later that these nanopipettes were used for delivery of liquids to a surface,[7,8] however not in a biological context. The demonstration of the ability to print patterns of proteins of submicrometer dimensions followed in 2002[9] and 2003.[10]

A nanopipette is filled with the liquid of choice, which is drawn to the tip of the probe by capillary

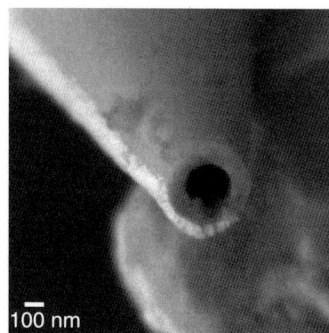

Figure 3. A nanopipette with a 200-nm-diameter aperture.

forces. Typically the liquid does not flow out of the pipette on its own because of the surface tension of the droplet that forms at its end. Flow occurs only upon contacting the pipette with a surface (Figure 4).

Two basic types of nanopipettes exist: the straight nanopipette and the cantilevered nano-pipette. The straight probes are controlled, just like in NSOM, using a "shear-force" detection mechanism. The other type can use all modes employed in AFM, that is, contact, tapping, and noncontact, owing to the fact that it is a bent, cantilevered probe. With the straight probe, though, it is difficult to write lines, because once in contact with the surface, the feedback signal vanishes (the oscillation amplitude is zero) and it is not possible to keep a constant force while translating the sample. The bent probe, however can easily write lines.

One more aspect of nano fountain pen nano-lithography (NFP) that is worth mentioning is that flow of the liquid out of the pipette to the surface is governed, among other parameters, by the wettability of the liquid-substrate system.

One advantage of NFP over DPN is its ability to print solutions of many types of molecules on many types of substrates (practically anything on anything), as we show in the subsequent text. Another, conceptual advantage (that has still to be demonstrated), is its ability to write "without pen lifting", assuming one can load the pipette with a train of various molecules that are written as they come out of the tip of the pipette.

The manufacture of nanopipettes is difficult, and they are apparently not amenable for either mass production or manufacture in arrays. However, work is under way in several groups, trying to combine the advantages of a fountain pen with the advantages of microfabrication, to end up with cheaper, parallel arrays of probes.

2.1.4   Scanning Near-field Photolithography (SNP)

Scanning near-field photolithography (SNP) uses the light emanating from an NSOM to induce light-assisted chemical reactions on a surface. Owing to the fact that NSOM illuminates an area with dimensions well under the wavelength of the light it uses, the chemical reactions are limited to that nanometer-sized region (Figure 5). In the context of bionanolithography, this method has been applied to immobilize proteins[11] and very recently DNA[12] with very high resolution. Although not purely "direct write", SNP is a very promising approach to bionanolithography.

2.1.5   Combinations and Variations

NPRW

Nanopen reader and writer (NPRW) is a combination of the nanografting technique and DPN technique.[14] Here a SAM monolayer is used as the "resist", like in nanografting, which is removed locally with the tip of an AFM probe. The AFM probe is precoated with a different molecule, thus it simultaneously "shaves" the resist SAM and deposits a new adsorbate. The same probe is then used to image the features produced this way (Figure 6). NPRW presents an advantage over DPN in that the resolution does not depend on the

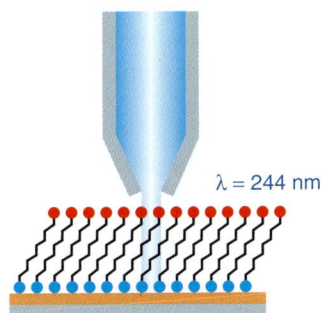

Figure 5. Schematic process of SNP. [Reproduced with permission Sun and Leggett[13] copyright 2002, American Chemical Society.]

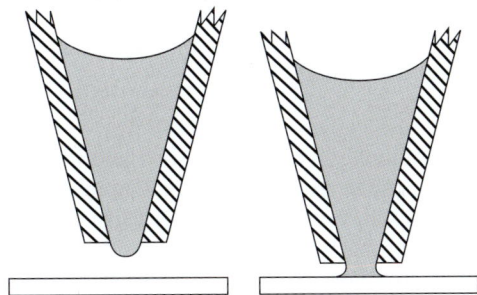

Figure 4. Schematic process of NFP writing.

NPRW

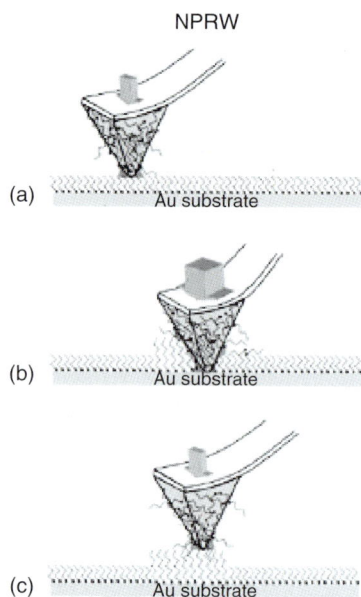

Figure 6. Schematic diagram of NPRW illustrating the three basic steps to produce and characterize a pattern under ambient laboratory conditions. [Reproduced with permission Amro et al.[14] copyright 2000, American Chemical Society.]

bare substrate or the humidity of the environment but on the tip–substrate contact force.

MFN

Unlike NPWR, in meniscus force nanolithography (MFN) the AFM probe is not coated with a SAM. Instead, it "swims" in a drop of solution containing the molecules to be written.[15] The surface tension of the drop is pressing the tip with high force on the substrate. This allows the feedback loop of the AFM to be disconnected and, furthermore, the high force flattens the underlying polycrystalline gold and strips it of the resist monolayer. Thus, MFN can proceed at very high writing speeds (Figure 7).

MFP

Micromachined fountain pen (MFP) is a modified AFM probe with integrated fluidic channels running over the cantilever beams (Figure 8), which can replace the pulled glass capillaries used in NFP.[16] The advantage it offers is the possibility of mass production of such probes, using standard technologies common in microelectronics.

Another variant of the AFM-based fountain pen probe is termed *nano fountain probe* (as opposed

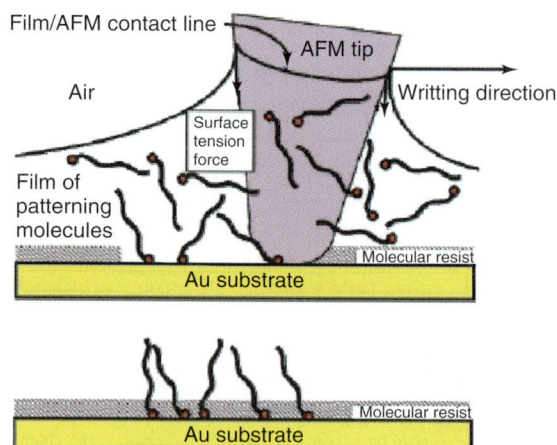

Figure 7. MFN principle. [Reprinted with permission Schwartz[15] copyright 2001, American Chemical Society.]

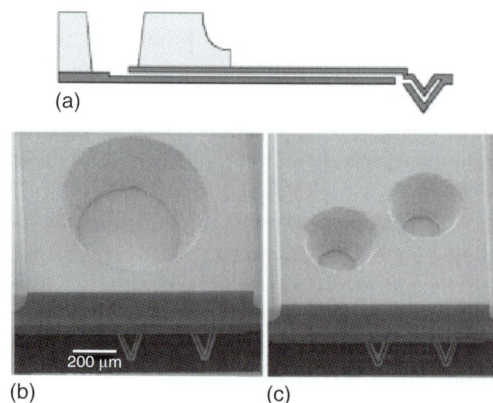

Figure 8. Micromachined fountain pen for AFM-based nano-patterning. [Reprinted with permission Deladi et al.[16] copyright 2004, American Institute of Physics.]

to nano fountain *pen*), and uses a "volcano"-shaped AFM probe (Figure 9). This development has apparently the capability of both writing sub-100-nm features and avoids the need for dipping the probe repetitively.[17]

E-DPN

Electrochemical AFM dip-pen nanolithography (E-DPN) is an extension of DPN, which makes use of the minuscule water meniscus formed between the AFM probe and the surface as a nanometer-sized electrochemical cell, in which metal salts can be dissolved, reduced into metals electrochemically, and deposited on the surface.[18] This is

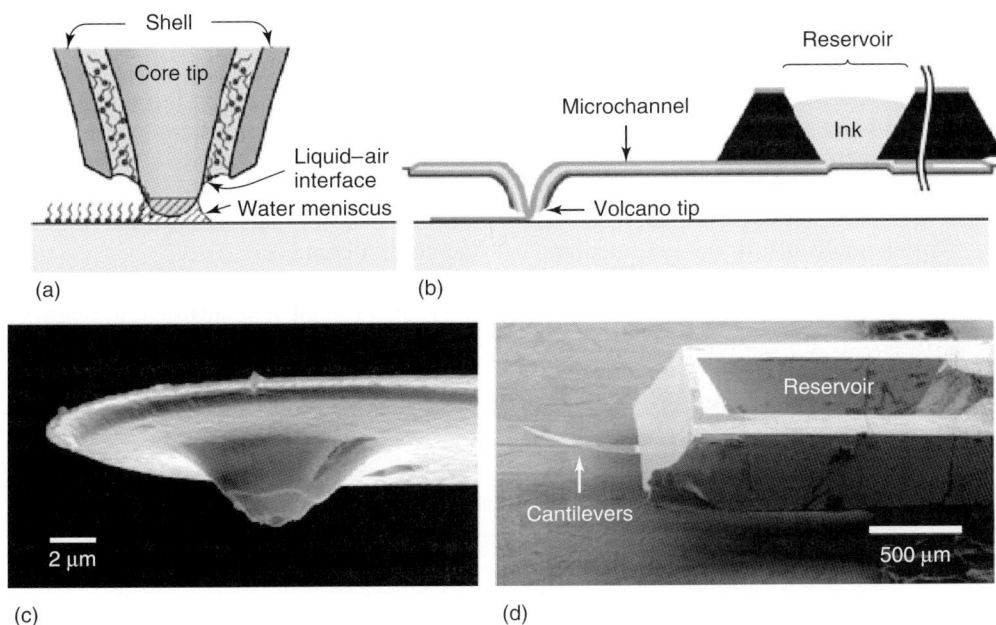

Figure 9. Writing mechanism of the volcano tip. [Reprinted with permission Kim et al.[17] copyright 2005, Wiley VCH.]

achieved by applying a voltage between the AFM tip and the substrate (Figure 10). This technique was used to write nanowires of metal, however it could conceivably be extended to biomolecules immobilized to these metallic features in a subsequent step.

E-NFP

Electrochemical nano fountain pen (E-NFP) uses a nanopipette, like NFP, but applies an electric field between two electrodes, one inserted into the nanopipette, and one inserted in the bath of ionic solution in which the pipette is immersed (Figure 11). The method offers the fine control

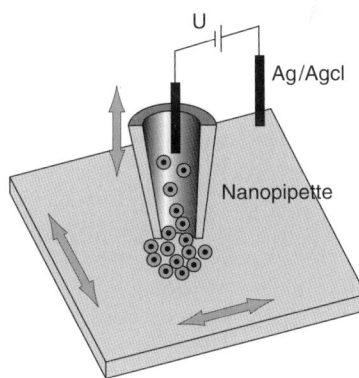

Figure 11. Schematic of the writing experiment. A voltage is applied between two Ag/AgCl electrodes, one inside the nanopipette, and one inserted into the bath of ionic solution. [Reprinted with permission Bruckbauer et al.[9] copyright 2002, American Chemical Society.]

of the delivery potentially down to the single-molecule level.[9]

2.2 Positive Nanobiolithography (Deposition of Material)

It is important to make a few distinctions before describing specific results.

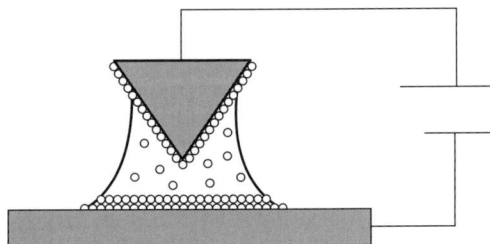

Figure 10. Schematic sketch of the E-DPN experimental setup. [Reprinted with permission Li et al.[18] copyright 2001, American Chemical Society.]

Direct write versus *indirect write*: By direct-write processes one means directly depositing the molecules of interest onto the substrate. For example, nanografting, as described in the preceding text, is not a direct-write technique. DPN, on the other hand, is a direct-write technique in some cases, and an indirect writing process in others, depending on the molecule to be patterned: if the directly written molecule is an intermediate stage, followed by additional steps to link the final molecule, then it is not considered "direct write". At this point, however, it is important to make the other distinction: *immobilization* of the patterned molecules *on the substrate*. The step of immobilization is important in order to achieve the goal of a functional biochip, and depending on the approach, may contribute to the definition of a lithography method as a direct or indirect one. If, for example, a technique is used to directly pattern a molecule that serves as the linker of the biomolecule to the substrate, then the technique is "direct write" as far as the linker is concerned, but indirect as far as the biomolecule is concerned. However, directly patterning the biomolecule on the substrate, even if possible, would not attain the goal of immobilization. Therefore, these definitions, if at all, should be used carefully and in the correct context.

One more point to make is that in some cases, fully operational nanoarrays (although very simplified, compared with microarrays) have been demonstrated, while in other cases only the patterning of the probe molecules has been presented. This is indicated in the brief description of the reported achievements in the subsequent text.

Several examples are described, grouped according to the nanopatterned biomolecules. These are also summarized in Table 1.

Table 1. A summary of biomolecules that have been patterned using nanobiolithography techniques

Molecule	Details	Method	References
DNA		MFN	15
		E-NFP	9
		Nanografting	19
		DPN	20
		SNP	12
Proteins	Lysozyme, IgG	Nanografting	1
	IgG	Nanografting	21
	Lysozyme	Nanografting	22
	Collagen	DPN	23
	BSA	SNP	11
	Streptavidin, BSA	DPN	24
	Protein G	E-NFP	9
	Lysozyme	DPN	25
	Lysozyme, IgG	DPN	26
	IgG, DNA	E-NFP	27
	His-tagged peptides	E-DPN	28
	IgG	DPN	29
	Enzyme: staphylococcal serine V8 protease	Immobilized DPN	30
	Protein G, GFP	NFP	10
	Enzyme: trypsin	NFP	31, 32
	IgG in nanowells	E-NFP	33
	Enzyme: DNase I	DPN	34
	IgG	DPN	35
	His6-ubiquitin, thioredoxin	DPN	36
	Avidin, BSA	c-AFM	37
	Avidin	DPN	38
Virus	CPMV	DPN, nanografting	39
	TMV	DPN	40
Single cell	*Escherichia coli*	DPN	41
Polymers	Photoresist	NFP	7
	PDMS	DPN	42
	TRIM	NFP	43

CPMV: cowpea mosaic virus; TMV: tobacco mosaic virus; PDMS: Polydimethylsiloxane; TRIM: trimethylolpropane trimethacrylate.

2.2.1 DNA

Patterning of DNA has been demonstrated with most of the techniques mentioned here. Nanografting,[19] MFN,[15] E-NFP,[9] DPN,[20] and recently SNP.[12] Importantly, ssDNA nanostructures have been proved to be amenable for hybridization with target complementary DNA fragments, in multifunctional arrays[20,27,44] (Figures 12 and 13).

2.2.2 Proteins

A large variety of proteins has been printed on submicrometer scale by virtually all the techniques discussed in the preceding text. *Lysozyme* was printed using *Nanografting*[1,22] and *DPN*.[25,26]

Immunoglobulin G (*IgG*) antibodies from various sources have been patterned using *Nanografting*,[1] for rabbit IgG[21] their function was proved with goat antirabbit IgG, *DPN*[26] where direct writing of rabbit IgG was demonstrated on a multicomponent (lysozyme) array and probed with anti-IgG, and antirabbit and antihuman IgG (Figure 14), probed with rabbit and human IgG, respectively[29] and anti-p24(human immunodeficiency virus (HIV)).[35] Rabbit IgG was also patterned using *E-NFP*, and was proved functional by probing with anti-IgG.[27,33]

Bovine serum albumin (*BSA*) has been patterned using *SNP*,[11] *DPN*,[24] and conductive AFM (*c-AFM*).[37] Another important protein, *avidin/streptavidin* has been patterned with various techniques: *DPN*,[24,38] *c-AFM*.[37] Protein G has been patterned with *E-NFP*[9] and *NFP*.[10] Other

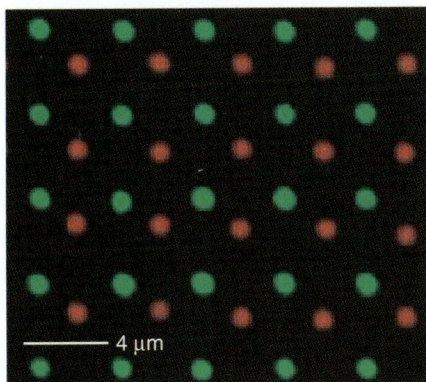

Figure 12. Combined red–green epifluorescence image of two different fluorophore-labeled sequences simultaneously hybridized to a two-sequence array deposited on an SiOx substrate by DPN. [Reprinted with permission Demers et al.[20] copyright 2002, AAAS.]

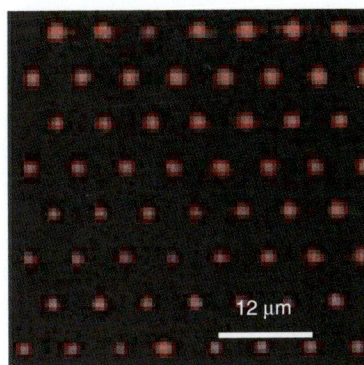

Figure 14. Array of dots consisting of antirabbit IgG (labeled with Alexa 594) on a negatively charged SiO2 surface. [Reprinted with permission Lim et al.[29] copyright 2003, Wiley VCH.]

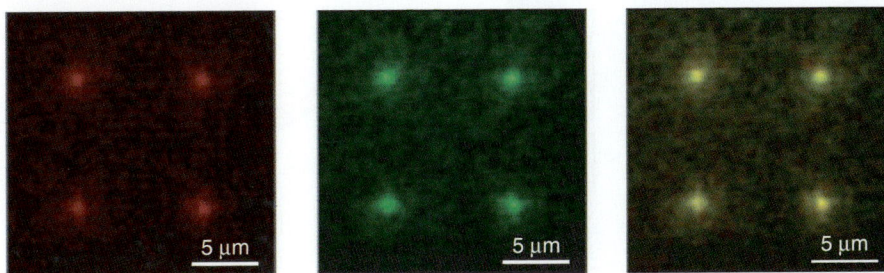

Figure 13. Spots of ssDNA labeled with biotin and Alexa Fluor 647 (Alexa 647 DNA, 35 mer) delivered by the nanopipette onto a streptavidin surface. Fluorescence images spotted DNA (red), hybridized complementary DNA (green), and combined image (yellow). The second oligonucleotide hybridizes selectively to the initial spotted oligonucleotide. [Reprinted with permission Bruckbauer et al.[27] copyright 2003, American Chemical Society.]

examples of proteins include his-tagged *peptides* with *E-DPN*,[28] his-tagged *ubiquitin* and *thioredoxin* with *DPN*[36] and *GFP* printed with *NFP*.[10] Also *collagen* was patterned using *DPN*.[23]

Enzymes have also been delivered to surfaces in one way or another, however they will be discussed subsequently in the context of enzyme-based negative lithography.

2.2.3 Virus

Entire viruses have been patterned using SPM-based nanobiolithography techniques. Two examples are (genetically modified) cowpea mosaic virus (CPMV) that has been patterned using nanografting and DPN[39] and tobacco mosaic virus (TMV), patterned using DPN.[40]

2.2.4 Polymers

Although polymers are not biological molecules, the ability to nanopattern them may find important applications in future nanobiochips. We mention here three examples: photoresist (a form of Poly(methyl methacrylate) (PMMA)) printed with NFP,[7] PDMS printed with DPN,[42] and trimethylolpropane trimethacrylate (TRIM) printed with NFP[43] (Figure 15).

2.3 Negative Nanobiolithography (Removal of Material)

Negative lithography, as opposed to positive lithography, is a process by which material is removed from surfaces in desired areas. Many nonbiological methods exist to perform negative lithography in general and nanolithography in particular. The methods can basically be divided into two groups: mechanical (such as indentation, scraping, etc.) and chemical (involving some type of etching agent).

The key element in biological negative lithography is the use of biological molecules that possess *specificity*. Unlike an acid attacking silicon, a metal, and so on, an enzyme will only recognize with high affinity a specific substrate.

Enzyme-based nanolithography uses the tools employed for positive nanolithography, to pattern enzymes on surfaces covered with the substrate of the enzyme.

So far three such examples have been demonstrated.

Staphylococcal serine V8 protease was immobilized to an AFM tip and scanned on a surface consisting of peptides immobilized on mica.[30] The protease recognizes either glutamic or aspartic acid residues in the peptide and digests the peptide's C

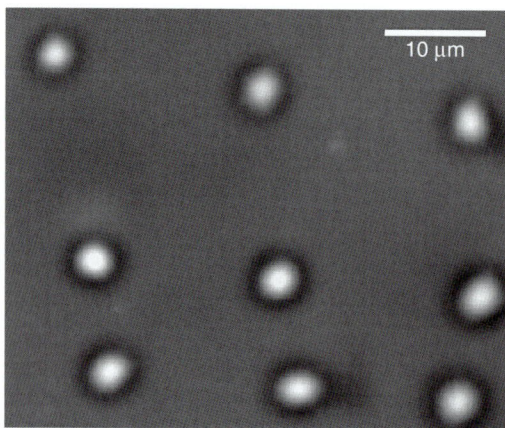

Figure 15. Dots of trimethylolpropane trimethacrylate (TRIM) deposited with NFP. [Reprinted with permission Sokuler and Gheber[43] copyright 2006, American Chemical Society.]

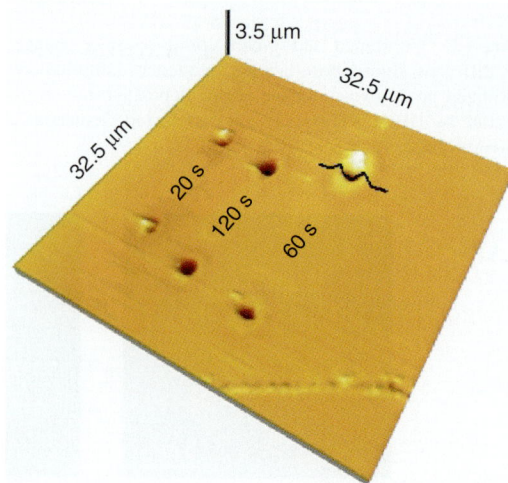

Figure 16. Nanowells created by delivering trypsin with NFP on a BSA-covered surface. [Reprinted with permission Ionescu et al.[31] copyright 2003 American Chemical Society.]

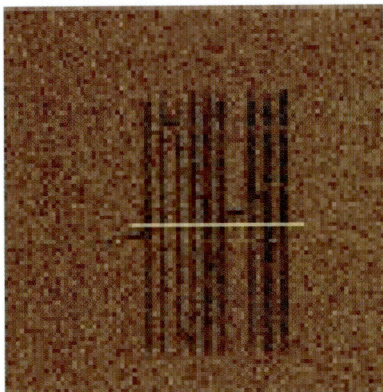

Figure 17. Nanotrenches in DNA surface using DNAse I and DPN. [Reprinted with permission Hyun et al.[34] copyright 2004, American Chemical society.]

terminus. It was demonstrated that the peptide containing glutamic acid was "etched" by the enzyme.

Trypsin, another protease, has been patterned using NFP on a BSA-covered surface, to create depressions. Trypsin recognizes lysine and/or arginine and hydrolyzes the peptide bond at these locations. BSA contains 86 lysine and arginine amino acids, therefore the cleavage by trypsin causes the collapse of the protein structure[31] (Figure 16). The same approach was used to create nanochannels in a BSA substrate.[32]

DNase I, a nonspecific endonuclease that digests double-stranded and single-stranded DNA into nucleotide fragments was patterned using DPN, onto a surface functionalized with an oligonucleotide SAM.[34] The enzyme digested the oligonucleotide substrate, leading to the creation of nanochannels (Figure 17).

3 OBSTACLES TO OVERCOME

3.1 Low Signal Intensity

Reduction of spot size by $\sim 10^3$ is expected to reduce fluorescence emission intensity by $\sim 10^6$ (like the reduction of the area) compared with signal intensities presently available in microarray technology, but hopefully less, due to improvement of surfaces and their binding capacity. This reduction in signal has to be compensated by more than just the transition to a high-NA objective, and

efforts are currently being invested by the scientific community to find solutions to this problem. For example, polymer microlenses that could significantly enhance the emitted light.[43]

3.2 Expensive Manufacture Methods

All the techniques described in the preceding text, based on SPM methods, are inherently expensive. The high price is due to the systems, the probes, and the fact that scanning techniques are serial, and thus slow.

Several proof-of-concept demonstrations show that DPN is in principle extendable to a parallel process,[5,45,46] which should speed up nanobiolithography processes.

3.3 Multiplicity of Molecules

Nanobioarrays demonstrated so far have been composed of two different molecules, to the best. These are obviously only the first steps, proofs of concept, however it is clear that if and when the previous problems are solved, the ability to nanopattern hundreds, if not thousands, of *different* biomolecules, will become a limiting step.

3.4 Label-free and Time-resolved Detection

One of the goals of nanobiolithography is to greatly improve the portability of biochips. Portable biochips can potentially be used for monitoring the quality of food, water, and the environment in general. However, the present biochip technology, based almost entirely on fluorescent labeling of the target, is incompatible with continuous monitoring. Therefore, label-free detection methods need to be developed, compatible with microarray and future nanoarray technology.

REFERENCES

1. K. Wadu-Mesthrige, S. Xu, N. A. Amro, and G. Y. Liu, Fabrication and imaging of nanometer-sized protein patterns. *Langmuir*, 1999, **15**, 8580–8583.
2. R. D. Piner, J. Zhu, F. Xu, S. H. Hong, and C. A. Mirkin, "Dip-pen" nanolithography. *Science*, 1999, **283**, 661–663.

3. C. A. Mirkin, S. H. Hong, and L. Demers, Dip-pen nanolithography: Controlling surface architecture on the sub-100 nanometer length scale. *Chemphyschem*, 2001, **2**, 37–39.

4. S. H. Hong, J. Zhu, and C. A. Mirkin, Multiple ink nanolithography: Toward a multiple-pen nano-plotter. *Science*, 1999, **286**, 523–525.

5. K. Salaita, S. W. Lee, X. F. Wang, L. Huang, T. M. Dellinger, C. Liu, and C. A. Mirkin, Sub-100 nm, centimeter-scale, parallel dip-pen nanolithography. *Small*, 2005, **1**, 940–945.

6. K. Lieberman and A. Lewis, Simultaneous scanning tunneling and optical near-field imaging with a micropipette. *Applied Physics Letters*, 1993, **62**, 1335–1357.

7. M. H. Hong, K. H. Kim, J. Bae, and W. Jhe, Scanning nanolithography using a material-filled nanopipette. *Applied Physics Letters*, 2000, **77**, 2604–2606.

8. A. Lewis, Y. Kheifetz, E. Shambrodt, A. Radko, E. Khatchatryan, and C. Sukenik, Fountain pen nanochemistry: Atomic force control of chrome etching. *Applied Physics Letters*, 1999, **75**, 2689–2691.

9. A. Bruckbauer, L. M. Ying, A. M. Rothery, D. J. Zhou, A. I. Shevchuk, C. Abell, Y. E. Korchev, and D. Klenerman, Writing with DNA and protein using a nanopipet for controlled delivery. *Journal of the American Chemical Society*, 2002, **124**, 8810–8811.

10. H. Taha, R. S. Marks, L. A. Gheber, I. Rousso, J. Newman, C. Sukenik, and A. Lewis, Protein printing with an atomic force sensing nanofountainpen. *Applied Physics Letters*, 2003, **83**, 1041–1043.

11. C. Philipona, Y. Chevolot, D. Leonard, H. J. Mathieu, H. Sigrist, and F. Marquis-Weible, A scanning near-field optical microscope approach to biomolecule patterning. *Bioconjugate Chemistry*, 2001, **12**, 332–336.

12. S. Q. Sun, M. Montague, K. Critchley, M. S. Chen, W. J. Dressick, S. D. Evans, and G. J. Leggett, Fabrication of biological nanostructures by scanning near-field photolithography of chloromethylphenyisiloxane monolayers. *Nano Letters*, 2006, **6**, 29–33.

13. S. Q. Sun and G. J. Leggett, Generation of nanostructures by scanning near-field photolithography of self-assembled monolayers and wet chemical etching. *Nano Letters*, 2002, **2**, 1223–1227.

14. N. A. Amro, S. Xu, and G. Y. Liu, Patterning surfaces using tip-directed displacement and self-assembly. *Langmuir*, 2000, **16**, 3006–3009.

15. P. V. Schwartz, Meniscus force nanografting: Nanoscopic patterning of DNA. *Langmuir*, 2001, **17**, 5971–5977.

16. S. Deladi, N. R. Tas, J. W. Berenschot, G. J. M. Krijnen, M. J. de Boer, J. H. de Boer, M. Peter, and M. C. Elwenspoek, Micromachined fountain pen for atomic force microscope-based nanopatterning. *Applied Physics Letters*, 2004, **85**, 5361–5363.

17. K. H. Kim, N. Moldovan, and H. D. Espinosa, A nanofountain probe with sub-100 nm molecular writing resolution. *Small*, 2005, **1**, 632–635.

18. Y. Li, B. W. Maynor, and J. Liu, Electrochemical AFM "Dip-Pen" nanolithography. *Journal of the American Chemical Society*, 2001, **123**, 2105–2106.

19. M. Z. Liu, N. A. Amro, C. S. Chow, and G. Y. Liu, Production of nanostructures of DNA on surfaces. *Nano Letters*, 2002, **2**, 863–867.

20. L. M. Demers, D. S. Ginger, S. J. Park, Z. Li, S. W. Chung, and C. A. Mirkin, Direct patterning of modified oligonucleotides on metals and insulators by dip-pen nanolithography. *Science*, 2002, **296**, 1836–1838.

21. J. R. Kenseth, J. A. Harnisch, V. W. Jones, and M. D. Porter, Investigation of approaches for the fabrication of protein patterns by scanning probe lithography. *Langmuir*, 2001, **17**, 4105–4112.

22. K. Wadu-Mesthrige, N. A. Amro, J. C. Garno, S. Xu, and G. Y. Liu, Fabrication of nanometer-sized protein patterns using atomic force microscopy and selective immobilization. *Biophysical Journal*, 2001, **80**, 1891–1899.

23. D. L. Wilson, R. Martin, S. Hong, M. Cronin-Golomb, C. A. Mirkin, and D. L. Kaplan, Surface organization and nanopatterning of collagen by dip-pen nanolithography. *Proceedings of the National Academy of Sciences of the United States of America*, 2001, **98**, 13660–13664.

24. J. Hyun, S. J. Ahn, W. K. Lee, A. Chilkoti, and S. Zauscher, Molecular recognition-mediated fabrication of protein nanostructures by dip-pen lithography. *Nano Letters*, 2002, **2**, 1203–1207.

25. K. B. Lee, S. J. Park, C. A. Mirkin, J. C. Smith, and M. Mrksich, Protein nanoarrays generated by dip-pen nanolithography. *Science*, 2002, **295**, 1702–1705.

26. K. B. Lee, J. H. Lim, and C. A. Mirkin, Protein nanostructures formed via direct-write dip-pen nanolithography. *Journal of the American Chemical Society*, 2003, **125**, 5588–5589.

27. A. Bruckbauer, D. J. Zhou, L. M. Ying, Y. E. Korchev, C. Abell, and D. Klenerman, Multicomponent submicron features of biomolecules created by voltage controlled deposition from a nanopipet. *Journal of the American Chemical Society*, 2003, **125**, 9834–9839.

28. G. Agarwal, R. R. Naik, and M. O. Stone, Immobilization of histidine-tagged proteins on nickel by electrochemical dip pen nanolithography. *Journal of the American Chemical Society*, 2003, **125**, 7408–7412.

29. J. H. Lim, D. S. Ginger, K. B. Lee, J. Heo, J. M. Nam, and C. A. Mirkin, Direct-write dip-pen nanolithography of proteins on modified silicon oxide surfaces. *Angewandte Chemie International Edition*, 2003, **42**, 2309–2312.

30. S. Takeda, C. Nakamura, C. Miyamoto, N. Nakamura, M. Kageshima, H. Tokumoto, and J. Miyake, Lithographing of biomolecules on a substrate surface using an enzyme-immobilized AFM tip. *Nano Letters*, 2003, **3**, 1471–1474.

31. R. E. Ionescu, R. S. Marks, and L. A. Gheber, Nanolithography using protease etching of protein surfaces. *Nano Letters*, 2003, **3**, 1639–1642.

32. R. E. Ionescu, R. S. Marks, and L. A. Gheber, Manufacturing of nanochannels with controlled dimensions using protease nanolithography. *Nano Letters*, 2005, **5**, 821–827.

33. A. Bruckbauer, D. J. Zhou, D. J. Kang, Y. E. Korchev, C. Abell, and D. Klenerman, An addressable antibody nanoarray produced on a nanostructured surface. *Journal of the American Chemical Society*, 2004, **126**, 6508–6509.

34. J. Hyun, J. Kim, S. L. Craig, and A. Chilkoti, Enzymatic nanolithography of a self-assembled oligonucleotide monolayer on gold. *Journal of the American Chemical Society*, 2004, **126**, 4770–4771.

35. K. B. Lee, E. Y. Kim, C. A. Mirkin, and S. M. Wolinsky, The use of nanoarrays for highly sensitive and selective detection of human immunodeficiency virus type 1 in plasma. *Nano Letters*, 2004, **4**, 1869–1872.

36. J. M. Nam, S. W. Han, K. B. Lee, X. G. Liu, M. A. Ratner, and C. A. Mirkin, Bioactive protein nanoarrays on nickel oxide surfaces formed by dip-pen nanolithography. *Angewandte Chemie International Edition*, 2004, **43**, 1246–1249.

37. J. H. Gu, C. M. Yam, S. Li, and C. Z. Cai, Nanometric protein arrays on protein-resistant monolayers on silicon surfaces. *Journal of the American Chemical Society*, 2004, **126**, 8098–8099.

38. Q. L. Tang, Y. X. Zhang, L. H. Chen, F. N. Yan, and R. Wang, Protein delivery with nanoscale precision. *Nanotechnology*, 2005, **16**, 1062–1068.

39. C. L. Cheung, J. A. Camarero, B. W. Woods, T. W. Lin, J. E. Johnson, and J. J. De Yoreo, Fabrication of assembled virus nanostructures on templates of chemoselective linkers formed by scanning probe nanolithography. *Journal of the American Chemical Society*, 2003, **125**, 6848–6849.

40. R. A. Vega, D. Maspoch, K. Salaita, and C. A. Mirkin, Nanoarrays of single virus particles. *Angewandte Chemie International Edition*, 2005, **44**, 6013–6015.

41. S. Rozhok, C. K. F. Shen, P. L. H. Littler, Z. F. Fan, C. Liu, C. A. Mirkin, and R. C. Holz, Methods for fabricating microarrays of motile bacteria. *Small*, 2005, **1**, 445–451.

42. D. L. Malotky and M. K. Chaudhury, Investigation of capillary forces using atomic force microscopy. *Langmuir*, 2001, **17**, 7823–7829.

43. M. Sokuler and L. A. Gheber, Nano fountain pen manufacture of polymer lenses for nano-biochip applications. *Nano Letters*, 2006, **6**, 848–853.

44. M. Z. Liu and G. Y. Liu, Hybridization with nanostructures of single-stranded DNA. *Langmuir*, 2005, **21**, 1972–1978.

45. S. H. Hang and C. A. Mirkin, A nanoplotter with both parallel and serial writing capabilities. *Science*, 2000, **288**, 1808–1811.

46. M. Zhang, D. Bullen, S. W. Chung, S. Hong, K. S. Ryu, Z. F. Fan, C. A. Mirkin, and C. Liu, A MEMS nanoplotter with high-density parallel dip-pen manolithography probe arrays. *Nanotechnology*, 2002, **13**, 212–217.

47

Nanosphere Lithography-Based Chemical Nanopatterns for Biosensor Design

Pascal Colpo, Andrea Valsesia, Patricia Lisboa and François Rossi

Institute for Health and Consumer Protection, Joint Research Centre, Ispra, Italy

1 INTRODUCTION

Patterning biomolecules on surfaces is a fundamental issue for many biosensor applications such as medical diagnostics, environment monitoring, food safety, or security applications.[1] For instance in genomics and proteomics areas, DNA and proteins are patterned in hundreds of micrometer spots, enabling thousands of analysis to be performed simultaneously on a small area. The step forward in terms of miniaturization leads naturally to submicron patterning. Nanoarrays represent a radical technology breakthrough that would provide an enormous performance enhancement compared to conventional technologies. They are intended to increase by several orders of magnitude the number of analyses in the same area and to lower the detection limits. Nanoarrays will be the technological basement of a new generation of miniaturized biochips for molecular diagnostics.

Many experiments are being performed worldwide to develop advanced sensing platforms having controlled surface chemistry with well-defined nanopatterns. The goal is to immobilize biomolecules on a surface in an active state, avoiding nonspecific adsorption. The main trends are to structure the surface in adhesive and nonadhesive zones to control the protein binding at the nanoscale. The surface densification of the recognition element results in an amplification of the recognition activity of the sensing surface, leading to an enhancement of the sensitivity and the specificity of the sensor. Another important consequence is the reduction of the analyte volume needed for the detection.

Many approaches are used to create chemical surface nanostructuring: for instance nanosoft lithography,[2,3] dip-pen lithography,[4–6] and nanofountain pen lithography.[7] These techniques allow the production of nanofeatures with typical dimensions of a few hundred nanometers. The approaches are based on the sequential dispensing of small (nano) quantities of functional molecules (polypeptides, oligonucleic acids, thiols) in specific locations, followed by the passivation of the remaining area using antiadhesion layers.

An alternative nanopatterning technique to fabricate nanofeatures is the so-called nanosphere lithography. This technique relies on the self-assembly of monodisperse nanoparticles that are used as a 2D nanomask during etching and deposition operations. This technique presents the advantage of being inexpensive and enables the production of nanotopography over large surfaces.[8]

This chapter describes a reliable technique to directly create a chemical nanopatterned surface using nanosphere lithography. The first part of the chapter provides some background information on colloidal lithography including techniques of deposition and phenomena involved in the colloidal mask formation. Then, two different strategies for surface chemical nanopatterning are presented: the first based on plasma deposited polymers and the second based on self-assembled monolayers (SAM).

2   NANOSPHERE LITHOGRAPHY

Colloidal-particle films have several technological and fundamental applications. They are employed as nanomasks for etching or deposition processes and as nanobuilding blocks for the nanostructuring of a surface. Whenever ordered in regular arrays (2D or 3D colloidal crystals), they can provide special optical properties (photonic crystal, enhanced plasmonic surface resonance, or surface-enhanced Raman scattering effects).[9]

Methods of *nanosphere lithography* used for nanofabrication differ as a function of the desired products. Metallic nanodots and nanoholes are, for instance, fabricated for localized surface plasmon resonance based sensors[10,11] and functional polymeric patterns, to create bioactive material with nanoscale resolution.[8]

The ultimate challenge is to set up a reliable method to control the organization of colloidal-particle films on a surface, which is driven by the particle–particle and particles–surface interactions.

In some applications, the interdistance between nanoparticles must be controlled.[12,13] The absorption of colloidal particles on surfaces is performed by means of electrostatic or specific chemical forces. If needed, the surface electrostatic charge is modified chemically by a positively charged layer, and the organization of colloidal layers (i.e., the density and interdistance of nanobeads) depends upon the concentration of the particles in the colloidal suspension and the ionic strength of the electrolytic solution.[14] When a well-ordered 2D crystal is used as a template for metal deposition over large areas, the efficiency of nanoparticle organization is often enhanced by using a surfactant to increase the wettability of the surface.

For both types of organization, deposition techniques such as dip-coating, Langmuir–Blodgett, or spin-coating can be used and optimized deposition parameters are needed to accurately control the number of particle layers created in the film.

To create nanobioactive surfaces, that is, surfaces with chemical contrast at the nanoscale, the nanoparticles are usually deposited on a polymeric layer that already has the functionality needed in the final product. The first functional layer used as a contact surface during the nanobead depositions is preferably used as deposited. Indeed, any additional surfactant that could alter its chemical function must be avoided. The spin-coating technique is widely used for colloidal film formation because of the simplicity of its implementation. A microdrop of polystyrene (PS) colloidal particles suspension is deposited on the hydrophilic surface with the spin-coater off (contact angle with pristine PS beads suspension <40°). The bulk volume of the drop is usually removed by a micropipette, in order to obtain a very thin layer of PS beads suspension on the surface.

The sample is then spun at a determined speed in order to accumulate the liquid and the PS beads at the boundary of the drop, leaving a monolayer of PS beads in the center. The evaporation rate of the liquid has to be maintained low enough to allow the organization of the PS beads in a hexagonal crystal lattice. In this way, a large area in the range $500 \times 500\,\mu m^2$ of homogenous PS beads crystal can be deposited on the surface.

3   CHEMICAL NANOPATTERNS

The surface chemical nanopatterning strategy combines colloidal lithography with surface functionalization techniques. The steps involved in the process are illustrated in Figure 1. First a layer rich in carboxylic functionalities is deposited on the substrate and is then covered by a monolayer of crystalline PS nanosphere (Figure 1a); the use of PS nanospheres is recommended since they have a good monodispersity factor (<10%) and they can be easily chemically etched by an O_2 plasma discharge. Moreover their surface chemistry can be adapted to the chemical properties of the surface in order to avoid the specific absorption of the spheres onto the surface. The postprocessing lift-off of the spheres can be done without using aggressive solvents, which could affect the stability of the carboxylic groups of the surface.

The nanomask pattern is then transferred to the carboxylic surface by means of O_2 plasma

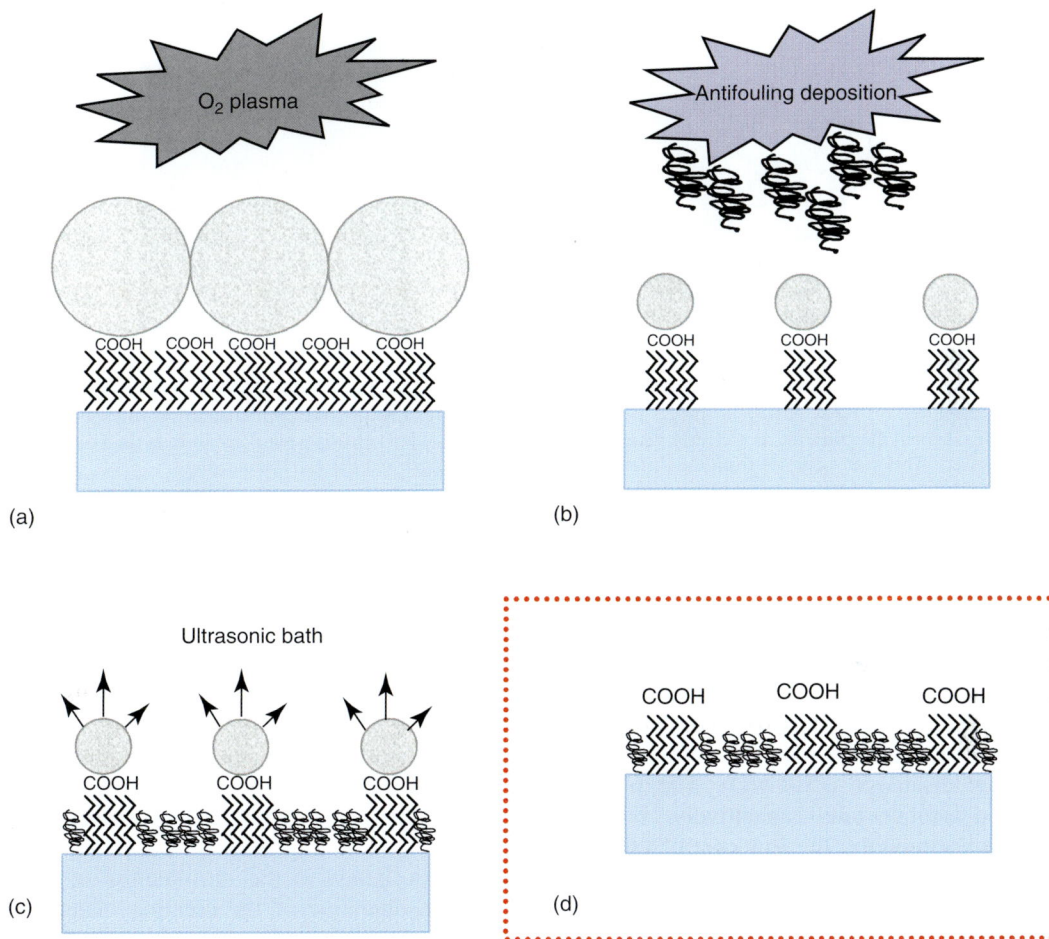

Figure 1. Scheme of the nanopatterning technique: (a) Plasma etching of PS-nano mask and creation of COOH nanostructures. (b) Deposition of antifouling (passivation) layer. (c) Residual PS beads are removed by ultrasonic bath. (d) Final nanostructured surface.

etching (Figure 1b). In this way, the PS spheres are reduced in diameter and the uncovered carboxylic groups of the surface are etched away. The etching time is accurately controlled in order to avoid the complete etching of the spheres protecting certain areas covered by the carboxylic functionalities. The residual etched nanospheres are used as a mask for the passivation (antifouling) of the remaining area (Figure 1c). The final step involves the lift-off of the residual nanosphere mask by an ultrasonic treatment of the surfaces in ultrapure water. The spheres are easily stripped by the mechanical forces induced by the ultrasound because of their weak interaction with the surface, leaving some nanoareas of carboxylic

functionalities surrounded by the antifouling layer (Figure 1d).

One of the advantages of this method is the compatibility with different surface functionalization techniques: both the carboxylic functionalized layer and the antifouling layer can be fabricated by plasma-enhanced chemical vapor deposition[15,16] (PE-CVD) or by classical wet chemistry (SAMs). Such a high flexibility allows the creation of chemical nanopatterns on different substrates and with different morphological properties, for example, 3D or 2D chemical nanopatterns.[a] Two representative examples are shown in Figure 2: Figure 2(a) shows the 3D structured morphology of polyacrylic acid (PAA) nanodomes (carboxylic

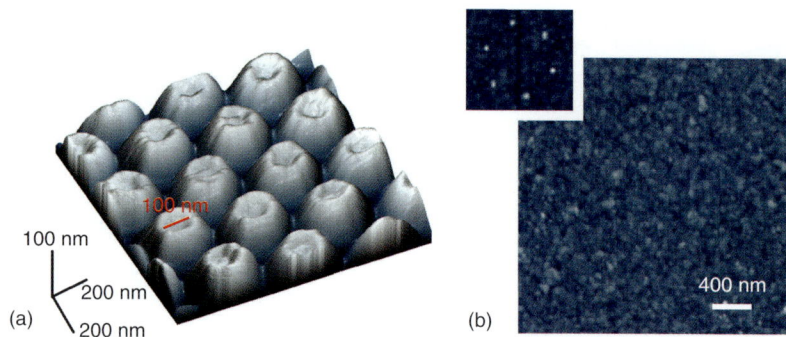

Figure 2. (a) AFM picture of the 3D nanopatterned surface obtained by the combination of the nanosphere lithography and PE-CVD surface functionalization. (b) AFM picture of the 2D nanopatterned surface obtained by the combination of nanosphere lithography and SAM. The hexagonal 2D pattern is not clearly visible from the picture but it is clearly seen if we look at the Fourier Transform of the picture (inset). This pattern arises from the different heights of the MHD and the HDT molecules.

functional) surrounded by a polyethylene glycol–like (antifouling) matrix,[8,17] while Figure 2(b) shows a SAM 2D patterned surface on gold constituted by 16-mercaptohexadodecanoic acid (MHD) nanospots surrounded by a hexadecanethiol (HDT) matrix.[18].

In the first example, the functional polymer films are deposited from a monomer vapor fragmented using a glow discharge (PE-CVD). The plasma reactors used are generally capacitively coupled, owing to their simplicity, the low cost of fabrication, and the mild conditions produced, avoiding excessive precursor fragmentation.[19] Plasma process parameters such as the working pressure, monomer flow, and the power forwarded to the discharge have to be controlled and optimized in order to minimize the monomer fragmentation in the plasma and to maintain high density of monomer functionality in the deposited films.

In the second example, the most delicate step is the preparation of the gold surface for the HDT (or any other passivation layer) self-assembly: in this case the O_2 plasma has the double role of removing the unmasked MHD SAM and preparing the surface for the HDT absorption.

As proof of the concept, labeled bovine serum albumin (BSA) proteins were incubated on a 3D nanopatterned surface where the carboxylic groups of the acrylic acid were activated using classical 1-ethyl-3-(3-dimethylaminopropil) carbodiimide/N-hydroxysuccinimide (EDC/NHS) chemistry.

Fluorescence contrast is clearly observed in Figure 3, consistent with the selective immobilization of BSA on the COOH-terminated nanoareas

(Figure 3c). The size of the fluorescence spots is diffraction limited to ~ 300 nm, indicating that the real size of the fluorescence features is smaller than this value, while the periodicity of the patterns is ~ 500 nm; moreover the same hexagonal surface pattern is confirmed from the fast Fourier transform (FFT) of the fluorescent image (inset of Figure 3c). These results confirm the selective immobilization of biomolecules on the activated COOH nanoareas with respect to the antifouling matrix.

As explained in the introduction of this chapter, the objective of the creation of such precise chemical nanopatterns is the surface densification of the recognition agents in the case of biosensing devices; one illustrative example is represented by the *immunosensor*, where the antibody is immobilized on a surface enabling the recognition of its antigen via the specific antibody–antigen reaction. In this case, the surface density of active "probes" is the surface density of antibodies immobilized with the antigen recognition sites available for the specific reaction (i.e., nondenatured).

In order to compare the number of active antibodies immobilized on a nonstructured, COOH-functionalized surface with respect to a 2D nanopatterned, COOH-functionalized surface,[b] a parallel enzyme-linked immunosorbent assay (ELISA) test has been performed. First a monoclonal antibody is immobilized on the surface. BSA is used in this case as a blocking agent for nonspecific absorption. Then, the sample is exposed to the antigen solution: only the antibodies correctly immobilized on the surface could recognize their antigen. A

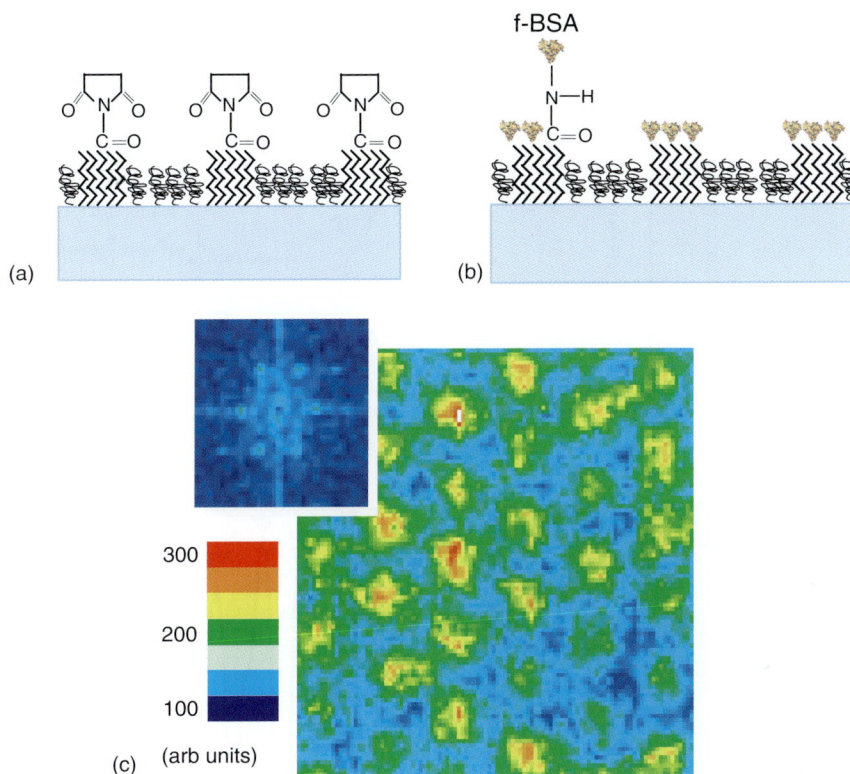

Figure 3. (a) COOH activation by EDC/NHS protocol. (b) Fluorescently labeled BSA (f-BSA) is covalently bound to the surface via its amino groups. (c) Typical fluorescence map of f-BSA immobilized on PAA nanocraters, in the inset, the FFT of the image: the hexagonal crystalline structure is clearly visible.

biotinated antibody then serves as a specific linker to the avidinated enzyme that catalyzes the chemiluminescent reaction with trifluorobenzene (TFB). The signal intensity of the final blue color (at 670 nm) of the solution is proportional to the "recognized" antigens, and hence to the amount of reactive antibodies. The transmittance spectra for the ELISA test for the nonpatterned and the nanopatterned surfaces are shown in Figure 4(a). The 670-nm blue signal of the ELISA-exposed nanopatterned sample was *four times larger* than the nonpatterned one, evidencing a significant increase of the number of recognition agents.

The morphology of the ELISA-exposed nanopatterns (Figure 4b) has a clear crystalline nature similar to the original chemical patterns as demonstrated by the FFT of the height function of the sample (Figure 4b inset). The height distribution of the clusters as measured by atomic force microscopy (AFM) is 54 ± 3 nm ($n = 63$) and is consistent with a theoretical value for the sum of the heights of the proteins involved in the ELISA test. As an example, a height profile along a crystalline spots line is shown in Figure 4(c).

The combination of the spectrometry results with the surface analysis on the ELISA test–exposed surfaces allows us to conclude that the nanopatterned surfaces are able to induce the immobilization of the antibodies (first step of the ELISA test) in such a way that the antibody–antigen reaction is more favorable than in the case of the nonpatterned ones.

4 OUTLOOK

Nanopatterning a surface at the nanoscale is a challenging task and is very promising for many fields of applications. Improving the capability of the functional surfaces to bind biomolecules

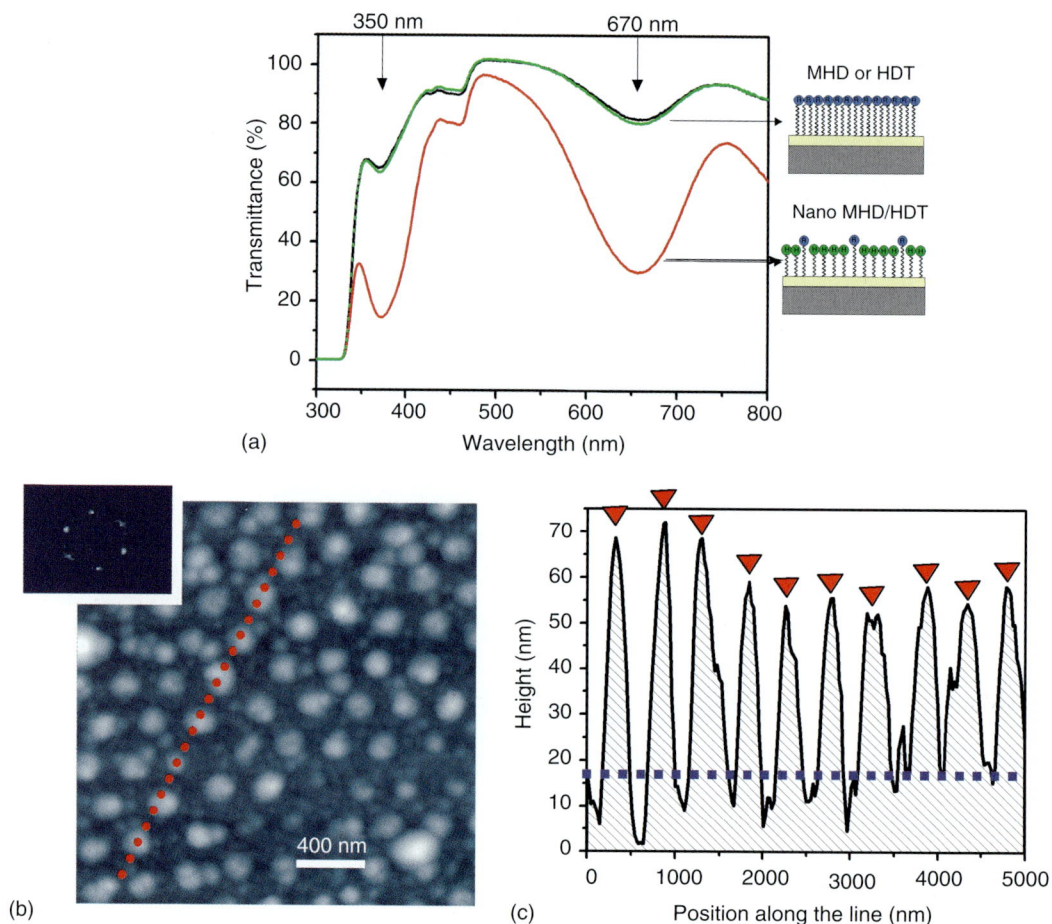

Figure 4. (a) Transmittance spectra for the solutions reacted with TFB for the nonpatterned and the nanopatterned samples. (b) AFM picture of the nanopatterned (COOH/antifouling) sample after the ELISA procedure (vertical scale: [0;106 nm]). The inset shows the FFT of the h(x, y) functions. (c) Height profile along the red line in Figure 3(a). The height of the spot (marked by red triangles) is calculated from the blue baseline.

in the active state, with a very low level of nonspecific absorption, is an enabling feature for the development of very sensitive devices that can be used as prescreening and early diagnostic tools. While nanopatterned biosensor surfaces are still the object of fundamental studies, interesting results are obtained in terms of selectivity and sensitivity of detection. Colloidal lithography is one of the techniques that can be implemented to produce these surfaces. It has many advantages such as low cost and parallel fabrication capability. Furthermore, this technique is compatible with classical surface functionalization techniques such as SAM and plasma deposition or functionalization. Nevertheless, compatibility with classical

sample preparation and handling systems still has to be implemented to realize the full potential of such nanodevices.

ACKNOWLEDGMENTS

This work has been supported by the NanoBioTech action of the Sixth Framework Program of the European Commission. The authors would like to acknowledge all the people involved in the NANOBIOTECH action, in particular: G. Ceccone, T. Sasaki, F. Bretagnol, and M. Lejeune, T. Meziani.

The authors are especially grateful to M. Garcia-Parajo and A. Bouma (MESA$^+$ Institute, The Netherlands) for the confocal microscopy measurements.

END NOTES

[a.] The pattern is considered 2D when the typical size of patterning in the plane of the surface is much greater than the height of the nanostructures.
[b.] These experiments have been performed on the 2D chemical nanopatterns.

REFERENCES

1. N. L. Rosi and C. A. Mirkin, Nanostructures in biodiagnostics. *Chemical Reviews*, 2005, **105**, 1547–1562.
2. J. P. Renault, A. Bernard, A. Bietsch, B. Michel, H. R. Bosshard, E. Delamarche, M. Kreiter, B. Hecht, and U. P. Wild, Fabricating arrays of single protein molecules on glass using microcontact printing. *The Journal of Physical Chemistry. B*, 2003, **107**(3), 703–711.
3. D. Falconnet, D. Pasqui, S. Park, R. Eckert, H. Schift, J. Gobrecht, R. Barbucci, and M. Textor, A novel approach to produce protein nanopatterns by combining nanoimprint lithography and molecular self-assembly. *Nano Letters*, 2004, **4**, 1909–1914.
4. K. B. Lee, S. J. Park, C. A. Mirkin, J. C. Smith, and M. Mrksich, Protein nanoarrays generated by dip-pen nanotithography. *Science*, 2002, **295**, 1702–1705.
5. K. B. Lee, E. Y. Kim, C. A. Mirkin, and S. M. Wolinsky, The use of nanoarrays for highly sensitive and selective detection of human immunodeficiency virus type 1 in plasma. *Nano Letters*, 2004, **4**, 1869–1872.
6. Y. S. Lee and M. Mrksich, Protein chips: from concept to practice. *Trends in Biotechnology*, 2002, **20**, S14–S18.
7. H. Taha, R. S. Marks, L. A. Gheber, I. Rousso, J. Newman, C. Sukenik, and A. Lewis, Protein printing with an atomic force sensing nanofountainpen. *Applied Physics Letters*, 2003, **83**, 1041–1043.
8. A. Valsesia, P. Colpo, M. M. Silvan, T. Meziani, G. Ceccone, and F. Rossi, Fabrication of nanostructured polymeric surfaces for biosensing devices. *Nano Letters*, 2004, **4**, 1047–1050.
9. B. G. Prevo and O. D. Velev, Controlled, Rapid Deposition of Structured Coatings from Micro- and Nanoparticle Suspensions. *Langmuir*, 2004, **20**, 2099–2107.
10. A. J. Haes, S. Zou, G. C. Schatz, and R. P. V. Duyne, Nanoscale optical biosensor: Short range distance dependence of the localized surface plasmon resonance of noble metal nanoparticles. *Journal of Physical Chemistry B*, 2004, **108**, 6961–6968.
11. T. Rindzevicius, Y. Alaverdyan, A. Dahlin, F. Hook, D. S. Sutherland, and M. Kall, Plasmonic sensing characteristics of single nanometric holes. *Nano Letters*, 2005, **5**(11), 2335–2339.
12. B. Kasemo, Biological surface science. *Surface Science*, 2002, **500**, 656–677.
13. M. Arnold, E. A. Cavalcanti-Adam, R. Glass, J. Blummel, W. Eck, M. Kantlehner, H. Kessler, and J. P. Spatz, Activation of integrin function by nanopatterned adhesive interfaces. *ChemPhysChem*, 2004, **5**, 383–388.
14. P. Hanarp, D. S. Sutherland, J. Gold, and B. Kasemo, Control of nanoparticle film structure for colloidal lithography. *Colloids and Surfaces A*, 2003, **214**, 23–36.
15. F. Bretagnol, L. Ceriotti, M. Lejeune, A. Papadopoulou-Bouraoui, M. Hasiwa, D. Gilliland, G. Ceccone, P. Colpo, and F. Rossi, Functional micropatterned surfaces by combination of plasma polymerization and lift-off processes. *Plasma Processes and Polymers*, 2006, **3**, 30–38.
16. A. Valsesia, M. M. Silvan, G. Ceccone, D. Gilliland, P. Colpo, and F. Rossi, Acid/base micropatterned devices for pH-dependent biosensors. *Plasma Processes and Polymers*, 2005, **2**, 334–339.
17. A. Valsesia, P. Colpo, T. Meziani, F. Bretagnol, M. Lejeune, T. Meziani, F. Rossi, A. Bouma, and M. Garcia-Parajo, Selective immobilization of protein clusters on polymeric nanocraters. *Advanced Functional Materials*, 2006, **16**(9), 1242–1246.
18. A. Valsesia, P. Colpo, P. Lisboa, M. Lejeune, T. Meziani, and F. Rossi, Immobilization of antibodies on biosensing devices by nanoarrayed self-assembled monolayers. *Langmuir*, 2006, **22**(4), 1763–1767.
19. H. K. Yasuda. *Plasma Polymerization*, Academic Press, London, 1985.

48

Quantum Dots: Their Use in Biomedical Research and Clinical Diagnostics

Stanley Abramowitz

Advanced Technology Group, Silver Spring, MD, USA

1 INTRODUCTION

Quantum dots are most useful in both biomedical research and for in vitro and in vivo clinical diagnostics. Quantum dots are semiconductor nanocrystals with dimensions varying from 1 to 10 nm. Because of the small dimensions, the physicochemical properties of quantum dots are dependent on their size. This is because of a phenomenon known as *quantum confinement*. Briefly the energy levels for nanocrystals are a function of the size of the quantum dot. This results in the emission of photons with small bandwidth, relatively long lifetime (about 10 ns) and, equally important, symmetric bandwidth. This is a potentially large advantage over the use of organic fluorophore dyes, which typically have fluorescence bands with a red tail emission and short lifetimes relative to quantum dots. Also, properly coated quantum dots are much more photostable even in biological systems than the commonly available fluorophores. A complete discussion of the electrical, optical, and magnetic properties of quantum dots together with the theory that predicts these properties can be found in several articles.[1-3] The wide use of quantum dots especially in research applications has been facilitated by the introduction of reliable and widely available technologies for the synthesis of quantum dots of reproducible size.[4] The bulk of the use of

quantum dots in medical research and diagnostics is based on their optical properties. Other advantages of quantum dots in comparison with the more commonly available organic fluorophores include their large quantum yields; ability to excite a wide variety of sizes of quantum dots, with emission ranging from the visible to the near infrared with one laser frequency or light source; the symmetric and narrow emission band width; and longer fluorescence lifetimes.[5] This obviates the need for fluorescence resonance enhancement transfer (FRET) dyes for multicolor labeling methodologies, as is used for labeling different parts of a cell and the four-color technology utilized in DNA analysis. The longer fluorescence lifetime permits the use of "gated detection technologies", thereby helping to avoid the natural short-time fluorescence present in biomaterials including tissue.[6] There are significant toxicology concerns when quantum dots are utilized for in vivo applications. There is a general concern for the toxicology of nanomaterials including quantum dots.[7] The toxicology of particular quantum dots are determined by their chemical make up, the stability of the chemical coatings that provide the means to solubilize the quantum dots into a colloidal solution, and the functionalization provided for interfacing with the biological moieties to be measured.[8] Typically a moiety is attached to the quantum dots for bonding to a variety of biological entities including DNA, proteins, antibodies, and so on. The toxicity

Handbook of Biosensors and Biochips. Edited by Robert S. Marks, David C. Cullen, Isao Karube, Christopher R. Lowe and Howard H. Weetall.
© 2007 John Wiley & Sons, Ltd. ISBN 978-0-470-01905-4.

is dependent upon several factors including the metals comprising the quantum dot, the chemical in vivo stability of the coating, and functionalization. The quantum dot and its higher-band-gap covering and the chemical species attached for solubilization into a colloidal suspension and functionalization in order to target biospecies is often referred to as the *quantum-dot conjugate*. In this chapter we use the term *quantum dot* to refer to this species that is utilized in biomedical research and clinical diagnostics.

2   PREPARATION AND SOURCES OF QUANTUM DOTS

Quantum dots have many applications. This chapter is limited to the preparation and sources of quantum dots for biology research and clinical diagnostics. Quantum dots are typically made up of semiconductor metals such as cadmium, selenium, indium, and so on. These metals can be extremely toxic and it is necessary to coat the quantum dot with another higher-band-gap material such as ZnS in order to minimize the phenomenon called *blinking*. A typical quantum dot will have a core of CdSe and a coating of a few molecules of ZnS. This coating will mitigate a blinking phenomenon that occurs because of a loss of energy through crystal imperfections and thereby increase the quantum yield. The quantum dot will then be coated with a layer that will enable its functionalization for biological uses and will permit the "solubilization" of the quantum dots into a colloidal solution or suspension. Quantum dots can be synthesized by heating pyrophoric materials including $Cd(CH_3)_2$ and elemental Se in trioctylphosphine oxide at temperatures of 250–350 °C for 24 h followed by a precipitation step that is size selective. It has been recently shown that less onerous Cd-containing species can be used including CdO, $Cd(OAc)_2$, and $CdCO_3$. A more convenient sonically driven synthesis that yields quantum dots of similar optical and magnetic properties has been published.[4] A summary of synthesis methodologies and citations to the thermally driven technologies can also be found in Ref. 4. This synthesis technology that utilizes sonic energy is more straightforward and simpler than the thermally driven methodologies and should be available to more laboratories.

The synthesized quantum dots are then coated with materials that enable its functionalization for biological use. These coatings include silica and other materials that can then be derivatized for attaching proteins, DNA, antibodies, and so on. Quantum dots that are coated and ready for biological use can be obtained from several companies including Invitrogen,[9] Evident Technologies,[10] and Crystalplex.[11] Much information concerning the use of quantum dots for biological research and the development of diagnostics can be found at the websites of these companies.

3   PHYSICOCHEMICAL PROPERTIES OF QUANTUM DOTS

Quantum dots are nanocrystals of semiconductor materials in the size range of 1–10 nm. The optical properties of quantum dots are of particular interest to biological and medical research. Colloidal semiconductor quantum dots are single crystals whose size can be controlled by the synthesis conditions. The various processes used yield quantum dots of a particular size that have distinctive absorption and emission spectra. The emission spectrum of a particular quantum dot is size dependent and varies from the near-infrared through the visible spectrum. Basically the absorption of a photon with energy above the semiconductor band gap results in an exciton or electron–hole pair. Unlike the commonly used organic fluorophores, the emission has a narrow bandwidth, often about 20–25 nm; is symmetric; has longer lifetimes, greater than 10 ns; can be excited by a single laser wavelength or other light source; and does not require the use of FRET dyes. This is shown in Figures 1 and 2 that are taken from Ref. 5.

The longer lifetime permits the use of timegated detection technologies. These properties have advantages over the commonly used fluorophores particularly with multicolor detection requirements, as is the case with some commonly utilized DNA sequence and resequencing applications. The nanocrystals are smaller than the Bohr exciton radius, which is typically about a few nanometers, and therefore the energy levels are quantized and related directly to the nanocrystal size, an effect commonly referred to as *quantum containment*. Surface defects in the quantum dot may lead to a blinking phenomenon and thereby,

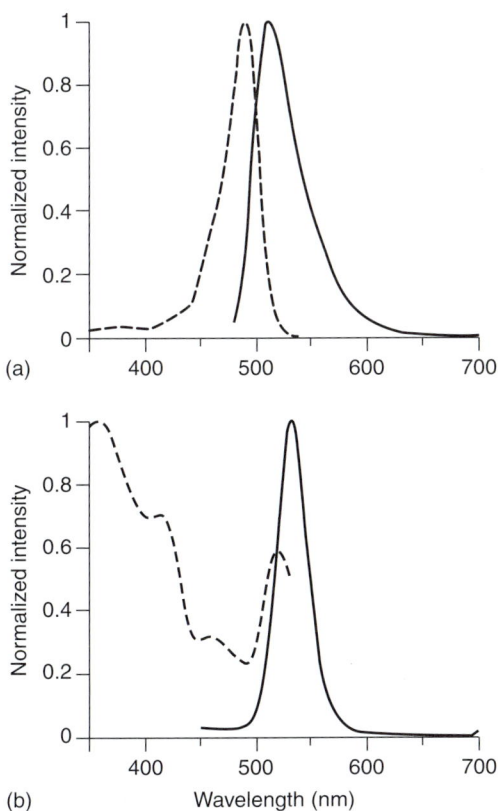

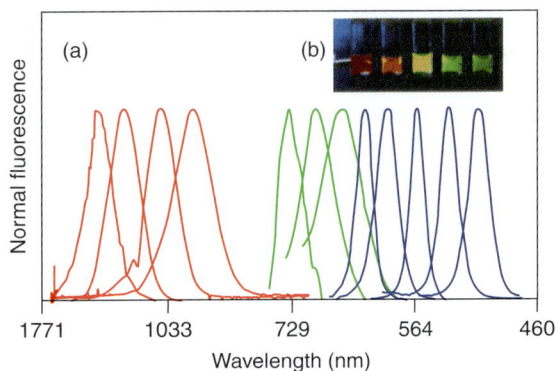

Figure 2. (a) Size- and material-dependent emission spectra of several surfactant-coated semiconductor nanocrystals in a variety of sizes. The blue series represents different sizes of CdSe nanocrystals (16) with diameters of 2.1, 2.4, 3.1, 3.6, and 4.6 nm (from right to left). The green series is of InP nanocrystals (26) with diameters of 3.0, 3.5, and 4.6 nm. The red series is of InAs nanocrystals (16) with diameters of 2.8, 3.6, 4.6, and 6.0 nm. (b) A true-color image of a series of silica-coated core (CdSe)-shell (ZnS or CdS) nanocrystal probes in aqueous buffer, all illuminated simultaneously with a handheld ultraviolet lamp.

Figure 1. Excitation (dashed) and fluorescence (solid) spectra of (a) fluorescein and (b) a typical water-soluble nanocrystal (NC) sample in PBS. The fluorescein was excited at 476 nm and the NC at 355 nm. Excitation spectra were collected with detection at 550 nm (fluorescein) and 533 nm (NC) because of the difference in emission spectra. The nanocrystals have a much narrower emission (32 nm compared with 45 nm at half maximum and 67 nm compared with 100 nm at 10% maximum), no red tail, and a broad, continuous excitation spectrum. [Reprinted with permission from M. Bruchez Jr, et al., Semiconductor nanocrystals as fluorescent biological labels, *Science*, 1998, **281**, 2013–2016. Copyright 1998 AAAS.]

dots and their useful lifetime in real biological systems including cells, cultures, and animals.

4 TOXICOLOGY

Nanomaterials are being used as constituents of many commercial products including filters, catalysts, paints, cosmetics, microelectronics, drug delivery vehicles, and for biological and medical research as discussed in this chapter. Therefore there is a general concern for their toxic potential though their interaction with the environment and biological systems. This chapter is only be concerned with the toxicity of quantum dots as they pertain to biological research, clinical studies, and in vivo and in vitro diagnostics. There have been several studies concerning the toxicology of quantum dots as they pertain to living systems. This is not surprising since some of the more common constituents of quantum dots, for instance, cadmium and selenium, are quite poisonous and readily dissolve in biological fluids. When quantum dots are covered with a layer of a higher-energy-gap material and then covered with a suitable chemical to allow functionalization and solubilization of the quantum dot into a colloidal

a loss of quantum yield. Basically the absorption of a photon with energy above the semiconductor band gap results in an exciton or electron–hole pair. Blinking can be mitigated by covering the quantum dot with a few atomic layers of a material with a higher band gap. Hence the covering of CdSe with ZnS, and the common description of this quantum dot as CdSe/ZnS. It is possible, with suitable covering, to obtain quantum yields approaching 90%. This covering and the functionalization of the quantum dot for biological use markedly increase the photostability of quantum suspension, they are not generally toxic as long as the coverings remain intact and do not allow bodily or biological fluids into the quantum-dot core. Also, as noted in the preceding text, the lifetimes of the quantum dots in biological media are sufficiently long to allow kinetic studies and the flow of at least the smaller quantum dots through the biological system studied. A summary of some of the more common methodologies for quantum-dot solubilization and functionalization is given in Ref. 12. A recent, complete study of the effects of high doses of CdSe/ZnS polyethylene glycol silanized quantum dots has indicated that they have a minimal impact on cells and are very promising for in vivo applications.[13]

5   IMAGING

A review of the use of nanocrystals including quantum dots for biological detection is a very useful primer to the use of nanocrystals for a wide variety of biological applications.[14] The breadth of the use of quantum dots for imaging can be illustrated by the following examples. Quantum dots are utilized for imaging in cells, tissues, and in living subjects. They have longer fluorescence life times in living subjects than the traditional fluorophores. A recent study of the fluorescence of quantum dots that have been designed to bind to the vasculature of human tumors in mice showed that the quantum dots were visible 20 min after injection and the fluorescence peaked at 6 h after injection, and the optical images clearly outlined the tumors showing that the quantum dots attached to the blood vessels growing in and around the tumors.[15] Another recent paper describes a methodology that allows for the fabrication of self-illuminating quantum dots, thereby alleviating the need for separate light sources for stimulating fluorescence. The quantum-dot conjugates are prepared with a mutant of the bioluminescent protein Renilla reniformis luciferase. The conjugates emit long-wavelength bioluminescent light in cells, animals, and tissues and are suitable for in vivo imaging. The system has been demonstrated in mice, tissue in living mice, and cells.[16] The use of quantum dots with suitable hydrodynamic diameters allows the passage of these quantum dots through the blood and lymphatic systems and they

can then be targeted to the moieties of interest. Recently quantum dots have been shown to be able to pass through the sequential lymph nodes and map the excavation (passing out of a fluid into the surrounding tissue) from the vasculature in a rat model. An InAs quantum dot with a ZnSe covering that was then coated with suitable coating for attachment to biological species of interest was utilized in this study. This quantum dot emits in the red region of the spectrum (750–920 nm), depending on the size of the InAs core.[17] When injected into Xenopus embryos, the quantum dots were stable, nontoxic, cell autonomous, and slow to photobleach. These properties allowed the fluorescence to be followed through animal development to the tadpole stage allowing the study of embryogenesis.[18] Semiconductor quantum dots have been used to label cancer markers such as Her2 and other cellular targets in living cells and pathology specimens. All the signals are specific for the intended targets and are reported to be brighter and considerably more photostable than the commonly available organic fluorophores. The use of quantum dots with different emission wavelengths targeted to different moieties in the cell has also been demonstrated.[19] Quantum dots are also being utilized for immunolabeling of membrane proteins and cells.[20]

6   DIAGNOSTICS

Quantum dots have been used for in vitro diagnostics. Their use has several potential advantages for fluorescence technologies, including the possibility of exciting many emission wavelengths, depending on the size of the quantum dot, with one laser excitation frequency, narrow, and symmetric emission bands that may make analysis simpler in the case of multicolor analyses, photostability, and high quantum yield. Quantum dots are being utilized in Fluorescence In Situ Hybridization (FISH), DNA analysis, targeting a variety of components of cells, and so on. A recent review outlining the utility of quantum dots in the probing of human chromosomes and DNA discusses the advantages of quantum dots and the challenges that need to be overcome to make them commercially viable.[21] Quantum dots have been used for FISH and molecular cytogenetics. They have been applied to karyotyping and to spectral karyotyping. They have also been used to determine the number of gene copies and their distribution among the chromosomes. A recent paper compares the fluorescence from quantum dots and conventional organic fluorophores when utilized for these applications. These researchers found the quantum dots to be brighter and more photo stabile than the organic fluorophores.[22] By using 10 intensity level and 6 colors, one can in principle encode for 1 million DNA or protein sequences. The quantum dots are embedded in beads and can be analyzed using flow cytometry.[23] Another study showed the use of multiplexed single nuclear polymorphism (SNP) genotyping using Qbead™ semiconductor quantum dots to code microspheres that are utilized in multiplex assays. By the use of combination mixtures of quantum dots with varying emission wavelengths, one can create useful spectral bar codes for complex DNA analysis.[24] There are several diagnostic companies, including Ventana, that are investigating the possibility of incorporating quantum-dot technology into their diagnostic platforms for anatomic pathology and cytology applications.[25] Quantum dots are especially useful where multicolor emission from several sites either within a tissue cell or diagnostic such as four-color DNA analysis are required. They have been used in the ParAllele analysis for phenotyping. Over 20 000 SNPs can be determined within a single assay. This is an inventive and efficient technology for the determination of a large number, up to 20 000, of SNPs in a single tube assay. The last step of the assay is the identification of a single DNA base that is assayed via a four-color assay. The targets are analyzed on an Affymetrix DNA array using quantum dots for the four-color assay. The advantages of this are the ability to efficiently excite the four colors corresponding to the four nucleotides with one laser frequency and the resultant emissions from the four bases to which the quantum dots are attached. The resultant emission bands are narrow and symmetric. This technology is scalable. At present 20 000 SNPs are analyzed via 4 probes for each of the DNA bases. (While it is in principle possible to analyze the SNPs using two probes, Affymetrix has decided to analyze for all four bases.) There is still much more room on the DNA chip for more probes to afford analysis of more than 20 000 SNPs. A discussion of this technology can be found at the Affymetrix website[26] and in recent archival publications.[27,28] Affymetrix, which is marketing this assay, is also working collaboratively with others on other DNA assays that may eventually be available as "off-the-shelf" DNA arrays that utilize quantum dots because of the advantages previously discussed. To my knowledge, the ParAllele DNA array that is available from Affymetrix is the only "off-the-shelf" commercial product utilizing quantum dots for in vitro clinical diagnostics. One can obtain quantum dots that are suitable for many of the applications discussed from quantum-dot suppliers.[9–11]

7 CONCLUSIONS

Quantum dots have great utility in the areas of biomedical research and clinical diagnostics. They are also being utilized for biosensors that can monitor REDOX reactions for both in vivo and in vitro applications.[29] This discussion of nanocrystals has been limited to quantum dots that have been functionalized for use in biomedical and diagnostic applications. It should be noted that other nanocrystals are also finding utility in these areas as well as other biomedical applications including drug delivery. Of special interest is a colloidal suspension of nanocrystal gold that has been utilized for several applications including DNA detection and imaging.[14,30] The National Cancer Institute of NIH maintains an excellent website that monitors the current literature in the biomedical and clinical uses of nanocrystals including quantum dots.[31] This website has abstracts of articles on the biomedical research and clinical uses of nanocrystals. Reports are posted weekly, and archives dating back to 2004 can be accessed. An excellent review of the potential and realized uses of nanocrystals, including quantum dots, can be found in Refs 12 and 14.

REFERENCES

1. A. P. Alivisatos, Semiconductor clusters, nanocrystals, and quantum dots. *Science*, 1996, **271**, 933–937.
2. Al. L. Efros and M. Rosen, The electronic structure of semiconductor nanocrystals. *Annual Review of Materials Science*, 2000, **30**, 475–421.

3. M. Nirmal and L. Brus, Luminescence photophysics in semiconductor nanocrystals. *Accounts of Chemical Research*, 1999, **32**, 407–414.

4. M. J. Murcia, D. L. Shaw, H. Woodruff, C. A. Naumann, B. A. Young, and E. C. Long, Facile sonochemical synthesis of highly luminescent ZnS-shelled CdSe quantum dots. *Chemistry of Materials*, 2006, **18**, 2219–2225.

5. M. Bruchez Jr, M. Moronne, P. Gin, S. Weiss, and A. P. Alivisatos, Semiconductor nanocrystals as fluorescent biological labels, *Science*, 1998, **281**, 2013–2016.

6. M. Dahan, T. Laurence, F. Pinaud, D. S. Chemla, A. P. Alivisatos, M. Sauer, and S. Weiss, Time gated biological imaging by use of colloidal quantum dots. *Optics Letters*, 2001, **26**, 825–827.

7. A. Nei, T. Xia, L. Madler, and N. Li, Toxic potential of materials at the nanolevel. *Science*, 2006, **311**, 622–627.

8. R. Hardman, A toxicologic review of quantum dots: toxicity depends on physicochemical and environmental factors. *Environmental Health Perspectives*, 2006, **114**, 165–172.

9. www.invitrogen.com/products/qdot/.

10. www.evident.tech.com.

11. www.crystalplex.com.

12. X. Michalet, F. F. Pinaud, L. A. Bentolila, J. M. Tsay, S. Doose, J. J. Li, G. Sundaresan, A. M. Wu, S. S. Gambhir, and S. Weiss, Quantum dots for live cells, in vivo imaging, and diagnostics. *Science*, 2005, **307**, 538–544.

13. T. Zhang, J. L. Stilwell, D. Gerion, L. Ding, O. Elboudwarej, P. A. Cooke, J. W. Gray, A. P. Alivisatos, and F. F. Chen, Cellular effect of high doses of silica-coated quantum dot profiled with high throughput gene expression analysis and high content cellomics measurements. *Nano Letters*, 2006, **6**, 800–808.

14. A. P. Alivisatos, The use of nanocrystals in biological detection. *Nature Biotechnology*, 2004, **22**, 46–52.

15. W. Cai, D.-W. Shin, K. Chen, O. Gheysens, Q. Cao, S. X. Wang, S. S. Gambhir, and X. Chen, Peptide-labeled near-infrared quantum dots for imaging tumor vasculature in living subjects. *Nano Letters*, 2006, **6**, 669–676.

16. M.-K. So, C. Xu, A. M. Loening, S. S. Gambhir, and J. Rao, Self-illuminating quantum dot conjugates for *In vivo* imaging. *Nature Biotechnology*, 2006, **24**, 339–343.

17. J. P. Zimmer, S. W. Kim, S. Ohnishi, E. Tanaka, J. V. Frangioni, and M. G. Bawendi, Size series of small indium arsenide-zinc selenide core-shell nanocrystals and their application to in vivo imaging. *Journal of the American Chemical Society*, 2006, **128**, 2526–2527.

18. B. Dubertret, P. Skourides, D. J. Norris, V. Noireaux, A. H. Brivanlou, and A. Libchaber, In vivo imaging of quantum dots encapsulated in phospholipid micelles. *Science*, 2002, **298**, 1759–1762.

19. X. Wu, H. Liu, J. Liu, K. N. Haley, J. A. Treadway, J. P. Larson, N. Ge, F. Peale, and M. P. Bruchez, Immunofluorescent labeling of cancer marker Her2 and other cellular targets with semiconductor quantum dots. *Nature Biotechnology*, 2003, **21**, 41–46.

20. A. Sukhanova, J. Devy, L. Venteo, H. Kaplan, M. Artemyev, V. Oleinikov, D. Klinov, M. Pluot, H. M. Cohen, and I. Nabiev, Biocompatible fluorescent nanocrystals for immunolabeling of membrane protein and cells. *Analytical Biochemistry*, 2004, **324**, 60–67.

21. Y. Xia and P. E. Barker, Semiconductor nanocrystal probes for human chromosomes and DNA. *Minerva Biotecnologica*, 2006, **16**, 281–288.

22. Y. Xiao and P. E. Barker, Semiconductor nanocrystal probes for human metaphase chromosomes. *Nucleic Acids Research*, 2004, **32**, e28.

23. M. Han, X. Gao, J. Z. Su, and S. Nie, Quantum dot tagged microbeads for multiplexed optical coding of biomolecules. *Nature Biotechnology*, 2001, **19**, 631–635.

24. H. Xu, M. Y. Sha, E. Y. Wong, J. Uphoff, Y. Xu, J. A. Treadway, A. Truong, E. O'Brien, S. Asquith, M. Stubbins, N. K. Spurr, E. H. Lai, and W. Mahoney, Multiplexed SNP genotyping using Qbead™ system: a quantum dot-encoded microsphere-based assay. *Nucleic Acids Research*, 2003, **31**, e43.

25. www.ventanamed.com.

26. www.affymetrix.com/technology/mip_technology.affx.

27. P. Hardenbol, J. Baner, M. Jain, M. Nilsson, E. A. Namsaraev, G. A. Karlin-Neumann, H. H. Fakrai-Rad, M. Ronaghi, T. D. Willis, U. Landegren, and R. W. Davis, Multiplexed genotyping with sequence-tagged molecular inversion probes. *Nature Biotechnology*, 2003, **21**, 673–678.

28. P. Hardenbol, F. Yu, J. Belmont, J. MacKenzie, C. Bruckner, T. Brudage, A. Boudreau, S. Chow, J. Eberle, A. Erbilgin, M. Falkowski, R. Fitzgerald, S. Ghose, O. Lartchouk, M. Jain, G. Karlin-Neumann, X. Lu, X. Miao, B. Moore, M. Moorhead, E. Namsaraev, S. Paternak, E. Prakash, K. Tran, Z. Wang, H. B. Jones, R. W. Davis, T. D. Willis, and R. A. Gibbs, Highly multiplexed molecular inversion probe genotyping: over 10,000 SNPs genotyped in a single tube assay. *Genome Research*, 2005, **15**, 269–275.

29. S. J. Clarke, C. A. Hollman, Z. Zhang, S. E. Bradford, N. M. Dimitrijevic, W. G. Minarik, and J. L. Naudeau, Photophysics of dopamine-modified quantum dots and effects on biological systems. *Nature Materials*, 2006, **5**, 409–417.

30. R. Elghanian, J. J. Storhoff, R. C. Mucic, R. L. Letsinger, and C. A. Mirkin, Selective colorimetric detection of polynucleotides based on the distance-dependent optical properties of gold nanoparticles. *Science*, 1997, **277**, 1078–1081.

31. http:nano.cancer.gov/news_center/anaotech_news.asp.

49

Manipulation and Detection of Magnetic Nanoparticles for Diagnostic Applications

Benjamin B. Yellen and Randall M. Erb

Department of Mechanical Engineering and Materials Science, Duke University, Durham, NC, USA

1 INTRODUCTION

Over the last few decades, the field of biosensors and biochips has flourished due to increasing applications in DNA and pathogen detection, clinical diagnostics, and the analysis of various biological and chemical materials.[1,2] Developing these micro total analysis systems (μTAS)[3] has required multidisciplinary collaboration between physicists, chemists, biologists, and engineers to accomplish various tasks; including integrating the microfluidics, immobilizing molecules on various surfaces, and detecting and manipulating desired components inside the fluid. A great deal of work has already been done on the microfluidics, leading to a diverse array of technologies[4,5] related to sample preparation, sample injection, reaction, separation, and detection.[6,7] In tandem, significant advances in material science has allowed for the development of stable monodisperse colloidal particles, which serve as an ideal candidate for loading sensor molecules onto mobile supports due to their high surface area-to-volume ratio. Strategies for the preparation of polymer and metal-based colloidal particles are provided in several recent review papers.[8–11] Significant work has also been conducted on synthesizing particles that emit a detectable signal through the incorporation of an optical, electrical, or magnetic material into the particles.[8,10–14] This feature allows for the positions of the particle to be detected by an array of sensors.

In comparison with optical and electrical detection schemes, magnetic detection systems tend to be highly sensitive, inexpensive, and compatible with most lab-on-a-chip applications.[15–22] *Magnetic* particles offer an additional advantage in that they can be manipulated by external magnets and transported to desired locations independently of other species inside the fluid. Although other approaches based on optical and electrical field manipulation also have the potential for dual functionality, the relative transparency of most biological materials to magnetic fields makes magnetism uniquely adapted for diagnostic and biomedical applications.

Here, the focus is to review recent work on magnetic detection and manipulation systems, and the mention of alternative approaches based on electric or optical fields is used only for comparison purposes. This paper is organized as follows. First, we review the basic principles governing magnetization in nanostructures and outline the types of magnetic materials used in biomedical applications. On the basis of this discussion, the criteria for manipulating magnetic particles in a fluidic system and its general scaling principles are outlined. Next, we present recent work on magnetic

Handbook of Biosensors and Biochips. Edited by Robert S. Marks, David C. Cullen, Isao Karube, Christopher R. Lowe and Howard H. Weetall.
© 2007 John Wiley & Sons, Ltd. ISBN 978-0-470-01905-4.

sensors used in biological detection, and outline the basic types of sensors used in magnetic assays as well as the principle of operation for the most commonly used variety, known as *spin valves*. Finally, this paper concludes with a discussion on the future outlook in the field of magnetic biosensors and biochips and a general comparison of magnetic versus optical and electrical strategies for manipulating and detecting biological materials.

2 MAGNETIZATION BEHAVIOR OF NANOSTRUCTURED MAGNETIC MATERIALS

To familiarize the reader with terms commonly found in magnetic material descriptions, this section begins with a discussion of how the magnetization M of a material relates to the external magnetic field, H. The magnetization behavior is typically depicted by a hysteresis loop, like those shown in Figure 1. Magnetic saturation, M_s, is a parameter, which characterizes the strength of a material's magnetization when all of its atomic spins are aligned with an external field. In this state, the material is *magnetized to the point of saturation*. When the external field is removed, any remaining magnetization is defined as the magnetic remanence, M_r. In materials that have magnetic remanence, opposing fields must be applied to decrease the material's remanent magnetization to zero. The critical field strength required to reverse the material's magnetization is defined

as the coercive field, H_c, also frequently called the *coercivity*.

There are several classes of magnetic materials, each of which is defined by common hysteretic traits. For the sake of simplicity, this discussion covers only the three most prominent groups, namely ferromagnetic, superparamagnetic, and diamagnetic materials. Ferromagnetism is a state describing materials that have some remaining magnetization after the external field is removed (i.e., they have magnetic remanence). This class of materials is known to exhibit hysteresis, which is a term implying that the material's magnetization is not uniquely defined by the strength and direction of the applied field, but it also depends on *the history of past exposure to a magnetic field*. A typical hysteresis curve for ferromagnetic materials is provided in Figure 1(a). A second class of materials, known as *superparamagnetic*, retains zero magnetization in the absence of external fields (i.e., no hysteresis). Owing to this behavior, the magnetization within superparamagnetic materials traces and retraces the same path on the hysteresis loop, which implies that its magnetization is not history dependent and is based solely on the strength and direction of the present external field, as shown in Figure 1(b). A third class of materials, known as *diamagnetic*, exhibits peculiar behavior in that the material's magnetization opposes the applied field. An example of a diamagnetic hysteresis loop is shown in Figure 1(c). Although the vast majority of biological materials (water, proteins, etc.) display slightly diamagnetic

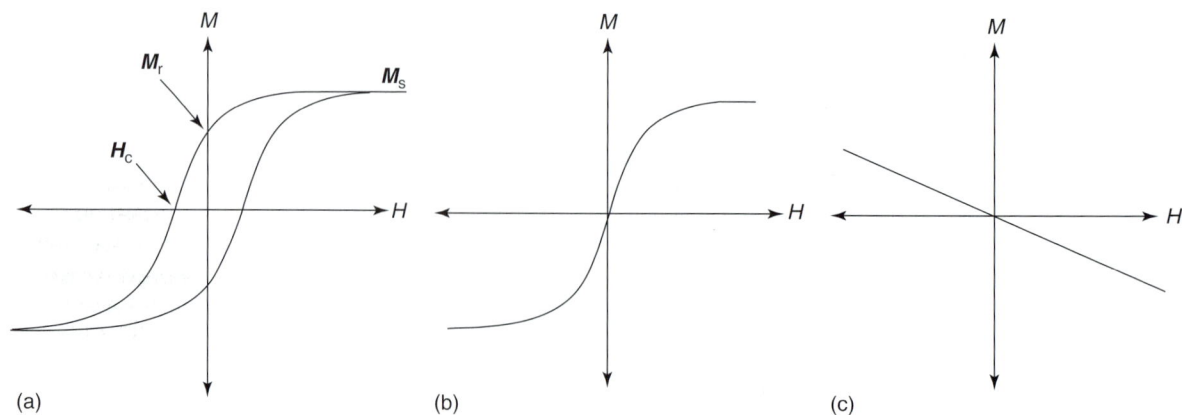

Figure 1. Characteristic hysteresis loops for (a) ferromagnetic, (b) superparamagnetic, and (c) diamagnetic materials.

properties, this effect is considered to be weak compared with ferromagnetic and superparamagnetic behavior of magnetic materials (i.e., Fe, Co, Ni, rare earth materials, and various alloys), and hence diamagnetic effects are often ignored in theoretical treatment.

With sufficient time, the magnetization within all magnetic materials will eventually relax to zero under the influence of randomizing thermal fluctuations. The rate at which the particle's magnetization relaxes is related to an energy barrier, ΔE, with respect to its equilibrium state, as given by: $\tau^{-1} = f_0 e^{-\Delta E / k_B T}$, where τ is the corresponding relaxation time constant. For materials displaying uniaxial anisotropy, the energy barrier is simply the product of the material's volume with its magnetic anisotropy constant: $\Delta E = KV$. The value f_0 has been determined experimentally to vary between 10^9 and 10^{13} s^{-1};[23] however, exact knowledge of the matching constant is not so crucial, since the exponential function dominates the behavior of the time constant. As an example, consider the behavior of a spherical iron nanoparticle with an anisotropy constant 10^{-6} erg cm^{-3} and with the following parameters ($f_0 = 10^9$ s^{-1}, $T = 298$ K), the magnetization within the material is predicted to survive for 0.1 ms for a particle with 10-nm diameter as compared to 10^9 s for a particle with 15-nm diameter. This analysis explains why <10-nm magnetic nanoparticles are the preferred magnetic material for biomedical applications. Below this critical size of roughly 10 nm, the particles become effectively superparamagnetic. The advantage of using superparamagnetic particles is that they readily respond to an applied magnetic field, yet the particles can be turned *off* when the external field is removed (i.e., they do not irreversibly agglomerate because they lack remanent magnetization).

In practice, the material is considered to be ferromagnetic if the remnant magnetization lasts longer than the experiment time (i.e., on the order of a few seconds). Magnetizations that relax more quickly are considered to be superparamagnetic because of the fact that their magnetization behaves more like a "super" large paramagnetic spin. Most often, the threshold of $\Delta E = KV \approx 25 k_B T$ is taken as the limit for superparamagnetism.[24] Below this limit, the time-averaged magnetization within an individual particle is described by the well-known Langevin's function:

$$\vec{M}(\vec{H})$$
$$= M_s L \left(\frac{\mu_0 M_s V H}{k_B T} \right) \frac{\vec{H}}{|H|}$$
$$= M_s \left(\coth \left(\frac{\mu_0 M_s V H}{k_B T} \right) - \frac{k_B T}{\mu_0 M_s V H} \right) \frac{\vec{H}}{|H|}$$

(1)

An approximation can be obtained in the low-field regime ($\mu_0 M_s V H \ll k_B T$) by using the Taylor series expansion of $\coth(\eta) \approx (1)/(\eta) + (\eta)/(3) + \ldots$, which results in the particle's magnetization reducing to a linear relationship with the applied field as: $\vec{M}(\vec{H}) = (\mu_0 M_s^2 V)/(3 k_B T)\vec{H} = \chi \vec{H}$, where the magnetic susceptibility $\chi = (\mu_0 V M_s^2)/(3 k_B T)$ is an experimentally determined parameter. In the high-field regime ($\mu_0 M_s V H \gg k_B T$), the particle's magnetization saturates according to the relationship: $1 - (k_B T)/(\mu_0 M_s V H)$. However, this circumstance is only reached in magnetic fields approaching 0.1 T.

3 MAGNETIC PARTICLES USED IN DIAGNOSTIC APPLICATIONS

Biomedical devices commonly employ magnetic particles for both in vivo and in vitro applications. In vivo applications encompass drug targeting, hyperthermia, and magnetic resonance imaging, while in vitro applications include separation/manipulation, detection, and magnetorelaxometry.[25] For these various applications, care should be taken to select the proper type of magnetic particles. Superparamagnetic particles come in two varieties: surfactant-stabilized ferrofluid and polymer-based micro/nanospheres. The key ingredient in both types of particles is nanoscale magnetic grains having dimensions tailored to promote a superparamagnetic response. Ferrofluid is a stabilized dispersion of superparamagnetic grains inside carrier fluids, such as water or hydrocarbon-based liquids.[26,27] The method for stabilizing these particles involves attaching long polymer chains or charge carrying molecules to the particle surfaces. These molecules induce repulsive electrostatic or steric interactions, which are used to overcome short-range attractive forces as well as attractive magnetic interactions, thereby allowing the particles to remain colloidally stable.

Polymer-based magnetic micro/nanospheres are formed by incorporating superparamagnetic grains into the particle matrix during microsphere formation. A number of techniques have already been developed for synthesizing microspheres from various biodegradable and nonbiodegradable polymers.[28,29] In most cases, the loading fraction of magnetic material within the microspheres is kept below 30% in order to preserve its superparamagnetic behavior by reducing magnetic interactions between the grains. Using straightforward surface modification techniques, functional groups can be attached to the particle's surface. These functional groups can be tailored to recognize specific proteins, DNA, and other biological substances through molecular affinity binding. The advantage of using magnetic microspheres is that the collective magnetic moment of the particle can be substantially increased due to the presence of a large number of encapsulated magnetic grains. Since magnetic force is proportional to the particle's magnetic moment, which in turn is dependent on the particle's volume, >100-nm magnetic particles are the preferred choice for biological separation.

4 MAGNETIC MANIPULATION SYSTEMS AND GENERAL SCALING PRINCIPLES

The aim of this section is to review recent progress on techniques used to manipulate superparamagnetic particles. First, the physical origin of magnetic forces on colloidal particles is provided, and it is followed by an overview of the types of magnetic field sources commonly employed in microsystems. The purpose of this discussion is to provide general design criteria for when a particle's motion is dominated by magnetic force as opposed to random Brownian motion.

Similar to the electrostatic treatment of polarizable materials, the net force on a superparamagnetic particle can be viewed as the force acting on equivalent magnetic charges distributed on the particle's surface. These charges arise because of the difference in the particle's magnetization $\vec{M}_p$ with respect to the magnetization of the surrounding fluid $\vec{M}_f$. The equivalent surface charge density

is represented by: $\sigma = \hat{n} \cdot (\vec{M}_p - \vec{M}_f)$, where $\hat{n}$ is a unit vector normal to the particle surface, leading to an expression for the net magnetic force as:

$$\vec{F} = \mu_0 \oiint_{S_p} \sigma \vec{H} \cdot \mathrm{d}S$$

$$= \mu_0 \oiint_{S_p} (\vec{M}_p - \vec{M}_f)\vec{H} \cdot \hat{n}\,\mathrm{d}S \qquad (2)$$

where $\vec{H}$ is the magnetic field at the location of the particle's surface, S_p. In applying Gauss's divergence theorem, (1) can be rewritten as:

$$\vec{F} = \mu_0 \iiint_{V_p} ((\vec{M}_p - \vec{M}_f) \cdot \nabla)\vec{H}\,\mathrm{d}V$$

$$= \mu_0((\vec{m}_p - \vec{m}_f) \cdot \nabla)\vec{H} \qquad (3)$$

where $\vec{m}_p$ and $\vec{m}_f$ are the magnetic moments of the particle and of the fluid volume that the particle displaces.

The most important conclusion that can be reached on the basis of the above expression is that the force is proportional to the *field gradient*, and it is simple to show that uniform magnetic fields will apply zero net magnetic force on a magnetic particle. Another conclusion that can be drawn from expression (3) is that both magnetic and nonmagnetic materials can be manipulated by magnetic force if a fluid of suitable magnetic susceptibility is chosen. Take, for example, a magnetic particle ($\vec{M}_p \neq 0$) that is situated inside a nonmagnetic carrier fluid, such as water ($\vec{M}_f \approx 0$). In this case, the force on the particle is simply a product of the particle's moment and the field gradient. A variety of systems currently employ this methodology in the field of magnetic separation.[30,31] If, on the other hand, the fluid contains a suspension of magnetic nanoparticles (i.e., ferrofluid), which are magnetized by an external field, then the average fluid magnetization becomes nonzero ($\vec{M}_f \neq 0$), thereby allowing even a nonmagnetic particle ($\vec{M}_p = 0$), which is placed inside the fluid, to acquire an *effective* net magnetic moment and be manipulated by field gradients. The latter phenomenon can be thought of as "negative magnetophoresis" as a corollary to the phenomenon of "negative dielectrophoresis", which occurs when electric fields are applied to manipulate weakly polar materials (e.g., plastic particles) immersed inside strongly polarized fluids (e.g., water). In the literature, this class of fluids is often called *inverse ferrofluids*, and work is ongoing on the design of manipulation systems around this effect.[32,33]

In the following analysis, it will be assumed that the particles are spherical and superparamagnetic, so that its magnetic dipole moment can be related linearly with the applied magnetic field. The dipole moment is given by: $\vec{m} = \bar{\chi} V_p \vec{H}$, where $\bar{\chi} = (3(\chi_p - \chi_f))/(3 + \chi_p + 2\chi_f)$ is a term denoting the effective susceptibility of the particle, and it includes both a shape factor, which accounts for demagnetizing fields within the spherical particle,[34] as well as the difference in the particle's susceptibility χ_p with respect to the susceptibility of the surrounding fluid χ_f. It is worth mentioning that if the particle's susceptibility is larger than that of the fluid (i.e., $\chi_p > \chi_f$), then the particle's moment is aligned parallel to the field and it will migrate toward the regions of magnetic field maxima. On the other hand, if the particle's susceptibility is less than that of the fluid (i.e., $\chi_p < \chi_f$), then the particle's moment is aligned antiparallel with the field and it will migrate toward the regions of magnetic field minima. It is a simple matter to demonstrate that the effective magnetic susceptibility for a spherical particle ranges within $-3/2 \leq \chi \leq 3$.

A variety of different magnetic field sources have been used to manipulate colloidal particles. Some are based on current sources, while others are based on the fields produced by magnetic materials, such as permanent or magnetizable magnets. It turns out that the most important feature for manipulating colloidal particles is not the type of source but rather its geometry. As a general rule, the field gradient varies inversely with the smallest dimension of the field producing structure. However, both large and small magnetic sources can play a significant role in microfluidic manipulation. Small magnetic sources have the advantage of producing stronger magnetic field gradients, but the field gradients are highly localized and, thus, can only manipulate particles in close proximity to the source. On the other hand, large magnetic structures produce weaker magnetic field gradients; however their fields and field gradients have the advantage of extending further into the fluid.

With this concept in mind, the following discussion attempts to outline the types of magnetic field sources used in micromanipulation and how their fields and field gradients decay with distance from the source.

Magnetic field–versus-distance relationships can be obtained for several different field producing geometries. For example, a semi-infinite current line produces fields that decay inversely with the first power of distance from the source (i.e., $H(\vec{r}) \propto r^{-1}$, $(\partial/(\partial r)H(\vec{r}) \propto r^{-2})$. This geometry has previously been used by several authors for manipulating magnetic particles, cells, and other materials suspended in fluids.[35,36] The edges of a large domain wall within a magnetic thin film, such as in iron garnet films,[37] also produce magnetic fields that decay inversely with the first power of distance from the source. The field produced by long wires magnetized perpendicularly to their long axes decays inversely with the square of distance away from the wire center (i.e., $H(\vec{r}) \propto r^{-2}$, $\partial/\partial r H(\vec{r}) \propto r^{-3})$. This type of structure has found applications in magnetic separation[38] and drug delivery.[39] The magnetic field also decays inversely with the square of distance away from one pole of a long-axially magnetized wire. Tightly wound current loops produce a magnetic field that decays inversely with the cube of distance from the source (i.e., $H(\vec{r}) \propto r^{-3}$, $(\partial/(\partial r)H(\vec{r}) \propto r^{-4})$. A similar type of field is produced by neighboring magnetic particles inside the fluid or by lithographically patterned small magnetic dots.[40,41]

Taking cues from the high-gradient magnetic separation (HGMS) community, multiple magnetic field sources are commonly used in magnetic manipulation systems.[38,39] This approach is taken because fundamental physics indicates that a single source of magnetic field cannot simultaneously produce strong fields and strong field gradients everywhere in space. The reason strong fields are also needed is that the moment of a superparamagnetic particle is proportional to the strength of the applied field. Since the field effectively contributes twice in the force equation (i.e., the force is the product of the magnetic moment and the field gradient), a common practice is to use at least two different types of magnetic field sources in colloidal manipulation: one source is typically a large magnet, which produces long-range fields for saturating the moments of all the particles in the fluid, while the other source is typically a small structure, which is used to supply strong local field gradients for attracting the particles. By combining the two types of sources, the particle moments and the field gradients are maximized and the effective reach of the magnetic manipulation system can be extended much further into the fluid.

An additional advantage to the two-source method is that programmable operations can be performed on particles inside the fluid when used in combination with programmable magnets such as those found in magnetic data storage substrates. In this approach, long-range fields can be used to bias the moment of the particle in either repulsive or attractive configurations with respect to the underlying magnetization pattern of the substrate. This phenomenon has been used to program the fluidic deposition of colloidal materials onto preselected sites on a substrate.[40]

Fluid drag and magnetic force are the basis for a class of magnetic manipulation models that are founded on modeling particle trajectories. The most typical approach is to ignore particle inertia (any particle acceleration happens over timescales that are minuscule compared to the timescale of particle movement) and to obtain particle velocity by equating the magnetic force in (3) with the Stoke's drag force: $\vec{F}_{\text{drag}} = 3\pi\eta d \cdot \vec{v}$, where d is the diameter of the particle, $\vec{v}$ is the particle's velocity, and η is the viscosity of the fluid. The time for a particle to reach the surface of the magnet S_{mag} from a given starting position X, therefore becomes:

$$\vec{v} = \frac{\partial x}{\partial t} = \frac{\vec{F}_{\text{mag}}}{3\pi\eta d} \quad t = \int_{S_{\text{mag}}}^{X} \frac{3\pi\eta d}{F_{\text{mag}}} \partial x \quad (4)$$

In the literature, one often finds statements that Brownian motion becomes an important concern for particles smaller than about 100 nm. Strictly speaking, such a general statement is invalid. A more accurate, although still approximate, criterion can be derived for when particle movement can be described in terms of deterministic trajectories as opposed to those based on random Brownian motion. This criterion is based on the idea that random deviation from a point on the trajectory by more than one particle diameter d should be unlikely. For magnetic manipulation, the likelihood of such deviation is small if the corresponding change of the particle's magnetic energy

during movement exceeds $k_{\text{B}}T$. Assuming that the magnetic force on the particle does not change appreciably as the particle moves within a sphere of diameter d, the change in magnetic energy is approximately equal to the magnetic force times d.

To get a feeling for the criteria mentioned in the preceding text, suppose that the force on a particle is being applied by a spherical magnet of diameter D, with magnetization M_0. If we assume the particle is magnetized by an externally applied field, H_0, then the force on the particle is given by the expression:

$$F_{\text{mag}} \cdot d = \frac{6\mu_0}{4\pi}\left(\frac{\pi}{6}\right)^2 \frac{\bar{\chi}H_0 M_0}{D} d^4 \frac{D^4}{R^4} \approx k_{\text{B}}T \quad (5)$$

From the above expression, it is clear that when the particle is located very far from the magnet (i.e., $R \gg D$), it experiences little magnetic force, and therefore random Brownian motion will certainly dominate. Let us, therefore, concentrate on some critical region surrounding the magnet, say $R = 10D$, *where D is again the characteristic dimension of the permanent magnet*. In the following analysis, we will use the magnetic particle properties ($\bar{\chi} \approx 1$), an applied bias field of $H_0 \approx 1\,\text{kOe}$, and assume that the magnetization of the spherical magnet is $M_0 \approx 10\,\text{kOe}$, all of which are common parameters in magnetic manipulation systems. When the characteristic size of the magnet is of the order of $1\,\mu\text{m}$, the above criterion implies that particles greater than $250\,\text{nm}$ move along reasonably well-defined trajectories when they are closer than about $10\,\mu\text{m}$ away from the magnet. Brownian motion dominates the trajectory for particles below this size or when the distance from the magnet is greater. In these cases, it is better to describe particle motion in terms of distribution functions, such as particle concentration. Clearly, 100-nm particle diameter is not some magical size below which particles cannot be manipulated. In fact, it is not difficult to show that even 50-nm particles can be manipulated along well-defined trajectories when they are within with about 2 diameters of the gradient-producing spherical magnet (i.e., $R = 2D$). This analysis implies that it is the size ratio between the particle and the magnetic manipulator that is the important criteria. Smaller particles can be manipulated without much regard for Brownian motion when the characteristic size of the gradient-producing magnet is also reduced. This finding provides additional motivation for using microfabrication technology, typically employed to create integrated circuits and microelectromechanical systems (MEMS), to miniaturize particle handling systems.

For particles below the critical threshold, detailed trajectory analysis is no longer meaningful and random Brownian motion dominates particle movement. Problems of this type occur in many other areas of fluid physics. The common feature of all these approaches is the use of particle distribution functions, which are usually interpreted in term of the probability of finding a particle within some finite region of space–time. The effect of Brownian motion on concentration is usually described by the diffusion flux density: $\vec{J}_{\text{diff}} = -D\nabla C(\vec{r}, t)$ where D is the diffusion coefficient and $C(\vec{r}, t)$ is the particle concentration. The effects of nonrandom particle motion can be incorporated through the average particle velocity $\vec{v}_{\text{p}}$ and the resulting drift flux density of particles is: $\vec{J}_{\text{drift}} = \vec{v}_{\text{p}}C(\vec{r}, t)$. The sum of the diffusion and drift flux densities constitutes the total particle flux density: $\vec{J} = \vec{J}_{\text{drift}} + \vec{J}_{\text{diff}}$. Evolution of the concentration can now be obtained through the conservation law:

$$\frac{\partial C}{\partial t} + \nabla \cdot \vec{J} = 0 \text{ or } \frac{\partial C}{\partial t}$$
$$= -\nabla \cdot (\vec{J}_{\text{diff}} + \vec{J}_{\text{drift}})$$
$$= \nabla \cdot (D\nabla C(\vec{r}, t)) - \nabla \cdot (\vec{v}_{\text{p}}C(\vec{r}, t))$$

$$(6)$$

Initial work along this direction was performed recently[42] and relatively simple analytical solutions were obtained for simple problems concerning the magnetic manipulation of ferrofluids.

5 MAGNETIC DETECTION SYSTEMS USED IN BIOSENSORS

Progress in magnetic biosensors has been driven by recent advances in techniques for fabricating and retrieving signals from miniature magnetic field sensors, which is a topic of great interest in magnetic disk drive technology. Leveraging past work in this field, magnetic field sensors have recently been adapted to detect binding of molecules onto a substrate. The basic sensing device consists of a magnetic field sensor, whose surface is chemically functionalized with probe molecules for recognizing target molecules of interest (i.e., bacteria, nucleic acids, proteins, viruses, cells, etc.). In the simplest case, the biomolecules to be detected (i.e., target molecules) are immobilized onto magnetic particles and they are subsequently exposed to an array of magnetic sensors, which are functionalized with either complementary or noncomplementary probe molecules. After sufficient time is provided for the magnetic labels to interact with the sensors through specific molecular recognition, the remaining magnetic labels are rinsed from the solution. Because magnetic labels are superparamagnetic, an external magnetic field is needed to magnetize the particles. A positive binding signal is therefore indicated by the perturbation of the external field signal by the stray dipole fields emanating from the magnetic label. An illustration of the general sensing mechanism is provided in Figure 2.

A variety of different sensors have been employed in biological detection. Low-field magnetic sensors, such as superconducting quantum interference devices (SQUID),[43] fluxgate magnetometers,[44] and induction coils, have previously been used in medical imaging, however owing to their large size they are not often employed in magnetic biosensors. Instead, the focus has been on solid-state magnetic field sensors which are amenable to batch fabrication techniques. The most common solid-state sensors used in magnetic biodetection are based on the Hall effect[15] or on the magnetoresistive effect, such as anisotropic magnetoresistance (AMR),[16] giant magnetoresistance (GMR)[17–20,22] or the planar Hall effect (PHE).[21] Although work is ongoing on the development of more sensitive cantilever-based sensors[45] and atomic magnetometers,[46] these developments will not be discussed in detail since the aim of this section is simply to outline the basic sensing mechanism used in magnetic biosensors. For this reason, the rest of this section is devoted to a discussion of GMR-based sensors, which to date are the most popularly employed device in magnetic biosensors.

GMR sensors, often called *spin valves*, consist of two layers of ferromagnetic material separated by a nonmagnetic spacer layer. The sensors are

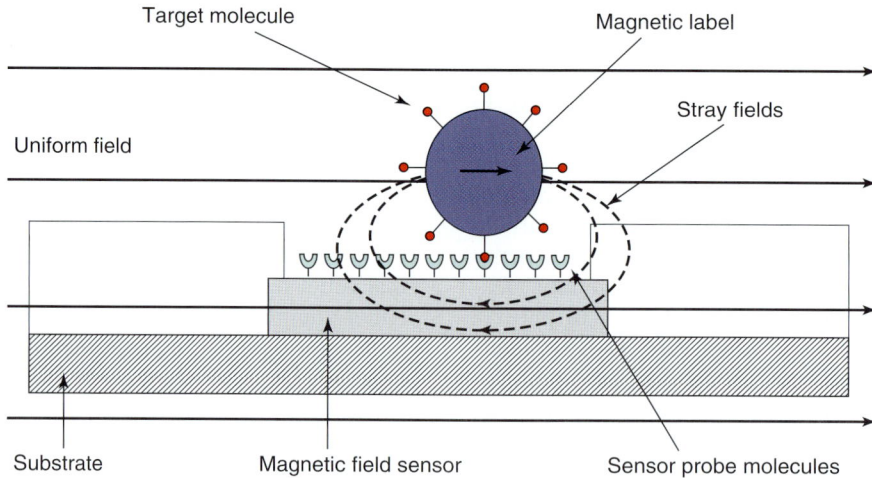

Figure 2. Magnetic labels immobilized with target molecules are exposed to a magnetic field sensor. Through specific molecular recognition, the magnetic label binds selectively to the probe molecules on the surface of the magnetic field sensor. The target molecule's presence is therefore detected by the stray fields the magnetic label produces in response to the uniform magnetizing field.

designed such that the magnetization of one layer is pinned by an antiferromagnetic layer, while the other layer's magnetization is free to rotate in response to an external field. In the typical design, the pinned layer is initially set such that its magnetization makes a 90° angle with the free layer's magnetization in the absence of external fields. The signal from the GMR sensor is sensitive to the sine of the angle between the pinned layer with respect to the free layer, and it is interpreted as a change in electrical resistance, ΔR, through the sensor material. Hence, the key to predicting the sensor signal is in determining the relative angle between the two layers, which is accomplished by modeling the magnetic energy within the free layer[47] as:

$$U = \frac{1}{2} H_K M_F \sin^2(\phi) - H_\perp M_F \cos(\phi)$$
$$- H_\parallel M_F \sin(\phi) \qquad (7)$$

where M_F is the magnetization of the free layer, $90° - \phi$ is the relative angle between the magnetization of the free and pinned layers, and H_K is the total anisotropy field in the free layer, which includes effects due to shape anisotropy and crystalline anisotropy; while $H_\parallel$ and $H_\perp$ are the external fields acting on the free layer, oriented either parallel or perpendicular to the pinned layer,

respectively. These external fields may result from bias fields, the field from the magnetic particles, and any other intrinsic fields in the sensor, associated with the sense current, interlayer coupling, and magnetostatic fields due to the pinned layer.

The equilibrium magnetization orientation within the spin valve is determined by minimizing the magnetostatic energy U of the free layer with respect to ϕ, which results in an approximate expression through the use of small angle approximation ($\phi < 25°$) as:

$$\sin(\phi) \cong \tanh\left(\frac{H_\parallel}{H_\perp + H_K}\right) \qquad (8)$$

This approximation is reasonable because the tilting angle of the free layer is often less than 10°.[48] The resistance in the spin valve can therefore be written as: $R = R_0 + (1)/(2)\delta R \sin(\phi)$, where R_0, is the resistance when the magnetizations of the free and pinned layers are perpendicular (i.e., $\phi = 0$), while δR is the maximum possible change in resistance of the spin valve when the magnetization of the free and pinned layers are oriented antiparallel. The magnetoresistance ratio is defined as $MR \approx (\delta R)/(R_0)$, and for a typical spin-valve sensor this ratio is of the order of 10%. Using (8), it is possible to write the change in resistance ΔR due to the presence of a magnetic particle as:

$$\Delta R = R_{\text{with}} - R_{\text{without}} = \frac{1}{2}\delta R \left[\tanh\left(\frac{H_{\parallel}^*}{H_{\perp}^* + H_K}\right) \right.$$
$$\left. - \tanh\left(\frac{H_{\parallel}}{H_{\perp} + H_K}\right) \right]$$
(9)

where $H_{\parallel}^*, H_{\perp}^*$ denotes the fields produced in the presence of a magnetic particle and includes its contribution to magnetizing the free layer, whereas $H_{\parallel}$, $H_{\perp}$ denotes the fields produced in the particle's absence when only the external fields and intrinsic sensor fields contribute to magnetizing the free layer.

While GMR sensors have been designed in a variety of shapes, the most common GMR sensor design is the rectangular spin-valve structure in which the pinned layer's magnetization is set along the sensor's minor axis by crystalline anisotropy, while the free layer is dominated by shape anisotropy and aligns along the sensor's major axis. A DC bias field is typically applied along the sensor's major axis, and a small-signal AC magnetic field is applied along the sensor's minor axis to generate a signal which can be easily detected with the use of a lock-in amplifier. On the basis of this design, it is possible to produce a voltage signal in the range of millivolts using modest external magnetic fields of <100 Oe, and a small sense current of only a few milliamperes.[15–22]

Magnetic biosensors are an ideal detection platform when only the presence or absence of a target molecule needs to be ascertained. However, quantitative comparison between different magnetic biosensor platforms has remained a difficult challenge due to the high nonlinearity of the field from magnetic labels. The sensor signal can be modeled to a first approximation as being proportional to the average of the magnetic particle's field across the sensor. Since the fields from magnetic particles are dipolar, the sensor signal depends strongly on the direction of the external field, the position of the particle, and the orientation of the sensor axis. Because of symmetry considerations, there are situations when a particle's signal is maximized; however, other situations permit a particle to be located on the sensor with no signal being obtained. Wherever possible, one must design the system to prevent these scenarios from occurring, since they lead to false-negative test results. Another problem encountered in magnetic biosensors, is that the particle's field can point in opposite directions at different points of the sensor, in effect destructively interfering with the sensor signal. This effect is more pronounced when the particle is much smaller than the sensor; however, this effect can be reduced by designing the sensor to be commensurate with the size of the magnetic label. In doing so, the particle's field becomes substantially more uniform across the sensor, thereby allowing a maximal signal to be obtained while simultaneously reducing its sensitivity to the particle's position above the sensor.

6 CONCLUSIONS AND OPEN QUESTIONS

Compared to competing technologies based on optical and electrical fields, the use of magnetic nanoparticles in biosensors and biochips has a number of advantages, which makes it uniquely suited for diagnostic applications. The advantage of using magnetic detection systems is that the sensor output is an electronic signal, which enables the entire sensor platform to be miniaturized without having to accommodate bulky optical sources. For this reason, magnetic biochips are ideally suited for portable applications used in the military and in health diagnostics. In addition, magnetic materials are more stable than their fluorescent counterparts (magnetic signals do not bleach), and if designed properly, magnetic biosensors can be used to quantitatively compare data between different biosensor experiments. Magnetic signals also benefit from a lower external noise level than their fluorescent counterparts.

Furthermore, magnetic manipulation systems allow for straightforward and potentially programmable handling of colloidal particles in fluids. This capability can be used to separate molecules, mix fluids, or effectively concentrate target molecules onto sensor devices. Compared with dielectrophoresis, which is the basis of optical or electrical techniques for manipulating neutrally charged particles, magnetophoresis has a number of advantages. For one, most biological materials are transparent to magnetic fields, which enables only the materials of interest to be manipulated without concern for heating or inducing chemical reactions. Moreover, basic physics indicates that magnetic forces are likely to be orders of magnitude stronger than forces produced by electrical or optical fields. The basis for this argument is that substantially higher spatial energy densities can be stored in magnetic fields than in electrical fields, which are limited by dielectric breakdown of roughly 10^6 V m^{-1} in most fluids. The energy density stored in a 1 T magnetic field, for example, is 3–4 orders of magnitude higher than the energy density stored in a corresponding 10^6 V m^{-1} electrical field. This analysis motivates future use of magnetic systems for manipulation of particles much smaller than is possible by electrical or optical fields.

Despite its recent success, several issues have limited the use of magnetic sensors more broadly in diagnostics. Compared with fluorescent sensors, which acquire signal through line of sight, the limitation in magnetic sensing is the highly nonlinear dipolar field produced by magnetic labels. As a result, magnetic signals are more sensitive to both the size and position of the label on the sensor than are fluorescent signals. This problem may be largely a technical issue, and one way to overcome this problem is to use sensors that are commensurate in size with the magnetic labels. Additionally, techniques can be devised to control the label's position on the sensor by magnetic force, thereby providing a more repeatable signal.

In conclusion, there is a bright future for magnetic biosensors. With increasing control over magnetic particle synthesis and with future advances in the design of smaller and more sensitive magnetic field sensors, magnetic biosensors may one day rival conventional methods and become more broadly adopted in the field of biosensors and biochips.

REFERENCES

1. D. H. Blohm and A. Guiseppi-Elie, New developments in microarray technology. *Current Opinion in Biotechnology*, 2001, **12**, 41–47.
2. J. Marx, DNA arrays reveal cancer in its many forms. *Science*, 2000, **289**, 1670–1672.
3. A. Manz, N. Graber, and H. M. Widmer, Miniaturized total chemical analysis systems: a novel concept for chemical sensing. *Sensors and Actuators, B Chemistry*, 1990, **1**(1–6), 244–248.
4. M. J. Madou, *Fundamentals of Microfabrication: The Science of Miniaturization*, 2nd Edn, CRC Press, Boca Raton, 2002.
5. D. R. Reyes, D. Iossifidis, P. A. Auroux, and A. Manz, Micro total analysis systems. 1. Introduction, theory, and technology. *Analytical Chemistry*, 2002, **74**(12), 2623–2636.
6. P. A. Auroux, D. Iossifidis, D. R. Reyes, and A. Manz, Micro total analysis systems. 2. analytical standard operations and applications. *Analytical Chemistry*, 2002, **74**(12), 2637–2652.
7. T. Vilkner, D. Janasek, and A. Manz, Micro total analysis systems. Recent developments. *Analytical Chemistry*, 2004, **76**(12), 3373–3385.
8. F. E. Kruis, H. Fissan, and A. Peled, Synthesis of nanoparticles in the gas phase for electronic, optical and magnetic applications-a review. *Journal of Aerosol Science*, 1998, **29**(5–6), 511–535.
9. H. Kawaguchi, Functional polymer microspheres. *Progress in Polymer Science*, 2000, **25**(8), 1171–1210.
10. S. Sun, H. Zeng, D. B. Robinson, S. Raoux, P. M. Rice, S. X. Wang, and G. Li, Monodisperse MFe$_2$O$_4$ (M = Fe, Co, Mn) Nanoparticles. *Journal of the American Chemical Society*, 2004, **126**, 273.
11. N. Nath and A. Chilkoti, Label free colorimetric biosensing using nanoparticles. *Journal of Fluorescence*, 2004, **14**(4), 377–389.
12. S. R. Nicewarner-Peña, R. G. Freeman, B. D. Reiss, L. He, D. J. Peña, I. D. Walton, R. Cromer, C. D. Keating, and M. J. Natan, Submicrometer metallic barcodes. *Science*, 2001, **294**, 137–141.
13. M. Han, X. Gao, J. Su, and S. Nie, Quantum-dot-tagged microbeads for multiplexed optical coding of biomolecules. *Nature Biotechnology*, 2001, **19**, 631–635.
14. D. H. Reich, M. Tanase, A. Hultgren, L. A. Bauer, C. S. Chen, and G. J. Meyer, Biological applications of multifunctional magnetic nanowires. *Journal of Applied Physics*, 2003, **93**, 7275–7280.
15. P. A. Besse, G. Boero, M. Demierre, V. Pott, R. Popovic, Detection of a single magnetic microbead using a miniaturized silicon Hall sensor. *Applied Physics Letters*, 2002, **80**, 4199–4201.
16. M. M. Miller, G. A. Prinz, S. F. Cheng, and S. Bounnak, Detection of a micron-sized magnetic sphere using a ring-shaped anisotropic magnetoresistance-based sensor: A model for a magnetoresistance-based biosensor. *Applied Physics Letters*, 2002, **81**(12), 2211–2213.
17. D. L. Graham, H. A. Ferreira, P. P. Freitas, and J. M. S. Cabral, High sensitivity detection of molecular recognition using magnetically labelled biomolecules and magnetoresistive sensors. *Biosensors and Bioelectronics*, 2003, **18**, 483–488.
18. D. L. Graham, H. A. Ferreira, and P. P. Freitas, Magnetoresistive-based biosensors and biochips. *Trends in Biotechnology*, 2004, **22**(9), 455–462.
19. M. Tondra, M. Porter, and R. J. Lipert, Model for detection of immobilized superparamagnetic nanosphere assay labels using giant magnetoresistive sensors. *Journal of Vacuum Science and Technology A*, 2000, **18**(4), 1125–1129.
20. G. Li, S. X. Wang, and S. Sun, Model and experiment of detecting multiple magnetic nanoparticles as biomolecular

labels by spin valve sensors. *IEEE Transactions on Magnetics*, 2004, **40**(4), 3000–3002.

21. L. Ejsing, M. F. Hansen, A. K. Menon, H. A. Ferreira, D. L. Graham, and P. P. Freitas, Planar Hall effect sensor for magnetic micro- and anobead detection. *Applied Physics Letters*, 2004, **84**(23), 4729–4731.

22. D. K. Wood, K. K. Ni, D. R. Schmidt, and A. N. Cleland, Submicron giant magnetoresistive sensors for biological applications. *Sensors and Actuators A*, 2005, **120**(1), 1201–1206.

23. D. P. E. Dickson, N. M. K. Reid, C. Hunt, H. D. Williams, M. El- Hilo, and K. O'Grady, Determination of f^0 for fine magnetic particles. *Journal of Magnetism and Magnetic Materials*, 1993, **125**(3), 345–350.

24. L. C. R. Néel, Influence of thermal fluctuations on the magnetization of ferromagnetic small particles. *Academy of Sciences*, 1949, **228**, 664–668.

25. P. Tartaj, M. D. P. Morales, S. Veintemillas-Verdauger, T. Gonzalez-Carreno, and C. J. Serna, The preparation of magnetic nanoparticles for applications in biomedicine. *Journal of Physics D: Applied Physics*, 2003, **36**, R182–R197.

26. R. Massart, Preparation of aqueous magnetic liquids in alkaline and acidic media. *IEEE Transactions on Magnetics*, 1981, **17**(2), 1247–1248.

27. S. W. Charles, *Magnetic Fluids (Ferrofluids)*, J. L. Dormann and D. Fiorani (eds), Elsevier Science Publishers, Amsterdam, 1992, pp. 267–276.

28. X. Li and Z. Sun, Synthesis of magnetic polymer microspheres and application for Immobilization of proteinase of balillus sublitis. *Journal of Applied Polymer Science*, 1995, **58**(11), 1991–1997.

29. H. P. Khng, D. Cunliffe, S. Davies, N. A. Turner, and E. N. Vulfson, The synthesis of sub-micron magnetic particles and their use for preparative purification of proteins. *Biotechnology and Bioengineering*, 1998, **60**(4), 419–424.

30. M. Uhlen, Magnetic separation of DNA. *Nature*, 1989, **340**(6236), 733–734.

31. C. H. Setchell, Magnetic separations in biotechnology-a review. *Journal of Chemical Technology and Biotechnology*, 1985, **35**(B), 175–182.

32. G. Helgesen, P. Pieranski, and A. T. Skjeltorp, Dynamic behavior of simple magnetic hole systems. *Physical Review A*, 1990, **42**, 7271–7280.

33. B. B. Yellen, O. Hovorka, and G. Friedman, Arranging matter by magnetic nanoparticle assemblers. *Proceedings of the National Academy of Sciences*, 2005, **102**, 8860–8864.

34. W. K. H. Panofsky and M. Phillips, *Classical Electricity and Magnetism*, 2nd Edn, Addison-Wesley, 1962.

35. C. S. Lee, H. Lee, and R. M. Westervelt, Microelectromagnets for the control of magnetic nanoparticles. *Applied Physics Letters*, 2001, **79**(20), 3308–3310.

36. D. L. Graham, H. Ferreira, J. Bernardo, P. P. Freitas, and J. M. S. Cabral, Single magnetic microsphere placement and detection on-chip using current line designs with integrated spin valve sensors: Biotechnological applications. *Journal of Applied Physics*, 2002, **91**(10), 7786.

37. L. E. Helseth, H. Z. Wen, R. W. Hansen, T. H. Johansen, P. Heinig, and T. M. Fischer, Assembling and Manipulating Two-Dimensional Colloidal Crystals with Movable Nanomagnets. *Langmuir*, 2004, **20**, 7323–7332.

38. F. J. Friedlaender, M. Takayasu, J. B. Rettig, and C. P. Kentzer, Particle flow and collection process in single wire HGMS studies. *IEEE Transactions on Magnetics*, 1978, **14**(6), 1158–1164.

39. B. B. Yellen, Z. G. Forbes, D. S. Halverson, G. Fridman, K. A. Barbee, M. Chorny, R. J. Levy, and G. Friedman, Targeted drug delivery to magnetic implants for therapeutic applications. *Journal of Magnetism and Magnetic Materials*, 2005, **293**(1), 647–654.

40. B. B. Yellen and G. Friedman, Programmable Assembly of Heterogeneous Colloidal Particle Arrays. *Advanced Materials*, 2004, **16**(2), 111–115.

41. B. B. Yellen, G. Fridman, and G. Friedman, Ferrofluid lithography. *Nanotechnology*, 2004, **15**, S562–S565.

42. O. Hovorka, B. B. Yellen, N. Dan, and G. Friedman, Self-consistent model of field gradient driven particle aggregation in magnetic fluids. *Journal of Applied Physics*, 2005, **97**, 10Q306.

43. S. H. Liao, S. C. Hsu, C. C. Lin, H. E. Horng, J. C. Chen, M. J. Chen, C. H. Wu, and H. C. Yang, High-Tc SQUID gradiometer system for magnetocardiography in an unshielded environment. *Superconductor Science and Technology*, 2003, **16**, 1426–1429.

44. P. Ripka, Review of fluxgate sensors. *Sensors and Actuators, A*, 1992, **33**, 129–141.

45. N. E. Jenkins, L. P. DeFlores, J. Allen, T. N. Ng, S. R. Garner, S. Kuehn, J. M. Dawlaty, and J. A. Marohn, Batch fabrication and characterization of ultrasensitive cantilevers with submicron magnetic tips. *Journal of Vacuum Science and Technology B*, 2004, **22**(3), 909–915.

46. P. D. D. Schwindt, S. Knappe, V. Shah, L. Hollberg, J. Kitching, L. A. Liew, and J. Moreland, Chip-scale atomic magnetometer. *Applied Physics Letters*, 2004, **85**(26), 6409–6411.

47. G. Li and S. X. Wang, Analytical and micromagnetic modeling for detection of a single magnetic microbead or nanobead by spin valve sensors. *IEEE Transactions on Magnetics*, 2003, **39**, 3313–3315.

48. Y. W. Tahk, K. J. Lee, and T. D. Lee, Spin tilting phenomenon in strongly coupled AFC media. *IEEE Transactions on Magnetics*, 2002, **38**(5), 2087–2089.

The Detection and Characterization of Ions, DNA, and Proteins Using Nanometer-Scale Pores

John J. Kasianowicz,[1] Sarah E. Henrickson,[1] Jeffery C. Lerman,[1] Martin Misakian,[1] Rekha G. Panchal,[2] Tam Nguyen,[2] Rick Gussio,[2] Kelly M. Halverson,[3] Sina Bavari,[3] Devanand K. Shenoy[4] and Vincent M. Stanford[5]

[1] *Electronics and Electrical Engineering Laboratory, National Institute of Standards and Technology, Gaithersburg, MD, USA,* [2] *Target Structure Based Drug Discovery Group, National Cancer Institute–SAIC, Frederick, MD, USA,* [3] *United States Army Medical Research Institute of Infectious Diseases, Frederick, MD, USA,* [4] *Naval Research Laboratory, Center for Bio/Molecular Science and Engineering, Washington, DC, USA and* [5] *Information Technology Laboratory, National Institute of Standards and Technology, Gaithersburg, MD, USA*

1 INTRODUCTION

Biology is controlled by interactions between different types of soft condensed matter including DNA, RNA, proteins, lipids, and carbohydrates.[1] Hydrogen bonding and other atom–atom interactions permit biological macromolecules to adopt three-dimensional structures that are robust over long enough timescales to perform useful work (e.g., the storage, transcription, transmission, and translation of information critical for the development, maintenance and propagation of life). In addition, they catalyze chemical reactions (e.g., the synthesis and degradation of other molecules) and act as inter- and intracellular transport machines. Here, we focus on the latter category of macromolecules because they have demonstrated potential for use in biosensor applications.

Ion channels are nanometer-scale pores formed by membrane-spanning proteins[2] that can catalyze the flow of up to $\sim 10^9$ ions/s. More than 50 years of research into the structure and function of ion channels demonstrates that a seemingly simple motif, a nanopore, performs many different tasks in cells and organelles. These include neuronal signal transmission,[3] antibiotic activity,[4] the transduction of signals within and between cells,[5] and the selective transport of ions and macromolecules. Failure of ion channels in vivo often leads to debilitating disease. For example, a defect in a chloride-selective ion channel is the molecular basis for cystic fibrosis.[6]

Biological nanopores also facilitate the transport of macromolecules in a wide variety of processes including protein translocation across membranes,[7] gene transduction between bacteria, and the transfer of genetic information from some viruses and bacteriophages to cells.[8] With the

Handbook of Biosensors and Biochips. Edited by Robert S. Marks, David C. Cullen, Isao Karube, Christopher R. Lowe and Howard H. Weetall.
Published in 2007 by John Wiley & Sons, Ltd. ISBN 978-0-470-01905-4. This chapter is in the public domain in the United States
but is copyright John Wiley & Sons Ltd. in the rest of the world.

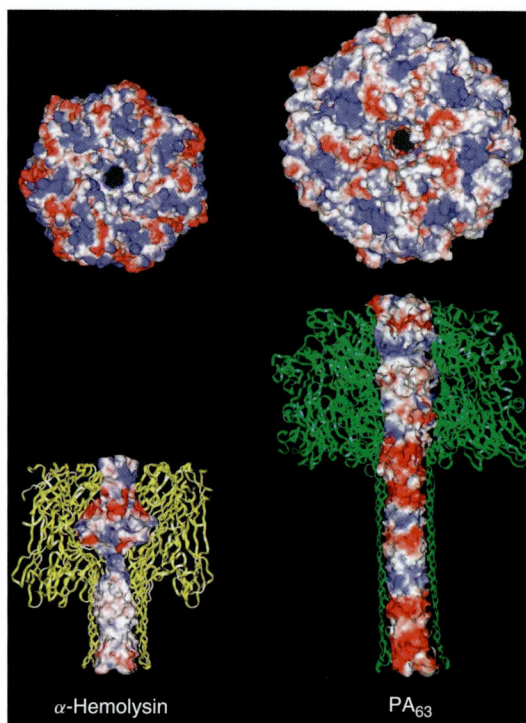

Figure 1. Structures of two protein nanopores formed by protein toxins used in the development of nanometer-scale biosensors. (Left) A crystal structure of the *Staphylococcus aureus* alpha-hemolysin (αHL) ion channel.[9] (Right) A model for the channel formed by *B. anthracis* protective antigen 63 $(PA_{63})_7$ is shown.[10] The αHL channel is approximately 10.5-nm tall.

goal of adapting protein nanopores for biosensor applications, we are studying how ions, DNA, and proteins bind to and alter the ionic current through two different channels formed by the bacterial toxins, *Staphylococcus aureus* alpha-hemolysin (αHL)[9] and *Bacillus anthracis* protective antigen 63 (PA_{63})[10] (Figure 1).

2   ION TRANSPORT THROUGH FULLY OPEN NANOSCALE PORES

2.1   Properties of *S. aureus* αHL, a Model Nanopore for Biosensing

α-Hemolysin is one of several toxins secreted by the bacteria *S. aureus*.[11] The 293–amino acid protein monomer has a molecular mass of 33.1 kDa, is water soluble, binds spontaneously to phospholipid membranes, and forms a

relatively large ion channel[12] from seven identical monomers.[9] The crystal structure of the channel shows a massive cap domain that extends beyond one of the membrane–solution interfaces and a relatively short β-barrel stem region that spans the membrane (Figures 1 and 2).

The αHL channel has several properties that make it ideal for use in biosensor systems. Like many other protein ion channels, the αHL nanopore gates (i.e., switches) between different conducting states.[13] Over 15 years ago, we found that the αHL channel can remain fully open for tens of minutes.[14] This permitted the development of the αHL channel as a fully open nanometer-scale test tube to study the reaction dynamics of ions binding to the pore,[15–17] of polymers partitioning into and binding to it,[18] and the transport of individual DNA molecules through it.[19] The results of these experiments are described below because they opened the possibility of using single nanopores for the detection and characterization of analytes.

Nanometer-scale pores are ideal for use in sensor applications for several reasons. First, they

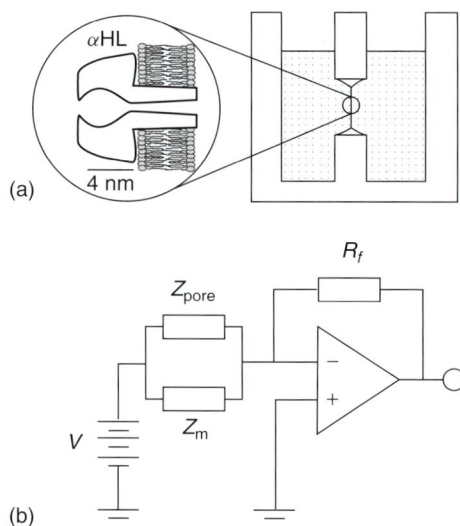

Figure 2. Experimental setup for measuring the ionic current through protein nanometer-scale pores. (a) A lipid membrane ($\sim$4-nm thick) is formed on a 100-μm diameter hole in a Teflon film that separates two identical Teflon chambers that each hold $\sim$2 ml of aqueous electrolyte solution. Ion channels are reconstituted into the membrane by adding protein to one chamber while stirring. (b) The impedance of the membrane and nanopore are measured using a low-noise voltage source and a high-impedance, high-bandwidth operational amplifier.

are commensurate with the size of the analytes of interest. Second, they are sufficiently small so that analytes can create a signal by a steric blockade of the pore or by changing the electrostatic potential near or inside the pore (i.e., the pore size is on the order of the Debye length in a typical electrolyte solution).[20] Third, the analyte does not necessarily need to be labeled (e.g., with a fluorescent probe) to be detected as a single entity. Fourth, the method is sufficiently sensitive (better than 1 nM).

2.2 Measurement of the Ionic Current through Protein Nanopores

Details of the experimental methods for reconstituting channels formed by αHL and similar proteins into planar lipid bilayer membranes are described elsewhere.[16] Briefly, a solvent-free diphytanoyl phosphatidylcholine lipid membrane is formed on a 20–100-μm diameter opening in a 25-μm thick polytetrafluoroethylene (PTFE, Teflon) partition that separates two halves of a Teflon chamber (Figure 2a). The two compartments contain identical aqueous solutions (e.g., 1 M KCl buffered to a constant pH value). The pore-forming protein is added to the aqueous phase in one compartment, herein called *cis*, which is stirred vigorously for $\sim$10 s. After the desired number of channels is formed, excess protein is removed from the chamber. An electric field is applied across the membrane by a matched pair of Ag–AgCl electrodes. The ionic current is converted to a voltage using a high-bandwidth amplifier (Figure 2b) after which the signal is passed through an analog low-pass filter and then digitized by an A/D converter. A negative applied potential drives anions from the cis side to the trans side.

If we assume the pore can be represented by a smooth circular cylinder filled with an electrolyte solution that has a conductivity equal to that of the bulk, σ_{bulk}, and that for small applied potentials Ohm's law is valid (i.e., $g = I/V$, where g is the single-channel conductance), a crude estimate for the diameter, d, of an ion channel can be obtained. If we also assume that the access resistance of the solution outside the pore[2] is zero, then the channel conductance is $g = \sigma A/l$, where A and l are the cross-sectional area and length of the pore, respectively, and σ is the conductivity of the solution inside the pore.

Thus $d = (4gl/\sigma\pi)^{1/2}$. For the αHL channel ($l \sim$ 10 nm) in 1 M KCl ($\sigma \sim 0.1$ S cm^{-1}) at pH 7.5, the single-channel conductance is $g \sim 1$ nS and the estimated diameter is $d \sim 1.12$ nm, slightly smaller than the limiting aperture obtained from the channel crystal structure (1.56 nm). Although naive, this simple calculation underscores the fact that ion channels are indeed nanometer-scale objects.

2.3 Effect of Electrostatics on the Current Flow in a Fully Open Channel

Figure 3(a) illustrates typical ionic current recordings for a single αHL channel that occur in response to different values of the applied potential.[21] Figure 3(b) shows the mean value of the current at each voltage, in the form of a current–voltage (I–V) relationship, for solutions with two different pH values. At low pH (pH 4.5), the I–V relationship is nearly ohmic. At higher pH (pH 7.5), the I–V relationship becomes slightly nonlinear and rectifying.[15,16,21] The difference in the I–V relationships at these two pH values is more striking at lower values of the ionic strength (*not shown*).

In principle, the binding of protons to amino acid side chains in the αHL channel could cause the I–V relationship to change by either altering the pore's geometry or the electrostatic potential profile along the pore axis. The latter effect can be described by a one-dimensional electrostatic model for the control of ion flow through the channel.[2,21] Figure 3(c) shows how the αHL channel is most likely situated in a 4-nm thick lipid bilayer membrane.[17] Figure 3(d) presents a simplified one-dimensional potential profile that identifies charged amino acid side chains inside the pore or near the pore entrances. The applied potential, V, is assumed to drop linearly along the channel length as in the Goldman–Hodgkin–Katz equation.[22,23] The well amplitudes change as the pH value of the electrolyte solution is varied (the positive barrier amplitudes remains fixed because of lysine's high pK value; the lysines are at positions 8, 131, and 147). Physically reasonable adjustments to the values of the well depths made it possible to obtain a good fit to the shape of the I–V relationships for pH 4.5 and 7.5. The change in the I–V relationship

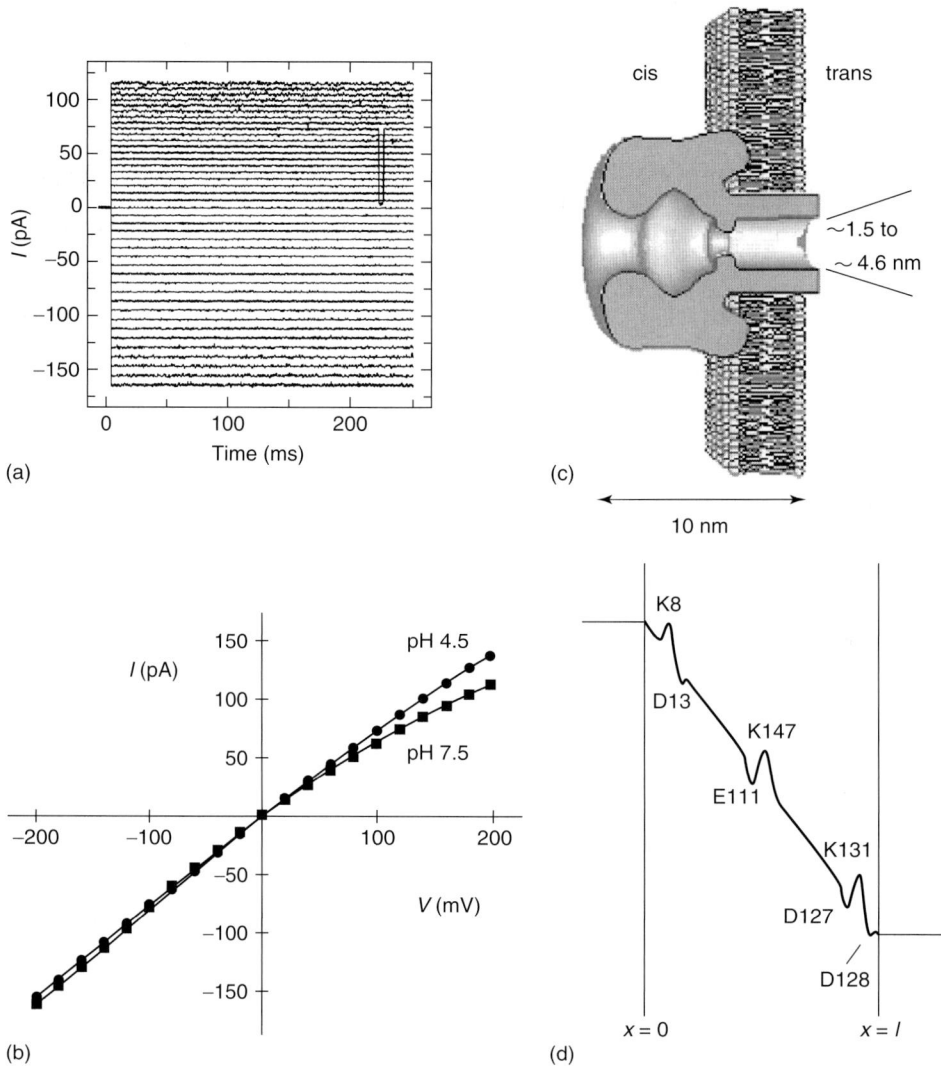

Figure 3. Experimental and theoretical description for ion flow through the fully open αHL channel. (a) Single-channel currents for applied potentials ranging from −200 mV to +200 mV in 10-mV increments. (b) Current–voltage relationship estimated from data similar to that in (a). The current is generally nonlinear in the applied potential, slightly rectifying, and pH dependent. (c) Cartoon illustration of the αHL channel in a lipid membrane. (d) A simplified one-dimensional electrical potential profile that includes the applied transmembrane voltage, and the barriers and wells caused by fixed charges inside the nanopore. Changes in the solution pH cause an asymmetric change in the amount of fixed charge in the αHL pore.[21] This property was used to demonstrate that single αHL channels could act as sensors for ions. [Reprinted from M. Misakian and J.J. Kasianowicz, Electrostatic influence on ionic current through the α-HL ion channel. *J. Membrane Biology*, 2003, **195**, 137–146, with permission from Springer.]

from nearly linear to nonlinear and rectifying most likely reflects the fact that ionizable side chains (e.g., D13, E111, D127, and D128) that reversibly bind protons under these conditions are asymmetrically distributed along the pore axis. Also, the pore diameter is commensurate with the Debye length.[20]

2.4 Single Nanopore-based Sensors: Analyte Detection Based on Reaction Dynamics

The results in Figure 3, and the simplified theoretical models that describe them, demonstrate how an ion can bind to a channel and leverage the flow of other ions through a nanometer-scale pore (i.e., much like the field-effect process in a transistor). Because ion binding to sites within the pore is generally a reversible process, one might expect that the channel conductance will fluctuate between two states as the ion associates with, and dissociates from the site. Moreover, because reaction kinetics are characteristic of the interaction between the analyte and the binding site, the dynamics of the current fluctuations will contain information about the analyte concentration and type.[24] Furthermore, the type of blockade (e.g., electrostatic control, steric blockade of the pore by analyte, etc.) provides additional details about the process. Therefore, the binding of the analyte to the channel could either cause the conductance to increase or decrease.

Let us suppose the pore has two conductance states that are characteristic of the binding site when it is never, or always, occupied (Figure 4, states 1 and 2). These two conditions could correspond to analyte concentrations that are much less than or much greater than the binding constant K in mol l^{-1}. K is defined by $K = k_{off}/k_{on}$ where k_{on} and k_{off} are the rate constants for the association and dissociation of the analyte to the binding site on the nanopore, and $1/k_{off}$ is the mean time that the analyte is bound to the site on the pore. When the analyte concentration is equal to the binding constant (i.e., $[A] = K$), the channel spends on average half the time in each of the two conductance states.

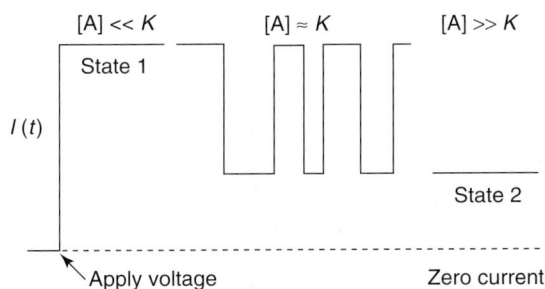

Figure 4. Schematic of nanopore-based method for analyte detection. Ionic current versus time for a pore challenged with three different analyte concentrations. First, in the absence of analyte ($[A] \ll K$), where K is the association constant for the binding of analyte to the pore, the current is in state 1. Second, for analyte concentration $[A] \approx K$, the pore conductance will fluctuate between two states (bound and unbound with analyte). Third, for high analyte concentration (i.e., $[A] \gg K$), the pore will be virtually always occupied with analyte and therefore be in state 2.

Figure 5(a) illustrates schematically how the ionic current appears as the analyte concentration varies. Figure 5(b) shows the mean ionic current variation if there is either only one analyte binding site on the channel or many independent binding sites with identical binding constants. Calibrating a given nanopore with known concentrations of a particular analyte allows the unique determination of the analyte concentration in a test solution by measuring the value of the mean ionic current.

The dynamic information in current fluctuations provides additional information that aids the estimation of the concentration and identification of the analyte species.[15,16,24] Spectral analysis is used to determine the frequency content of a time series.[25] For a random telegraph, two-state system, the spectral density of the current noise is described by:

$$S(f) = \frac{4(\Delta i_{1,2})^2 \tau^2}{((\tau_1 + \tau_2)(1 + (2\pi f \tau)^2))} \quad \text{or}$$

$$S(f) = \frac{S(0)}{(1 + (f/f_c)^2)}$$

where τ_1 and τ_2 are the mean times spent in states 1 and 2, respectively; $1/\tau = 1/\tau_1 + 1/\tau_2$ or $\tau = \tau_1 \tau_2/(\tau_1 + \tau_2)$ (in seconds); f is the frequency (in Hz); and $\Delta i_{1,2}$ is the difference in current between states 1 and 2.[15,26] Figure 5(c) is a theoretical plot of the spectral density as a function of frequency for a given analyte type and concentration (given values for k_{on} and k_{off}). At low frequencies, $S(f)$ approaches a constant value, $S(0)$. At greater frequencies, $S(f)$ decreases as $1/f^2$. At the characteristic corner frequency $f_c = 1/(2\pi \tau)$, the spectral density drops twofold from its value at $f = 0$ (i.e., $S(f_c) = S(0)/2$).

A least-squares fit of the expression above for $S(f)$ to the experimental ion current spectral density data provides estimates for $S(0)$ and τ. Figure 5(d) illustrates how these two parameters vary with analyte concentration. At both extremes of the concentration, the current fluctuations, and therefore $S(0)$ (solid line), are virtually nil because the binding site is either always unoccupied or fully occupied. $S(0)$ rises to a maximum value near the analyte concentration $[A] \sim K$, because the site is occupied or unoccupied about half the time.

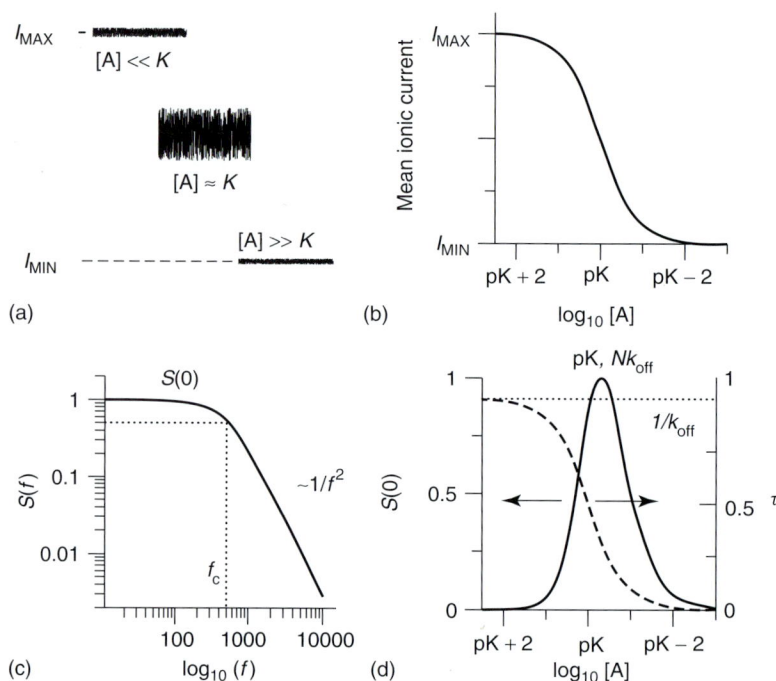

Figure 5. Spectral analysis of analyte-induced current fluctuations provides estimates for the binding constant and kinetic rate constants for the reactions. (a) Hypothetical ionic current versus time recordings for three different analyte concentrations. The fluctuations in the ionic current about the mean value are caused by the reversible reaction between the analyte and a site or sites on a single nanopore. They are minimal at the extremes of analyte concentration and maximal at $[A] \approx K$. (b) For increasing analyte concentration, the mean ionic current decreases monotonically from the maximum to the minimum current values. (c) Relative power spectral density (PSD) for the ionic current fluctuations. The low-frequency PSD value, $S(0)$, and the corner frequency, f_c, provide information about the binding constant and kinetic rate constants. For reactions that are Markovian, the PSD decreases as $1/f^2$ for $f > f_c$. (d) At the extreme values of analyte concentration, $S(0)$ (black line) is minimal; for analyte concentrations $[A] \approx K$, it rises to a maximal value. The characteristic relaxation time for the interaction between the analyte and the nanopore, $\tau = 1/2\pi f_c$, decreases monotonically with increasing analyte concentration (dashed line). In the limit of zero analyte concentration, $\tau \sim 1/k_{off}$.

The plot in Figure 5(d) also illustrates that at low analyte concentration, the characteristic lifetime τ (dashed line) equals $1/k_{off}$ because the dominant timescale is the mean time the analyte is bound to a site on the channel. As the analyte concentration increases, a second timescale, the time it takes the analyte to bind and react with the site, contributes and τ decreases monotonically to zero. The variation of $S(0)$ with analyte concentration depends on both the pK, the product of the number of binding sites, N, and k_{off}. Thus, if the pK and k_{off} are determined from a calibration of the mean current (Figure 5b) and τ (Figure 5d), the number of binding sites can be estimated from the values of $S(0)$.[15]

This method discriminates particularly well between different analytes that bind to the nanopore because it makes use of both the thermodynamic (pK) and kinetic (k_{on} and k_{off}) information. For example, it was demonstrated that a single αHL channel can distinguish between isotopic ion species (i.e., aqueous hydronium and deuterium ions).[15,16] The spectral method was subsequently applied to genetically engineered versions of the αHL channel tuned to bind divalent cations.[17]

3 NEUTRAL POLYMER AND DNA TRANSPORT IN A SINGLE NANOPORE

3.1 Neutral Polymer Probes of Channel Structure Interact with the Pore

Krasilnikov and colleagues developed a method to use nonelectrolyte polymers of poly(ethylene glycol) (PEG) to estimate the diameter of ion channels.[27,28] It is well known that PEG decreases the bulk conductivity of ionic solutions. Thus, PEGs that are sufficiently small enter the pore and decrease the channel conductance. Polymers larger than the diameters of the two channel mouths rarely partition into the pore and therefore have little or no effect on the conductance. The dependence of channel conductance on the PEG molecular weight determines the pore's PEG molecular mass cutoff, and by inference, the pore diameter.

The single-channel recordings in Figure 6(a) illustrate the effect of different molecular mass PEGs on the αHL channel conductance. Note that the relatively large PEG 8000 rarely decreases the conductance. In contrast, PEG 200 significantly reduces the mean current. The ratio of the conductance in the presence of PEG to that in the absence of the polymer demonstrate that PEGs with molecular mass less than 3000 partition into the channel (Figure 6b). The diameter of the αHL channel is estimated from these data and the measured values of PEG hydrodynamic radii[29] are indicated on the plot.

The single-channel current recordings in the presence of PEG 2000 (Figure 6a) are noisy. Ion current fluctuations should indeed occur when the polymer randomly partitions into and out of the pore. However, the observed noise is orders of magnitude greater than expected based on the calculated residence time for the polymer diffusion inside the pore. Specifically, the one-dimensional diffusion equation $\langle \Delta x \rangle^2 = 2D\tau_D$[30] suggests the polymer should diffuse the length of the channel in a time $\tau_D \sim 10$ ns. However, the current recordings shown in Figure 6(a), which were filtered to 1-kHz bandwidth, indicate otherwise. Indeed, the mean residence time for PEG in the αHL pore, deduced from the excess current noise, was $\sim 100 \mu s$.[18]

3.2 Detecting Individual Polynucleotides that Thread through a Single αHL Channel

The previous result (Figure 6a) indicated that the αHL nanopore can interrogate a polymer for a time much greater than the time taken for the polymer to diffuse through the channel. Because of this property and because the αHL channel can remain fully open for long times, an opportunity is provided to study the details of DNA transport in a highly confined space. Specifically, we demonstrated that individual molecules of single-stranded DNA are

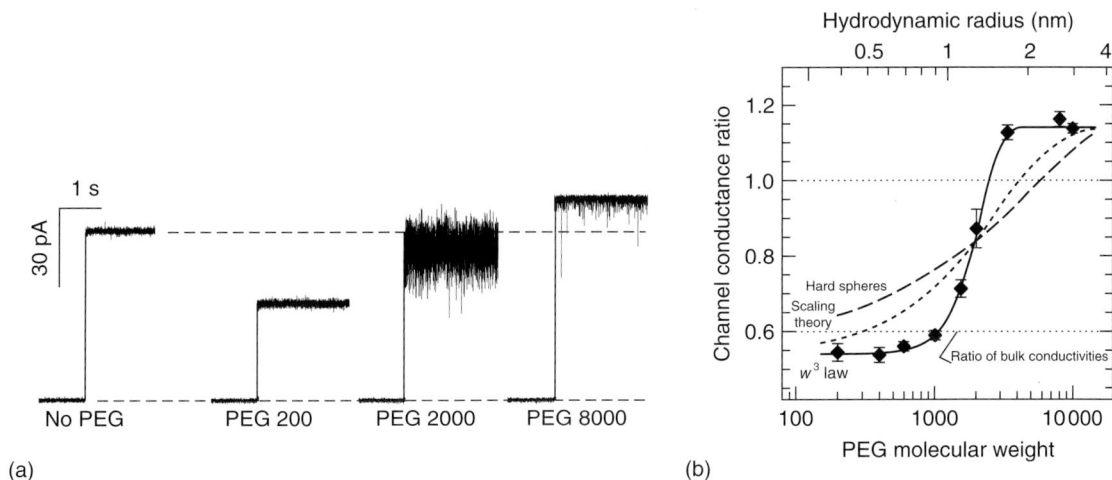

Figure 6. Estimating the size of the αHL channel with nonelectrolyte polymers. Polymers of poly(ethylene glycol), PEG, reduce the bulk conductivity of an electrolyte solution. (a) Sufficiently small PEGs partition into the solitary channel and reduce the current of spontaneously forming channels. (b) The dependence of the single-channel conductance on the polymer Molecular Weight (MW) is used to estimate the PEG MW cutoff, and thus provides an estimate for the diameter of the aqueous-filled channel pore.

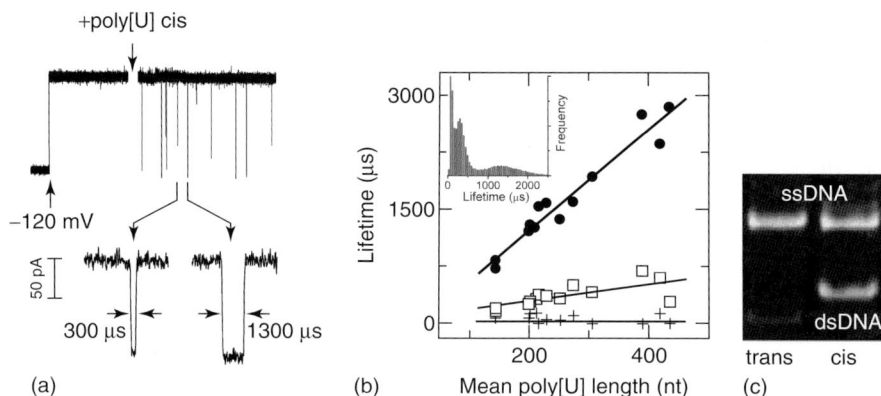

Figure 7. Polynucleotides are driven into a single αHL channel by a transmembrane potential difference. (a) Single-channel recordings in the absence and presence of single-stranded RNA show transient blockades. (b) Polymer-induced blockade lifetimes (inset: histogram of blockade lifetimes for a given length poly[U] RNA show 3 characteristic lifetime values). The polynucleotide-induced lifetimes for the two slowest blockade types are proportional to the polymer length, which suggests the polynucleotide threads completely through the nanopore. (c) PCR demonstrates that single-, but not double-stranded DNA is transported through the αHL channel from the cis to the trans side.

driven electrophoretically into and through a single αHL ion channel.[19,31] Because the mobility of negatively charged polynucleotides is less than that of monovalent ions and the polymer occupies space that small mobile ions normally would, polynucleotides decrease the channel conductance when they are inside the pore (Figure 7a). The lifetime of the polymer-induced current blockades is proportional to the contour length for polynucleotides comparable to or longer than the pore length, (Figure 7b). Polymerase chain reaction technology (PCR) confirmed that single-stranded, but not blunt-ended double-stranded, DNA was transported through the pore. The latter two results strongly suggest that the polymer threads through the pore as a linear rod.[19] In that report, we hypothesized that a single nanopore could rapidly sequence DNA if each base caused a unique current blockade level. We subsequently demonstrated that different homopolymers cause distinctly different ion current blockade signatures.[32,33] Because DNA sequencing with a single nanopore is yet to be realized, intense investigation continues.[34–41]

3.3 Sensor Technologies Based on DNA–Nanopore Interactions

For relatively short polynucleotides that interact with the αHL channel, the time-averaged

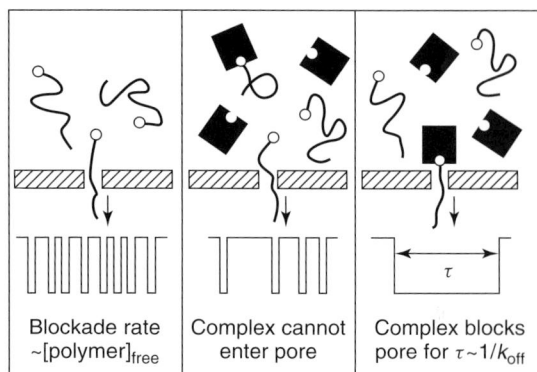

Figure 8. Sensor models based on the interaction between a single nanopore and polymers. It is assumed that polymers with binding sites for analytes have unfettered access to the nanopore. The entry or transport of individual polymers causes a transient decrease in the ionic current (left). Addition of analytes that bind to the polymer change the ability of the latter to partition into (center) and/or transport through the pore (right). In the latter model, the analyte : polymer complex blocks the pore for a time that corresponds to the mean lifetime of the complex (i.e., $\tau \sim 1/k_{off}$). [Adapted from Kasianowicz, et al.,[42] 2001.]

blockade rate increases linearly with the polymer concentration.[31] This, in part, permitted simultaneous multiple analyte detection with polymers and a single nanopore (Figure 8a–c).[42] Briefly, a binding site for a specific analyte is attached to an αHL channel–permeant polymer. Analyte binding to this site alters the ability of the polymer to thread through the pore. In the first case, the analyte concentration is deduced from the decrease in the mean number of blockades per unit time (Figure 8b). For the second detection scheme (Figure 8c), the analyte concentration is estimated from the mean time for nanopore occlusion by the analyte/polymer complex after the electric field is applied. Because different polymer types cause different current blockade patterns (see below), a single nanopore can be used to simultaneously detect different analytes. This method is particularly useful because it does not require the analytes to be labeled. In addition, changing the applied potential permits the force on the bond between the analyte and polymer to be varied, which may help identify a particular analyte in the presence of molecules with similar, but not identical properties.

4 ADVANCED SIGNAL PROCESSING METHODS: READING INFORMATION ENCODED IN POLYMERS

4.1 Polynucleotide-induced Current Blockades Characteristic of Polymer Type

As shown in Figure 7, the transport of individual polynucleotides through a single ion channel is easily observed electronically because the polymers occlude the channel and thereby reduce the flow of ions through the nanopore. The length of a polymer can be estimated from the mean current blockade lifetime (Figure 7b). Can information encoded in the polymer be read from the electronic signals?

The current recordings in Figure 9(a) illustrate the blockades caused by identical length poly[dT], poly[dA], and poly[dC] molecules. Note that the lifetime and substate patterns for a given homopolynucleotide are clearly distinguishable from those caused by the others. Nevertheless, the variation in individual blockade patterns for a given polymer type requires a stochastic analysis of the signals.

The current blockades depend on the characteristics of the polynucleotide (e.g., base composition, secondary structure, and interactions between polymer subunits) and the nanopore (e.g., pore geometry and local electric field in the lumen). The

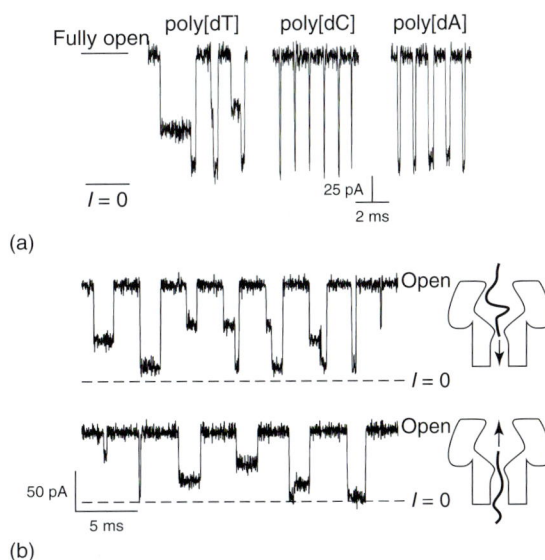

Figure 9. Polynucleotides threading through a single αHL channel cause transient ionic current blockades that are characteristic of the polymer. (a) Blockades caused by individual 100-nt long poly[thymine] (poly[dT]), poly[cytosine] (poly[dC]), and poly(adenosine) (poly[dA]). (b) The blockade patterns for poly[dT] depend upon the side to which the polymer is added.

signals caused by 100-base-long homopolymers of thymine (i.e., poly[dT]$_{100}$) are discussed in greater detail because their rich state structure reveals characteristics of the pore geometry. Additionally, the signals provide an example of how information encoded in a polymer might be read electronically via a nanopore.

The current blockades shown in Figure 9(b) for poly[dT]$_{100}$ entering the cis entrance of the αHL channel, as indicated by the cartoon inset, often show a characteristic shallow-then-deep blockade pattern. In contrast, when the polymer is driven from the opposite direction (i.e., trans side), the blockade pattern is reversed (i.e., a deep-then-shallow pattern prevails).

Figure 10 illustrates many blockades induced by poly[dT]$_{100}$. This "forest from the trees" representation obscures the details of any individual blockades (Figure 9b) but it aids the visual identification of the most probable occluded current states. For example, the three darkest bands in the current recording (Figure 10a, left) and the largest peaks in the current amplitude histogram for $\sim 10^4$ blockades (Figure 10a, right) correspond to the fully open state and two most probable occluded current

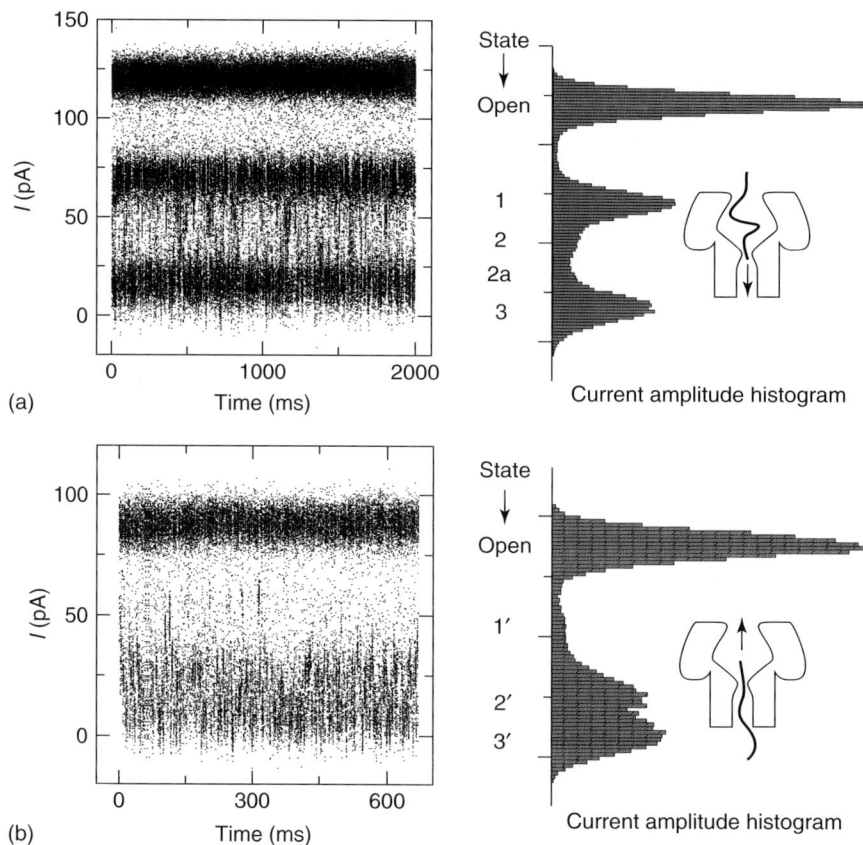

Figure 10. Ionic current time series for poly[dT]$_{100}$ transit events with lifetimes between 40 µs and 2 ms (left) and current amplitude distribution (right) for polymers entering the pore from either the (a) cis or (b) trans pore entrances for applied potentials of V = −120 or +120 mV, respectively. The two time series depict the open channel current band and bands corresponding to the three most probable occluded states. The limited number of occluded states may represent the negotiation of the polymer through the various-diameter segments of the pore.

states (states 1 and 3). The speckle between the lower two dark bands (Figure 10b, left) represents a less frequently occurring, but statistically significant state (state 2) and an even less probable state (state 2a). We ignore the latter state in subsequent discussion. The technique for determining the state sequences of individual events relies on Viterbi decoding[43] of the dwell times within states of Gaussian mixtures fit to the observed amplitude densities. For the range of event lifetimes under consideration (≤ 2 ms), three blockade states and one open channel state are statistically adequate to describe the amplitude distribution.

Using these population distributions, the state sequences in current flow levels from individual molecules as they entered the ~1.5-nm diameter

opening in the channel were decoded. Under these experimental conditions only hundreds of ions per microsecond flow through the pore when it is partially occluded by a polynucleotide. Thus, statistical techniques are fundamental to characterizing the blockade states. For relatively short lifetime current blockades, the states and state sequences suggest that the polymer–pore interaction is simple. Moreover, it can be demonstrated that the state parameters and their sequence depend on the direction of transit, and how long the polymer and pore interact with each other.

A statistical analysis of the current amplitude histogram in Figure 10(a, right), shows that three occluded states result in a good fit to the data. The cis open channel state mean current

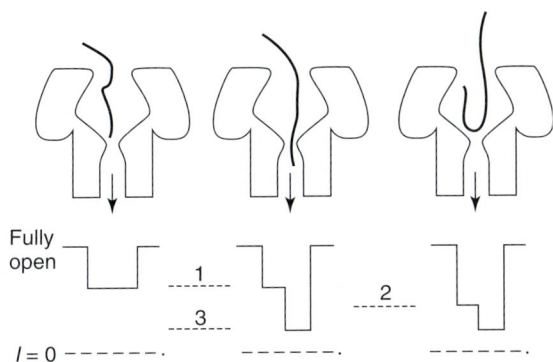

Figure 11. A cartoon that illustrates a possible mechanism for the three most probable transient current blockades caused by poly[dT]$_{100}$.

($\sim$120 pA), and the occluded state mean current values ($\sim$70, $\sim$46, and $\sim$17 pA), have probability weights of 0.62, 0.2, 0.03, and 0.15 respectively. Similar results are obtained for events caused by poly[dT]$_{100}$ molecules entering the trans entrance (Figure 10b). In that case, the trans fully open single-channel current average ($\sim$88 pA), and the three most probable occluded states ($\sim$53, $\sim$26, and $\sim$8 pA) occur with probability weights 0.53, 0.04, 0.19, and 0.24, respectively. Interestingly, the ratios of mean current values for each of the three most probable occluded states to the respective mean open channel current for poly[dT]$_{100}$ entering the pore from the cis side (Figure 10a) do not appear to correspond to those for the three most likely occluded states for polymers threading the pore in the opposite direction (Figure 10b).

One interpretation of the latter result, illustrated in Figure 11, suggests that the degree of ionic current blockade correlates with the amount of poly[dT] mass in either the pore vestibule (shallow blockade), or the smallest channel aperture (deeper blockade). This simple hypothesis is consistent with the blockade patterns caused by poly[dT] transport in either direction and the lifetime distributions of poly[dT]-induced blockades. These results, and others shown here, demonstrate that DNA can be used to probe the geometry of αHL channel. By inference, this technique may prove useful for probing the structures of other nanometer-scale pores, including those made in solid-state materials.

4.2 Blockade State Sequences Evolve with the Event Lifetime

Because the physical length of the homopolymer is constant for the poly[dT]$_{100}$ experiment, differences in the blockade event lifetime and the state sequences allow us to characterize the physical properties of the nanopore.

Figure 12 illustrates how the morphology of poly[dT]$_{100}$-induced current blockades evolves with increasing event lifetime over the range from $60 \leq \Delta\tau_{cis} \leq 600\,\mu s$ (i.e., for polymer added to the cis side). Here, the single-channel current time series for ensembles of events at three representative lifetimes (i.e., $\Delta\tau_{cis} = 60$, 140, and 600 μs) are aligned at the onset of each channel blockade. The colors indicate the frequency of current values; that is, the closer to red end of the spectrum, the greater the number of occurrences. Because few 60-μs events exhibit the deep blockade state (state 3), the polymer most likely only entered the pore vestibule and did not thread completely through the pore. The 140-μs ensemble shows a bifurcation of the event set into shallow blockades, which are qualitatively similar to the 60-μs events, and deep blockades in which the homopolymer most likely was driven past the narrowest diameter of the channel and

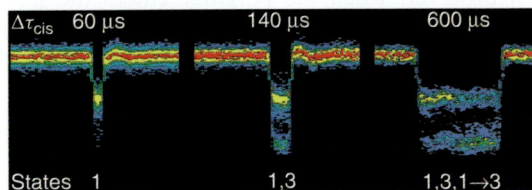

Figure 12. Event time-amplitude histograms showing characteristic ionic current signatures for increasing blockade duration. Poly[dT]$_{100}$-induced current blockades stratified by duration show that the event structure evolves in a simple manner. Sixty microsecond long events show a state 1 conductance level; 140 μs long events show a bimodal conductance morphology (state 1 and state 3); 600 μs long events show state 1 and increasingly frequent state 3 blockades. The 600 μs blockades also show the emergence of state 1 to state 3 transitions within the events. These signatures most likely represent the progress of the polymer through the various limiting apertures of the nanopore. In these experiments, the polynucleotide entered the αHL channel from the cis pore mouth. The color scale is adjusted to the declining frequency of longer events to best visualize the common event morphologies.

threaded through the pore (see Figure 11 center). The 600-μs event ensemble shows a third event morphology, a shallow-then-deep blockade, that is composed of three event types: shallow blockades, deep blockades, and shallow-then-full blockades. Figure 9(a, left) shows three such individual events.

4.3 Extracting Information Encoded in Polymers

Figure 13 illustrates typical single-channel current blockades for poly[dT]$_{100}$ entering from the cis entrance of the channel, and the Maximum A Posteriori probability estimate of the state sequence (black), subject to the constraint imposed by a persistent Hidden Markov Model (HMM), which applies a penalty for state transitions, superimposed on the actual time series data (gray). For polymers entering the pore from the cis side (Figure 13a), the blockades show the three common patterns of Figure 12; i.e., state 1 only, state 3 only, and the intra-event transition from state 1 to state 3. Also included is an example of a relatively rare event comprised of a transition from state 2 to state 3. A corresponding class of events is observed when the polymer enters the trans side (Figure 13b). However, note that the transitions are from a more occluded state to a less occluded one, with the opposite two-step pattern observed when the polymer enters the pore from the cis entrance. The relatively small number of current blockade states and patterns also

suggests that a simple description of the blockade mechanism at these short event lifetimes may be valid. Figure 13(b) shows individual events illustrating shallow blockades, deep-then-shallow blockades, and events proceeding directly to a deep blockade. A mirror symmetrical morphology is observed in trans-to-cis event ensembles of this duration range: the deep-then-partial blockade. These results most likely reflect the fact that the channel cross section is asymmetric: the narrower segments are closer to the trans pore entrance.

Analysis of a large number ($\sim 2 \times 10^5$) of poly[dT]$_{100}$-induced blockade events, from both the cis and trans directions suggest an interesting and potentially important use of polymers as molecular rulers for the ion channel. The wide range of blockade durations, from 20 μs to over 2 full seconds, also suggests that the very long events may represent ssDNA that are folded in ways requiring substantial time periods to unfold and thereby access states that allow transit. The state structure of the amplitude distribution for longer events, characterized by components of a Gaussian mixture, evolves substantially over logarithmic increments of event duration beyond the 60–600-μs range just discussed, and is shown in Figure 14. Progressively greater proportions of long-duration events are spent in deep blockade states and require many more Gaussian mixture components to model adequately than do the deep blockade portions of the short-duration events. We have resolved up to 39 states with the aid of digital

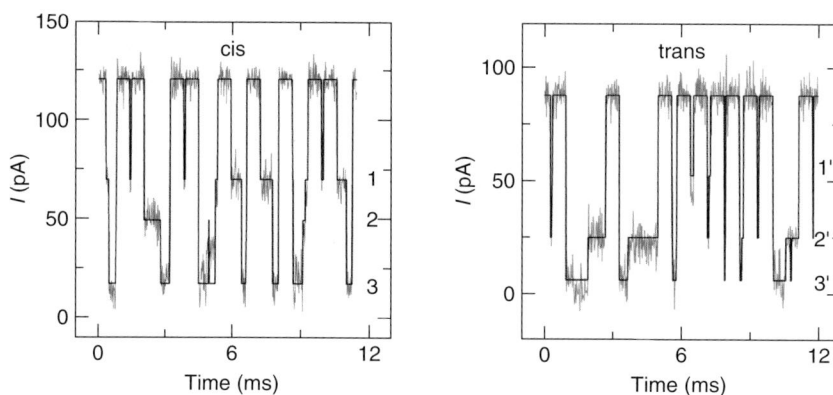

Figure 13. Typical single-channel current blockades for poly[dT]$_{100}$ entering from the cis entrance (a) or trans entrance (b) and the maximum-likelihood estimate of the amplitude state sequence (black) superimposed on the data (gray). The technique for determining the state sequences from the data relies on Viterbi decoding of the dwell times.

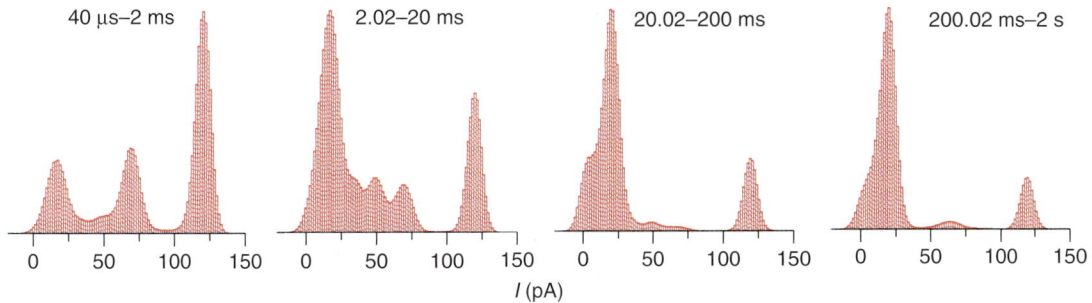

Figure 14. Blockade event lifetime histograms for poly[dT]$_{100}$ molecules that enter the cis side of a single αHL nanopore. For blockade lifetimes over a wide range (40 μs–2 s), virtually all of the events (99%) have lifetimes <5.5 ms. The lifetimes for each order of magnitude have clearly defined and differing Gaussian amplitude states. This representation of the data provides a fingerprint of the molecule as it interacts with the nanopore. We suggest that the ability to unambiguously decode the many Gaussian states caused by polymer–nanopore interactions provides an "electronic spectrogram" that could be used to uniquely identify polymers of interest.

filtering techniques and were able to estimate the required parameters.

Clarification of the actual nanopore cross section and its functional interaction with polymers in transit suggests that the most fully blocked states will provide the best coign of vantage for higher-resolution measurements of the ssDNA molecules themselves. Possibly, this is because the more complete blockade levels represent the interaction of the polymer with the narrowest point of constriction in the nanopore channel. If true, our state decoding technique, or an equivalent one, will be needed to extract information encoded in polymers.

4.4 Statistical Methods

It was previously demonstrated that statistical signal processing could be used to estimate and measure substates and their lifetimes in the DNA-induced current blockades.[19,33,44] In many blockade events, the single-channel current is piecewise stationary, with states that overlap in amplitude but not in time. This signal structure prompted modification of the classical HMM of Baum[45] to make a maximum a posteriori state sequence estimate. Statistical characterizations of the DNA-induced αHL channel current blockade states were based on Gaussian mixture models (GMMs) estimated using an expectation maximization (EM) procedure[46] rather than the classical supervised forward–backward iteration.

This provides a set of Gaussian components and mixture weights from a large population of events. The amplitudes of individual piecewise stationary segments, and their lifetimes, can then be estimated using Viterbi decoding,[43] which provides a MAP state sequence as the most probable generating function for the observed time series.

While the maximum a posteriori state sequences are globally optimal, given the correct output distributions, it is well known that the EM/GMM estimator can arrive at local likelihood maxima that are globally suboptimal. However, a good statistical criterion can mitigate the practical impact of this limitation, with random initial conditions being explored until no further improvements in the likelihood or goodness-of-fit criteria are forthcoming. The well-known Kolmogorov–Smirnov (KS) statistic offers such a stopping criterion. The KS statistic is computed for each candidate amplitude mixture model, and additional components added if the p-value of the model fit could be rejected at the 0.05 level.

Because the GMM components usually overlap, simply assigning each point to the highest-likelihood component results in many physically implausible transitions between states that overlap in amplitude. To model the time coherence in the signal implied by the passage of polynucleotides through the pore, the GMM components were employed as the output distributions of an ergodic HMM with a transition matrix favoring state persistence, meaning it has a dominant diagonal with much smaller nonzero probabilities off the main diagonal. Also, an unconstrained maximum-likelihood GMM can result in distribution states that are highly heteroscedastic, which can therefore result in unrealistic decoding of the state sequences. A solution was to constrain the variances to equality during the GMM/EM estimation procedure. This offers a better approximation to the actual physical phenomenon of a nanometer-scale pore that is occluded by a comparable-sized polymer than does unconstrained variances for the states.

In the pioneering work applying HMM processing to channel conductance data described by Chung and colleagues,[47,48] they state "Perhaps the most subjective part of the HMM processing method, like any parameter estimation scheme, is finding the state dimension—or the number of conductance states in our example—in hidden Markov chain processes." In contrast, the state-identification technique described above for quantifying current blockades[44] is fully adaptive to the number of channel current states, and results in maximum a posteriori estimates of the state parameters and state sequences within the events.[33,44] This newer method uses the KS goodness-of-fit criterion as a stopping rule for signal-amplitude states. It stops when the model cannot be rejected at the p value of 0.05, rather than using an ad hoc estimate of the point of diminishing returns. Once an adequate channel amplitude state model has been estimated using the EM algorithm for mixtures, a maximum a posteriori Viterbi algorithm decodes the entire event.

The state sequences, like those shown in Figure 13, also provide lifetime and amplitude measurements for further analyses, such as discriminators for molecule types or automated extraction of structural information from individual molecules, and for analysis of the kinetics of encounters between the αHL nanopore and single DNA molecules.

To summarize, the measurement algorithm discussed above estimates the number of states, the mixture weights, state means, a pooled estimate of the state variance, and a state sequence for the entire ensemble blockade events. Thus, it can be reasonably characterized as being fully data-driven or unsupervised, in contrast to other systems, which require supervised training or enforce arbitrary assumptions on the number of states, state means, and state variances.

5 DETECTING ANTHRAX TOXINS WITH A SINGLE NANOPORE

5.1 Properties of the *B. anthracis* PA_{63} Ion Channel

The bacterium *Bacillus anthracis* secretes three proteins that cause late-stage anthrax infection. The properties of these toxins, and the steps leading to cell intoxication by them, have been outlined elsewhere.[49] Protective antigen 83 (PA_{83}, 83 kDa) binds to target cells, is cleaved by furin-like proteases, and the 63-kDa fragment that remains associated with the membrane surface (PA_{63}) oligomerizes into a heptameric prepore ($PA_{63})_7$. Lethal factor (LF, 90 kDa) and edema factor (EF, 89 kDa) subsequently bind to ($PA_{63})_7$, and the complexes LF:($PA_{63})_7$ and EF:($PA_{63})_7$ become associated with endocytotic vesicle membranes. Acidification of the endosome interior occurs and is believed to cause the heptameric ($PA_{63})_7$ prepore to form a fully functional transmembrane ion channel that facilitates the entry of LF and EF into the cytoplasm. Once inside the cell, LF and EF interfere with MAPKK signaling proteins and increase cAMP production, respectively, leading to cell death by two different mechanisms.

5.2 Nanopore-based Assay for Toxin Detection and Characterization

Figures 15(a–d) demonstrate that full-length LF and EF convert the ($PA_{63})_7$ channel $I–V$ relationship from slightly nonlinear to highly rectifying.[50] The data in Figure 15(d) show that the extent to which LF blocks the channels depends on the LF concentration. The ratio of the current in the presence of LF to that in the absence of the LF is shown in Figure 15(e). A simple analysis of this data suggests that LF binds with high affinity to the ($PA_{63})_7$ channel (K $\sim$ 50 pM). Similar results were obtained with EF. The ability to detect LF and EF in trace quantities with the ($PA_{63})_7$ channel provides a significant advantage for biosensor applications (e.g., the sensitive detection of LF and EF). Unlike surface plasmon resonance technique,

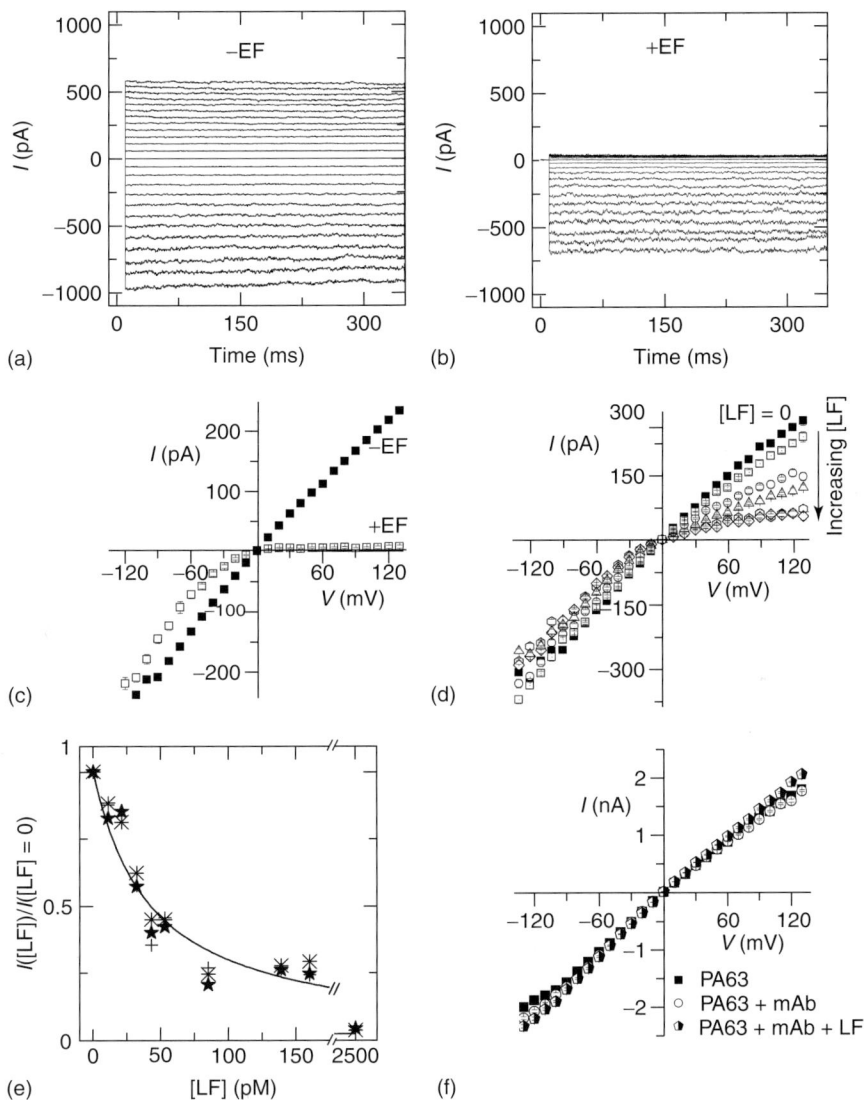

Figure 15. The binding of edema factor (EF) and lethal factor (LF) to the PA_{63} channel provides a novel method for anthrax toxin detection and anthrax therapeutic screening applications. (a) and (b) The current recordings for PA_{63} ion channels without and with EF present, respectively. (c) The instantaneous $I–V$ relationship of PA_{63} channels in the absence (filled squares) and presence of EF. (d) LF also affects the $I–V$ relationship of PA63 channels in an asymmetric manner. The concentration of LF was increased from zero to 160 pM. (e) The ratio of the ionic current in the presence of LF to that in its absence as a function of LF concentration for three different values of the applied potential. The solid line through the data is the result of a least-squares fit of an equation to the data that assumes 1:1 binding between LF and a PA_{63} channel and a 40 pM binding constant. (f) The $I–V$ relationships of PA_{63} channels alone and in the presence of either a monoclonal antibody (1G3-1-1) or the mAb with and 3 nM LF are virtually identical.

which is also used to characterize the binding of LF and EF to PA_{63}, the electronic method provides a functional test of PA_{63}, LF, and EF. In addition, results can be obtained in minutes rather than in days as is required by cell-based methods.

5.3 Potential for High-throughput Screening of Therapeutic Agents against Anthrax

Another promising prospect includes identifying and screening compounds and/or antibodies that disrupt toxin function or interaction.[50] To test this hypothesis, we used a previously characterized monoclonal antibody (mAb), 1G3-1-1, which prevents the binding of LF to PA_{63} through steric hindrance.[51] The mAb 1G3-1-1-neutralized LF toxin both in vivo and in vitro. Figure 15(f) demonstrates that the antibody had virtually no effect on the $I-V$ relationship of the PA_{63} channel. However, it completely inhibited LF-induced blockade. The results suggest that the $(PA_{63})_7$ channel can function as a biosensor for rapidly screening anthrax therapeutics. Agents that bind to the PA_{63} channel or to either LF or EF and that do not cause current blockades by themselves would inhibit LF- or EF-induced rectification of the $(PA_{63})_7$ channel $I-V$ relationship.

5.4 Detection of Anthrax Lethal Toxin from Infected Animals

The strikingly distinct and sensitive electrophysiological signature produced by the binding of LF and/or EF to PA_{63} suggests its application as a possible clinical diagnostic assay. Recent work provided evidence for this.[50] The first study demonstrated that the anthrax lethal toxin complex is present as a complex in the serum of infected animals.[50] Both the in vitro purified lethal toxin complex and the complex from the serum exhibited enzymatic activity. As the next step to show the potential of the method as a diagnostic tool, the anthrax lethal toxin complex was isolated from the serum of infected animals using size-exclusion chromatography. This process was necessary to prevent possible interference from cellular and bacterial proteins. Addition of a small aliquot of sample (purified in vivo $(PA_{63})_7$:LF complex) containing only $\sim$30 pM initial LF concentration to the cis chamber with an applied potential of -50 mV resulted in the reconstitution of strongly rectifying nanopores. This strong rectification is seen not only in these samples from infected animals, but also with PA_{63} channels after the addition of LF and a purified $(PA_{63})_7$:LF complex formed in vitro. These results suggest not only a clinical diagnostic biosensor, but also a sensitive assay that lends credence to previous results that showed complexed $(PA_{63})_7$:LF is present in infected animals' blood.

6 EMERGING TECHNOLOGIES

6.1 Planar Lipid Bilayer Technology: Limitations and Future Directions

The examples described in this chapter provide evidence that protein nanopores offer significant potential as components of biosensors. However, ion channels currently only function in lipid bilayer membranes, which currently are not a robust and practical matrix for real-world sensor applications.

There are two approaches that might resolve this issue. First, channels have been reconstituted into membranes affixed to a gold electrode on a solid support. For example, Cornell and colleagues demonstrated a robust and functional gramicidin channel–based sensor system tethered to a solid support.[52] That impressive method has yet to demonstrate single-channel activity in such matrices. Until then, detailed kinetic information inherent in the reversible binding of the analyte to the nanopore will be lost to population averaging. Second, the possibility of reconstituting single channels into membranes that can be hardened after the channels have formed has been explored. For example, it was recently shown that αHL and PA_{63} channels can be reconstituted into membranes formed by photopolymerizable lipids.[53] In addition, upon exposure to UV light, the physical properties of those membranes (e.g., monolayer surface pressure and bilayer membrane specific capacitance) change, as one would expect if the lipids polymerize with their neighbors. Moreover, the function of single αHL channels is unaffected by the polymerization. However, it is not yet clear how robust these membranes are or even what fraction of the lipids are polymerized. These questions need to be addressed before polymerized lipids can be used as a stable platform for nanopore-based sensors.

6.2 Solid-state Nanopores

There are other methods being developed to make robust nanopores. Because lipid membranes are fragile, several groups are trying to determine if nanopores formed in synthetic solid materials can provide the functionality of ion channels. Several methods have been developed to produce solid-state nanopores. In one technique, micrometer-thick plastic is bombarded with a weak α-particle beam. The substrate is then chemically etched to form a single pore that is nanometer-scale in diameter at one end.[54,55] However, these pores are micrometer scale in length, which limits their utility in some applications. More recently, Golovchenko's group and others made nanometer-scale pores in solid supports (e.g., silicon nitride) by etching a pit in the material on one side and then bombarding the opposite side with a carefully controlled ion beam.[56–58] The hole is then "sculpted" to ~2-nm diameter with an electron beam. These solid-state portals have demonstrated capability of detecting DNA molecules and should find their way into practical applications if the cost of producing the nanopores can be reduced.

6.3 Theory and Modeling of Nanopores

Theoretical analyses of polymer partitioning into simple model geometries are now providing valuable insight into the physics of DNA confinement in structures with biologically relevant length scales.[35,59–61,62–74] Theory and simulation of how nanopores function and how analytes might react with them will clearly enable the rational design of nanopore-based sensors.

7 CONCLUSIONS

Proof of concept for nanopore-based detection and quantitation of a wide variety of analytes, including ions, polynucleotides, and lethal toxins, has been demonstrated. However, there are several barriers that need to be overcome to make single nanopores practical and useful in sensor applications. First, they need to be made robust. Second, for DNA sequencing applications, single nanopores need to better discriminate between closely similar polynucleotides. Both goals are within reach and will hopefully be demonstrated soon.

ACKNOWLEDGMENTS

The work performed in our laboratories was supported in part by the National Institute of Standards and Technology (NIST) Advanced Technology Program, the NIST "Single Molecule Manipulation and Measurement" program, the National Science Foundation (NIRT grant CTS-0304062), the Medical Biological Defense Research Program, the US Army Medical Research Institute of Infectious Diseases (USAMRIID Research Plan 02-4-2C-012), the National Cancer Institute, National Institutes of Health, under contract N01-CO-1240, in part by the Developmental Therapeutics Program in the Division of Cancer Treatment and Diagnosis of the National Cancer Institute and the Defense Advanced Research Projects Agency. The identification of commercial materials and their sources is made to adequately describe the experimental results. This identification neither implies recommendation by the NIST or USAMRIID nor does it imply that the material is the best available.

REFERENCES

1. L. Stryer, *Biochemistry*, 3rd Edn, W.H. Freeman, San Francisco, 1988.
2. B. Hille, *Ionic Channels of Excitable Membranes*, 2nd Edn, Sinauer Associates, Sunderland, 1992.
3. A. L. Hodgkin and A. F. Huxley, Currents carried by sodium and potassium ions through the membrane of the giant axon of *Loligo*. *Journal of Physiology (London)*, 1952, **116**, 449–472.
4. M. Leippe, J. Andra, and H. J. Mullereberhard, Cytolytic and antibacterial activity of synthetic peptides derived from amoebapore, the pore-forming peptide of entamoeba-histolytica. *Proceedings of the National Academy of Sciences of the United States of America*, 1994, **91**, 2602–2606.
5. A. Harris, Emerging issues of connexin channels: biophysics fills the gap. *Quarterly Reviews of Biophysics*, 2001, **34**, 325–472.
6. M. Welsh, Cystic fibrosis. *Scientific American*, 1995, **273**(6), 52–59.
7. S. M. Simon and G. Blobel, A protein-conducting channel in the endoplasmic reticulum. *Cell*, 1991, **65**, 371–380.
8. R. V. Miller, Bacterial gene swapping in nature. *Scientific American*, 1998, **278**(1), 66–71.
9. L. Z. Song, M. R. Hobaugh, C. Shustak, S. Cheley, H. Bayley, and J. E. Gouaux, Structure of Staphylococcal alpha-hemolysin, a heptameric transmembrane pore. *Science (USA)*, 1996, **274**, 1859–1866.
10. T. L. Nguyen, Three-dimensional model of the pore form of anthrax protective antigen. Structure and biological implications. *Journal of Biomolecular Structure and Dynamics*, 2004, **22**, 253–266.
11. S. Bhakdi, R. Fussle, and J. Tranum-Jensen, Alpha-toxin of *Staphylococcus aureus*. *Microbiological Research*, 1991, **55**, 733–751.

12. G. Menestrina, Ionic channels formed by *Staphylococcus aureus* alpha-toxin: voltage dependent inhibition by divalent and trivalent cations. *The Journal of Membrane Biology*, 1986, **90**, 177–190.

13. O. Beckstein, P. C. Biggin, P. Bond, J. N. Bright, C. Domene, A. Grottesi, J. Holyoake, and M. S. P Sansom, Ion channel gating: insights via molecular simulations. *FEBS Letters*, 2003, **555**, 85–90.

14. J. J. Kasianowicz, Nanopores. Flossing with DNA. *Nature Materials*, 2004, **3**, 355–356.

15. S. M. Bezrukov and J. J. Kasianowicz, Current noise reveals protonation kinetics and number of ionizable sites in an open protein ion channel. *Physical Review Letters*, 1993, **70**, 2352–2355.

16. J. J. Kasianowicz and S. M. Bezrukov, Protonation dynamics of the α-toxin ion channel from spectral analysis of pH dependent current fluctuations. *Biophysical Journal*, 1995, **69**, 94–105.

17. J. J. Kasianowicz, D. L. Burden, L. Han, S. Cheley, and H. Bayley, Genetically engineered metal ion binding sites on the outside of a channel's transmembrane β-barrel. *Biophysical Journal*, 1999, **76**, 837–845.

18. S. M. Bezrukov, I. Vodyanoy, R. A. Brutyan, and J. J. Kasianowicz, Dynamics and free energy of polymers partitioning into a nanoscale pore. *Macromolecules*, 1996, **29**, 8517–8522.

19. J. J. Kasianowicz, E. Brandin, D. Branton, and D. W. Deamer, Characterization of individual polynucleotide molecules using a membrane channel. *Proceedings of the National Academy of Sciences of the United States of America*, 1996, **93**, 13770–13773.

20. S. G. A. McLaughlin, Electrostatic potentials at membrane-solution interfaces. *Current Topics in Membranes and Transport*, 1977, **9**, 71–144.

21. M. Misakian and J. J. Kasianowicz, Electrostatic influence on ionic current through the α-HL ion channel. *The Journal of Membrane Biology*, 2003, **195**, 137–146.

22. D. E. Goldman, Potential, impedance, and rectification in membranes. *The Journal of General Physiology*, 1943, **27**, 37–60.

23. A. L. Hodgkin and B. Katz, The effect of sodium ions on the electrical activity of the giant axon of the squid. *The Journal of Physiology (London)*, 1949, **108**, 37–77.

24. G. Feher and M. Weissman, Fluctuation spectroscopy: determination of chemical reaction rates from the frequency spectrum of fluctuations. *Proceedings of the National Academy of Sciences of the United States of America*, 1973, **70**, 870–875.

25. L. J. deFelice, *Membranes and Current Noise*, Plenum Press, New York, 1989.

26. S. Machlup, Noise in semiconductors: spectrum of a two-parameter random signal. *Journal of Applied Physics*, 1954, **25**, 341–343.

27. O. V. Krasilnikov, R. Z. Sabirov, V. I. Ternovsky, P. G. Merzliak, and J. N. Muratkodjaev, A simple method for the determination of the pore radius of ion channels in planar bilayer membranes. *FEMS Microbiology Immunology*, 1992, **105**, 93–100.

28. O. V. Krasilnikov, *Sizing Channels with Neutral Polymers*, in *NATO Advanced Research Workshop. Structure and Dynamics of Confined Polymers*, J. J. Kasianowicz, M. S. Z. Kellermayer, and D. W. Deamer (eds), Kluwer Press, Dordrecht, 2002, pp. 97–115.

29. S. Kuga, Pore size distribution analysis of gel substances by size exclusion chromatography. *Journal of Chromatography*, 1981, **206**, 449–461.

30. A. Einstein, *Investigations on the Theory of Brownian Movement*, Dover Press, New York, 1956.

31. S. E. Henrickson, M. Misakian, B. Robertson, and J. J. Kasianowicz, Driven asymmetric DNA transport in a nanometer-scale pore. *Physical Review Letters*, 2000, **85**, 3057–3060.

32. M. Akeson, D. Branton, J. J. Kasianowicz, E. Brandin, and D. W. Deamer, Microsecond time-scale discrimination between polycytidylic acid and polyadenylic acid segments within single RNA molecules. *Biophysical Journal*, 1999, **77**, 3227–3233.

33. J. J. Kasianowicz, S. E. Henrickson, M. Misakian, H. H. Weetall, B. Robertson, and V. Stanford, *Physics of DNA Threading Through a Nanometer Pore and Applications to Simultaneous Multianalyte Sensing*, in *NATO Advanced Research Workshop. Structure and Dynamics of Confined Polymers*, J. J. Kasianowicz, M. S. Z. Kellermayer, and D. W. Deamer (eds), Kluwer Press, Dordrecht, 2002, pp. 141–163.

34. M. Prashant and R. K. Lal, The dnaSET: a novel device for single-molecule DNA sequencing. *IEEE Transactions on Electron Devices*, 2004, **51**, 2004–2012.

35. M. Muthukumar, Theory of sequence effects of DNA translocation through protein and nanotubes. *Electrophoresis*, 2002, **23**, 1417–1420.

36. J. Lagerqvist, M. Zwolak, and M. Di Ventra, Fast DNA sequencing via transverse electronic transport. cond-mat/0601394. *Nano Letters*, 2006 **6**, 779–782.

37. W. Vercoutere, S. Winters-Hilt, H. Olsen, D. Deamer, D. Haussler, and M. Akeson, Rapid discrimination among individual DNA hairpin molecules at single-nucleotide resolution using an ion channel. *Nature Biotechnology*, 2001, **19**, 248–252.

38. J. Kling, Ultrafast DNA sequencing. *Nature Biotechnology*, 2003, **21**, 1425–1427.

39. S. Winters-Hilt, W. Vercoutere, V. S. DeGuzman, D. Deamer, M. Akeson, and D. Haussler, Highly accurate classification of Watson-Crick base pairs on termini of single DNA molecules. *Biophysical Journal*, 2003, **84**, 967–976.

40. V. M. Stanford and J. J. Kasianowicz, *Transport of DNA Through a Single Nanometer-Scale Pore: Evolution of Signal Structure*, IEEE Workshop on Genomic Signal Processing and Statistics, Baltimore, 2004.

41. M. Zwolak and M. Di Ventra, Electronic signature of DNA nucleotides. *Nano Letters*, 2005, **5**, 421–424.

42. J. J. Kasianowicz, S. E. Henrickson, H. H. Weetall, and B. Robertson, Simultaneous multianalyte detection with a nanopore. *Analytical Chemistry*, 2001, **73**, 2268–2272.

43. A. J. Viterbi, Error bounds for convolutional codes and an asymmetrically optimum decoding algorithm. *IEEE Transactions on Information Theory*, 1967, **IT-13**, 260–267.

44. V. Stanford and J. J. Kasianowicz, Using HMMs to quantify signals from DNA driven through a nanometer-scale pore, *Workshop on Genomic Signal*

Processing and Statistics (GENSIPS), Raleigh, NC, 2002, CP1-03.

45. L. Baum, An inequality and associated maximization technique in statistical estimation of probabilistic functions of a Markov process. *Inequalities*, 1972, **3**, 1–8.

46. R. Redner and H. Walker, Mixture densities, maximum likelihood and the EM algorithm. *SIAM Review*, 1984, **26**, 195–239.

47. S. H. Chung, J. B. Moore, L. Xia, L. S. Premkumar, and P. W. Gage, Characterization of single channel currents using digital signal processing techniques based on Hidden Markov models. *Philosophical Transactions of the Royal Society of London Series B: Biological Sciences*, 1990, **329**, 265–285.

48. S.-H. Chung and P. W. Gage, *Signal Processing Techniques for Channel Current Analysis Based on Hidden Markov Models*. in *Methods in Enzymology: Ion Channels, Part B*, P. M. Conn (ed), Academic Press, New York, 1998.

49. P. Ascenzi, P. Visca, G. Ippolito, A. Spallarossa, M. Bolognesi, and C. Montecucco, Anthrax toxin: a tripartite lethal combination. *FEBS Letters*, 2002, **531**, 384–388.

50. K. M. Halverson, R. G. Panchal, T. L. Nguyen, R. Gussio, S. F. Little, M. Misakian, S. Bavari, and J. J. Kasianowicz, Asymmetric blockade of anthrax protective antigen ion channel by lethal factor: anthrax therapeutic sensor. *The Journal of Biological Chemistry*, 2005, **280**, 34056–34062.

51. S. F. Little, J. M. Novak, J. R. Lowe, S. H. Leppla, Y. Singh, K. R. Klimpel, B. C. Lidgerding, and A. M. Friedlander, Characterization of lethal factor binding and cell receptor binding domains of protective antigen of *Bacillus anthracis* using monoclonal antibodies. *Microbiology*, 1996, **142**, 707–715.

52. B. A. Cornell, V. L. BraachMaksvyti, L. G. King, P. D. J. Osman, B. Raguse, L. Wieczorek, and R. J. Pace, A biosensor that uses ion-channel switches. *Nature (London)*, 1997, **387**, 580–583.

53. D. K. Shenoy, W. Barger, A. Singh, R. G. Panchal, M. Misakian, V. M. Stanford, and J. J. Kasianowicz, Functional reconstitution of ion channels in polymerizable lipid membranes. *Nano Letters*, 2005, **5**, 1181–1185.

54. T. K. Rostovtseva, C. L. Bashford, G. M. Alder, G. N. Hill, C. McGiffert, P. Y. Apel, G. Lowe, and C. A. Pasternak, Diffusion through narrow pores: movement of ions, water and nonelectrolytes through track-etched PETP membranes. *Journal of Membrane Biology*, 1996, **151**, 29–43.

55. Z. Siwy and A. Fulinski, Fabrication of a synthetic nanopore ion pump. *Physical Review Letters*, 2002, **89**, 198103.

56. J. Li, D. Stein, C. McMullan, D. Branton, M. J. Aziz, and J. A. Golovchenko, Ion-beam sculpting at nanometre length scales. *Nature (London)*, 2001, **412**, 166–169.

57. A. J. Storm, J. H. Chen, X. S. Ling, H. W. Zandbergen, and C. Dekker, Electron-beam-induced deformations of SiO$_2$ nanostructures. *Journal of Applied Physics*, 2005, **98**, 014307.

58. J. B. Heng, C. Ho, T. Kim, R. Timp, A. Aksimentiev, Y. V. Grinkova, S. Sligar, K. Schulten, and G. Timp, Sizing DNA using a nanometer-diameter pore. *Biophysical Journal*, 2004, **87**, 2905–2911.

59. A. Aksimentiev, J. B. Heng, G. Timp, and K. Schulten, Microscopic kinetics of DNA translocation through synthetic nanopores. *Biophysical Journal*, 2004, **87**, 2086–2097.

60. M. Muthukumar, Modeling of polynucleotide translocation through protein pores and nanotubes. *Electrophoresis*, 2002, **23**, 2697–2703.

61. E. Slonkina and A. B. Kolomeisky, Polymer translocation through a long nanopore. *Journal of Chemical Physics*, 2003, **118**, 7112–7118.

62. W. Sung and P. J. Park, Polymer translocation through a pore in a membrane. *Physical Review Letters*, 1996, **77**, 783–786.

63. D. K. Lubensky and D. R. Nelson, Driven polymer translocation through a narrow pore. *Biophysical Journal*, 1999, **77**, 1824–1838.

64. M. Muthukumar, Polymer translocation through a hole. *Journal of Chemical Physics*, 1999, **111**, 10371–10374.

65. P. deGennes, Passive entry of a DNA molecule into a small pore. *Proceedings of the National Academy of Sciences of the United States of America*, 1999, **96**, 7262–7264.

66. K. K. Kumar and K. L. Sebastian, Adsorption-assisted translocation of a chain molecule through a pore. *Physical Review E*, 2000, **62**, 7536–7539.

67. K. L. Sebastian and A. K. R. Paul, Kramers problem for a polymer in a double well. *Physical Review E*, 2000, **62**, 927–939.

68. M. Muthukumar, Translocation of a confined polymer through a hole. *Physical Review Letters*, 2001, **86**, 3188–3191.

69. S. K. Lee and W. Sung, Barrier crossing of a semiflexible ring polymer. *Physical Review E*, 2001, **64**, 041801.

70. C. Y. Kong and M. Muthukumar, Modeling of polynucleotide translocation through protein pores and nanotubes. *Electrophoresis*, 2002, **23**, 2697–2703.

71. W. Sung and P. J. Park, *The polymer barrier problem, In: NATO Advanced Research Workshop, Structure and Dynamics of Confined Polymers, Kluwer Press*. Eds. J. J. Kasianowicz, M. S. Z. Kellermayer, and D. W. Deamer. pp. 261–280. Dordrecht, The Netherlands. 2002.

72. A. Troisi, A. Nitzan, and M. A. Ratner, A rate constant expression for charge transfer through fluctuating bridges. *Journal of Chemical Physics*, 2003, **119**, 5782–5788.

73. M. Muthukumar, Polymer escape through a nanopore, *Journal of Chemical Physics*, 2003, **118**, 5174–5184.

74. C. Y. Kong and M. Muthukumar, Polymer translocation through a nanopore. II. Excluded volume effect. *Journal of Chemical Physics*, 2004, **120**, 3460–3466.

FURTHER READING

L. E. Baum and T. Petrie, Statistical inference for probabilistic functions of finite state Markov chains. *Annals of Mathematical Statistics*, 1966, **37**, 1559–1563.

R. Blaustein and A. Finkelstein, Voltage-dependent block of anthrax toxin channels in planar phospholipid bilayer membranes by symmetric tetraalkylammonium ions. *The Journal of General Physiology*, 1990, **96**, 905–919.

B. Katz, *Nerve, Muscle, and Synapse*, McGraw-Hill, New York, 1966.

X.-M. Wang, R. Wattiez, M. Mock, P. Falmagne, J.-M. Ruysschaert, and V. Cabiaux, Structure and interaction of PA63 and EF (edema toxin) of *Bacillus anthracis* with lipid membrane. *Biochemistry*, 1997, **36**, 14906–14913.

L. Xu and M. Jordan, On convergence properties of the EM algorithm for Gaussian mixtures. *Neural Computation*, 1996, **8**, 129–151.

S. Zhang, E. Udho, Z. Wu, R. J. Collier, and A. Finkelstein, Protein translocation through anthrax toxin channels formed in planar lipid bilayers. *Biophysical Journal*, 2004, **87**, 3842–3849.

Conducting Polymer Nanowire-Based Biosensors

Adam K. Wanekaya,[1] Wilfred Chen,[2] Nosang V. Myung[2] and Ashok Mulchandani[2]

[1] Chemistry Department, Missouri State University, Springfield, MO, USA and [2] Department of Chemical and Environmental Engineering and Center for Nanoscale Science and Engineering, University of California, Riverside, CA, USA

1 INTRODUCTION

Conducting polymer nanowires (CP NWs) are part of one-dimensional (1D) nanostructured materials that include materials such as nanotubes, nanosprings, nanobelts, and other nanowires. These materials are the smallest-dimension structures that can be used for efficient transport of electrons and are thus critical to the function and integration of high-density nanoscale devices. Consequently, they are the focus of intensive research in sensing, optoelectronics, and other applications due to their unique properties. Because of their high surface-to-volume ratio and tunable electron transport properties due to quantum confinement effect, their electrical properties are strongly influenced by minor perturbations. This property provides an avenue for the much desired rapid, label-free, and direct electrical detection to the point that single-molecule detection is possible. The advantages of label-free detection include a simple homogenous assay format without separation and washing and rapid near-real-time response. This has the potential to impact biological research as well as screening in medical, environmental, and homeland security applications.

Field effect transistor (FET)-type devices have many advantages such as small size and weight, fast response, high reliability, and on-chip integration with low-cost mass production. Basically, a FET consists of two electrodes, a source and a drain, connected by a semiconducting channel. The current can flow through the semiconductor only when appropriate voltage is applied to the gate electrode. The modulation of the applied gate potential has a tremendous effect on the conductivity of the semiconducting channel. This phenomenon can thus be used to amplify electrical signals resulting from the interaction of charged species with the semiconducting channel. The dependence of the conductance on gate voltage makes FETs natural candidates for electrically based sensing. Although the concept of sensing with FETs was introduced several decades ago, the limited sensitivity of planar devices precluded them from having a large impact.

FETs based on 1D nanostructures are based on a similar framework but are more sensitive because, unlike planar FETs, they avoid the reduction in conductance changes caused by lateral current shunting. When used as the gate of a FET device, 1D nanostructures offer significant advantages over 2D thin-film planar gates. First, binding

Handbook of Biosensors and Biochips. Edited by Robert S. Marks, David C. Cullen, Isao Karube, Christopher R. Lowe and Howard H. Weetall.
© 2007 John Wiley & Sons, Ltd. ISBN 978-0-470-01905-4.

to the surface of 1D nanostructures can lead to depletion or accumulation of carriers in the "bulk" of the nanometer-diameter structure versus only the surface region of a planar device, giving rise to large resistance/conductance changes, to the point that single-molecule detection is possible. Second, the direct conversion of chemical information into an electronic signal can take advantage of the existing microelectronic technology and lead to miniaturized sensor devices. Third, the sizes of biological molecules (BM), such as proteins and nucleic acids are comparable to nanoscale building blocks. Therefore, any interaction between such molecules should induce significant changes in the electrical properties of 1D nanostructures. Finally, the small size of the nanostructures makes development of high-density arrays of individually addressable nanostructures for simultaneous analysis of different substances possible. Also, 1D nanostructures, such as CP NWs are extremely attractive for nanoelectronics because they can function both as devices and as the wires that access them.

Until recently the advancement of 1D nanostructures was slow because of the difficulties associated with the synthesis and fabrication of these nanostructures with well-controlled dimensions, morphology, phase purity, and chemical composition. Three classes of 1D nanostructured materials namely carbon nanotubes (CNTs), silicon nanowires (Si NWs), and lately CP NWs have shown profound performance in device fabrication in general and in label-free detection technology in particular. A major advantage of conducting polymers (CPs) is that they can be functionalized before polymerization, during polymerization, and after polymerization. This makes them very flexible materials because optimal conditions can be used for each step, to obtain optimum polymer conductivity and orientation of the functionalized moiety. Also, the conductivities of CPs can be modulated up to 15 orders of magnitude by changing the dopant and monomer/dopant ratios. It has been further demonstrated that their conductivity can be further modulated by controlling their oxidation state.

In this chapter we discuss the fabrication, functionalization, and the sensing applications based on CP NWs. We highlight the successes and limitations of various methods from different laboratories and how we and other researchers have attempted to address these limitations. We conclude by outlining some successes and challenges that are associated with the CP NWs.

2 FABRICATION OF CP NWS

CP NWs can be fabricated by a variety of methods including lithography, mechanical stretching, electrospinning, template-directed synthesis, and templateless alignment methods. Whatever the method, a good fabrication procedure should enable the simultaneous control of morphology, dimensions, and properties. The objective of this section is to highlight some of the methods that have been used in the fabrication of CP NWs.

2.1 Electrospinning

This method was originally developed for generating ultrathin polymer fibers.[1] It uses electrical forces to produce polymer fibers with nanometer-scale diameters. A microfabricated scanned tip is used as an electrospinning source. The tip is dipped in a polymer solution to gather a droplet. A voltage applied to the tip causes the formation of a Taylor cyclone, and at sufficiently high voltages, a polymer jet is extracted from the droplet. By moving the source relative to a surface that acts as a counter electrode, oriented nanofibers can be deposited and integrated with microfabricated surface structures. The morphology of the fibers depends on the solution concentration, applied electric strength, and the feeding rate of the precursor solution. Uniform nanofiber depositions have thus been fabricated by the electrospinning method.[2–5]

2.2 Lithography

CP NWs can also be fabricated by lithographic methods. The major advantage of this method is the fact that CPs can be precisely patterned based on different lithographic principles. For example, dip-pen nanolithography (DPN) is a scanning probe nanopatterning technique in which an atomic force microscope (AFM) tip is used to deliver molecules to a surface via a solvent meniscus. The ionically charged CPs are used as the "ink" on oppositely charged substrates providing a significant driving force for the generation of stable patterns on substrates. Using this technique, polyaniline,[6] polypyrrole (PPY),[6] and poly(3,4-ethylenedioxythiophene)[7] nanowire lines down to 310-, 290-, and 30-nm widths, respectively, have been reported.

Recently, Fuchs and coworkers demonstrated the lithographic fabrication of CP NW patterns by a copolymer strategy.[8] In the first step of the procedure, a pattern is defined on a photoresist using e-beam lithography. Second, a copolymer film is deposited by the chemical oxidation of the appropriate monomers. Finally, the resist is lifted-off leaving patterns of CP NWs as shown in Figure 1.

2.3 Mechanical Stretching

CPs are electrochemically polymerized on a scanning tunneling microscope (STM) tip that is insulated except for a few square nanometers, thus localizing the growth of the polyaniline (PANI)

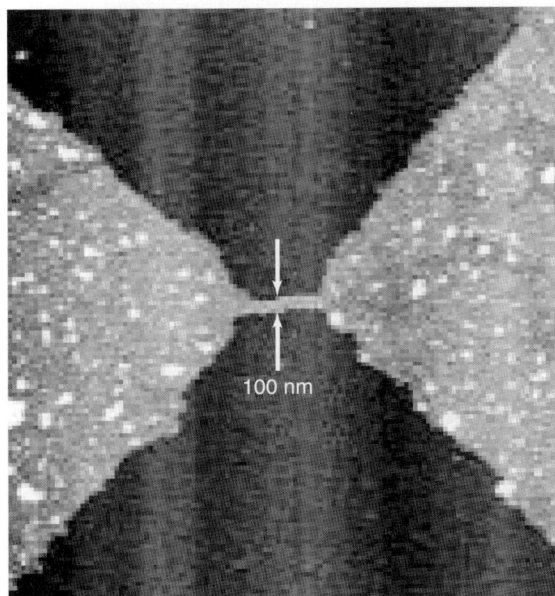

Figure 1. A 100-nm-wide PPY NW bridging the ends of two PPY microelectrodes. [Reprinted with permission Dong et al.[8] copyright 2005, Wiley VCH.]

NW at the tip end. After the formation of the CP NW, the diameter of the nanowire may be further reduced by stretching it with the STM tip. A highly conductive PANI NW with a diameter of about 20 nm has been fabricated using this method.[9]

2.4 Template-directed Methods

This method entails the synthesis of the desired material within the pores of a nanoporous template membrane. The membranes employed have cylindrical pores of uniform diameter. Porous anodic aluminum oxide (AAO), track-etched polymer membranes, and mica are the three types of membranes that are commonly used. The template-directed method is an attractive procedure in the fabrication of CP NWs. The CP NWs may be synthesized within the pores by chemical or electrochemical deposition procedures.[10,11] Electrochemical deposition is accomplished by coating one face of the template with an inert conducting material and using it as the anode. The template is then suspended in a solution containing the appropriate monomer and dopant. Application of electric current on the template induces anodic polymerization of the monomer within the template pores to form CP NWs. The length of the nanowires is determined by the current density and deposition time. The diameter of the nanowire is determined by the pore diameter of template. Chemical template synthesis can be similarly accomplished by simply immersing the template into a solution of the desired monomer and its oxidizing agent. Following the CP NW deposition, the inert conducting film is dissolved using appropriate acids or bases. Organic solvents may be used to dissolve polymer templates.

DNA has also been used as a template in the assembly and polymerization of aniline to form PANI nanowires. For example, He and coworkers[12] synthesized polyaniline nanowires by stretching, aligning, and immobilizing double-stranded λ-DNA on a thermally oxidized Si chip by the molecular combing method.[13,14] Then the DNA templates were incubated in protonated aniline monomer solution to emulsify and organize the aniline monomers along the DNA chains. Finally, the aligned aniline monomers were polymerized enzymatically by adding horseradish peroxidase (HRP) and H_2O_2 successively to form polyaniline/DNA nanowires. Simmel and coworkers also synthesized polyaniline nanowires templated by DNA using three methods.[15] They found that DNA templating worked best for polyaniline formed by oxidative polymerization of aniline with ammonium persulfate, both in solution and on templates immobilized on a chip. DNA was also a good template for polyaniline formed by enzymatically catalyzed polymerization utilizing HRP. However, immobilization of these structures between contact electrodes was compromised by extensive protein adsorption to the surface.

2.5 Templated Polymerization within Lithographically Defined Nanoscale Channels

Recently, template polymerization within lithographically defined nanoscale channels has proved to be very attractive in the fabrication of CP NWs[16,17] and CP nanoribbons.[18] The nanoscale channels have built-in electrical contacts that enable the application of electrical potentials for polymerization of the monomers within the channels. The built-in electrical contacts thus remain and become interconnects to the array components. This procedure avoids the harsh and cumbersome template dissolution processes that are necessary in the conventional template-directed polymerization and are thus not suitable for biological applications. Further, any postsynthesis manipulation and electrical contacting processes that are required in the conventional template-directed polymerization are not necessary. Figure 2 shows a scanning electron microscope (SEM) image of a 100-nm-wide and 3-μm-long PANI NW grown within lithographically defined nanoscale channels. The nanowire was continuous, well defined, and dendrite-free and exhibited precise dimensionality. The ability to fabricate individually addressable nanowires with a high aspect ratio was demonstrated by synthesizing two 200-nm-wide by 2.5-μm-long PPY NWs (Figure 3).

A complementary approach to the ones reported above is based on CP nanojunctions formed by bridging two electrodes separated with a gap of 1–100 nm.[19] The gap between the electrodes is

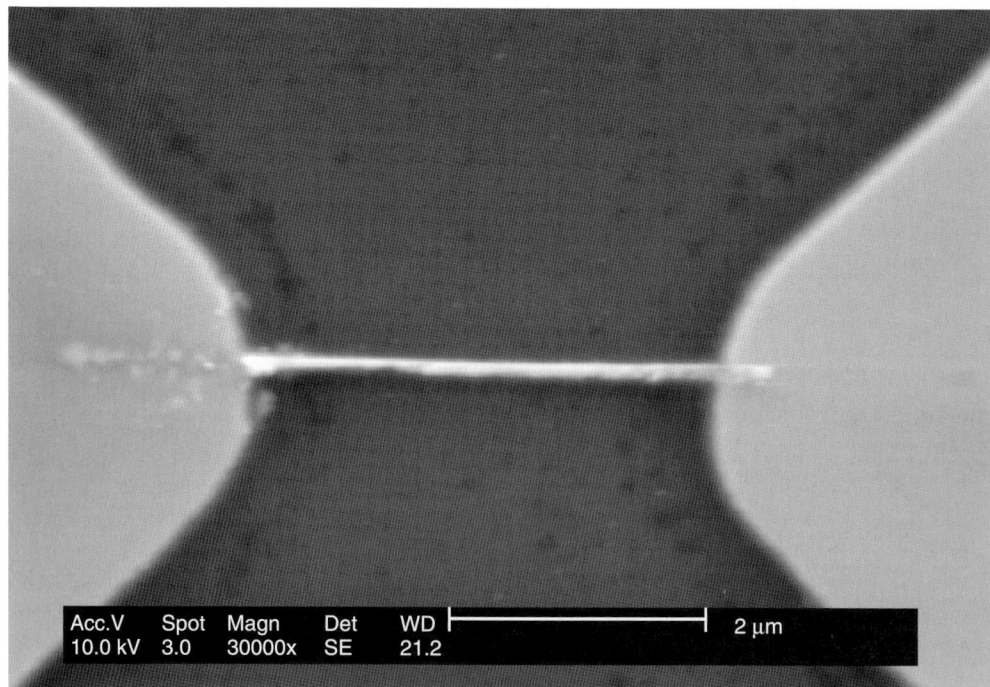

| Acc.V | Spot | Magn | Det | WD | | 2 μm |
| 10.0 kV | 3.0 | 30000x | SE | 21.2 | | |

Figure 2. SEM image of a 100-nm-wide by 3-μm-long PANI nanowire. [Reprinted with permission from Ramanathan et al.[16] © 2004 American Chemical Society.]

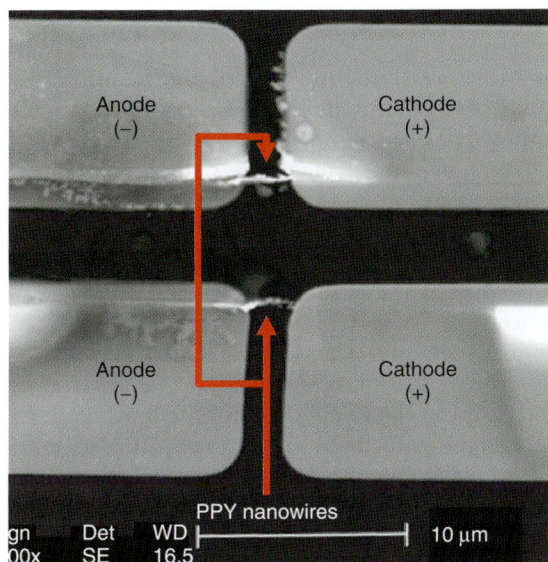

Figure 3. SEM image of two 200-nm-wide by 2.5-μm-long PPY nanowires separated by 10 μm, deposited one at a time. [Reprinted with permission from Ramanathan et al.[16] © 2004 American Chemical Society.]

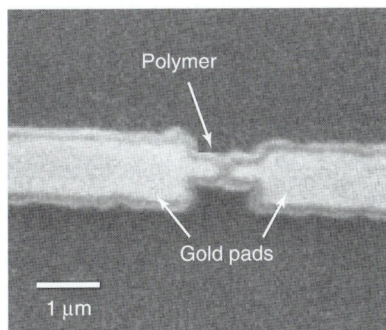

Figure 4. SEM image of polyaniline-polyacrylic acid films deposited on gold pads with 20–60 nm gaps. [Reprinted with permission from Forzani et al.[19] © 2004 American Chemical Society.]

reduced down to ∼1 nm by first electrochemically depositing Au onto the electrodes thus causing an increase in the current flowing across the gap due to quantum tunneling effect. Au atoms may be etched away from the electrodes to optimize the gap by taking advantage of the reversibility of the electrochemical process. The gaps are then bridged with CPs by the cycling potential of the nanoelectrodes in solutions of the corresponding monomers (Figure 4). However, unlike the polymerization within lithographically defined nanoscale channels, the length-to-width ratio of the nanojunction is not well defined and controlled and the aspect ratio is small. On the other hand, since the conductance path is much smaller, the nanojunction approach is particularly suitable for poorly conducting polymers or polymers that loose much of their conductivity upon attachment of receptor groups.

2.6 Templateless Polymerization of Aligned CP NWs

To date, template-directed polymerization using porous membranes has been the most preferred method for the fabrication of oriented CP NWs.

However, there have been reports that oriented 1D nanostructure CP structures could only be obtained for wires or rods with a large diameter.[20] Dissolution of the membrane supports for CP NWs with a diameter of less than 100 nm caused the nanowires to collapse without preferred orientation.[20] As a means of solving this problem, some researchers have devised a method of direct templateless electropolymerization of large arrays of oriented CP NWs with a diameter much smaller than 100 nm. This technique involves a three-step electrochemical deposition procedure where a high current density is applied in the first step followed by the application of reduced current densities in the subsequent steps.[21,22] The hypothesis is that at high current densities, CP nanoparticles are formed that are then used as nucleation sites to grow the extended CP NWs at reduced current densities in the second and third steps. A similar technique has been used for the fabrication of CP NW junction arrays.[23]

3 FUNCTIONALIZATION OF CP NWS

Functionalization of CP NWs is often necessary for their functionality and biocompatibility. Therefore, the interface between BM and CP NWs is critical because the unique electrical properties of these nanoscale materials when utilized in conjunction with the remarkable biomolecular recognition capabilities, could lead to miniature bioelectronic devices. The ultimate goal in functionalization is to retain, as much as possible, the properties of the biological molecule and the CP

NW. The biological molecule should be stable and should be able to retain its biorecognition properties. Most of the research in the functionalization of CPs has been based on thin films. However, the same techniques should also be applicable in the functionalization of CP NWs. The functionalization of CPs can be carried out in three ways: before, during, and after the polymerization process. The fourth procedure is an entrapment technique where the target material is immobilized during the polymerization processes. These processes are summarized in Scheme 1 with pyrrole as the model monomer.

The first technique involves the covalent linkage of a specific biological molecule to the starting monomer prior to electropolymerization. This method is good only if the biological molecule is stable during polymerization. Good applications of this procedures include the fabrication of a peptide chip by the electro-copolymerization of pyrrole and pyrrole-peptide, [24] the electropolymerization of oligonucleotide (ODN)-functionalized pyrrole,[25-27] and the electropolymerization of N-biotinylated pyrrole.[28,29] The disadvantage of N-substituted pyrroles is the slower rate of polymerization and a marked drop in the conductivity of the polymer matrix due to nonplanarity of the PPY chain induced by the substituents. An alternative to the N-substitution is derivatization of the pyrrole ring in the 3 or 4 position, which normally requires lower anodic potentials and exhibits higher conductivities.[30]

The second route involves the specific incorporation of target molecules as dopants during the electropolymerization procedure. Therefore, functionalization is achieved by the irreversible capture of a negatively charged target molecule within the polymer matrix. A good example is the recent demonstration of the incorporation of PPY NWs with CNTs via a template-directed procedure.[31] The negatively charged CNTs acted as dopants and served as charge-balancing counterions. It was further demonstrated that BM could be easily adsorbed on the CNTs prior to polymerization.[32,33] Such simultaneous incorporation of CNTs and BM impart electrocatalytic and biocatalytic properties, respectively, thus improving the sensing performance. BM have also been similarly incorporated into PPY in the absence of supporting electrolyte.[34]

The third procedure is often termed as *postpolymerization functionalization*. In this case, the functionalization with the biological molecule is performed after the polymerization. An appropriate functional group in the polymer is allowed to covalently bind to another functional group of the targeted biological molecule. This approach, therefore, requires the synthesis of CPs having reactive entities that are used as anchoring points to graft the appropriate functional groups on the biological molecule. The major advantage of this sequential procedure is the possibility of using optimal conditions for each step to obtain optimum polymer conductivity and precise localization and orientation of the biological molecule. A good example is the recent postpolymerization functionalization of PPY films with biotin entities.[35] In this case, the β-ferrocene ethylamine used as redox probe was immobilized via a coupling reaction on the surface of a preformed PPY film bearing activated ester groups onto which biotin entities were immobilized. Other examples include ODN immobilization by postfunctionalization of PPY-bearing easy leaving groups[36] and PPY-bearing carboxylic acid groups that enable easy covalent binding to biomolecules.[37]

The fourth procedure involves entrapment of target molecule within CPs. It involves the electropolymerization of the monomer in a solution containing the dopant and the target biomolecule. The major advantage of this procedure is that entrapment of the biomolecules occurs without any chemical reaction that could affect the activity of the biological molecule. Also, previous functionalization of the biological molecule by specific labels via chemical modification and genetic engineering is not necessary. Disadvantages associated with this procedure include the possibility of reduced accessibility, catalytic activity, and flexibility of the biological molecule as a result of random immobilization and steric hindrances within the CP matrix.[38,39] We have successfully used this procedure to entrap avidin within the matrix of a 100-nm-wide PPY NW. In this process, avidin was entrapped during the electropolymerization of PPY NW in a solution containing an avidin-conjugated quantum dot, the pyrrole monomer dopant in a single step within lithographically defined nanoscale channels (Figure 5a).[17] The immobilization of the avidin was confirmed by energy dispersive X-ray (EDX) spectrum analysis of the PPY which showed the presence of a Cd peak in the PPY entrapped with the avidin-conjugated quantum dot and no Cd peak in the control sample (Figure 5b

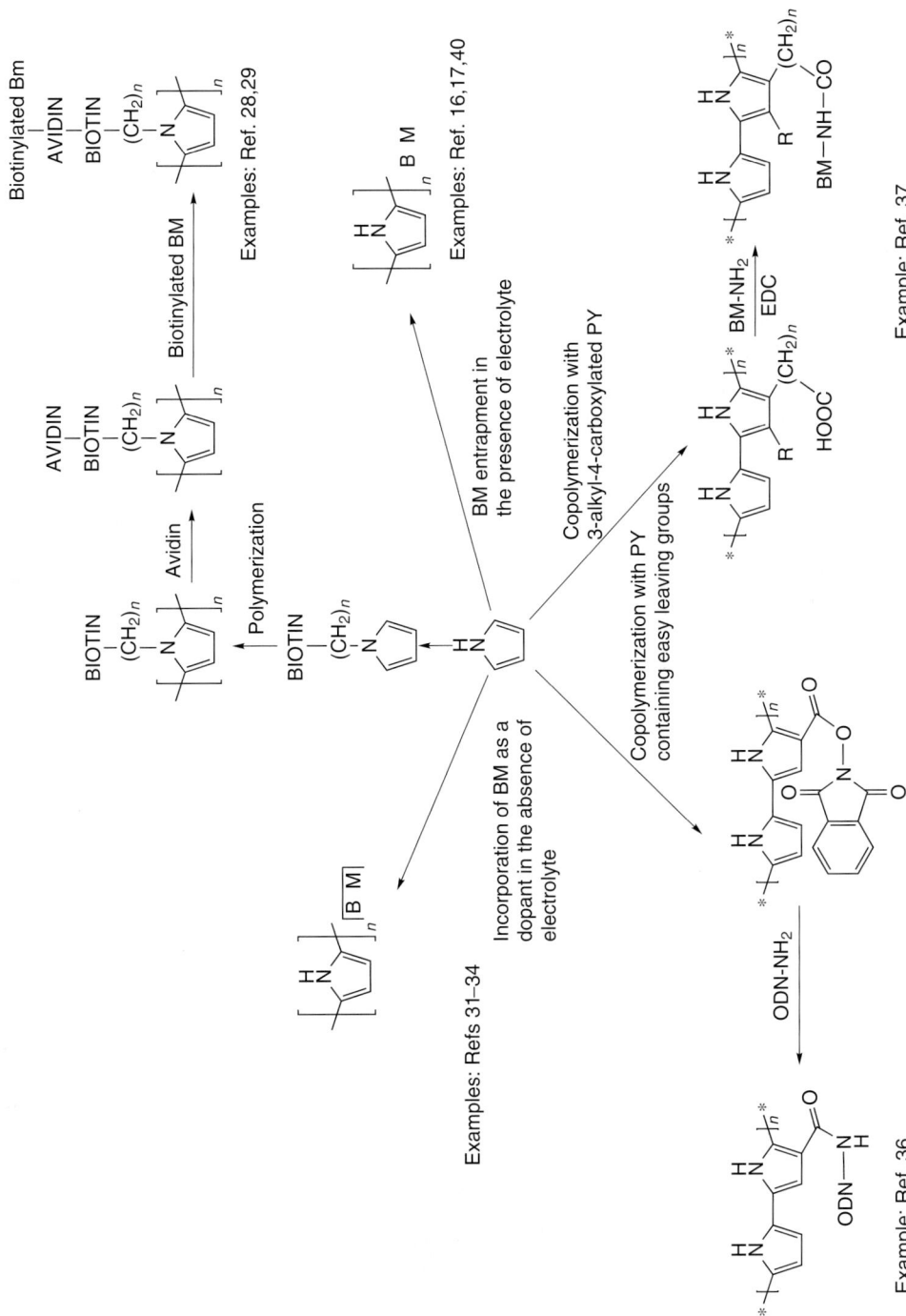

Scheme 1. Polymerization of conducting polymer with pyrrole as the model monomer. BM: biomolecule; EDC: 1-ethyl-3-(3-dimethylaminopropyl)-carbodiimide; ODN: oligonucleotide; PY: pyrrole

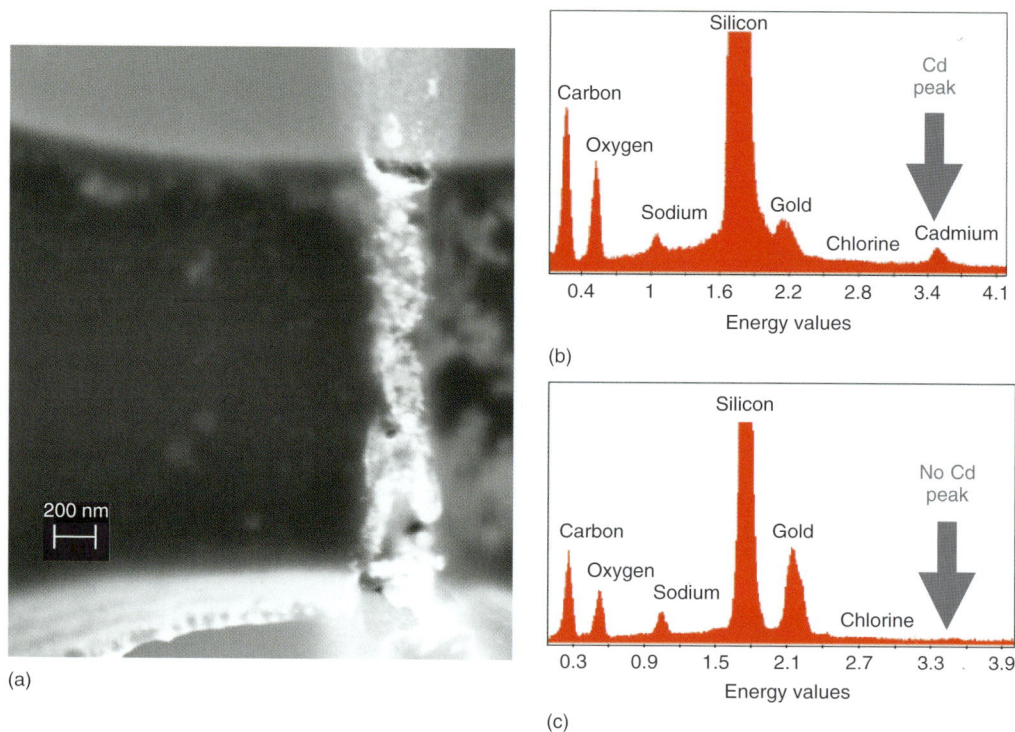

(a)

(b)

(c)

Figure 5. (a) SEM image of an Aqd-embedded polypyrrole nanowire (200-nm wide). The EDX analysis of polypyrrole nanowire with (b) and without (c) avidin-conjugated quantum dot. [Reprinted with permission from Ramanathan et al.[17] © 2005 American Chemical Society.]

and 5c). AFM phase imaging analysis further confirmed the presence of avidin on the polymer surface (Figure 6). The higher contrast on the avidin-functionalized PPY surface compared with the control PPY reflected the difference in variation and hardness in the two polymers due to the presence of quantum dots that were conjugated to avidin. Similarly, Mallouk and coworkers recently reported the fabrication of gold-capped, avidin-entrapped PPY NWs using the conventional template-directed electrochemical deposition.[40]

Finally, the emerging technique of imprinting CP NW with BM is very promising.[41,42] In this procedure, the imprint molecule may be immobilized on the pore walls of silica-treated nanoporous alumina membranes followed by the template-directed CP NW deposition.[42] The alumina membranes are then dissolved leaving the CP NWs with biological molecule binding sites on the surface. This technique may be a good alternative to the immobilization of bioreceptors for affinity sensor development. However, the use

of electropolymerized molecularly imprinted polymers for sensor development is still very young.

4 ASSEMBLY/ALIGNMENT OF CP NWS

Alignment of CP NWs into practical, functional sensors is sometimes necessary depending on the method that was used for their fabrication. For example CP NWs fabricated via template-directed synthesis and electrospinning methods require some kind of alignment/assembly procedures to form functional sensor devices. In principle, procedures that have been used in the alignment of other 1D nanostructures like CNTs and Si NWs can potentially be used in the alignment of CP NWs. For example, capping CP NWs with nickel ends and using applied magnetic field to orient and align between nickel electrodes as was recently demonstrated for other 1D nanostructures materials.[43,44] Tao and coworkers demonstrated a magnetic-field-assisted method to assemble an

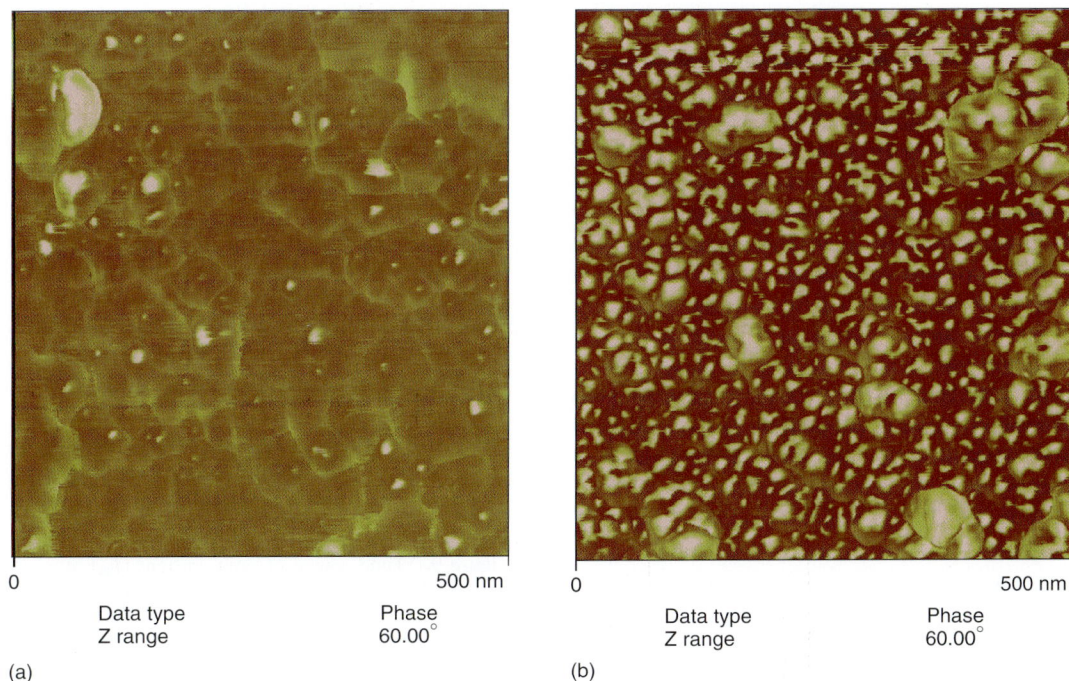

Data type   Phase
Z range   60.00°

(a)

Data type   Phase
Z range   60.00°

(b)

Figure 6. AFM phase images for (a) PPY and (b) PPY with avidin-conjugated quantum dots. [Reprinted with permission from Ramanathan et al.[17] © 2005 American Chemical Society.]

array of electrically wired CP junctions.[45] An aqueous solution containing CP-coated Au/Ni/Au metallic bars was introduced onto an array of parallel microelectrodes. In the presence of a magnetic field, the polymer-coated magnetic bars are aligned perpendicular to the microelectrodes, thus enabling metal/polymer/metal junctions between the two microelectrodes.

On the other hand, some CP NW fabrication methods totally avoid the alignment process altogether. Good examples are the template-directed synthesis within lithographically defined nanoscale channels,[16,17] dip-pen lithography,[6,7] other lithographic methods,[8] and some forms of templateless polymerization.[23]

5 SENSING APPLICATIONS BASED ON CP NWS

Figure 7 presents the real-time responses of avidin-functionalized PPY NW to different concentration of biotin conjugated to a 20-mer DNA oligonucleotide, elucidating the application of biologically

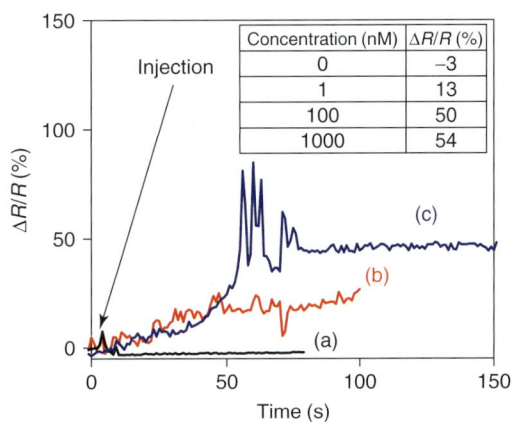

Concentration (nM)	$\Delta R/R$ (%)
0	−3
1	13
100	50
1000	54

Figure 7. Electrical responses of an unmodified nanowire (a) to 100 nM biotin-DNA (single stranded) and avidin-embedded polypyrrole (200 nm) nanowires to 1 nM (b) and 100 nM (c) biotin-DNA. The responses were recorded on two separate polypyrrole-avidin nanowires. Polypyrrole nanowires containing entrapped avidin were grown using 25 nM pyrrole in 10 mM NaCl and avidin. [Reprinted with permission from Ramanathan et al.[17] © 2005 American Chemical Society.]

functionalized CP NWs for label-free sensing.[17] The sensing was very rapid and concentrations

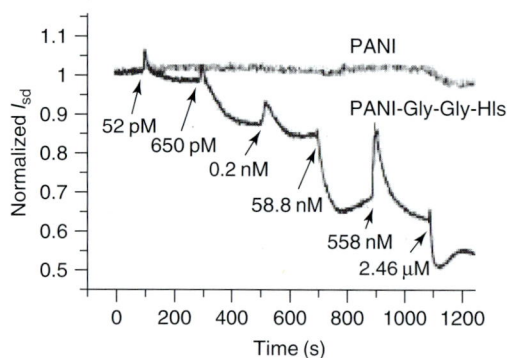

Figure 8. Simultaneous monitoring of the time course of normalized drain current (I_{sd}) on peptide-functionalized PANI (poly(GGH-ANI)) and regular PANI nanojunctions at drain potential (V_{sd}) of 0.4 V upon the injection of Cu(NO$_3$)$_2$ solutions. Numbers indicate Cu^{2+} final concentrations. [Reused with permission from Alvaro Díaz Aguilar et al.[46] © 2005.]

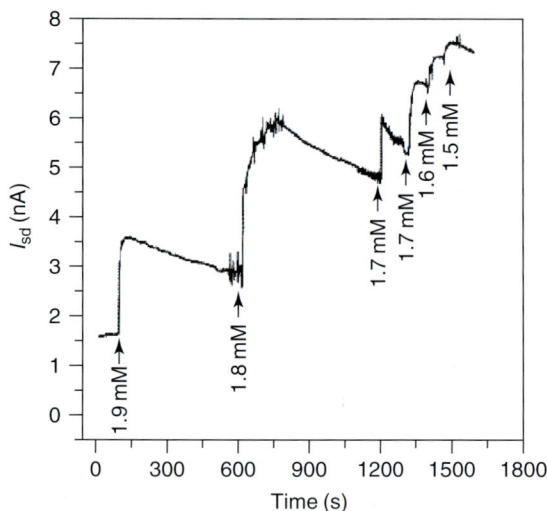

Figure 9. Time course of drain current (I_{sd}) at a gate voltage of 35 mV versus SCE (drain voltage = −20 mV) for PANi/GOx nanojunction in McIlvaine's buffer upon successive additions of 40 mM of glucose. [Reprinted with permission from Forzani et al.[19] © 2004 American Chemical Society.]

of biotin-DNA of up to 1 nM were detected in a few seconds. Similar response was observed with anti-avidin IgG. A sensitivity of potentially single-molecule detection is possible by adjusting the nanowire's conductivity to a value closer to the lower end of a semiconductor. Another demonstration of label-free sensing using biologically functionalized CPs was given by Tao, where a polyaniline nanojunction functionalized with gly–gly–his, hexa-his, and glucose oxidase (GOx) was used for the detection of Cu^{2+}, Ni^{2+} (Figure 8),[46] and glucose (Figure 9),[19] respectively. While the former involved a label-free detection due to the interaction of the metal ions with the peptide-functionalized polyaniline nanojunction, the detection of glucose involved its catalytic oxidation by the enzyme followed by the reoxidation of the GOx by O$_2$ in solution, which produces H$_2$O$_2$. The H$_2$O$_2$ then oxidized polyaniline, thus triggering an increase in the polyaniline conductivity. The increase in polyaniline conductivity constituted the analytical signal. Owing to the small size of the nanojunction sensor, the GOx was regenerated naturally without the need of redox mediators. Therefore, responses were very fast (<200 ms) and a minimal amount of oxygen was consumed.

6 SUMMARY AND OUTLOOK

Compared to other 1D nanostructured materials like Si NWs and CNTs, CP NWs are relatively new in the chemical and biosensing arena. However, they offer distinct and very attractive advantages compared to other 1D nanostructured materials. For example, CP NWs can easily be synthesized using benign reagents at ambient conditions through well-known chemical and electrochemical procedures. Their conductivities can be modulated by up to 15 orders of magnitude by changing the dopant and monomer/dopant ratios. They can be functionalized before, during, and after synthesis. Also functional BM can be incorporated into the CP NWs in a one-step procedure within built-in electrical contacts thus avoiding the cumbersome alignment and positioning manipulations. These are major advantages of CP NWs over other 1D nanostructured materials like Si NWs and CNTs.

One of the disadvantages of CP NWs is that they are mechanically weak and are more likely to break easily. Also, unlike the fabrication of CP NWs within defined nanoscale channels, the template-directed synthesis of CP NWs still requires postsynthetic alignment and positioning. This is one area that needs improvement. While several novel sensing concepts based on CP NWs have been demonstrated, incorporating these materials into routine functional integrated devices remains a challenge. Therefore, advances in the capabilities of assembling larger and more complex CP NW arrays and integrating them with nanoscale electronics may lead to wider applications in health, environmental, and homeland security sectors.

REFERENCES

1. D. H. Reneker and I. Chun, Nanometre diameter fibres of polymer, produced by electrospinning. *Nanotechnology*, 1996, **7**, 216–223.

2. J. Kameoka and H. G. Craighead, Fabrication of oriented polymeric nanofibers on planar surfaces by electrospinning. *Applied Physics Letters*, 2003, **83**, 371–373.

3. J. Kameoka, R. Orth, Y. N. Yang, D. Czaplewski, R. Mathers, G. W. Coates, and H. G. Craighead, A scanning tip electrospinning source for deposition of oriented nanofibres. *Nanotechnology*, 2003, **14**, 1124–1129.

4. J. Kameoka, R. Orth, B. Ilic, D. Czaplewski, T. Wachs, and H. G. Craighead, An electrospray ionization source for integration with microfluidics. *Analytical Chemistry*, 2002, **74**, 5897–5901.

5. H. Q. Liu, J. Kameoka, D. A. Czaplewski, and H. G. Craighead, Polymeric nanowire chemical sensor. *Nano Letters*, 2004, **4**, 671–675.

6. J. H. Lim and C. A. Mirkin, Electrostatically driven dip-pen nanolithography of conducting polymers. *Advanced Materials*, 2002, **14**, 1474–1477.

7. B. W. Maynor, S. F. Filocamo, M. W. Grinstaff, and J. Liu, Direct-writing of polymer nanostructures: poly(thiophene) nanowires on semiconducting and insulating surfaces. *Journal of the American Chemical Society*, 2002, **124**, 522–523.

8. B. Dong, D. Y. Zhong, L. F. Chi, and H. Fuchs, Patterning of conducting polymers based on a random copolymer strategy: toward the facile fabrication of nanosensors exclusively based on polymers. *Advanced Materials*, 2005, **17**, 2736–2741.

9. H. X. He, C. Z. Li, and N. J. Tao, Conductance of polymer nanowires fabricated by a combined electrodeposition and mechanical break junction method. *Applied Physics Letters*, 2001, **78**, 811–813.

10. C. R. Martin, Nanomaterials—a membrane-based synthetic approach. *Science*, 1994, **266**, 1961–1966.

11. C. R. Martin, Template synthesis of electronically conductive polymer nanostructures. *Accounts of Chemical Research*, 1995, **28**, 61–68.

12. Y. F. Ma, J. M. Zhang, G. J. Zhang, and H. X. He, Polyaniline nanowires on Si surfaces fabricated with DNA templates. *Journal of the American Chemical Society*, 2004, **126**, 7097–7101.

13. A. Bensimon, A. Simon, A. Chiffaudel, V. Croquette, F. Heslot, and D. Bensimon, Alignment and sensitive detection of DNA by a moving interface. *Science*, 1994, **265**, 2096–2098.

14. H. Nakao, H. Hayashi, T. Yoshino, S. Sugiyama, K. Otobe, and T. Ohtani, Development of novel polymer-coated substrates for straightening and fixing DNA. *Nano Letters*, 2002, **2**, 475–479.

15. P. Nickels, W. U. Dittmer, S. Beyer, J. P. Kotthaus, and F. C. Simmel, Polyaniline nanowire synthesis templated by DNA. *Nanotechnology*, 2004, **15**, 1524–1529.

16. K. Ramanathan, M. A. Bangar, M. H. Yun, W. F. Chen, A. Mulchandani, and N. V. Myung, Individually addressable conducting polymer nanowires array. *Nano Letters*, 2004, **4**, 1237–1239.

17. K. Ramanathan, M. A. Bangar, M. Yun, W. Chen, N. V. Myung, and A. Mulchandani, Bioaffinity sensing using biologically functionalized conducting-polymer nanowire. *Journal of the American Chemical Society*, 2005, **127**, 496–497.

18. C. Y. Peng, A. K. Kalkan, S. J. Fonash, B. Gu, and A. Sen, A "grow-in-place" architecture and methodology for electrochemical synthesis of conducting polymer nanoribbon device arrays. *Nano Letters*, 2005, **5**, 439–444.

19. E. S. Forzani, H. Q. Zhang, L. A. Nagahara, I. Amlani, R. Tsui, and N. J. Tao, A conducting polymer nanojunction sensor for glucose detection. *Nano Letters*, 2004, **4**, 1785–1788.

20. J. Duchet, R. Legras, and S. Demoustier-Champagne, Chemical synthesis of polypyrrole: structure-properties relationship. *Synthetic Metals*, 1998, **98**, 113–122.

21. L. Liang, J. Liu, C. F. Windisch, G. J. Exarhos, and Y. H. Lin, Direct assembly of large arrays of oriented conducting polymer nanowires. *Angewandte Chemie International Edition*, 2002, **41**, 3665–3668.

22. J. Liu, Y. H. Lin, L. Liang, J. A. Voigt, D. L. Huber, Z. R. Tian, E. Coker, B. Mckenzie, and M. J. Mcdermott, Templateless assembly of molecularly aligned conductive polymer nanowires: a new approach for oriented nanostructures. *Chemistry- A European Journal*, 2003, **9**, 605–611.

23. M. M. Alam, J. Wang, Y. Y. Guo, S. P. Lee, and H. R. Tseng, Electrolyte-gated transistors based on conducting polymer nanowire junction arrays. *Journal of Physical Chemistry B*, 2005, **109**, 12777–12784.

24. T. Livache, H. Bazin, P. Caillat, and A. Roget, Electroconducting polymers for the construction of DNA or peptide arrays on silicon chips. *Biosensors and Bioelectronics*, 1998, **13**, 629–634.

25. N. Lassalle, A. Roget, T. Livache, P. Mailley, and E. Vieil, Electropolymerisable pyrrole-oligonucleotide: synthesis and analysis of ODN hybridisation by fluorescence and QCM. *Talanta*, 2001, **55**, 993–1004.

26. N. Lassalle, P. Mailley, E. Vieil, T. Livache, A. Roget, J. P. Correia, and L. M. Abrantes, Electronically conductive polymer grafted with oligonucleotides as electrosensors of DNA—preliminary study of real time monitoring by in situ techniques. *Journal of Electroanalytical Chemistry*, 2001, **509**, 48–57.

27. P. Guedon, T. Livache, F. Martin, F. Lesbre, A. Roget, G. Bidan, and Y. Levy, Characterization and optimization of a real-time, parallel, label-free, polypyrrole-based DNA sensor by surface plasmon resonance imaging. *Analytical Chemistry*, 2000, **72**, 6003–6009.

28. S. Cosnier and A. Lepellec, Poly(pyrrole-biotin): a new polymer for biomolecule grafting on electrode surfaces. *Electrochimica Acta*, 1999, **44**, 1833–1836.

29. L. M. Torres-Rodriguez, A. Roget, M. Billon, and G. Bidan, Synthesis of a biotin functionalized pyrrole and its

electropolymerization: toward a versatile avidin biosensor, *Chemical Communications*, 1998, 1993–1994.

30. D. Delabouglise, J. Roncali, M. Lemaire, and F. Garnier, Control of the lipophilicity of polypyrrole by 3-Alkyl substitution. *Journal of the Chemical Society, Chemical Communications*, 1989, 475–477.

31. J. Wang, J. H. Dai, and T. Yarlagadda, Carbon nanotube-conducting-polymer composite nanowires. *Langmuir*, 2005, **21**, 9–12.

32. J. Wang and M. Musameh, Carbon-nanotubes doped polypyrrole glucose biosensor. *Analytica Chimica Acta*, 2005, **539**, 209–213.

33. H. Cai, Y. Xu, P. G. He, and Y. Z. Fang, Indicator free DNA hybridization detection by impedance measurement based on the DNA-doped conducting polymer film formed on the carbon nanotube modified electrode. *Electroanalysis*, 2003, **15**, 1864–1870.

34. J. Wang and M. Jiang, Toward genolelectronics: nucleic acid doped conducting polymers. *Langmuir*, 2000, **16**, 2269–2274.

35. M. L. Calvo-Munoz, B. E. A. Bile, M. Billon, and G. Bidan, Electrochemical study by a redox probe of the chemical post-functionalization of N-substituted polypyrrole films: application of a new approach to immobilization of biotinylated molecules. *Journal of Electroanalytical Chemistry*, 2005, **578**, 301–313.

36. H. KorriYoussoufi, F. Garnier, P. Srivastava, P. Godillot, and A. Yassar, Toward bioelectronics: specific DNA recognition based on an oligonucleotide-functionalized polypyrrole. *Journal of the American Chemical Society*, 1997, **119**, 7388–7389.

37. A. I. Minett, J. N. Barisci, and G. G. Wallace, Immobilisation of anti-Listeria in a polypyrrole film. *Reactive and Functional Polymers*, 2002, **53**, 217–227.

38. L. Cocheguerente, A. Deronzier, P. Mailley, and J. C. Moutet, Electrochemical immobilization of glucose-oxidase in poly(amphiphilic pyrrole) films and its application to the preparation of an amperometric glucose sensor. *Analytica Chimica Acta*, 1994, **289**, 143–153.

39. S. Cosnier and C. Innocent, A novel biosensor elaboration by electropolymerization of an adsorbed amphiphilic pyrrole tyrosinase enzyme layer. *Journal of Electroanalytical Chemistry*, 1992, **328**, 361–366.

40. R. M. Hernandez, L. Richter, S. Semancik, S. Stranick, and T. E. Mallouk, Template fabrication of protein-functionalized gold-polypyrrole-gold segmented nanowires. *Chemistry of Materials*, 2004, **16**, 3431–3438.

41. Y. Li, H. H. Yang, Q. H. You, Z. X. Zhuang, and X. R. Wang, Protein recognition via surface molecularly imprinted polymer nanowires. *Analytical Chemistry*, 2006, **78**, 317–320.

42. H. H. Yang, S. Q. Zhang, F. Tan, Z. X. Zhuang, and X. R. Wang, Surface molecularly imprinted nanowires for biorecognition. *Journal of the American Chemical Society*, 2005, **127**, 1378–1379.

43. S. Niyogi, C. Hangarter, R. M. Thamankar, Y. F. Chiang, R. Kawakami, N. V. Myung, and R. C. Haddon, Magnetically assembled multiwalled carbon nanotubes on ferromagnetic contacts. *Journal of Physical Chemistry B*, 2004, **108**, 19818–19824.

44. A. K. Bentley, J. S. Trethewey, A. B. Ellis, and W. C. Crone, Magnetic manipulation of copper-tin nanowires capped with nickel ends. *Nano Letters*, 2004, **4**, 487–490.

45. H. Q. Zhang, S. Boussaad, N. Ly, and N. J. J. Tao, Magnetic-field-assisted assembly of metal/polymer/metal junction sensors. *Applied Physics Letters*, 2004, **84**, 133–135.

46. A. D. Aguilar, E. S. Forzani, X. L. Li, N. J. Tao, L. A. Nagahara, I. Amlani, and R. Tsui, Chemical sensors using peptide-functionalized conducting polymer nanojunction arrays. *Applied Physics Letters*, 2005, **87**, 193108–193111.

Biosensors Based on Single-Walled Carbon Nanotube Near-Infrared Fluorescence

Paul W. Barone,[1] Esther S. Jeng,[1] Daniel A. Heller[2] and Michael S. Strano[1]

[1] *Department of Chemical Engineering, Massachusetts Institute of Technology, Cambridge, MA, USA and* [2] *Department of Chemistry, Massachusetts Institute of Technology, Cambridge, MA, USA*

The optical characteristics of individual single-walled carbon nanotubes (SWNTs) make them excellent candidates for optical sensors, as has been shown previously.[1–3] SWNTs can be visualized as sheets of graphene rolled into seamless structures. Depending on the angle of rolling (chirality) and the diameter of the nanotube, semiconducting, or metallic tubes are formed.[4] Individual semiconducting nanotubes photoluminesce with discrete bands between 900 and 1600 nm. This range in the near-infrared (nIR) region is in the "tissue-transparent window" where the scattering, autofluorescence,[5] and absorption of optically dense biologically relevant molecules such as blood, water, and tissue[6–8] are low, allowing for transmission through this otherwise highly scattering media[6,7] (Figure 1a). In particular, Rayleigh scattering, which constitutes the majority of the scattering intensity, is inversely proportional to the wavelength raised to the fourth power, resulting in dramatically reduced scattering in the nIR.[9] Therefore, SWNTs have the potential to be used in biological applications and environments, including live cells and tissues. Also SWNTs are sensitive to molecular events at their surface[1,4] and are uniquely photostable.[10] Figure 1(b) compares the rate of photobleaching of an organic fluorophore, quantum dots, and carbon nanotubes, showing that nanotubes excited at high laser fluence for 10 h show no diminution in emission.[10] This section presents the use of individually dispersed HiPCO SWNT (Rice Research Reactor run 107, length 400–1000 nm) as fluorescent biosensors. Fluorescent and Raman scattering measurements are done simultaneously using a Kaiser Optical Holospec f/1.8 imaging spectrograph with excitation at 785 nm.

1 SINGLE-WALLED CARBON NANOTUBE GLUCOSE SENSORS

Implantable biosensors are a major focus of ongoing biomedical research,[11–17] because of their many advantages over the current analyte detection techniques.[17] Such sensors would have an immediate impact on diseases such as diabetes, where continuous monitoring along with intensive therapy is often necessary to prevent long-term complications.[18] SWNTs present one avenue

Handbook of Biosensors and Biochips. Edited by Robert S. Marks, David C. Cullen, Isao Karube, Christopher R. Lowe and Howard H. Weetall.
© 2007 John Wiley & Sons, Ltd. ISBN 978-0-470-01905-4.

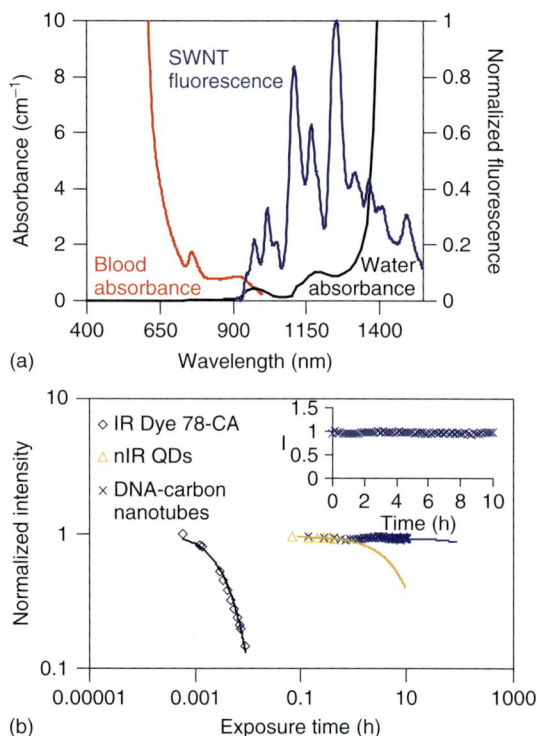

(a)

(b)

Figure 1. (a) "Tissue-transparent window" composed of the absorption spectra of blood (red) and water (black). Fluorescence spectrum of SWNT (blue). (b) Photobleaching curve of a near-IR organic dye, near-infrared quantum dots (QDs), and carbon nanotubes. The nanotubes show no photobleaching over 10 h of excitation. [Reprinted with permission Heller et al.[10] copyright 2005, Wiley VCH.]

toward the creation of implantable biosensors.[1–3] The combination of the nanotube's sensitivity

to surface adsorption events[4,19,20] and their nIR fluorescence[7] makes them promising candidates for fluorescence-based devices.

One method to create a SWNT sensor that is both sensitive and selective, is to link the actions of enzymes or proteins to the nanotube and its fluorescence. The idea being that, when the enzymatic reaction or binding event occurs, a measurable change in nanotube fluorescence would be observed. In the case of enzymes, a reaction intermediary is necessary to mediate the nanotube fluorescence in response to the enzymatic reaction.[1] Figure 2 shows a schematic of a glucose sensor using glucose oxidase (GOx) as the enzyme and potassium ferricyanide ($K_3Fe(CN)_6$) as the intermediary. The enzyme is adsorbed to the side of the nanotube where it catalyzes the oxidation of β-D-glucose, producing gluconolactone and hydrogen peroxide. Ferricyanide ions adsorb to the side of the nanotube, which quench SWNT emission, then react with the hydrogen peroxide resulting in a recovery of SWNT fluorescence.

To assemble the sensor, GOx is first adsorbed to the surface of the SWNT. Nanotubes are initially solubilized using sodium cholate.[1,3,7] The GOx is self assembled on the surface of the nanotube by removing the cholate coating the SWNT through dialysis. Such a process allows the GOx to retain its functionality. Figure 3 shows SWNT fluorescence spectra before and after GOx assembly via a 20 h dialysis. The adsorption of GOx to the surface of the nanotube causes the fluorescence emission to decrease in energy, as compared to cholate suspended nanotubes. Such a shift is indicative of a more porous coating on

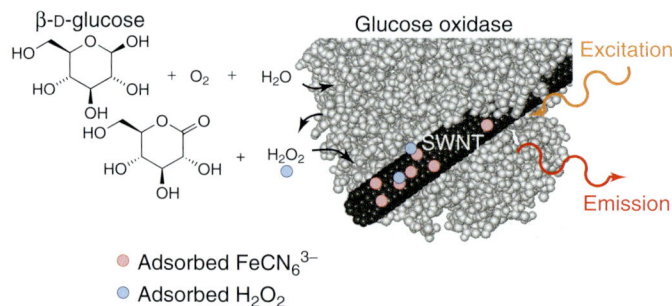

Figure 2. Schematic of glucose sensing using enzyme–nanotube complex and a reaction mediator ($Fe(CN)_6^{3-}$). Glucose oxidase adsorbed to the nanotube catalyzes the oxidation of glucose, producing hydrogen peroxide. The hydrogen peroxide then reacts with the adsorbed ferricyanide causing a fluorescence restoration. [Reprinted with permission Barone et al.[1] copyright 2005, Nature Publishing Group.]

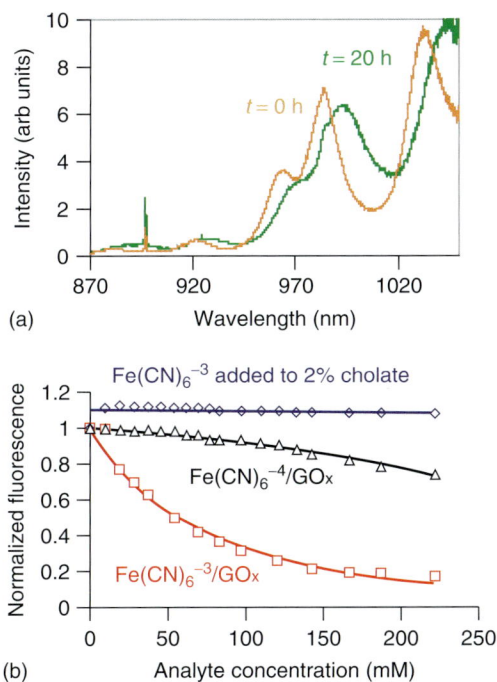

(a)

(b)

Figure 3. (a) Single-walled carbon nanotube fluorescence before (orange) and after (green) adsorption of GOx from a 20 h dialysis. (b) GOx suspended nanotube fluorescence decreases up to 83% when titrated with ferricyanide (red). Titration with the redox partner ferrocyanide (black) results in a lesser degree of fluorescence attenuation. The addition of ferricynide to sodium cholate suspended nanotubes does not alter the fluorescence intensity (blue). [Reprinted with permission Barone et al.[1] copyright 2005, Nature Publishing Group.]

the surface of the nanotube.[21] The porous enzyme coating makes permeation by the ferricyanide

possible. Addition of ferricyanide to the GOx suspended SWNT solution results in ferricyanide adsorption to the nanotube and a measurable fluorescence decrease (Figure 3b). The attenuation in SWNT fluorescence is concentration dependent with nanotube fluorescence decreasing by a maximum of 83% at 225 mM ferricyanide. If potassium ferricyanide's redox partner, potassium ferrocyanide ($K_4Fe(CN)_6$) is instead titrated into the GOx suspended SWNT solution, the attenuation of SWNT fluorescence is less. Interestingly, the adsorption of ferricyanide is partially irreversible. Figure 4 shows fluorescence from GOx suspended nanotubes contained in a dialysis cassette immersed in standard Tris buffer at pH 7 and 37 °C. Additions of 10 and 120 mM ferricyanide to the buffer cause SWNT fluorescence to attenuate. Cycling to reagent free buffer causes only a partial restoration of nanotube fluorescence indicating irreversible adsorption.

Addition of β-D-glucose to the GOx–ferricyanide–SWNT sensing complex results in nanotube fluorescence restoration. Figure 5(a) shows the response of nanotube fluorescence as 62.5 mM ferricyanide is added to GOx suspended SWNT followed by subsequent additions of 1.4, 2.8, and 4.1 mM total glucose. The fluorescence recovery after glucose additions is rapid, on the order of tens of seconds. The total fluorescence recovery can then be mapped to glucose added to the sample (Figure 5b). The resulting response function follows a Type 1 Langmuir adsorption isotherm with a $K_m = 0.91$ mM^{-1} and a calculated sensitivity of 37.4 µM. Conceivably, this type of sensing system

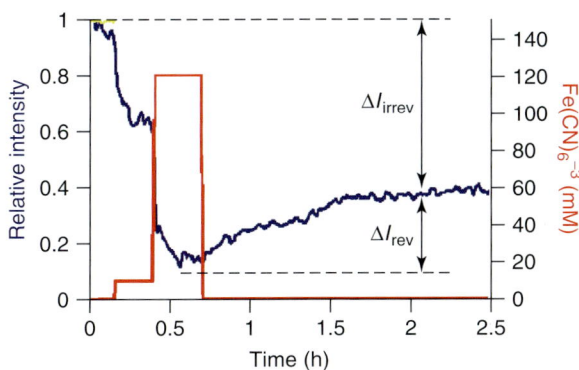

Figure 4. GOx suspended SWNT fluorescence (blue) decreases upon addition of 10 and 120 mM ferricyanide (red). Removal of free ferricyanide results in only a partial fluorescence recovery, indicating irreversible adsorption to the nanotube. [Reprinted with permission Barone et al.[1] copyright 2005, Nature Publishing Group.]

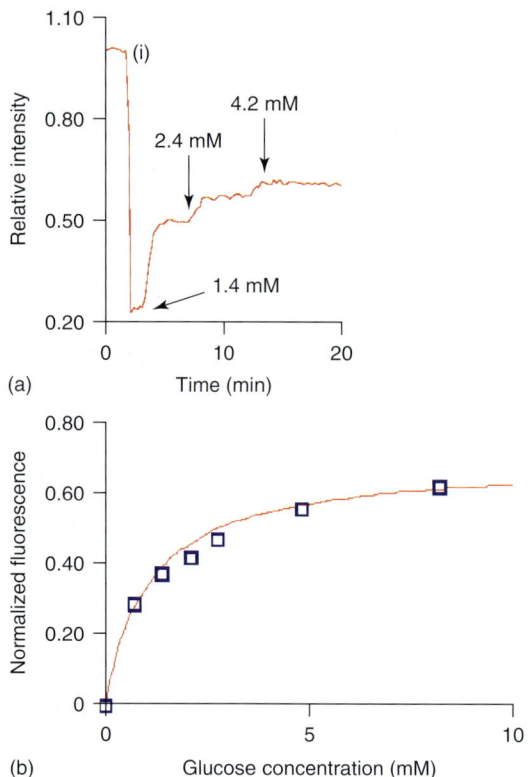

Figure 5. (a) GOx suspended SWNT fluorescence decreases upon addition of 62.5 mM ferricyanide and recovers upon subsequent additions of glucose. (b) The response function follows a Type 1 Langmuir adsorption isotherm. [Reprinted with permission Barone et al.[1] copyright 2005, Nature Publishing Group.]

could be implanted inside the body through the use of a small dialysis capillary.[1,22] Such a capillary would allow glucose to diffuse in, but would retain the sensing medium since the GOx–SWNT complex would be too large to diffuse through and the ferricyanide is irreversibly adsorbed to the surface of the nanotube. This sensing device is analogous to the flux-based electrochemical sensors currently being devised for implantation.

This work demonstrates the viability of carbon nanotubes as components in implantable biomedical sensors. Passive, optically responsive implantable sensors have clear advantages over electrochemical[23] and photo-electrochemical[24,25] devices. Additionally, the sensing mechanism outlined above is not relegated to just GOx, but could also be used with other redox enzymes that create hydrogen peroxide as a by-product, such as lactate oxidase or glutamate oxidase.

2 SINGLE-WALLED CARBON NANOTUBES AS DNA HYBRIDIZATION SENSORS

Detection of specific DNA sequences has many applications in the medical, life, and environmental sciences.[26–34] The use of fluorescence detection is advantageous because of the sensitivity and selectivity of the technique,[35] as well as ease of use. SWNT can be individually suspended by adsorbing molecules to its surface in solution, allowing for fluorescent label-free and dye-free[1,21,36,37] systems. Earlier work has shown that some DNA sequences adsorb strongly to the surface of nanotubes resulting in a stable suspension.[37,38] The use of SWNT led to the first photobleaching resistant, nanoparticle system that could detect DNA hybridization through the modulation of an nIR fluorescence signal from the DNA–SWNT complex.[3] This system allowed for optical detection of selective hybridization of DNA with its complementary strand directly on the surface of SWNT, and therefore opens possibilities for new types of nanotube-based biosensors.

The DNA–SWNT used for the hybridization sensors is assembled through a two-step dialysis technique (Figure 6a). In the first step a sodium cholate–SWNT solution[1,3] is dialyzed against standard Tris buffer pH 7.6 in the presence of the oligonucleotide strands through a membrane with a 12–14 kDa molecular weight cutoff (MWCO). A random sequence DNA1 (5′-TAG CTA TGG AAT TCC TCG TAG GCA-3′) was used to illustrate the concept of the sensor.[3] The sodium cholate dialyzes out through the membrane while the larger oligonucleotide strands assemble on the SWNT surface. The remaining free DNA strands are removed through a second dialysis (100 kDa MWCO). Combined photoluminescence and Raman spectra of the initial cholate-SWNT and final DNA1–SWNT solutions shows a bathochromic shift (Figure 6b). The shifting of SWNT fluorescence energy has been attributed to a change in the coverage of the SWNT surface in water.[21] The small cholate molecules are able to pack the SWNT surface more densely than the oligonucleotide strands. An atomic force microscopy (AFM) image (Figure 6c) shows that the DNA strands are adsorbed randomly to the surface of the SWNT and uneven heights suggest that each strand is folded on itself.

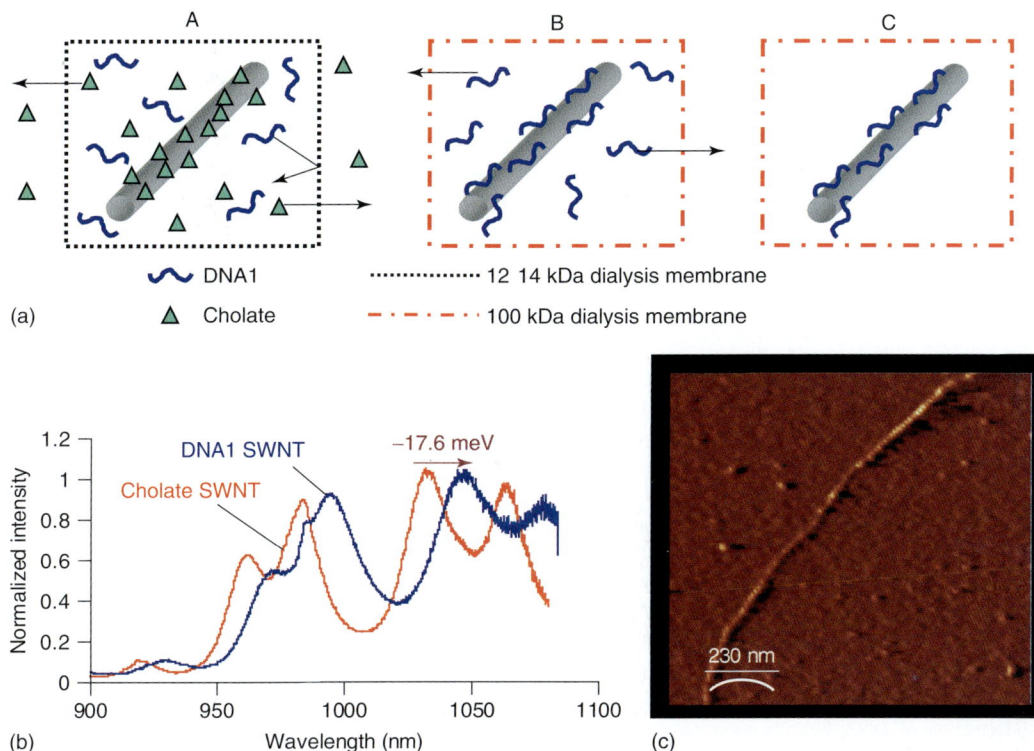

Figure 6. Assembly of DNA1 on SWNT (a) (A) Cholate–SWNT is dialyzed (12–14 kDa membrane) in the presence of DNA1, resulting in cholate removal and DNA1 adsorption to SWNT. (B) Free DNA1 is removed through a second dialysis step (100 kDa membrane). (C) Final DNA1–SWNT sample. (b) Comparison of initial cholate–SWNT and final DNA1–SWNT photoluminescence spectra reveals a bathochromic shift of 17.6 meV in the (7,5) nanotube with DNA assembly. (c) AFM image of DNA suspended SWNT with thicker sections denoting nonuniform adsorption of DNA on the SWNT with varying heights.

Hybridization of DNA1–SWNT with the complementary deoxyribonucleic acid strand (cDNA 5′-GCC TAC GAG GAA TTC CAT AGC T-3′) is transduced through a hypsochromic shift in the (6,5) nanotube fluorescence. An illustration of the detection scheme and the hypsochromic shift are shown in Figure 7(a) and (b). The (6,5) SWNT was excited with a 785-nm laser while the fluorescence ($\lambda_{max} = 994$ nm) was monitored. The DNA1–SWNT was incubated with cDNA and noncomplementary deoxyribonucleic acid (nDNA 5′-TCG ATA CCT TAA GGA GCA TCC G-3′) for 48 h to ensure that steady state was reached. The addition of cDNA caused a fluorescence energy increase of up to 2.02 ± 0.07 meV while additions of nDNA at the same concentrations resulted in negligible energy changes (Figure 7c). The energy increased with the concentration of cDNA added until (cDNA) reached 400 nM, where

the shift remained 2.02 meV. This energy shift can be attributed to the change in the dielectric environment of the SWNT, which changes the exciton binding energy. The dielectric constant, ε, at the SWNT surface is determined to be

$$\varepsilon = \alpha \varepsilon_{DNA} + (1 - \alpha)\varepsilon_{H_2O} \qquad (1)$$

using a local dielectric medium approximation where α is the fraction of SWNT surface area covered by DNA, and $\varepsilon_{DNA} = 2.1$, $\varepsilon_{H_2O} = 88$ represent the dielectric constants of DNA and water, respectively. The exciton binding energy is a function of the dielectric constant according to the following equation:

$$E = A\mu^{n-1} r_t^{n-2} \varepsilon^{-n} \qquad (2)$$

where A and n are constants (24.1 and 1.4 eV) found by fitting nanotubes with diameters of

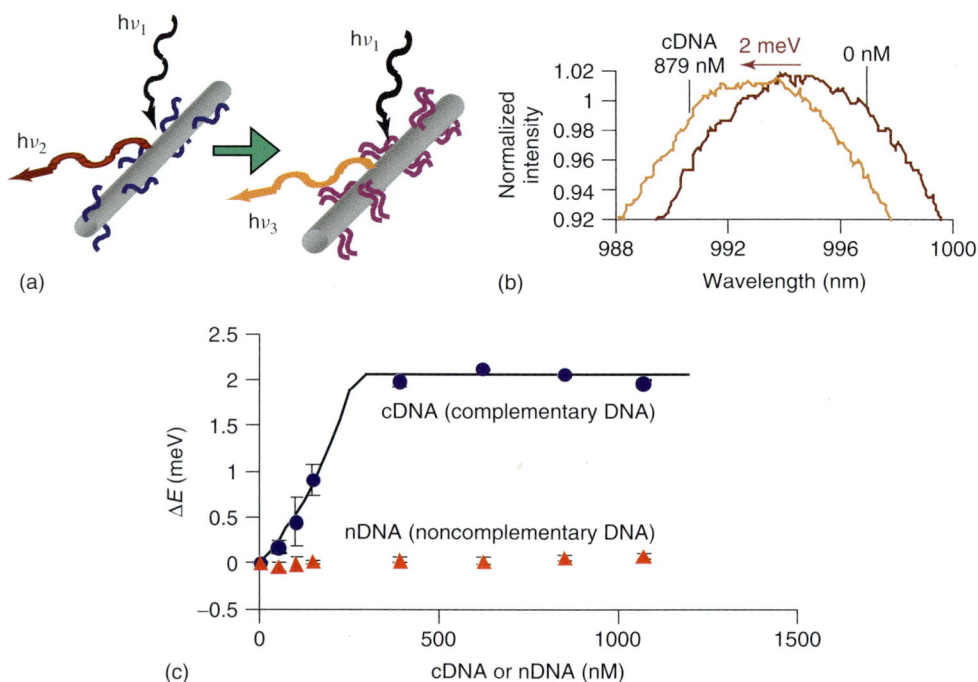

Figure 7. Detection of DNA hybridization on the SWNT surface (a) Illustration of detection mechanism. Incident laser light (energy $h\nu_1$) causes SWNT to fluoresce at a given energy ($h\nu_2$). After hybridization the same incident light energy causes the SWNT to fluoresce at a different energy ($h\nu_3$). (b) The 2 meV hypsochromic shift of (6,5) SWNT fluorescence energy after addition of complementary DNA (cDNA) is shown. (c) The steady-state energy increase of the SWNT fluorescence is shown (blue circles) as a function of cDNA added to the system. In contrast, the nDNA causes a negligible change in energy (red triangles). The solid line represents the calculated exciton binding energy change that is theorized to occur as the dielectric environment around the SWNT is changed. [Reprinted with permission Jeng et al.[3] copyright 2006, American Chemical Society.]

1–2.5 nm, $\mu = 0.068$ is the reduced effective mass of the (6,5) nanotube,[6] and $\beta = 0.0529$ nm is the Bohr radius constant. Approximating that hybridization doubles the surface area coverage of SWNT and using the energy increase of 2.02 meV, the surface area coverage is predicted to be 3.5–7%. In this model, Perebeinos et al. solved the Bethe–Salpeter equation to find an exciton binding energy scaling relationship.[39] The model is shown as the solid line in Figure 7(c). Hybridization on the SWNT surface was confirmed using Forster resonance energy transfer (FRET) of fluorophore labeled DNA (5′-TAG CTA TGG AAT TCC TCG TAG GCA-3′-6-FAM™ and TAMRA™ NHS Ester–5′- GCC TAC GAG GAA TTC CAT AGC T-3′).[3]

The hybridization kinetics on SWNT were studied through transient measurements of the fluorescence energy shift. Unlike the hybridization of free DNA, the DNA1–SWNT system

cannot be adequately described with a second-order reaction.[40] Instead the hybridization can be modeled using a two-step Langmuir isotherm followed by a first-order reaction (Figure 8a), similar to previous modeling.[41] The equilibrium number of occupied sites on the SWNT surface, $A\theta$, is fixed by the adsorption step as shown below:

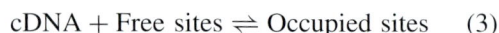

$$A + \theta \rightleftharpoons A\theta$$

$$\text{cDNA} + \text{Free sites} \rightleftharpoons \text{Occupied sites} \quad (3)$$

The number of occupied sites is determined by combining a total site (θ_T) balance and the equilibrium constant, K.

$$\theta_T = A\theta + \theta \quad (4)$$

$$K = \frac{[A\theta]}{[A](\theta_T - [A\theta])} \quad (5)$$

(a) Type 1 Langmuir isotherm   First-order kinetic reaction

(b)

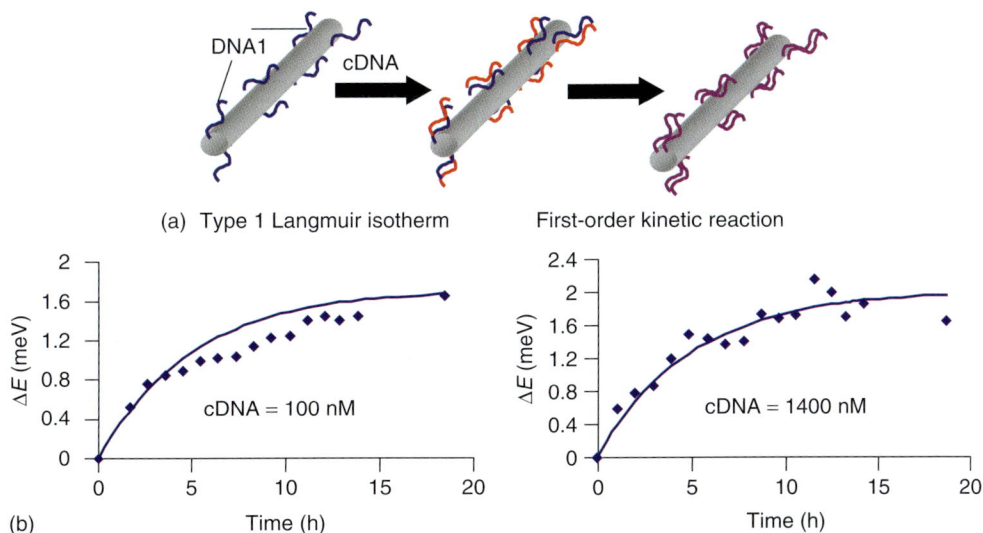

Figure 8. A two-step model of adsorption followed by reaction is fitted to the measured energy shifts caused by addition and hybridization of complementary cDNA (22 bases) with DNA1 (24 bases) on SWNT. (a) An illustration of the two-step model. (b) The change in SWNT energy with time at various concentrations (100–1400 nM) of complementary DNA are fitted using the same kinetic constant $k = 5.57 \times 10^{-5} \mathrm{s}^{-1}$, total binding sites on SWNT θT = 120 nM, and equilibrium constant $K = 5 \times 10^{7} \mathrm{M}^{-1}$. The measured and fitted energy shifts of the lowest and highest concentrations of cDNA are shown. [Reproduced from Jeng, E. S., Barone, P. W., Nelson, J. D., and Strano, M. S. *Small* (2007) submitted.]

The total concentration of hybridization sites in solution is θ_T (M), $A\theta$ is the concentration of occupied sites (M), K is the equilibrium constant (M^{-1}), and A is the concentration of cDNA strands (M). An occupied site concentration is determined from coupling of these two equations.

$$[A\theta_0] = \frac{K[A_0][\theta_T]}{1 + K[A_0]} \tag{6}$$

A slower and irreversible hybridization reaction of two complementary strands follows the adsorption step.

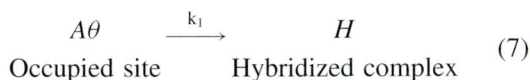

$$\begin{array}{ccc} A\theta & \xrightarrow{k_1} & H \\ \text{Occupied site} & & \text{Hybridized complex} \end{array} \tag{7}$$

H is the concentration of the hybridized complex (M) and k_1 is the kinetic constant in s^{-1}M^{-1}. The reaction is modeled as a simple first-order reaction

$$\frac{d[A\theta]}{dt} = -k_1[A\theta] = -\frac{d[H]}{dt} \tag{8}$$

with a hybridized complex formation of

$$H = (1 - e^{-k_1 t}) \frac{K[A_0][\theta_T]}{1 + K[A_0]} \tag{9}$$

The hybridization reaction that is transduced through the energy shift of the (6,5) SWNT, is fitted using the method of least squares as shown in Figure 8(b). At 25 °C the kinetic constant, k_1, is $5.57 \times 10^{-5} \mathrm{s}^{-1}$, and the equilibrium constant, K, is $5 \times 10^{7} \mathrm{M}^{-1}$. The total number of binding sites is 120 nM, which is the amount of DNA adsorbed to the SWNT surface. The process of hybridization on SWNT takes about 3.4 h to reach half of the steady-state energy shift observed for detection.

Hybridization DNA–SWNT sensors hold promise for the detection of specific DNA sequences in many applications. Although this new technology can still be improved through a decrease in detection time and an enhancement in the transduction signal, this label-free detection mechanism has the potential to be used for nonphotobleaching detection in scattering biological media and even inside cells.

3   SINGLE-WALLED CARBON NANOTUBE ION SENSORS

SWNTs functionalized with DNA respond to divalent metal ions of the type that induce the structural conformational transition of DNA from the native B form to the Z form.[42] Nanotube fluorescence shows a reversible concentration-dependent red shift of up to 16 meV upon introduction of metal cations.

Oligonucleotides containing certain sequences will transit from the right-handed B form to the left-handed Z form upon exposure to divalent metal cations.[42–44] This change in DNA secondary structure is caused preferentially by ions that bind to the DNA bases. The structural transformation was exploited for sensing by adsorbing 30-mer oligonucleotides onto the nanotube sidewall in a tight, helically-wrapped configuration. The method of encapsulation, developed by Zheng et al.[38,45] forms a colloidally stable complex (Figure 9a).[2]

Complexes of $d(GT)_{15}$ oligonucleotides and HiPCO carbon nanotubes (GT–SWNT) were prepared via ultrasonication. Single-stranded DNA

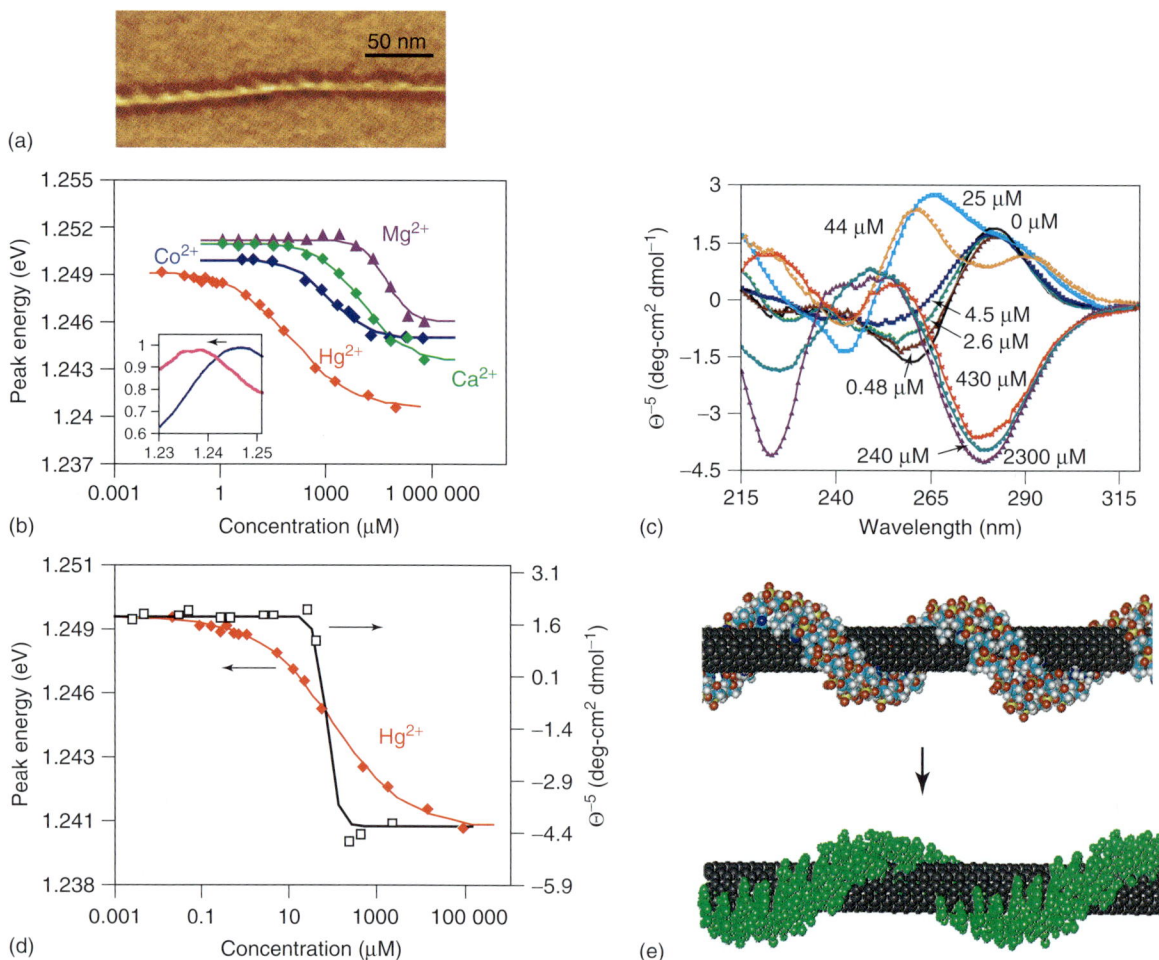

Figure 9. (a) AFM image of GT-DNA-encapsulated carbon nanotube complexes synthesized by ultrasonication. [Reproduced from Zheng, M. et al. *Science*, **302**(2003):1545–1548.] (b) Metal ion-induced spectral changes of the GT-SWNT complex show a monotonic energy decrease. One nanotube species is used for all measurements. Inset: Nanotube spectra at starting and ending concentrations of Hg^{2+}. (c) Circular dichroism spectra of the free $d(GT)_{15}$ oligonucleotide with increasing concentrations of Hg^{2+}. (d) Nanotube emission energy (red) and 285 nm ellipticity from CD spectra (black) versus Hg^{2+} concentration. (e) Representation of B–Z transition on carbon nanotube. [Reprinted with permission Heller et al.[2] copyright 2006, AAAS.]

with the GT sequence is postulated to form a double-stranded structure on the nanotube via weak pairing of G to T. The complexes were probed via nIR spectrofluorometry in a pH 7.4 buffer. The nanotube emission displayed a distinct, monotonic red shift with concentration of divalent metal cations (Figure 9b). The relative sensitivity to the ions added was dictated by the relative tendency for the ions to bind to DNA.[46] Removal of the metal ions restored the emission energy, indicating a reversible thermodynamic transition.

Circular dichroism (CD) spectroscopy, performed under identical conditions, confirmed that the unbound oligonucleotide of the same sequence undergoes a conformational change from the B to the Z form seen in the inversion of the 285 nm peak, which indicates a reversal of helicity (Figure 9c). We compare the ellipticity of the 285-nm CD peak versus Hg^{2+} concentration with the fluorescent emission energy from the nanotube, under identical conditions (Figure 9d). The overlapping points of inflection indicate that the difference in the free energy (ΔG) changes for the DNA on and off the nanotube is quite small ($\Delta(\Delta G) \sim 0.05\,k_B T$ per phosphate, where $k_B T$ is the thermal energy). Thus, the transitions for DNA in solution or adsorbed on the SWNT can be considered thermodynamically identical.

It is interesting to note that the slopes at the inflection show a distinct difference. Pohl[42] describes the B–Z transition, which requires a double-stranded helix to separate, change helicity and re-form, as a process of nucleation and propagation in series. The dsDNA strand initially separates with a ratio of rate constants β_B/β_Z while propagation proceeds as a series of equilibrium steps proportional to the number of base-pairs, N, as the dislocation proceeds down the chain. The expression for the fractional transition K,[42,47] contains a scaling factor C_0 which is the ion concentration (C) where K is independent of oligonucleotide length (N).

$$K = \left(\frac{C}{C_0}\right)^{aN} \left(\frac{\beta_B}{\beta_Z} + \left(\frac{C}{C_0}\right)^{aN}\right)^{-1} \quad (10)$$

Pohl used this equation to describe his observed changes in slope with oligonucleotide length or the propagation length, aN. Regression of the data in Figure 9(d) suggests that, given a length of 30

nucleotides on the free DNA, the nanotube-bound DNA exhibits an effective length of 5 nucleotides. In other words, the transition proceeds through only one-sixth the number of transitions as in the case of the free strand. As expected, β_B/β_Z, which is associated with the initiation of the event, is similar for the cases on and off the nanotube (1.21 and 1.04, respectively). The model suggests that the transition propagates in small steps and requires about $2\pi/3$ rads of the strand to unravel for propagation down the nanotube (Figure 9e).

Detection of divalent ions is also possible inside live cells. DNA-nanotube complexes (GT–SWNT) entered mammalian cells via endocytosis and reside in a perinuclear orientation within the cells over long time intervals.[10] Nanotubes were detected after 3 months in cell culture. Fluorescence area maps composed of multiple raster-scanned spectra show nanotube emission accumulated near the nucleus in the cytoplasm surrounded by membrane bound vesicles (Figure 10a).[10]

The GT–SWNT complexes function as single-cell sensors for ion concentration upon addition of Hg^{2+} into the cell media. Murine 3T3 fibroblasts containing GT–SWNT complexes were perfused with various concentrations of $HgCl_2$ in the extracellular buffer space and were monitored via photoluminescence. The SWNT emission, although shifted by 3 meV already upon uptake within the cell, red shifts additively with increasing Hg^{2+} concentration. After correcting for the initial shift caused by the new environment, the response of cell-bound DNA–SWNT fits the model curve created by the same complexes in pure buffer (Figure 10b).

Detection is also possible in media, such as muscle tissue and blood, which possess strong visible absorption. A GT–SWNT-filled dialysis capillary was inserted into whole blood and mammalian tissue and probed via photoluminescence. Also, nanotube complexes were added to a black dye solution (optical density >4). Detection of mercuric chloride was possible through these highly absorptive materials (Figure 10c). The nIR fluorescence of GT–SWNT in the dye solution exhibited the same response as GT–DNA in pure buffer. In whole blood and tissue, the presence of interfering absorbers of Hg^{2+} (free DNA, proteins, etc.) shift the observed sensitivity to larger values ($C_0 = 3500\,\mu M$ in blood and $8000\,\mu M$ in tissue), however the complexes still provided a measure

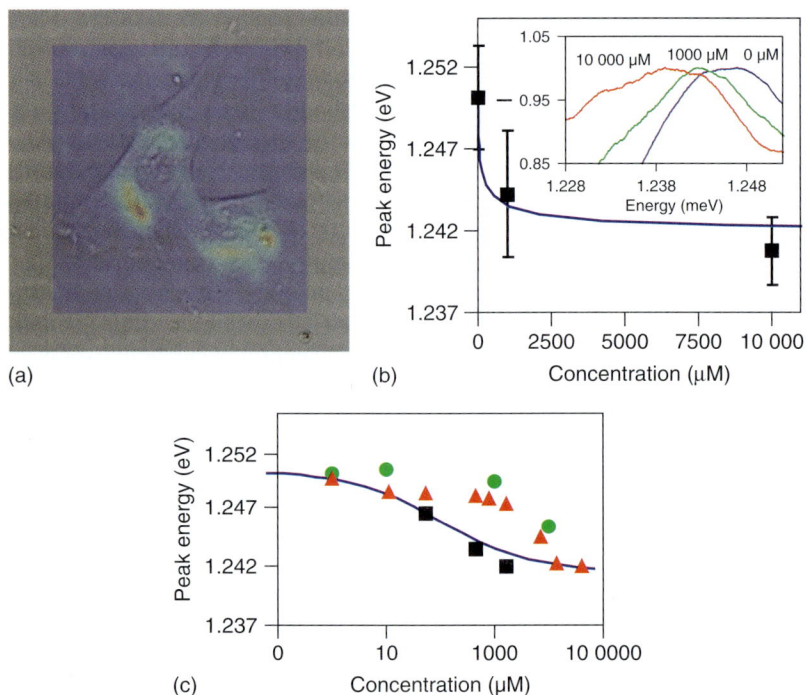

(a)

(b)

(c)

Figure 10. (a) Area map of nanotube emission. (b) Emission energy of GT–SWNT fluorescence upon addition of Hg^{2+}. Blue curve obtained from original GT–SWNT shift taken outside of cells. Inset: Individual spectra obtained at each Hg^{2+} concentration. (c) Emission energy of GT–SWNT in a solution of black ink (black squares), chicken muscle tissue (green circles), and whole rooster blood (red triangles) shown on blue curve obtained from the pristine GT–SWNT shift. [Reprinted with permission Heller et al.[2] copyright 2006, AAAS.]

of the residual ions that are locally bound to the complex in these heterogeneous media.

This work reveals the use of SWNT as biosensors for ions in live cells, opening an avenue for single-cell detection in a host of applications.

4 FUTURE DIRECTION OF SINGLE-WALLED CARBON NANOTUBE BIOSENSORS

The work outlined in this chapter shows the promising beginnings of a new type of sensor based on SWNT nIR photoluminescence. While we have demonstrated the ability to detect a wide range of analytes, from nucleic acids to small molecules, there is still much work left to do. A fundamental understanding of signal transduction to the nanotube is necessary for the rational design of any SWNT-based device. Future research in this area will not only focus on the application of current detection mechanisms for other analytes,

such as an enzymatic sensor for glutamate or the detection of single-nucleotide polymorphisms in DNA, but will also focus on the discovery and explanation of new detection modalities.

REFERENCES

1. P. W. Barone, S. Baik, D. A. Heller, and M. S. Strano, Near-infrared optical sensors based on single-walled carbon nanotubes. *Nature Materials*, 2005, **4**(1), 86–92.

2. D. A. Heller, E. S. Jeng, T. K. Yeeung, B. M. Martinez, A. E. Moll, J. B. Gastala, and M. S. Strano, Optical detection of DNA conformational polymorphism on single-walled carbon nanotubes. *Science*, 2006, **311**(5760), 508–511.

3. E. S. Jeng, A. E. Moll, AA. C. Roy, J. B. Gastala, and M. S. Strano, Detection of DNA hybridization using the near-infrared band-gap fluorescence of single-walled carbon nanotubes. *Nano Letters*, 2006, **6**(3), 371–375.

4. R. Saito, G. Dresselhaus, and M. S. Dresselhaus, *Physical Properties of Carbon Nanotubes*, Imperial College Press, London, 1998.

5. R. Weissleder and V. Ntziachristos, Shedding light onto live molecular targets. *Nature Medicine*, 2003, **9**(1), 123–128.

6. S. M. Bachilo, M. S. Strano, C. Kittrell, R. H. Hauge, R. E. Smalley, and R. B. Weisman, Structure-assigned optical spectra of single-walled carbon nanotubes. *Science*, 2002, **298**(5602), 2361–2366.

7. M. J. O'Connell, S. M. Bachilo, C. B. Huffman, V. C. Moore, M. S. Strano, E. H. Haroz, K. L. Rialon, P. J. Boul, W. H. Noon, C. Kittrell, J. P. Ma, R. H. Hauge, R. B. Weisman, and R. E. Smalley, Band gap fluorescence from individual single-walled carbon nanotubes. *Science*, 2002, **297**(5581), 593–596.

8. S. Wray, M. Cope, D. T. Delpy, J. S. Wyatt, and E. O. R. Reynolds, Characterization of the near-infrared absorption-spectra of cytochrome-Aa3 and hemoglobin for the non-invasive monitoring of cerebral oxygenation. *Biochimica et Biophysica Acta*, 1988, **933**(1), 184–192.

9. C. F. Bohren and D. R. Huffman, *Absorption and Scattering of Light by Small Particles*, Wiley, New York, 1983, p. 530.

10. D. Heller, S. Baikk, T. E. Eurell, and M. S. Strano, Single-walled carbon nanotube spectroscopy in live cells: towards long-term labels and optical sensors. *Advanced Materials*, 2005, **17**(23), 2793–2799.

11. L. L. E. Salins, R. A. Ware, C. M. Ensor, and S. Daunert, A novel reagentless sensing system for measuring glucose based on the galactose/glucose-binding protein. *Analytical Biochemistry*, 2001, **294**(1), 19–26.

12. H. Fang, G. Kaur, and B. H. Wang, Progress in boronic acid-based fluorescent glucose sensors. *Journal of Fluorescence*, 2004, **14**(5), 481–489.

13. M. Shichiri, N. Asakawa, Y. Yamasaki, R. Kawamori, and H. Abe, Telemetry glucose monitoring device with needle-type glucose sensor - a useful tool for blood-glucose monitoring in diabetic individuals. *Diabetes Care*, 1986, **9**(3), 298–301.

14. K. W. Johnson, J. J. Mastrototaro, D. C. Howey, R. L. Brunelle, P. L. Burdenbrady, N. A. Bryan, C. C. Andrew, H. M. Rowe, D. J. Allen, B. W. Noffke, W. C. McMahan, R. J. Morff, D. Lipson, and R. S. Nevin, Invivo evaluation of an electroenzymatic glucose sensor implanted in subcutaneous tissue. *Biosensors and Bioelectronics*, 1992, **7**(10), 709–714.

15. B. J. Gilligan, M. C. Shults, R. K. Rhodes, and S. J. Updike, Evaluation of a subcutaneous glucose sensor out to 3 months in a dog-model. *Diabetes Care*, 1994, **17**(8), 882–887.

16. W. K. Ward, E. S. Wilgus, and J. E. Troupe, Rapid detection of hyperglycemia by a subcutaneously-implanted glucose sensor in the rat. *Biosensors and Bioelectronics*, 1994, **9**(6), 423–428.

17. D. A. Gough, J. C. Armour, and D. A. Baker, Advances and prospects in glucose assay technology. *Diabetologia*, 1997, **40**, S102–S107.

18. H. Shamoon, H. Duffy, N. Fleischer, et al., The effect of intensive treatment of diabetes on the development and progression of long-term complications in insulin-dependent diabetes-mellitus. *New England Journal of Medicine*, 1993, **329**(14), 977–986.

19. T. Durkop, S. A. Getty, E. Cobas, and M. S. Fuhrer, Extraordinary mobility in semiconducting carbon nanotubes. *Nano Letters*, 2004, **4**(1), 35–39.

20. R. J. Chen, S. Bangsaruntip, K. A. Drouvalakis, N. W. S. Kam, M. Shim, Y. M. Li, W. Kim, P. J. Utz, and H. J. Dai, Noncovalent functionalization of carbon nanotubes for highly specific electronic biosensors. *Proceedings of the National Academy of Sciences of the United States of America*, 2003, **100**(9), 4984–4989.

21. M. S. Strano, V. C. Moore, M. K. Miller, M. J. Allen, E. H. Haroz, C. Kittrell, R. H. Hauge, and R. E. Smalley, The role of surfactant adsorption during ultrasonication in the dispersion of single-walled carbon nanotubes. *Journal of Nanoscience and Nanotechnology*, 2003, **3**(1–2), 81–86.

22. R. Ballerstadt and J. S. Schultz, A fluorescence affinity hollow fiber sensor for continuous transdermal glucose monitoring. *Analytical Chemistry*, 2000, **72**(17), 4185–4192.

23. A. Guiseppi-Elie, C. H. Lei, and R. H. Baughman, Direct electron transfer of glucose oxidase on carbon nanotubes. *Nanotechnology*, 2002, **13**(5), 559–564.

24. R. Tantra, R. S. Hutton, and D. E. Williams, A biosensor based on transient photoeffects at a silicon electrode. *Journal of Electroanalytical Chemistry*, 2002, **538**, 205–208.

25. L. M. Peter, Dynamic aspects of semiconductor photoelectrochemistry. *Chemical Reviews*, 1990, **90**(5), 753–769.

26. F. S. Nolte, B. Metchock, J. E. McGowan, A. Edwards, O. Okwumabua, C. Thurmond, P. S. Mitchell, B. Plikaytis, and T. Shinnick, Direct-detection of mycobacterium-tuberculosis in sputum by polymerase chain-reaction and DNA hybridization. *Journal of Clinical Microbiology*, 1993, **31**(7), 1777–1782.

27. S. J. Hamiltondutoit, G. Pallesen, M. B. Franzmann, J. Karkov, F. Black, P. Skinhoj, and C. Pedersen, Aids-related lymphoma - histopathology, immunophenotype, and association with epstein-barr-virus as demonstrated by insitu nucleic-acid hybridization. *American Journal of Pathology*, 1991, **138**(1), 149–163.

28. K. Senda, Y. Arakawa, K. Nakashima, H. Ito, S. Ichiyama, K. Shimokata, N. Kato, and M. Ohta, Multifocal outbreaks of metallo-beta-lactamase-producing pseudomonas aeruginosa resistant to broad-spectrum beta-lactams, including carbapenems. *Antimicrobial Agents and Chemotherapy*, 1996, **40**(2), 349–353.

29. F. J. Louws, D. W. Fulbright, C. T. Stephens, and F. J. Debruijn, Specific genomic fingerprints of phytopathogenic xanthomonas and pseudomonas pathovars and strains generated with repetitive sequences and PCR. *Applied and Environmental Microbiology*, 1994, **60**(7), 2286–2295.

30. S. Juretschko, G. Timmermann, M. Schmid, K. H. Schleifer, A. Pommerening-Roser, H. P. Koops, and M. Wagner, Combined molecular and conventional analyses of nitrifying bacterium diversity in activated sludge: nitrosococcus mobilis and nitrospira-like bacteria as dominant populations. *Applied and Environmental Microbiology*, 1998, **64**(8), 3042–3051.

31. Y. Wei, J. M. Lee, C. Richmond, F. R. Blattner, J. A. Rafalski, and R. A. LaRossa, High-density microarray-mediated gene expression profiling of escherichia coli. *Journal of Bacteriology*, 2001, **183**(2), 545–556.

32. L. D. Kuykendall, B. Saxena, T. E. Devine, and S. E. Udell, Genetic diversity in Bradyrhizobium-japonicum Jordan 1982 and a proposal for Bradyrhizobium-elkanii Sp-Nov. *Canadian Journal of Microbiology*, 1992, **38**(6), 501–505.

33. M. J. Heller, DNA microarray technology: devices, systems, and applications. *Annual Review of Biomedical Engineering*, 2002, **4**, 129–153.

34. A. C. Pease, D. Solas, E. J. Sullivan, M. T. Cronin, C. P. Holmes, and S. P. A. Fodor, Light-generated oligonucleotide arrays for rapid DNA-sequence analysis. *Proceedings of the National Academy of Sciences of the United States of America*, 1994, **91**(11), 5022–5026.

35. M. U. Kumke, G. Li, L. B. McGown, G. T. Walker, and C. P. Linn, Hybridization of fluorescein-labeled DNA oligomers detected by fluorescence anisotropy with protein-binding enhancement. *Analytical Chemistry*, 1995, **67**(21), 3945–3951.

36. V. C. Moore, M. S. Strano, E. H. Haroz, R. H. Hauge, R. E. Smalley, J. Schmidt, and Y. Talmon, Individually suspended single-walled carbon nanotubes in various surfactants. *Nano Letters*, 2003, **3**(10), 1379–1382.

37. M. Zheng, A. Jagota, M. S. Strano, A. P. Santos, P. Barone, S. G. Chou, B. A. Diner, M. S. Dresselhaus, R. S. McLean, G. B. Onoa, G. G. Samsonidze, E. D. Semke, M. Usrey, D. J. Walls, Structure-based carbon nanotube sorting by sequence-dependent DNA assembly. *Science*, 2003, **302**(5650), 1545–1548.

38. M. S. Strano, M. Zheng, A. Jagota, G. B. Onoa, D. A. Heller, P. W. Barone, and M. L. Usrey, Understanding the nature of the DNA-assisted separation of single-walled carbon nanotubes using fluorescence and Raman spectroscopy. *Nano Letters*, 2004, **4**(4), 543–550.

39. V. Perebeinos, J. Tersoff, and P. Avouris, Scaling of excitons in carbon nanotubes. *Physical Review Letters*, 2004, **92**(25), 257402.

40. J. G. Wetmur and N. Davidson, Kinetics of renaturation of DNA. *Journal of Molecular Biology*, 1968, **31**(3), 325–633.

41. D. Erickson, D. Q. Li, and U. J. Krull, Modeling of DNA hybridization kinetics for spatially resolved biochips. *Analytical Biochemistry*, 2003, **317**(2), 186–200.

42. F. M. Pohl, Salt-induced transition between 2 double-helical forms of oligo(Dc-Dg). *Cold Spring Harbor Symposia on Quantitative Biology*, 1982, **47**, 113–117.

43. T. M. Jovin, D. M. Soumpasis, and L. P. Mcintosh, The transition between B-DNA and Z-DNA. *Annual Review of Physical Chemistry*, 1987, **38**, 521–560.

44. A. Rich, The biology of left-handed Z-DNA. *Journal of Biological Chemistry*, 1996, **271**(20), 11595–11598.

45. M. Zheng, A. Jagota, E. D. Semke, B. A. Diner, R. S. Mclean, S. R. Lustig, R. E. Richardson, and N. G. Tassi, DNA-assisted dispersion and separation of carbon nanotubes. *Nature Materials*, 2003, **2**(5), 338–342.

46. J. Duguid, V. A. Bloomfield, J. Benevides, and G. J. Thomas, Raman spectral studies of nucleic-acids. 44. Raman-spectroscopy of DNA-metal complexes. 1. Interactions and conformational effects of the divalent-cations - Mg, Ca, Sr, Ba, Mn, Co, Ni, Cu, Pd, and Cd. *Biophysical Journal*, 1993, **65**(5), 1916–1928.

47. F. M. Pohl and T. M. Jovin, Salt-induced co-operative conformational change of a synthetic DNA: equilibrium and kinetic studies with poly(dG-dC). *Journal of Molecular Biology*, 1972, **67**, 375–396.

FURTHER READING

E. S. Jeng, P. W. Barone, J. D. Nelson, M. S. Strano, Hybridization kinetics and thermodynamics of DNA adsorbed to individually dispersed single walled carbon nanotubes. *Small*, 2007 (submitted).

Part Six

Array Technologies

Array Technologies

Isao Karube

School of Bionics, Tokyo University of Technology, Tokyo, Japan

Two score years have passed since my first paper on biosensors appeared in a scientific journal. Since then, the discoveries in the field of biochemistry and molecular biology have been breathtaking. As represented by the completion of the sequence of the human genome, the developments in biotechnology have provided a deeper understanding of our lives. As a result, we know not only what is occurring within whole cells but also when, where, and how the reaction occurs at the distinct location of the cells or tissues. These latest scientific findings brought us new developments in biotechnology and biosensor research that are now indispensable and versatile tools for various studies. Biosensors already provide many benefits to the public and are now vital for patients who have diabetes (e.g. glucose sensors), and medical testing (e.g. immune sensors). Today, biosensor research finds wide-ranging applications, and has become a part of the studies ranging from the detection of fundamental reaction to environment and high-tech health care.

Due to natural phenomena and the complexity of the biological process, introducing a one-to-one correspondence of only a single molecule became insufficient to cover a wide-range, panoramic view. The request for a simultaneous read-out of many different components contributed to the development of multiple-sensing tools. These demands were considered to be critical in the development of multifunctional biosensing devices. Recent developments in array technologies such as DNA arrays allowed remarkable progress in genomics. The incremental advances in medical science acknowledged the importance of proteomics. Therefore, the research has now focused on taking advantage of using semiconductor technologies, as it turned out, to perform several tasks at the same time on extremely minute scales. The array technology now offers views in a systematic way to survey the variation of proteins and has now lead to further development.

This section is designed to assimilate knowledge of existing biosensor array technologies of both academic and industrial interest that are accessible to scientists, engineers, students and newcomers to the subject. It begins with an introductory article on nucleic acid arrays (*see* **Nucleic Acid Arrays**). The technical foundations of array technology as well as recent research and development are discussed. With considerable effort and tightening of experimental conditions, the researchers developed an automated protein chip system along with the detection system. A fully-automated, two-dimensional electrophoresis (2DE) system for proteomic analysis system is described (*see* **Protein Chips and Detection Tools**). This 2DE system has one huge advantage compared with conventional laboratory work; it can be controlled by machines without the influence of humans (researchers). A decrease in the reproducible error is expected by using this system.

The specific, characterized proteins with different binding conditions on the solid surface allow simplification of the workflow and improve analytical sensitivity. The technology used in the array

Handbook of Biosensors and Biochips. Edited by Robert S. Marks, David C. Cullen, Isao Karube, Christopher R. Lowe and Howard H. Weetall.
© 2007 John Wiley & Sons, Ltd. ISBN 978-0-470-01905-4.

of mass spectrometry (MS) detection is described. The use of MS in the characteristic manner of proteins has an advantage in comparing particular proteins across samples (*see* **Surface-Enhanced Laser Desorption/Ionization (SELDI) Technology**). The optical fiber array (*see* **Fiber-Optic Array Biosensors**) and surface plasmon resonance array technologies (*see* **Surface Plasmon Resonance Array Devices**) have contributed to the advance of biosensor research. The array detection technology, by multiplexing assays, generates thousands of signals simultaneously, with a small dimension.

Electronic devices have been used for the electrochemical detection of toxicity in water (*see* **An Electrochemical Biochip Sensor for the Detection of Pollutants**), DNA and protein detection (*see* **Label-Free Gene and Protein Sensors Based on Electrochemical and Local Plasmon Resonance Devices**) in a label-free format. Many approaches have been reported in electrochemical analysis on biosensor research to monitor chemical and biological agents in nature. The chapters in this section focus on applying electrochemical analysis by integrating other technologies, such as micro-fabrication, carbon nanotubes, and optical probes, etc. The label-free detection of biomolecules has also been discussed by the use of the microcantilever array devices (*see* **Microcantilever Array Devices**). These devices are used to track the location where the surface stress changes. The advantage of this technique is its small size and use in identifying local information in small sample volumes.

New technical challenges for the artificial biochemical sniffers have been discussed for the environmental assessment and human odor analysis (*see* **Biosniffers (Gas-Phase Biosensors) as Artificial Olfaction**). This research development has a potential for integration with a portable instrument to be used in the biosensor sensor system, and expected to have potential for growth in new industry by the integration of array technology.

This section introduces and discusses leading-edge research and development of array technologies. Although the technologies described here are quite new, systems for the detection of chemical agents and biological molecules are in demand in medical treatment, the food industry, environmental science, etc. Because certain technologies described in this section have become established in only a few laboratories, the tools remain prohibitively expensive, owing to complicated handling procedures and costs. However, these problems are likely to be solved by social need, economy of scale, market competition, and rapid implementation of analysis. The new generation of array technologies is still in its infancy, but will meet demand in the near future.

To understand the complicated nature and biological processes, simultaneous and accurate detection by a sensor system is required. Therefore, instead of typical single-function biosensors, multiple-function sensor devices with small configuration and small sample volumes using array technology will form the next approach for biosensor research.

53

Nucleic Acid Arrays

Hirotaka Miyachi

School of Bionics, Tokyo University of Technology, Tokyo, Japan

1 INTRODUCTION

Over the past years, the completion of the genome project over a variety of organisms such as human,[1–3] mouse,[4,5] birds, fishes,[6] yeast, plants, fruit fly,[7] bacteria,[8] marsupial species,[9] and in the other species, has generated a large amount of genomic information. The sequencing data from such diverse organisms has led to a comprehensive functional analysis. These advantages facilitated the demands to understand biological mechanisms in a comprehensive genome complexity. Without a research tool known as *nucleic acid array* (also referred to as *DNA array, DNA microarray,* or *DNA chip*) a survey of a large number of genes may not have been possible.[10] DNA array offers tremendous advantages over traditional methods compared to the conventional single gene analysis methods such as northern hybridization[11,12] and reverse transcriptase-polymerase chain reaction (RT-PCR). These days, DNA array technology has become one of the principal technologies for the high-throughput analysis in biological studies, and is capable of determining gene expression patterns and DNA sequence information of thousands of genes in a single experiment.

This review is aimed to summarize the general aspects and to discuss the latest developments in the different application areas of nucleic acid array–based technology.

2 TECHNICAL FOUNDATIONS OF DNA ARRAY

The utilization of sequence complementarity of DNA duplex is the principal method in DNA array. Traditionally, labeled nucleic acid molecules were used to determine the location of a particular sequence of DNA within a complex mixture. The detection relies on a hybridization procedure called *Southern blotting.*[13] In the Southern blot procedure, DNA fragments digested by DNA restriction endonucleases are separated by electrophoresis, and denatured by soaking the gel in an alkaline solution. The fragments of single-stranded DNAs were then blotted onto a nylon membrane, reproducing the distribution of DNA fragments in the gel. This membrane is immersed in a solution containing a radioactively labeled DNA probe. By exposing X-ray film to the membrane, the results displayed on autoradiography, reveal the fragment to which the probe hybridizes. With the advent of recombinant technology, this technique is used for gene cloning and mapping, DNA fingerprinting, and restriction fragment length polymorphisms (RFLPs). The technique relies on membrane-based analysis introducing particular sequence correspondence between clones and hybridization signals.

In every newly sequenced genome, it was almost impossible to describe relationships to known genes. The researchers established approaches to hybridize mRNA to cDNA libraries spotted on nylon filters for the expression analysis. This

Handbook of Biosensors and Biochips. Edited by Robert S. Marks, David C. Cullen, Isao Karube, Christopher R. Lowe and Howard H. Weetall.
© 2007 John Wiley & Sons, Ltd. ISBN 978-0-470-01905-4.

analytical method seems to be more relevant to the form of the DNA array technology today, but modification of the immobilization method of probes is yet to be improved.

The technical foundation for the development of DNA array revealed after the use of the solid support as an immobilization matrix. The improvement has also been made to the probe immobilization procedure by the use of robotic equipment called spotter[14] or photolithographic techniques[15] adapted from the semiconductor technology.

The main distinction between DNA arrays and dot blots is the use of an impermeable solid substrate, such as glass or silicon. These types of substances have a number of advantages compared to the porous membrane used in dot blots, where the effect of the diffusion is negligible; because liquid solution cannot pass through the support, it results in direct access to the target DNA with probes. The flatness of the support is also advantageous in visualizing fluorescent microscopic images, improving reproducibility without the effect of interference of the support.[16]

As a result of the genome project, there has been an explosion in the amount of information available on the DNA sequence and a huge number of novel genes have been identified. Consequently, this feature greatly facilitated the study to organize and catalog the huge amount of DNA sequence information into the assignment of cellular functions to newly identified genes.

3   THE APPLICATION OF DNA ARRAY

Although the most common use of DNA arrays is gene expression profiling,[17] this analytical tool has been used for multiple applications including genotyping, large-scale sequencing, copy number analysis, novel gene identification, diagnostics,[18] and DNA–protein interactions. In expression profiling, DNA array is used to estimate the amount of mRNA transcribed across different tissues at the different cell stages,[19,20] response to drugs, and comparative expression research between normal and diseased states.[21] These studies are likely to help us to understand and explore the molecular physiological changes. The effect of the transcription and signal transduction by the various compounds are useful to define target molecules for

drug discovery.[22] Alternatively, it could help us to clarify the specificity of the effects of inhibitors in cells, organs, and tissues.

DNA array can also be used to read the particular sequence of a genome. SNP (single-nucleotide polymorphisms) DNA array is one of the examples used to identify genetic variation in individuals.[23] The genetic variations are thought to be responsible for the susceptibility to genetically caused diseases in individuals; therefore the rapid detection and discovery of genetic predisposition to disease using DNA arrays are of importance to health care and medical treatment for selecting drugs and effective treatments. The microarrays for the determination of SNPs are being used to profile somatic mutations in cancer. Alternatively, DNA arrays to resequence portions of the genome in individuals have also been developed.[24–27]

4   DNA ARRAY FABRICATION

In general, oligonucleotides can be deposited on nylon membranes,[28] controlled pore glass beads,[29] activated dextran,[30] avidin-coated polystyrene beads,[31] nitrocellulose,[32] polystyrene matrix,[33] acrylamide gel,[34] silicon, and glass.[35] DNA arrays are usually made using an impermeable, rigid substrate, such as glass microscope slide and silicon. Therefore, target nucleic acids can be accessed by the probe without the effect of diffusion into pores.

The oligonucleotides can be printed, spotted, or synthesized directly onto the support. DNA arrays can be classified into two groups depending on their fabrication method: (i) solid-phase synthesis, and (ii) by spotting presynthesized materials onto the solid substrates.

4.1   Solid-phase Synthesis

The first solid-phase synthesis utilizing photolithographic technology was applied to synthesize polypeptides. Photolithography, a method derived from semiconductor technology, possesses the ability of producing oligonucleotides with independent positions on the solid substrate.[36–39] Phosphoramidite chemistry, which can be activated by light irradiation at appropriate positions and respond to characteristic patterns of lithographic

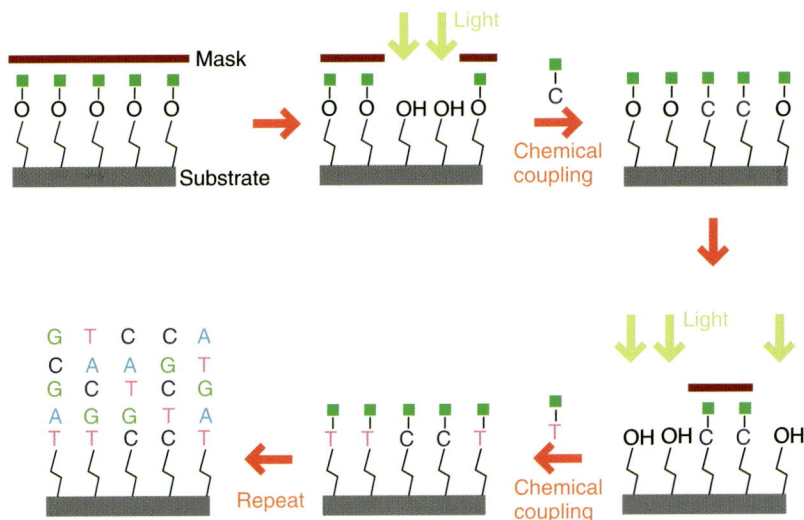

Figure 1. Photolithographic DNA synthesis.

masks used to either transmit or block light onto the surface, is used (Figure 1). Mechanical and optical instrumentation are adopted and only distinct areas are illuminated. The coupling occurs only where that has been deprotected. The repetition of the removal of the photoprotecting groups and the phosphoramidite coupling steps synthesize probe DNAs on appropriate locations on the substrate by immersing in a solution containing adenine, thymine, cytosine, or guanine (GeneChip® array technology by Affymetrix).

New array fabrication formats are being developed. The use of a mechanical system similar to computer overhead projection systems and a Digital Micromirror Device (DMD) similar to Texas Instruments' Digital Light Processor (DLP), have been reported for the replacement of the physical chromium masks used in Affymetrix array fabrication methods.[40,41] The Maskless Array Synthesizer (MAS) technology (NimbleGen) has flexibility in creating a custom array.

4.2 Immobilization of Presynthesized Oligonucleotides

An alternative method is to place DNA directly on a solid surface. DNA fragments from the known genes, polymerase chain reaction (PCR) products, and oligonucleotides can be placed by using robotic devices, known as *spotters*, that accurately deposit the DNA solution onto the support (Figure 2). Many thousands of spots are deposited in nanoliter quantities on a small surface area. The data generated from a single array experiment can mount up easily, where some DNA array experiments can contain up to tens of thousands of probe immobilized spots.

4.3 DNA Array Fabricated by Semiconductor Technologies

CombiMatrix has developed circuits containing arrays of individually addressable microelectrodes on the semiconductor device. Each microelectrode has the ability to selectively direct chemical reaction, facilitate the in situ synthesis of oligonucleotides onto each spot on the surface. The technique for the oligonucleotide synthesis reaction is called *virtual flask technology* where the reaction occurs only within the layer above each electrode.[42,43]

Using semiconductor technology, Nanogen has developed NanoChip® Electronic Microarray for the rapid movement and concentration of a negatively charged DNA and/or RNA molecule by applying a positive electric current to designated electrodes on the device. The technology is applicable not only to electronically addressing nucleic

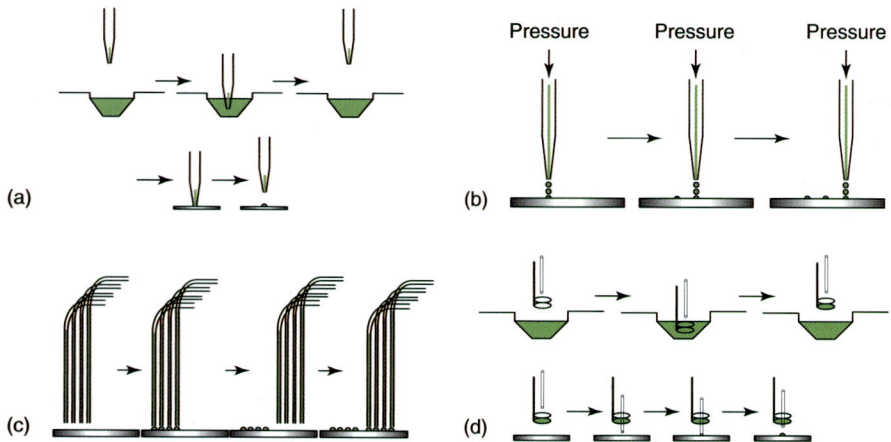

Figure 2.   Examples of the spotter. (a) Pin, (b) ink-jet, (c) capillary, and (d) pin and ring type.

acid solutions and hybridization but can also apply stringency to remove nonspecifically bounded target DNAs after the hybridization process.[44–46] The characteristic feature to concentrate and remove molecules on the device may offer advantages in efficiency and obtaining accurate results.

4.4   Other DNA Array Fabrication Technologies

Mitsubishi Rayon has created fibrous DNA array called *Genopal*™, developed with the slicing method (http://www.mrc.co.jp/genome/e/index. html). Oligonucleotides are immobilized in the hollow fiber and put together in a "block", and the blocks of DNA arrays are sliced. In this method, many DNA arrays can be created with unified specification, and are suited for the experiments for large quantities that are interested in the same sequences. Each oligonucleotide is immobilized in a 3D structured spot, said to be suitable for the effective hybridization and has high reproducibility, because equal quantities of the same sequences are thought to be immobilized in duplicate to each DNA array.

5   TARGET HYBRIDIZATION AND IMAGE ANALYSIS

A DNA array can be prepared from any known DNA sequences from any source, and positioned on a solid substrate. DNA array can be hybridized with other fluorescently labeled target nucleic acids. In general, mRNA (messenger RNA) samples are collected from cells at two different cell stages, types of cells, or tissue samples. Two target DNAs are made from the cDNA, labeled with nucleotides that fluorescence in the different colors for each sample.[47] After the addition of labeled cDNAs to a DNA array, cDNA hybridize to complementary probe sequences at appropriate temperatures and target DNA are washed to remove unhybridized target DNA.

After the hybridization step, DNA array will be placed in a confocal laser scanning microscope. The target hybridized spots are excited by the laser, while spots that do not hybridize to a target are not. The fluorescence intensities are detected with each fluorescent by analyzing the digital image of the DNA array. Background data for each spot are calculated to create the ratios of the fluorescence intensities for every spot. Each spot on a DNA array is associated with a particular gene. Therefore, by comparing the fluorescence intensities for each spots, we may conclude which cells express genes of interest more than the other. The location and intensity of a fluorescent will also provide information of whether the mutation is present in the genome used as a target DNA. The computer program creates a table of the ratios labeled in the different colors.[48] For example, when two fluorescent intensities were at nearly equal intensities, the computer may conclude that both cells express gene at the same level.

The analysis of DNA arrays poses a large number of statistical problems, including standardization of the data for every spot on the array and this cause difficulties in evaluating all the data on every single gene.

6 POLYMORPHISM ANALYSIS

DNA arrays are used to detect polymorphisms in a gene sequence. In the analysis, probe sequences immobilized on support differs from that of the other spots in the same DNA array, sometimes by only one or a few sequences. This type of the analysis is called *single-nucleotide polymorphisms*, or SNPs (pronounced "snips"); small genetic differences of approximately 1 bp (base pair) in every 1000 bp that can occur between human population.

The SNPs DNA array technology is used to survey a risk of developing particular diseases. The genomic DNA taken from an individual is hybridized to DNA array loaded with various labeled SNP sequences. The spots derived from the fluorescence scanning on the DNA array will show greater intensity when the target DNA hybridizes to a specific probe. SNP analysis can be used to test whether the individual may have or is at risk for developing a particular disease.

7 DNA ARRAY FABRICATED BY PLASMA-POLYMERIZATION TECHNOLOGY

Sequence-specific discrimination is the most important issue in DNA–DNA hybridization based gene analysis. Recent developments of DNA array technology have brought us a platform of already working daily routine systems and are widely applied to characterize and monitor the genome sequencing, expression analysis, disease diagnosis, and biological response to genotoxic contaminants. Although there are many promising new fields and techniques, some problems in selectivity still remain. To improve the specificity, one important aspect of the study will be the modification of the immobilization technique of the probe DNA on the solid support. Several technical issues

must be considered in investigating an immobilization technique including the chemical stability, the amenability to chemical modification, the surface area and loading capacity, and the degree of nonspecific binding.

A method to decrease nonspecific binding of a target DNA has been described by the use of hydrophobic properties of thin films deposited using plasma-polymerization (PP) technique.[49] Plasma processing is a standard industrial method for the deposition of thin polymeric films and modification of material surfaces (Figure 3). The films can be formed free of pinholes, will adhere strongly to a wide range of materials, and are highly resistant to chemical and physical treatments.[50,51] PP technique is attractive because it is easy to control film thickness, the procedure can be performed in a dry condition, and the surface properties of the film can be controlled easily by changing monomer gas.

A plasma-polymerized film (PPF) of hexamethyldisiloxane (HMDS: $(CH_3)_3SiOSi(CH_3)_3$) was used[52,53] to immobilize streptavidin on a glass substrate. Another layer of HMDS was additionally plasma-polymerized to the absorbed streptavidin on the substrate. As a result, the streptavidin was "embedded" between the two PPFs of HMDS, enabling them to capture biotinylated molecules such as end-primed biotinylated oligonucleotides (Figure 4). The second PPF-layer of approximately 30–45 Å of PPF was sufficient to capture biotinylated molecules, while thicknesses of more than 90 Å significantly hindered the

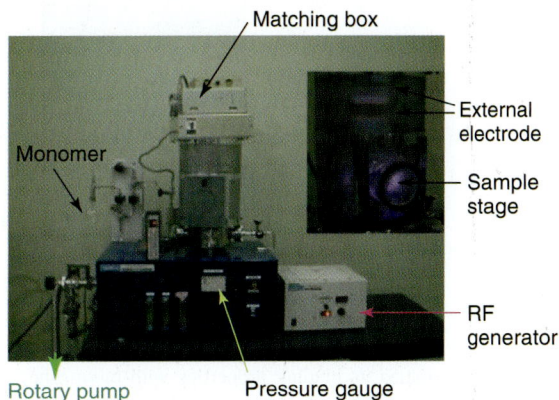

Figure 3. An overview of the plasma-polymerization apparatus used to deposit thin film.

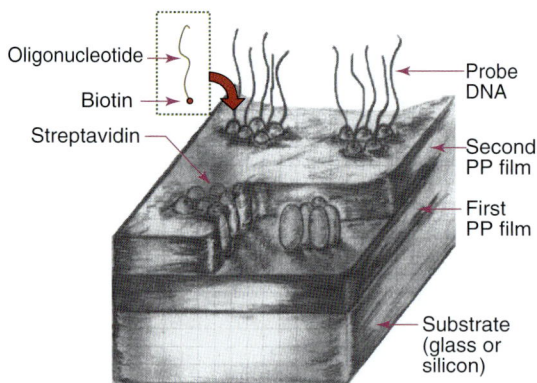

Figure 4. The illustration of the expected dimension of the immobilized protein.

streptavidin–biotin interactions (Figure 5). Fluorescence analysis revealed that the absence of either of the HMDS-PP layers, which is used for immobilizing streptavidin molecules, resulted in a decrease in biotin binding.[49] This immobilization technique was used to bind biotinylated oligonucleotides in sequence-specific DNA–DNA hybridization. The hydrophobic properties of the HMDS-PPF (Figure 6) decreased nonspecific DNA binding to the substrate compared with a conventional poly-L-lysine-coated array.[49] The method described here shows remarkable adaptability for the DNA hybridization analysis.

Antibodies, too, can be immobilized on the solid support using PP technology. Antibodies are spotted and the second PPF-layer of approximately 60 Å was sufficient to define presence or absence of proteins in a sample. Antibody array has been applied for the elucidation of interaction partners (Figure 7).

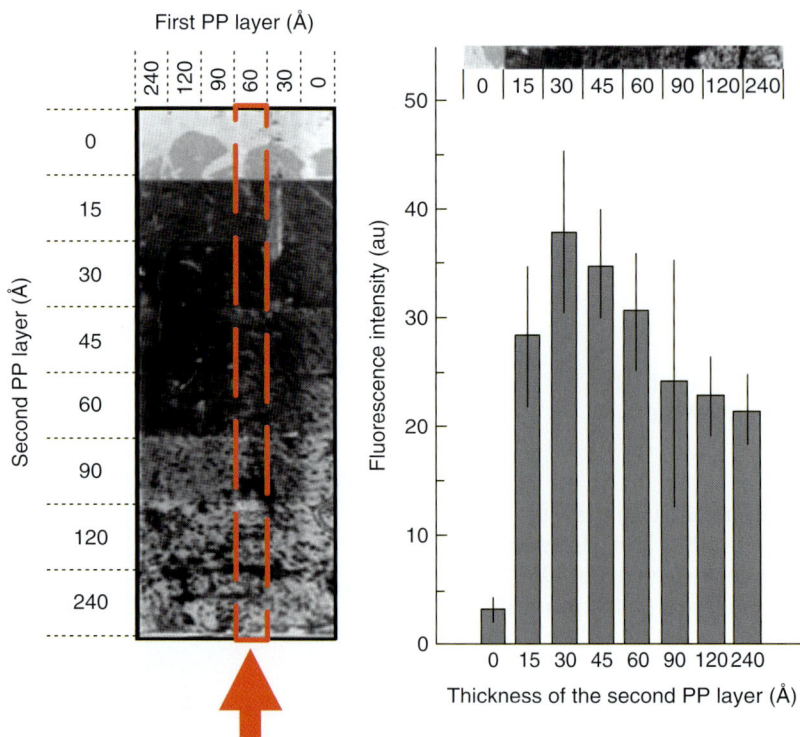

Figure 5. The result of the fluorescent intensities on the different thicknesses of the PPF. In order to examine whether immobilized streptavidin retained its binding properties with biotin, the streptavidin-immobilized support was immersed in fluorescein-labeled biotin solution and screened for fluorescence. The result showed that the highest fluorescence intensity is observed at 30 Å for the second plasma-polymerization layer. The fluorescence intensity dropped dramatically upon application of more than 30 Å. The result indicates that the PP layer of greater thickness have an adverse effect on streptavidin–biotin interactions. We assume that the streptavidin binding site was buried after deposition of second film greater than 30 Å.

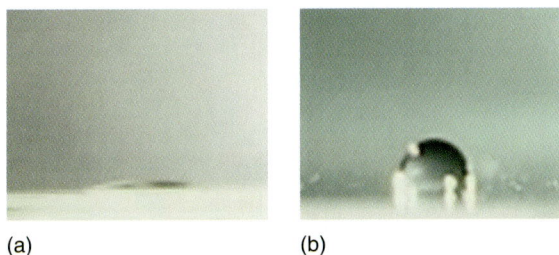

Figure 6. The hydrophobic properties of HMDS-PPF. A side view of 2 μl of water delivered (a) glass, (b) 6-nm of HMDS plasma-polymerized film.

8 THE EFFECT OF THE SURFACE PROPERTIES

In our previous work, PPF of HMDS was applied to devise an immobilization method as a decent DNA-immobilization matrix.[49] Compared to the conventional poly-lysine matrix, nonspecific binding of target DNAs onto the surface was reduced.[54] This may be attributed to the hydrophobicity of the HMDS-PP layer. Though probe-surrounding properties may influence DNA interactions, there have been only a few systematic studies of the factors affecting hybridization behavior.

To improve the overall efficiency of hybridization on the DNA array, we studied the hybridization behavior by changing the surface hydrophobicity and determined whether probe-surrounding properties are essential factors for the specific DNA hybridization on the DNA array.[55] This study was done by immobilizing probe DNAs on a hydrophobic and/or hydrophilic surface PPF based on the PP–DNA array fabrication method. The sequences of two polymorphic loci were used as a probe on the array and the effect on single-nucleotide mismatch detection has been studied with the unique structure of the target DNA.[56]

To investigate whether or how hybridization is affected by the surface properties, 5'-biotinylated oligonucleotides were immobilized by streptavidin–biotin binding on the surface-modified support. The probe sequences were derived from positions within the human SNPs site, *ApoE* gene.[57–61] Among three common ApoE isoforms; ApoE2 has Cys at amino acids 112 and 158; ApoE3, Cys at amino acid 112 and an Arg at 158; ApoE4, Arg at both positions. The *ApoE* genotypes were tested at the two different allele positions (at 112 and 158). Eight types of biotinylated oligonucleotides were immobilized, and hybridized with the labeled target complementary oligonucleotide.[55] The highest signal was obtained with the fully complementary oligonucleotides compared with those with the single-base mismatch or having unrelated sequences. The results also showed that the fluorescence signals obtained using the HMDS-PP (surface hydrophobic) array was higher than that from the acetonitrile–PP (surface hydrophilic) array (Figure 8).

A comparison of time dependency of hybridization between two surface properties was also examined. The results of the hybridization analysis show that the fluorescence signal increased rapidly when HMDS-PP (hydrophilic support surface) was

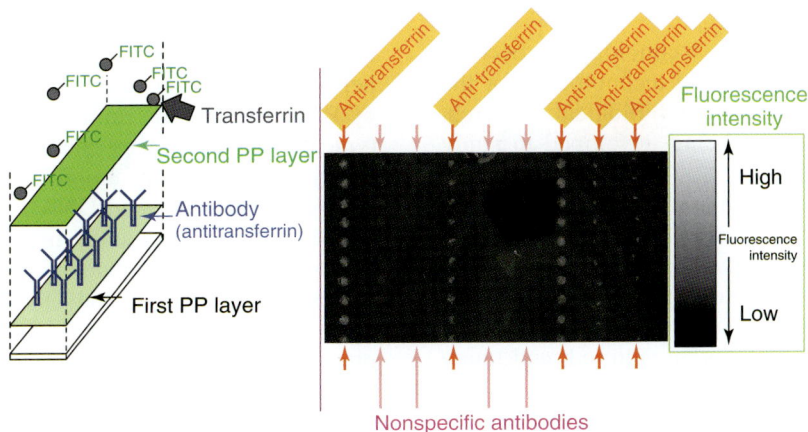

Figure 7. Antibody array fabricated by PP technology.

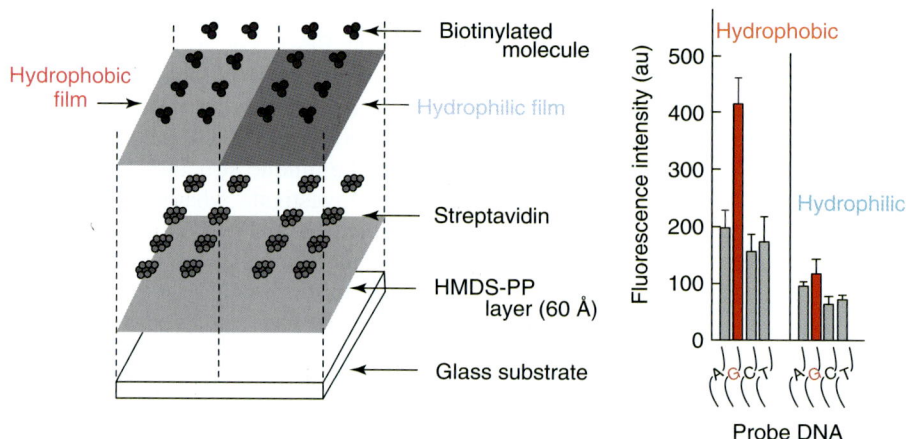

Figure 8. Hybridization experiment on hydrophilic and hydrophobic DNA array surfaces.

used as an immobilization matrix. It is also an important factor to carefully control the surface charge of the DNA array. The differences of the hybridization behavior may be because the probe DNAs were detached from the PP matrix and that contributed to an improvement in the signal-to-noise ratio.

These results suggest that the surface properties of DNA array are one of the crucial factors to obtain reliable results and significantly influence the environment in which the DNA hybridization takes place.

9 OUTLOOK OF DNA ARRAY TECHNOLOGY

Over the years, DNA array technology has become one of the principal platform technologies for the high-throughput analysis in biological research. New array formats are being developed to study biological functions.

Cell-based array is one of the platforms of the DNA array–based technique to elucidate the biological function of each gene function.[62,63] In cell-based array technology, plasmids or RNAs are spotted on the substrate and used to transfect cultured cells. Cells were cultured, and transfected with the localized probes. This technique allows the rapid generation of a desired phenotype, screening of specific proteins, and assays on transient phenotypes of living cells in real time

with limited adherent cells for easily transfectable cells.

A new platform has been proposed for the genomewide location analysis. ChIP-chip (also known as *ChIP-on-chip*) is a technique that joins DNA array (DNA chip) with chromatin immuno-precipitation (ChIP) technology. ChIP-chip is used to investigate interactions of the specific DNA-binding proteins (DBPs) within a particular region on the genomic DNA. In ChIP-chip assay, intact cells are fixed and stabilized with formaldehyde in vivo to preserve protein–DNA interactions. The cross-linked chromatin-associated proteins to DNA are then conjugated with specific antibodies associated with the protein of interest. Chromatin immunoprecipitation is performed by immunoprecipitating antibody–protein–DNA complexes. The DNA is eluted by the reversal of cross-links and the removal of proteins. DNA fragments are then performed blunting and ligation steps followed by PCR; because the immunoprecipitated DNA is typically present in very small quantities, it is necessary to perform an amplification step prior to subsequent labeling and DNA array analysis. The results of ChIP-chip assay indicate the functions of transcriptional regulators including promoters, repressors, enhancers, and DNA replication, and so on,[64–66] in the development and/or disease progression. Any DBP can be analyzed with this technique, without the knowledge of the potential binding sites and therefore the applications of ChIP-chip assay gives genomewide distribution of the binding sites of many DBPs, promoter regions, transcription factors, chromatin-associated proteins, DNA replication, recombination, histone modifications, and chromatin structure.

Fukumori et al. reported a rapid and efficient detection of DBPs using a single-stranded stem-loop structured DNA as a probe.[67] To demonstrate the validity of detecting DBPs on the solid support, Cro protein[68] derived from λ phage was used as a model DBP associated with two commonly used modification enzymes; exonuclease III and *Taq* DNA polymerase. By combining these two enzymes and by the use of fluorescently labeled deoxynucleotide as one of the substrates in the extension step, DBPs were to be detected by a counterpart of fluorescence incorporated signal visualized as a "spot" on DNA array. The assay depends on inhibition of *Exo* III reaction, the same principle known as *Exo* III stop assay[69–71] and strand elongation by a thermostable *Taq* DNA polymerase, commonly used in PCR. The unique feature of this method is that it uses a stem-loop structured probe, because fluorescently labeled dUTP is incorporated covalently to the digested probe DNA allowing extreme washing without any effect on the resulting fluorescence signals. Therefore, the errors caused by the nonspecific adsorption can be avoided. This reaction can also be applied to detect DBPs present in the nucleus. NF-κB[72] in the Jurkat cell extract has been detected in the same way. In this method, any DBPs can be detected within 30 min using a stem and loop ds-DNA array without labeling proteins or the use of the antibodies.

Obtaining an insight into entire genomes for the acquisition of complete and reliable information on cells, tissues, or the entire organisms, that previously would have taken years, has been shortened to a matter of weeks by the use of DNA array technology. A new method of applying other technologies with a DNA array is still in its infancy but growing rapidly due to the interest in gaining a wider understanding of the biological processes. These new developments may become a new platform for the high-throughput screening technology in the future.

REFERENCES

1. International Human Genome Sequencing Consortium, Finishing the euchromatic sequence of the human genome. *Nature*, 2004, **431**(7011), 931–945.

2. J. C. Venter, J. C. Venter, M. D. Adams, E. W. Myers, P. W. Li, R. J. Mural, G. G. Sutton, H. O. Smith, M. Yandell, C. A. Evans, R. A. Holt, J. D. Gocayne, P. Amanatides, R. M. Ballew, D. H. Huson, J. R. Wortman, Q. Zhang, C. D. Kodira, X. H. Zheng, L. Chen, M. Skupski, G. Subramanian, P. D. Thomas, J. Zhang, G. L. Gabor Miklos, C. Nelson, S. Broder, A. G. Clark, J. Nadeau, V. A. McKusick, N. Zinder, A. J. Levine, R. J. Roberts, M. Simon, C. Slayman, M. Hunkapiller, R. Bolanos, A. Delcher, I. Dew, D. Fasulo, M. Flanigan, L. Florea, A. Halpern, S. Hannenhalli, S. Kravitz, S. Levy, C. Mobarry, K. Reinert, K. Remington, J. Abu-Threideh, E. Beasley, K. Biddick, V. Bonazzi, R. Brandon, M. Cargill, I. Chandramouliswaran, R. Charlab, K. Chaturvedi, Z. Deng, V. Di Francesco, P. Dunn, K. Eilbeck, C. Evangelista, A. E. Gabrielian, W. Gan, W. Ge, F. Gong, Z. Gu, P. Guan, T. J. Heiman, M. E. Higgins, R. R. Ji, Z. Ke, K. A. Ketchum, Z. Lai, Y. Lei, Z. Li, J. Li, Y. Liang, X. Lin, F. Lu, G. V. Merkulov, N. Milshina, H. M. Moore, A. K. Naik, V. A. Narayan, B. Neelam, D. Nusskern, D. B. Rusch, S. Salzberg, W. Shao, B. Shue, J. Sun, Z. Wang, A. Wang, X. Wang, J. Wang, M. Wei, R. Wides, C. Xiao, C. Yan, A. Yao, J. Ye, M. Zhan, W. Zhang, H. Zhang, Q. Zhao, L. Zheng, F. Zhong, W. Zhong, S. Zhu, S. Zhao, D. Gilbert, S. Baumhueter, G. Spier, C. Carter, A. Cravchik, T. Woodage, F. Ali, H. An, A. Awe, D. Baldwin, H. Baden, M. Barnstead, I. Barrow, K. Beeson, D. Busam, A. Carver, A. Center, M. L. Cheng, L. Curry, S. Danaher, L. Davenport, R. Desilets, S. Dietz, K. Dodson, L. Doup, S. Ferriera, N. Garg, A. Gluecksmann, B. Hart, J. Haynes, C. Haynes, C. Heiner, S. Hladun, D. Hostin, J. Houck, T. Howland, C. Ibegwam, J. Johnson, F. Kalush, L. Kline, S. Koduru, A. Love, F. Mann, D. May, S. McCawley, T. McIntosh, I. McMullen, M. Moy, L. Moy, B. Murphy, K. Nelson, C. Pfannkoch, E. Pratts, V. Puri, H. Qureshi, M. Reardon, R. Rodriguez, Y. H. Rogers, D. Romblad, B. Ruhfel, R. Scott, C. Sitter, M. Smallwood, E. Stewart, R. Strong, E. Suh, R. Thomas, N. N. Tint, S. Tse, C. Vech, G. Wang, J. Wetter, S. Williams, M. Williams, S. Windsor, E. Winn-Deen, K. Wolfe, J. Zaveri, K. Zaveri, J. F. Abril, R. Guigó, M. J. Campbell, K. V. Sjolander, B. Karlak, A. Kejariwal, H. Mi, B. Lazareva, T. Hatton, A. Narechania, K. Diemer, A. Muruganujan, N. Guo, S. Sato, V. Bafna, S. Istrail, R. Lippert, R. Schwartz, B. Walenz, S. Yooseph, D. Allen, A. Basu, J. Baxendale, L. Blick, M. Caminha, J. Carnes-Stine, P. Caulk, Y. H. Chiang, M. Coyne, C. Dahlke, A. Mays, M. Dombroski, M. Donnelly, D. Ely, S. Esparham, C. Fosler, H. Gire, S. Glanowski, K. Glasser, A. Glodek, M. Gorokhov, K. Graham, B. Gropman, M. Harris, J. Heil, S. Henderson, J. Hoover, D. Jennings, C. Jordan, J. Jordan, J. Kasha, L. Kagan, C. Kraft, A. Levitsky, M. Lewis, X. Liu, J. Lopez, D. Ma, W. Majoros, J. McDaniel, S. Murphy, M. Newman, T. Nguyen, N. Nguyen, M. Nodell, S. Pan, J. Peck, M. Peterson, W. Rowe, R. Sanders, J. Scott, M. Simpson, T. Smith, A. Sprague, T. Stockwell, R. Turner, E. Venter, M. Wang, M. Wen, D. Wu, M. Wu, A. Xia, A. Zandieh, and X. Zhu, The sequence of the human genome. *Science*, 2001, **291**(5507), 1304–1351.

3. E. S. Lander, L. M. Linton, B. Birren, C. Nusbaum, M. C. Zody, J. Baldwin, K. Devon, K. Dewar, M. Doyle, W. FitzHugh, R. Funke, D. Gage, K. Harris, A. Heaford, J. Howland, L. Kann, J. Lehoczky, R. LeVine, P. McEwan, K. McKernan, J. Meldrim, J. P. Mesirov, C. Miranda, W. Morris, J. Naylor, C. Raymond, M. Rosetti, R. Santos, C. Sheridan, C. Sougnez, N. Stange-Thomann, N. Stojanovic, A. Subramanian, D. Wyman, J. Rogers, J. Sulston, R. Ainscough, S. Beck, D. Bentley, J. Burton, C. Clee, N. Carter, A. Coulson, R. Deadman, P. Deloukas, A. Dunham, I. Dunham, R. Durbin, L. French, D. Grafham, S. Gregory, T. Hubbard, S. Humphray, A. Hunt, M. Jones, C. Lloyd, A. McMurray, L. Matthews, S. Mercer, S. Milne, J. C. Mullikin, A. Mungall, R. Plumb, M. Ross, R. Shownkeen, S. Sims, R. H. Waterston, R. K. Wilson, L. W. Hillier, J. D. McPherson, M. A. Marra, E. R. Mardis, L. A. Fulton, A. T. Chinwalla, K. H. Pepin, W. R. Gish, S. L. Chissoe, M. C. Wendl, K. D. Delehaunty, T. L. Miner, A. Delehaunty, J. B. Kramer, L. L. Cook, R. S. Fulton, D. L. Johnson, P. J. Minx, S. W. Clifton, T. Hawkins, E. Branscomb, P. Predki, P. Richardson, S. Wenning, T. Slezak, N. Doggett, J. F. Cheng, A. Olsen, S. Lucas, C. Elkin, E. Uberbacher, M. Frazier, R. A. Gibbs, D. M. Muzny, S. E. Scherer, J. B. Bouck, E. J. Sodergren, K. C. Worley, C. M. Rives, J. H. Gorrell, M. L. Metzker, S. L. Naylor, R. S. Kucherlapati, D. L. Nelson, G. M. Weinstock, Y. Sakaki, A. Fujiyama, M. Hattori, T. Yada, A. Toyoda, T. Itoh, C. Kawagoe, H. Watanabe, Y. Totoki, T. Taylor, J. Weissenbach, R. Heilig, W. Saurin, F. Artiguenave, P. Brottier, T. Bruls, E. Pelletier, C. Robert, P. Wincker, D. R. Smith, L. Doucette-Stamm, M. Rubenfield, K. Weinstock, H. M. Lee, J. Dubois, A. Rosenthal, M. Platzer, G. Nyakatura, S. Taudien, A. Rump, H. Yang, J. Yu, J. Wang, G. Huang, J. Gu, L. Hood, L. Rowen, A. Madan, S. Qin, R. W. Davis, N. A. Federspiel, A. P. Abola, M. J. Proctor, R. M. Myers, J. Schmutz, M. Dickson, J. Grimwood, D. R. Cox, M. V. Olson, R. Kaul, C. Raymond, N. Shimizu, K. Kawasaki, S. Minoshima, G. A. Evans, M. Athanasiou, R. Schultz, B. A. Roe, F. Chen, H. Pan, J. Ramser, H. Lehrach, R. Reinhardt, W. R. McCombie, M. de la Bastide, N. Dedhia, H. Blöcker, K. Hornischer, G. Nordsiek, R. Agarwala, L. Aravind, J. A. Bailey, A. Bateman, S. Batzoglou, E. Birney, P. Bork, D. G. Brown, C. B. Burge, L. Cerutti, H. C. Chen, D. Church, M. Clamp, R. R. Copley, T. Doerks, S. R. Eddy, E. E. Eichler, T. S. Furey, J. Galagan, J. G. Gilbert, C. Harmon, Y. Hayashizaki, D. Haussler, H. Hermjakob, K. Hokamp, W. Jang, L. S. Johnson, T. A. Jones, S. Kasif, A. Kasprzyk, S. Kennedy, W. J. Kent, P. Kitts, E. V. Koonin, I. Korf, D. Kulp, D. Lancet, T. M. Lowe, A. McLysaght, T. Mikkelsen, J. V. Moran, N. Mulder, V. J. Pollara, C. P. Ponting, G. Schuler, J. Schultz, G. Slater, A. F. Smit, E. Stupka, J. Szustakowski, D. Thierry-Mieg, J. Thierry-Mieg, L. Wagner, J. Wallis, R. Wheeler, A. Williams, Y. I. Wolf, K. H. Wolfe, S. P. Yang, R. F. Yeh, F. Collins, M. S. Guyer, J. Peterson, A. Felsenfeld, K. A. Wetterstrand, A. Patrinos, M. J. Morgan, P. de Jong, J. J. Catanese, K. Osoegawa, H. Shizuya, S. Choi, Y. J. Chen, and J. Szustakowski, Initial sequencing and analysis of the human genome. *Nature*, 2001, **409**(6822), 860–921.

4. C. M. Wade, E. J. Kulbokas III, A. W. Kirby, M. C. Zody, J. C. Mullikin, E. S. Lander, K. Lindblad-Toh, and M. J. Daly, The mosaic structure of variation in the laboratory mouse genome. *Nature*, 2002, **420**(6915), 574–578.

5. R. H. Waterston, K. Lindblad-Toh, E. Birney, J. Rogers, J. F. Abril, P. Agarwal, R. Agarwala, R. Ainscough, M. Alexandersson, P. An, S. E. Antonarakis, J. Attwood, R. Baertsch, J. Bailey, K. Barlow, S. Beck, E. Berry, B. Birren, T. Bloom, P. Bork, M. Botcherby, N. Bray, M. R. Brent, D. G. Brown, S. D. Brown, C. Bult, J. Burton, J. Butler, R. D. Campbell, P. Carninci, S. Cawley, F. Chiaromonte, A. T. Chinwalla, D. M. Church, M. Clamp, C. Clee, F. S. Collins, L. L. Cook, R. R. Copley, A. Coulson, O. Couronne, J. Cuff, V. Curwen, T. Cutts, M. Daly, R. David, J. Davies, K. D. Delehaunty, J. Deri, E. T. Dermitzakis, C. Dewey, N. J. Dickens, M. Diekhans, S. Dodge, I. Dubchak, D. M. Dunn, S. R. Eddy, L. Elnitski, R. D. Emes, P. Eswara, E. Eyras, A. Felsenfeld, G. A. Fewell, P. Flicek, K. Foley, W. N. Frankel, L. A. Fulton, R. S. Fulton, T. S. Furey, D. Gage, R. A. Gibbs, G. Glusman, S. Gnerre, N. Goldman, L. Goodstadt, D. Grafham, T. A. Graves, E. D. Green, S. Gregory, R. Guigó, M. Guyer, R. C. Hardison, D. Haussler, Y. Hayashizaki, L. W. Hillier, A. Hinrichs, W. Hlavina, T. Holzer, F. Hsu, A. Hua, T. Hubbard, A. Hunt, I. Jackson, D. B. Jaffe, L. S. Johnson, M. Jones, T. A. Jones, A. Joy, M. Kamal, E. K. Karlsson, D. Karolchik, A. Kasprzyk, J. Kawai, E. Keibler, C. Kells, W. J. Kent, A. Kirby, D. L. Kolbe, I. Korf, R. S. Kucherlapati, E. J. Kulbokas, D. Kulp, T. Landers, J. P. Leger, S. Leonard, I. Letunic, R. Levine, J. Li, M. Li, C. Lloyd, S. Lucas, B. Ma, D. R. Maglott, E. R. Mardis, L. Matthews, E. Mauceli, J. H. Mayer, M. McCarthy, W. R. McCombie, S. McLaren, K. McLay, J. D. McPherson, J. Meldrim, B. Meredith, J. P. Mesirov, W. Miller, T. L. Miner, E. Mongin, K. T. Montgomery, M. Morgan, R. Mott, J. C. Mullikin, D. M. Muzny, W. E. Nash, J. O. Nelson, M. N. Nhan, R. Nicol, Z. Ning, C. Nusbaum, M. J. O'Connor, Y. Okazaki, K. Oliver, E. Overton-Larty, L. Pachter, G. Parra, K. H. Pepin, J. Peterson, P. Pevzner, R. Plumb, C. S. Pohl, A. Poliakov, T. C. Ponce, C. P. Ponting, S. Potter, M. Quail, A. Reymond, B. A. Roe, K. M. Roskin, E. M. Rubin, A. G. Rust, R. Santos, V. Sapojnikov, B. Schultz, J. Schultz, M. S. Schwartz, S. Schwartz, C. Scott, S. Seaman, S. Searle, T. Sharpe, A. Sheridan, R. Shownkeen, S. Sims, J. B. Singer, G. Slater, A. Smit, D. R. Smith, B. Spencer, A. Stabenau, N. Stange-Thomann, C. Sugnet, M. Suyama, G. Tesler, J. Thompson, D. Torrents, E. Trevaskis, J. Tromp, C. Ucla, A. Ureta-Vidal, J. P. Vinson, A. C. Von Niederhausern, C. M. Wade, M. Wall, R. J. Weber, R. B. Weiss, M. C. Wendl, A. P. West, K. Wetterstrand, R. Wheeler, S. Whelan, J. Wierzbowski, D. Willey, S. Williams, R. K. Wilson, E. Winter, K. C. Worley, D. Wyman, S. Yang, S. P. Yang, E. M. Zdobnov, M. C. Zody, and E. S. Lander, Initial sequencing and comparative analysis of the mouse genome. *Nature*, 2002, **420**(6915), 520–562.

6. S. Aparicio, J. Chapman, E. Stupka, N. Putnam, J.-M. Chia, P. Dehal, A. Christoffels, S. Rash, S. Hoon, A. Smit, M. D. S. Gelpke, J. Roach, T. Oh, I. Y. Ho, M. Wong, C. Detter, F. Verhoef, P. Predki, A. Tay, S. Lucas,

P. Richardson, S. F. Smith, M. S. Clark, Y. J. K. Edwards, N. Doggett, A. Zharkikh, S. V. Tavtigian, D. Pruss, M. Barnstead, C. Evans, H. Baden, J. Powell, G. Glusman, L. Rowen, L. Hood, Y. H. Tan, G. Elgar, T. Hawkins, B. Venkatesh, D. Rokhsar, and S. Brenner, Whole-genome shotgun assembly and analysis of the genome of Fugu rubripes. *Science*, 2002, **297**(5585), 1301–1310.

7. M. D. Adams, S. E. Celniker, R. A. Holt, C. A. Evans, J. D. Gocayne, P. G. Amanatides, S. E. Scherer, P. W. Li, R. A. Hoskins, R. F. Galle, R. A. George, S. E. Lewis, S. Richards, M. Ashburner, S. N. Henderson, G. G. Sutton, J. R. Wortman, M. D. Yandell, Q. Zhang, L. X. Chen, R. C. Brandon, Y. H. Rogers, R. G. Blazej, M. Champe, B. D. Pfeiffer, K. H. Wan, C. Doyle, E. G. Baxter, G. Helt, C. R. Nelson, G. L. Gabor, J. F. Abril, A. Agbayani, H. J. An, C. Andrews-Pfannkoch, D. Baldwin, R. M. Ballew, A. Basu, J. Baxendale, L. Bayraktaroglu, E. M. Beasley, K. Y. Beeson, P. V. Benos, B. P. Berman, D. Bhandari, S. Bolshakov, D. Borkova, M. R. Botchan, J. Bouck, P. Brokstein, P. Brottier, K. C. Burtis, D. A. Busam, H. Butler, E. Cadieu, A. Center, I. Chandra, J. M. Cherry, S. Cawley, C. Dahlke, L. B. Davenport, P. Davies, B. de Pablos, A. Delcher, Z. Deng, A. D. Mays, I. Dew, S. M. Dietz, K. Dodson, L. E. Doup, M. Downes, S. Dugan-Rocha, B. C. Dunkov, P. Dunn, K. J. Durbin, C. C. Evangelista, C. Ferraz, S. Ferriera, W. Fleischmann, C. Fosler, A. E. Gabrielian, N. S. Garg, W. M. Gelbart, K. Glasser, A. Glodek, F. Gong, J. H. Gorrell, Z. Gu, P. Guan, M. Harris, N. L. Harris, D. Harvey, T. J. Heiman, J. R. Hernandez, J. Houck, D. Hostin, K. A. Houston, T. J. Howland, M. H. Wei, C. Ibegwam, M. Jalali, F. Kalush, G. H. Karpen, Z. Ke, J. A. Kennison, K. A. Ketchum, B. E. Kimmel, C. D. Kodira, C. Kraft, S. Kravitz, D. Kulp, Z. Lai, P. Lasko, Y. Lei, A. A. Levitsky, J. Li, Z. Li, Y. Liang, X. Lin, X. Liu, B. Mattei, T. C. McIntosh, M. P. McLeod, D. McPherson, G. Merkulov, N. V. Milshina, C. Mobarry, J. Morris, A. Moshrefi, S. M. Mount, M. Moy, B. Murphy, L. Murphy, D. M. Muzny, D. L. Nelson, D. R. Nelson, K. A. Nelson, K. Nixon, D. R. Nusskern, J. M. Pacleb, M. Palazzolo, G. S. Pittman, S. Pan, J. Pollard, V. Puri, M. G. Reese, K. Reinert, K. Remington, R. D. Saunders, F. Scheeler, H. Shen, B. C. Shue, I. Sidén-Kiamos, M. Simpson, M. P. Skupski, T. Smith, E. Spier, A. C. Spradling, M. Stapleton, R. Strong, E. Sun, R. Svirskas, C. Tector, R. Turner, E. Venter, A. H. Wang, X. Wang, Z. Y. Wang, D. A. Wassarman, G. M. Weinstock, J. Weissenbach, S. M. Williams, T. Woodage, K. C. Worley, D. Wu, S. Yang, Q. A. Yao, J. Ye, R. F. Yeh, J. S. Zaveri, M. Zhan, G. Zhang, Q. Zhao, L. Zheng, X. H. Zheng, F. N. Zhong, W. Zhong, X. Zhou, S. Zhu, X. Zhu, H. O. Smith, R. A. Gibbs, E. W. Myers, G. M. Rubin, and J. C. Venter, The genome sequence of Drosophila melanogaster. *Science*, 2000, **287**(5461), 2185–2195.

8. F. R. Blattner, G. Plunkett, C. A. Bloch, N. T. Perna, V. Burland, M. Riley, J. Collado-Vides, J. D. Glasner, C. K. Rode, G. F. Mayhew, J. Gregor, N. W. Davis, H. A. Kirkpatrick, M. A. Goeden, D. J. Rose, B. Mau, and Y. Shao, The complete genome sequence of Escherichia coli K-12. *Science*, 1997, **277**(5331), 1453–1474.

9. T. S. Mikkelsen, M. J. Wakefield, B. Aken, C. T. Amemiya, J. L. Chang, S. Duke, M. Garber, A. J. Gentles, L. Goodstadt, A. Heger, J. Jurka, M. Kamal, E. Mauceli,

S. M. Searle, T. Sharpe, M. L. Baker, M. A. Batzer, P. V. Benos, K. Belov, M. Clamp, A. Cook, J. Cuff, R. Das, L. Davidow, J. E. Deakin, M. J. Fazzari, J. L. Glass, M. Grabherr, J. M. Greally, W. Gu, T. A. Hore, G. A. Huttley, M. Kleber, R. L. Jirtle, E. Koina, J. T. Lee, S. Mahony, M. A. Marra, R. D. Miller, R. D. Nicholls, M. Oda, A. T. Papenfuss, Z. E. Parra, D. D. Pollock, D. A. Ray, J. E. Schein, T. P. Speed, K. Thompson, J. L. VandeBerg, C. M. Wade, J. A. Walker, P. D. Waters, C. Webber, J. R. Weidman, X. Xie, M. C. Zody, Broad Institute Genome Sequencing Platform; Broad Institute Whole Genome Team Assembly, J. Baldwin, A. Abdouelleil, J. Abdulkadir, A. Abebe, B. Abera, J. Abreu, S. C. Acer, L. Aftuck, A. Alexander, P. An, E. Anderson, S. Anderson, H. Arachi, M. Azer, P. Bachantsang, A. Barry, T. Bayul, A. Berlin, D. Bessette, T. Bloom, T. Bloom, L. Boguslavskiy, C. Bonnet, B. Boukhgalter, I. Bourzgui, A. Brown, P. Cahill, S. Channer, Y. Cheshatsang, L. Chuda, M. Citroen, A. Collymore, P. Cooke, M. Costello, K. D'Aco, R. Daza, G. De Haan, S. DeGray, C. DeMaso, N. Dhargay, K. Dooley, E. Dooley, M. Doricent, P. Dorje, K. Dorjee, A. Dupes, R. Elong, J. Falk, A. Farina, S. Faro, D. Ferguson, S. Fisher, C. D. Foley, A. Franke, D. Friedrich, L. Gadbois, G. Gearin, C. R. Gearin, G. Giannoukos, T. Goode, J. Graham, E. Grandbois, S. Grewal, K. Gyaltsen, N. Hafez, B. Hagos, J. Hall, C. Henson, A. Hollinger, T. Honan, M. D. Huard, L. Hughes, B. Hurhula, M. E. Husby, A. Kamat, B. Kanga, S. Kashin, D. Khazanovich, P. Kisner, K. Lance, M. Lara, W. Lee, N. Lennon, F. Letendre, R. LeVine, A. Lipovsky, X. Liu, J. Liu, S. Liu, T. Lokyitsang, Y. Lokyitsang, R. Lubonja, A. Lui, P. MacDonald, V. Magnisalis, K. Maru, C. Matthews, W. McCusker, S. McDonough, T. Mehta, J. Meldrim, L. Meneus, O. Mihai, A. Mihalev, T. Mihova, R. Mittelman, V. Mlenga, A. Montmayeur, L. Mulrain, A. Navidi, J. Naylor, T. Negash, T. Nguyen, N. Nguyen, R. Nicol, C. Norbu, N. Norbu, N. Novod, B. O'Neill, S. Osman, E. Markiewicz, O. L. Oyono, C. Patti, P. Phunkhang, F. Pierre, M. Priest, S. Raghuraman, F. Rege, R. Reyes, C. Rise, P. Rogov, K. Ross, E. Ryan, S. Settipalli, T. Shea, N. Sherpa, L. Shi, D. Shih, T. Sparrow, J. Spaulding, J. Stalker, N. Stange-Thomann, S. Stavropoulos, C. Stone, C. Strader, S. Tesfaye, T. Thomson, Y. Thoulutsang, D. Thoulutsang, K. Topham, I. Topping, T. Tsamla, H. Vassiliev, A. Vo, T. Wangchuk, T. Wangdi, M. Weiand, J. Wilkinson, A. Wilson, S. Yadav, G. Young, Q. Yu, L. Zembek, D. Zhong, A. Zimmer, Z. Zwirko, Broad Institute Whole Genome Team Assembly, D. B. Jaffe, P. Alvarez, W. Brockman, J. Butler, C. Chin, S. Gnerre, I. MacCallum, J. A. Graves, C. P. Ponting, M. Breen, P. B. Samollow, E. S. Lander, and K. Lindblad-Toh, Genome of the marsupial Monodelphis domestica reveals innovation in non-coding sequences. *Nature*, 2007, **447**(7141), 167–177.

10. A. Marshall and J. Hodgson, DNA chips: an array of possibilities. *Nature Biotechnology*, 1998, **16**(1), 27–31.

11. J. C. Alwine, D. J. Kemp, and G. R. Stark, Method for detection of specific RNAs in agarose gels by transfer to diazobenzyloxymethyl-paper and hybridization with DNA probes. *Proceedings of the National Academy of Sciences of the United States of America*, 1977, **74**(12), 5350–5354.

12. J. C. Alwine, D. J. Kemp, B. A. Parker, J. Reiser, J. Renart, G. R. Stark, and G. M. Wahl, Detection of specific RNAs or specific fragments of DNA by fractionation in gels and transfer to diazobenzyloxymethyl paper. *Methods in Enzymology*, 1979, **68**, 220–242.

13. E. M. Southern, Detection of specific sequences among DNA fragments separated by gel electrophoresis. *Journal of Molecular Biology*, 1975, **98**(3), 503–517.

14. V. G. Cheung, M. Morley, F. Aguilar, A. Massimi, R. Kucherlapati, and G. Childs, Making and reading microarrays. *Nature Genetics*, 1999, **21**(Suppl. 1), 15–19.

15. R. J. Lipshutz, S. P. A. Fodor, T. R. Gingeras, and D. J. Lockhart, High density synthetic oligonucleotide arrays. *Nature Genetics*, 1999, **21**(Suppl. 1), 20–24.

16. E. Southern, K. Mir, and M. Shchepinov, Molecular interactions on microarrays. *Nature Genetics*, 1999, **21**(Suppl. 1), 5–9.

17. M. Schena, D. Shalon, R. W. Davis, and P. O. Brown, Quantitative monitoring of gene expression patterns with a complementary DNA microarray. *Science*, 1995, **270**(5235), 467–470.

18. J. G. Hacia, Resequencing and mutational analysis using oligonucleotide microarrays. *Nature Genetics*, 1999, **21**(Suppl. 1), 42–47.

19. V. R. Iyer, M. B. Eisen, D. T. Ross, G. Schuler, T. Moore, J. C. Lee, J. M. Trent, L. M. Staudt, J. Hudson, M. S. Boguski, D. Lashkari, D. Shalon, D. Botstein, and P. O. Brown, The transcriptional program in the response of human fibroblasts to serum. *Science*, 1999, **283**(5398), 83–87.

20. P. T. Spellman, G. Sherlock, M. Q. Zhang, V. R. Iyer, K. Anders, M. B. Eisen, P. O. Brown, D. Botstein, and B. Futcher, Comprehensive identification of cell cycle-regulated genes of the yeast Saccharomyces cerevisiae by microarray hybridization. *Molecular Biology of the Cell*, 1998, **9**(12), 3273–3297.

21. J. Marx, Medicine. DNA arrays reveal cancer in its many forms. *Science*, 2000, **289**(5485), 1670–1672.

22. C. Debouck and P. N. Goodfellow, DNA microarrays in drug discovery and development. *Nature Genetics*, 1999, **21**(Suppl. 1), 48–50.

23. A. Chakravarti, Population genetics–making sense out of sequence. *Nature Genetics*, 1999, **21**(Suppl. 1), 56–60.

24. Z. Strezoska, T. Paunesku, D. Radosavljević, I. Labat, R. Drmanac, and R. Crkvenjakov, DNA sequencing by hybridization: 100 bases read by a non-gel-based method. *Proceedings of the National Academy of Sciences of the United States of America*, 1991, **88**(22), 10089–10093.

25. N. E. Broude, T. Sano, C. L. Smith, and C. R. Cantor, Enhanced DNA sequencing by hybridization. *Proceedings of the National Academy of Sciences of the United States of America*, 1994, **91**(8), 3072–3076.

26. R. Drmanac, S. Drmanac, I. Labat, R. Crkvenjakov, A. Vicentic, and A. Gemmell, Sequencing by hybridization: towards an automated sequencing of one million M13 clones arrayed on membranes. *Electrophoresis*, 1992, **13**(8), 566–573.

27. R. Drmanac, I. Labat, I. Brukner, and R. Crkvenjakov, Sequencing of megabase plus DNA by hybridization: theory of the method. *Genomics*, 1989, **4**(2), 114–128.

28. J. Meinkoth and G. Wahl, Hybridization of nucleic acids immobilized on solid supports. *Analytical Biochemistry*, 1984, **138**(2), 267–284.

29. S. S. Ghosh and G. F. Musso, Covalent attachment of oligonucleotides to solid supports. *Nucleic Acids Research*, 1987, **15**, 13 5353–5372.

30. T. R. Gingeras, D. Y. Kwoh, and G. R. Davis, Hybridization properties of immobilized nucleic acids. *Nucleic Acids Research*, 1987, **15**(13), 5373–5390.

31. V. Lund, R. Schmid, D. Rickwood, and E. Hornes, Assessment of methods for covalent binding of nucleic acids to magnetic beads, Dynabeads, and the characteristics of the bound nucleic acids in hybridization reactions. *Nucleic Acids Research*, 1988, **16**(22), 10861–10880.

32. R. K. Saiki, P. S. Walsh, C. H. Levenson, and H. A. Erlich, Genetic analysis of amplified DNA with immobilized sequence-specific oligonucleotide probes. *Proceedings of the National Academy of Sciences of the United States of America*, 1989, **86**(16), 6230–6234.

33. S. R. Rasmussen, M. R. Larsen, and S. E. Rasmussen, Covalent immobilization of DNA onto polystyrene microwells: the molecules are only bound at the $5'$ end. *Analytical Biochemistry*, 1991, **198**(1), 138–142.

34. E. Fahy, G. R. Davis, L. J. DiMichele, and S. S. Ghosh, Design and synthesis of polyacrylamide-based oligonucleotide supports for use in nucleic acid diagnostics. *Nucleic Acids Research*, 1993, **21**(8), 1819–1826.

35. U. Maskos and E. M. Southern, Oligonucleotide hybridizations on glass supports: a novel linker for oligonucleotide synthesis and hybridization properties of oligonucleotides synthesised in situ. *Nucleic Acids Research*, 1992, **20**(7), 1679–1684.

36. S. P. Fodor, J. L. Read, M. C. Pirrung, L. Stryer, A. T. Lu, and D. Solas, Light-directed, spatially addressable parallel chemical synthesis. *Science*, 1991, **251**(4995), 767–773.

37. G. McGall, J. Labadie, P. Brock, G. Wallraff, T. Nguyen, and W. Hinsberg, Light-directed synthesis of high-density oligonucleotide arrays using semiconductor photoresists. *Proceedings of the National Academy of Sciences of the United States of America*, 1996, **93**(24), 13555–13560.

38. M. H. Caruthers, Gene synthesis machines: DNA chemistry and its uses. *Science*, 1985, **230**(4723), 281–285.

39. A. C. Pease, D. Solas, E. J. Sullivan, M. T. Cronin, C. P. Homes, and S. P. Fodor, Light-generated oligonucleotide arrays for rapid DNA sequence analysis. *Proceedings of the National Academy of Sciences of the United States of America*, 1994, **91**(11), 5022–5026.

40. E. F. Nuwaysir, W. Huang, T. J. Albert, J. Singh, K. Nuwaysir, A. Pitas, T. Richmond, T. Gorski, J. P. Berg, J. Ballin, M. Mccormick, J. Norton, T. Pollock, T. Sumwalt, L. Butcher, D. Porter, M. Molla, C. Hall, F. Blattner, M. R. Sussman, R. L. Wallace, F. Cerrina, and R. D. Green, Gene expression analysis using oligonucleotide arrays produced

by maskless photolithography. *Genome Research*, 2002, **12**(11), 1749–1755.

41. S. Singh-Gasson, R. D. Green, Y. J. Yue, C. Nelson, F. Blattner, M. R. Sussman, and F. Cerrina, Maskless fabrication of light-directed oligonucleotide microarrays using a digital micromirror array. *Nature Biotechnology*, 1999, **17**(10), 974–978.

42. A. L. Ghindilis, M. W. Smith, K. R. Schwarzkopf, K. M. Roth, K. Peyvan, S. B. Munro, M. J. Lodes, A. G. Stover, K. Bernards, K. Dill, and A. McShea, CombiMatrix oligonucleotide arrays: genotyping and gene expression assays employing electrochemical detection. *Biosensors and Bioelectronics*, 2007, **22**(9–10), 1853–1860.

43. R. H. Liu, T. Nugyen, K. Schwarzkopf, H. S. Fuji, A. Petrova, T. Siuda, K. Peyvan, M. Bizak, D. Danley, and A. McShea, Fully integrated miniature device for automated gene expression DNA microarray processing. *Analytical Chemistry*, 2006, **78**(6), 1980–1986.

44. C. F. Edman, D. E. Raymond, D. J. Wu, E. Tu, R. G. Sosnowski, W. F. Butler, M. Nerenberg, and M. J. Heller, Electric field directed nucleic acid hybridization on microchips. *Nucleic Acids Research*, 1997, **25**(24), 4907–4914.

45. P. N. Gilles, D. J. Wu, C. B. Foster, P. J. Dillon, and S. J. Chanock, Single nucleotide polymorphic discrimination by an electronic dot blot assay on semiconductor microchips. *Nature Biotechnology*, 1999, **17**(4), 365–370.

46. M. J. Heller, A. H. Forster, and E. Tu, Active microelectronic chip devices which utilize controlled electrophoretic fields for multiplex DNA hybridization and other genomic applications. *Electrophoresis*, 2000, **21**(1), 157–164.

47. P. O. Brown and D. Botstein, Exploring the new world of the genome with DNA microarrays. *Nature Genetics*, 1999, **21**(Suppl. 1), 33–37.

48. D. E. Bassett Jr, M. B. Eisen, and M. S. Boguski, Gene expression informatics–it's all in your mine. *Nature Genetics*, 1999, **21**(Suppl. 1), 51–55.

49. H. Miyachi, A. Hiratsuka, K. Ikebukuro, K. Yano, H. Muguruma, and I. Karube, Application of polymer-embedded proteins to fabrication of DNA array. *Biotechnology and Bioengineering*, 2000, **69**(3), 323–329.

50. F. F. Shi, Recent advances in polymer thin films prepared by plasma polymerization Synthesis, structural characterization, properties and applications. *Surface and Coatings Technology*, 1996, **82**, 1–15.

51. S. Kurosawa, N. Kamo, D. Matsui, and Y. Kobatake, Gas sorption to plasma-polymerized copper phthalocyanine film formed on a piezoelectric crystal. *Analytical Chemistry*, 1990, **62**, 353–359.

52. G. Akovali, D. G. Mamedov, and Z. M. O. Rzaev, Plasma surface modification of polyethylene with organosilicon and organotin monomers. *European Polymer Journal*, 1996, **32**, 375–383.

53. D. J. Li, F. Z. Cui, H. Q. Gu, and J. Zhao, Ion-beam-induced biomedical behavior of hexamethyldisiloxane films on deposited polyetherurethanes. *Applied Surface Science*, 1998, **126**, 1–10.

54. H. Miyachi, G.-T. Nomizo, A. Hiratsuka, K. Ikebukuro, H. Muguruma, and I. Karube *Fabrication of DNA Sensing Array Using Plasma Polymerization*, The Sixth World Congress on Biosensors. San Diego, USA, 24–26 May 2000, ELSEVIER.

55. H. Miyachi, K. Ikebukuro, K. Yano, H. Aburatani, and I. Karube, Single nucleotide polymorphism typing on DNA array with hydrophobic surface fabricated by plasma-polymerization technique. *Biosensors and Bioelectronics*, 2004, **20**(2), 184–189.

56. H. Miyachi, K. Yano, K. Ikebukuro, M. Kono, S. Hoshina, and I. Karube, Application of chimeric RNA-DNA oligonucleotides to the detection of pathogenic microorganisms using surface plasmon resonance. *Analytica Chimica Acta*, 2000, **407**(1–2), 1–10.

57. R. W. Mahley, Apolipoprotein E: cholesterol transport protein with expanding role in cell biology. *Science*, 1988, **240**(4852), 622–630.

58. E. H. Corder, A. M. Saunders, W. J. Strittmatter, D. E. Schmechel, P. C. Gaskell, G. W. Small, A. D. Roses, J. L. Haines, and M. A. Pericak-Vance, Gene dose of apolipoprotein E type 4 allele and the risk of Alzheimer's disease in late onset families. *Science*, 1993, **261**(5123), 921–923.

59. Y. K. Paik, D. J. Chang, C. A. Reardon, G. E. Davies, R. W. Mahley, and J. M. Taylor, Nucleotide sequence and structure of the human apolipoprotein E gene. *Proceedings of the National Academy of Sciences of the United States of America*, 1985, **82**(10), 3445–3449.

60. A. D. Roses, Alzheimer diseases: a model of gene mutations and susceptibility polymorphisms for complex psychiatric diseases. *American Journal of Medical Genetics*, 1998, **81**(1), 49–57.

61. A. M. Saunders, W. J. Strittmatter, D. Schmechel, P. H. St George-Hyslop, M. A. Pericak-Vance, and S. H. Joo, Association of apolipoprotein E allele epsilon 4 with late-onset familial and sporadic Alzheimer's disease. *Neurology*, 1993, **43**(8), 1467–1472.

62. J. Ziauddin and D. M. Sabatini, Microarrays of cells expressing defined cDNAs. *Nature*, 2001, **411**(6833), 107–110.

63. D. Vanhecke and M. Janitz, High-throughput gene silencing using cell arrays. *Oncogene*, 2004, **23**(51), 8353–8358.

64. J. D. Lieb, X. Liu, D. Botstein, and P. O. Brown, Promoter-specific binding of Rap1 revealed by genome-wide maps of protein-DNA association. *Nature Genetics*, 2001, **28**(4), 327–334.

65. B. Ren, F. Robert, J. J. Wyrick, O. Aparicio, E. G. Jennings, I. Simon, J. Zeitlinger, J. Schreiber, N. Hannett, E. Kanin, T. L. Volkert, C. J. Wilson, S. P. Bell, and R. A. Young, Genome-wide location and function of DNA binding proteins. *Science*, 2000, **290**(5500), 2306–2309.

66. V. R. Iyer, C. E. Horak, C. S. Scafe, D. Botstein, M. Snyder, and P. O. Brown, Genomic binding sites of the yeast cell-cycle transcription factors SBF and MBF. *Nature*, 2001, **409**(6819), 533–538.

67. T. Fukumori, H. Miyachi, and K. Yokoyama, Exo-Taq-based detection of DNA-binding protein for homogeneous and microarray format. *Journal of Biochemistry (Tokyo)*, 2005, **138**(4), 473–478.

68. W. F. Anderson, D. H. Ohlendorf, Y. Takeda, and B. W. Matthews, Structure of the cro repressor from

bacteriophage lambda and its interaction with DNA. *Nature*, 1981, **290**(5809), 754–758.

69. C. Wu, An exonuclease protection assay reveals heat-shock element and TATA box DNA-binding proteins in crude nuclear extracts. *Nature*, 1985, **317**(6032), 84–87.

70. B. Han, H. Hamana, and T. Shinozawa, Sensitive detection of the binding of E2F to its promoter by exonuclease III- and BssHII-protection PCR assays. *Journal of Biochemical and Biophysical Methods*, 1999, **39**(1–2), 85–92.

71. J. K. Wang, T. X. Li, Y. F. Bai, and Z. H. Lu, Evaluating the binding affinities of NF-kappaB p50 homodimer to the wild-type and single-nucleotide mutant Ig-kappaB sites by the unimolecular dsDNA microarray. *Analytical Biochemistry*, 2003, **316**(2), 192–201.

72. F. Chen, V. Castranova, X. Shi, and L. M. Demers, New insights into the role of nuclear factor-kappaB, a ubiquitous transcription factor in the initiation of diseases. *Clinical Chemistry*, 1999, **45**(1), 7–17.

54

Protein Chips and Detection Tools

Kenji Yokoyama, Atsunori Hiratsuka, Hideki Kinoshita, Keisuke Usui and Yoshio Suzuki

Research Center of Advanced Bionics, National Institute of Advanced Industrial Science and Technology (AIST), Tsukuba, Japan

1 INTRODUCTION

Proteome research comprehensively analyzing proteins occupies a special place in the methods of analyzing vital phenomena incapable of being elucidated by genome and transcriptome analyses. It has been a long time since the word *proteome* began to prevail. At the same time, substantial progress in protein analysis hardly seems real. One of the factors impeding the progress of proteome analysis is delay in the development of analytical instruments and techniques. Two-dimensional (2D) electrophoresis[1,2] and mass spectrometry (MS) as conventional proteome analysis methods come readily to mind—proteins are separated by 2D electrophoresis and identified with a mass spectrometer. The process of conventional 2D electrophoretic analysis, however, is yet to be automated. So far, only fully trained personnel have been able to obtain reproducible results. Even with a small gel, the electrophoresis process takes 20 h or longer; for a large gel, it takes about 3 days, thereby lowering the throughput extremely.

Another method of proteome analysis is liquid chromatography (LC)/MS shotgun protein analysis,[3] with the combined use of LC and MS. In this method, protein samples are analyzed according to the following steps: (i) the samples are digested with trypsin and (ii) the resultant peptides are separated by LC and then subjected to MS to determine each of the peptides. The advantages of this method are that the analysis of huge proteins, including membrane proteins, is convenient because of digestion performed before separation, and the analytical process can be easily automated; however, a number of peptide fragments produced from a protein are repeatedly analyzed, thereby producing low efficiency. This method is not suitable for quantitative analysis of posttranslational modification. Furthermore, isoelectric point and molecular weight, which have been conventionally used by researchers, are not available as well as the enormous database for 2D electrophoresis. Thus, 2D electrophoresis is a useful method and has many advantages in proteome analysis.

Our research group is currently developing a system allowing fully automated 2D electrophoresis as well as the completion of analysis and detection within 1 h. In this chapter, we introduce the protein microarray typical of a chip for proteome analysis and the high-throughput fully automated 2D electrophoresis system. A novel fluorescent dye for staining proteins in a 2D electrophoresis gel is also described.

2 PROTEIN MICROARRAY

A DNA microarray, on which a lot of DNAs used as probes are arrayed on a solid substrate, is used for identifying DNA or RNA. A protein

Handbook of Biosensors and Biochips. Edited by Robert S. Marks, David C. Cullen, Isao Karube, Christopher R. Lowe and Howard H. Weetall.
© 2007 John Wiley & Sons, Ltd. ISBN 978-0-470-01905-4.

microarray is used for arraying proteins in the place of DNAs. Proteins are typically analyzed, in some cases, nucleic acids, sugars, and low-molecular-weight compounds are also used. Like the DNA microarray, the protein microarray has been attracting attention because a high through-put of comprehensive biochemical analyses has become necessary.

The typical preparation process and the use of a protein microarray are as follows: (i) a target protein is spotted on a glass slide or silicon substrate for immobilization to prepare a protein microarray; (ii) then, a sample protein solution labeled with a fluorescent dye is contacted on the substrate surface, reacting within a given time to bind a sample protein matching the capturing protein (in the case of proteins to be measured). Quantitative determination of the protein or its affinity with a capturing protein can be studied by visualizing the binding behavior with a fluorometer with a positional resolution, that is fluorescence microscope or fluorescence image scanner, or by measuring the fluorescence intensities. If it is difficult to label a sample protein with a fluorescent dye, fluorescent detection can be achieved by combining the binding sample protein with the corresponding labeled antibody.

Recently, substrates coated with a three-dimensional gel are often used in place of a simple glass substrate: the surface of the glass substrate is coated with cross-linked hydrophilic polymers having functional groups forming covalent linkages with amino acids including N-hydroxy-succinimide esters. This substrate on which the capturing protein is immobilized has the advantage of ensuring the stability of the analysis. As the capturing protein is immobilized in gel, the reaction is similar in conditions to an aqueous solution system. In addition, capturing proteins are immobilized three dimensionally; a number of proteins can be immobilized, thereby enabling highly sensitive detection of the target substances. Glass substrates coated with three-dimensional gel are commercially available.

Unlike DNA microarrays offering stable immobilization of probes, the protein microarrays in general have a problem with stability because of the proteins immobilized on the substrate. Thus, the protein microarray is partially marketed, although its market scale is smaller than that of the DNA microarray. To solve the problem on stability, in vitro probe protein synthesis on each spot was proposed.[4]

3 AUTOMATED 2D ELECTROPHORESIS SYSTEM

3.1 On-chip 2D Electrophoresis System

The protein microarray is of advantage exclusively in identifying known proteins; the conventional 2D electrophoresis method is suitable for identifying unknown proteins or evaluating protein processing or post-translational modification. As mentioned above, however, the 2D electrophoretic analysis of proteins requires considerable time as well as a high level of skill. Therefore, we are aiming at developing a system that allows shorter running time for the 2D electrophoretic analysis, with simple operation. Initially, our research group designed a system for performing first-dimensional isoelectric focusing (IEF) and second-dimensional sodium dodecyl sulfate-polyacrylamide gel electrophoresis (SDS-PAGE) on a common substrate: a 2D electrophoresis chip integrated with IEF and SDS-PAGE has been developed.[5]

A schematic illustration of the chip is shown in Figure 1. Specific features of the structures were fabricated into the chip. Details of the chip layout are as follows. The chip was constructed of three different parts, a polymer plate, a glass cover, and a solution inlet. An immobilized pH gradient (IPG) gel strip and a polyacrylamide gel were fabricated as the first and second separation regions, respectively. The two regions were arranged perpendicular to each other and spatially separated by a junction structure. Grooves on both sides of the chip were fabricated as solution reservoirs for second-dimension electrophoresis. A pair of electrodes was connected via the reservoirs.

The solution inlet was fabricated and equipped with an IPG gel strip. The details of the structures are illustrated in Figure 2. The solution inlet comprised a cover plate, a slit plate, an IPG gel strip, and a pair of electrodes. The slit plate covered the junction structure. A capillary structure was formed by the cover plate, slit plate, and IPG gel strip. The solutions were loaded into the IPG gel by capillary action. The cathode and anode were

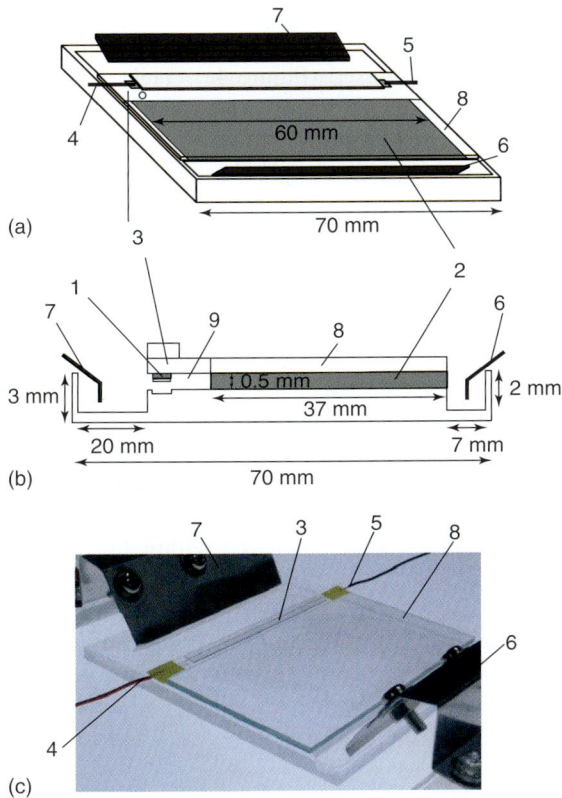

Figure 1. Chip layout. (a) Perspective view, (b) cross section, and (c) photograph. 1: IPG gel strip; 2: polyacrylamide gel; 3: solution inlet; 4, 5, 6, 7: electrodes; 8: glass cover; 9: junction structure.

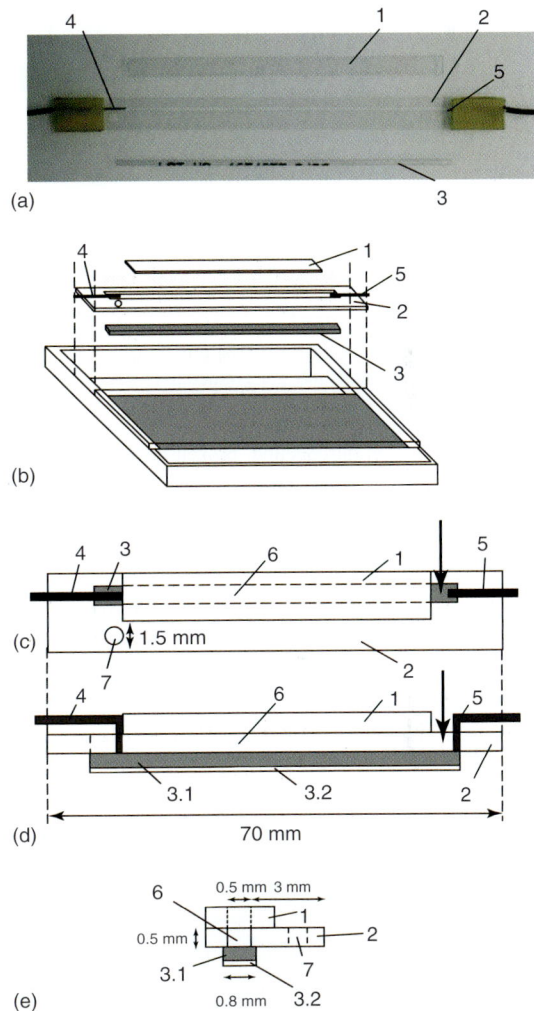

Figure 2. Structure of solution inlet. (a) Photograph of decomposed structure, (b) schematic drawing of the structure, (c) overhead view, (d) cross-section front view, (e) cross-section side view. Arrow: sample solution inlet; 1: cover plate; 2: slit plate; 3: IPG gel strip (3.1: IPG gel, 3.2: backing plate); 4: anode; 5: cathode; 6: capillary structure; 7: agarose inlet.

fabricated on the slit plate and contacted to both sides of the IPG gel. The entrance of the agarose solution formed on top of the junction structure.

The electrophoresis procedure for the solution inlet and the junction structure is illustrated in Figure 3. Initially, a sample solution was loaded into the dry IPG gel strip from the opening of the capillary. The capillary structure on the IPG gel prevented leakages of the solution. The IPG gel was rehydrated and expanded downward. The backing plate was fitted on a gutter, which was fabricated on the chip substrate. The IEF was then performed. After performing the IEF, the agarose solution was introduced from the entrance into the junction structure. Thus, the IPG and polyacrylamide gels were connected. Subsequently, the samples were transferred from the IPG gel to the polyacrylamide gel for the second separation.

The results of 2D electrophoresis are shown in Figure 4. Before performing the SDS-PAGE, the protein samples were observed in the IPG gel region (Figure 4a). When the SDS-PAGE was performed, the protein bands moved from the IPG gel region to the junction region within 3 min (Figure 4b). After 6 min, the band was getting narrow in the junction region (Figure 4c). These results showed that the agarose gel region played a role on a stacking gel concentrating proteins. After

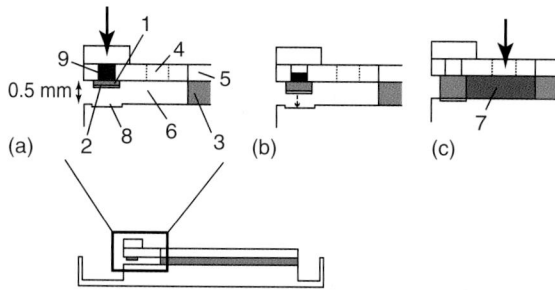

Figure 3. Cross section of solution inlet and junction structure. (a) Sample introduction, (b) rehydration and IEF, (c) agarose introduction and gel connection. The entire chip structure is shown at the bottom. The thick line is around the solution inlet and the junction structure. Solid arrow: entrance for solution introduction; broken arrow: direction of gel inflation; 1: IPG gel; 2: backing plate; 3: polyacrylamide gel; 4: agarose inlet; 5: glass cover; 6: junction structure; 7: agarose; 8: gutter; 9: sample solution.

the protein bands reached the SDS-polyacrylamide gel, the second separation was started along the direction of the potential (Figure 4d). When the lowest-molecular-weight protein reached the terminus of the gel, four kinds of proteins were separated as a diagonal line across the second-dimensional gel; as the pI increased, the molecular weight decreased (Figure 4e). This separation result was identical to the result using the conventional 2D gel separation method. The precedent smear band in the basic area was Cy5 bound with lysine (Figure 4c,d). This material was produced during the process of preparing the Cy5-labeled proteins. These results showed that the junction structure efficiently transferred the focused samples to the second-dimension gel without fractionation, and the samples were separated depending on their molecular weights by SDS-PAGE. These techniques including the function of the solution inlet, junction structure, miniaturized gel, and the simultaneous detection system achieved a 2D gel electrophoresis in a chip system. In addition, all the procedures were employed within 1 h.

Figure 5 shows the results of the 2D separation of the extracted proteins. These results were achieved using the chip and commercially available minigel system, respectively. The proteins were separated and the protein spots were visualized over the entire gel region. Figure 5(a,b)

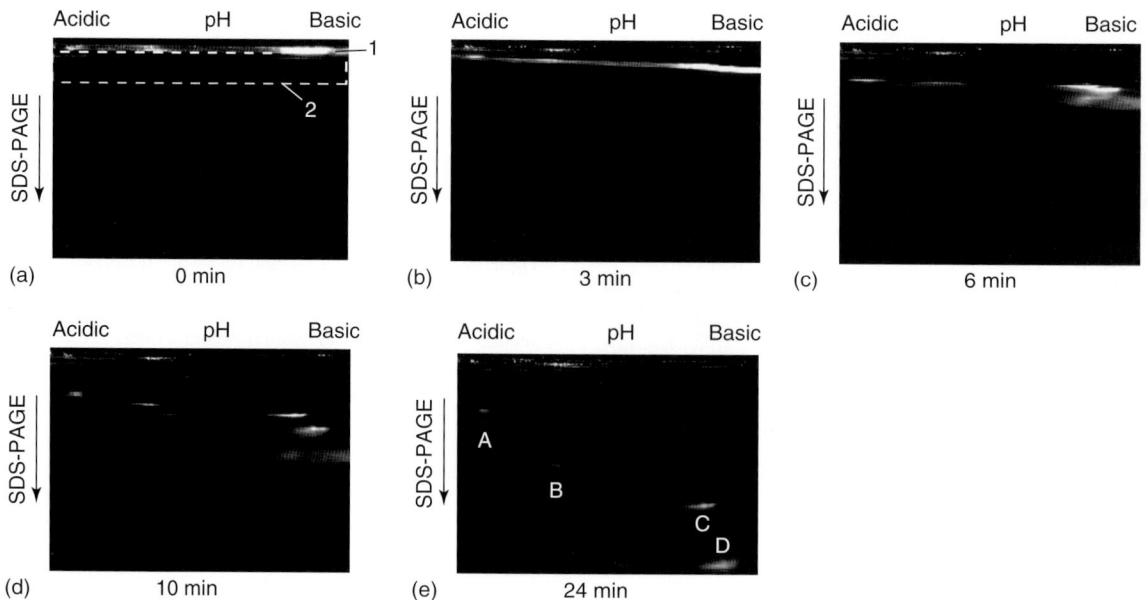

Figure 4. Continuous images of the second separation results of 2D protein markers. 2D protein markers (2 μg) labeled with Cy5 were applied to the chip. The images were continuously taken by a CCD camera while performing the SDS-PAGE. Images were taken according to the time course. The times when taking the images are shown below the images. (A) *Aspergillus niger* amyloglucosidase; (B) ovalbumin; (C) human erythrocytes carbonic anhydrase; (D) horse heart myoglobin. 1: IPG gel; 2: junction structure.

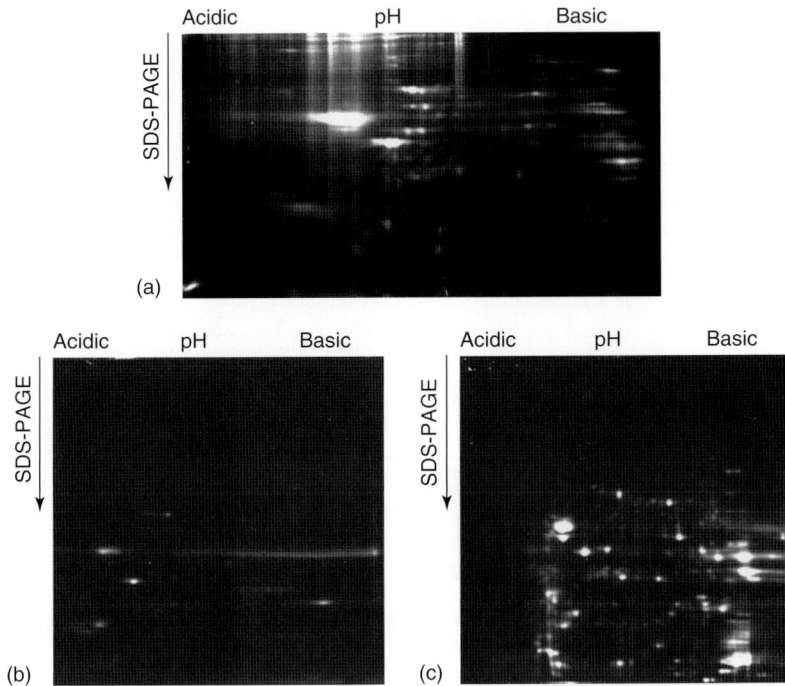

Figure 5. 2D separation result of extracted proteins from mouse brain. The extract proteins were labeled with Cy5. 2D results were achieved using the chip and a commercially available minigel system (ZOOM IPGRunner system, Invitrogen Corporation; IPG gel strip: ZOOM Strip pH 3–10 NL, Cat. No. ZM0011; PAG: Nupage 4–12% Bis-Tris ZOOM Gel, Cat. No. NP0330BOX). The chip system (a,b) 5 μg protein applied, minigel system and (c) 40 μg protein applied.

shows the results from 5 μg of the protein sample applied for comparing numbers of the spot Migration length and gel thickness for SDS-PAGE of the minigel are 80.0 and 1.0 mm, respectively. Meanwhile, the dimension of the chip is almost half (37 mm long and 0.5 mm thick). This miniaturization of the chip may cause the short-time migration and the low protein diffusions. Hence, there were lesser fluorescence reductions and a large amount of the proteins were visualized on the chip. Figure 5(c) shows the result obtained by applying 40 μg of the sample using the minigel system. A number of protein spots were visualized by increasing the amount of the sample. Although the chip visualized lesser spots than the minigel using 40 μg of the sample, more than 100 spots were clearly visualized using the chip. Accordingly, the chip system still showed the utility and the resolving power to separate complex biological samples.

In the normal 2D electrophoresis, the entire process takes 10 h or longer and manual transfer of gel is required. On the contrary, use of the chip reduced the time required for obtaining 2D electrophoresis images to 60 min including 5 min elapsing from the loading of the protein sample to the loading of the agarose, accomplishing our goal time of less than 1 h. However, there are problems preventing commercialization, including the loading of agarose between the first-dimensional and second-dimensional electrophoretic analysis. We have made the transition to a separate model of the fully automated 2D electrophoresis system.

3.2 High-throughput Fully Automated 2D Electrophoresis System

Development personnel are eager to develop the 2D electrophoretic system in which the IEF system is integrated with the SDS-PAGE system, which does not necessarily appeal to users. The current, better choice is the electrophoretic system only requiring smaller sample quantities and a providing high-throughput, automated output of analytical results, irrespective of whether the components composing the system are built into a single piece. The separate model has fewer factors crucial to development than an integrated model, seemingly facilitating the development. The authors separated IEF, staining, washing, and SDS-PAGE units from the integrated model, and

chose in its place an automated transfer system that can carry samples.

Figure 6 shows the system diagram and photograph of the fully automated 2D electrophoresis system. This system is designed to sequentially convey the IEF chip containing the IEF gel strip fixed on the support. Figure 7 shows photographs of the IEF, the reaction solution, and the

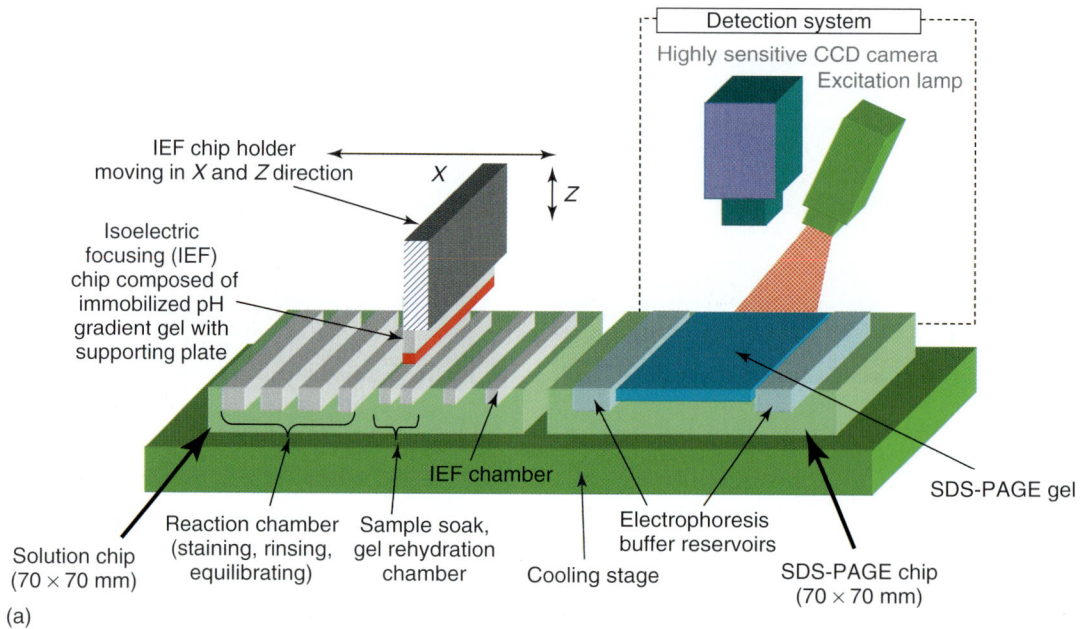

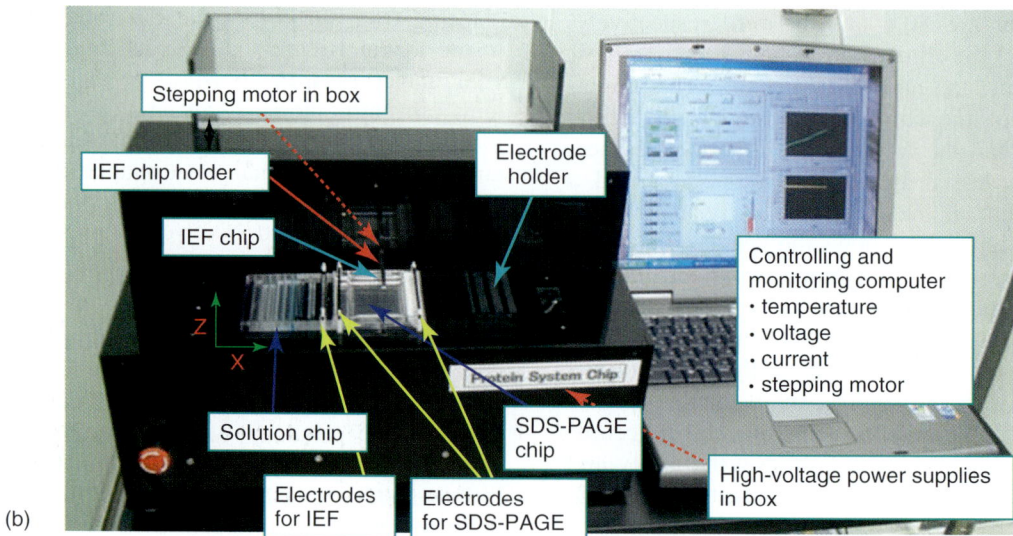

Figure 6. Fully automated 2D electrophoresis system.

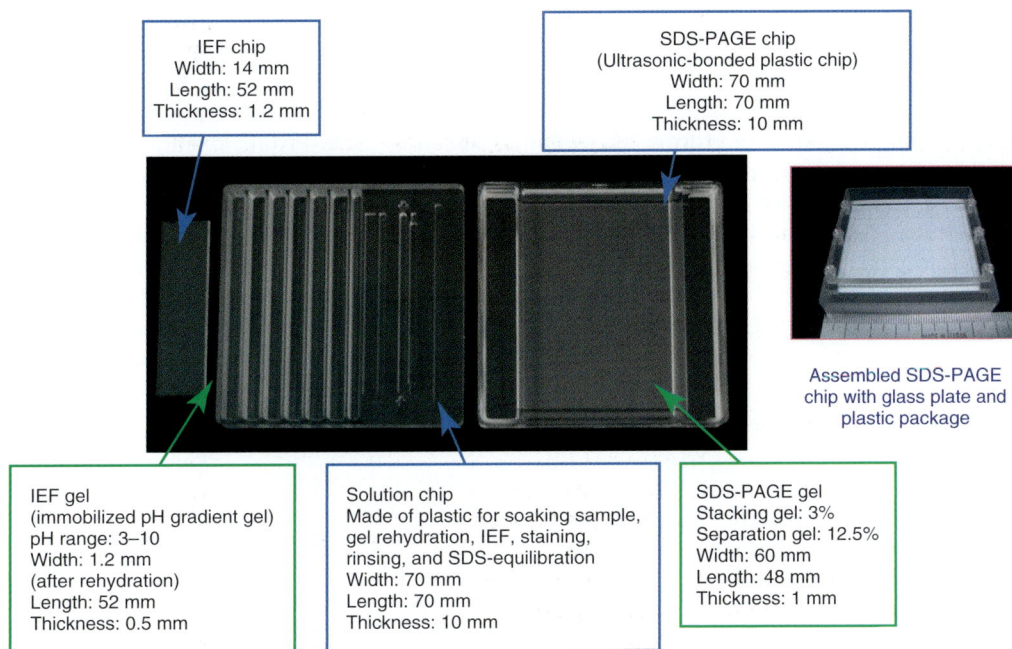

Figure 7. Chips of fully automated 2D electrophoresis system.

SDS-PAGE chips. The operation procedure is as follows: First, the IEF chip holder catches the dried IEF chip and immerses it in a protein sample solution, and then transfers it to the IEF gel swelling solution chamber. Then, the IEF chip is transferred to the IEF chamber, in which a certain voltage is applied to the IEF gel. On completion of the IEF process, the IEF chip is transferred sequentially to the washing chamber followed by the staining chamber, modifying proteins with a fluorescent dye including Cy5. Next, the excessive dye is washed away from the chip, equilibrating with SDS. Subsequently, the IEF chip is transferred to the starting point of the second-dimensional SDS-PAGE gel to contact with the IEF gel, initiating the SDS-PAGE process. Equipped with a charge coupled device (CCD) camera, this system can visualize, in real time, the separation process during the SDS-PAGE process, thereby achieving a reduction of the conventional operating time of 20 h to 60–90 min: the loading of protein samples and gel swelling, 10 min; IEF, 20–30 min; staining/washing/equilibration, 10–20 min; and SDS-PAGE/detection, 20–30 min. Whereas the conventional method requires the manual transfer of gel at every process step, this system can do everything fully automatically from the loading of a protein sample to detection. Also, the system is reduced in size to one-fourth or less, compared with the conventional 2D electrophoresis system, saving installation space.

The performance of the system was evaluated using solubilized liver proteins as a protein sample. As for intermediate staining with the SDS-PAGE chip assembled with a glass slide, the results from intermediate staining (Cy5) and post-SDS-PAGE staining (SYPRO Ruby) were satisfied. On the contrary, with a chip welded with a plastic substrate by supersonic bonding, noise was caused in a low-molecular-weight region possibly by the adhesion of Cy5 to the plastic surface; coating of the plastic substrate is under review. Reproducibility of this system was compared with a commercialized manual small gel system, and the results showed a high reproducibility in both systems. Failure in the evaluation of reproducibility may be attributed to the unparalleled skills of our research personnel conducting the comparative test. Therefore, results from many other research personnel with different skills should be evaluated for ensuring an accurate comparison. Then, IEF and SDS-PAGE were evaluated for their resolution;

and the resolution was comparable to the small gel, although it did not reach that of the large gel, about $200\,mm^2$.

As mentioned above, this system, including reproducibility and resolution, performs as well as or better than commercialized systems equal in size. In addition, the automated acquisition of the 2D electrophoretic data in a shorter time seemingly proves that the method using this system is apparently superior compared with conventional methods.

4 FLUORESCENT DYE FOR STAINING PROTEINS

We have reported a novel fluorescent dye (Dye 1) based on the intramolecular charge transfer for the highly sensitive detection of a protein whose structure is shown in Figure 8.[6] Dye 1 has the following characteristics: (i) Dye 1 itself indicates a weak fluorescence, whereas a strong red emission is observed upon binding to a protein molecule, (ii) a high molar extinction coefficient and high fluorescent quantum yield upon binding to protein, (iii) low protein-to-protein variation, (iv) low interference from nonprotein substances (inorganic salts, detergents, reductants, organic solvents, etc.), (v) high chemical and photophysical stabilities, and (vi) low-cost starting materials and easy synthetic method (one-step synthesis). Moreover, the successful demonstration of protein staining in minigels after SDS-PAGE was performed using Dye 1.

In this chapter, we described a study to develop the high-performance staining of proteins in the gel using Dye 1 as a further application to carry out the high-throughput protein analysis. The normal protein staining protocol for Coomassie brilliant blue (CBB) staining or SYPRO Ruby staining, which is "SDS-PAGE, fixation, staining, washing, and detection", was simplified while maintaining sensitivity for proteins and the sharp band resolution of proteins in the gel. The experimental results finally showed that protein staining using Dye 1 can simultaneously be carried out together with SDS-PAGE separation by dissolving Dye 1 into the electrophoresis buffer solution. Moreover, it was successful in shortening the handling time of protein staining after SDS-PAGE (general SYPRO Ruby method takes about 18 h, whereas the procedure presented here takes 45 min) to obtain the experimental results without the loss of the sensitivity for proteins in the gel, and it was found that Dye 1 had significantly better characteristics than commercially available dyes for protein staining.

To demonstrate the staining of proteins in the gel using Dye 1, proteins after the separation using the 1D SDS-PAGE minigels were fixed in the gel by incubation in the fixing solution for 30 min and then immersed in the staining solution containing Dye 1 for 30 min. After washing of the gels by the fixation solution for 30 min, the bands of the proteins on the gel were scanned using the image analysis systems. Figure 9 shows the typical gel images of the bovine serum albumin (BSA) after Dye 1 staining. The association of Dye 1 with BSA and the visual examination of the staining response were successful. The Dye 1 staining indicated a good linear relationship between the integrated volume of the densitometry units for scanned bands of BSA in the gel and BSA concentration as shown in Figure 9(b). The detection limit was 7.0 ng/band of BSA, which is as sensitive as the short-protocol silver staining methods.[7] To examine the reproducibility of the calibration graph, reproducibility tests ($n = 3$) were carried out. The relative standard deviation of the response was found to be within 2.0%. Other proteins at the various concentrations (chymotrypsinogen A, transferrin, immunoglobulin G (IgG)) were stained by Dye 1 after SDS-PAGE, and their calibration graphs were constructed as shown in Figure 9(b). From this result, this protein staining using Dye 1 indicated only slight protein-to-protein variation, which makes it possible to accurately compare the protein expression levels and to monitor the correct

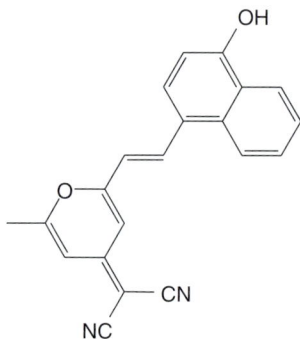

Figure 8.   Chemical structure of Dye 1.

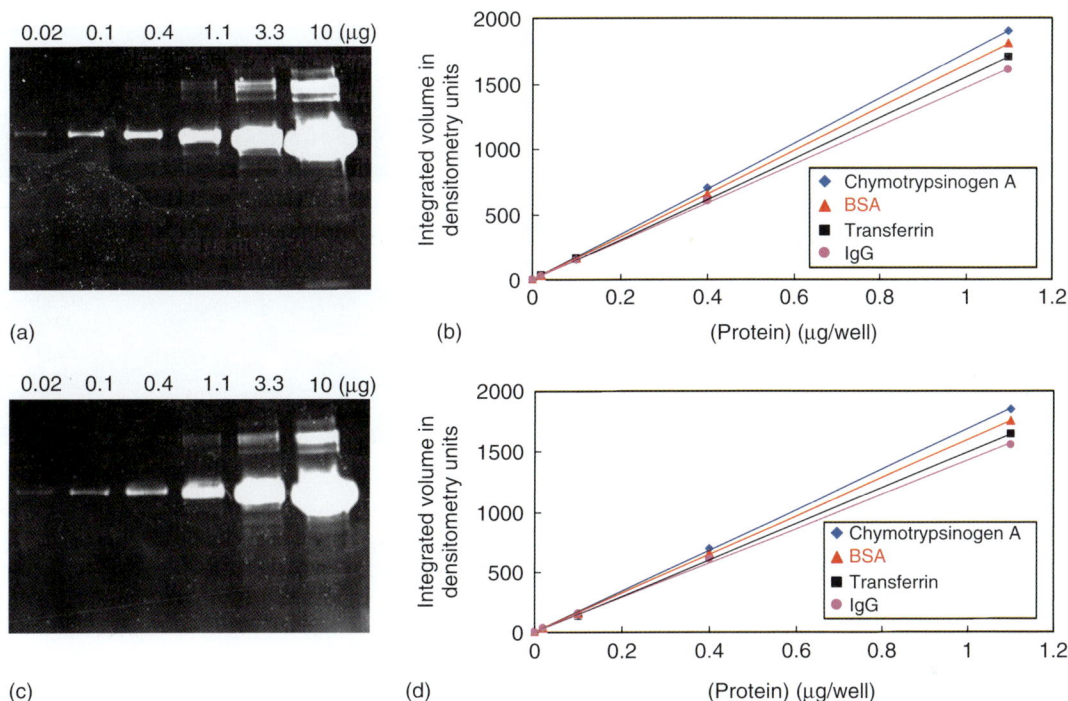

Figure 9. Fluorescent staining and densitometry of proteins with Dye 1. (a) Typical gel image of BSA stained by Dye 1 using conventional staining protocol. (b) The relationship between the integrated volume in densitometry units for scanned bands and various protein concentrations (BSA, chymotrypsinogen A, transferrin, IgG) in gels using conventional staining protocol. (c) Gel image of BSA stained by Dye 1 using rapid staining protocol. (d) The relationship between the integrated volume in densitometry units for scanned bands and various protein concentrations (BSA, chymotrypsinogen A, transferrin, IgG) in gels by 1D SDS-PAGE using rapid staining protocol. The numerical values in (a) and (c) are the amounts of BSA applied to each gel well.

protein concentration using only one calibration graph such as commercially available fluorescent dyes (SYPRO Ruby and SYPRO Orange).

For the SDS-PAGE measurement, it takes a long time to obtain the results after separation of the proteins, because of the multiple steps and intensive labor for protein staining in the gel using the conventional staining method, such as SYPRO Ruby staining or CBB staining. The staining protocol of SYPRO Ruby after SDS-PAGE and each treatment time is as follows: fixation (30 min), staining (overnight), washing (30 min), and detection (it takes a total time of about 18 h).

Dye 1 was useful for protein staining in the gel, and the proteins were completely stained by Dye 1 within 30 min, a time that is much shorter than that for the SYPRO Ruby staining and CBB staining. Therefore, it was thought that Dye 1 might reduce the protein staining time, and each treatment in

the general protocol for the protein staining was simplified.

In our previous report, the fixation procedure was omitted in order to investigate whether proteins were able to be stained by Dye 1 in the presence of an excess amount of sodium dodecyl sulfate (SDS).[6] The association of Dye 1 with BSA and IgG and a visual examination of the staining response were successful in the presence of an excess amount of SDS. After washing of the gel for 30 min, clear protein bands were obtained owing to reduction of the background. Under the same experimental conditions, SYPRO Ruby and CBB could not stain the proteins, because the excess amount of SDS interfered with the binding of these dyes to the proteins. On the other hand, it was recognized that the binding between Dye 1 and the proteins was not affected by the excess amount of SDS in the solution in the previous report, which contributed to the success of the protein staining by Dye 1 independent of the SDS removal.

As the next step to reduce the staining protocol time, the SDS-PAGE experiment was carried out together with the protein staining. An electrophoresis buffer solution containing Dye 1 was prepared under the optimum experimental conditions, and Dye 1 was bound to the protein during the SDS-PAGE experiment. Figure 9(c) shows the gel images of the various concentrations of the BSA using Dye 1 staining during a 1D SDS-PAGE experiment after washing the gels. The association of Dye 1 with proteins during the SDS-PAGE experiment was successful and the staining response was clearly observed by visual examination. For this staining method, it was successful in greatly reducing the staining time and intensive handling from several hours for the general method to 15 min. The integrated volume in densitometry units for the scanned protein bands in the gel had a linear relationship with the concentration of the proteins ($r^2 > 0.995$), and little protein-to-protein variation was observed as shown in Figure 9(d). The detection limit of the proteins using this staining method was 7.0 ng/band of BSA (signal-to-noise ratio was 3.0) when washing the gel, which is as sensitive as the general staining methods using Dye 1, and it was noted that the sensitivity for proteins was maintained regardless of reducing the staining procedures and time.

Other proteins at various concentrations (chymotrypsinogen A, transferrin, IgG) were stained by Dye 1 under the same experimental conditions, and their calibration graphs were constructed as shown in Figure 9(d). As a result, this protein staining using Dye 1 indicated only a slight protein-to-protein variation similar to the result of the general staining method.

Under the same experimental conditions, 2D SDS-PAGE was carried out for the separation of 40 µg of mouse brain lysates, and well-resolved gel images were observed. These gel images are shown in Figure 10. The fluorescent image of proteins with some nonspecific spots and small background was observed despite the SDS removal and excess dyes in the gel as shown in Figure 10(a). By washing the gel, the background and nonspecific spots in the gel disappeared, and a clearer gel image was observed as shown in Figure 10(b). As a result, the binding of Dye 1 to proteins during the

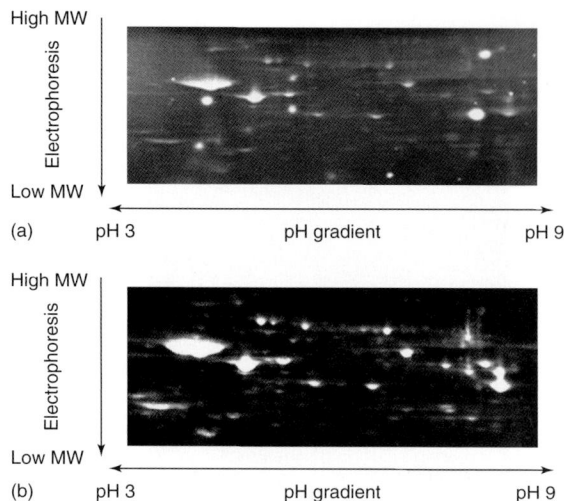

Figure 10. Rapid staining with Dye 1 for 2D PAGE of mouse brain lysate (40 µg) before (a) and after (b) washing gel.

SDS-PAGE experiment was successful regardless of the 1D or 2D SDS-PAGE gel separation.

SYPRO Orange is a commercially available fluorescent protein staining dye which makes it possible to stain proteins in the presence of SDS and is a simple and rapid method.[8,9] Although this stain works well with the 1D SDS-PAGE gel, its performance with 2D gels was inconsistent and failed to achieve the sensitivity levels obtained by the 1D SDS-PAGE.[10] Moreover, no proteins were stained by SYPRO Orange under the same experimental conditions, because the movement of SYPRO Orange into the gel was prevented by the 1D strip gel. Both Dye 1 and SYPRO Orange could bind to proteins using a simple and rapid method, whereas Dye 1 succeeded in the staining of proteins not only in 1D SDS-PAGE gels, but also in 2D gels with a high sensitivity. From this viewpoint, it was demonstrated that Dye 1 had significantly improved characteristics than the commercially available staining dyes.

Our procedures significantly reduced protein staining times for the SDS-PAGE compared to the general method for the SYPRO Ruby staining of 18 h and for CBB at 105 min. For this study, it took 15 min (or 45 min in the case of washing of the gel after SDS-PAGE together with staining). Moreover, the gel image analysis to detect the protein bands in this experiment could be directly performed after the SDS-PAGE experiment without the labor-intensive treatments such as fixation, washing, and lengthy staining of the general staining protocol using SYPRO Ruby and CBB. Although the detection limit of the protein for Dye 1 (7 ng/band) was slightly lower than that for SYPRO Ruby (1–2 ng/band), the detection limit of the proteins in this study was as sensitive as the short-protocol silver staining methods (7 ng/band), and was much higher than that of CBB (64 ng/band). The SDS-PAGE experiment using Dye 1 made it possible to carry out the high-throughput protein analysis and highly sensitive detection of proteins, which satisfies the requirements in the rapidly growing field of proteomics.

The present study demonstrated the high-performance staining for 1D and 2D SDS-PAGE using the novel protein-binding fluorophore, Dye 1. The proteins in the gel could be stained by Dye 1 during the SDS-PAGE experiment by preparation of the electrophoresis buffer solution containing Dye 1 under optimum conditions and by the binding to proteins in the gel during the SDS-PAGE experiment. This method for SDS-PAGE significantly simplified the staining protocols without any loss of the protein-to-protein variation and sensitivity.

Recently, proteomic analysis has become an important field and it is increasingly important to refine the techniques generating proteomic data. This highly sensitive, rapid, and easy handling staining method using Dye 1 should be widely applicable and convenient for multiple scientific disciplines including biochemistry, medicine, and pharmacology.

5 PROSPECTS FOR THE FUTURE

The fully automated 2D electrophoresis system developed by the authors is believed to be an epoch-making system that allows anyone to conduct a simple, rapid 2D electrophoretic analysis conventionally limited to a few highly trained researchers. This makes protein analysis more familiar to us; it may be used for laboratory tests in the near future. The global market scale of products related to the fully automated 2D electrophoresis, including instruments and consumable items, reached US$3.13 million in 2003.[11] If a chip is developed that allows post-translational modifications including glycosylation and phosphorylation, the market scale will greatly expand in the future.

ACKNOWLEDGMENTS

The development of the automated 2D electrophoresis system was financially supported as the High-throughput Biomolecule Analysis System Project by the New Energy and Industrial Technology Development Organization (NEDO), Japan. Authors thank cooperative researchers: Y. Unuma, Y. Maruo, T. Matsushima, K. Takahashi, M. Mieda, M. Nakamura from Sharp corporation, K. Sakairi, C. Hayashida, M. Kano, K. Ueyama from Toppan Printing, I. Namatame, K. Yodoya, Y. Ishii, T. Shibata, H. Inamochi, Y. Nakada, Y. Ogawa, H. Marusawa, T. Komatsu, Y. Saito from Astellas Pharma, K. Yano, S. Akutsu, and I. Karube from Tokyo University of Technology.

REFERENCES

1. G. A. Scheele, Two-dimensional gel analysis of soluble proteins. Charaterization of guinea pig exocrine pancreatic proteins. *Journal of Biological Chemistry*, 1975, **250**, 5375.
2. P. H. O'Farrell, High resolution two-dimensional electrophoresis of proteins. *Journal of Biological Chemistry*, 1975, **250**, 4007.
3. D. A. Wolters, M. P. Washburn, and J. R. Yates III, An Automated Multidimensional Protein Identification Technology for Shotgun Proteomics. *Analytical Chemistry*, 2001, **73**, 5683.
4. N. Ramachandran, E. Hainsworth, B. Bhullar, S. Eisenstein, B. Rosen, A. Y. Lau, J. C. Walter, and J. LaBaer, Self-Assembling Protein Microarrays. *Science*, 2004, **305**, 86.
5. K. Usui, A. Hiratsuka, K. Shiseki, Y. Maruo, T. Matsushima, K. Takahashi, Y. Unuma, K. Sakairi, I. Namatame, Y. Ogawa, and K. Yokoyama, A self-contained polymeric 2-DE chip system for rapid and easy analysis. *Electrophoresis*, 2006, **27**, 3635.
6. Y. Suzuki and K. Yokoyama, Design and Synthesis of Intramolecular Charge Transfer-Based Fluorescent Reagents for the Highly-Sensitive Detection of Proteins. *Journal of the American Chemical Society*, 2005, **127**, 17799.
7. F. Gharahdaghi, C. R. Weinberg, D. A. Meagher, B. S. Imai, and S. M. Mische, Mass spectrometric identification of proteins from silver-stained polyacrylamide gel: a method for the removal of silver ions to enhance sensitivity. *Electrophoresis*, 1999, **20**, 601.
8. T. Steinberg, L. Jones, R. P. Haugland, and V. Singer, SYPRO orange and SYPRO red protein gel stains:

one-step fluorescent staining of denaturing gels for detection of nanogram levels of protein. *Analytical Biochemistry*, 1996, **239**, 223.

9. L. Steinberg, R. P. Haugland, and V. Singer, Applications of SYPRO orange and SYPRO red protein gel stains. *Analytical Biochemistry*, 1996, **239**, 238.

10. J. P. Malone, M. R. Radabaugh, R. M. Leimgruber, and G. S. Gerstenecker, Practical aspects of fluorescent staining for proteomic applications. *Electrophoresis*, 2001, **22**, 919.

11. *World 2 D Gel Electrophoresis Market*, Frost & Sullivan, Palo Alto, 2004.

55

Surface-Enhanced Laser Desorption/ Ionization (SELDI) Technology

Lee O. Lomas[1] and Scot R. Weinberger[2]

[1] *Ciphergen Biosystems Inc., Fremont, CA, USA and* [2] *GenNext Technologies Inc., Montara, CA, USA*

1 OVERVIEW OF SELDI BIOCHIP TECHNOLOGY

Surface enhanced laser desorption/ionization mass spectrometry (SELDI-MS) was first conceptualized in the early 1990s when Hutchens and Yip demonstrated that chromatographic affinity probes used to specifically enrich fragments of lactoferrin could then be directly presented to a laser desorption/ionization source for mass spectrometric detection.[1] The capability of the probe to actively participate in the extraction of the analyte and subsequent removal of sample components that interfere with or suppress ionization was a logical improvement over the classical matrix-assisted laser desorption/ionization (MALDI) applications where the sample probe surface plays a passive role in the analytical scheme and merely presents the sample to the mass spectrometer for analysis; in order to produce usable MALDI-MS signal, crude samples must first be fractionated and purified of any ionization suppressants such as salts, chaotropic agents, detergents, and so on.

SELDI, as originally defined by Hutchens and Yip, consists of two subsets of technology: surface enhanced affinity capture (SEAC) and surface enhanced neat desorption (SEND).[1] By far the SELDI array technology showing the most utility to date is SEAC and as such is generally referred to as SELDI in the published domain. In this format, the probe surface plays an active role in the extraction, concentration, and presentation of the analyte and elimination of ionization suppressants. Figure 1 depicts the elements of a SELDI biochip and associated variety of chemical and biochemical SELDI array surfaces used in differential protein expression applications. Chemical surface arrays are derivatized with classic chromatographic separation ligands such as reverse phase, ion exchange, immobilized metal affinity capture (IMAC), and normal phase media. Such surfaces, with broad binding properties, are typically used for general protein profiling and de novo biomarker discovery, where large populations of proteins are compared (e.g., from diseased vs normal samples) with the goal of elucidating differentially expressed elements. Biomolecules bind to these surfaces through hydrophobic, electrostatic, coordinate covalent bond or Lewis acid–base interaction, the strengths of which can be directly modulated by modifying the binding and/or washing buffer compositions.

Biochemical arrays are created by immobilizing bait molecules upon the surface of preactivated SELDI surfaces via covalent attachment using either primary amines or hydroxyl groups. In this way, specific protein-interaction arrays of virtually any content may be created, including antibodies,

Handbook of Biosensors and Biochips. Edited by Robert S. Marks, David C. Cullen, Isao Karube, Christopher R. Lowe and Howard H. Weetall.

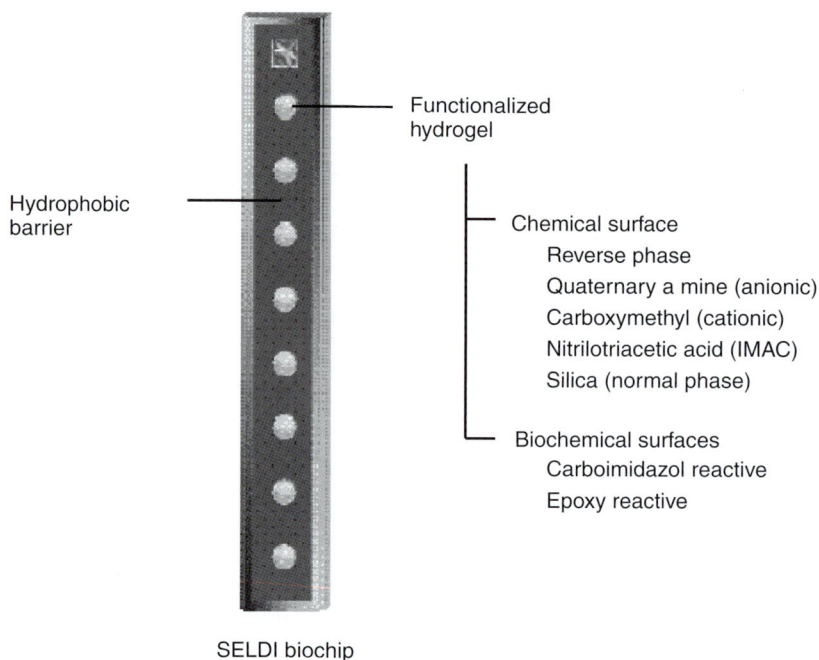

Figure 1. Elements of a SELDI biochip. The array consists of a glass-coated aluminum strip that displays discrete affinity locations or spots. Each spot incorporates a hydrogel that is functionalized with classical chromatographic ligands, such as C9–C12 aliphatic chains (reverse phase), quaternary amines (strong anionic), carboxymethyl (weak cationic), nitrilotriacetic acid (IMAC) and silicon oxide (normal phase). The biochips also incorporate a hydrophobic barrier that surrounds the spot locations and prevents sample movement between spots. Dimensions and spot locations are such that 12 strips side-by side provide the standard 96-well microtiter footprint and sample processing can be achieved using standard robotics.

receptors, enzymes, DNA, small molecules, ligands, and lectins. In contrast with standard chromatographic media, these biochemical surfaces provide a greater degree of enrichment of captured analytes, because of the high specificity of biomolecular interactions. Because specific biochemical interaction motifs demonstrate high affinity and low equilibrium disassociation constants, biochemical surfaces facilitate a vast array of microscale experiments that facilitate the analysis of very low sample volumes. Such experiments include SELDI immuno assays, targeted protein identification and/or purification, ligand binding domain analysis, epitope-mapping experiments, post–translational modification detection, as well as reliable quantitative studies, even when fishing for target proteins within a complex biological sample. When compared to MALDI, SELDI-based arrays have demonstrated not only a simplified workflow, but also improved analytical sensitivity and associated mass detection limit. The latter is attributed to a marked reduction in sample loss

inherent in combining and miniaturizing sample processing workflow.

2 USE OF SELDI IN PROTEOMICS

2.1 Differential Protein Display and Biomarker Discovery

Over the course of the last decade, SELDI technology has mostly been applied to challenges of proteomics research. Among today's most popular proteomic research activities is differential protein display or expression monitoring. Differential protein display is a comparative technique that contrasts protein profiles between different organisms, individuals, pathogenic and/or metabolic conditions, and phenotypic response to environmental or chemical challenges. Unlike differential studies of transcription, differential protein display studies are not easily enabled by amplification strategies such as reverse polymerase chain reaction (PCR). In this manner, differential protein studies require approaches that isolate and enrich both major as well as minor protein constituents from a complex biological mixture, with specific attention to preanalytical and analytical biases that may compromise conclusional integrity. The proteomic community performs differential protein display for biomarker discovery in two formats; the so-called top down and bottom up approaches. SELDI has been used most extensively in the top down approach whereby protein mass signatures within a biological sample are detected and compared without any further reduction to peptide fragments by post sample collection methods such as global tryptic digestion. The implication of this is that proteins and protein modifications due to the biology of the system are preserved through to detection as they may be indicative of the biological question to be answered. For such experiments, the workflow is performed in two phases; a scouting phase whereby a large number of chromatographic separations are performed using multiple array chemistries and binding/washing conditions, and a validation phase whereby select differentially expressed protein candidates are validated using only the chromatographic surface and binding/washing condition that gave the initial differential profile. As an example, the process of SELDI-based sample preparation using four different chemical surfaces is demonstrated in Figure 2. A series of orthogonal SELDI surfaces including reverse phase, anionic, cationic, and IMAC loaded with a transitional metal such as Cu^{2+}, Ni^{2+}, or Zn^{2+} are arranged in plate format. A complex biological sample is deposited upon every chemically active "spot" of each array. After binding, the spots on each array are washed with appropriate buffers in gradient manner. Within a given array, subsequent spots experience a greater degree of stringency, removing analytes with comparatively weaker surface interaction potential and enriching for those of strong surface affinity. In some cases, particularly when using specific biomolecular interactions, purification to almost-complete homogeneity is possible, without substantial loss of analyte.

Benefits of this workflow also reveal insight into the physical–chemical properties of the analyte retained on the surfaces such as hydropathicity, charge, pI, and post–translational modifications such as phosphorylation and/or glycosylation. Such de facto knowledge can be further exploited particularly in analyte purification whereby such knowledge can be used to design efficient scale-up purification strategies by matching the array chemistry and binding/washing conditions used for discovery with that of a more conventional chromatographic bead chemistry in a column format. This is particularly useful when additional purified material is needed for the purpose of protein identification and further characterization.

In contrast to research-based proteomics, clinical proteomic studies endeavor to follow the progress of disease within an individual or a small population with the ultimate goal of finding biomarkers potentially useful as diagnostic agents or new drug targets; this topic is now known best as "translational medicine". Under such circumstances, sample or tissue availability is limited and the dependence upon highly efficient, small-scale techniques is becoming more and more essential. Typically protein populations between groups are compared using univariate and/or multivariate statistical analysis schemes with the ultimate goal of elucidating a protein or groups of proteins whose expression levels correlate with a given clinical condition.[2] Automation requirements here are primarily focused upon running many samples in a massively parallel manner and only proteins of interest are further characterized to provide insight into identification and post–translational modifications, or to provide insight into the disease mechanism or host response to disease. Because of its reproducibility, throughput, and starting material requirements, SELDI-MS has gained acceptance as a tool of choice for clinical proteomic studies.

3 SELDI CLINICAL PROTEOMIC STUDIES

Clinical proteomics aims to scan the realm of expressed proteins to identify biomarkers that can answer specific clinical questions. The most obvious are markers that can be used for diagnosis or prognosis. Another important issue that clinical proteomics promises to help resolve, is the ability to predict a patient's response to a specific drug. For example, diagnostic markers can themselves be candidates for drug targets and pharmaceutical companies pursue clinical proteomics to identify markers that predict toxicity of candidate drugs. The overriding determinant of the success of a clinical proteomics program is the choice

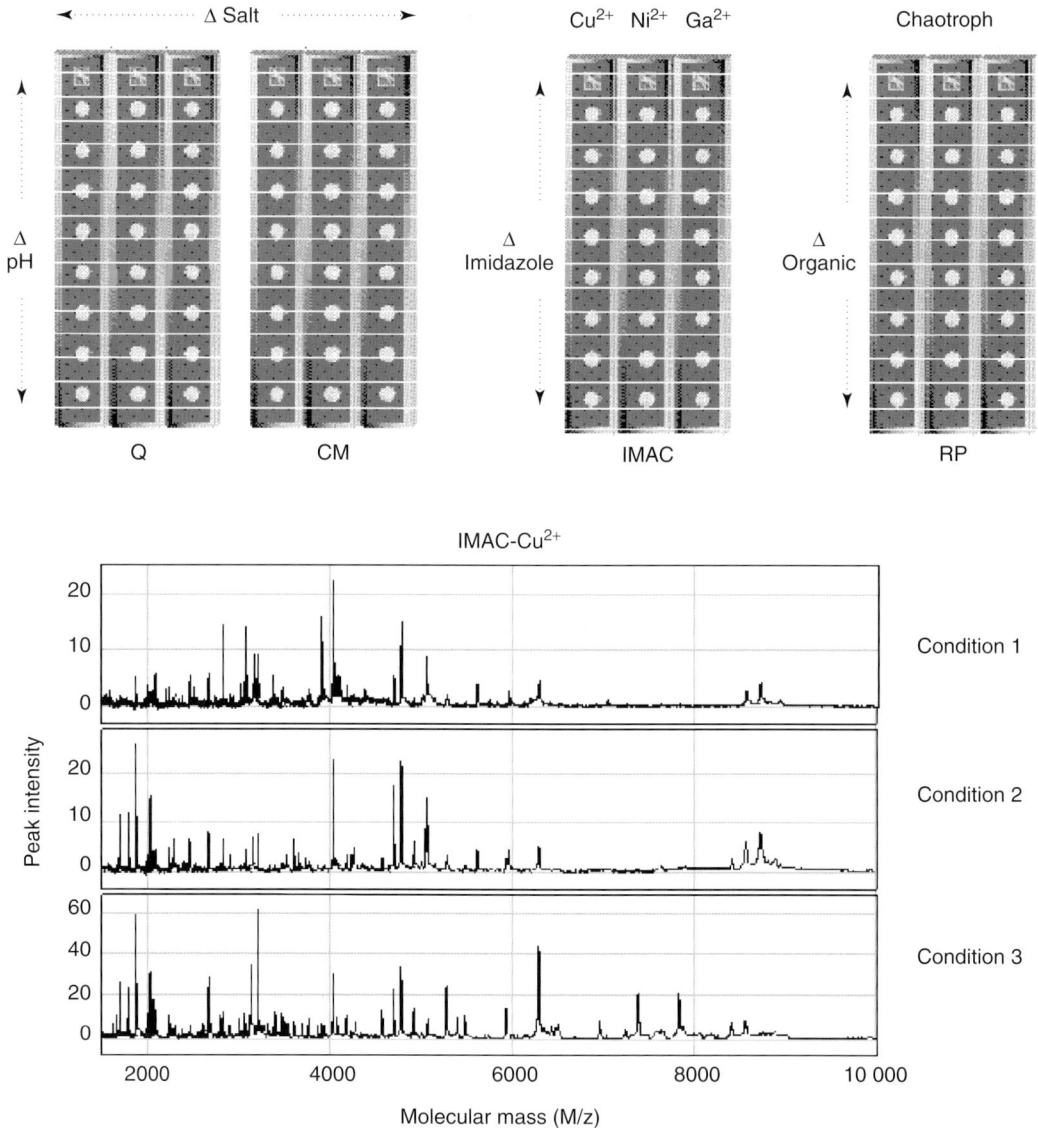

Figure 2. Typical workflow for a discovery phase using SELDI biochips. A series of arrays of different chromatographic characteristics are screened, such as anionic (Q), cationic (CM), IMAC, and reverse phase. Samples are incubated on chemistries under a number of different binding/washing conditions to modulate the specific population of proteins retained by the surface. After final wash to remove any buffer components that may interfere with MS detection, MS spectra are generated that represent a protein fingerprint of that sample under the specific chemistry and binding conditions; which can then be compared across samples such as control versus disease to identify differentially expressed proteins.

of clinical question followed by careful study design and implementation.[3] The underlying clinical question will drive the decision on the choice of proteomics technique, the success criteria, how many samples to examine, how to analyze the data, and ultimately, whether the clinical proteomics

program is a success. Therefore, the initial task is to balance the advantages of simplifying the clinical study with the practical utility of the outcome. Obviously the more limited the population on which an outcome is based, the more limited the population the outcome can be applied to, unless additional work is performed to demonstrate the validity of the biomarkers for a more general population. The importance of proper study design cannot be more exemplified than by a study published by Petricoin et al.[4] that described the grail of clinical proteomics, the capability of a panel of biomarkers to distinguish early-stage ovarian cancer with 100% sensitivity, 95% specificity and a positive predictive value of 94%. Although this study energized the research community, subsequent scrutiny indicated a number of study design flaws that ultimately provided a conclusory bias that is now generally acknowledged.[5] Although this study polarized the proteomic community with respect to the utility of biomarkers, it did solidify for all the importance of all aspects of study design including sampling methods, sample collection, preparation and data generation, candidate biomarker selection based on appropriate statistical analysis, biomarker validation, identification, and finally quantitative assay generation. Today, clinical proteomics is becoming a concerted effort that involves biostatisticians, clinicians, and the bio-analytical core facilities. During the course of the last two years, researchers have used SELDI array technology to perform biomarker discovery research in a variety of diseases including infectious disease,[6–8] Alzheimer's disease,[9–13] and cancer.[14–17] Further information regarding SELDI array technology and clinical proteomics research is found in the following recent reviews.[18–23]

As with research proteomics, once the samples are defined and appropriately collected, a typical SELDI clinical proteomic study begins with a discovery phase, in which assay conditions are tested on a relatively small number of samples. Usually, profiling proceeds with at least 30 samples in each classification group (e.g., disease vs healthy or treated vs untreated). This number of samples is usually enough to yield greater than 90% statistical confidence in single markers with p values <0.01 and to allow the use of some forms of multivariate analysis. The samples themselves are another critical parameter. Because the initial sample set size is relatively small, inherent biological variability always threatens the ability to conclude that the differences seen are consequences specific to the disturbance under study. Therefore, it is imperative that the study includes well-chosen samples (e.g., patients of the same age group or

all of a single sex) and, equally important, appropriately chosen controls. Naturally, in vitro studies show less variability than do animal studies, which in turn show less variability than human studies.

Finding single proteins responsible for differentiating disease versus normal or treated versus untreated is a natural first step in analyzing expression data.[24] SELDI analysis generally employs nonparametric statistical methods since one cannot assume that peak intensity data conforms to a normal distribution and SELDI studies often have a small sample size. These methods include the Mann-Whitney, the nonparametric equivalent of the student's t-test, and the Kruskal-Wallis, the nonparametric equivalent of analysis of variance (ANOVA) thus eliminating any assumption on the distribution of the peak intensity data.[25] Essentially, these nonparametric tests sort the peak intensities and their corresponding ranks are used in the p-value calculation. The p-value results of these tests help identify potential markers in conjunction with data visualization tools.

These statistical tests simply give an indication of group mean differences, which may not always be helpful if there is a very large spread in the distribution of data. Simply increasing the sample size can improve p values while discrimination between groups may remain poor.[26] This is especially important when attempting to use the biomarker in a diagnostic assay as a biomarker's p value may have no correlation to its clinical value. Furthermore, one may not find single biomarkers with acceptable p values. At this point, it is prudent to turn to other analytical methods that are both multivariate in nature and lend themselves toward developing a clinical assay. There are a number of different analysis tools that identify and use multivariate patterns in the expression profile data for the purposes of identifying groupings and/or classifying groupings. These methods can generally be lumped into one of the two categories—unsupervised learning in the form of cluster analysis, or supervised learning in the form of classification methods. A number of these (only a fraction of existing literature is cited here) have been applied to gene expression data including clustering and visualization,[27,28] self-organizing maps,[29] and support vector machines.[30] Most SELDI clinical proteomic studies have relied upon regression tree-based methods,[31,32] and for the most part, use the algorithms embodied in Ciphergen's Biomarker PatternsS™oftware. Similar to the way a clinician makes his/her diagnosis by correlating and integrating various findings from a patient's physical examination with laboratory test results, the classification tree creates similar rules based on peak intensity such as, "peak intensity at 13 979 Da <5.169, and at 3272 Da >12.283—therefore this sample belongs to the disease group". In addition, for each peak cluster that is determined to be a good classifier, Biomarker Patterns Software determines the intensity value that serves as a threshold above or below which a given classification is assigned.

A number of characteristics of classification trees make it an attractive tool for protein expression studies. The model is easy to interpret compared to "black-box" classifiers such as neural networks and nearest neighbor classifiers. The protein peaks used in the model are easily attainable by examination of the rules, and these rules are easily validated by examination of the spectra. This openness of the tree-based model is an attractive feature for researchers wanting both a diagnostic assay as well as potential therapeutic targets. In addition, classification trees can sift through all the input variables and select the subset to use in the tree. As such, it alleviates some of the burden of performing feature selection up front. Regardless of study design, once a protein of interest has been detected, protein characterization efforts often ensue. Proteins are characterized by identifying post-translational modifications, providing primary sequence information, and, ultimately, by elucidating protein identity.

4   ASSAY DEVELOPMENT AND VALIDATION

Once a candidate biomarker or panel of markers has been identified, a phase of assay design and validation ensues. This involves optimizing the methods for the routine detection and quantitation of these markers in a massively parallel process. In situations where the analyte to be quantified is a single member of a single family, methods such as immunoassay become an obvious choice for this purpose. However, in instances where the analyte is a specific isoform or protein fragment from a larger protein family, the use of MS

offers an excellent opportunity to provide highly specific analyte detection. As discussed in the preceding text, the incorporation of activated surfaces onto the SELDI-MS probes have overcome many of the difficulties associated with extracting analytes from small sample volumes. Additionally, SELDI biochips are specifically formatted in individual strips of eight spots per array. When 12 such arrays are combined together, the resultant footprint resembles the industry standard 96-well microtiter plate format and allows many options for automation using conventional fluidic robots. Utilization of such robotics not only yields a marked improvement in analytical throughput, but also demonstrates superior qualitative and quantitative reproducibility when compared to manual processing of arrays. For serum profiling experiments, typical protein quantitative profiles demonstrate %coefficient of variation (CVs) of less than 20%. Reducing %CV below this value, however, continues to be challenging due to the laser desorption/ionization process that is inherent in MALDI and SELDI-MS. For example, the laser desorption process is a complex and poorly understood event that depends on the interaction of a laser ionization source and co-crystals of sample and matrix. Variability in the total protein composition between samples can influence the matrix co-crystallization and influence relative ion abundance formations with respect to different analytes to be detected and during the laser desorption/ionization event, ionization of one analyte cannot occur independent of any other analyte exposed at the same time to the ionization source. The best opportunity to define a reliable method of normalization is to make all parameters as similar as possible, however, by definition, calibrants differ in analyte concentration in a predictable manner while samples may differ unpredictably. When such differences are dramatic, even common methods to normalize using total ion current normalization are not sufficient. Most recently, we have developed new methods that have generally resolved much of the normalization difficulties associated with analyte: matrix co-crystallization and laser desorption/ionization. This method relies on many of the principles of displacement chromatography[33] and is particularly well suited for use with SELDI biochips. The method involves spiking the sample with a foreign matrix of proteins that contain members of high affinity to the chosen array

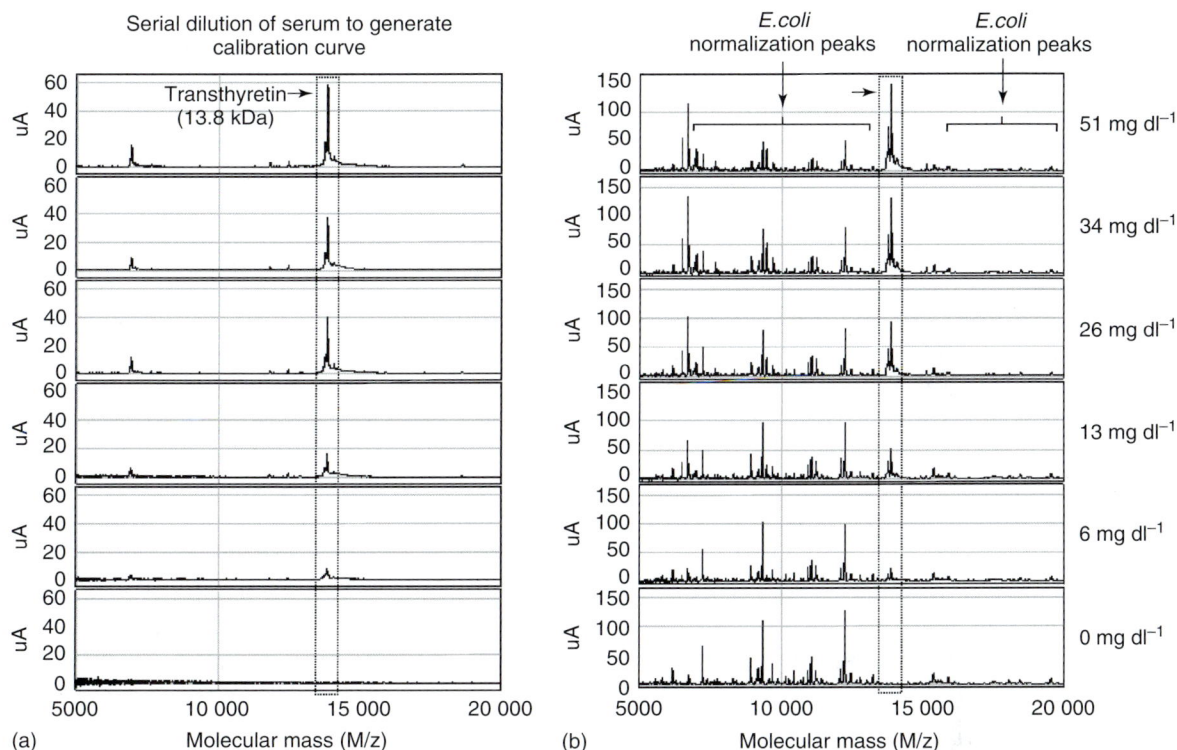

Figure 3. Normalization by addition of a complex protein matrix. MS spectra of a decreasing calibration series of cysteinylated transthyretin (from top to bottom) generated from serial dilution of serum into buffer only (a) or into buffer containing a constant amount of *E. coli* extract (b). Dilution of serum into buffer only cannot be adequately normalized by total ion current methods to compensate for assay variability because of the lack of sufficient unchanging background. Dilution of serum into buffer containing a constant amount of *E. coli* protein provides both sufficient unchanging background for normalization and also acts as an internal standard that can also compensate for variability throughout the assay. Using such a method, %CV across assays (day-to-day, operator-to-operator) are 5–10%.

surface. As an example, Ciphergen is currently developing a seven-protein marker assay panel that may aid in the stratification of women with a pelvic mass.[34] One such protein marker, a cysteinylated form of transthyretin, is particularly difficult to quantitate in serum because it is an abundant protein and standard curves are generally prepared by serial dilution of control serum into an appropriate assay buffer. Thus, the overall protein background and subsequent mass spectra generated from the highest concentration calibrant is significantly different from the mass spectra generated from the lowest concentration calibrant (Figure 3) and normalization methods using total ion currents are not valid. Supplementing such samples with a foreign material, in this case an *Escherichia coli* extract, provides an overall consistency in mass spectra between calibrants and

samples. Additionally, the individual mass signatures associated with *E. coli* specific proteins allows for an internal normalization process that also accounts for variability across the entire sample handling process and has resulted in absolute quantitation with high reproducibility (CV < 10%).

5   FURTHER SEPARATION OF COMPLEX MIXTURES

The ability to detect low abundant species remains a critical challenge in deciphering an entire proteome and correlating proteome changes with metabolic events for diagnostic purposes. Anderson and Anderson[35] have thoroughly described the topic of protein concentration dynamic range in serum and the obstacles in detecting low-abundance proteins. Depletion methodologies are frequently used to remove the most abundant species, however, this removal not only fails significantly to enrich trace proteins, but it may also nonspecifically and nonreproducibly deplete them due to their interactions with the removed high-abundance proteins. Albuminomics, for example, specifically looks at only biomolecules that are co-depleted during albumin removal because of their specific interaction with albumin.[36] Although this is an interesting approach for the detection of specific and potentially low-abundance molecules, its clinical utility has yet to be established.

Recently, we reported a new methodology that reduces the protein concentration range of a complex mixture, like whole serum, through the simultaneous dilution of high-abundance proteins and the concentration of low-abundance proteins. The principle of this novel strategy is based on the selective adsorption of proteins on a solid-phase combinatorial ligand library under capacity-overloading binding conditions. The spatial arrangement of amino acids within a protein defines its physicochemical properties, for example, isoelectric point, charge density and hydrophobicity index, and conformation. The latter determines the ability of a protein to interact in vivo with other molecules having complementary structures and forms the basis of protein separation by affinity chromatography where the interacting molecule (ligand) is chemically attached to a solid carrier.[37] The complementary proteins to the immobilized ligands are captured from complex mixtures up to the saturation of the available ligand. With sufficient diversity of ligands, it is theoretically possible to have a ligand to every protein in a complex mixture, ensuring that each is adsorbed. When a biological extract like serum is exposed to such a ligand library under specific capacity-constrained conditions, an abundant protein will quickly saturate all of its available high affinity ligands and the vast majority of the same protein will remain unbound. In marked contrast, a trace protein will not saturate all its high affinity ligands and the majority of the same protein will be bound. Thus, based on the saturation-overloading principle, a combinatorial solid-phase library will enrich for trace proteins relative to their abundant counterparts. Following

washing to remove unbound or weakly bound proteins, elution of the adsorbed proteins from the beads will result in a solution with a narrower dynamic range of protein concentrations while still representing all proteins present in the original material.

To be effective, the library must meet three criteria: (i) a sufficient reproducible diversity of ligands must be present to reliably bind each protein in the mixture; (ii) dissociation constants of the ligands and proteins must be compatible with the protein concentration; and (iii) the ligand and its support must be compatible with the unfractionated test sample and have a binding capacity high enough to capture sufficient protein to be detected by current methods. The technology is founded upon libraries of peptide ligands on which proteins can be adsorbed. On the basis of the work of Merrifield[38] on solid-phase synthesis using the "split, couple and recombine" method, libraries of peptide ligands are synthesized on resin beads.[39–41] Each bead has millions of copies of a single, unique ligand, and each bead potentially has a different ligand. Using only the 20 natural amino acids, a library of linear hexapeptides contains 20^6, or 64 million different ligands. The addition of unnatural amino acids and D-enantiomers into branched, linear, or circular ligands generates a potential library diversity that is, for all practical purposes, unlimited and may contain a ligand to every protein present in a biological sample.

In the current format, a given volume of affinity beads, typically $100\,\mu l$ to 1 ml, is incubated with at least a 10-fold excess of biological sample for about 1–3 h. Once ligands bind their corresponding proteins, the beads are washed to eliminate unbound or weakly bound proteins. Adsorbed proteins are subsequently released by means of classical elution methods used in chromatography. The eluted protein mixture can then be analyzed by standard methods such as 1-dimensional electrophoresis (DE) and 2-DE and/or SELDI-MS. As reviewed by Righetti et al.[42] such enrichment was evidenced in a variety of very different biological samples, including human sera, low concentration cell culture supernatant, and chicken egg white. Thus, the method largely addresses the problem of the dynamic range in clinical proteomic analyses.

6 CONCLUDING REMARKS

During the course of the last decade, SELDI protein array technology has demonstrated utility in the analysis of protein complexes, the discovery of relevant biomarkers, and the creation of assays for clinical, drug discovery, or basic biological research. With the more recent global proteomic focus on biomarkers and their utility in human health, the amount of time and resources directed to this topic will likely increase at a pace not seen earlier. Irrespective of the detection method, a fundamental problem of enriching for the disproportionately large number of low-abundance proteins in a background of a few dozen high-abundance proteins is the current challenge. Currently, the clear direction is for the depletion of the abundance proteins followed by the concentration of the rare species. Whether this is accomplished serially by abundance protein depletion followed by subsequent remaining analyte concentration and/or liquid chromatographic steps, or simultaneously via combinatorial approaches such as that used in Ciphergen's equalizer bead technology, the ultimate decision will be driven by the reproducibility of the methods and how this translates to the biological question. Greatest efforts will also be directed to adequately work with samples at ever-decreasing volumes. Considerable efforts are already directed toward micro and nanofluidics separation techniques that can be interfaced with conventional detection methods such as fluorescence or MS. In this area, such capabilities of chromatographic separation or equalizer bead enrichment followed by secondary cleanup on SELDI arrays is a logical approach to this problem. In terms of SELDI surface chemistry, continued efforts in developing novel affinity surfaces to further extract unique populations of proteins from a biological sample will provide an additional level of specificity and will ultimately lead toward the design of specific surfaces for the routine quantitative analyses of analytes demonstrated to be statistically correlated with a biological question.

Improvements in chip reader technology is also expected as work toward the creation of MS devices with simultaneous high detection sensitivity for proteins and mass resolving power for peptides and proteins is enabled by implementing new laser optic, ion optic, and mass analyzer approaches. Additionally, it is expected that the next generation of chip readers will not only demonstrate improved analytical performance when compared with their predecessors, but also provide a more robust, easier-to-use, and quantitative platform for the purpose of facilitating translational proteomic studies as well as the performance of in vitro diagnostic tests.

REFERENCES

1. T. W. Hutchens and T. T. Yip, New desorption strategies for the mass spectrometric analysis of macromolecules. *Rapid Communications in Mass Spectrometry*, 1993, **7**, 576–580.
2. S. R. Weinberger, E. A. Dalmasso, and E. T. Fung, Current achievements using ProteinChip array technology. *Current Opinion in Chemical Biology*, 2002, **6**, 86–91.
3. E. T. Fung, Strategies in clinical proteomics: from discovery to assay. *Preclinica*, 2004, **2**, 253–258.
4. E. F. Petricoin, A. M. Ardekani, B. A. Hitt, P. J. Levine, V. A. Fusaro, S. M. Steinberg, G. B. Mills, C. Simone, D. A. Fishman, E. C. Kohn, and L. A. Liotta, Use of proteomic patterns in serum to identify ovarian cancer. *Lancet*, 2002, **359**, 572–577.
5. D. F. Ransohoff, Lessons from controversy: ovarian cancer screening and serum proteomics. *Journal of the National Cancer Institute*, 2005, **97**, 315–319.
6. J. L. Hess, L. Blazer, T. Romer, L. Faber, R. M. Buller, and M. D. P. Boyle, Immunoproteomics. *Journal of Chromatography B*, 2005, **815**, 65–75.
7. D. Barzaghi, I. D. Jenefir, L. P. Kimberly, and T. L. Born, Use of surface-enhanced laser desorption/ionization -time of flight to explore bacterial proteomes. *Proteomics*, 2004, **4**, 2624–2628.
8. X. D. Zhu, W. H. Zhang, C. L. Li, Y. Xu, W. J. Liang, and P. Tien, New serum biomarkers for detection of HBV-induced liver cirrhosis using SELDI protein chip technology. *World Journal of Gastroenterology*, 2004, **10**, 2327–2329.
9. E. Head, K. Moffat, P. Das, F. Sarsoza, W. W. Poon, G. Landsberg, et al., Beta-amyloid deposition and tau phosphorylation in clinically characterized aged cats. *Neurobiology of Aging*, 2005, **26**, 749–763.
10. O. Yalkinoglu, G. Koenig, D. F. Hochstrasser, J. C. Sanchez, O. Carrette, *SELDI-TOF MS Detection and Identification of Protein Biomarkers for Diagnosing Alzheimer's Disease*, Vol. Application: *WO*, Bayer Healthcare A.-G, Germany, 2004, pp. 38.
11. A. S. Maddalena, A. Papassotiropoulos, C. Gonzalez-Agosti, A. Signorell, T. Hegi, T. Pasch, et al., Cerebrospinal fluid profile of amyloid β peptides in patients with alzheimer's disease determined by protein biochip technology. *Neurodegenerative Diseases*, 2004, **1**, 231–235.
12. H. D. Lewis, D. Beher, D. Smith, L. Hewson, N. Cookson, D. S. Reynolds, et al., Novel aspects of accumulation dynamics and Ab composition in transgenic models of AD. *Neurobiology of Aging*, 2004, **25**, 1175–1185.

13. P. Lewczuk, H. Esselmann, W. T. Groemer, M. Bibl, J. M. Maler, P. Steinacker, et al., Amyloid beta peptides in cerebrospinal fluid as profiled with surface enhanced laser desorption/ionization time-of-flight mass spectrometry: evidence of novel biomarkers in Alzheimer's disease. *Biological Psychiatry*, 2004, **55**, 524–530.

14. L. Le, K. Chi, S. Tyldesley, S. Flibotte, D. L. Diamond, M. A. Kuzyk, and M. D. Sadar, Identification of serum amyloid a as a biomarker to distinguish prostate cancer patients with bone lesions. *Clinical Chemistry*, 2005, **51**, 695–707.

15. K. Junker, J. Gneist, C. Melle, D. Driesch, J. Schubert, U. Claussen, and F. Von Eggeling, Identification of protein pattern in kidney cancer using ProteinChip arrays and bioinformatics. *International Journal of Molecular Medicine*, 2005, **15**, 285–290.

16. Y. F. Wong, T. H. Cheung, K. W. K. Lo, V. W. Wang, C. S. Chan, T. B. Ng, et al., Protein profiling of cervical cancer by protein-biochips: proteomic scoring to discriminate cervical cancer from normal cervix. *Cancer Letters (Amsterdam, Netherlands)*, 2004, **211**, 227–234.

17. C. Melle, R. Kaufmann, M. Hommann, A. Bleul, D. Driesch, G. Ernst, and F. Von Eggeling, Proteomic profiling in microdissected hepatocellular carcinoma tissue using protein chip technology. *International Journal of Oncology*, 2004, **24**, 885–891.

18. K. R. Coombes, Analysis of mass spectrometry profiles of the serum proteome. *Clinical Chemistry*, 2005, **51**, 1–2.

19. S. J. Walker and A. Xu, Biomarker discovery using molecular profiling approaches. *International Review of Neurobiology*, 2004, **61**, 3–30.

20. T. D. Veenstra, L. R. Yu, M. Zhou, and T. P. Conrads, Diagnostic proteomics: serum proteomic patterns for the detection of early stage cancers. *Disease Markers*, 2003, **19**, 209–218.

21. N. Tang, P. Tornatore, and S. R. Weinberger, Current developments in SELDI Affinity technology. *Mass Spectrometry Reviews*, 2003, **23**, 34–44.

22. E. F. Petricoin and L. A. Liotta, Clinical applications of proteomics. *Journal of Nutrition*, 2003, **133**, 2476S–2484S.

23. H. J. Issaq, T. P. Conrads, D. A. Prieto, R. Tirumalai, and T. D. Veenstra, SELDI-TOF MS for diagnostic proteomics. *Analytical Chemistry*, 2003, **75**, 148A–155A.

24. A. D. Long, H. J. Mangalam, B. Y. Chan, L. Tolleri, G. W. Hatfield, and P. Baldi, Improved statistical inference from DNA microarray data using analysis of variance and a Bayesian statistical framework: analysis of global gene expression in *Escherichia coli* K12. *Journal of Biological Chemistry*, 2001, **276**, 19937–19944.

25. R. Sprinthall, *Basic Statistical Analysis*, Prentice-Hall, New Jersey, 1997.

26. T. F. Pajak, G. M. Clark, D. J. Sargent, L. M. McShane, and M. E. Hammond, Statistical issues in tumor marker studies. *Archives of Pathology and Laboratory Medicine*, 2000, **124**, 1011–1015.

27. A. Ben-Dor, R. Shamir, and Z. Yakhini, Clustering gene expression patterns. *Journal of Computational Biology*, 1999, **6**, 281–297.

28. M. B. Eisen, P. T. Spellman, P. O. Brown, and D. Botstein, Cluster analysis and display of genome-wide expression patterns. *Proceedings of the National Academy of Sciences of the United States of America*, 1998, **95**, 14863–14868.

29. T. R. Golub, D. K. Slonim, P. Tamayo, C. Hurard, M. Gaasenbeek, J. P. Mesirov, et al., Molecular classification of cancer: class discovery and class prediction by gene expression monitoring. *Science*, 1999, **286**, 531–537.

30. E. J. Moler, M. L. Chow, and L. S. Mian, Analysis of molecular profile data using generative and discriminative methods. *Physiological Genomics*, 2000, **4**, 109–126.

31. L. Breiman, J. Friedman, R. Olshen, C. Stone, *Classification and Regression Trees*, Wadsworth, Pacific Grove, California, 1984.

32. H. Zhang, C. Y. Yu, B. Singer, and M. Xiong, Recursive partitioning for tumor classification with gene expression microarray data. *Proceedings of the National Academy of Sciences of the United States of America*, 2001, **98**, 6730–6735.

33. H. Kalasz, Editorial: displacement chromatography. *Journal of Chromatographic Science*, 2003, **41**, 281–283.

34. Z. Zhang, et al., Three biomarkers identified from serum proteomic analysis for the detection of early stage ovarian cancer. *Cancer Research*, 2004, **64**, 5882–5890.

35. N. L. Anderson and N. G. Anderson, The human plasma proteome: history, character, and diagnostic prospects. *Molecular and Cellular Proteomics*, 2002, **1**, 845–867.

36. M. S. Lowenthal, et al., Analysis of albumin-associated peptides and proteins from ovarian cancer patients. *Clinical Chemistry*, 2005, **51**, 1933–1945.

37. M. Wilchek, T. Miron, and J. Kohn, Affinity chromatography. *Methods in Enzymology*, 1981, **104**, 3–55.

38. R. B. Merrifield, Automated synthesis of peptides. *Science*, 1965, **150**, 178–185.

39. K. S. Lam, et al., A new type of synthetic peptide library for identifying ligand-binding activity. *Nature*, 1991, **354**, 82–84.

40. A. Furka and F. Sebetyen, General methods for rapid synthesis of multi-component peptide mixtures. *International Journal of Peptide and Protein Research*, 1991, **37**, 487–493.

41. A. D. Watts, N. H. Hunt, B. D. Hanbly, and G. Chaudhri, Separation of tumor necrosis factor alpha isoforms by two-dimensional polyacrylamide gel electrophoresis. *Electrophoresis*, 1997, **18**, 1086–1091.

42. P. G. Righetti, E. Boschetti, L. Lomas, A citterio, protein equalizer™ technology: the quest for a 'democratic proteome'. *Proteomics*, 2006, **6**, 3980–3992.

56

Fiber-Optic Array Biosensors

Rahela Gašparac and David R. Walt

Department of Chemistry, Tufts University, Medford, MA, USA

1 INTRODUCTION

Biosensors are chemical sensors that consist of three basic elements: a receptor (biological recognition element), transducer (physical component), and a separator (membrane or coating).[1–3] Optical biosensors are a class of biosensors that use light to detect the presence of a chemical or biochemical. An optical fiber biosensor is a biosensor that employs either a single optical fiber or an optical fiber array as the biosensing platform.

A typical single optical fiber is made of glass or plastic. Each fiber consists of an inner glass called the *core* and an outer glass called the *cladding*. Both core and cladding are encased in a buffer coating that protects the fiber from moisture and damage.[4] Optical fibers operate via a process called *total internal reflection*, which occurs when the core refractive index is higher than the clad refractive index (Figure 1a). Several optical fibers can be fused and drawn into a coherent bundle, creating an optical fiber array (Figure 1b).[5] Optical fiber arrays containing between 3000 and 100 000 individual optical fibers fused into a total diameter of 0.2–2 mm have been reported.[6] Fiber-optic arrays possess both advantages and disadvantages for biosensing.[7–11] Some important advantages include increased sensitivity, amenability to miniaturization, relatively low cost, electromagnetic immunity, and geometric flexibility.[9]

Different sensing chemistries may be attached to the end of the optical fiber array, allowing spatial and spectral resolution. Thus, optical fiber arrays have become extremely important in medical research,[12,13] clinical diagnostics,[14,15] food quality control,[14,16,17] and pharmacology and pharmaceuticals development.[18] Biomolecules commonly immobilized on optical fiber arrays include oligonucleotides, polymerase chain reaction (PCR) products, enzymes, and antibodies. Living cells have also been loaded in microwell optical fiber arrays, allowing live whole-cell biosensing. Fiber-optic array biosensors are classified on the basis of the nature of the biological recognition element used for sensing. In this review, we discuss nucleic acid, enzyme, whole-cell, and immunoassay (antibody/antigen) fiber array–based biosensors.

1. Nucleic acid biosensing relies on immobilization of a single-stranded deoxyribonucleic acid (ssDNA) on one end of the fiber-optic array surface. Detection is based on hybridization between ssDNA and a fluorescently labeled complementary ssDNA sequence.
2. Enzymes are one of the most commonly used biological recognition elements for fiber-optic biosensors. The enzyme acts as a catalyst, facilitating a highly specific and sensitive reaction. The products of the enzymatically catalyzed reaction are usually detected either directly or indirectly upon interaction with an indicator.
3. Whole-cell fiber-optic array biosensors commonly employ genetically engineered whole cells as the biological recognition elements. A wide variety of live mammalian cell types such

Handbook of Biosensors and Biochips. Edited by Robert S. Marks, David C. Cullen, Isao Karube, Christopher R. Lowe and Howard H. Weetall.
© 2007 John Wiley & Sons, Ltd. ISBN 978-0-470-01905-4.

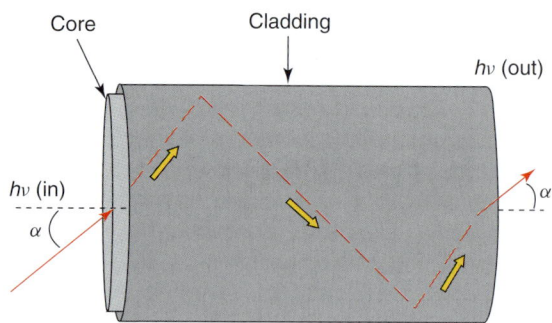

(a)

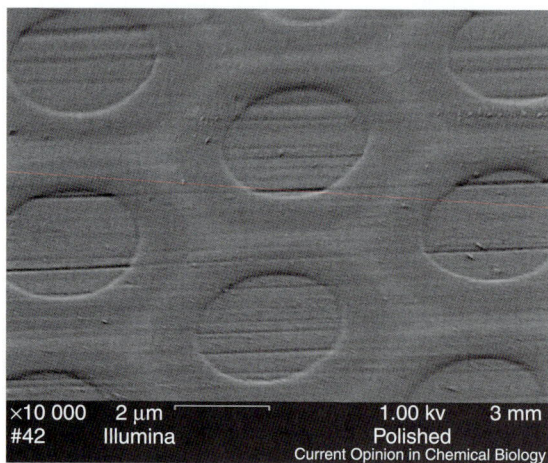

(b)

Figure 1. (a) Schematic of an optical fiber. Light propagates in the fiber along the entire length because of the refractive index differences of the core and clad materials. (b) Scanning electron micrograph (SEM) of a region of an optical imaging fiber array. The circles are the fiber cores that carry light signals. The darker gray regions are the cladding that confines the light within the cores. [Reprinted with permission Walt[5] copyright 2002, Elsevier.]

as neurons, cardiomyocytes, hepatocytes, and immune cells, as well as viruses, yeasts, and bacteria, have been used to fabricate whole-cell-based biosensors.

4. Immunoassay fiber-optic array biosensors report on the specific binding between an antibody as the biological recognition element and an antigen. Immunoassays are usually performed in one of three modes: direct, competitive, and sandwich. The specific binding of an antibody to an antigen allows for the detection of sample analytes by a variety of immunoassay modes.

2 FIBER-OPTIC ARRAYS AS NUCLEIC ACID BIOSENSORS

Over the past decade, there has been a significant increase in the development and use of nucleic acid–based fiber-optic arrays. This increase is, in part, due to the completion of the Human Genome Project and availability of genetic information from additional sequencing efforts.[19,20] Nucleic acid optical fiber arrays have been applied to pathogen detection,[21,22] medical diagnostics,[12,13] new drug discovery,[18] genetic analysis for detecting single-nucleotide polymorphisms (SNPs),[23–25] and gene expression analysis[26–28] with the potential for whole-genome screening.[29,30] Nucleic acid biosensing relies on the immobilization of an ssDNA called the *probe* on one end of the fiber-optic array surface. The detection method is based on hybridization between the probe and a fluorescently labeled complementary sequence of ssDNA called the *target*. The target of interest may first need to be amplified using PCR. The optimization of the nucleic acid immobilization strategy, the choice of fluorophore, and the hybridization temperature may improve selectivity, signal intensity, and signal-to-noise ratio.[13,22,31–33] The following section is a review of the recent advances in nucleic acid–based biosensing on optical fiber array platforms.

A versatile optical fiber array platform has been fabricated by chemically etching the distal end of a polished optical fiber bundle (typically containing between 6000 and 50 000 individual fibers). Chemical etching agents such as hydrofluoric acid or diluted hydrochloric acid are typically employed. Chemical etching allows selective etching of the fiber's core relative to the cladding.[10,34–36] Since the etching rate is faster for the core material, an array of identical microwells is created with diameters corresponding to the individual fiber cores (Figure 2a). Microwells of different depths can be tailored depending on the etching-agent concentration and the exposure time.[34,35] Figure 2(a) shows an atomic force micrograph (AFM) image of an etched optical fiber bundle containing ~3.6-μm-diameter microwells that are ~3-μm deep.[37] Complementary-sized latex or silica microspheres with different sensing chemistries can be loaded into the microwells either by applying a small aliquot of microsphere suspension directly to the

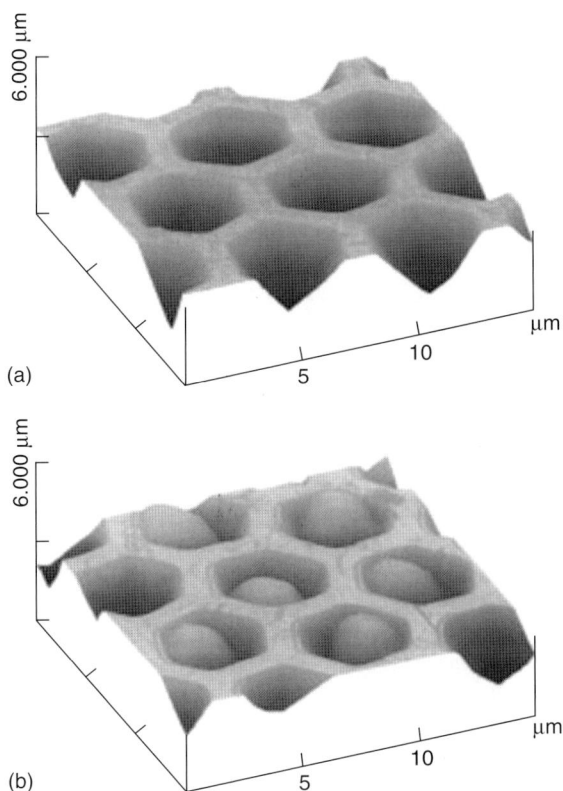

Figure 2. (a) Atomic force micrograph (AFM) image of ~3.6-μm-diameter microwells created by chemically etching a polished, 1000-μm-diameter imaging fiber. The distal end of the imaging fiber was submerged into a buffered hydrofluoric acid solution for 80 s. The microwells produced are ~3-μm deep. (b) AFM image of ~3.6-μm-diameter microwells containing a single 3.1-μm-diameter microsphere in each microwell. [Reprinted with permission Michael et al.[37] copyright 1998, American Chemical Society.]

optical fiber end or by dipping the etched optical fiber into a microsphere-containing solution (Figure 2b). Figure 2(b) shows an AFM image of ~3.6-μm-diameter microwells containing a single 3.1-μm microsphere in each microwell. It is important to note that the microspheres match the size of the etched microwell and remain firmly in the microwells via electrostatic interactions. Since each microwell is connected to its own optical fiber, each microwell can be interrogated as an individual sensor.

Nucleic acid optical fiber arrays can be prepared by attaching oligonucleotide probes to the microspheres. Oligonucleotide probes are usually attached to microspheres via an amine-linker, but other attachment strategies have been reported.[38] Each microsphere carries a specific DNA probe sequence. Microspheres of different DNA probe sequences are mixed together into a library and randomly distributed onto the etched optical fiber array surface (Figure 3a).[10,27,35,36,39,40] The position of the individual microspheres on the fiberoptic array cannot be predetermined. Thus, it is necessary to have a method for resolving the location of the microspheres in the array. Two methods for microsphere registration are employed.

In the first method, a unique combination of fluorescent dyes is used to encode different types of microspheres either before or after DNA probe attachment. Fluorescent dyes used for encoding must have different excitation and emission wavelengths that allow a unique signature, that is, an "optical bar code" for each microsphere type to be determined. The encoding step is important because it enables the positional registration of every microsphere in the fiber array.[11,39,41] Each microsphere with its unique optical bar code can then be easily decoded by simply collecting a series of fluorescence images with a charge-coupled device (CCD) camera at different excitation and emission wavelengths, and analyzing the resultant intensities of each microsphere with imaging software (Figure 3b). Once the positional registration of each encoded microsphere with its specific DNA probe in the array is determined, the platform is then exposed to a fluorescently labeled target DNA solution, producing signals only on those microspheres bearing the DNA probes complementary to the target DNA in solution. It is important to note, that fluorescent dyes used for microsphere encoding must have different excitation and emission wavelengths than fluorescent dyes used for target DNA labeling to prevent spectral overlap.

The second approach to array decoding uses a combinatorial decoding scheme that allows each microsphere type in the optical fiber array to be identified.[39,42] In this approach, each DNA probe-functionalized microsphere is identified using an algorithm that involves sequential hybridization of fluorescently labeled target DNA or decoders complementary to probe DNA sequences (Figure 4). Gunderson et al. were able to decode thousands of different DNA sequences with only a few fluorescent labels (called *states*) and several sequential hybridizations (called *stages*).[42]

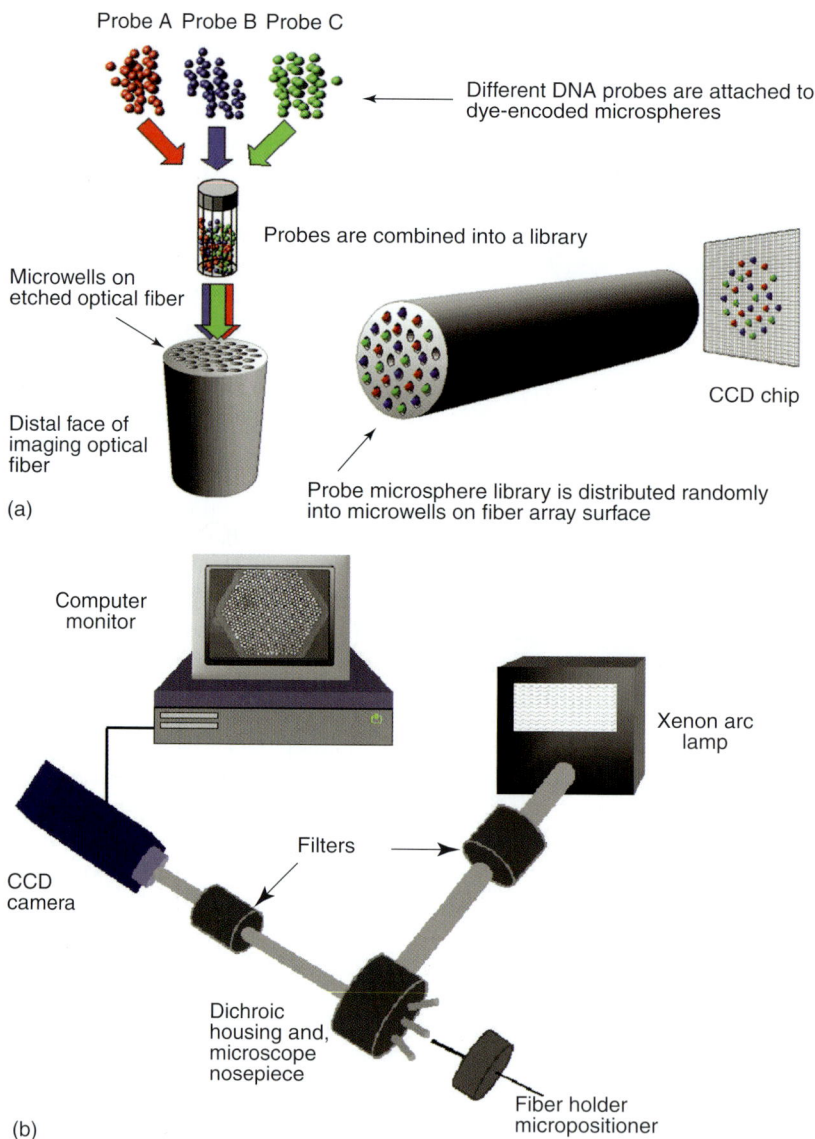

Figure 3. (a) Decoding scheme using a dye-encoded-microspheres approach. Three different DNA probes are attached to fluorescently encoded microspheres, combined into a probe microsphere library, and randomly distributed on the distal end of the optical fiber end. Each microsphere with its unique optical bar code is decoded by collecting a series of fluorescence images with a charge-coupled device (CCD) camera at different excitation and emission wavelengths. (b) The general setup of the custom-built imaging system consisting of computer-controlled modified epifluorescence microscope, a xenon arc lamp, excitation and emission filters, microscope objectives, and a CCD camera. Fluorescence images are processed using imaging software.

Figure 4 illustrates an example of decoding eight different microsphere types using sequential hybridization of fluorescently labeled target DNA in a two-state, three-stage system. Target DNA sequences complementary to DNA probes attached to the microspheres were synthesized and mixed in different combinations, creating unique decoder pools. The decoder pools were then used in sequential hybridizations with the microsphere array, generating a unique combinatorial code for each microsphere type. This combinatorial code identifies each microsphere type. With more

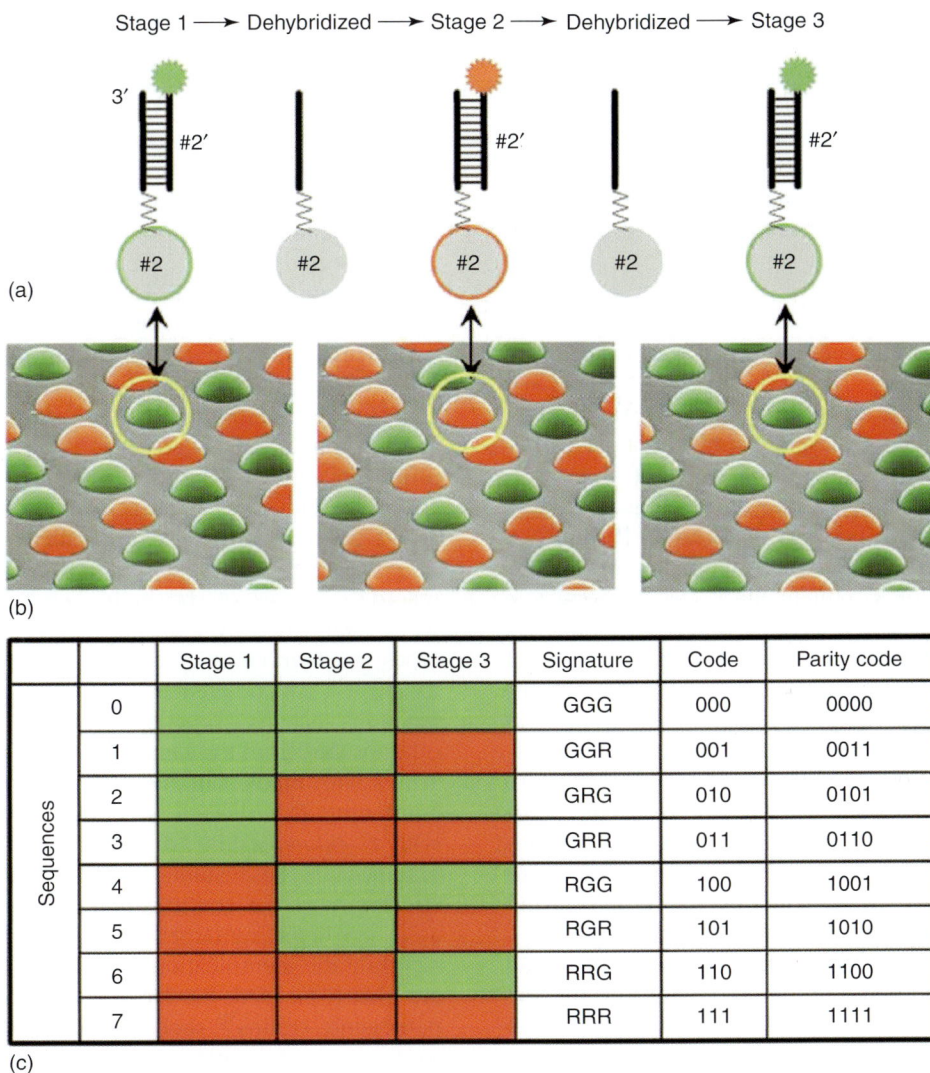

Figure 4. Combinatorial decoding process. (a) The sequential hybridization process is illustrated for a single microsphere, of microsphere type 2. In stage 1, a complementary decoder hybridizes to the oligonucleotide probe that is attached to the microsphere. The decoder is labeled with a fluorophore (green in stage 1, red in stage 2, and green in stage 3). The fluorescent signal is read by imaging the entire array. The array is then dehybridized, and the process is repeated for two more stages. (b) A scanning electron micrograph of an array of microspheres, false colored to represent three sequential hybridization stages. The images, taken collectively, reveal a combinatorial code for each microsphere. Note that the microsphere circled in yellow has the color signature GRG or code 010. (c) Colors or states are assigned to individual decoder sequences in each stage to produce a unique combination across stages. This signature, or code, identifies each microsphere type. As indicated in the parity code column, an extra decoding stage can be performed to provide an error checking parity bit. After three stages of decoding, all the microspheres are uniquely identified by their color. [Reprinted with permission Gunderson et al.[42] copyright 2004, Cold Spring Harbour Laboratory Press.]

dyes and more stages, the combinatorial decoding scheme enables the identification of up to 1792 different microsphere types[42] and was used in applications such as SNP genotyping[24,43,44] and gene expression profiling.[27,28]

Epstein et al. reported zeptomole detection limits (DLs) (~600 molecules) of three different target DNA sequences using this microsphere fiber array platform.[45] The DNA probe sequences used for this study belonged to interleukin 2 (IL2),

IL6, and the F508C mutation in the cystic fibrosis gene. The mixture of probe DNA–functionalized microspheres was randomly distributed on the 3.1-μm-diameter fiber array etched distal face. Hybridization with fluorescently labeled target DNA was allowed to proceed for up to 12 h. Satisfactory hybridization signals were achieved by averaging 10 microspheres of each probe type, enabling an acceptable signal with 7% coefficient of variation. In addition, target DNA concentrations of 1 fM gave 100% reproducible results, while 100 aM concentrations had a reproducibility of 80%.

Ferguson et al. reported a 10 fM DL of fluorescently labeled DNA targets using an optical encoding scheme.[46] Yeakley et al. used the microwell optical fiber array platform for profiling messenger ribonucleic acid (mRNA) isoforms.[47] The mRNA transcripts were analyzed without prior RNA purification or complementary deoxyribonucleic acid (cDNA) synthesis. The detection of mRNA isoforms was achieved from 10 to 100 pg of total cellular RNA. The reported sensitivity may be suitable for monitoring gene expression in tissues and the fiber-optic array matrix format permits parallel analysis of different samples. Steemers et al. were able to screen unlabeled genomic cystic fibrosis–related DNA targets using molecular beacons (MBs) immobilized on microspheres.[48] Thrombin detection was achieved by Lee et al.[49] using immobilized aptamers on microspheres.

Bowden et al. developed a fiber-optic array integrated with a microfluidic platform for the detection of the biological warfare agent (BWA) simulant *Bacillus thuringiensis Kurstaki* (*BTK*) capable of attomolar target DNA detection.[50] BWA detection draws lots of attention these days in order to minimize human casualties caused by a biological attack, improve public health, and enhance homeland security.[51] The detection of BWAs is very challenging because it needs to be specific, rapid, and extremely low concentrations of harmful substances need to be identified. In these studies 50-mer probe oligonucleotides were designed. The microfluidic fiber-optic array platform is depicted in Figure 5. The target DNA was passed across the fiber array surface via pressure-driven flow in a microfluidic channel, ensuring laminar flow. Using this system, the hybridization of 10 aM target DNA occurred in less then 15 min at a flow rate of 1 μ l min^{-1}. Comparison studies of static (no flow) and microfluidic measurements showed two significant advantages of the microfluidic setup: faster hybridization and lower DLs (1000-fold lower over static setup). It has been also shown that dehybridization in the microfluidic setup were both very rapid and very reproducible, enabling regeneration of the fiber-optic array.

In continuing research on BWA detection, Song et al. reported the use of the fiber-optic array platform for multiplexed BWA detection.[52,53] Eighteen different 50-mer oligonucleotide sequences were designed for *Bacillus anthracis, Yersinia pestis, Francisella tularensis, Brucella melitensis, Clostridium botulinum, Vaccinia* virus, and one BWA simulant—*BTK*. A cyanuric chloride coupling procedure was used to enhance DNA probe coupling efficiency to the encoded microspheres.[46] A DL of 10 fM was reported for *B. anthracis, Y. pestis, Vaccinia* virus, and *BTK*. The cross-reactivity assays of the multiplexed DNA array were established by exposing single-microsphere types to their complementary DNA targets and to the 17 noncomplementary ones. Some cross-reactions occurred but usually were limited to other probes from the same organism. Multiplexed arrays with mixed samples were also prepared by combining *B. anthracis* with *BTK*, and *Y. pestis* with *F. tularensis* in different ratios. The results obtained in these studies clearly indicated high specificity of the amplification and hybridization assay for detecting target BWAs.

While the identification and detection of BWAs are clearly needed, the classification of emerging pathogenic strains[54] such as SARS (severe acute respiratory syndrome),[55] West Nile virus,[56] and a virulent strain of *Vibrio cholerae* O139[57] are of great importance as well. In the effort to develop a fiber-optic array for bacterial detection, Shepard et al. used multilocus sequence typing (MLST) methodology[58–60] on a fiber-optic array platform to characterize 12 *Escherichia coli* strains at five loci: *ycgW, yaiN, osmB, galS*, and *serW*.[54] This group specifically selected a set of probe sequences for each locus allowing them to simplify the response to a binary signal/no signal response upon hybridization to each of the probes. Using this approach, 6 different oligonucleotide probes were enough to rapidly classify 12 different *E. coli* strains. A patterned response method was able to classify an "unknown" strain as wild-type nonpathogenic *E. coli* strain ECOR-44.

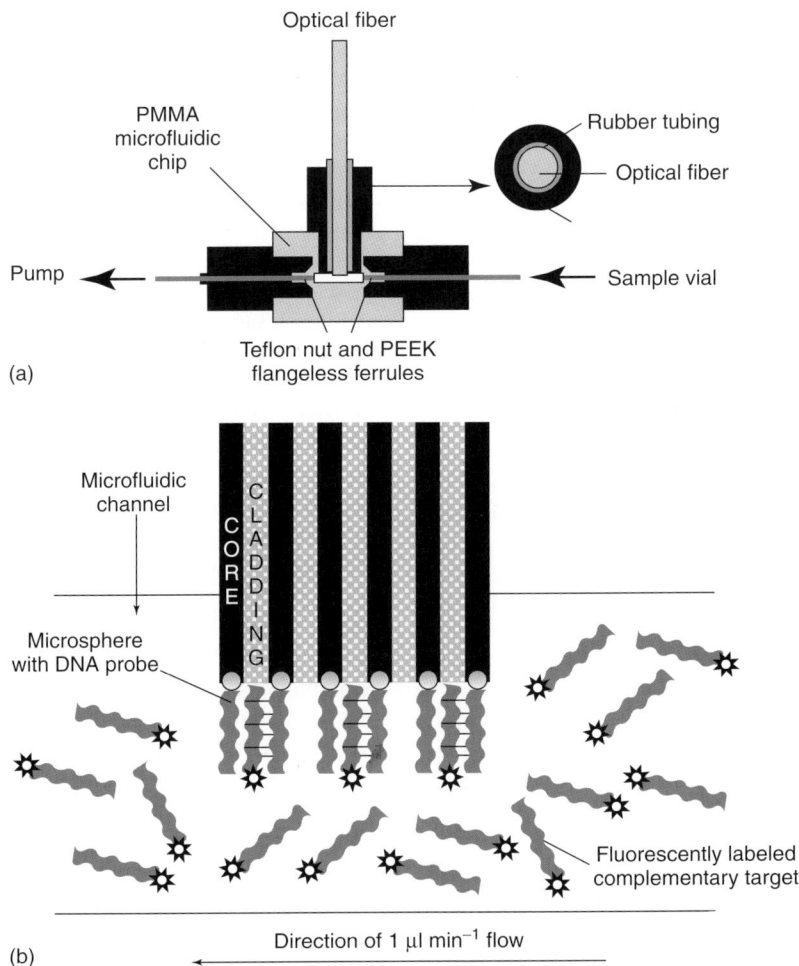

Figure 5. (a) A cross-sectional schematic diagram showing the microfluidic T-junction layout. The poly(methyl methacrylate) (PMMA) microfluidic chip was fabricated in a Tefzel tee consisting of a 1.25-mm inner diameter channel and a dead volume of <20 μl. The fiber bundle is positioned perpendicular to the flowing stream. The optical fiber array was jacketed by Teflon tubing and attached to the flow chamber by a PEEK nut and ferrule. (b) Schematic diagram of the optical imaging fiber bundle showing DNA hybridization in the microfluidic manifold. DNA probe microspheres that are complementary to the fluorescently labeled DNA targets fluoresce upon illumination through the imaging fiber proximal end. [Reprinted with permission Bowden et al.[50] copyright 2005, American Chemical Society.]

One of the challenges in the development of nucleic acid biosensors is SNP detection.[61] This process is often cumbersome—it requires amplification, followed by time-consuming enzymatic digestion or extension, and separation of the resulting products. Several authors reported the use of a commercially available microsphere-based optical fiber array platform for high-throughput SNP genotyping.[24,43,44] This platform not only combines multiplexed analysis but is also highly accurate, robust, and cost-effective, giving the opportunity to associate genes with phenotypes of interest. Shen and coworkers showed that high-throughput SNP genotyping can be performed on a universal microsphere fiber-optic array platform using the Illumina GoldenGate™ genotyping assay.[43] This assay detects up to 1536 SNPs in a single DNA sample (minimum 250 ng at 50 ng μl^{-1}). Gunderson et al. developed a whole-genome genotyping (WGG) assay on the microsphere-based optical fiber array platform that was used for genotyping hundreds of previously characterized SNPs.[44] A high signal-to-noise ratio was achieved by combining hybridization of relatively high concentrations (~2–3 pM) of WGA genomic deoxyribonucleic acid (gDNA) to arrayed 50-mer oligonucleotide probes with allele-specific primer extension (ASPE) and signal amplification. A high specificity was achieved in the presence of the entire genome.

In the continuing research of the WGG assay technology, Steemers et al. described an improved version of the WGG assay.[29] This version of the WGG assay includes single-base extension (SBE) rather than the ASPE step. The details of the ASPE and SBE steps in the WGG assay can be found elsewhere.[30] The genotyping performance was demonstrated by resequencing homozygous loci from the 100 k Human-1 Genotyping Bead-Chip, showing accurate and reproducible SNP genotyping assay data. Finally, Brogan et al. gave a detailed overview of the use of the microsphere-functionalized fiber-optic array platform for SNP, genotyping, and quantitative gene expression profiling.[21]

In addition, several recent publications by the Krull research group described nucleic acid–based optical fibers that did not require traditional labels.[31,62–64] The intercalating fluorescent dye thiazole orange (TO) was covalently linked to a 5′-oligonucleotide probe (nonpathogen *Erwinia herbicola* gene fragment)-modified fused silica optical fiber.[63] Hybridization with the complementary oligonucleotide sequence occurred in a sample solution and was monitored by detection of fluorescence intensity changes. It was shown that TO exhibits enhanced fluorescence intensity upon hybridization to a complementary DNA sequence in solution.[62] A sixfold fluorescence intensity enhancement has been reported upon hybridization with the complementary DNA sequence. Hence, the fluorescence intensity was decreased upon hybridization with the SNP target sequence, suggesting the potential application of DNA TO-based sensing in SNP analysis.

3 FIBER-OPTIC ARRAYS AS ENZYME BIOSENSORS

Enzymes are the most abundantly used biological recognition elements for fiber-optic biosensors.

In these systems, the enzyme acts as a catalyst facilitating a highly specific and sensitive reaction. The products of the enzymatically catalyzed reaction can be detected in direct fashion or indirectly upon interaction with an indicator (reagent).[8,65] Direct detection mode is possible when the substrates, products, or intermediates are spectroscopically active (e.g., they exhibit absorbance, fluorescence, or chemiluminescence (CL)). In this case, the optical transducer directly detects the optical changes. Indirect detection via an indicator is necessary when the substrates, products, or intermediates of the enzymatic reaction possess no intrinsic optical properties (e.g., O_2, CO_2, NH_3, pH, NADH, H_2O_2). In this case, the reduction or production of one of these species is detected by its reaction with the indicator.[7,9,66,67]

In the last decade, a plethora of papers have been published about enzyme optical fiber biosensors based on different types of transduction mechanisms.[7–9,68] Enzymes are employed in either pure form or contained on or within a biological matrix such as a cell or vesicle. This review focuses mainly on the fiber-optic array platform when substrates, products, or intermediates of the enzymatic reaction exhibit optical properties (fluorescence, CL, or electrogenerated chemiluminescence or electrochemiluminescence (ECL)).

3.1 Fluorescence Methods

Durrieu et al. developed an enzymatic whole-cell fiber-optic biosensor for the detection of freshwater pollutants such as Pb^{2+} and Cd^{2+}.[69,70] *Chlorella vulgaris* cells, which display alkaline phosphatase (AP) on the external surface of their cell membranes, were immobilized on a glass microfiber filter. This filter was placed in front of an optical fiber bundle inside a flow cell. The AP-catalyzed reaction produces two reaction products, phosphate and fluorescent methylumbelliferone (MUF) from methylumbelliferoyl phosphate (MUP). It has been shown that the AP catalytic activity is strongly inhibited in the presence of heavy metals, resulting in a decrease of MUF fluorescence intensity. Comparisons between a fiber-optic biosensor and a microplate reader were similar. Approximately 35% AP inhibition was achieved for 0.01 mg l^{-1} concentrations of Cd^{2+} and Pb^{2+}.

Doong and Tsai developed a fluorescence solgel-based fiber-optic biosensor for the determination of acetylcholine.[71] This approach employed the fluorescent dye fluorescein isothiocyanate (FITC)-dextran encapsulated with acetylcholinesterase (AChE) in a solgel network. The solgel matrix was immobilized on the distal end of an optical fiber bundle. The enzymatic reaction of AChE and acetylcholine forms acetic acid, resulting in a fluorescence intensity change of the pH-sensitive fluorescent FITC-dextran dye. This reversible and reproducible fiber-optic biosensor showed linearity between 0.5 and 20 mM of acetylcholine. These authors also reported the use of this fiber-optic biosensor for detecting organophosphorous pesticides such as paraoxon. A 30% AChE inhibition was obtained when 152 ppb paraoxon was added into the system, resulting in increased fluorescence intensity (Figure 6).

A fluorescence optical fiber biosensor based on *E. coli* cytoplasmic membranes as the source of enzyme was described by Ignatov et al.[72] *E. coli* cells were cultured in the presence of sodium lactate to initiate formation of the enzyme L-lactate oxidase. The L-lactate oxidase enzyme catalyzes the conversion of L-lactate to pyruvate. During this process, molecular oxygen is consumed and peroxide is liberated. An oxygen-sensitive dye, ruthenium tris(diphenylphenoanthroline) $(Ru(ph)_2phen)_3^{2+}$, was encapsulated in a thin polysiloxane film on the surface of an optical fiber bundle. When the enzyme is active, oxygen is consumed in the catalytic reaction, causing a fluorescence increase of the Ru-based dye. Interferences with other substances (glucose, fructose, and glutamic acid in 50-mM concentrations) were not observed, demonstrating good selectivity of the sensor.

Issberner et al. reported an enzyme-based fiber-optic array biosensor for L-glutamate release and reuptake from the foregut plexus of the tobacco hornworm *Manduca sexta*.[73] A gel composed of the enzyme L-glutamate oxidase (GLOD), a pH-sensitive fluorescent dye SNAFL (5- and 6-carboxy seminaphthofluorescein), and poly(acrylamide-*co*-*N*-acryloxysuccinimide) (PAN) was attached to the distal face of an optical fiber array by spin-coating. When L-glutamate reacts with GLOD, NH_3 is released. The increased ammonia concentration reduces the fluorescence intensity of SNAFL. The DL of L-glutamate was 10–$100 \mu M$. Interferences with other amino acids both in vitro and in vivo were not observed, demonstrating good specificity of the L-glutamate sensor. This paper also demonstrated the ability to visualize L-glutamate release in vivo from the nerves of the foregut plexus with a relatively high temporal resolution. This platform is distinguished from other enzyme-based array approaches because L-glutamate release can be localized using the spatial resolution of the array to provide both chemical and visual images of the sample.

A dual-enzyme fiber-optic array biosensor for pyruvate has been developed by Zhang et al.[74] An optical fiber bundle consisting of 14 individual fibers was used for these studies. Lactate oxidase (LOX) and lactate dehydrogenase (LDH) were immobilized at the optical fiber distal end. Pyruvate detection was accomplished by monitoring the consumption of the fluorophore NADH during the enzymatic reactions:

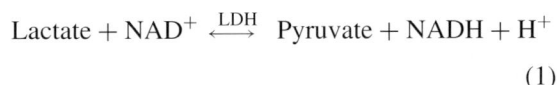

$$Lactate + NAD^+ \xrightarrow{LDH} Pyruvate + NADH + H^+ \quad (1)$$

$$Lactate + O_2 \xrightarrow{LOX} Pyruvate + H_2O_2 \quad (2)$$

By adjusting the pH of the sample solution the equilibrium of the first reaction can be controlled. NADH and pyruvate diffuse from the bulk solution

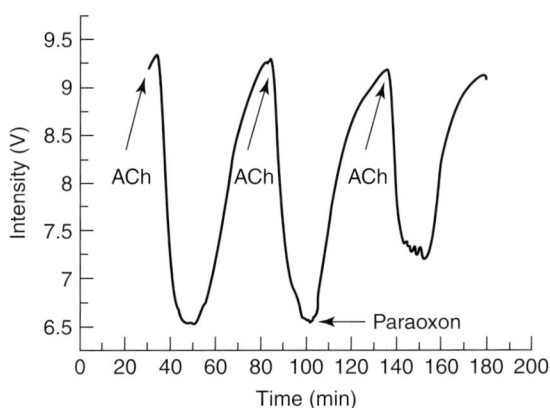

Figure 6. Time–response curve of an acetylcholinesterase-based fiber-optic biosensor before and after exposure to 0.54 mM (152 ppb) paraoxon. Paraoxon was added at the time indicated by the arrow. [Reprinted with permission Doong and Tsai.[71] Copyright 2001, Elsevier.]

toward the dual-enzyme (LDH and LOX) layer immobilized at the optical fiber end. At low pH, LDH catalyzes the formation of lactate and NAD^+. As the reaction proceeds, pyruvate is reduced and NADH is consumed at the fiber end. LOX then regenerates pyruvate as shown in the second reaction, allowing more NADH to react thereby making the measurement more sensitive. A steady-state NADH concentration is established when the rate of NADH consumption is counterbalanced with the NADH diffusion from bulk solution. The sensor was tested in response to the change of several key parameters: pH dependence, the pyruvate concentration, the NADH concentration, and amount of LOX applied to the optical fiber end. This dual-enzyme biosensor for pyruvate showed eightfold and fivefold improved sensitivity and DL, respectively, compared to single-enzyme fiber-optic biosensors. In addition, the biosensor works well when large amounts of NADH are added to the sample solution. Thus, it cannot be used in vivo where the levels of NADH and oxygen cannot be controlled.

Single-molecule enzymology is the ultimate in ultrasensitive biosensing.[75] In an effort to achieve this type of sensitivity, Rissin et al. developed a fiber-optic array–based biosensor capable of single-enzyme molecule detection.[76] The activity of single β-galactosidase molecules were monitored in femtoliter sized microwells that were fabricated by etching one end of a fiber-optic array (Figure 2a).[34] Each individual optical fiber monitored a reaction in which fluorescent products were built up from single-enzyme molecule catalysis. The fluorescence intensity of the enzymatic product resorufin was monitored across the array by β-galactosidase catalysis of the substrate resorufin-β-D-galactosidase. The collective monitoring of many single molecules over a large area, combined with Poisson statistics, allowed the enzyme concentration in the bulk solution to be calculated. The number of single β-galactosidase molecules predicted from the Poisson distribution correlated well with the observed number of molecules in the array measurements (Table 1). This sensor was capable of detecting β-galactosidase concentrations of 72 fM.

3.2 Chemiluminescence and Electrogenerated Chemiluminescence Methods

CL is the generation of visible light by the release of energy from a chemical reaction. ECL is a form of CL generated by an electrochemically initiated reaction.[77] Both methods offer numerous advantages over traditional optical techniques (fluorescence, absorbance, etc.). The background in CL and ECL is significantly reduced due to the lack of an optical excitation source. As an analytical technique, ECL offers greater control over a reaction because the initiation trigger can be easily modulated. Additionally, ECL typically offers better selectivity and sensitivity over CL. A few recently published CL and ECL fiber-optic array biosensors based on an enzymatic reaction are described here.

Navas Diaz et al. developed an enzyme-based fiber-optic biosensor for the detection of hydrogen

Table 1. Digital readout from the arrays[(a)]

Digital readout of enzyme concentrations			Experiments			Actual % active	
Enzyme-to-well ratio	Concentration	Poisson % active	t1	t2	t3	Average	Stand dev
1:5	7.20×10^{-12}	18.2	14.87	15.44	14.56	14.96	0.45
1:10	3.60×10^{-12}	9.5	11.46	14.96	6.92	11.11	4.03
1:20	1.80×10^{-12}	4.9	5.57	4.78	5.73	5.36	0.70
1:40	9.00×10^{-13}	2.5	3.47	2.80	1.92	2.73	0.78
1:80	4.50×10^{-13}	1.2	1.46	1.26	1.31	1.34	0.11
1:100	3.60×10^{-13}	1.0	1.10	1.11	1.70	1.30	0.34
1:200	1.80×10^{-13}	0.5	0.25	0.42	0.74	0.47	0.25
1:500	7.20×10^{-14}	0.2	0.10	0.15	0.09	0.11	0.03

[Reprinted with permission Rissin and Walt[76] copyright 2006, American Chemical Society.]
a The actual percentages of chambers exhibiting activity, in comparison to the expected percentage calculated from the Poisson distribution, are listed for the various concentrations analyzed.

peroxide (H_2O_2) as well as the phenol derivatives *p*-iodophenol, *p*-coumaric acid, and 2-naphthol.[78] The enzyme horseradish peroxidase (HRP) was encapsulated in a solgel matrix and immobilized on the distal end of an optical fiber bundle. The reaction of luminol and H_2O_2 forms luminol radicals that initiate CL light emission. When the CL enhancers *p*-iodophenol, *p*-coumaric acid, and 2-naphthol were added to the luminol–H_2O_2–HRP system enhanced CL occurred.[79,80] The DLs for *p*-iodophenol, *p*-coumaric acid, and 2-naphthol were reported to be 0.83 μM, 15 nM, and 48 nM,

respectively, showing that *p*-coumaric acid is the best CL enhancer of the luminol–H_2O_2–HRP system. The CL enhancement of the phenol derivatives on H_2O_2 detection was established as well. The best CL enhancement was achieved with *p*-coumaric acid when a DL of 1.6 μM H_2O_2 was obtained.

Chovin et al. demonstrated the used of a nanoaperture fiber-optic array platform[4,81–87] for ECL detection of NADH in the presence of tris(2,2′-bipyridine) ruthenium (II) chloride ($Ru(bpy)_3Cl_2$).[88] Figure 7 is a schematic

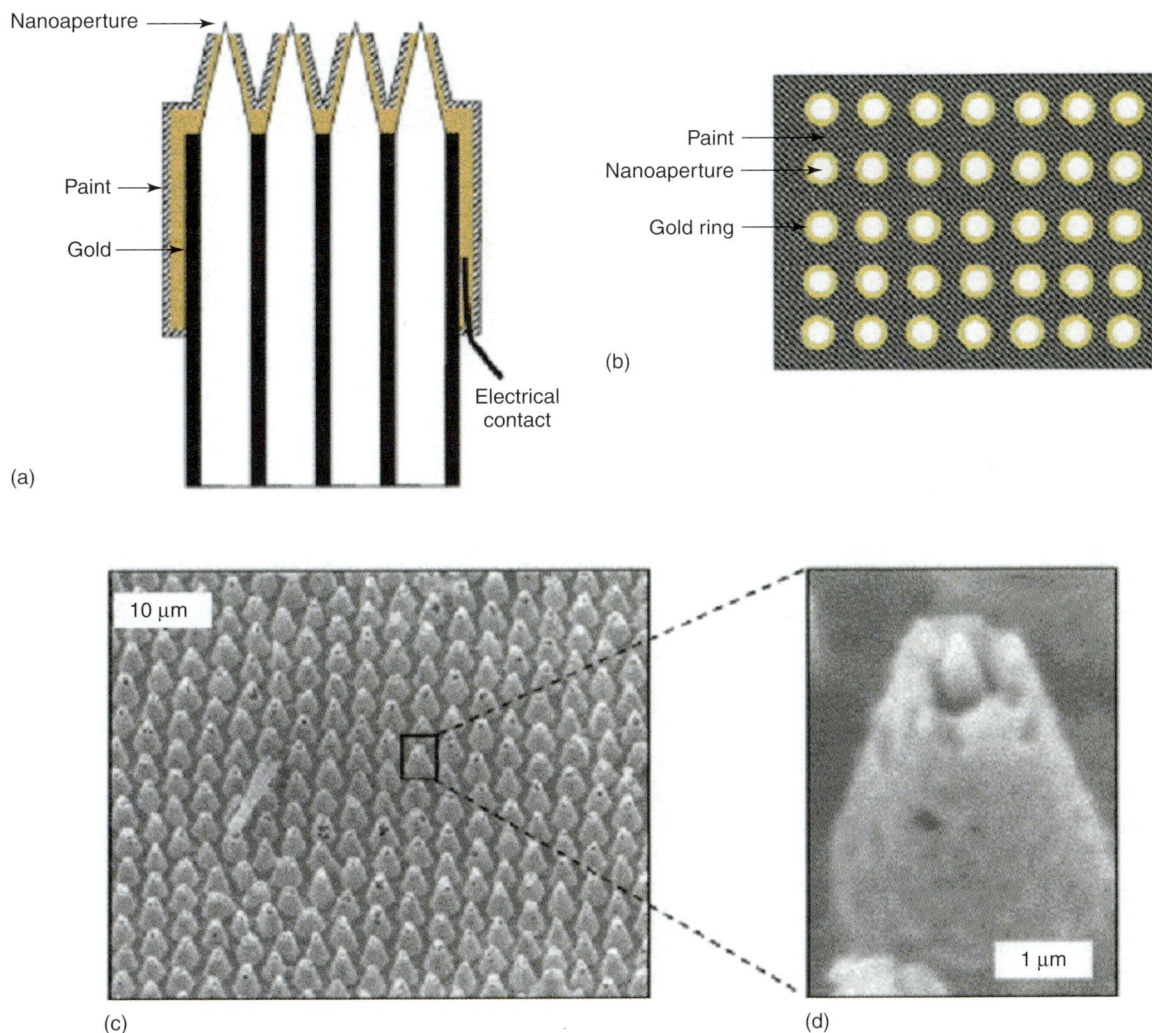

Figure 7. Schematic illustration of the nanoaperture array, (a) side view and (b) top view. (c) Scanning electron micrograph of the distal face of the nanoaperture array. (d) High magnification of a single nanoaperture in the array. [Reproduced by permission of Elsevier from Chovin et al.[88]]

illustration showing the nanoaperture fiber-optic array which comprised ~6000 individual optical fibers. The nanoaperture fiber array was sputter-coated with gold, creating a gold ring that acted as an electrode. Additionally, each nanoaperture confines light within each optical fiber core. The $Ru(bpy)_3^{2+}$ complex mediates the NADH oxidation, generating ECL light at the distal end of the nanoaperture array by each gold ring (radius ~200 nm). A fraction of the collected ECL light is then transmitted through the optical fiber nanoaperture array and imaged with a CCD camera. Thus, this optical fiber array platform combines imaging and electrochemical properties. The nanoaperture array was characterized by cyclic voltammetry, showing good electrochemical performance. Stable and reproducible ECL intensities showed linear behavior in the NADH concentration range of 0.5–15 mM.

4　FIBER-OPTIC ARRAYS AS WHOLE-CELL BIOSENSORS

Another trend in fiber-optic array biosensing is employing living cells as biorecognition elements. Whole-cell-based sensing is unique, because live cells generate a specific physiological or genetic signal in response to a wide variety of biologically active compounds or environmental conditions. These biosensors are often considered functional biosensors because they only respond when the analyte of interest can access and bind to its receptor(s) and cause downstream effects. Cell biosensors provide information about bioavailability and biological relevance in the context of a living cell. In contrast, conventional sensors and biosensors only report on *binding*.

Whole-live-cell biosensing has recently found applications for water quality testing,[89] clinical diagnostics,[90] and metal monitoring.[91] A wide variety of mammalian cell types such as neurons, cardiomyocytes, hepatocytes, immune cells, as well as viruses, yeasts, and bacteria have been used to fabricate whole-cell-based biosensors.[92–94] Methods for analyzing cell responses include cytometry,[95–97] electrochemistry,[98,99] capillary electrophoresis,[100–102] and various separation techniques.[103] These methods are usually invasive, disturbing the cell contents, require complex experimental setups, and are limited to short-term measurements. To overcome these limitations, new techniques offering small-volume sampling, noninvasiveness, and repetitive measurements of whole cells are needed.

Taylor et al. demonstrated that fluorescent dye-encoded NIH 3T3 mouse fibroblast cells can be loaded onto fiber-optic microwell arrays and optically interrogated.[104] An optical fiber array containing ~6750 individual fibers (diameter = 7 μm) was etched, creating a 3-μm-deep microwell. Each microwell can host a single cell that preferentially attaches to the well bottom, eliminating the need for an entrapment matrix. Three distinct lipophilic dyes (PKH26, PKH67, and $DiIC_{18}$) were used to encode the cells to enable them to be distinguished from one another. The locations of the cells within the array were detected by a CCD camera by collecting the fluorescence from the encoding dye upon excitation (Figure 3b). The encoding fluorescent dyes were rapidly incorporated into the cell and exhibited no toxic effects on the cell. In fact, cells retained their viability over 20 h without disturbing the cell contents, making them suitable for practical screening applications.

The whole-cell sensor arrays have become an attractive approach for high-throughput screening (HTS) applications.[92,105,106] Walt and coworkers used the microwell fiber-optic array platform for simultaneous monitoring of large numbers of individual cells from different strains or cell lines.[107–109] The use of the fiber-optic array platform was demonstrated using the yeast (*Saccharomyces cerevisiae*) two hybrid (Y2H) system[107] and genetically modified bacterial (*E. coli*) strains[108,109] for monitoring single-cell expression. Fluorescence detection was employed to monitor gene expression using *lacZ*, EGFP, ECFP, and DsRed reporter genes. A Y2H system was constructed by transforming yeast cells with two plasmids in order to detect protein–protein interactions in vivo. If two proteins of interest interact, the reporter gene transcription is enabled, allowing these reporter genes to be detected.[110,111] Figure 8 shows a scanning electron micrograph (SEM) image of a single yeast cell and an array of yeast cells in a fiber-optic array.[108] Yeast cells were encoded with fluorescent dyes, enabling them to be identified. Fluorescence signals obtained from individual cells of 3 yeast strains showed that 33% of cells in a positive strain, 5% in a negative strain, and none in a wild-type strain

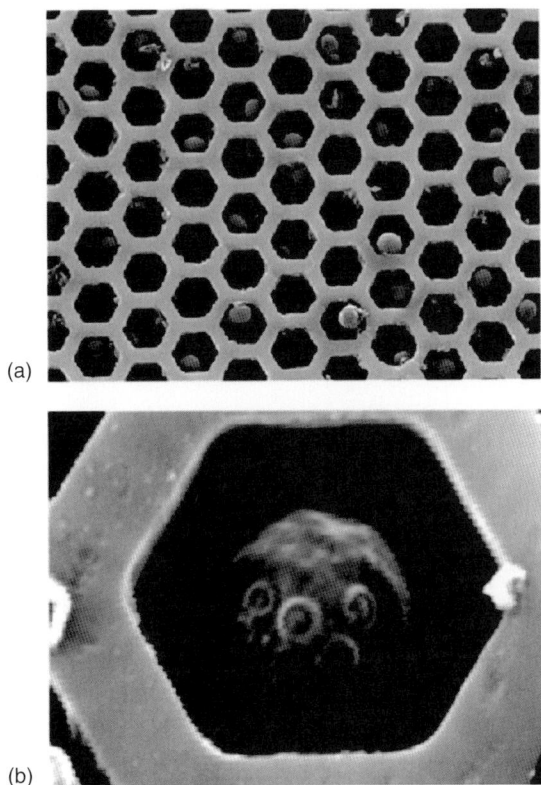

(a)

(b)

Figure 8. Loading cells into wells. Scanning electron micrograph (SEM) of (a) an array of yeast cells in microwells and (b) a single yeast cell in a microwell. Each microwell has a diameter of $6\,\mu m$. [Reprinted with permission Biran and Walt[107] copyright 2002, American Chemical Society.]

exhibited a fluorescence increase over 100 units from the background after 4 h.[107] In another study, bacteria were genetically engineered to express reporter genes coding for optically detectable proteins. Three plasmids were used, each encoding for red, green, or cyan fluorescent protein. This genetic encoding scheme creates a unique optical signature for individual cells, allowing positional registration of each cell in the array.

Many different sensing cell lines have been constructed for the detection of toxic metals,[112–115] aromatic compounds,[116] genotoxins,[117,118] and drug discovery.[119] Biran et al. incorporated an E. coli mercury-sensing strain on the microwell fiber-optic array platform.[108] The bacterial strain E. coli RBE27-13 containing lacZ fused to zntA was used. Single-cell lacZ expression was monitored as a function of mercury concentration

$(0–5\,\mu M)$. As low as $100\,nM\,Hg^{2+}$ was detected within 1 h, demonstrating the possibility of using the system for mercury ion detection. Ikariyama et al. used E. coli pTSN316 strain carrying a xylR-xylS::lux fusion for benzene-related aromatics detection.[120] These authors immobilized a genetically modified E. coli strain on a fiber-optic bundle composed of 75 individual fibers (diameter $= 0.3\,mm$). A DL of $5\,\mu M\,l^{-1}$ for m-xylene was achieved, showing promise for using this platform for detecting environmental contamination by benzene derivatives.

Kuang et al. used a high-density optical fiber array platform for genotoxin detection including the compounds mitomycin C (MMC), N-methyl-N-nitro-N-nitrosoguanidine, hydrogen peroxide, nalidixic acid, and formaldehyde.[109] E. coli cells (strain MG1655) carrying a toxin-sensitive recA promoter fused to a fluorescent gfp reporter gene (recA::gfp) were placed on the etched end of an optical fiber bundle and used as sensing components. This whole-cell biosensor exhibited highly sensitive genotoxic detection during short incubation times ($1\,ng\,ml^{-1}$ MMC after 90 min). With longer incubation times, it should be possible to detect even lower MMC concentrations. A shelf lifetime of two weeks and active sensing lifetime of more than 6 h were reported.

Whole-cell–based fiber-optic microwell arrays have also been used for fundamental cell biology studies such as monitoring gene expression rates and the dynamics of genetic noise.[121] A single living cell is a complex system in which many biochemical processes occur simultaneously, exhibiting inherent stochasticity or noise in the process of gene expression. Genetic noise provides information about the structure of gene regulatory circuits[122,123] and helps in predicting and designing gene network functions.[124–126] Monitoring gene expression allows investigation of genetic noise among individual cells in a cell population.[127] Two populations of cells (E. coli strains carrying pSC101 plasmids with lacZ::gfp or recA::gfp fusions) were used to simultaneously monitor genetic activity in single cells over 80 min (Figure 9). Hundreds of individual E. coli cells were employed to study lacZ and recA gene expression kinetics and dynamic changes in cell population noise. Fluorescence signal intensities were plotted for individual recA::gfp cells

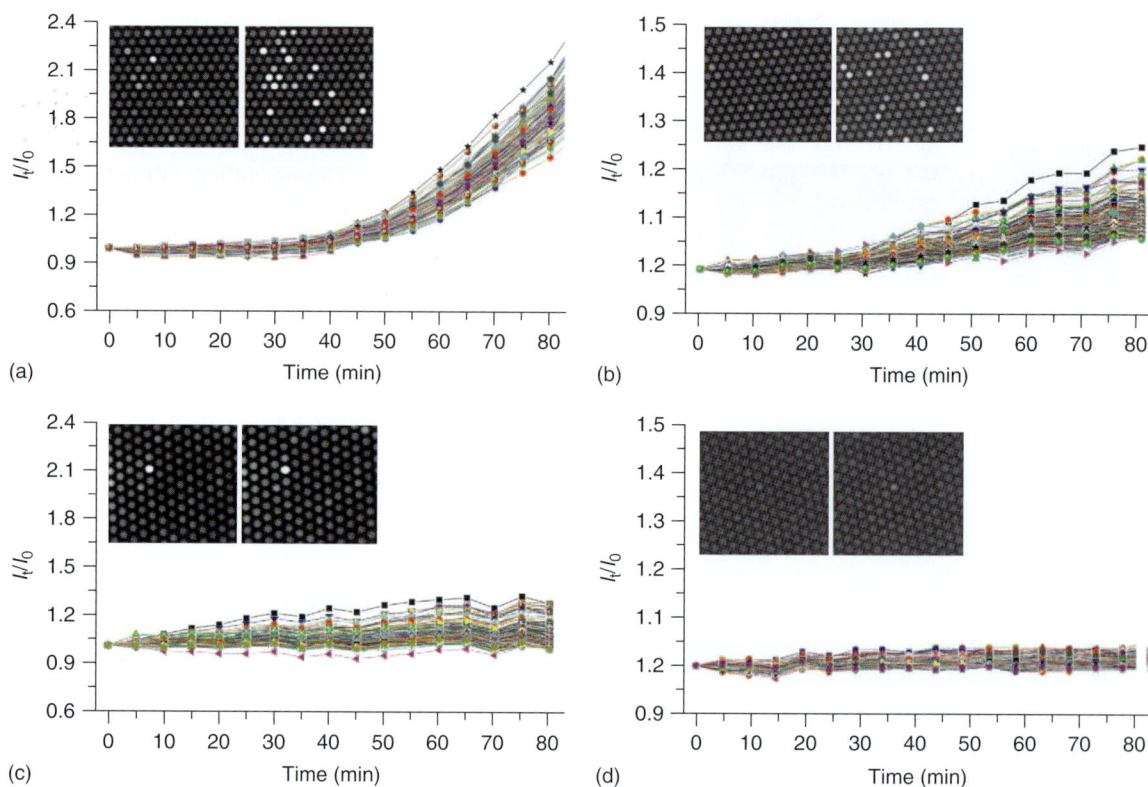

Figure 9. Monitoring *lacZ* and *recA* promoter activity in multiple individual cells. Fluorescence signals are expressed as the fluorescence intensity increase (I_t/I_0). (a) Individual *recA::gfp* cells from an array exposed to $10 \,\mu g \, ml^{-1}$ MMC. (b) *recA::gfp* cells in the control array incubated only with medium. (c) Individual *lacZ::gfp* cells in an array exposed to 5 mM IPTG. (d) *lacZ::gfp* cells in the control array incubated only with medium. Approximately 200 cells from each array are shown. Fluorescence images from a small portion of the imaging fiber-based cell array corresponding to each experiment are shown in the insets. In each inset, the left panel shows the fluorescence image at time 0 and the right panel shows the image after 80 min. [Reprinted with permission Kuang et al.[121] copyright 2004, American Chemical Society.]

in the control array incubated only with M9 medium, for individual *recA::gfp* cells in the array exposed to M9 medium supplemented with $10 \,\mu g \, ml^{-1}$ MMC, and for individual *lacZ::gfp* cells that were induced with 5 mM isopropyl β-D-1-thiogalactopyranoside (IPTG) (Figure 9). Gene expression was measured by acquiring fluorescence images from the array (Figure 9 insets) every 5 min. In addition, gene noise values for both *recA* and *lacZ* showed that expression levels and noise of noninduced *recA* cells were higher than those of *lacZ* cells (recA noise 40.33 ± 16.4, lacZ noise 8.5 ± 5.8). The fully induced *lacZ* gene was noisier than the fully induced *recA* gene, in spite of the more complex gene regulation mechanism of *lacZ*.

Recently, a nonetched fiber-optic array platform was used to monitor whole-cell and subcellular migration.[128] Cell migration analysis is of great importance in cancer proliferation. In many cases, malignant cancer cells migrate from the original tumor site and create metastases in other parts of the body.[128] The authors deposited fluorescently encoded fibroblasts onto a polished nonetched optical fiber array.[128] Cell migration was monitored with and without exposure of the whole array to different concentrations of the antimigratory drug nocodazole. Fluorescence intensity spikes were recorded over time when cells migrated over an individual fiber. This group showed that when nocodazole was applied, the whole-cell and subcellular migration velocity decreases, causing each cell to spend more time over an individual fiber. In the whole-cell studies, the cell velocity decreased ~6 times upon 300 nM nocodazole exposure, and in subcellular studies the cell velocity decreased ~3 times for the same nocodazole exposure concentration. This array platform can be easily used to detect changes in other cell types by altering the drugs and fluorescent encoding dyes.

5 FIBER-OPTIC ARRAYS AS IMMUNOASSAY BIOSENSORS

Immunoassay biosensors are a class of biosensors that report on the specific binding between an antibody as the biological recognition element and an antigen (analyte). Antibodies possess high specificity and affinity for antigens of interest. An antigen can be a naturally fluorescent compound or can be fluorescently labeled. A primary antibody is typically immobilized on the surface of a transducer.[129] Immunoassays are usually performed in one of four modes: direct, competitive, binding inhibition, and sandwich (Figure 10).[9,130–132]

Figure 10 shows assay formats commonly employed in immunosensing. In a direct assay (Figure 10a), the fluorescence signal is generated when a naturally fluorescent antigen and immobilized antibody form a complex. In this format, the signal is directly proportional to the amount of antigen. The sensitivity of this assay depends on the number of immobilized antibody molecules available for complex formation with the antigen. In a competitive assay format, the unlabeled analyte in the sample and fluorescently labeled antigen compete for immobilized antibody binding sites (Figure 10b). In this assay, the fluorescence intensity of the labeled antigen is measured. The unlabeled analyte blocks binding by the labeled antigen. Thus, the measured signal intensity is inversely related to the concentration of the unlabeled analyte: the higher the fluorescence signal, the lower the analyte concentration. Finally, the sandwich assay is based on the formation of a complex, the "sandwich", between two highly specific antibodies—a primary antibody that is immobilized on the transducer and a second fluorescently labeled antibody that binds to a second antigenic site (Figure 10c). The signal in sandwich immunoassays is generated only if the analyte binds to both the primary antibody and the fluorescently labeled antibody. This assay is highly specific with the ability to detect specific analytes at low concentrations, enabling the reduction of false positives.[133,134]

Immunosensing signal transduction can be achieved by a variety of different techniques such as electrochemical, piezoelectric, optical, or calorimetric ones.[132,135–138] In recent years considerable research effort has been devoted to the development of fiber-optic immunoassay biosensors for the detection of specific analytes,[16,116,131,139–142] harmful food-borne pathogens,[17,139,143] clinically relevant drugs,[144] proteins,[18,133,145–148] and viruses.[135,149] Several studies have focused on the optical fiber probe immunoassay biosensors,[150–152] and some of them have focused on the development of the optical fiber array immunoassay biosensors.[153–156] The following section is a review of the recent advances in immunoassay-based biosensing on optical fiber array platforms.

Szurdoki et al. used an etched optical fiber bundle (Figure 2a) to simultaneously detect the clinically important drugs digoxin and theophylline.[144] Two sets of amine-functionalized microspheres (3.15 μm in diameter) were subjected to different encoding dyes and attachment strategies.

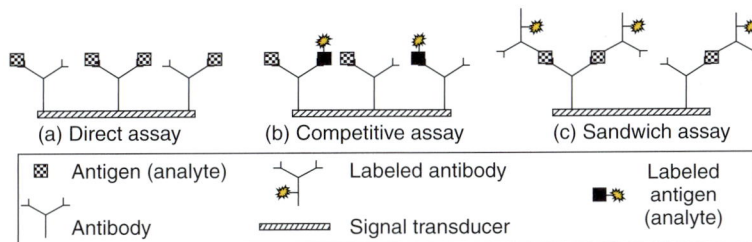

Figure 10. Assay formats commonly employed in immunosensors: (a) direct assay, (b) competitive assay, and (c) sandwich assay.

The amine-functionalized microspheres for the theophylline assay were encoded with BODIPY 493/503 dye. These encoded microspheres were activated with glutaraldehyde and derivatized with polyethyleneimine to increase the number of antigen binding sites on the microspheres. The functionalized microspheres were then coated with a theophylline–casein conjugate, and treated with casein to minimize nonspecific binding. Microspheres for the digoxin assay were encoded with amine reactive Cy 5.5 monofunctional dye. Digoxin was attached to microspheres by the periodate/borohydrate method and treated with casein to minimize nonspecific binding. The competitive immunoassay was performed with HRP-labeled polyclonal rabbit antitheophylline or antidigoxin in the presence of the desired analyte. The generated fluorescence signal was amplified by a catalyzed reporter deposition method.[157–159] This fiber-optic array platform allowed a semiquantitative analysis and provided a proof of principle. Microspheres analyzed on microscope slides showed high sensitivity for both assays. The IC_{50} was ~8 ppb for theophylline, and ~0.5 ppb for digoxin.

Recently, Rissin et al. used a high-density optical fiber array–based duplexed sandwich immunoassay to detect two salivary immune system proteins: immunoglobulin A (IgA) and lactoferrin.[160] Methylstyrene/divinylbenzene microspheres with unique optical bar codes were functionalized with monoclonal primary antibodies to IgA and lactoferrin. After loading the microspheres onto an optical fiber array, incubation with $10\,\mu g\,ml^{-1}$ lactoferrin and $100\,\mu g\,ml^{-1}$ IgA fluorescently labeled secondary monoclonal antibodies was performed, followed by rinsing the unbound antibodies with blocking buffer. The custom-built epifluorescence imaging system shown in Figure 3(b) was used to acquire all fluorescence images. The working concentration range of IgA was between 700 pM and 100 nM, while the working concentration range for lactoferrin was between 385 pM and 10 nM (Figure 11). Lactoferrin exhibited a smaller concentration range compared to IgA, probably due to the higher specificity of lactoferrin antibodies for their target. This fiber-optic array showed enough specificity for possible useful measurements on diluted human saliva samples.

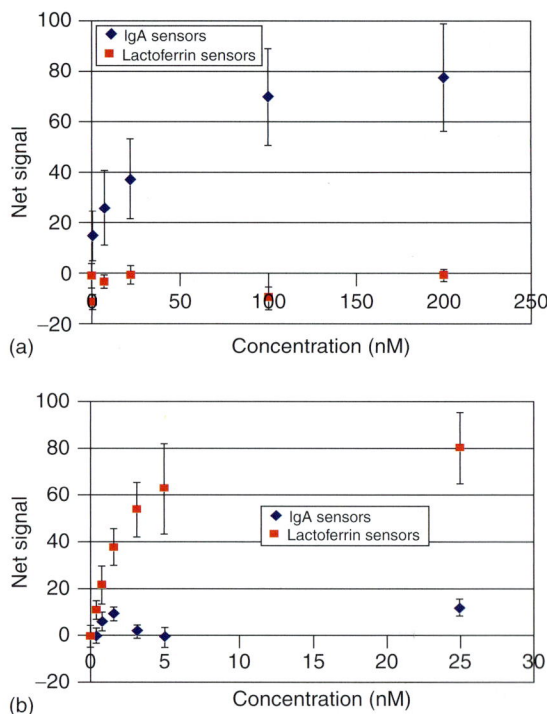

Figure 11. Dose–response curve for (a) IgA detection and (b) lactoferrin detection showing probe specificity and working concentration range. [Reprinted with permission Rissin and Walt[160] copyright 2006, Elsevier.]

6 CONCLUSIONS

Optical fiber array technology has advanced substantially in the last decade. The marriage of this detection technology with multiplexed specific and sensitive assays results in a powerful, cost-effective detection system for genotyping, gene expression studies, pharmaceutical research, cell-based biosensing, enzyme, and immunoassay high-content analysis. An optical fiber microsphere array platform generates thousands of signals simultaneously, allowing for the identification of a specific oligonucleotide sequence on every microsphere in the array, offering multiplexed genotyping, and gene expression analysis with the potential for whole-genome screening. The ability to reuse DNA arrays is a major advantage of this platform. Its small dimensions enable small sample volumes to be used, making it amenable for integration with a portable instrument.

Enzyme-based fiber-optic array biosensors are the most extensively used among the biosensors described in this chapter. These biosensors typically employ enzymes in either pure form or contained on or within a biological matrix such as a cell or vesicle. Enzyme-based fiber-optic arrays have been described for the detection of freshwater pollutants, organophosphorous pesticides, and neurotransmitters. Microsphere-based fiber-optic arrays demonstrated the ability to image the localized chemical releases. Single-molecule enzymology studies on optical fiber bundles are becoming another attractive area toward ultrasensitive biosensing, revealing molecular behaviors at the single-molecule level.

The development of versatile whole-cell fiber-optic arrays provides a means to perform screening of large populations of cells, allowing environmental, toxicological, gene expression kinetics, single-cell genetic noise monitoring, genotoxin monitoring, and drug discovery. This technology demonstrates significant potential for subcellular imaging, and multiorganelle interactions.

The specificity of antibodies covalently attached to the fiber-optic array platform makes them suitable for reliable drug detection, food quality control, and disease diagnostics. Antibodies can be used for the recognition of an antigen from complex samples such as blood, urine, or saliva. Despite improvements in immunoassay development, improvements in antibody production and in understanding immunoreaction mechanisms and kinetics are required.

ACKNOWLEDGMENTS

The DOE, NIH, NSF, NCI, and DARPA supported the work discussed in this chapter.

REFERENCES

1. D. R. Thevenot, K. Toth, R. A. Durst, and G. S. Wilson, Electrochemical biosensors: recommended definitions and classification. *Biosensors and Bioelectronics*, 2001, **16**, 121–131.
2. M. Mehrvar and M. Abdi, Recent developments, characteristics, and potential applications of electrochemical biosensors. *Analytical Sciences*, 2004, **20**, 1113–1126.
3. M. Mehrvar, C. Bis, J. M. Scharer, M. Moo-Young, and J. H. Luong, Fiber-optic biosensors–trends and advances. *Analytical Sciences*, 2000, **16**, 677–692.
4. J. M. Tam, S. Szunerits, and D. R. Watt, Optical fibers for nanodevices. *Encyclopedia of Nanoscience and Nanotechnology*, 2004, **8**, 167–177.
5. D. R. Walt, Imaging optical sensor arrays. *Current Opinion in Chemical Biology*, 2002, **6**, 689–695.
6. D. R. Walt, Fiber optic imaging sensors. *Accounts of Chemical Research*, 1998, **31**, 267–278.
7. D. J. Monk and D. R. Walt, Optical fiber-based biosensors. *Analytical and Bioanalytical Chemistry*, 2004, **379**, 931–945.
8. B. Kuswandi, R. Andres, and R. Narayanaswamy, Optical fibre biosensors based on immobilised enzymes. *The Analyst*, 2001, **126**, 1469–1491.
9. M. D. Marazuela and M. C. Moreno-Bondi, Fiber-optic biosensors–an overview. *Analytical and Bioanalytical Chemistry*, 2002, **372**, 664–682.
10. J. R. Epstein and D. R. Walt, Fluorescence-based fiber optic arrays: a universal platform for sensing. *Chemical Society Reviews*, 2003, **32**, 203–214.
11. J. R. Epstein, A. P. K. Leung, K.-H. Lee, and D. R. Walt, High-density, microsphere-based fiber optic DNA microarrays. *Biosensors and Bioelectronics*, 2003, **18**, 541–546.
12. D. R. Walt, Chemistry: miniature analytical methods for medical diagnostics. *Science*, 2005, **308**, 217–219.
13. P. A. E. Piunno and U. J. Krull, Trends in the development of nucleic acid biosensors for medical diagnostics. *Analytical and Bioanalytical Chemistry*, 2005, **381**, 1004–1011.
14. X. J. Liu and W. H. Tan, Development of an optical fiber lactate sensor. *Mikrochimica Acta*, 1999, **131**, 129–135.
15. M. Bibikova, Z. Lin, L. Zhou, E. Chudin, E. W. Garcia, B. Wu, D. Doucet, N. J. Thomas, Y. Wang, E. Vollmer, T. Goldmann, C. Seifart, W. Jiang, D. L. Barker, M. S. Chee, J. Floros, and J. B. Fan, High-throughput DNA methylation profiling using universal bead arrays. *Genome Research*, 2006, **16**, 383–393.
16. S. Ko and S. A. Grant, A novel fret-based optical fiber biosensor for rapid detection of Salmonella typhimurium. *Biosensors and Bioelectronics*, 2006, **21**, 1283–1290.
17. D. R. DeMarco, E. W. Saaski, D. A. McCrae, and D. V. Lim, Rapid detection of Escherichia Coli O157:H7 in ground beef using a fiber-optic biosensor. *Journal of Food Protection*, 1999, **62**, 711–716.
18. D. Yu, B. Blankert, J. C. Vire, and J.-M. Kauffmann, Biosensors in drug discovery and drug analysis. *Analytical Letters*, 2005, **38**, 1687–1701.
19. P. Janssen, B. Audit, I. Cases, N. Darzentas, L. Goldovsky, V. Kunin, N. Lopez-Bigas, J. M. Peregrin-Alvarez, J. B. Pereira-Leal, S. Tsoka, and C. A. Ouzounis, Beyond 100 genomes. *Genome Biology*, 2003, **4**, 402–404.
20. P. Janssen, L. Goldovsky, V. Kunin, N. Darzentas, and C. A. Ouzounis, Genome coverage, literally speaking. *EMBO Reports*, 2005, **6**, 397–399.
21. K. L. Brogan and D. R. Walt, Optical fiber-based sensors: application to chemical biology. *Current Opinion in Chemical Biology*, 2005, **9**, 494–500.

22. A. Almadidy, J. Watterson, P. A. E. Piunno, S. Raha, I. V. Foulds, P. A. Horgen, A. Castle, and U. Krull, Direct selective detection of genomic DNA from coliform using a fiber optic biosensor. *Analytica Chimica Acta*, 2002, **461**, 37–47.

23. J. H. Watterson, S. Raha, C. C. Kotoris, C. C. Wust, F. Gharabaghi, S. C. Jantzi, N. K. Haynes, N. H. Gendron, U. J. Krull, A. E. Mackenzie, and P. A. E. Piunno, Rapid detection of single nucleotide polymorphisms associated with spinal muscular atrophy by use of a reusable fiber-optic biosensor. *Nucleic Acids Research*, 2004, **32**, e18-1–e18-9.

24. J. B. Fan, A. Oliphant, R. Shen, B. G. Kermani, F. Garcla, K. L. Gunderson, M. Hansen, F. Steemers, S. L. Butler, P. Deloukas, L. Galver, S. Hunt, C. McBride, M. Bibikova, T. Rubano, J. Chen, E. Wickham, D. Doucet, W. Chang, D. Campbell, B. Zhang, S. Kruglyak, D. Bentley, J. Haas, P. Rigault, L. Zhou, J. Stuelpnagel, and M. S. Chee, Highly parallel SNP genotyping. *Cold Spring Harbor Symposia on Quantitative Biology*, 2003, **68**, 69–78.

25. B. G. Healey, R. S. Matson, and D. R. Walt, Fiberoptic DNA sensor array capable of detecting point mutations. *Analytical Biochemistry*, 1997, **251**, 270–279.

26. M. Bibikova, D. Talantov, E. Chudin, J. M. Yeakley, J. Chen, D. Doucet, E. Wickham, D. Atkins, D. Barker, M. Chee, Y. Wang, and J.-B. Fan, Quantitative gene expression profiling in formalin-fixed, paraffin-embedded tissues using universal bead arrays. *American Journal of Pathology*, 2004, **165**, 1799–1807.

27. J.-B. Fan, J. M. Yeakley, M. Bibikova, E. Chudin, E. Wickham, J. Chen, D. Doucet, P. Rigault, B. Zhang, R. Shen, C. McBride, H.-R. Li, X.-D. Fu, A. Oliphant, D. L. Barker, and M. S. Chee, A versatile assay for high-throughput gene expression profiling on universal array matrices. *Genome Research*, 2004, **14**, 878–885.

28. K. Kuhn, S. C. Baker, E. Chudin, M.-H. Lieu, S. Oeser, H. Bennett, P. Rigault, D. Barker, T. K. McDaniel, and M. S. Chee, A novel, high-performance random array platform for quantitative gene expression profiling. *Genome Research*, 2004, **14**, 2347–2356.

29. F. J. Steemers, W. H. Chang, G. Lee, D. L. Barker, R. Shen, and K. L. Gunderson, Whole-genome genotyping with the single-base extension assay. *Nature Methods*, 2006, **3**, 31–33.

30. Illumina Homepage. (2006). Information on the SNP genotyping and whole-genome genotyping; http://www.illumina.com/products/snp/whole_genome_genotyping.ilmn.

31. A. Almadidy, J. Watterson, P. A. E. Piunno, I. V. Foulds, P. A. Horgen, and U. Krull, A fiber-optic biosensor for detection of microbial contamination. *Canadian Journal of Chemistry*, 2003, **81**, 339–349.

32. J. H. Watterson, P. A. E. Piunno, C. C. Wust, S. Raha, and U. J. Krull, Influences of non-selective interactions of nucleic acids on response rates of nucleic acid fiber optic biosensors. *Fresenius Journal of Analytical Chemistry*, 2001, **369**, 601–608.

33. J. Zeng, A. Almadidy, J. Watterson, and U. J. Krull, Interfacial hybridization kinetics of oligonucleotides immobilized onto fused silica surfaces. *Sensors and Actuators, B: Chemical*, 2003, **B90**, 68–75.

34. P. Pantano and D. R. Walt, Ordered nanowell arrays. *Chemistry of Materials*, 1996, **8**, 2832–2835.

35. D. D. Bernhard, S. Mall, and P. Pantano, Fabrication and characterization of microwell array chemical sensors. *Analytical Chemistry*, 2001, **73**, 2484–2490.

36. D. R. Walt, Techview: molecular biology: bead-based fiber-optic arrays. *Science*, 2000, **287**, 451–452.

37. K. L. Michael, L. C. Taylor, S. L. Schultz, and D. R. Walt, Randomly ordered addressable high-density optical sensor arrays. *Analytical Chemistry*, 1998, **70**, 1242–1248.

38. G. Steinberg, K. Stromsborg, L. Thomas, D. Baker, and C. Zhao, Strategies for covalent attachment of DNA to beads. *Biopolymers*, 2004, **73**, 597–605.

39. J. R. Epstein, J. A. Ferguson, K.-H. Lee, and D. R. Walt, Combinatorial decoding: an approach for universal DNA array fabrication. *Journal of the American Chemical Society*, 2003, **125**, 13753–13759.

40. J. M. Tam, L. Song, and D. R. Walt, Fabrication and optical characterization of imaging fiber-based nanoarrays. *Talanta*, 2005, **67**, 498–502.

41. J. R. Epstein, I. Biran, and D. R. Walt, Fluorescence-based nucleic acid detection and microarrays. *Analytica Chimica Acta*, 2002, **469**, 3–36.

42. K. L. Gunderson, S. Kruglyak, M. S. Graige, F. Garcia, B. G. Kermani, C. Zhao, D. Che, T. Dickinson, E. Wickham, J. Bierle, D. Doucet, M. Milewski, R. Yang, C. Siegmund, J. Haas, L. Zhou, A. Oliphant, J.-B. Fan, S. Barnard, and M. S. Chee, Decoding randomly ordered DNA arrays. *Genome Research*, 2004, **14**, 870–877.

43. R. Shen, J.-B. Fan, D. Campbell, W. Chang, J. Chen, D. Doucet, J. Yeakley, M. Bibikova, E. Wickham Garcia, and C. McBride, High-throughput Snp genotyping on universal bead arrays. *Mutation Research-Fundamental and Molecular Mechanisms of Mutagenesis*, 2005, **573**, 70–82.

44. K. L. Gunderson, F. J. Steemers, G. Lee, L. G. Mendoza, and M. S. Chee, A genome-wide scalable SNP genotyping assay using microarray technology. *Nature Genetics*, 2005, **37**, 549–554.

45. J. R. Epstein, M. Lee, and D. R. Walt, High-density fiber-optic genosensor microsphere array capable of zeptomole detection limits. *Analytical Chemistry*, 2002, **74**, 1836–1840.

46. J. A. Ferguson, F. J. Steemers, and D. R. Walt, High-density fiber-optic DNA random microsphere array. *Analytical Chemistry*, 2000, **72**, 5618–5624.

47. J. M. Yeakley, J.-B. Fan, D. Doucet, L. Luo, E. Wickham, Z. Ye, M. S. Chee, and X.-D. Fu, Profiling alternative splicing on fiber-optic arrays. *Nature Biotechnology*, 2002, **20**, 353–358.

48. F. J. Steemers, J. A. Ferguson, and D. R. Walt, Screening unlabeled DNA targets with randomly ordered fiber-optic gene arrays. *Nature Biotechnology*, 2000, **18**, 91–94.

49. M. Lee and D. R. Walt, A fiber-optic microarray biosensor using aptamers as receptors. *Analytical Biochemistry*, 2000, **282**, 142–146.

50. M. Bowden, L. Song, and D. R. Walt, Development of a microfluidic platform with an optical imaging microarray capable of attomolar target DNA detection. *Analytical Chemistry*, 2005, **77**, 5583–5588.

51. D. R. Walt and D. R. Franz, Biological warfare detection. *Analytical Chemistry*, 2000, **72**, 738A–746A.

52. L. Song, S. Ahn, and D. R. Walt, Detecting biological warfare agents. *Emerging Infectious Diseases*, 2005, **11**, 1629–1632.

53. L. Song, S. Ahn, and D. R. Walt, Fiber-optic microsphere-based arrays for multiplexed biological warfare agent detection. *Analytical Chemistry*, 2006, **78**, 1023–1033.

54. J. R. E. Shepard, Y. Danin-Poleg, Y. Kashi, and D. R. Walt, Array-based binary analysis for bacterial typing. *Analytical Chemistry*, 2005, **77**, 319–326.

55. K. V. Holmes and L. Enjuanes, Virology: the sars coronavirus: a postgenomic era. *Science*, 2003, **300**, 1377–1378.

56. M. Enserink, Infectious disease: west Nile's surprisingly swift continental sweep. *Science*, 2002, **297**, 1988–1989.

57. R. R. Colwell, Global climate and infectious disease: the cholera paradigm. *Science*, 1996, **274**, 2025–2031.

58. C. C. Stuart, Nucleotide sequence-based typing of bacteria and the impact of automation. *BioEssays*, 2002, **24**, 858–862.

59. M. C. J. Maiden, J. A. Bygraves, E. Feil, G. Morelli, J. E. Russell, R. Urwin, Q. Zhang, J. Zhou, K. Zurth, D. A. Caugant, I. M. Feavers, M. Achtman, and B. G. Spratt, Multilocus sequence typing: a portable approach to the identification of clones within populations of pathogenic microorganisms. *Proceedings of the National Academy of Sciences of the United States of America*, 1998, **95**, 3140–3145.

60. E. R. Packard, R. Parton, J. G. Coote, and N. K. Fry, Sequence variation and conservation in virulence-related genes of bordetella pertussis isolates from the UK. *Journal of Medical Microbiology*, 2004, **53**, 355–365.

61. V. N. Kristensen, D. Kelefiotis, T. Kristensen, and A.-L. Borresen-Dale, High-throughput methods for detection of genetic variation. *BioTechniques*, 2001, **30**, 318–332.

62. X. Wang and U. J. Krull, Tethered thiazole orange intercalating dye for development of fiber-optic nucleic acid biosensors. *Analytica Chimica Acta*, 2002, **470**, 57–70.

63. X. Wang and U. J. Krull, A self-contained fluorescent fiber optic DNA biosensor. *Journal of Materials Chemistry*, 2005, **15**, 2801–2809.

64. X. Wang and U. J. Krull, Synthesis and fluorescence studies of thiazole orange tethered onto oligonucleotide: development of a self-contained DNA biosensor on a fiber optic surface. *Bioorganic and Medicinal Chemistry Letters*, 2005, **15**, 1725–1729.

65. O. S. Wolfbeis, Fiber-optic chemical sensors and biosensors. *Analytical Chemistry*, 2002, **74**, 2663–2677.

66. I. del Villar, I. R. Matias, F. J. Arregui, and R. O. Claus, Fiber-optic hydrogen peroxide nanosensor. *IEEE Sensors Journal*, 2005, **5**, 365–371.

67. I. Del Villar, I. R. Matias, F. J. Arregui, J. Echeverria, and R. O. Claus, Strategies for fabrication of hydrogen peroxide sensors based on electrostatic self-assembly (ESA) method. *Sensors and Actuators, B: Chemical*, 2005, **B108**, 751–757.

68. A. Kishen, M. S. John, C. S. Lim, and A. Asundi, A fiber optic biosensor (FOBS) to monitor mutans streptococci in human saliva. *Biosensors and Bioelectronics*, 2003, **18**, 1371–1378.

69. C. Durrieu, C. Chouteau, L. Barthet, J.-M. Chovelon, and C. Tran-Minh, A Bi-enzymatic whole-cell algal biosensor for monitoring waste water pollutants. *Analytical Letters*, 2004, **37**, 1589–1599.

70. C. Durrieu and C. Tran-Minh, Optical algal biosensor using alkaline phosphatase for determination of heavy metals. *Ecotoxicology and Environmental Safety*, 2002, **51**, 206–209.

71. R. A. Doong and H. C. Tsai, Immobilization and characterization of sol-gel-encapsulated acetylcholinesterase fiber-optic biosensor. *Analytica Chimica Acta*, 2001, **434**, 239–246.

72. S. G. Ignatov, J. A. Ferguson, and D. R. Walt, A fiber-optic lactate sensor based on bacterial cytoplasmic membranes. *Biosensors and Bioelectronics*, 2001, **16**, 109–113.

73. J. P. Issberner, C. L. Schauer, B. A. Trimmer, and D. R. Walt, Combined imaging and chemical sensing of l-glutamate release from the foregut plexus of the lepidopteran, manduca sexta. *Journal of Neuroscience Methods*, 2002, **120**, 1–10.

74. W. Zhang, H. Chang, and G. A. Rechnitz, Dual-enzyme fiber optic biosensor for pyruvate. *Analytica Chimica Acta*, 1997, **350**, 59–65.

75. Y. Rondelez, G. Tresset, K. V. Tabata, H. Arata, H. Fujita, S. Takeuchi, and H. Noji, Microfabricated arrays of femtoliter chambers allow single molecule enzymology. *Nature Biotechnology*, 2005, **23**, 361–365.

76. D. M. Rissin and D. R. Walt, Digital concentration readout of single enzyme molecules using femtoliter arrays and poisson statistics. *Nano Letters*, 2006, **6**, 520–523.

77. K. A. Fahnrich, M. Pravda, and G. G. Guilbault, Recent applications of electrogenerated chemiluminescence in chemical analysis. *Talanta*, 2001, **54**, 531–559.

78. M. C. Ramos, M. C. Torijas, and A. N. Diaz, Enhanced chemiluminescence biosensor for the determination of phenolic compounds and hydrogen peroxide. *Sensors and Actuators, B: Chemical*, 2001, **B73**, 71–75.

79. A. Navas Diaz, F. Garcia Sanchez, and J. A. Gonzalez Garcia, Phenol derivatives as enhancers and inhibitors of luminol-H2o2-horseradish peroxidase chemiluminescence. *Journal of Bioluminescence and Chemiluminescence*, 1998, **13**, 75–84.

80. A. Navas Diaz, M. C. Ramos Peinado, and M. C. Torijas Minguez, Sol-gel horseradish peroxidase biosensor for hydrogen peroxide detection by chemiluminescence. *Analytica Chimica Acta*, 1998, **363**, 221–227.

81. S. Szunerits, J. M. Tam, L. Thouin, C. Amatore, and D. R. Walt, Spatially resolved electrochemiluminescence on an array of electrode tips. *Analytical Chemistry*, 2003, **75**, 4382–4388.

82. S. Szunerits and D. R. Walt, The use of optical fiber bundles combined with electrochemistry for chemical imaging. *Chemphyschem*, 2003, **4**, 186–192.

83. Y. H. Liu, T. H. Dam, and P. Pantano, A Ph-sensitive nanotip array imaging sensor. *Analytica Chimica Acta*, 2000, **419**, 215–225.

84. A. Chovin, P. Garrigue, G. Pecastaings, H. Saadaoui, I. Manek-Hoenninger, and N. Sojic, Microarrays of

near-field optical probes with adjustable dimensions. *Ultramicroscopy*, 2006, **106**, 57–65.

85. A. Chovin, P. Garrigue, L. Servant, and N. Sojic, Electrochemical modulation of remote fluorescence imaging at an ordered opto-electrochemical nanoaperture array. *Chemphyschem*, 2004, **5**, 1125–1132.

86. A. Chovin, P. Garrigue, and N. Sojic, Electrochemiluminescent detection of hydrogen peroxide with an imaging sensor array. *Electrochimica Acta*, 2004, **49**, 3751–3757.

87. A. Chovin, P. Garrigue, P. Vinatier, and N. Sojic, Development of an ordered array of optoelectrochemical individually readable sensors with submicrometer dimensions: application to remote electrochemiluminescence imaging. *Analytical Chemistry*, 2004, **76**, 357–364.

88. A. Chovin, P. Garrigue, and N. Sojic, Remote NADH imaging through an ordered array of electrochemiluminescent nanoapertures. *Bioelectrochemistry*, 2006, **69**, 25–33.

89. K. D. Hughes, D. L. Bittner, and G. A. Olsen, New fluorescence tools for investigating enzyme activity. *Analytica Chimica Acta*, 1995, **307**, 393–402.

90. N. Zurgil, S. Gerbat, P. Langevitzh, M. Tishler, M. Ehrenfeld, and Y. Shoenfeld, Intracellular fluorescence polarization measurements by the cellscan system: detection of cellular activity in autoimmune disorders. *Israel Journal of Medical Sciences*, 1997, **33**, 273–279.

91. S. Ramanathan, W. Shi, B. P. Rosen, and S. Daunert, Sensing antimonite and arsenite at the subattomole level with genetically engineered bioluminescent bacteria. *Analytical Chemistry*, 1997, **69**, 3380–3384.

92. S. Belkin, Microbial whole-cell sensing systems of environmental pollutants. *Current Opinion in Microbiology*, 2003, **6**, 206–212.

93. A. Baeumner, Biosensors for environmental pollutants and food contaminants. *Analytical and Bioanalytical Chemistry*, 2003, **377**, 434–445.

94. A. M. Karlsson, K. Bjuhr, M. Testorf, P. A. Oberg, E. Lerner, I. Lundstrom, and S. P. S. Svensson, Biosensing of opioids using frog melanophores. *Biosensors and Bioelectronics*, 2002, **17**, 331–335.

95. L. J. Dietz, R. S. Dubrow, B. S. Manian, and N. L. Sizto, Volumetric capillary cytometry: a new method for absolute cell enumeration. *Cytometry*, 1996, **23**, 177–186.

96. H. N. Zhou, A. Xu, J. A. Gillispie, C. A. Waldren, and T. K. Hei, Quantification of Cd59(-) mutants in human-hamster hybrid (a(L)) cells by flow cytometry. *Mutation Research-Fundamental and Molecular Mechanisms of Mutagenesis*, 2006, **594**, 113–119.

97. M. Burmolle, L. H. Hansen, and S. J. Sorensen, Use of a whole-cell biosensor and flow cytometry to detect AHL production by an indigenous soil community during decomposition of litter. *Microbial Ecology*, 2005, **50**, 221–229.

98. L. Huang and R. T. Kennedy, Exploring single-cell dynamcis using chemically-modified microelectrodes. *Trends in Analytical Chemistry*, 1995, **14**, 158–163.

99. K. A. Marx, Quartz crystal microbalance: a useful tool for studying thin polymer films and complex biomolecular systems at the solution-surface interface. *Biomacromolecules*, 2003, **4**, 1099–1120.

100. M. Moini and H. L. Huang, Application of capillary electrophoresis/electrospray ionization-mass spectrometry to subcellular proteomics of escherichia coli leribosomal proteins. *Electrophoresis*, 2004, **25**, 1981–1987.

101. F. E. Regnier, D. H. Patterson, and B. J. Harmon, Electrophoretically-mediated microanalysis (EMMA). *Trends in Analytical Chemistry*, 1995, **14**, 177–181.

102. W. Wang, S. B. Han, and W. R. Jin, Determination of lactate dehydrogenase isoenzymes in single rat glioma cells by capillary electrophoresis with electrochemical detection. *Journal of Chromatography. B, Analytical Technologies in the Biomedical and Life Sciences*, 2006, **831**, 57–62.

103. J. B. Shear, H. A. Fishman, D. Garigan, R. N. Zare, N. L. Allbritton, and R. H. Scheller, Single cells as biosensors for chemical separations. *Science*, 1995, **267**, 74–77.

104. L. C. Taylor and D. R. Walt, Application of high-density optical microwell arrays in a live-cell biosensing system. *Analytical Biochemistry*, 2000, **278**, 132–142.

105. F. Becker, N. Costa-Roldán, C. Kaufmann, U. Hanke, C. Degenhart, S. Baumann, W. Wallner, A. Huber, S. Dedier, S. Dill, S. Meier-Ewert, N. Kley, K. Murthi, C. Smith, J. Come, D. Kinsman, M. Hediger, N. Bockovich, and A. F. Kluge, A three-hybrid approach to scanning the proteome for targets of small molecule kinase inhibitors. *Chemistry and Biology*, 2004, **11**, 211–223.

106. D. Vanhecke and M. Janitz, High-throughput gene silencing using cell arrays. *Oncogene*, 2004, **23**, 8353–8358.

107. I. Biran and D. R. Walt, Optical imaging fiber-based single live cell arrays: a high-density cell assay platform. *Analytical Chemistry*, 2002, **74**, 3046–3054.

108. I. Biran, D. M. Rissin, E. Z. Ron, and D. R. Walt, Optical imaging fiber-based live bacterial cell array biosensor. *Analytical Biochemistry*, 2003, **315**, 106–113.

109. Y. Kuang, I. Biran, and D. R. Walt, Living bacterial cell array for genotoxin monitoring. *Analytical Chemistry*, 2004, **76**, 2902–2909.

110. S. Fields and O. K. Song, A novel genetic system to detect protein protein interactions. *Nature*, 1989, **340**, 245–246.

111. P.-O. Vidalain, M. Boxem, H. Ge, S. Li, and M. Vidal, Increasing specificity in high-throughput yeast two-hybrid experiments. *Methods*, 2004, **32**, 363–370.

112. K. Hakkila, P. Leskinen, T. Green, R. Marks, A. Ivask, and M. Virta, Detection of bioavailable heavy metals in eilatox-oregon samples using whole-cell luminescent bacterial sensors in suspension or immobilized onto fibre-optic tips. *Journal of Applied Toxicology*, 2004, **24**, 333–342.

113. S. Leth, S. Maltoni, R. Simkus, B. Mattiasson, P. Corbisier, I. Klimant, O. S. Wolfbeis, and E. Csöregi, Engineered bacteria based biosensors for monitoring bioavailable heavy metals. *Electroanalysis*, 2002, **14**, 35–42.

114. K. B. Riether, M. A. Dollard, and P. Billard, Assessment of heavy metal bioavailability using Escherichia Coli Zntap::lux and copap::lux-based biosensors. *Applied Microbiology and Biotechnology*, 2001, **57**, 712–716.

115. I. Biran, R. Babai, K. Levcov, J. Rishpon, and E. Z. Ron, Online and in situ monitoring of environmental pollutants: electrochemical biosensing of cadmium. *Environmental Microbiology*, 2000, **2**, 285–290.

116. A. A. Wise and C. R. Kuske, Generation of novel bacterial regulatory proteins that detect priority pollutant

phenols. *Applied and Environmental Microbiology*, 2000, **66**, 163–169.

117. Y. Davidov, R. Rozen, S. Belkin, D. R. Smulski, T. K. Van Dyk, A. C. Vollmer, D. A. Elsemore, and R. A. LaRossa, Improved bacterial SOS promoter::lux fusions for genotoxicity detection. *Mutation Research-Genetic Toxicology and Environmental Mutagenesis*, 2000, **466**, 97–107.

118. B. Polyak, E. Bassis, A. Novodvorets, S. Belkin, and R. S. Marks, Bioluminescent whole cell optical fiber sensor to genotoxicants: system optimization. *Sensors and Actuators, B: Chemical*, 2001, **74**, 18–26.

119. Y. Kuang and D. R. Walt, Monitoring "promiscuous" drug effects on single cells of multiple cell types. *Analytical Biochemistry*, 2005, **345**, 320–325.

120. Y. Ikariyama, S. Nishiguchi, T. Koyama, E. Kobatake, M. Aizawa, M. Tsuda, and T. Nakazawa, Fiber-optic-based biomonitoring of benzene derivatives by recombinant E. Coli bearing luciferase gene-fused TOL-plasmid immobilized on the fiber-optic end. *Analytical Chemistry*, 1997, **69**, 2600–2605.

121. Y. Kuang, I. Biran, and D. R. Walt, Simultaneously monitoring gene expression kinetics and genetic noise in single cells by optical well arrays. *Analytical Chemistry*, 2004, **76**, 6282–6286.

122. M. D. Levin, Noise in gene expression as the source of non-genetic individuality in the chemotactic response of Escherichia Coli. *FEBS Letters*, 2003, **550**, 135–138.

123. J. M. G. Vilar, H. Y. Kueh, N. Barkai, and S. Leibler, Mechanisms of noise-resistance in genetic oscillators. *Proceedings of the National Academy of Sciences of the United States of America*, 2002, **99**, 5988–5992.

124. F. J. Isaacs, J. Hasty, C. R. Cantor, and J. J. Collins, Prediction and measurement of an autoregulatory genetic module. *Proceedings of the National Academy of Sciences of the United States of America*, 2003, **100**, 7714–7719.

125. J. Paulsson, Summing up the noise in gene networks. *Nature*, 2004, **427**, 415–418.

126. M. L. Simpson, C. D. Cox, and G. S. Sayler, Frequency domain analysis of noise in autoregulated gene circuits. *Proceedings of the National Academy of Sciences of the United States of America*, 2003, **100**, 4551–4556.

127. P. Cluzel, M. Surette, and S. Leibler, An ultrasensitive bacterial motor revealed by monitoring signaling proteins in single cells. *Science*, 2000, **287**, 1652–1655.

128. C. DiCesare, I. Biran, and D. R. Walt, Individual cell migration analysis using fiber-optic bundles. *Analytical and Bioanalytical Chemistry*, 2005, **382**, 37–43.

129. W. Schramm, S. H. Paek, and G. Voss, Strategies for the immobilization of antibodies. *ImmunoMethods*, 1993, **3**, 93–103.

130. M. C. Moreno-Bondi, J. Mobley, J. P. Alarie, and T. Vo-Dinh, Antibody-based biosensor for breast cancer with ultrasonic regeneration. *Journal of Biomedical Optics*, 2000, **5**, 350–354.

131. T. Vo Dinh, M. J. Sepaniak, G. D. Griffin, and J. P. Alarie, Immunosensors: principles and applications. *ImmunoMethods*, 1993, **3**, 85–92.

132. J. Cruz Helder, C. Rosa Carla, and G. Oliva Abel, Immunosensors for diagnostic applications. *Parasitology Research*, 2002, **88**, S4–S7.

133. D. M. Rissin and D. R. Walt, Duplexed sandwich immunoassays on a fiber-optic microarray. *Analytica Chimica Acta*, In press.

134. G. MacBeath, Protein microarrays and proteomics. *Nature Genetics*, 2002, **32**, 526–532.

135. S. Herrmann, B. Leshem, S. Landes, B. Rager-Zisman, and R. S. Marks, Chemiluminescent optical fiber immunosensor for the detection of anti-west nile virus igg. *Talanta*, 2005, **66**, 6–14.

136. M. Liu, Q. X. Li, and G. A. Rechnitz, Gold electrode modification with thiolated hapten for the design of amperometric and piezoelectric immunosensors. *Electroanalysis*, 2000, **12**, 21–26.

137. G. Sakai, T. Saiki, T. Uda, N. Miura, and N. Yamazoe, Evaluation of binding of human serum albumin (HSA) to monoclonal and polyclonal antibody by means of piezoelectric immunosensing technique. *Sensors and Actuators, B: Chemical*, 1997, **B42**, 89–94.

138. E. Zacco, M. I. Pividori, X. Llopis, M. del Valle, and S. Alegret, Renewable protein a modified graphite-epoxy composite for electrochemical immunosensing. *Journal of Immunological Methods*, 2004, **286**, 35–46.

139. S. A. Grant, M. E. Pierce, D. J. Lichlyter, and D. A. Grant, Effects of immobilization on a fret immunosensor for the detection of myocardial infarction. *Analytical and Bioanalytical Chemistry*, 2005, **381**, 1012–1018.

140. L. L. E. Salins, S. K. Deo, and S. Daunert, Phosphate binding protein as the biorecognition element in a biosensor for phosphate. *Sensors and Actuators, B: Chemical*, 2004, **B97**, 81–89.

141. R. S. Marks, A. Novoa, D. Thomassey, and S. Cosnier, An innovative strategy for immobilization of receptor proteins on to an optical fiber by use of poly(pyrrole-biotin). *Analytical and Bioanalytical Chemistry*, 2002, **374**, 1056–1063.

142. T. Konry, A. Novoa, S. Cosnier, and R. S. Marks, Development of an "electroptode" immunosensor: indium tin oxide-coated optical fiber tips conjugated with an electropolymerized thin film with conjugated cholera toxin B subunit. *Analytical Chemistry*, 2003, **75**, 2633–2639.

143. T. Geng, M. T. Morgan, and A. K. Bhunia, Detection of low levels of listeria monocytogenes cells by using a fiber-optic immunosensor. *Applied and Environmental Microbiology*, 2004, **70**, 6138–6146.

144. F. Szurdoki, K. L. Michael, and D. R. Walt, A duplexed microsphere-based fluorescent immunoassay. *Analytical Biochemistry*, 2001, **291**, 219–228.

145. I. Balcer Heath, J. Kwon Hyun, and A. Kang Kyung, Assay procedure optimization of a rapid, reusable protein C immunosensor for physiological samples. *Annals of Biomedical Engineering*, 2002, **30**, 141–147.

146. J. Glökler and P. Angenendt, Protein and antibody microarray technology. *Journal of Chromatography. B, Analytical Technologies in the Biomedical and Life Sciences*, 2003, **797**, 229–240.

147. U. B. Nielsen and B. H. Geierstanger, Multiplexed sandwich assays in microarray format. *Journal of Immunological Methods*, 2004, **290**, 107–120.

148. S. Aoyagi and M. Kudo, Development of fluorescence change-based, reagent-less optic immunosensor. *Biosensors and Bioelectronics*, 2005, **20**, 1680–1684.

149. T. Konry, A. Novoa, Y. Shemer-Avni, N. Hanuka, S. Cosnier, A. Lepellec, and R. S. Marks, Optical fiber immunosensor based on a poly(pyrrole-benzophenone) film for the detection of antibodies to viral antigen. *Analytical Chemistry*, 2005, **77**, 1771–1779.

150. R. Kapoor, N. Kaur, E. T. Nishanth, S. W. Halvorsen, E. J. Bergey, and P. N. Prasad, Detection of trophic factor activated signaling molecules in cells by a compact fiber-optic sensor. *Biosensors and Bioelectronics*, 2004, **20**, 345–349.

151. J. H. Watterson, P. A. E. Piunno, C. C. Wust, and U. J. Krull, Controlling the density of nucleic acid oligomers on fiber optic sensors for enhancement of selectivity and sensitivity. *Sensors and Actuators, B: Chemical*, 2001, **B74**, 27–36.

152. D. R. Fry and D. R. Bobbitt, Hapten immobilization for antibody sensing using a dynamic modification protocol. *Talanta*, 2001, **55**, 1195–1203.

153. M. C. Moreno-Bondi, J. P. Alarie, and T. Vo-Dinh, Multi-analyte analysis system using an antibody-based biochip. *Analytical and Bioanalytical Chemistry*, 2003, **375**, 120–124.

154. K. Blank, A. Lankenau, T. Mai, S. Schiffmann, I. Gilbert, S. Hirler, C. Albrecht, H. Clausen-Schaumann, M. Benoit, and H. E. Gaub, Double-chip protein arrays: force-based multiplex sandwich immunoassays with increased specificity. *Analytical and Bioanalytical Chemistry*, 2004, **379**, 974–981.

155. G. MacBeath and S. L. Schreiber, Printing proteins as microarrays for high-throughput function determination. *Science*, 2000, **289**, 1760–1763.

156. M. Pawlak, E. Schick, M. A. Bopp, M. J. Schneider, P. Oroszlan, and M. Ehrat, Zeptosens' protein microarrays: a novel high performance microarray platform for low abundance protein analysis. *Proteomics*, 2002, **2**, 383–393.

157. M. N. Bobrow, G. J. Litt, K. J. Shaughnessy, P. C. Mayer, and J. Conlon, The use of catalyzed reporter deposition as a means of signal amplification in a variety of formats. *Journal of Immunological Methods*, 1992, **150**, 145–149.

158. M. N. Bobrow, K. J. Shaughnessy, and G. J. Litt, Catalyzed reporter deposition, a novel method of signal amplification. 2. Application to membrane immunoassays. *Journal of Immunological Methods*, 1991, **137**, 103–112.

159. M. N. Bobrow, T. D. Harris, K. J. Shaughnessy, and G. J. Litt, Catalyzed reporter deposition, a novel method of signal amplification–application to immunoassays. *Journal of Immunological Methods*, 1989, **125**, 279–285.

160. D. M. Rissin and D. R. Walt, Duplexed sandwich immunoassays on a fiber-optic microarray. *Analytica Chimica Acta*, 2006, **564**, 34–39.

57

Surface Plasmon Resonance Array Devices

Masayasu Suzuki,[1] Yasunori Iribe[1] and Tatsuya Tobita[2]

[1] Departmemt of Electric and Electronic Engineering, University of Toyama, Toyama, Japan and
[2] NTT Advanced Technology Corp., Atsugi, Japan

1 SURFACE PLASMON RESONANCE (SPR) AND SPR IMMUNOSENSORS

Many types of immunosensors have been developed, but only surface plasmon resonance (SPR) immunosensors, e.g., the BIAcore™ system, have achieved commercial success. Surface plasmon oscillation is a localized wave that propagates along the interface between the metal and the ambient medium, and this wave is very sensitive to changes in the refractive index near the metal surface. Figure 1 shows the principle of SPR measurement based on Kretschmann prism arrangement.[1] This is based on total internal reflection in a glass prism onto which a thin metal film is deposited. Under the condition that total internal reflection occurs, a part of the light, which is called the *evanescent field*, penetrates outside the prism. When the metal film is sufficiently thin (e.g., ca. 50 nm) the evanescent field can penetrate the metal film and set up a surface plasmon at the metal–ambient interface. This surface plasmon wave propagates along the interface between the metal and the ambient medium. When the propagation vector of the evanescent wave, $k_e(=k_p \sin\theta$, k_p is propagation vector of incident light), equals that of the surface plasmon wave, $k_{sp}(= (\omega/c)[\varepsilon n^2/(\varepsilon + n)]^{-(1/2)})$, the resonance occurs, and most of the incident light is transferred into the surface plasmon wave and a sharp minimum in the reflected light intensity is observed. This condition is expressed as follows:

$$k_p \sin\theta = (\omega/c)[\varepsilon n^2/(\varepsilon + n)]^{-(1/2)} \quad (1)$$

This incident angle, θ, is called the *SPR angle*. If the light source (angular frequency ω, velocity c), metal (dielectric constant ε), and prism are not changed, the SPR angle depends only on the refractive index of the ambient medium near the metal, n.

Although the SPR phenomenon was initially used for the characterization of thin metal films, the group from Linkoping University have applied SPR to gas and biochemical sensing.[2] These studies were linked to the development of the BIAcore™ system,[3] which enabled label-free, real-time affinity monitoring of biochemical species such as antibodies, DNA, receptors, and so on. Although SPR-based immunosensors also have many advantages, e.g., multichannel integration with enzyme sensors, conventional SPR immunosensor systems require large instruments. For this reason the miniaturization of the SPR sensor system was investigated, and Texas Instruments developed and commercialized a fully integrated, miniaturized, manufacturable SPR sensor, the Spreeta™.[4] This sensor chip costs only

Handbook of Biosensors and Biochips. Edited by Robert S. Marks, David C. Cullen, Isao Karube, Christopher R. Lowe and Howard H. Weetall.
© 2007 John Wiley & Sons, Ltd. ISBN 978-0-470-01905-4.

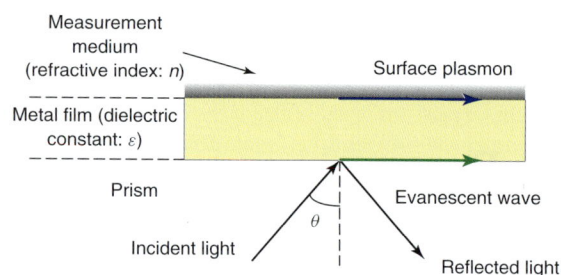

Figure 1. Principle of surface plasmon resonance sensor.

approximately US$50, and includes all the necessary components for SPR measurements, namely, light source, polarizer, prism, sensing area, and SPR angle detector. Suzuki et al.[5] have developed a miniature SPR immunosensor using this sensor chip.

2 TWO-DIMENSIONAL SPR IMAGERS AND THEIR APPLICATIONS

In order to monitor immunoreactions occurring in a large number of spots on microarrayed protein chips or cell chips, conventional SPR sensors could not be applied. The two-dimensional SPR imaging technique is suitable for this purpose. Theoretically, two-dimensional SPR imaging could be realized by adding another dimension to a conventional SPR sensor. Actually such instruments were constructed by several groups, but the instruments became very large and expensive. The first practical two-dimensional SPR imager was reported by Corn et al.[6] They achieved two-dimensional SPR imaging with the conventional Kretschmann configuration instrument by using a collimated beam as incident light. This SPR imager has already been commercialized by GWC technologies (USA).[7] Traditional SPR sensors determine the shift in "SPR angle" (angle of minimum reflectivity) when the material adsorbs to the surface. But angle scanning requires a long time to determine the SPR angle. Therefore, commercial "one-dimensional" SPR sensors, such as BIAcore™, employ a charge coupled device (CCD) line sensor as a detector. In two-dimensional SPR sensors, angle scanning is unavoidable in determining the SPR angle. GWC's SPR imager takes SPR measurement at a fixed angle of incidence,

and collects the reflected light with a CCD camera. This enables real-time monitoring of SPR images on the surface of DNA or protein arrays. Since SPR sensor is an affinity sensor, it can be applied to immunoreactions, DNA–DNA or DNA–RNA interaction, receptor assay, and so on. Many applications using this SPR imager have been reported. Lee et al.[8] applied this SPR imager as a detector of DNA microarrays. They created one- or two-dimensional DNA microarrays from microfluidic channels on a thin gold film (SPR sensor membrane). This two-dimensional DNA array was used to detect a 20 fmol sample of in vitro transcribed RNA from the *uidA* gene of a transgenic *Arabidopsis thaliana* plant. Wegner et al.[9] applied the SPR imager to peptide array. They studied the kinetics of protein adsorption/desorption onto peptide microarrays by using a SPR imager. Real-time measurement with SPR sensor is a powerful tool for kinetics studies. They also applied it to the study of the surface enzymatic activities of the protease factor Xa. Further application could be found in GWC's website.[7]

The same type of SPR imagers have been commercialized by Toyobo Co. Ltd. (Japan). They also supply a spotting machine for preparing DNA or protein arrays suitable for observation with the SPR imager. Kyo et al.[10] applied this SPR imager to label-free detection of proteins with antibody arrays. They measured the expression of eight proteins in the mouse brain. The detection limit of the antibody array was approximately 30 ng ml^{-1} in crude cell lysate. Inamori et al.[11] applied this system to detect and quantify on-chip phosphorylated peptides using a phosphate to capture molecules. The peptide probes, which were phosphorylated on the surface by protein kinase A, could be detected and quantified by this SPR imager. The advantage of such prism-coupled SPR is its high sensitivity compared with grating-coupled SPR described in the next paragraph.

The SPR imager based on grating-coupled SPR was also reported.[12] The advantages of grating-based SPR sensing include the fact that a prism is not necessary to excite surface plasmons and optical quality of the substrate is not crucial. Furthermore, inexpensive and disposable plastic gratings can be used as substrates. This type of SPR imagers were originally commercialized by Applied Biosystems and are now marketed as BIAcore Flexchip.

Localized SPR based on a nanoparticle layer is another principle for realizing SPR imaging. The advantage of this SPR imager is that it is free of SPR image deformation, which is unavoidable in the SPR imager based on Kretschmann configuration. But the sensor performance is strongly affected by the formation of nanoparticles. Improvement of reproducibility might be necessary for this type of SPR imagers.

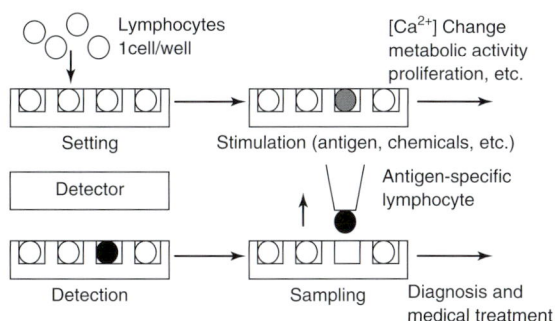

Figure 2. Microarray-based antigen-specific B lymphocytes screening.

3 SINGLE-CELL-BASED MICROARRAY CHIP SYSTEM

In this section, we describe the importance of microarrayed well chips and microarrayed immunosensors. Recently, much attention has been focused on cell chips. Typical cell chips are toxicity test chips using hepatic cells and microarray chips, whose well size is over hundreds of micrometers. On the other hand, studies on single-cell analysis are also increasing. After the completion of human genome sequencing, it is now necessary to develop means to observe the functions of biomolecules in single cells as minimal units of living cells. Therefore the First International Workshop on Approaches to Single-Cell Analysis was held at Uppsala University, Sweden, in 2006.[13] In Japan, the "Lifesurveyor" project (supported by MEXT, Japan) was started in 2005.[14] This project aims at the development of tools and methods for analyzing single cells based on an accurate quantitative and digital analysis of molecules.

We have developed a single-cell-based microarray chip system for rapid screening of antigen-specific B lymphocytes.

Detection and collection of antigen-specific B lymphocytes for the target antigen is quite important for the development of antibody medicines which are expected as "future medicines". We are, in Toyama Medical-Bio Cluster Projects, developing the antigen-specific B lymphocytes screening system based on the microarrayed lymphocyte chip on which a quarter million of 10-μm microwells, which are approximately the same size as human lymphocytes, are arranged.[15] Figure 2 shows the procedure of the microarray-based, antigen-specific B lymphocytes screening.

This screening process consists of the process of seeding the cell onto the microarrayed cell chips, the stimulation process with target antigens, the

Figure 3. Single-cell-based automatic cell screening system.

detection process for target cells, and the collection process of the target cells. We have developed the automatic single-cell screening system based on the microarrayed cell chips. The system consists of a cell seeding unit for microwell array chips, a high-resolution detection unit, and an automatic single-cell collection unit. Figure 3 shows the photograph of the developed cell screening system.

We engaged in the development of the novel sensor systems for microwell array chips. In the single-cell-based microarrayed cell chips, the applicable sensing technologies are very limited

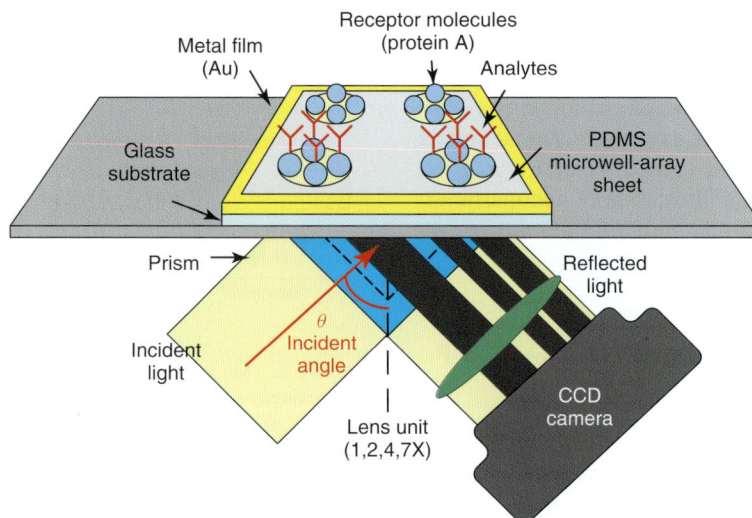

Figure 4. Structure of the two-dimensional SPR imaging sensor.

because of their small size and large number of spots. High-resolution imaging techniques are required. We developed the microarrayed optical pH and oxygen sensor systems using the high-resolution laser confocal scanner for cell activity monitoring,[16] and the high-resolution, two-dimensional SPR imaging affinity sensor system for protein detection.[17]

4   CONSTRUCTION OF THE MICROSCOPIC SPR IMAGER

Among the various types of SPR imagers, the SPR imager based on Kretschmann configuration might be the best choice for realizing high-resolution imaging. Therefore we employed the SPR imager based on Kretschmann configuration (2DSPR-04A, NTT advanced technologies). Figure 4 shows the structure of our SPR imager. A collimated beam, generated by an LED source ($\lambda = 660$, 740, and 840 nm) illuminates the metal sensor surface, which is in contact with the sample solution, through a coupling prism. The incident beam is transverse magnetically (TM) polarized and the reflected light is imaged onto the cooled CCD camera via a microscopic long working distance (WD) lens (1X, 2X, 4X, and 7X). Figure 5 shows the photograph of our SPR imager.

The disadvantage of the SPR imager based on Kretschmann configuration is deformation of the

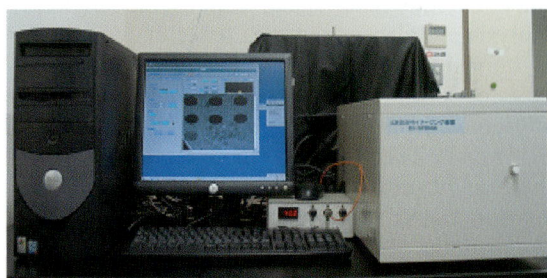

Figure 5. Two-dimensional SPR imaging sensor.

SPR image because the image is given a sidelong glance. This problem might be serious when the image is small as in the present study. In order to decrease this deformation, a high-refractive optical system was employed. The conventional SPR sensor employs BK7 for prism because its refractive index is similar to glass. In our SPR imager, SF6, a high-refractive material, was used as the prism and as a substrate for the sensor chip. Furthermore, a cooled CCD camera was employed in order to suppress the entropy noise. In the conventional Kretschmann type SPR imager, the SPR sensor chip was set vertically because of its optical system structure. This is not suitable for microarrayed well chips. In our SPR imager, the SPR sensor chip can be set horizontally by employing an optical fiber cable.

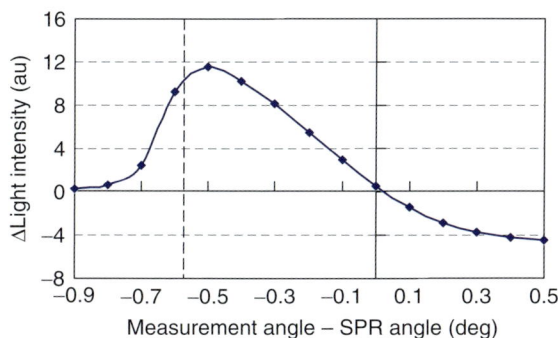

Figure 6. Effect of measurement angle on SPR sensitivity, measured with gold film and 770-nm light source.

5 OPTIMIZATION OF SPR SENSING

In order to improve the sensitivity of the SPR sensor, measurement conditions of the sensor were optimized. Sensitivity could be greatly improved by optimization of the measurement angle, exposure time, light source intensity, and so on. For example, the maximum response was obtained when the measurement angle was set at SPR angle: $-0.5°$, as shown in Figure 6.

By optimization of the measurement angle and exposure time of CCD camera, SPR sensitivity was improved by over 20 times, as shown in Figure 7 (at $20\,\text{mg ml}^{-1}$ glucose). Under this condition, immunoglobulin G (IgG) was measured in 10- and 30-μm microwells. In 30-μm microwells, sensor response to $0.1\,\text{mg ml}^{-1}$ mouse IgG was improved by five times. Under the same condition, $0.01\,\text{mg ml}^{-1}$ IgG also could be detected. In 10-μm microwells, $0.1\,\text{mg ml}^{-1}$ IgG could be successfully detected with this microarrayed 2D-SPR sensor (Figure 8). These results show the possibility of real-time monitoring for antibody production by a single B lymphocyte in a microwell.

6 RESOLUTION OF SPR IMAGE AND METALS OF SPR SENSOR CHIPS

Improvement of the resolution of SPR images (sharpness of images) is very important for two-dimensional SPR sensing. SPR phenomena are affected by the dielectric constant (ε) of thin metal film on the prism. Dielectric constants of metals

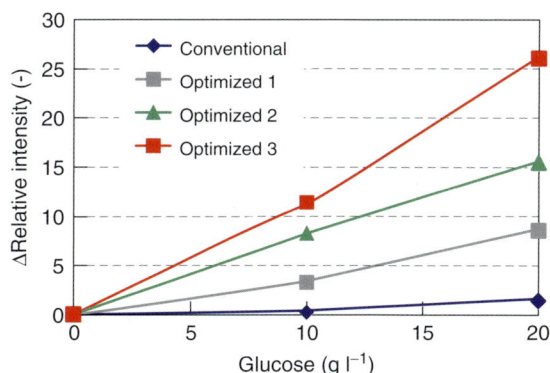

Figure 7. Effect of exposure time and measurement angle on SPR sensitivity. Exposure time and measurement angle were 2 s and SPR angle (conventional), 20 s and SPR angle (optimized 1), 2 s and (SPR angle: $-0.5°$) (optimized 2), and 20 s and (SPR angle: $-0.5°$) (optimized 3), respectively.

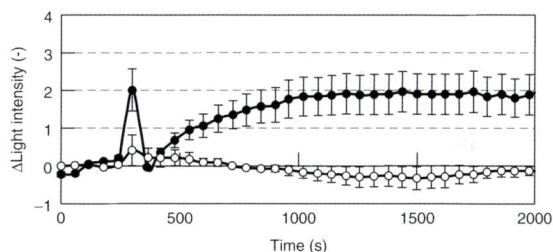

Figure 8. IgG detection in 10-μm wells $0.1\,\text{g l}^{-1}$ mouse IgG was measured under the optimized condition 3. Gold film and 770-nm light source were used.

are expressed as a complex number. For example, dielectric constant of gold is $-21.7 + 1.36\text{i}$, and that of silver is $-28.7 + 0.34\text{i}$ when the wavelength of light source is 770 nm and temperature is 20 °C. The real part of the dielectric constant (ε_r) denotes reflection, and the imaginary part (ε_i) reflects absorption. Therefore, a larger $|\varepsilon_i/\varepsilon_r|$ means higher energy absorption. SPR curves may become broader and SPR sensitivity might be decreased in the case of higher $|\varepsilon_i/\varepsilon_r|$. For example, $|\varepsilon_i/\varepsilon_r|$ of gold, silver, and aluminum are 0.063, 0.012, and 0.573, respectively. This means silver film might be a suitable material for highly sensitive SPR measurement.

Therefore the relationship between the brightness gradient at the boundary of well images and the kind of metal film was investigated, and the resolution of 2D-SPR imaging was improved.

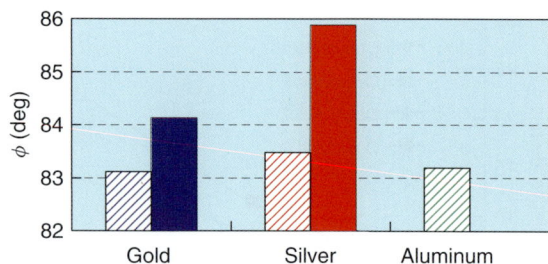

Figure 9. Comparison of resolution with various metal films. Metal film was prepared with vacuum vapor deposition (open column) or ion-sputtering (closed column). Distilled water in 10-μm wells was measured by using 770-nm light source.

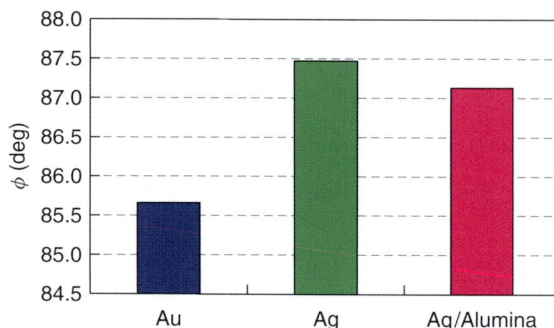

Figure 10. Effect of alumina coating on resolution. Distilled water in 10-μm wells was measured by using 670-nm light source.

By using a 10-μm microarray, the effect of the metal layer on the resolution of SPR images was investigated. As shown in Figure 9, the clearest image was obtained with silver film. Approximately similar results were obtained with gold and aluminum films.

The effects of the preparation methods of metal layers on the resolution were also investigated. Silver and gold layers were prepared by the ion-sputtering method and vacuum vapor deposition method. Clearer and higher-resolution SPR images were obtained with metal layers prepared by the ion-sputtering method as shown in Figure 9.

The results show that the highest resolution was obtained by the silver layer prepared by the ion-sputtering method.

The problem with using a silver layer as an SPR sensor membrane is its instability in buffer solution. Actually, the prepared silver layer was destroyed when dipped in 0.05 M Hepes-HCl buffer (pH 7.5) containing 4.46 g l^{-1} NaCl for 2 h. In order to improve the stability of the silver film, Zhu et al.[18] reported the effects of protection layers, such as Al_2O_3, Cr_2O_3, and Nb_2O_5, on the stability of silver layer for SPR biosensors. We employed a thin layer of alumina as a protection layer for the silver film.

The effect of alumina layer as a protection layer was investigated. One nanometer of aluminum layer was prepared by vacuum vapor deposition on the silver layer of the SPR sensor chip. Then the sensor chip was baked at 100 °C for 30 min.

The stability of the silver layer coated with the thin film of alumina was investigated by dipping it in 0.05 M Hepes-HCl buffer (pH 7.5) containing 4.46 g l^{-1} NaCl for 2 h. Even after 2 h, no destruction of silver film was observed.

In the same way, the resolution of the SPR image was compared by using a 10-μm microarray on alumina-coated silver, bared silver, and gold. As shown in Figure 10, a sharp image was obtained with alumina-coated silver film as in bared silver film.

Although silver film is unstable in buffer solution, alumina-coated silver film is stable in buffer solution, and shows approximately similar characteristics as SPR sensor membrane.

7 SENSITIVITY DECREASE OF SPR SENSING IN 10-μM-ORDER AREA

Figure 11 shows the relationship between SPR sensitivity and well size. SPR sensitivity was evaluated by glucose responses. With 30-μm wells, SPR sensitivity was the same as in the case without wells (bared gold film). But the SPR sensitivity was drastically decreased with 10- and 8.5-μm microwells. This sensitivity decrease was also observed with the SPR immunosensing for mouse IgG. Since SPR is a phenomenon of surface waves, it may depend on the wavelength of incident light if the area is extremely small as in 10-μm microwells. Therefore a shorter wavelength would be suitable for SPR measurement in 10-μm microwells. But the wavelength of incident light affects the dielectric constant of metal film. Figure 12 shows the relationship between wavelength and the ratio of the real (ε_r) and imaginary parts (ε_i) of the dielectric constant, $|\varepsilon_i/\varepsilon_r|$, of gold

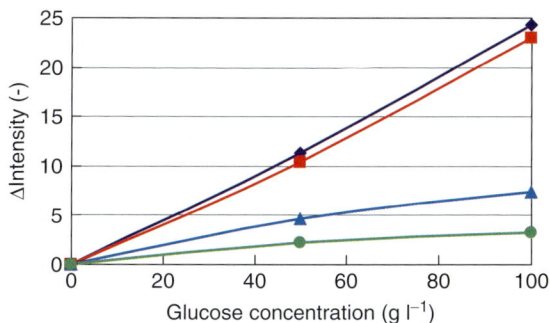

Figure 11. SPR sensitivity and well size. ♦: without wells, ■: 30 μm wells, ▲: 10 μm wells, •: 8.5 μm wells. Measured with gold film and 670-nm light source.

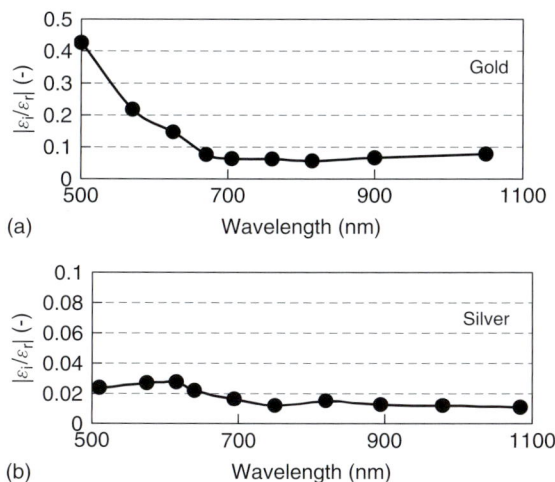

Figure 12. Wavelength of light source and dielectric constant.

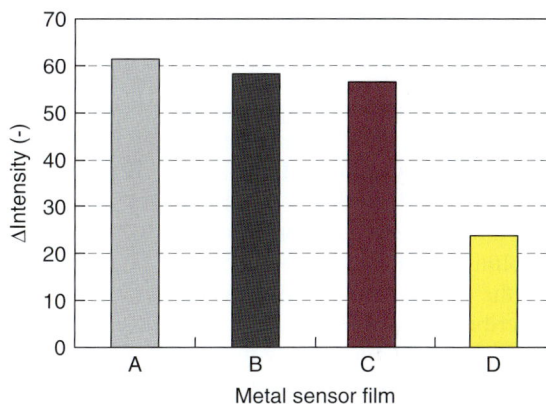

Figure 13. Comparison of SPR sensitivity with various metal films. Metal sensor film: (A) Ag (55 nm), (B) Ag (55 nm)/alumina (1 nm), (C) Ag (55 nm)/alumina (1 nm)/Au (2 nm), and (D) Au (65 nm); $100 \, g \, l^{-1}$ glucose was measured with 670-nm light source.

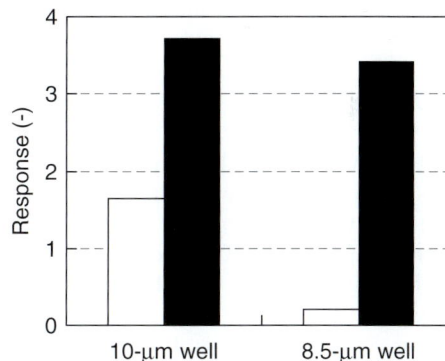

Figure 14. Detection of immunoglobulin G with various metal films. Metal sensor film D (Au (65 nm)) (open column) or metal sensor film C (Ag (55 nm)/alumina (1 nm)/Au (2 nm)) (closed column) was used to measure $0.1 \, g \, l^{-1}$ mouse IgG. A 670-nm light source was used for SPR measurement.

(Figure 12a) and silver (Figure 12b). In the case of gold sensor film, $|\varepsilon_i/\varepsilon_r|$ was drastically increased when the wavelength of light source was under 700 nm. Therefore SPR curves became broad and the sensitivity of SPR sensor became lower. But in the case of the silver film, $|\varepsilon_i/\varepsilon_r|$ was stable over 500 nm. Therefore with the use of a silver sensor film and shorter-wavelength light source, more sensitive SPR measurement could be possible even in a small area in the order of under 10 μm.

Figure 13 shows the comparison of SPR sensitivity with various metal films. The sensitivity was evaluated with the response to $100 \, g \, l^{-1}$ glucose. For sensor film C, called multilayered metal film,

gold film (2 nm) was prepared on alumina-coated silver film to immobilize affinity ligands, by using bifunctional molecules with thiols. High SPR sensitivity was also obtained with three kinds of silver-based sensor films.

Finally, mouse IgG was detected by using the multilayered metal film as sensor film. As shown in Figure 14, greater response to $0.1 \, g \, l^{-1}$ mouse IgG could be obtained with the multilayered metal film, especially with the 8.5-μm microwells, where the response was approximately 16 times higher than the gold sensor film.

8 CONCLUSION

In this chapter, two-dimensional SPR imaging techniques and their application to microarray technologies have been described. After the general survey of two-dimensional SPR imaging techniques, including commercially available instruments, our recent studies on the high-resolution SPR imager which may be applicable to the single-cell-based microwell array were described. The SPR sensor has wide application fields. Not only the affinity measurement described in this chapter, but direct measurement of cell dynamics may also be possible. Giebel et al.[19] have reported the imaging of cell/substrate contacts of living cells with the SPR imager. The problems of the SPR imager at the present stage may be because of low sensitivity. In order to achieve high sensitivity as in BIAcore™, effective control measures which eliminate the effects of environmental fluctuations may be necessary.

ACKNOWLEDGMENTS

Most of the work described in this chapter was done as a project study organized by Toyama Medical-Bio Cluster Projects supported by the Ministry of Education, Culture, Sports, Science and Technology of Japan, and Toyama prefecture.

The work was partially supported by the Grant-in-Aid for Scientific Research on Priority Areas "Lifesurveyor" and "Bio-Manipulation" from the Ministry of Education, Culture, Sports, Science and Technology of Japan.

REFERENCES

1. E. Kretschmann and H. Raether, Radiative decay of non-radiative surface plasmons excited by light. *Zeitschrift fur Naturforschung*, 1968, **23**, 2135.
2. B. Liedberg, C. Nylander, and I. Lundstroem, Surface. Plasmon resonance for gas detection and biosensing. *Sensors and Actuators*, 1983, **4**, 299–304.
3. R. L. Earp and R. E. Dessy, *Surface Plasmon Resonance*, in *Commercial Biosensors*, G. Ramsay (ed), John Wiley & Sons, 1998, pp. 99–164.
4. J. Melendez, R. Carr, D. U. Bartholomew, K. Kukanskis, L. Elkind, S. Yee, C. Furlong, and R. Woodbury, A commercial solution for surface plasmon sensing. *Sensors and Actuators*, 1996, **B35–B36**, 212–216.
5. M. Suzuki, F. Ozawa, W. Sugimoto, and S. Aso, Miniature surface-plasmon resonance immunosensors–rapid and repetitive procedure. *Analytical and Bioanalytical Chemistry*, 2002, **372**, 301–304.
6. A. G. Frutos and R. M. Corn, SPR of ultrathin organic films. *Analytical Chemistry*, 1998, **70**, 449A–455A.
7. SPR imager II array instrument, GWC Technologies, www.gwctechnologies.com/gwcSPRimager.htm, 2007.
8. H. J. Lee, T. T. Goodrich, and R. M. Corn, SPR imaging measurements of 1D and 2D DNA microarrays in PDMS microfluidic channels on gold thin films. *Analytical Chemistry*, 2001, **73**, 5525–5531.
9. G. J. Wegner, A. W. Wark, H. J. Lee, E. Codner, T. Saeki, S. Fang, and R. M. Corn, Real time SPR imaging measurements for the multiplexed determination of protein adsorption/desorption kinetics and surface enzymatic reactions on peptide microarrays. *Analytical Chemistry*, 2004, **76**, 5677–5684.
10. M. Kyo, K. Usui-Aoki, and H. Koga, Label-free detection of proteins in a crude cell lysate with antibody arrays by surface plasmon resonance imaging technique. *Analytical Chemistry*, 2005, **77**, 7115–7121.
11. K. Inamori, M. Kyo, Y. Nishiya, Y. Inoue, T. Sonoda, E. Kinoshita, T. Koike, and Y. Katayama, Detection and quantification of on-chip phosphorylated peptides by surface plasmon resonance imaging techniques using a phosphate capture molecule. *Analytical Chemistry*, 2005, **77**, 3979–3985.
12. B. K. Singh and A. C. Hiller, Surface plasmon resonance imaging of biomolecular interactions on a grating-based sensor array. *Analytical Chemistry*, 2006, **78**, 2009–2018.
13. *The First International Workshop on Approaches to Single-Cell Analysis*, Uppsala, Sweden, June, 2006, www.conference.slu.se/single cell, 2006.
14. Lifesurveyor Project, www.tuat.ac.jp/~surveyor/Eng/Eng-index.htm, 2007.
15. S. Yamamura, S. R. Rao, Y. Omori, Y. Tokimitsu, S. Kondo, H. Kishi, A. Muraguchi, Y. Takamura, and E. Tamiya, High-throughput screening and analysis for antigen specific single-cell using microarray. *Micro Total Analysis Systems*, 2004, **1**, 178–180.
16. M. Suzuki, H. Nakabayashi, Y. Jing, and M. Honda, Optical pH and oxygen sensing for micro-arrayed cell chips. *Micro Total Analysis Systems*, 2005, **2**, 1482–1484.
17. M. Suzuki, S. Hane, T. Ohshima, Y. Iribe, and T. Tobita, Detection of antibodies in 10 μm wells on micro-arrayed cell chips by 2D-SPR affinity imaging. *Micro Total Analysis Systems*, 2006, **1**, 756–758.
18. X.-M. Zhu, P.-H. Lin, P. Ao, and L. B. Sorensen, Surface treatment for surface plasmon resonance biosensors. *Sensors and Actuators*, 2002, **B84**, 106–112.
19. K.-F. Giebel, C. Bechinger, S. Herminghaus, M. Riedel, P. Leiderer, U. Weiland, and M. Bastmeyer, Imaging of cell/substrate contacts of living cells with surface plasmon resonance microscopy. *Biophysical Journal*, 1999, **76**, 509–516.

58

Label-Free Gene and Protein Sensors Based on Electrochemical and Local Plasmon Resonance Devices

Kagan Kerman,[1] Tatsuro Endo[2] and Eiichi Tamiya[3]

[1] *School of Materials Science, Japan Advanced Institute of Science and Technology (JAIST), Ishikawa, Japan,* [2] *Department of Mechano-Micro Engineering, Tokyo Institute of Technology, Yokohama, Japan and* [3] *Department of Applied Physics, Osaka University, Osaka, Japan*

1 INTRODUCTION

Since its discovery in early 1950s by Watson and Crick,[1] DNA has never ceased to fascinate the researchers in many diverse fields. Double helix is formed with the coupling of two strands with complementary base sequences. The codes behind the order of the four nucleotides have made DNA the most attractive biomaterial in scientific research. The detection of specific DNA base sequences has a significant impact in various fields of life sciences. For this task, DNA biosensors using several detection principles; such as electrochemical,[2–5] piezoelectric,[6,7] and optical[8,9] transducers have been developed. Recently, micro- or nanometer-scale DNA biosensors and biochips have become a major interest with the significant advances in fabrication technologies.

2 LABEL-FREE ELECTROCHEMICAL GENE SENSORS BASED ON THE OXIDATION OF DNA BASES

The majority of the label-free electrochemical DNA biosensors are based on the determination of purine oxidation current signals, mainly the guanine oxidation signal. This is due to the fact that guanine has the lowest oxidation potential of all DNA bases.[10–12] The main oxidation product, 8-oxo-7,8-dihydroguanine[13] is considered the biomarker of DNA damage by oxidative stress[14] and can be quantified using its oxidation peak current intensity.[15,16]

The detailed studies on the electrochemical behavior of purine and pyrimidine bases were reported. The majority of these reports were about the electrochemical reduction of purine and pyrimidine derivatives on mercury electrodes. The reduction peak current signal of guanine, adenine, and cytosine could be observed; however, no polarographic wave was reported for thymine on the mercury electrodes.[10] Carbon electrodes were extensively used for the detection of oxidation peak current signals derived from the purine bases.[17,18] The electroactivity of pyrimidine-derivative compounds at solid electrodes was reported.[19] Brett and coworkers reported that guanine and adenine were more easily detected than thymine and cytosine.[20] The well-defined oxidation signal of guanine has been utilized for the detection of DNA

Handbook of Biosensors and Biochips. Edited by Robert S. Marks, David C. Cullen, Isao Karube, Christopher R. Lowe and Howard H. Weetall.
© 2007 John Wiley & Sons, Ltd. ISBN 978-0-470-01905-4.

hybridization.[21–25] Guanine oxidation signal was also monitored to detect the interaction mechanisms between several types of drugs and DNA.[26–28] Toxic molecules that interact with DNA could also be detected by monitoring the changes in guanine oxidation signal in connection with electrochemical DNA biosensors.[29–31]

The use of inosine-substituted probes and the appearance of a guanine signal upon hybridization with the native target DNA molecule provided new challenges in electrochemical gene sensors.[32,33] Ariksoysal et al.[33] described a label-free electrochemical DNA biosensor protocol that the detection of hybrid formation was performed by using the guanine oxidation signal of the target DNA. A pencil-based renewable biosensor for label-free electrochemical detection of hybridization was also reported by Kara et al.[21] Magnetocomposite electrodes have recently been developed for the label-free electrochemical detection of DNA hybridization using magnetic beads.[34]

Carbon nanotubes (CNTs) possess unique properties that are amenable to biosensor applications. CNTs are one-dimensional structures that are extremely sensitive to electronic perturbations, readily functionalized with biorecognition layers, and compatible with many semiconducting manufacturing processes.[35,36] To date, there have been several reports on the electrochemical detection of DNA hybridization using CNT-modified electrodes.[37,38] Whereas electrochemical methods rely on electrochemical behavior of the labels, measurement of direct electron transfer between CNTs and DNA molecules paves the way for label-free DNA detection. Single-stranded deoxyribonucleic acid (ssDNA) has been recently demonstrated to interact noncovalently with single-walled carbon nanotubes (SWNTs).[39] The ssDNA forms a stable complex with individual SWNTs by wrapping around them by means of the aromatic interactions between nucleotide bases and SWNT sidewalls. Double-stranded DNA molecules have also been proposed to interact with SWNTs as major groove binders.[40]

Figure 1 shows the differential pulse voltammograms for the detection of polynucleotides. The voltammetric oxidation peaks of 10-ppm poly-adenine (polyA) and poly-guanine (polyG) were obtained at a carbon paste electrode (CPE). CPE was prepared by mixing graphite powder (Fisher) and mineral oil (Acheson 38) in a 70 : 30

(w/w) ratio. The carbon paste was then tightly packed into a Teflon tube (3 mm i.d.). SWNT-paste electrode was prepared by mixing SWNT powder and mineral oil in a 60 : 40 (w/w) in a similar fashion as reported by Palleschi and coworkers.[41] About eightfold amplification in the oxidation signal of polyA was observed at the SWNT-paste electrode (Figure 1a). The oxidation peak of polyA shifted 0.15 V toward lower potential values, which was attributed to the good electron transfer characteristics of SWNTs. PolyG oxidation signal was also amplified threefold, and its oxidation peak shifted about 0.25 V toward lower potential values at SWNT-paste CPE (Figure 1b).

We anticipate that the development of DNA sensors based on the intrinsic electrochemical signals will continue to evolve and bring new and exciting findings and a closer look into the structure and interaction mechanisms of DNA with small ligands.

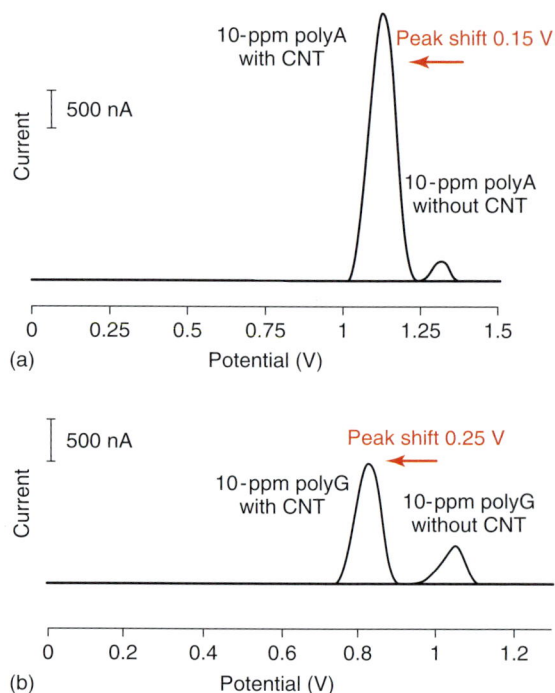

Figure 1. Differential pulse voltammograms for the label-free electrochemical detection of polynucleotides; (a) 10-ppm poly-adenine (polyA) at a bare carbon paste electrode (CPE), 10-ppm polyA at SWNT-paste electrode, (b) 10-ppm poly-guanine (polyG) at a bare CPE, 10-ppm polyG at SWNT-paste electrode in 0.50 M acetate buffer solution (pH 4.5).

3 LABEL-FREE GENE AND PROTEIN SENSORS BASED ON FIELD-EFFECT TRANSISTORS

A field-effect transistor (FET) device uses an electric field to control the shape, and hence the conductivity of a "channel" in a semiconductor material. FETs are sometimes employed as voltage-controlled resistors. FET is simpler in concept than the bipolar transistor. The channel region of any FET is either doped to produce n-type semiconductor, producing an "N channel", or with p-type to produce a "P channel". The doping determines the polarity of the gate operation.

FET devices can be constructed from a wide range of materials. FETs made of semiconductor nanowires (NWs) and SWNTs show promising potential for biosensor applications. The depletion or accumulation of charge carriers, which are caused by the binding of charged biological macromolecules on the surface of NWs or SWNTs, greatly affects the entire cross-sectional conduction pathway of these nanostructures.

Electrical detection of biomolecular interactions with metal–insulator–semiconductor diodes has recently been reported by Estrela et al.[42] They have also reported the detection of biomolecular interactions using FET devices; such as metal-oxide-semiconductor field-effect transistor (MOSFET) capacitor structures and polycrystalline silicon thin-film transistors (Poly-Si TFTs).[43]

FET-based biosensors in connection with semiconducting Si NWs are promising candidates, since the doping type and concentration can be controlled, which allows tuning of the sensitivity without the need for an external gate. Since the report of Lieber and coworkers[44] about the ultrasensitive detection of biological and chemical species by exploring nanoscale Si NW-based FETs, there has been an intensive amount of research on FET-based biosensors with Si NWs. NW arrays were modified with antibodies for influenza A, and discrete conductance changes were observed due to the binding and unbinding events in the presence of influenza A but not paramyxovirus or adenovirus.[45] Lieber and coworkers also reported the electrical detection of small-molecule inhibitors of ATP binding to Abl, a protein tyrosine kinase, by using Si NW-FET devices.[46] NW arrays also allowed highly selective and sensitive multiplexed detection of prostate-specific antigen (PSA), PSA-α_1-antichymotrypsin, carcinoembryonic antigen, and mucin-1 with a detection limit of $0.9\,pg\,ml^{-1}$ in undiluted serum samples.[47] Recently, Lieber and coworkers[48] demonstrated that Ge/Si core/shell NWs had long carrier mean free paths at room temperature. They concluded that the performance of Ge/Si NW-FETs was comparable to similar-length CNT-FETs and substantially exceeded the length-dependent scaling of planar silicon MOSFETs.

CNT-based FETs also have excellent operating characteristics,[49,50] and they have already been explored for highly sensitive electronic detection of gases.[51–53] We developed back-gate CNT-FET nanosensor arrays, where the biomolecules could easily be immobilized onto the gold-coated back gate (Figure 2a and b). We have employed these back-gate devices based on CNT-FETs for the

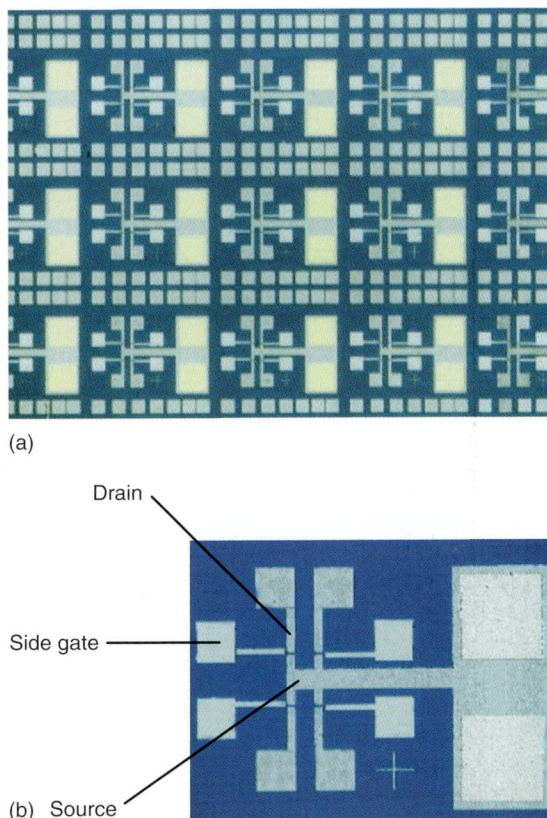

(a)

(b)

Figure 2. (a) Microelectrode array based on carbon nanotube field-effect transistor (CNT-FET) devices and (b) close-up image of a single CNT-FET device.

ultrasensitive detection of hybridization between peptide nucleic acid (PNA) probes and DNA target strands. PNA is a synthetic DNA analog, in which both the phosphate and the deoxyribose of the DNA backbone are replaced by a polypeptide.[54] Even so, PNA retains the ability to hybridize with complementary DNA or RNA sequences. An additional advantage of using PNA instead of DNA in sensing systems is its neutral polypeptide backbone, which leads to an improved PNA–DNA base-pairing free of ionic strength influence, and resistance to degradation by proteases and nucleases.[55]

A self-assembled monolayer (SAM) of PNA probes related to the tumor necrosis factor alpha gene (TNF-α) was prepared on the gold electrode at the reverse side of the CNT-FET device as illustrated in Figure 3(a). A time-dependent conductance increase was monitored upon flowing negatively charged DNA target molecules through the microfluidic channel of a poly(dimethylsiloxane) (PDMS) chip as shown in Figure 3(b). The high selectivity of PNA probes only toward

the full-complementary DNA samples enabled the rapid and simple discrimination against single-nucleotide polymorphism (SNP) or noncomplementary (NC) DNA.[56] Concentration-dependent measurements indicated a limit of detection of 6.8 fM target DNA.[56]

4 CONDUCTING POLYMERS FOR LABEL-FREE ELECTROCHEMICAL GENE SENSORS

Conducting polymers (CP) are attractive substrates for gene sensors, because they can act as an electronic transducer for the charged species binding to the surface. No redox indicator is necessary, because the CP itself can directly report the hybridization event. An additional advantage of CP is that a bioelement that serves as a probe can be covalently grafted to the polymer backbone, and a variety of organic synthetic methods are available for this purpose.[57] Several methods for the immobilization of DNA probes onto CP have been reported.[58,59]

Initial approaches to the construction of CP-based gene sensors included the direct adsorption of the oligonucleotides onto oxidized polypyrrole (PPy) films by electrostatic attractions[60] or incorporation of the oligonucleotides into the polymer film as a macro-counterion.[61] However, these methods suffered from effects arising from constraints on the orientation of the oligonucleotides resulting in high steric and kinetic barriers to hybridization, and the possibility of oxidative damage to the oligonucleotide probes. To overcome such steric constraints, several groups have electrocopolymerized, N-position-substituted pyrroles.[62] Mugweru and Rusling[63] adsorbed poly(4-vinylpyridine) (PVP) with attached Ru(bpy)$_2^{2+}$ onto pyrolytic graphite electrodes for the development of reusable DNA sensors. Recently, Komarova et al.[64] developed a label-free DNA sensor using ultrathin films of PPy doped with an oligonucleotide probe related to a biowarfare virus, *Variola major*. Piro et al.[65] reported the characterization of a new bifunctional electroactive polymer, poly(5-hydroxy-1,4-naphthoquinone (juglone)-co-5-hydroxy-3-thioacetic acid-1,4-naphthoquinone), which was used for the label-free electrochemical detection of DNA hybridization.

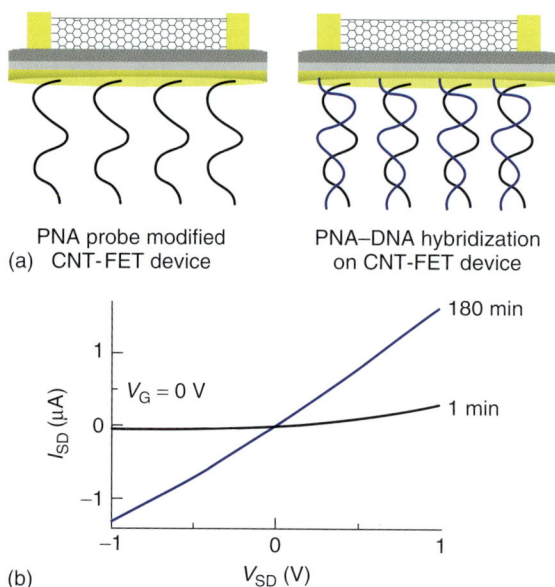

Figure 3. (a) Schematic illustration of a back-gate carbon nanotube field-effect transistor (CNT-FET) device. Peptide nucleic acid (PNA) probes were attached onto the gold surface using covalent modification. (b) As the hybridization with the negatively charged target DNA molecules took place, the source–drain current response versus source–drain potential increased.

Peng et al.[66] reported synthesis of a new precursor monomer, (4-(3-pyrrolyl) butanoic acid), which was used to form a copolymer of poly(pyrrole-co-4-(3-pyrrolyl) butanoic acid) by electropolymerization. The relatively long butanoic acid side chains positioned the oligonucleotide probe away from the copolymer backbone and facilitated efficient hybridization. The hybridization with target DNA was detected in connection with cyclic voltammetry and the AC impedance spectroscopy. Label-free DNA detection has recently been performed using modified conducting PPy films at microelectrodes.[67]

The use of PPy in conjunction with bioaffinity reagents is a powerful method that has expanded the range of applications of electrochemical detection and its future development is expected to continue. The use of a wide range of counterions will provide significant improvements in affinity at the PPy ion-exchange sites. The application of CNTs will also be beneficial to the development of PPy-based biosensors at the nanoscale.

5 IMPEDANCE SPECTROSCOPY FOR THE LABEL-FREE GENE AND PROTEIN SENSORS

Electrochemical impedance spectroscopy (EIS) (sometimes also called *AC impedance*) is an electrochemical technique, in which a low-amplitude alternating potential (or current) wave is imposed on top of a direct current potential (often the corrosion potential and zero imposed current). EIS has been explored in detail as the dominant method in label-free biosensors owing to its high sensitivity in the detection of small proteins; such as interferon γ.[68] Impedimetric immunosensors have been reviewed in depth by Katz and Willner.[69] Moreover, conjugated biomolecules with gold nanoparticles have recently been used for the amplification of impedance and capacitance signals.[70] The comparison of label-free and impedance-amplifying label-based systems was performed by Ma et al.[71] The sensing of DNA binding drugs using gold substrates modified with gold nanoparticles was reported by Li et al.[72] DNA hybridization was detected using EIS investigation of conducting properties of a functionalized polythiophene matrix.[73] Yang et al.[74] reported

a new antibody immobilization strategy based on electrodeposition of nanometer-sized hydroxyapatite for label-free capacitive immunosensors. Oligonucleotide-functionalized PPy was also applied to the impedimetric detection of DNA hybridization.[75] Interdigitated ultramicroelectrodes using electrochemical redox probes were developed for impedimetric detection of DNA.[76] An impedance array biosensor was fabricated by Yu et al.[77] for the detection of multiple antibody–antigen interactions.

Möller et al.[78] have recently utilized the resistance in the gap between the microfabricated electrodes for monitoring DNA hybridization. Capture DNA was immobilized in the gap. Then, the biotin-modified target DNA hybridized with its complementary capture DNA. Afterward, the chip was incubated with either streptavidin-modified gold nanoparticles or a streptavidin–peroxidase polymer, which was bound to the biotin modification. Silver enhancement produced a conductive layer to bridge the gap between the electrodes, so that a significant drop in the electrical resistance could be detected.

Aptamers, derived from the Latin word, "aptus" meaning "to fit", are artificial nucleic acid ligands that can be generated against amino acids, drugs, proteins, and many other molecules.[79] They are isolated from combinatorial libraries of synthetic nucleic acids by a process called *SELEX* (*systematic evolution of ligands by exponential enrichment*) involving adsorption, recovery, and re-amplification.[80] Aptamers, first reported in the early 1990s, are still attractive in the areas of therapeutics and diagnostics and offer themselves as ideal candidates for use as biocomponents in biosensors (aptasensors), possessing many advantages over the other affinity sensors.[81] Recently, aptamers have been applied as the recognition layer in impedimetric biosensors. Aptamer-based array electrodes were developed by Xu et al.[82] for immunoglobulin detection. An aptamer-based impedance measurement assay was also developed for thrombin with a detection limit of 0.1 nM.[83] Radi et al.[84] developed reusable impedimetric aptasensors. Wang and coworkers reported the recognition-induced switching of the surface charge for aptasensors.[85] Surely, the potential of EIS as a biosensor is immense, and this exciting research area is on the brink of exponential growth.

6   LABEL-FREE ELECTROCHEMISTRY OF PROTEINS

Although many advanced transducers have been developed in the past decades, the electronic transduction of protein interactions has been a major challenge in electrochemistry, because the proteins have been considered to be "electroinactive". Thus, a major interest has been focused on "metalloproteins", which are chemical combinations of protein atoms with ions of metals such as iron, calcium, copper, and zinc. The first reports on the direct electron transfer between the metal ion center of the metalloprotein and an electrode surface were published in 1977, when Eddowes et al.[86] and Yeh et al.[87] independently showed that cytochrome *c* on bipyridyl-modified gold and tin-doped indium oxide electrodes, respectively, went through "bioelectrocatalysis" with reversible cyclic voltammograms. The electrocatalytic voltammetry of metalloproteins continues to be the subject of intense investigation[88,89] to the present day.

The catalytic evolution of hydrogen on dropping mercury electrodes in weakly alkaline solutions was a milestone discovery of Heyrovsky, which had caused a series of publications to investigate the possibility of protein electrochemistry.[90] Especially, the Bridcka reaction, which represents the catalytic lowering of hydrogen overvoltage in the presence of cobalt ions has been the main means of protein polarography.[90] The first introduction of the direct oxidation of tyrosine (Tyr), tryptophan (Trp), and cysteine (Cys) residues on carbon electrodes, about two decades ago, created a new notion in electroanalysis: that not only metalloproteins, but also all proteins that contained either Tyr, Trp, and/or Cys could be detected directly on carbon electrodes.[91,92] Oxidation of Tyr and Trp at a wax-impregnated spectroscopic graphite electrode was reported to be a two-electron transfer process.[93] The alanine-substituted indole ring of Trp is susceptible to electrochemical oxidation from its heteroatom.[91–93] Several groups reported the electrochemical detection of clinically important proteins and amino acids.[94–96] Recently, the electrochemical oxidation signal of amyloid-β peptides related to Alzheimer's disease was employed as the indicator of fibril formation.[97]

The electrical responses of some representative proteins and amino acids are shown in Figure 4. Bovine serum albumin (BSA) showed a wide electrochemical peak of ~208 nA at 500 pg ml^{-1} (Figure 4C), but the signal was not as high as one would have expected from a protein with 21 Tyr and 3 Trp residues. The structure of BSA with about 607 amino acids might have masked the exposure, and eventually the contact of these electroactive amino acids with the carbon electrode surface. As expected, the electroactive amino acids, Tyr (Figure 4A) and Trp (Figure 4B) alone gave the highest electrical responses. The proteins with small molecular weights, such as human chorionic gonadotropin (hCG) (Figure 4D), and PSA (Figure 4E), gave high responses. This trend in their signals was attributed to the steric proximity of the electroactive residues of these proteins to the electrode surface and their less complex structure owing to their small molecular weight.

The detection of electroactive amino acids was assessed on the bare Pt and SWNT-modified Pt electrodes in connection with differential pulse voltammetry (DPV).

SWNT-Pt electrodes were prepared by dispersing acid-treated SWNTs in chitosan solution. A 2% (w/v) chitosan solution was prepared in 1% (v/v) acetic acid. SWNTs were treated in concentrated nitric acid for 5 h, filtered and

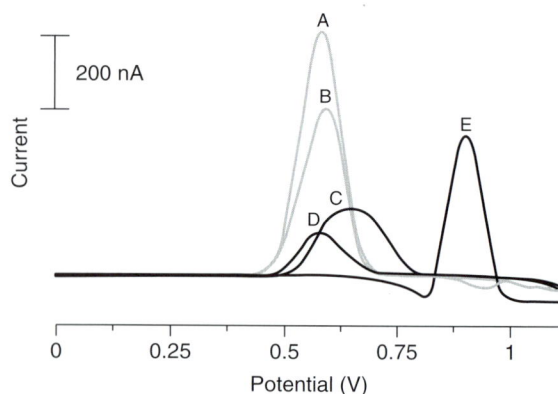

Figure 4. Differential pulse voltammograms of proteins and amino acids on a carbon paste electrode (CPE), (A) 250 µg ml^{-1} tyrosine (Tyr), (B) 250 µg ml^{-1} tryptophan (Trp), (C) 500 µg ml^{-1} bovine serum albumin (BSA), (D) 300 µg ml^{-1} human chorionic gonadotropin hormone (hCG), (E) 0.250 µg ml^{-1} prostate-specific antigen (PSA) in 50-mM phosphate buffer solution (pH 7.4).

washed with ultrapure water until the filtrate was neutral, and finally dried under vacuum. Then, a desired amount of CNTs ($1\,mg\,ml^{-1}$) was dissolved in chitosan solution under 1-h sonication. Chitosan–CNT solution ($20\,\mu l$) was spread uniformly on the Pt electrode surface. After the solution was dried in air, the SWNT-modified electrode was immersed in 0.1 M NaOH for 30 min to make the film more stable. Then, the electrode was rinsed with ultrapure water and air-dried. The scanning electron microscopy (SEM) images of the SWNT–chitosan matrices are shown in Figure 5(a) and (b). The differential pulse voltammograms for the detection of Tyr, Trp, and Cys are shown in Figure 6(a–c), respectively. The electrochemical oxidation signals were clearly observed from the SWNT-modified electrodes (red line). On the other hand, low signals were obtained from the devices with bare Pt-arrayed electrodes (blue line). Moreover, the peak current intensities for the amino acids obtained from SWNT-modified electrodes were about 20-fold higher than those obtained from bare Pt, which was attributed to the remarkable enhancement in total surface area of the working electrode with the efficient electron transfer characteristics of SWNTs.

Further research about the intrinsic electroactivity of proteins will surely provide extensive data regarding their structure, and protein–protein, protein–nucleic acid, or protein–drug interaction mechanisms. Label-free electrochemistry provides a powerful method for the challenge of developing effective biosensors for identifying, quantifying, and characterizing proteins.

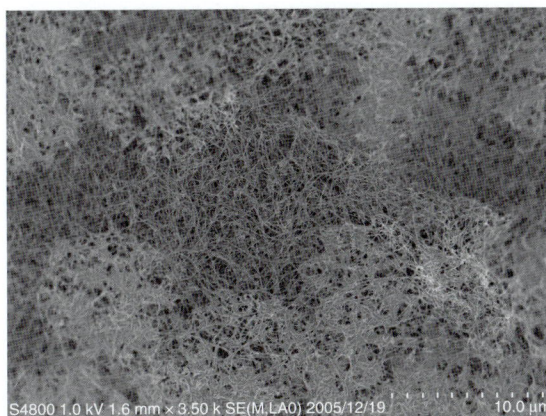

(a)

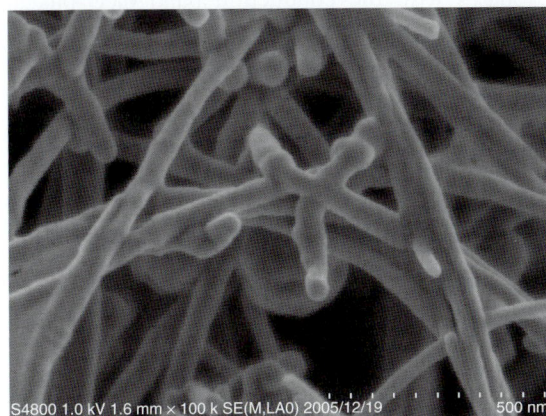

(b)

Figure 5. (a) Scanning electron microscopy (SEM) image of a single-walled carbon nanotube (SWNT)-dispersed chitosan matrix with the close-up of the same SEM image (b).

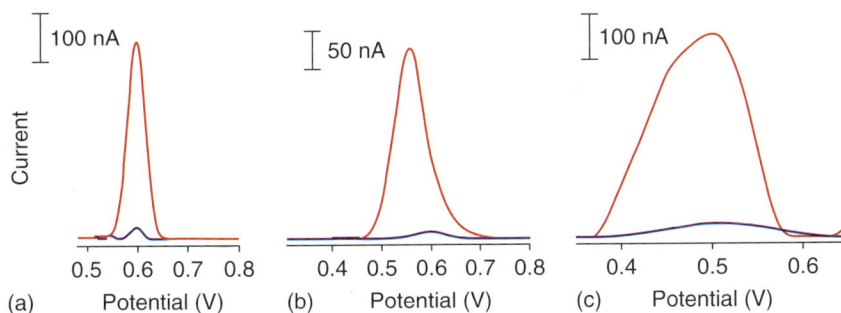

(a) (b) (c)

Figure 6. Differential pulse voltammograms of amino acids at SWNT-Pt (red line) and bare Pt electrodes (blue line), (a) $10\,\mu g\,ml^{-1}$ Tyr, (b) $5\,\mu g\,ml^{-1}$ Tyr, (c) $5\mu g\,ml^{-1}$ Cys, in 50 mM phosphate buffer solution (pH 7.4).

7 LOCALIZED SURFACE PLASMON RESONANCE

Colloidal gold and silver have been attractive materials for architects, scientists, and artists for centuries because of their unique optical properties. Such properties are strongly dependent on the size, shape, and local environment of their nanostructures, as described by Mie theory.[98] Briefly, as the size of a metal structure decreases from the bulk scale (meters to micrometers) to the nanoscale (<100 nm), the movement of electrons through the internal metal framework becomes restricted. As a result, metal nanoparticles display extinction bands in their UV–visual spectra, when the incident light resonates with the conduction band electrons at their surfaces. These charge-density oscillations are simply defined as *local surface plasmon resonance* (*LSPR*). The excitation of LSPR by light at an incident wavelength, where resonance would occur, results in the appearance of surface plasmon (SP) absorption bands. The intensity and position of the SP absorption bands are characteristic of the type of the nanomaterial, its diameter, and its distribution. LSPR is also highly sensitive to the changes of its surrounding environment.[99,100] Thus, LSPR enables the detection of an immediate increase in thickness of a biomolecular layer on the surface of a substrate caused by a reaction between the solution component under study and the receptor layer immobilized on the surface.

The high sensitivity of LSPR has been utilized to design biosensors for the label-free detection of biomolecular interactions between various ligands.[101–104] However, almost all of the LSPR-based optical biosensors were designed in connection with gold and silver nanoparticles. The synthesis of the nanoparticles and the control of their diameters have been the most challenging problems. Since LSPR has extreme sensitivity toward size, a uniform layer of nanoparticles with the exact same size had to be built for reproducible results. Moreover, monitoring small shifts in the peak wavelength obtained from these nanoparticles was a difficult task for the complicated applications to real samples. Therefore, we focused our attention on the fabrication of gold-capped nanoparticle layer substrates.[105] The gold-capped nanoparticle substrate brings several advantages along with its cost-effective and easy fabrication procedure. In addition, the optical characteristics of the gold-capped nanoparticle layer substrate depend on the dielectric nanoparticle diameter and the thickness of the gold layer, which is deposited on the glass substrate surface. These characteristics may be attributed to a similar phenomenon described for the plasmon resonance spectra of the nanoshell-structured nanoparticles. As reported by Petit et al.[106] the variations in the intensity of the plasmon resonances depended on the cluster size. The electronic relaxation after the electromagnetic excitation is accelerated due to the decrease in damping with increasing particle size.[107]

8 LSPR-BASED GENE AND PROTEIN SENSORS

Advanced chemical and biological nanosensors for the detection of single molecules are under development.[108] Chau et al.[109] have recently reported a novel reflection-based LSPR fiber-optic probe for chemical and biochemical sensing. The sensor is based on intensity measurement of the internal reflected light at a fixed wavelength from an optical fiber where the extinction cross-section of self-assembled gold nanoparticles on the unclad portion of the optical fiber changes with the different refractive index of the environment near the gold surface. They have achieved the detection of "spectroscopically silent" Ni^{2+} ions and the detection of streptavidin and staphylococcal enterotoxin B at the pM level. Silver particles generated by the evaporation–condensation method were deposited on the solgel-derived silica film or the silica glass substrate to form LSPR sensors.[110]

We constructed the LSPR-based nanochip with a unique core-shell-structured nanoparticle layer (Figure 7a), the surface-modified silica nanoparticle (particle diameter: 100 nm) was used as the *core*, and the *shell* was applied with the top (30 nm) and the bottom (40 nm) gold layers that were deposited using thermal deposition. When the top gold layer was deposited on the nanoparticle layer–modified substrates using thermal evaporator, we could observe the LSPR band at 520 nm with a simple optical probe. The multiarray ($20 \times 60 \, mm^2$) of core-shell-structured nanoparticle layer substrate was used for biosensor applications (Figure 7b).

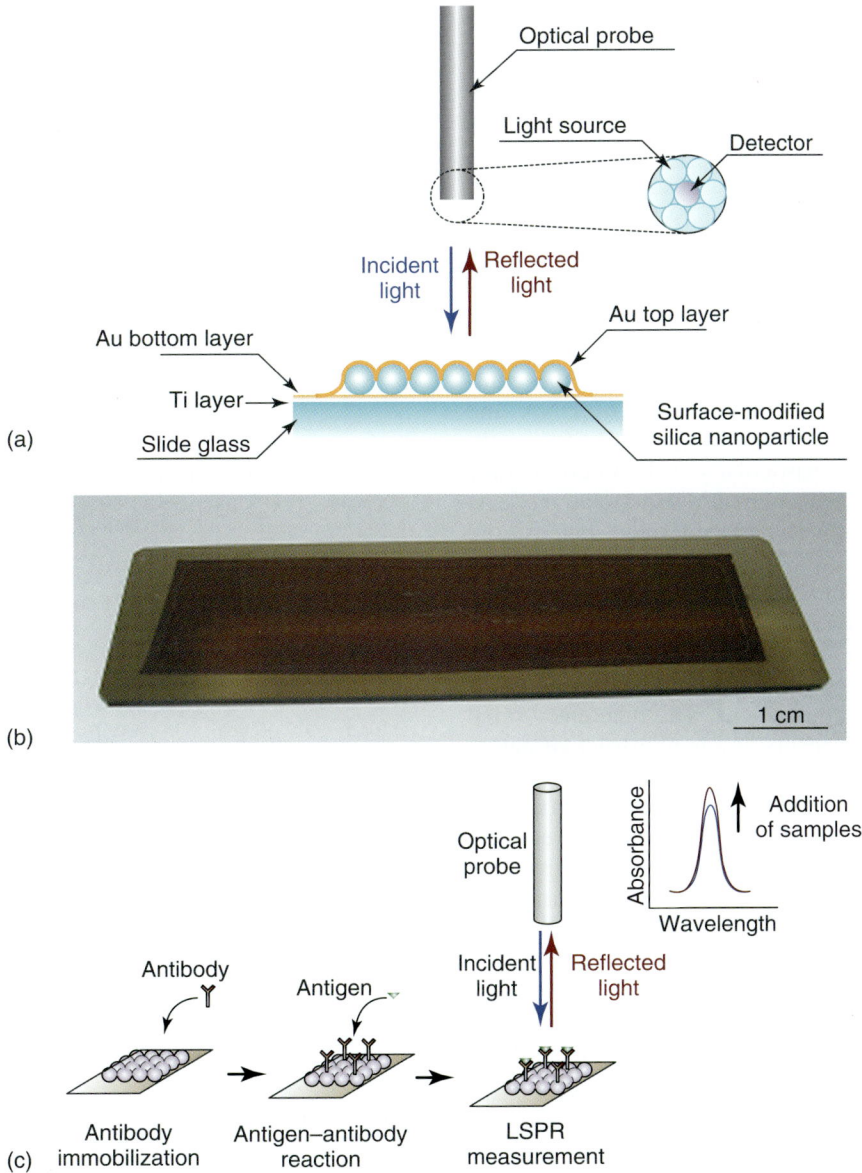

Figure 7. (a) Construction of the multiarray LSPR-based nanochip. The surface-modified silica nanoparticles were aligned onto the gold-deposited glass substrate surface. Consecutively, the top gold layer was evaporated onto the silica nanoparticle layer, (b) photograph of the nanoparticle layer substrate, (c) illustration for the experimental flow of the LSPR-based immunosensor application.

We used PNAs as the biorecognition layer on our core-shell nanoparticle layer substrate and detected PNA–PNA and PNA–DNA hybridization reactions.[111] Moreover, the nanoparticle layer substrate was applied as an immunosensor. The antifibrinogen antibody was immobilized onto the nanoparticle layer substrate surface (Figure 7c). Different concentrations of fibrinogen were introduced to the antifibrinogen antibody-immobilized nanoparticle layer substrate surface, and the change in the absorption spectrum, caused by the antigen–antibody reaction, was observed.

By using this antifibrinogen antibody-immobilized nanoparticle layer substrate, the detection limit for fibrinogen was $10 \, ng \, ml^{-1}$.

9 CONCLUSIONS

Here, we would like to quote *Label-free Methods Are Not Problem Free* from Emma Hitt.[112] On the other hand, there have been many significant and promising advances to solve the problems of label-free methods and improve them since the publication of her article.[112] Especially for the commonly sample-limited applications of DNA and protein analysis, electrochemical and LSPR methods have become particularly well suited, as they can be miniaturized and multiplexed relatively easily in connection with nanomaterials; such as gold nanoparticles and CNTs. However, the superiority of label-free electrochemistry would be due to its ability to provide new and complementary information about the oxidation/reduction process.

LSPR biosensors have a promising potential in a wide variety of applications. The inherently small size of the nanosensors and nondestructive nature of the optical measurement are significant advantages over macroscale biosensors or electrochemical schemes that usually do not permit the recovery of the target analyte. Eventually, we anticipate the development of credit card–sized sensor arrays in connection with electrochemical or LSPR-based methods in the near future.

REFERENCES

1. J. Watson and F. Crick, A structure for deoxyribose nucleic acid. *Nature*, 1953, **171**, 737–738.
2. E. Palecek, Past, present and future of nucleic acids electrochemistry. *Talanta*, 2002, **56**, 809–819.
3. T. G. Drummond, M. G. Hill, and J. K. Barton, Electrochemical DNA sensors. *Nature Biotechnology*, 2003, **21**, 1192–1199.
4. K. Kerman, M. Kobayashi, and E. Tamiya, Recent trends in electrochemical DNA biosensor technology. *Measurement Science and Technology*, 2004, **15**, R1–R11.
5. L. Murphy, Biosensors and bioelectrochemistry. *Current Opinion in Chemical Biology*, 2006, **10**, 177–184.
6. C. K. O'Sullivan and G. G. Guilbault, Commercial quartz crystal microbalances–theory and applications. *Biosensors and Bioelectronics*, 1999, **14**, 663–670.
7. K. A. Marx, Quartz crystal microbalance: a useful tool for studying thin polymer films and complex biomolecular systems at the solution-surface interface. *Biomacromolecules*, 2003, **4**, 1099–1120.
8. R. Karlsson, SPR for molecular interaction analysis: a review of emerging application areas. *Journal of Molecular Recognition*, 2004, **17**, 151–161.
9. P. Pattnaik, Surface plasmon resonance: applications in understanding receptor-ligand interaction. *Applied Biochemistry and Biotechnology*, 2005, **126**, 79–92.
10. E. Palecek, M. Fojta, F. Jelen, and V. Vetterl, *Electrochemical Analysis of Nucleic Acids*, in *Encyclopedia of Electrochemistry, Bioelectrochemistry*, A. J. Bard, M. Stratmann, and G. S. Wilson (eds), John Wiley & Sons, Weinheim, Germany, 2002, Vol. 9, pp. 365–429.
11. S. Steenken, Purine bases, nucleosides, and nucleotides: aqueous solution redox chemistry and transformation reactions of their radical cations and e- and OH adducts. *Chemical Reviews*, 1989, **89**, 503–520.
12. A. M. Oliveira Brett, S. H. P. Serrano, J. A. P. Piedade, *Electrochemistry of DNA*, in *Comprehensive Chemical Kinetics, Applications of Kinetic Modelling 37*, R. G. Compton and G. Hancock (eds), Elsevier, Oxford, 1999, pp. 91–119.
13. W. Saenger and C. R. Cantor, *Principles of Nucleic Acid Structure*, Springer, New York, 1984.
14. B. Halliwell and J. M. C. Gutteridge, *Free Radicals in Biology and Medicine*, 3rd Edn, Oxford University Press, New York, 1999.
15. A. M. Oliveira-Brett, J. A. P. Piedade, and S. H. P. Serrano, Electrochemical oxidation of 8-oxoguanine. *Electroanalysis*, 2000, **12**, 969–973.
16. I. A. Rebelo, J. A. P. Piedade, and A. M. Oliveira-Brett, Development of an HPLC method with electrochemical detection of femtomoles of 8-oxo-7,8-dihydroguanine and 8-oxo-7,8-dihydro–2′-deoxyguanosine in the presence of uric acid. *Talanta*, 2004, **63**, 323–331.
17. C. M. A. Brett, A. M. Oliveira Brett, and S. H. P. Serrano, On the adsorption and electrochemical oxidation of DNA at glassy carbon electrodes. *Journal of Electroanalytical Chemistry*, 1994, **366**, 225–231.
18. A. M. Oliveira Brett, L. A. Silva, and C. M. A. Brett, Adsorption of guanine, guanosine, and adenine at electrodes studied by differential pulse voltammetry and electrochemical impedance. *Langmuir*, 2002, **18**, 2326–2330.
19. A. M. Oliveira Brett and F.-M. Matysik, Voltammetric and sonovoltammetric studies on the oxidation of thymine and cytosine at a glassy carbon electrode. *Journal of Electroanalytical Chemistry*, 1997, **429**, 95–99.
20. A. M. Oliveira-Brett, V. Diculescu, and J. A. P. Piedade, Electrochemical oxidation mechanism of guanine and adenine using a glassy carbon microelectrode. *Bioelectrochemistry*, 2002, **55**, 61–62.
21. P. Kara, D. Ozkan, A. Erdem, K. Kerman, S. Pehlivan, F. Ozkinay, D. Unuvar, G. Itirli, and M. Ozsoz, Detection of achondroplasia G380R mutation from PCR amplicons by using inosine modified carbon electrodes based on electrochemical DNA chip technology. *Clinica Chimica Acta*, 2003, **336**, 57–64.
22. K. Kerman, Y. Morita, Y. Takamura, and E. Tamiya, Label-free electrochemical detection of DNA hybridization on gold electrode. *Electrochemistry Communications*, 2003, **5**, 887–891.

23. B. Meric, K. Kerman, D. Ozkan, P. Kara, and M. Ozsoz, Indicator-free electrochemical DNA biosensor based on adenine and guanine signals. *Electroanalysis*, 2002, **14**, 1245–1250.

24. K. Kerman, D. Ozkan, P. Kara, A. Erdem, B. Meric, P. E. Nielsen, and M. Ozsoz, Label-free bioelectronic detection of point mutation by using peptide nucleic acid probes. *Electroanalysis*, 2003, **15**, 667–670.

25. K. Kerman, Y. Morita, Y. Takamura, M. Ozsoz, and E. Tamiya, DNA-directed attachment of carbon nanotubes for enhanced label-free electrochemical detection of DNA hybridization. *Electroanalysis*, 2004, **16**, 1667–1672.

26. A. Erdem and M. Ozsoz, Electrochemical DNA biosensors based on DNA–drug interactions. *Electroanalysis*, 2002, **14**, 965–974.

27. J. A. P. Piedade, I. R. Fernandes, and A. M. Oliveira-Brett, Electrochemical sensing of DNA–adriamycin interactions. *Bioelectrochemistry*, 2002, **56**, 81–83.

28. B. Meric, K. Kerman, D. Ozkan, P. Kara, A. Erdem, O. Kucukoglu, E. Erciyas, and M. Ozsoz, Electrochemical biosensor for the interaction of DNA with the alkylating agent 4,4′-dihydroxy chalcone based on guanine and adenine signals. *Journal of Pharmaceutical and Biomedical Analysis*, 2002, **30**, 1339–1346.

29. K. Kerman, B. Meric, D. Ozkan, P. Kara, A. Erdem, and M. Ozsoz, Electrochemical DNA biosensor for the determination of benzo[α]pyrene–DNA adducts. *Analytica Chimica Acta*, 2001, **450**, 45–52.

30. F. Lucarelli, A. Kicela, I. Palchetti, G. Marrazza, and M. Mascini, Electrochemical DNA biosensor for analysis of wastewater samples. *Bioelectrochemistry*, 2002, **38**, 113–118.

31. M. Ozsoz, A. Erdem, P. Kara, K. Kerman, and D. Ozkan, Electrochemical biosensor for the detection of interaction between arsenic trioxide and DNA based on guanine signal. *Electroanalysis*, 2003, **15**, 613–619.

32. J. Wang and A. N. Kawde, Amplified label-free electrical detection of DNA hybridization. *Analyst*, 2002, **127**, 383–386.

33. D. O. Ariksoysal, H. Karadeniz, A. Erdem, A. Sengonul, A. A. Sayiner, and M. Ozsoz, Label-free electrochemical hybridization genosensor for the detection of hepatitis B virus genotype on the development of lamivudine resistance. *Analytical Chemistry*, 2005, **77**, 4908–4917.

34. A. Erdem, M. I. Pividori, A. Lermo, A. Bonani, M. del Valle, and S. Alegret, Genomagnetic assay based on label-free electrochemical detection using magneto-composite electrodes. *Sensors and Actuators B*, 2006, **114**, 591–598.

35. J. Wang, Carbon-nanotube based electrochemical biosensors: a review. *Electroanalysis*, 2004, **17**, 7–14.

36. E. Katz and I. Willner, Biomolecule-functionalized carbon-nanotubes: applications in nanobioelectronics. *ChemPhysChem*, 2004, **5**, 1084–1104.

37. J. Li, H. T. Ng, A. Cassell, W. Fan, H. Chen, Q. Ye, J. Koehne, J. Han, and M. Meyyappan, Carbon nanotube nanoelectrode array for ultrasensitive DNA detection. *Nano Letters*, 2003, **3**, 597–602.

38. K. Kerman, Y. Morita, Y. Takamura, and E. Tamiya, Escherichia coli single-strand binding protein-DNA interactions on carbon-nanotube-modified electrodes from a label-free electrochemical hybridization sensor. *Analytical and Bioanalytical Chemistry*, 2005, **381**, 1114–1121.

39. M. Zheng, A. Jagota, E. D. Semke, B. A. Diner, R. S. Mclean, S. R. Lustig, R. E. Richardson, and N. G. Tassi, DNA-assisted dispersion and separation of carbon nanotubes. *Nature Materials*, 2003, **2**, 338–342.

40. M. Zheng, A. Jagota, M. S. Strano, A. P. Santos, P. Barone, S. G. Chou, B. A. Diner, M. S. Dresselhaus, R. S. Mclean, G. B. Onoa, G. G. Samsonidze, E. D. Semke, M. Usrey, and D. J. Walls, Structure-based carbon nanotube sorting by sequence-dependent DNA assembly. *Science*, 2003, **302**, 1545–1548.

41. F. Valentini, A. Amine, S. Orlanducci, M. L. Terranova, and G. Palleschi, Carbon nanotube purification: preparation and characterization of carbon nanotube paste electrodes. *Analytical Chemistry*, 2003, **75**, 5413–5421.

42. P. Estrela, P. Migliorato, H. Takiguchi, H. Fukushima, and S. Nebashi, Electrical detection of biomolecular interactions with metal-insulator-semiconductor diodes. *Biosensors and Bioelectronics*, 2005, **20**, 1580–1586.

43. P. Estrela, A. G. Stewart, F. Yan, and P. Migliorato, Field effect detection of biomolecular interactions. *Electrochimica Acta*, 2005, **50**, 4995–5000.

44. Y. Cui, Q. Wei, H. Park, and C. M. Lieber, Nanowire nanosensors for highly sensitive and selective detection of biological and chemical species. *Science*, 2001, **293**, 1289–1292.

45. F. Patolsky, G. Zheng, O. Hayden, M. Lakadamyali, X. Zhuang, and C. M. Lieber, Electrical detection of single viruses. *Proceedings of the National Academy of Sciences of the United States of America*, 2004, **101**, 14017–14022.

46. W. U. Wang, C. Chen, K. H. Lin, Y. Fang, and C. M. Lieber, Label-free detection of small-molecule-protein interactions by using nanowire nanosensors. *Proceedings of the National Academy of Sciences of the United States of America*, 2005, **102**, 3208–3212.

47. G. Zheng, F. Patolsky, Y. Cui, W. U. Wang, and C. M. Lieber, Multiplexed electrical detection of cancer markers with nanowire sensor arrays. *Nature Biotechnology*, 2005, **23**, 1294–1301.

48. J. Xiang, W. Lu, Y. Hu, Y. Wu, H. Yan, and C. M. Lieber, Ge/Si nanowire heterostructures as high-performance field-effect transistors. *Nature*, 2006, **441**, 489–493.

49. S. J. Tans, A. R. M. Verschueren, and C. Dekker, Room-temperature transistor based on a single carbon nanotube. *Nature*, 1998, **393**, 49–52.

50. P. Avouris, Molecular electronics with carbon nanotubes. *Accounts of Chemical Research*, 2002, **35**, 1026–1034.

51. J. Kong, N. R. Franklin, C. Zhou, M. G. Chapline, S. Peng, K. Cho, and H. Dai, Nanotube molecular wires as chemical sensors. *Science*, 2000, **287**, 622–625.

52. P. G. Collins, K. Bradley, M. Ishigami, and A. Zettl, Extreme oxygen sensitivity of electronic properties of carbon nanotubes. *Science*, 2000, **287**, 1801–1804.

53. R. J. Chen, S. Bangsaruntip, K. A. Drouvalakis, N. Wong Shi Kam, M. Shim, Y. Li, W. Kim, P. J. Utz, and H. Dai, Noncovalent functionalization of carbon nanotubes for highly specific electronic biosensors. *Proceedings of the*

National Academy of Sciences of the United States of America, 2003, **100**, 4984–4989.

54. P. E. Nielsen, M. Egholm, R. H. Berg, and O. Buchardt, Sequence-selective recognition of DNA by strand displacement with a thymine-substituted polyamide. *Science*, 1991, **254**, 1497–1500.

55. P. Wittung, P. E. Nielsen, O. Buchardt, M. Egholm, and B. Norden, DNA-like double helix formed by peptide nucleic acid. *Nature*, 1994, **368**, 561–563.

56. K. Kerman, Y. Morita, Y. Takamura, E. Tamiya, K. Maehashi, and K. Matsumoto, Peptide-nucleic acid-modified carbon nanotube field-effect transistor for ultra-sensitive real-time detection of DNA hybridization. *Nanobiotechnology*, 2005, **1**, 65–70.

57. J. Janata and M. Josowicz, Conducting polymers in electronic chemical sensors. *Nature Materials*, 2003, **2**, 19–24.

58. N. Lassalle, P. Mailley, E. Vieil, T. Livache, A. Roget, J. P. Correia, and L. M. Abrantes, Electronically conductive polymer grafted with oligonucleotides as electrosensors of DNA preliminary study of real time monitoring of by in situ techniques. *Journal of Electroanalytical Chemistry*, 2001, **509**, 48–57.

59. H. Korri-Youssoufi, F. Garnier, P. Srivastava, P. Godillot, and A. Yassar, Toward bioelectronics: specific DNA recognition based on an oligonucleotide-functionalized polypyrrole. *Journal of the American Chemical Society*, 1997, **119**, 7388–7389.

60. D. S. Minehan, K. A. Marx, and S. K. Tripathy, Kinetics of DNA binding to electrically conducting polypyrrole films. *Macromolecules*, 1994, **27**, 777–783.

61. J. Wang, M. Jiang, A. Fortes, and B. Mukherjee, New label-free DNA recognition based on doping nucleic-acid probes within conducting polymer films. *Analytica Chimica Acta*, 1999, **402**, 7–12.

62. J. Wang and M. Jiang, Toward genolelectronics: nucleic acid doped conducting polymers. *Langmuir*, 2000, **16**, 2269–2274.

63. A. Mugweru and J. F. Rusling, Catalytic square-wave voltammetric detection of DNA with reversible metallopolymer-coated electrodes. *Electrochemistry Communications*, 2001, **3**, 406–409.

64. E. Komarova, M. Aldissi, and A. Bogomolova, Direct electrochemical sensor for fast reagent-free DNA detection. *Biosensors and Bioelectronics*, 2005, **21**, 182–189.

65. B. Piro, J. Haccoun, M. C. Pham, L. D. Tran, A. Rubin, H. Perrot, and C. Gabrielli, Study of the DNA hybridization transduction behavior of a quinone-containing electroactive polymer by cyclic voltammetry and electrochemical impedance spectroscopy. *Journal of Electroanalytical Chemistry*, 2005, **577**, 155–165.

66. H. Peng, C. Soeller, N. Vigar, P. A. Kilmartin, M. B. Cannell, G. A. Bowmaker, R. P. Cooney, and J. Travas-Sejdic, Label-free electrochemical DNA sensor based on functionalized conducting polymer. *Biosensors and Bioelectronics*, 2005, **20**, 1821–1828.

67. C. dos Santos Riccardi, H. Yamanaka, M. Josowicz, J. Kowalik, B. Mizaikoff, and C. Kranz, Label-free DNA detection based on modified conducting polypyrrole films at microelectrodes. *Analytical Chemistry*, 2006, **78**, 1139–1145.

68. M. Bart, E. C. A. Stigter, H. R. Stapert, G. J. de Jomg, and W. P. van Bennekom, On the response of a label-free interferon-γ immunosensor utilizing electrochemical impedance spectroscopy. *Biosensors and Bioelectronics*, 2005, **21**, 49–59.

69. E. Katz and I. Willner, Probing biomolecular interactions at conductive and semiconductive surfaces by impedance spectroscopy: routes to impedimetric immunosensors, DNA-sensors, and enzyme biosensors. *Electroanalysis*, 2003, **15**, 913–947.

70. J. Wang, J. A. Profitt, M. J. Pugia, and I. Suni, Au nanoparticle conjugation for impedance and capacitance signal amplification in biosensors. *Analytical Chemistry*, 2006, **78**, 1769–1773.

71. K.-S. Ma, H. Zhou, J. Zoval, and M. Madou, DNA hybridization detection by label-free versus impedance amplifying label with impedance spectroscopy. *Sensors and Actuators B*, 2006, **114**, 58–64.

72. C.-Z. Li, Y. Liu, and J. H. T. Luong, Impedance sensing of DNA binding drugs using gold substrates modified with gold nanoparticles. *Analytical Chemistry*, 2005, **77**, 478–485.

73. C. Gautier, C. Cougnon, J.-F. Pilard, and N. Casse, Label-free detection of DNA hybridization based on EIS investigation of conducting properties of functionalized polythiophene matrix. *Journal of Electroanalytical Chemistry*, 2006, **587**, 276–283.

74. L. Yang, W. Wei, X. Gao, J. Xia, and H. Tao, A new antibody immobilization strategy based on electrodeposition of nanometer-sized hydroxyapatite for label-free capacitive immunosensor. *Talanta*, 2005, **68**, 40–46.

75. C. Tlili, H. Korri-Youssoufi, L. Ponsonnet, C. Martelet, and N. J. Jaffrezic-Renault, Electrochemical impedance probing of DNA hybridization on oligonucleotide-functionalized polypyrrole. *Talanta*, 2005, **68**, 131–137.

76. V. Dharuman, T. Grunwald, E. Nebling, J. Albers, L. Blohm, and R. Hintsche, Label-free impedance detection of oligonucleotide hybridisation on interdigitated ultramicroelectrodes using electrochemical redox probes. *Biosensors and Bioelectronics*, 2005, **21**, 645–654.

77. X. Yu, R. Lv, Z. Ma, Z. Liu, Y. Hao, Q. Li, and D. Xu, An impedance array biosensor for detection of multiple antibody-antigen interactions. *Analyst*, 2006, **131**, 745–750.

78. R. Möller, R. D. Powell, J. H. Hainfeld, and W. Fritzsche, Enzymatic control of metal deposition as key step for a low-background electrical detection for DNA chips. *Nano Letters*, 2005, **5**, 1475–1482.

79. C. Tuerk and L. Gold, Systematic evolution of ligands by exponential enrichment: RNA ligands to bacteriophage T4 DNA polymerase. *Science*, 1990, **249**, 505–510.

80. A. D. Ellington and J. W. Szostak, *In vitro* selection of RNA molecules that bind specific ligands. *Nature*, 1990, **346**, 818–822.

81. A. E. Radi, J. L. Acero Sanchez, E. Baldrich, and C. K. O'Sullivan, Reagentless, reusable, ultrasensitive electrochemical molecular beacon aptasensor. *Journal of the American Chemical Society*, 2006, **128**, 117–124.

82. D. Xu, D. Xu, X. Yu, Z. Liu, W. He, and Z. Ma, Label-free electrochemical detection of aptamer-based array electrodes. *Analytical Chemistry*, 2005, **77**, 5107–5113.

83. H. Cai, T. M.-H. Lee, and I.-M. Hsing, Label-free protein recognition using an aptamer-based impedance measurement assay. *Sensors and Actuators B*, 2006, **114**, 433–437.

84. A. E. Radi, J. L. Acero Sanchez, E. Baldrich, and C. K. O'Sullivan, Reusable impedimetric aptasensor. *Analytical Chemistry*, 2005, **77**, 6320–6323.

85. M. C. Rodriguez, A.-N. Kawde, and J. Wang, Aptamer biosensor for label-free impedance spectroscopy detection of proteins based on recognition-induced switching of the surface charge. *Chemical Communications*, 2005, **34**, 4267–4269.

86. M. J. Eddowes and A. O. H. Hill, Novel method for the investigation of the electrochemistry of metalloproteins: cytochrome c. *Journal of the Chemical Society, Chemical Communications*, 1977, **21**, 771–772.

87. P. Yeh and T. Kuwana, Reversible electrode reaction of cytochrome c. *Chemistry Letters*, 1977, **10**, 1145–1148.

88. A. Christenson, E. Dock, L. Gorton, and T. Ruzgas, Direct heterogeneous electron transfer of theophylline oxidase. *Biosensors and Bioelectronics*, 2004, **20**, 176–183.

89. Z. Dai, X. Xu, and H. Ju, Direct electrochemistry and electrocatalysis of myoglobin immobilized on a hexagonal mesoporous silica matrix. *Analytical Biochemistry*, 2004, **332**, 23–31.

90. M. Heyrovsky, Early polarographic studies on proteins. *Electroanalysis*, 2004, **16**, 1067–1073.

91. J. A. Reynaud, B. Malfoy, and A. Bere, The electrochemical oxidation of three proteins: RNAase A, bovine serum albumin and concanavalin A at solid electrodes. *Journal of Electroanalytical Chemistry*, 1980, **116**, 595–606.

92. V. Brabec and I. Schindlerova, Electrochemical behavior of proteins at graphite electrodes. Part III. The effect of protein adsorption. *Bioelectrochemistry and Bioenergetics*, 1981, **8**, 451–458.

93. V. Brabec and V. Mornstein, Electrochemical behavior of proteins at graphite electrodes. II. Electrooxidation of amino acids. *Biophysical Chemistry*, 1980, **12**, 159–165.

94. E. Palecek, F. Jelen, C. Teijeiro, V. Fucik, and T. M. Jovin, Biopolymer-modified electrodes in the voltammetric determination of nucleic acids and proteins at the submicrogram level. *Analytica Chimica Acta*, 1993, **273**, 175–186.

95. L. Moreno, A. Merkoçi, S. Alegret, S. Hernández-Cassou, and J. Saurina, Analysis of amino acids in complex samples by using voltammetry and multivariate calibration methods. *Analytica Chimica Acta*, 2004, **504**, 251–257.

96. E. Palecek, M. Masarik, R. Kizek, K. Kuhlmeier, J. Hassmann, and J. Schülein, Sensitive electrochemical determination of unlabeled MutS protein and detection of point mutations in DNA. *Analytical Chemistry*, 2004, **76**, 5930–5936.

97. M. Vestergaard, K. Kerman, M. Saito, N. Nagatani, Y. Takamura, and E. Tamiya, A rapid label-free electrochemical detection and kinetic study of Alzheimer's amyloid beta aggregation. *Journal of the American Chemical Society*, 2005, **127**, 11892–11893.

98. G. Mie, Grundlagen einer theorie der materie. *Annalen Der Physik*, 1908, **25**, 377–445.

99. R. Jin, Y. Cao, C. A. Mirkin, K. L. Kelly, G. C. Schatz, and J. G. Zheng, Photoinduced conversion of silver nanospheres to nanoprisms. *Science*, 2001, **294**, 1901–1903.

100. N. P. Prasad, *Nanophotonics*, John Wiley & Sons, New York, 2004.

101. A. J. Haes and R. P. Van Duyne, A unified view of propagating and localized surface plasmon resonance biosensors. *Analytical and Bioanalytical Chemistry*, 2004, **379**, 920–930.

102. N. Nath and A. Chilkoti, Label-free biosensing by surface plasmon resonance of nanoparticles on glass: optimization of nanoparticle size. *Analytical Chemistry*, 2004, **76**, 5370–5378.

103. A. J. Haes, L. Chang, W. L. Klein, and R. P. Van Duyne, Detection of a biomarker for Alzheimer's disease from synthetic and clinical samples using a nanoscale optical biosensor. *Journal of the American Chemical Society*, 2005, **127**, 2264–2271.

104. D. A. Stuart, A. J. Haes, C. R. Yonzon, E. M. Hicks, and R. P. Van Duyne, Biological applications of localised surface plasmonic phenomenae. *IEEE Proceedings: Nanobiotechnology*, 2005, **152**, 13–32.

105. T. Endo, S. Yamamura, N. Nagatani, Y. Morita, Y. Takamura, and E. Tamiya, Localized surface plasmon resonance based optical biosensor using surface modified nanoparticle layer for label-free monitoring of antigen–antibody reaction. *Science and Technology of Advanced Materials*, 2005, **6**, 491–500.

106. C. Petit, P. Lixon, and M. P. Pileni, Synthesis of cadmium sulfide in situ in reverse micelles. 2. Influence of the interface on the growth of the particles. *Journal of Physical Chemistry*, 1990, **94**, 1598–1603.

107. T. S. Ahmadi, S. L. Logunov, M. A. El-Sayed, in *Nanostructured Materials: Clusters, composites, and thin films*, V. M. Shalaev and M. Moskovits (eds), American Chemical Society, Washington, 1997, pp. 125–140.

108. C. R. Yonzon, D. A. Stuart, X. Zhang, A. D. McFarland, C. L. Haynes, and R. P. Van Duyne, Towards advanced chemical and biological nanosensors-an overview. *Talanta*, 2005, **67**, 438–448.

109. L.-K. Chau, Y.-F. Lin, S.-F. Cheng, and T.-J. Lin, Fiber-optic chemical and biochemical probes based on localized surface plasmon resonance. *Sensors and Actuators B*, 2006, **113**, 100–105.

110. N. Hashimoto, T. Hashimoto, T. Teranishi, H. Nasu, and K. Kamiya, Cycle performance of sol-gel optical sensor based on localized surface plasmon resonance of silver particles. *Sensors and Actuators B*, 2006, **113**, 382–388.

111. T. Endo, K. Kerman, N. Nagatani, Y. Takamura, and E. Tamiya, Label-free detection of peptide nucleic acid-DNA hybridization using localized surface plasmon resonance based optical biosensor. *Analytical Chemistry*, 2005, **77**, 6976–6984.

112. E. Hitt, Label-free methods are not problem free. *Drug Discovery and Development*, 2004, **9**, 1–6.

59

An Electrochemical Biochip Sensor for the Detection of Pollutants

Rachela Popovtzer,[1] Yosi Shacham-Diamand[1] and Judith Rishpon[2]

[1] *Department of Physical Electronics, Tel Aviv University, Ramat-Aviv, Israel and* [2] *Department of Molecular Microbiology and Biotechnology, Tel Aviv University, Ramat-Aviv, Israel*

1 INTRODUCTION

In the present era, the increasing concern regarding the effect of environmental pollution on the public human health and well-being have led to the necessity of developing sensors to monitor air, wastewaters, and drinking water. The threat of chemical warfare terrorism and the pollution of ground water due to rapid industrialization stimulated numerous studies and investigations of methods to detect the toxicity in water. Both acute and chronic toxicity are important, although they pose different problems. In this chapter we focus mainly on acute toxicity detection, as it requires shorter measurement time and simpler decision-making process than chronic toxicity detection. Acute toxicity detection also poses severe immediate damage and has significant implications to society.

The use of living organisms as active components in electronic devices is an innovative and challenging area combining recent progress in molecular biology and micro technology. Bacterial-based biosensor chips can be defined as measurement electronic devices that use living cells as sensing elements. They can detect the signal and also record, process, and analyze the obtained information. The significance of these "whole-cell" biosensors lies in their capability to provide functional physiological information regarding the effect of chemicals on living systems. The conventional sensors such as molecular, enzymes, or DNA sensors are based on specific interactions and structural recognition within the biological materials; thus, they can only identify and quantify anticipated chemicals. Whole-cell-based biosensor chips can be applied in many fields including health care and medical applications, pharmaceutical screening, and environmental monitoring.

The detection mechanism combines a biologically sensitive element with a physical or chemical transducer to detect the presence of specific or nonspecific compounds in a given external environment. There are many detection methods used in biosensors, sensors and biochips, while the most common are mechanical, optical, and electrochemical. The sensors are expected to provide highly sensitive, accurate, and rapid results. In addition, monitoring environmental pollution in the field requires robust, miniaturized, and portable sensors.

In this study we utilized the electrochemical detection method, which proved to be simple, sensitive, provides fast response time and can be performed using compact and mobile equipment.[1] The portability of such a system is essential for on-site environmental monitoring, and for various medical applications. Miniaturization of electrochemical devices is highly important

Handbook of Biosensors and Biochips. Edited by Robert S. Marks, David C. Cullen, Isao Karube, Christopher R. Lowe and Howard H. Weetall.
© 2007 John Wiley & Sons, Ltd. ISBN 978-0-470-01905-4.

because of portability, convenient handling, and reduced requirement of reagents.

2   ELECTROCHEMICAL NANOBIOCHIP FOR TOXICITY DETECTION IN WATER

An electrochemical nanobiochip for water toxicity detection is presented. In this chapter we describe chip design, fabrication, and performance. Bacteria, which have been genetically engineered to respond to environmental stress, act as a sensor element and trigger a sequence of processes, which leads to generation of electrical current. The specific design and process of the electrochemical "lab on a chip", which integrates the recombinant bacteria provides highly accurate, sensitive, and rapid detection of acute toxicity in water.

In the first part of this chapter, we will describe the design and fabrication of the silicon chip. Then the general concept of the bacterial genetic engineering and the cascade of mechanisms by which the bacteria generate a current signal will be explained. The second part of the chapter contains experimental results of toxicity detection in water; followed by extensive discussion of the efficiency of the present method in comparison to other available methods. Finally the possible use of the electrochemical "lab on a chip"-based whole-cell sensors for different applications is described.

3   DEVICE FABRICATION

The silicon-based biochip has been designed and fabricated using standard microsystem-technology (MST) methods. Its architecture includes an array of miniaturized electrochemical cells. The nanovolume chambers (i.e., the electrochemical cells) contain the bacteria, which are brought into contact with the water to be examined. The cylindrical chambers hold 100 nl volume each. All arrays include positive and negative control chambers.

The device is manufactured in two parts: (i) a disposable chip—with the nanochambers containing the bacteria, and (ii) a reusable chip—with an interface to electronic circuitry that includes a multiplexer, potentiostat, temperature control, and a pocket PC for sensing and data analysis. This setting allows continuous reusing for multiple measurements.

The chip was produced from silicon, and contains an array of eight miniaturized electrochemical cells. Each electrochemical cell consists of three circular-shaped electrodes, surrounded by an insulating silicon nitride layer: (i) gold working electrode, (ii) gold counter electrode and (iii) Ag/AgCl reference electrode. The electrodes are made by gold sputtering, microlithography, and by selectively depositing Ag and anodizing it in a chloride-containing solution for the reference electrode. The chambers walls were constructed from photo-polymerized polyimide (SU-8) as shown in Figure 1. The silicon chip was wire bonded to

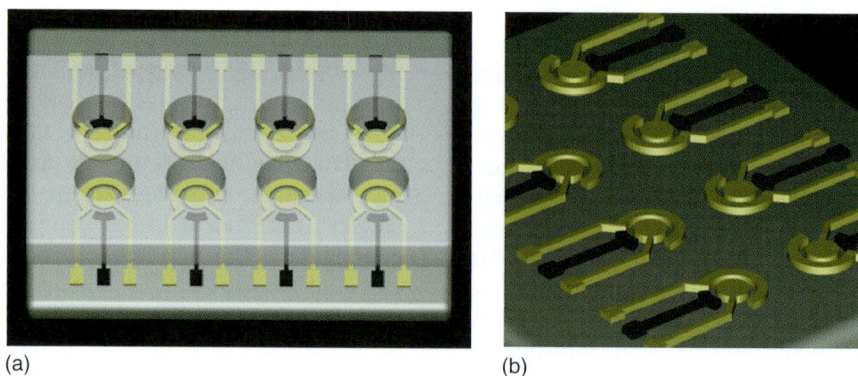

(a)   (b)

Figure 1. Image of the electrochemical chip. (a) Silicon chip contains an array of eight miniaturized electrochemical cells with external pads. (b) The electrodes without the top layer (SU-8). Each electrochemical cell consists of three circular-shaped electrodes: gold working and counter electrodes and Ag/AgCl reference electrode.

a plastic chip that was interfacing the electronic circuit.

The signal sensing is performed by the handheld palm-potentiostat with an interface to electronic circuitry for electrode signal regulation and detection (Palm Instruments BV-2004). The electronics consists of eight independent duplicate circuits of electrochemical cells, which are temperature controlled. A potential is applied between a working and a reference electrode in each electrochemical cell, and the output current is measured.

4 BACTERIAL-BASED BIOCHIPS

The most important reason for utilizing living cells as biosensors lies in their capability to provide functional information, that is, information about the effect of a stimulus on living organisms. The aim of the present biochip is to offer the unique "functionality" sensing capability. It answers the question "Is the water toxic?" and it does not intend to perform chemical analysis or to identify the nature of the toxicant. These whole-cell sensing systems can be visualized as an environmental switch, which is turned on in the presence of toxins or stressful conditions. In many cases, functional rather than analytical information is ultimately desired. In other applications, it complements and adds a special feature to a system by emulating the behavior of living entities.

5 GENETICALLY ENGINEERED BACTERIA

Whole-cell-based biosensors are intact, living unicellular organisms that have been genetically engineered to produce a measurable signal in response to a specific chemical or physical perturbation in their environment. They contain two essential genetic elements: a promoter gene and a reporter gene. The promoter gene is turned on (transcribed) when the toxic chemical, or target agent, is present in the cell's environment. In a normal cell, the promoter gene is coupled to number of genes that are transcribed and then translated into proteins that help the cell in either combating or adapting to the toxicant. In the case of a recombinant whole-cell biosensor, this promoter gene is coupled to his normal reporter genes, and additionally, to a specific reporter gene that was chosen because of

his ability to be easily assayed and to produce a detectable signal. Therefore, when the promoter gene senses a toxicant, he activates the reporter genes. Activation of the reporter gene leads to production of reporter proteins that ultimately generate some type of detectable signal. Therefore, the presence of a signal indicates that the biosensor has sensed chemical interference in its environment. This work was carried out using *Escherichia coli* (MC1061), which is the most convenient tool for genetic engineering, and a suitable host cell for carrying a variety of promoter–reporter fusions.

6 PROMOTERS FOR CELL-BASED BIOSENSORS

Three different recombinant strains of *E. coli* that contain the *fabA* (responsive to membrane damage-related stress), *dnaK* and *grpE* (both responsive to protein damage-related stress) promoters, were fused to the bacterial *lacZ* gene. Using different types of promoters that respond to a variety of stresses, mechanisms can react in response to a broad spectrum of chemicals.

7 REPORTERS FOR CELL-BASED BIOSENSORS

Reporter genes are nucleic acid sequences that can be easily assayed. In recombinant cell-based biosensors, the reporter gene is fused to a promoter of a gene whose expression is preferred to be monitored. There are several commonly used reporter proteins such as green fluorescent protein (GFP), bacterial luciferase (Lux), and β-galactosidase. In this work, the different promoters were fused to the *lacZ* reporter gene, which was monitored using an electrochemical assay of β-galactosidase activity. The activity of β-galactosidase is determined by using the substrate p-aminophenyl-β-D-galactopyranoside (PAPG).

8 BIOELECTROCHEMISTRY: BACTERIAL CURRENT GENERATION

Several detection methods for β-galactosidase are available including colorimetric histochemical, fluorescent, luminescent, and electrochemical. These detection strategies are dependent on the substrates used. The most common substrates employed for β-Gal are o-nitrophenyl-β-D-galactopyranoside (ONPG) for colorimetric detection, 5-bromo-4-chloro-3-indolyl-β-D-galactoside (X-gal) for histochemical detection, 4-methylumbelliferyl-β-D-galactopyranoside (MUG) for fluorometry, 1,2-dioxetane substrates for luminescence, and PAPG for electrochemical analysis. In our work we utilize the electrochemistry detection method that requires the addition of an external substrate (PAPG), but it does not require lysis or permeabilization of the cells and thus is ideal for on-line continuous measurement of enzymatic activity.[2] Moreover, this detection strategy can be performed in turbid solutions and under anaerobic conditions.

Recombinant bacteria, which were genetically engineered to respond to environmental stress, act as a sensing element and trigger a cascade mechanism: (i) In the presence of a toxin, this promoter is activated, and induces the production of the reporter enzyme β-galactosidase. (ii) As the enzyme is produced, the activity of β-galactosidase is determined by using the substrate PAPG. (iii) The product of the enzymatic reaction, p-aminophenol (PAP), is oxidized on the working electrode. This oxidation current is monitored and visualized on the PC screen.

9 BACTERIAL CULTURES

E. coli strains were grown to early log phase at 30 °C in 100 ml of Luria broth (LB) medium with aeration by shaking. Ampicillin, at a final concentration of $100 \mu g \, ml^{-1}$, was added to ensure plasmid maintenance. Cultures at 3×10^7 cells per ml were used for all experiments.

10 WATER TOXICANTS MEASUREMENTS

The analytical performance of the biochip was determined with various chemicals, in order to exemplify its ability to detect acute toxicity in water. Toxic chemicals were introduced to the recombinant E. coli cultures. The measurement setup includes two units: the silicon chip attached to the printed circuit board (PCB) that was replaced in every experiment, and the reusable

electronic unit for sensing and data analysis. A fixed potential was applied between the working and the reference electrode in each electrochemical cell in the array on the chip by the potentiostat, and the output current signal was measured.

In these experiments we used recombinant E. coli bacteria bearing plasmid with one of the following promoters: dnaK, grpE, or fabA. These promoters were fused to the reporter enzyme β-galactosidase. The toxic chemical, in each experiment, was introduced to the bacterial samples in the presence of the substrate PAPG. Immediately after (~1 s), the suspensions were placed in the electrochemical cells. The response of the bacteria to the different toxic chemicals was measured on line by applying a potential of 220 mV. The substrate, PAPG, was added to a final concentration of $0.8 \, mg \, ml^{-1}$ (100 nl total volume). The product of the enzymatic reaction (PAP) was monitored by its oxidation current using the amperometric technique. Additional measurements in the absence of the bacteria were performed to exclude the possibility of electroactive species in the LB medium, in the substrate, or in the substrate and the LB medium mixture, which can contribute to the current response.

10.1 Ethanol Detection

Ethanol, an efficient inducer of the heat shock proteins, was introduced to E. coli cultures harboring the promoters grpE and dnaK. The samples of the E. coli cultures (3×10^7 cells per ml) with the PAPG substrate were brought into contact with samples containing increasing concentrations of ethanol, and were immediately (~after 1 min) placed into the electrochemical chambers. The induced β-galactosidase activity was monitored electrochemically and the results are shown in Figure 2.

These results show the bacterial ability to generate a cascade of mechanisms that leads to current signal in the biochip system. For example for the grpE promoter, after 600 s, a current signal of 120 nA is produced in response to 2% ethanol, and a current of 20 nA is produced in response to 0.5% ethanol. These current signals are well above the background signal, which is defined as the measured current of the bacteria, bacterial medium, and the substrate without ethanol. For

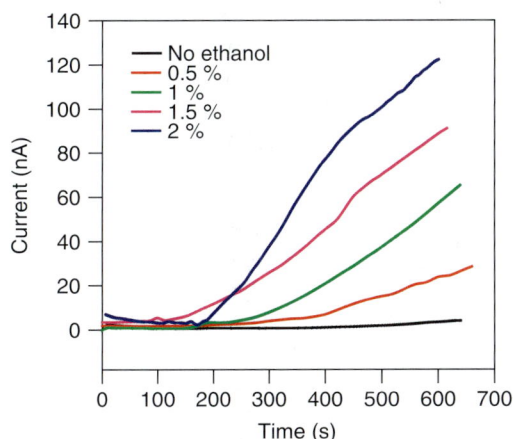

Figure 2. Amperometric response curves for on-line monitoring of ethanol using the electrochemical silicon chip. The recombinant *E. coli* containing a promoterless *lacZ* gene fused to promoter *grpE* exposed to 0.5–2% concentration of ethanol. The bacteria cultures with the substrate PAPG and the ethanol were placed into the 100 nl volume electrochemical cells on the chip immediately after the ethanol addition (~1 min) and were measured using the amperometric technique at 220 mV.

Figure 3. Amperometric response curves for on-line monitoring of different *E. coli* reporters in response to the addition of 1% ethanol, using the electrochemical silicon chip. The different *E. coli* reporters are *fabA*, *dnaK*, and *grpE*. Measurement performed immediately after the ethanol addition (~1 min) at 220 mV working potential versus Ag/AgCl reference electrode. The LB curve represents the bacterial response to the LB medium with the substrate PAPG without ethanol.

the *dnaK* promoter, after 600 s, a current signal of 70 nA is produced in response to 2.5% ethanol, and a current of 10 nA is produced in response to 1% ethanol.

The results indicate that there is a direct correlation between the currents signals and the toxicants concentrations.

In order to prevent false alarms, all arrays can include positive and negative control chambers. In the positive control chamber, pure water is to be added besides adding the tested sample with the unknown chemicals to the bacterial solution. In case a current signal was generated, it is a false alarm. A negative control chamber may include w.t. (MG1655) *E. coli* bacteria that constitutively expresses β-galactosidase, thus, current should be generated in all cases. When no current is generated, measurement is incorrect, because of bacterial death caused by adding highly toxic chemicals. However, chemicals can produce only constant DC current signal, while the enzymatic reaction acts as an intrinsic amplifier, and generates increasing current signal.

Real-time detection of the response of recombinant *E. coli* bacteria, with one of the promoters *dnaK*, *grpE*, or *fabA*, to ethanol is shown in Figure 3.

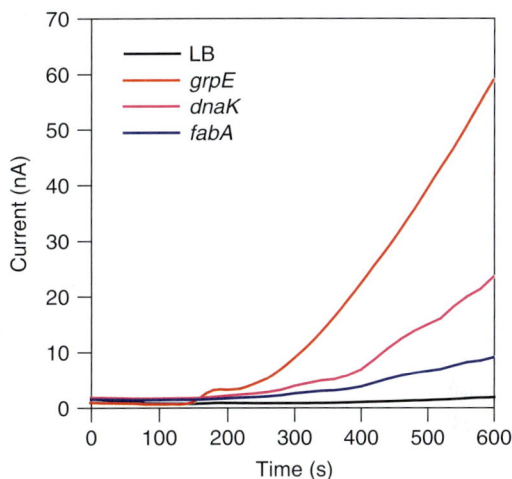

The results show that for all different promoters, concentration of 1% ethanol could be detected within 10 min.

10.2 Phenol Detection

Phenol was introduced to the *E. coli* bacteria culture harboring the promoter *grpE* under the same conditions as the ethanol. The samples of the *E. coli* cultures with the PAPG substrate were brought into contact with samples containing increasing concentrations of phenol, and were immediately (approx. after 1 min) placed into the electrochemical chambers. The induced β-galactosidase activity was monitored electrochemically and the results are shown in Figure 4. As in the previous experiment, the results demonstrate the direct correlation between the currents signals and the toxicants concentrations, and the fast response time.

Real-time detection of the responses of *E. coli* bacteria, with one of the promoters *dnaK*, *grpE*, or *fabA*, to phenol are shown in Figure 5. The results show that concentrations as low as 1.6 ppm of phenol could be detected within 6 min.

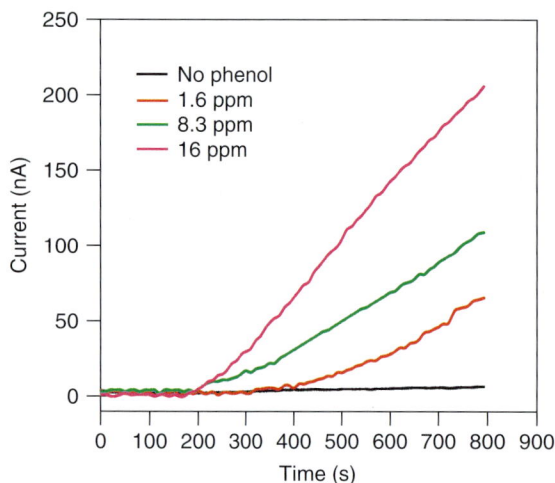

Figure 4. Amperometric response curves for on-line monitoring of phenol using the electrochemical silicon chip. The recombinant *E. coli* containing a promoterless *lacZ* gene fused to promoter *grpE* exposed to 1.6–16 ppm concentration of phenol. The bacteria cultures with the substrate PAPG and the phenol were placed into the 100 nl volume electrochemical cells on the chip immediately after the phenol addition (~1 min) and were measured at 220 mV.

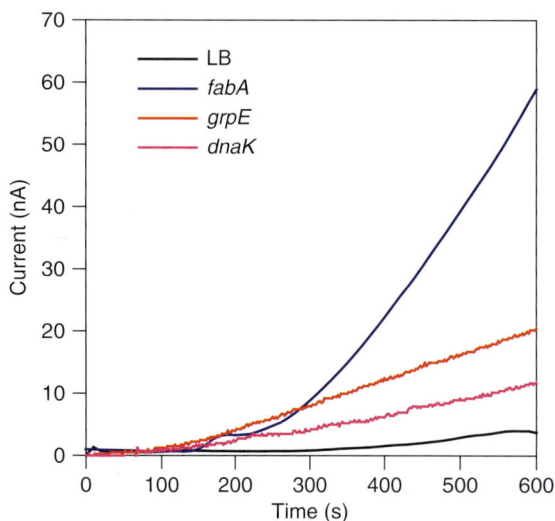

Figure 5. Amperometric response curves for on-line monitoring of different *E. coli* reporters in response to the addition of 1.6 ppm phenol, using the electrochemical silicon chip. The different *E. coli* reporters are *fabA*, *dnaK*, and *grpE*. Measurement performed immediately after the ethanol addition (~1 min) at 220 mV working potential versus Ag/AgCl reference electrode. The LB curve represents the bacterial response to the LB medium with the substrate PAPG without phenol.

In all measurements, an induction period appears before the current emerges since there is a response time at the genetic level.

11 DISCUSSION

The results that were obtained from all measurements indicate that this bacterial-based sensor system is highly sensitive and provides rapid results. Concentrations as low as 0.5% of ethanol and 1.6 ppm of phenol could be detected in less than 10 min of exposure to the toxic chemical (Figures 2 and 4). In comparison to other bacterial-based studies, which utilize optical detection methods for water toxicity detection, these results proved to be more sensitive and produce faster response time. Recent study,[3] which utilized bioluminescent *E. coli* sensor cells, detected 0.4 M (2.35%) ethanol after 220 min. An additional study[1] based on fluorescent reporter system (GFP), enabled detection of 6% ethanol and 295 ppm phenol after more than 1 h. Cha et al.[4] used optical detection methods of fluorescent GFP proteins, detected 1 g of phenol per liter (1000 ppm) and 2% ethanol, after 6 h. Other studies,[5] could not be directly compared due to different material used, however their time scale for chemicals identification is hours.

Different intensity response of the various bacterial sensors, *dnaK*, *grpE*, and *fabA*, to ethanol (Figure 3) and phenol (Figure 5) is due to the specific activation of each promoter to the type of the toxicant. The promoters *dnaK* and *grpE* are sensitive to protein damage (SOS system), thus, they were induced in response to ethanol which is known as *protein damage agent*.[6] *grpE* showed high induction activity in response to ethanol, *dnaK* showed reduced enzyme activity, and *fabA* was only slightly induced above the background level. *fabA* promoter is sensitive to membrane damage, and thus, reacts to phenol exposure, which is known membrane damage chemical. As expected, *grpE* and *dnaK* promoters were less activated by phenol (Figure 5).

The results that were obtained from measuring the bacterial harboring the *grpE* and *dnaK* promoters response to ethanol (Figures 2 and 6), are in agreement with the results of previous studies,[4,7] which illustrate that the response current signals

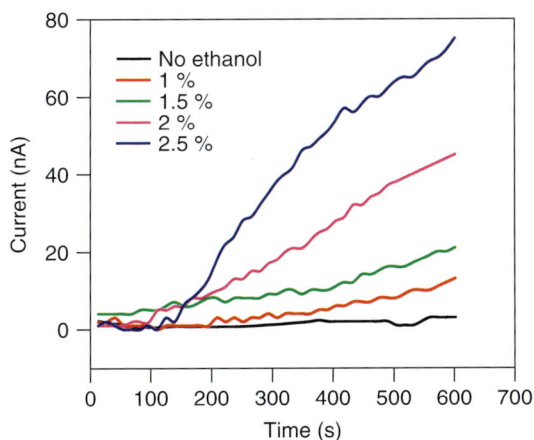

Figure 6. Amperometric response curves for on-line monitoring of ethanol using the nanobiochip. The recombinant *E. coli* containing a promoterless *lacZ* gene fused to promoter *dnaK* exposed to 0.5–2% concentration of ethanol. The bacteria cultures with the substrate PAPG and the ethanol were placed into the 100 nl volume electrochemical cells on the chip immediately after the ethanol addition (∼1 min) and were measured using the amperometric technique at 220 mV.

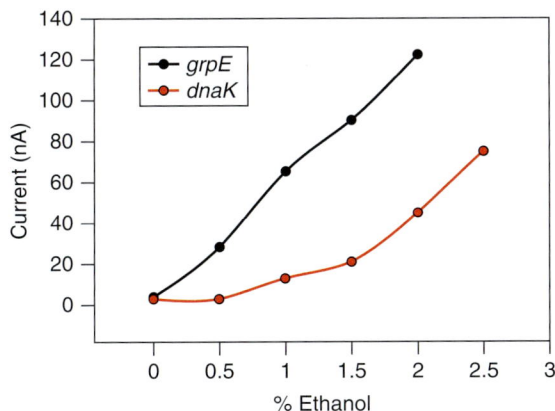

Figure 7. Comparison between the *grpE* and *dnaK* promoters response to ethanol detection. The two experiments were performed under the same conditions. Current points were sampled after 600 s.

generated by the *dnaK* promoter were typically lower than that of the *grpE* promoter (Figure 7).

Different intensity response of the bacterial sensors *fabA, grpE*, and *dnaK* to phenol (Figure 5) is due to the specific activation response of the promoter to the type of the toxicant. *fabA* promoter is sensitive to membrane damage, while

the promoters *dnaK* and *grpE* are more sensitive to protein damage.[6] Therefore, *E. coli* bacteria harboring the *fabA* promoter showed high induction activity in response to phenol exposure, which is a known membrane damage chemical. As expected, *grpE* and *dnaK* promoters were less activated by phenol.

There is a fundamental difference between nonenzymatic microbial sensing systems (e.g., GFP), in which the concentration of the reporter protein is determined, and enzymatic systems, which are based on measuring enzyme activity (e.g., β-galactosidase and alkaline phosphatase). Biochemical process that intends to produce a measurable signal has a great benefit while utilizing enzymatic activity. Since enzymes form continuously, and each enzyme reacts with many substrate molecules successively, this enzymatic mechanism serves as an intrinsic amplifier; consequently, the signal produces faster responses, is more sensitive, and increasing with time.[8] For example, for the *fabA* promoter, after 300 s, a current signal of 10 nA is produced in response to 1.6 ppm phenol, and a current of 14 nA is produced in response to 3.3 ppm phenol (Figure 8). The current signal increases significantly with time, since the current results from an enzymatic reaction.[9] The enzymes are continuously generated due to a sustained exposure of the bacteria to the toxicant. Therefore, after 600 s, a current signal of 60 nA is produced in response to 1.6 ppm phenol, and 80 nA in response to 3.3 ppm phenol.

Figure 8 shows that current intensity is linearly proportional to the toxicant concentration, while the toxicant concentration is above the detection limit. The dashed line represents a linear interpolation between the minimal detection concentration (1.6 ppm) and 0 phenol concentration. When the phenol concentration was lower than 1.6 ppm, no significant reaction was measured; the current was approximately 0 during the measurement time.

Electrochemical detection methods are also highly sensitive, since the measured current signal, that is, the number of transferred electrons that result from an enzymatic reaction can be accurately quantified. Furthermore, electrochemical monitoring can be carried out in turbid solutions or even under anaerobic conditions, as it does not require oxygen, in contrast to the light-emitting biosensors.[10] Combining enzymatic system with

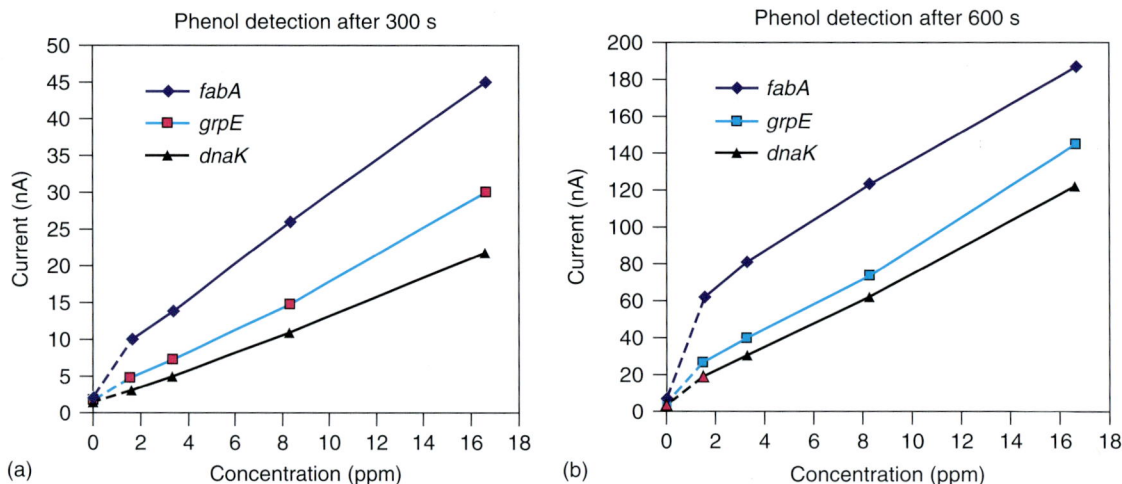

Figure 8. Recombinant bacterial current response to increasing concentrations of phenol, (a) after 300 s and (b) after 600 s. The *E. coli* reporters are *fabA, grpE*, and *dnaK*. Amperometric measurements were performed immediately after phenol addition ($\sim$1 min) at 220 mV working potential versus Ag/AgCl reference electrode. The dashed line represents a linear interpolation between the minimal detection concentration (1.6 ppm) and 0 phenol concentration.

electrochemical detection methods enables measurements in turbid solutions and under anaerobic conditions.[11]

This new design of nanochambers array on chips allows a broad band of measurements. We can simultaneously test eight different toxicant types with the general stress responsive promoter by introducing to each chamber a different toxicant, or, in order to specify unknown aqueous sample, we can test the sample with eight different stress responsive promoters, thus we obtain an indication of the toxicant type. In addition, the array configuration enables the addition of positive and negative control chambers for each experiment.

12   CONCLUSIONS

A novel integration between microelectronic device and living organisms for electrochemical detection of toxicity in water has been successfully developed and tested. This electrochemical "lab on a chip" that integrates bacteria as sensors cells, provides rapid and sensitive real-time electrochemical detection of acute toxicity in water. A clear signal is produced within less than 10 min of exposure to various concentrations of toxicants, or to stress conditions, with a direct correlation between

the toxicant concentration and the induced current. The small volume (100 nl) of the electrochemical cells allows minimum interaction with the water flow, resulting in increased speed and sensitivity due to the small diffusion length of the analytes to the electrode surface. The construction of an array of nanochambers on one chip leads to high throughput in addition to the capability of performing multi experiments simultaneously and independently. During the measurement period the bacteria remained active and were capable of performing cellular gene expression and enzymatic activity, which demonstrates the chip biocompatibility.

These results emphasize the advantages of merging electrochemical detection methods with adjusted design and process of nanovolume electrochemical cells array on silicon chip, which results in small sampling requirement and fast response time.

The electrochemical detection method has been shown to be sensitive, simple, and can be performed even in turbid solutions and under anaerobic conditions. By using compact and portable device, we are capable of measuring water toxicity in the field.

It is expected that this technology can be most beneficial to the health care and medical industry, for environmental monitoring and for control of chemical or pharmaceutical industry processing.

REFERENCES

1. S. Belkin, Microbial whole-cell sensing systems of environmental pollutants. *Current Opinion in Microbiology*, 2003, **6**, 206–212.
2. I. Biran, L. Klimentiy, R. Hengge-Aronis, E. Z. Ron, and J. Rishpon, On-line monitoring of gene expression. *Microbiology*, 1999, **145**, 2129–2133.
3. J. R. Premkumar, O. Lev, R. S. Marks, B. Polyak, R. Rosen, and S. Belkin, Antibody-based immobilization of bioluminescent bacterial sensor cells. *Talanta*, 2001, **55**, 1029–1038.
4. H. J. Cha, R. Srivastava, V. M. Vakharia, G. Rao, and W. E. Bentley, Green fluorescent protein as a noninvasive stress probe in resting Escherichia coli cells. *Applied and Environmental Microbiology*, 1999, **65**, 409–414.
5. D. E. Nivens, T. E. McKnight, S. A. Moser, S. J. Osbourn, M. L. Simpson, and G. S. Sayler, Bioluminescent bioreporter integrated circuits: potentially small, rugged and inexpensive whole-cell biosensors for remote environmental monitoring. *Journal of Applied Microbiology*, 2004, **96**, 33–46.
6. S. Daunert, G. Barrett, J. S. Feliciano, R. S. Shetty, S. Shrestha, and W. Smith-Spencer, Genetically engineered whale-cell sensing systems: coupling biological recognition with reporter genes. *Chemical Reviews*, 2000, **100**, 2705–2738.
7. S. Rupani, K. Konstatninov, P. Dhurjati, T. Vandyk, W. Majarian, and R. Larossa, Online monitoring of recombinant Escherichia-coli in batch and continuous cultures using a grpe promoter bioluminescence reporter gene system. *Abstracts of Papers of the American Chemical Society*, 1994, **207**, 127–BIOT.
8. E. Sagi, N. Hever, R. Rosen, A. J. Bartolome, J. R. Premkumar, R. Ulber, O. Lev, T. Scheper, and S. Belkin, Fluorescence and bioluminescence reporter functions in genetically modified bacterial sensor strains. *Sensors and Actuators. B, Chemical*, 2003, **90**, 2–8.
9. R. Popovtzer, T. Neufeld, N. Biran, E. Z. Ron, J. Rishpon, and Y. Shacham-Diamand, Novel integrated electrochemical nano-biochip for toxicity detection in water. *Nano Letters*, 2005, **5**, 1023–1027.
10. I. Biran, R. Babai, K. Levcov, J. Rishpon, and E. Z. Ron, Online and in situ monitoring of environmental pollutants: electrochemical biosensing of cadmium. *Environmental Microbiology*, 2000, **2**, 285–290.
11. Y. Paitan, D. Biran, I. Biran, N. Shechter, R. Babai, J. Rishpon, and E. Z. Ron, On-line and in situ biosensors for monitoring environmental pollution. *Biotechnology Advances*, 2003, **22**, 27–33.

60

Microcantilever Array Devices

Daniel Haefliger and Anja Boisen

Department of Micro- and Nanotechnology, Technical University of Denmark, Kongens Lyngby, Denmark

1 BACKGROUND

Micrometer-sized diving boards, also called *cantilevers*, started to be used for sensing purposes shortly after the invention of the atomic force microscope (AFM) in 1986.[1] Basically, the AFM functions as a miniaturized phonograph where images are obtained by scanning the surface with an AFM probe. The probe consists of a sharp tip mounted on a cantilever, which deflects due to the forces between tip and sample.

Usually, the deflection of the AFM probe is detected by optical leverage where a laser is reflected off the backside of the cantilever onto a position-sensitive diode. However, for operations in nontransparent liquids and for the realization of compact AFMs, other readout techniques are needed. Since 1989, research groups have been reporting on self-sensing AFM probes where the cantilever deflection is sensed at the cantilever by using different integrated transducer principles. Among the promising principles that have been developed are piezoresistive,[2] piezoelectric,[3] and capacitive readout.[4]

The AFM probe needs to have micrometer dimensions in order to achieve a relatively high resonant frequency (in the kHz regime) and a low spring constant (in the N/m regime). Typically the cantilevers are 100-μm long, 50-μm wide, and 0.5-μm thick. The high resonant frequency makes the probe less sensitive to external vibrations and the low spring constant improves the force sensitivity of the probe. Because of the small dimensions microfabrication is necessary. The first micromachined cantilevers with integrated tips were realized in 1990 by Tom Albrecht and coworkers in the group at Stanford University, CA, USA,[5] and by the group of O.Wolter at IBM Sindelfingen, Germany.[6] These microfabricated AFM probes soon became commercially available through the companies Park Scientific Inc. and NanoWorld AG. Today, several companies offer AFM probes in a large span of spring constants and resonant frequencies. These values are tuned by the choice of cantilever material(s) and geometry.

Possibly inspired by some of the often encountered annoyances in AFM work, Thomas Thundat and coworkers at Oakridge National Laboratory, TN, USA, in 1994, started to explore the cantilever's possible potential as a physical and chemical sensor.[7] Their initial work demonstrated that an AFM probe which is coated on one side with a metal layer for improved reflection of the laser beam will be prone to bimorph effects. This can cause severe drift problems in the AFM setup. However, the effect can also be used as a simple principle for a very sensitive thermometer, where a static cantilever deflection can be related to a given temperature change. Moreover, it was demonstrated that a metal-coated cantilever kept at constant temperature responds reproducibly with a constant bending to humidity changes and to exposure of other vapors such as mercury. Also, it was shown that the amount of adsorbates could be

Handbook of Biosensors and Biochips. Edited by Robert S. Marks, David C. Cullen, Isao Karube, Christopher R. Lowe and Howard H. Weetall.
© 2007 John Wiley & Sons, Ltd. ISBN 978-0-470-01905-4.

estimated by monitoring the shift in the resonant frequency of the cantilever.

The change in the static deflection of the cantilever is related to the difference in surface stress of the two faces of the cantilever, as will be discussed later. A related technique to study surface-stress changes is called the *bending plate technique* and was first observed and applied on wafers of III–V compounds in the 1960s.[8] By introducing the cantilever technique this measurement principle was significantly improved. Using the bending plate technique, surface-stress changes of 1 N m^{-1} were normally reported, whereas the first measurements of surface-stress changes with the micrometer-sized cantilevers were of the order of 10^{-3} N m^{-1}.

The change in the resonant frequency of the cantilevers is caused by a change in mass and/or stiffness of the cantilever. For the micrometer-sized AFM probes people initially reported mass changes in the picogram regime. The mass sensitivity has continuously been improved as discussed in the subsequent text and has been pushed, by researchers like Michael Roukes at the California Institute of Technology, Pasadena, CA, USA, to the point where single molecule detection is possible.

Soon after the cantilever-based sensing principle was initiated, researchers started to apply surface-stress change measurements in liquid environment. For example, the unspecific binding of proteins to a hydrophobic surface was reported in 1996,[9] and studies of electrochemical processes were also initiated in 1996.[10] The cantilever sensors were further used for online monitoring of the self-assembly of alkanethiol monolayers on gold-coated cantilevers.[11] The surface stress, in this work, was seen to be related to the length of the alkane chain.

Most of the early cantilever-based sensing work was performed in standard AFM setups using normal AFM probes. In a normal AFM setup, it is not possible to measure on more than one cantilever at a time. This makes it impossible to simultaneously perform control measurements on a reference cantilever (RC). The AFM probes used in these experiments typically had tips although for sensor applications this was not necessary. One of the first specifically designed large scale sensor systems was developed by Christoph Gerber and coworkers at the IBM Research Laboratory in Zurich, Switzerland. An array of polymer-coated Si cantilevers was used to simultaneously detect the adsorption-induced change in the surface stress of the cantilevers. The system could readout from eight cantilevers simultaneously using an optical leverage technique and different alcohols could be detected due to their different adsorption rates in the polymers.[12] In these measurements uncoated cantilevers were employed for reference.

The potential of the cantilever-based sensing in the field of diagnostics was highlighted in 2000, when it was demonstrated that a pair of cantilevers coated with two short strands of DNA oligos that only differ by a single base can be used for single nucleotide polymorphism (SNP) detection.[13] From the bioapplication point of view, these data were the onset of many studies related to the specific recognition of DNA, proteins, and macromolecules.

On the technology side cantilevers with integrated readout started to appear in the late 1990s. As for the AFM, the integrated readout facilitates the realization of compact devices that can operate in even nontransparent environments. For example, in 2000, piezoresistive cantilevers for surface-stress sensing were presented demonstrating the possibility of realizing a dense array of sensors with a compact readout system.[14] Even highly advanced systems with integrated electronics were soon demonstrated.[15] The field of self-sensing cantilevers has developed rapidly. Today several groups around the world are investigating different readout techniques for the cantilever, and the technological solutions typically address either the surface-stress detection or the mass detection.

Today (2006) cantilever-based sensors are developed and commercialized by several companies including, NanoNord A/S in Aalborg, Denmark, Protiveris Inc. in Rockville, MD, USA, Concentris GmbH in Basel, Switzerland, and SIS GmbH in Herzogenrath, Germany. NanoNord A/S develops cantilevers with piezoresistive readout whereas the three other companies use different types of optical detection. The company SIS GmbH provides a solution where the deformation of the complete cantilever is monitored simultaneously.

2   MEASUREMENT PRINCIPLES

In general, three different operation modes are employed in cantilever arrays as illustrated in Figure 1: *static mode*, *dynamic mode*, and *heat*

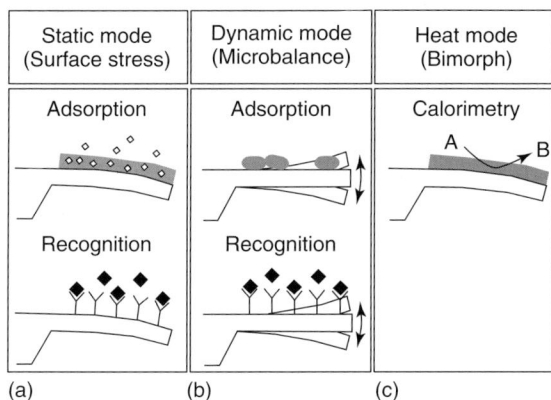

Figure 1. Operation modes.

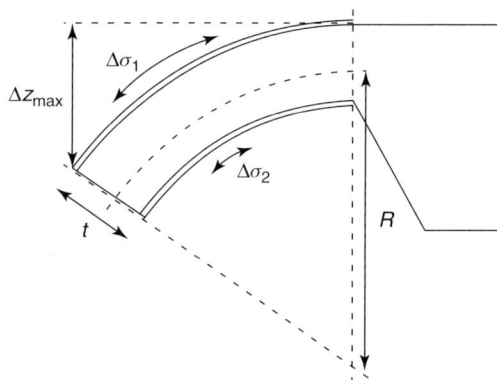

Figure 2. Lateral view of a thin cantilever of thickness t loaded by surface-stress changes $\Delta\sigma_1$ and $\Delta\sigma_2$. The cantilever bends around a neutral plane with a constant radius R.

mode. Common to all modes is the fact that the cantilever's surface is usually functionalized in such a way that one surface is rendered chemically active while the other surface is passivated by a chemically inactive substance. Chemical and physical reactions on the active cantilever surface can then be observed by the temporal evolvement of the cantilever's response. The reaction partner immobilized on the cantilever surface is usually a liquid or solid such as a polymer film or a self-assembled monolayer (SAM) of biomolecules. The other partner, which is to be detected via the reaction with the immobilized substance, can be a gas, vapor, biomolecule, or microorganism.

In the *static mode*, the chemical or physical reaction occurring on the cantilever surface is transduced into a nanomechanical bending of the cantilever. The reaction provokes a change in surface stress on the chemically active surface. The passivated surface remains inactive yielding no change in surface stress. Hence, due to the asymmetric surface property of the cantilever, an asymmetric surface stress is produced upon chemical or physical reaction. This asymmetric loading causes the cantilever to bend. The detection of the reaction is finally achieved by measuring the nanomechanical bending by means of optical or electrical sensors.

The bending of commonly used cantilevers of a few 100-μm lengths is typically between a few nanometers and a few hundred nanometers. This minute deflection imposes particularly high demands on the detection systems. The most widely used techniques are described in Section 3.

From a mechanical point of view, the deflection of the cantilever upon a change in surface stress is usually described by the simple classical relationships that Stoney derived and applied to electroplated metal films at the beginning of the twentieth century.[16] Stoney's equation provides a first order approximation for the radius of curvature, R, and the deflection of the cantilever tip, Δz_{max}, in relation to the differential surface-stress change $\Delta(\sigma_1 - \sigma_2) = \Delta\sigma_1 - \Delta\sigma_2$ (Figure 2):

$$\frac{1}{R} = \frac{6(1-\nu)}{Et^2}(\Delta\sigma_1 - \Delta\sigma_2) \qquad (1)$$

$$\Delta z_{max} \cong \frac{3(1-\nu)L^2}{Et^2}(\Delta\sigma_1 - \Delta\sigma_2) \qquad (2)$$

where $\Delta\sigma_1$ and $\Delta\sigma_2$ are the surface-stress changes on the active and the passivated surfaces, respectively. t and L denote the thickness and the length of the cantilever, respectively, E is the Young's modulus, and ν the Poisson's ratio of the material. Stoney's formula assumes that the thickness of the functionalizing coatings on the active and passivated cantilever surfaces is much thinner than the thickness of the cantilever. It thus provides good estimations for cantilevers coated with very thin films. Measured surface-stress changes typically range between 0.001 and 1 N m^{-1}.

The physical origin of the surface stress is discussed in the following, separately for unspecific and specific reactions. In the case of unspecific reactions, one cantilever surface is commonly activated by a thin polymer coating which can adsorb vapor from, for example, water or alcohols.[12,14] The polymer film swells upon diffusion of the vapor molecules into the polymer and bends the cantilever away from the coated surface. The bending is caused by compressive stress induced onto the cantilever due to a volumetric change of the active film (Figure 1). The interpretation of the origin of surface stress is, however, more complex to interpret in the case of specific reactions such as DNA hybridization and antibody–antigen binding. In this case, one cantilever surface is commonly activated by covalent binding of a receptor via thiol–gold chemistry. The cantilever is coated with a 20–100-nm-thick gold film on which a SAM of thiolated receptors is immobilized. The other surface of the cantilever is usually passivated by adsorption of bovine serum albumine (BSA) to avoid unspecific binding of the analyte. Several different interactions occur when the analyte reacts with the receptor. The reaction changes the size of the molecules forming the SAM on cantilever and it may alter conformation and charge distribution of the species. These modifications affect the interaction between the molecules via electrostatic, dispersion, steric, osmotic, and solvation forces, which induce a net change in surface stress.[13,17] To which extent each force is contributing to the surface stress depends heavily on the involved species and the specific reaction. Recently, theoretical calculations revealed that interactions between the molecules and the gold film used for immobilization can also significantly contribute to a surface-stress change.[18] In this case, the stress originates from surface charge redistribution in the gold film in response to the charges in the molecules.

Microcantilevers operating in *dynamic mode* are essentially mechanical oscillators that are driven at resonance. The resonance frequency f_0 for a rectangular cantilever can be expressed as

$$f_0 = 0.323 \sqrt{\frac{k}{m^*}} \qquad (3)$$

where k is the spring constant and m^* the effective mass of the cantilever. The resonance frequency of microcantilevers typically ranges in the kilohertz to megahertz regime. These vibrating cantilevers are used as microbalances capable of measuring minute mass changes of a few atograms (10^{-18} g). By adsorption of a mass Δm on the surface of the cantilever its effective mass will be increased. As

shown by Equation (3), a change in the effective mass translates into a shift of the resonant frequency. In the case of mass adsorption the resonance will decrease. Using appropriate electronics to track the resonance shift, Δf, the adsorbed mass can be determined by

$$\Delta m = 0.104k \left(\frac{1}{(f_0 - \Delta f)^2} - \frac{1}{f_0^2} \right) \qquad (4)$$

This interpretation of the frequency shift assumes that changes in the cantilever's spring constant k during mass adsorption are negligible. This is the case for adsorption of single particles such as microorganisms[19] (Figure 1b) and molecules at low concentrations. However, from Equation (3), it is clear that changes in resonance frequency can result from both a change of the effective mass m^* and variations in the spring constant k. The spring constant relates directly to the stiffness of the cantilever, which is determined by the cantilever material and any applied surface stress. As seen in static mode measurements, adsorption of molecules by diffusion or biological recognition can provoke considerable surface-stress changes. At high concentration, these stress changes affect the stiffness of the cantilever and have to be taken into consideration when interpreting the measured frequency shift. The resonance shift can then be approximated by[20]

$$\Delta f = \frac{1}{2} f_0 \left(\frac{\Delta k}{k} - \frac{\Delta m^*}{m^*} \right) \qquad (5)$$

where Δk denotes the surface-stress change and Δm^* the change in specific mass. Equation (5) is valid as long as Δk and Δm^* are much smaller than the stiffness and mass of the cantilever, respectively. In most experiments, changes in k and m^* are small and these requirements are fulfilled. To decouple the influence of mass loading and spring constant variations, a simultaneous measurement of bending and resonant frequency shifts is required.[21]

The *heat mode* employs the bimorph effect to measure the heat adsorbed or released by an endothermic or exothermic reaction on the cantilever surface. The cantilever is coated asymmetrically on one surface with a layer of different thermal expansion properties than those of the cantilever material (Figure 1c). Some of the most popular coatings are metals such as gold, platinum, or aluminum. If such a cantilever is subjected to temperature changes, it will bend to the mismatch in thermal expansion coefficients of the cantilever material and the coating. Detecting the bending via optical or electronic means allows for the measurement of temperature changes in the microkelvin range.

3   INSTRUMENTATION

The measurement of cantilever deflections or mass change in the nanometer or atogram range imposes high demands on the detection systems, and various intriguing setups have been developed. We concentrate here on an overview of the most widely used techniques in the field of static and dynamic mode biosensors.

The readout of the cantilever deflection in *static mode* operation is predominantly achieved by the optical lever method originally developed for atomic force microscopy.[22] As illustrated in Figure 3(a), a laser beam of typically few milliwatts power is focused onto the free end of the cantilever. The cantilever surface acts as a mirror reflecting the beam onto a position-detection diode. Depending on the degree of cantilever bending, the reflected laser beam is tilted away from its initial position resulting in a shift of the laser spot impinging onto the diode. The travel distance of the laser spot is proportional to the cantilever tip deflection and can be extracted electronically from the position-detection diode. A microcantilever array commonly exists of at least two cantilevers. One cantilever is functionalized with the receptor and acts as the active measurement cantilever (MC) while the other exhibits an inert surface and is used for reference. Numerous measurements have shown that a RC is vital to ensure reliable signal interpretation. Cantilever sensors are prone to various artefacts induced by variations in ambient temperature, pressure, humidity, and composition of biological solutions (pH, ionic strength, etc.).[7] By implementing a reference, such crosstalk can be cancelled out to a large extent. Increasing the number of cantilevers in an array opens up the possibility of exposing several MCs and RCs in a single experiment under identical conditions in terms of temperature, flow rate, and pressure. The MCs can thereby be functionalized

with different receptors enabling fast screening for multiple analytes, including negative controls (Figure 3a).

Figure 3(b) shows a cantilever array made in silicon used together with optical readout. The deflection readout of each individual cantilever is achieved in one of the following ways: (i) in parallel fashion by an array of laser diodes and position sensors aligned to the cantilever array,[12] (ii) sequentially by one laser diode scanned over the cantilever array,[23] or (iii) in parallel by flood illumination of the complete array and digital image analysis to track the spot of light reflected off each individual cantilever.[24] Cantilever arrays are commonly fabricated in silicon or silicon nitride using conventional processes known from the semiconductor industry. Recently, arrays have also been fabricated in polymers as shown in Figure 3(c).[25,26] Polymers are much softer than silicon since they can exhibit up to 100 times lower Young's modulus, E. As is clear from Equation (2), a low E

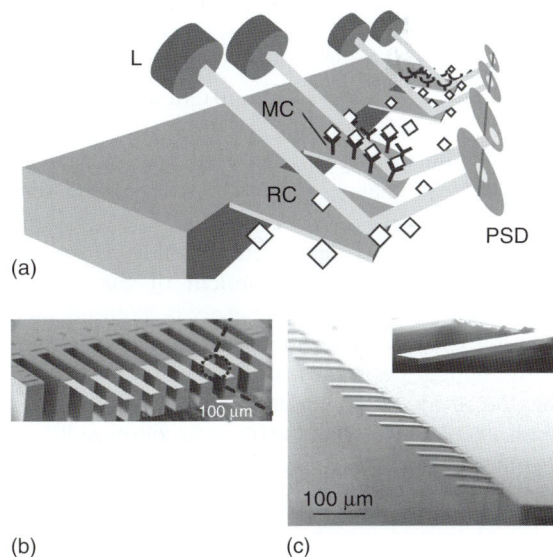

(a)

(b)   (c)

Figure 3. Microcantilever array with optical lever readout. (a) Principle with laser (L) and position-sensitive diode (PSD). The array shows two pair of sensors with a measurement cantilever (MC) and a reference cantilever (RC). Note the beam shift on the PSD of the deflected MC. (b) Scanning electron microscopy image of a silicon cantilever array fabricated at the IBM Zurich Research Laboratory. The cantilevers are 1-μm thick, 500-μm long, and 100-μm wide. (c) Polymeric cantilever array fabricated in epoxy resist (Epon SU-8). The cantilevers are 4.5-μm thick, 200-μm long and 20-μm wide.

(a)

(b)   2 mm

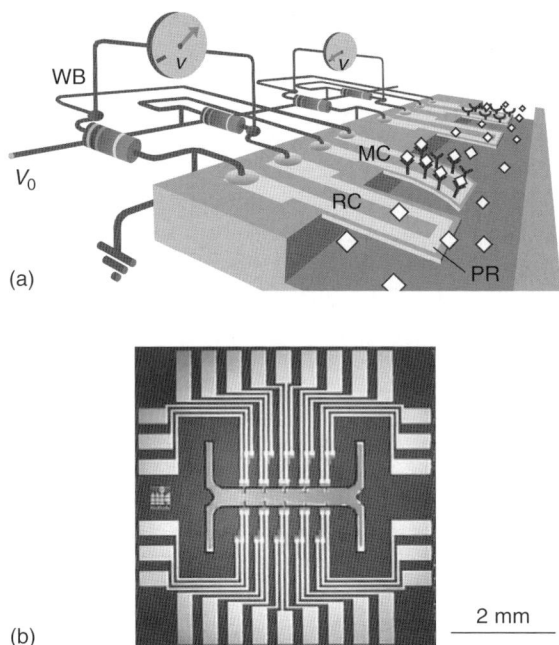

Figure 4. Cantilever array with piezoresistive readout. (a) Principle with Wheatstone bridge (WB), piezoresistor (PR), measurement cantilever (MC), and reference cantilever (RF). (b) Optical microscopy image of an integrated piezoresistive cantilever sensor chip.

translates into a larger deflection z_{max} for a given surface-stress change and thus increases the sensor's sensitivity.

Optical readouts, however, suffer from several disadvantages: tedious alignment of the optical components to the cantilever array is required, susceptibility to optical artefacts arising from changes in the optical properties of the medium surrounding the array (e.g., variation in refractive index upon exchange of the medium) is observed, and operation in nontransparent solutions is impossible. These problems are generally circumvented by using cantilever arrays with piezoresistive deflection readouts (Figure 4a). In this configuration a thin, stress-sensitive film is directly integrated into the cantilever material. This film consists of a piezoresistive material such as doped silicon that changes electrical conductivity when it is strained by the cantilever bending. The change in conductivity can be measured via a simple Wheatstone bridge. Unlike the optical lever system, the integrated piezoresistive readout is very compact allowing for complete on-chip integration

of the electronic circuitry and a microfluidic handling system useful for biochemical diagnostics. Figure 4(b) shows a silicon chip with an array of ten cantilevers equipped with encapsulated metal wires suitable for operation in a liquid environment.[27] Such chips bear high potential in portable, point-of-care diagnostic devices. The main disadvantage of piezoresistive cantilevers is, however, to date, imposed by the high intrinsic noise level, which affects the resolution and sensitivity.

For *dynamic mode* sensors, the readout inherently demands for an external actuation to drive the cantilevers in resonance. Actuation of microstructures is widely used in the field of microelectromechanical systems (MEMS) and a large variety of mechanisms have been developed. Most actuation schemes use piezoelectrically, electrostatically, magnetically, or thermally induced forces for excitation. The detection of the cantilever oscillation can be performed by employing the same type of forces for passive sensors. Moreover, detection of the vibration by the optical lever technique or piezoresistors (PRs) as discussed in the preceding text is often used. In the simplest readout scheme, the cantilever is mounted on a millimeter-sized driving piezo and the cantilever oscillation is detected by an optical lever. Due to the large size of the piezoelectric element, the excitation, however, stimulates not only the cantilever but also the complete measurement setup. The noise jeopardizes the detection of the cantilever resonance when working in liquids. To confine the stimulation to the cantilever, beams with integrated piezoelectric elements have been fabricated.[28] The piezos can be used simultaneously for detection of the cantilever resonance. Such integrated readout schemes enable very compact sensors design. Dynamic microcantilevers can be hooked up to integrated circuitry and microprocessors on a single substrate using MEMS and complementary metal oxide semiconductor (CMOS) technology. Figure 5 shows a microcantilever sensor monolithically integrated into a CMOS chip designed to detect volatile organic compounds.[15] The cantilever is excited by a thermally actuated bimorph element; the response is acquired by integrated PRs. The chip combines the cantilever sensor with a capacitive and calorimetric sensor to improve the wealth of information about the analyte.

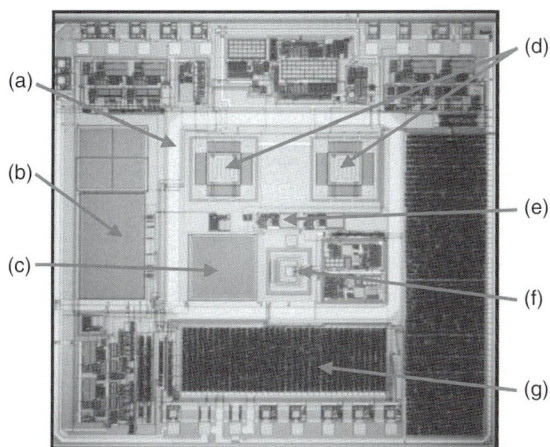

Figure 5. Optical microscopy image of a gas microsensor chip (size: 7×7 mm). The different components include: (a) flip-chip frame, (b) reference capacitor, (c) sensing capacitor, (d) calorimetric sensor, (e) temperature sensor, (f) dynamic cantilever sensor, and (g) digital interface. [Reprinted with permission Hagleitner et al.[15] copyright 2001, Nature Publishing Group.]

The need of dynamic mode devices as pure microbalances drives the development of ultrahigh sensitive mass sensors for trace analysis. One route to achieve high mass sensitivity is offered by downscaling the cantilever structure. The smaller the cantilever, the smaller its mass becomes. Intuitively, a given minute mass added to a small cantilever exerts higher influence on the resonance behavior than if it is adsorbed on a bigger and heavier structure. Mathematically, the resolution $\Delta m / \Delta f$ of a resonating rectangular cantilever can be derived from Equation (4):

$$\frac{\Delta m}{\Delta f} \cong 6.172 m \sqrt{\frac{m}{k}} = 0.208 \frac{k}{f_0^3} \,(\text{kg/Hz}) \quad (6)$$

which provides further evidence of increased mass sensitivity on reducing the size and therefore mass of a cantilever. Figure 6 shows a resonating nanocantilever microfabricated in polysilicon.[29] The actuation and detection for readout is performed by electrostatic electrodes. The mass of such a nanocantilever is about 10 000 times smaller than for conventional silicon cantilevers as shown in Figure 3(b). It provides an ultrahigh mass resolution of a few atogram per Hertz. The readout signal for such minute mass changes is very low and requires direct integration of the nanosensor

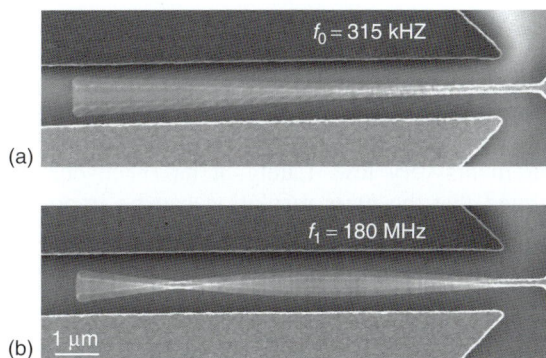

Figure 6. Scanning electron microscopy image of an electrostatically actuated nanocantilever made of polysilicon. (a) Fundamental mode, f_0 and (b) first mode, f_1. The cantilever is 800-nm wide, 2-μm thick, and 40-μm long.

into CMOS circuitry to reduce the noise level and improve measurement reliability.

The functionalization of all kinds of cantilevered sensors with receptors can be achieved by self-assembly of thiolated molecules on a gold coating. For this, the gold is sputter- or evaporation-deposited on the cantilever structure in vacuum equipment. A top-down approach for functionalization is offered by depositing the receptor via ink-jet printing or by dipping the cantilever array into a conformal microcapillary array filled with the receptor solution.

4 APPLICATIONS

The microcantilever array devices are generally employed for label-free detection of biomolecules using both static and dynamic mode operations. Moreover, the fundamentals behind the signal generation can be used to track the evolution of local surface-stress changes for fundamental studies. Such measurements are difficult to achieve by any other technology. Also, the cantilever is inherently a very sensitive thermometer, and owing to its small size, can be used for local temperature measurements or for calorimetry on very small sample volumes. In terms of mass detection, the ultimate limit is the detection of a single molecule and the use of cantilevers in mass spectroscopy is being pursued.

The specific recognition of biomolecules has mainly been performed using static mode. Such measurements are generally performed in liquid environment, which imposes high experimental challenges for dynamic mode operation. The liquid damps the oscillation of the cantilever rendering the precision of the resonant frequency measurements very low. Lately, it has been demonstrated that the mass resolution can be improved by operating the device at higher order modes (as shown in Figure 6b), where the damping loss is lower.[30] Specific recognition has been demonstrated for mainly DNA,[31] proteins,[32] bacteria,[19] and antigens.[33] For specific recognition, the cantilever surface needs to be coated with a "detector" layer that binds only to one type of molecules. For DNA detection, the common technology is to immobilize short strands of thiolated DNA on a gold surface. The discussion in cantilever sensing today is: "what is the optimal surface for generation of a large surface-stress signal?" Here the data obtained from standard characterization techniques such as fluorescents and radio labelling cannot be directly transferred to the cantilever surface. A high density of target DNA on a surface does not automatically give a large change in surface stress. Instead, it seems that the molecules should be placed as close to the surface as possible and the sensor signal will be strongly affected by the pH and the salt concentration of the solution.[17] For bacteria detection, recent results indicate that the largest results are obtained by using Fab fragments on the surface, hereby bringing the molecules close to the cantilever surface.[34]

Ongoing work by the group of Martin Hegner at the University of Basel, Switzerland has demonstrated that it is possible to detect DNA oligos in concentrations of pM, which indicates that the specific recognition can be performed without polymerase chain reaction (PCR) amplification. It has even been possible to detect specific DNA oligos in a background mixture of many DNA fragments. Moreover, the kinetics of DNA immobilization and hybridization can be investigated. In these experiments, the quality of the surface plays an essential role. For example, the cleanliness of the gold has been demonstrated to have a huge impact on the size of the surface-stress signal and the adsorption rate of the thiols.[35] By immobilizing 25-mer DNA oligos at different concentrations on a gold-coated cantilever, the surface free energy of the thiol-modified DNA-oligo adsorption on gold is found to be 32.4 kJ mol^{-1}. The adsorption

experiments also indicate that first a single layer of DNA oligos is assembled on the gold surface after which a significant unspecific adsorption takes place on top of the first DNA-oligo layer.[36]

Pathogens such as *Escherichia coli* (*E. coli*) have been detected employing dynamic mode. The cantilever is coated with anti-*E. coli* antibodies and then dipped into a solution containing the bacteria.[19] After surface adsorption of *E. coli*, the cantilever is dried and the resonant frequency is compared to the resonant frequency prior to the exposure to the bacteria. The resonant frequency revealed to drop as expected due to the added mass of the adsorbed *E. coli*. The system enabled the detection of a single *E. coli* cell. A calculation according to Equation (4) revealed that the cell weighed 665 fg, which is consistent with other reports. By adding a nutritious agarose film onto the cantilever, the biosensor has been employed to detect the growth of the pathogen. Sensitivity in the pg/Hz regime has been achieved, which is seen to be three orders of magnitude higher than for common quartz crystal microbalances (QCM). The high sensitivity enabled rapid cell-growth detection within less than an hour, about 10 times faster than with QCM.[37]

Cantilevers coated with proteins for specific fungus recognition can be used as small Petri dishes for investigation of fungus growth.[38] Initially, the cantilevers are placed in a solution of fungus and afterwards kept in a humid environment. The presence of fungus and mycelium growth are monitored as a change in the mass of the cantilever using dynamic mode operation and the mass of a single fungal spore can be detected. Vital spores can be detected in situ within 4 h, which is approximately 10 times faster than the procedures applied today in fungal detection. The biosensor could detect the target fungi in a range of 10^3–10^6 colony-forming units per milliliter. Potentially, such measurements can be used in, for example, water and food quality control.

The ability to measure very small changes in surface stress is a unique feature of the cantilever device operated in the static mode. For example, conformational changes in proteins can be monitored. Protein folding is hard to measure by other simple techniques and could potentially be used, for example, in drug screening. As a test system, conformational changes of bacteriorhodopsin have quantitatively been monitored.[39] The results indicate that it is possible to measure ligand unbinding from membrane proteins based on the intrinsic nanomechanical changes of the receptor.

The most impressive mass sensing achieved to date has been presented by Michael Roukes group at California Institute of Technology, in Pasadena, CA, USA. Their current generation of devices is sensitive to added mass at the level of a few zeptograms.[40] In their experiments, this represents about thirty xenon atoms which is the typical mass of an individual protein molecule. The high mass resolution is primarily achieved by reducing the cantilever dimensions to the nanometer regime. The sensor system is operated at temperatures below 50 K.

Metal-coated cantilevers can be used as calorimeters. One of the first applications was the investigation of surface catalytic reactions. A cantilever was coated with a thin Pt layer, which is used to catalyze the exothermic conversion of H_2 and O_2 to form water vapor.[41] Lately, the exothermic reaction caused by the deflagration of a small amount of trinitrotoluene (TNT) has been monitored using heated cantilevers.[42] Such micrometer-sized devices might hold great potential for handheld explosive sniffers.

5 OUTLOOK

The field of microcantilever array devices is expanding on the technology as well as the application side. More players are joining the game and more advanced devices with integrated fluidics, readout, and actuation are being developed. Moreover, large arrays are now being built and there seems to be no real practical limit to the number of cantilevers that can be operated in parallel. Just think of the millipede project from IBM Research Laboratory in Zurich, Switzerland,[43] where more than 1000 cantilevers are operated simultaneously for reading and writing on a polymer surface for data storage. Polymer cantilevers are emerging and these devices will probably become widely used because they are more sensitive than Si-based devices and cheaper to fabricate. They offer high prospects for the field of single-use, disposable devices. Recently, the principle of chemically induced change in surface stress has been explored for direct actuation of micromechanical structures

such as valves. The valves are self actuated upon a chemical trigger, enabling new possibilities in the development of autonomous devices for use in, for example, drug delivery.[44]

Also, the surface chemistry developed and the applications that are being reported become more advanced in terms of detection limit and complexity of the receptor-ligand schemes. However, there are still several fundamental questions to be addressed. For example, the origin of surface-stress changes is still not fully understood and several models have been proposed for, for example, the generation of signals from DNA hybridization. Nowadays researchers start to perform thorough and detailed studies on rather simple systems such as alkanethiols. They verify these results against computational simulations, taking advantage of recent developments in the field of modeling of intra- and intermolecular interactions.

It seems that no real commercial breakthrough has yet occurred for cantilever sensors, but it can be expected soon. The technology is in a maturing phase where some of the unanswered questions are now being addressed. One important experimental factor is the use of RCs as controls and to filter unspecific signals stemming from variations in temperature and buffer solutions. Moreover, the chemical blocking of the backside of the cantilever is crucial for static mode measurements in order to guarantee an unambiguous signal interpretation. It is thus crucial to conduct the measurements under very well-controlled conditions. For proof-of-concept of a novel sensor application, it is important to perform control measurements by other supplementing techniques such as X-ray photoelectron spectroscopy (XPS), surface plasmons resonance, fluorescence detection, and Fourier transform infra red spectroscopy (FTIR). Such hybrid setups are starting to emerge.

The cantilever sensors have picked up speed and the research society now seems to address the fundamental questions such as signal generation and optimal sensing principles. At the same time, more sensitive devices and highly advanced applications such as detection of conformational changes in proteins and the detection of the mass of a single protein are appearing at a high pace. In conclusion, the cantilever sensors have demonstrated unencumbered mass sensitivity and surface-stress sensitivity and it seems very likely that those features will be utilized for future products in, for example, diagnostics and drug discovery. The cantilever field is exciting and dynamic. The limit for detection is constantly challenged and just the imagination sets the limit to new applications.

REFERENCES

1. C. F. Quate, G. Binnig, and C. Gerber, Atomic force microscope. *Physical Review Letters*, 1986, **5**, 930–932.
2. M. Tortonese, H. Yamada, R. C. Barrett, and C. F. Quate, *Atomic Force Microscopy using Piezoresistive Cantilever*, In *Proceedings of Transducers '91 Conference*, San Francisco, CA, USA, 1991, 448–451.
3. T. Itoh and T. Suga, Development of a force sensor for atomic fore microscopy using piezoelectric thin films. *Nanotechnology*, 1993, **4**, 218–224.
4. J. Brugger, R. A. Buser, and N. F. de Rooij, Micromachined atomic force microprobe with integrated capacitive read-out. *Journal of Micromechanics and Microengineering*, 1992, **2**, 218–220.
5. T. R. Albrecht, S. Akamine, T. E. Carver, and C. F. Quate, Microfabrication of cantilever styli for the atomic force microscope. *Journal of Vacuum Science and Technology A*, 1990, **8**, 3386–3396.
6. O. Wolter, T. Bayer, and J. Greschner, Micromachined silicon sensors for scanning force microscopy. *Journal of Vacuum Science and Technology B*, 1991, **9**, 1353–1357.
7. T. Thundat, R. J. Warmack, G. Y. Chen, and D. P. Allison, Thermal and ambient-induced deflections of scanning force microscope cantilevers. *Applied Physics Letters*, 1994, **64**, 2894–2896.
8. H.-J. Butt and R. Raiteri, *Measurements of the Surface Tension and Surface Stress of Solids*, in *Surface Characterization Methods*, Marcel Dekker, 1999.
9. H.-J. Butt, A sensitive method to measure changes in the surface stress of solids. *Journal of Colloid and Interface Science*, 1996, **180**, 251–260.
10. S. J. O'Shea, M. E. Welland, T. A. Brunt, A. R. Ramadan, and T. Rayment, Atomic force microscopy stress sensors for studies in liquids. *Journal of Vacuum Science and Technology B*, 1996, **14**, 1383–1385.
11. R. Berger, E. Delamarche, H. P. Lang, C. Gerber, J. K. Gimzewski, E. Meyer, and H.-J. Güntherodt, Surface stress in the self-assembly of alkanethiols on gold. *Science*, 1997, **276**, 2021–2024.
12. H. P. Lang, R. Berger, F. Battiston, J. P. Ramseyer, E. Meyer, C. Andreoli, J. Brugger, P. Vettiger, M. Despont, T. Mezzacasa, L. Scandella, H. J. Güntherodt, C. Gerber, and J. K. Gimzewski, A chemical sensor based on a micromechanical cantilever array for the identification of gases and vapors. *Applied Physics A*, 1998, **66**, 561–564.
13. J. Fritz, M. K. Baller, H. P. Lang, H. Rothuizen, P. Vettiger, E. Meyer, H. J. Güntherodt, C. Gerber, and J. K. Gimzewski, Translating biomolecular recognition into nanomechanics. *Science*, 2000, **288**, 316–318.
14. A. Boisen, J. Thaysen, H. Jensenius, and O. Hansen, Environmental sensors based on micromachined cantilevers with integrated read-out. *Ultramicroscopy*, 2000, **82**, 11–16.
15. C. Hagleitner, A. Hierlemann, D. Lange, A. Kummer, N. Kerness, O. Brand, and H. Baltes, Smart single-chip gas sensor microsystem. *Nature*, 2001, **414**, 293–296.
16. G. G. Stoney, The tension of metallic films deposited by electrolysis. *Proceedings of the Royal Society of London Series A*, 1909, **82**, 172–177.
17. G. Wu, H. Ji, K. Hansen, T. Thundat, R. Datar, R. Cote, M. F. Hagan, A. K. Chakraborty, and A. Majumdar, Origin of nanomechanical cantilever motion generated from biomolecular interactions. *Proceedings of the National Academy of Sciences of the United States of America*, 2001, **98**, 1560–1564.
18. P. Grutter, M. Godin, V. Tabbard-Cosa, H. Bourque, T. Monga, Y. Nagai, and R. B. Lennox, *Cantilever-Based Sensing: Origins of Surface Stress*, In *Proceedings of International Workshop on Nanomechanical Sensors*, Copenhagen, Denmark, May 7-10, 2006, 36–37.
19. B. Ilic, D. Czaplewski, M. Zalalutdinov, H. G. Craighead, P. Neuzil, C. Campagnolo, and C. Batt, Single cell detection with micromechanical oscillators. *Journal of Vacuum Science and Technology B*, 2001, **19**, 2825–2828.
20. G. Y. Chen, T. Thundat, E. A. Wachter, and R. J. Warmack, Adsorption-induced surface stress and its effects on resonance frequency of microcantilevers. *Journal of Applied Physics*, 1995, **77**, 3618–3622.
21. F. M. Battiston, J. P. Ramseyer, H. P. Lang, M. K. Baller, C. Gerber, J. K. Gimzewski, E. Meyer, and H. J. Güntherodt, A chemical sensor based on a microfabricated cantilever array with simultaneous resonance-frequency and bending readout. *Sensors and Actuators, B*, 2001, **77**, 122–131.
22. G. Meyer and N. M. Amer, Novel optical approach to atomic force microscopy. *Applied Physics Letters*, 1988, **53**, 1045–1047.
23. M. Alvarez and J. Tamayo, Optical sequential readout of microcantilever arrays for biological detection. *Sensors and Actuators, B*, 2005, **106**, 687–690.
24. M. Yue, H. Lin, D. E. Dedrick, S. Satyanarayana, A. Majumdar, A. S. Bedekar, J. W. Jenkins, and S. Sundaram, A 2-D microcantilever array for multiplexed biomolecular analysis. *Journal of Microelectronic Systems*, 2004, **13**, 290–299.
25. M. Calleja, J. Tamayo, M. Nordström, and A. Boisen, Low-noise polymeric nanomechanical biosensors. *Applied Physics Letters*, 2006, **88**, 113901.
26. D. Haefliger and A. Boisen, Three-dimensional microfabrication in negative resist using printed masks. *Journal of Micromechanics and Microengineering*, 2006, **16**, 951–957.
27. J. Thaysen, R. Marie, and A. Boisen, *Cantilever-Based Bio-Chemical Sensor Integrated in a Microfluidic Handling System*, In *Proceedings of IEEE MEMS Conference*, Interlaken, Switzerland, 2001, 401–404.
28. J. H. Lee, K. S. Hwang, J. Park, K. H. Yoon, D. S. Yoon, and T. S. Kim, Immunoassay of Prostate-Specific Antigen (PSA) using resonant frequency shift of piezoelectric nanomechanical microcantilever. *Biosensors and Bioelectronics*, 2005, **20**, 2157–2162.
29. Z. J. Davis, G. Abadal, B. Helbo, O. Hansen, F. Campabadal, F. Pérez-Murano, J. Esteve, E. Figueras, J. Verd, N. Barniol, and A. Boisen, Monolithic integration of mass

sensing nano-cantilevers with CMOS circuitry. *Sensors and Actuators, A*, 2003, **105**, 311–319.

30. M. K. Ghatkesar, V. Barwich, T. Braun, A. H. Bredekamp, U. Drechsler, M. Despont, H. P. Lang, M. Hegner, and C. Gerber, *Real Time Mass Sensing by Nanomechanical Resonators in Fluid*, *Proceedings of IEEE Sensors*, Vienna, Austria, 2004, 1060–1063.

31. R. McKendry, J. Zhang, Y. Arntz, T. Strunz, M. Hegner, H. P. Lang, M. K. Baller, U. Certa, E. Meyer, H. J. Guntherodt, and C. Gerber, Multiple label-free biodetection and quantitative DNAbinding assays on a nanomechanical cantilever array. *Proceedings of the National Academy of Sciences of the United States of America*, 2002, **99**, 9783–9788.

32. Y. Arntz, J. E. Seelig, J. Zhang, H. P. Lang, P. Hunziker, J. P. Ramseyer, E. Meyer, M. Hegner, and C. Gerber, A label-free protein assay based on a nanomechanical cantilever array. *Nanotechnology*, 2003, **14**, 86–90.

33. G. Wu, R. H. Datar, K. M. Hansen, T. Thundat, R. J. Cote, and A. Majumdar, Bioassay of prostatespecific antigen (PSA) using microcantilevers. *Nature Biotechnology*, 2001, **19**, 856–860.

34. N. Backmann, C. Zahnd, F. Huber, A. Bietsch, A. Plückthun, H. P. Lang, H. J. Güntherodt, M. Hegner, and C. Gerber, A label-free immunosensor array using single-chain antibody fragments. *Proceedings of the National Academy of Sciences of the United States of America*, 2005, **102**, 14587–14592.

35. A. G. Hansen, M. W. Mortensen, J. E. T. Andersen, J. Ulstrup, A. Kühle, J. Garnæs, and A. Boisen, Stress formation during self-assembly of alkanethiols. *Probe Microscopy*, 2001, **2**, 139–150.

36. R. Marie, A. Boisen, J. Thaysen, H. Jensenius, and C. B. Christensen, Adsorption kinetics and mechanical properties of thiol-modified DNA -oligos on gold investigated by microcantilever sensors. *Ultramicroscopy*, 2002, **91**, 29–36.

37. A. J. Detzel, G. A. Campbell, and R. Mutharasan, Rapid assessment of Escherichia coli by growth rate on piezoelectric-excited millimeter-sized cantilever (PEMC) sensors. *Sensors and Actuators, B*, 2006, in press.

38. N. Nugaeva, K. Y. Gfeller, N. Backmann, H. P. Lang, M. Düggelin, and M. Hegner, Nanomechanical cantilever array sensors for selective fungal immobilization and real-time growth detection. *Biosensors and Bioelectronics*, 2005, **21**, 849–856.

39. T. Braun, N. Backmann, M. Vögtli, A. Bietsch, A. Engel, H. P. Lang, C. Gerber, and M. Hegner, Conformational change of bacteriorhodopsin quantitatively monitored by microcantilever sensors. *Biophysical Journal*, 2006, **90**, 2970–2977.

40. Y. T. Yang, C. Callegari, X. L. Feng, K. L. Ekinci, and M. L. Roukes, Zeptogram-scale nanomechanical mass sensing. *Nano Letters*, 2006, **6**, 4583–4586.

41. J. K. Gimzewski, C. Gerber, E. Meyer, and R. R. Schlitter, Observation of a chemical reaction using a micromechanical sensor. *Chemical Physics Letters*, 1994, **217**, 589–594.

42. L. A. Pinnaduwage, A. Wig, D. L. Hedden, A. Gehl, D. Yi, T. Thundat, and R. T. Lareau, Detection of trinitrotoluene via deflagration on a microcantilever. *Journal of Applied Physics*, 2004, **95**, 5871–5875.

43. A. Knol, P. Bächtold, J. Bonan, G. Cherubini, M. Despont, U. Drechsler, U. Dürig, B. Gotsmann, W. Häberle, C. Hagleitner, D. Jubin, M. A. Lantz, A. Pantazi, H. Pozidis, H. Rothuizen, A. Sebastian, R. Stutz, P. Vettiger, D. Wiesmann, and E. S. Eleftheriou, Integrating nanotechnology into a working storage device. *Microelectronic Engineering*, 2006, **83**, 1692–1697.

44. D. Haefliger, R. Marie, and A. Boisen, *Self-Actuated Polymeric Valve for Autonomous Sensing and Mixing*, *Digest of Technical Papers Transducers '05 Conference*, Seoul, Korea, 2005, 1569–1572.

61

Biosniffers (Gas-Phase Biosensors) as Artificial Olfaction

Kohji Mitsubayashi

Institute of Biomaterials and Bioengineering, Tokyo Medical and Dental University, Tokyo, Japan

1 A BIOCHEMICAL SNIFFER FOR A GAS-PHASE BIOASSAY

The detection and quantification of gaseous substances with high sensitivity and selectivity are required in many different areas such as environmental assessment, food process control, fire detection, anti-terrorism measures, etc. In addition, some substances in the gas phase can be related to human health and behavior. For example, chemical malodors can affect human mental and physical conditions.[1] In several reports, analysis of the volatile components in the expiratory gas of patients can help in the diagnosis of diseases such as oral ailments,[2] hepatocirrhosis, and cancer of the lung,[3,4] head, and neck regions. Expiratory gas analysis would provide a noninvasive, convenient, and safe method of diagnosing and monitoring disease states.

A way for measuring gaseous substances with high sensitivity and selectivity is therefore required. Many types of gas sensors have been investigated and developed. Considerable effort has gone into improving the ethanol selectivity and sensitivity of semiconductor-type gas sensors.[5–7] Semiconductor sensors are still inadequate for sensing multianalyte samples such as expiratory gas, because the sensor response is based only on changes in the electrical conductivity of the device, following adsorption of gaseous substances.[1,5–8] Although an improvement in semiconductor sensors and quartz crystal microbalance (QCM) devices (including electronic nose) has been reported, it is still far from the selectivity achievable using biological recognition systems such as enzymes.

For the measurement of chemical substances in the liquid phase, biosensors have been extensively researched. Biosensors use biologically derived materials such as enzymes, microorganisms, antigen and antibody, and organelles, which possess high specificity for their substrates.[9] It is difficult to use biosensors for detection of gaseous chemicals, however, because of deactivation of the biological component in the gas phase.

Some kinds of biochemical gas sensors (biosniffers) as gas-phase bioassay have been developed in many fields such as environmental assessment, clinical and dental measurement, and so on. The biosniffers for continuous monitoring have been constructed using a reaction unit with gas and liquid cells. Stick-type biosniffers as disposable devices for a batch gas measurement have been also fabricated by microelectromechanical systems (MEMS) or screen-printing process. A drug metabolizing system and an enzyme inhabitation mechanism are also used as biological recognition materials. In this chapter, the diaphragm-type and the stick-type biosniffers are introduced with their

Handbook of Biosensors and Biochips. Edited by Robert S. Marks, David C. Cullen, Isao Karube, Christopher R. Lowe and Howard H. Weetall.
© 2007 John Wiley & Sons, Ltd. ISBN 978-0-470-01905-4.

sensor performances. Applications of the biosniffers (i.e., physiological analysis, dental diagnosis, environmental analysis) are also explained.

2 BIOSNIFFERS WITH A DIAPHRAGM REACTION UNIT

2.1 Alcohol Biosniffer

Expiratory gas analysis would provide a noninvasive, convenient, and safe method of diagnosing and monitoring disease states. One of the major applications of breath analysis is in the quantification and detection of ethanol in expiratory gas after alcohol consumption. Blood ethanol concentration can be determined from its concentration in breath, a blood-breath alcohol partition ratio of 2000 having been widely adopted.[10] The breath alcohol level can be related to the degree of intoxication (i.e., haziness: $0.5–1\,g\,l^{-1}$ in blood (130–260 ppm in breath); slight drunkenness: $1–1.5\,g\,l^{-1}$ (260–390 ppm); drunkenness: $1.5–2.5\,g\,l^{-1}$ (390–650 ppm))[11] and used as a measure of drunken driving.

In certain industrial fields, such as fermentation and distillation, the ethanol vapor concentration can reach toxic levels, causing inflammation of the nasal mucous membrane and conjunctiva, irritation of the skin, and, at high levels, even alcohol poisoning. The maximum permitted concentration of ethanol vapor in the workplace as defined by American Conference of Governmental Industrial Hygienists (ACGIH) is 1000 ppm.

Since gaseous ethanol is a malodorous substance for humans (the ethanol selection and detection limits for the human sense of smell have been reported to be 6.1 and 0.36 ppm,[12] respectively), The continuous monitoring of ethanol concentration in the gas phase is significant for assessing the human health and behavior related to ethanol vapor from a physiological point of view.

An alcohol biosensor for the liquid phase has also been investigated and applied widely for the measurement of ethanol concentration in fermentation processes. Alcohol oxidase (AOD, EC 1.1.3.13) is commonly used in the construction of alcohol biosensors, catalyzing the oxidation of lower primary alcohols according to the reaction:

$$R–CH_2OH + O_2 \xrightarrow{\text{AOD}} R–CHO + H_2O_2$$

A biosniffer was constructed using the AOD catalytic reaction for the measurement of gaseous substances.

2.1.1 Construction of the Alcohol Biosniffer

Figure 1 illustrates the structure of the biosniffer for measurement of ethanol vapor.[13] The sensor consisted of an enzyme electrode, a reaction cell with two compartments for liquid and gas phases, and a porous diaphragm membrane between these compartments. The enzyme electrode was constructed using a commercially available Clark-type dissolved oxygen electrode with an enzyme membrane.

AOD (EC 1.1.3.13) was used for constructing the enzyme electrode. For enzyme immobilization, AOD was mixed with acrylamide gel monomer solution in a volume ratio of 1 : 4, then sandwiched between 2 glass plates, and irradiated with a UV lamp (254 nm) for 5 min in order to photocrosslink the monomer solution and immobilize the enzyme. The enzyme electrode was constructed by cutting the enzyme membrane to the required dimensions using a scalpel and placing it onto the sensing area of the dissolved oxygen electrode.

Two hollow polytetrafluoroethylene (PTFE) tubes (OD: 40 mm, ID: 20 mm) were cut to a length of 20 mm and the surfaces of the tubes were polished. Four PTFE tube connectors (OD: 6 mm, ID: 2 mm) were screwed into the outside of all of the tapped holes. A porous PTFE membrane was sandwiched between the two tube blocks, which were then held together firmly using a mechanical clamp. In this way, the membrane acted as a separating diaphragm between the two hollow compartments of the tubes.

A cylindrical rubber stopcock (OD: 20.5 mm) was inserted into the large hole of one of the tubes, thus forming the gas compartment of the reaction cell. The tip of the enzyme electrode was immersed in the liquid compartment filled with phosphate buffer (0.067 M, pH 7) and adjusted so as to directly touch the surface of the diaphragm membrane. Gas and phosphate buffer solution could be flowed individually through each inlet tube connector to the gas and liquid compartments of the reaction cell, respectively. Gaseous substances in the gas compartment could diffuse through into the liquid compartment of the reaction cell through the PTFE diaphragm membrane.

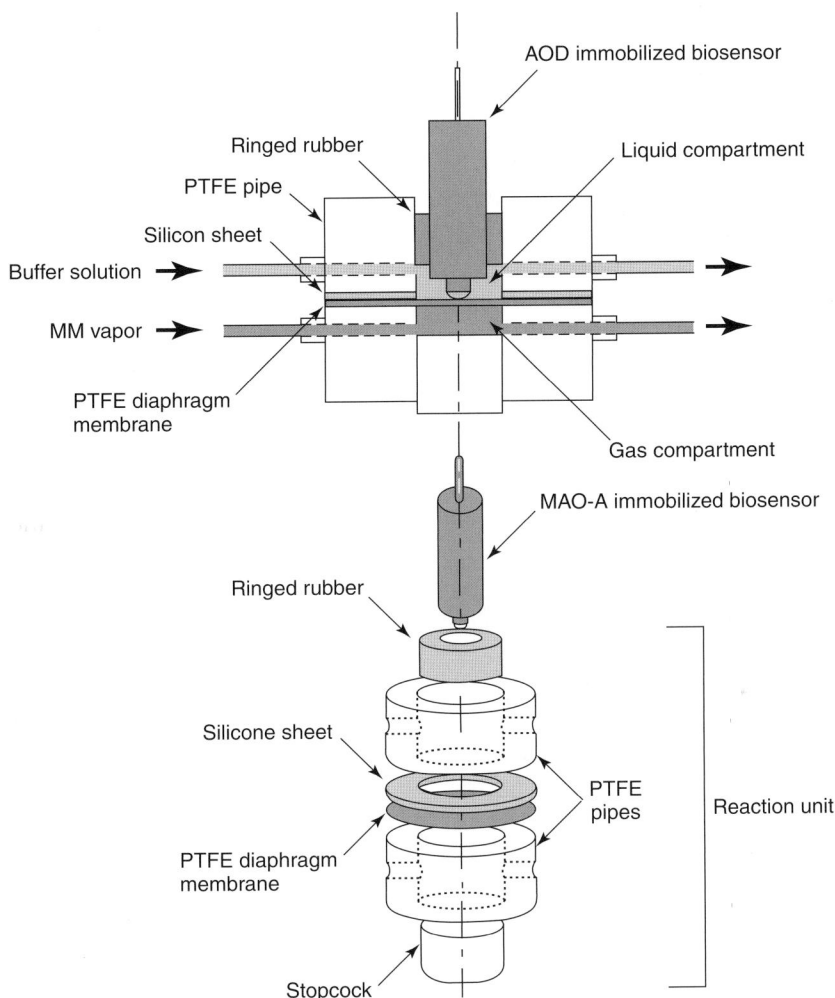

Figure 1. Structure of an alcohol biosniffer with a diaphragm reaction unit. [T. Minamide, et al., *Sensors and Actuators, B*, 2005, **108**, 639–645.]

2.1.2 Characteristics of the Alcohol Sniffer

Figure 2 shows the current response of the biosniffer to ethanol in the gas phase at a concentration of 41.5 ppm for 5 min with and without buffer flow ($1.9 \, \text{l} \, \text{h}^{-1}$) through the liquid compartment. As the figure indicates, experiments both with and without buffer flow gave rectangular curves, in which the current decreased rapidly following application of ethanol, followed by a steady state current which gradually increased to the initial output following standard air application. The decrease in output current relates to the concentration of ethanol in the gas phase, since ethanol that diffuses through the enzyme membrane is oxidized by AOD using oxygen as electron acceptor, causing a decrease in the concentration of dissolved oxygen (see the following equations).

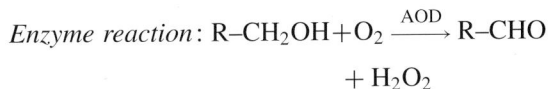

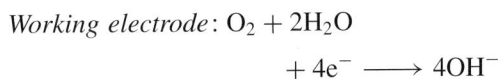

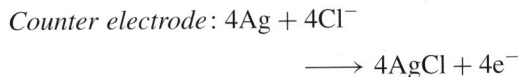

$$Enzyme\ reaction: \text{R–CH}_2\text{OH} + \text{O}_2 \xrightarrow{\text{AOD}} \text{R–CHO} + \text{H}_2\text{O}_2$$

$$Working\ electrode: \text{O}_2 + 2\text{H}_2\text{O} + 4\text{e}^- \longrightarrow 4\text{OH}^-$$

$$Counter\ electrode: 4\text{Ag} + 4\text{Cl}^- \longrightarrow 4\text{AgCl} + 4\text{e}^-$$

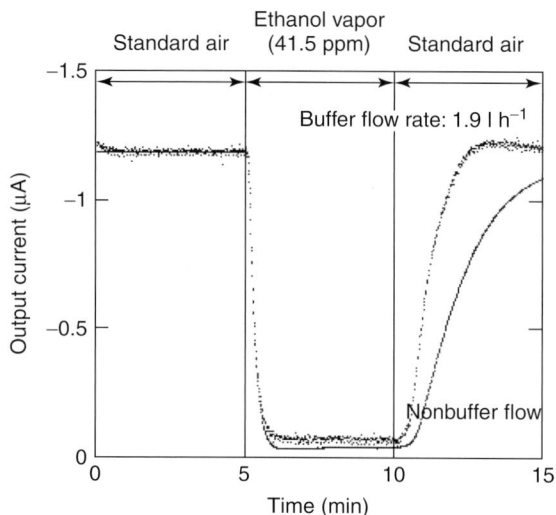

Figure 2. Typical responses for ethanol measurement in the gas phase using the biosniffer with buffer flow and without buffer flow. [Reprinted with permission Mitsubayashi et al.[13] copyright 1994.]

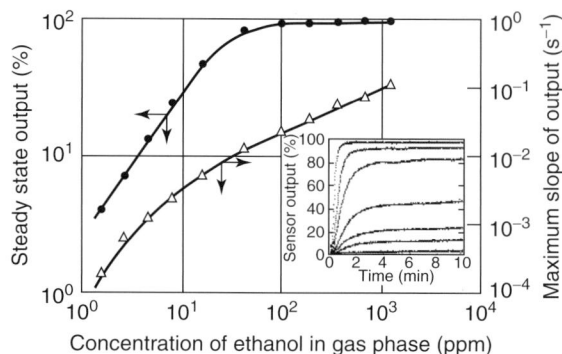

Figure 3. Calibration curves of biosniffer using steady state output (filled circles) and maximum response slope (open triangles). Inset: typical response to varying concentrations of ethanol in the gas phase (1.57, 4.49, 7.86, 15.7, 41.5, 98.2, and 680 ppm). [Reprinted with permission Mitsubayashi et al.[13] copyright 1994.]

Comparison of the curves showed that the sensor current without buffer flow had lower noise and slightly higher sensitivity than with buffer flow. With buffer flow, however, there was a more rapid recovery of the initial current following application of standard air.

The steady state current value and the maximum slope calibration curves for ethanol in the gas phase are shown in Figure 3. The inset figure shows the typical response of the biosniffer to gaseous ethanol at various concentrations. In these figures, the steady state output signal is represented as a percentage of the current value for standard air. As Figure 3 indicates, the steady output and the slope of the response were related to the concentration of ethanol in the gas phase over the range of 1.57–41.5 ppm and 15.7–1242 ppm, respectively, deduced from exponential regression analysis of the log–log plot by a method of least squares according to the following equations:

$$\text{Output}(\%) = 3 \times [\text{gaseous ethanol (ppm)}]^{0.943} \quad (1)$$

$$\text{Slope}(s^{-1}) = 0.00091 \times [\text{gaseous ethanol (ppm)}]^{0.685} \quad (2)$$

This calibration range using the maximum slope covers the maximum permissible concentration of ethanol vapor in the workplace (1000 ppm) and the alcohol levels encountered in breath (over 130 ppm)[11] and can be utilized to determine the level of intoxication as described in the preceding text. The biosniffer was calibrated using the steady state output, to give a detection limit of 1.57 ppm for the sensor (diaphragm membrane pore size: 1–2 μm, thickness: 0.25 mm), which is lower than the ethanol selective detection limit (6.1 ppm), but higher than ethanol detection limit (0.36 ppm) for the human sense of smell.[12]

2.1.3 Gas Selectivity of the Alcohol Biosniffer

The selectivity of both the biosniffer and the commercially available semiconductor gas sensor for several gases and blends of gases is shown in Figure 4. The semiconductor gas sensor responded to all of the applied gases, confirming the extremely poor selectivity of such devices as generally known.[8] The biosniffer using an immobilized AOD electrode, however, gave negligible responses to all the chemicals other than ethanol. The gas-phase sensor output using a mixture of ethanol and n-pentane was the same as the response to ethanol alone with a confidence level of 97.5%. The biosniffer thus possessed much greater selectivity and accuracy than the semiconductor gas sensor.

Applied gas substances
(concentration: ppm)

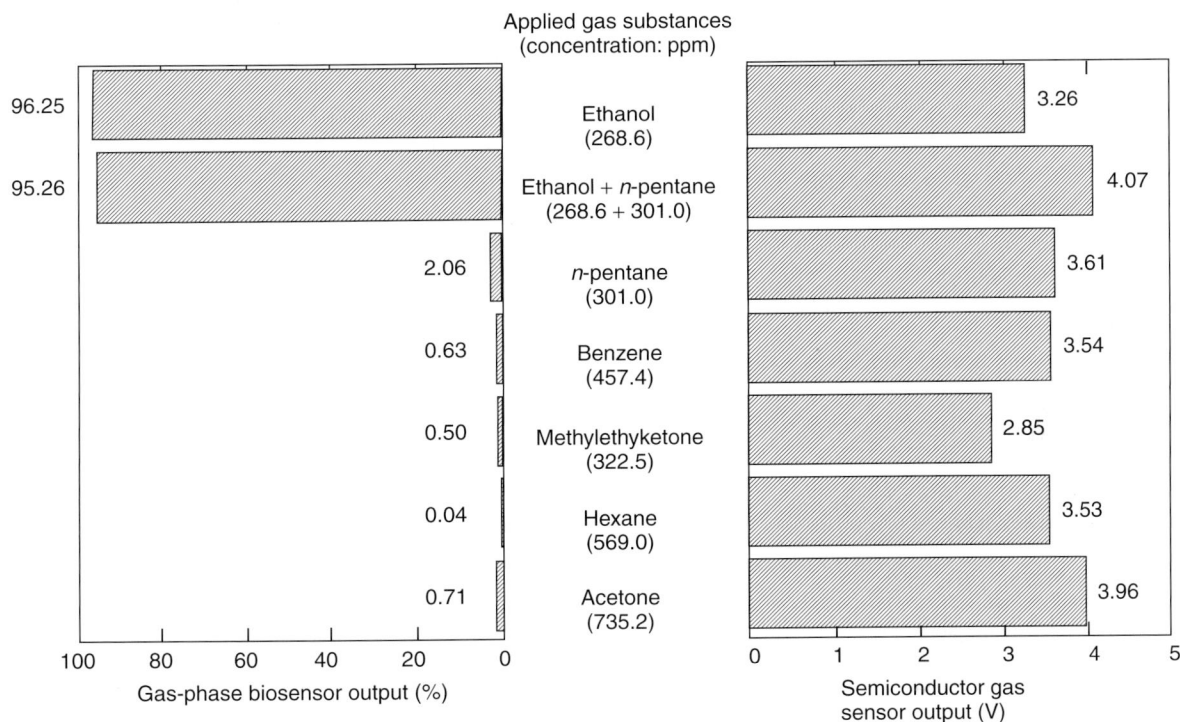

Figure 4. Gas selectivity of the biosniffer and semiconductor gas sensor for various substances in the gas phase. [Reprinted with permission Mitsubayashi et al.[13] copyright 1994.]

2.2 Acetaldehyde Biosniffer

Gaseous acetaldehyde is also one of the chemical malodorous substances. The maximum permitted concentrations of gaseous acetaldehyde as defined by ACGIH and by the Environment Agency in Japan are 100 and 50 ppm, respectively.[14,15] The convenient measurement of acetaldehyde concentration in the gas phase is required in the fields of alcohol fermentation process (brewery, winery) and physiological research of alcohol metabolism as is well known. The bioelectronic sniffer for acetaldehyde vapor was also constructed using the diaphragm reaction unit with the porous diaphragm membrane.[16]

2.2.1 Acetaldehyde Biosniffer with the Diaphragm Reaction Unit

As mentioned in the previous section, the aldehyde dehydrogenase (ALDH) immobilized biosensor was incorporated into the reaction cell with

both gas and liquid phase compartments separated by the PTFE diaphragm membrane,[13] thus obtaining the bioelectronic sniffer for acetaldehyde vapor. Acetaldehyde is dehydrogenated by ALDH using oxidized β-Nicotinamide-adenine dinucleotide (NAD$^+$, oxidized form) as electron acceptor. Then NADH (reduced form) is dehydrogenated by diaphorase using potassium ferricyanide (1 mmol l^{-1}) as an electrochemical mediator, thus obtaining the oxidizing current of its reduction form at the Pt-electrode[17] as shown in Figure 5. A fixed voltage of 81 mV (vs a Pt-wire counterelectrode) was applied to the Pt working electrode coated with the hydrophilic polytetrafluoroethylene (H-PTFE) membrane.[18]

The sensor output was related to the concentration of acetaldehyde in the gas phase over the range 0.525–20 ppm using the 1–2 μm pore size and 0.105–5.25 ppm using the 20–30 μm pore size, respectively, deduced from exponential regression analysis of the log–log plot by a method of least squares according to the following

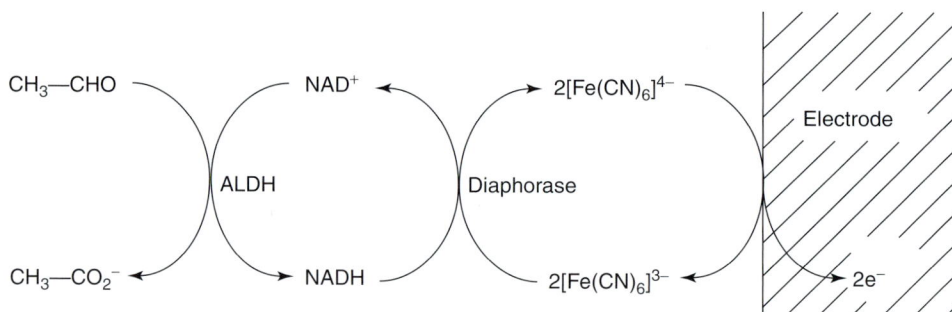

Figure 5. Enzymatic and electrochemical reactions for measuring acetaldehyde. [Reprinted with permission Mitsubayashi et al.[16] copyright 2003.]

equation:

Pore size: $1–2 \, \mu m$

Output current (nA) =

$$97.8[\text{acetaldehyde (mmol l}^{-1})]^{0.935} \quad (3)$$

Pore size: $20–30 \, \mu m$

Output current(nA) =

$$226[\text{acetaldehyde (mmol l}^{-1})]^{0.979} \quad (4)$$

By use of the more porous type of diaphragm membrane, it is possible to improve the detection limit from 0.525 to 0.105 ppm of gaseous acetaldehyde. The detection limit of the sniffer is equal to the sensitive level 3 (0.15 ppm) for sense of smell in humans.[12] The calibration range of the biosniffer covered the maximum permitted concentrations of acetaldehyde vapor in the United States (100 ppm) and Japan (50 ppm) as described in the preceding text.

3 ODOR ANALYSIS WITH DRUG METABOLIZING SYSTEM FOR ODOR CHEMICALS

3.1 Trimethylamine (Fish-odor) Sniffer with Flavin-containing Monooxygenase

Trimethylamine (TMA, fish-odor substance), is one of the volatile organic compounds (VOCs) and the specified malodorous substances as defined by ACGIH and the Environment Agency, Government of Japan. In certain industrial areas, such as a chemical plant and a canning factory for marine products, the concentration of TMA fume can reach toxic levels, causing inflammation of the nasal mucous membrane, conjunctiva irritation of the skin, and, at high levels, even a spontaneous combustion (autoignition temperature: 190 °C) and a gas explosion (explosive limit: 2–11.6 vol%).[19] The maximum permissible concentration of TMA vapors in the workplace is 5 ppm ($12 \, \text{mg m}^{-3}$, time-weighted average (TWA) concentration) and 15 ppm ($36 \, \text{mg m}^{-3}$, short-term exposure limit (STEL) concentration) as defined by ACGIH.[19]

In humans, TMA is metabolized exclusively to trimethylamine N-oxide (TMAO) with up to $60 \, \text{mg day}^{-1}$ excreted in the urine of healthy people and less than 5% excreted as the parent compound.[20,21] The reaction is widely considered to be catalyzed by flavin-containing monooxygenase (FMO, EC 1.14.13.8) as one of the xenobiotic metabolizing enzymes,[22] which can decompose and detoxicate most chemicals in vivo, even the inhaled VOC. Trimethylaminuria or "fish-odor syndrome" is a human disorder characterized by an impaired ability to convert the malodorous TMA to its odorless N-oxide.[23] A mutation in the FMO gene, which decreases TMA metabolism, has been described recently.[24] FMO3 (type 3 of the polymorphic FMO family enzymes) has been recognized to be the major hepatic form in adults and catalyzes the majority of TMAO formation from TMA[25] in vivo with the following reaction:

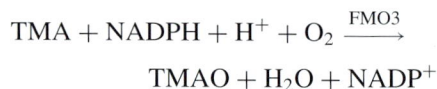

$$\text{TMA} + \text{NADPH} + \text{H}^+ + \text{O}_2 \xrightarrow{\text{FMO3}}$$
$$\text{TMAO} + \text{H}_2\text{O} + \text{NADP}^+$$

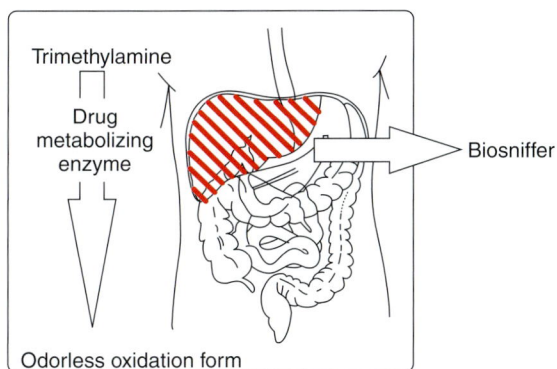

Figure 6. Drug metabolizing system for trimethylamine in human liver. [K. Mitsubayashi, *Annales de Chimie-Science des Materiaux*, 2004, **29**(6), 103–114.]

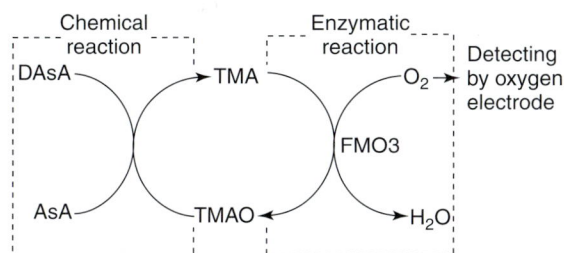

Figure 7. Principle of cyclic reaction for signal-amplified biosensor for trimethylamine (TMA) using FMO3 enzyme and ascorbic acid (AsA) as reducing reagents. TMAO: trimethylamine *N*-oxide; DAsA: dehydroascorbic acid. [Reprinted with permission Mitsubayashi and Hashimoto[26] copyright 2002, IEEE.]

The TMA biosensor was constructed by cutting the FMO3 immobilized membrane to the required dimensions using a cutter and placing it onto the sensing area of the dissolved oxygen electrode (Figure 6).[26] In order to amplify the output signal of the biodevices, the substrate regeneration cycle was applied for measuring TMA in the liquid and gas phases by coupling with the reducing reagent system of ascorbic acid (AsA)[27] and FMO3 enzyme, respectively (Figure 7).

3.1.1 FMO Immobilized Biosniffer

Figure 8 illustrates the output response of the FMO3 biosniffer with 10 mmol l⁻¹ AsA in phosphate buffer to TMA vapor varying the concentration. The sensor output is represented by the

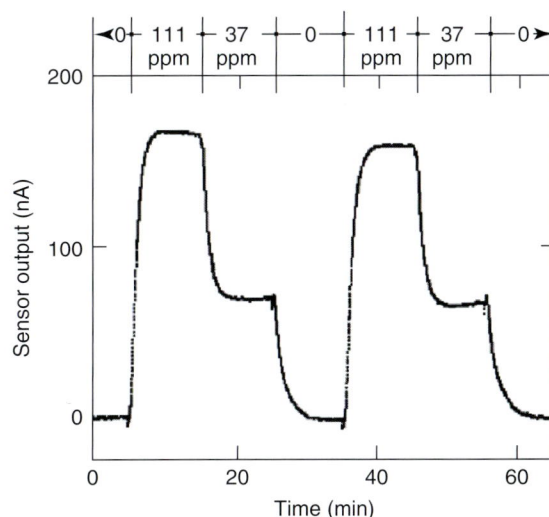

Figure 8. Typical response of the biosniffer to varying concentrations of trimethylamine in the gas phase. [Reprinted with permission Mitsubayashi and Hashimoto[26] copyright 2002, IEEE.]

absolute value of the difference between the initial and the present current value for standard purified air. As the figure indicates, the output increased rapidly following application of TMA vapor, followed by a steady state output that decreased as a result of lowering the concentration of TMA and then approached the initial output following standard air application (TMA free); and this was successfully repeated for the TMA measurement. On the basis of these results, it was possible to use the biosniffer, with good response, for continuous monitoring of the change in the concentration of TMA vapor. The increase in the sensor output relates to the concentration of gaseous TMA, since TMA that diffuses through the enzyme membrane is oxidized by FMO3 using oxygen as an electron acceptor, causing a decrease in the concentration of dissolved oxygen.

The sensor output was related to the concentration of TMA vapor over the range of 0.52–105 ppm, deduced from exponential regression analysis of the log–log plot by the least-squares method according to the following equation:

$$\text{Sensor output(nA)} =$$

$$2.13[\text{gaseous TMA(ppm)}]^{0.89} \quad (5)$$

This calibration range of the sniffer with FMO3 covers the maximum permissible concentration of TMA vapor in the workplace (TWA: 5 ppm, STEL: 15 ppm by ACGIH), thus allowing the determination of the level of intoxication. As described in the preceding text, the TMA sensing level 5 (3 ppm) for the human sense of smell was also encountered to the sniffer calibration range.

4 HUMAN ODOR ANALYSIS BY THE BIOSNIFFERS

4.1 Stick-type Biosniffers for Ethanol and Acetaldehyde in Breath Air after Drinking

Expiratory gas analysis would provide a convenient and safe noninvasive method of diagnosing and monitoring disease states. One of the major applications of breath analysis is in the quantification and detection of ethanol and acetaldehyde (as metabolic products of ethanol) in expiratory gas after alcohol consumption. Blood ethanol concentration can be determined from its concentration in breath, a blood-breath alcohol partition ratio of 2000 having been widely adopted.[5,11,28]

It has also been reported that breath acetaldehyde is related to its blood concentration as an intoxicant substance, especially for aldehyde dehydrogenase type 2 (ALDH2) deficiency type. Nearly half of the Japanese people are negative for ALDH2. In this section, stick-type biosniffers for the convenient measurement of gaseous ethanol and acetaldehyde in breath air are described.

4.1.1 Construction of Stick-type Biosniffers for Ethanol and Acetaldehyde

Both biosniffers, for ethanol and acetaldehyde, were constructed in sandwich configurations with a stick shape as shown in Figure 9 (a: ethanol, b: acetaldehyde).[29]

For the ethanol sniffer, carbon and Ag/AgCl electrodes were formed by printing the respective pastes onto each sides of a filter membrane, respectively. By applying AOD, the oxidation current of hydrogen peroxide being produced by AOD enzyme reaction could be measured by amperometric analysis between carbon and Ag/AgCl electrodes. The electrode-coated membrane was cut using a scalpel into the shape of a stick. In order

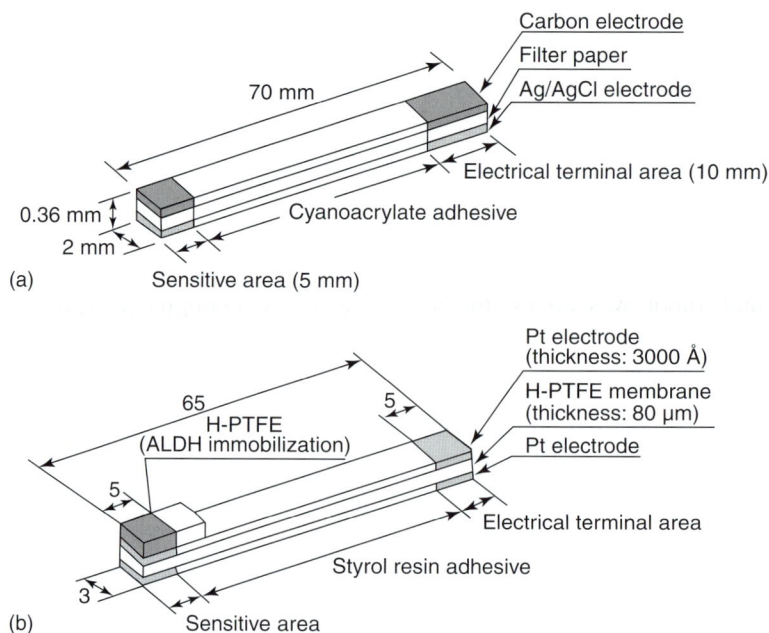

Figure 9. Structure of bioelectronic sniffers for gaseous ethanol (a) and acetaldehyde (b). [Reprinted with permission Mitsubayashi et al.[29] copyright 2005, Elsevier.]

to isolate the sensitive area, a cyanoacrylate adhesive was applied to the middle part of the electrode membrane. The stick membrane was thus separated into three discrete areas: sensitive area, lead area, and electrical terminal area as illustrated in Figure 9(a). AOD was immobilized in the sensitive area of the stick membrane.

Figure 9(b) illustrates the structure of the biosniffer for acetaldehyde. The acetaldehyde sniffer was constructed according to a similar method. The platinum electrodes were formed by sputter deposition on both sides of a H-PTFE membrane. The oxidation current of NADH (reduced form) produced by ALDH from the enzyme reaction with NAD^+ (oxidized form) could be measured by amperometric analysis between the two Pt electrodes.[30] ALDH (EC 1.2.1.5) was immobilized on the additional H-PTFE membrane with PVA-SbQ.

4.1.2 Performances of Stick-type Biosniffers in the Gas Phase

The enzyme electrodes were used as sniffer devices for assessing gaseous substances. A gas-sampling bag (880 ml) was filled with various concentrations of gaseous substances supplied from the gas generator. The sensitive area of the stick-type sniffers, wetted with $10\,\mu l$ of phosphate buffer (pH 8, 100 mM, with or without $2\,mmol\,l^{-1}NAD^+$), was quickly inserted into the opening mouth of the sampling bag containing the gaseous substance.

The AOD sniffer responded to standard ethanol vapor and gave a steady state output, which was related to the applied ethanol concentration. The response time to reach 90% of the steady current after applying ethanol vapor was approximately 40 s, which is slower than that in the liquid phase. The steady output was related to the concentration of ethanol in the gas phase over the range 1–500 ppm, deduced from logarithmic regression analysis of the semilog plot by a method of least squares according to the following equation:

$$Output\ current(\mu A) = 0.141$$
$$\times 1.09\log[gaseous\ ethanol(ppm)] \quad (6)$$

This calibration range covered the alcohol levels encountered in breath (over 130 ppm) after drinking. The detection limit of 1 ppm is lower than the ethanol selective detection limit for the human sense of smell (6.1 ppm).[11]

The ALDH sniffer was also used to measure acetaldehyde from 0.11 to 10 ppm, deduced from exponential regression analysis of the log–log plot by a method of least squares according to the following equation:

$$Output\ current(nA) =$$
$$108[gaseous\ acetaldehyde(ppm)]^{0.430} \quad (7)$$

Upon testing with various gas substances at 2 ppm, the ALDH biosniffer did not respond to any of the chemicals apart from acetaldehyde, thus indicating high gas selectivity attributable to the substrate specificity of ALDH, and similar to that of the AOD sniffer.

4.1.3 Breath Analysis with Biosniffers after Drinking

For physiological applications, the concentrations of ethanol and acetaldehyde in expired gas after drinking were measured for healthy male subjects aged 21–35. Figure 10 illustrates the measurement processes of the breath analysis (ethanol and acetaldehyde) by the biosniffers and gas detector tubes, respectively. The expired gas was collected in the sampling bag at 15 or 20 min intervals after the subjects took 350 ml of beer (5.5% alcohol), and was then used for gas measurement with the same method as described in the preceding text. The sniffer devices were used repeatedly after cleaning the sensitive area with phosphate buffer solution.

Both the sniffers for ethanol and acetaldehyde were applied in the analysis of breath air collected from the ALDH2 [+] and [−] subjects after drinking. Figure 11 illustrates the comparisons of the ethanol (a) and acetaldehyde (b) concentration changes with time in breath ethanol between the ALDH2 [+] (open square) and ALDH2 [−] (filled circle) subjects, which had been classified by the ethanol patch test. The concentration values of ethanol in the figure are represented as average of the measured concentrations from the subjects by the biosniffer.

As Figure 11(a) indicates, both the concentrations of breath ethanol from the ALDH2 [+] and [−] subjects increased following alcohol drinking,

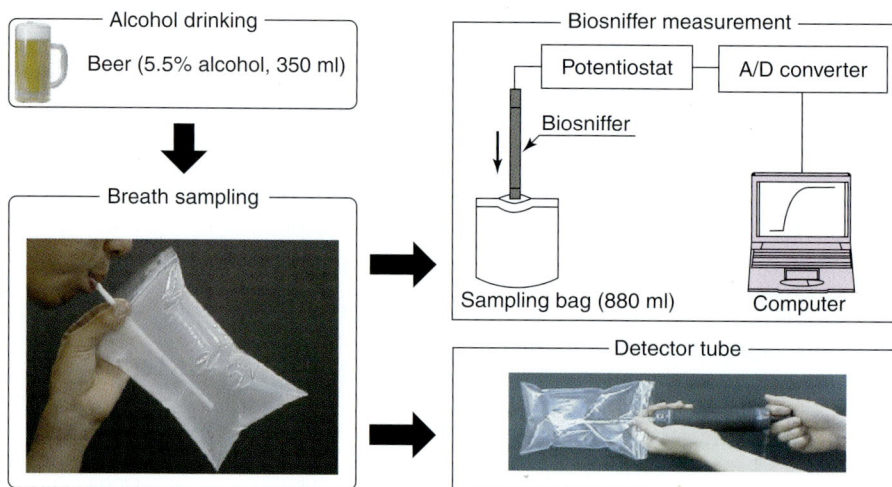

Figure 10. Measurement processes of breath analysis by the biosniffers and gas detector tubes for ethanol and acetaldehyde after drinking (350 ml of beer: 5.5% alcohol). [Reprinted with permission Mitsubayashi et al.[29] copyright 2005, Elsevier.]

with the peak value at 30 min after drinking, and then decreased gradually over a period of hours. These peak values for the ALDH2 [+][5] and [−] subjects were evaluated as alcohol haziness or slight drunkenness levels after drinking (even 350 ml of beer: 5.5% alcohol), and reached the human sense of smell level 3 (100 ppm) for ethanol vapor as noted in the preceding text. It was possible to use the biosniffer repeatedly by rinsing the sensor tip, in contrast with the disposable detector tube. The sniffer device is convenient to use for the noninvasive analysis of ethanol metabolism condition using expired air. As Figure 11(a) also indicates, the breath ethanol concentration in the ALDH2 [−] subjects was higher (approx. twofold) than that in the ALDH2 [+] subjects, from beginning to end of the examination. As a previous paper has reported,[31] the lower activity of ALDH2 induced an adverse effect on ethanol metabolism, with the result that ethanol and acetaldehyde remain in the human body for a long time.

Breath acetaldehyde as an intoxicant substance was also analyzed by the ALDH sniffer for the ALDH2 [+] and [−] subjects, similar to the breath ethanol measurement. The comparison of the concentration changes in breath acetaldehyde with time between the ALDH2 [+] and ALDH2 [−] subjects is illustrated in Figure 11(b). Like the results of breath ethanol, both the concentrations of breath acetaldehyde from the ALDH2 [+] and

[−] subjects increased following alcohol drinking, with a peak value at 30 min after drinking, and then decreased gradually over a period of hours. The peak concentrations in breath air of the ALDH2 [+] and [−] reached the human sense of smell level 3.5 (0.46 ppm) and 4 (1.4 ppm), respectively, for gaseous acetaldehyde. For example, breath acetaldehyde is a sharper odorous chemical for the human olfactory nerve system than ethanol in breath after drinking.

In addition, the acetaldehyde concentration in the ALDH2 [−] subject breath air was much higher (approx. 10-fold) than that of the ALDH2 [+] subjects. The significant difference of breath acetaldehyde concentrations between ALDH2 [+] and [−] subjects is clarified by a T-test. As the results indicated, the lower activity of ALDH2 induced an adverse effect on the ethanol metabolism, thus acetaldehyde remained in the human body, even in human expired air. The biosniffer with high gas selectivity would be an effective and convenient noninvasive approach to evaluate the condition and rate of ethanol metabolism using expired air, even with halitosis.

4.2 Optical Biosniffer for Methyl Mercaptan in Halitosis

An optical biosniffer for methyl mercaptan (MM), one of the major odorous chemicals in halitosis

(a)

(b)

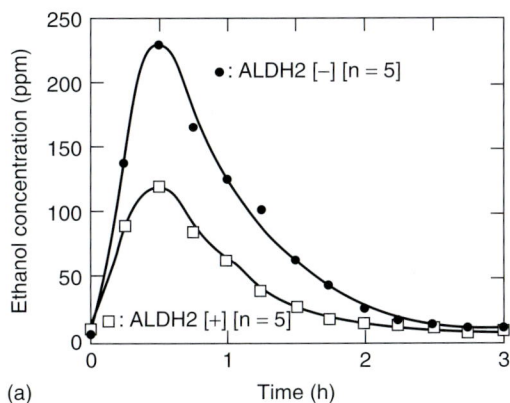

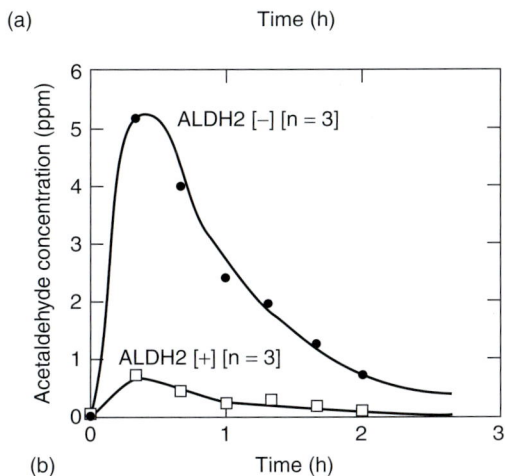

Figure 11. Comparisons of the ethanol and acetaldehyde concentration changes with time in the expired air between ALDH2 [−] (filled circle) and ALDH2 [+] (open square) subjects after drinking. [Reprinted with permission Mitsubayashi et al.[29] copyright 2005, Elsevier.]

(bad breath) was also constructed by immobilizing monoamine oxidase type A (MAO-A) onto a tip of a fiber-optic oxygen sensor (OD: 1.59 mm) with an oxygen-sensitive ruthenium organic complex (excitation: 470 nm, fluorescent: 600 nm). After evaluating the sensor characteristics using a gas flow measurement system with a gas generator, the optical biosniffer was applied to expired gases from healthy male volunteers for halitosis analysis as a physiological application.

4.2.1 Construction of an Optical Biosniffer for MM

Figure 12 illustrates the structure of the optical biosniffer for MM vapor.[32] The sniffer device

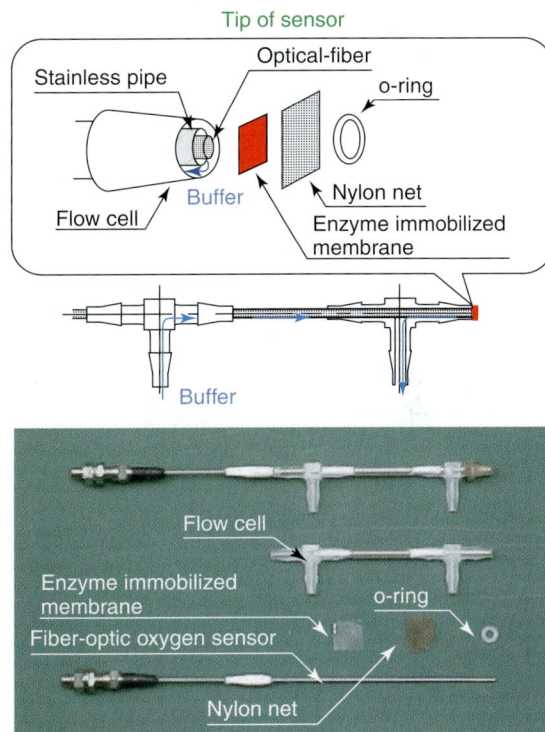

Figure 12. Structure and photograph of an optical biosniffer with a sensor tip cleaning system in the gas phase and the device components. [Reprinted with permission Mitsubayashi et al.[33] copyright 2006, Elsevier.]

consisted of an enzyme membrane, a reaction unit, and an oxygen-sensitive optical fiber with an AOD membrane. The optical fiber was coated by solgel process with ruthenium organic complex, which generates optical quenching to the existence of oxygen molecule in both the liquid and gas phases. MAO-A (EC 1.4.3.4.) was used as the MM recognition material for the halitosis biosniffer.[33]

The performance of the biosniffer for MM vapor was assessed with the batch flow measurement system. The output increased rapidly following application of MM vapor and reached a steady state output within 3 min because of the reduction of optical quenching of the ruthenium organic complex emission induced by the consumption of oxygen (as electron acceptor) which is the result of the MAO-A catalytic reaction with MM.

The calibration curve of the optical biosniffer for MM in the gas phase is illustrated in Figure 13.[34] An arrow sign in this figure indicates a threshold (200 ppb MM) of pathologic halitosis. As the

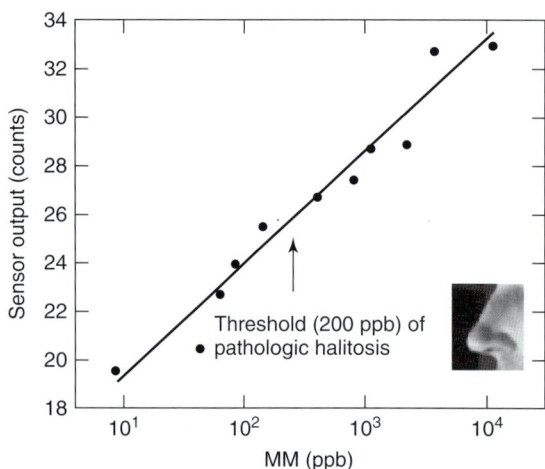

Figure 13. Calibration curve of the optical biosniffer for methyl mercaptan in the gas phase (arrow: 200 ppb MM as threshold of pathologic halitosis). [Reprinted with permission Mitsubayashi et al.[33] copyright 2006, Elsevier.]

figure illustrates, the changes in output of the biosniffer were found to be related to the MM concentrations in the gas phase because gaseous MM that diffuses through the enzyme membrane was oxidized by MAO-A using oxygen, causing a decrease in the concentration of dissolved oxygen. The calibration range of the biosniffer for MM vapor was from 8.7 to 11 500 ppb including an MM threshold (200 ppb) of pathologic halitosis and the human sense of smell level 3.5 (10 ppb). The sensor output deduced from logarithmic regression analysis of the semilog plot by a method of a least squares was according to the following equation:

$$\text{Sniffer output(counts)} =$$
$$14.54 + 4.61 \log[\text{MM(ppb)}] \quad (8)$$
$$(\text{with} 10 \, \text{mmol} \, \text{l}^{-1} \, \text{AsA})$$

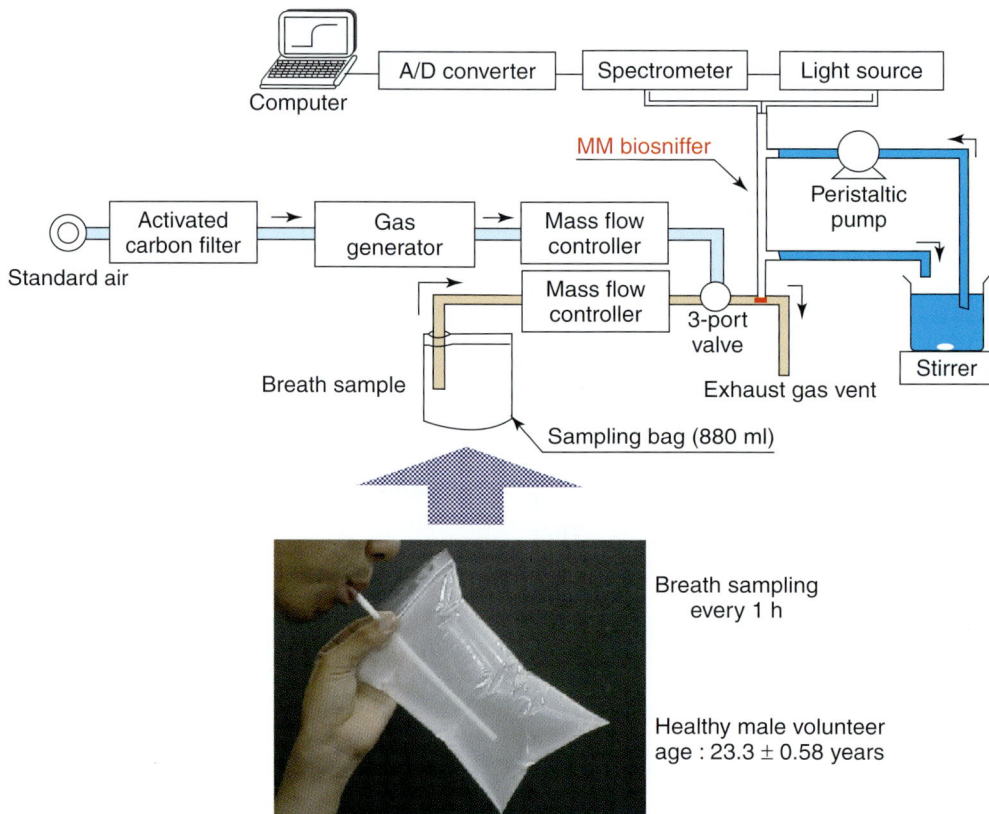

Figure 14. Breath analysis procedure and system by the optical biosniffer with MAO-A for methyl mercaptan in the gas phase. [Reprinted with permission Mitsubayashi et al.[33] copyright 2006, Elsevier.]

4.2.2 Halitosis Monitoring with the Optical Sniffer

For a physiological application, breath samples from healthy male volunteer subjects were applied to the gas measurement system (Figure 14) with the optical biosniffer for monitoring halitosis (expired air). No subjects (23.3 ± 0.58 years) had oral or dental disease. As shown in Figure 14, the expired gas was collected in the sampling bag (800 ml) every 1 h from 7 am to 4 pm in a day, and then used for gas measurement with the same method as described in the preceding text.

As the physiological application, the breath samples from healthy male volunteer subjects were provided to the gas measurement system with the optical biosniffer for monitoring halitosis. Figure 15 shows the change of the average and the standard deviation in the biosniffer output corresponding to breath samples from the subjects during the daytime. As the figure indicates, the high output of biosniffer on awakening decreased significantly after breakfast and increased gradually by noon. The elevated sensor signal declined quickly after lunch and increased toward evening again. The sniffer signal would be found to take a regular fluctuation with low deviations dependent on food intakes. As reported in the

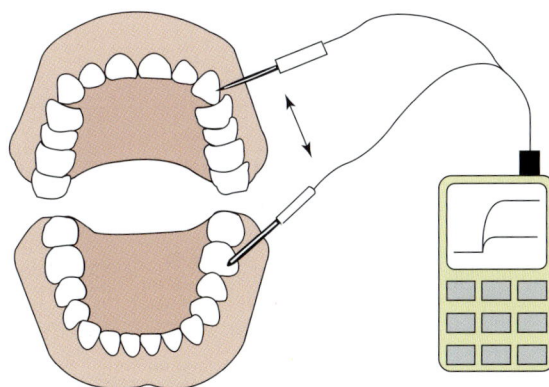

Figure 16. A halitosis local analysis with the fiber-optic sniffer to detect an odor source site. [K. Mitsubayashi, et al., *Analytica Chimica Acta,* 2006, **20**, 573–574, 75–80.]

previous papers, human halitosis changes throughout the day, decreasing after food consumption and increasing over the course of time relative to saliva flow,[35,36] thus resulting in bad breath on awakening because of reduced saliva secretion during sleep. In this study, the sensor signal to breathe air from young male volunteers had not reached the threshold value (approx. 25.2 counts) calculated from the calibration equation using the MM threshold (200 ppb) of pathologic halitosis.

The potential applications of the thin biosniffer include not only the batch analysis of expired air for diagnosing and monitoring disease states in the respiratory and digestive systems, but also a local (spot) analysis to search an odorous source in the mouth cavity (Figure 16), and so on.

5 POTENTIAL APPLICATIONS OF THE BIOSNIFFERS AS ARTIFICIAL OLFACTION

The detection and quantification of gaseous substances, such as odors, toxic and combustible gases, with high sensitivity and selectivity are required in many different areas. The biosniffers have good gas selectivity because of the biological recognition materials such as enzyme catalytic reaction, drug metabolizing system, and enzyme inhabitation mechanism. The biochemical sniffers measure harmful chemicals (such as VOCs, nerve gases, illegal drugs), not only for environmental assessment but also for antiterrorism measures in place of sniffer dogs. In addition, some substances

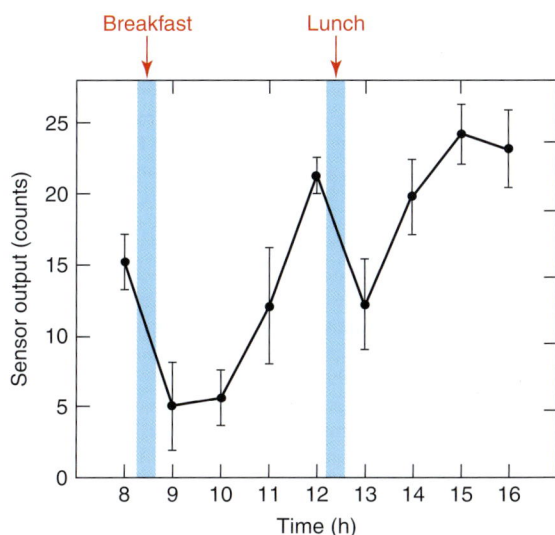

Figure 15. Change in the sniffer output to breath sample from healthy male volunteers during daytime. [Reprinted with permission Mitsubayashi et al.[33] copyright 2006, Elsevier.]

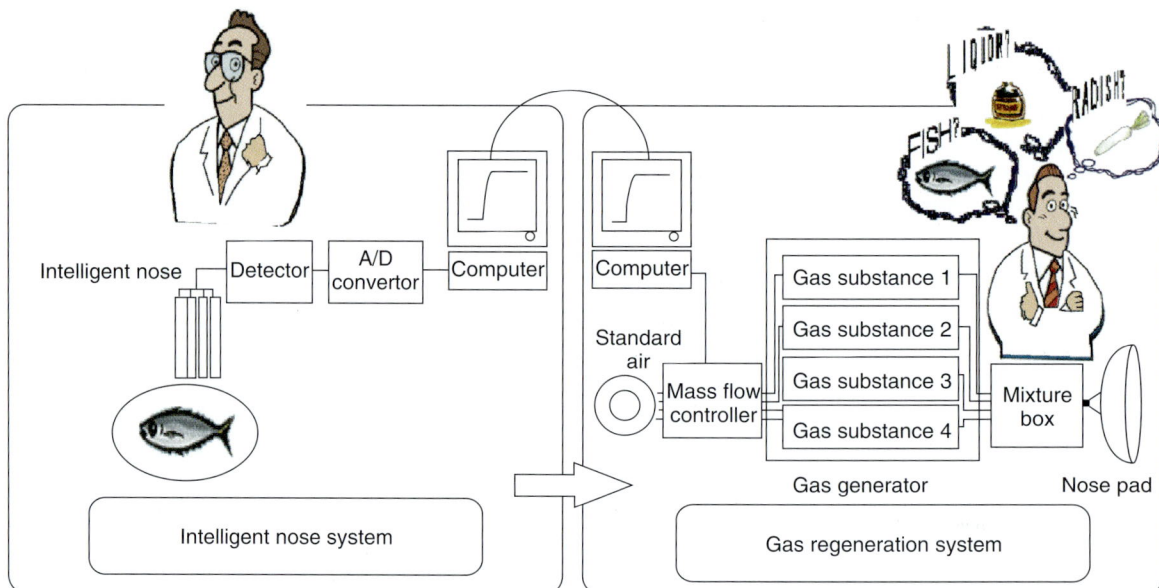

Figure 17. Schematic diagram of a smell communication system with an artificial olfaction.

in the gas phase can be related to human health and behavior. The expiratory gas and human odor analysis by the biosniffer would provide a noninvasive, convenient, and safe method of diagnosing and monitoring disease states, and also a detection of disaster victims.

An artificial olfaction was developed with the integrated array system of the fiber-optic biosniffers with biological materials for analyzing and monitoring various kinds of chemicals in the gas phase. A novel chemical code system with nonodorous vapors and a chemical biometrics using human odor pattern have also constructed the selective sniffers because they can recognize and identify gaseous chemicals as digital information in the computer. The intelligent olfaction device would be used for the development of a smell communication system (Figure 17) by integrating with a gas generator, thus improving the quality of chemical information in the gas phase (smell informatics).

REFERENCES

1. M. Tonoike, Current status and future view of development for odorant sensors. *Bulletin of the Electrotechnical Laboratory*, 1988, **52**(5), 63–79.

2. J. G. Kostelc, G. Preti, P. R. Zelson, L. Brauner, and P. Baehmi, Oral odors in early experimental gingivitis. *Journal of Periodontal Research*, 1984, **19**, 303–312.

3. M. Watanabe, Unten-tekisei (Driver's Attitude for Safety Driving. J. *Japan Association of Traffic Medicine & Engineering*, 1992, 39–46.

4. L. Thomas (ed), *Alcohol and Nutrition: The Energy Value of Alcohol in the Diets of Alcoholics*, Pamphlet of Monell Chemical Senses Center, 1986.

5. Y. Hiranaka, T. Abe, and H. Murata, Gas-dependent response in the temperature transient of SnO_2 gas sensors. *Sensors and Actuators, B*, 1992, **9**, 177–182.

6. T. Maekawa, N. Miura, and N. Yamazoe, Development of SnO_2-based ethanol gas sensor. *Sensors and Actuators, B*, 1992, **9**, 63–69.

7. Z. Chen and K. Colbow, MgO-doped Cr_2O_3: solubility limit and the effect of doping on the resistivity and ethanol sensitivity. *Sensors and Actuators, B*, 1992, **9**, 49–53.

8. X. Wang, S. Yee, and P. Carey, An integrated array of multiple thin-film metal oxide sensors for quantification of individual components in organic vapor mixtures. *Sensors and Actuators, B*, 1993, **13–14**, 458–461.

9. I. Karube, H. Matsuoka, S. Suzuki, E. Watanabe, and K. Toyama, Determination of fish freshness with an enzyme sensor system. *Journal of Agricultural and Food Chemistry*, 1984, **32**, 315.

10. A. W. Jones, Variability of the blood: breath ratio in vivo. *Journal of Studies on Alcohol*, 1978, **39**, 1931–1939.

11. S. Sato, K. Kitagawa, in *The Portable Alcohol Testerin*, *Traffic Safety Series*, Japan Association of Traffic Medicine & Engineering, © MIZUHO printing, Nagoya (Japan), 1993, pp. 26–30.

12. Japan Environmental & Sanitary Center, *A Report of Chemical Malodor Analysis*, *(A Study for The Japan Environmental Agency)*, 1980, pp. 248–250.

13. K. Mitsubayashi, K. Yokoyama, T. Takeuchi, and I. Karube, Gas-phase biosensor for ethanol. *Analytical Chemistry*, 1994, **66**, 3297–3302.

14. T. Yoshida, *Revision and Trend Measures of Preventive a Bad Smell Code*, N.T.S Co., Ltd, Tokyo, 1996, p. 16.

15. Special Pollution Section, *Conservation of the Atmosphere Bureau, Environmental Agency, Preventive a bad Smell Code*, Gyosei Co., Ltd, 1993, p. 18.

16. K. Mitsubayashi, H. Amagai, H. Watanabe, and Y. Nakayama, Bioelectronic sniffer with a diaphragm flow-cell for acetaldehyde vapor. *Sensors and Actuators, B*, 2003, **95**, 303–308.

17. T. Noguer and J. L. Marty, High sensitive bienzymic sensor for the detection of dithiocarbamate fungicides. *Analytica Chimica Acta*, 1997, **347**, 63–70.

18. T. Noguer and J. L. Marty, An amperometric bienzyme electrode for acetaldehyde detection. *Enzyme and Microbial Technology*, 1995, **17**, 453–459.

19. *International Chemical Safety Cards (ICSCs)*, The International Program on Chemical Safety & the Commission of the European Communities, 1998, pp. ICSC # 0206. Prepared in the context of cooperation between the International Programme on Chemical Safety & the Commission of the European Communities © IPCS CEC 1993. http://www.ilo.org/public/english/protection/safework/cis /products/icsc/dtasht/_icsc02/icsc0206.htm.

20. A. Q. Zhang, S. C. Mitchell, R. Ayesh, and R. L. Smith, Determination of trimethylamine and relate aliphatic amines in human urine by head-space gas chromatography. *Journal of Chromatography*, 1992, **584**, 141–145.

21. M. Al-Waiz, S. C. Mitchell, J. R. Idle, and R. L. Smith, The metabolism of 14C-labelled trimethylamine and its N-oxide in man. *Xenobiotica*, 1987, **17**, 551–558.

22. D. M. Ziegler, Recent studies on the structure and function of. multisubstrate flavin-containing monooxygenases. *Annual Review of Pharmacology and Toxicology*, 1993, **33**, 179–199.

23. S. C. Mitchell, The fish-odor syndrome. *Perspectives in Biology and Medicine*, 1996, **39**, 514–526.

24. D. H. Lang, C. K. Yeung, R. M. Peter, C. Ibarra, R. Gasser, K. Itagaki, R. M. Philpot, and A. E. Rettie, Isoform specificity of trimethylamine N-oxygenation by human flavin containing monooxygenase (FMO) and P450 enzymes - Selective catalysis by FMO3. *Biochemical Pharmacology*, 1998, **56**, 1005–1012.

25. C. T. Dolphin, A. Janmohamed, R. L. Smith, E. A. Shepard, and I. R. Philips, Missense mutation in flavin-containing mono-oxygenase 3 gene, FMO3, underlies fish-odour syndrome. *Nature Genetics*, 1997, **17**, 491–494.

26. K. Mitsubayashi and Y. Hashimoto, Bioelectronic sniffer device for trimethylamine vapor using flavin containing monooxygenase, *IEEE Sensors Journal*, 2002, **2**(3), 133–139.

27. Y. Hasebe, K. Oshima, O. Takise, and S. Uchiyama, Chemically amplified kojic acid responses of tyrosinase-based biosensor, based on inhibitory effect to substrate recycling driven by tyrosinase and L-ascorbic acid. *Talanta*, 1995, **42**, 2079–2085.

28. T. Maekawa, J. Tamaki, N. Miura, and N. Yamazoe, Development of SnO_2-based ethanol gas sensor. *Sensors and Actuators, B*, 1992, **9**, 63–69.

29. K. Mitsubayashi, H. Matsunaga, G. Nishio, S. Toda, and Y. Nakanishi, Bioelectronic sniffers for ethanol and acetaldehyde in breath air after drinking. *Biosensors and Bioelectronics*, 2005, **20**, 1573–1579.

30. J. K. Park, H. J. Yee, K. S. Lee, W. Y. Lee, M. C. Shin, T. H. Kim, and S. R. Kim, Determination of breath alcohol using a differential-type amperometric biosensor based on alcohol dehydrogenase. *Analytica Chimica Acta*, 1999, **390**, 83–91.

31. N. Enomoto, S. Takase, N. Takada, A. Takada, A. Alcoholic liver disease in heterozygotes of mutant and normal aldehyde dehydrogenase-2 genes. *Hepatology*, 1991, **13**(6), 1071–1075.

32. K. Mitsubayashi, T. Kon, and Y. Hashimoto, Optical bio-sniffer for ethanol vapor using an oxygen-sensitive optical fiber. *Biosensors and Bioelectronics*, 2003, **19**, 193–198.

33. K. Mitsubayashi, T. Minamide, K. Otsuka, H. Kudo, and H. Saito, Optical bio-sniffer for methyl mercaptan in halitosis. *Analytica Chimica Acta*, 2006, **573–574**, 75–80.

34. T. Minamide, K. Mitsubayashi, N. Jaffrezic-Renault, K. Hibi, H. Endo, and H. Saito, Bioelectronic detector with monoamine oxidase for halitosis monitoring. *The Analyst*, 2005, **130**, 1490–1494.

35. E. L. Attia and K. G. Marshall, Halitosis. *Canadian Medical Association Journal*, 1982, **126**, 1281–1285.

36. M. Rosenberg, Clinical assessment of bad breath: Current concepts. *The Journal of the American Dental Association*, 1996, **127**, 475–481.

Part Seven

Data Analysis, Conditioning, and Presentation

Data Analysis, Conditioning, and Presentation

David C. Cullen

Cranfield Health, Cranfield University, Silsoe, UK

Within the range of disciplines and components that go together to make a biosensor device, the relative effort given to handling of the basic data output of a device is often small. This is especially true of the majority of reports in the research literature. Often, little more than a simple calibration graph with some error bars is given. This situation obviously overlooks the key role that data analysis, conditioning, and presentation play in real world applications of biosensor technology. For a commercial device, these topics are crucial for a successful device. This is true at all stages of the life of a product, from the R&D phase, through regulatory approval, quality assurance and control, and of course for a final analytical measurement.

The following section covers three key areas concerning data handling, plus an emerging approach. Firstly, with glucose sensors dominating the commercial biosensor sector, this represents a mature field concerning data handling and offers a good example of the current state of the art. Given the apparent simplicity of the basic electrochemical glucose sensing approach, the details and amount of effort and resources devoted to the data handling aspects may come as a surprise to some readers (see **Design of Data Algorithms for Blood Glucose Biosensors** by John Rippeth and Wah Ho). Secondly, the biochip and micro-array community has from its earliest embodiment had to deal with significant levels of raw data and to consider how this can be converted into easily interpretable and useful information. This is encompassed in the ballooning discipline of bioinformatics that focuses on data analysis, conditioning and presentation. Consideration of this area with a focus on micro-array techniques will hopefully offer insight into approaches that may be transferred to other array-based biosensor devices that are being considered for applications within point-of-care, environmental and other non-laboratory locations (see **Microarray Analysis Software and its Applications** by Conrad Bessant). Thirdly, and more generally, consideration of the quality of data arising from analytical analyses and the relationship of these to the diagnostic value of measurements, for example, taking into account the end application of the information, will be presented (see **Data Validation and Interpretation**, by Ursula Spichiger-Keller). Lastly, Bayesian theory is a topic that is being applied to an ever-increasing range of applications. Biosensor development and data processing is no different and it is therefore time to consider how this topic can benefit the biosensor community (see **Introduction to Bayesian Methods for Biosensor Design** by Edmund Jackson and William Fitzgerald).

Handbook of Biosensors and Biochips. Edited by Robert S. Marks, David C. Cullen, Isao Karube, Christopher R. Lowe and Howard H. Weetall.
© 2007 John Wiley & Sons, Ltd. ISBN 978-0-470-01905-4.

Design of Data Algorithms for Blood Glucose Biosensors

John J. Rippeth and Wah O. Ho

Research and Development, Hypoguard Ltd., Woodbridge, UK

1 INTRODUCTION

Blood glucose biosensor test kits (blood glucose meters, BGMs), are primarily used by diabetics at home or in the workplace. Users are usually people without formal training in clinical diagnostic testing with diverse educational backgrounds and an age range from the young to the elderly. This dictates to the manufacturers that glucose testing meters and strips must be as simple and intuitive as possible with safeguards to minimize the risk of erroneous results. Diabetes is a disease in which the body does not produce or properly use insulin. Insulin is a hormone produced by the pancreas in response to an increase in circulating glucose concentration, for example, after a meal or a sugary drink, that signals cells to absorb glucose and convert this to energy. Diabetics can be classified into two types. Broadly, around 10% of diabetics are Type 1, who are insulin dependent and rely on injections of insulin to control their blood glucose concentrations to within acceptable limits. It is generally recommended that they test at least four times a day. The remaining 90% of diabetics are Type 2, who are noninsulin-dependent and are usually able to manage their blood glucose concentrations through diet, drugs, and exercise. These people usually test less frequently. Three good sources of information with regard to diabetes and diabetes care can be found on the websites from the American Diabetes Association, Diabetes UK, and the World Health Organization.[1–3] Accuracy and precision in glucose measurement is critical in blood glucose biosensor development, as an inaccurate measurement can lead to contradictory treatment to that actually required. The biosensor and meter system must be designed so that the user is unable to affect the result, for example, by placing an insufficient amount of blood onto the sensor strip, or by attempting to test outside the specified operating temperature range. If the BGM is misused, the meter should detect this error state and notify the user with a warning or error.

2 MARKET AND REGULATORY REQUIREMENTS

Various regulatory organizations and market requirements dictate the performance safety limits within which BGMs are expected to perform. These naturally form part of any BGM design input. Documents such as ISO 15197:2003 detail the minimum requirements of BGMs in terms of precision and accuracy.[4] Repeatability or precision is evaluated by making 100 replicate measurements at 5 glucose concentrations over specified intervals within the range of 30–400 mg dl^{-1} (1.7–22.2 mmol l^{-1}) (Table 1). Generally, the precision is evaluated by calculating

Handbook of Biosensors and Biochips. Edited by Robert S. Marks, David C. Cullen, Isao Karube, Christopher R. Lowe and Howard H. Weetall.

Table 1. Glucose concentration intervals for precision (or repeatability) evaluation

Interval	Glucose concentration	
	(mmol l^{-1})	(mg dl^{-1})
1	1.7–2.8	30–50
2	2.9–6.1	51–110
3	6.2–8.3	111–150
4	8.4–13.9	151–250
5	14.0–22.2	251–400

[Reproduced by permission of the European Committee for Standardization.]

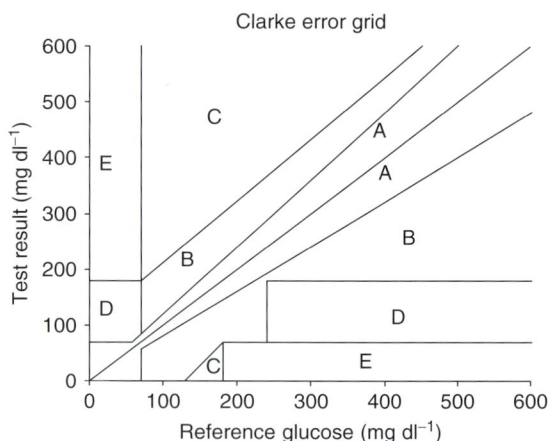

Figure 1. Clarke error grid. (A) <20% bias, clinically accurate result, correct action or inaction; (B) >20% bias but would lead to benign or no treatment; (C) leads to unnecessary corrective treatment; (D) potentially dangerous failure to detect and treat hyper- or hypoglycemia; (E) erroneous treatment zone, corresponding treatments would be opposite to that required to correct hypo- or hyperglycemia.

the average, standard deviation (sd), and the coefficient of variation (CV) (by dividing the sd by the average and multiplying by 100) for each glucose concentration. For concentrations greater than 75 mg dl^{-1}, a CV value is used which should be less than 5%, and for concentrations less than 75 mg dl^{-1}, an SD value is used which should be less than 9 mg dl^{-1}.

Accuracy is evaluated by obtaining results from at least 100 donors over a stratified glucose range from <50 to >400 mg dl^{-1} (<2.8 to >22.2 mmol l^{-1}) (Table 2). Acceptable accuracy is defined as having 95% of results falling within 0.83 mmol l^{-1} (15 mg dl^{-1}) of the reference glucose value for glucose concentrations <4.2 mmol l^{-1} (<75 mg dl^{-1}) and within ±20% of the reference glucose value for glucose concentrations ≥ 4.2 mmol l^{-1} (>75 mg dl^{-1}). Reference glucose concentration values are typically obtained from benchtop analyzers such as the YSI 2300 STAT Plus™ from Yellow Springs Instruments (YSI) Inc.[5]

One commonly used method to assess the accuracy of a system or to compare different systems is to use a Clarke error grid plot, which compares the glucose concentration result obtained from a reference analyzer to the result obtained from a meter and strip test system from the same sample.[6] The grid clearly illustrates the consequence of an inaccurate reading and how it can lead to inappropriate treatment (Figure 1). Results falling within zone A are considered clinically accurate, that is, it would lead the user either to a course of action to correct a hypo- or hyperglycemic state or to one of inaction if the glucose result is within the normal range.

One of the key areas on the Clarke error grid is at low glucose levels (<70 mg dl^{-1}), where a positive bias in meter result can lead to a result in zone D, or worse, in E, which implies a hyperglycemic treatment (e.g., an insulin injection) for a hypoglycemic subject resulting in gross hypoglycemia leading to coma and potentially death.

3 DRIVING FORCES FOR BGM DEVELOPMENT

The three driving forces for the development of a BGM system are as follows: (i) An ability to

Table 2. Glucose concentrations of samples for system accuracy evaluation and the ideal distribution of the total number of samples at each glucose interval

Interval	Percentage of samples (%)	Glucose concentration	
		(mmol l^{-1})	(mg dl^{-1})
1	5	<2.8	<50
2	15	2.8–4.3	50–80
3	20	4.4–6.7	80–120
4	30	6.7–11.1	120–200
5	15	11.2–16.6	201–300
6	10	16.7–22.2	301–400
7	5	>22.2	>400

[Reproduced by permission of the European Committee for Standardization.]

fit the lifestyle of the user. These systems will not be used under laboratory conditions of controlled temperature and humidity; neither will the actual blood testing be conducted in a controlled laboratory fashion with any sample preparation or use a specialized sample delivery. The test will usually be performed by placing a drop of blood acquired from a simple fingerprick onto the strip. (ii) Any new system must be competitive in performance and specifications and offer advantages over existing systems in the market or anticipated for market. (iii) Cost, which is continually falling due to competition with more products entering the market from low-cost geographical regions, and increased simplicity of meter and strip design.

The above requirements have led to challenges in system development as the market demands continuous performance improvements and additional features, such as glucose trend analysis, smaller blood volumes, shorter time to display result, all at continual lower costs.

Advanced strip design and inexpensive bulk manufacturing techniques, such as screenprinting, have reduced costs and improved the quality of BGMs. Improved algorithm design has enabled these biosensors to be interrogated in a manner to meet new marketing specifications. The following is a list of the minimum expected requirements for a new BGM:

1. sample volume $<1\,\mu l$, preferably $<500\,nl$;
2. read time of $<10\,s$, preferably $<5\,s$;
3. $20\text{--}600\,mg\,dl^{-1}$ $(0.5\text{--}33.3\,mmol\,l^{-1})$ glucose range (hypoglycemia is recognized $<40\,mg\,dl^{-1}$);
4. $30\text{--}55\%$ hematocrit, preferably $20\text{--}70\%$ hematocrit range, to have $<10\%$ deviation in final result;
5. operating temperature $15\text{--}35\,°C$, preferably $10\text{--}40\,°C$;
6. precision: CV $<5\%$ in the glucose range $75\text{--}600\,mg\,dl^{-1}$ $(4.2\text{--}33.3\,mmol\,l^{-1})$;
7. accuracy: 95% of results within 20% of a reference glucose value;
8. adequate blood fill detection (insufficient volume of a sample would give a low estimate of the glucose level);
9. blood misdose detection (e.g., blood placed on wrong part of strip, adding a second drop during measurement of the first drop);

10. download capability (PC software to analyze glucose trends to assess how well glucose concentration has been controlled);
11. simple to use (single button, calibration free);
12. large display (diabetics may have poor vision);
13. robust (withstand transport in a pocket or bag);
14. discrete (small, easy to carry).

4 BLOOD GLUCOSE METER – SOFTWARE REQUIREMENTS

The algorithms used in biosensor meters can be considered to be split into two functions, the first is the actual control of the meter and how it arrives at the final result and the second is the user interface software. The former should ensure that the electrochemical measurement taken arrives at an accurate final blood glucose value rapidly, using the meter calibration algorithms and any error detection; the latter should be easy and intuitive for a user to make a measurement.

The algorithms performed by a particular BGM are very much dependent on the intrinsic performance of the biosensor itself and would be optimized as such. The simplest and most widespread measurement for an amperometric BGM is a single endpoint current measurement, which is converted to a glucose value through an algorithm. Arriving at a correct final glucose value requires BGM algorithms to have the necessary fail-safe mechanisms for simple and reliable use. The flowchart (Figure 2) outlines how such algorithms may operate.

5 INITIAL METER AND STRIP CHECKS

The first part of the meter algorithm is a check of the meter hardware to ensure that all the components are working correctly or are the correct components (i.e., the calibration chip). Meters and calibration chips are often assigned an original equipment manufacturer (OEM) number, which needs to match and is usually done for commercial reasons to separate different geographical markets (which may have different pricing structures) or if a meter system is sold by more than one branded distributor. It can also be used as method of controlling the introduction of an upgraded meter or biosensor strip.

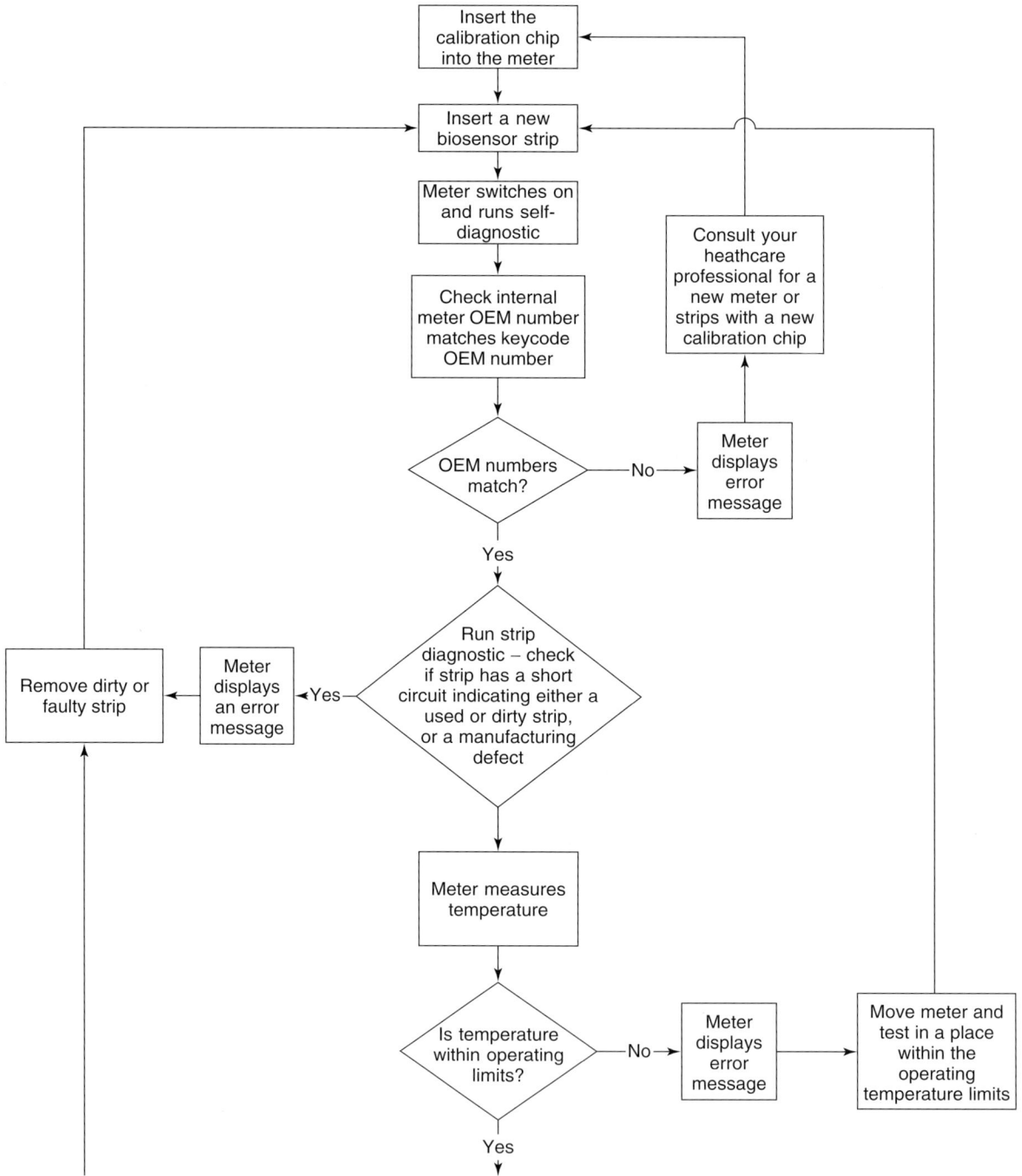

Figure 2. Flowchart illustrating the operational steps taken by users and meter algorithms in obtaining a glucose reading from a typical blood glucose monitor.

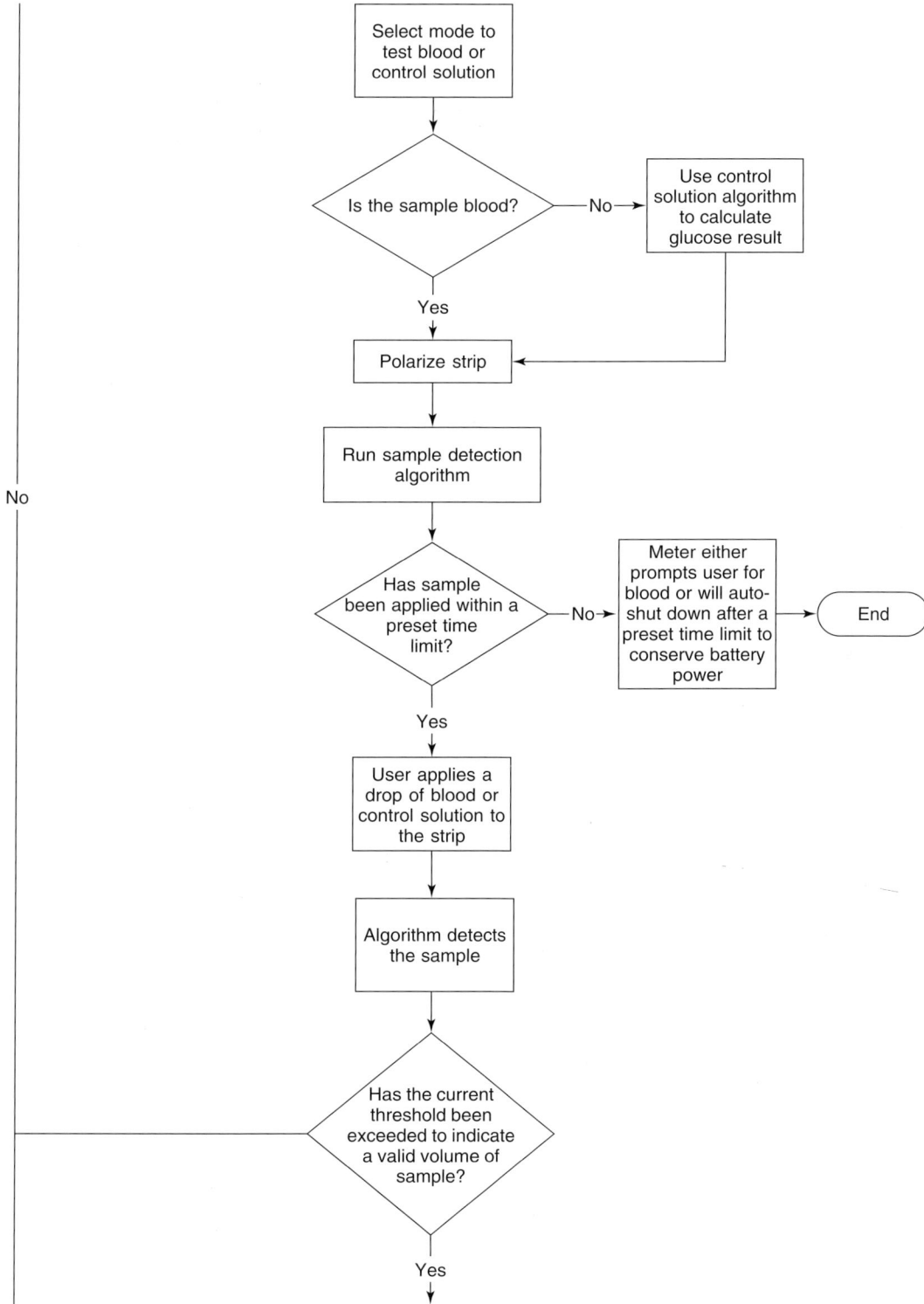

Figure 2. (*continued*).

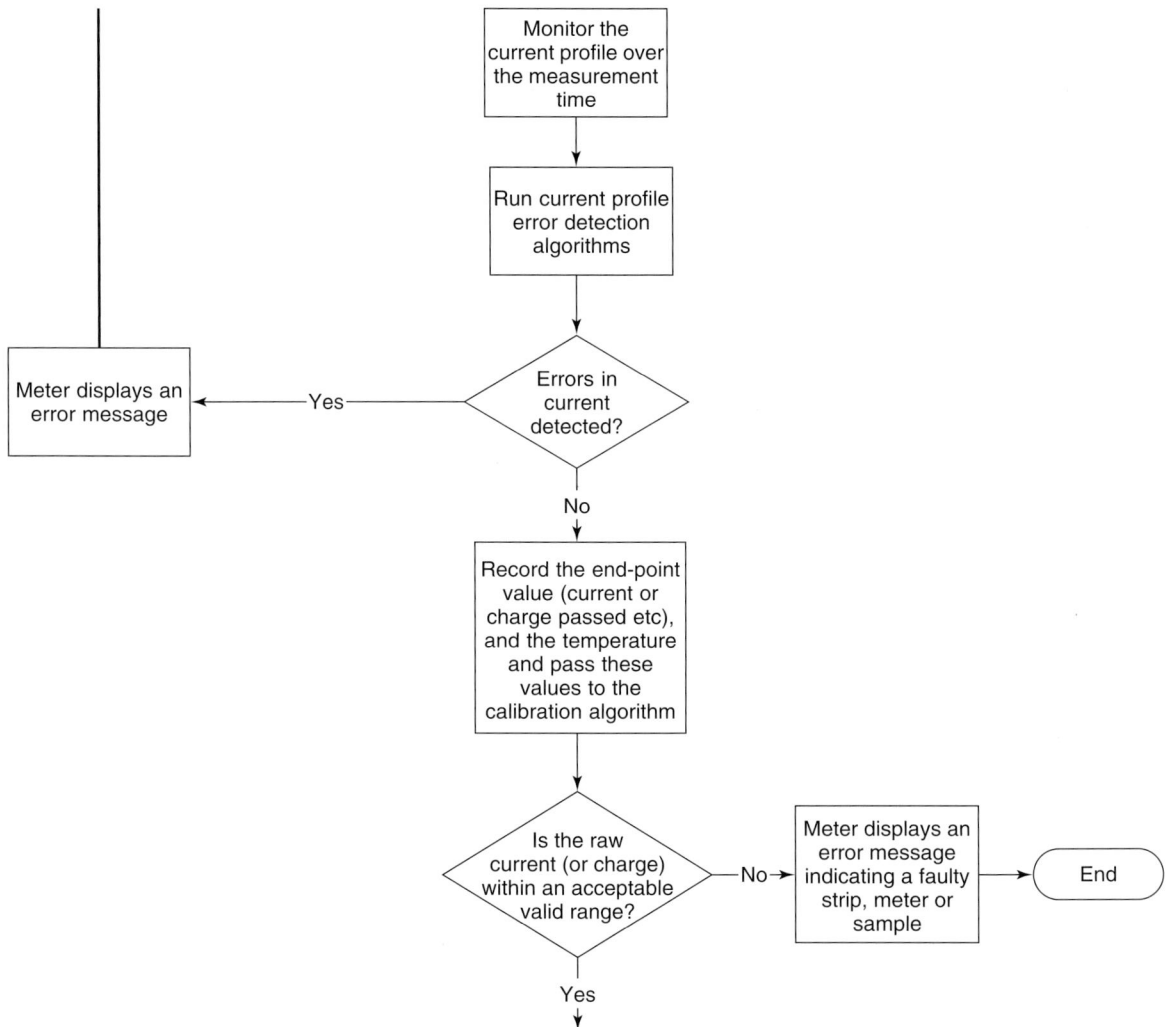

Figure 2. (*continued*).

The second part of the algorithm is a check on the biosensor and other conditions prior to the test being performed. As BGM biosensors are for single use, the meter must be able to determine if a strip has been used or not. Often this is done by simply checking if there is a leakage current between the working and reference electrodes and if this exceeds a set value, an error is indicated; this method can also detect manufacturing defects, for example, short circuiting between electrodes.

The next stage is the biosensor measurement. Initially the strip is polarized at the required potential and the current is continuously sampled.

When a sample such as blood enters the strip, the meter detects the current change, and if this exceeds a predetermined threshold, the meter starts to monitor the current being produced over the time of measurement. The initial sampling rate is fast which ensures that the reading is started at an accurate time point but can then be lowered to allow the data to be processed by inexpensive electronic components.

During the time between the introduction of a sample and the final displayed meter result, a whole host of processing techniques can be employed. The simplest technique is a single

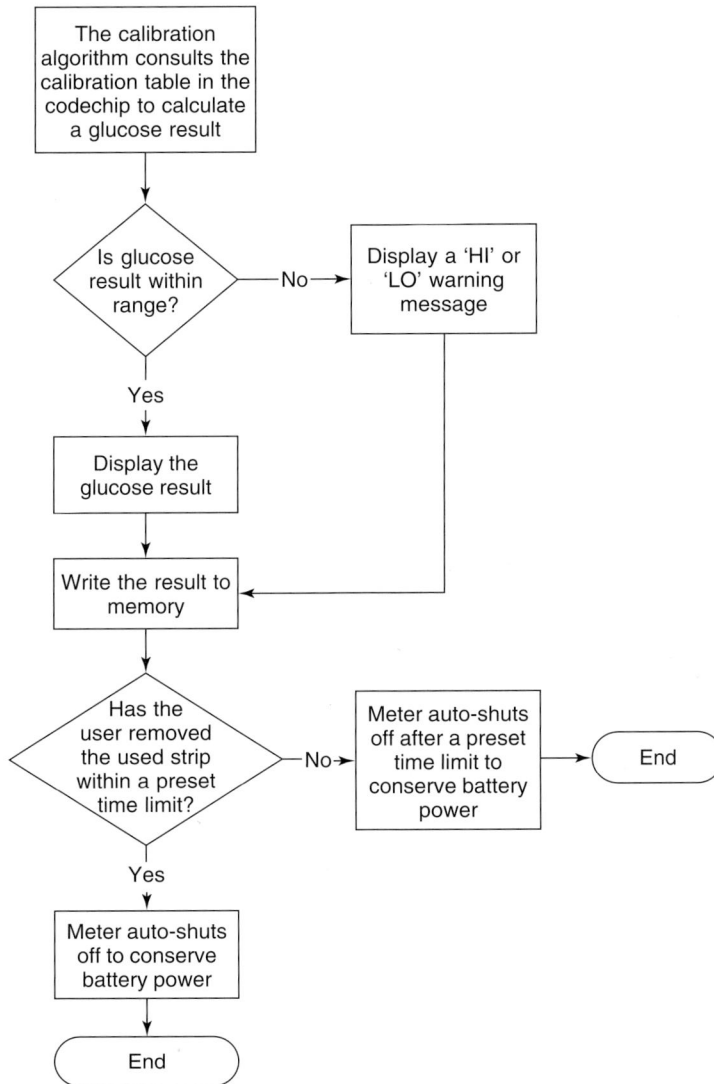

Figure 2. (*continued*).

record of the current at the end of the measurement time. This single current value is compared to calibration information and converted by the algorithm to a glucose value.

More sophisticated processes can switch the meter to an inactive potential during a pre-incubation step which allows reagents to dissolve and react with glucose in the sample to produce a buildup of electroactive species such as hydrogen peroxide or a reduced form of a mediator.[7] At the end of the incubation period, the meter changes the polarizing potential to an active value which causes oxidation of the electroactive species and produces a current proportional to the glucose concentration. An advantage with this technique is that given enough reagent, all or most of the glucose can be converted to an electroactive product during the incubation period. The final analytical process is much simplified because there is no need to consider the combined reaction kinetics of both the biochemical reaction between glucose and reagents (i.e., enzymes, mediators) and the electrochemical reaction of either hydrogen peroxide or the reduced mediator at the electrode. The final current is thus read under a constant concentration of electroactive analyte, which leads to a more precise determination of the current and thus a more precise final glucose result.

6 ERROR DETECTION

Once the current profile of a biosensor has been characterized, then detection of errors in the current measurement profile can be identified and addressed by procedures in the algorithm. Amperometric biosensors are in general expected to follow Cottrell-type current decay behavior, and one strategy for error detection can be to determine deviation from this and thus detect abnormalities.[8] This requires the meter to monitor current throughout the measurement period and test for such deviations, which could occur in real time or at the end of measurement. Processing the passing current in real time would mean the meter could flag an error immediately when it is detected, otherwise the meter continues until the end before flagging an error. The choice would be determined by memory capacity and commercial decisions. The first method requires no great memory storage and thus has the potential to be cheaper to implement and probably slightly quicker to display a result, good or otherwise.

This current profile analysis must be sensitive enough to detect abnormalities but must not be so sensitive to be triggered by any acceptable background noise or by signal deviations that do not significantly affect the final result. There would be an extremely limited amount of tolerance in the market for a BGM that continuously gave error messages and quickly used up expensive pots of strips.

Any error detection must cover current response profiles over the entire glucose response range as well as consider the range of operating temperatures. Therefore, any threshold settings must be validated for the algorithm to operate correctly within these limits.

Figure 3 is an example of a typical current response from a BGM with the current measured at various points that could be used as the basis of detecting abnormal current responses. If the current (i_0) differs positively or negatively from its immediate predecessor (i_{-1}) by more than a

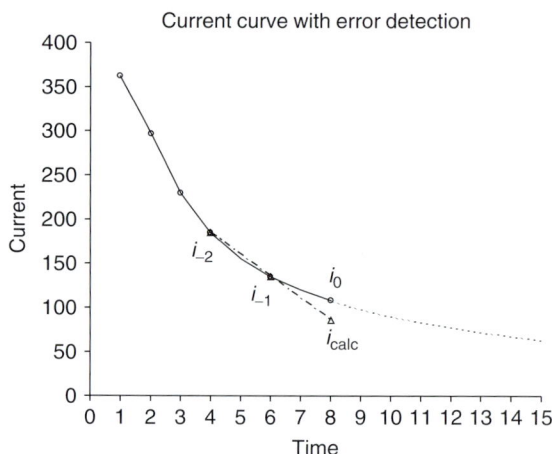

Figure 3. A typical current profile from an amperometric biosensor strip with error detection. The measured current (i_0) is compared to a calculated current (i_{calc}) which is the extrapolated value from the previous two current values (i_{-1} and i_{-2}). If the difference between i_0 and i_{calc} is within some preset limit (e.g., 5%), then no error is generated. This method is useful in detecting sudden current spikes where the current curve departs sharply beyond the preset limit from the expected continued current path (shown as the dashed line continuing from the solid line). The dashed line represents a prediction of current values over time based on a Cottrell current decay. Significant departures from the model could indicate an error in the glucose measurement.

preset value or percentage, then an error is flagged, otherwise the algorithm continues to compare the next measured current to the previous current until the end of the measurement time. This is useful for situations where the current profile grossly deviates over a relatively long period from its expected rate of change of current profile, which could lead to a gross under- or overestimation of the glucose concentration.

Another strategy, which could run in parallel with the first error detection above, would be for a second algorithm to compare the measured current (i_0) to a value of i_{calc} that has been calculated from a linear extrapolation between i_{-1} and i_{-2}. If the values of i_0 and i_{calc} differ by more than a predetermined limit, then the measurement can be stopped and an error flagged. This is useful for situations where there is a current spike that could cause the current profile to read an unexpectedly high current, leading to a false high-glucose result.

A more sophisticated treatment of the current profile error analysis could use the first few seconds of the current to predict the remainder of the current profile based on the Cottrell model (Figure 3, dashed line). Significant deviations from the model could indicate an error.

7 BLOOD GLUCOSE DETERMINATION

The majority of BGM biosensors are sold in pots or packets of 25–50 strips, which are packaged with an electronic calibration chip or strip that contains calibration information for that specific batch. This is in contrast to older systems where a meter would have a store of predetermined calibration datasets and the user has to enter into the meter a code read off the strips pot. Failure to do so could lead to erroneous results and thus to inappropriate treatment or inaction. This reliance on the user to enter the correct batch specific code was removed when calibration chips were introduced. It was expected that user compliance to change calibration information with a new pot of strips would increase due to the simpler operation of removing an old calibration chip and inserting the new one.

This calibration chip is inserted into the BGM and the information is either copied into the meter memory when the meter is turned on or read directly off the chip during meter operation.[9] The business model of BGM systems is to create a return on the investment made on strip and meter development and costs of promotion. Fierce competition keeps prices low, and manufacturers and distributors commonly resort to providing users and institutions (e.g., diabetic clinics, hospitals, long-term care homes) with the initial meter and strip package at below manufacturing cost or for free. The return in investment is made by the sale of further strips although competition keeps profit margins low; therefore several packs of strips need to be sold in order to recoup the cost of each meter placed. This means that a meter life cycle is usually at least 3 years during which time controlled changes to the associated test strip may be made. This may be the use of different materials for cost reduction or performance improvement.

The advantage of using a swappable chip to provide the calibration information to the meter means that any changes to the performance of the strip, either through product enhancement or variations within manufacturing tolerances, can immediately be accounted for to enable the same meter to deliver a consistently precise and accurate result. This flexibility can be further extended to using strips for analytes other than glucose in the same meter as long as the correct calibration chip is inserted into the meter, for example, the Abbott Medisense® Optium Xceed™ system which also measures blood ketones (ß-hydroxybutyric acid).

Some established biosensor products now have such stable manufacturing processes that the same calibration chip can be used for each new batch of strips. The obvious commercial benefit of this is that such meters and strips are advertised as "calibration free" systems, for example, the Abbott Medisense ExacTech® RSG™ system. Users insert a strip into the meter and immediately test; there are no requirements for plugging in separate calibration chips or entering a calibration code, and thus such a system is very simple to use. This is an older biosensor system that has been relaunched as a new meter system that has the calibration built into the meter; such systems are usually sold in low-cost, third-world markets as they often do not offer the performance benefits that the Western market demands.

8 CALCULATION OF THE FINAL GLUCOSE VALUE

This subject is one of the most complex and critical parts of the algorithm where a raw signal such as current with another key factor such as temperature is passed through a calculation process to arrive at an accurate and precise estimate of the concentration of glucose in a blood or control solution sample as judged by the Clarke error grid (Figure 1). Essentially the relationship between glucose concentration, temperature, and current is a three-dimensional surface in which the dependent variable is current. Potentially, the equation used to describe this can be quite complex and can have at least four parameters. The process of developing the algorithm to calculate the glucose concentration may be quite long and complex. However much of the complexity and effort of deriving the descriptive equation for this three-dimensional relationship can be reduced and simplified to a two-dimensional calibration table of glucose concentration, temperature, and current.

Usually, the data in the calibration table is created by a running series of calibration curves for blood or control solution over a wide range of glucose concentrations and temperatures, which is discussed in detail below (control solution is a glucose solution(s) that is used to check if the meter and strip system is working correctly as a quality control check). A calibration table is then independent of any complex calibration calculation, and allows a large amount of flexibility in the dimensions of calibration as this can all be done prior to writing a table as well, ensuring a large degree of future proofing against any changes that may occur.

9 BIOSENSOR CALIBRATION

Glucose determination is done on the basis of biosensor calibration, and it should dictate correctly to the user either a course of action (i.e., inject insulin or ingest glucose) or inaction (glucose concentration acceptable, no adjustment necessary). An incorrect glucose value could lead the user to inappropriate action as exemplified by the Clarke error grid (Figure 1).

Biosensor strip manufacture is primarily a batch process. The material of the strip will most likely be supplied on a roll or precut sheets and made in batches, inks are made in batches, and reagents such as enzymes and stabilizers are supplied from batches. Traditionally, the most common, cheapest, and fastest method of manufacturing single-use disposable test strips is the screenprinting of one or more inks onto a polymer substrate. This process has a start and an end as inks are loaded, printed, dried, and repeated until the strip has been built up, and the number of sheets to be printed has been achieved. This may be followed by the application of the biochemical reagents, typically glucose oxidase or glucose dehydrogenase, with a mediator such as potassium ferricyanide ferrocene or hexaammineruthenium (III) chloride. Common application methods include dropcoating, dipcoating, screenprinting, and ink-jetting, depending on the viscosity of the reagent and quantity.

Through this process, batch-to-batch variations of strips arise, since the conditions, or batches of material, printing temperature and humidity, print thickness, and print quality may vary. This could give rise to slightly different responses in current to the same glucose concentration, hence the need to characterize each batch through a calibration process. This process would also serve to identify where the process has failed, that is, a quality control process in which the product would show an unacceptable calibration (e.g., currents too low, not linear enough, etc.).

Calibration usually involves running a series of venous blood samples over a range of glucose concentrations slightly beyond the glucose measurement range claimed by the product, typically 10–900 mg dl^{-1}. The type of hardware used to operate the strip and record the data is worthy of mention.

Commercial or bespoken electrochemical workstations may be used for the initial development of any strip, as these have the advantage of being flexible enough to carry out a wide variety of electroanalytical procedures such as amperometry, coulometry, potentiometry, and impedimetry. The final choice of method is made on the basis of which gives the most precise and accurate result in a reasonably short amount of time for many different batches of strips, and also ideally does not infringe upon the intellectual property (i.e., patents) of other companies. Once a method has been decided, the hardware and software from these electrochemical workstations must be reduced and simplified to the point of a handheld, battery operated, blood glucose monitor operated by a diabetic at home.

Blood glucose monitors for home testing comprise a biosensor test strip and an accompanying meter that work together as "the test system". It is unlikely that a strip from one manufacturer would work correctly, if at all, in another manufacturer's meter. At worst, a glucose result would be displayed that could be wildly inaccurate and certainly should not be used to indicate any action or inaction. In order to have a high degree of confidence in the results of calibration, the current data should ideally be collected with the same meter used by the user of the strip, so that the data accounts for any hardware characteristics of the meter that may influence the final calculation of the glucose concentration. Using the meter for calibration purposes also saves the need for any time-consuming and expensive validation and verification exercises to qualify the use of such a parallel system.

In practice, these meters may be enhanced by having more memory to collect all of the current response, storing this in memory for later download and analysis, although the essential hardware elements responsible for polarizing the strip and measuring the current would be the same. Using meters has the advantage that many meters can be used in parallel under a variety of temperatures to collect replicate measurements for statistical treatment of the data to estimate the accuracy and precision of the batch of strips. The current is recorded for each glucose concentration at the claimed read time of the product (e.g., 5 or 10 s).

The nature of the test samples for deriving the calibration for a batch of strips is worthy of discussion. The usual intended sample for BGMs is fresh capillary blood obtained by a finger prick with a typical volume of 1–10 µl. However, much greater volumes (milliliters) of blood are required in order to fulfill all of the various testing required to derive the calibration data for the batch, and to carry out other quality control checks such as effects of drugs and endogenous substances found in blood that have the potential to influence the glucose result. The use of capillary blood also necessitates the need to run calibration trials at diabetic clinics in order to gain the required information over the entire glucose range which would be costly and time consuming; using healthy volunteers even through a glucose tolerance test would only satisfy a narrow range of what is needed. A glucose tolerance test is where a volunteer imbibes a meal high in sugar to increase their blood glucose; for a healthy individual, this has the transient effect of temporarily increasing the blood glucose, but for a diabetic, the blood glucose will not recover quickly to a normal baseline concentration. To satisfy this requirement for greater volumes of test blood, venous blood is used. Venous blood may be collected into any suitable preservative or anticoagulant such as heparin or potassium ethylenediaminetetraacetic acid (K-EDTA). This blood may then be spiked with glucose from a concentrated aqueous stock solution to the various glucose concentrations required for calibration. The calibrations obtained from venous blood must either match that of fresh capillary blood or be shown to match after a correction factor has been applied. Where venous blood samples may give different responses to capillary samples, issues such as the presence of the

anticoagulant, hemolysis, differing oxygen tension, and the age of the blood need to be considered.

Once a dataset of currents has been collected through many measurements of different glucose concentrations at a variety of temperatures, the mathematical process of fitting a calibration line through the data can begin. Initially, a typical dataset would show a family of current response against glucose concentration curves at each temperature as illustrated in Figure 4.

This calibration at its simplest and most preferable would be a linear relationship. This can be fitted with a simple linear equation. Deviations from linearity add a complication, although it may be possible to fit a second-order polynomial or other types of equations such as logarithmic plots and power or exponential fits. Depending on the equation or model chosen, the complexity of using the calibration model to convert any current value to a glucose value will become evident. The complexity further increases when factors other than glucose can affect the current measured. One obvious factor is temperature. The temperature at which the current is measured should also be recorded in order to provide an accurate estimate of glucose over a range of permitted temperatures. Following the determination of the calibration, it is often converted into a simple calibration table containing the three critical factors current (or an equivalent measure, e.g., total charge passed, Cottrell-type measurements), temperature, and glucose. It is this table that is stored in the calibration chip and read by the calibration algorithm in the BGM to convert its raw measurement

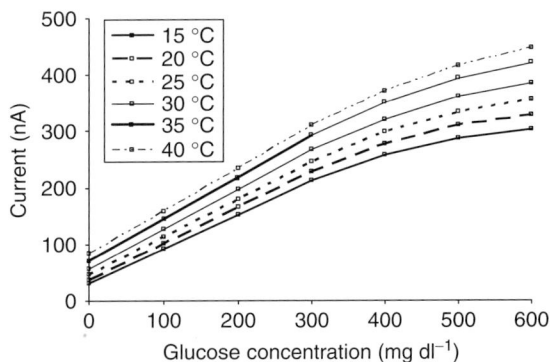

Figure 4. Glucose calibration curves obtained over a range of temperatures used to derive the calibration table shown in Table 3.

Table 3. Calibration table representing the calibration curves shown in Figure 4. Interpolation by the calibration algorithm between the current values in the table at a measured temperature enables the value of the glucose concentration in a blood sample to be estimated. For example, a strip returning a current of 207 nA measured at a temperature of 23 °C would give a glucose value of 250 mg dl^{-1}. Values beyond the highest glucose value in the table can be linearly extrapolated from the last pair of currents in the table, that is, from the currents, at 500 and 600 mg dl^{-1}, at each temperature

		Temperature (°C)					
		15	20	25	30	35	40
	0	30	38	47	58	70	83
	100	91	102	114	127	142	159
	200	152	165	180	197	215	235
Glucose (mg dl^{-1})	300	212	229	247	267	288	311
	400	258	277	298	321	345	371
	500	288	311	335	361	388	417
	600	303	329	356	385	415	447

data (i.e., current, charge, voltage) into a glucose value.

The amperometric method used to measure the current response from a glucose strip is used along with the calibration table to arrive at a glucose value, this measurement may simply be the value of the current at a defined endpoint, for example, the current after 10 s, to be used by the calibration algorithm to calculate the glucose concentration. The ambient temperature is also measured to allow any temperature correction as the kinetics of the biosensor will be influenced by temperature. The simplest forms of the calibration table contain values which are points or nodes selected from calibration curves of the biosensor glucose–current response at each temperature. The cells in the calibration table (Table 3, Figure 4) contain the value of the current at each pair of glucose concentration and temperature. When a meter has been used with a sample, the final current (or coulometric value, etc.) is recorded together with the temperature. The calibration algorithm compares the recorded current value to the values in the calibration table. If there are no exact matches, then the algorithm can interpolate between neighboring cells to derive the glucose concentration value that will be displayed to the user. Values of current beyond the table limits may, for simplicity, be linearly extrapolated from the edge of the table and the glucose value displayed. An added safety check by the algorithm would be that if the current exceeds the currents in the table beyond a preset limit, then this may indicate

an error state either with the meter or strip and an error message could be indicated to the user; this is an important safety issue as a fault that would give a current that equates to a very high or low glucose value may lead the user to instigate the incorrect treatment on a false diagnosis, whereas an error message would merely lead to a retest. Owing to the nature of utilizing a relatively small number of nodes to represent the full calibration curve, very low or very high glucose concentrations may become increasingly inaccurate, especially if the calibrations curves are not linear over a wide range (ideally 10–900 mg dl^{-1}); therefore the choice of node positions and the number of nodes needs to address this issue to achieve an acceptable degree of accuracy at the limits of the calibration curve. The use of a calibration table allows a great deal of flexibility in the glucose calibration, so, complicated mathematical models can be used to characterize the blood glucose response and can then be described by a relatively simple table, without the need to build such models into the meter algorithms, resulting in a saving in the cost of expensive processors.

10 BLOOD GLUCOSE METER USER INTERFACE

This is the part of any meter algorithm that the user will see and control; it must be designed so that the user cannot influence the final blood glucose result however the meter is used. Meter hardware and software design is often done by consulting groups of meter users, so-called focus groups, to gauge the needs of real diabetic users, to ensure that what is designed and developed fits those needs. After the meter, and its software, has been developed, it undergoes a series of consumer and manufacturer testing to ensure its operation does not lead to undesirable consequences through use; this is essentially a series of challenge tests, that is, making sure all of the documented features of the meter actually work as intended, such as results being stored correctly, alarms working, screens displaying all of the digits and icons correctly, drop tests, vibration tests, and so on, to locate any potential hardware and software errors or oversights.

BGMs are designed to be small and portable but not too small as many diabetics have dexterity problems and poor eyesight, so the meter ergonomics needs to be designed for easy handling as well as having a large and clear display for the user to read the result. However, the needs of different customers do differ with medical facilities wanting larger, durable units that can withstand heavy use, while self-testers usually prefer something more discrete so that they can carry the device in a pocket or a small bag. The display must be designed such that any errors or user flags are easily observed by the user, often as extra icons, which in turn must have an obvious symbol; such flag icons can include a thermometer to indicate a high or low temperature, a battery to indicate a low battery, or a word like "Ketone" to indicate the danger of increased ketone concentration associated with a high glucose reading, and hypo- or hyperglycemic flags; these types of flags can also use an audible signal to alert the user. The display is also controlled so that a user knows it is in action during a measurement such as displaying a countdown or a series of diminishing flashing bars.

The meter will be designed with a minimum number of buttons, both to remove complexity of use and also to reduce cost. Many meters only have a single button for control; however, extended features in a meter would obviously require extra buttons. The display must be designed in such a way that only the displayed result and not the time or date be recognized as a result leading to a misreading. Similarly the units of measure (either $mmol\,l^{-1}$ or $mg\,dl^{-1}$) must be clearly displayed so that an incorrect diagnosis cannot be made. The units of measure are often changeable on a meter, so depending upon which market a particular meter is destined for, this is often set prior to dispatch so that the user does not have to make any change. The units of measure are usually distinguished by having a decimal point with $mmol\,l^{-1}$ and never having a decimal point with $mg\,dl^{-1}$ results (e.g., $3.3\,mmol\,l^{-1}$ or $60\,mg\,dl^{-1}$).

Most meters feature a memory function that can access previous results for the user to manage their condition; this memory function must distinguish between blood glucose results and any control solution results. Control solutions are supplied with BGM kits so that the user can run a check on the meter and strip prior to use to ensure both are functioning within acceptable limits. Any averaging of blood glucose results should therefore not include control solution results. The averaging is often time based with a number of different averages given such as the average over the last 24 h, 7 days, and 14 days. Some meters even provide a graphing function, so trends can be easily observed (e.g., Lifescan® One Touch® Ultrasmart®). The results in the meter are often accessible by a personal computer, so a user or clinician can easily download the patient's data to keep a record and monitor a patient's condition.

While each manufacturer will have their own systems for measuring glucose or any other analyte, the basic principle of converting the raw data (i.e., current (amperometry), charge (coulometry), voltage (potentiometry), resistance, impedance, optical reflectance, or absorbance) obtained from measuring the response from a blood sample in a test strip with a meter, must still pass through a calibration process that correlates the raw data or signal to glucose concentration. Algorithms similar to those described earlier are then able to use this calibration data to present the user with a glucose concentration value. Thus this chapter illustrates that algorithms controlling data and meter responses within blood glucose monitors are the essential interface between the measurement technology and the user, helping to guide millions of diabetics toward good control of their glucose concentration and thus their long-term health.

REFERENCES

1. *American Diabetes Association Homepage*, http://www.diabetes.org/, 2006.
2. *Diabetes UK Homepage*, http://www.diabetes.org.uk/, 2006.
3. World Health Organization, *Diabetes Program*, http://www.who.int/diabetes/, 2006.
4. International Standard, EN ISO 15197, 2003, *In vitro Diagnostic Test Systems–Requirements for Blood-Glucose Monitoring Systems for Self-Testing in Managing Diabetes Mellitus*.
5. *Yellow Springs Instrument Incorporated Homepage*, http://www.ysi.com/, 2006.
6. W. L. Clarke, D. Cox, L. A. Gonder-Frederick, W. Carter, and S. L. Pohl, Evaluating clinical accuracy of systems for self-monitoring of blood glucose. *Diabetes Care*, 1987, **10**, 622–628.
7. N. J. Szuminsky, J. Jordan, P. A. Pottgen, and J. L. Talbott, *Method and Apparatus for Amperometric Diagnostic Analysis*, US Patent RE36268.
8. F. G. Cottrell, Der Reststrom bei galvanischer Polarisation, betrachtet als ein Diffusionsproblem. *Zeitschrift fur Physikalische Chemie*, 1902, **42**, 385.
9. B. E. White, R. A. Parks, P. G. Ritchie, and T. A. Beaty, *Biosensing Meter with Pluggable Memory Key*, US Patent 5,366,609.

63

Microarray Analysis Software and its Applications

Conrad Bessant

Cranfield Health, Cranfield University, Silsoe, UK

1 INTRODUCTION

As discussed in detail elsewhere in this book, array technology permits multiple biological entities to be monitored in a single experiment. Today the term *arrays* is most associated with gene expression microarrays, but there are many other types of bioarrays. For example, a collection of enzyme-mediated amperometric biosensors for different metabolites might be used for the determination of fruit ripeness or a chip spotted with antibodies may be used for the detection of a range of proteins. Whatever the array, the way in which the data acquired from it is processed and interpreted requires some thought. It is possible to treat array data simply as a collection of individual assays, but this looses valuable relationships between the individual components of the data, and anyway becomes intractable when dealing with arrays with more than a handful of elements. A more advanced, *multivariate*, approach to the interpretation of the acquired data is therefore required.

The aim of this chapter is to introduce the techniques used to analyze data acquired from bioarrays, focusing primarily on spotted cDNA microarrays as a case study. This focus has been chosen because the data analysis approaches for such arrays are well established, and spotted arrays are the typical format used by researchers generating novel array platforms due to the relative ease of fabrication. The chapter focuses on generic principles, rather than on specific software packages, hardware platforms, or applications as the details of these change rapidly over time. However, lists of software available at the time of writing are provided as a starting point for those wishing to implement the techniques discussed.

Processing of microrray data is best considered in two specific stages. Firstly, it is necessary to determine the quantity of material bound to each spot by processing the image scanned from the array. For gene expression microarrays, this ultimately results in a list of the expression levels of each gene represented on the array. The second stage of analysis is to extract biologically relevant information from this table of expression levels. The exact way in which this latter stage is done depends on the particular biological question being asked, but typically involves some kind of the comparison between the data from different samples. The remainder of this chapter describes these two main analysis stages.

2 IMAGE PROCESSING

2.1 Nature of Data

Before considering how the image data is processed, it is important to appreciate the nature of the data involved. This can vary considerably according to the specific array platform. In the case

Handbook of Biosensors and Biochips. Edited by Robert S. Marks, David C. Cullen, Isao Karube, Christopher R. Lowe and Howard H. Weetall.
© 2007 John Wiley & Sons, Ltd. ISBN 978-0-470-01905-4.

of a spotted cDNA array, the starting point is two high-resolution scanned images of the array—one for each of the fluorescent compounds (typically Cy3 and Cy5) used to label the cDNA from the samples being analyzed. The level of fluorescence intensity of a pixel in the image is proportional to the amount of material bound at that particular location. Often the two images are combined (following normalization as described in Section 2.2.2) in such a way that the red component of each pixel represents the intensity at that point of the label associated with the sample of interest, and the green component indicates the intensity of the control label. Therefore, spots associated with genes that are upregulated compared to the control appear red and those that are downregulated appear green.

Microarray images are necessarily of high resolution and therefore large (typically at least 1500 × 3500 pixels). These images cannot be compressed using methods of high compression ratio such as JPEG, as these introduce artifacts into the image that degrade the quality of the results.

2.2 Aim of Image Processing

The ultimate aim of the image processing is to produce a list of the relative expression levels for each individual gene. This requires several steps:

- Identify spots and separate them from the background (often referred to as feature extraction).
- Relate each spot to its associated gene.
- Check that each spot is of the required quality, and detect spotting anomalies.
- Normalize the data to account for interimage variability, background fluorescence, and non-specific hybridization.
- Determine the relative amount of material bound to each spot.

Relating spots to genes is not scientifically challenging, as the designer of the array will have known which fragment of cDNA they were spotting in each position, and therefore what gene each spot corresponds to. However, there is an information management issue in that for any analysis there needs to be a record of the array design, listing the genes represented on the array and the

positions at which they were spotted. Microarray analysis software requires that this information be provided in a suitably formatted file that can be loaded prior to analysis. For commercially available arrays, such files are provided by the array vendor, but for purpose-built arrays these obviously need to be generated by the array designer.

2.2.1 Alignment of Spots to Grid, Determination of Spot Quality, and Errors

Spots on an array tend to be deposited in a simple grid formation, but inherent errors in spotting precision mean that spots are not always on a perfect linear alignment. Furthermore, scanning can introduce a slight angle or displacement to the grid layout. A step is therefore required to rotate and shift the image so that it aligns with the grid design expected for the specific microarray. A grid is then superimposed on the image, in which each spot should fall into a particular cell within that grid. Each cell can then be linked to the relevant gene using the microarray design information mentioned earlier. All of this is fairly straightforward in computational terms and is built into currently available software packages (see Section 2.4) as an automatic process.

Sometimes, a spot will not fall fully within a square on the grid. This is an indication of a problem with the array, either regarding the spotting process or contamination. A spot may be out of position or merged with another, or a piece of dust may have fallen on the array, creating a feature in the image that spans multiple grid squares. In these cases the cell in question is usually flagged as being unreliable and therefore excluded from subsequent analysis. If the array is large enough and well designed, there should be repetitions of the spot elsewhere on the array to mitigate this type of problem.

Once a spot has been assigned to a grid square, it is defined as a shape to separate it from the background. The purpose of this is both to define which pixels are to be used to calculate the expression level of the gene associated with that spot, and to determine the background level of fluorescence across the array. Assigning a boundary between spot and background can also be used as a quality measure—ideally each spot should be circular, so calculating how well a circle fits to the spot outline is one way of approximating spotting quality. This is a fairly trivial computational process and so can be carried out rapidly in software, even on a microarray with thousands of spots. Distribution of fluorescence across a spot is also an indicator of spotting quality—generally we are seeking uniform distribution across the spot. However, spots sometimes form a concave shape, giving them a donut appearance in the scanned image, or they have a diminishing level of material across the width of the spot.

2.2.2 Normalization

Aside from method development studies, it is unlikely that an array image will be considered in isolation. Real experiments will more typically involve the analysis of multiple samples and/or multiple time points, resulting in multiple array images for a given experiment. To make this type of analysis possible, it is essential to equalize the fluorescence intensity across the group of images so that they can be sensibly considered in the same analysis. This process is referred to as normalization, and essentially involves scaling the intensity of the microarray images such that they are directly comparable across an experiment. Indeed, it is even necessary to normalize the intensities of the two different labels from a single array, as they may be subject to a systematic error, which prevents the intensity of the two labels being directly compared. Such a disparity between labels can be caused by factors such as the scanner, the topology of the array, or the labeling or binding processes.

Because of the many potential sources of systematic error, the exact normalization method depends on the type of array and scanning platform being used, and sources of errors can be tackled collectively or in turn. For example, a simple way of addressing inconsistencies in scanning intensity across a set of individual arrays would be to scale the intensities of every pixel in the image such that the background intensity is consistent across the arrays. More often the array will have been designed with a selection of control spots, typically housekeeping genes whose expression level is expected to be consistent across the experiments. The intensity values from each array can then be scaled so that the values for these control spots are consistent across the experiment. There are many more advanced normalization techniques, built on solid statistical foundations, which space does not allow us to include here. Suffice to say, normalization is an essential part of microarray data analysis and it has a significant impact on all the analysis that follows.

2.2.3 Calculation of Expression Levels

Once the above steps have been carried out, determination of the expression levels is straightforward, essentially taking the average intensity of pixels within the defined spot area. These intensities are then stored alongside the spot information in a results file. In cases where intensity is not uniform across the spot, as mentioned in Section 2.2.1, something more advanced than a simple average will be needed to determine a representative spot intensity. Normally, the raw intensities are transformed by taking logs to base 2, as this results in a more convenient distribution of data for most microarray data sets.

2.3 MGED Reporting Standards

To enable sharing of unambiguous data between different researchers, the Microarray Gene Expression Data (MGED) Society produced the MIAME[1] (Minimum Information About a Microarray Experiment) standard for reporting microarray results. MIAME captures normalized gene expression data, as well as the raw data from which it was derived, and other pertinent information such as experimental procedures, experimental design, and the design of the microarray. The microarray and gene expression markup language (MAGE-ML) is the primary embodiment of the MIAME standard, providing an extensible markup language (XML)-based data format for convenient exchange of data. Most of the microarray image processing software packages are capable of producing MAGE-ML output, and it is often used as a starting point for the data interpretation described in the next part of this chapter. Many scholarly journals now require that researchers submitting papers involving microarray analysis deposit MIAME compliant data supporting their work into a public repository

Table 1. Examples of software for microarray image processing

Software	Provider	License	Web site
BlueFuse	BlueGnome	Commercial	4
GenePix Pro	Molecular Devices	Commercial	5
ScanAlyze	Michael Eisen, University of California at Berkeley	Free for academic use	6
TIGR Spotfinder (part of TM4)	The Institute for Genomic Research	Open source	7

such as Gene Expression Omnibus (GEO)[2] or ArrayExpress.[3]

2.4 Microarray Image Processing Software

Various software packages are available for microarray image processing. A selection of some of the most cited software available at the time of writing is listed in Table 1. As software specifications can change rapidly, it is recommended that the reader consults the providers' web sites for the latest information regarding functionality.

3 DATA INTERPRETATION

3.1 Aims

Quantifying the level of expression of each gene on each array is only the first stage in microarray data processing. In most experiments, the researcher will analyze multiple samples, and will want to compare the results from these different samples. Typical aims of the data interpretation step will be identification of differentially expressed genes, clustering of molecular entities according to behavior, clustering of samples according to their profile, or identification of genes whose behavior changes over time. The exact nature of the data analysis will vary according to the biological study being conducted, but analysis of the raw data by eye alone is obviously out of the question when the expression of thousands of genes has been monitored across tens or hundreds of samples. Even for small arrays, the application of multivariate data interpretation techniques can allow more information to be extracted from the available data as it considers the data as a whole, rather than piece by piece. Several of the most common multivariate data interpretation techniques are described below.

3.2 Nature of Data

Before discussing how data can be analyzed, it is important to consider exactly what data we are starting with and how this is most conveniently organized. In essence, all data collected from a series of array experiments can be represented by some kind of data matrix. In what follows, we will concentrate on gene expression microarrays for simplicity, but the techniques described are equally appropriate to data from other types of array. In a microarray experiment, the gene expression levels for each sample can be represented by a vector of expression measurements, denoted as x_i, where i is the sample number. Within this vector, each element x_{ij} represents the expression level of gene j on the microarray. The expression level of each gene can be considered to be a variable, of which there are many, hence array data is clearly highly multivariate and multivariate data analysis techniques are required to interpret it.

The vectors representing the gene expression levels from individual samples can be amalgamated into an $I \times J$ data matrix, X, where I is the total number of samples considered and J is the number of genes per array. Each row of the X matrix therefore represents an individual array, while each column indicates the expression level of each specific gene over all samples. A single data matrix is therefore sufficient to describe all the samples analyzed in a given series of experiments. Figure 1 shows how such a data matrix is constructed.

An important issue with microarray data is that missing values are possible due to spots failing the quality tests described earlier, in Section 2.2.1. Most multivariate data analysis algorithms are

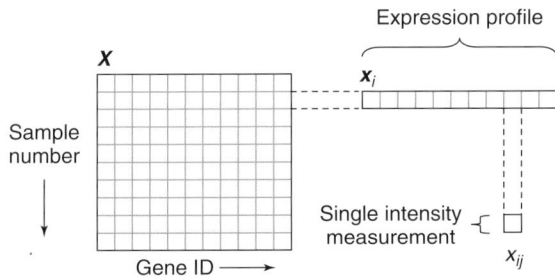

Figure 1. Organization of gene expression data into a data matrix. For data from a gene expression microarray, the row vector x_i would be the gene expression profile over the samples analyzed. Each column of the matrix represents the expression profile of an individual gene over all sample.

unable to deal directly with missing values, so these need to be replaced or removed. The most elegant solution is to estimate the missing value on the basis of data from the rest of the matrix, although it is important to remember that the new value is only an estimate and there is obviously a limit to the number of values that can be substituted in this way. An alternative approach is to simply remove all the data associated with a particular gene (an entire column) or sample (entire row).

Once a data matrix has been constructed, there are essentially two main data interpretation activities that can be of interest: data exploration and sample classification. Approaches for tackling these tasks are described below, along with examples of where these approaches have been used.

3.3   Exploratory Data Analysis

The aim of data exploration techniques is to provide a way of visualizing variation within large multivariate data sets. This is sometimes an end in itself, but it is also a useful way of evaluating whether the data is of sufficient quality or sufficient information content to warrant further analysis. Clearly, there is no point in expending effort attempting to classify samples into different groups according their gene expression profiles if initial exploration of the data shows that there is no sign of correlation between the data acquired and the sample types analyzed.

3.3.1   Scatter Plots

One of the simplest, yet most effective, forms of exploratory analysis is the construction of scatter plots. By plotting, for each gene, a point on a graph at coordinates (a_j, b_j), where a_j is the expression level of gene j in sample A and b_j is the expression level in sample B, genes which show substantially different expression levels between the two samples can be clearly seen. Typically, the expression values are plotted on log scales to provide more clarity to the figure. Genes with similar expression levels fall along a diagonal line across the plot. Genes that fall more than a specified distance from this line can be considered to exhibit a significant difference in expression between the two samples. The definition of a significant difference varies depending on the application and according to the general level of noise in the data, but typically a twofold change in expression would be considered significant. Lines marked on the scatter plot representing a twofold change can be superimposed so that the genes of interest can be clearly seen. The scatter plot shown in Figure 2 is an example of this. Most gene expression analysis software permits the creation of these plots, and allow the user to click on the points of interest to see what genes they relate to. Also, through a simple comparative test, the software is able to automatically generate a list

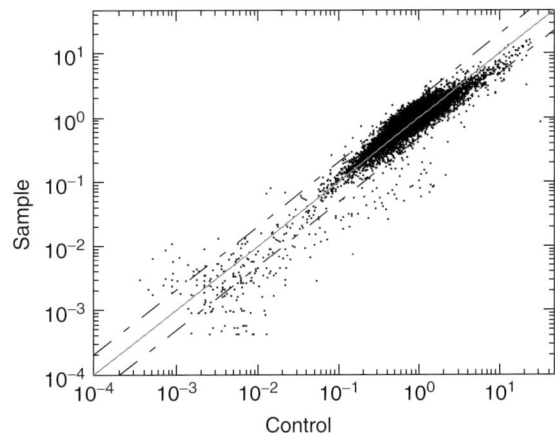

Figure 2. Comparison of data from two microarrays for samples taken from two different positions in mouse brain. The solid line indicates identical expression between the two samples. The dotted lines denote a twofold expression change either way. [Reprinted with permission Brown et al.[8] copyright 2002, Coldspring Harbour Press.]

of such genes, which the researcher would then investigate further.

3.3.2 Principal Components Analysis

A significant limitation of the scatter plot approach is that it is limited to pairwise comparisons, with just two samples in any one plot. If we want to compare data from more than two samples or compare the actual expression profiles of multiple genes, then more advanced techniques are required. One such technique is principal components analysis (PCA).

PCA is a way of reducing a large multivariate data matrix into a matrix with a much smaller number of variables, without loosing important information within the data. The principle behind PCA is that the multivariate data can be decomposed by linear projections onto a new coordinate system. The new axes, known as *principal components* (*PCs*), are orientated such that the first PC captures the largest amount of common variance. The next PC is orthogonal (meaning totally uncorrelated) to the first and captures common variance in its direction. The maximum number of PCs is limited to the number of variables in the original multivariate data set but, due to the reorientation of the coordinate system to maximize common variance, most data variance can be captured by a much smaller number of PCs due to correlation and redundancy in the original data. In mathematical terms, PCA is the reduction of the data matrix, X, into two smaller matrices, the scores, T, and loadings, P. The product of the two, plus a residual matrix, E, gives the original data matrix (equation 1).

$$X = T \cdot P + E \qquad (1)$$

There are a number of algorithms for calculating T and P, the most common being nonlinear iterative partial least squares (NIPALS) and singular value decomposition (SVD). In both cases, T is a matrix of column eigenvectors, with the columns in order of largest variance first, hence PCs are delivered in order of their information content. The scores matrix T is determined by multiplying X by the matrix of loadings, P, as shown in Figure 3. In simple terms, this means that the scores for a particular sample are weighted sums of the original

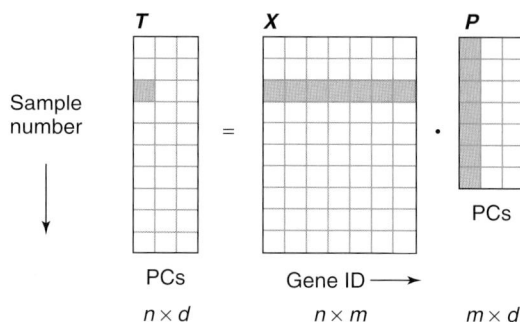

Figure 3. Relationship between the data matrix (X), scores matrix (T), and loadings matrix (P) in PCA. In this simple example, the number of samples, n, is 10, the number of measured variables (e.g., genes), m, is 7, and the number, d, of PCs considered is 3. The highlighted row in X and column in P show what is required to generate first PC score for sample 3.

variables. For example, the first PC score for the third sample in the data matrix shown in Figure 3 would be calculated as follows:

$$t_{3,1} = x_{3,1}p_{1,1} + x_{3,2}p_{2,1} + x_{3,3}p_{3,1} + x_{3,4}p_{4,1}$$
$$+ x_{3,5}p_{5,1} + x_{3,6}p_{6,1} + x_{3,7}p_{7,1} \qquad (2)$$

In many cases just the first two or three components are sufficient to capture the bulk of the variance in a given data set. Each sample can then be plotted on a simple two- or three-dimensional graph at the position dictated by its first two or three PCA scores. The relative position of the samples in this plot indicates the relative similarities between samples, with similar samples appearing at similar positions within the graph. As PCA uses linear projection, random noise does not fit the model so is relegated to the latter PCs. For this reason PCA lends itself to compression of linear data with simultaneous filtering of random noise. PCA can be used to compress nonlinear data; however, more PCs have to be kept, which would normally be rejected as being noise, as the nonlinear element of the data will be contained within these PCs.

An example of a PCA scores plot generated from microarray data is shown in Figure 4. In this study, cells were exposed to the carcinogen benzo(a)pyrene (BaP) and its noncarcinogenic isomer benzo(e)pyrene (BeP). Using this plot of the scores of the first two PCs it can clearly be seen that the cells respond differently to the different

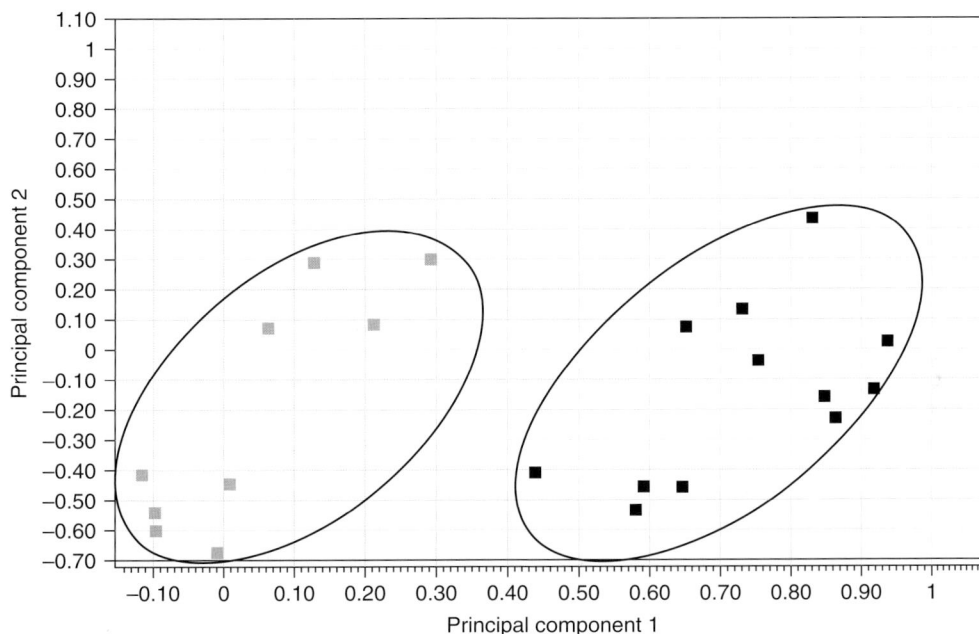

Figure 4. A PCA scores plot generated from a data matrix comprising 165 selected genes, monitored over 21 samples. Samples exposed to BaP (black dots) exhibit clearly different behavior to those exposed to BeP (grey dots). [Adapted from Hockley et al., 2006.[9]]

compounds, which provides strong evidence for more detailed investigation.

3.3.3 Hierarchical Cluster Analysis

Hierarchical cluster analysis (HCA) is another exploratory data analysis technique which, like PCA, is designed to reveal relationships between samples or between the molecular entities (e.g., genes) being studied. The result of HCA is a tree diagram, or dendrogram, in which each sample is represented by a branch, and the distance between branch tips indicates the level of similarity between samples.

The dendrograms are created by a recursive process in which the pairwise similarity between every sample and every other sample is calculated. The samples representing the two most similar samples are then joined using branches whose length is related to the level of similarity between the samples. The process is then repeated, with the two samples already accounted for being agglomerated in such a way that they can be considered as a single sample. This process is repeated until all samples have been joined together. An example of the result of this process for a fairly small data set is shown in Figure 5. This method is clearly capable of displaying the relation between entities in a data set, and unlike PCA it is easily extended to very large data sets without cluttering the plot or loosing information. It is therefore very popular in microarray analysis, and many examples of its use can be found in the literature.

All hierarchical clustering follows the general approach set out above, but there are a lot of variations in how the similarity between samples is calculated, and how samples are joined together. The primary method of determining the level of similarity between two samples is by calculating the distance between them in the multidimensional space of the measured variables (e.g., the gene expression values). The process is easy to visualize for two measured variables, as shown in Figure 6, but is equally applicable to any number of variables. Taking the two-dimensional case in the figure as an example, the most intuitive distance measure is the euclidean distance—the shortest distance between the two points. This

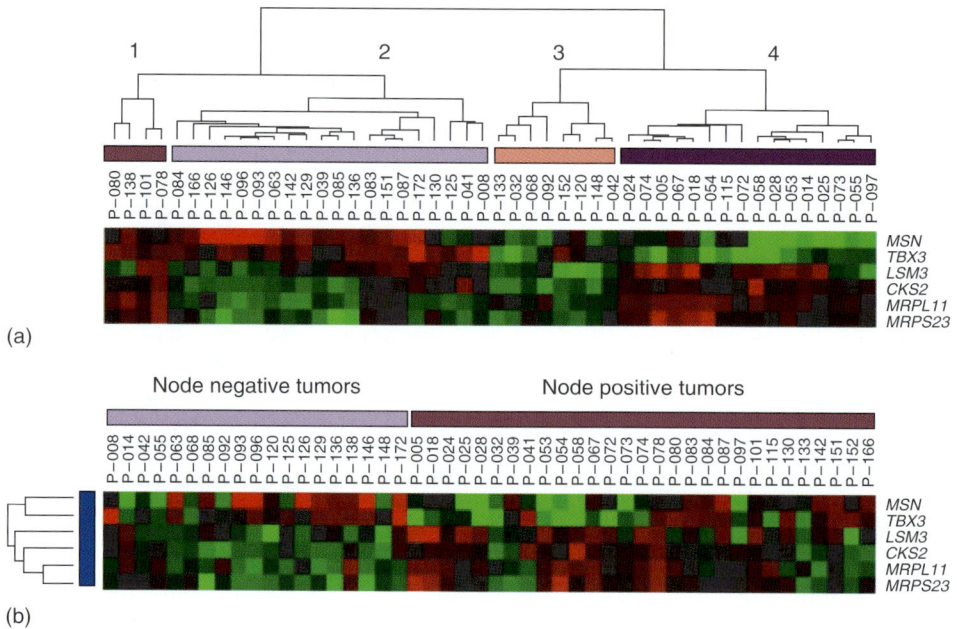

(a)

(b)

Figure 5. Example output when performing HCA on gene expression data. In both cases, a data matrix comprising the expression levels of 6 genes measured in 48 samples has been used. These have then been clustered according to (a) the expression profile of each sample and (b) the behavior of each gene. The blocky heat map represents the gene expression data matrix (equivalent to Figure 1), in which the rows and columns have been rearranged in light of the HCA results. [Adapted from Lyng et al., 2006.[22]]

distance, d, is trivially calculated using Pythagoras' theorem

$$d_{A,B} = \sqrt{(A_1 - B_1)^2 + (A_2 - B_2)^2} \qquad (3)$$

Extending this to further variables simply involves adding the squared differences for the other variables within the square root. For the case of N variables, the calculation for each sample would

be as follows:

$$d_{A,B} = \sqrt{\sum_{n=1}^{N} (A_n - B_n)^2} \qquad (4)$$

However, the euclidean distance is not the only measure. If we want to particularly emphasize samples that are markedly different from others, we can amplify the distance by squaring it. For the two-dimensional example, the squared euclidean distance is simply equation (3) with the square root removed.

If we want to emphasize the difference between samples according the value of the largest difference between values of a single variable, regardless of what that variable is, we can use the Chebychev distance:

$$d_{A,B} = \max |A_n - B_n| \qquad (5)$$

Just as there is a choice of method for calculating the distance, or similarity, between two samples, so there is a range of linkage algorithms

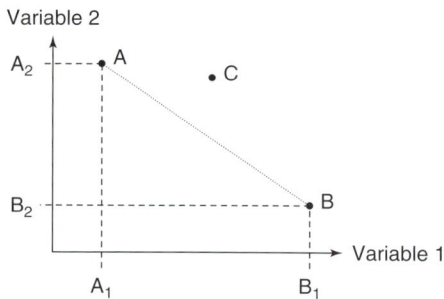

Figure 6. Illustration of distance between samples in variable space. In this case, we consider two samples, A and B, with two measured variables. There are many ways in which the distance between the samples can be calculated.

for joining clusters together as the clustering process progresses. Essentially a linkage algorithm defines which point in a cluster is used to represent that cluster when the distances are calculated. The most obvious approach is the weighted average, where each cluster is represented by the average position in the variable space of the samples that make up the cluster—this essentially represents the "center of gravity" of the cluster. Other popular methods include complete and single linkage. Using complete linkage the distance between two clusters is calculated using the largest distance between individual points in those clusters—this promotes tight clusters over those with more variance. Single linkage is the opposite, where the distance is measured according to the closest two points in the two clusters—this allows clusters to be joined on the basis of just two similar samples, regardless of the spread across the variable space that each cluster exhibits.

A more advanced linkage algorithm, called *Ward's method*, moves away from simple geometric solutions and joins clusters not just on simple distance measures but according to which of the agglomerated clusters will have the least variance. This approach has the benefit of promoting tight clusters, but does not suffer the sensitivity to outliers found in complete linkage. For this reason, it is often used as the linkage algorithm of choice in microarray work.

Sometimes, if noise, correlation, and redundancy are expected to be significant, PCA is applied prior to cluster analysis, with the distance between samples being calculated according to the positions of the samples in PC scores space instead of in the original variable space.

Clearly there is a wide range of possible combinations of parameters for performing HCA, and experience shows that these can result in markedly different dendrograms, leading to potentially different interpretations of the data set. It is therefore very important to ensure that the particular distance measures and linkage algorithms used are appropriate, either by considering in detail how each approach works and how this relates to the particular data set being analyzed, or by following the best practice described in the literature for similar data sets. It is also important to consider the robustness of the results obtained—if a particular clustering behavior is observed only in the dendrogram created by a very specific set of HCA parameters then it may not be wise to assume that the clusters genuinely represent the relationships between the samples.

3.4 Classification

In many applications, we are particularly interested in being able to classify samples according to their analytical response. For example, many papers have been published showing how gene expression profiles can be used to classify biological samples into "healthy" and "diseased" states for particular diseases. This is important because it raises the possibility of detecting diseases according to the behavior of multiple biomarkers, rather than a single biomarker as has traditionally been the case. This has the potential to improve diagnosis accuracy, simply because it takes into account more biological factors. Such classification could be done by looking at the output of an exploratory technique such as HCA or PCA, but we really want an automated computational method if we are to ensure objectivity and high data throughput.

Multivariate classification is the name given to the data analysis approach used to achieve this. It involves building a classification model from a data matrix acquired from samples of known class. The model is effectively a mathematical transformation relating the measured variable to a number indicating the class of sample (e.g., 0 for healthy, 1 for diseased). Crucially, a separate matrix of data from samples of known class is collected and used to test the resulting classification model. The performance of the calibration model can therefore be quoted using easily understood quantitative measures such as the proportion of test samples that are correctly identified by the model. Alternatively, the performance can be specified in terms of the specificity and sensitivity of the model, which are derived individually from the proportion of correctly identified positive samples and correctly identified negative samples.

There exists a plethora of methods for constructing a classification model—far more than can be dealt with in detail here, but the details can be found in chemometrics textbooks.[10,11] Provided samples from similar classes cluster well in a PC scores plot, one of the easiest solutions is to divide the scores plot into sections using

Table 2. Software for microarray data interpretation

Software	Provider	License	Web site
ArrayMiner	Optimal Design	Commercial	13
Bioconductor	Fred Hutchinson Research Center	Open source	14
GeneSpring GX	Agilent Technologies	Commercial	15
Partek Genomics Suite	Partek	Commercial	16
Rosetta Resolver	Rosetta Biosoftware	Commercial	17
TM4	The Institute for Genomic Research	Open source	18
Vector Xpression	Invitrogen	Commercial	19

a collection of linear boundaries. New samples are then identified according to which side of the boundaries they fall on. In the example in Figure 4 the two response types could easily be defined by inserting a vertical boundary at the point where the PC1 score equals 0.35. This approach is referred to as linear discriminant analysis (LDA). LDA is capable of automatically generating the boundaries using fairly simple mathematics, and the technique can be extended to multiple dimensions—in three dimensions the boundary becomes a two-dimensional plane, and in higher dimensions it is a hyper plane. This means that LDA can be used on the original data matrix as well as on PCA scores, regardless of the number of variables measured.

In more complex data sets, where there are many classes of sample, or classes of sample which cluster in an unusual shape or with a lot of variance, it is not always possible to separate classes using simple linear features defined by LDA. One solution to this problem is soft independent modeling of class analogies (SIMCA), which works by creating individual PC models for each class, using different numbers of PCs for each class if necessary. Classification of an unknown sample can then be performed by projecting it into each PCA model to look for the best fitting class. If the classes still cannot be defined, it is necessary to look to machine learning approaches such as artificial neural networks (ANNs).[12] ANNs are an extremely powerful tool, based on a rough approximate simulation of the neural systems that make up the brain. Given sufficient data, it is theoretically possible to train an ANN to model any relationship that exists between the measured data and sample class, regardless of its complexity. In reality, however, correctly training an ANN is a time consuming and difficult task, and careful validation of the models produced is essential if the network

is not to become over fitted to the test set and therefore unable to classify new samples that are presented to it.

3.5 Software

The are many general-purpose data analysis packages available that have implementations of the algorithms above and can therefore be used to interpret microarray data. Such packages include Matlab, R, and Statistica. As such general packages always carry an overhead in terms of adapting them to a specific task, several packages have been produced specifically for microarray analysis. A selection of some of the most widely used software at the time of writing is listed in Table 2. As with the image analysis software, updates are provided frequently so the reader is encouraged to consult the providers' web sites. Most of these web sites provide downloadable trial versions of the software, allowing its fitness for purpose to be determined prior to purchase.

4   CONCLUSIONS

This chapter intended to provide an awareness of current approaches for the analysis of data from sensor arrays. Clearly, it has not been possible in the space available to cover all the possible techniques, but those described represent the core of common practice. There is now an ever-increasing selection of textbooks available, which explain the approaches covered here in much more detail.[20,21] Hopefully, it is clear from this chapter that powerful techniques for the analysis of data from bioarrays are well established. Now that these techniques have been implemented in user-friendly software packages they are within easy reach of the laboratory researcher, but it is important to realize that the processes underlying these analyses are reasonably complex, and that an appreciation of what is happening is essential if we are to have confidence in the information extracted.

Looking ahead, bioinformatics is very much an ongoing research discipline, and new algorithms and software are being generated all the time. This will undoubtedly lead to further advances in the information, which can be extracted from gene expression data, and the confidence that we can associate with that data. To give just one example, much work is currently under way to reliably extract gene regulatory networks from gene expression data. The interested reader is encouraged to monitor journals such as *Bioinformatics, BMC Bioinformatics*, and *PLoS Computational Biology* to see the directions in which research is progressing.

REFERENCES

1. http://www.mged.org/Workgroups/MIAME/miame.html.
2. http://www.ncbi.nlm.nih.gov/geo/.
3. http://www.ebi.ac.uk/arrayexpress/.
4. http://www.cambridgebluegnome.com, 2007.
5. http://www.moleculardevices.com, 2007.
6. http://rana.lbl.gov/EisenSoftware.htm, 2007.
7. http://www.tm4.org/spotfinder.html, 2007.
8. V. M. Brown, A. Ossadtchi, A. H. Khan, S. Yee, G. Lacan, W. P. Melega, S. R. Cherry, R. M. Leahy, and D. J. Smith, Multiplex three dimensional brain gene expression mapping in a mouse model of Parkinson's disease. *Genome Research*, 2002, **12**(6), 868–884.
9. S. L. Hockley, V. M. Arlt, D. Brewer, I. Giddings, and D. H. Phillips, Time- and concentration-dependent changes in gene expression induced by benzo(a)pyrene in two human cell lines, MCF-7 and HepG2. *BMC Genomics*, 2006, **7**, 260.
10. M. Otto, *Chemometrics: Statistics and Computer Application in Analytical Chemistry*, John Wiley & Sons, 1998.
11. R. Brereton, *Chemometrics: Data Analysis for the Laboratory and Chemical Plant*, John Wiley & Sons, 2003.
12. H. B. Demuth, M. H. Beale, and M. T. Hagan, *Neural Network Design*, PWS Publishing, 1996.
13. http://www.optimaldesign.com, 2007.
14. http://www.bioconductor.org, 2007.
15. http://www.genespring.com, 2007.
16. http://www.partek.com/, 2007.
17. http://www.rosettabio.com, 2007.
18. http://www.tm4.org/, 2007.
19. http://www.invitrogen.com, 2007.
20. D. Stekel, *Microarray Bioinformatics*, Cambridge University Press, 2003.
21. G. J. McLachlan, K. A. Do, and C. Ambroise, *Analyzing Microarray Gene Expression Data*, John Wiley & Sons, 2004.
22. H. Lyng, R. S. Brøvig, D. H. Svendsrud, R. Holm, O. Kaalhus, K. Knutstad, H. Oksefjell, K. Sundfør, G. B. Kristensen, and T. Stokke, Gene expressions and copy numbers associated with metastatic phenotypes of uterine cervical cancer. *BMC Genomics*, 2006, **7**, 268.

64

Data Validation and Interpretation

Ursula E. Spichiger-Keller

Centre for Chemical Sensors and Chemical Information Technology, Swiss Federal Institute of Technology (ETH), Zurich, Switzerland

1 INTRODUCTION

The development of analytical tools for diagnostic purposes is having an increasing impact in analytical and clinical chemical research. The results of a chemical test contribute to deciding how to treat an individual or a source and are, therefore, economically and socially relevant. Two parameters, the quality of the chemical analysis and the biological uncertainty, together contribute to the information that can be drawn from the specimen that was investigated by chemical analysis. The value of a test result is established using a validation procedure where information from both processes, the chemical analysis and the biological processes, are involved in the final allocation of a result to the population of "affected" or "unaffected" sources. For the qualitative assessment of a diagnostic test, the biological uncertainty[1] (intraindividual variation, cv_{intra}, relative to the interindividual variation, cv_{inter},) has to be related to the analytical variation (cv_a relative to cv_{intra}). In a very useful diagram, Keller[2–4] related these criteria to the allowable limits of uncertainty given by CAP[5] for a number of diagnostic parameters. The analytical value of a test result is established using an analytical validation process (see Section 3), and the diagnostic value of a test result is established by investigating the biological processes and their uncertainties.

1.1 Communication, Validation, and Information Yield

A general drawback of automation with high-throughput screening and array technologies is the production of large amounts of data, which are never transformed into information.[4] Screening tests are run to collect information on decisive characteristics of a representative sample of a natural source (individuals, animals, air, rivers, etc.) or a product. Source and receiver (investigator, customer) are connected by a communication process described by Shannon and Weaver.[6] They distinguish three planes of communication, which contribute to the uncertainty of an information process: the technical plane (which affects the reliability of the communication process), the plane of semantics (which has to do with the real content of the information contained in the message), and the plane of effectiveness (which concerns the effect of a message on the receiver). By validating the method that is used to yield information, the sources of uncertainty are supposed to be reduced or at least elucidated. If both the analytical and biological validation processes are omitted, however, information is as good as lost! As Gabrieli[7] says: "Data per se are lifeless." Data validation is absolutely necessary to interpret data and to yield information on the status of a source. In this respect, information technology is essential in stringently regulated fields of analytical chemistry such as

Handbook of Biosensors and Biochips. Edited by Robert S. Marks, David C. Cullen, Isao Karube, Christopher R. Lowe and Howard H. Weetall.
© 2007 John Wiley & Sons, Ltd. ISBN 978-0-470-01905-4.

environmental chemistry, food chemistry, forensic analytical chemistry, and clinical chemistry.[8]

The standard procedure is for an analytical method to be validated by the provider and the user, who normally follow different protocols (see Section 3). In addition to the purely technical and statistical validation of results, the plane of *effectiveness* of the information yielded should be investigated in a second phase as well as how to interpret data (see Section 2). In this second phase, the biological uncertainty of a test result is investigated as well as the impact of a particular information on the receiver. It is important to carry out this phase not only for medical diagnosis but in other fields of analytical chemistry as well, since a misinterpreted analytical result may have serious consequences in both domains.

1.2 Data Allocation and the Discriminating Power of a Test

Handling a request addressed to an analytical laboratory induces formulation of a diagnostic question comparable to a medical decision. Such questions could be: "What is the phosphate level of the water of a river or lake?", "Is the hormone level too high or not?", or "What is the reason for the 'fish acute toxicity syndrome (FATS)' in a fish pool?".[9,10] FATS is a respiratory and cardiovascular disease of fish caused by toxic agents in water. The cutoff values which are, as an example, accepted for environmental pollution (the allowable concentration limits), in food technology, and forensic analytical chemistry are regularly based on scientist's recommendations, which then become fixed in law. However, those working in the field already had extensive validation processes running long before such recommendations were published.

The intention of each chemical analysis is to clearly allocate a specimen to the "healthy, unpolluted, unaffected" population or, alternatively, to the "affected, polluted, diseased" one. The discriminating power of a test is defined by its *efficiency*[11] and *efficacy*[4] (see Table 1). In the course of this allocation procedure, a set of numerical data is reduced to a binary scale, and the final decision is independent of the distance of a result from the center of gravity (median, mean) of its population. Remarkably, it is usually easier to make a single analysis highly efficient and effective than it

Table 1. Terminology used to describe the diagnostic performance of a test:[4] (i) characteristics of the source[a] and (ii) interpretation of the analytical test result (tr) referred to a fixed discriminator position[b]

Diagnostic sensitivity	TP/DIS	Fraction of correctly allocated DIS, TP results among all DIS
Diagnostic specificity	TN/REF	Fraction of correctly allocated REF, TN results among all REF
Positive predictive value, pV+, pVpos	TP/ALL POS	Fraction of true positive test results among all positive test results
Negative predictive value, pV−, pVneg	TN/ALL NEG	Fraction of true negative test results among all negative test results
Efficiency	(TP + TN)/ALL	Fraction of correctly allocated specimens among all classified
Efficacy	TP/ALL	Fraction of correctly allocated positive tr among all classified tr
Prevalence	DIS/ALL	Number of affected sources in the population investigated.

[Reprinted from Spichiger-Keller,[4] with permission from Wiley-VCH.]
[a] DIS: diseased, affected; REF: nondiseased, nonaffected population; ALL: REF + DIS; n: number of subjects, sources, or specimens under study.
[b] POS: positive test result; TP: true positive test result (POS and DIS); FP: false positive test result (α-error) (POS and REF); NEG: negative test result; TN: true negative test result (NEG and REF); FN: false negative test result (β-error) (NEG and DIS); tr: analytical test result; X_i, class of values of the ith test.

is to achieve this for an analytical array test since each additional test field in an array statistically decreases the odds of receiving a clearly discriminating result.[12,13] If the fields in a test array are responding to independent qualities of a sample, these odds decrease exponentially with the product of the number of test fields. In summary, the probability of receiving a "normal" result decreases with the number of simultaneously evaluated test parameters and depends on the biological interrelationship between test results.[12]

Example: Between 1979 and 1983, the IFCC published a series of recommendations on the theory of reference values.[14] By convention the reference interval of an analyte was defined as the central 0.95 fraction of all values from samples collected from a reference population. If these samples were analyzed by 23 different laboratory tests and if all test results were independent of each other, the probability of finding all test results within the reference range would be 0.95[15] which is equal to 0.3 or 30%. Contrarily, if all tests were totally redundant, this probability would be 1.0 and 100%; it would have been sufficient to investigate a sample by only one test. The truth lies mostly in between. A typical example is shown in Figure 1 where, the data of 196 men and 96 women out of 601 male and 528 female blood donors represented the reference range after elimination of partially correlating extreme values.

2 THE DISCRIMINATING POWER OF A TEST AND INTERPRETATION OF ANALYTICAL RESULTS

2.1 The Quality of Screening Tests and Decision Making (The Discriminators)

When screening technology became available in 1970s with high-throughput analysis of biochemical profiles, multiple testing became popular among physicians and hospitals, even though it was not necessarily more cost-effective or did not

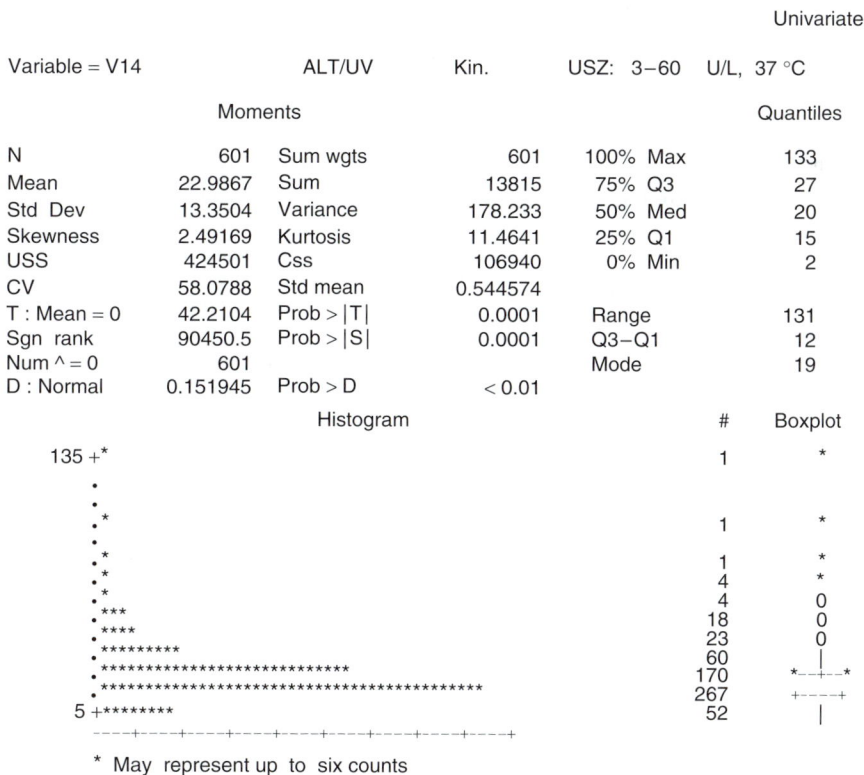

Univariate

Variable = V14		ALT/UV	Kin.	USZ: 3–60	U/L, 37 °C

Moments					Quantiles

N	601	Sum wgts	601	100% Max	133
Mean	22.9867	Sum	13815	75% Q3	27
Std Dev	13.3504	Variance	178.233	50% Med	20
Skewness	2.49169	Kurtosis	11.4641	25% Q1	15
USS	424501	Css	106940	0% Min	2
CV	58.0788	Std mean	0.544574		
T : Mean = 0	42.2104	Prob > |T|	0.0001	Range	131
Sgn rank	90450.5	Prob > |S|	0.0001	Q3−Q1	12
Num ^ = 0	601			Mode	19
D : Normal	0.151945	Prob > D	< 0.01		

```
                    Histogram                              #    Boxplot
      135 +*                                               1       *
          .
          .
          .*                                               1       *
          .
          .*                                               1       *
          .*                                               4       *
          .*                                               4       0
          .***                                            18       0
          .****                                           23       0
          .*********                                      60       |
          .***************************                   170    *--+--*
          .*********************************************  267    +----+
        5 +********                                        52       |
          ---+----+----+----+----+----+----+----+----+
```

* May represent up to six counts

Figure 1. Frequency distribution of the results of a diagnostic screening test, the alanine aminotransferase (ALT) in the heparinated blood plasma of 601 male blood donors (compare Table 2). To get the reference values for "healthy subjects", the extreme values were eliminated by an iterative process over 38 laboratory tests. The samples of four hundred and five male blood donors were eliminated by matching the central 0.98 fraction of the results of all laboratory tests. The final reference values for ALT were set to $<50\,U\,l^{-1}$ for men and women (University Hospital Zurich, older values $<60\,U\,l^{-1}$). These values change with the analytical method applied (optimized kinetic UV test at 37 °C), the population studied, and the sex and age of the subjects investigated. The cutoff value between the undiseased and diseased population is relevant to distinguish the two groups. Data of univariate statistics as presented by the SAS statistics software (SAS Statistics Institute, CA).[12,13] [Reprinted from Spichiger and Vonderschmitt[13] with permission from the American Association for Clinical Chemistry.]

always produce more reliable diagnostic results than single analyses. In the 1980s and 1990s, guidelines were published by the American College of Pathologists and the US Preventive Services Task Force, which increased the awareness regarding the prevalence and incidence of specific diseases and the effectiveness of screening tests and preventive examinations.[16,17]

2.2 The Screening Test as Used in Medical Applications

A diagnostic process in modern medicine is a structured process that starts with the history (anamnesis) of a "case" and ends with the treatment. A screening test is normally used as the initial step of a sequential process to analyze a risk. Screening in medicine is an attempt to detect disease in asymptomatic persons where the prevalence (see Table 1) is normally small. In preventive medicine, a screening test is typically used to detect a risk (e.g., pollution) before widespread damage occurs. In medical care, a screening test is applied to answer diagnostic questions highly relevant to society and public health such as the early detection of myocardial infarction,[18] liver disease,[19] diabetes,[20] and genetic predispositions,[21] but also in preventive veterinary medicine.[22,23] For an example, see Table 2.[15]

Table 2. The relationship between the diagnostic parameters and the prevalence of an event. Two hundred and seventeen patients subjected to chronic hemodialysis were investigated with a test for alanine aminotranferase (ALT), an enzyme immunoassay to detect hepatitis C virus (HCV) antibodies and a nested PCR procedure to genotype the HCV/RNA. The objective was to adjust the cutoff value of the ALT test in order to identify patients with HCV infection. The diagram shows an example for the calculations of the diagnostic specificity and sensitivity related to the prevalence of the HCV infection in the selected population. Seventeen infected patients were identified by HCV/RNA positive results. This population was assumed to be truly diseased (DIS). By reducing the cutoff level of the ALT screening test, one more subject was detected as truly positive (TP), which results in an increased sensitivity of the test. Simultaneously, six subjects showed positive ALT values. These subjects have to be additionally investigated by the immunoassay and, in a second step, the HCV/RNA test. Test 1, test 2, allocation of subjects by the ALT test before and after reducing the cutoff value to 50%.[15]

Previous prevalence, united kingdom		6.9% = 0.069	(15 / 217)

Total population:

Test 1

Number	average age (years)		Test	Results			
		Status	NEG	POS		Test parameters	
		Affected	5	10	0.67	Sensitivity	67%
217	51.2 y	Unaffected	168	34	0.83	Specificity	83%
		All	173	44			
		pV neg	: 0.971	0.227 :	pV pos		

After   test 2

	Test	results			
Status	NEG	POS		Test parameters	
Affected	4	11	0.73	Sensitivity	73%
Unaffected	162	40	0.80	Specificity	80%
All	166	51			
pV neg	: 0.976	0.216 :	pV pos		

The success of a screening procedure manifests itself in the ratio of individuals who are correctly allocated to the group of "affected" and "unaffected" individuals relative to those who are incorrectly allocated (see Table 1). The parameters that describe this ratio for both groups are the *efficiency* and, the more economically relevant parameter, the *efficacy* of a test. These two parameters summarize the quality of a screening procedure, however, they rely on a number of more basic parameters.[2,4,11,12,24,25] These basic parameters involve the *biological uncertainty* described by factors such as the *intra-* and *interindividual variation* of test results for healthy and diseased individuals;[1,4,12] the typical frequency distribution of the test results for the two populations "unaffected" and "affected", which include the biological variations; the overlap of the test results for two populations; and the detection limit, which is characteristic for the analytical method. In addition, the analytical uncertainties contribute to the overall uncertainty of the individual allocation (see Section 3).

In summary, several factors are related to the efficiency and efficacy of a test:

- The *prevalence* of affected sources, which is the basis of the Bayes Theorem and the Bayes equations (a priori and a posteriori probability of answer A and B) (see Section 7.2. in this volume and Refs 4 and 9)
- The position of a *cutoff limit* (truncation limit) and the distance of an individual value from the cutoff limit, which divides the diagnostic results of two overlapping populations into two entities
- The *analytical quality* of a diagnostic tool, which should allow to distinguish two populations (analytical uncertainty of a result)
- The position of the *detection* and *determination limits* in a laboratory test referred to the frequency distribution of the "unaffected" (reference values) and the "affected" population
- As the sum of all these performance standards, the diagnostic *specificity*, and *sensitivity* of a diagnostic tool
- The *positive and negative predictive value* of tests and the *likelihood ratio*[2,11]

These factors provide a basis for interpreting the diagnostic result. They allow conclusions

to be drawn about the accuracy of the information yielded and the results to be quantified (probability of an event).[26] In the following part, these performance criteria are explored in more detail.

2.3 The Frequency Distribution and the Cutoff Value

The diagnostic *sensitivity* and *specificity* of a test depend on the position of the cutoff value that separates the two populations, "unaffected" and "affected" (see Table 1). The *sensitivity* is the response to the question "What percentage of all diseased or affected species are identified by the test result?" (i.e., the fraction of true positive test results). The *specificity* refers to the percentage of all nonaffected species that show a negative test result (the fraction of true negative results). Both discriminators are linked to a reference population of samples drawn from unaffected sources[12,27] or to a cutoff limit that allows the results to be referred to a borderline value (which is typically the case in legal metrology). Unlike *analytical* sensitivity and specificity, *diagnostic* sensitivity and specificity refer to the power of the test to separate two populations from each other. The investigator normally is interested in the *positive event* since this "case" will be investigated further with a set of more specific tests and then normally treated.

The *positive predictive value* of a diagnostic test defines the percentage of species with positive test results that are indeed affected (see Table 1). The *negative predictive value* describes the opposite situation and answers the question: "What percentage of species with negative test results are really not affected?". The *positive and negative predictive values* are influenced by the *prevalence* of a specific event or "case" in a population (see Table 2). The rarer a case, the higher the uncertainty that a specific test will be able to identify this case. The positive predictive value becomes low. This is the situation with tests to identify many genetic disorders and was the case in tests for identifying HIV in the mid 1980s. For DNA tests, the uncertainty, in the best case, can be compensated for by increasing the number of redundant and dependent results in the nucleic acid profile. Generally redundant tests are not recommended.

They raise costs without providing any additional benefit.[15,22,23]

2.4 The Likelihood Ratio

In medical diagnostics, the terms considered in the Section 2.3 specify the power of a test to identify a "case" in an epidemiological situation. The *diagnostic sensitivity* and *specificity* of a test are regularly specified for new tests and provide valuable information if novel assays are compared to older ones. Gambino,[28] however, warned in a critical comment that laboratories often fail to recognize that sensitivity and specificity vary with the strength of the signal observed. A signal and result far higher than the cutoff value is more likely to indicate "disease" than a value just beyond the cutoff is. However, both values are reported as positive. This problem is linked to the final step in chemical analysis of data reduction where data are reduced from a continuous numerical scale to a binary decision. In view of this problem, the *likelihood ratio* is specified.[2,11] In order to evaluate the likelihood ratio, a relevant range, where the values of the affected and not affected population are overlapping, is divided into classes with different "weights". In each class, the ratio between negative results for those unaffected and positive results for affected sources is calculated. This procedure allows the distance (weight) of a class of results from the cutoff position along a scale of numerical data (x axis) to be taken into account.

2.5 The Influence of the Detection and Determination Limit on Decision Making

The performances of screening tests and point-of-care tests (POCT)[29] that are economically attractive are often of limited analytical quality. It is not primarily the uncertainty of the test that is critical but the detection limit. In order to distinguish between "polluted" sources and natural ones and to sort out "affected" people from a population, a decisive cutoff value has to be set. This cutoff value is a reference limit that allows two different conditions to be distinguished, that is, healthy and diseased or natural and polluted.

Example: For populations suffering from metabolic disorders where a low level of the target analyte as well as a high level are critical (for blood glucose or hormones, for instance), two cutoff values have to be set. These discriminate between three populations namely a "normal" status, a status linked to "low" results and a status linked to "high" results. Such two-sided statistical distributions of test results involving two cutoff values are relatively rare. A similar situation, however, arises in the quality assessment of production lines where vitamins, catalysts, enzymes, dyes, and so on, are added to a product and the quantity of the added compound can be "just appropriate", "low" or "high" or "out of range" depending on the distance from the center of gravity on both sides of the frequency distribution.

Skewed distributions (see Figure 1) are the rule for tests where the detection limit of the test is close to the cutoff value and for environmental specimens where two populations are not clearly separated. The degree of pollution results in a long right-hand tailing of the frequency distribution of the analytical results. In the first case, the distribution of test results is distorted artificially owing to the detection and determination limit of the tests.[30]

For the design and quality of a screening test, the detection limit of the analytical procedure is most relevant. An inappropriate detection limit will not enable the results to be allocated correctly. The performance of a test may, nevertheless, be sufficient to sort out extreme cases and to identify highly polluted sources. A similar situation arises if the analytical selectivity of a test is insufficient. In contrast, the uncertainty contributes primarily to the rigor with which two events are separated from each other. As a consequence, the diagnostic quality of the test is impaired.

2.6 Point-of-care Testing (POCT)

Emergency tests and a relatively small number of screening tests are preferentially executed on site and near the patient—the so-called "POCT".[29] It is then possible to obtain an answer immediately and to tightly coordinate diagnostics and treatment (remediation). POC tests should therefore be simple, have a reasonable price, and demonstrate the appropriate discriminating power.

3 THE VALIDATION OF AN ANALYTICAL PROCESS

3.1 Introduction

The analytical chemist is supposed to answer a specific question about the composition of a specimen by measuring one or several target analytes (or measurands). In many cases in chemical analysis, the *measurand* will be the concentration of an analyte. However, chemical analysis is used to measure other quantities additionally, for example, color, pH or temperature and therefore the term *measurand* is more generally used in quality assessment guides.

In order to receive the required information, the specialist selects, develops, and uses chemical or physicochemical methods which suit the purpose and provide the necessary quality, which may not necessarily be the best one, but adequate for investigating the specimen. According to Kellner et al.,[31] the *validation of an analytical method* means having "to identify all possible sources of errors which might affect its performance". The validation of an analytical method is the process that shows whether the results produced by this method are reliable and reproducible, and whether the method is suitable for the intended application.

A more substantial and more stringent definition is given by the international organizations such as IUPAC, which refer the definition to reference materials and to quantities defined by the SI (Système International d'Unités) (see subsequent text). However, independent of the definition, the results of every analytical method must meet the goals of discriminating between at least two different conditions of a specimen and two different populations of results. In the preanalytical phase, this discrimination task has to be decided on and the expected efficiency and efficacy of the analytical method must be identified. A result subjected to a high uncertainty is of no value.

In parallel, the term *model validation* describes the "assessment of the extent to which a diagnosis/an assumption/a model is well founded, is tractable, and fulfills the purpose for which it is formulated" (Ref. 11, p. 109). As a basis for the development of multiple tests and diagnostic arrays, a model of the chemical/biochemical background of a diagnostic question must be formulated. This model must be validated on the basis of analytical investigations before it is used daily to solve diagnostic challenges. Depending on how the model performs, it will either be confirmed or rejected.

In the first part of this section, the focus was on the diagnostic impact of an analytical method and its uncertainties. In the second part, the technical sources of the analytical uncertainty are considered.

The analytical uncertainty is referred to: the choice of the analytical method, the sampling and pretreatment of a specimen, sample storage and transport, the determination process (manual, automatic, highly complex, etc.), calibration and quality of reference specimens, calculations, choice of quantities, presentation of results, validation, and comments.

3.2 Quality Assessment is the Pathway to Validated Results

The basics of classical quality assessment guidelines are described in Ref. 31 The guidelines of international organizations are specified for different applications and purposes and documented in various documents published by IFCC, ISO, and NIST,[30,32–35] and in volumes such as the WHO *Guidelines for Drinking-Water Quality*.[36] Moreover, the quality assessment protocols in the EU and the standardization bodies of every country have their own documents on their homepages for downloading.

Here, more specific questions concerning the applications of biosensors and sensor arrays and the *quality of information* are addressed.

The procedure for providing a valid diagnosis and a "best practice" treatment is similar worldwide, wherever the standard of medical care is comparable. Therefore, diagnostic tools and treatments can be commercialized globally. The circle can be considered closed, if the analytical methods and results are of comparable quality around the globe. This last goal can be met by implementing the recommendations for the quality assessment of analytical methods and tools of international organizations such as the Cooperation on International Traceability in Analytical Chemistry (CITAC),[37] the World Health Organization (WHO),[36] the International Federation of Clinical Chemistry (IFCC),[30] the International Union of Pure and Applied Chemistry (IUPAC),[38] and others. Each of these organizations has a number of committees and subcommittees with specialists who are involved in global surveys and drawing up recommendations.

One such organization is the ISO organization in Geneva,[32] which publishes international recommendations and standards, known as *ISO guides and ISO standards*, devoted to a range of specific topics (see Refs 32–35).

3.3 Method Validation and Quantifying Uncertainty

Generally *validation of a method* means to *trace back* the results of an analytical investigation to internationally accepted quantities and to the values found in a reference material by a method that achieves the best possible performance linked to the lowest possible uncertainty for a biological sample.

The relevant ISO guidelines 15194:2002(E) are entitled "In vitro diagnostic medical devices—Measurement of quantities in samples of biological origin—Description of reference materials". Section 5.9 of this ISO guide is devoted to the validation process for an analytical method, where the steps involved are:

5.9.1 Planning of the experimental design
5.9.2 Assessment of homogeneity
5.9.3 Statistical evaluation of results
5.9.4 Assessing the stability
5.9.5 Value assignment
5.9.6 Value and uncertainty of measurements assigned by one measurement procedure in one laboratory
5.9.7 Regional recognition (of the reference material).

These subtitles give us an idea of the content and complexity of the "validation process", which has to be performed not only for traditional wet-chemical screening tests but in as much for analytical procedures by chemical and biochemical sensors such as ion-selective electrodes, immunoassays etc. (see Refs 12, 13, 15, 39, 40).

3.4 The Terminology Used in Chemical Analysis and Diagnostics

The *International Vocabulary of Basic and General Terms in Metrology* (VIM, International vocabulary of basic and general terms in metrology. ISO, Geneva (1993)) defines a number of statistical terms such as *uncertainty, standard deviation,* and *traceability.*

3.4.1 Uncertainty and Standard Deviation

The definition of the term *uncertainty* (of measurement) used in this protocol is taken from the version adopted in the current International Vocabulary of Basic and General Terms in Metrology:

"A parameter associated with the result of a measurement that characterizes the dispersion of the values that could reasonably be attributed to the measurand."

This parameter may be assigned, for example, a *standard deviation* (or a given multiple of it) or the width of a confidence interval.

The uncertainty of measurement comprises, in general, many components. Some of these components may be evaluated from the statistical distribution of the results of a series of measurements and can be characterized by standard deviations. The other components, which can also be characterized by standard deviations, are evaluated from model probability distributions (model distributions, reference values) based on experience or other information. The ISO Guide refers to these different cases as Type A and Type B estimations, respectively.

Every experimentally evaluated, measured value is subjected to *uncertainties* inherent in the physicochemical procedure and in the handling of a sample. This uncertainty of the value reduces the discriminating power of the result and the performance of the method. The validation process investigates and characterizes the performance of an analytical method and the discriminating power of its results. Simultaneously, the process provides information about the efficiency of the method for the purpose for which the method was selected. One EU document, the EURACHEM/CITAC guide,[41] (2nd edition), is devoted to "Quantifying Uncertainty in Analytical Measurement".

3.4.2 Traceability and Comparability

The VIM-guide, ISO, Geneva (1993) defines *traceability* as the "property of the result of a measurement or the value of a standard whereby it can be related to stated references, usually national, or international standards, through an unbroken chain of comparisons all having stated uncertainties".

This definition implies a need for effort at the national and international level to provide widely accepted reference standards, and at the individual laboratory level to demonstrate the necessary links to these standards. At the national and international level, the *comparability* between national measurement systems is being continually improved by intercomparison of measurement standards at the National Metrology Institute (NMI) level. A multilateral mutual recognition agreement was signed in 1999 by the member nations of the Meter Convention in response to the need for an open, transparent, and comprehensive scheme to give users reliable quantitative information on the comparability of national metrology systems.

3.5 Standard Reference Materials (SRMs)

These are to a large extent provided by (European) Institute for Reference Materials and Measurements (IRMM) [42] and the (US) National Institute for Standardization (NIST) [43]: http://ts.nist.gov/ts/htdocs/230/232/232.htm)

The National Institutes for Standardization such as NIST support accurate and compatible measurements by certifying and providing Standard Reference Materials (SRMs) with well-characterized composition or properties, or both. The mission of the IRMM is to promote a common and reliable European measurement system in support of EU policies. The prime objective of the IRMM is to build up confidence in the comparability of measurements by producing and disseminating internationally accepted quality assurance tools. These tools include validated methods, reference materials, reference measurements, interlaboratory comparisons, and training.

NIST provides more than 1100 products. These are used to perform instrument calibrations in units as part of the overall quality assurance programs, to verify the accuracy of specific measurements, and to support the development of new measurement methods. Industry, academia, and government use reference materials such as NIST SRMs "to facilitate commerce and trade and to advance research and development. NIST SRMs are also one mechanism for supporting measurement traceability in the United States".[43]

NIST SRMs are currently available for use in areas such as microanalysis, health, and industrial hygiene (e.g., DNA profiling), forensics (e.g., drug abuse, DNA profiling), ion activity (pH, pNa, pK, pCl, pF), and food and agriculture (e.g., vitamins, fat, proteins). Reference materials (RM) can be ordered on-line (https://srmors.nist.gov/).

The products are labeled as "*Certified Reference Material (CRM), Reference Material (RM), NIST Standard Reference Material*® *(SRM), NIST Reference Material, NIST Traceable Reference Material*™ (NTRM™)". The homepage[43] provides a link where definitions of these materials can be found. Similar to the label of the materials, the values of the amount or concentration of a target analyte is defined by the terms *Certified Value* and *Reference Value*. The certificates for the content and specification of the products are shipped together with the product. The certificates define and confirm the quality of the product.

3.6 "Abuse and Misuse of Quantities"[31]

The first edition of the "Manual of Symbols and Terminology for Physicochemical Quantities and Units" was prepared for publication on behalf of the Physical Chemistry Division of IUPAC by M.L. Glashan in 1969. He described the objective of his contribution in the preface to that first edition as being "to secure clarity and precision, and wider agreement in the use of symbols, by chemists in different countries, among physicists, chemists, and engineers, and by editors of scientific journals". (Ref. 38). By convention physical quantities are organized in a multidimensional system built upon seven base quantities (length, mass, time, electric current, thermodynamic temperature, amount of substance, and luminous intensity). All other physical quantities are called *derived quantities*. The dimensions of these quantities are algebraically derived from the seven base quantities by multiplication and division. Units, which do not

The process of chemical analysis: analytical and diagnostic validation of data

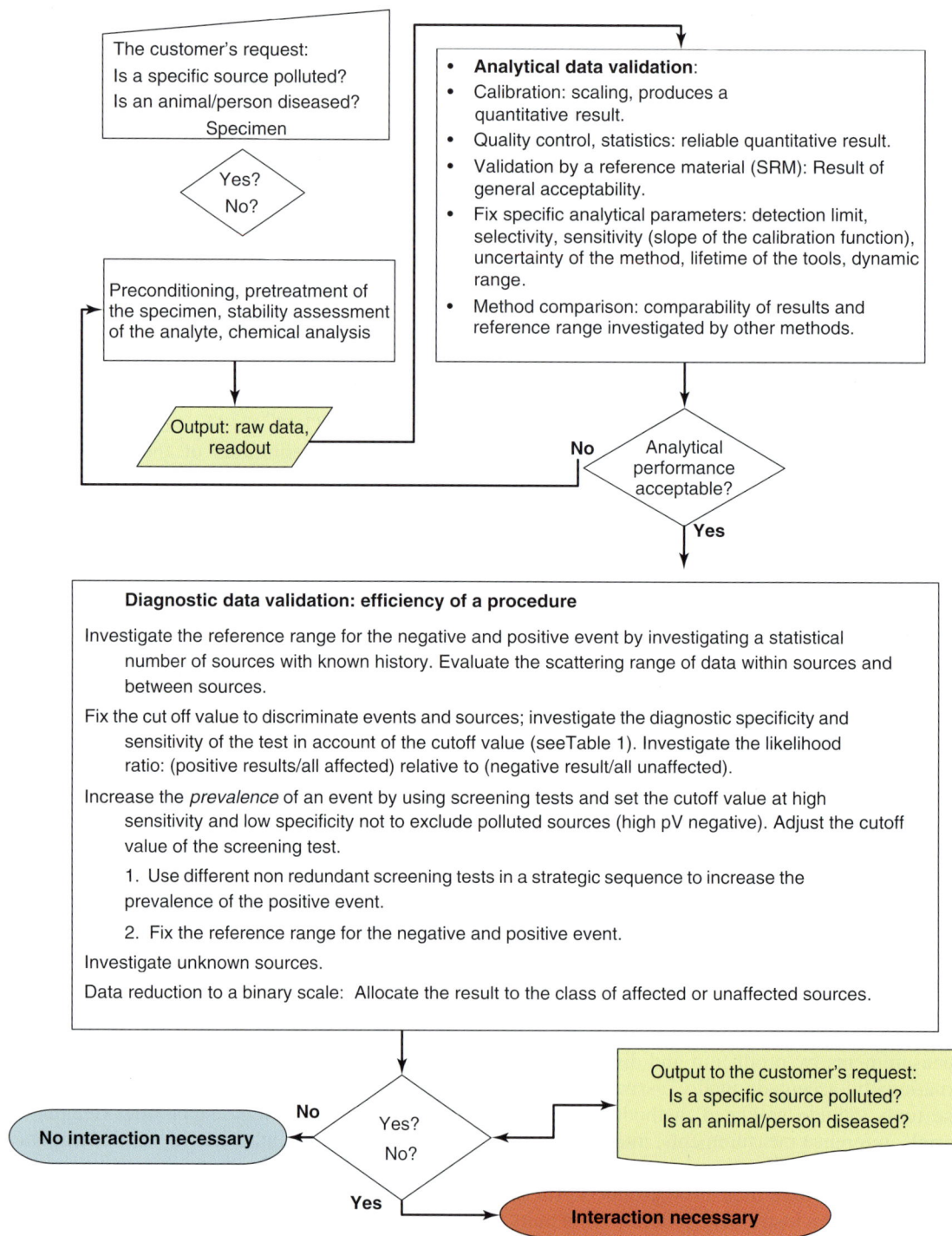

The customer's request:

Is a specific source polluted?

Is an animal/person diseased?

Specimen

Yes?
No?

Preconditioning, pretreatment of the specimen, stability assessment of the analyte, chemical analysis

Output: raw data, readout

- **Analytical data validation**:
- Calibration: scaling, produces a quantitative result.
- Quality control, statistics: reliable quantitative result.
- Validation by a reference material (SRM): Result of general acceptability.
- Fix specific analytical parameters: detection limit, selectivity, sensitivity (slope of the calibration function), uncertainty of the method, lifetime of the tools, dynamic range.
- Method comparison: comparability of results and reference range investigated by other methods.

No   Analytical performance acceptable?

Yes

Diagnostic data validation: efficiency of a procedure

Investigate the reference range for the negative and positive event by investigating a statistical number of sources with known history. Evaluate the scattering range of data within sources and between sources.

Fix the cut off value to discriminate events and sources; investigate the diagnostic specificity and sensitivity of the test in account of the cutoff value (seeTable 1). Investigate the likelihood ratio: (positive results/all affected) relative to (negative result/all unaffected).

Increase the *prevalence* of an event by using screening tests and set the cutoff value at high sensitivity and low specificity not to exclude polluted sources (high pV negative). Adjust the cutoff value of the screening test.

1. Use different non redundant screening tests in a strategic sequence to increase the prevalence of the positive event.

2. Fix the reference range for the negative and positive event.

Investigate unknown sources.

Data reduction to a binary scale: Allocate the result to the class of affected or unaffected sources.

No   Yes?
No?

No interaction necessary

Output to the customer's request:
Is a specific source polluted?
Is an animal/person diseased?

Yes   Interaction necessary

Figure 2. The process of chemical analysis: analytical and diagnostic validation of data.

conform to these fundamental guidelines, are not recommended.

Given the SI dimensions (Système International), quantities such as ppm, ppb, mmHg, and g/100 ml are not recommended and should not be used anymore. It is absolutely essential to state which quantity is measured by the analytical method. A typical example is that of measuring the active molality of substrates and electrolytes by biosensors and ion-selective electrodes, which is reported in molar units (mol^{-1} instead of osmol or active molality per mass of solvent (in mol per kg water)). The active molality is a quantity defined on a purely thermodynamic basis and, at the same time, it is the quantity measured by sensors and biosensors in a specimen directly. This quantity is independent of the amount of proteins, lipids, and water in a specimen. Unfortunately, most reference materials also do not yet routinely report these quantities.

4 CONCLUSIONS

The technical validation of methods and procedures which are applied to investigate the substance concentration of decisive chemical compounds (analytes, measurands) is absolutely necessary in order to be able to interpret the results of laboratory tests. The technical validation has to be executed globally on the basis of generally valid international recommendations and rules for the results to be globally comparable. The goal of the technical validation of analytical procedures is to minimize the technical uncertainty of laboratory results.

In addition to the technical uncertainties, biological variations, and shifts of the value of a measurand contribute to the overall uncertainty of a test result. In order to investigate these natural uncertainties, the technical uncertainties must be known. Figure 2 provides an overview on the subsequent validation processes involved. The procedure and rules are the basis to validate the analytical data, to reduce data to a binary decision, and to draw conclusions in environmental control, health monitoring, quality assessment of food, and others. The whole process is the fundament of actions to be initiated.

REFERENCES

1. C. G. Fraser, *Interpretation of Clinical Chemistry Laboratory Data*, Blackwell Scientific, Oxford, 1986.
2. H. Keller, Clinical and laboratory investigation. *Scandinavian Journal of Clinical and Laboratory Investigation*, 1986, **46**, Suppl. 181, 1–74.
3. H. Keller, *Klinisch-Chemische Labordiagnostik für die Praxis*, Georg Thieme Verlag, Stuttgart, New York, 1986.
4. U. E. Spichiger-Keller, *Chemical Sensors and Biosensors for Medical and Biological Applications*, Wiley-VCH, Weinheim, 1998.
5. College of American Pathologists (CAP), Analytical, *Goals in Clinical Chemistry*, in *Proceedings of the 1976 Aspen Conference on Analytical Goals in Clinical Chemistry*, F. R. Elevitch (ed), British Library, Boston, Special Issue 6842–48635, 1978.
6. C. E. Shannon and W. Weaver, *The Mathematical Theory of Communication*, University of Illinois Press, Chicago, 1949.
7. E. R. Gabrieli, *Clinically Oriented Documentation of Laboratory Data*, Academic Press, London, 1972, pp. 9–36.
8. K. Eckschlager and K. Danzer, *Information Theory in Analytical Chemistry*, in *Chemical Analysis*, J. D. Wineforder (ed), John Wiley & Sons, New York, 1994, Vol. 128.
9. http://www.epa.gov/Region5/defs/html/tsca.htm EPA Region 5, Toxic Substances Control Act of 1976.
10. S. P. Bradbury, T. R. Henry and R. W. Carlson, in *Practical Applications of Quantitative Structure-Activity Relationship (QSAR) in Environmental Chemistry and Toxicology*, Chemical and Environmental Science Series W. Karcher, and J. Devillers (eds), Kluwer Academic Publisher, Dodrecht, 1990, Vol. 1, pp. 295–315.
11. H. Keller and Ch. Trendelenburg, *Data Presentation, Interpretation*, in *Clinical Biochemistry*, H. Ch. Curtius and M. Roth (eds), Walter de Gruyter, Berlin, New York, 1989, Vol. 2.
12. U. Spichiger-Keller, *A Selfconsistent Set of Reference Values*, Ph.D. Thesis Nr. 8830, Swiss Federal Institute of Technology, 1989.
13. U. E. Spichiger and D. Vonderschmitt, A self-consistent set of reference values for 23 clinical chemical analytes. *Clinical Chemistry*, 1989, **35**, 448–452.
14. H. E. Solberg, IFCC, 1985/5, The theory of reference values, Part 5. Statistical treatment of collected reference values. Determination of reference limits. *Journal of Clinical Chemistry and Clinical Biochemistry*, 1983, **21**, 749–760; IFCC, International Federation of Clinical Chemistry. http://www.ifcc.org.
15. E. P. A. Lopes, E. C. Gouveia, A. C. C. Albuquerque, L. H. B. C Sette, L. A. Mello, R. C. Moreira and M. R. C. D. Coelho, Determination of the cut-off value of serum alanine aminotransferase in patients undergoing hemodialysis, to identify biochemical activity in patients with hepatitis C viremia, *Journal of Clinical Virology*, 2006, **35**, 298–302.
16. G. H. Tabas and M. S. Vanek, *Is 'routine' Laboratory Testing a Thing of the Past?*, 2006, www.postgradmed.com/issues/1999/03_99/tabas.htm.

17. M. I. Evans, R. S. Galen, and D. W. Britt, *Principles of Screening, Seminars in Perinatology*, Elsevier, 2005, 364–366, Evans@CompreGen.com.

18. A. H. B. Wu, Analytical and clinical evaluation of new diagnostic tests for myocardial damage. *Clinica Chimica Acta*, 1998, **272**, 11–21.

19. D. Thabut, S. Naveau, F. Charlotte, J. Massard, V. Ratziul, F. Imbert-Bismut, D. Cazals-Hatem, A. Abella, D. Messous, F. Beuzen, M. Munteanu, J. Taieb, R. Moreau, D. Lebrec and T. Poynard, The diagnostic value of biomarkers (AshTest) for the prediction of alcoholic steato-hepatitis in patients with chronic alcoholic liver disease. *Journal of Hepatology*, 2006, **44**, 1175–1185.

20. F. Guerrero-Romero and M. Rodríguez-Morán, Lowered criterion for normal fasting plasma glucose: impact on the detection of impaired glucose tolerance and metabolic syndrome. *Archives of Medical Research*, 2006, **37**, 140–144.

21. M. K. Schwartz, Genetic testing and the clinical laboratory improvement amendments of 1988: present and future. *Clinical Chemistry*, 1999, **45**, 739–745.

22. D. Jordan and S. A. McEwen, Herd-level test performance based on uncertain estimates of individual test performance, individual true prevalence and herd true prevalence. *Preventive Veterinary Medicine*, 1998, **36**, 187–209.

23. R. D. Smith and B. D. Slenning, Decision analysis: dealing with uncertainty in diagnostic testing. *Preventive Veterinary Medicine*, 2000, **45**, 139–162.

24. R. Galen and S. Gambino, *Beyond Normality: The Predictive Value and Efficiency of Medical Diagnoses*, John Wiley & Sons, New York, 1975.

25. R. S. Galen and S. R. Gambino, *Norm und Normabweichung klinischer Daten*, Gustav Fischer Verlag, Stuttgart, 1979.

26. C. Heusghem, A. Albert, and E. S. Benson, *Advanced Interpretation of Clinical Laboratory Data*, in *Clinical and Biochemical Analysis*, M. K. Schwartz (ed), Marcel Dekker, New York and Basel, 1982, Vol. 2.

27. H. F. Martin, B. J. Gudzinowicz, and H. Fanger, *Normal Values in Clinical Chemistry, A Guide to Statistical Analysis of Laboratory Data*, in *Clinical and Biochemical Analysis*, M. K. Schwartz (ed), Marcel Dekker, New York and Basel, 1975, Vol. 2.

28. R. Gambino, The misuse of predictive value—or why you must consider the odds. *Annalidell Istituto Superiore Di Sanita*, 1991, **27**(3), 395–399 PMID: 1809058 [PubMed—indexed for MEDLINE].

29. Ch. P. Price and J. M. Hicks, *Point-of-Care Testing*, The American Association for Clinical Chemistry, AACC-Press, Washington, 1999 (jhicks@cnmc.org; c.p.price@mds.qmw.ac.uk).

30. IFCC/NCCLS, EP17-A, *Protocols for Determination of Limits of Detection and Limits of Quantitation, Approved Guidelines*, NCCLS, Wayne, 2005, 24/34; IFCC, International Federation of Clinical Chemistry, homepage: http://www.ifcc.org/ifcc.asp.

31. R. Kellner, J.-M. Mermet, M. Otto, and H. M. Widmer, *Analytical Chemistry*, Wiley-VCH, Weinheim, 1998.

32. ISO, *Capability of Detection—Part 1 to 4*, ISO 11843–1 to 11843–4, ISO, Geneva, 1997 and 2000, International Organization for Standardization, Geneva, Homepage and ISO documents to order, http://www.iso.org/iso/en/ISO Online.frontpage; http://www.iso.org/iso/en/isostore.

33. ISO, *Accuracy (Trueness and Precision) of Measurements and Results—Part 1. General Principles and Definitions*, ISO 5725–1, Geneva, 1994.

34. ISO 3822-1; 1999, Measurement uncertainty, and ISO/TS 21749:2005, http://store.iso.org/isostore/public.

35. ISO/TS 21748, Geneva, 2004, *Guide to the Expression of Uncertainty in Measurement*, ISO, Geneva, 1993. International Organization for Standardization, Geneva, Homepage and ISO documents to order, http://www.iso.org/iso/en/ISOOnline.frontpage; http://www.iso.org/iso/en/isostore.

36. World Health Organization (WHO), *Guidelines for Drinking-Water Quality*, 2nd Edn, WHO, Geneva, 1993, Vol. 1, (ISBN 92 4 154460); 1996, Vol. 2, (ISBN 92 4 154480 5); 1997 Vol. 3, (ISBN 92 4 154503 8), homepage: www.who.int/water_sanitation_health/dwq/gdwq2v1/en/print.html.

37. CITAC, *Co-Operation on International Traceability in Analytical Chemistry*, 2006, http://www.citac.ws.

38. IUPAC, International Union of Pure and Applied Chemistry, 2006, http://www.iupac.org/index_to.html.

39. U. E. Spichiger, *Basic Principles in Magnesium Assessment. The Relationship Between the Diagnostic Value of a Laboratory Determination and the Quality of the Analytical Procedure*, in *Magnesium, A Relevant Ion*, B. Lasserre and J. Durlach (eds), John Libbey & Company Limited, London, 1991, pp. 217–226.

40. H. W. Bucher, U. E. Spichiger, and E. Jenny, *Total Magnesium Concentrations in Serum and Erythrocytes Before and After Treatment with Magnesium l-Aspartate Hydrochloride: Reference to Clinical Symptoms*, in *Magnesium 1993*, S. Golfe, D. Dralle and L. Vecchiet (eds), John Libbey & Company Limited, London, 1993, pp. 155–161.

41. CITAC, *Quantifying Uncertainty in Analytical Measurement (QUAM)*, 2nd Edn, 2000 P1, homepage: http://www.citac.ws, and http://www.measurementuncertainty.org. The guide has been produced primarily by a joint EURACHEM/CITAC Working Group in collaboration with representatives from AOAC International and EA. Production of the guide was in part supported under the contract with the UK Department of Trade and Industry as part of the National Measurement System Valid Analytical Measurement (VAM) Programme.

42. IRMM, Institute for Reference Materials and Measurements, Homepage, http://www.irmm.jrc.be; Contact: European Commission; Directorate-General Joint Research Centre, Institute for Reference Materials and Measurements; B-2440 Geel, Belgium. For reference materials sales, contact the sales department, 2006, E-mail: jrc-irmm-rm-sales@ec.europa.eu.

43. NIST, (US) National Institute for Standardization, homepage: http://ts.nist.gov/ts/htdocs/230/232/232.htm); program questions: Public Inquiries Unit: (301) 975-NIST (6478), TTY (301) 975–8295, NIST, 100 Bureau Drive, Stop 1070, Gaithersburg, MD 20899–1070, 2006.

Introduction to Bayesian Methods for Biosensor Design

Edmund S. Jackson and William J. Fitzgerald

Department of Engineering, University of Cambridge, Cambridge, UK

1 INTRODUCTION

Biosensors are an exciting technology, rapidly growing in sophistication and expanding in their application. One factor is common to all biosensors: an a priori knowledge of the entity to be measured by the sensor. It is in discovering this entity, be it a molecule, molecular concentration or relationship between such concentrations, or any other indicator of state, that the chapters shall be concerned with.

A typical workflow resulting in a biosensor is the collection of both control and condition samples, high sensitivity screening of those samples to discover the properties that discriminate between control and condition, verification of these findings and finally the design of a biosensor. The philosophy is that the initial high sensitivity screening, based on sensitive, high dynamic range technologies such as NMR or mass spectrometry, is able to measure a wide variety of substances and hence provide the greatest probability of finding interesting substances. However, due to their expense and complicated operation these technologies are not widely available. Hence, it is desired to identify a small subset of these interesting substances against which a specific and inexpensive biosensor may be designed, for wide deployment.

Hence the work of the Bayesian analyst is in the *mining* of high throughput data to identify this subset of interesting targets. However this classification must be very reductionist in nature. It is clearly of no value to construct a classifier, regardless of performance, that requires a great many inputs over a wide dynamic range, for then the original instrument is required to perform the classification, whereas the goal is to classify using a smaller, inexpensive device. Hence, the designed classifier must have "acceptable" performance, but the primary design goal is the minimizing of the number of inputs. This is often thought of as "finding biomarkers", for with a few such markers a specific, and hence inexpensive, device may be designed against each in order to perform the classification.

Unfortunately, the Bayesian paradigm in inclined toward the opposite, with complex classifiers, such as model averaging methods, as the norm. Nevertheless, it is entirely possible to proceed fruitfully and in this chapter we shall explore such avenues. Bearing in mind the nonspecialist nature of our audience we shall provide an introduction to each idea, presented simply with references to more thorough treatments for the adventurous.

This chapter is organized linearly, presenting all the required concepts and components and arriving finally at the desired classifier. In the Section on 2 we introduce Bayesian inference and its two component tasks, parameter estimation and model selection. In order to achieve these tasks on real data we require techniques of numerical integration which we describe in the Section on 3.

Handbook of Biosensors and Biochips. Edited by Robert S. Marks, David C. Cullen, Isao Karube, Christopher R. Lowe and Howard H. Weetall.
© 2007 John Wiley & Sons, Ltd. ISBN 978-0-470-01905-4.

Finally, in the Section on 4 we present a Bayesian classification algorithm tuned specifically to the problem of biosensor design.

2   BAYESIAN INFERENCE

We begin our exploration with an introduction to Bayesian inference and its two constituents *parameter estimation* and *model selection*. It will be seen that prediction and classification, are particular cases of the ideas given here.

2.1   Parameter Estimation

At the first level it is assumed that one of the models within a chosen set, is the correct (or appropriate) model with which to interpret the data and the problem of inference at this level consists of extracting values for the free parameters of the model, given the observed data.[1]

Bayes' theorem may be used to express the posterior probability of the parameters as:

$$P(\boldsymbol{\theta}_k|\mathbf{D}, \mathcal{M}_k) = \frac{P(\mathbf{D}|\boldsymbol{\theta}_k, \mathcal{M}_k)\, P(\boldsymbol{\theta}_k \mid \mathcal{M}_k)}{P(\mathbf{D} \mid \mathcal{M}_k)} \quad (1)$$

where $\mathcal{M}_k$ represents the kth model, $\mathbf{D}$ is a vector of observed data, $\boldsymbol{\theta}_k$ is a vector of model parameters. $P(\boldsymbol{\theta}_k|\mathbf{D}, \mathcal{M}_k)$ is the *posterior probability* and $P(\mathbf{D}|\boldsymbol{\theta}_k, \mathcal{M}_k)$ is the *likelihood*. $P(\boldsymbol{\theta}_k \mid \mathcal{M}_k)$ is the *prior probability* of the parameters $\boldsymbol{\theta}_k$ before the data were observed and $P(\mathbf{D} \mid \mathcal{M}_k)$ is a quantity called the *evidence* which at this level is a constant.

The evidence plays an essential role in *model selection*. All of the probabilities and probability density functions have to be conditioned on $\mathcal{M}_k$, which represents the particular *model structure* being considered. For example, if we were considering spectral estimation and we were interested in the number of sinusoids present in the observed data, the *model structures*, $\mathcal{M}_k$, would be $\mathcal{M}_1$, $\mathcal{M}_2$, $\mathcal{M}_3$, and so on, corresponding to one, two or three sinusoids being consistent with the data.[2]

In many problems it may not be required to estimate all the elements of the model parameter vector $\boldsymbol{\theta}_k$ and parameters which are of no interest are known as *nuisance parameters*. A powerful feature of the Bayesian framework is that the *nuisance*

parameters may be removed from consideration by integration, a process called *marginalization*.

The parameter vector may be partitioned into two sets of parameters: $\boldsymbol{\theta}_{k,1}$, which are the parameters of interest and $\boldsymbol{\theta}_{k,2}$, which are the *nuisance parameters*.

The posterior probability can now be rewritten as:

$$P(\boldsymbol{\theta}_k|\mathbf{D}, \mathcal{M}_k) \equiv P(\boldsymbol{\theta}_{k,1}, \boldsymbol{\theta}_{k,2}|\mathbf{D}, \mathcal{M}_k) \quad (2)$$

And the posterior marginal density of the parameters of interest, $\boldsymbol{\theta}_{k,1}$, may be obtained from

$$P(\boldsymbol{\theta}_{k,1})|\mathbf{D}, \mathcal{M}_k) = \int_{\mathcal{R}^k} P(\boldsymbol{\theta}_{k,1}, \boldsymbol{\theta}_{k,2}|\mathbf{D}, \mathcal{M}_k)\, \mathrm{d}\boldsymbol{\theta}_{k,2} \quad (3)$$

In an estimation problem one assumes that the model is true for some unknown values of the model parameters, and one explores the constraints imposed on the parameters by the data, using Bayes' theorem. The hypothesis space for an estimation problem is therefore the set of possible values of the parameter vector $\boldsymbol{\theta}_k$, for a fixed, assumed *model structure*, $\mathcal{M}_k$, and it is this vector that will form the *hypothesis* that will be used in Bayes' theorem. The data form the sample space, and both the hypothesis space and the sample space may be either discrete or continuous.

Any data which may be described in terms of a linear combination of basic functions with an additive Gaussian noise component satisfies the *general linear model*. Suppose the observed data may be described by a model of the form:

$$d(n) = \sum_{q=1}^{Q} b_q g_q(n) + e(n) \quad \text{for} \quad 1 \leq n \leq N \quad (4)$$

where $g_q(n)$ is the value of a time dependent model function $g_q(t)$ evaluated at time t_n, represented by integers, n, for uniform sampling.

This can be written in the form of a matrix equation:

$$\mathbf{D} = \mathbf{G}\,\mathbf{b} + \mathbf{e} \quad (5)$$

where: $\mathbf{D}$ is an $N \times 1$ vector of data points, $\mathbf{e}$ is an $N \times 1$ vector of noise samples, $\mathbf{G}$ is an $N \times Q$ matrix whose columns are the basic functions evaluated at each point in the time series and $\mathbf{b}$ is a $Q \times 1$ linear coefficient vector.

Many of the *standard* signal processing model structures can be represented with the general linear model: the sinusoidal model, the autoregressive (AR) model, the autoregressive with external input (ARX) model, the nonlinear autoregressive (NAR) model, the Volterra model, the radial basis function model and so on.[1]

Under certain simplifying assumptions, for example, uniform and Jeffreys' priors for certain parameters,[1] and assuming Gaussian noise statistics, the linear parameters in the *general linear model* and the noise variance can be integrated out of the posterior distribution giving the marginal posterior for the remaining parameters, $\boldsymbol{\theta}_Q$, of interest:[1]

$$P\left(\boldsymbol{\theta}_Q \mid \mathbf{D}, \mathcal{M}_Q\right)$$

$$\times \propto \frac{\left[\mathbf{D}^\mathsf{T}\mathbf{D} - \mathbf{D}^\mathsf{T}\mathbf{G}\left(\mathbf{G}^\mathsf{T}\mathbf{G}\right)^{-1}\mathbf{G}^\mathsf{T}\mathbf{D}\right]^{\frac{-(N-Q)}{2}}}{\sqrt{\det\left(\mathbf{G}^\mathsf{T}\mathbf{G}\right)}} \quad (6)$$

Note that this is a function of $\boldsymbol{\theta}_Q$ only. This means that there is no need to know about the standard deviation nor the values of the linear parameters in order to infer the values of $\boldsymbol{\theta}_Q$. Here the integrals have been done analytically so the dimensionality of the parameter space was reduced for each parameter integrated out. This reduction of the dimensionality is a property of Bayesian marginal estimates and can be a major advantage in many applications.

2.2 Model Selection

There are numerous statistical inference problems that require model selection with respect to a set of competing models, $\{\mathcal{M}_k\}$, $k = 1, \ldots, K$. This is the basis of the second level of inference. The Bayesian solution to the model-selection problem is to choose one that fits the prior knowledge and data the best, that is, one that has the highest posterior probability. This is however, only optimal in the sense of maximizing the expected utility function if one uses a zero-one utility function and the choice of a model is the main objective rather than as the basis of a decision problem.[2,3]

This is of course rather nice, but the question starts to become a problem if you take it too far.

One must always keep in mind what it is that one is requiring from a model selection criteria—the "true" model might not be within the set chosen, one might just be interested in how well the model predicts future data and the concept of a "true" model might not be appropriate—or a "true" model might not even exist.

In many cases, the optimal model choice is taken as the one that maximizes $\arg\max_k P(\mathcal{M}_k \mid \mathbf{D})$.

Since

$$P(\mathcal{M}_k \mid \mathbf{D}) \propto P(\mathbf{D} \mid \mathcal{M}_k)P(\mathcal{M}_k) \quad (7)$$

one can readily see that the central element of the Bayesian model selection procedure is the evaluation of the quantity $P(\mathbf{D} \mid \mathcal{M}_k)$, referred to as the *marginal likelihood*, *integrated likelihood* or *model evidence*.

Denoting $\boldsymbol{\theta}_k$ to be a vector of parameters under model $\mathcal{M}_k$, $P(\mathbf{D} \mid \mathcal{M}_k)$ can be written as

$$P(\mathbf{D} \mid \mathcal{M}_k) = \int P(\mathbf{D} \mid \boldsymbol{\theta}_k, \mathcal{M}_k)P(\boldsymbol{\theta}_k \mid \mathcal{M}_k)\,\mathrm{d}\boldsymbol{\theta}_k \quad (8)$$

It can also be expressed through a straightforward application of Bayes' theorem as

$$P(\mathbf{D} \mid \mathcal{M}_k) = \frac{P(\mathbf{D} \mid \boldsymbol{\theta}_k, \mathcal{M}_k)P(\boldsymbol{\theta}_k \mid \mathcal{M}_k)}{P(\boldsymbol{\theta}_k \mid \mathbf{D}, \mathcal{M}_k)} \quad (9)$$

which gives a new interpretation to model evidence as the normalizing constant that makes the product of the prior and the likelihood the density of the posterior distribution.

Note that equation (9) holds for all $\boldsymbol{\theta}_k$. It follows that provided there exists one point $\boldsymbol{\theta}_k^*$ in the entire parameter space for which normalized values of the prior density, likelihood and posterior density are available, the evidence can be computed using this identity. One might thus be inclined to think that this is a far easier way of evaluating the evidence than the integral approach. Nevertheless, though it is often trivial to compute the prior density and the likelihood function, to obtain the normalized posterior density is generally as difficult a task as computing a complex multidimensional integral. Equation (9) does suggest however that in addition to numerical integration methods, one can also employ multivariate density estimation techniques to compute the model evidence.[4]

The Bayes' factor $BF_{a,b}(\mathbf{D})$ for model $\mathcal{M}_a$ against model $\mathcal{M}_b$, which is used extensively in Bayesian hypothesis testing and model selection, is defined to be the ratio of the respective evidences of the two models, namely,

$$BF_{a,b}(\mathbf{D}) \triangleq \frac{P(\mathbf{D} \mid \mathcal{M}_a)}{P(\mathbf{D} \mid \mathcal{M}_b)} \qquad (10)$$

The Bayes' factor can be interpreted as providing a measure of how much the data $\mathbf{D}$ have increased or decreased the odds on $\mathcal{M}_a$ relative to $\mathcal{M}_b$, and is given by,

$$BF_{a,b}(\mathbf{D}) = \frac{P(\mathcal{M}_a \mid \mathbf{D})}{P(\mathcal{M}_b \mid \mathbf{D})} \bigg/ \frac{P(\mathcal{M}_a)}{P(\mathcal{M}_b)} \qquad (11)$$

$BF_{a,b}(\mathbf{D}) > 1$ signifies that in relative terms model $\mathcal{M}_a$ is more plausible in the light of $\mathbf{D}$. The posterior odds ratio $P(\mathcal{M}_a \mid \mathbf{D})/P(\mathcal{M}_b \mid \mathbf{D})$ provides a summary of the evidence for $\mathcal{M}_a$ against $\mathcal{M}_b$. From the preceding text, it is clear that the model evidence depends on the corresponding prior density $P(\boldsymbol{\theta}_k \mid \mathcal{M}_k)$ as well as the likelihood function. In a general inference setting involving model uncertainty, two levels of prior knowledge are required, $P(\mathcal{M}_k)$ and $P(\boldsymbol{\theta}_k \mid \mathcal{M}_k)$. While the impact of the former on the end result is clear, the latter affects the result in a way that is rather subtle, but certainly not insignificant.

3 INTRODUCTION TO MONTE CARLO SIMULATION

In the last sections it has been shown how to conduct inference in situations where the integrals required to perform marginalization have been analytically tractable. However, in more realistic situations the integrations have to be performed numerically and this requirement has lead to a full-scale investigation of such integration methods. It is now clear that methods based on Monte Carlo techniques hold the greatest scope and in this paper the ideas concerning Markov Chain Monte Carlo methods will be introduced. References 1, 5–7 deal with other schemes.

Three areas of Bayesian analysis that require evaluation of integrals are as follows:

- **Evidence evaluation** Evidence is a real number whose value is used as a merit index for comparing the performance of different models at describing the data. We can either calculate *absolute* evidence,

$$P(\mathbf{D} \mid \mathcal{M}_k)$$
$$= \int P(\mathbf{D} \mid \boldsymbol{\theta}_k, \mathcal{M}_k) P(\boldsymbol{\theta}_k \mid \mathcal{M}_k) \, d\boldsymbol{\theta}_k \quad (12)$$

which is the normalizing factor for the product of likelihood and prior, or calculate *relative evidence*, which is the ratio of the evidence of one model compared to another.

- **Marginalization** A marginal density is a function of one or more parameters that is used to compare the plausibility of different parameter values, and is as such used for parameter estimation. The shape of the marginal density is more important than its size, and in fact, we almost always dispense with the computation of the normalization factor. Because the marginal density is a curve we must compute enough sample points for it to be possible to make an accurate plot of the density. Typically, about thirty sample points are required for a good representation of a one dimensional marginal density. This means evaluating no less than thirty integrals!

$$P(\boldsymbol{\theta}_{k,1}) \mid \mathbf{D}, \mathcal{M}_k)$$
$$= \int_{\mathcal{R}^k} P(\boldsymbol{\theta}_{k,1}, \boldsymbol{\theta}_{k,2} \mid \mathbf{D}, \mathcal{M}_k) \, d\boldsymbol{\theta}_{k,2} \quad (13)$$

- **Moments and expectation values** Computing moments and expectations of functions is not significantly different to computing marginal densities or evidence, depending on whether the result is in the form of a curve or a single real number. However, unlike marginalization, it is important not to discard normalization constants.

$$\mathbb{E}_\pi(f) = \int f(\boldsymbol{\theta}) \pi(\boldsymbol{\theta}) \, d\boldsymbol{\theta} \qquad (14)$$

For most statistical applications, the numerical aspect of the problem can be cast in the form of computing the expected value of some function of interest $f(\boldsymbol{\theta})$, with respect to a target probability density $\pi(\boldsymbol{\theta})$, where $\boldsymbol{\theta}$ is a vector of parameters of interest. In Bayesian inference, π is typically taken as a posterior distribution. For scalar θ, the usual way to evaluate an analytically intractable integral will be via *deterministic* numerical integration. As the dimensionality increases, however, there are problems associated with applying deterministic techniques. In contrast, Monte Carlo integration is as easy in 10 dimensions as in 1 dimension.

Suppose $\theta^1, \ldots, \theta^n$ are an iid sample set from π and that we evaluate

$$\hat{f} = \frac{1}{n} \sum_{i=1}^{n} f(\theta^i) \qquad (15)$$

Then $\hat{f}$ is said to be a Monte Carlo estimator of $\mathbb{E}_\pi(f)$. Since

$$\mathbb{E}\hat{f} = \frac{1}{n} \sum_{i=1}^{n} \mathbb{E}f(\theta^i) = \mathbb{E}_\pi(f) \qquad (16)$$

$\hat{f}$ is clearly an unbiased estimator. and the variance is given by

$$\mathrm{var}(\hat{f}) = \frac{1}{n} \int [f(\theta) - \mathbb{E}_\pi(f)]^2 \pi(\theta)\, d\theta \qquad (17)$$

implying that the error (i.e., the standard deviation) of a Monte Carlo estimator is $O(n^{-1/2})$, which can be contrasted with some deterministic methods which may be of $O(n^{-4})$ or even $\exp(-n)$.

The basic problem that Monte Carlo addresses is how to generate random samples from a given probability distribution π on some state space. The theoretical underpinning of Monte Carlo is the duality between samples and the distributions from which they are generated: given a distribution, there exist, in principle, ways to generate samples from it; and given samples, one can, at least approximately, recreate the distribution.

There are broadly two classes of Monte Carlo methods: *static* and *dynamic*.

Static methods generate a sequence of iid random numbers from the desired distribution π, an example being rejection sampling in the one-dimensional case. In theory, a static simulation method always exists, because of the following identity for a k-component parameter vector,

$$\pi(\theta) = \pi(\theta_1, \ldots, \theta_k) = \prod_{j=1}^{k} \pi(\theta_j \mid \theta_{j-1}, \ldots, \theta_1) \qquad (18)$$

However, the univariate distributions on the right-hand side are rarely all available in a form that is suitable for simulation purposes. Therefore static simulation methods are typically not applicable to complex problems.

Dynamic Monte Carlo simulates a stochastic process that has π as its unique equilibrium distribution (by which it is meant that π is the distribution to which a process converges irrespective of the starting point). Once equilibrium has been reached, under *ergodicity* one can substitute sample path (temporal) averages for ensemble averages. The crucial part of a dynamic Monte Carlo method is how to invent the best ergodic stochastic time evolution for the underlying system that converges to the desired distribution. In practice, one invariably uses a first-order Markov process to accomplish this, hence the term *Markov Chain Monte Carlo* (MCMC). In physical terms, different MCMC algorithms simply correspond to different invented *dynamics*. High-order Markov processes or non-Markovian processes are significantly more difficult to analyze in terms of their theoretical convergence properties.

3.1 The Theory of Markov Chains

A Markov chain with state space E is a sequence of E-valued random variables $\{\theta_i\}$ such that the probability of one state given previous states is $P(\theta_i \mid \theta_{i-1}, \ldots, \theta_0) = P(\theta_i \mid \theta_{i-1})$. The elements of E are thought of as the possible states of a system, θ_i representing the state at time i. To predict the future state of a Markovian system, one only needs knowledge of the present. Knowledge of the past is not required—it has already been fully accounted for by the present state.[5]

A Markov chain is defined by two components: the initial distribution $P(\theta_0)$ and the transition kernel, $T(\theta, A) = P(\theta_{i+1} \in A \mid \theta_i = \theta)$,[a] where $\mathcal{E}$ denotes a σ-algebra on E, $\forall \theta \in E$ and $\forall A \in \mathcal{E}$. $T(\theta, \cdot)$ is a probability measure on $(E, \mathcal{E})$: $E \times \mathcal{E} \to [0, 1]$.

3.2 Definitions of Key Concepts

- **Equilibrium distribution** Let $T^n(\boldsymbol{\theta}_0, A)$ represent the conditional distribution $P(\boldsymbol{\theta}_n \in A \mid \boldsymbol{\theta}_0)$. π is an *equilibrium* distribution for the chain if for π-almost[b] all $\boldsymbol{\theta}_0$,

$$\lim_{n \to \infty} T^n(\boldsymbol{\theta}_0, A) = \pi(A), \forall A \qquad (19)$$

- **Invariant distribution** π is an *invariant* or *stationary* distribution for a Markov chain if the transition kernel T is such that $\pi(A) = \int \pi(d\boldsymbol{\theta}) T(\boldsymbol{\theta}, A), \forall A$. It is possible for a Markov chain to have more than one invariant distribution.
- **Irreduciblity** A Markov chain is π-*irreducible* on $(E, \mathcal{E})$ if for a σ-finite measure π, $\forall \boldsymbol{\theta} \in E$, $\forall A$ with $\pi(A) > 0$, $\exists n$, such that $T^n(\boldsymbol{\theta}, A) > 0$.
- **Aperiodicity** A Markov chain is periodic if, informally, there are portions of the state space it can only visit at certain regularly spaced times; otherwise, it is *aperiodic*.
- **Reversibility and detailed balance** A Markov chain with transition kernel T is reversible with respect to distribution π if and only if $\forall B, C \in \mathcal{E}$, the *detailed balance* condition

$$\int_B \pi(d\boldsymbol{\theta}) T(\boldsymbol{\theta}, C) = \int_C \pi(d\boldsymbol{\theta}) T(\boldsymbol{\theta}, B) \qquad (20)$$

is satisfied. Equivalently, detailed balance holds if and only if

$$\pi(d\mathbf{x}) T(\mathbf{x}, d\mathbf{y}) = \pi(d\mathbf{y}) T(\mathbf{y}, d\mathbf{x}), \quad \forall \mathbf{x}, \mathbf{y} \in E \qquad (21)$$

The two sides of this identity are measures on $\mathcal{E} \otimes \mathcal{E}$ and detailed balance means these two measures are identical. If T is reversible with respect to π, π is invariant for T. A reversible Markov chain has the property that for any function f,

$$\int f(\mathbf{y}) \pi(d\mathbf{x}) T(\mathbf{x}, d\mathbf{y}) = \int f(\mathbf{y}) \pi(d\mathbf{y}) T(\mathbf{y}, d\mathbf{x})$$

$$= \int f(\mathbf{y}) \pi(d\mathbf{y}) \qquad (22)$$

It follows that detailed balance is a sufficient condition for having an invariant distribution, the former being more amenable to analysis.

- **Transience, persistence, and recurrence** For a discrete state space, a state $\mathbf{x}$ is *persistent* if a system starting at $\mathbf{x}$ is certain to return to it sometime in the future. Transience of state $\mathbf{x}$ is equivalent to $P(\boldsymbol{\theta}_n = \mathbf{x} \; io \mid \boldsymbol{\theta}_0 = \mathbf{x}) = 0$, where io means *infinitely often*, and to $\sum_n T^n(\mathbf{x}, \mathbf{x}) < \infty$; and persistence of state $\mathbf{x}$ is equivalent to $P(\boldsymbol{\theta}_n = \mathbf{x} \; io \mid \boldsymbol{\theta}_0 = \mathbf{x}) = 1$ and to $\sum_n T^n(\mathbf{x}, \mathbf{x}) = \infty$. In the discrete case, if the Markov chain is irreducible, the states are either all transient or all persistent. The chain itself can accordingly be called either *transient* or *persistent*, respectively.

  A crucial concept in convergence theory is *recurrence*. The definition given in Ref. 8 for a discrete state space is that a π-irreducible chain with invariant distribution π is recurrent if, $\forall B$ with $\pi(B) > 0$, $P(\boldsymbol{\theta}_n \in B \; io \mid \boldsymbol{\theta}_0 = \mathbf{x}) > 0$, $\forall \mathbf{x}$, and $P(\boldsymbol{\theta}_n \in B \; io \mid \boldsymbol{\theta}_0 = \mathbf{x}) = 1$, for π-almost all $\mathbf{x}$. It is *Harris recurrent* if $P(\boldsymbol{\theta}_n \in B \; io \mid \boldsymbol{\theta}_0 = \mathbf{x}) = 1$, $\forall \mathbf{x}$.
- **Convergence** It is shown in Ref. 8 that if a chain is π-irreducible and has π as an invariant distribution, then it must be positive recurrent, where the term positive refers to the total mass of the measure being finite,[c] and π is its *unique* invariant distribution. If the chain is also aperiodic, $\forall \boldsymbol{\theta}$ π-almost,

$$||T^n(\boldsymbol{\theta}, \cdot) - \pi(\cdot)|| \to 0 \qquad (23)$$

where $|| \cdot ||$ denotes the total variation distance.[d] If the chain is Harris recurrent, the convergence occurs $\forall \boldsymbol{\theta}$.

The main point about convergence is whether or not the effect of the initial state wears off. Having an invariant distribution does not guarantee convergence to it. A much stronger condition for convergence is positive recurrence, which is a corollary of irreducibility and having a proper invariant distribution. With positive recurrence, convergence holds for *almost* every starting point. Harris recurrence ensures convergence from *every* starting point.

If $\{\theta^1, \ldots, \theta^n, \ldots\}$ is a realization of a Markov chain, for the sample path average $\frac{1}{n}\left\{\sum_{i=1}^{n} f(\theta^i)\right\}$ to converge to $\mathbb{E}_\pi(f)$, one needs positive recurrence with invariant distribution π. The condition for ergodicity is positive Harris recurrence and aperiodicity. Limiting properties of sample path averages however do not depend on aperiodicity.

An ergodic Markov chain with invariant distribution π is said to be *geometrically* ergodic if there exists a function h such that $\int |h(\boldsymbol{\theta})| \pi(\mathrm{d}\boldsymbol{\theta}) < \infty$ and a positive constant $r < 1$ such that

$$||T^n(\boldsymbol{\theta}, \cdot)|| \leq h(\boldsymbol{\theta})r^n \qquad (24)$$

There are other forms of ergodicity and the convergence rate can also be characterized by studying how quickly the sample path average of an arbitrary square π-integrable function approaches its expectation under the invariant distribution.

3.3   The Metropolis–Hastings Algorithm

The Metropolis–Hastings algorithm[9] is a method of constructing a *reversible* Markov transition kernel with a specified invariant distribution. A Metropolis–Hastings kernel on $(E, \mathcal{E})$ can be expressed as,[1]

$$T(\mathbf{x}, \mathrm{d}\mathbf{y}) = Q(\mathbf{x}, \mathrm{d}\mathbf{y})A(\mathbf{x}, \mathbf{y}) + I(\mathbf{x} \in \mathrm{d}\mathbf{y})$$
$$\times \int [1 - A(\mathbf{x}, \boldsymbol{\theta})]Q(\mathbf{x}, \mathrm{d}\boldsymbol{\theta}) \qquad (25)$$

where I is the indicator function; $A(\mathbf{x}, \mathbf{y})$ is a function: $E \times E \to [0, 1]$; $Q(\mathbf{x}, \mathrm{d}\mathbf{y})$ is a transition kernel that generates a candidate $\mathbf{y}$ for the next state conditional on the current state $\mathbf{x}$. The candidate is accepted with probability $A(\mathbf{x}, \mathbf{y})$.

The Metropolis–Hastings kernel satisfies detailed balance if and only if

$$\pi(\mathrm{d}\mathbf{x})Q(\mathbf{x}, \mathrm{d}\mathbf{y})A(\mathbf{x}, \mathbf{y}) = \pi(\mathrm{d}\mathbf{y})Q(\mathbf{y}, \mathrm{d}\mathbf{x})A(\mathbf{y}, \mathbf{x}) \qquad (26)$$

If with transition kernel Q and invariant distribution π, $r(\mathbf{x}, \mathbf{y})$ represents the rate of transitions from $\mathbf{x}$ to $\mathbf{y}$ relative to those from $\mathbf{y}$ to $\mathbf{x}$, it is shown in Ref. 10 that in order to satisfy the detailed balance condition, A must satisfy

$$\frac{A(\mathbf{y}, \mathbf{x})}{A(\mathbf{x}, \mathbf{y})} = r(\mathbf{x}, \mathbf{y}) \qquad (27)$$

If there exists a *common dominating measure* with respect to which $\pi(\mathrm{d}\mathbf{x})$ has density $\pi(\mathbf{x})$ and $Q(\mathbf{x}, \mathrm{d}\mathbf{y})$ has density $q(\mathbf{x}, \mathbf{y})$,

$$r(\mathbf{x}, \mathbf{y}) = \frac{\pi(\mathbf{x})q(\mathbf{x}, \mathbf{y})}{\pi(\mathbf{y})q(\mathbf{y}, \mathbf{x})} \qquad (28)$$

The standard Metropolis–Hastings acceptance probability can be written as

$$A_{\mathrm{MH}}(\mathbf{x}, \mathbf{y}) = \min\left[1, \frac{\pi(\mathbf{y})q(\mathbf{y}, \mathbf{x})}{\pi(\mathbf{x})q(\mathbf{x}, \mathbf{y})}\right] \qquad (29)$$

It is worthy of note that there exist several alternative forms of the acceptance probability function, but the Hastings version is optimal for a quite extensive range of choices chiefly because it rejects proposals less frequently than the others.

For distributions of high dimensions, it is often preferable to update only a subset of the target vector at a time so as to avoid low acceptance rates. A general version of the Metropolis–Hastings algorithm is given below:

1. Partition $\boldsymbol{\theta}$ into m blocks such that $\boldsymbol{\theta} = (\boldsymbol{\theta}_1, \ldots, \boldsymbol{\theta}_m)$; let $\boldsymbol{\theta}_{-j}$ be a subvector containing all the components of $\boldsymbol{\theta}$ not included in $\boldsymbol{\theta}_j$.
2. At iteration $t + 1$, select a subvector $\boldsymbol{\theta}_j$ for updating by a deterministic or random mechanism that ensures each component of $\boldsymbol{\theta}$ should be visited with nonzero probability. Let $\boldsymbol{\theta}_{-j}^{(t)}$ be the most recent update of $\boldsymbol{\theta}_{-j}$. Update the jth block by generating a candidate $\boldsymbol{\theta}_j'$ from a proposal density $q_j(\boldsymbol{\theta}_j', \boldsymbol{\theta}_j \mid \boldsymbol{\theta}_{-j})$.
3. Accept it with probability

$$\min\left[1, \frac{\pi(\boldsymbol{\theta}_j', \boldsymbol{\theta}_{-j}^{(t)})q_j(\boldsymbol{\theta}_j^{(t)}, \boldsymbol{\theta}_j' \mid \boldsymbol{\theta}_{-j}^{(t)})}{\pi(\boldsymbol{\theta}_j^{(t)}, \boldsymbol{\theta}_{-j}^{(t)})q_j(\boldsymbol{\theta}_j', \boldsymbol{\theta}_j^{(t)} \mid \boldsymbol{\theta}_{-j}^{(t)})}\right]$$

or else $\boldsymbol{\theta}_j^{(t+1)} = \boldsymbol{\theta}_j^{(t)}$.
4. Increment t and go to step 2.

Step 1 can be modified to allow for random partitioning of $\boldsymbol{\theta}$ subject to the condition that each component may be visited with positive probability while maintaining irreducibility of the Markov chain.

3.4   Reversible Jump MCMC

An MCMC algorithm that jumps between parameter spaces of variable dimensions can be used and is referred to as the *reversible jump* MCMC sampler.[11] In this approach, the state space is defined to be the union of the parameter spaces of all the entertained models. The Metropolis–Hastings algorithm is generalized to allow for moves between subspaces with different dimensions by using proposals that match the differing degrees of freedom.

The reversible jump MCMC framework enables an algorithm to be constructed which samples from $p(k, \theta^{(k)}|\mathbf{D})$, where k indexes the models of differing dimensionality, $\theta^{(k)}$ are the parameters of the kth model and $\mathbf{D}$ is the data. We need to construct the transition kernel $T(\mathbf{x}, \mathrm{d}\mathbf{x}')$ to maintain detailed balance when moving between the different subspaces. This is in general a difficult problem, drawing on measure theory, however, we only require the simplest case in our application.

When the current state is $\mathbf{x}$, choose to make a move of type m which will take the current state to $\mathrm{d}\mathbf{x}'$ with probability $q_m(\mathbf{x}, \mathrm{d}\mathbf{x}')$ (this includes the probability of choosing this move type). The acceptance probability can be derived as

$$A_m(\mathbf{x}, \mathbf{x}') = \min\left(1, \frac{\pi(\mathrm{d}\mathbf{x}')q_m(\mathbf{x}', \mathrm{d}\mathbf{x})}{\pi(\mathrm{d}\mathbf{x})q_m(\mathbf{x}, \mathrm{d}\mathbf{x}')}\right) \quad (30)$$

The main condition when using this in practice is that the proposal density must be constructed such that $\pi(\mathrm{d}\mathbf{x})q_m(\mathbf{x}, \mathrm{d}\mathbf{x}')$ has finite density with respect to a symmetric measure. This can be done by constructing the proposals such that $\pi(\mathrm{d}\mathbf{x}')q_m(\mathbf{x}', \mathrm{d}\mathbf{x})$ and $\pi(\mathrm{d}\mathbf{x})q_m(\mathbf{x}, \mathrm{d}\mathbf{x}')$ are of the same dimensionality (even though $\mathbf{x}$ and $\mathbf{x}'$ are of differing dimensionality), by making the q_m's functions of a number of new random variables, which are introduced to match the dimensions.

The argument is complex in general, however it may be greatly simplified in our case. Following Ref. 12, we impose that the variable in the space of unknown dimension are discrete, and only allow transdimensional moves that add or remove a single dimension without altering the value of the variables in the other dimensions. These are termed *birth* and *death* steps respectively. In addition, a within-dimension step called a *move* is allowed, which alters the values of the parameters in one dimension, leaving the rest unmodified. These moves occur with independent probability b_k, d_k, and $1 - b_k - d_k$ respectively, when the current dimension is k.

In this case a Metropolis–Hastings algorithm, using independence sampling, will have an acceptance probability similar to equation (28) of Ref. 12

$$A_m(\mathbf{x}, \mathbf{y}) = \min\left[1, \frac{\pi(\mathbf{y})q(\mathbf{x}, \mathbf{y})}{\pi(k)q(\mathbf{y}, \mathbf{x})} \cdot R\right] \quad (31)$$

$$= \min\left[1, BF(\mathbf{x}, \mathbf{y}) \cdot R\right] \quad (32)$$

Here R accounts for the change of dimension, and is

$$R = \begin{cases} \frac{d_{k+1}}{b_k}, & \text{Birth} \\ \frac{b_{k-1}}{d_k}, & \text{Death} \\ 1, & \text{Move} \end{cases} \quad (33)$$

By setting $b_k = d_k = 1$ in all cases except at the dimensionality borders ($k = 0$ or $k = k_{\max}$), $R = 1$. For further details see Refs. 3, 11–14.

4 BAYESIAN CLASSIFICATION

Having begun in general by defining Bayesian inference as the process of model selection and parameter estimation in Section 2, and then having shown in Section 3 how to use Monte Carlo methods to perform the analytically intractable integrals that result, it remains only to provide the specifics of models to be used for our problem.

Our goal is classification: based on a feature set of measurements, $X \in \mathbb{R}^{n \times p}$, to correctly predict a class $Y \in \mathcal{C}^n, \mathcal{C} \in [1, q] \in \mathbb{Z}$, where n is the number of samples, p is the number of features extracted from each and q is the number of classes (typically two). In addition, given our goal of biosensor production, we wish to minimize p.

4.1 Bayesian Regression

We shall present a framework, drawing heavily on Ref. 12 (which gives an extremely good treatment), that affects classification based on regression. Consider the basic linear regression model

$$Y = \boldsymbol{\beta}X + \boldsymbol{\varepsilon} \quad (34)$$

$$y_j = \sum_{i=1}^{p} \beta_i x_{i,j} + \epsilon, \quad \forall j \in [1, n] \quad (35)$$

where $\epsilon_i \sim \mathcal{N}(0, \sigma^2)$, and hence $Y \sim \mathcal{N}(\boldsymbol{\beta}X, \boldsymbol{\varepsilon})$. Thus the likelihood may be written[12]

$$p\left(X|\boldsymbol{\beta}, \sigma^2\right) = \left(2\pi\sigma^2\right)^{-n/2}$$
$$\times \exp\left(-\frac{(Y - X\boldsymbol{\beta})'(Y - X\boldsymbol{\beta})}{2\sigma^2}\right)$$

(36)

Our inference is thus over $\boldsymbol{\beta}$ and σ^2, and hence we must place priors over these quantities. The standard approach is to use a normal-inverse Γ distribution for the joint distribution as it is the conjugate prior[12]

$$p(\boldsymbol{\beta}, \sigma^2)$$
$$= p(\boldsymbol{\beta}|\sigma^2)p(\sigma^2)$$
$$= \mathcal{N}(\mathbf{m}, \sigma^2, \mathbf{V})\mathcal{IG}(a, b)$$
$$= \frac{b^a}{(2\pi)^{k/2}|\mathbf{V}|^{1/2}\Gamma(a)}(\sigma)^{-2(a+(k/2)+1)}$$
$$\times \exp\left[\frac{-(\boldsymbol{\beta} - \mathbf{m})'\mathbf{V}^{-1}(\boldsymbol{\beta} - \mathbf{m}) + 2b}{(2\sigma^2)}\right]$$

(37)

Utilizing equations (36) and (37) the parameter posterior is then[12]

$$p(\boldsymbol{\beta}, \sigma^2|X)$$
$$= \frac{p(\boldsymbol{\beta}, \sigma^2)p(X|\boldsymbol{\beta}, \sigma^2)}{p(X)}$$
$$= \frac{(b^*)^{a^*}(\sigma)^{-2(a^*+(k/2)+1)}}{(2\pi)^{k/2}|\mathbf{V}|^{1/2}\Gamma(a)}$$
$$\times \exp\left[\frac{-(\boldsymbol{\beta} - \mathbf{m}^*)'\mathbf{V}^{*-1}(\boldsymbol{\beta} - \mathbf{m}^*) + 2b^*}{(2\sigma^2}\right]$$

(38)

where

$$\mathbf{m}^* = \left(\mathbf{V}^{-1} + X'X\right)^{-1}\left(\mathbf{V}^{-1}\mathbf{m} + \mathbf{X}'Y\right)$$

$$\mathbf{V}^* = \left(\mathbf{V}^{-1} + X'X\right)^{-1}$$

$$a^* = a + n/2$$

$$b^* = b + \frac{1}{2}\left(\mathbf{m}'\mathbf{V}^{-1}\mathbf{m} + Y'Y - \mathbf{m}^{*'}\mathbf{V}^{*-1}\mathbf{m}^*\right)$$

These equations are sufficient for parameter estimation, however for model selection we require the Bayes' factor between two models. In this case the models will simply be linear regressions where the parameter matrix, $\boldsymbol{\beta}$, is of different dimensions, corresponding to the number of features included in the regression.

Using the preceding equations it can be shown that the evidence for a model with i features included in the regression[12] is

$$p(X|M_i) = \frac{p(X|\boldsymbol{\beta}_i, \sigma^2)p(\boldsymbol{\beta}_i, \sigma^2)}{p(\boldsymbol{\beta}_i, \sigma^2|X)}$$
$$= \frac{|\mathbf{V}_i^*|^{1/2}b^a\Gamma(a_i^*)}{|\mathbf{V}_i|^{1/2}\pi^{n/2}\Gamma(a)}(b_i^*)^{-a_i^*} \quad (39)$$

and hence the Bayes' factor in favor of a model of dimension i over another of dimension j is[12]

$$BF(M_i, M_j) = \frac{|\mathbf{V}_j|^{1/2}|\mathbf{V}_i^*|^{1/2}(b_j^*)^{a^*}}{|\mathbf{V}_i|^{1/2}|\mathbf{V}_j^*|^{1/2}(b_i^*)^{a^*}} \quad (40)$$

4.2 Bayesian Classification

Having examined Bayesian regression, we now turn to classification, which we cast as a regression problem. The regression framework described in the preceding text relies on $Y \sim \mathcal{N}(\cdot, \cdot)$, whereas in (two-class) regression $Y \sim \mathcal{BR}$. In order to proceed we propose, following Ref. 12 closely once again, introducing a latent variable, Z, associated with Y, which is normally distributed and against which we may thus perform regression.

Consider equation (34) once again, but substituting a latent variable, z for y.

$$Z = \boldsymbol{\beta}X + \boldsymbol{\varepsilon} \quad (41)$$

$$z_j = \sum_{i=1}^{p}\beta_i x_{i,j} + \epsilon, \quad \forall j \in [1, n] \quad (42)$$

where in this case $\epsilon_i \sim \mathcal{N}(0, 1)$. If regression is performed on z, similar results to the general regression described above result, but for the knowledge of that $\sigma^2 = 1$. Thus

$$p\left(X|\boldsymbol{\beta}, \sigma^2\right) = (2\pi)^{-n/2}$$
$$\times \exp\left(-\frac{(Z - X\boldsymbol{\beta})'(Z - X\boldsymbol{\beta})}{2}\right)$$

(43)

$$p(\sigma^2) = \delta(1) \tag{44}$$

$$p(\boldsymbol{\beta}|X) = \mathcal{N}(\mathbf{m}^*, \mathbf{V}^*) \tag{45}$$

$$= \frac{1}{(2\pi)^{k/2}|\mathbf{V}|^{1/2}}$$

$$\times \exp\left[\frac{-(\boldsymbol{\beta}-\mathbf{m}^*)'\mathbf{V}^{*-1}(\boldsymbol{\beta}-\mathbf{m}^*)}{2}\right] \tag{46}$$

$$p(X|Z) \propto \frac{|\mathbf{V}^*|^{1/2}\exp(-b^*)}{|\mathbf{V}|^{1/2}(2\pi)^{n/2}} \tag{47}$$

where

$$\mathbf{m}^* = \left(\mathbf{V}^{-1} + X'X\right)^{-1}\left(\mathbf{V}^{-1}\mathbf{m} + \mathbf{X}'Z\right)$$

$$\mathbf{V}^* = \left(\mathbf{V}^{-1} + X'X\right)^{-1}$$

$$b^* = \frac{1}{2}\left(\mathbf{m}'\mathbf{V}^{-1}\mathbf{m} + Z'Z - \mathbf{m}^{*'}\mathbf{V}^{*-1}\mathbf{m}^*\right)$$

Finally, the Bayes' factor is

$$BF(M_i, M_j) = \frac{|\mathbf{V}_j^*|^{1/2}|\mathbf{V}_1^*|^{1/2}}{|\mathbf{V}_1|^{1/2}|\mathbf{V}_2^*|^{1/2}}\exp(b_j^* - b_i^*) \tag{48}$$

It remains to associate Z with Y, which is done by assuming (see Ref. 12 for justification) for each feature j that

$$p(y_j = 1|z_j, \boldsymbol{\beta}) = \begin{cases} 1, & \text{if } z_j > 0 \\ 0, & \text{else} \end{cases} \tag{49}$$

in which case, by marginalization of z_j, (Section 2.1)

$$p(y_j = 1|\boldsymbol{\beta}) = \int_{-\infty}^{\sum_{i=1}^{p}\beta_i x_{i,j}} \frac{1}{\sqrt{2\pi}}\exp\left(\frac{-a^2}{2}\right)da \tag{50}$$

which is simply the normal distribution function (CDF), which is simple to sample from and hence perform a numerical integration over.

A measure of the predictive performance of a set of parameters $\boldsymbol{\beta}$ is thus the Euclidean distance of the predicted labels, from the true labels.

Thus we require a numerical method in order to estimate the distribution (46) and another numerical integration to marginalize the equation (49). Fortunately the two may be elegantly combined such that we sample z for its marginalization, and for each such sample perform a single step of a RJMCMC (Section 3.4) integration, using the linear regression model. This is described in Algorithm 1 in the subsequent text.

The fully Bayesian solution is to perform model averaging for prediction, that is, in order to label an unseen datum $\mathbf{x}$ to use each of the $\beta^{(t)}$ imputed in the preceding text[12]

$$p(y = 1|\mathbf{x}, X) \approx \frac{1}{N}\sum_{t=0}^{T} p(y = 1|\mathbf{x}, \beta^{(t)})$$

$$= \frac{1}{N}\sum_{t=0}^{T}\int_{-\infty}^{\sum_{i=1}^{p}\beta_i^{(t)}x_{i,j}} \frac{1}{\sqrt{2\pi}}$$

$$\times \exp\left(\frac{-a^2}{2}\right)da \tag{52}$$

This is unacceptable for our purposes as biosensors are not equipped with either the memory or the computational wherewithal to perform such a calculation. Instead we must compromise. The usual solution in such cases is the *maximum a posteriori* solution, however it is not directly available to us as the parameter posterior $\text{argmax}_{\boldsymbol{\beta}}\, p(X|\boldsymbol{\beta})$ is a function of the latent variable, z, requiring its integration for use. This is the purpose of $\gamma^{(t)}$ in 1, we can approximate the MAP parameter by

$$\hat{\boldsymbol{\beta}} = \boldsymbol{\beta}^{(t)} : t = \text{argmax}_t \gamma^{(t)} \tag{53}$$

which is the $\boldsymbol{\beta}$ that performs the most accurate prediction on our training set.

Another improper notion from the Bayesian perspective that is required by pragmatism, is the stipulation of a $k_{\max}$, the maximum number of predictors to allow in the model. Specification of this quantity is arbitrary and hence a design time

$$\gamma = \sqrt{\sum_{j=1}^{n}\left(Y_j - \left(\int_{-\infty}^{\sum_{i=1}^{p}\beta_i^{(t)}x_{i,j}} \frac{1}{\sqrt{2\pi}}\exp\left(\frac{-a^2}{2}\right)da\right)\right)^2} \tag{51}$$

Algorithm 1 Bayesian RJMCMC Classifier

---

for $t = 0$ to Num Iterations **do**

$\quad Z_j \leftarrow \mathcal{N}(\sum_{i=1}^{p} \beta_i^{(t)} x_{i,j}), \forall j$ %Sample Latent Variables

$\quad\quad Z_j \leftarrow N(\sum_{i=1}^{p} \beta_i^{(t)} x_{i,j}), \forall j$ %Sample Latent Variables

$\quad \boldsymbol{\beta}^{(t)} \leftarrow \mathcal{N}(\mathbf{m}^*, \mathbf{V}^*)$ %Sample Regression Parameters

$\quad\quad \boldsymbol{\beta}^{(t)} \leftarrow N(\mathbf{m}^*, \mathbf{V}^*)$ %Sample Regression Parameters

$\quad$ %Propose move:

$\quad$ %Sample Move Type

$\quad$ **if** $k == 1$ **then**

$\quad\quad u_1 \leftarrow U(0, b_k)$ %Force Birth Move

$\quad$ **else if** $k == k_{\max}$ **then**

$\quad\quad u_1 \leftarrow U(b_k, 1)$ %Death or Move Move

$\quad$ **else**

$\quad\quad u_1 \leftarrow U(0, 1)$ %Any Move

$\quad$ %Propose Parameter

$\quad$ **if** $u_1 \leq b_k$ **then**

$\quad\quad \beta_b \leftarrow p(\beta)$ %**Birth Move:** Sample Variable

$\quad\quad \boldsymbol{\beta}' = [\boldsymbol{\beta}, \beta_b]$ %Add to Set

$\quad$ **else if** $u_1 \leq b_k + d_k$ **then**

$\quad\quad r \leftarrow U[1, k]$ %**Death Move:** Sample Index

$\quad\quad \boldsymbol{\beta}' = \boldsymbol{\beta}^{(t)} \backslash \beta_r$ %Remove Variable

$\quad$ **else**

$\quad\quad r \leftarrow U[1, k]$ %Death Move

$\quad \boldsymbol{\beta}'' = \boldsymbol{\beta}^{(t)} \backslash \beta_r$

$\quad\quad \beta_b \leftarrow p(\beta)$ %Birth Move

$\quad \boldsymbol{\beta}' = [\boldsymbol{\beta}'', \beta_b]$

$\quad$ **end if**

$\quad$ %Accept or Reject Proposed Move

$\quad u_2 \leftarrow U[0, 1]$ %Sample Acceptance Probability

$\quad$ **if** $u_2 < \min\left\{1, BF(\boldsymbol{\beta}', \boldsymbol{\beta}^{(t)}) \cdot R\right\}$ **then**

$\quad\quad \boldsymbol{\beta}^{(t+1)} = \boldsymbol{\beta}'$ %Accept Proposed Move

$\quad$ **else**

$\quad\quad \boldsymbol{\beta}^{(t+1)} = \boldsymbol{\beta}^{(t)}$ %Reject Proposed Move

$\quad$ **end if**

$\quad\quad \gamma^{(t)} = \gamma$ %Record Performance

end for

---

decision, motivated either through the eventual limitations of the biosensor technology, or tuned iteratively by observing the predictive performance over a range of $k_{\max}$ and choosing the best compromise of performance and simplicity.

A final critical step is the estimation of predictive performance on unseen data. We advocate the process of cross-validation. The data is split into two nonoverlapping groups, termed the *training and testing data*. The algorithm is run on the training set and the performance on the testing set is noted. This is then repeated with different training and testing sets and the average performance is taken as an estimate of prediction accuracy. This is a robust estimate of the classifier performance on unseen data.

5 CONCLUSION

In this chapter we have attempted to indicate the main role that Bayesian statistics plays in the realm of biosensors. Given the nature of the problem, we have argued that main contribution of the Bayesian methods is in classification, specifically the identification of features to design the biosensors. We have argued that as a large majority of biosensors are noncomputing and operate by sensing the presence, absence, or large concentration differences, of molecules (or other features), the Bayesian method must produce a classifier amenable to such implementation. Specifically, this requires a small feature set and a noncomputing discriminant. We have endeavored to produce a complete discussion of just such a system, with an introduction to all the necessary components.

We have presented an introduction to Bayesian classification specifically for biosensor design with the minimum necessary accompaniment of detail. Initially we introduced Bayesian inference and indicated the two main divisions of labor: parameter estimation and model selection. These problems highlighted the need for numerical integration techniques which we introduced briefly, giving specific attention to the Metropolis–Hastings algorithm and its transdimensional generalization, the reversible jump algorithm. With this framework in hand, we then introduced a popular Bayesian classification algorithm, and made specialization for the application area of biosensor design.

We have only touched briefly on each area, and recommend further reading on the area. Specifically in Bayesian statistics we recommend Ref. 15 and for a more philosophical outlook Ref. 16. As regards numerical techniques Ref. 7 is outstanding and as our numerous citations indicate, Ref. 12 is a prime source for Bayesian techniques for classification. The Bayesian approach is not the only one for classification and Ref. 17 provides an excellent overview of the plethora of alternatives.

END NOTES

[a.] Sometimes in the definition of the Markov chain, the transition kernel is allowed to depend on i.
[b.] A property that holds for θ outside a set of measure 0 under π is said to hold π-almost everywhere, or for π-almost all θ. The symbols $\forall$ and $\exists$ mean "for all" and "there exists" respectively.
[c.] There is a possibility that the invariant measure is null.
[d.] For a measure ϕ on $(E, \mathcal{E})$, the total variation distance $||\phi|| \triangleq \sup_{A \in \mathcal{E}} \phi(A) - \inf_{A \in \mathcal{E}} \phi(A)$. There

are several other distance measures that can be used to measure closeness to the target distribution.

REFERENCES

1. J. J. K. O'Ruanaidh and W. J. Fitzgerald, *Numerical Bayesian Methods Applied to Signal Processing*, Springer Verlag, New York, 1996.
2. J. M. Bernardo and A. F. M. Smith, *Bayesian Theory*, John Wiley & Sons, New York, 1994.
3. C. Andrieu, P. M. Djuric, and A. Doucet, Model Selection by MCMC Computation. Signal Processing, unpublished.
4. J. Besag, A candidate's formula: a curious result in Bayesian prediction. *Biometrika*, 1989, **76**(1), 183.
5. W. R. Gilks, S. Richardson, and D. J. Spieglehalter, *Markov Chain Monte Carlo in Practice*, Chapman Hall, New York, 1996.
6. A. E. Gelfand and A. F. M. Smith, Sampling based approaches to calculating marginal densities. *Journal of the American Statistical Association*, 1990, **410**(85), 398–409.
7. C. P. Robert and G. Casella, *Monte Carlo Statistical Methods*, Springer Verlag, 1999.
8. L. Tierney, Markov chains for exploring posterior distributions (with discussion). *Annals of Statistics*, 1994, **22**(4), 1701–1762.
9. W. K. Hastings, Monte Carlo sampling methods using Markov chains and their applications. *Biometrika*, 1970, **57**, 97–109.
10. L. Tierney, A note on metropolis-hastings kernels for general state spaces, School of Statistics, University of Minnesota, no. 606, June, 1995.
11. P. J. Green, Reversible jump Markov chain Monte Carlo computation and Bayesian model determination. *Biometrika*, 1995, **82**, 711–732.
12. D. G. Denison, C. C. Holmes, B. K. Mallick, and A. F. Smith, *Bayesian Methods for Nonlinear Classification and Regression*, John Wiley & Sons, 2002.
13. R. Morris and W. J. Fitzgerald, *A Sampling Based Approach to Line Scratch Removal From Motion Picture Frames*, In: Proc. IEEE International Conference on Image Processing, Lausanne, Switzerland, 1996, September.
14. C. P. Robert and G. Casella, *Monte Carlo Statistical Methods*, 2nd Edn, Springer, 2004.
15. J. M. Bernardo and A. F. M. Smith, *Bayesian Theory*, Wiley, 2004.
16. E. Jaynes, *Probability Theory: The Logic of Science*, Cambridge University Press, 2003.
17. T. Hastie, R. Tibshirani, and J. Friedman, *The Elements of Statistical Learning*, Springer, 2001.

Part Eight

Areas and Examples of Biosensor Applications

Areas and Examples of Biosensor Applications

Institute of Biotechnology, University of Cambridge, Cambridge, UK

There is a greater demand for chemical and biological intelligence now than there has ever been in the history of analytical chemistry. Such demand stems from the desire to gain a better understanding of the major issues facing the planet in the future, the human genome, the impact of infectious diseases, bioterrorism, bioenergy, global warming and climate change, and the consequent requirement for making measurements of key analytes in a variety of samples and circumstances. Nowhere more important are these measurements than in clinical medicine and the pharmaceutical industry, where recent understanding of the genome, proteome, and metabolome is likely to lead to new drug targets, personalized medicine, and diagnostics based on novel biomarkers. In principle, biosensors could be applied to disease predisposition, pre-symptomatic diagnosis, prognosis, drug and patient stratification, compliance, and adverse drug reactions. Jacobs describes genetic and other DNA-based biosensor applications, which may find application as pre-disposition monitors for a variety of genetics-based disorders and, when combined with suitable dietary or lifestyle indicators, as prognostic indicators. More recently, two significant developments in biomedical diagnostics have been reported. First, the use of sophisticated hyphenated mass spectrometry techniques has allowed the identification of trace peptide and protein biomarkers in readily accessible fluids with diagnostic potential. In this section,

Mascini picks up this theme (*see* **Examples of Biosensors for the Measurement of Trace Medical Analytes**) and describes examples of biosensors for the measurement of trace medical analytes. The second major development concerns the migration of analytical procedures from the well-serviced central laboratory to the physician's office, operating theatre, ward, or home. Lateral flow immunochromatographic assays are described by Davies (*see* **Lateral-Flow Immunochromatographic Assays**) and constitute a major contribution to user-friendly, zero-power devices for monitoring pregnancy hormones, heart disorders, drugs of abuse, and other analytes in the home, workplace, and roadside. The notion of field operable biosensors is elaborated on by Lobel (*see* **Field-Operable Biosensors for Tropical Dispatch**) and demonstrates how far this technology has come since the first introduction of biosensors, some three decades ago. In principle, making the technology accessible to lay users was one of the original goals of the then nascent biosensor community, but has proven difficult to implement without the introduction of inexpensive measurement and manufacturing processes.

However, many of these developments are in the realm of future activity, and despite considerable speculation on the value of biosensor technologies, and a substantial worldwide research activity, the reality is that there are very few biosensors actually in the market. The exception, and an archetypal example of a home-use sensor, highlighted by Hall, is glucose measurement via conventional electrochemical biosensors, which constitutes a $4 billion market of about 7 billion tests per annum for diabetes management (*see* **Glucose Measurement Within Diabetes via "Traditional" Electrochemical Biosensors**). However, as reviewed by Newman and Pickup, even this market is beginning to change with newer approaches aimed at real-time monitoring of glucose concentrations via a variety of minimally-invasive and non-invasive techniques being developed (*see* **Biosensors for Monitoring Metabolites in Clinical Medicine**). Other sensors for debilitating neurological and psychiatric disorders and infectious diseases facing modern and aging societies are described by Kornguth (*see* **Biosensors for Neurological Disease**) and Baril (*see* **Need for Biosensors in Infectious Disease Epidemiology**).

Biosensor, bioarray, and biochip technologies are increasingly being exploited in non-medical diagnostics. For example, the pharmaceutical industry has desires to increase its throughput for drug discovery by coupling a variety of these approaches to miniaturized combinatorial chemistry to allow synthesis and bioassay to be linked in the same high throughput system. Legge engages this theme and describes the utility of biosensors in the pharmaceutical industry (*see* **Utility of Biosensors in the Pharmaceutical Industry**). Similarly,

other sectors have not been slow to appreciate the value of these platform technologies. For example, Rogers describes their application in environmental monitoring (*see* **Chip-Based Biosensors for Environmental Monitoring**), Karube for environmental biological oxygen demand (BOD) and other applications (*see* **Environmental Biochemical Oxygen Demand and Related Measurement**), Gauglitz for monitoring trace pollutants (*see* **Optical Biosensor for the Determination of Trace Pollutants in the Environment**), Terry for agriculture, horticulture, hydroponics and related applications (*see* **Agriculture, Horticulture, and Related Applications**) and Schnerr for the food and beverage industries (*see* **Food and Beverage Applications of Biosensor Technologies**).

Finally, a significant advantage of this raft of the technologies is that they are lightweight and useable in remote environments with limited requirements for power. Such characteristics make them ideal for extraterrestrial applications on the International Space Station, as described by Baumstark-Khan (*see* **From Earth to Space: Biosensing at the International Space Station**), for planetary exploration for life forms, as noted by Cullen (*see* **Life Detection within Planetary Exploration: Context for Biosensor and Related Bioanalytical Technologies**), and potentially in the future for truly exo-galaxy investigation.

66

Genetic and Other DNA-Based Biosensor Applications

Wim Laureyn,[1] Tim Stakenborg[1,2] and Paul Jacobs[3]

[1] NEXT-Nano-Enabled Systems, Leuven, Belgium, [2] Veterinary Research Institute, Brussels, Belgium and [3] Innogenetics NV, Zwijnaarde, Belgium

1 INTRODUCTION

The interest and research effort into the development of DNA diagnostic tools has increased dramatically. Without any doubt, the development of the polymerase chain reaction (PCR) heralded a new era in molecular analysis: the specific analysis of extremely small quantities of DNA suddenly became feasible. As a consequence, applications in which the sensitivity extends far beyond the sensitivity for protein analysis, down to below 100 targets per ml, could be addressed. Not only did this open the way to sensitive detection of organisms such as bacteria or viruses but it also produced sufficient material that enabled further study of other structural elements of the detected nucleic acids and to expand insights into structure–function relationships. The discovery of numerous genetic loci related to monogenic diseases and complex polygenic diseases, the rapid spread of (newly emerging) infectious agents, the problems associated with drug resistance, the increasing quality standards for food production and consumption, and the necessity of detecting genetically modified organisms (GMO) are just a few examples of what has led to completely new challenges in molecular diagnostics. Now, not only the speed of detection but also the specific detection of single nucleotide

polymorphisms (SNPs) or combinations of alleles in a particular sample has become essential. The advent of microarray technology provided an analytical tool that allows genome-wide expression profiling,[1,2] which has proved to be an increasingly important methodology. An example can be found in the field of cancer research where arrays may be used for early disease detection, prediction of therapy, and follow-up of treatment outcome.[3,4]

Biosensors, in combination with instruments that can detect, analyze, and quantify these new targets, are expected to revolutionize healthcare, particularly in the field of molecular diagnostics. Systems are awaited that are especially user-friendly (i.e., no hands-on manipulations, fast and easy interpretation of results) and that can be used preferably as portable instruments for point-of-care (POC) applications.[5,6] However, according to the strict definition, a biosensor can be defined merely as a device containing a biological sensing element coupled to a transducer that generates a detectable signal.[7] In this perspective, a DNA-based biosensor is a device incorporating oligonucleotides either intimately connected to or integrated within a transducer. This exact translation of the biosensor definition to DNA-based biosensors is particularly important to delineate the scope of this chapter. Recent developments

Handbook of Biosensors and Biochips. Edited by Robert S. Marks, David C. Cullen, Isao Karube, Christopher R. Lowe and Howard H. Weetall.
© 2007 John Wiley & Sons, Ltd. ISBN 978-0-470-01905-4.

in miniaturization have resulted in a plethora of technologies and assay formats that are referred to as biosensors although they do not fulfill the strict requirement of integration or intimate contact with a transducer. As a consequence, microarrays or DNA chips as such are not considered as biosensors and therefore not discussed in this chapter, nor are the many examples of miniaturized capillary separation-based systems that are often associated with biochips or lab-on-a-chip devices.

In this chapter, a number of application fields for DNA-based biosensors are described, together with some examples to illustrate the existing state of the art, without pursuing complete coverage of all current research. Nonetheless, the given examples highlight the specific nature of DNA-based diagnostics, its impact on biosensor requirements, and how specific challenges in this field can be addressed. An overview, illustrated with some of our original work, provides aspects of a potential route toward affordable DNA-sensor arrays with a particular focus on multiplexing and compatibility with polymer integration technology.

2　OVERVIEW OF APPLICATIONS

2.1　Infectious Diseases

Worldwide, one death in three is the result from a communicable or infectious disease.[8] Millions of people suffer from acute and chronic disorders caused by infectious agents (e.g., influenza, hepatitis C, hepatitis B, human papilloma and human immunodeficiency viruses, *Mycobacterium tuberculosis*, *Plasmodium* species, etc.), some of which are implicated in cancer development.[9] As a consequence, research has been expressly directed toward developing newer methods to detect pathogenic agents and treat the diseases they cause. Shorter detection times may lead to faster treatment, examining the molecular epidemiology may yield information for prevention or vaccination, and improvements in food and water quality may result in better sanitary practices. Unsurprisingly, most research in the area of biosensors is directly or indirectly linked to the diagnosis of infectious diseases.

Numerous examples of biosensors optimized for the detection of infectious agents are available.[10] The simultaneous detection of these agents for some applications (e.g., food-borne diseases) is particularly demanding as sample preprocessing steps are far from universal and certainly not all available in a standardized format.[11,12] Also, the choice of target genes may be debatable. Often, species-specific genes, or more generic type sequences such as rRNA sequences, are used as targets.[13–15] The latter sequences have the advantage that they are relatively stable and are present in high amounts in bacterial cells. On the other hand, they are largely conserved, even among different taxonomic units, and the specificity must be well assessed before using them as a marker. In that respect, the spacer region between the genes coding for rRNA may form a good alternative. It can likewise be amplified with universal primers, but in addition shows sufficient variability to allow adequate differentiation and specificity.[16]

Besides bacteria, DNA-sensor technologies have also been reported for some of the most important human viruses such as human papillomaviruses,[17] hepatitis viruses,[18] and human immunodeficiency viruses.[19,20] In the case of bacteria, a surface plasmon resonance (SPR) fluorescence sensor used PCR-generated fragments from *Neisseria gonorrhoeae* and *Chlamydia trachomatis* as target sequences. Under experimental conditions, the technique could detect as low as 10^6 DNA copies and was also shown to be useful for the detection of SNPs.[21] Moreover, in parasitology, DNA-based biosensors for the detection of, or the sensitivity to, eukaryotic parasites are currently being developed.[22] A DNA chip has already been used to monitor the sensitivity of children toward *Plasmodium* spp., the causative agents of malaria.[23]

DNA sensors have not been developed only for diseases that are important for humans. A sensor based on a quartz crystal microbalance (QCM) was described that could specifically detect 1-ng RNA originating from two different orchid viruses. A 10-fold lower detection limit was observed with crude orchid samples.[24]

2.2　Human Genetic Testing

2.2.1　Genetic Disorders

Since the completion of the Human Genome Project, an increasing understanding of the human genetic code has led to the discovery of many new hereditary genetic disease loci. Base mutations, deletions, insertions, chromosomal aneuploidy, or triplet and repeat expansions may all relate to genetic disorders. To better detect and understand such monogenetic or complex diseases (e.g., Alzheimer's disease, sickle cell anemia, schizophrenia, depression, attention-deficit/hyperactivity disorder (ADHD), etc.), the mapping of different genotypes has received worldwide attention (www.hapmap.org).

Cystic fibrosis is a monogenic disease for which already close to 1000 mutations have been identified in the cystic fibrosis transmembrane conductance regulator (*CFTR*) gene. Persons with mutations in this gene may suffer from the accumulation of mucus, especially in the lungs. Although cystic fibrosis is widespread, the distribution of these mutations in the gene and their relative frequency varies markedly between geographic and ethnic populations. However, early diagnosis of the disease allows the timely initiation of selected treatments in order to retard irreversible damage and improve the quality of life and life expectancy of the patient.[25]

The numerous multiethnic variants and the move toward neonatal or even prenatal detection of these genetic disorders require sensitive biosensor systems able to detect multiple factors simultaneously on samples of minimal size. In early work on electrochemical DNA biosensors, the selective detection of the common ΔF508 deletion mutation of the *CFTR* gene was used as a demonstrator.[26] More recently, ΔF508 has also been used as a demonstrator for the ultrasensitive (<10 fM) and reproducible electrical detection of sequence variations using nanowire-based sensors.[27] Advances in large-scale assembly of nanowires should allow high-throughput multiplex DNA detection.[28]

Cystic fibrosis could be the first diagnostic field to really benefit from DNA biosensor systems. The eSensor™ (Clinical Micro Sensors, Inc., a former business unit of Motorola, acquired by Osmetech Inc., Roswell, GA, USA) is a commercially available system detecting a panel of 24 mutations recommended by the American College of Medical Genetics and American College of Obstetricians and Gynecologists (ACMG/ACOG). The assay is based on sandwich hybridization of the single-stranded target to a capturing probe immobilized on a gold electrode. Added reporter probes labeled with ferrocene are detected upon hybridization by

charge transfer through the molecular wires. Free, nonhybridized redox material does not interfere because of the blocking insulator molecules.[29] The construction of the electrode array is based on established printed-circuit board technology. The moderate ambitions with respect to probe site density will most probably result in an affordable product.

DNA-based biosensors are currently also under investigation for other genetic diseases. Using impedance spectroscopy, an oligonucleotide mutation related to the lethal Tay–Sachs disease was examined.[30] Microgravimetric detection of specific hybridization using QCMs has been reported for Tay–Sachs disease[31] as well as for the rapid detection of β-thalassemia disease.[32] In addition, an optical detection technique was used to quantify SNPs associated with spinal muscular atrophy.[33] DNA probes, specific for the *SMN1* and *SMN2* gene fragments, were immobilized on the surface of fused silica optical fiber segments. Labeled DNA targets were subsequently measured by means of an epifluorescence scanning instrument in a total internal reflection fluorescence format. It is worthwhile to mention that interfacial thermal denaturation profiles were collected by applying a temperature ramp (25–85 °C, using a rate of 0.3 °C min^{-1}), yielding extra information about the hybridization event.

2.2.2 Oncology

The number of persons diagnosed annually with cancer, and the established finding that earlier detection almost invariably leads to better clinical outcome, highlights the importance in developing new treatment and diagnostic tools in this area. Although risk factors of various origins have been documented, the mapping of genetic mutations and, increasingly, expression profiling, provide direct links to cancer. Mutations may activate oncogenes or repress tumor-suppressor antigens and may eventually lead to the development of cancers. One such example is the *P53* tumor-suppression gene that encodes a nuclear protein with an essential role in the regulation of the cell cycle. Mutation of this gene, leading to deficiency or absence of the P53 protein, is believed to contribute to the majority of cancers and already a number of SPR and QCM analyses were performed to detect such *P53* mutations.[34,35] *BRCA1* and *BRCA2* are two other tumor-suppressor genes, several genotypes of which are linked to an increased risk for breast cancer. Certain mutations of the *BRCA1* gene were successfully determined using a biochip.[36] Such research may well allow for predictive testing in the field of cancer diagnosis.[6,37] Although genomic analysis is primarily a research tool to study vast amounts of data, many of the complex patterns can be represented by subsets of more manageable numbers of markers, thus paving the way for practical multiparametric assays to allow simultaneous disease detection and prognosis, or to provide information on preferred treatment options.[38]

There is important evidence that free circulating DNA in plasma may be useful as a marker for disease prognosis as well as for therapy monitoring. Indeed, circulating cell-free DNA appears to be the result of cell death, either caused by apoptosis/necrosis or associated with disease and malignant cells. The mere increase of circulating DNA plasma levels has been studied as a prognostic marker for long-term survival of hepatocellular carcinoma patients.[39] A more specific approach used the *hTERT* gene as a marker in plasma to study its significance in lung cancer.[40] Even detection of DNA in urine has been mentioned as a possible cancer diagnostic, but in this case certain precautions need to be taken into account.[41] It is surprising that little or no work on DNA-based biosensors for these targets in plasma and urine is in progress, since it appears to offer a unique condition in which the complexity of DNA extraction is expected to be significantly lower than that required for cells or virus particles. Furthermore, the levels of detection ($1–10\,\text{ng ml}^{-1}$) do not seem to be particularly demanding.

2.2.3 Tissue Typing for Transplantation

Successful organ or tissue transplantation relies upon sufficient control of the host immune response to induce tolerance toward the foreign material that is introduced. In this respect, the human leukocyte antigen (HLA) system plays a pivotal role. The HLA system enables the body to distinguish "self" from "nonself" material and to react accordingly by mounting an adequate immune response. To this end, HLA molecules are presented at the cell surface, forming a discrete recognition pattern, unique for every person (with the exception of monozygotic

twins). As a consequence, the identification of donor material that matches the recipient HLA type as closely as possible is of major importance to minimize the risk of rejection. Taking into account the variety of alleles that have been discovered (and still continue to be discovered), a typing methodology based on sequence-specific oligonucleotide hybridization must be able to address hundreds of specific sites simultaneously. Although microarray platforms offer sufficient multiplexing, classical solid-phase arrays require tremendous development and optimization efforts for such a complex set of oligonucleotide probes. The bead-based LabMap® system (Luminex Corp., Austin, TX, USA) offers a suspension array platform that is being used for this purpose, but is still limited to below 100 probes per reaction. The PamChip™ technology (PamGene, 's-Hertoghenbosch, The Netherlands), on the other hand, offers significant advantages for the development of complex typing assays such as the one presented by HLA typing.[42] This unique microarray system makes use of nanoporous alumina membranes and allows efficient mixing and dynamic monitoring of melting curves without the need for complex microfluidics.[43] However, of the currently available systems, neither LabMap nor the Pam-Gene technology can be regarded strictly as biosensors. Apart from fundamental binding studies on the Biacore® biosensor system, little work on biosensors specifically for tissue typing can be found. This may be explained by the complexity of the topic and the demanding requirements with respect to multiplexing.

2.2.4 Forensics

Genomic profiling of individuals is increasingly linked to forensic science or the search for evidence in crime scenes. Residual traces of DNA can be used for genetic fingerprinting or profiling and may help to convict suspects or to exonerate prisoners. Currently, a PCR of 13 common heterologous short tandem repeat (STR) sequences or base pair differences in mitochondrial DNA is used for individual profiling.[44] These data are subsequently stored in national databases (e.g., combined DNA index system (CODIS) and national DNA database (NDNAD)). Although the different STR fragments are now analyzed after electrophoresis, DNA arrays or biosensors may be useful to simultaneously hybridize various fragments. However, as current techniques are sufficiently robust and only specialized accredited laboratories are allowed to carry out these investigations, little effort is being made to develop portable sensing systems for these kinds of applications. It might even be questioned whether less stringent systems are desirable in view of the impact of mistyping.

2.3 Agriculture

As DNA sensors are still under full investigation, much less research has been reported for applications not related to human diseases. Nevertheless, many potential applications can be found in the area of agriculture.

Molecular marker-assisted selection (MAS) may greatly improve plant selection[45] and animal breeding.[46] Already, many alleles linked to bio-technological benefits (e.g., higher yield, better resistance to disease and/or stress factors, modified nutritional properties, etc.) have been discovered, but the use of MAS is rather limited. In particular, for inbreeding and/or minor crops, the price involved for the selection of different traits is still too high and cheaper alternatives are needed.[47]

Genetic engineering has also gained attention in food production. The use of GMO may result in higher yields and better food safety.[48] On the other hand, the effects of large-scale implementation on environmental safety, public opinion, intellectual property rights, and internal quality parameters may not be overlooked. For these reasons, performance monitoring of GMO needs to be assured.

At present, few research groups are developing DNA sensors for either the screening of GMO or MAS. Nevertheless, some data have been published. Selection markers of genetically modified plants were used to investigate the usefulness of a piezoelectric sensor for label-free detection,[49,50] while an optical sensor was described that monitors hybridization events with fluorescent-labeled target sequences in real time.[51] In another study, an electrochemical detector was optimized using a specific ssDNA target of transgenetic corn.[52]

2.4 Bioterrorism

Bioterrorism has existed for decades,[53] but has received much attention since packages containing anthrax were posted in 2001. It has driven the need for a fast, sensitive, and accurate testing of unknown, potentially biohazardous samples. Detectors for biological warfare agents should be decentralized and networked.[54] In addition, their obligatory use in or near the area of attack requires portable systems that allow rapid, yet sensitive diagnosis. Because of sensitivity needs, many systems for biological warfare agents are based on real-time PCR.[55,56] In order to ensure specificity and statistical confidence in the result of an assay for very low copy numbers (e.g., 100 copies ml^{-1}), a minimum volume of a few hundred microliters must be processed. Integration and interfacing with sample pretreatment, therefore, requires advanced microfluidic technologies to bridge the gap between relatively large sample volumes and microstructures used for rapid analysis.[57] A biosensor, based on an isothermal amplification step in combination with an optical detection, was validated for the detection of viable *Bacillus anthracis* spores.[58] For further examples, readers are referred to a comprehensive review on this issue.[59]

It must be emphasized that the challenge remains to detect any of the possible pathogenic microorganisms that could be used in an attack, taking into account the huge background of similar nonpathogenic microorganisms constantly present in the environment and the desired speed of detection of such organisms. Arrays of sensitive sensors with rapid readout are therefore desirable in order to realize an adequate identification and decision system.

2.5 Personalized Medicine

Personalized medicine is increasingly mentioned in connection with multiparameter and biosensor applications. Indeed, several monogenic genotypic variants (alleles) have been correlated with phenotypic variations, for example, adverse drug reactions or even serious adverse events. In 2004, rofecoxib, a drug used widely at that time to treat patients with arthritis, was voluntarily withdrawn from the market, after some patients displayed signs of an increased risk for cardiovascular disease. It must, however, be noted that, the side effects and the therapeutic efficacy of drugs usually vary from person to person, making individually adapted treatments preferable. A first step toward such personalized drug delivery was enabled by Food and Drug Administration (FDA) approval of the AmpliChip® CYP450 (Roche Diagnostics, Basel, Switzerland) test. In this test, the genotype of two genes of the cytochrome P450 can be determined, which gives an indication of the patient's ability to metabolize and clear certain medical compounds. A second example is the UGT1A1 Invader® assay (Third Wave Technologies, Madison, WI, USA) for detection of irinotecan-associated toxicity in the treatment of metastatic colorectal cancer.

3 REQUIREMENTS IMPOSED ON DNA-BASED BIOSENSORS

Despite impressive progress in the development of various promising types of biosensors (Figure 1), some challenges are still very difficult to overcome. Consequently, very few sensors have been validated with real samples and little more than a proof of concept has been reported. The examples given in the preceding text show that developers of DNA biosensors face two important challenges: the necessity for a combination of extreme sensitivity and selectivity required for detection, plus the ever-increasing complexity of the clinical question to be addressed, which calls for the combination of multiple polymorphisms.

The success of the glucose sensor has led to POC diagnostics often being put forward as the ideal biosensor application. From a medical point of view, detection of glucose offers an ideal scenario: the assay is monoparametric, it can be measured using an enzyme with a high catalytic turnover, the time-frame for decision-making is favorable, and the outcome leads to a clear action with immediate benefit for the patient. However, bringing DNA molecular tests to the POC imposes a number of challenging requirements that are not met by current DNA-based biosensors. The target DNA is most often embedded in a complex matrix, which, depending on its origin, can be very heterogeneous. As a consequence, extraction and further purification of the DNA is mandatory before its

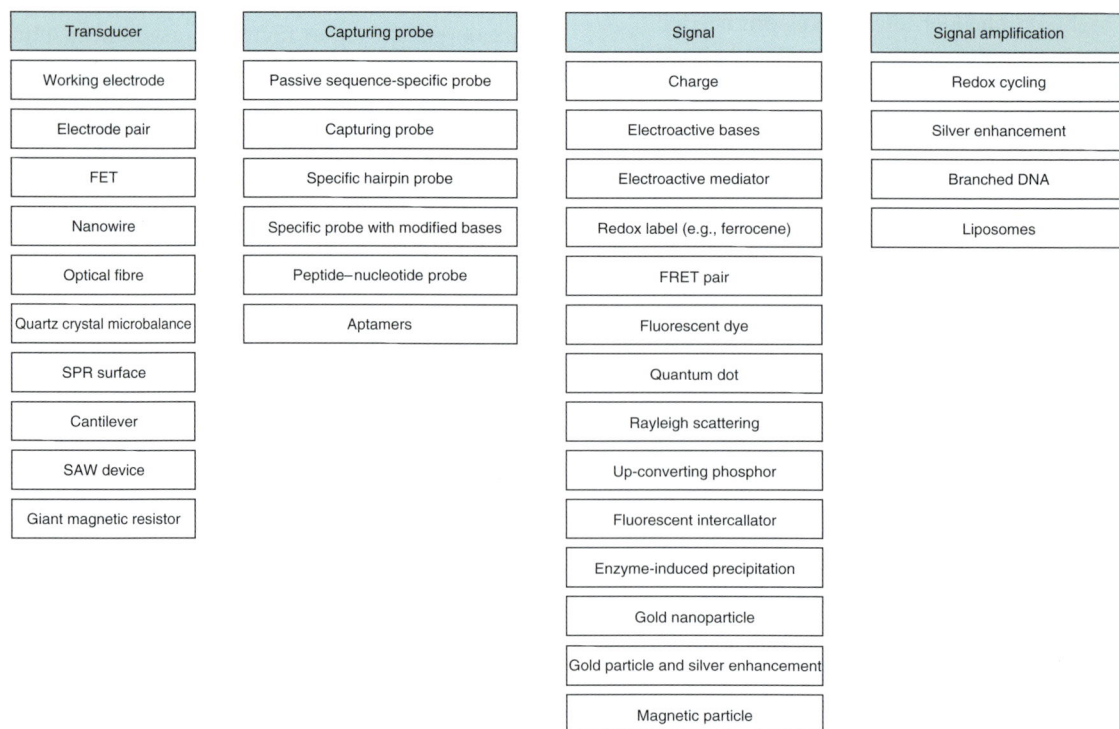

Transducer	Capturing probe	Signal	Signal amplification
Working electrode	Passive sequence-specific probe	Charge	Redox cycling
Electrode pair	Capturing probe	Electroactive bases	Silver enhancement
FET	Specific hairpin probe	Electroactive mediator	Branched DNA
Nanowire	Specific probe with modified bases	Redox label (e.g., ferrocene)	Liposomes
Optical fibre	Peptide–nucleotide probe	FRET pair	
Quartz crystal microbalance	Aptamers	Fluorescent dye	
SPR surface		Quantum dot	
Cantilever		Rayleigh scattering	
SAW device		Up-converting phosphor	
Giant magnetic resistor		Fluorescent intercallator	
		Enzyme-induced precipitation	
		Gold nanoparticle	
		Gold particle and silver enhancement	
		Magnetic particle	

Figure 1. A schematic overview of the variety of reported DNA-based sensing schemes based on the basic transduction principle, the type of capturing molecules used, the signal-generating component, and possible signal-amplification strategies.

actual detection. A POC system should therefore couple a DNA-based biosensor with highly sensitive, multiplex-specific detection to some device that preprocesses the sample. In many cases, the yield of target DNA is low, down to a few copies per milliliter of biological fluid. Biosensors with sufficient sensitivity to reach these targets have not yet been developed. Even when cultivating cells or bacteria under optimal conditions, below picomolar detection limits must be reached.[60] The strategy that has been explored in many different variants is that of signal amplification. It is, however, questionable whether it is reasonable to assume that the required specificity can be guaranteed in vitro on the basis of the hybridization specificity and kinetics alone.[61] A certain level of nonspecific reaction will not influence the clinical outcome when there is an abundance of specific material or where the hybridization kinetics are sufficiently discriminative. However, in view of the increased complexity of many applications, it is desirable to safeguard options in order to enhance their robustness. One way to achieve this is by using additional information of the intrinsic melting behavior of the formed duplexes, a feature that is jeopardized if precipitation reactions are used as a signal-amplification strategy.

Enzymatic target-amplification reactions such as PCR allow the selective enrichment of the particular genetic loci to be analyzed up to an abundant level far above that of the background. Without any doubt PCR-based tests are currently indispensable in the field of molecular diagnostics. Since PCR requires sequential thermal cycling, miniaturization offers the benefit of increased speed.[62] In combination with real-time fluorescence methods to monitor the progress of the PCR cycle by cycle, a major step toward integration has been taken. This has already lead to the development of POC systems such as the GeneXpert® (Cepheid, Sunnyvale, CA, USA) and Liat™ Analyzer (IQUUM Inc., Marlborough, MA, USA). A major drawback of this approach is the limitation on the number of mutations or polymorphisms that can be analyzed simultaneously. DNA-based biosensors could offer such multiplexing capabilities.

An alternative approach could be envisaged using miniaturized transponders.[63] A transponder is, in principle, a device that combines a transmitter and a responder circuit. The PharmaSeq microtransponders (PharmaSeq Inc., New York, NY,

USA) are highly miniaturized and are activated by light. The chips, measuring only $0.25 \times 0.25 \times 0.1$ mm, contain a photocell that, upon illumination with a laser-beam, powers a radio transmitter circuit. The activated chip then transmits its hard-coded 64-bit identification code. This code can be linked to the type of probe that has been immobilized on the microtransponder. Interrogation with a second laser system is used to determine the presence of fluorescent-labeled target molecules in a flow-cytometric setup. It might be feasible to further refine this type of chip to hold an electronic detection element connected to the logic of the identification transmitter, activating the chip only if hybridization has occurred. This would simplify the setup significantly, and, instead of integrating the wet-chemical process steps into fluidic cartridges, the electronic detection elements would be fully immersed in any homogeneous reaction system, reporting the binding events only upon activation with light. Despite the already extreme miniaturization, the combination of a high number of such chips in a small sample volume poses some challenges in keeping the chips suspended and consequently individually addressable.

Another revolutionary development, specifically focused on the challenge of multiparameter testing, could emerge from single-molecule sequencing. One promising approach uses 50-nm-wide waveguides to restrict the effective observation volume to the vicinity of a single immobilized DNA polymerase molecule. When this polymerase incorporates a fluorescent-tagged nucleotide, the tag is excited and the emission is detected. Using four dyes with distinct spectral characteristics, each sequentially incorporated nucleotide can theoretically be identified.[64]

Besides the attractive detection features, many sensing devices may offer additional functionality. Although not exploited as a biosensor system, electrophoretic attraction and repulsion has been used in the NanoChip® system (Nanogen Inc., San Diego, CA, USA) for rapid hybridization and increased stringency.[65] Electronically addressed in situ synthesis is used by Combimatrix Corp. (Mukilteo, WA, USA) for making custom arrays[66] for which an electrochemical readout system was developed and is marketed as ElectraSense™. Magnetic bead-based systems offer the advantage that manipulation can easily be implemented and combined with sensitive detection and even evaluation of the binding force.[67]

In summary, the DNA-based biosensor in the strict sense will only be a part of a fully integrated system or a so-called lab-on-a-chip device.[68] Ideally, such a device would integrate multiple bioanalytical functions into an easy-to-use, small, and portable instrument with further potential benefits including low reagent consumption, low time-to-result, and minimal waste.[6] In most cases, an array of sensing elements will be required to address multiple parameters (SNPs, mutations, etc.) simultaneously.

4 A POLYMER ARRAY OF INTERDIGITATED ELECTRODE DNA SENSORS

4.1 Introduction

Electrochemical biosensors offer great potential for miniaturization and integration. This aspect is of particular interest for DNA-based biosensors due to the necessity of analyzing multiple genetic loci simultaneously. As mentioned earlier, some sensor arrays are currently finding their way to commercialization, such as the GeneLyzer™ (Toshiba Corporation, Tokyo, Japan) or the eSensor™ system. With 20 and 36 sensing sites respectively, the multiplexing capability, although better than that of real-time fluorescence systems, still falls short in accommodating an ever-increasing number of applications. In this regard, some researchers investigated the capability of combining integrated circuit sensor arrays such as photodiode arrays[69] or a charge-coupled device (CCD)-element[70] with oligonucleotide-capturing probe arrays.

The polyanionic nature of DNA has been exploited in a number of sensing schemes that assess the charge density in the case of hybridization to a solid-phase immobilized probe (Table 1). A simple way of assessing charge is by evaluating

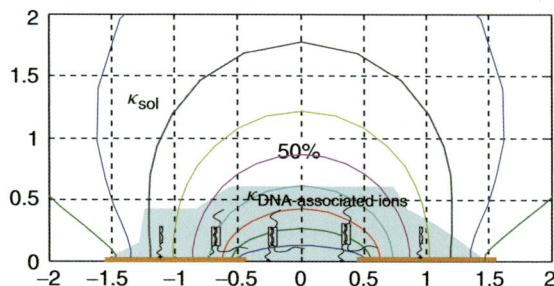

Figure 2. Contour plot indicating the reach of the field lines if an electric field is applied between two planar electrodes of width 1 arb unit and with an equal interelectrode spacing of 1 arb unit. Assuming that a typical probe–target duplex of about 30 bp measures about 10 nm, and a typical target DNA sequence would comprise 300 or more bases, submicron electrode width and spacing is required in order to bring the assessed portion of the field lines within the area where the binding event is monitored. As the associated ion distribution is highly influenced by the ionic strength, the associated diffuse charge layer will expand further away from the solid phase as low-ionic-strength buffers are used (light-blue shaded area). In that case, the change in local conductivity ($\kappa_{DNA-associated\,ions}$) versus the background electrolytic conductivity of the measurement solution (κ_{sol}) can be optimized.

the local electrolytic impedance. Most impedance-based work addresses the electrochemical interfacial impedance, since probes are easily immobilized on electrode surfaces in a self-assembled monolayer. An alternative approach is to treat the microenvironment between two electrodes as a conductivity cell, in which the concentration of mobile ions determines the local electrolytic conductivity. However, to obtain a sufficient signal-to-noise ratio in such a setup, the electrical field lines need to be confined as close as possible to the molecules to be assessed (Figure 2). As a consequence, micron-sized or even submicron-sized electrode structures are needed.

Since cost is always a major driver, this section describes attempts made to develop a scalable process that results in such micron-sized thin-film

Table 1. Different detection methods using nucleic acid–associated charge

Transducer	Effect	Method	References
Silicon oxide	Interfacial impedance	Electrical impedance spectroscopy	71
	Field effect	Surface potential	72
	Interfacial impedance	Optoelectrochemical impedance spectroscopy	73
Si nanowires	Field effect		27
Gold electrode	Capacitance		74
Electrode pair	Interdigitated electrode impedance	Reduction in associated mobile counterions	75

microelectrode structures, for which an innovative manufacturing approach has been evaluated in combination with the assessment of label-free hybridization detection. This route has been researched by a European consortium within the 5th Framework, coordinated by Innogenetics NV (Ghent, Belgium) to assess its potential use in future markets in the in vitro diagnostics field. It is based on building awareness that multiparameter diagnostics will gain in importance in routine analysis, and the assumption that dedicated technologies for detection should be compatible with low-cost manufacturing (http://cordis.europa.eu/nanotechnology/src/pressroom-pub.htm).

4.2   Directional Metal Deposition on 3D Structures for Microelectrode Production

Until recently, micron-sized electrode structures could only be obtained with sophisticated and expensive lithographic techniques. Furthermore, this type of manufacturing is mainly used in silicon foundries. The use of silicon as a passive carrier material is expensive, but it offers the best substrate to realize highly accurate micron-sized structures by micromachining. If combined with state-of-the-art micromolding techniques, these micron-sized structures can be replicated in polymers resulting in very inexpensive microstructures.

It is known that, under certain conditions, metal deposition by evaporation suffers from a shadowing effect and will only deposit metal on the surfaces that are "seen" from the standpoint of the evaporation source. This effect has been exploited for the realization of arrays of nanowires.[76] Elaborating further on this principle, an appropriate design of a three-dimensional microstructure in a substrate could result in a self-contained shadowing effect that enables the creation of electrically disconnected interdigitated electrodes (IDEs). Such a three-dimensional design is shown in Figure 3. The combination of channels and hills results in the proper shadow regions where no metal is deposited, while the exposed parts form the electrodes of the interdigitated pairs.

An attractive feature of IDE structures is that they show a very reproducible and well-controlled impedance characteristic. Furthermore,

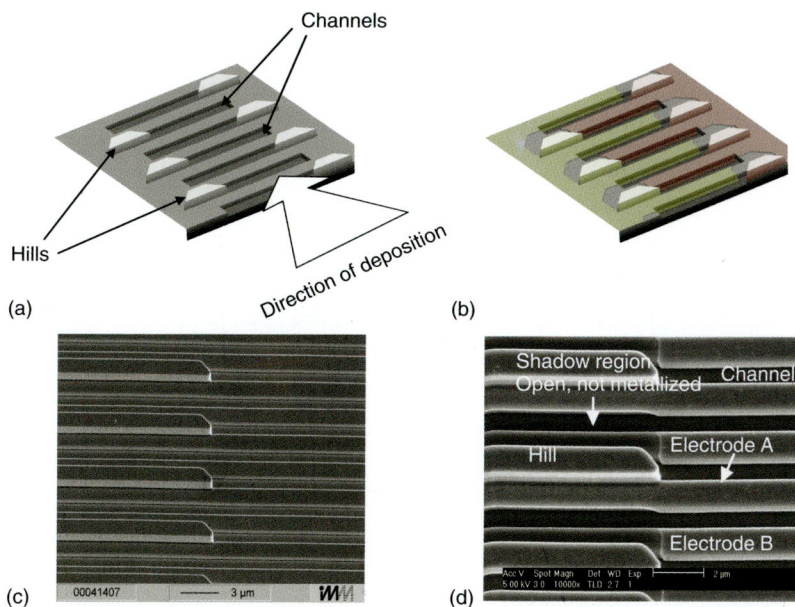

Figure 3. An appropriate combination of parallel channels and hills forms the basis for making interdigitated electrode structures by a single directional metal deposition: (a) before metal deposition and (b) after metal deposition. Scanning electron microscope images of realized three-dimensional structures as they were designed to show the desired shadowing effect: (c) detail of a silicon master structure and (d) the result after metal deposition.

this characteristic can be described very accurately using a fairly simple electrical equivalent model.[77] A mathematical fit of this model with the experimental data allows easy interpretation (Figure 4).

A plot of the impedance characteristics after metal deposition on the silicon master structures, which can be used as the basis for polymer replication, is given in Figure 5. The impedance characteristics on the resulting three-dimensional interdigitated structures show a similar profile as the silicon-based planar interdigitated structures that were used as a reference. In the low-frequency range (1–10 kHz), the characteristic is almost purely capacitive. Between 10 kHz and a few hundred kHz, a clear resistive plateau reflecting the bulk electrolyte resistance is observed.

At the higher end of the tested frequency range (1 MHz), the parasitic capacitances dominate the impedance.

Using state-of-the-art micromolding techniques, the silicon microstructures can be replicated in polymers. The combination with directional metal deposition leads to an elegant, two-step production process: In a first step, replicas of the appropriate three-dimensional structures are produced by injection molding. In the second step, metal is deposited by directional deposition, which can be processed in the same fashion, leading to affordable IDEs.

The mold inserts to be used in the molding tools can be derived from the silicon microstructures by galvanoplating (Figure 6). Multicavity molding

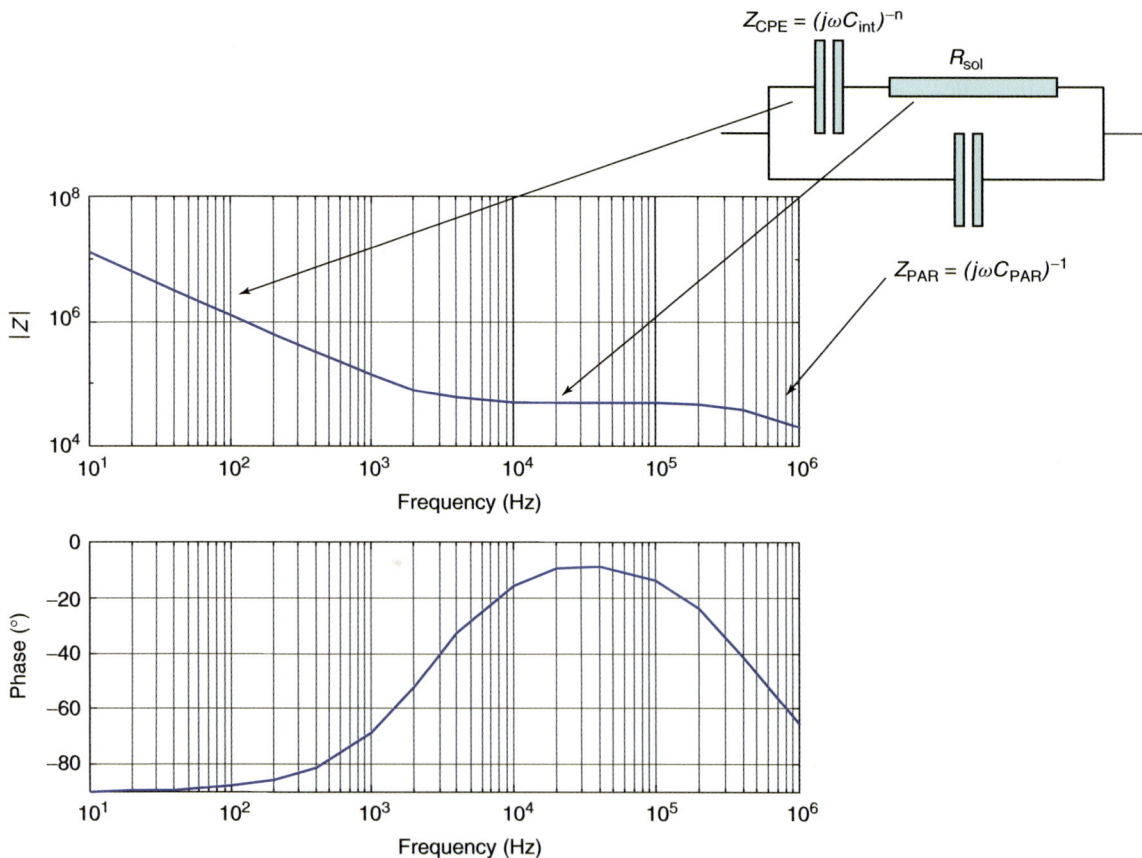

Figure 4. Typical impedance plots (modulus and phase) of low-signal impedance when using interdigitated electrodes in an electrolytic solution. Inset illustrates a simple equivalent model that fits this characteristic. Z_{CPE} is a so-called constant phase element that approximates the interface impedance. As the interdigitated electrodes are driven in the linear region (low driving signal typically 5–10 mV$_{rms}$), the value of n is between 0.9 and 1, being close to purely capacitive. Z_{PAR} is the parasitic capacitance. The parameter of interest is the electrolytic conductivity in the close proximity of the electrodes, which is reflected in the R_{sol}.

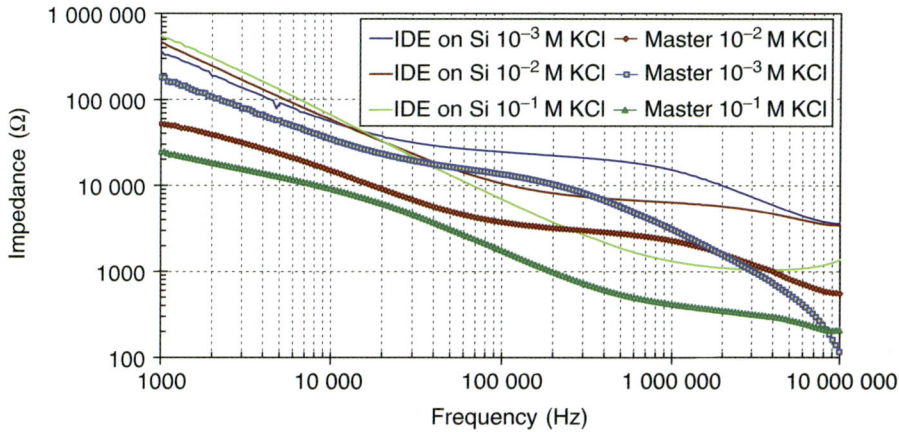

Figure 5. Illustration of the similarity of the electronic impedance measured with state-of-the-art silicon-based 1-μm planar interdigitated electrode structures (IDE, lines) and structures obtained by directional metal deposition on the proposed three-dimensional structures as they were realized in silicon. A clear resistive region at decreasing levels as electrolyte concentration increases is present in both cases. In the low-frequency region, a flattening of the slope for the three-dimensional structures can be observed. This is due to a secondary resistive effect between the contact pads and the electrolyte solution. Careful encapsulation removes this effect as is proved in the case of the standard planar electrodes.

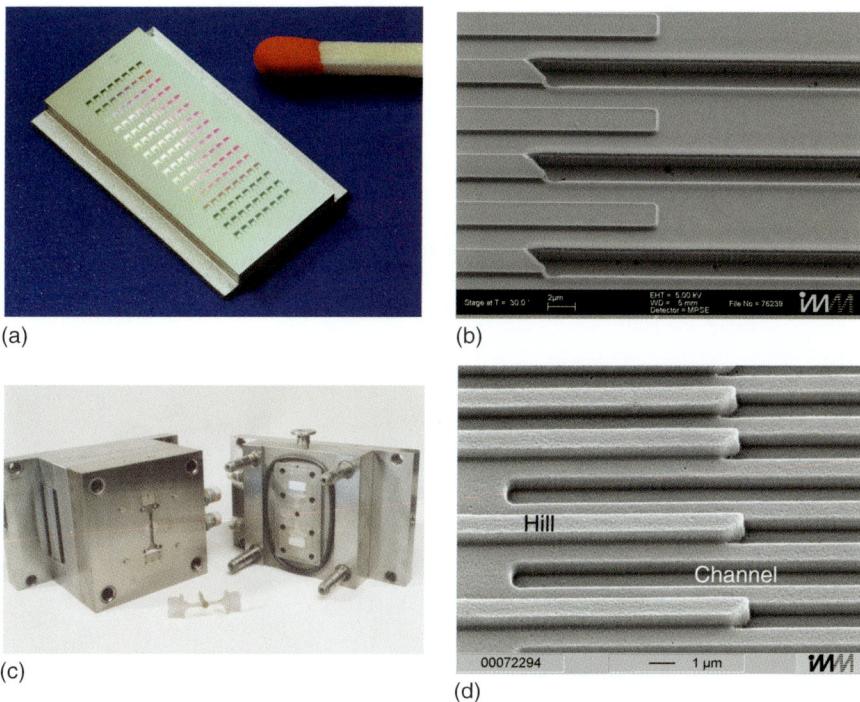

(a)

(b)

(c)

(d)

Figure 6. Several master structures for arrays of interdigitated electrode structures have been tested. Using these silicon master structures, mold inserts were fabricated by galvanoforming. (a) A prototype for a 128-*sensel* array. (b) SEM image showing the detailed negative. The mold inserts were used in standard injection molding equipment to make polymer replicas. (c) The molding chamber in which two cavities containing the mold inserts are addressed simultaneously. Each molding shot therefore results in two replicas still fixed to each other by the runner polymer. The resulting pair of devices can be seen in the foreground. (d) SEM image of a detail of the resulting molded replica showing the accurate reproduction of the channels and hills that form the basis of the required three-dimensional structures.

tools are often used to further reduce the cost. For reasons of minimizing additional process steps to realize interconnections to each of the electrodes, a physical shadow mask can be used during the metal deposition step to form a common interconnect line to one electrode of each of the interdigitated pairs in a column. To avoid the need for a multilayer process, access to the counterelectrode of each individual pair can be provided by contact pins through the housing of the fluidic chip after assembly (Figure 7).

4.3 Assessing the DNA-associated Electrolytic Conductivity

The goal of this work was to make an array of detection elements for the specific detection of hybridization of amplified DNA targets to immobilized probe oligonucleotides based on the assessment of electric charge accumulation. The phosphate backbone of DNA provides a negative charge to the molecule at neutral pH. These negative charges are neutralized by cationic counterions. The counterions are associated to the phosphate backbone in a condensed layer and/or a diffuse layer, depending on the ionic strength of the solution.

Figure 8 shows the impedance characteristics of a 1-μm silicon-based planar IDE structure before and after probe immobilization and after hybridization. There is a clear effect of probe immobilization in the capacitive region as well as in the resistive plateau. After hybridization with an 87-nucleotide HLA-DQB synthetic target, a further distinct decrease in the equivalent resistance was observed, while the capacitive component remained unchanged. The corresponding

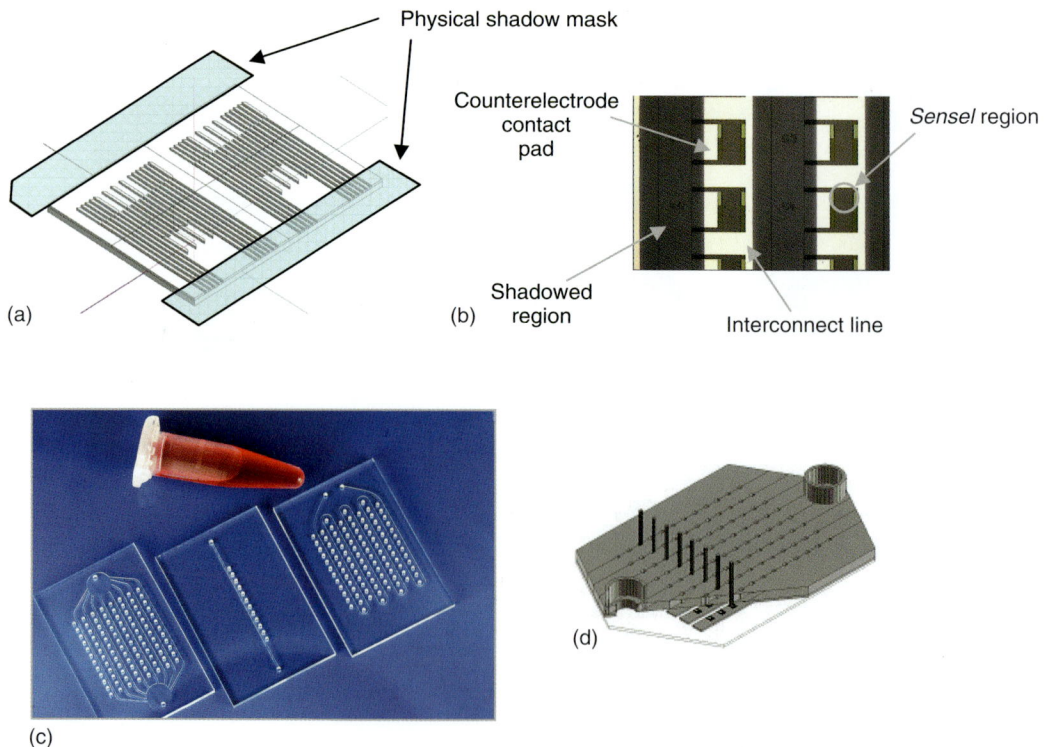

Figure 7. During the directional metal deposition step, an additional physical shadow mask was aligned on top of the structures to define the column addressing interconnection lines while electrically separating the different columns (a). At this stage, an array of 96 interdigitated sensing elements or *sensels* was formed in an 8-column by 12-row format. One electrode of each sensing element in one column is already interconnected and the interconnect lines reach the edge of the arrays (b). A polymer cover defines the fluidic path over the different sensing elements and contains perforations that allow contact between the counterelectrodes and contact pins. A number of prototype fluidic lids have been made (c). A diagram of the 96-element assembly is shown (d).

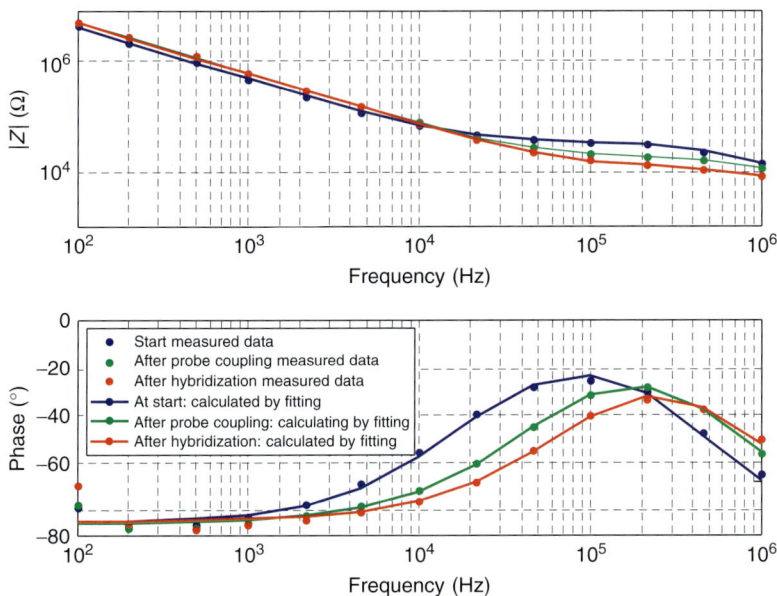

Figure 8. Measured impedance characteristic for a 1-μm interdigitated *sensel* at different stages. The discrete dots are the measured raw data points. The lines are the result of the calculated data based on the best fit of the equivalent model to the measured data. The resulting extracted parameters are summarized in Table 2.

equivalent model parameters calculated by curve fitting, are given in Table 2.

In a first investigation of the effect of decreasing the electrode dimensions on the impedance shift upon accumulation of DNA, planar IDE structures with varying electrode width and spacing were tested. The 15 *sensels* (defined as *sens*ing *el*ement in analogy to the definition of "pixel") of one single chip were activated for oligonucleotide coupling by application of a mercaptosilane film. A great deal of attention was paid to the optimization of this activation step since it is critical to obtain stable impedances throughout the whole process. The impedance characteristics of each individual *sensel* were measured in the frequency range of 1 kHz–1 MHz. A 10-μM oligonucleotide solution was manually spotted on 12 of the 15 *sensels*. Three types of thiol (SH)-modified

oligonucleotides were used in this instance: (i) a 16-nucleotide HLA-DQB probe with a single hexaethylene glycol (HEG) spacer molecule between the probe and the SH-group; (ii) a variant of this same probe, with a five-unit HEG spacer; and (iii) a 25-nucleotide *Mycobacterium* probe with a single HEG-unit spacer but preceded by 41 adenines before the actual specific probe (66 nucleotides in total). After probe coupling and washing, the impedance characteristics were recorded again in the same measurement buffer. The results showed that a low-conductivity buffer is preferable for this purpose. A 1-mM Tris or a 50-mM glycine buffer was the buffer of choice since they both conserved the duplexes formed after the hybridization reaction. For interpretation, the change in resistance before and after the probe coupling was determined as:

$$\text{Relative } \Delta R = \frac{(R_{\text{sol before}} - R_{\text{sol after}})}{R_{\text{sol before}}} \quad (1)$$

The graph in Figure 9 summarizes the results. Each bar represents the average of the change of impedance of four *sensels*. It can be seen that the change in resistance increases with decreasing electrode dimensions. In addition, the decrease

Table 2. Calculated equivalent model parameters after fitting

	C_{int}	n	$R_{\text{sol}}(\Omega)$	C_{PAR} (pF)
Start	465×10^{-12}	0.94	37 323	9.9
Modification	363×10^{-12}	0.94	20 825	11.0
Hybridization	398×10^{-12}	0.93	13 534	14.0

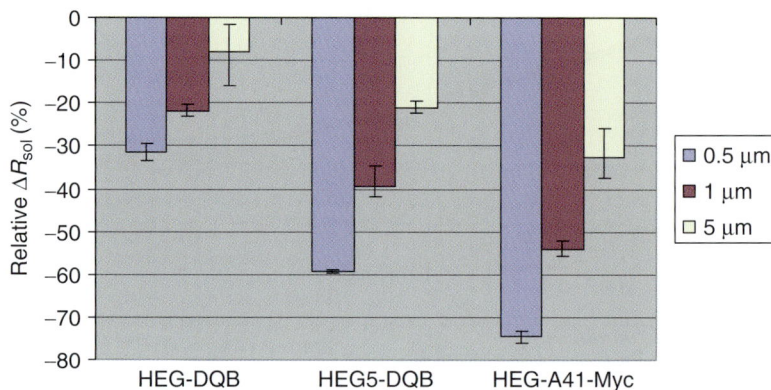

Figure 9. Change in R_{sol} extracted from the measured data by curve fitting for the three types of interdigitated structures that have been investigated (0.5, 1, and 5 μm), comparing the value after silanization, preparing the *sensels* for immobilization, and after probe coupling. The experiment summarizes the data of 12 *sensels* of a 15-*sensel* array. Each type of probe was deposited on a group of four *sensels* Error bars indicate full range (minimum and maximum out of the four observations).

is systematically higher for longer oligonucleotide probes (25 vs 66 nucleotide probes). The effect of the longer ethylene glycol spacer on the resistance change might be due to the increased hydrophilicity near the surface.

4.4 Detection of Sequence-specific Hybridization

In a subsequent step, the response to hybridization with synthetic complementary oligonucleotides was studied in more detail. Using planar prototype electrode structures, a further analysis of the effect of the reduction of dimensions on hybridization detection was performed. Although a somewhat increased response was evident in some cases for the electrode structures with the smallest dimensions, the reproducibility was lower, mainly due to the instability of the submicron-sized gold electrodes even when produced using the standard liftoff process (data not shown). Therefore, only 1-μm IDE structures were used in further experiments.

To challenge the specificity of detection, three variants of HLA-DQB probes were used on a single chip: in addition to the DQB probe used earlier, a probe with three nucleotides difference in the center of the probe-sequence (DQB3MM) and one with only a single nucleotide difference (DQB1MM) were immobilized on a 15-*sensel*

array. Firstly, a synthetic 80-base oligonucleotide with a perfectly matching complementary region for the DQB probe was incubated over the 15-*sensel* array at a concentration of 10 nM. Subsequently, the hybrids formed were denatured on chip and incubated again with 10-nM synthetic target, but in this case with a perfectly complementary region for the DQB3MM probe. The respective results are shown in Figure 10. When probes with the same spacer are used, a signal-to-noise ratio (perfect match signal versus mismatch signal) of at least 4 is obtained.

In a further set of experiments, the DQB probe and the DQB1MM probe were used on a single chip and hybridized with a 261-nucleotide PCR fragment of DQB (~1 nM). As one of the primers in the PCR reaction contained a 5′-phosphate group, single-stranded material could be obtained before incubation by digestion with λ exonuclease. It can be seen that using such a long target the effect of the spacer length for the specific probe becomes more significant. Again comparing the probes with the same spacer, a signal-to-noise ratio of higher than 3 was achieved.

These results show that specific post-PCR detection of multiple targets using this approach is feasible without any additional signal-amplification strategy. Additional functionality, such as on-chip localized heating, has also been demonstrated (data not shown), opening the way to multiplex melting analysis. The integration of the *sensels* in a microfluidic system that would allow dynamic

Probe	Sequence
HEG3-DQB	3′HS-(CH$_2$)$_6$-(HEG)$_3$ -CAGGTGG**C**TGCGGGCC-5′
HEG5-DQB	3′HS-(CH$_2$)$_6$-(HEG)$_5$ -CAGGTGG**C**TGCGGGCC-5′
HEG3-DQB1MM	3′HS-(CH$_2$)$_6$-(HEG)$_3$ -CAGGTGG**G**TGCGGGCC-5′
HEG3-DQB3MM	3′HS-(CH$_2$)$_6$-(HEG)$_3$ -CAGGTG**TGC**GCGGGCC-5′

Target	Sequence
80 nt	5′CT**GTCCACCGACGCCCGG**GCCCCCTCCAGGACTTCCTTCT GGCTGTTCCAGTACTCGGCGCTAGGCCGCCCCTGCGGCGT 3′
87 nt	5′GCACACCCT**GTCCACACGCGCCCGG**GCCCCCTCCAGGACTTCC TTCTGGCTGTTCCAGTACTCGGCGCTAGGCCGCCCCTGCGGCGT 3′

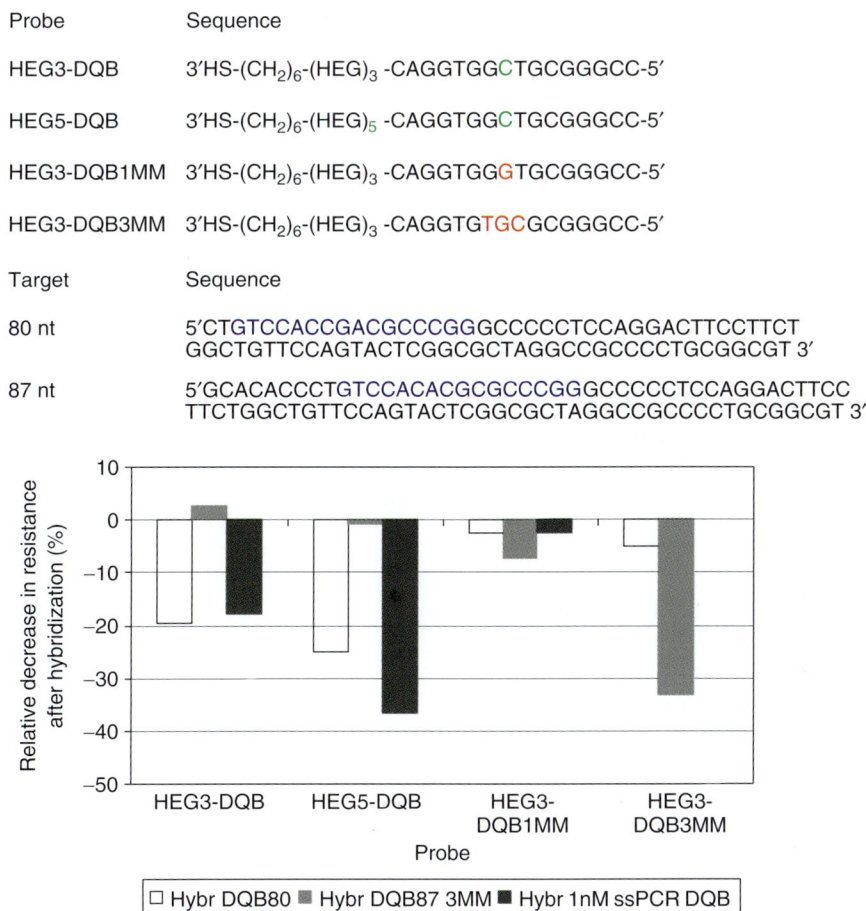

Figure 10. Change in R_{sol} extracted from the measured data by curve fitting for 1-μm interdigitated structures, comparing the value before and after hybridization. The graph summarizes the data for three experiments: the first target was an 80-nucleotide synthetic complement for the DQB probe, the next was a synthetic complement of 87 nucleotides matching the DQB3MM, and, finally, a 261-nucleotide ssPCR product of DQB was hybridized at a concentration of approximately 1 nM.

incubation and improve mass transfer is a straightforward next step.

5 CONCLUSIONS AND FUTURE PERSPECTIVES

An enormous potential may be expected from future developments in DNA-based biosensors in view of the ever-increasing number of applications that require the decoding of specific nucleic acid targets in a single sample. Arrays of biosensors offer the potential of efficient and simultaneous detection of multiple hybridization events without the need for sophisticated instrumentation. As

discussed, it will be quite a challenge to attain the required sensitivity and specificity if the combination with a target-amplification step is not realized. In addition, for most real-life applications, the DNA target molecules are embedded in a heterogeneous complex matrix. Integration with at least the extraction and purification steps is required, preferably even in closed-tube systems, if the full benefit of biosensor-based systems is to be appreciated in easy-to-use, handheld instruments.

Integration of biochemical processes such as cell lysis, DNA extraction and purification, and enzymatic target amplification, preferably with arrays of sensors, brings enormous challenges with respect to microfluidics, electronic interconnection, and interfacing. A unique approach that could circumvent a number of these challenges makes use of extremely miniaturized transponders. The development of these extremely small chips, known as "smart dust", may offer a completely new perspective to sensing and associated instrumentation.

ACKNOWLEDGMENTS

The work on the Polymer Array of Interdigitated Electrode DNA Sensors was carried out in the framework of two projects funded by the Flemish Government (IWT 960096, IWT 030059) and one EC-funded project (BRPR-CT98-0770). The contributions of our colleagues at Innogenetics (Georges Van Reybroeck, Walter Hofer, Wim Tachelet), IMEC Vzw (Peter Vangerwen, Andrew Campitelli, Jan Suls), IMM GmbH (Peter Detemple, Marion Bär), Cranfield University (Jeff Newman, David Cullen) and Biodot Ltd. (Chris Flack, Anthony Peloe) are greatly appreciated. Finally we would like to thank André Van de Voorde for the many constructive discussions and valuable input and Anne Farmer for the critical review of the manuscript.

REFERENCES

1. J. D. Hoheisel, Microarray technology: beyond transcript profiling and genotype analysis. *Nature Reviews Genetics*, 2006, **7**(3), 200–210.
2. J. Khan, L. H. Saal, M. L. Bittner, Y. D. Chen, J. M. Trent, and P. S. Meltzer, Expression profiling in cancer using cDNA microarrays. *Electrophoresis*, 1999, **20**(2), 223–229.
3. Y. Pawitan, J. Bjohle, L. Amler, A. L. Borg, S. Egyhazi, P. Hall, X. Han, L. Holmberg, F. Huang, S. Klaar, E. T. Liu, L. Miller, H. Nordgren, A. Ploner, K. Sandelin, P. M. Shaw, J. Smeds, L. Skoog, S. Wedren, and J. Bergh, Gene expression profiling spares early breast cancer patients from adjuvant therapy: derived and validated in two population-based cohorts. *Breast Cancer Research*, 2005, **7**(6), R953–R964.
4. B. K. Zehentner, D. C. Dillon, Y. Q. Jiang, J. C. Xu, A. Bennington, D. A. Molesh, X. Q. Zhang, S. G. Reed, D. H. Persing, and R. L. Houghton, Application of a multigene reverse transcription-PCR assay for detection of mammaglobin and complementary transcribed genes in breast cancer lymph nodes. *Clinical Chemistry*, 2002, **48**(8), 1225–1231.
5. T. M. H. Lee and I. M. Hsing, DNA-based bioanalytical microsystems for handheld device applications. *Analytica Chimica Acta*, 2006, **556**(1), 26–37.
6. S. A. Soper, K. Brown, A. Ellington, B. Frazier, G. Garcia-Manero, V. Gau, S. I. Gutman, D. F. Hayes, B. Korte, J. L. Landers, D. Larson, F. Ligler, A. Majumdar, M. Mascini, D. Nolte, Z. Rosenzweig, J. Wang, and D. Wilson, Point-of-care biosensor systems for cancer diagnostics/prognostics. *Biosensors and Bioelectronics*, 2006, **21**(10), 1932–1942.
7. A. Brecht and G. Gauglitz, Optical probes and transducers. *Biosensors and Bioelectronics*, 1995, **10**(9–10), 923–936.
8. World Health Organization, *Scaling up the Response to Infectious Diseases. A Way Out of Poverty*, Report on infectious diseases, 2002 WHO/CDS/2002.7 (Geneva: WHO, 2002). link: www.who.int/infectious-disease-report/index.html.
9. S. A. Hoption, J. P. van Netten, and C. van Netten, Acute infections as a means of cancer prevention: opposing effects to chronic infections. *Cancer Detection and Prevention*, 2006, **30**(1), 83–93.
10. B. Pejcic, R. D. Marco, and G. Parkinson, The role of biosensors in the detection of emerging infectious diseases. *Analyst*, 2006, **131**(10), 1079–1090.
11. P. D. Patel, Overview of affinity biosensors in food analysis. *Journal of AOAC International*, 2006, **89**(3), 805–818.
12. A. Rasooly and K. E. Herold, Biosensors for the analysis of food- and waterborne pathogens and their toxins. *Journal of AOAC International*, 2006, **89**(3), 873–883.
13. A. K. Deisingh and M. Thompson, Sequences of *E. coli* O157:H7 detected by a PCR-acoustic wave sensor combination. *Analyst*, 2001, **126**(12), 2153–2158.
14. J. C. Liao, M. Mastali, V. Gau, M. A. Suchard, A. K. Moller, D. A. Bruckner, J. T. Babbitt, Y. Li, J. Gornbein, E. M. Landaw, E. R. McCabe, B. M. Churchill, and D. A. Haake, Use of electrochemical DNA biosensors for rapid molecular identification of uropathogens in clinical urine specimens. *Journal of Clinical Microbiology*, 2006, **44**(2), 561–570.
15. B. P. Nelson, M. R. Liles, K. B. Frederick, R. M. Corn, and R. M. Goodman, Label-free detection of 16S ribosomal RNA hybridization on reusable DNA arrays using surface plasmon resonance imaging. *Environmental Microbiology*, 2002, **4**(11), 735–743.
16. G. Jannes and V. D. De, A review of current and future molecular diagnostic tests for use in the microbiology laboratory. *Methods in Molecular and Cellular Biology*, 2006, **345**, 1–21.
17. J. H. Wang, W. L. Fu, M. H. Liu, Y. Y. Wang, Q. Xue, J. F. Huang, and Q. Y. Zhu, Multichannel piezoelectric genesensor for the detection of human papilloma virus. *Chinese Medical Journal*, 2002, **115**(3), 439–442.
18. M. Chen, M. H. Liu, L. L. Yu, G. R. Cai, Q. H. Chen, R. Wu, F. Wang, B. Zhang, T. L. Jiang, and W. L. Fu, Construction of a novel peptide nucleic acid piezoelectric gene sensor microarray detection system. *Journal of Nanoscience and Nanotechnology*, 2005, **5**(8), 1266–1272.
19. K. Kukanskis, J. Elkind, J. Melendez, T. Murphy, G. Miller, and H. Garner, Detection of DNA hybridization using the TISPR-1 surface plasmon resonance biosensor. *Analytical Biochemistry*, 1999, **274**(1), 7–17.

20. J. Wang, X. H. Cai, G. Rivas, H. Shiraishi, P. A. M. Farias, and N. Dontha, DNA electrochemical biosensor for the detection of short DNA sequences related to the human immunodeficiency virus. *Analytical Chemistry*, 1996, **68**(15), 2629–2634.

21. B. Campbell, J. Lei, D. Kiaei, D. Sustarsic, and S. El Shami, Nucleic acid testing using surface plasmon resonance fluorescence detection. *Clinical Chemistry*, 2004, **50**(10), 1942–1943.

22. A. Logrieco, D. W. Arrigan, K. Brengel-Pesce, P. Siciliano, and I. Tothill, DNA arrays, electronic noses and tongues, biosensors and receptors for rapid detection of toxigenic fungi and mycotoxins: a review. *Food Additives and Contaminants*, 2005, **22**(4), 335–344.

23. X. B. Zhong, L. Leng, A. Beitin, R. Chen, C. McDonald, B. Hsiao, R. D. Jenison, I. Kang, S. H. Park, A. Lee, P. Gregersen, P. Thuma, P. Bray-Ward, D. C. Ward, and R. Bucala, Simultaneous detection of microsatellite repeats and SNPs in the macrophage migration inhibitory factor (MIF) gene by thin-film biosensor chips and application to rural field studies. *Nucleic Acids Research*, 2005, **33**(13), e121.

24. A. J. C. Eun, L. Q. Huang, F. T. Chew, S. F. Y. Li, and S. M. Wong, Detection of two orchid viruses using quartz crystal microbalance-based DNA biosensors. *Phytopathology*, 2002, **92**(6), 654–658.

25. J. Zielenski and L. C. Tsui, Cystic fibrosis: genotypic and phenotypic variations. *Annual Review of Genetics*, 1995, **29**, 777–807.

26. K. M. Millan and S. R. Mikkelsen, Sequence-selective biosensor for DNA-based on electroactive hybridization indicators. *Analytical Chemistry*, 1993, **65**(17), 2317–2323.

27. J. Hahm and C. M. Lieber, Direct ultrasensitive electrical detection of DNA and DNA sequence variations using nanowire nanosensors. *Nano Letters*, 2004, **4**(1), 51–54.

28. D. Whang, S. Jin, Y. Wu, and C. M. Lieber, Large-scale hierarchical organization of nanowire arrays for integrated nanosystems. *Nano Letters*, 2003, **3**(9), 1255–1259.

29. R. M. Umek, S. W. Lin, J. Vielmetter, R. H. Terbrueggen, B. Irvine, C. J. Yu, J. F. Kayyem, H. Yowanto, G. F. Blackburn, D. H. Farkas, and Y. P. Chen, Electronic detection of nucleic acids: a versatile platform for molecular diagnostics. *Journal of Molecular Diagnostics*, 2001, **3**(2), 74–84.

30. A. Bardea, F. Patolsky, A. Dagan, and I. Willner, Sensing and amplification of oligonucleotide-DNA interactions by means of impedance spectroscopy: a route to a Tay-Sachs sensor. *Chemical Communications*, 1999, **1**, 21–22.

31. A. Bardea, A. Dagan, I. Ben-Dov, B. Amit, and I. Willner, Amplified microgravimetric quartz crystal-microbalance analyses of oligonucleotide complexes: a route to a Tay-Sachs biosensor device. *Chemical Communications*, 1998, **7**, 839–840.

32. X. C. Zhou, L. Q. Huang, and S. F. Y. Li, Microgravimetric DNA sensor based on quartz crystal microbalance: comparison of oligonucleotide immobilization methods and the application in genetic diagnosis. *Biosensors and Bioelectronics*, 2001, **16**(1–2), 85–95.

33. J. H. Watterson, S. Raha, C. C. Kotoris, C. C. Wust, F. Gharabaghi, S. C. Jantzi, N. K. Haynes, N. H. Gendron,

U. J. Krull, A. E. Mackenzie, and P. A. E. Piunno, Rapid detection of single nucleotide polymorphisms associated with spinal muscular atrophy by use of a reusable fibre-optic biosensor. *Nucleic Acids Research*, 2004, **32**(2), e18.

34. T. Jiang, M. Minunni, P. Wilson, J. Zhang, A. P. Turner, and M. Mascini, Detection of TP53 mutation using a portable surface plasmon resonance DNA-based biosensor. *Biosensors and Bioelectronics*, 2005, **20**(10), 1939–1945.

35. J. Wang, P. E. Nielsen, M. Jiang, X. H. Cai, J. R. Fernandes, D. H. Grant, M. Ozsoz, A. Beglieter, and M. Mowat, Mismatch sensitive hybridization detection by peptide nucleic acids immobilized on a quartz crystal microbalance. *Analytical Chemistry*, 1997, **69**(24), 5200–5202.

36. D. Stipp, Gene chip breakthrough. *Fortune*, 1997, **135**(6), 56.

37. D. P. Malinowski, Molecular diagnostic assays for cervical neoplasia: emerging markers for the detection of high-grade cervical disease. *Biotechniques*, 2005, **38**(4), S17–S23.

38. A. Rasooly and J. Jacobson, Development of biosensors for cancer clinical testing. *Biosensors and Bioelectronics*, 2006, **21**(10), 1851–1858.

39. N. Ren, Q. H. Ye, L. X. Qin, B. H. Zhang, Y. K. Liu, and Z. Y. Tang, Circulating DNA level is negatively associated with the long-term survival of hepatocellular carcinoma patients. *World Journal of Gastroenterology*, 2006, **12**(24), 3911–3914.

40. G. Sozzi, D. Conte, M. E. Leon, R. Cirincione, L. Roz, C. Ratcliffe, E. Roz, N. Cirenei, M. Bellomi, G. Pelosi, M. A. Pierotti, and U. Pastorino, Quantification of free circulating DNA as a diagnostic marker in lung cancer. *Journal of Clinical Oncology*, 2003, **21**(21), 3902–3908.

41. Y. H. Su, M. J. Wang, D. E. Brenner, A. Ng, H. Melkonyan, S. Umansky, S. Syngal, and T. M. Block, Human urine contains small, 150 to 250 nucleotide-sized, soluble DNA derived from the circulation and may be useful in the detection of colorectal cancer. *Journal of Molecular Diagnostics*, 2004, **6**(2), 101–107.

42. I. De Canck, G. Mersch, I. Dossche, P. Adams, R. Bray, H. Gebel, and G. Parish, Testing a new technology platform for typing HLA alleles (4-MAT (TM)). *Tissue Antigens*, 2006, **67**(6), 557.

43. R. van Beuningen, H. van Damme, P. Boender, N. Bastiaensen, A. Chan, and T. Kievits, Fast and specific hybridization using flow-through microarrays on porous metal oxide. *Clinical Chemistry*, 2001, **47**(10), 1931–1933.

44. M. Schanfield, Forensic science. *Croatian Medical Journal*, 2001, **42**(3), 217–352.

45. R. K. Varshney, D. A. Hoisington, and A. K. Tyagi, Advances in cereal genomics and applications in crop breeding. *Trends in Biotechnology*, 2006, **24**(11), 490–499.

46. P. K. Basrur and W. A. King, Genetics then and now: breeding the best and biotechnology. *Revue Scientifique Et Technique De L Office International Des Epizooties*, 2005, **24**(1), 31–49.

47. R. K. Varshney, A. Graner, and M. E. Sorrells, Genomics-assisted breeding for crop improvement. *Trends in Plant Science*, 2005, **10**(12), 621–630.

48. V. Garcia-Canas, A. Cifuentes, and R. Gonzalez, Detection of genetically modified organisms in foods by DNA amplification techniques. *Critical Reviews in Food Science and Nutrition*, 2004, **44**(6), 425–436.

49. I. Mannelli, M. Minunni, S. Tombelli, and M. Mascini, Quartz crystal microbalance (QCM) affinity biosensor for genetically modified organisms (GMOs) detection. *Biosensors and Bioelectronics*, 2003, **18**(2–3), 129–140.

50. M. Minunni, S. Tombelli, J. Fonti, M. M. Spiriti, M. Mascini, P. Bogani, and M. Buiatti, Detection of fragmented genomic DNA by PCR-free piezoelectric sensing using a denaturation approach. *Journal of the American Chemical Society*, 2005, **127**(22), 7966–7967.

51. C. Peter, M. Meusel, F. Grawe, A. Katerkamp, K. Cammann, and T. Borchers, Optical DNA-sensor chip for real-time detection of hybridization events. *Fresenius Journal of Analytical Chemistry*, 2001, **371**(2), 120–127.

52. Y. Ren, K. Jiao, G. Y. Xu, W. Sun, and H. W. Gao, An electrochemical DNA sensor based on electrodepositing aluminum ion films on stearic acid-modified carbon paste electrode and its application for the detection of specific sequences related to bar gene and CP4 Epsps gene. *Electroanalysis*, 2005, **17**(23), 2182–2189.

53. M. D. Phillips, Bioterrorism: a brief history. *Northeast Florida Medicine Journal*, 2005, **56**(1), 32–35.

54. D. Ivnitski, D. J. O'Neil, A. Gattuso, R. Schlicht, M. Calidonna, and R. Fisher, Nucleic acid approaches for detection and identification of biological warfare and infectious disease agents. *Biotechniques*, 2003, **35**(4), 862–869.

55. C. A. Bell, J. R. Uhl, T. L. Hadfield, J. C. David, R. F. Meyer, T. F. Smith, and F. R. Cockerill, Detection of *Bacillus anthracis* DNA by lightcycler PCR. *Journal of Clinical Microbiology*, 2002, **40**(8), 2897–2902.

56. A. R. Hoffmaster, R. F. Meyer, M. P. Bowen, C. K. Marston, R. S. Weyant, K. Thurman, S. L. Messenger, E. E. Minor, J. M. Winchell, M. V. Rassmussen, B. R. Newton, J. T. Parker, W. E. Morrill, N. McKinney, G. A. Barnett, J. J. Sejvar, J. A. Jernigan, B. A. Perkins, and T. Popovic, Evaluation and validation of a real-time polymerase chain reaction assay for rapid identification of *Bacillus anthracis*. *Emerging Infectious Diseases*, 2003, **9**(4), 511.

57. P. Belgrader, M. Okuzumi, F. Pourahmadi, D. A. Borkholder, and M. A. Northrup, A microfluidic cartridge to prepare spores for PCR analysis. *Biosensors and Bioelectronics*, 2000, **14**(10–11), 849–852.

58. A. J. Baeumner, B. Leonard, J. McElwee, and R. A. Montagna, A rapid biosensor for viable B. Anthracis spores. *Analytical and Bioanalytical Chemistry*, 2004, **380**(1), 15–23.

59. S. S. Iqbal, M. W. Mayo, J. G. Bruno, B. V. Bronk, C. A. Batt, and J. P. Chambers, A review of molecular recognition technologies for detection of biological threat agents. *Biosensors and Bioelectronics*, 2000, **15**, 549–578.

60. T. T. Goodrich, H. J. Lee, and R. M. Corn, Direct detection of genomic DNA by enzymatically amplified SPR imaging measurements of RNA microarrays. *Journal of the American Chemical Society*, 2004, **126**(13), 4086–4087.

61. A. C. Syvanen and H. Soderlund, DNA sandwiches with silver and gold. *Nature Biotechnology*, 2002, **20**(4), 349–350.

62. N. A. Northrup, C. Gonzalez, D. Hadley, R. F. Hills, P. Landre, S. Lehew, R. Saw, and R. Watson, *A MEMS-Based Miniature DNA Analysis System*. In: *Proceedings of the IEEE International Conference on Solid-State Sensors and Actuators*, Stockholm, Sweden, 1995, 764–767.

63. W. Mandecki, *Three-Dimensional Arrays of Microtransponders Derivatized with Oligonucleotides*, D&MD Library Series Publication, Southborough, 1999, pp. 179–187.

64. J. Korlach, M. Levene, S. W. Turner, H. G. Craighead, and W. W. Webb, Single molecule analysis of DNA polymerase activity using zero-mode waveguides. *Biophysical Journal*, 2002, **82**(1), 507A.

65. R. Sosnowski, M. J. Heller, E. To, A. H. Forster, and R. Radtkey, Active microelectronic array system for DNA hybridization, genotyping and pharmacogenomic applications. *Psychiatric Genetics*, 2002, **12**(4), 181–192.

66. R. H. Liu, T. Nguyen, K. Schwarzkopf, H. S. Fuji, A. Petrova, T. Siuda, K. Peyvan, M. Bizak, D. Danley, and A. McShea, Fully integrated miniature device for automated gene expression DNA microarray processing. *Analytical Chemistry*, 2006, **78**(6), 1980–1986.

67. G. U. Lee, L. A. Chrisey, and R. J. Colton, Direct measurement of the forces between complementary strands of DNA. *Science*, 1994, **266**(5186), 771–773.

68. J. Wang, From DNA biosensors to gene chips. *Nucleic Acids Research*, 2000, **28**(16), 3011–3016.

69. T. Vo-Dinh, J. P. Alarie, N. Isola, D. Landis, A. L. Wintenberg, and M. N. Ericson, DNA biochip using a phototransistor integrated circuit. *Analytical Chemistry*, 1999, **71**(2), 358–363.

70. J. B. Lamture, K. L. Beattie, B. E. Burke, M. D. Eggers, D. J. Ehrlich, R. Fowler, M. A. Hollis, B. B. Kosicki, R. K. Reich, S. R. Smith, R. S. Varma, and M. E. Hogan, Direct-detection of nucleic-acid hybridization on the surface of a charge-coupled-device. *Nucleic Acids Research*, 1994, **22**(11), 2121–2125.

71. W. Cai, J. R. Peck, D. W. van der Weide, and R. J. Hamers, Direct electrical detection of hybridization at DNA-modified silicon surfaces. *Biosensors and Bioelectronics*, 2004, **19**(9), 1013–1019.

72. J. Fritz, E. B. Cooper, S. Gaudet, P. K. Sorger, and S. R. Manalis, Electronic detection of DNA by its intrinsic molecular charge. *Proceedings of the National Academy of Sciences of the United States of America*, 2002, **99**(22), 14142–14146.

73. E. Souteyrand, C. Chen, J. P. Cloarec, X. Nesme, P. Simonet, I. Navarro, and J. R. Martin, Comparison between electrochemical and optoelectrochemical impedance measurements for detection of DNA hybridization. *Applied Biochemistry and Biotechnology*, 2000, **89**(2–3), 195–207.

74. C. Guiducci, C. Stagni, G. Zuccheri, A. Bogliolo, L. Benini, B. Samori, and B. Ricco, DNA detection by integrable electronics. *Biosensors and Bioelectronics*, 2004, **19**(8), 781–787.

75. M. Gheorghe and A. Guiseppi-Elie, Electrical frequency dependent characterization of DNA hybridization. *Biosensors and Bioelectronics*, 2003, **19**(2), 95–102.

76. J. Jorritsma, M. A. M. Gijs, J. M. Kerkhof, and J. G. H. Stienen, General technique for fabricating large arrays of nanowires. *Nanotechnology*, 1996, **7**(3), 263–265.

77. P. Van Gerwen, W. Laureyn, W. Laureys, G. Huyberechts, M. O. De Beeck, K. Baert, J. Suls, W. Sansen, P. Jacobs, L. Hermans, and R. Mertens, Nanoscaled interdigitated electrode arrays for biochemical sensors. *Sensors and Actuators B-Chemical*, 1998, **49**(1–2), 73–80.

FURTHER READING

A. Rangel-Lopez, R. Maldonado-Rodriguez, M. Salcedo-Vargas, J. M. Espinosa-Lara, A. Mendez-Tenorio, and K. L. Beattie, Low density DNA microarray for detection of most frequent TP53 missense point mutations. *BMC Biotechnology*, 2005, **5**.

67

Examples of Biosensors for the Measurement of Trace Medical Analytes

Maria Minunni, Sara Tombelli, Sonia Centi and Marco Mascini

Department of Chemistry, University of Florence, Florence, Italy

1 INTRODUCTION

Analyses in the clinical laboratory concern metabolites in human blood and urine in the millimolar concentration range. However, a better understanding of several diseases requires the measurements of different analytes, such as those reported in Table 1, in the micro- and submicromolar range.

Steroids, drugs and their metabolites, hormones, and protein factors represent the molecular targets of many clinical analyses for elucidating several diseases. The concentration of these molecules is in the range 10^{-11}–10^{-6} mol l^{-1}. Conventional approaches are based mainly on immunochemical assays, coupled to enzymatic or fluorescent labels. These methods assure very high sensitivity and specificity and short analysis time (few hours). However, there is a need for label-free approaches, which should be more rapid (from many hours to minutes).

An interesting approach for the detection of trace clinical analytes is based on affinity biosensors.

The biological element in an affinity biosensor can be an antibody, a receptor (natural or synthetic), a nucleic acid or an aptamer. In all of these interactions, the binding between the target analyte and the immobilized biomolecule on the transduction element is governed by an affinity interaction like the antigen–antibody (Ag–Ab), the

DNA–DNA or the protein–nucleic acid binding. The specificity of the biosensor system is given by the immobilized molecule.

Many examples related to the analysis of clinical relevant analytes by immunosensing have been reported in the last 20 years when this approach first started.[1–4] The most relevant applications are related to the detection of hormones, drugs, and protein factors such as immunoglobulin and their subclass identification.[5–9]

More recently, in the last decade, DNA-based sensing has appeared for real applications to clinical diagnostics to detect the presence of pathogenic species responsible of infections, to identify genetic polymorphisms[10,11] and to detect point mutations.[12–14]

In DNA-based sensors the affinity interaction is the hybridization reaction between the DNA probe immobilized on the transducer surface and the complementary sequence in solution. This reaction is fast and specific. The specificity relies on the probe sequence, which is chosen on the basis of the target sequence (analyte) when the DNA-based sensor is developed. When the hybridization reaction occurs, a complex consisting of double stranded DNA (dsDNA) is formed at the sensor surface. This complex can be revealed using different transduction principles. Some of them require labeling of reagents some others are label free. So far most of the reported samples consist in amplified DNA, mainly by polymerase chain

Handbook of Biosensors and Biochips. Edited by Robert S. Marks, David C. Cullen, Isao Karube, Christopher R. Lowe and Howard H. Weetall.
© 2007 John Wiley & Sons, Ltd. ISBN 978-0-470-01905-4.

Table 1. Concentrations of clinically relevant analytes

Concentration (mol l^{-1})	Drugs	Steroids, aminoacids	Proteins, polypeptides	Antibodies
10^{-6}		Cortisol, estradiol	Placental lactogen	
10^{-7}				Specific IgG
10^{-8}	Digoxin	Corticosterone		Rubella
10^{-9}		Progesterone	Prolactin, insulin, hCG	
10^{-10}			hGH, luteinizing hormone	Total IgE
10^{-11}			TSH, oxytocin	

hCG: human chorionic gonadotropin; hGH: human growth hormone; TSH: thyroid stimulating hormone.

reaction (PCR), although some attempts in detecting target sequences directly into genomic DNA are reported.[15–19]

Very recently a new category of relevant receptors, nucleic acid sequences selected in vitro named aptamers, has found an interesting application in the so-called *aptasensors*, which represent a most interesting and innovative emerging field in biosensor research. So far different aptamers have been selected[20–33] and a list of analytes of clinical interest is reported in Table 2.

All the mentioned affinity sensors have been coupled to a variety of transduction principles.

In this chapter we report some examples of the work carried out over the last 10 years by our group, starting from the development of immunosensors for progesterone detection by electrochemical sensing. Then DNA-based sensing for point mutation detection in the β-globin gene, occurring in β-thalassemia, will be reported and

Table 2. Selected protein-binding aptamers

Target	Type of aptamer	References
Thrombin	DNA	20
IgE	DNA	21
NF-kB	RNA	22
Lysozyme	DNA	23
HIV-1 tat protein	RNA	24
VEGF	RNA	25
CD4 antigen	RNA	26
HIV-1 gag protein	RNA	26
L-selectin	DNA	28
IRP	RNA	29
HIV-1 rev protein	RNA	30
PDGF	DNA	31
TTF1	DNA	32
Acetylcholine receptor	RNA	33

VEGF: vesicular endothelial growth factor; IRP: iron regulatory protein; PDGF: platelet-derived growth factor; TTF1: thyroid transcription factor.

finally aptasensing for thrombin detection will be discussed. The last two examples deal with piezoelectric and electrochemical sensing.

Some considerations are important at this regard: when developing a biosensor for clinical diagnostics, a relevant aspect is the simplification of operation, and test strips are at present still superior.[34] Their mass production allowed the development of very cheap devices, which are user friendly; the well-known glucose electrode is a typical example. Nowadays it is also possible to print many different working electrodes by screen printing technology allowing multianalyte detection in one measurement shot. However, the dilemma of a disposable or multiuse sensing is still open: savings provided by reusable sensors should not be exceeded by the expenses of necessary maintenance.[34] The choice of the device to be used depends on the final user: test strips are mainly employed in home monitoring, in the doctors consulting room, and in small clinical laboratories, while reusable devices are better employed in clinical laboratories (as bench instruments) or in bed side instrumentation in hospitals.

Here we choose to give examples of both approaches: disposable sensing using screen-printed electrodes (SPEs) while reusable sensors are based on piezoelectric sensing.

2 IMMUNOSENSORS FOR HORMONE DETECTION

The first reported application refers to the detection of progesterone by a disposable immunosensor, based on the use of screen-printed electrodes SPEs coupled to modern electrochemical techniques.

The measurement of this hormone is important in women life.

Progesterone, a C21 steroid secreted by the corpus luteum, promotes the development of the endometrial lining. Serum levels of progesterone rise during the luteal phase of the menstrual cycle. If conception occurs, during pregnancy, levels increase from the end of the first trimester to term. Because progesterone is required for the continuation of pregnancy, low levels are associated with luteal phase defect, ectopic gestation, and miscarriage.

The main role of such a hormone is to make the uterine mucosa able to accept the fertilized egg and during the pregnancy it is necessary for the placenta maintenance. In Table 3 the reference concentration range of progesterone in plasma is reported.

Immunoassays are commonly used for the determination of progesterone in serum, urine, or saliva. Most of these were radioimmunoassays (RIAs), which involved the handling of radioactive materials.[35] Given the inherent problems, different nonradioactive methods have been developed for measuring progesterone.[36–39]

Among the large number of possible immunoassay techniques, the enzyme-linked immunosorbent assays (ELISA) combined with a colorimetric measurement are the most widely used for measuring hormone concentrations.[40–42] Another interesting approach is the use of immunosensors. Both electrochemical and optical biosensors have been reported, using screen printing technology coupled to cyclic voltammetry[43,44] and chronoamperometry.[45] For optical sensing surface plasmon resonance (SPR) has been used.[46]

SPEs have been used in immunosensor development for progesterone detection as a solid phase for the competitive immunoassay as described in Figure 1.

An enzymatic label is used to estimate the extent of the affinity reaction and in this case alkaline phosphatase was used as label and 1-naphthyl phosphate as enzymatic substrate. The addition of the enzymatic substrate gives rise to the formation of an electroactive product (1-naphthol), which can be detected by an electrochemical

Table 3. Progesterone levels

Sex	Phase	Progesterone ($ng\ ml^{-1}$)
Women	Follicular	0.1–1.5
	Luteal	3.8–28.0
Men		<0.6

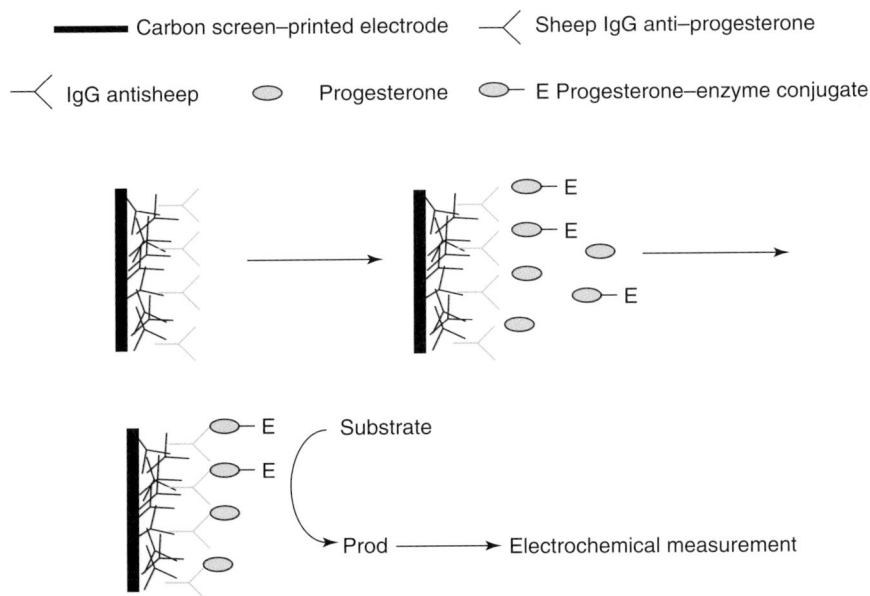

Figure 1. Electrochemical immunosensor based on a competitive immunoassay scheme for progesterone detection.

measurement using differential pulse voltamme-try (DPV). In our approach IgG specific for sheep $(10 \mu g \, ml^{-1})$ were bound to the carbon surface through passive adsorption, in order to favor the oriented immobilization of antibodies against progesterone from sheep.

Using optimized parameters, a dose–response curve for progesterone was performed. The experiments were carried out incubating for 30 min the sensor surface modified by adsorption of antisheep IgG and then by immobilization of antiprogesterone IgG with competition solutions containing progesterone-AP and different concentrations of progesterone (in the range $0.01–100 \, ng \, ml^{-1}$). After a washing step, the enzymatic substrate was added to the sensor surface and after 5 min of incubation time the electrochemical measurement was performed.

The calibration curve was obtained reporting the height of the peaks against the progesterone concentration and it is shown in Figure 2. A signal decrease was observed with the lowest current measured for $100 \, ng \, ml^{-1}$ concentration. The EC_{50} value (the analyte concentration necessary to displace 50% of the enzyme label) was calculated as $2 \, ng \, ml^{-1}$, whereas the limit of detection (LOD), calculated by evaluation of the mean of the blank solution response (containing the tracer only) minus two times the standard deviation, was estimated as $32 \, pg \, ml^{-1}$. The reproducibility of the assay, evaluated as average coefficient of variation (CV) on three measurements, was 10%.

Nonspecific adsorption was evaluated comparing the signals measured in presence and in

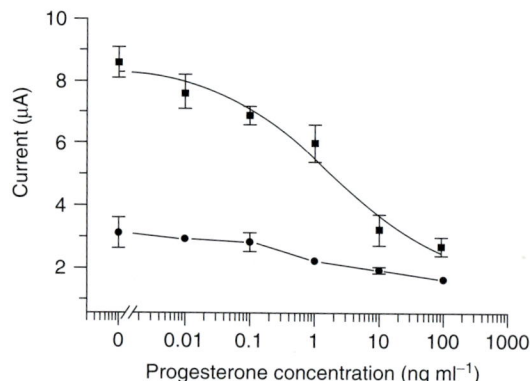

Figure 3. Comparison between the signals measured in presence (■) and in absence (●) of IgG antiprogesterone.

absence of antibody against progesterone. For this purpose, the experiments were carried out incubating the sensor surface modified by immobilization of antisheep rIgG with the competition solution containing progesterone-AP and different concentrations of progesterone (in the range $0.01–100 \, ng \, ml^{-1}$). Figure 3 indicates a residual signal due to nonspecific adsorption estimated as 40% lower than the specific binding.

This system represents an interesting approach for the detection of this analyte in human specimens as well as in cow milk for application to food analysis and to detect metabolite levels in veterinary testing and animal husbandry.[47]

Nowadays different commercially available devices for electrochemical immunoassays are available with automated liquid handling, sample dispensing, and so on, which make these approaches competitive respect to traditional immunoassays.[48,49]

3 DNA-BASED BIOSENSORS FOR GENETIC DISEASES INVESTIGATION

Nucleic acid–based biosensors represent a promising tool for gene sequence analysis and for mutation detection.[50] In this case the immobilization of single-stranded oligonucleotides (probe) on the surface of the transducing element is necessary and the variations of the transducer signal caused by the hybridization between the probe and the complementary strand in solution (target) are recorded. In particular, these devices can be applied to the

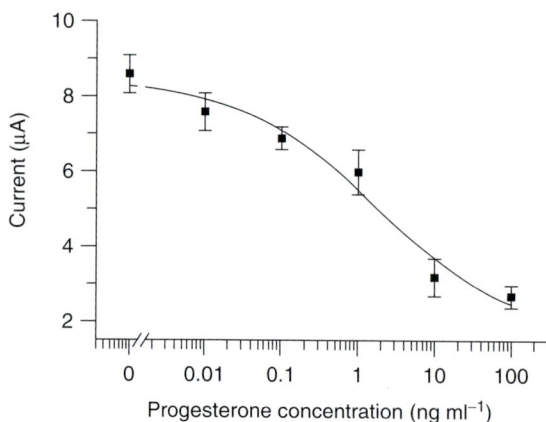

Figure 2. Calibration curve for progesterone.

detection of point mutations: when the matching with the complementary target is complete the biosensor reports the maximum response, whereas a decrease in the signal is observed when a mismatch occurs.

Several types of nucleic acid–based biosensors have been developed over recent years. In particular, piezoelectric biosensors offer the possibility of monitoring the hybridization reaction in real time and without the use of any label[51–54] and they have been presented as alternatives to gel electrophoresis and to traditional methods of specific DNA sequences detection, where labeled probes are required. DNA piezoelectric biosensors have been already realized[55–62] and applied with success to clinical analysis.[11,53,63,64]

3.1 Application to Point Mutations Detection: the Case of β-Thalassemia

A DNA piezoelectric biosensor has been developed for the determination of a mutation which is frequent in β-thalassemia syndromes in the Mediterranean area.[65] The biosensor was used to detect the extent of the hybridization between the immobilized probe and PCR-amplified DNA samples extracted from human blood of healthy, β-thalassemia healthy carriers, and thalassemic patients.

β-thalassemia is an autosomal recessive disease characterized by reduced or absent β-globin gene expression resulting from gene deletions or mutations that affect the transcription or stability of mRNA products.[66] β-thalassemias are caused by mutations in and around the structural gene of the adult hemoglobin (HbA) β-chain. More than a hundred mutations have been characterized around the world. In the Mediterranean population, this disease is caused mainly by the nonsense C → T substitution at codon 39 (west Mediterranean area).[67]

The determination of codon 39 mutation by the use of a DNA piezoelectric biosensor, which is able to detect the hybridization reaction both using synthetic 25-mer oligonucleotides or real PCR-amplified DNA samples and to distinguish between sequences differing in only one base in both cases, has been proposed as an alternative method.

A probe internal to the codon 39 has been identified, carrying the base involved in the mutation in the center of the sequence (5'-biotin-CAA AGA ACC TCT GGG TCC AAG GGT A-3').

Piezoelectric quartz crystals of 10 MHz with gold electrodes were modified with the probe using the strong affinity between biotin and streptavidin. The probe was biotinylated and streptavidin was previously immobilized following an optimized protocol.[11]

Experiments with complementary, noncomplementary and mismatched oligonucleotides (25-mer) were performed by using appropriate oligonucleotide solutions (100 μl).

In Figure 4 the responses of fully complementary (100%), mismatch (100%), and a mixture of the two (50 + 50%) are compared. In the presence of the mismatch a decreased signal is always obtained (conc. >1 ppm), comparing the responses at the same concentration. The decrease in the sensor response in the case of mismatch is 30% when the oligonucleotides are tested at the same concentration. Also the solution of the two oligonucleotides together, mimicking the healthy carrier (50% complementary and 50% mismatch), could be distinguished by evaluating the frequency shift, which is about 15% lower than that obtained with the complementary strand.

The reproducibility of the system is good (CV%$_{av}$ = 9%, calculated over three repeated measurements for each concentration on the same

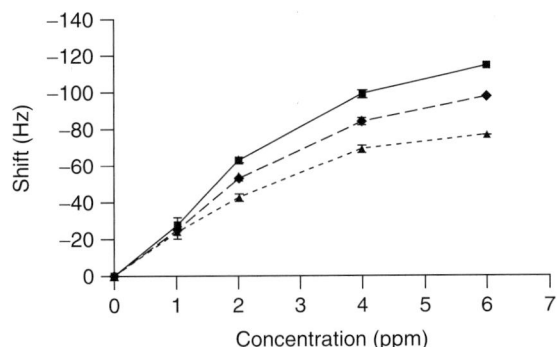

Figure 4. Calibration curves obtained with probe 2 immobilized on the crystal. Concentration range: 1–6 ppm. Solutions prepared in hybridization buffer. Frequency recorded after 10 min of interaction with the probe. (———) complementary; (- - - -) 50% complementary +50% mismatch; (·····) mismatch.

crystal). The reproducibility among different crystals ($n = 10$) is also evaluated resulting in an average CV% < 10%.

It is important to note that the measurements obtained with a negative control (noncomplementary strand) did not result in a frequency shift, showing the high specificity of the system.

For the analysis of real samples, DNA was extracted from peripheral blood of healthy, β-thalassemia healthy carrier and thalassemic patients. All the samples are tested with a standard method, which included DNA amplification by PCR using four 5'-biotinylated primers, amplifying two fragments of different length (771 and 577 bp).

After PCR and purification, the 771 bases amplicons were denatured by heating at 95 °C for 5 min followed by cooling in ice for 1 min and 100 µl of the solution was then added to the cell well.

The hybridization reaction proceeds for 20 min, and then the crystal is washed with the hybridization buffer. In this application, all the samples were diluted to 0.15 µM in order to compare the results and to have a significant frequency shift.

To evaluate the specificity of the system, not only the PCR blanks (PCR reagents without DNA template) are used, but two other samples (negative controls) were also tested: PCR-amplified DNA extracted from plants (negative control) and PCR-amplified DNA extracted from β-thalassemia healthy carrier having a mutation in another region of DNA, not in codon 39 were chosen. In these samples the sequence of codon 39 corresponds to the one of healthy patients DNA.

As reported in Figure 5, the biosensor is able to distinguish between samples from healthy (100% CAG), healthy carrier (50% CAG + 50% TAG) and thalassemic patients (100% TAG).

Moreover, the system results are very specific as the negative control (PCR-amplified DNA from plants) resulted in a negligible signal and samples from β-thalassemia healthy carriers carrying a mutation in different region, shows the same behavior of samples from healthy patients.

When real samples are employed it is possible to perform up to 10 measurements on the same crystal, regenerating between cycles with an acid treatment.

The results demonstrated that piezoelectric biosensor could be successfully applied to the analysis of real clinical samples of DNA amplified by PCR for β-thalassemia detection. Rapid assignment of the samples to different genotype groups could be possible based on such a quick method.

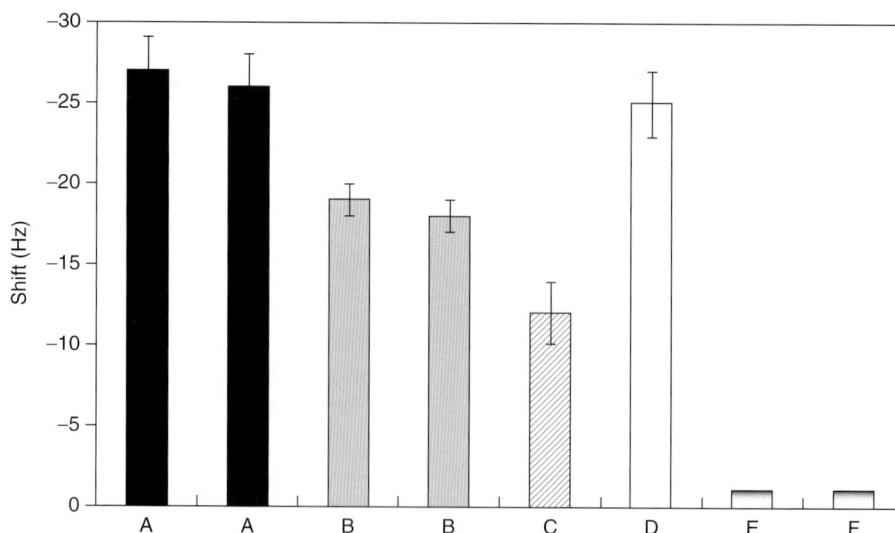

Figure 5. Results with purified real samples. All the samples are diluted to 80 ppm. Blanks consisted of PCR mixture without DNA template. The negative control corresponds to PCR-amplified DNA from plants. The mean value estimated over three replicate measurements. (A) Healthy patients; (B) healthy carrier patients; (C) β-thalassemic patients; (D) healthy carrier patients with mutation not in codon 39; (E) PCR blank; (F) negative control.

The detection principle has a wide range of applicability, since the specificity of the system relies on the particular choice of the probe. It could be applied to the detection of many diseases based on gene mutations and also to characterize different genotypes (i.e., polymorphism investigation).

Biosensors were demonstrated to be a powerful tool for fast screening of clinical samples, providing fast and cheap responses. For these reasons they represent an interesting approach for a preliminary screening of samples, eventually further characterized by more expensive and time consuming, well established nonbiosensor approaches (e.g., real-time PCR).

4   APTAMER-BASED BIOSENSORS FOR PROTEINS DETECTION

Aptamers are nucleic acid ligands that have been designed through an in vitro selection process called *SELEX* (*Systematic Evolution of Ligands by Exponential Enrichment*).[68,69] These DNA/RNA ligands have been selected to bind non-nucleic acid targets to which they bind through their particular structures and not via a hybridization reaction. The high affinity of aptamers for their targets is given by their capability of folding upon binding their target molecule: they can incorporate small molecules into their nucleic acid structure or integrate into the structure of larger molecules such as proteins. Because of the high diversity of molecular shapes of all possible nucleotide sequences, aptamers have been selected for a wide array of targets, including proteins, carbohydrates, lipids, or small molecules. Aptamers offer several advantages over antibodies as they can be completely engineered in vitro, produced and labeled by chemical synthesis, and possess good storage properties.[70,71]

In addition, aptamers receptors, due to their production by chemical synthesis, have a number of advantages that make them very promising in analytical applications. Aptamers have been recently used in analytical chemistry as immobilized ligands or in homogeneous assays.[71] A very attractive application is the exploitation of aptamers as biorecognition elements in biosensors.

The application of aptamers as biocomponents in biosensors offers classical affinity sensing methods mainly based on antibodies, a multitude of advantages, such as the possibility of easy regeneration of the function of immobilized aptamers, their homogeneous preparation, and the possibility of using different detection methods due to easy labeling.[70,71]

Quartz crystals biosensors have been developed with aptamers immobilized on the sensing surface.[72,73] Liss et al.[72] compared the behavior of aptamers as receptor molecules with that of conventional antibodies.

Regarding the application to trace level analytes the detection of IgE (Table 1) could represent an interesting example. Using an IgE-specific aptamer and an antiIgE antibody as model system, with a piezoelectric transduction, the aptasensor resulted in an increased stability with respect to the relative immunosensor (several weeks) and presented a more complete and reversible surface regeneration. The sensitivity of the two systems was comparable (minimum detectable IgE concentration, $100\,\mu g\,l^{-1}$) but the one having the aptamer as recognition element presented an extended linear measurement range (up to $10\,mg\,l^{-1}$). A similar comparison between an aptamer- and an antibody-based quartz crystal biosensor has been presented using HIV-1 Tat protein as target molecule.[73] The developed piezoelectric aptasensor, proved to be specific, reproducible, and reusable in the detection of Tat protein. Both receptors, aptamer and antibody, detected Tat at a minimum concentration of $0.01\,\mu M$ with a comparable reproducibility in terms of coefficient of variation (CV% = 6%).

Aptamers have been used as biorecognition element in optical sensors, both labeled (fluorescence-based methods) or in systems not requiring labels (SPR). A fluorescence aptamer-based biosensor array has been developed using aptamers specific for several proteins with relevance to cancer, inosine monophosphate dehydrogenase (IMPDH), VEGF and basic fibroblast growth factor (bFGF).[74] The array was proposed for multiplex analysis of proteins of clinical interest present at trace levels (ppt).

The possible use of unlabeled aptamers in optical biosensors has been evidenced mainly in methods based on SPR. The SPR-based device Biacore™ (Biacore, Uppsala, Sweden) has been widely used to study the affinity of the selected aptamers for their target molecules, such as HIV-1 Rev protein,[75] CD4 antigen[26] and TTF1.[32]

A label free electrochemical method has been presented by Kawde et al.[76] The detection of lysozyme at submicromolar levels has been achieved by combining the protein specific aptamer immobilized onto magnetic particles and the electrochemical measurement of the captured protein (oxidation of tyrosine and tryptophan residues). This reagent-less label-free detection cannot be accomplished with traditional immunoassays due to the presence of the electroactive residues both in the target protein and in the antibody. Also this method has been presented as an alternative technique for the development of protein biochips to be applied in clinical analysis.

4.1 Application to Thrombin Detection

Thrombin (factor IIa) is the last enzyme protease involved in the coagulation cascade and it converts fibrinogen to insoluble fibrin that forms the fibrin gel either in physiological conditions or in a pathological thrombus.[77] The concentration of thrombin in blood can vary considerably: thrombin, not present in blood under normal conditions, can reach low micromolar concentrations during the coagulation process but low levels (low nanomolar) of thrombin generated early in hemostasis are

also important to the overall process.[78] Thrombin represents an interesting clinical target in trace level.

The thrombin-binding aptamer (15-mer, 5′-GGTTGGTGTGGTTGG-3′) has been the first DNA aptamer selected in vitro, specific for a protein without nucleic acid's binding properties[20] and it has been studied as an anticlotting therapeutic tool. This aptamer has been used as a model system, coupled to different transduction principles to demonstrate the wide applicability of aptamers as bioreceptors in biosensors.[79–82] The thrombin-binding aptamer has been extensively investigated: its G-quartet structure has been established[83,84] and the binding site has been studied and determined.[85]

We report here an example of an aptamer-based assay for the detection of thrombin directly into human plasma.[86] The assay is based on electrochemical transduction coupled to magnetic particles, modified with the aptamer. With the addition of a second aptamer after the interaction of the analyte with the modified beads, similar to immunoassays, this approach can be thought as a sandwich method (Figure 6). The work aims to couple a simple target capturing step, by employing magnetic beads, to the high sensitivity requested for thrombin analysis, assured by the electrochemical detection with DPV on SPEs. A sandwich assay format is chosen following an approach similar to the one reported by Ikebukuro et al.[81] who employed

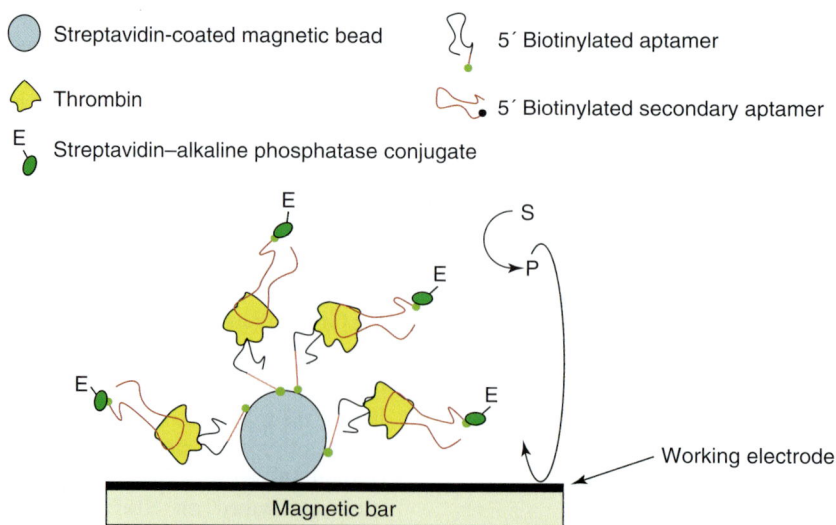

Figure 6. Scheme of the electrochemical sandwich assay coupled to magnetic beads.

an electrochemical detection using a conventional gold electrode and limited to standard solutions of the analyte down to a concentration of 10 nM. In both approaches, two selected aptamers binding thrombin in two different, non-overlapping, sites are used. The protein captured by the first aptamer is detected after the addition of the second biotinylated aptamer and of streptavidin labeled with an enzyme (alkaline phosphatase). Detection of the product generated by the enzymatic reaction was achieved by DPV onto SPEs. This approach based on the use of two different aptamers overcomes some of the drawbacks related to the use of antibodies.

Using this novel design, an electrochemical biosensor recognizing thrombin with high affinity, sensitivity, and specificity was obtained, opening the possibility of a real application to diagnostics or medical investigation.

The optimized conditions resulted in an immobilized biotinylated aptamer, named primary aptamer, with a polyT tail to ensure a certain flexibility in the binding (1 μM 15-mer: 5′-biotin-TT TTT TTT TTT TTT TTT TTT GGT TGG TGT GGT TGG-3′) and a biotinylated 29-mer as secondary aptamer to be used in the sandwich (5′-biotin-AGTCCGTGGTAGGGCAGGTTGG-GGT GACT-3′) (0.1 μM, incubation time 15′).

A dose-response curve for thrombin is reported in Figure 7. The height of the peaks obtained by DPV measurements for different concentrations of thrombin increases with the increase of thrombin concentration showing a typical behavior of a sandwich assay. In the figure, the signals are reported as peak current against log of thrombin concentration. A signal increase was observed for thrombin concentrations greater than 0.1 nM and the highest current was measured at a protein concentration of 100 nM. The reproducibility, expressed as the average CV is 8% ($n = 5$ in the range 0–100 nM). The detection limit (DL) is 0.45 nM. Human serum albumin (HSA) at physiological concentration in human plasma (HSA 72 μM (5000 ppm)) was used to test the specificity of the aptamer, which was found to be very high since a negligible signal in presence of HSA or of the blank solution was observed.

The applicability of the assay to plasma samples was evaluated by investigating the presence of any matrix effect. Plasma, after fibrinogen precipitation, diluted 10 times was tested alone or

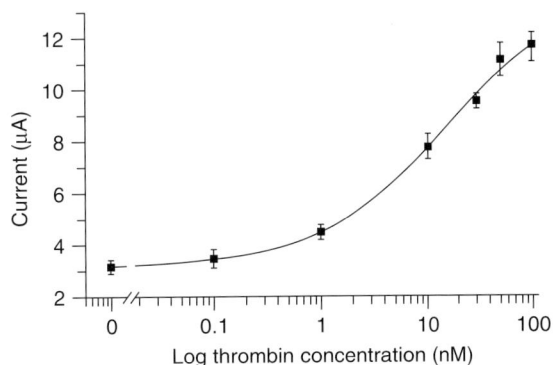

Figure 7. Electrochemical assay for thrombin coupled to magnetic beads: dose–response curve for thrombin.

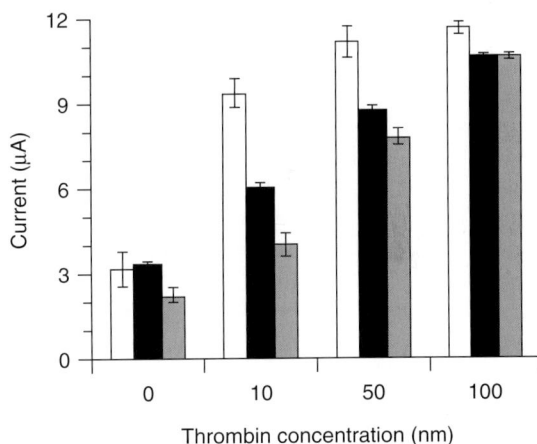

Figure 8. Results obtained with serum samples spiked with thrombin (black histograms) and comparison with the same concentrations tested in buffer (white histograms) and in plasma after fibrinogen precipitation (gray histograms).

spiked with thrombin (in the concentration range 0–100 nM), and the results were compared with thrombin standard buffer solutions. Figure 8 shows that comparable responses were found for buffer, and plasma: addition of thrombin to the sample resulted in protein concentration-dependent signal. The samples were incubated with the beads (coated with primary aptamer) and then the assay was carried out by adding the secondary aptamer and the conjugate. For plasma a weak matrix effect was found considering the lower currents (average decrease ~80%) measured with respect to buffer. Probably this decrease is due to a reduced concentration of thrombin available for the binding caused, by the interaction with some matrix components.

These results demonstrate the applicability of aptamer-based assays to the detection of molecules of clinical interest present in trace level, directly into plasma samples.

5 CONCLUSIONS

The detection of clinical important molecules at micromolar or submicromolar concentration, employing different strategies in affinity sensing has been reported. The examples span from an electrochemical immunosensor for progesterone using disposable devices, to nucleic acid–based sensors, for the detection of point mutations, which are based on the hybridization reaction, revealed in real time by label free piezoelectric sensing. Finally, a new emerging category of receptors called *aptamers* have been introduced and discussed in the application to thrombin detection.

Both single use and multiuse sensing have been shown to be suitable for trace analysis in clinical chemistry. Both approaches possess sensitivity, reproducibility, and analysis times compatible with the final application. However, while strips are widely commercialized and applied to the analysis of clinical analytes by nonskilled personnel, for piezoelectric sensing much less commercial investments and diffusion are seen at the moment.

REFERENCES

1. P. B. Luppa, L. J. Sokoll, and D. W. Chan, Immunosensor-principles and applications to clinical chemistry. *Clinica Chimica Acta*, 2001, **314**, 1–26.
2. W. M. Mullett, E. P. C. Lai, and J. M. Yeung, Surface plasmon resonance-based immunoassays. *Methods*, 2000, **22**, 77–91.
3. J. Lin and H. Ju, Electrochemical and chemiluminescent immunosensors for tumor markers. *Biosensors and Bioelectronics*, 2005, **20**, 1461–1470.
4. D. Purvis, O. Leonardova, D. Farmakovsky, and V. Cherkasov, An ultrasensitive and stable potentiometric immunosensor. *Biosensors and Bioelectronics*, 2003, **18**, 1385–1390.
5. B. Zhang, Q. Mao, X. Zhang, T. Jiang, M. Chen, F. Yu, and W. Fu, A novel piezoelectric quartz micro-array immunosensor based on self-assembled monolayer for determination of human chorionic gonadotropin. *Biosensors and Bioelectronics*, 2004, **19**, 711–720.
6. G. Lillie, P. Payne, and P. Vadgama, Electrochemical impedance spectroscopy as a platform for reagentless bioaffinity sensing. *Sensors and Actuators B: Chemical*, 2001, **78**, 249–256.
7. B. S. Attili and A. A. Suleiman, A piezoelectric immunosensor for the detection of cocaine. *Microchemical Journal*, 1996, **54**, 174–179.
8. A. de la Escosura-Muñz, M. B. González-García, and A. Costa-García, Aurothiomalate as an electroactive label for the determination of immunoglobulin M using glassy carbon electrodes as immunoassay transducers. *Sensors and Actuators B: Chemical*, 2006, **114**, 473–481.
9. X. Su, F. T. Chew, and S. F. Y. Li, Self-assembled monolayer-based piezoelectric crystal immunosensor for the quantification of total human IgE. *Analytical Biochemistry*, 1999, **273**, 66–72.
10. G. Marrazza, G. Chiti, M. Mascini, and M. Anichini, Detection of human apolipoprotein E genotypes by DNA electrochemical biosensor coupled with PCR. *Clinical Chemistry*, 2000, **46**, 31–37.
11. S. Tombelli, M. Mascini, L. Braccini, M. Anichini, and A. P. F. Turner, Coupling of a DNA piezoelectric biosensor and polymerase chain reaction to detect apolipoprotein E polymorphisms. *Biosensors and Bioelectronics*, 2000, **15**, 363–370.
12. D. Dell'Atti, S. Tombelli, M. Minunni, and M. Mascini, Detection of clinically relevant point mutations by a novel piezoelectric biosensor. *Biosensors and Bioelectronics*, 2006, **21**, 1876–1879.
13. T. Jiang, M. Minunni, P. Wilson, J. Zhang, A. P. F. Turner, and M. Mascini, Detection of *TP53* mutation using a portable surface plasmon resonance DNA-based biosensor. *Biosensors and Bioelectronics*, 2005, **20**, 1939–1945.
14. F. Patolsky, A. Lichtenstein, and I. Willner, Electronic transduction of DNA sensing processes on surfaces: amplification of DNA detection and analysis of single-base mismatches by tagged liposomes. *Journal of the American Chemical Society*, 2001, **123**, 5194–5205.
15. M. Minunni, S. Tombelli, J. Fonti, M. M. Spiriti, and M. Mascini, Detection of genomic DNA by PCR-free piezoelectric sensing. *Journal of American Chemical Society*, 2005, **127**, 7966–7967.
16. A. Almadidy, J. Watterson, P. A. E. Piunno, S. Raha, I. V. Foulds, P. A. Horgen, A. Castle, and U. Krull, Direct selective detection of genomic DNA from coliform using a fibre optic biosensor. *Analytica Chimica Acta*, 2002, **461**, 37–41.
17. T. T. Goodrich, H. J. Lee, and R. Corn, Direct detection of genomic DNA by enzymatically amplified SPR imaging measurements of RNA microarrays. *Journal of the American Chemical Society*, 2004, **126**, 4086–4087.
18. F. Patolsky, A. Lichtenstein, M. Kotler, and I. Willner, Electronic transduction of polymerase or reverse transcriptase induced replication processes on surfaces: highly sensitive and specific detection of viral genomes. *Angewandte Chemie International Edition*, 2001, **40**, 2261–2265.
19. R. B. Towery, N. C. Fawcett, P. Zhang, and J. A. Evans, Genomic DNA hybridizes with the same rate constant on the QCM biosensor as in homogeneous solution. *Biosensors and Bioelectronics*, 2001, **16**, 1–8.
20. L. C. Bock, L. C. Griffin, J. A. Latham, E. H. Vermaas, and J. J. Toole, Selection of single-stranded DNA molecules that bind and inhibit human thrombin. *Nature*, 1992, **355**, 564–566.

21. T. W. Wiegand, P. B. Williams, S. C. Dreskin, M. H. Jouvin, J. P. Kinet, and D. Tasset, High-affinity oligonucleotide ligands to human IgE inhibit binding to Fc epsilon receptor I. *Journal of Immunology*, 1996, **157**, 221–230.

22. L. L. Lebruska and L. L. Mather, Selection and characterization of an RNA decoy for transcription factor NF-kappa B. *Biochemistry*, 1999, **38**, 3168–3174.

23. J. C. Cox and A. D. Ellington, Automated selection of anti-protein aptamers. *Bioorganic and Medicinal Chemistry*, 2001, **9**, 2525–2531.

24. R. Yamamoto, M. Katahira, S. Nishikawa, T. Baba, K. Taira, and P. K. R. Kumar, A novel RNA motif that binds efficiently and specifically to the Tat protein of HIV and inhibits the trans-activation by Tat of transcription *in vitro* and *in vivo*. *Genes to Cells*, 2000, **5**, 371–388.

25. D. W. Drolet, L. Moon-Mcdermott, and T. S. Romig, An enzyme-linked oligonucleotide assay. *Nature Biotechnology*, 1996, **14**, 1021–1025.

26. E. Kraus, W. James, and A. N. Barclay, Novel RNA ligands able to bind CD4 antigen and inhibit CD4+ T lymphocyte function. *Journal of Immunology*, 1998, **160**, 5209–5212.

27. M. A. Lochrie, S. Waugh, D. G. Pratt, J. Clever, T. G. Parslow, and B. Polisky, In vitro selection of RNAs that bind to the human immunodeficiency virus type-1 gag polyprotein. *Nucleic Acids Research*, 1997, **25**, 2902–2910.

28. B. J. Hicke, S. R. Watson, A. Koenig, C. K. Lynott, R. F. Bargatze, and Y. F. Chang, DNA aptamers block L-selectin function *in vivo*. *The Journal of Clinical Investigation*, 1996, **98**, 2688–2692.

29. F. Lisdat, D. Utepbergenov, R. F. Haseloff, I. E. Blasig, W. Stocklein, F. W. Scheller, and R. Brigelius-Flohe, An optical method for the detection of oxidative stress using protein-RNA interaction. *Analytical Chemistry*, 2001, **73**, 957–962.

30. W. Xu and A. D. Ellington, Anti-peptide aptamers recognize amino acid sequence and bind a protein epitope. *Proceedings of the National Academy of Sciences of the United States of America*, 1997, **94**, 7475–7480.

31. X. Fang, Z. Cao, T. Beck, and W. Tan, Molecular aptamer for real-time oncoprotein PDGF monitoring by fluorescence anisotropy. *Analytical Chemistry*, 2001, **73**, 5752–5757.

32. M. B. Murphy, S. T. Fuller, P. M. Richardson, and S. A. Doyle, An improved method for the *in vitro* evolution of aptamers and applications in protein detection and purification. *Nucleic Acids Research*, 2003, **31**, e110–e118.

33. H. Ulrich, J. E. Ippolito, O. R. Pagan, V. A. Eterovic, R. M. Hann, H. Shi, J. T. Lis, M. E. Eldefrawi, and G. P. Hess, *In vitro* selection of RNA molecules that displace cocaine from the membrane-bound nicotinic receptor. *Proceedings of the National Academy of Sciences of the United States of America*, 1998, **95**, 14051–14056.

34. F. Sheller and F. Schubert, *Biosensors, Technique and Instrumentation in Analytical Chemistry Series*, Elsevier, 1992, pp. 291–322.

35. J. Reinsberg, E. Jost, and H. Van Der Ven, Performance of the fully automated progesterone assays on the Abbott. *Clinical Biochemistry*, 1997, **30**, 469–471.

36. S. A. Miller, M. S. Morton, and A. Turkes, Chemiluminescence immunoassay for progesterone in plasma incorporating acridinium ester labelled antigen. *Annals of Clinical Biochemistry*, 1988, **25**, 27–34.

37. S. E. Kakabakos and M. J. Khosravi, Direct time-resolved fluorescence immunoassay of progesterone in serum involving the biotin-streptavidin system and the immobilized-antibody approach. *Clinical Chemistry*, 1992, **38**, 725–730.

38. M. J. Choi, J. Choi, D. Y. Yoon, J. Park, and S. A. Eremin, Fluorescence polarization immunoassay of progesterone. *Biological and Pharmaceutical Bulletin*, 1997, **20**, 309–314.

39. J. Y. Hong and M. J. Choi, Development of one-step fluorescence polarization immunoassay for progesterone. *Biological and Pharmaceutical Bulletin*, 2002, **25**, 1258–1262.

40. A. Basu, T. G. Shrivastav, and S. K. Maitra, A direct antigen heterologous enzyme immunoassay for measuring progesterone in serum without using displacer. *Steroids*, 2006, **71**, 222–230.

41. L. A. Singer, M. S. A. Kumar, W. Gavin, and S. L. Ayres, Determining the onset of parturition in the goat using CITE progesterone tests. *Small Ruminant Research*, 2004, **52**, 203–209.

42. R. W. Claycomb, M. J. Delwiche, C. J. Munro, and R. H. BonDurant, Rapid enzyme immunoassay for measurement of bovine progesterone. *Biosensors and Bioelectronics*, 1998, **13**, 1165–1171.

43. Y. F. Xu, M. Velasco-Garcia, and T. T. Mottram, Quantitative analysis of the response of an electrochemical biosensor for progesterone in milk. *Biosensors and Bioelectronics*, 2005, **20**, 2061–2070.

44. R. M. Pemberton, J. P. Hart, P. Stoddard, and J. A. Foulkes, A comparison of 1-naphthyl phosphate and 4 aminophenyl phosphate as enzyme substrates for use with a screen-printed amperometric immunosensor for progesterone in cows' milk. *Biosensors and Bioelectronics*, 1999, **14**, 495–503.

45. J. P. Hart, R. M. Pemberton, R. Luxton, and R. Wedge, Studies towards a disposable screen-printed amperometric biosensor for progesterone. *Biosensors and Bioelectronics*, 1997, **12**, 1113–1121.

46. E. H. Gillis, J. P. Gosling, J. M. Sreenan, and M. Kane, Development and validation of a biosensor-based immunoassay for progesterone in bovine milk. *Journal of Immunological Methods*, 2002, **267**, 131–138.

47. M. N. Velasco-Garcia and T. Mottram, Biosensors in the livestock industry: an automated ovulation prediction system for dairy cows. *Trends in Biotechnology*, 2001, **19**, 433–434.

48. Web Page: http://www.alderonbiosciences.com/.

49. Web Page: http://www.diagnoswiss.com., 2006.

50. C. N. Jayarajah and M. Thompson, Signaling of transcriptional chemistry in the on-line detection format. *Biosensors and Bioelectronics*, 2002, **17**, 159–171.

51. S. Tombelli, M. Mascini, C. Sacco, and A. P. F. Turner, A DNA piezoelectric biosensor assay coupled with a polymerase chain reaction for bacterial toxicity determination in environmental samples. *Analytica Chimica Acta*, 2000, **418**, 1–9.

52. L. M. Furtado and M. Thompson, Hybridization of complementary strand and single-base mutated oligonucleotides detected with an on-line acoustic wave sensor. *Analyst*, 1998, **123**, 1937–1945.

53. X. C. Zhou, L. Q. Huang, and S. F. Y. Li, Microgravimetric DNA sensor based on quartz crystal microbalance: comparison of oligonucleotide immobilization methods and the application in genetic diagnosis. *Biosensors and Bioelectronics*, 2001, **16**, 85–95.

54. A. K. Deisingh and M. Thompson, Sequences of *E. coli* 0157:H7 detected by a PCR-acoustic wave sensor combination. *Analyst*, 2001, **126**, 2153–2158.

55. S. Yamaguchi, T. Shimomura, T. Tatsuma, and N. Oyama, Adsorption, immobilization, and hybridization of DNA studied by the use of quartz crystal oscillators. *Analytical Chemistry*, 1993, **65**, 1925–1927.

56. H. Su and M. Thompson, Rheological and interfacial properties of nucleic acid films studied by thickness—shear mode sensor and network analysis. *Canadian Journal of Chemistry*, 1996, **74**, 344–358.

57. F. Caruso, E. Rodda, D. Furlong, K. Niikura, and Y. Okahata, Quartz crystal microbalance study of DNA immobilization and hybridization for nucleic acid sensor development. *Analytical Chemistry*, 1997, **69**, 2043–2049.

58. I. Mannelli, M. Minunni, S. Tombelli, and M. Mascini, Quartz crystal microbalance (QCM) affinity biosensor for genetically modified organisms (GMOs) detection. *Biosensors and Bioelectronics*, 2003, **18**, 129–140.

59. Y. Okahata, M. Kawase, K. Nukura, F. Ohtake, H. Furusawa, and Y. Ebara, Kinetic measurements of DNA hybridisation on an oligonucleotide-immobilised 27-MHz quartz crystal microbalance. *Analytical Chemistry*, 1998, **70**, 1288–1296.

60. H. Sun, Y. Zhang, and Y. Fung, Flow analysis coupled with PQC/DNA biosensor for assay of *E. coli* based on detecting DNA products from PCR amplification. *Biosensors and Bioelectronics*, 2006, **22**, 506–512.

61. F. He and S. Liu, Detection of *P. aeruginosa* using nano-structured electrode-separated piezoelectric DNA biosensor. *Talanta*, 2004, **62**, 271–277.

62. M. Minunni, I. Mannelli, M. M. Spiriti, S. Tombelli, and M. Mascini, Detection of highly repeated sequences in non-amplified genomic DNA by bulk acoustic wave (BAW) affinity biosensor. *Analytica Chimica Acta*, 2004, **526**, 19–25.

63. X. Zhou, L. Liu, M. Hu, L. Wang, and J. Hu, Detection of hepatitis B virus by piezoelectric biosensor. *Journal of Pharmaceutical and Biomedical Analysis*, 2002, **27**, 341–345.

64. C. K. O'Sullivan and G. G. Guilbault, Commercial quartz crystal microbalances—theory and applications. *Biosensors and Bioelectronics*, 1999, **14**, 663–670.

65. M. Minunni, S. Tombelli, R. Scielzi, I. Mannelli, M. Mascini, and C. Gaudiano, Detection of β-thalassemia by a DNA piezoelectric biosensor coupled with polymerase chain reaction. *Analytica Chimica Acta*, 2003, **48**, 55–64.

66. S. R. Chern and C. P. Chen, Molecular prenatal diagnosis of thalassemia in Taiwan. *International Journal of Gynecology and Obstetrics*, 2000, **69**, 103–106.

67. L. Rigoli, A. Meo, M. R. Miceli, K. Alessio, R. A. Caruso, M. A. La Rosa, D. C. Salpietro, M. Ricca, and I. Barberi,

68. A. D. Ellington and J. W. Szostak, In vitro selection of RNA molecules that bind specific ligands. *Nature*, 1990, **346**, 818–822.

69. C. Tuerk and L. Gold, Systematic evolution of ligands by exponential enrichment: RNA ligands to bacteriophage T4 DNA polymerase. *Science*, 1990, **249**, 505–510.

70. C. K. O'Sullivan, Aptasensors-the future of biosensing. *Analytical and Bioanalytical Chemistry*, 2002, **372**, 44–48.

71. S. Tombelli, M. Minunni, and M. Mascini, Analytical applications of aptamers. *Biosensors and Bioelectronics*, 2005, **20**, 2424–2434.

72. M. Liss, B. Petersen, H. Wolf, and E. Prohaska, An aptamer-based quartz crystal protein biosensor. *Analytical Chemistry*, 2002, **74**, 4488–4495.

73. M. Minunni, S. Tombelli, A. Gullotto, E. Luzi, and M. Mascini, Development of biosensors with aptamers as bio-recognition element: the case of HIV-1 Tat protein. *Biosensors and Bioelectronics*, 2004, **20**, 1149–1156.

74. T. G. McCauley, N. Hamaguchi, and M. Stanton, Aptamer-based biosensor arrays detection and quantification of biological macromolecules. *Analytical Biochemistry*, 2003, **319**, 244–250.

75. D. I. Van Ryk and S. Venkatesan, Real-time kinetics of HIV-1 rev-rev response element interactions. *Journal of Biological Chemistry*, 1999, **274**, 17452–17463.

76. A. Kawde, M. C. Rodriguez, T. M. H. Lee, and J. Wang, Label-free bioelectronic detection of aptamer-protein interactions. *Electrochemistry Communications*, 2005, **7**, 537–540.

77. C. A. Holland, A. T. Henry, H. C. Whinna, and F. C. Church, Effect of oligodeoxynucleotide thrombin aptamer on thrombin inhibition by heparin cofactor II and antithrombin. *FEBS Letters*, 2000, **484**, 87–91.

78. M. A. Shuman and P. W. Majerus, The measurement of thrombin in clotting blood by radioimmunoassay. *The Journal of Clinical Investigation*, 1976, **58**, 1249–1258.

79. E. Baldrich, A. Restrepo, and C. K. O'Sullivan, Aptasensor development: elucidation of critical parameters for optimal aptamer performance. *Analytical Chemistry*, 2004, **76**, 7053–7063.

80. T. Hianik, V. Ostatná, E. Zajacová, E. Stoikova, and G. Evtugyn, Detection of aptamer-protein interactions using QCM and electrochemical indicator methods. *Bioorganic and Medicinal Chemistry Letters*, 2005, **15**, 291–295.

81. K. Ikebukuro, C. Kiyohara, and K. Sode, Electrochemical sensing of protein using two aptamers in sandwich manner. *Biosensors and Bioelectronics*, 2005, **20**, 2168–2172.

82. T. M. A. Gronewold, S. Glass, E. Quandt, and M. Famulok, Monitoring complex formation in the blood-coagulation cascade using aptamer-coated SAW sensors. *Biosensors and Bioelectronics*, 2005, **20**, 2044–2052.

83. R. F. Macaya, P. Schultze, F. W. Smith, J. A. Roe, and J. Feigon, Thrombin-binding DNA aptamer

Molecular analysis of β-thalassaemia patients in a high incidence area of southern Italy. *Clinical and Laboratory Haematology*, 2001, **23**, 373–378.

forms a unimolecular quadruplex structure in solution. *Proceedings of the National Academy of Sciences*, 1993, **90**, 3745–3749.

84. I. Smirnov and R. H. Shafer, Effect of loop sequence and size on DNA aptamer stability. *Biochemistry*, 2000, **39**, 1462–1468.

85. L. R. Paborsky, S. N. McCurdy, L. C. Griffin, J. J. Toole, and L. L. Leung, The single-stranded DNA aptamer-binding site of human thrombin. *Journal of Biological Chemistry*, 1993, **268**, 20808–20811.

86. S. Centi, S. Tombelli, M. Minunni, and M. Mascini, *Analytical Chemistry*, 2007, **79**, 1466–1473.

Biosensors for Monitoring Metabolites in Clinical Medicine

John C. Pickup

Metabolic Unit, King's College London School of Medicine, London, UK

1 INTRODUCTION

Metabolites are relatively small molecular weight intermediates or products of metabolic reactions. They are measured in clinical medicine for several reasons including their use as biomarkers of disease presence and severity (e.g., glucose in diabetes), as an index of organ or tissue functioning (e.g., urea and creatinine in renal function tests), or as risk factors for the future development of disease (e.g., cholesterol and atherosclerosis).

Until the last few decades, measurements in clinical chemistry have been centered on large central laboratories in hospitals and the discrete analysis of samples, usually blood or urine. However, the emergence of biosensor technology in recent years, particularly in the form of miniature probe-type devices, has provided the clinician and patient with three new and valuable capabilities: near-patient testing at the bedside, in the doctor's office, or in the home and workplace; the possibility of also monitoring analyte concentrations in the body itself (in vivo) rather than in samples of fluid extracted from the body; and the ability to sense analyte fluctuations continuously or near-continuously throughout the day.[1]

The most important metabolite for sensing, both clinically and commercially, is undoubtedly glucose in the context of diabetes, which is reviewed elsewhere in this volume. Here, I discuss a number of other clinically important analytes where biosensing technologies are increasingly playing a role.

2 MONITORING RENAL FUNCTION: UREA AND CREATININE

Since the ammonia produced by oxidative deamination is toxic to the body, ammonia is formed into urea in the liver and excreted by the kidney; creatinine is the breakdown product of creatine phosphate in muscle. Serum concentrations of urea and creatinine are routinely used as measures of glomerular function in the kidney. If the glomerular filtration rate (GFR) falls because of kidney disease or restriction of the blood supply to the kidney, waste products of metabolism like urea and creatinine are retained in the blood. A further clinical use is the on-line or intermittent assessment of the adequacy of renal dialysis by measuring urea or creatinine in the dialysate fluid.[2–4]

The reference range for urea is about 2.5–8.0 mM but levels can reach 50 mM in serious renal disease. Creatinine levels are normally about 40–130 μM but begin to increase when the GFR has fallen to half its normal value, reaching several hundred micromolar in serious kidney disease. Generally, creatinine is preferred over urea

Handbook of Biosensors and Biochips. Edited by Robert S. Marks, David C. Cullen, Isao Karube, Christopher R. Lowe and Howard H. Weetall.
© 2007 John Wiley & Sons, Ltd. ISBN 978-0-470-01905-4.

as a measure of renal function because it is less affected by short-term dietary changes.

Urea biosensors are based on immobilized urease, which catalyzes the hydrolysis of urea to ammonia and carbon dioxide:

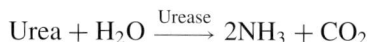

$$\text{Urea} + H_2O \xrightarrow{\text{Urease}} 2NH_3 + CO_2$$

Numerous urea sensors have been described where the products of this reaction are detected electrochemically or optically.[1] Ammonia is hydrolyzed to NH_4^+, which can be sensed potentiometrically by a change in pH at an H^+ ion–selective electrode or by detecting NH_4^+ at an ammonium ion–selective electrode incorporating an ionophore such as nonactin.[5] Miniaturized sensors can be manufactured using ion-selective field-effect transistors (ISFETs) with the enzyme immobilized at the gate.[6] However, a significant problem with NH_4-sensitive electrodes is interference by Na^+ and K^+ ions due to the limited selectivity of the ionophore. For example, in a disposable biosensor for urea there was excellent agreement between sensor-determined urea and a reference method in blood from healthy subjects, but with samples from the renal laboratory with expected electrolyte disturbances there was significant error.[5]

Alternatively, the sensor can be based on measuring the charged products at a conductivity electrode. Here, an alternating potential is applied between two electrodes and the increase in ionic strength in the solution mediates an increase in current between the electrodes. This technology has been used for continuous monitoring of urea in the dialysate fluid when patients are undergoing hemodialysis. The dialysate is allowed to flow through an enzyme column of immobilized urease on porous glass[2] or acrylic beads[3] and then to a conductivity electrode.

A recently described optical urea sensor uses pH-sensitive transducer reflection holograms manufactured by embedding interference fringe patterns comprising layers of ultrafine silver grains within thin polymer hydrogel films.[7] When acidic or basic monomers are incorporated within the polymer, pH changes induce ionization and cause the grating to swell, increasing fringe separation and thus changing the diffraction wavelength, and therefore the color. Urease was immobilized within the polymer film and the initial rate of

change in diffraction wavelength increased nonlinearly, with saturation above 50 mM.

The first biosensor for creatinine, a potentiometric enzyme electrode, was described by Meyerhoff and Rechnitz in 1976.[8] Of the many creatinine sensors subsequently reported,[9] the most common technology has been an amperometric three-enzyme system consisting of immobilized creatinine amidohydrolase, creatine amidohydrolase, and sarcosine oxidase, which sequentially converts creatinine to creatine, sarcosine, and thence to hydrogen peroxide, glycine, and formaldehyde:

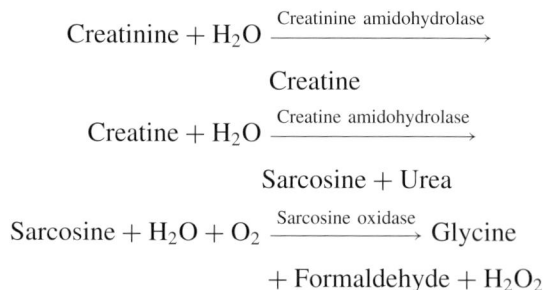

$$\text{Creatinine} + H_2O \xrightarrow{\text{Creatinine amidohydrolase}}$$
$$\text{Creatine}$$
$$\text{Creatine} + H_2O \xrightarrow{\text{Creatine amidohydrolase}}$$
$$\text{Sarcosine} + \text{Urea}$$
$$\text{Sarcosine} + H_2O + O_2 \xrightarrow{\text{Sarcosine oxidase}} \text{Glycine}$$
$$+ \text{Formaldehyde} + H_2O_2$$

The final reaction has usually been monitored by electrochemical detection of H_2O_2, though measurement of O_2 consumption is also possible. Endogenous creatine is a potential interferent.

Potentiometric creatinine biosensors are generally based on the immobilization of creatinine iminohydrolase (deiminase) at the surface of an NH_4^+-sensitive ion-selective electrode:[9,10]

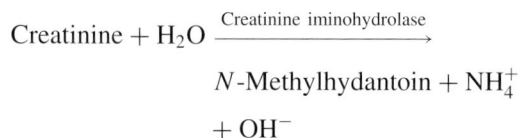

$$\text{Creatinine} + H_2O \xrightarrow{\text{Creatinine iminohydrolase}}$$
$$N\text{-Methylhydantoin} + NH_4^+$$
$$+ OH^-$$

Although such systems are simpler than the three-enzyme amperometric creatinine sensors and avoid interference from creatine, there can be interference from NH_4^+ in blood and urine.

A recently described optical system known as *intelligent polymerized crystalline colloidal array* (IPCCA) technology, somewhat reminiscent of the hologram sensor for urea described in the preceding text, has been adapted for creatinine sensing.[11] Here, an array of colloidal polystyrene latex particles that diffract light was embedded in a pH-responsive polyacrylamide hydrogel, forming a "photonic crystal", and also containing creatinine deiminase. Metabolism of creatinine and release of OH^- swells the gel and red shifts the diffraction according to the creatinine concentration.

3 MONITORING TISSUE OXYGENATION: LACTIC ACID

Lactate monitoring is important in critical care[12] and sports medicine.[13] Lactic acidosis is usually seen in medicine in seriously ill patients as a result of systemic hypoperfusion and/or tissue hypoxia, for example, cardiogenic, septic, or hypovolemic shock, severe anemia, and severe asthma. Lactic acidosis is less commonly associated with a number of illnesses where there is no obvious evidence of tissue hypoxia, such as liver disease, malignancy, and drug and toxin exposure. The clinical importance of this analyte can be seen, for example, from the fact that the initial arterial lactate concentration in patients in intensive care units correlates negatively with the probability of survival.[12] Blood lactate accumulation is also a predictor of exercise performance[13] and lower lactate accumulation per unit of work reflects a higher training status.[14]

Numerous discrete lactate biosensors have been described, and several commercial analyzers are available. A widely used example is the amperometric enzyme electrode incorporated in the Yellow Springs Instruments (YSI) device, which is based on lactate oxidase and a technology identical to the YSI glucose analyzer (save for the use of glucose oxidase in the latter), that is, oxidation of lactate producing pyruvate and hydrogen peroxide, which can be detected electrochemically by the positively charged base electrode:

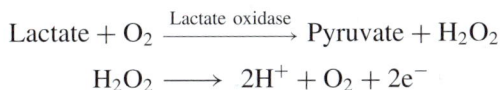

$$\text{Lactate} + O_2 \xrightarrow{\text{Lactate oxidase}} \text{Pyruvate} + H_2O_2$$

$$H_2O_2 \longrightarrow 2H^+ + O_2 + 2e^-$$

In this instrument, lactate oxidase is immobilized between an outer polycarbonate membrane controlling lactate diffusion and an inner cellulose acetate membrane, which restricts interfering electroactive substances, though a dilution step is also used. Modifications to the lactate oxidase technology that have been used to minimize interfering species in unprocessed blood samples that might

be oxidized at the high potential required to detect H_2O_2 and/or render the device less dependent on oxygen levels include lowering the required electrode overpotential by its modification with Prussian blue[15] and "wiring" the enzyme to the electrode via an osmium-based redox polymer, so that electrons flow directly from the enzyme's prosthetic group, $FADH_2$, to the electrode, rather than to oxygen.[16]

The alternative enzyme, lactate dehydrogenase, has been less used in lactate sensors because of the need to co-immobilize the cofactor NAD:

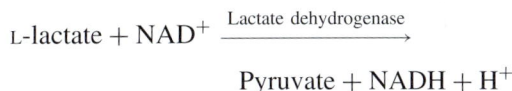

$$\text{L-lactate} + NAD^+ \xrightarrow{\text{Lactate dehydrogenase}}$$
$$\text{Pyruvate} + NADH + H^+$$

One example is an optical lactate sensor where lactate dehydrogenase and its cofactor were encapsulated in a silica solgel and generated NADH, measured by its fluorescence at 460 nm or its absorbance.[17] Nevertheless, there was some leaching of the enzyme and cofactor from the matrix. Lactate dehydrogenase has also been employed in a sensor described by D'Auria et al.,[18] but here the dye 8-anilino-1-naphthalene was noncovalently bound to the enzyme as a reporter, displaying a 40% decrease in fluorescence emission on binding to lactate.

Moving away from enzymes and naturally occurring protein receptors, Looger et al.[19] have recently used a computer-based method to change the binding specificities of bacterial periplasmic proteins, leading to "virtually designed" receptors, including one for lactate. After mutating the wild-type gene as specified in the program, protein designs were expressed, modified with fluorescent dyes conjugated to cysteine residues introduced by mutation, and shown by change in fluorescence with binding of lactate.

Continuous (as opposed to discrete) lactate sensing is receiving increasing attention, and it is probably second only to glucose as an analyte that favors such a monitoring strategy. Technologies include continuous blood sampling via a heparinized double-lumen cannula and flow through a chamber containing an amperometric lactate oxidase–based electrode.[20] Several microdialysis-based continuous lactate sensors have been described; in a recent example,[21] the GlucoDay (Menarini) continuous in vivo glucose analyzer was modified with a sensor employing immobilized lactate oxidase and microdialysis probes implanted subcutaneously in rabbits and a human volunteer. A twisted-wire dual glucose and lactate sensor has also been evaluated for continuous sensing of these analytes, implanted in the subcutaneous tissue of in rats, and has promise for human in vivo monitoring.[22]

4 MONITORING NUCLEIC ACID METABOLISM: URIC ACID

Uric acid is primarily the end product of purine metabolism, that is, the nucleic acids adenine and guanine. Because urate is relatively insoluble, elevated levels can cause precipitation, for example in the urinary tract as stones and in the joints as gout. Causes of hyperuricemia include reduced excretion by the kidney and increased nucleic acid turnover, for example, in malignancy.

Enzyme electrodes for urate are based on uricase:

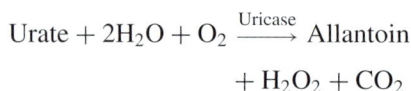

$$\text{Urate} + 2H_2O + O_2 \xrightarrow{\text{Uricase}} \text{Allantoin} + H_2O_2 + CO_2$$

Although early urate sensors detected O_2 consumption in the preceding reaction, more typically the H_2O_2 produced is monitored, an example being a device with uricase cross-linked to albumin, with an organosilane-treated polycarbonate outer membrane to extend the linear range and diminish biofouling, and an inner cellulose acetate membrane to exclude electroactive interferents.[23]

Fluorescence-based urate detection has some advantages in that it is extremely sensitive and can be done in highly dilute samples, thus reducing the risk of interference. A system has been described[24] incorporating uricase and peroxidase coencapsulated in solgel, with (initially nonfluorescent) amplex red detection of produced H_2O_2:

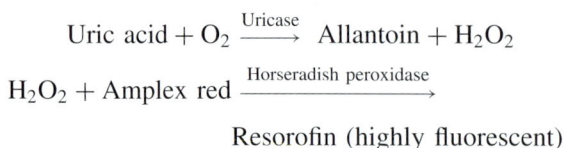

$$\text{Uric acid} + O_2 \xrightarrow{\text{Uricase}} \text{Allantoin} + H_2O_2$$

$$H_2O_2 + \text{Amplex red} \xrightarrow{\text{Horseradish peroxidase}}$$

$$\text{Resorofin (highly fluorescent)}$$

Hydrogen peroxide production by oxidase enzymes such as uricase can also be monitored by an electrochemiluminescent method, where luminol is oxidized at an electrode surface such as glassy carbon which catalyzes its chemiluminescence when reacting with H_2O_2.[25] This methodology is extremely sensitive and may offer some advantages for operating in complex solutions such as blood without interference.

5 MONITORING LIPID METABOLISM: CHOLESTEROL, TRIGLYCERIDE, AND KETONES

Lipid and lipoprotein disorders are some of the commonest metabolic diseases. Cholesterol, particularly the low-density lipoprotein (LDL) cholesterol subfraction, is the best established risk factor for the development of atherosclerosis. Cholesterol levels can be elevated because of primary genetic disorders (e.g., familial hypercholesterolemia), or secondary to some other condition such as kidney disease, or most commonly because of a combination of many genes producing a tendency to elevated cholesterol in the presence of environmental influences such as diet. Lowered concentrations of high-density lipoprotein (HDL) cholesterol are also related to arterial disease. Triglyceride and its carrier lipoprotein, very low density lipoprotein (VLDL), are typically elevated in type 2 diabetes in association with low HDL, a combination known as *dyslipidemia* and associated with increased cardiovascular risk. Hypertriglyceridemia also occurs in some primary genetic defects such as lipoprotein lipase deficiency; triglyceride excess is associated with pancreatitis.

Cholesterol biosensors frequently make use of the enzyme cholesterol oxidase, measuring oxygen consumption at an oxygen electrode or amperometric H_2O_2 production:[26,27]

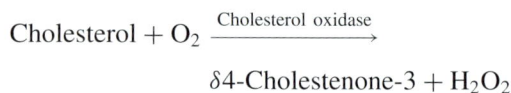

$$\text{Cholesterol} + O_2 \xrightarrow{\text{Cholesterol oxidase}}$$

$$\delta 4\text{-Cholestenone-3} + H_2O_2$$

This reaction has been adapted to fiber-optic technology for free cholesterol sensing by monitoring oxygen consumption via the fluorescence of a dye such as decacyclene that is quenched by molecular oxygen.[28] Since about 70% of the serum cholesterol is esterified with fatty acids, it is necessary to co-immobilize cholesterol oxidase with cholesterol esterase for blood total cholesterol detection:[29]

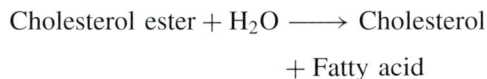

$$\text{Cholesterol ester} + H_2O \longrightarrow \text{Cholesterol} \\ + \text{Fatty acid}$$

An alternative optical approach to H_2O_2 production is the color development of an added dye in the presence of peroxidase[29] (therefore, not strictly a biosensor).

Cholesterol is one of the analytes that have been successfully detected by biosensors employing the relatively new electrode material of carbon nanotubes, which promote electron transfer and sensitivity and decrease the overvoltage for H_2O_2 detection, for example, a screen-printed device where carbon paste is modified with nanotubules.[30]

An alternative enzyme for cholesterol biosensors is the mono-oxygenase cytochrome P450scc which catalyzes cholesterol side chain cleavage:

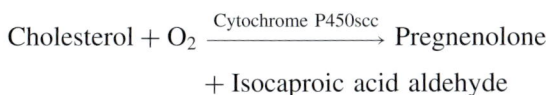

$$\text{Cholesterol} + O_2 \xrightarrow{\text{Cytochrome P450scc}} \text{Pregnenolone} \\ + \text{Isocaproic acid aldehyde}$$

Riboflavin can be used as an electrochemical mediator to transfer electrons from an electrode to the enzyme, reducing the cytochrome heme iron.[31]

Triglyceride sensing has received comparatively little attention, perhaps because its clinical value is less certain than that of cholesterol. One example uses the enzyme lipase immobilized on oxidized porous silicon, the spongelike matrix of which has a large surface area and great absorptive properties:[32]

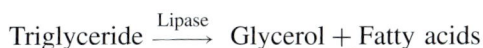

$$\text{Triglyceride} \xrightarrow{\text{Lipase}} \text{Glycerol} + \text{Fatty acids}$$

The resultant change in pH is the basis of a potentiometric measurement of triglyceride.

Ketones form in the body as the result of the breakdown of fats under conditions of insulin deficiency, such as in undiagnosed or uncontrolled type 1 diabetes. The three ketone bodies are acetone, which is volatile; 3-hydroxybutyric acid; and acetoacetic acid. The standard clinical test for ketones is based on a color reaction with

nitroprusside, but this is relatively insensitive, not reacting with the quantitatively most important of the ketones, 3-hydroxybutyrate. There is thus a clinical need for an improved and more sensitive ketone monitoring technology.

A recently developed commercial device for sensing 3-hydroxybutyrate in blood[33,34] is an adaptation of the successful MediSense glucose sensor technology, where a ferrocene mediator transfers electrons from the flavoprotein glucose oxidase to the base electrode.[35] The electrode of the ketone sensor incorporates 3-hydroxybutyrate dehydrogenase (HBDH), with the mediated detection of the NADH which is generated by the oxidation of the ketone:

$$\text{3-Hydroxybutyrate} + NAD^+ \xrightarrow{\text{HBDH}} \text{Acetoacetic} \\ \text{acid} + NADH$$

Ferrocene is not suitable as an NADH mediator, and amongst various quinoid NADH mediators that have been studied, 1,10-phenylanthroline quinone does not inhibit the enzyme and has good performance in the commercial biosensor MediSense Optium.

However, a bienzyme system for ketone detection is claimed to have improved operation;[36] a recent report employs immobilized HBDH and salicylate hydroxylase on a Clark-type oxygen electrode (to monitor consumed O_2), with the NADH produced enabling the decarboxylation and hydroxylation of the salicylate:

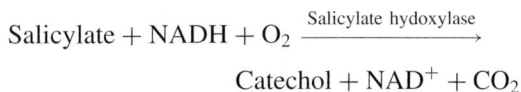

$$\text{Salicylate} + NADH + O_2 \xrightarrow{\text{Salicylate hydoxylase}} \\ \text{Catechol} + NAD^+ + CO_2$$

Note that the operating buffer requires added NAD and salicylate, so the device is again not a biosensor by conventional definition.

6 MONITORING JAUNDICE: BILIRUBIN

Jaundice is the typical yellow discoloration of the skin due to excessive blood levels of bilirubin; it is apparent when serum levels are greater than about $22\,\mu M$. Bilirubin is the breakdown product of the heme group found in hemoglobin and cytochromes and is normally eliminated by conjugation with glucuronic acid in the liver and excretion in the bile. The main causes of raised bilirubin levels and jaundice are hemolysis (the increased breakdown of blood cells and hemoglobin), liver disease causing failure of the conjugating mechanism, and obstruction of the biliary system (e.g., gall stones). Bilirubin itself is potentially toxic, particularly in newborns, and is an important analyte in neonatal monitoring,[37] but is also used as a general indicator of the liver and blood disorders in all patients.

Bilirubin biosensors have received rather little attention, however. Bilirubin oxidase potentially can be incorporated in enzyme electrodes:

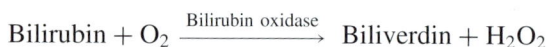

$$Bilirubin + O_2 \xrightarrow{\text{Bilirubin oxidase}} Biliverdin + H_2O_2$$

However, two significant problems are that the enzyme is comparatively unstable and easily undergoes denaturation, and bilirubin itself is electrochemically active and is oxidized in association with the formation of a dark-colored, eventually insulating polymer at the base electrode.[38] One approach to a sensor with improved functioning is a multilayer of enzyme covalently attached to a (gold) base electrode (which has increased stability and excludes bilirubin), with electron transfer mediated by a ferrocene derivative.[39]

Another approach is to make use of the pseudoperoxidase activity of hemoglobin, whereby hemoglobin catalyzes the conversion of bilirubin to biliverdin in the presence of H_2O_2. The H_2O_2 can be generated in situ by co-immobilizing hemoglobin with glucose oxidase in a polyvinyl alcohol membrane, with detection of biliverdin spectrophotometrically at 450 nm:[39]

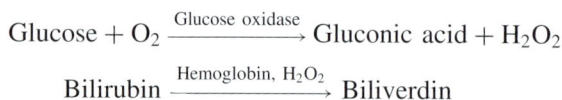

$$Glucose + O_2 \xrightarrow{\text{Glucose oxidase}} Gluconic\ acid + H_2O_2$$

$$Bilirubin \xrightarrow{\text{Hemoglobin, } H_2O_2} Biliverdin$$

7 CONCLUSIONS

The development and description of new biosensor technologies for clinically important analytes in recent years is impressive but has far exceeded their take up into clinical practice. The impact of metabolite sensors is still modest in most diseases, excepting the measurement of glucose in diabetes. This is partly because many altered analyte levels

in disease and risk states, apart from glucose, are relatively constant throughout the day and do not need to be monitored frequently and rapidly. The true impact of biosensors in clinical chemistry may come from the potential to construct cheap and robust multianalyte sensors on a single miniature device—the so-called "lab on a chip". Armed in the future with this type of technology at the bedside, in the home and workplace, and in developing countries where access to conventional hospital laboratories is limited, we will likely see a major biosensor-led improvement in clinical care.

REFERENCES

1. P. D'Orazio, Biosensors in clinical chemistry. *Clinica Chimica Acta*, 2003, **334**, 41–69.
2. P. Thavarungkul, H. Håkanson, O. Holst, and B. Mattiasson, Continuous monitoring of urea in blood during dialysis. *Biosensors and Bioelectronics*, 1991, **6**, 101–107.
3. L. D. Ciana and G. Caputo, Robust, reliable biosensor for continuous monitoring of urea during dialysis. *Clinical Chemistry*, 1996, **42**, 1079–1085.
4. B. Tombach, J. Schneider, F. Matzkies, R. M. Schaefer, and G. C. Chemnitius, Amperometric creatinine biosensor for hemodialysis patients. *Clinica Chimica Acta*, 2001, **312**, 129–134.
5. C. Eggenstein, M. Borchardt, C. Diekman, B. Gründig, C. Dumschat, K. Cammann, M. Knoll, and F. Spener, A disposable biosensor for urea determination in blood based on an ammonia-sensitive transducer. *Biosensors and Bioelectronics*, 1999, **14**, 33–41.
6. A. P. Soldatkin, J. Montorial, W. Sant, C. Martlet, and N. Jaffrezic-Renault, A novel urea sensitive biosensor with extended dynamic range based on recombinant urease and ISFETs. *Biosensors and Bioelectronics*, 2003, **19**, 131–135.
7. A. J. Marshall, D. S. Young, J. Blyth, S. Kabilan, and C. R. Lowe, Metabolite-sensitive holographic biosensors. *Analytical Chemistry*, 2004, **76**, 1518–1523.
8. M. Meyerhoff and G. A. Rechnitz, An activated enzyme electrode for creatinine. *Analytica Chimica Acta*, 1976, **85**, 277–285.
9. A. J. Killard and M. R. Smith, Creatinine biosensors: principles and designs. *Trends in Biotechnology*, 2000, **18**, 433–437.
10. G. G. Guilbault and P. R. Coulet, Creatinine-selective enzyme electrodes. *Analytica Chimica Acta*, 1983, **152**, 223–228.
11. A. C. Sharma, T. Jana, R. Kesavamoorthy, L. Shi, M. A. Virji, D. N. Finegold, and S. A. Asher, A general photonic crystal sensing motif: creatinine in bodily fluids. *Journal of the American Chemical Society*, 2004, **126**, 2971–2977.
12. B. A. Mizock and J. L. Falk, Lactic acidosis in critical illness. *Critical Care Medicine*, 1992, **20**, 80–93.
13. I. Jacobs, Blood lactate. Implications for training and sports performance. *Sports Medicine*, 1986, **3**, 10–25.

14. K. E. Friedl, The promise of lactic acid monitoring in ambulatory individuals. *Diabetes Technology and Therapeutics*, 2004, **3**, 402–404.

15. R. Garjonyte, Y. Yigzaw, R. Meskys, A. Malinauskas, and L. Gorton, Prussian Blue-and lactate oxidase-based amperometric biosensor for lactic acid. *Sensors and Actuators, B*, 2001, **79**, 33–38.

16. I. Katakis and A. Heller, L-α-glycerophosphate and L-lactate electrodes based on the electrochemical 'wiring' of oxidases. *Analytical Chemistry*, 1992, **64**, 1008–1013.

17. C.-I. Li, Y.-H. Lin, C.-L. Shih, J.-P. Tsaur, and L.-K. Chau, Sol-gel encapsulation of lactate dehydrogenase for optical sensing of L-lactate. *Biosensors and Bioelectronics*, 2002, **17**, 323–330.

18. S. D'Auria, Z. Gryczynski, I. Gryczynski, M. Rossi, and J. R. Lazowicz, A protein biosensor for lactate. *Anal Biochem*, 2000, **283**, 83–88.

19. L. L. Looger, M. A. Dwyer, J. J. Smith, and H. W. Hellinga, Computational design of receptor and sensor proteins with novel functions. *Nature*, 2003, **423**, 185–190.

20. C. Meyerhoff, F. Bischof, F. J. Mennel, F. Sternberg, J. Bican, and E. F. Pfeiffer, On line continuous monitoring of blood lactate in men by a wearable device based upon an enzymatic amperometric lactate sensor. *Biosensors and Bioelectronics*, 1993, **8**, 409–414.

21. A. Poscia, D. Messeri, D. Moscone, F. Ricci, and F. Valgimigli, A novel continuous lactate monitoring system. *Biosensors and Bioelectronics*, 2005, **20**, 2244–2250.

22. W. K. Ward, J. L. House, J. Birck, E. M. Anderson, and L. B. Jansen, A wire-based dual-analyte sensor for glucose and lactate: in vitro and in vivo evaluation. *Diabetes Technology and Therapeutics*, 2004, **3**, 389–401.

23. F. H. Keedy and P. Vadgama, Determination of urate in undiluted whole blood by enzyme electrode. *Biosensors and Bioelectronics*, 1991, **6**, 491–499.

24. D. Martinez-Pérez, M. Ferrer, and C. R. Mateo, A reagent less fluorescent sol-gel biosensor for uric acid detection in biological fluids. *Analytical Biochemistry*, 2003, **322**, 238–242.

25. C. A. Marquette, A. Degiuli, and L. J. Blum, Electrochemiluminescent biosensors array for the concomitant detection of choline, glucose, glutamate, lactate, lysine and urate. *Biosensors and Bioelectronics*, 2003, **19**, 433–439.

26. I. Satoh, I. Karube, and S. Sazuki, Enzyme electrode for free cholesterol. *Biotechnology and Bioengineering*, 1977, **19**, 1095–1099.

27. M. Mascini and G. G. Guilbault, Clinical uses of enzyme electrode probes. *Biosensors*, 1986, **2**, 147–172.

28. W. Trettnack and O. S. Wolfbeiss, A fiberoptic cholesterol biosensor with an oxygen optrode as the transducer. *Analytical Biochemistry*, 1990, **184**, 124–127.

29. A. Krug, A. A. Sulerman, and G. G. Guilbault, Enzyme-based fiber-optic device for the determination of total and free cholesterol. *Analytica Chimica Acta*, 1992, **256**, 263–268.

30. G. Li, J. M. Liao, G. Q. Hu, N. Z. Ma, and P. J. Wu, Study of carbon nanotube modified biosensor for monitoring total cholesterol in blood. *Biosensors and Bioelectronics*, 2005, **20**, 2140–2144.

31. V. Shumyantseva, G. Deluca, T. Bulko, S. Carrara, C. Nicolini, S. A. Usanov, and A. Archakov, Cholesterol amperometric biosensor based on cytochrome P450. *Biosensors and Bioelectronics*, 2004, **19**, 971–976.

32. R. R. K. Reddy, A. Chadha, and E. Bhattacharya, Porous silicon based potentiometric triglyceride biosensor. *Biosensors and Bioelectronics*, 2001, **16**, 313–317.

33. N. J. Forrow, G. S. Sanghera, S. J. Walters, and J. L. Watkin, Development of a commercial amperometric biosensor electrode for the ketone D-3-hydroxybutyrate. *Biosensors and Bioelectronics*, 2005, **20**, 1617–1625.

34. R. W. K. Chiu, C. S. Ho, S. F. Tong, K. F. Ng, and C. W. Lam, Evaluation of a new handheld biosensor for point-of-care testing of whole blood beta-hydroxybutyrate concentration. *Hong Kong Medical Journal*, 2002, **8**, 172–176.

35. A. E. G. Cass, G. Davis, G. D. Francis, H. A. O. Hill, W. J. Aston, I. J. Higgins, E. V. Plotkin, L. D. L. Scott, and A. P. F. Turner, Ferrocene-mediated enzyme electrode for the amperometric determination of glucose. *Analytical Chemistry*, 1984, **56**, 667–671.

36. R. C. H. Kwan, P. Y. T. Hon, W. C. Mak, L. Y. Law, J. Hu, and R. Renneberg, Biosensor for rapid determination of 3-hydroxybutyrate using bienzyme system. *Biosensors and Bioelectronics*, 2006, **21**, 1101–1106.

37. I. Murkovic, M. D. Steinberg, and B. Murkovic, Sensors in neonatal monitoring: current practice and future trends. *Technology and Health Care*, 2003, **11**, 399–412.

38. B. Shoham, Y. Migron, I. Willner, and B. Tartakovsky, A bilirubin biosensor based on a multilayer network enzyme electrode. *Biosensors and Bioelectronics*, 1995, **10**, 341–352.

39. M. M. Vidal, I. Delgadillo, M. H. Gil, and J. Alonso-Chamarro, Study of an enzyme coupled system for the development of fibre optical bilirubin sensors. *Biosensors and Bioelectronics*, 1996, **11**, 347–354.

69

Need for Biosensors in Infectious Disease Epidemiology

Laurence Baril

Unit of Infectious Disease Epidemiology, Institut Pasteur, Dakar, Sénégal

1 WHAT IS INFECTIOUS DISEASE EPIDEMIOLOGY?

Infectious disease epidemiology deals with three main issues: causes, distribution, and control of infectious diseases occurring in the human population. This is based on the observation that most infectious diseases do not occur randomly, but are related to environmental, economic, societal, and personal characteristics. In the past decades, increased development, deforestation, and other environmental changes have brought people into contact with animals (recent epidemic of SARS, and Ebola and Marburg hemorrhagic fevers) or insects (vectorborne diseases such as human infections with the dengue, West Nile, or Chikungunya viruses) that harbor diseases only rarely encountered before in humans.[1–4] In addition, the recent threefold increase in international travel as well as the greater importation of fresh foods across national borders are other examples of potential exposures with growing numbers of people who have no or weakened immunity to these new pathogens (see Annex 1).

The etymology of *epidemiology* comes from the Greek *epi* = among, *demos* = people, and *logos* = science, whereas the definition of infectious disease is related to a communicable disease. Communicable disease was initially defined as a disease that can be transmitted from person to person through a microbial agent. However, of the around 1400 recognized species of human pathogens (prions, viruses, bacteria, fungi, protozoa, and helminths) almost 60% are zoonotic, that is, able to infect other host species.[5] These recognized zoonotic pathogens have been, at least once, successfully introduced from a nonhuman source. In infectious disease epidemiology, the identification of the pathogen (the disease-causing organism) is a pivotal step in understanding infectious disease etiology.[6,7] Then, it is necessary to study the factors that influence the occurrence of this disease and its distribution in the human population. Therefore, infectious disease epidemiology examines epidemic (excess or new) and endemic (always present) infectious diseases. It is a cornerstone methodology of public health research for identifying risk factors for infectious diseases and determining optimum treatment or preventive measures. Infectious diseases are still the leading cause of death worldwide. It is estimated that around 180 human pathogens are regarded as emerging or reemerging among which two-thirds are known to be zoonotic.[5] However, most zoonotic pathogens are not highly transmissible within human populations and do not cause major epidemics but only outbreaks.

In infectious diseases, the etiologic agent and the biological interaction between host and pathogen are the major components leading to disease. Small

Handbook of Biosensors and Biochips. Edited by Robert S. Marks, David C. Cullen, Isao Karube, Christopher R. Lowe and Howard H. Weetall.
© 2007 John Wiley & Sons, Ltd. ISBN 978-0-470-01905-4.

changes in the nature of the host–pathogen interaction can lead to the emergence or reemergence of an epidemic. There is a *joint action* between the biological events including genetic factors and the societal determinants in transmission of infectious diseases in the human population. The AIDS pandemic and the reemergence of vectorborne diseases could illustrate this causal coaction of biological and societal or environmental factors. The transmission of AIDS was dramatic in the intravenous drug user and homosexual communities, both stigmatized by society in highly developed countries, and in the heterosexual population living in developing countries, abandoned by the global economical system. For vectorborne diseases such as malaria, the reemergence of dengue, or the recent Chikungunya epidemic, the slow-down of the vector-control measures, the impact of warmed-up climate, the growing suburban population are certainly among the causative factors that promote the high burden of those diseases in tropical areas.

Historically, the epidemiologic approach has helped to explain the transmission of communicable diseases such as cholera and measles; this was done by discovering what exposures or host factors were shared by individuals who became sick. The founding event in infectious disease epidemiology was when John Snow, in 1854, removed a public water pump in London and ended an outbreak of cholera. Nowadays, the identification of the pathogens using newly developed biological tools has changed the epidemiological approach to infectious diseases. In this way, pathogens were quite rapidly identified in the recent epidemics of West Nile virus, SARS coronavirus, and Ebola and Marburg hemorrhagic fevers (see Annex 1). However, in terms of public health measures in response to epidemics, a rapid recognition of emerging infections requires the core capacity to assure almost an "on-the-field" detection. This detection implies the use of newly developed diagnostic tools, ideally a lab on a chip or, at least, some portable (easy to use) tools such as optical fiber biosensors.[8–10] In addition, on-the-field, noninvasive remote diagnosis tools are necessary for safety reasons. Such diagnostic tools to identify existing and possibly new pathogens could be essential for determining control and prevention efforts during the investigation of a new outbreak.

2 METHODS IN INFECTIOUS DISEASE EPIDEMIOLOGY

The epidemiologists attempt to answer the following questions:

- Who is prone to a particular infectious disease?
- What exposure do the patients have in common?
- Where is the risk highest?
- When is the disease most likely to occur and what is its trend over time?
- By how much does the risk increase given a specific exposure?
- How many cases of the disease could be avoided by eliminating the exposure?

However, many diseases (diarrheal or respiratory diseases, meningitis, etc.) can be caused by more than one species of pathogen. This explains that most of the time the first epidemiological approach is a *syndromic approach* and emphasizes how much the rapid identification of the pathogen helps the epidemiologist to carry out the first control and prevention measures.

During the second half of the twentieth century, different methods have been developed to understand factors that influence the risk of onset and severity of infectious diseases:

- outbreak investigation to identify the component *causes* of a specific infectious disease, its incubation period, and its mode of transmission;
- surveillance system with *notifiable* infectious diseases at national or international levels of reporting;[11,12]
- cohort studies to understand incidence, determinants, and occurrence patterns of a particular disease (mainly applicable to chronic infectious diseases such as HIV infection or viral hepatitis);[13]
- clinical research to improve prevention measures or therapy;
- modeling method to contribute to the formulation of public health policies and support public health decision-makers.[14]

The different outcome measures used by the epidemiologists are summarized in Annex 2.

3 A MULTIDISCIPLINARY APPROACH IS ESSENTIAL

Epidemiological studies in humans offer the advantage of producing results relevant to people, but have the disadvantage of not always allowing perfect control of study conditions. Unlike experimental science, budget and societal concerns impose some limitations on epidemiological research. One major difficulty in epidemiological research is the expensive effort needed to gather the necessary information. In addition, epidemiology research is sometimes viewed as a collection of data analyzed using some statistical tools! However, the epidemiologists use gathered data on a broad range of biomedical domains in an interactive way to generate or expand theory, to test hypotheses, and to make assertions about which relationships are causal and about how they are causal. Thus, one essential concept in epidemiology is to keep in mind that people are classified according to causal indicators or etiologic agents. Some causal components could remain hidden despite efforts to understand a complex causal mechanism. Moreover, the organization and the logistics required to obtain biological samples for etiologic diagnostics and to transport these samples to the laboratory of reference are difficult. This is why it is essential that diagnostic tools be developed on the basis of biosensor, lab-on-a-chip, or biochip technologies that would enable the epidemiologists to carry out on-the-field tests.

Therefore, a multidisciplinary approach is essential to investigate emerging or reemerging infectious diseases. Not only infectious disease epidemiologists or public health authorities but also researchers from other fields should be rapidly recruited, such as virologists and immunologists for the study of microbial pathogenesis and host responses, researchers in biotechnologies to develop novel diagnostics tools and even new therapeutic or vaccine approaches, and researchers in the field of vector or animal control. This multidisciplinary approach at an early phase of the investigation will also allow collaborators to obtain *well-characterized* samples that are essential to study disease severity biomarkers and test newly developed diagnostics tools that require proper statistical analysis for validation (see Annex 3).

4 COMMUNICATIONS AND REGULATIONS AT AN INTERNATIONAL LEVEL

One other challenge for the epidemiologists is to translate surveillance or epidemics information into disease prevention and control activities and also to optimize the exploitation of the results. Several means of information sharing are now available through the web at national or international levels (available from the websites of the World Health Organization (WHO), the US Centers for Disease Control and Prevention (CDC), the ProMED society, etc.). Emphasis has been placed on the critical importance of communicating alerts about clusters of illness, data on disease trends, and guidelines for disease prevention. In this regard, the recent SARS epidemic was quite a success in terms of the collaborative international effort to contain the epidemic. In addition, WHO has set up a legal instrument to notify the outbreak of diseases of international importance: the International Health Regulations (IHR, available from http://www.who.int/csr/ihr). The most recent IHR version emphasizes the immediate notification of all disease outbreaks of urgent international importance. This should facilitate alert and appropriate international response while awaiting laboratory verification.

5 PREPAREDNESS FOR EPIDEMICS

Preparedness for epidemics (including deliberate epidemics) is becoming a major issue for national and international public health organizations. For instance, it has been estimated that only 0.4% of all extant bacterial species have been identified despite the development of different molecular methods such as DNA microarray technology.[2]

We have highlighted how important laboratory assessment is for identifying the cause of an infectious disease, but what would epidemiologists *ideally* need on the field from an efficient collaboration with the partners from other disciplines?

- coordination of the respective activities and sharing of the information: this includes the elaboration of a written protocol and standard operational procedures;

- previous laboratory management skills (including biosafety issues as defined by WHO[15,16]) in the field conditions;
- rapid, reproducible, high-sensitivity, and specific bioassays (see Annex 3) to detect antigens for early diagnostic, and to detect antibodies (IgM and IgG) for confirmation of the diagnostic during the early convalescent phase;
- the laboratory assessment should allow for syndromic approach (i.e., taking into account differential diagnosis) and for the detection of subclinical cases to monitor the spread of the disease;
- the bioassays should be able to perform the analysis using small amounts of different biological samples (blood, urine; stools, cerebrospinal fluid, tissue biopsy, etc.);
- a compatible electronic data system to be able to perform comprehensive data analysis and easily diffuse information.

Adequate funding and institutional and community support are also critical to perform an effective investigation and to implement control measures.

6 CONCLUSION

Like novel biotechnologies, infectious disease epidemiology is still an *emerging* discipline. This domain requires a strong interdisciplinary approach; the epidemiologists face the problem of obtaining cooperation from numerous people to perform epidemiological investigation or research essential in efforts to understand, prevent, control, and respond to new or reemerging infectious diseases. Innovations in biotechnology have an increasingly important role in identifying pathogens. Novel laboratory testing tools such as optical biosensors will improve the quality of the epidemiological information. In developing countries with resource constrains where most of the recent new viral epidemics have occurred in the past decades (see Annex 1), such innovative diagnosis tools are necessary to improve the feasibility of epidemiological research. This is not a paradox when we look at what happened in the telecom and informatics domains where cell phones and microcomputers have been easily implemented and taken up by the population all over the world.

7 ANNEX 1: EXAMPLES OF INFECTIOUS DISEASE SURVEILLANCE SYSTEMS AND RECENTLY DIAGNOSED VIRAL DISEASES

1. Examples of disease-specific surveillance:
   (a) tuberculosis
   (b) HIV/AIDS
   (c) sexually transmitted infections
   (d) foodborne/waterborne diseases (*Campylobacter, Clostridium, Escherichia coli* O157:H7, *Listeria*, nontyphoid *Salmonell a*, or *Staphylococcus aureus*, etc.)
   (e) respiratory diseases
   (f) vectorborne/zoonotic diseases (malaria, dengue, Japanese encephalitis, yellow fever, etc., in some tropical areas or as imported diseases)
   (g) drug resistance (malaria, tuberculosis, bacterial community-acquired pneumonia, etc.)
   (h) vaccine-preventable diseases (including rabies)
   (i) diseases transmitted by blood transfusions or blood products (hepatitis C, for instance)
   (j) other emerging infectious diseases (agents of viral hemorrhagic fevers, Lyme disease, borreliae, bartonellae, babesiae, rickettsiae, etc.)
   (k) parasitic diseases (trypanosomiasis, etc.).

   The disease-specific surveillance systems to improve the prevention and control of emerging or reemerging infectious diseases differ from country to country.[11,12] In addition, public health risks may change over time; therefore, priorities are reviewed periodically by the public health authorities.

2. Examples of recently diagnosed viral diseases and causes of death in human population are shown in Figure 1.

8 ANNEX 2: LEXICON OF THE CURRENT DEFINITIONS USED IN EPIDEMIOLOGY

Definitions adapted from Refs. 13 and 14.

Descriptive measures: these are used for identifying populations and subgroups at high or low risk of infectious diseases and for monitoring time trends for specific infectious diseases.

Analytic measures: these are used for identifying specific factors that increase or decrease the

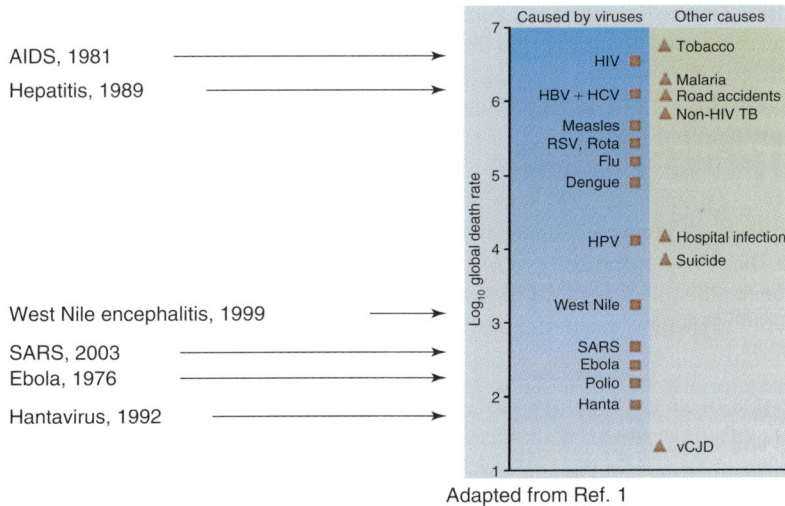

Adapted from Ref. 1

Figure 1. Recently diagnosed viral diseases and causes of death in human population.

risk of infectious disease and for quantifying the associated risk.

Hypothesis in statistical testing: Assumptions to be tested in a statistical procedure, the *null hypothesis* (noted H_0) is often in the form "no relationship exists between x and y". When the statistical test cannot disprove the null hypothesis, it is termed *failure to reject the null hypothesis* rather than acceptance. In statistical testing, a hypothesis is accepted if a sample contains sufficient evidence to reject the null hypothesis. In most cases, the alternative hypothesis (H_1) is the expected conclusion (why the test was completed in the first place).

P value: It is the probability of obtaining a finding at least as impressive as that obtained by assuming the null hypothesis is true, so that the finding was the result of chance alone. The fact that P values are based on this assumption is crucial to their correct interpretation. Therefore, in practical terms, the higher the P value, the higher the probability that the observations you are studying are just chance. If the null hypothesis (H_0) is true, the significance level (α) is the probability that it will be rejected in error (a decision known as a *Type 1 error*). P values used to decide the *statistical significance* represents the level of belief in the null hypothesis (choice of level often set at $\alpha = 0.05$).

Power of a statistical test: It is the probability that the test will reject a false H_0, or in other words,

that it will not make a Type II error (probability β). As power increases, the chances of a Type II error decrease, and vice versa. Therefore, power is equal to $1 - \beta$.

Case fatality rate: number of deaths over a period per 1000 persons at risk.

Incidence density: number of new cases over a period per 100 000 persons at risk.

Cumulative incidence: number of new cases divided by the studied persons over a period.

Prevalence: number of existing cases at a given time per 100 persons at risk.

Odds ratio (OR): The ratio of the odds of an event occurring in one group to the odds of it occurring in another group. These groups might be an experimental group and a control group, or any other dichotomous classification. If the probabilities of the event in each of the groups are p (first group) and q (second group), then the odds ratio is $OR = p(1-q)/q(1-p)$. An OR of 1 indicates that the condition or event under study is equally likely in both groups. An OR greater than 1 indicates that the condition or event is more likely in the first group and an OR less than 1 indicates that the condition or event is less likely in the first group.

Relative risk (RR): It is the risk of an event (or of developing a disease) relative to exposure; it is used frequently in the statistical analysis of binary outcomes where the outcome of interest has relatively low probability: $RR = p_{exposed}/p_{control}$. RR is different from OR, although it asymptotically approaches it for small probabilities.

An *RR* of 1 means there is no difference in risk between the two groups. An *RR* less than 1 means the event is less likely to occur in the experimental group than in the control group. An *RR* greater than 1 means the event is more likely to occur in the experimental group than in the control group.

Hazard ratio: the effect of an explanatory variable on the hazard or risk of an event (used in survival analysis).

Confidence interval (CI): It is an interval between two numbers with an associated probability p, which is generated from a random sample of an underlying population, such that if the sampling was repeated numerous times and the confidence interval recalculated from each sample according to the same method, a proportion p of the confidence intervals would contain the population parameter. Confidence intervals are the most prevalent form of interval estimation.

Infectivity (R_0): The possible magnitude of an infectious disease outbreak is related to the basic reproduction number R_0. For pathogens that are minimally transmissible within human populations $(R_0$ close to 0), outbreak size is determined largely by the number of introductions from the reservoir. For pathogens that are highly transmissible within human populations $(R_0 > 1)$, outbreak size is determined largely by the size of the susceptible population. For pathogens that are moderately transmissible within human populations (corresponding to $R_0 \sim 1$), notable outbreaks are possible (for instance, if multiple introductions occur), but the scale of these outbreaks is very sensitive to small changes in R_0.

9 ANNEX 3: LEXICON OF THE PERFORMANCE OF NEWLY DEVELOPED DIAGNOSTIC TOOLS

Some indicators should help to estimate the performances of laboratory methods and to support their comparisons (see Table 1). For instance, to determine whether a patient has a certain infectious disease or not, a newly developed diagnostic tool should be compared to a reference method through the following indicators:

Sensitivity (Se): number of true positives/ (number of true positives + number of false negatives).

Specificity (Sp): number of true negatives/(number of true negatives + number of false positives).

Positive predictive value: number of true positives/(number of true positives + number of false positives). It is the proportion of patients who present the disease, among the patients diagnosed by the method.

Negative predictive value: number of true negatives/(number of true negatives + number of false negatives). It is the proportion of patients who do not present the disease, among the patients not diagnosed by the method.

Positive predictive value is called *precision*, and sensitivity is known as *recall*. F measure can be used as a single measure of the performance of the test, it is calculated as follows: $F = 2 \times$ precision $\times$ recall/(precision + recall).

The *receiver operating characteristic* (ROC) is a graphical plot of the sensitivity versus 1-specificity for a binary testing system as its discrimination threshold is varied. The ROC curve can also be represented equivalently by plotting the fraction of true positives versus the fraction of false positives.

Reproducibility refers to the ability of a laboratory method to be accurately reproduced, or replicated.

Table 1. Link between statistical tests and performance of laboratory tests

	H$_1$	H$_0$		Disease	No disease
Statistical significance	$1 - \beta$	α	Positive test	*Se*	$1 - Sp$ (False positive)
No statistical significance	β	$1 - \alpha$	Negative test	$1 - Se$ (False negative)	*Sp*

The *calibration curve* is a plot of how the instrumental response changes with changing concentration (for instance) of substance to be measured (adapted from Ref. 8). The operator will create a series of standards across a range of concentrations near the expected unknown concentration. Analyzing each of the standards using the chosen technique will produce a series of readings. The results could follow different types of equations (linear, logarithmic, sigmoid, etc.) and the curve also indicates the precision of the results. Moreover, if there is evidence of a relationship, the technique of correlation ($r =$ *correlation coefficient*) allows one to test the statistical significance of the association. However, a significant correlation only shows that two factors vary in a related way but does not necessarily demonstrate a causal relationship.

The lower detection limit is important to report; it describes the lower analyte dilution that can be efficiently distinguished by the detecting system. For a definite dilution, the difference of the signal level between the experimental points with positive or control biological samples is modulated by the standard deviation of the control (adapted from Ref. 17).

Some recommendations before publications of laboratory testing methods are available from http:// www.consort-statement.org/ stardstatement. htm.[17,18]

REFERENCES

1. R. A. Weiss and A. J. McMichael, Social and environmental risk factors in the emergence of infectious diseases. *Nature Medicine*, 2004, **12**, 70–76.
2. A. S. Fauci, New and re-emerging diseases: the importance of biomedical research. *Emerging Infectious Diseases*, 1998, **4**, 374–378.
3. N. D. Wolfe, A. A. Escalante, W. B. Karesh, A. Kilbourn, A. Spielman, and A. A. Lal, Wild primate in emerging infectious disease research: the missing link? *Emerging Infectious Diseases*, 1998, **4**, 149–158.
4. J. Lederberg, Emerging infections: an evolutionary perspectives. *Emerging Infectious Diseases*, 1998, **4**, 366–370.
5. M. E. Woolhouse and S. Gowtage-Sequeria, Host range and emerging and re-emerging pathogens. *Emerging Infectious Diseases*, 2005, **11**, 1842–1847, see Appendix available on http://www.cdc.gov/ncidod/eid/vol11no12/05-0997_app.htm).
6. P. Houpikian and D. Raoult, Traditional and molecular techniques for the study of emerging bacterial diseases: one laboratory's perspective. *Emerging Infectious Diseases*, 2002, **8**, 122–131.
7. C. A. Cummings and D. A. Relman, Using DNA microarrays to study host-microbe interactions. *Emerging Infectious Diseases*, 2000, **6**, 513–525.
8. S. Herrmann, B. Leshem, S. Landes, B. Rager-Zisman, and R. S. Marks, Chemiluminescent optical fiber immunosensor for the detection of anti-West Nile virus IgG. *Talanta*, 2005, **66**, 6–14.
9. T. Konry, A. Novoa, Y. Shemer-Avni, N. Hanuka, S. Cosnier, A. Lepellec, and R. S. Marks, Optical fiber immunosensor based on a poly(pyrrole-benzophenone) film for the detection of antibodies to viral antigen. *Analytical Chemistry*, 2005, **77**, 1771–1779.
10. S. Cosnier, R. E. Ionescu, S. Herrmann, L. Bouffier, M. Demeunynck, and R. S. Marks, Electroenzymatic polypyrrole-intercalator sensor for the determination of West Nile virus cDNA. *Analytical Chemistry*, 2006, **78**, 7054–7057.
11. World Health Organization. *Settings Priorities in Communicable Surveillance*, 2006, (available from www.who.int/csr/labepidemiology).
12. World Health Organization. *Developing Laboratory Partnerships to Detect Infections and Prevent Epidemics*, 2005, (available from www.who.int/csr/labepidemiology).
13. K. J. Rothman and S. Greenland, *Modern Epidemiology*, 2nd Edn, Lippincott—Raven Publishers, Philadelphia, 1998.
14. R. M. Anderson and R. M. May, *Infectious Diseases of Humans: Dynamics and Control*, Oxford University Press, Oxford, 2002.
15. World Health Organization. *Laboratory Biosafety Manual*, 3rd Edn, 2004, (available from http://www.who.int/csr/resources:publications/biosafety).
16. World Health Organization. *Guidance on Regulations for the Transport of Infectious Substances*, 2005, (available from http://www.who.int/csr/resources/publications/transport).
17. N. Smidt, A. W. Rutjes, D. A. van der Windt, R. W. Ostelo, P. M. Bossuyt, J. B. Reitsma, L. M. Bouter, and H. C. de Vet, Reproducibility of the STARD checklist: an instrument to assess the quality of reporting of diagnostic accuracy studies. *BMC Medical Research Methodology*, 2006, **6**, 12.
18. P. M. Bossuyt, L. Irwig, J. Craig, and P. Glasziou, Comparative accuracy: assessing new tests against existing diagnostic pathways. *British Medical Journal*, 2006, **332**, 1089–1092.

70

Biosensors for Neurological Disease

Kathryn M. Bell and Steven E. Kornguth

Center for Strategic and Innovative Technologies, University of Texas at Austin, Austin, TX, USA

1 INTRODUCTION

Biosensors are capable of rapid, sensitive, and accurate identification of metabolic, proteomic, and genomic biomarkers of neurological disease (ND). Many neurological disorders can be identified by testing the activity of a single enzyme (e.g., mental retardation in phenylketonuria is associated with reduced levels of phenylalanine hydroxylase) or genomic element (e.g., a large number of deoxynucleotide triplet repeats in Huntington's disease, HD). For other diseases the analysis of a single biomarker may provide a tentative diagnosis that is burdened by a high number of false positives or negatives. The ability to generate a multitarget profile of disease probability or "signature" of disease is therefore of major interest. As an example, autoantibody signatures for prostate cancer were developed from markers present in prostate cancer tissues, a technique that can be adapted to nervous system cancers.[1] Using iterative biopanning of a phage display library derived from the cancer tissues, a 22-phage peptide panel had 88.2% specificity and 81.6% sensitivity in discriminating the prostate cancer group from the control group. In an analogous manner, a robust sensor array that could detect multiple genomic and proteomic ND markers from saliva, urine, serum, or cerebrospinal fluid (CSF) could increase the probability of accurate diagnosis.

For most neurological disorders there is no one test that can provide a definitive diagnosis. In addition to the collection of full medical history and multiple neurological examinations, tests such as magnetic resonance imaging, electromyography, and electroencephalography are performed. Where established, tests for a single, or in some cases a limited set, of biochemical or genetic biomarker(s) are performed. Conventional biochemical tests include immunoassay (e.g., enzyme-linked immunosorbent assay, ELISA, for amyloid-β peptides in Alzheimer's disease, AD) and enzymatic assay (e.g., hexosaminidase A assay for Tay–Sachs). Using conventional genetic tests, gene mutation is determined using PCR (e.g., allele-specific PCR for Tay–Sachs) or RT-PCR (e.g., retroviral detection). The disadvantages of conventional immunoassay include poor precision, time required, and difficulty automating the techniques. Though highly sensitive, one disadvantage of PCR-based methods is the reliance upon polymerase enzyme function, which can be costly and reduce reliability of the system.[2] Immunosensors, microarrays, and nonenzymatic amplification techniques are expected to markedly reduce these disadvantages in diagnosing disease, including ND. Not only do biosensors currently serve as diagnostics, they also enable tissue-derived and genome-based biomarker discovery that can identify the components of complex disease signatures.

Handbook of Biosensors and Biochips. Edited by Robert S. Marks, David C. Cullen, Isao Karube, Christopher R. Lowe and Howard H. Weetall.
© 2007 John Wiley & Sons, Ltd. ISBN 978-0-470-01905-4.

For the health care provider, the process of ND determination is complex. Measurements of analytes in biological fluids are routinely performed at locations outside of the physician's office or hospital. Conventional laboratory testing would be complemented by rapid sensor-based methods. For the patient, availability of biosensors could greatly facilitate disease management. In many cases, biomarkers exist in the blood and urine that can indicate state of progression, recovery, or response to medication.[3] Unlike glucose, pregnancy, and HIV home test kits, there are currently no products on the market that allow a patient to track the state of neurological health. Likewise, there is no health care industry standard for clinical biomarker profiling of ND.

2   DISCUSSION

2.1   Biosensors for Neurological Disease

Biosensors are analytical devices composed of a high-affinity recognition probe of biological origin and a transducer, which converts chemical signals to electrical or optical outputs. In general, all sensors are designed to achieve one or more of the following goals: lowering the limits of detection, parallel detection of multiple analytes or signals (multiplexing), and signal amplification by several orders of magnitude.

Current disease biosensors, including several platforms that have been used for ND analyte sensing, are capable of binding genetic targets or biochemical analytes in saliva, urine, blood, CSF, or tissue of interest. The concept of creating disease biomarker signatures involves testing for multiple targets, some of which may hold high correlation to a specific disease and some of which may indicate general nervous system involvement but do not identify a specific illness. Negative controls must be included on the signature platform to rule out false positives. Table 1 is a selected list of potential targets for disease signatures. Some of the targets are already used in ND testing, while others have been identified but are not yet incorporated into clinical testing.

There is a current requirement for separate platforms for genomic and proteomic or biochemical biosensing. The specificity and sensitivity of genomic sensing is dependent upon the thermal dissociation of the hydrogen bonding between the bases of nucleic acid. The specificity and sensitivity of biochemical sensing in the case of immunological-based sensing is dependent upon the dissociation constant, or K_d, of the probe–analyte complex. This complex is dependent not only upon hydrogen bonding but also upon electrostatic forces, π-bonding, and van der Waals forces.

Patient sensitivity to and tolerance of treatment are crucial to selecting the proper course of care for individuals with NDs associated with significant pain and inflammation. Understanding the mechanisms behind cancer and ND symptoms and correlating biomarkers to pain, stress, and inflammation can advance current treatment modalities. Many biomarkers for pain susceptibility and chronic inflammation have been identified including interleukins (ILs), interferon-α, and tumor necrosis factor (TNF).[14,56,57] Activation of these pathways is involved in the pathogenesis of many complex diseases including NDs, cancer, and cardiovascular disease.

Until recently, the genes associated with cytokine production were not well understood. In addition to the long list of cytokines that were known to be involved in pain and inflammation, a recent study has identified a gene associated with cytokine production. The selenoprotein S (SEPS1) gene (also called *SELS*, *SELENOS*, or *VIMP*) was identified as a mediator of inflammation.[58] SEPS1 polymorphisms were identified and strongly associated (multivariate $p = 0.0000002$) with elevated levels of the cytokines IL-6, IL-1-β, and TNF-α in plasma. This knowledge will allow for sensors to determine gene polymorphisms associated with pain and inflammation, elucidating the risk of stress susceptibility. Table 1 includes a brief list of known stress susceptibility markers. Pain susceptibility when combined with ND biomarker data can provide a powerful tool for creating individual patient susceptibility profiles. Personalized medicine can be realized on diverse, multiplexed platforms with technology that is commercially available.

2.2   Biomarker Overlap

Supporting the concept of disease signatures is the common event in which any one biomarker exists

Table 1. Biomarker targets for neurological disease, and stress (pain/inflammation) susceptibility

General category of disorder, incidence	Primary cause of neurological disorder, incidence	Biochemical biomarker	Genetic biomarker	Sample source
Degenerative	AD (CNS), 120/100 000, over 60 years of age[4]	Tau protein, phospho-tau[5]		CSF
		ApoE[5]		Serum
		Aggregated β-amyloid peptide[5,6]		Serum, CSF
			APP mutations[5]	Serum
			Presenilin-1 and -2 mutations[5]	Serum
	Parkinson's Disease (CNS), 100/100 000 over 65 years of age[4]	8-hydroxydeoxyguanosine[7]		Urine
		α-synuclein contained in Lewy bodies[5]		Brain histology
		Altered DAT[5]		Brain histology
		DARP[8]		CSF
			α-synuclein mutations[5]	Serum
			Parkin gene mutations[5]	Serum
			UCH-L1 mutations[5]	Serum
			NR4A2 mutations[5]	Serum
	Fragile X syndrome (CNS) 6/100 000 males[4]		FMR1 CGG repeats[9]	Serum
	HD (CNS) 0.5–5/100 000[4]		CAG repeats[5]	Serum
Autoimmune	Lupus (CNS, PNS), 122/100 000[a),10]	ANAb[11]		Serum
		Cardiolipin[11]		Serum
		IgG[12]		CSF in mice
		TRAIL[13]		Serum

(continued overleaf)

Table 1. (continued)

General category of disorder, incidence	Primary cause of neurological disorder, incidence	Biochemical biomarker	Genetic biomarker	Sample source
	MS (CNS), 10–80/100 000, dependent on geography, ethnicity, and environmental factors[4]	Myelin basic protein[15,16]		CSF, urine, cervical lymph nodes
		Oligoclonal IgG[17]		CSF
		Glycoprotein[17]		CSF
		Autoantibodies to the 20S proteasome[18]		Serum, CSF
		Anti-hnRNP-B1 antibody[19,20]		Serum, CSF
			HLA-DR mutations[21]	CSF, CNS, and thymic tissue
	Guillain–Barré syndrome (PNS), 1.5/100 000[4]	Anti-GQ1b ganglioside antibody in Miller–Fisher syndrome[22]		Serum
	Myasthenia gravis (PNS) 0.3/100 000[4]	Autoantibodies to AchRs[19]		Serum, CSF
		Autoantibodies to MuSK[19]		Serum, CSF
	PND (CNS)	ANA-Ma2[23] (for SCLC[19]), ANA-1 (anti-Hu), ANA-2 (anti-Ri), ANA-3[23]		Serum, CSF, histology
		PCA-1, PCA-2, PCA-Tr, CRMP5-IgG, amphiphysin IgG[23]		Serum, CSF
		Anti-Rc antibody, for SCLC[24]		Serum, histology
		Progelatinase B/proMMP-9 correlating with CRP[25]		Pleural fluids
Cardiovascular accident, 100–300/100 000	Stroke/impeded vascular flow (CNS)	CK[26]		Serum
		cTnl[27]		Serum
	Myocardial infarction	CRP[28]		Serum
		IL-1Ra[29]		Serum
Seizure disorders, 44/100 000	Pyridoxine-dependent seizures (CNS)	Pipecolic acid[30]		Serum, CSF
Primary brain tumors, 15/100 000[31]	Astrocytoma (CNS), 6-12/100 000[31,32]	VEGF[33]		CSF
			ASCL1 (Drosophila; HASH 1)[34]	Serum, histology

Category	Disease	Biomarker	Sample source
	Glioblastoma multiforme (CNS), 7.5/100 000[31]	Cathepsin D[35]	Serum
		IGF-1[36]	Serum
		ANX7 for survivorship[37]	Histology
		MAPK, PKC, PLC-γ[38]	Histology
		Polysialic acid NCAM[39,40]	CSF, histology
		Antirecoverin[33]	Serum, CSF
	Medulloblastoma (CNS), 0.3/100 000 in adult[31]	Survivin[41]	CSF, histology
		BMI-1 oncogene pathway[42,43]	Histology
Infectious	Meningioma (CNS), 3/100 000[31]	Polysialylated NCAM[44] (PML)	Histology
	PML (CNS), 5000/100 000 in AIDS	JC virus[46,47]	CSF, histology
	Creutzfeldt-Jakob disease (CNS), 0.2/100 000	PrP[45]	Histology; CSF, urine[45]
		NSE, tau protein, amyloid-β, 14-3-3 protein[6]	CSF
Inborn errors of metabolism	Glycogen storage diseases (CNS)	PRKAG2 gene[48]	Serum
		LAMP2 gene[48]	Serum
	Tay–Sachs (CNS), 170/100 000 in Ashkenazi jews	HEXA gene variants[49]	Serum
Channelopathies, 1/100 000[31]	Hyperkalemic periodic paralyses (PNS)	(Hyperkalemic) CACNA1s, SCN4A genes[51]	Serum
	Chloride channel myopathy	(Chloride channel myopathy) CLCN1 gene[52]	Serum
	MD (PNS)	CK MM isoforms[50](MD)	Serum
Stress susceptibility	Manifested as inflammation, pain	IL-2, −6, −8, −10[53]	Serum
		TNF-α[39]	Serum
		IFN-γ[54]	Serum
		CRP[55]	Serum
		SEPS1 gene (also called SELS, SELENOS)[53]	Serum

ApoE: apolipoprotein E; APP: amyloid precursor protein; DAT: dopamine transporter; DARP: dopamine releasing protein; UCH-L1: ubiquitin carboxy-terminal hydrolase L1; NR4A2: nuclear receptor family; FMR1: fragile X mental retardation gene 1; ANAb: antineuronal nuclear antibody b; TRAIL: tumor necrosis factor–related apoptosis inducing ligand; HLA-DR: histocompatibility leukocyte antigen; AchRs: acetylcholine receptors; MuSK: muscle-specific receptor tyrosine kinase; PND: paraneoplastic neurological disorder; PCA-1: Purkinje cell cytoplasmic Ab type 1, anti-Yo; CRMP5: collapsin-response mediator protein; Anti-Rc: antirecoverin; CRP: C-reactive protein; CK: creatine kinase; cTn1: cardiac troponin 1; IL-1Ra: interleukin 1 receptor agonist; VEGF: vascular endothelial growth factor; ASCL1: Achaete–scute complex-like 1; IGF-1: insulin-like growth factor 1; ANX7: annexin VII; MAPK: mitogen-activated protein kinase; NCAM: neural cell adhesion molecule; PML: progressive multifocal leukoencephalopathy; PrP: prion protein; NSE: neuron-specific enolase; PRKAG2: associated membrane protein–adenosine monophosphate γ 2; LAMP2: lysosome-associated membrane protein 2; MD: muscular dystrophy; IFN: interferon.
(a)Excluding drug-induced lupus.

in multiple diseases or syndromes. For example, a tumor marker might be found across several different cancer etiologies. Interleukins are general markers for inflammation but are also indicators of pain susceptibility. Myelin basic protein is released during traumatic neural injury, but is also a strong indicator for multiple sclerosis (MS). These are examples where a single marker is not pathognomonic for a specific disease. The use of a multiplexed platform in conjunction with a robust statistical database can generate a profile of susceptibility to specific diseases.[1,59] Multiplexed sensors can be used to address the issue of biomarker overlap. Table 2 demonstrates selected examples where a particular biomarker appears changed in concentration across multiple diseases, including ND.

2.3 Currently Available Biosensor Technology

The function of biosensors for ND is twofold; they can be used for biomarker discovery and as diagnostic tools. The sensors discussed here include a broad range of technologies including commercially available kits and experimental proofs of principle. These sensors exhibit levels of detection comparable to or better than the conventional assay. Many procedures currently used for the diagnosis of ND in clinical laboratories involve multistep immunoassays. Because the immunoassay relies upon antibody (Ab)–antigen (Ag) binding, such tests are readily being converted to operate in biosensor systems such as Biacore® SPR (surface plasmon resonance), Roche Diagnostics' handheld therapeutic drug monitor, and other advanced Ab-based diagnostic tools. Extensive biomarker discovery is ongoing with the use of microarrays from companies such as Affymetrix® and Agilent (gene based) and Ciphergen (protein based). A brief list of biosensor companies and their platforms is included in Table 3. Many of the platforms listed in the table are discussed in more detail here.

2.3.1 Immunosensors

Immunosensors detect the binding event between Ab and Ag. Immobilized Abs or immunoglobulins are used to detect the presence of specific Ags,

Table 2. Biomarker overlap with corresponding diseases

Biomarkers	Corresponding diseases
α-synuclein contained in Lewy bodies	Parkinson's disease, DLB, multiple system atrophy[5]
Antirecoverin	Paraneoplastic neurological disorder (cancer-associated retinopathy,[60] SCLC[24]), glioma[33]
Cathepsin D	Glioblastoma,[35] colorectal cancer[61]
CRP	General marker of inflammation,[55] stroke,[28] multiple tumor etiologies[25]
IGF-1	Reduced levels in AD;[62] elevated levels in colorectal,[61] breast,[63] prostate,[64] lung[65] cancer
IgG and albumin in CSF	Breached blood–brain barrier due to CNS damage during systemic lupus erythematosus-like disease,[12] traumatic neural injury, MS[66]
MAPK family: PKC, PLC	Pancreatic cancer,[67] non-SCLC[68]
Myelin basic protein	MS, HIV, PML, HTLV-1 associated myelopathy,[69] traumatic neural injury
Polysialic acid NCAMS	Diseased muscles,[70] multiple tumors including medulloblastoma,[40] meningioma,[44] retinal cell carcinoma,[71] and choroid plexus tumors[72]
Survivin	Elevated in medulloblastoma[41] and pancreatic cancer[73]—negative prognostic marker of survival
VEGF	Elevated in many cancers including glioma,[33] colorectal, and lung adenocarcinoma[74]

HTLV-1: Human T-cell Lymphotropic Virus; DLB: dementia with Lewy bodies.

or conversely immobilized immunogenic peptides are used to detect the presence of specific Abs in biological fluids. These sensors minimize, and in some cases eliminate, time-consuming separation and washing steps and utilize the convenience of various physical transducers. Many immunosensors can be regenerated after immersion in acid (e.g., HCl, H_2SO_4), base (e.g., tetraethylamine), denaturant (e.g., urea), or high salt. The regenerated probes can be reused cost-efficiently.

A large class of immunosensors utilizes SPR for detection and quantitation of neurological biomarkers. For AD detection, a nanoscale optical biosensor based on SPR has been used for quantitative assay of the interaction between

Table 3. Biosensor platforms and technology

Company	Platform probe/technology	Target analyte type
Affymetrix®	GeneChip®: DNA-based microarray	Genetic
Agilent	DNA-based microarray, uses 60-mer probes	Genetic
Applied Biosystems	DNA-based microarray and protein biomarker discovery	Genetic, proteomic
Biobarcode	Nonenzymatic amplification	Genetic, proteomic, immunogenic, small molecules, complex targets
Cepheid	DNA-based microarray	Genetic
Ciphergen	ProteinChip: protein, peptide, and Ab microarrays	Proteomic, immunogenic
Luminex	xMAP technology: cDNA or peptide/protein/Ab-coated beads	Genetic, proteomic, immunogenic
Plexigen	GeneCube™, geneCard™: genetic-, proteomic-, and immunogenic-based	Genetic, proteomic, immunogenic
Roche Diagnostics	TDM technology: Ab-based handheld device	Drug analytes, potentially any Ag–Ab pair
Third Wave Technologies	Invader® technology: nucleic acid quantitation	Genetic

the amyloid-β-derived diffusible ligands (ADDLs) and specific anti-ADDL Abs.[75] Biacore® SPR systems have been used to characterize normal and expanded polyglutamine tracts to determine pathologic threshold for HD.[76] The HD trait marker is readily identifiable, but additional markers are needed for a better understanding of age of onset and course of the disease.

SPR is also widely used for detection of cancer biomarkers. Many examples exist for the detection of the prostate cancer marker, prostate-specific antigen (PSA), at nano- to femtomolar levels.[77,78] There are additional examples for microcantilever and amperometric detection of PSA.[79–81] Another cancer diagnostic model is the piezoelectric immunosensor array for clinical immunophenotyping of acute leukemia.[82] Regeneration of the probe was performed through denaturation of target in high-molar urea. In another example, picomolar levels of IL-8 were detected in human saliva in oropharyngeal cancer patients with less than 3% coefficient of variation using SPR.[83] IL-8 is an indicator of inflammation and is a general marker for cancer and stress susceptibility.

Ultrasonic radiation was tested for inexpensive regeneration of an Ab-based immunosensor for breast cancer Ag.[84] Capable of dissociating the Ab–Ag pair for reuse of the probe, ultrasonic radiation does not require use of strong acids or bases. The same research group has done work on Ab-based multianalyte systems, which performed simultaneous detection of multiple targets.[85] Often termed *biochips*, many versions of this technology exist. One example is a fluorescence-based nine-analyte array sensor that detected both pathogenic

bacteria and bacteria-derived Ags on one platform "without significant cross-reactivity".[86]

SPR has been used to detect complex targets as well. Efficient and selective capture of peptide-labeled metastatic epithelial cancer cells from flowing blood was recently demonstrated using SPR.[87] When immobilized on appropriate surfaces, these peptides could be used in both in vivo and ex vivo cell separation devices. Selective peptides that bind viruses, bacteria, or cells involved in the ND process could be incorporated into the SPR system. Conceivably, highly selective cost-effective aptamers could replace peptides in this application.

Roche Diagnostics currently offers an FDA-approved, patented electrochemical immunosensor technology for therapeutic drug monitoring (TDM) in point-of-care testing (POCT) (http://www.fuentek.com/technologies/TDM2.htm).[88] Designed as a plug-in for a handheld device, immobilized Abs are used to detect drug concentrations in biological samples (e.g., saliva, urine, serum). Because the detection is Ab-based, the Roche TDM technology is readily adaptable to Ab detection of ND biomarkers.

2.3.2 Microarrays

Protein- or peptide-based chips have been used for biomarker discovery. Ciphergen Biosystems, Inc. offers the ProteinChip platform for biomarker discovery and analysis. Extensive biomarker panels have been identified for determination of AD using this platform.[89,90] Ciphergen protein arrays have identified a novel panel of early-stage AD biomarkers from CSF samples using surface enhanced laser desorption/ionization time-of-flight (SELDI-TOF) mass spectrometry. Twenty-seven total markers were characterized using five different solid phase chemistries.[90] A multimarker prototype assay of four markers was able to distinguish mild AD from control individuals with high sensitivity and specificity in a blinded test set.

Luminex bead-based array systems allow for multiplexed detection and quantification of analytes using both nucleic acid (e.g., cDNA) and protein (e.g., Ab) complexed probes. Research using Luminex systems characterized 756 sera from cervical cancer patients. There are more than 100 different human papillomaviruses (HPVs), many of which cause proliferative diseases such as cervical cancer. In this study, simultaneous detection of Abs against 27 in situ affinity-purified recombinant HPV proteins was accomplished.[91] The Luminex xMAP® technology specifications indicate that the assay has the capacity for testing up to 100 different Ags in parallel per well of reaction. In May of 2005, Luminex Corporation announced that it has teamed with Zeus Scientific to offer multiplex autoimmune disease test kits.

SuperArray Bioscience Corporation provides high-throughput gene expression profiling featuring Plexigen's geneCube™ technology, which utilizes stacked detection arrays. The arrays can be functionalized with nucleic acids, proteins, or Abs and can be used to screen for a wide variety of targets in biological fluids. The technology can analyze hundreds of thousands of samples in days. For example, 30 geneCubes can measure the expression of "200 genes in 10 000 samples in less than 2 months" (http://www.plexigen.com/technology. htm).[92]

Affymetrix® microarrays are the dominant nucleic acid–based array used for multigene expression profiling. A compelling example of how current array capabilities can provide genetic profiling for cancer was demonstrated by Glinsky et al.[59] The research group identified an 11-gene probability-of-cancer signature based on the BMI-1 oncogene pathway. An Affymetrix® microarray was used to test for the expression of these cancer-related genes. When combined with mathematical analysis of patient therapy outcome data, the authors were able to correlate the 11-gene signature to probability of survival. Furthermore,

they were able to do so across 11 different types of cancer. This same strategy is being applied to AD and other NDs to identify disease signatures. The Affymetrix® GeneChip® was used to identify gene expression changes following traumatic spinal cord injury in rats.[93] With the help of The MathWorks, Inc. (MATLAB) algorithms to analyze the extensive data sets, a large panel of genes including caspase, multiple cathepsins, and IL-6 was identified.

2.3.3 Amplification and Quantification Technologies

One problem challenging conventional detection of early disease states is the low concentration of biomarkers in biological fluids. Many of the techniques discussed here are highly sensitive and selective, yielding impressive levels of detection (nano- to femtomolar). Directly addressing the issue of detection levels is the biobarcode system. Essentially a nonenzymatic amplification method, the biobarcode assay developed by Chad Mirkin et al. at Northwestern, has shown promise as a fast, sensitive, inexpensive, enzyme-free method for detecting tiny amounts of nucleic acid, protein, small molecules, and metal ions.[94] While previous less-invasive detection of pathogenic AD markers in CSF has not been possible due to low biomarker concentrations (<1 picomolar), the barcode assay was able to detect the amyloid-β-derived AD marker in CSF at 100 fM–100 aM. Notably, the biobarcode assay when combined with a PCR step to amplify signal can detect PSA at 6 orders of magnitude higher sensitivity than conventional ELISA (3 aM vs 3 pM, respectively).[95] The barcode assay has also been used for detection of the AD biomarker tau protein. Biochemical markers previously analyzed through invasive tumor biopsy or postmortem brain histology (e.g., ADDLs, prion protein PrP) might now be detectable in urine, blood, or CSF as a result of more sensitive detection like the biobarcode assay and genomics (e.g., SEPS1 gene for cytokine expression).[58,75,94]

Quantitation of RNAs can determine viral titer, disease progression, and therapeutic efficacy.[96] The Third Wave Technologies Invader® assay has been developed for detection of nucleic acids. The RNA Invader uses fluorescence resonance energy transfer (FRET)-based signal amplification for detecting RNA.[97] The assay can detect fewer than 100 copies of RNA per reaction and discriminate among 95% homologous sequences (1 in $\geq 20\,000$). In biplex reaction format, the RNA Invader can analyze simultaneous expression of two genes in the same sample. The DNA Invader can be used for genotyping and gene expression monitoring without the target amplification steps required in conventional sequencing and other methods. The Invader assay has been used to monitor gene expression of detoxification enzymes including several cytochrome P450 variants and multidrug resistant enzyme (MDR1). Expression of these drug metabolizers and transporters can cause drug–drug interactions or determine efficacy of a drug treatment.[98] Rapid methods for determining drug potential have been explored with the Invader assay by examining neuronal differentiation biomarkers of treated pheochromocytoma cells.[99] The assay has also been used to examine expression levels of inflammatory biomarkers including IL-10 and TNF-α[100] and degenerative disease biomarkers including the AD-associated apolipoprotein E.[101] The Invader technology is offered as "accurate, simple to use, scalable, and cost-effective".[102]

One of the concerns with sensor technology is the time lapse between acquisition of sample and data output. All of the platforms described here can provide data output within 1 h from the application of sample onto the sensor device. Generally, the most time-consuming step is sample preparation for the device, amplification of genomic sequences, or releasing the ligand to be detected from the sample. This process may require several hours. Additional time is needed for interpretation of the data output from the device and statistical analysis. This interpretation is based upon comparing the signature elements with a database correlating a signature with a differential diagnosis. In addition, the data has to be correlated with the clinical presentation of the patient and with prior medical history. We believe that a diagnosis cannot be determined based on a single laboratory test, but must involve correlations with the clinical presentation.

2.4 Market Demand

Biosensors in general and disposable sensors in particular offer utility over traditional clinical laboratory testing methods by permitting (i) rapid indication of clinical state, (ii) determination of the progression of illness, (iii) determination of a patient's vulnerability/resilience to disease and corresponding treatment, and (iv) private self-assessment by a patient. All of these factors affect the market demand for medical sensors. Sensors that meet market potential include glucose, pregnancy, and HIV tests. The costs of the devices range from $2 (pregnancy test strip) to around $100 (glucose monitor). An individual glucose test costs only about 10¢. Not only are the sensor products available over the counter but they also provide rapid information of clinical state to the patient and health care professional. Similar to glucose monitors, ND sensors could complement disease management by monitoring effects of diet, exercise, and medication. Ideally, the tests would require only microliters of blood or small urine sample, limiting discomfort and allowing for easy assessment in POCT. Even in cases where a CSF sample is required for detection of ND analytes, the process can be faster and more sensitive than conventional testing.

Using the diabetes/glucose monitor as a comparison template, an estimated 6–12 million persons in the United States have type II diabetes.[103] An estimated 360 000 people have AD and over 300 000 people have Parkinson's in the United States alone (Table 1). Many of these degenerative NDs have late onsets (over 50 years of age) and the baby boom generation now fills this age-group. This represents a significant economic driver for investment in biosensors for NDs. The market funding for pharmaceutical research and development in the central nervous system (CNS) disorder area approximates the funding for cancer and endocrine disorders and is estimated to become $15 billion by 2007.[104] These areas receive about double the funding of cardiovascular disorders and infectious diseases.

Factors affecting investment include the large number of persons affected by ND; the cost of testing; the cost of technology which involves the availability of reusable or disposable devices; the need for disease detection, diagnosis, and management (POCT); the need for personalized medicine (e.g., sensors for distinguishing proper morphine dosage due to the great variance in individual sensitivity); and continued discovery of biomarkers and development of disease signatures.

3 CONCLUSION

Several neurological disorders, whether autoimmune or degenerative, have a variable course during progression of the disease. MS can follow a course of exacerbations and remissions, each with differing lengths of time; it can also follow a steadily progressive course. Signature elements can distinguish these patterns. HD exhibits a variable age of onset depending, in part, upon the number of triplet repeats in the genome. The progression of AD and Parkinson's is highly variable. Parkinsonism exhibits the most variability in symptoms, rate of progression, and etiology. Sensor systems based on biomarkers highly correlated to disease states would assist in prognosis of and clinical care for these illnesses.

The decoding of the human genome now permits an analysis of the relationship between phenotypic expressions of ND and the genomic correlates of that disease. Similarly correlates between individual responses to drugs and effectiveness of the drugs in treating specific diseases are now possible. For example, genes that encode detoxifying enzymes (e.g., cytochrome P450, glutathione S-transferase) and multidrug resistant pumps have been shown to affect patient response to specific therapeutic drugs and stress. Biosensor systems allow for the definition of genomic and proteomic signatures necessary for improved determination of disease susceptibility, prognosis, and treatment. Proper application of available technologies can allow for management of ND and aid in the selection of therapeutics based upon the individual genome. The age distribution of the US population in the coming two decades will significantly increase the need for ND management and treatment, and therefore the need for sensor platforms.

REFERENCES

1. X. Wang, J. Yu, A. Sreekumar, S. Varambally, R. Shen, D. Giacherio, R. Mehra, J. E. Montie, K. J. Pienta, M. G. Sanda, P. W. Kantoff, M. A. Rubin, J. T. Wei, D. Ghosh, and A. M. Chinnaiyan, Autoantibody signatures in prostate cancer. *New England Journal of Medicine*, 2005, **353**(12), 1224–1235. Comment in: *New England Journal of Medicine*, 2005, **353**(12), 1288–1290.

2. J. Lin and H. Ju, Electrochemical and chemiluminescent immunosensors for tumor markers. *Biosensors and Bioelectronics*, 2005, **20**(8), 1461–1470.

3. A. Szilvas, A. Blazovics, G. Szekely, E. Dinya, J. Feher, and G. Mozsik, Comparative study between the free radicals and tumor markers in patients with gastrointestinal tumors. *Journal of Physiology-Paris*, 2001, **95**, 247–252.

4. M. Victor and A. H. Ropper, *Principles of Neurology*, 7th Edn, Mcgraw-Hill, 2001.

5. V. Rachakonda, T. H. Pan, and W. D. Le, Biomarkers of neurodegenerative disorders: how good are they? Review. *Cell Research*, 2004, **14**(5), 347–358.

6. B. Van Everbroeck, J. Boons, and P. Cras, Cerebrospinal fluid biomarkers in Creutzfeldt-Jakob disease. Review. *Clinical Neurology and Neurosurgery*, 2005, **107**(5), 355–360, Epub 2005 Jan 12.

7. S. Sato, Y. Mizuno, and N. Hattori, Urinary 8-hydroxydeoxyguanosine levels as a biomarker for progression of Parkinson disease. *Neurology*, 2005, **64**(6), 1081–1083.

8. L. Verdugo-Diaz, C. Morgado-Valle, G. Solis-Maldonado, and R. Drucker-Colin, Determination of dopamine-releasing protein (DARP) in cerebrospinal fluid of patients with neurological disorders. *Archives of Medical Research*, 1997, **28**(4), 577–581.

9. Y. Curlis, C. Zhang, J. J. Holden, P. K. Loesch, and R. J. Mitchell, Haplotype study of intermediate-length alleles at the fragile X (FMR1) gene: ATL1, FMRb, and microsatellite haplotypes differ from those found in common-size FMR1 alleles. *Human Biology*, 2005, **77**(1), 137–151.

10. K. M. Uramoto, C. J. Michet Jr, J. Thumboo, J. Sunku, W. M. O'Fallon, S. E. Gabriel, Trends in the incidence and mortality of systemic lupus erythematosus, 1950–1992. *Arthritis and Rheumatism*, 1999, **42**(1), 46–50.

11. R. A. Sinico, B. Bollini, E. Sabadini, L. Di Toma, A. Radice, The use of laboratory tests in diagnosis and monitoring of systemic lupus erythematosus. Review. *Journal of Nephrology*, 2002, **15**(Suppl. 6), S20–S27.

12. M. M. Sidor, B. Sakic, P. M. Malinowski, D. A. Ballok, C. J. Oleschuk, and J. Macri, Elevated immunoglobulin levels in the cerebrospinal fluid from lupus-prone mice. *Journal of Neuroimmunology*, 2005, **165**(1–2), 104–113.

13. V. Rus, V. Zernetkina, R. Puliaev, C. Cudrici, S. Mathai, and C. S. Via, Increased expression and release of functional tumor necrosis factor-related apoptosis-inducing ligand (TRAIL) by T cells from lupus patients with active disease. *Clinical Immunology*, 2005, **117**(1), 48–56.

14. C. L. Raison, M. Demetrashvili, L. Capuron, and A. H. Miller, Neuropsychiatric adverse effects of interferon-alpha: recognition and management. *CNS Drugs*, 2005, **19**(2), 105–123.

15. J. Palace, Clinical and laboratory characteristics of secondary progressive MS. Review. *Journal of the Neurological Sciences*, 2003, **206**(2), 131–134.

16. B. O. Fabriek, J. N. Zwemmer, C. E. Teunissen, C. D. Dijkstra, C. H. Polman, J. D. Laman, and J. A. Castelijns, In vivo detection of myelin proteins in cervical lymph nodes of MS patients using ultrasound-guided fine-needle aspiration cytology. *Journal of Neuroimmunology*, 2005, **161**(1–2), 190–194.

17. S. A. McMillan, G. V. McDonnell, J. P. Douglas, and S. A. Hawkins, Evaluation of the clinical utility of cerebrospinal fluid (CSF) indices of inflammatory markers in multiple sclerosis. *Acta Neurologica Scandinavica*, 2000, **101**(4), 239–243.

18. A. Storstein, A. Knudsen, and C. A. Vedeler, Proteasome antibodies in paraneoplastic cerebellar degeneration. *Journal of Neuroimmunology*, 2005, **165**(1–2), 172–178.

19. K. Endo, Advances in neuroimmunological laboratory studies on neuromuscular diseases. Review Japanese. *Rinsho Byori*, 2004, **52**(9), 759–766.

20. E. Sueoka, M. Yukitake, K. Iwanaga, N. Sueoka, T. Aihara, and Y. Kuroda, Autoantibodies against heterogeneous nuclear ribonucleoprotein B1 in CSF of MS patients. *Annals of Neurology*, 2004, **56**(6), 778–786.

21. E. Prat, U. Tomaru, L. Sabater, D. M. Park, R. Granger, N. Kruse, J. M. Ohayon, M. P. Bettinotti, and R. Martin, HLA-DRB5*0101 and -DRB1*1501 expression in the multiple sclerosis-associated HLA-DR15 haplotype. *Journal of Neuroimmunology*, 2005, **167**(1–2), 108–119.

22. K. Paparounas, Anti-GQ1b ganglioside antibody in peripheral nervous system disorders: pathophysiologic role and clinical relevance. *Archives of Neurology*, 2004, **61**(7), 1013–1016.

23. S. J. Pittock, T. J. Kryzer, and V. A. Lennon, Paraneoplastic antibodies coexist and predict cancer, not neurological syndrome. *Annals of Neurology*, 2004, **56**(5), 715–719.

24. A. V. Bazhin, M. S. Savchenko, O. N. Shifrina, S. A. Demoura, S. Y. Chikina, G. Jaques, E. A. Kogan, A. G. Chuchalin, and P. P. Philippov, Recoverin as a paraneoplastic antigen in lung cancer: the occurrence of anti-recoverin autoantibodies in sera and recoverin in tumors. *Lung Cancer*, 2004, **44**(2), 193–198.

25. J. Kotyza, M. Pesek, V. Bednarova, M. Terl, and B. Werle, Pleural fluids associated with metastatic lung tumors are rich in progelatinase B/proMMP-9. *Neoplasma*, 2005, **52**(5), 388–392.

26. T. Q. Kong, C. J. Davidson, S. N. Meyers, J. T. Tauke, M. A. Parker, and R. O. Bonow, Prognostic implication of creatine kinase elevation following elective coronary artery interventions. *JAMA*, 1997, **277**(6), 461–466.

27. E. Di Angelantonio, M. Fiorelli, D. Toni, M. L. Sacchetti, S. Lorenzano, A. Falcou, M. V. Ciarla, M. Suppa, L. Bonanni, G. Bertazzoni, F. Aguglia, and C. Argentino, Prognostic significance of admission levels of troponin I in patients with acute ischaemic stroke. *Journal of Neurology Neurosurgery and Psychiatry*, 2005, **76**(1), 76–81.

28. L. Masotti, E. Ceccarelli, S. Forconi, and R. Cappelli, Prognostic role of C-reactive protein in very old patients with acute ischaemic stroke. *Journal of Internal Medicine*, 2005, **258**(2), 145–152.

29. G. Patti, S. Mega, V. Pasceri, A. Nusca, G. Giorgi, E. M. Zardi, A. D'Ambrosio, A. Dobrina, and G. Di Sciascio, Interleukin-1 receptor antagonist levels correlate with extent of myocardial loss in patients with acute myocardial infarction. *Clinical Cardiology*, 2005, **28**(4), 193–196.

30. B. Plecko, C. Hikel, G. C. Korenke, B. Schmitt, M. Baumgartner, F. Baumeister, C. Jakobs, E. Struys, W. Erwa, and S. Stockler-Ipsiroglu, Pipecolic acid as

31. a diagnostic marker of pyridoxine-dependent epilepsy. *Neuropediatrics*, 2005, **36**(3), 200–205.

31. C. G. Goetz and E. J. Pappert, *Textbook of Clinical Neurology*, ISBN 0-7216-6423-7, Philadelphia, WB Saunders, 1999.

32. S. H. Landis, T. Murray, S. Bolden, and P. A. Wingo, Cancer statistics, 1999. *CA: A Cancer Journal for Clinicians*, 1999, **49**(1), 8–31, 1.

33. P. Sampath, C. E. Weaver, A. Sungarian, S. Cortez, L. Alderson, and E. G. Stopa, Cerebrospinal fluid (vascular endothelial growth factor) and serologic (recoverin) tumor markers for malignant glioma. *Cancer Control*, 2004, **11**(3), 174–180.

34. K. Somasundaram, S. P. Reddy, K. Vinnakota, R. Britto, M. Subbarayan, S. Nambiar, A. Hebbar, C. Samuel, M. Shetty, H. K. Sreepathi, V. Santosh, A. S. Hegde, S. Hegde, P. Kondaiah, and M. R. Rao, Upregulation of ASCL1 and inhibition of Notch signaling pathway characterize progressive astrocytoma. *Oncogene*, 2005, **24**(47), 7073–7083.

35. M. E. Fukuda, Y. Iwadate, T. Machida, T. Hiwasa, Y. Nimura, Y. Nagai, M. Takiguchi, H. Tanzawa, A. Yamaura, and N. Seki, Cathepsin D is a potential serum marker for poor prognosis in glioma patients. *Cancer Research*, 2005, **65**(12), 5190–5194.

36. L. A. Trojan, P. Kopinski, M. X. Wei, A. Ly, A. Glogowska, J. Czarny, A. Shevelev, R. Przewlocki, D. Henin, and J. Trojan, IGF-I: from diagnostic to triple-helix gene therapy of solid tumors. Review. *Acta Biochimica Polonica*, 2002, **49**(4), 979–990.

37. K. S. Hung and S. L. Howng, Prognostic significance of annexin VII expression in glioblastomas multiforme in humans. *Journal of Neurosurgery*, 2003, **99**(5), 886–892.

38. C. Mawrin, S. Diete, T. Treuheit, S. Kropf, C. K. Vorwerk, C. Boltze, E. Kirches, R. Firsching, and K. Dietzmann, Prognostic relevance of MAPK expression in glioblastoma multiforme. *International Journal of Oncology*, 2003, **23**(3), 641–648.

39. C. Dubois, D. Figarella-Branger, G. Rougon, and C. Rampini, Polysialylated NCAM in CSF, a marker for invasive medulloblastoma. French. *Comptes Rendus des Seances de la Societe de Biologie et de ses Filiales*, 1998, **192**(2), 289–296.

40. C. Dubois, A. Okandze, D. Figarella-Branger, C. Rampini, and G. Rougon, A monoclonal antibody against meningococcus group B polysaccharides used to immunocapture and quantify polysialylated NCAM in tissues and biological fluids. *Journal of Immunological Methods*, 1995, **181**(1), 125–135.

41. J. Pizem, A. Cort, L. Zadravec-Zaletel, and M. Popovic, Survivin is a negative prognostic marker in medulloblastoma. *Neuropathology and applied Neurobiology*, 2005, **31**(4), 422–428.

42. C. Leung, M. Lingbeek, O. Shakhova, J. Liu, E. Tanger, P. Saremaslani, M. Van Lohuizen, and S. Marino, Bmi1 is essential for cerebellar development and is overexpressed in human medulloblastomas. *Nature*, 2004, **428**(6980), 337–341.

43. G. V. Glinsky, O. Berezovska, and A. B. Glinskii, Microarray analysis identifies a death-from-cancer signature predicting therapy failure in patients with multiple

types of cancer. *Journal of Clinical Investigation*, 2005, **115**(6), 1503–1521.

44. D. Figarella-Branger, P. H. Roche, L. Daniel, H. Dufour, N. Bianco, and J. F. Pellissier, Cell-adhesion molecules in human meningiomas: correlation with clinical and morphological data. *Neuropathol Appl Neurobiol.*, 1997, **23**(2), 113–122.

45. H. Seeger, M. Heikenwalder, N. Zeller, J. Kranich, P. Schwarz, A. Gaspert, B. Seifert, G. Miele, and A. Aguzzi, Coincident scrapie infection and nephritis lead to urinary prion excretion. *Science*, 2005, **310**(5746), 324–326.

46. J. R. Berger andI. J. Koralnik, Progressive multifocal leukoencephalopathy and natalizumab-unforeseen consequences. *New England Journal of Medicine*, 2005, **353**(4), 414–416, Epub 2005 Jun 9.

47. E. O. Major, K. Amemiya, C. S. Tornatore, S. A. Houff, and J. R. Berger, Pathogenesis and molecular biology of progressive multifocal leukoencephalopathy, the JC virus-induced demyelinating disease of the human brain. Review. *Clinical Microbiology Reviews*, 1992, **5**(1), 49–73.

48. J. C. Roos and T. M. Cox, Glycogen storage diseases and cardiomyopathy. *New England Journal of Medicine*, 2005, **352**(24), 2553 author reply 2553. No abstract available.

49. M. Karpati, E. Gazit, B. Goldman, A. Frisch, R. Colombo, and L. Peleg, Specific mutations in the HEXA gene among Iraqi Jewish Tay-Sachs disease carriers: dating of founder ancestor. *Neurogenetics*, 2004, **5**(1), 35–40, Epub 2003 Nov 27.

50. X. Zhao, C. Bi, and Z. Yang, Studies on the creatine kinase MM isoforms of normal and Duchenne muscular dystrophic patients. *Chinese Medical Journal*, 1998, **111**(1), 75–77.

51. W. Y. Ng, K. F. Lui, A. C. Thai, and J. S. Cheah, Absence of ion channels CACN1AS and SCN4A mutations in thyrotoxic hypokalemic periodic paralysis. *Thyroid*, 2004, **14**(3), 187–190.

52. B. J. Simpson, T. A. Height, G. Y. Rychkov, K. J. Nowak, N. G. Laing, B. P. Hughes, and A. H. Bretag, Characterization of three myotonia-associated mutations of the CLCN1 chloride channel gene via heterologous expression. *Human Mutation*, 2004, **24**(2), 185.

53. J. E. Curran, J. B. Jowett, K. S. Elliott, Y. Gao, K. Gluschenko, J. Wang, D. M. Azim, G. Cai, M. C. Mahaney, A. G. Comuzzie, T. D. Dyer, K. R. Walder, P. Zimmet, J. W. Maccluer, G. R. Collier, A. H. Kissebah, and J. Blangero, Genetic variation in selenoprotein S influences inflammatory response. *Nature Genetics*, 2005, **37**(11), 1234–1241, Epub 2005 Oct 9.

54. T. Ek, M. Jarfelt, L. Mellander, and J. Abrahamsson, Proinflammatory cytokines mediate the systemic inflammatory response associated with high-dose cytarabine treatment in children. *Medical and Pediatric Oncology*, 2001, **37**(5), 459–464.

55. A. Y. Wang, Prognostic value of C-reactive protein for heart disease in dialysis patients. *Current Opinion in Investigational Drugs*, 2005, **6**(9), 879–886.

56. X. Chevalier, B. Giraudeau, T. Conrozier, J. Marliere, P. Kiefer, and P. Goupille, Safety study of intraarticular injection of interleukin 1 receptor antagonist in patients with painful knee osteoarthritis: a multicenter study. *Journal of Rheumatology*, 2005, **32**(7), 1317–1323.

57. S. Kopp, P. Alstergren, S. Ernestam, S. Nordahl, P. Morin, and J. Bratt, Reduction of temporomandibular joint pain after treatment with a combination of methotrexate and infliximab is associated with changes in synovial fluid and plasma cytokines in rheumatoid arthritis. *Cells Tissues Organs*, 2005, **180**(1), 22–30.

58. J. E. Curran, J. B. M. Jowett, K. S. Elliott, Y. Gao, K. Gluschenko, J. Wang, D. M. A. Azim, G. Cai, M. C. Mahaney, A. G. Comuzzie, T. D. Dyer, K. R. Walder, P. Zimmet, J. W. MacCluer, G. R. Collier, A. H. Kissebah, and J. Blangero, Genetic variation in selenoprotein S influences inflammatory response. *Nature Genetics*, 2005, **37**(11), 1234–1241.

59. G. V. Glinsky, O. Berezovska, and A. B. Glinskii, Microarray analysis identifies a death-from-cancer signature predicting therapy failure in patients with multiple types of cancer. *Journal of Clinical Investigation*, 2005, **115**(6), 1503–1521.

60. C. E. Thirkill, A. M. Roth, and J. L. Keltner, Cancer-associated retinopathy. *Archives of Ophthalmology*, 1987, **105**(3), 372–375.

61. E. Skrzydlewska, M. Sulkowska, A. Wincewicz, M. Koda, and S. Sulkowski, Evaluation of serum cathepsin B and D in relation to clinicopathological staging of colorectal cancer. *World Journal of Gastroenterology*, 2005, **11**(27), 4225–4229.

62. T. Watanabe, A. Miyazaki, T. Katagiri, H. Yamamoto, T. Idei, and T. Iguchi, Relationship between serum insulin-like growth factor-1 levels and Alzheimer's disease and vascular dementia. *Journal of the American Geriatrics Society*, 2005, **53**(10), 1748–1753.

63. J. V. Vadgama, Y. Wu, G. Datta, H. Khan, and R. Chillar, Plasma insulin-like growth factor-I and serum IGF-binding protein 3 can be associated with the progression of breast cancer, and predict the risk of recurrence and the probability of survival in African-American and hispanic women. *Oncology*, 1999, **57**(4), 330–340.

64. A. Wolk, S. O. Andersson, C. S. Mantzoros, D. Trichopoulos, and H. O. Adami, Can measurements of IGF-1 and IGFBP-3 improve the sensitivity of prostate-cancer screening? *Lancet*, 2000, **356**(9245), 1902–1903.

65. D. Olchovsky, I. Shimon, I. Goldberg, T. Shulimzon, A. Lubetsky, A. Yellin, C. Pariente, A. Karasik, and H. Kanety, Elevated insulin-like growth factor-1 and insulin-like growth factor binding protein-2 in malignant pleural effusion. *Acta Oncologica*, 2002, **41**(2), 182–187.

66. K. P. Johnson, Cerebrospinal fluid and blood assays of diagnostic usefulness in multiple sclerosis. Review. *Neurology*, 1980, **30**(7 Pt 2), 106–109.

67. J. M. Summy, J. G. Trevino, C. H. Baker, and G. E. Gallick, c-Src regulates constitutive and EGF-mediated VEGF expression in pancreatic tumor cells through activation of phosphatidyl inositol-3 kinase and p38 MAPK. *Pancreas*, 2005, **31**(3), 263–274.

68. K. T. Chang, C. M. Tsai, Y. C. Chiou, C. H. Chiu, K. S. Jeng, and C. Y. Huang, IL-6 induces neuroendocrine dedifferentiation and cell proliferation in non-small cell lung cancer cells. *The American Journal of Physiology—Lung Cellular Molecular Physiology*, 2005, **289**(3), L446–L453, Epub 2005 May 13.

69. M. Ohta, K. Ohta, M. Nishimura, and T. Saida, Detection of myelin basic protein in cerebrospinal fluid and serum from patients with HTLV-1-associated myelopathy/tropical spastic paraparesis. *Annals of Clinical Biochemistry*, 2002, **39**(Pt 6), 603–605.

70. D. Figarella-Branger, J. Nedelec, J. F. Pellissier, J. Boucraut, N. Bianco, and G. Rougon, Expression of various isoforms of neural cell adhesive molecules and their highly polysialylated counterparts in diseased human muscles. *Journal of the Neurological Sciences*, 1990, **98**(1), 21–36.

71. L. Daniel, C. Bouvier, B. Chetaille, J. Gouvernet, A. Luccioni, D. Rossi, E. Lechevallier, X. Muracciole, C. Coulange, and D. Figarella-Branger, Neural cell adhesion molecule expression in renal cell carcinomas: relation to metastatic behavior. *Human Pathology*, 2003, **34**(6), 528–532.

72. D. Figarella-Branger, H. Lepidi, C. Poncet, D. Gambarelli, N. Bianco, G. Rougon, and J. F. Pellissier, Differential expression of cell adhesion molecules (CAM), neural CAM and epithelial cadherin in ependymomas and choroid plexus tumors. *Acta Neuropathologica*, 1995, **89**(3), 248–257.

73. M. A. Lee, G. S. Park, H. J. Lee, J. H. Jung, J. H. Kang, Y. S. Hong, K. S. Lee, D. G. Kim, and S. N. Kim, Survivin expression and its clinical significance in pancreatic cancer. *BMC Cancer*, 2005, **5**, 127.

74. R. Cressey, O. Wattananupong, N. Lertprasertsuke, and U. Vinitketkumnuen, Alteration of protein expression pattern of vascular endothelial growth factor (VEGF) from soluble to cell-associated isoform during tumourigenesis. *BMC Cancer*, 2005, **5**, 128.

75. A. J. Haes, L. Chang, W. L. Klein, and R. P. Van Duyne, Detection of a biomarker for Alzheimer's disease from synthetic and clinical samples using a nanoscale optical biosensor. *Journal of the American Chemical Society*, 2005, **127**(7), 2264–2271.

76. M. J. Bennett, K. E. Huey-Tubman, A. B. Herr, A. P. West Jr, S. A. Ross, and P. J. Bjorkman, Inaugural article: a linear lattice model for polyglutamine in CAG-expansion diseases. *Proceedings of the National Academy of Sciences of the United States of America*, 2002, **99**(18), 11634–11639.

77. L. Huang, G. Reekmans, D. Saerens, J. M. Friedt, F. Frederix, L. Francis, S. Muyldermans, A. Campitelli, and C. V. Hoof, Prostate-specific antigen immunosensing based on mixed self-assembled monolayers, camel antibodies and colloidal gold enhanced sandwich assays. *Biosensors and Bioelectronics*, 2005, **21**(3), 483–490.

78. F. Yu, B. Persson, S. Lofas, and W. Knoll, Surface plasmon fluorescence immunoassay of free prostate-specific antigen in human plasma at the femtomolar level. *Analytical Chemistry*, 2004, **76**(22), 6765–6770.

79. J. H. Lee, K. S. Hwang, J. Park, K. H. Yoon, D. S. Yoon, and T. S. Kim, Immunoassay of prostate-specific antigen (PSA) using resonant frequency shift of piezoelectric nanomechanical microcantilever. *Biosensors and Bioelectronics*, 2005, **20**(10), 2157–2162.

80. K. W. Wee, G. Y. Kang, J. Park, J. Y. Kang, D. S. Yoon, J. H. Park, and T. S. Kim, Novel electrical detection of label-free disease marker proteins using piezoresistive self-sensing micro cantilevers. *Biosensors and Bioelectronics*, 2005, **20**(10), 1932–1938.

81. P. Sarkar, P. S. Pal, D. Ghosh, S. J. Setford, and I. E. Tothill, Amperometric biosensors for detection of the prostate cancer marker (PSA). *International Journal of Pharmaceutics*, 2002, **238**(1–2), 1–9.

82. H. Wang, H. Zeng, Z. Liu, Y. Yang, T. Deng, G. Shen, and R. Yu, Immunophenotyping of acute leukemia using an integrated piezoelectric immunosensor array. *Analytical Chemistry*, 2004, **76**(8), 2203–2209.

83. C. Y. Yang, E. Brooks, Y. Li, P. Denny, C. M. Ho, F. Qi, W. Shi, L. Wolinsky, B. Wu, D. T. Wong, and C. D. Montemagno, Detection of picomolar levels of interleukin-8 in human saliva by SPR. *Lab on a Chip*, 2005, **5**(10), 1017–1023.

84. M. C. Moreno-Bondi, J. Mobley, J. P. Alarie, and T. Vo-Dinh, Antibody-based biosensor for breast cancer with ultrasonic regeneration. *Journal of Biomedical Optics*, 2000, **5**(3), 350–354.

85. M. C. Moreno-Bondi, J. P. Alarie, and T. Vo-Dinh, Multi-analyte analysis system using an antibody-based biochip. *Analytical and Bioanalytical Chemistry*, 2003, **375**(1), 120–124.

86. C. R. Taitt, G. P. Anderson, B. M. Lingerfelt, M. J. Feldstein, and F. S. Ligler, Nine-analyte detection using an array-based biosensor. *Analytical Chemistry*, 2002, **74**(23), 6114–6120.

87. S. Aggarwal, S. Janssen, R. M. Wadkins, J. L. Harden, and S. R. Denmeade, A combinatorial approach to the selective capture of circulating malignant epithelial cells by peptide ligands. *Journal of Biomaterials*, 2005, **26**(30), 6077–6086.

88. Roche Diagnostics biosensor for therapeutic drug monitoring (TDM) in point-of-care testing (POCT); http://www.fuentek.com/technologies/TDM2.htm# techdet, 2006.

89. L. H. Choe, M. J. Dutt, N. Relkin, and K. H. Lee, Studies of potential cerebrospinal fluid molecular markers for Alzheimer's disease. *Electrophoresis*, 2002, **23**(14), 2247–2251.

90. O. Carrette, I. Demalte, A. Scherl, O. Yalkinoglu, G. Corthals, P. Burkhard, D. F. Hochstrasser, and J. C. Sanchez, A panel of cerebrospinal fluid potential biomarkers for the diagnosis of Alzheimer's disease. *Proteomics*, 2003, **3**(8), 1486–1494.

91. T. Waterboer, P. Sehr, K. M. Michael, S. Franceschi, J. D. Nieland, T. O. Joos, M. F. Templin, and M. Pawlita, Multiplex human papillomavirus serology based on in situ-purified glutathione s-transferase fusion proteins. *Clinical Chemistry*, 2005, **51**(10), 1845–1853.

92. Plexigen geneCube™ Technology, North Carolina, US; http://www.plexigen.com/technology.htm, 2006.

93. J. B. Aimone, J. L. Leasure, V. M. Perreau, M. Thallmair, and The Christopher Reeve Paralysis Foundation Research Consortium, Spatial and temporal gene expression profiling of the contused rat spinal cord. *Experimental Neurology*, 2004, **189**(2), 204–221.

94. D. G. Georganopoulou, L. Chang, J. M. Nam, C. S. Thaxton, E. J. Mufson, W. L. Klein, and C. A. Mirkin, Nanoparticle-based detection in cerebral spinal fluid of a soluble pathogenic biomarker for Alzheimer's disease.

Proceedings of the National Academy of Sciences of the United States of America, 2005, **102**(7), 2273–2276.

95. J. M. Nam, S. I. Stoeva, and C. A. Mirkin, Bio-barcode-based DNA detection with PCR-like sensitivity. *Journal of the American Chemical Society*, 2004, **126**(19), 5932–5933.

96. K. Ariyoshi, S. Jaffar, A. S. Alabi, N. Berry, M. Schim van der Loeff, S. Sabally, P. T. N'Gom, T. Corrah, R. Tedder, and H. Whittle, Plasma RNA viral load predicts the rate of CD4 T cell decline and death in HIV-2-infected patients in West Africa. *AIDS*, 2000, **14**(4), 339–344.

97. P. S. Eis, M. C. Olson, T. Takova, M. L. Curtis, S. M. Olson, T. I. Vener, H. S. Ip, K. L. Vedvik, C. T. Bartholomay, H. T. Allawi, W. P. Ma, J. G. Hall, M. D. Morin, T. H. Rushmore, V. I. Lyamichev, and R. W. Kwiatkowski, An invasive cleavage assay for direct quantitation of specific RNAs. *Nature Biotechnology*, 2001, **19**(7), 673–676. Erratum in: *Nature Biotechnology*, 2002, **20**(3), 307.

98. J. B. Mills, K. A. Rose, N. Sadagopan, J. Sahi, and S. M. de Morais, Induction of drug metabolism enzymes and MDR1 using a novel human hepatocyte cell line. *Journal of Pharmacology and Experimental Therapeutics*, 2004, **309**(1), 303–309.

99. P. Roux, I. Menguy, S. Soubigou, J. Chinn, S. Ricard, S. Williams, J. D. Guitton, T. Tian, S. Singh, and C. Grepin, Direct measurement of multiple mRNAs in nerve growth factor-induced PC12 cells using electrophoretic tags to monitor biomarkers of neuronal differentiation in 96-well format. *Assay and Drug Development Technologies*, 2004, **2**(6), 637–646.

100. P. Agarwal, M. C. Oldenburg, J. E. Czarneski, R. M. Morse, M. R. Hameed, S. Cohen, and H. Fernandes, Comparison study for identifying promoter allelic polymorphism in interleukin 10 and tumor necrosis factor alpha genes. *Diagnostic Molecular Pathology*, 2000, **9**(3), 158–164.

101. R. W. Kwiatkowski, V. Lyamichev, M. de Arruda, and B. Neri, Clinical, genetic, and pharmacogenetic applications of the invader assay. *Molecular Diagnosis*, 1999, **4**(4), 353–364.

102. M. de Arruda, V. I. Lyamichev, P. S. Eis, W. Iszczyszyn, R. W. Kwiatkowski, S. M. Law, M. C. Olson, and E. B. Rasmussen, Invader technology for DNA and RNA analysis: principles and applications. *Expert Review of Molecular Diagnostics*, 2002, **2**(5), 487–496.

103. CDC, Surveillance for Diabetes Mellitus—United States, 1980–1989. *Morbidity and Mortality Weekly Report*, 1993, **42**(SS-2), 1–20.

104. R. Steinbrook, Wall street and clinical trials. *New England Journal of Medicine*, 2005, **353**(11), 1091–1093.

FURTHER READING

K. Blennow, CSF biomarkers for Alzheimer's disease: use in early diagnosis and evaluation of drug treatment. *Expert Review of Molecular Diagnostics*, 2005, **5**(5), 661–672.

http://www.emedicine.com/emerg/topic334.htm.

http://www.emedicine.com/med/topic2692.htm.

http://www.emedicine.com/neuro/topic624.htm.

71

Utility of Biosensors in the Pharmaceutical Industry

Trevor Chapman,[1] Coulton Legge[2] and Ash Patel[3]

[1] *Drug Discovery (Neurodegeneration Research), GlaxoSmithKline, Harlow, UK*
[2] *Pharmaceutical Development, GlaxoSmithKline, Ware, UK and* [3] *Drug Discovery (Biopharmaceuticals), GlaxoSmithKline, Beckenham, UK*

1 INTRODUCTION

The pharmaceutical industry has undergone a rapid transformation over the last decade, as a result of many factors and influences. Much of this change has resulted from a change in the R&D process for discovering new drug molecules that is now more akin to scientific investigation than serendipitous good fortune; all of the low-hanging fruit drug molecules have been picked. While one discouraging consequence of this is that the average cost to bring a new drug to the market has now escalated to be greater than $1 billion and takes on the order of 10 years to reach the market, our understanding of the way in which diseases function is growing at an unprecedented rate. The process of discovering drugs, while complicated, can be broken down into several key development areas (Figure 1), each of which requires access to specific knowledge and tools. The range of scientific disciplines required through this process is extensive, requiring both multi- and interdisciplinary skills. So, while biological knowledge space is very much the new science of the pharmaceutical industry, it requires access to all other disciplines. This was evident, for example, in the speed with which the coding sequence of human DNA was unraveled. It relied not just on genetic understanding, but also on analytical instrumentation development to speed up the decoding of DNA segments. Also, high-throughput screening (HTS) developed through the introduction of high-density microtiter plates and breakthroughs in the ability to dispense low volumes of sample into them and then detect the assay results through recent developments in imaging and spectroscopy. Much of these advances can be directly attributed to the instrumentation, which enabled increased productivity and gave scientists tools that would no longer be the bottleneck in the process of drug discovery.

Biosensors are one of the main tools that are assisting our knowledge of disease and how molecules can interact with targets to solicit a positive pharmacological response in a label-free environment. They can play a part in many aspects of drug discovery (Figure 1), but they are of most immediate importance in the areas of secondary screening and in the monitoring of biopharmaceutical product development (red chevrons in Figure 1), both of which will be discussed in detail in this chapter. It should be noted that although much of the current focus of the use of biosensors in an industrial setting is in the later parts of drug discovery, this should not indicate that they are of little use in early research. The fact is that the most value can be obtained in their current setting, but

Handbook of Biosensors and Biochips. Edited by Robert S. Marks, David C. Cullen, Isao Karube, Christopher R. Lowe and Howard H. Weetall.
© 2007 John Wiley & Sons, Ltd. ISBN 978-0-470-01905-4.

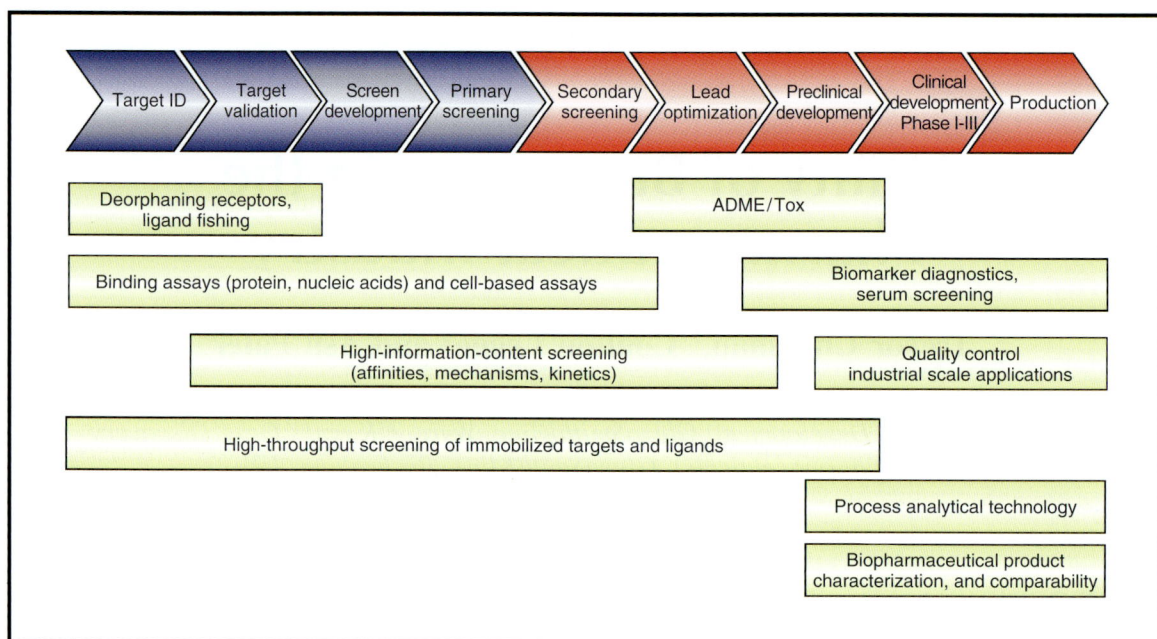

Figure 1. Overview of the drug discovery process within the pharmaceutical industry and where biosensors can be applied. The chevrons colored red represent the part of the drug discovery process that is discussed in this chapter. It should be noted that this does not indicate any scale in terms of time, as some processes can last several times longer than others.

with time, development, and success in their current application fields, they should become more utilized across the whole process. For example, biosensors could be developed for molecular interaction analysis in supporting pathway genomics, deorphaning receptors, and for characterizing the functionality of reagents within primary screens. These applications will only gain more support and uptake once biosensor technology becomes more mature in its current applications.

Primary (high-throughput) screening is performed to throw huge libraries of potential drug-like molecules against targets for disease. This results in a number of compounds generating a signal that can be considered an indication of action. In most cases this signal is derived from a fluorescent label, which while having a major impact on the productivity of screening, does add steps to the process. While this approach is very much qualitative in nature, it allows the number of molecules for subsequent analysis to be reduced from hundreds of thousands (even millions) to several thousands. This reduced set can then be investigated further to increase our quantitative data and knowledge in

what is referred to as *secondary screening*. By nature, secondary screening involves more focused work on the action between target and molecule and thus requires access to more detailed analytical tools, biosensors forming an essential part of them. An important feature in this step is to utilize label-free systems to ensure that the measured response is truly indicative of the biological system.

Another important growth area in the pharmaceutical industry has been the development of biological drug molecules, which are distinguished by being large and complex molecules rather than the more traditional small molecular weight drugs.

The manufacture of biopharmaceuticals has been a major development within the pharmaceutical industry over the last decade, requiring large-scale processing equipment to generate comparatively small quantities of end product. The impact of on-line and at-line analysis of these processes, which can vary dramatically under certain changes, has resulted in the application of biosensors to process analytical technology (PAT) to ensure quality production.

The following sections will detail the current requirements of secondary screening and biopharmaceutical manufacture and highlight areas of current biosensor application within them. Conclusions will be drawn at the end of this chapter on the current shortcomings of biosensors, to give the reader an indication of where future developments will need to be met to expand their application field throughout the drug discovery process.

2 BIOSENSORS IN DRUG DISCOVERY SECONDARY SCREENING

The mainstay for biosensor use in pharmaceutical discovery research has traditionally been in the field of antibody characterization and interaction analysis for larger biomolecules rather than small molecular weight druglike compounds. The expansion in biopharmaceutical research in recent years has secured the value of biosensors. As an example, the Biacore A100 has been used to screen for potential therapeutic antibodies with high affinity and potency (http://www.biacore.com). This relatively new array base instrument can monitor four unpurified monoclonal antibody supernatants individually over four antigen surfaces and a control surface at the same time, so reasonable numbers of samples can be screened fairly quickly. Maximum binding responses and off-rates are used for ranking potentially useful monoclonals, since the purity and concentration of the antibodies is unknown at an early stage in screening for monoclonal supernatants. After purification of selected clones an instrument like the Biacore S100 can be used to obtain full and comparative kinetic and affinity information. Furthermore, one can check that affinity is preserved after any modifications to the antibody, such as humanization of a mouse sequence to modify the Fc region to reduce immunogenicity and toxicity. Finally epitope mapping and mutational analysis can be performed on biosensors to back up patent claims or aim to further enhance a selected antibody's properties.

In recent years the sensitivity of certain optical biosensors has improved to the extent that small molecular weight druglike molecules, typically at around 400–500 Da, can now be detected when binding to immobilized therapeutic targets without the need for labels. Prior to this, drug binding was only monitored on label-free optical biosensors in less direct competition assays, where their interference on interactions between larger binding partners could be monitored, similar to the antagonist screen recently used by Vassilev et al.[1] Instruments such as the Biacore S51, and the earlier Biacore 3000, have been marketed toward drug screening applications.[2,3] However, it is not only sensitivity that is important in drug screening. Assay throughput is also crucial when testing large libraries of compounds. The above-mentioned systems utilize a flow delivery system across 2–4 channels, and therefore compounds are serially tested with a throughput of up to 384 tests per day. This figure is dependent on each compound test cycle time and may be reduced, for example, if time-consuming surface regeneration steps are required. Clearly this throughput is insufficient for current random compound HTS campaigns that may run over 1 million compounds. With an example hit rate of 1%, even full secondary screening would be unfeasible for 10 000 compounds. However, the information-rich data provided by screening compounds with biosensors such as the Biacore S51 (affinities, kinetics, stoichiometries of binding) can provide added value in the selection of the best candidates for further study from a more manageable number of random primary hits. Alternatively, and more usually the case, biosensors are applied to screening project compounds generated by chemists exploring structure–activity relationships around a particular chemical template.

A number of other biosensor manufacturers are beginning to achieve the sensitivity obtained by Biacore and required for compound binding detection. For example, the SRU BIND™ (http://www.srubiosystems.com) and the Corning Epic™ (http://www.corning.com) are two optical biosensors that are plate based and have the potential of testing up to 384 compounds in parallel. Certainly, early work on the SRU BIND™ system has shown promise in providing compound binding affinities from equilibrium data.[4] Akubio (http://www.akubio.com) have also been developing a multichannel biosensor that can detect compound binding and relies on an acoustic rather than optical technology for detection. It can be expected that other label-free biosensor technologies will emerge in the near future, aimed at the lucrative yet competitive drug screening market.

All these systems, however, will need good automated data handling packages for correct and fast interpretation of results.

If biosensors are to replace more conventional technologies used in drug discovery and compound secondary screening, then they must possess a number of advantages that will drive their increased use. One advantage that biosensor assays often have stems from their simplicity. They are often label-free assays with few components. Without the need to design and test relevant detection labels they can often be relatively quick to set up and modify if necessary. Another advantage they hold is in the elimination of false positive compound "hits" from primary HTS screens. These can originate from compound color quenching in fluorescent assays, for example. Biophysical detection systems used in many biosensors are often less susceptible to these false positives. False positives can also originate from nonspecific binding or even chemical interference with HTS assay labels and secondary reagents. Biosensor binding data can often eliminate these false positives and be used to select specific binders to target proteins rather than to other components of the primary assay. It is often possible to mimic a primary screen assay on a biosensor surface with the minimum of components, and at the same time, by exploiting the multichannel nature of many biosensors, include off-target components to which binding may be undesirable. Off-target components may for example, be nucleic acid used in a nucleic acid–associated protein target primary screen. Another off-target activity that could be excluded in parallel biosensor screening is drug binding to similar protein family members, which may lead to a toxicity liability. Indeed the same similar protein can provide an indication of a nonspecific binding mode of action and be used to eliminate detergents and "promiscuous" inhibitors. Biosensors directly measuring mass of binding components can often distinguish between "clean" interactions with a binding stoichiometry of 1 : 1 and "dirty" interactions from multisite or adsorbing nonspecific binders, detergents, DNA intercalators, and even insoluble compound aggregates coming out of solution. The later interactions can be identified by unexpectedly high maximum binding responses and complex kinetic data.

A flow system biosensor, such as the Biacore S51, can provide affinity and kinetic information on drug binding and this can be applied to build up a quantitative structural and functional relationship (QSAR) within a single compound template series under test, as used by Markgren et al.[5] for HIV protease inhibitors. However, affinity data is usually the driver in selecting the most potent compounds for further development, and occasionally, kinetic information may become increasingly important. For example, if the pharmacokinetic data suggests a short half-life for a compound series, then a long off-rate may be key to success. In reality, the evidence to support such use of kinetic data alone is not usually available at such an early stage in drug discovery. Recently thermodynamic data was also used by Shuman et al.[6] to characterize HIV protease inhibitors, but in a similar way the necessary structural information is rarely available for newly identified drug targets.

Published examples of the use of biosensors for drug screening are increasing and have recently been reviewed by Myszka,[7] Rich,[8] and Comley.[9] Although these examples are often on more stable targets and soluble compound series, the information provides an insight into the diversity of drug target assays that can be developed. Inhibitors to enzymes such as carbonic anhydrase II, HIV protease, and thrombin have been studied.[4,10,11] The success of kinase assays is increasing with careful immobilization strategies.[12,13] Screens involving proteins or drugs directly binding to nucleic acids are also possible.[14–16] Receptor protein assays have been possible with some purified proteins[1,17] and a few brave attempts have been made to measure direct compound binding to receptors expressed in their natural membrane environment.[18] Lastly, a number of papers have focused on measuring drugs binding to the serum proteins albumin and $\alpha(1)$-acid glycoprotein[19] and on drugs penetrating lipid membranes,[20] although biosensors are yet to routinely replace existing methods, except when additional, structural, or kinetic information is needed.

As yet biosensors do not play a dominant role in the secondary drug screening environment and this is because of a number of limitations with present technology. For the most developed systems such as the Biacore S51 that have achieved the sensitivity required for small molecule work, automation for compound screening and the comparative cost in protein and compound use per assay is not a hurdle. Limited throughput is a problem and this has been referred to earlier. Serial testing of compounds over the same target protein surface can create a second problem. With random compound sets under test, it is often inevitable that one nonspecific binder will irreversibly damage the test surface and spoil the remainder of a screen run. Array or plate-based approaches to screening compounds in parallel is more preferable and it is hoped that systems like the SRU BIND™ or Corning Epic™ continue to be developed. For such systems, a 384- and 1536-well-plate-based compound delivery system would be beneficial and in-well referencing for solvent effects and nonspecific binding would be a great advantage. Kinetic ranking of compounds may not be possible in a plate-based format unless a flow or mixing system were introduced. FortéBio have developed one such mixing system in the Octet™ instrument (http://www.fortebio.com).

The correct immobilization of target proteins to a biosensor surface is critical in maintaining structure and activity for drug recognition and binding. Certain targets are difficult to assemble on biosensors, for example, membrane spanning proteins such as 7TM receptors, multicomponent enzymes and certain unstable proteases. Sufficient immobilized protein is also needed in order to detect direct binding of small molecular weight compounds in present biosensors with their current sensitivity. Poor developments in immobilization strategies can adversely affect the likely success of biosensors in drug discovery applications. This has been compounded by a high level of intellectual property in this area, which restricts competitors from marketing similar and successful immobilization strategies.

A typical drug may need to pass through cell membranes or the blood–brain barrier before binding to a hydrophobic pocket on its target protein. For this reason, many compounds in drug screening libraries are hydrophobic by nature and require solvents such as dimethyl-sulphoxide (DMSO) to aid solubility. In fact, as a matter of routine, compounds submitted for screening in large pharma organizations often arrive as DMSO stock solutions and should be filtered after addition to running buffers or have their solubilities checked prior to test on biosensors. Biosensors employed in drug discovery need to be robust to test solutions containing at least 1% DMSO. For optical biosensors affected by refractive index, it is vital that controls and calibrations are run to take into account solvent effects between samples. Biosensor technologies less affected by solvent effects, such as acoustic devices, may hold advantages in this area.

There is much debate between advocates of drug-binding assays as opposed to target activity–modulation assays. Biosensor assays that measure binding only may fall into the trap of detecting too many compounds that bind but do not modulate activity. It is a possibility that too many of these false positives can make the selection of candidate drugs a very difficult task.

Finally, biosensors often require expert users to be skilled in experimental design and data interpretation: biosensors are not currently a technology platform that could be considered for general use and are a long way from being considered as a "black-box". Consequently, researchers involved in screening require detailed training, which may also be a factor in preventing biosensors playing a dominant role in drug screening.

In conclusion of this section, biosensors are being used as platforms for information-rich but low throughput secondary drug screens, to confirm the activities of "hits" and prioritize leads for further development. Parallel rather than serial screening technologies are required to increase throughput and robustness if biosensors are ever to be used for high-throughput random compound screening.

3 BIOSENSORS AS A TOOL FOR BIOPHARMACEUTICAL MANUFACTURE

Biopharmaceuticals are therapeutic agents produced by recombinant DNA technology in living systems and include recombinant monoclonal antibodies, recombinant proteins, therapeutic vaccines, and gene therapy products. The development of biopharmaceuticals, compared to that of small drug molecules, is an extremely complex entity that poses its own unique challenges.

Some major challenges in the industry include speed-to-market, scale up, product differentiation, and more emphasis toward outsourcing due to limited manufacturing capacity. Ultimately, the focus is toward providing patients with innovative medicines but requires creating a positive regulatory environment to facilitate timely approval of products by meeting high quality, efficacy and safety standards. Researchers must identify candidate molecules for disease targets with sufficiently high binding affinity that they can be translated into lower doses of the drug to meet safe and acceptable dosing regime for patients. Cell line development, maximizing fermentation/purification for production efficiency, scale up from laboratory to pilot/manufacturing scale, addressing capacity issues, formulation for stability, release testing to strict quality standards for safety and efficacy together with testing safety in preclinical studies are just some of the significant issues faced prior to starting a program of clinical studies.

The industry is therefore reliant on new and emerging technologies to address such challenges and biosensors can fulfil some of these requirements. A selection of biosensors are either currently available or under development, and those most commonly used are based on optical surface plasmon resonance (SPR) principles. Biosensors provide an exquisite binding specificity for biomolecules with specific ligand partners. Such interactions cause changes in physical parameters (e.g., mass, enthalpy, conformation change, etc.), measured by transducers to produce a digital signal, which is directly proportional to the interaction.

Following identification and selection of a candidate biopharmaceutical molecule for manufacture, recombinant DNA is transfected into a suitable cell line for expression of the relevant protein. Regulatory guidance[21] requires the selection of cell clones prior to the manufacture of a cell bank, which is both labor intensive and time consuming. Thousands of potential clones require screening for high but stable expression of the therapeutic product, requiring aseptic removal of 50–100 µl supernatants from 96 well plates. However, the protein concentrations in many of the samples are too low (low µg ml^{-1} range) to be detected using current methods. Screening therefore tends to be biased toward high productivity samples, but rarely takes into account the quality of the product. This may be achieved through the use of biosensors. For example, Hellwig et al.[22]

have monitored clones after scale up into shake-flasks to produce sufficient fusion proteins to enable detection on a Biacore 2000 using ligand-coated sensor surfaces. A careful understanding of cellular function during cell line development can thus prove to be useful in monitoring cell line stability. Bionas (http://www.bionas.de) recently launched an in vitro silicon sensor chip used as an analogue for a cell culture plate to profile metabolic activity of living cells using a label-free and noninvasive method by simultaneous measurement of oxygen consumption, acidification, and even adhesion of living cells, without compromising sterility. Similarly, ACEA Bioscience Inc. (http://www.aceabio.com) introduced a revolutionary microelectronic cell sensor system, based on microelectronic cell sensor arrays integrated into the bottom of microtiter plates. The system provides quantitative measurements of the biological status of the cells (e.g., cell growth, apoptosis, and changes in cell morphology) by using electrical impedance in real-time cell electronic sensing (RT-CES).

Such technologies afford a continuous monitoring of metabolic activities of cloned cells in culture plates, but as a potential application for PAT, they lack an ability to measure the quantity of expressed proteins, essential for the selection of optimal clones for manufacture of biopharmaceuticals.

A key driver for the pharmaceutical (including biopharmaceutics) industry has now been introduced by the Food and Drug Administration (FDA) in the shape of a draft guidance document on "PAT – A Framework for Innovative Pharmaceutical Manufacturing and Quality Assurance",[23] to enable manufacturers to develop and implement new efficient tools for use during pharmaceutical development, manufacturing, and quality assurance, while maintaining or improving the current level of product quality assurance. PAT is considered a paradigm shift in the manufacture and control of pharmaceuticals by demonstrating process consistency. It involves timely measurements of raw materials, intermediates, and process parameters and control of key critical process and quality parameters ("risk management") with an aim to ensure product quality. It offers a number of advantages for example, optimization of product development, improvement in product quality, and product knowledge. By adopting biosensors for monitoring bioprocesses, the industry will benefit from reducing the number of batch failures, through a rationalization of in-process controls and either a reduction or elimination of batch testing of the finished product.

Adapting cell lines for culturing in suspension, so that they can be grown in batch and fed-batch bioreactors with consistent media formulation and the ability to scale up is considered the most variable step in the manufacture of biopharmaceuticals and the industry can benefit from early implementation of PAT. Small but subtle changes to the culture conditions can dramatically alter post–translational modification and alter the impurity profile of the product. Although productivity of mammalian cells in bioreactors has reached the gram per liter range, opportunities still exist for improving mammalian cell systems through further advancements in production systems as well as through vector and host cell engineering and appropriate post–translational modifications such as glycosylation. More recently, a number of manufacturers have started to use spectroscopic probes (e.g., UV, IR, and fluorescence), together with biosensors and applying multivariate data analysis to better understand fermentation processes for a robust and more reliable monitoring and control strategy. However, estimation of on-line or at-line monitoring of product titers is still an issue that requires binding of specific ligands to sensor surfaces. The Spreeta chip (Sensata Technologies, http://www.sensata.com), based on SPR, coupled to a Flow Injection Analyser (FIA system) offers an opportunity to utilize biosensors for at-line monitoring of product concentration in production vessels. Bracewell et al.[24] have shown that an SPR sensor with automated sample handling only requires 30 s to perform an analysis compared with an enzyme-linked immunosorbent assay (ELISA) which requires approximately 3–24 h to determine product concentration. Utility of SPR sensors for at-line monitoring is not only limited to quantitation of proteins but also offers an opportunity to determine binding kinetics of the product to its ligand. For recombinant monoclonal antibodies, immobilization of the molecule via the Fc domain (using either anti-Fc antibodies or Protein A), followed by the antigen offers a unique ability for at-line monitoring of quality through measurements of the immunoreactive fraction for a typical fermentation batch analysis. Additionally, immobilization of suitable lectins with binding specificity for specific glycoforms offers a unique opportunity for continuous monitoring of post–translational modifications in cell culture systems. For off-line monitoring, recombinant proteins have traditionally relied on labor-intensive and time-consuming methodologies such as immunoassays, rapid chromatography, and nephelometry. The use of optical biosensors has been directly compared against traditional methods and offers some significant advantages in terms of speed, automation, and assay variability by Baker et al.[25]

In bioreactors, glucose sensors are very important, especially for fed-batch processes, to monitor the carbon source as a feeding strategy to meet metabolic demands of cells in culture. Other components such as amino acids, especially lysine and glutamine, are important markers of cell growth and protein secretion in mammalian cell cultures. Therefore, continuous monitoring of these components in media to prevent depletion of these vital components is important. Ammonia, a byproduct of mammalian culture is toxic and can inhibit cell growth. Designing of suitable biosensors to monitor such components during a typical fermentation process is essential. In this respect, De Lorimier et al.[26] reported the use of bacterial periplasmic binding proteins (bPBPs), with specificity for a wide variety of small molecules. These binding proteins undergo a large, ligand-mediated conformational change, which is linked to reporter functions to monitor ligand concentrations. This provides an opportunity to engineer sensitive molecules for reagentless optical biosensors for a diverse range of specific ligands (e.g., sugars, amino acids, anions, cations, and dipeptides). On-line biosensors consisting of a combination of a continuous flow microcalorimeter with a dielectric spectroscope (to monitor viable cell mass) have also been developed by Guan et al.[27] Specific heat flow rate and enthalpy changes are stoichiometrically related to the net specific rates of substrates, products, and indeed to specific growth rate, and therefore a direct reflection of metabolic rate. For a batch culture of Chinese hamster ovary cells, producing recombinant human interferon-γ (IFN-γ), total metabolic rate decreased with increasing time in batch culture, coincident with the decline in the two major substrates, glucose and glutamine, and the accumulation of the by-products, ammonia and lactate. Ultrasonic measurements have also been used for on-line monitoring of biomass by Zips.[28]

Formulation is pivotal for a successful biopharmaceutical product. Careful control of excipients, salt concentration, and pH is required for long-term storage of the product while maintaining safety and efficacy of the molecule. Changes in salt concentrations may either have little or a dramatic impact on the binding affinity for monoclonal antibodies. Reduced sensitivity to salt concentrations suggests that electrostatic interactions may be relatively less important in the overall binding affinity. Similarly, changes in pH can significantly influence binding characteristics (association and dissociation rates) and affinity binding by 8–20-fold within a pH range 5.5–8.5.[29] Biosensors offer an opportunity to rapidly analyze numerous buffer conditions in parallel with antibody characterization and aid in the selection of optimal formulation conditions. They also offer an ability to screen stability when binding of the product to its ligand may be the only relevant bioassay, which closely mimics mode of action. Recently, Biacore C was specifically designed for rapid concentration measurements and meets high standards of method validation and good manufacturing principals (GMP) demands by regulatory agencies.

A significant challenge for the biopharmaceutical industry occurs during scale up, which is currently being addressed through the "Design for Manufacture" initiative throughout the pharmaceutical industry. Changes in scale must be carefully monitored to ensure that the product purity, protein structure, concentration, and potency remain comparable. Products may contain new impurities with an altered quality, safety, stability, and efficacy profile. This frequently requires a detailed physicochemical, conformational, and biological characterization of the product before and after scale up. Characterization of proteins is considered far more challenging than that of small molecules. Large molecules tend to be complex with varied structures and are frequently sensitive to temperature, pH, and isotonicity, changes in which can result in loss of higher order structure or formation of aggregated structures under suboptimal conditions. Glycosylation also contributes toward the heterogeneity of the molecule. Good and robust

analytical data is required to demonstrate pharmaceutical equivalence between products from different batches. Such supporting data is essential to underwrite changes to the process during development and to provide a high degree of confidence to the regulatory authorities. This requires sophisticated analytical methods to demonstrate comparability and a detailed characterization strategy, which requires new tools, technologies, and more complex assays. Measurements of immunoreactive fraction, affinity, association and dissociation rates, and enthalpy changes are powerful characterization techniques offered by optical biosensors.

Additionally, determination of potency is a key measure of the final active structure, which cannot always be determined using physicochemical characterization methods. Effective potency assays often require cell-based assays to monitor functional biological activity. During a comparability study, it is essential to determine changes in the native structure of the protein, which can have a direct effect on the biological activity of the molecule. The technology commercialized by Biacore offers a potential to identify changes in the native structure of the molecule through epitope mapping and aid in comparability studies. While utility of biosensors for potency determination is less extensive, ACEA's RT-CES technology (mentioned earlier) may be useful and is noninvasive and harmless for cells. Cells cultured in individual microtiter plates for bioassays can be monitored continuously and provide a quantitative measurement of functional activity (e.g., cell proliferation, cytotoxicity profiling, cell adhesion, functional analysis of growth factor/death receptor, etc.). In some situations, the key mode of action may simply be a binding event (i.e., monoclonal antibody binding to specific proteins in serum). Under this circumstance, determination of immunoreactive fraction analysis alone may be considered as a true functional activity assay. Routine immunogenicity testing is also used as a quality control test to demonstrate batch-to-batch consistency for traditional vaccines (tetanus and diphtheria toxoids). Although regulatory authorities expect animal testing to demonstrate potency, the biosensor offers an attractive opportunity for in vitro testing of immunogenicity.

Protein therapeutics inevitably induce an immune response and while such a response is undesirable, it can reduce efficacy and in extreme cases lead to a development of a life threatening immune reaction. Emphasis is therefore placed on testing of immunogenicity at early stages of product development. This involves early detection of antibodies (either binding and or neutralization) against the therapeutic protein in preclinical and clinical studies. Measurement of the antibody response and biomarker is now a regulatory requirement during the development of the therapeutics and post–marketing surveillance of immunogenicity of protein therapeutic is an FDA requirement. Swanson et al.[30] reported a comparison between radioimmune precipitation, ELISA, biosensor immunoassay, and biological assay for the detection, quantitation, and characterization of antibodies against recombinant human Erythropoietin in patient samples. A combination of biosensor-based immunoassay (for detection of low-affinity antibodies) together with the bioassay (to monitor neutralizing activity) was recommended. More recently, introduction of the Biacore T100/A100 systems to detect an immune response by characterizing serum antibodies has made this strategy more practical.

In conclusion, while the utility of biosensors to aid biopharmaceutical product development has been extensive, there are still some areas (e.g., detectors for continuous purification process with feedback controls) that need to be explored and further developed.

4 CONCLUSIONS

The pharmaceutical industry is currently undergoing a great deal of transformation across many areas of research and development. The drivers for this are multitude, but are generally related to the costs associated with discovery and development of new drug candidates and how these are passed on to the end customer, who could be the patient, the doctor prescribing the medication, or the healthcare practice and/or insurance company who pays the bill. The R&D process is undergoing an industrialization in order to establish improved productivity through the use of new technologies. However, all this has to be done while still maintaining equivalent safety concerns for the final patient—the regulatory bodies police the industry and will continue to push pharmaceutical firms to engage any new methods that will

ensure patient safety and drug efficacy. With this in mind, biosensors, which have been explored for a number of years, offer the potential to bring quantitative assessment to what has traditionally been an experience-based qualitative process. This leads to improved understanding and increased knowledge of molecular interactions in biological systems. However, current biosensor platforms have suffered a number of drawbacks that have limited their increased uptake.

Two intimately linked variables, that are yet to be fully explored in the development of biosensors, relate to speed of analysis and throughput. While many biosensor systems operate much more rapidly, in comparison to most incumbent techniques, they have always suffered when they are expanded to high throughputs. Higher throughputs are required because even if speeds of analysis are reduced, one still has to overcome the practical situation of preparing samples and being ready for the eventuality that a sensor is poisoned or requires regeneration. Thus, the need to combine these two factors is very strong and once accomplished will undoubtedly open up greater scope for biosensor implementation. It is also important to note that in doing this, one must maintain the unique advantage that biosensors have of being quantitative. Should this be lost then they may simply become a new screening platform, in competition with others. The unique property of generating knowledge should not be lost, but through the development of speed and throughput, biosensors should make the transition from secondary to primary screening. It also follows from this that current biosensor platforms come in a wide variety of different platforms and do not operate on any standard, in comparison to HTS where the microtiter plate is evident. This reduces the flexibility of biosensors when considering the aspects of multiplexing between competing technologies.

Standardization can be even more of a problem in the PAT arena where the concept of plug-and-play is very much the requirement for analytical systems. The same issue used to be true in conventional screening where one had to invest in different instruments for different analysis techniques, for example, fluorescence and absorbance, but now this is not the case—a single instrument can be bought which not only incorporates multiple optical sensing schemes but can also include integrated liquid dispensing to aid in kinetic studies. The biosensor community needs to be aware of this and consider the possibility of developing standard platforms for implementation. In addition to this, instrumentation and control will have to be further developed that will enable a reduction in the skills required to operate—the concept of the black-box instrument is an essential requirement for future uptake across increased applications and use by multiple users, rather than individual experts in support groups.

On a final note, it is essential to remember that virtually all biosensor platforms require a degree of materials processing. These processes are an essential component of the system, often requiring the most development for successful implementation. Material linkage to whichever transduction technique is used can often be the difference between success and failure, and as such has attracted a great deal of intellectual property protection through patents. Thus, further developments in novel materials will be required that enable close coupling and analysis of biological responses. These will have to take into account the real life application scenarios for biosensors; samples *will* poison the sensors, requiring regeneration and solvents, and impurities *will* be present that were previously needed to be eliminated. Overcoming these issues would normally require an additional process in the already time-consuming sample preparation step, but biosensor platforms that are developed and can overcome these limitations will have increased scope in their application.

The previous sections have given an insight into how biosensors are currently being targeted in the two areas of drug discovery that can establish the most value from their current use. And, with further advancements in their current limitations, identified and discussed in the preceding text, their scope will widen and biosensors will become more mainstream within the industry. Biosensors hold a great deal of potential, and given their intimate link to the understanding of disease and ability to monitor biological processes, they should continue to expand across the industry and become common tools of use, rather than expert-driven instrumentation systems.

REFERENCES

1. L. T. Vassilev, B. T. Vu, B. Graves, D. Carvajal, F. Podlaski, Z. Filipovic, N. Kong, U. Kammlott, C. Lukacs, C. Klein, N. Fotouhi, and E. A. Liu, In vitro activation of the p53 pathway by small-molecule antagonists of MDM2. *Science*, 2004, **303**, 844–848.

2. D. G. Myszka, Analysis of small-molecule interactions using Biacore S51 technology. *Analytical Biochemistry*, 2004, **329**, 316–323.

3. D. G. Myszka and R. L. Rich, Implementing surface plasmon resonance biosensors in drug discovery. *Pharmaceutical Science and Technology Today*, 2000, **3**(9), 310–317.

4. B. T. Cunningham, P. Li, S. Schulz, B. Lin, C. Vaird, J. Gerstenmaier, C. Genick, F. Wang, E. Fine, and L. Laing, Label-free assays on the BIND system. *Journal of Biomolecular Screening*, 2004, **9**, 481–490.

5. P. O. Markgren, M. T. Lindgren, K. Gertow, R. Karlsson, M. Hamalainen, and U. H. Danielson, Determination of interaction kinetic constants for HIV-1 protease inhibitors using optical biosensor technology. *Analytical Biochemistry*, 2001, **291**, 207–218.

6. C. F. Shuman, M. D. Hamalainen, and U. H. Danielson, Kinetic and thermodynamic characterization of HIV-1 protease inhibitors. *Journal of Molecular Recognition*, 2004, **17**, 106–119.

7. D. G. Myszka and R. L. Rich, SPR's high impact on drug discovery. *Drug Discovery World*, 2003, Spring, 49–55. SPR impact in drug discovery.

8. R. L. Rich and D. G. Myszka, Why you should be using more SPR biosensor technology. *Drug Discovery Today. Technologies*, 2004, **1**, 301–308.

9. J. Comley, Label free detection: new biosensors facilitate broader range of drug discovery applications. *Drug Discovery World*, 2005, **6**, 63–74.

10. Y. S. Day, C. L. Baird, R. L. Rich, and D. G. Myszka, Direct comparison of binding equilibrium, thermodynamics, and rate constants determined by surface- and solution-based biophysical methods. *Protein Science*, 2002, **11**, 1017–1025.

11. R. Karlsson, M. Kullman-Magnusson, M. D. Hamalainen, A. Remaeus, K. Andersson, P. Borg, E. Gyzander, and J. Deinum, Biosensor analysis of drug-target interactions: direct and competitive binding assays for investigation of interactions between thrombin and thrombin inhibitors. *Analytical Biochemistry*, 2000, **265**, 340–350.

12. D. Casper, M. Bukhtiyarova, and E. B. Springman, A Biacore biosensor method for detailed kinetic binding analysis of small molecule inhibitors of p38 α mitogen-activated protein kinase. *Analytical Biochemistry*, 2004, **325**, 126–136.

13. H. Nordin, M. Jungnelius, R. Karlsson, and O. P. Karlsson, Kinetic studies of small molecule interactions with protein kinases using biosensor technology. *Analytical Biochemistry*, 2005, **340**, 359–368.

14. R. L. Chapman, T. B. Stanley, R. Hazen, and E. P. Garvey, Small molecule modulators of HIV Rev/Rev response element interaction identified by random screening. *Antiviral Research*, 2002, **54**, 149–162.

15. S. H. Verhelst, P. J. Michiels, G. A. van der Marel, C. A. van Boeckel, and J. H. van Boom, SPR evaluation of various aminoglycoside-RNA hairpin interactions reveal low degree of selectivity. *Chembiochem*, 2004, **5**, 937–942.

16. R. Gambari, G. Feriotto, C. Rutigliano, N. Bianchi, and C. Mischiati, Biospecific interaction analysis of low-molecular weight DNA-binding drugs. *Journal of Pharmacology and Experimental Therapeutics*, 2000, **294**, 370–377.

17. P. Debnam, P. Huxley, I. R. Matthews, D. Thrige, and J. Abery, Screening lead CD80 inhibitors. *Current Drug Discovery*, 2004, 25–29 January issue www.liebertonline.com/doi/pdf/10.1089/adt.2004.2.407.

18. M. A. Cooper, Advances in membrane receptor screening and analysis. *Journal of Molecular Recognition*, 2004, **17**, 286–315.

19. S. Cimitan, M. T. Lindgren, C. Bertucci, and U. H. Danielson, Early absorption and distribution analysis of antitumor and anti-AIDS drugs: lipid membrane and plasma protein interactions. *Journal of Medicinal Chemistry*, 2005, **48**, 3536–3546.

20. Y. N. Abdiche and D. G. Myszka, Probing the mechanism of drug/lipid membrane interactions using Biacore. *Analytical Biochemistry*, 2004, **328**, 233–243.

21. International Conference on Harmonisation: Quality Guideline ICH-Q5b – Quality Of Biotechnological Products: Analysis Of The Expression Construct In Cells Used For Production Of R-DNA Derived Protein Products, which can be accessed through the ICH website at http://www.ich.org.

22. S. Hellwig, F. Robin, J. Drossard, N. P. G. Raven, C. Vaquero-Martin J. E. Shively, and R. Fischer, Production of carcinoembryonic antigen (CEA) N-A3 domain in Pichia pastoris by fermentation. *Biotechnology and Applied Biochemistry*, 1999, **30**, 267–275.

23. This guidance document can be accessed through the FDA Center for Drug Evaluation and Research website at http://www.fda.gov/CDER/GUIDANCE/6419fnl.htm.

24. D. G. Bracewell, R. A. Brown, A. Gill, and M. Hoare, Monitoring and control of bioproducts from conception to production in real-time using an optical biosensor. *Chemical Engineering and Technology*, 2001, **24**(7), 25–31.

25. K. N. Baker, M. H. Rendall, A. Patel, P. Boyd, M. Hoare, R. B. Freedman, and D. C. James, Rapid monitoring of recombinant protein products: a comparison of current technologies. *Trends in Biotechnology*, 2002, **20**(4), 149–156.

26. R. M. De Lorimier, J. J. Smith, M. A. Dwyer, L. L. Looger, K. M. Sali, C. D. Paavola, S. S. Rizk, S. Sadigov, D. W. Conrad, L. Loew, and H. W. Hellinga, Construction of a fluorescent biosensor family. *Protein Science*, 2002, **11**(11), 2655–2675.

27. Y. Guan, P. M. Evans, and R. B. Kemp, Specific heat flow rate: An on-line monitor and potential control variable of specific metabolic rate in animal cell culture that combines microcalorimetry with dielectric spectroscopy. *Biotechnology and Bioengineering*, 1998, **58**(5), 464–477.

28. A. Zips and U. Faust, Determination of biomass by ultrasonic measurements. *Applied and Environmental Microbiology*, 1989, **55**, 1801–1807.

29. Early kinetic screening of hybridomas for confident antibody selection using Biacore A100: An application note (number 84) provided through the Biacore website at http://www.biacore.com/lifesciences/technology/application_notes/index.html.

30. S. J. Swanson, J. Ferbas, P. Mayeux, and N. Casadevall, Evaluation of methods to detect and characterize antibodies against recombinant human erythropoietin, *Nephron Clinical Practice*, 2004, **96**, c88–c95.

72

Glucose Measurement Within Diabetes via "Traditional" Electrochemical Biosensors

Elizabeth A. H. Hall

Institute of Biotechnology, University of Cambridge, Cambridge, UK

1 INTRODUCTION[1-8]

Elevated blood glucose levels caused either by an inability to produce sufficient insulin or by failure to use insulin in regulation of glucose affect an increasing proportion of the population in the developed and developing world. The condition, known as *diabetes mellitus*, is diagnosed in >17 million people in the United States. Type 2 diabetes has reached epidemic levels in Asia, where younger people are being affected, and 194 million people worldwide have diabetes. Causes attributed to this increase include poor nutrition and fast food culture, obesity, and sedentary lifestyle. By 2025, the International Diabetes Foundation estimates 333 million people will have diabetes, equivalent to the rate currently seen in the United States repeated worldwide (Table 1), making diabetes one of the most important international health issues.

The inability to control blood glucose leads to hypoglycemia (low blood glucose) or hyperglycemia (excessive blood sugar levels). The former causing mental confusion, convulsions, or even coma and death whereas the latter results in a wide range of long-term microvascular and neuropathic complications due to abnormally high levels of protein glycosylation.

The discovery of insulin, in 1922, and the development of a self-monitoring glucose test kit in 1978 mark events that enabled significant advances in the management of diabetes mellitus in the twentieth century. Self-monitoring of blood glucose levels, allowed diabetics to adjust treatment regimens and achieve more normal blood glucose levels and as a result of data from different trails in the early 1990s (e.g., the Diabetes Control and Complications Trial, released in 1993[9]) in which intensive insulin therapy and self-monitoring of blood glucose were shown to give better glycemic control, self-monitoring of blood glucose levels has become the standard of care.

The original measurement technology available for home use was a urine glucose test strip, limited to diagnosing hyperglycemia on the basis of the chromogenic reduction of a copper solution by glucose and other reducing sugars. This was superseded by a glucose biosensor that has traditionally used the enzyme *glucose oxidase* (*GOx*). The key feature of GOx, enabling it to be used in an *electrochemical biosensor* is that it is a *redox enzyme*. Redox enzymes are so named because they contain a cofactor or prosthetic group that contains a redox system. A common group of redox proteins are the *flavoproteins* and there are about 80 different enzymes containing: flavin adenine dinucleotide

Handbook of Biosensors and Biochips. Edited by Robert S. Marks, David C. Cullen, Isao Karube, Christopher R. Lowe and Howard H. Weetall.
© 2007 John Wiley & Sons, Ltd. ISBN 978-0-470-01905-4.

Table 1. Incidence of diabetes assuming the current rate of incidence in the United States is projected worldwide

Region	Incidence	Population
North America	25 360 167	431 122 873
Central America	1 171 369	19 913 300
Caribbean	229 291	3 897 960
South America	17 705 226	300 988 901
Northern Europe	1 171 072	19 908 270
Western Europe	9 102 581	154 743 914
Central Europe	9 142 561	155 202 055
Eastern Europe	12 310 147	209 272 527
SW Europe	3 727 481	63 367 202
Southern Europe	4 041 470	68 705 006
SE Europe	3 011 101	51 188 794
Northern Asia	161 841	2 751 314
Central Asia	2 856 803	48 565 676
Eastern Asia	89 833 064	1 527 162 188
Southwestern Asia	4 052 583	68 893 918
Southern Asia	83 306 574	1 416 211 830
Southeastern Asia	29 834 507	507 186 716
Middle East	10 630 621	180 720 662
Northern Africa	7 111 597	120 897 168
Western Africa	4 293 681	67 817 637
Central Africa	4 696 275	79 836 729
Eastern Africa	10 299 076	175 084 338
Southern Africa	4 290 175	81 933 044
Oceania	1 725 130	29 327 241
TOTAL	**340 064 393**	**5 784 699 263**

(FAD) and flavin mononucleotide (FMN). The flavin unit is very strongly associated with the protein: covalently bound to the amino acids of the protein or held by H-bonding and weaker forces.

The electrochemistry of the redox group from these enzymes can be shown by examining the voltammetric $i-V$ (current–voltage) curves for their respective prosthetic groups. For example, for FMN (Figure 1) the process occurs via a semiquinoid free radical intermediate. Depending on pH and measurement conditions, the $i-V$ curve can reveal the charge transfer (CT) as one two-electron step or two one-electron steps. Embedded within the flavin enzymes, the redox potential for the FMN or FAD may vary by more than 500 mV from one enzyme to another, according to the effect of the association forces with the protein.

The class of enzymes known as *oxidoreductases*, which include GOx, are electron transfer agents that participate in CT pathways culminating in molecular oxygen. GOx (glucose 1-oxidase or β-D-glucose:oxygen-1-oxidoreductase, EC 1.1.3.4), a globular protein 2.5 × longer than its diameter is a flavin enzyme with molecular mass of 150–180 kDa, and has been extensively used to monitor blood glucose levels in diabetics due to its catalytic specificity for the oxidation of glucose:

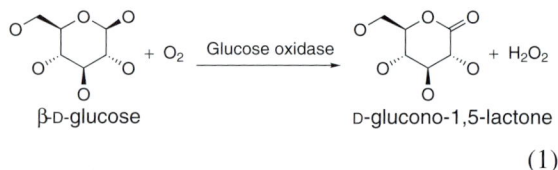

β-D-glucose + O₂ → (Glucose oxidase) → D-glucono-1,5-lactone + H₂O₂

$$\tag{1}$$

GOx was discovered in 1928, the year before penicillin was isolated and is probably the most investigated and used enzyme for biosensor applications. In principle GOx is ideally suited for measurement of glucose, since the number of electrons used in the reduction of FAD during the glucose oxidation process is directly related to the amount of glucose consumed or the concentration of glucose present (Figure 2).

However, the FAD group of GOx is deeply embedded within a protective protein matrix (Figure 3), so that the matrix or glycoprotein shell surrounding the redox site creates an effective kinetic barrier for electron transfer. Direct electron transfer between protein and an electrode is controlled mainly by three factors[10]:

- reorganization energies,
- potential differences and orientations of the redox-active sites involved in each oxidation state, and
- distances between redox-active sites and CT agent/mediator.

Thus, glucose oxidase cannot get close enough to the electrode for direct electron transfer to occur. The protein is in the way.

This dilemma was solved first by Leyland Clark.

2   CLARK OXYGEN AND GLUCOSE ELECTRODES[11–13]

An important early glucose biosensor was based on the Clark oxygen electrode developed in 1950s primarily for blood-gas measurements. This exploits the reduction of oxygen at a cathode:

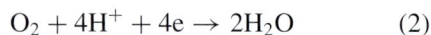

$$O_2 + 4H^+ + 4e \rightarrow 2H_2O \tag{2}$$

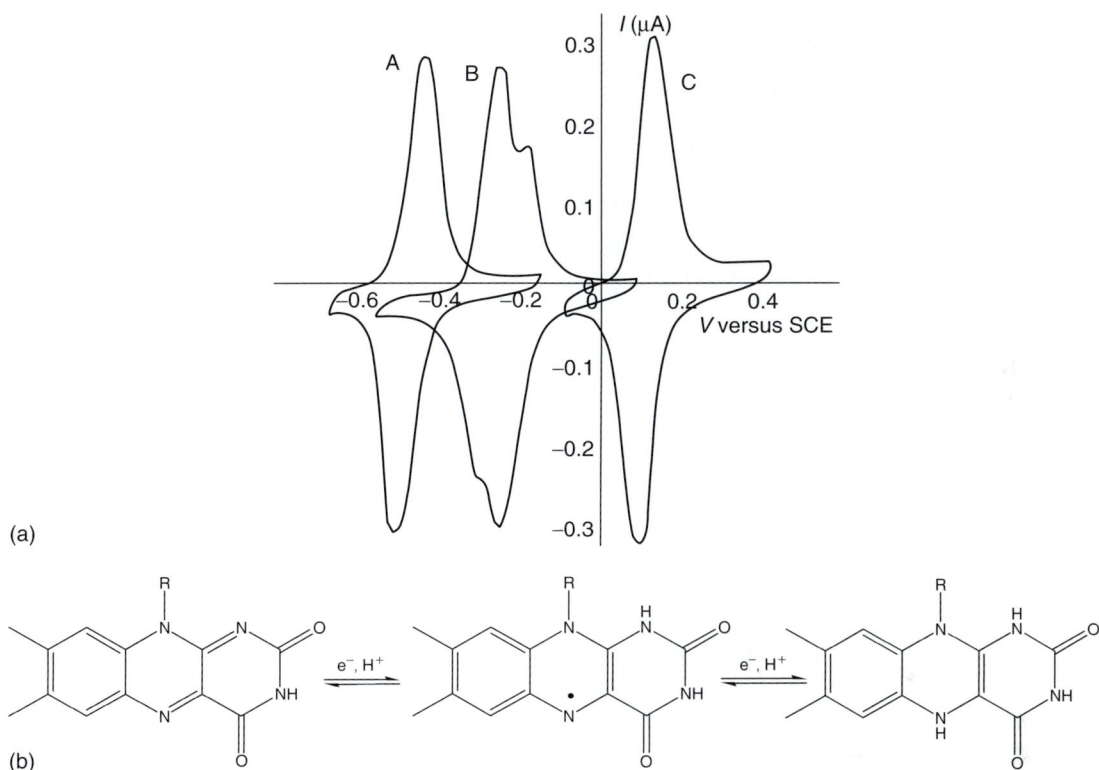

Figure 1. (a) FMN electrochemistry (A) pH13.4; (B) pH 8.0; (C) pH 1.0; (b) FMN redox pathways.

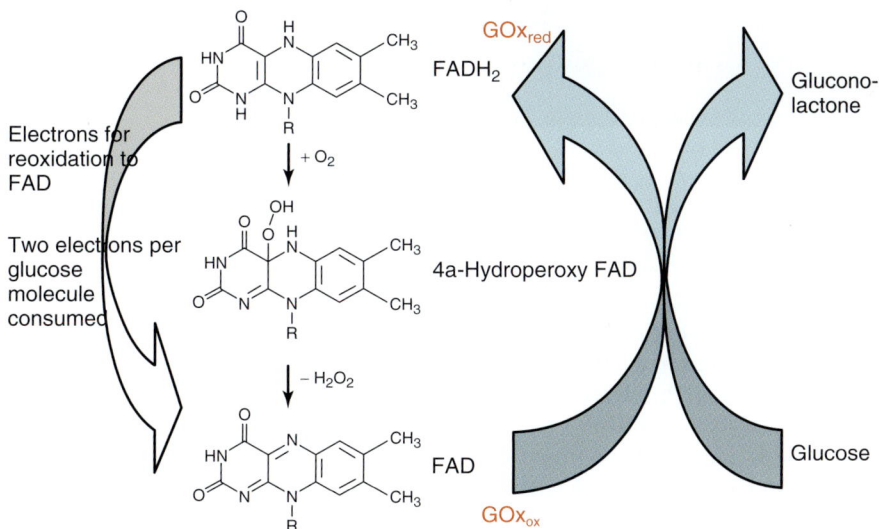

Figure 2. Redox pathway in GOx associated with glucose oxidation. The electrochemical glucose biosensor needs to be able to monitor the number of electrons used in the FAD/FADH$_2$ redox process.

(a)

(b)

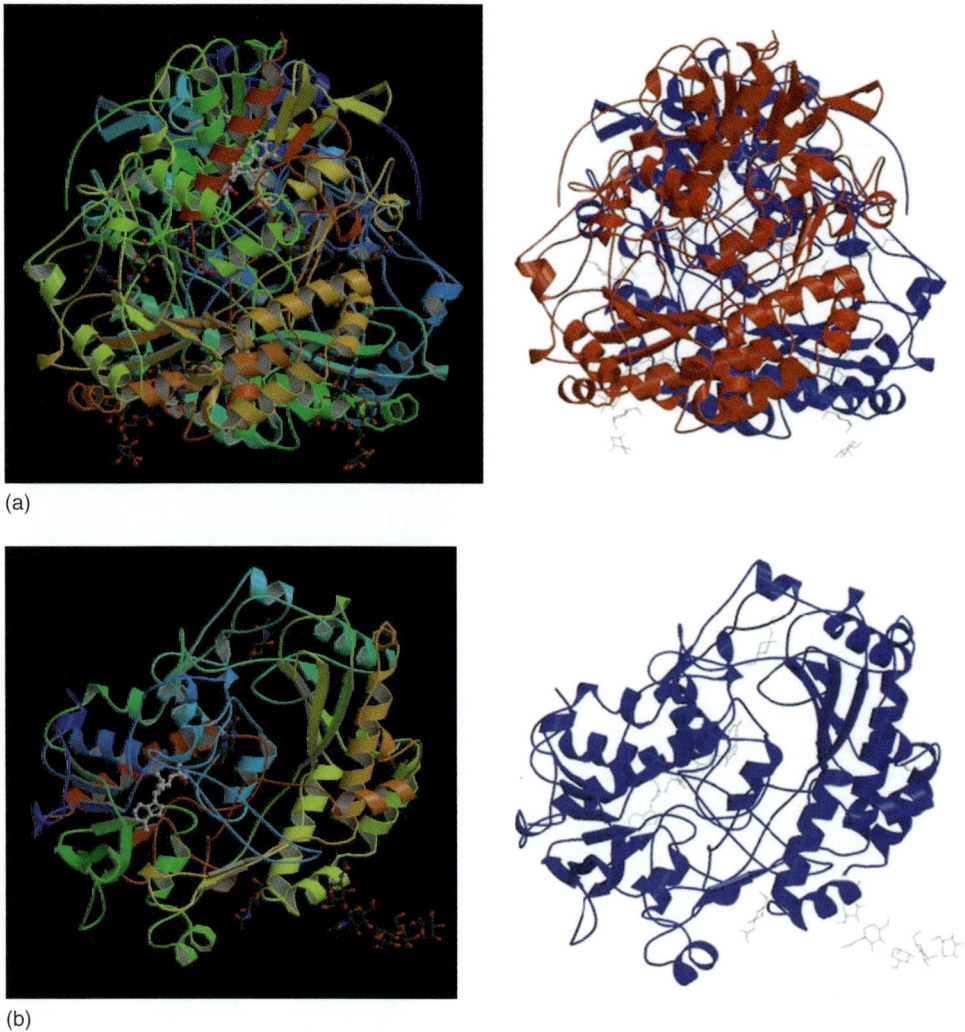

Figure 3. The glucose oxidase (GOx) E.C.1.1.3.4 (a) homo dimer from data in the protein data bank (PDB); 3D image of structure of PDB code 1gpe (b) one GOx protein monomer showing the deeply embedded FAD site surrounded by the protein shell. From data in the PDB 3D image of structure of PDB code 1gal.

As can be seen in Figure 4 the four-electron reduction ($n = 4$) can be measured on the diffusion-controlled plateau of a Pt electrode at a potential of approximately ~ -650 mV relative to Ag/AgCl. The Clark electrode is immersed in a simple electrolyte solution (e.g., KCl) and separated from the sample by a semipermeable membrane that allows diffusion of oxygen. Since blood contains many other electroactive species that could interfere with the measurement, this semipermeable membrane also acts to separate the sample from the internal electrolyte solution. A steady-state current, i, for the electrode is obtained, which depends on the partial pressure of oxygen (P_{O_2}) given by:

$$i = nF \frac{P_m}{b} P_{O_2} \tag{3}$$

where F is Faraday's constant, P_m is the permeability of O_2 in the membrane, and b is the membrane thickness.

From this oxygen electrode, Clark and Lyons went on to describe the first biosensor—the

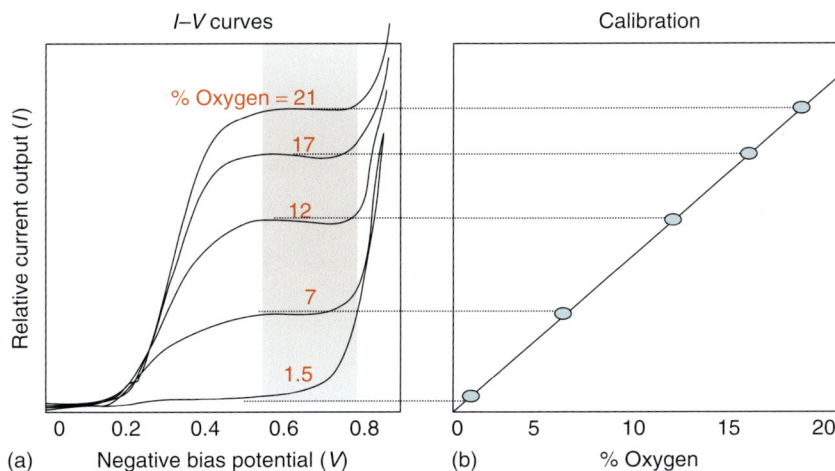

Figure 4. Oxygen reduction waves measured by linear sweep ($i–V$) voltammetry (a) and the resultant calibration plot (b). Potentials measured versus Ag/AgCl.

glucose enzyme electrode. This contained an embedded oxygen electrode within a concentrated solution of the enzyme glucose oxidase and was used to measure glucose as part of a study into continuous patient monitoring in the operating room during surgery. The principle involved is already demonstrated by examination of equation (1): the enzyme-catalyzed oxidation of glucose consumes oxygen. By placing GOx, in the electrolyte solution between the membrane and the oxygen electrode, Clark found that the concentration of glucose could be monitored by measuring the decrease in oxygen tension resulting from the enzyme reaction (equation 1). In the early designs, GOx was held between two semipermeable membranes, wrapped around the end of a cylinder, which contained an internal solution with the oxygen sensing and reference electrodes.

Alternatively, equation (1) also reveals that glucose concentration could be correlated with production of hydrogen peroxide, which conveniently can be oxidized at a potential of approximately +0.6 V versus Ag/AgCl. This enables a current measurement for hydrogen peroxide to be calibrated for glucose concentration. After refinement, this latter approach became the basis for the first commercial instrument developed by Yellow Springs Instruments (YSI) in the early 1970s. YSI instruments have retained this modus operandi and their glucose measurement still follows this principle. The membranes contain three layers: for

example, the outer porous polycarbonate, limiting the diffusion of glucose from the sample into the middle enzyme layer, preventing the reaction from becoming enzyme limited. The third internal layer, for example cellulose acetate, permits hydrogen peroxide diffusion to the electrode, but blocks many electrochemically active compounds that could interfere with the measurement (Figure 5). This remains a successful laboratory-based instrument with autocalibration, reuse of the enzyme membrane, and capability for high-throughput batch measurement. It has not been designed for the diabetic wishing to monitor glucose levels several times a day in their own home or discretely at work or in a more public place. It is certainly not portable.

Furthermore, interference effects—like other electroactive species arriving at the working electrode contributing to the observed current, giving an erroneous result—demand the careful choice of membrane, electrode material, and applied potential. The YSI multilayer membrane is thus not a viable solution for a disposable, noncalibrating instrument designed for the self-testing market. The required self-test system needs to be *small*, *cheap*, and *portable*, but the fundamental and essential requirement of any self-test assay method is accuracy. System calibration by the user is eliminated, but the sensor signal must still indicate the blood glucose concentration within a specified error. The reported concentration must be reliable

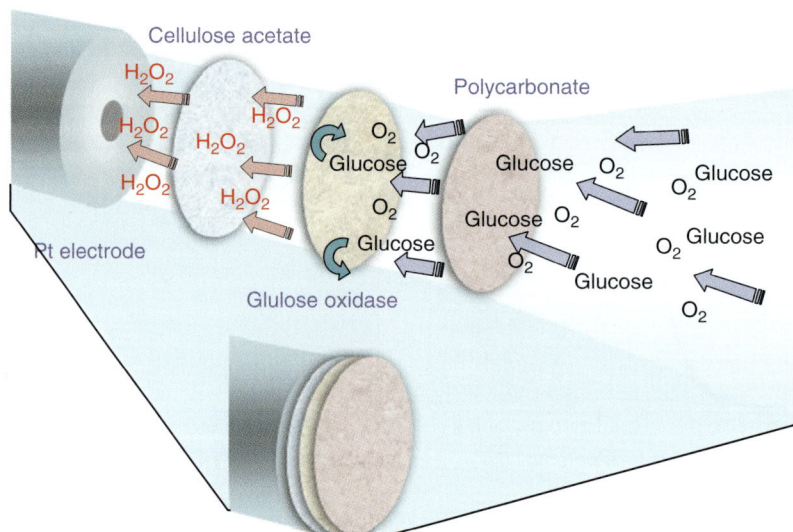

Figure 5. Construction of the YSI glucose amperometric biosensor.

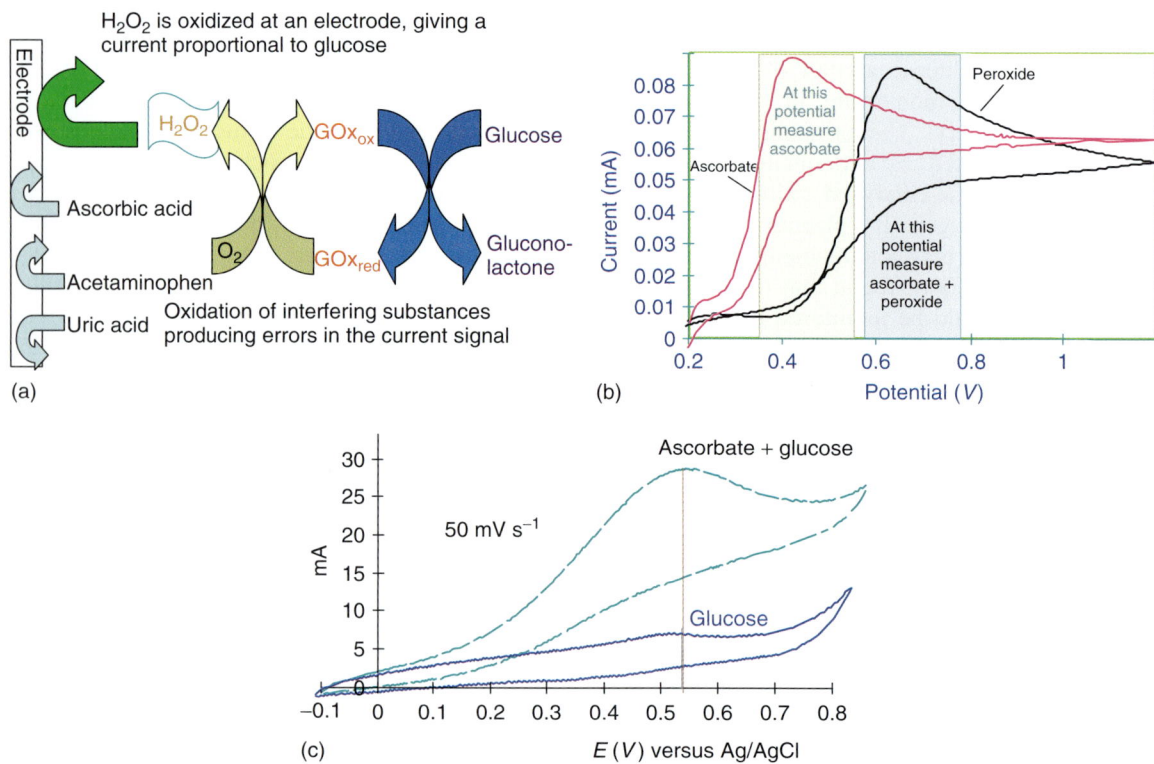

Figure 6. (a) Interference at the measuring potential for hydrogen peroxide by common electroactive species present in blood samples. i–V cyclic voltammograms for (b) ascorbate and hydrogen peroxide showing how ascorbate can be separated from peroxide, but peroxide cannot be separated from ascorbate. (c) Measurement of glucose using a membrane GOx electrode in the presence and absence of ascorbate.

for the user and the user must not be required to make further judgments about its accuracy before acting on the result.

Figure 6 demonstrates the problem of electrochemical interference by some common electroactive species. The issue is clear: hydrogen peroxide is oxidized at a higher potential than many interferents and thus at the potential required to perform the peroxide assay, the current measured (Figure 6b and c) potentially leads to falsely high readings. Accordingly, a key consideration is to lower the measuring potential below that for the interferents. Clearly, this creates a problem if hydrogen peroxide is the measurand, but the redox potential for FMN (Figure 1) suggests that it should be possible to reoxidize the *enzyme* at a lower potential, if the problem of achieving direct electron transfer to the protein can be overcome. There are several ways of achieving this.

3   ALTERNATIVE OXIDIZING SUBSTRATES/MEDIATORS[14-24]

The action of GOx is sometimes called a *ping-pong mechanism* that involves two molecules of FAD per GOx. This describes the mechanism by which an oxidase enzyme moves from the fully oxidized state to the fully reduced form and back to an oxidized state in a catalytic cycle (Figure 6a), and for GOx has been described according to Michaelis–Menten kinetics:

$$\frac{\mathrm{d}[S]_t}{\mathrm{d}t} = \frac{k_2[E][S]_t}{[S]_t + K_m} \qquad (4)$$

where $[S]$ is the substrate (glucose), $[E]$ enzyme concentration, and K_m the Michaelis–Menten constant.

The reduced GOx normally forms a complex with an electron acceptor (such as O_2, Figure 8a), regenerating the active, oxidized form of the enzyme. However, there is precedent for other oxidizing substrates in addition to oxygen (Figure 7), such as quinones (pH 5.6) and indophenols (acidic pH), diamines, ferrocenes, ferricyanide, methylene blue, meldola blue, phenoxazines, tetrathiafulvalene, tetracyanoquinodimethane and benzyl viologen (pH 7.5), tris(2,2'-bipyridine) cobalt(III) perchlorate, and so on, so the possibility of using an alternative electron acceptor or so-called mediator offers a clear advantage in a glucose test strip, if the measuring potential can be lowered. An enzyme *mediator* is thus a redox couple that gives efficient and rapid electron transfer to or from the enzyme. However, it has to be remembered

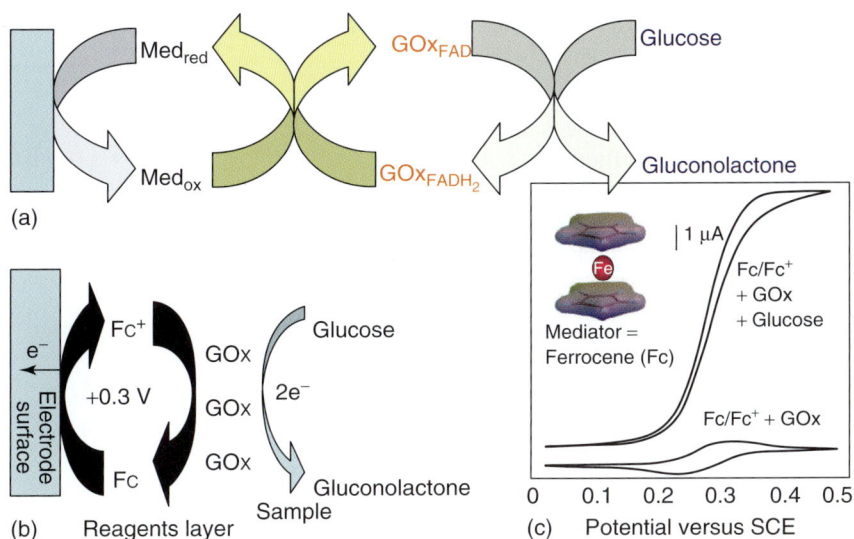

Figure 7. (a) The glucose oxidase redox system coupled with a mediator to shuttle electrons between enzyme and electrode. (b) Compound electrode components with layer on the electrode containing both enzyme and mediator. (c) Cyclic voltammogram of ferrocene/ferrocenium redox couple, recorded in the presence of GOx and in the absence and presence of glucose.

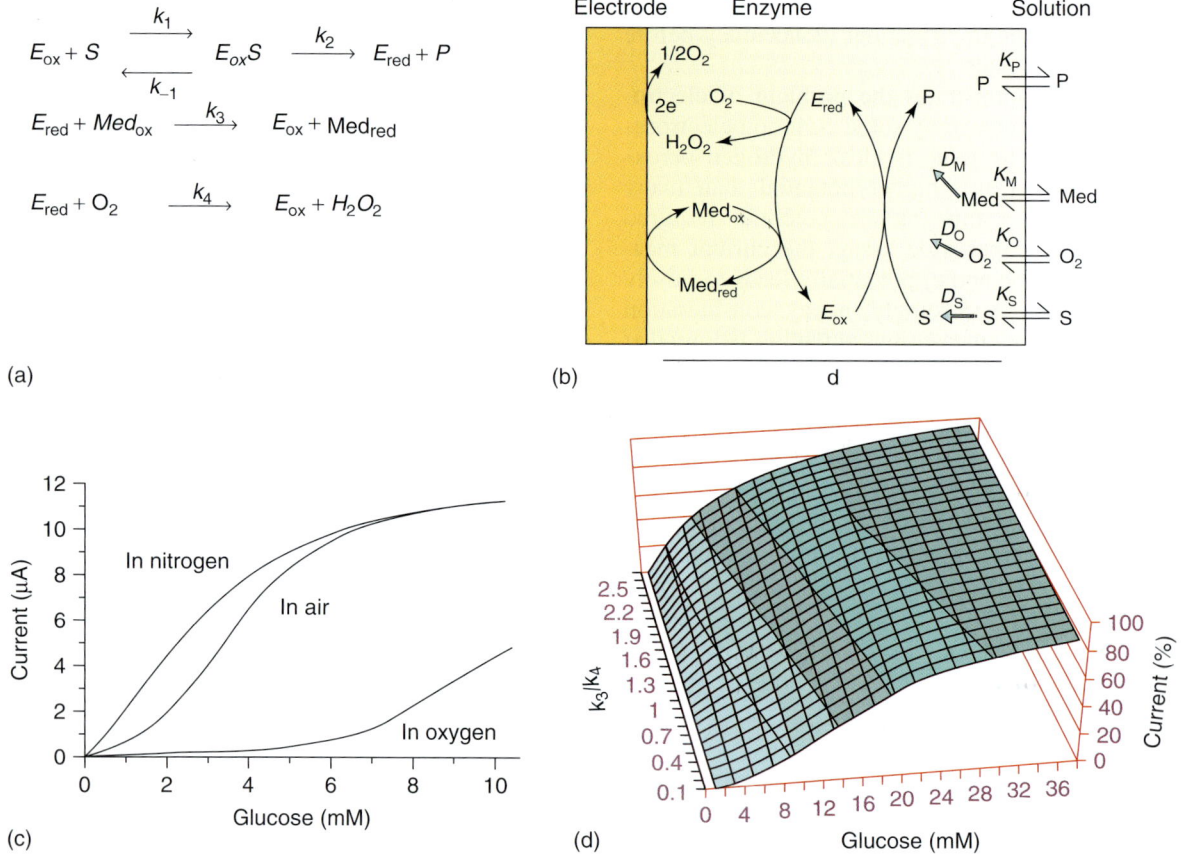

$$E_{ox} + S \xrightarrow[k_{-1}]{k_1} E_{ox}S \xrightarrow{k_2} E_{red} + P$$

$$E_{red} + Med_{ox} \xrightarrow{k_3} E_{ox} + Med_{red}$$

$$E_{red} + O_2 \xrightarrow{k_4} E_{ox} + H_2O_2$$

(a)

(b)

(c)

(d)

Figure 8. (a) Enzyme reaction pathways; (b) scheme of an enzyme layer biosensor with diffusion into the layer of thickness d; (c) calibration curve for glucose using a mediator with rate constant $10^5\,\mathrm{l\,mol^{-1}\,s^{-1}}$ reaction with glucose oxidase; and (d) effect on the calibration curve for glucose depending on k_3/k_4.

that oxygen may still be present in the sample, so that the mediator must compete efficiently with the oxygen to reoxidize the enzyme, else the current measured at the mediator's oxidation potential will not represent the total glucose concentration (Figure 8c).

3.1 Oxygen–mediator Competition

We can understand this by examining the kinetics of the reactions described in Figure 8(a). In the sensor, the reaction becomes a function of time and space with diffusion into the enzyme layer of thickness $y = d$. Taking the simplest case of one-dimensional diffusion obeying Fick's laws:

$$D_O \frac{d^2[O_2]}{dy^2} = k_4[E_{red}][O_2]$$

$$= \frac{k_2 k_1}{k_{-1} + k_2}[E_{ox}][S] \quad (5)$$

$$D_M \frac{d^2[Med_{ox}]}{dy^2} = k_3[E_{red}][Med_{ox}]$$

$$= \frac{k_2 k_1}{k_{-1} + k_2}[E_{ox}][S] \quad (6)$$

and

$$D_S \frac{d^2[S]}{dy^2} = k_1[E_{ox}][S] - k_{-1}[ES]$$

$$= \left(k_1 - \frac{k_{-1} k_1}{k_{-1} + k_2}\right)[E_{ox}][S] \quad (7)$$

where D_O, D_M, and D_S are the diffusion coefficients of oxygen, mediator, and the glucose within the enzyme layer. The electrode current is directly proportional to the first differential of the concentration of electroactive species present at the electrode surface. The model can be solved numerically to predict the calibration curve and Figure 8(c,d) reveals the mediator/oxygen competition for the enzyme (k_3/k_4), particularly at lower glucose concentration. The importance of maximizing k_3/k_4 is evident. The P_{O_2} of an average venous blood sample is approximately 40 mm Hg (equivalent to $\sim$0.06 mmol l^{-1} dissolved O_2). In contrast, for an arterial sample, P_{O_2} of approximately 110 mm Hg is typical, or 0.15 mmol l^{-1} dissolved oxygen, with capillary samples a little lower. Therefore, measurements of glucose taken at different sites may be expected to produce different results, resulting in difficulty for the user in interpretation.

3.2 Mediator Candidates

The Fe(III)/(II) redox couple is a well known and efficient, fast electron transfer reversible system. Among the substrates listed above, ferrocenes show efficient electron acceptance properties at an optimum pH (Table 2). This system was the basis for the blood glucose sensor first developed by MediSense (1987) and represented a major breakthrough in self-test technology. Test strips were manufactured by screen-printing a series of layers on to a poly(vinyl chloride) strip. It used a three-electrode cell with a silver/silver chloride electrode (SCE) employed as the combined reference/counterelectrode. There were two working electrodes, the first containing enzyme and mediator and the second containing mediator. The latter electrode's function was to provide a "blank", quantifying the level of interfering substances that may be present in the sample. Thus, the glucose concentration was measured by subtracting the current obtained at the second electrode from that obtained at the first electrode.

Adopting the same principle, other redox couples can also be used, and have become varyingly successful. For example, one particular favorite is Prussian blue:

$$2Fe(CN)_6{}^{3-} + 2e^- \leftrightarrow 2Fe(CN)_6{}^{4-} \quad (8)$$

3.3 Improving CT Between Enzyme and Mediator

However, one aspect of all these mediators is that $k_3/k_4 < 1$. Another approach is to return to the problem described by Marcus' theory of CT and to consider how to improve the CT success rate. Remembering that only a certain number of collisions between the freely diffusing mediator and the protein macromolecule will be anywhere near the redox site and have the right orientation to allow CT, any method that improves this collision success rate is likely to have some impact on the CT kinetics. An obvious way to improve this is by chemical modification of amino acid groups on the protein, so that they become derivatized by redox groups. Keeping in mind that the chemistry must be amenable to mass production, the main candidates for derivatization are lysine and cysteine groups, since their derivatization is most facile. There are two ways of approaching this: in the first case, what might be called the *brute force method* of opportunist modification of all convenient surface groups (Figure 9a); this produces a highly derivatized enzyme (e.g., GOx has 24 accessible lysine residues per monomer, 3 lying in the funnel between the monomers making up the dimer) whereas the second method gives a so-called intelligent site-directed modification, close to the redox site of the CT pathway through the protein. The latter is only realistically achieved by introducing convenient amino acid groups (most likely cysteine) at the appropriate position by site-directed mutagenesis. Unlike the wild-type enzyme available off the shelf for the brute force method, the mutant enzyme has to be cloned, expressed, and produced in quantities suitable for mass production of biosensors. Although very high CT rates can be achieved, the cost effectiveness of this has to be proved for GOx, although the method is demonstrated for other enzymes in research use.

In contrast, the more general surface modification method has been adopted in self-test glucose sensors. Foulds and Lowe were among the first to demonstrate the potential of enzyme modification in the 1980s, using a ferrocene–pyrrole conjugate modified GOx that was subsequently linked into a conducting polypyrrole polymer "wiring matrix" (Figure 9b), providing CT pathway control all the way from the redox site in the enzyme to the electrode. Their early work identified two core

Table 2. Redox mediators, giving their redox potentials in aqueous solution, and second-order rate constants (k) for their reduction by reduced GOx from *Aspergillus niger* (pH 5.5 or 7)

Mediator redox potential (mV) versus SCE	Structure	k_s (M^{-1} s^{-1})
Ferrocene 210		0.26×10^{-5}
Fc(Me)$_2$ 109		0.77×10^{-5}
Fc-CH$_2$-N-(CH$_3$)$_2$ 370		5.5×10^{-5}
Fc(COOH) 290		2.0×10^{-5}
Fc(COOH)$_2$ 403		0.26×10^{-5}
p-Ferrocenylaniline 245		Not reported
Benzoquinone 275		1.97×10^{-5}
1,4-Bis(*N*,*N*-dimethylamino) benzene 450		1.2×10^{-5}
4,4-Dihydroxybiphenyl 320		3.0×10^{-5}
Methylene blue 30		1.86×10^{-5}
Tetrathiafulvalene 150		1.28×10^{-5}
[Os(Me$_2$ bpy)$_2$ Cl$_2$]$_5$ 150		1.2×10^{-5}
[Ru(bpy)$_2$ Cl$_2$] 300		1.8×10^{-5}
[Ru(2-phenylimidazole) (1,10-phenanthroline)]PF$_6$ 280		0.75×10^7
[Os(phenylpyridine)(1,10-phenanthroline)$_2$] PF$_6$ 55		1.1×10^7

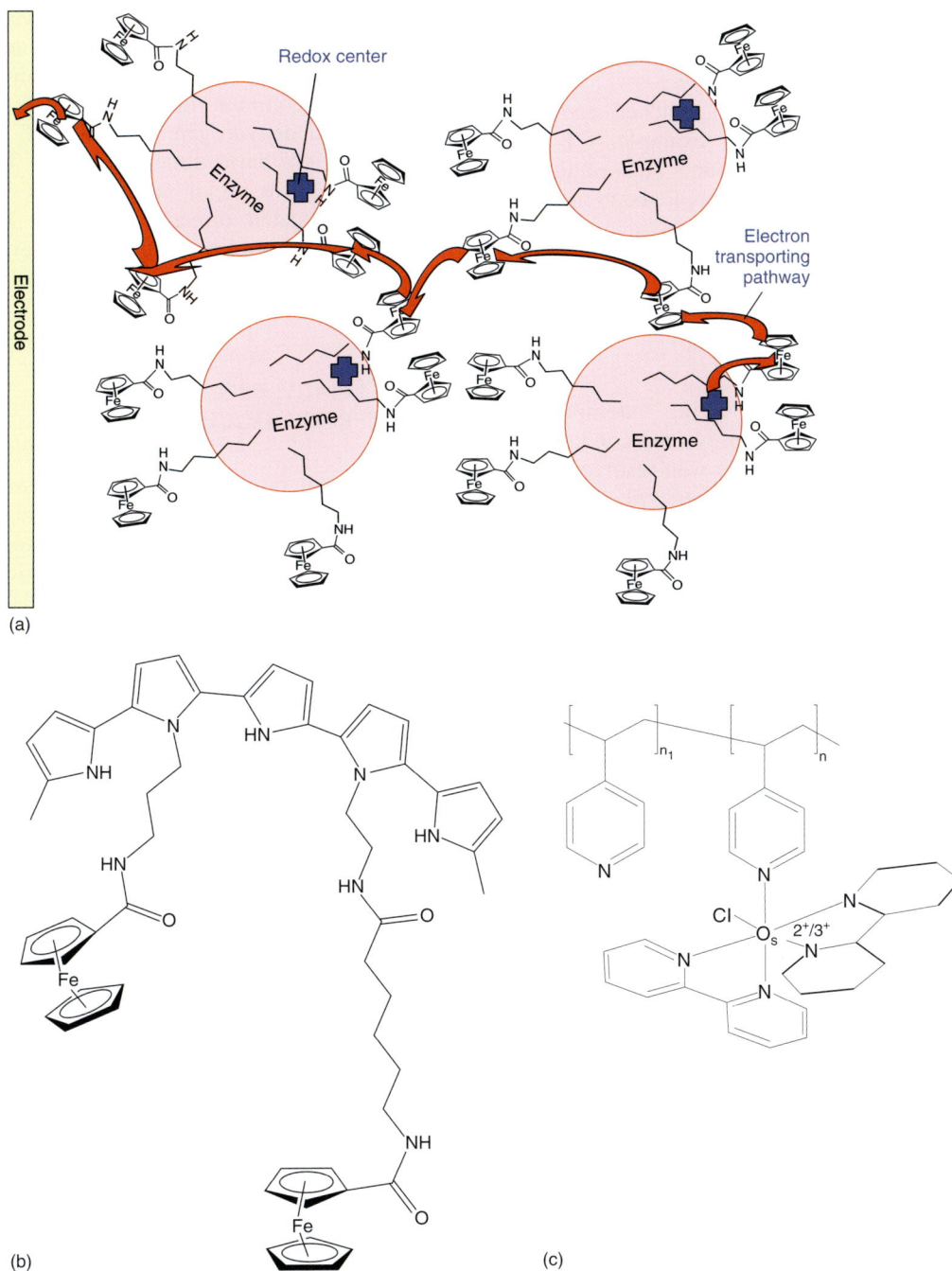

(a)

(b)   (c)

Figure 9. (a) Covalently modified with ferrocene electron-relay groups. Covalent attachment of electron mediators units at the protein periphery yields short electron transfer distances and provides an electron transporting pathway. Electrical contacting of immobilized enzymes through (b) a redox modified conducting polymer, for example shown for ferrocene and polypyrrole and (c) incorporation of mediators as complexes with polycationic redox polymers containing $Os(bpy)_2Cl$ groups attached to a poly(vinylpyridine) backbone. The Os site can be switched from Os(II) to Os(III) upon application of a potential and can also exchange electrons with the flavin site on the enzyme.

parameters that have been optimized in subsequent developments:

- The CT between the group attached and the enzyme must be fast. For example, CT for pyrrole is slow and so this cannot be used alone. A fast redox couple must be employed to modify the enzyme to ensure fast kinetics.
- A CT "wire" between the modified enzyme and electrode must be created (it is not sufficient to modify the surface of the enzyme without also making the link to the electrode).

A weakness of this early "wiring" attempt was the kinetics of the mixed redox system–conducting polymer hybrid. However, this was resolved during the subsequent decade by others, who adopted redox polymers rather than conducting polymers to move charge to the electrode. Heller led one of the research efforts synthesizing GOx "wiring" electron conducting hydrogels. These were cross-linked, water-soluble polymers containing, osmium poly pyridine–based fast redox centers (Figure 9c). By cross-linking nonprecipitating electrostatic adducts of polyanionic GOx with an excess of the polycationic osmium complex–derived, polyvinyl pyridine–based, polymer, the GOx became "wired" to the electrode, and the gel was highly permeable to sample (glucose) uptake. This system therefore also has many attractive features for GOx sensor fabrication, including the achievement of $FADH_2$ oxidation via the osmium redox couple at a potential of approximately—0.1 V versus Ag/AgCl.

3.4 Alternative Enzyme Choice

Even with fast wiring, the opportunity for short-circuiting the CT pathway with oxygen remains a possibility, since it is the natural cosubstrate for GOx. Alternative enzymes have therefore also been explored. Quinoproteins utilize quinone species as cofactors and prosthetic groups. In most bacterial quinoproteins pyrroloqinoline quinone (PQQ, Figure 10c) is tightly but noncovalently bound. Most of the quinoproteins are dehydrogenases, and often do not require any further soluble cofactors, the natural electron acceptor being a copper protein or membrane-bound ubiquinone.

These enzymes can be used in the same way as described for oxidases and have an easily accessible redox potential, that is lower than most of the common interferents. The soluble PQQ–glucose dehydrogenase (GDH EC 1.1.5.2, Figure 10a) is a basic dimeric enzyme. The glucose binding site is a wide solvent accessible cavity, located directly above the $PQQH_2$. Routine assays employ artificial electron acceptors without danger of susceptibility to oxygen interference.

4 ACCURACY, PRECISION, AND REPRODUCIBILITY[25–31]

In practice, in producing the self-test strips, the electrode is coated with the mediator and enzyme giving disposable, single-use minielectrodes that simply "plug and play" into the handheld instrument. Each of the test strips now available performs reasonably well as a home use test for glucose, but it has been a challenge to equal the YSI test in terms of accuracy and precision. Figure 11(a and b) shows an early (1999) comparison between some test strips from different manufacturers and highlights the potential for variation between and within strips. The clear improvement that has emerged over a 5-year period is also evident (Figure 11c). There are several factors needing consideration that can influence the measurement and must be accommodated in the calibration:

For example, in whole blood glucose is distributed between plasma and the red cells. By using a dilution step, such as that employed in the original YSI, a 25 times dilution into isotonic buffer ensures that virtually all of the glucose in the red cells is transferred to the solution phase. However, for a rapid single-step self-test kit, dilution is not an option. The tests can suffer from a significant hematocrit (HCT) bias. Normal levels of hematocrit are *ca* 45% and thus it would be rational to calibrate the test for this level, but if 21% change per percent HCT bias is observed, the glucose estimate will be 10% low when the hematocrit is 55%, and 10% high when the hematocrit is 35%.

Models of enzyme electrodes have allowed variations in the device parameters to be investigated. Parameters such as dissolution and sample wicking

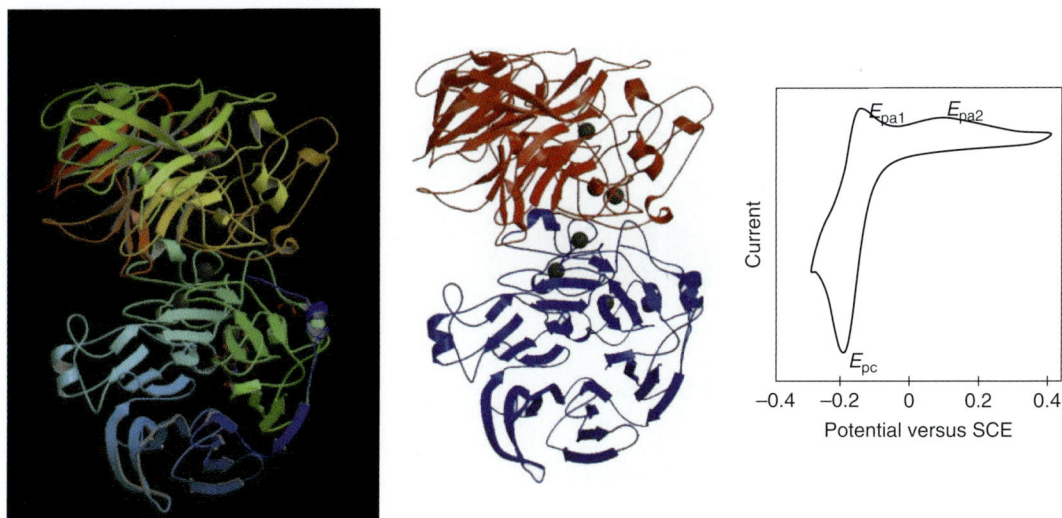

(a)

(b)

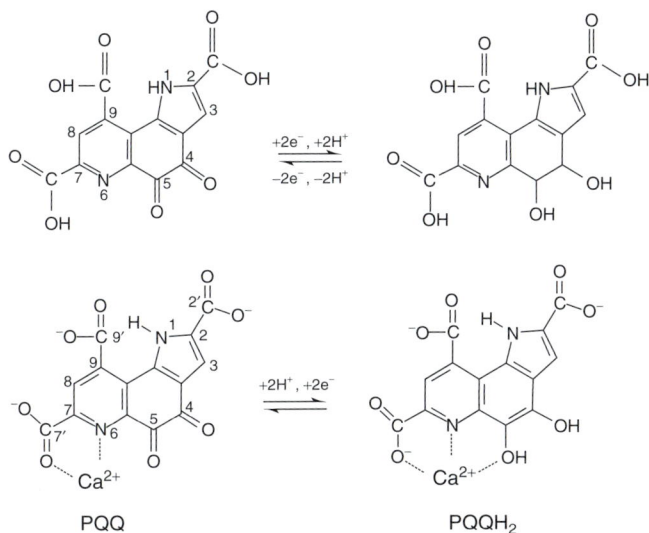

(c)

Figure 10. (a) PQQ–glucose dehydrogenase structure, from data in the protein data bank (PDB); 3D image of structure of PDB code 1c9u. (b) Cyclic voltammogram of the PQQ redox group (C electrode, 0.2 mM in phosphate buffer pH 7.2). (c) Electron and proton transfer equilibria, with stabilization of the carboxylate group through cation binding.

or integrated control and measurement techniques (steady state, dynamic, end point, etc.) can all be investigated producing very valuable models to aid in the design process.[32,33] Even from a simplified analysis of the amperometric steady-state Leyland Clark geometry, an understanding of the early GOx electrodes can be obtained. The reaction scheme for an oxidase enzyme linked assay, monitoring hydrogen peroxide or mediator, results in nonlinear second-order differential equations, from which the electrode current can be predicted, since it is related to the flux of electroactive material to the electrode surface, and hence the first differential of the concentration at $x = 0$ (Figure 12a). Applying boundary conditions, equations (5–7) can be nondimensionalized, giving the respective Thiele moduli which can be evaluated and are indicative of the reaction/diffusion balance

(a)

(b)

Comparison with reference method (YSI) for a selection of test strips ($n = 2500$).

	Difference, Mean ± SD%
Newer (2004 data)	
Test strip A	−0.6 ± 6.3
Test strip B	1.6 ± 1.1
Test strip C	0.9 ± 0.8
Test strip D	2.1 ± 1.2
Early (1999 data)	
Test strip I	−17.6 ± 19.6
Test strip II	10.8 ± 10.0
Test strip III	5.8 ± 4.7
Test strip IV	−13.5 ± 10.6

(c)

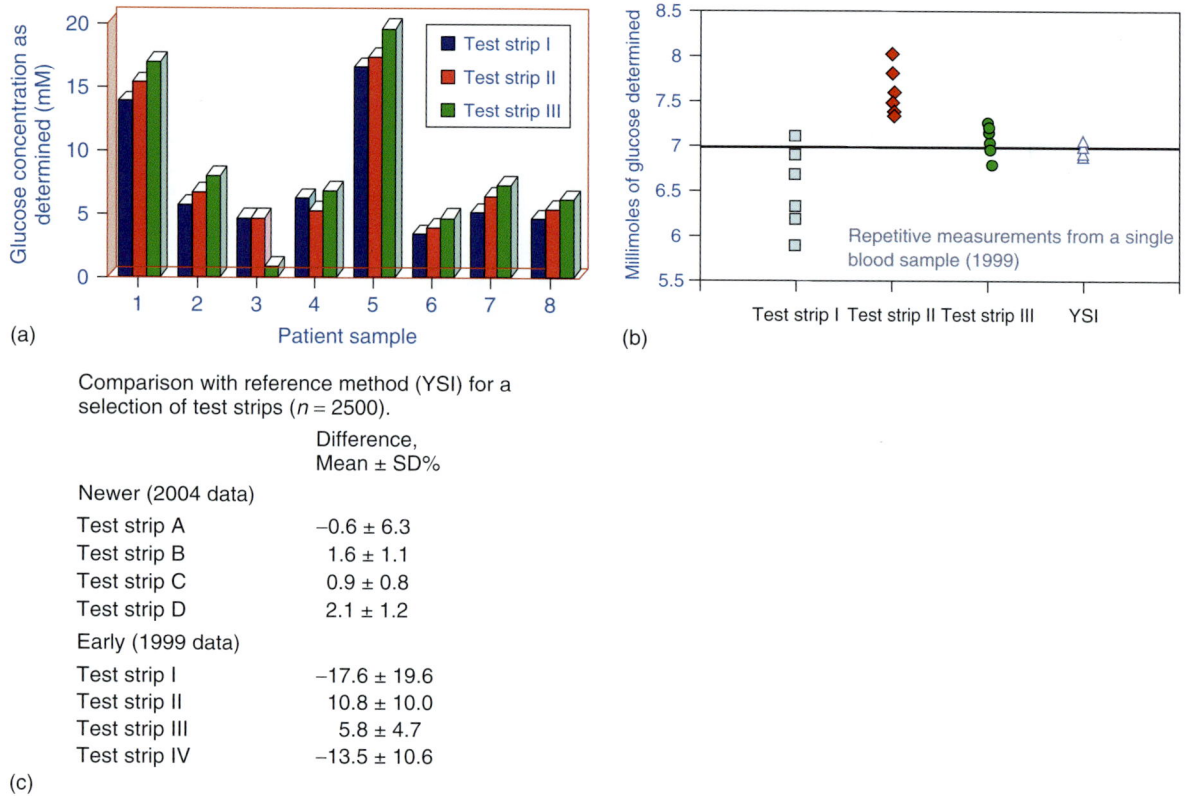

Figure 11. Comparative data collected from test strips (data compiled from multiple studies). (a) Three different test strips used to compare samples from eight patients. (b) The same sample repeatedly measured using three different test strips and compared with the YSI glucose measurement. (c) Data from different test strips in 1999 and 2004 compared with YSI reference measurement.

characteristic of the enzyme matrix of the designed biosensor:

$$\Phi_O^2 = \left(\frac{d^2 k_2 [E_T]}{D_O [O_2]_b} \right) \quad (9)$$

$$\Phi_M^2 = \left(\frac{d^2 k_2 [E_T]}{D_M [S]_b} \right) \quad (10)$$

$$\Phi_S^2 = \left(\frac{d^2 k_2 [E_T]}{D_S [S]_b} \right) \quad (11)$$

The thickness, d and enzyme loading $[E_T]$, are the two parameters within Φ^2 which are controlled in the fabrication of the biosensor. Matching Φ^2 for electrodes is important in obtaining matched output. The membrane thickness is especially sensitive, since its effect appears as a squared term.

In the field of electronics, deposition and definition is performed to precise limits of accuracy. In this field, thick- and thin-film technologies have become versatile and cheap tools in the fabrication of miniaturized electronic circuits. Biosensor fabrication has been compared with chip manufacturing, but biosensor devices differ from integrated circuit (IC) manufacture because they require a greater diversity of materials (mainly organic with potentially shorter shelf-lives) and initially lower volumes per product. This creates an obvious need for a nonsilicon approach to inexpensive disposable chemical and biological sensors and sensing systems. However, by merging conventional manufacturing technologies with IC-type fabrication methodologies, a variety of manufacturing options more suited to this type of biomedical application has been developed. Combined with the use of more cost-effective materials,

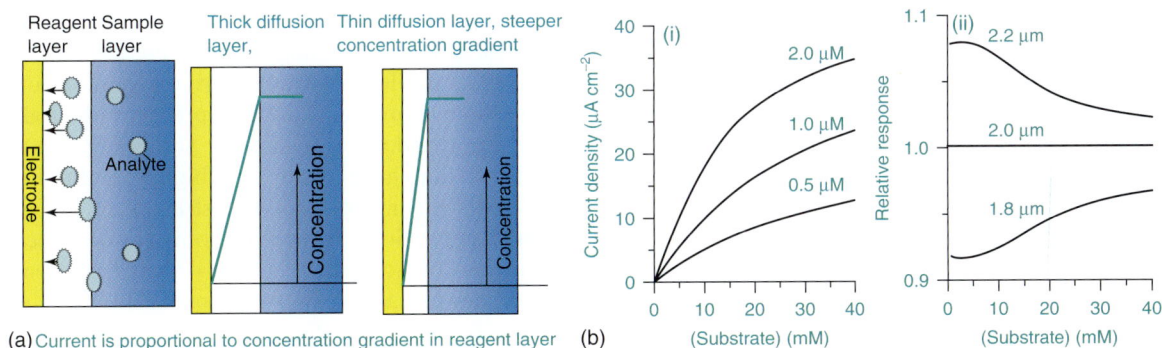

Figure 12. (a) Schematic of "thick" and "thin" enzyme layer sensors. (b) (i) Comparison of calibration curves at different enzyme layer thicknesses and (ii) effect on relative current response for $2 \pm 0.2 \, \mu m$ enzyme layer.

like polymers and plastics, this finally achieves a viable reality for the economical application of biosensor devices. In particular, deposition technologies based on a solution, ink, or paste have been successfully adapted for biosensor fabrication, but in the 1980–1990s these options were still limited in their resolution and replay accuracy, thereby also impacting the performance data (Figure 11c) and revealing one factor in understanding the early variability in test strips.

To further understand the challenges of accuracy and reproducibility, it is necessary to consider the impact of these fabrication processes and the deposition of the layers on the electrodes and their subsequent dissolution in the sample or, in the case of a nonsoluble immobilized layer, the diffusion of the sample through the layers.

Thick-film screen-printing

$100 \, \mu m$ lateral resolution; thickness control of approximately $500 \, \mu m$.

Ink-jet printing

1–500 000 drops per second; 0.5–1.5 nl per drop; 1–5-μm resolution.

Photolithography

Conventional resolution of approximately 250 nm with a 350–450-nm light source; with proteins typically 400–1800 μm; thickness control $125 \pm 10 \, \mu m$.

Figure 12(b)(ii) shows the effect of a change of just $\pm 0.2 \, \mu m$ on the deposited thickness for a

2D layered geometry as depicted in Figure 12(a). It is most striking that the variation in the relative response is also dependent on the substrate (analyte) concentration and can become a complex algorithm for the calibration process. Thus, the rewards for overcoming these issues through improved fabrication precision *or* changes in geometry *or* measurement technique are substantial.

In the decade since these first test strips were manufactured, different deposition technologies which are widely used tools for the fabrication of electronic circuits have been further adapted in their own right for biosensors. Screen-printing has become available with a number of special pastes with lower firing temperatures, leading to more options for biological sensor fabrication. Using this thick-film technology for the whole device (all components: electrodes to enzyme) is cheaper and it becomes possible to produce disposable sensors at a rather low price for a low- to medium-scale production.

A so-called soft lithography printing process, called *microcontact printing*, is also showing potential for printing biomolecules. The process prints directly from a patterned elastomeric stamp. The elastomer can compensate for some degree of local surface-roughness up to approximately 1 μm, depending on the material properties, whereas macroscopic warp ($>100 \, \mu m$) is compensated by the flexibility of the backplane. In the past, printing and soft lithography have been used only for single component transfer but new contact printing allows precise alignment on the final substrate. Proteins (e.g., enzymes such as GOx) adsorb spontaneously to the hydrophobic surface of a stamp (e.g., polydimethylsiloxane (PDMS) with a high Young's modulus) by incubation with an aqueous solution and the main advantage is the capability of processing large surfaces in one step with arbitrary patterns. There are two ways to print different components onto a single substrate using soft-lithographic techniques: (i) successive, iterative inking and printing (Figure 13) and (ii) parallel inking of a stamp followed by a single printing step.

However, perhaps the biggest impact in deposition technologies for GOx sensors so far, has

come in ink-jet printing technology. Over the past 20 years, this technology has been a dominant player in the low-cost color printer market but it has also become accepted as a precision microdispensing technology. There are some key characteristics that have made it a highly useful fluid microdispensing tool for industrial, medical, and other applications. It requires no masks or screens, it is noncontact and it is data-driven (processing information can be written directly in the CAD package and stored digitally).

Manufacturing of devices for biomedical diagnostics using ink-jet technology dates to the 1980s,

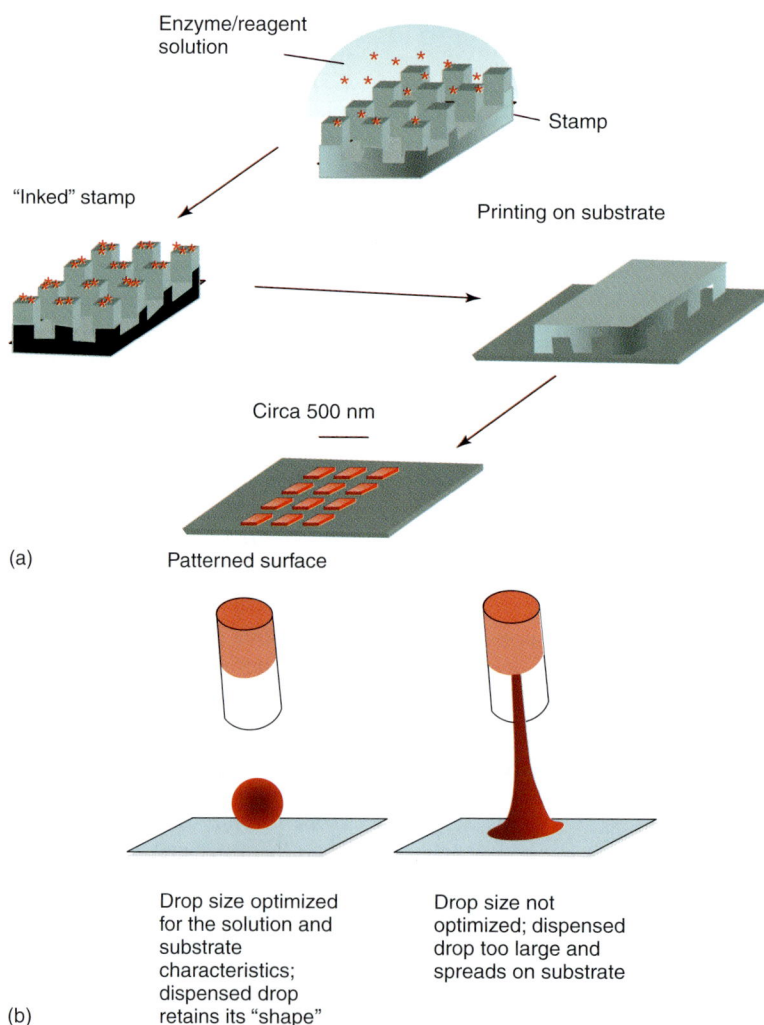

Figure 13. (a) Contact printing process applied for enzyme printing. (b) Impact of metered volume on the dispensed drop, depending on the substrate properties (e.g., hydrophobicity).

with the fabrication of glucose sensors being one of the first examples, but the billion-dollar pregnancy test kit market predominantly driving the early technology. Precision *micro*dispensing based on ink-jet technology has been used in medical diagnostics since the early 1990s. It can reproducibly dispense spheres of fluid with diameters of 15–100 μm (2 pl–5 nl). Perhaps a key driver toward better resolution and repeatability came in the effort to miniaturize biological assays and to conduct many assays in parallel for gene chip technology. This required high-density resolution of DNA probes, say better than 75 μm spots on <200 μm centers.

To be able to use ink-jet methods, the working fluid must be low viscosity, nearly Newtonian, and free of particles of the size of the orifice diameter. Controlled sample dispensing is then the critical-step and precise metering and application of the dispensed component (droplet) is essential for all quantitative measurements.

The dispensing process is optimized according to the characteristics of the solution (ink) *and* the substrate materials used for the wetted surfaces (Figure 13b) setting different limitations for proteins to those established in DNA printing. Proteins adhere to most surfaces but can loose their functionality (denature), so that controlling the solution and wetted surface area is critical.

The general trend toward miniaturization (irrespective of the deposition technology employed) offers several advantages. In small volumes, the ratio of surface area to enclosed volume is considerably greater. This results in the surface effects and capillary and adsorption phenomena dominating instead of the volume effects at higher dimensions. Electrochemical double layers have considerable influence, since their thickness, 10–100 nm, becomes a significant proportion of the total enzyme thickness and mass transport by diffusion becomes fast, so that mixing and chemical reactions can be accelerated accordingly. This means that the impact on Φ^2 of variation in deposition is lessened and, of course, the measurement result is obtained faster. Furthermore, the components are used more efficiently, and lower volumes of samples or reagents are required. As the sample size is reduced to a minimum, the *total* amount of glucose within the sample can become consumed within the measurement period. If the blood volume used can be accurately sampled, a coulometric assay then offers a viable alternative to steady-state current measurement.

In conclusion, for the successful self-test glucose biosensor, the test must be simple to run and convenient for use by a layperson as part of the normal daily routine. Future refinements will no doubt further increase the reliability and precision of each of the tests, so that the evolution of glucose assay methods will continue to be central to the development of therapies for diabetes mellitus. This chapter has given a flavor of a few of the issues that have been considered in the history of the glucose test strip. However, blood glucose assay with this methodology, still requires frequent blood sampling with the associated degree of inconvenience. Traditionally for practical reasons, blood is collected from the fingertips, a sensitive part of the body, so the sampling should be as painless as possible, invade a minimum sampling site (frequent testing leads to heavily calloused hands) and require a small volume (<2–3 μl).

Studies suggest that some people with diabetes may choose not to monitor their glucose levels and strive for close blood glucose control because of the obtrusiveness of blood sampling. The frequency of blood sampling that is acceptable to most patients is still less than the frequency of blood glucose fluctuations and does not detect spontaneous hypoglycemia or other rapid blood glucose imbalances. As useful as the present glucose assay technologies have become, an important feature that is not part of a disposable self-test strip is continuous monitoring.[34–39] This would permit routine therapeutic intervention. Continuous glucose sensing provides a basis for insulin administration at more appropriate dosages and timing, or for automatic insulin delivery from a pump and could also be employed beneficially in parallel with other existing or potential forms of insulin replacement, such as transplantation or hybrid islet devices. Nevertheless, even though the clinical benefit of continuous measurement is easy to defend, continuous sensing technologies still have to be unobtrusive and more convenient than discrete sensing approaches to achieve widespread application and replacement of the disposable self-test kit.

REFERENCES

1. A. E. G. Cass, *Biosensors, A Practical Approach*, IRL Press at Oxford University Press, 1990.

2. E. A. H. Hall, *Biosensors*, Open University Press/Wiley, 1990.

3. E. Magner, Trends in electrochemical biosensors. *Analyst*, 1998, **123**, 1967–1970.

4. L. S. Kuhn, Biosensors: blockbuster or bomb? electrochemical biosensors for diabetes monitoring. *The Electrochemical Society Interface*, 1998, **7**(4), 26–31.

5. N. R. Stradiotto, H. Yamanaka, and M. V. B. Zanoni, Electrochemical sensors: a powerful tool in analytical chemistry. *Journal of the Brazilian Chemical Society*, 2003, **14**(2), 159–173.

6. D. C. Klonoff, Continuous glucose monitoring roadmap for 21st century diabetes therapy. *Diabetes Care*, 2005, **28**, 1231–1238.

7. J. D. Newman and A. P. F. Turner, Home blood glucose biosensors: a commercial perspective. *Biosensors and Bioelectronics*, 2005, **20**(12,), 2435–2453.

8. S. P. Mohanty and E. Kougianos, Biosensors: a tutorial review. *IEEE Potentials*, 2006, **25**(2), 35–40.

9. The Diabetes Control and Complications Trial Research Group, The effect of intensive treatment of diabetes on the development and progression of long-term complications in insulin-dependent diabetes mellitus. *New England Journal of Medicine*, 1993, **329**(14), 977–986.

10. R. A. Marcus and N. Sutin, *Electron Transfer Reactions in Chemistry: Theory and Experiment*, Nobel Lecture Dec, 1985, http://nobelprize.org/nobel_prizes/chemistry/laureates/.

11. K. H. Mancy, D. A. Okun, and C. N. Reilley, A galvanic cell oxygen analyzer. *Journal of Electroanalytical Chemistry*, 1962, **4**, 65–92.

12. L. C. Clark and C. Lyons, Electrode systems for continuous monitoring in. cardiovascular surgery. *Annals of the New York Academy Sciences*, 1962, **102**, 29–45.

13. S. J. Updike and G. P. Hicks, The enzyme electrode. *Nature*, 1967, **214**, 986–988.

14. A. E. Cass, G. Davis, G. D. Francis, H. A. Hill, W. J. Aston, I. J. Higgins, E. V. Plotkin, L. D. Scott, and A. P. Turner, Ferrocene-mediated enzyme electrode for amperometric determination of glucose. *Analytical Chemistry*, 1984, **56**, 667–671.

15. C. Bourdillon, C. Demaille, J. Moiroux, and J. M. Saveant, New insights into the enzymic catalysis of the oxidation of glucose by native and recombinant glucose oxidase mediated by electrochemically generated one-electron redox cosubstrates. *Journal of the American Chemical Society*, 1993, **115**, 2–10.

16. J. Kulys, T. Buch-Rasmussen, K. Bechgaard, V. Razumas, J. Kazlauskaite, J. Marcinkeviciene, J. B. Christensen, and H. E. Hansen, Study of the new electron transfer mediators in glucose oxidase catalysis. *Journal of Molecular Catalysis*, 1994, **91**, 407–420.

17. V. N. Goral and A. D. Ryabov, Reactivity of the horseradish peroxidase compounds I and II toward organometallic substrates. A stopped-flow kinetic study of oxidation of ferrocenes. *Biochemistry and Molecular Biology International*, 1998, **45**, 61–71.

18. A. D. Ryabov, V. S. Kurova, V. N. Goral, M. D. Reshetova, J. Razumiene, R. Simkus, and V. Laurinavicius, p-Ferrocenylaniline and p-ferrocenylphenol: promising materials for analytical biochemistry and bioelectrochemistry. *Chemistry of Materials*, 1999, **11**, 600–604.

19. A. D. Ryabov, Y. N. Firsova, A. Y. Ershov, and I. A. Dementiev, Spectrophotometric kinetic study and analytical implications of the glucose oxidase-catalyzed reduction of [MIII(LL)$_2$Cl$_2$] + complexes by D-glucose (M = Os and Ru, LL = 2, 2′-bipyridine and 1,10-phenanthroline type ligands). *Journal of Biological Inorganic Chemistry*, 1999, **4**, 175–182.

20. A. D. Ryabov, Y. N. Firsova, V. N. Goral, V. S. Sukharev, A. Y. Ershov, C. Lejbolle, M. J. Bjerrum, and A. V. Eliseev, Horseradish peroxidase-catalyzed oxidation of cis-[RuII(LL)2XY] complexes by hydrogen peroxide (LL = 2, 2-bipyridine and 1,10-phenanthroline): equilibria, kinetics, mechanism, and active site reassembly. *Inorganic Reaction Mechanisms*, 2000, **2**, 343–360.

21. N. Anicet, A. Anne, C. Bourdillon, C. Demaille, J. Moiroux, and J.-M. Saveant, Electrochemical approach to the dynamics of molecular recognition of redox enzyme sites by artificial cosubstrates in solution and in integrated systems. *Faraday Discussions*, 2000, **116**, 269–279.

22. A. D. Ryabov, V. S. Sukharev, L. Alexandrova, R. Le Lagadec, and M. Pfeffer, New synthesis and new bio-application of cyclometalated ruthenium(II) complexes for fast mediated electron transfer with peroxidase and glucose oxidase. *Inorganic Chemistry*, 2001, **40**, 6529–6532.

23. C. Loechel, A. Basran, J. Basran, N. J. Scrutton, and E. A. H. Hall, Using triethylamine dehydrogenase in an enzyme linked amperometric electrode. Rational design engineering of a 'wired' mutant. *Analyst*, 2003, **128**, 889–898.

24. A. W. Bott, Investigation of enzyme-mediated electron transfer using DigiSim®. *Current Separations*, 2004, **20**, 4–8.

25. J. C. N. Chan, R. Y. M. Wong, C.-K. Cheung, P. Lam, C.-C. Chow, V. T. F. Yeung, E. C. Y. Kan, K.-M. Loo, M. Y. L. Mong, and C. S. Cockram, Accuracy, precision and user-acceptability of self blood glucose monitoring machines. *Diabetes Research and Clinical Practice*, 1997, **36**, 91–104.

26. L.-J. Cartier, P. Leclerc, M. Pouliot, L. Nadeau, G. Turcotte, and B. Fruteau-de-Laclos, Toxic levels of acetaminophen produce a major positive interference on glucometer elite and accu-chek advantage glucose meters. *Clinical Chemistry*, 1998, **44**, 893–894.

27. R. Weitgasser, B. Gappmayer, and M. Pichler, Newer portable glucose meters—analytical improvement compared with previous generation devices? *Clinical Chemistry*, 1999, **45**, 1821–1825.

28. Z. Tang, J. H. Lee, R. F. Louie, and G. J. Kost, Effects of different hematocrit levels on glucose measurements with handheld meters for point-of-care testing. *Archives of Pathology and Laboratory Medicine*, 2000, **124**, 257–266.

29. J. C. Boydand and D. E. Bruns, Quality specifications for glucose meters: assessment by simulation modeling of errors in insulin dose. *Clinical Chemistry*, 2001, **47**(2), 209–214.

30. P. B. Böhme, M. Floriot, M.-A. Sirveaux, D. Durain, R. O. Ziegler, P. Drouin, and B. Guerci, Evolution of analytical

performance in portable glucose meters in the last decade. *Diabetes Care*, 2003, **26**, 1170–1175.

31. W. L. Clarke, S. Anderson, L. Farhy, M. Breton, L. Gonder-Frederick, D. Cox, and B. Kovatchev, Evaluating the clinical accuracy of two continuous glucose sensors using continuous glucose—error grid analysis. *Diabetes Care*, 2005, **28**(10), 2412–2417.

32. N. Martens and E. A. H. Hall, Model for an immobilized oxidase enzyme electrode in the presence of two oxidants. *Analytical Chemistry*, 1994, **66**, 2763–2770.

33. M. E. G. Lyons, Modelling the transport and kinetics of electroenzymes at the electrode/solution interface. *Sensors*, 2006, **6**, 1765–1790.

34. D. Moatti-Sirat, G. Velho, and G. Reach, Evaluating in vitro and in vivo the interference of ascorbate and acetaminophen on glucose detection by a needle-type glucose sensor. *Biosensors and Bioelectronics*, 1992, **7**, 345–352.

35. W. Kerner, M. Kiwit, B. Linke, F. S. Keck, H. Zier, and E. F. Pfeiffer, The function of a hydrogen peroxide-detecting electroenzymatic glucose electrode is markedly impaired in human subcutaneous tissue and plasma. *Biosensors and Bioelectronics*, 1993, **8**, 473–482.

36. D. Gough and J. C. Armour, Development of the implantable glucose sensor: what are the prospects and why is it taking so long? *Diabetes*, 1995, **44**, 1005–1009.

37. J. Pickup, Technology advances in glucose monitoring. *Pediatric Diabetes*, 2002, **B**, 125–126.

38. V. Bozzetti, M. Viscardi, R. Bonfanti, A. Azzinari, F. Meschi, E. Bognetti, and G. Chiumello, Results of continuous glucose monitoring by GlucoWatch® biographer in a cohort of diabetic children and adolescents under real-life conditions. *Pediatric Diabetes*, 2003, **4**, 57–58.

39. I. M. Wentholt, J. B. Hoekstra, and J. H. Devries, A critical appraisal of the continuous glucose–error grid analysis. *Diabetes Care*, 2006, **29**(8), 1805–1811.

Field-Operable Biosensors for Tropical Dispatch

Rodica E. Ionescu,[1] Victoria Yavelsky,[2] Tamar Amir,[2] Natalie Gavrielov[2] and Leslie Lobel[2]

[1] *Centre de Genie Electrique de Lyon, Ecole Centrale de Lyon, Lyon, France and* [2] *Department of Virology, Ben-Gurion University of the Negev, Beer-Sheva, Israel*

1 INTRODUCTION – INFECTIOUS DISEASES

Emerging infectious diseases generally arise from infectious agents that circulate in their respective reservoir before making a jump to humans. Although many of these emerging infectious agents can and have caused significant morbidity and mortality, only recently has there been a concerted effort for rapid pathogen identification. The National Institutes of Health of the United States has prioritized pathogens in large part owing to the recent awareness of the potential use of these dangerous infectious diseases as agents of bioterror or biowarfare (Table 1).[1] Nonetheless, for successful prevention and treatment of infectious agents that emerge from either the environment or result from nefarious plots, multiple detection technologies need to be developed to serve different fields, such as bioscience research, medical diagnostics, analytical screening for food processing, and environmental testing.

In tropical areas, especially equatorial regions of the world, pathogens such as bacteria and viruses are responsible for grave respiratory diseases (such as measles, respiratory syncytial virus, tuberculosis, TB) and sexually transmitted diseases (such as gonorrhea and human immunodeficiency virus, HIV). In addition, many diseases can be spread through contaminated water and food resources, since clean water and sanitary conditions are often a luxury. One prime example is *Typhoid fever* (or enteric fever) caused by *Salmonella* Typhi that leads to about 16.6 million cases and 600 000 deaths annually with the vast majority of cases occurring in southeast Asia, Africa, and South America.[2]

Insects are major vectors of emerging diseases worldwide as they transmit deadly tropical diseases by picking up the pathogen from an infected person or animal and transmitting it to uninfected organisms during the feeding process. This is the case for the single-stranded RNA viruses of the *Flaviviridae* family, which replicate in the cytoplasm of infected cells. The members of this family consist of very small (50 nm) spherical enveloped virions and are transmitted to humans by mosquitoes predominantly in Africa, the Americas, and Australia. Representative viruses of the *Flaviviridae* family are West Nile virus (WNV), dengue virus (DEN), yellow fever (YF) virus, and Japanese encephalitis virus.[3]

The importance of studying such viruses can not be overstated since it has been reported, for example, that a rapid spread of WNV may pose a significant public health problem in the coming

Handbook of Biosensors and Biochips. Edited by Robert S. Marks, David C. Cullen, Isao Karube, Christopher R. Lowe and Howard H. Weetall.
© 2007 John Wiley & Sons, Ltd. ISBN 978-0-470-01905-4.

Table 1. Biological pathogens

National institute of allergy and infectious diseases (NIAID) category A, B, and C priority pathogens

Category A	Category B
Bacillus anthracis (anthrax)	*Burkholderia pseudomallei*
Clostridium botulinum (botulinum neurotoxin)	*Coxiella burnetii* (Q fever)
Yersinia pestis (plague)	*Brucella* species (brucellosis)
Variola major (smallpox) and other pox viruses	*Burkholderia mallei* (glanders)
Viral hemorrhagic fevers	Ricin toxin (from *Ricinus communis*)
Arenaviruses	Epsilon toxin of *Clostridium perfringens*
LCM, Junin virus,	Staphylococcus enterotoxin B
Machupo virus	Typhus fever (*Rickettsia prowaxekii*)
Guanarito virus,	
Lassa fever	Food and waterborne pathogens
Bunyaviruses	Bacteria
Hantaviruses	Diarrheagenic *E. coli*
Rift Valley fever	Pathogenic *Vibrios*
Flaviviruses	*Shigella* species
Dengue	*Salmonella*
Filoviruses	*Listeria monocytogenes*
Ebola	*Campylobacter jejuni*
Marburg	*Yersinia enterocolitica*
	Viruses (Caliciviruses)
Category C	Protozoa
Tickborne hemorrhagic fever viruses	*Cryptosporidium parvum*
	Cyclospora cayatanensis
Crimean–Congo	*Giardia lamblia*
Hemorrhagic fever virus	*Entamoeba histolytica*
Tickborne encephalitis viruses	*Toxoplasma*
Yellow fever	Microsporidia
Multidrug-resistant TB	Additional viral encephalitides
Influenza	West Nile virus
Other rickettsias	LaCrosse
Rabies	California encephalitis
	VEE
	EEE
	WEE
	Japanese encephalitis virus
	Kyasanur Forest virus

http://www.niaid.nih.gov/biodefense/bandc_priority.htm.

years.[4] Thus far, no licensed human WNV vaccine is available to protect at-risk populations from WNV-induced illness. On the other hand, YF virus causes a viral hemorrhagic fever in humans with a fatality rate that exceeds 50%,[5] affecting predominantly the human population in sub-Saharan Africa and tropical South America despite the availability of a safe and effective vaccine.

Dengue is the most common mosquito-borne viral disease in humans; however, viral pathogenesis has been difficult to examine because there are no laboratory or animal models of the disease. Nonetheless, indirect evidence suggests that dengue viruses differ in virulence, including their pathogenesis in humans.[6] It was experimentally demonstrated that dengue serotype 2 viruses causing fatal dengue hemorrhagic fever[7] epidemics (southeast Asian genotype) could outcompete viruses that cause dengue fever only (American genotype). This implies that the southeast Asian genotype will continue to displace other strains.[8]

The origin in nature of another deadly virus, Ebola, remains a mystery and humans are easily infected by close contact with patients. Since Ebola infections have mortality rates of 72% on average,[9] rapid laboratory diagnosis of Ebola hemorrhagic fever is very important for preventing the spread of infection in a population. Indeed, much research has focused on the rapid diagnosis of this and other deadly viruses, as the potential decimation of a large civilian or military population is apparent in the absence of a rapid response and quarantine based on accurate information. In this same vein, one of the most dramatic diseases that has created a global pandemic, AIDS, has been characterized by the rapid emergence and devastating spread of human immunodeficiency virus type 1 (HIV-1) accounting for more than 56% of all global infection.[10] Major outbreaks are occurring in every country of southern Africa, with some regions reporting adult prevalence rates as high as 40%.[11] The rapid spread of HIV has occurred largely in the absence of appropriate diagnostic and monitoring systems. Countries that have deployed diagnostic and monitoring for HIV along with treatment, such as the United States and interestingly Uganda, have made significant progress in the battle against AIDS.

To fight viral and bacterial diseases, numerous techniques of direct or indirect viral or bacterial detection are being or have been developed. These include both immunological and nucleic acid–based analytics. Most immunological detection methods used for bacterial pathogen identification are also applicable for the detection of viruses with nearly identical sample preparation formats.[12] However, sample preparation for nucleic acid–based detection of viruses, such as separation and isolation of viral DNA or RNA, is much simpler and faster than that for bacterial detection. Nonetheless, immunological detection of viruses is more difficult than that of bacteria. Viral antigens are often changed or altered allowing the virus to escape detection by immunological methods. Since most detection methods target either viral antigens displayed on the surface of the host cells or host antibodies produced as a result of immune response to these antigens, such methods are not very useful for detecting viral infection at its onset. However, sensitive immunological methods have been developed to detect a variety of viruses for example, herpes simplex viruses (HSV)[13] and hepatitis A virus.[14]

Generally speaking, traditional detection methods such as plating, culturing, staining, and biochemical tests are often time-consuming, requiring extensive training in microbiology and lengthy time for obtaining results. The situation is somewhat improved with automated systems, which use similar routine procedures but have minimal human input. Automated systems incorporating polymerase chain reaction (PCR),[15] DNA and RNA probe techniques,[16] flow cytometry,[17] immunomagnetic separation,[18] as well as quartz crystal microbalance (QCM),[19] surface plasmon resonance (SPR),[20] and microelectromechanical devices[21] can identify microorganisms in few hours, although they require expensive equipment and specialized reagents. Furthermore, they are either too costly for wide applicability or have difficulty discriminating false-positive and false-negative results.

For direct pathogen detection, various molecular methods have been introduced over the years, like classical PCR, real-time PCR, multiplex PCR, improved in situ hybridization assays, and in situ PCR.[22] Techniques that are still under development include PCR robotics, PCR for protein detection (DNA tags), novel nucleic acid hybridization methods, amplification without thermocycling, and macro- and microarrays. With respect to indirect detection, recombinant proteins and novel monoclonal antibodies are being employed for immunochromatographic lateral flow assays, enzyme-linked immunosorbent assays (ELISA), time-resolved fluorescence (TRF), immunomagnetic separation–electrochemiluminescence (ECL) and biosensors.[23]

As previously mentioned, conventional methods for pathogen detection require time-consuming steps, although they yield reproducible and reliable data. As such, methods including ELISA and PCRs are still actively employed for pathogen detection. In addition, labor-intensive techniques such as electron microscopy are also a useful tool and have two advantages over ELISA and nucleic acid–based tests. After a simple and fast negative stain preparation, "open view" of electron microscopy allows morphologic sample identification (10 min) and differential diagnosis of the various agents contained in the specimen.[24] Electron microscopy can be applied to many types of samples (e.g., DEN, Rift valley virus,[25]) and can also hasten routine cell-culture diagnosis. However, to fully exploit and control the potential of pathogen diagnostics using electron microscopy, this technique must be run in parallel with other standard laboratory assays.[26] Furthermore, it is too expensive and complicated for routine use.

For speed and sensitivity in direct and indirect pathogen detection, portable, rapid, and miniaturized medical and environmental diagnostic kits must be developed. To this end, biosensors have become pivotal as efficient detectors of local bacteria or viral infectious agents and require few "particles" to stimulate signal generation. For practical applications, biosensors can detect much fewer than 1000 microorganisms per ml of suspension. In terms of experimental timing, whereas ELISA and PCR tests are performed over 10–28 and 4–6 h, respectively, biosensors can make accurate determinations within 2 h under working conditions and with enhanced sensitivity.[27,28]

2 BIOSENSOR CAPABILITIES FOR FIELD OPERATIONS

Owing to their unique features, biosensors have been pursued by the military as field-operable, real-time instruments to detect and identify pathogenic microorganisms in contaminated areas. The origin of biosensors is in the union of molecular biology with electronic information technology. In essence, they make use of specific technologies to qualify or quantify parameters of biomolecule–analyte reactions. As such, they can be classified according to their bioreceptor or transducer type. Examples of various systems in operation and development are listed in Table 2.

Bioreceptors are key for the specificity of biosensor technology and are responsible for binding the analyte of interest to the sensor for measurement. They can take many forms, and the different bioreceptors that have been used are as numerous as the different analytes that have been monitored using biosensors. However, bioreceptors can generally be classified into five major categories: (i) antibody/antigen, (ii) enzymes, (iii) nucleic acids, (iv) cellular structures/cells, and (v) biomimetic.

Depending on the method of signal transduction, a number of detection technologies are being investigated and validated for microorganism detection such as electrochemical-, piezoelectric-, optical-, acoustic-, and thermal-based systems.[29] In addition, biosensors can be classified into sensors for direct or indirect (labeled) analyte detection. Direct biosensors measure physical changes induced by immunocomplex formation, whereas the indirect ones are generally based on detection of products resulting from a biochemical reaction. Another biosensor classification distinguishes between biocatalytic and bioaffinity sensors. The biocatalytic biosensor uses primarily enzymes as the biological material, catalyzing a biochemical reaction that emits a signal. The bioaffinity biosensors use specific binding proteins (antibodies), nucleic acids, or whole cells and are designed to monitor the binding phenomena. If antibodies or antibody fragments are the biological elements the device is called an *immunosensor*. With respect to bacteria, biosensors can also be grouped into sensors operating in batch (intermittent) and continuous (monitoring) mode.

Overall, biosensors are efficient diagnostic tools, predominantly in point-of-care testing, based on molecular recognition technologies that include antibodies, peptides, aptamers, and nucleic acids that are specific for target pathogens. Additional strategies are still under development, which will finally transform biosensor technologies into field-operable instruments to provide rapid, real-time pathogen detection, reducing substantially the labor, time, and cost as compared to classical laboratory testing.

2.1 Immunosensors

Currently, there are three popular types of immunosensors including electrochemical (potentiometric, amperometric, or conductometric/capacitive), optical, and microgravimetric. Electrochemical devices have some advantages over optical sensors in that they can operate in turbid media, offer comparable instrumental sensitivity, and are more amenable to miniaturization. Modern electroanalytical techniques have very low detection limits (down to 10^{-9} M), which can even be achieved using small volumes (1–20 μl) of samples.[30]

The immunosensors can either be developed as direct (nonlabeled) or indirect (labeled) devices. Whereas the direct sensors detect the physical changes during formation of an immunocomplex, the indirect ones use signal-generating labels when incorporated into the complex. Labels that are used include a spectrum of compounds ranging from enzymes such as peroxidase, glucose oxidase, catalase, or alkaline phosphatase to electroactive

Table 2. Examples of devices useful for developing biodefense programs

Type	Features
Bacillus microchip	Detects *Bacillus anthracis*, and identifies it from among other generic members such as *Bacillus thuringiensis*, *Bacillus subtilis*, and *Bacillus cereus*
BIDS	Biological integrated detection system detects through a laser-based sensor large areas under biological attack. Also functions as a warning system. BIDS is also capable of speeding-up treatment of biowarfare casualties by narrowing down the range of identities of specific biological agents used as bioweapons. Variations of the system allow for the detection of between to four and eight biological warfare agents in lees than an hour. The system is transportable for use by vehicle and laboratory-designed aircraft
CRP	Critical reagent program designed to provide a ready available resources of antibodies, antigens, and gene probes for use in field detection and neutralization of biological warfare agents
IBAD	Interim biological detector designed as a manual handheld assay for use on ships with links to aural and visual alarms, IBAD provides advance warning of the presence of biological warfare agents through immunochromatographic analysis
IOTA	Voltametric instrument comprising miniaturized electrodes for optional use with antibodies, enzymes, organic dyes, and molecules for detection of heavy metals in body fluids, microorganisms, pesticide contaminants in foods and potable water, and so on, accompanied by graphic computation
JBPDS	Joint biological point detection system is designed for use in protecting ports, naval ships, airfields, and as a portable warning system in conjunction with meteorological data. Automatic detection and identification of up to 10 biological warfare agents in less than 30 h is feasible. Enhanced versions of the systems focus on providing rapid facilities for the identification of 25 biological warfare agents thus speeding-up choice of treatment of casualties
LRBSDS	Long-range biological standoff detection system possesses a detection range of 50 km and through a laser eye distinguishes between artificial and natural aerosol clouds. The system has also been designed for complementary use with BIDS
LIBRA	Comprises quartz crystal resonators coated with optional layers of antibodies, enzymes, and so on, for use in identification of microorganisms, pesticides, and other dangerous organic molecules and chemical gases with computer prints
MAGIChip	Microarray of gel-immobilized compounds that identify simultaneously numerous biological agents through reliance on microbe-specific gene sequences and microbe-specific sequences of ribosomal ribonucleic acids (*r*RNAs)
PAB	Biosensor system with potentiometric alternating biosensing silicon chip, which interacts with a biological element such as cells, enzymes, and so on, with measured pH rates or redox potential variation. Used in determining metabolic variations in bacterial cells in response to presence of pollutants, drugs, hormones, pesticides, and so on, with graphic computation
Portal Shield	Used in the southeast Asian region for the protection of harbor and airfields, this biodefense system facilitates biological detection and identification, decontamination of biosensor equipment, and reduction of casualties

species such as ferrocene and fluorescent material (fluorescein, ruthenium complexes, etc.).

2.1.1 Electrochemical Pathogen Detection

Potentiometric Immunosensors
All potentiometric sensors such as ion-selective electrodes, transmembrane potential sensors, and field-effect transistors measure alterations in surface potential at near-zero current flow. The main advantages of these sensors are their simplicity and small electrode dimensions, which facilitates miniaturization. However, potentiometric methods in general suffer from the lack of high sensitivity and occasional nonspecificity. One application

that uses a light-addressable potentiometric sensor is the detection of *Venezuelan* equine encephalitis virus where the pathogen low level of detection (LOD) limit is 30 ng ml^{-1}.[31] Moreover, a similar approach was used for rapid detection of *Newcastle* disease virus with LOD varying from 400 ng ml^{-1} to 1.3 ng ml^{-1} with an antibody incubation step ranging from 1 to 60 min. The assay format is suitable for both virus and protein antigens. Furthermore, new assays can be developed and optimized readily, often within 1 day.[32]

Amperometric Immunosensors
Amperometric immunosensors are based on measurement of a current flow generated by an electrochemical reaction at constant voltage. For direct sensing there are only a few applications available

since most analytes (proteins) are not able to act as redox partners for the electrochemical reaction. Therefore, electrochemical species such as oxygen or hydrogen peroxide used either as substrates or as reaction products are introduced for sensing the electrochemical reaction of the analyte. This indirect testing approach has excellent sensitivity. In one application that combines flow-through immunofiltration with amperometry, bacteria *Escherichia coli O157:H7* detection is reported.[33] The authors obtained LOD of 100 cells per ml in an approximately 30-min experimental run. By using an electrogenerated biotinylated polypyrrole film copolymerized with pyrrole-lactobionamide monomer it was possible to amperometrically detect cholera antitoxin in the presence of the horse radish peroxidase (HRP) marker with a LOD down to 50 ng ml^{-1}.[34] With two other enzymatic markers based on biotinylated polyphenol oxidase (PPO-B) and biotinylated glucose oxidase (GOX-B) the LOD was 100 and 1 μg ml^{-1}, respectively.[35] The same research team designed amperometric immunosensors for the detection of photochemical grafted T7 bacteriophage displaying a specific WNV epitope with the low LOD for the antibody titer being 1:10^6. For bacteriophage anchorage, a novel copolymeric film was used based on pyrrole-benzophenone and tris(bipyridine pyrrole) ruthenium (II) units.[36]

An amperometric sensor for detection of antibodies to *Salmonella* Typhi in the serum of patients was also developed.[37] This involved use of screen-printed electrodes and a recombinant flagellin fusion protein. An indirect ELISA was used for detection of antibodies to *Salmonella* Typhi in the patient serum. The time taken for the detection by this electrochemical method is 1 h and 15 min, as compared to the time taken by other analytic means being 18 h.

Conductometric and Capacitive Immunosensors
These immunosensors measure conductivity changes at fixed voltage as a result of biochemical reactions, when biological entities are immobilized onto noble metals (such as Au or Pt) electrodes. Since biological media has high ionic strength, making it difficult to sense small conductivity changes, an ion-channel conductance immunosensor was developed.[38] As an application an *E. coli O157:H7*–monitoring conductometric biosensor was designed.[39] The LOD was approximately 7.9×10 colony-forming units per ml within a 10-min process. By changing the specificity of the antibodies, this biosensor can become a platform technology for sensitive detection of other types of pathogens too.

2.1.2 *Optical Immunosensors*

For bioanalytics the most frequently used transducers are optical immunosensors due to their durability despite frequent manipulations, rapid signal generation, and ease of data interpretation. In addition, the potential use of fiber-optic bundles with this technology can considerably speed-up multiple-pathogen identification in a clinical laboratory setting.[40,41] The optical technology can provide information regarding the status of several phenomena such as luminescence, refractive index, fluorescence, and optical surface modification. They have been employed in multiple applications over the years; in optical detection of either the direct label-free immunological reaction of labeled immunoentites or of the indirect product(s) of enzymatic reactions. Fluorescent labels have been the most popular but the bio- and chemiluminescence labels are used frequently too and are rapidly becoming just as popular.

A rapid assay for cholera toxin (CT) detection using a fluorescence-based biosensor has been developed[42] using an optical detection sensor. This sensor was capable of analyzing six samples simultaneously for CT in 20 min with few manipulations required by the operator. The biochemical assays utilized a ganglioside "capture" for capture of analyte, immobilized on discrete locations of the surface of an optical waveguide. Limits of detection for CT were 200 ng ml^{-1} in direct assays and 40 ng ml^{-1} and 1 μg ml^{-1} in sandwich-type assays performed using rabbit and goat tracer antibodies, respectively. This assay claimed to be the first description of a non-antibody-based recognition system in a multispecific planar array sensor. The CT B subunit molecules were also detected using indium tin oxide–coated fiber-optics modified with electroploymerized biotinylated polypyrrole films and used for tethering molecular recognition probes through avidin–biotin interactions. The obtained LOD of the optical immunosensor was 100 ng ml^{-1}.[43]

An ELISA-based optical fiber methodology was also developed for the detection of anti-WNV IgG antibodies using a silanization technique for anchoring biological moieties. The lower antibody titer obtained was 1:10[6].[44] More recently, the same research team developed an optical immunosensor for Ebola virus detection by using an electrogenerated poly(pyrrole-benzophenone) film deposited upon an indium tin oxide–modified fiber-optic for tethering Ebola virus antigen.[45] The lowest detectable titers measured with this optical sensor for anti-Ebola IgG for subtypes Zaire and Sudan were 1:960 000 and 1:1 000 000, respectively. Overall, fiber-optic immunosensors are highly sensitive and specific when compared with conventional immunoassays such as colorimetric ELISA systems.

2.1.3 Microgravimetric Sensors

Microgravimetric sensors are divided into QCM devices that apply a thickness-shear mode and devices applying a surface acoustic wave as the detection principle. The development of acoustic technology for the field of biology was recently reviewed.[46] Thus far, comparison of results suggests that[47] piezoimmunosensors may be powerful alternatives to optical sensors.

QCM technology was used for the development of an immunosensor for detection of HIV.[48] The sensor was based on the immobilization of recombinant viral peptides on the surface of the transducer and direct detection of anti-HIV antibodies in human sera. Recently, detection of Ebola glycoprotein using a QCM immunosensor was presented.[49] The immunosensors for the sensitive and selective detection of Ebola glycoprotein from the Zaire and Sudan/Gulu strains were assembled using four different antibody samples. Monoclonal IgG1, IgG2a, IgM, and polyclonal IgG were immobilized on QCM gold electrodes using a variety of antibody capture proteins. These sensor assemblies allowed for the determination of specificity of one antibody sample for a given Ebola strain and for a determination of the detection limits. Initial detection limits were in the range of 10–50 nM of Ebola glycoprotein. Antibody-based biosensors that are sensitive and selective for a given Ebola strain, as well as inexpensive and portable,

have tremendous potential for early environmental Ebola detection.

Impedimetric Immunosensors
A group of sensitive pathogen detection strategies using electrical characterization, which scans the detection volume with an electrical frequency sweep over a range of frequencies is electrical impedance spectroscopy (EIS). The technique inspects impedances typically from tens of hertz to the megahertz range. EIS systems are simpler to construct than optical systems and more compatible with microtechnology, as well as more reliable. Applications have been reported on microfabricated EIS prototypes intended for use in miniaturized systems with microfluidics for direct biological species detection, which are based on changes in solution conductivity.[50] The total fluidic path volume in the device is on the order of 30 nl. Flow fields in the closed chip were mapped by particle image velocimetry. Electrical impedance measurements of suspensions of the live microorganism *Listeria innocua* injected into the chip demonstrate an easy method for detecting the viability of a few bacterial cells. By-products of bacterial metabolism modify the ionic strength of a low-conductivity suspension medium, significantly altering its electrical characteristics.

2.2 Immunosensor Applications

Immunosensors for the detection of bacterial pathogens are rapidly evolving with improved sensitivity and specificity. Many companies such as Roche, Morningstar, and Diagnology have developed simple dipstick/dot blot tests for rapid detection of enteric pathogens, while others are developing sensitive, immunosensors that selectively detect antigenic targets for example, Molecular Devices' Threshold System, ILA based on sandwich ELISA immunological detection of antigen. Although, immunosensors have greatly decreased the time of sample analysis, they are not as sensitive as nucleic acid–based ones (detection limit about 102 cfu).[12]

2.3 DNA Sensors for Microorganism Detection

Clinical applications of nucleic acid (gene) probes represent an intensive area of research.[51] The nucleic acid probe describes a segment of nucleic acid that specifically recognizes and binds to a nucleic acid target. The recognition is dependent upon the formation of a stable duplex between the two nucleic acid strands. This is different from the antigen–antibody complex formation where hydrophobic, ionic, and hydrogen bonds are involved. Moreover, the nucleic acid coupling takes place at regular (nucleotide) intervals along the chain of the nucleic acid duplex, whereas antigen–antibody linkage occurs only at a few specific sites (epitopes). Although nucleic acid recognition is very stable, the main advantage of DNA-based sensors is that they can be easily regenerated through melting of the duplex by controlling buffer concentration and other variables.[52] The detection of specific nucleic acid sequences provides the basis for detecting a wide variety of bacterial and viral pathogens.

Conventional detection methods of a specific DNA use either radiolabeled probes[53] or gene amplification (the PCR).[54] Since both techniques suffer from the short shelf life of labeled probes, high cost, hazards, disposal problems of the radioactive residues, and from relatively long time for analysis, new DNA-biosensor technology was introduced, which can operate under special conditions.[51,55] With respect to the nature of the physical transducer used, the DNA sensors can be classified as electrochemical, optical, or gravimetric.

2.3.1 Electrochemical DNA Biosensors

Recently, detection of DNA sequences to the waterborne pathogen *Cryptosporidium* have been developed.[56] The sensor relies on the immobilization of a specific oligonucleotide to *Cryptosporidium* onto the carbon-paste transducer and employs a highly sensitive chronopotentiometric transduction mode for monitoring the hybridization event. After a 20–30-min hybridization time very sensitive detection of *Cryptosporidium* was achieved. Similar hybridization-chronopotentiometric schemes were developed for detection of pathogens such as *Giardia*, *E. coli*, and *Mycobacterium tuberculosis* (MTB).[56,57] In particular, a biosensor for the determination of short sequences from MTB has been described. The sensor relies on modification of the carbon-paste transducer with

27- or 36-mer oligonucleotide probes and their hybridization to complementary strands from the MTB DNA direct repeat region. Chronopotentiometry is employed to transduce the hybridization event. Short (5–15 min) hybridization periods permit quantitation of nanogram per milliliter levels of MTB target DNA to transduce the hybridization event.[56]

2.3.2 Optical DNA Biosensors

A fiber-optic DNA sensor array is capable of simultaneously monitoring multiple hybridization events. Typically, DNA probes are immobilized in an acrylamide-based polymer matrix on the surface of an optical fiber. Sample DNA is amplified by PCR and real-time hybridization of 5′-fluorescein isothiocyanate–labeled target oligonucleotides to the array is monitored by following fluorescence emission at 530 nm. Detection of labeled target oligonucleotides in the range of 0.2–196 nmol l^{-1} is possible with these systems and identification of a point mutation in the *H-ras* oncogene PCR product is a good proof of concept.[58]

SPR and evanescent wave sensing using optical waveguides can be used to enhance the fluorescence emission of labeled oligonucleotides bound to surface-attached probes. SPR is efficient for the study of association and dissociation kinetics as well as affinity constants for binding of complimentary target strands in solution. The lowest detectable concentration of target oligonucleotides (9.2 nmol l^{-1}) compares favorably with other biosensor methods that require labels. Regeneration of the surface-immobilized probe is also possible, allowing reuse of the sensor without significant loss of hybridization activity.[59,60]

2.3.3 Gravimetric Biosensors

Gravimetric biosensor technology is based on thin-film bulk acoustic wave resonators on silicon and the feasibility of detecting DNA and protein molecules has been demonstrated. The detection principle of these sensors is label-free and relies on a resonance frequency shift caused by mass loading of an acoustic resonator, a principle very well known from QCM. Integrated zinc oxide bulk acoustic wave resonators with resonance frequencies around 2 GHz have been fabricated, employing an acoustic mirror for isolation from the silicon substrate. DNA oligos have been thiol-coupled to the gold electrode by on-wafer dispensing. In a further step, samples have either been hybridized or alternatively a protein has been coupled to a receptor. Measurements have demonstrated that the new biosensor is capable of both protein detection as well as DNA hybridization without using a label. Owing to the substantially higher oscillation frequency, these sensors have demonstrated much higher sensitivity and resolution compared to QCM.[61]

3 NANOTECHNOLOGY AND BIOSENSORS

New technologies such as "nanotechnology" hold considerable promise for advances in the development of immunosensors for pharmaceutical and medical diagnostic applications (particularly in proteomics and cellomics).[62] Nanotechnology involves the scaling down to microfluidic and nanofluidic biochips and the design and construction of platforms from the bottom up. The result of ongoing nanodiagnostic research will be improvements in the sensitivity and extent of the present limits of molecular diagnostics. Miniaturization is key for this technology, focusing on the integration of all steps of an analytical process into a single-device (known as "lab on chip"), which will be partly disposable.[63]

Some of the earliest applications of nanotechnology reported for molecular diagnostics involve the use of nanoparticles; however, there is some concern about their potential effects on the human body and the environment. Nonetheless, nanotechnology has great potential for the production of in-dwelling controlled-release devices with autonomous operation that is responsive to individual needs. This has led to much research focused on the development of implantable, inexpensive immunosensors[64] for clinical applications.

3.1 Nanotechnological Methods for the Detection of Pathogenic Microorganisms and Viruses

The detection of microorganisms and viruses using a "portable" atomic force microscope (AFM)[65]

have been reported. The method is based on the creation of an immunochip with highly specific antibodies and a miniaturized scanning probe microscope performs the detection of the captured material. The AFM has the ability to sense tiny bumps on the surface at the nanometer scale. Such a technique is very sensitive and does not destroy the biological entities, so they can be further analyzed. The typical immunochip surface is produced with the Langmuir–Blodgett (LB) method[66] using amphiphilic polyelectrolytes and includes (i) production of an affinity surface (antibodies, receptors, and DNA (RNA) probes can be employed); (ii) performance of specific adsorption of the substance bearing the analyte; (iii) scanning of the surface; and (iv) image analysis and identification of the biological agent using pattern-recognition techniques. This method was demonstrated for analysis of *Coxiella burnetii*, *vaccinia virus*, and *Yersinia pestis* with a high sensitivity of down to a few bound entities.[67] At this stage there are no practical devices for field use due to size constraints and cost.

3.2 Nanobiosensor Applications

Another application that utilizes the capabilities of the AFM technology for virus detection is based on the ViriChip device,[68] a small silicon chip about 6 mm across that accommodates tiny droplets of antibodies on the surface. The chip can be printed with hundreds of different antibodies that can then be used for analysis in record time of different viral infectious agents simultaneously, using just a single drop of blood from the patient. These antibodies serve as receptor capture surfaces for viruses, which attach themselves selectively to them. Once the viruses have landed on a particular droplet, they can be detected using the AFM technology. Researchers have demonstrated a proof of principle for this technique by detecting six different strains of Coxsackie B virus, a virus that causes symptoms ranging from mild cold to death and is one of the key factors leading to the failure of heart transplants. The ability to detect it in a rapid and sensitive way will save thousands of lives by allowing physicians to determine in real time if a donor heart is infected.

A unique example of nanotechnology implemented for viral detection (e.g., influenza A) in real time and for impure samples is a detector based on nanowire transistors. Such a detector uses antibodies attached to aldehyde-coated nanowires to capture the individual virus particles causing conductance changes of the nanowire, which signals the presence of virus. The researchers found that the duration of the bind-and-release cycle depends on the density of the antibody proteins on the nanowires. The cycle averaged just over 1 s at a low concentration of antibody proteins, about 20 s at a moderate density, and 5–10 min at a high density. This chip detector can be useful for studying how viruses bind to receptors by determining which viruses bind to which receptors, how long virus particles bind to receptors, and what substances block or disrupt binding. The device can also eventually be used to detect individual biomolecules, including viral DNA and proteins.

4 LAB-ON-CHIP TECHNOLOGIES

Microfluidic devices integrating sample handling, reagent mixing, separation, and detection processes are of considerable interest. Since the early 1990s, when the modern concept of the micro total analysis system (μTAS, or lab on a chip) was proposed by Manz et al.[69] considerable effort was made toward the development of highly miniaturized analytical systems. Miniaturization revealed many potential benefits including low consumption of costly reagents (nanoscale volumes) and power, minimized handling of dangerous materials, short reaction times, portability, and versatility in design, and capability for parallel operations. Such "miniaturized sensors" can be ideal tools for monitoring personal exposure to individual compounds and pathogens.[28,70] In order to prove the principle, many examples of bioassays and biological procedures have been miniaturized into a chip format including PCR,[71] DNA analysis and sequencing,[72] electrophoresis,[73] immunoassays,[74] and intra- and intercellular analysis.[75] The main challenges in micro- and nanofluidics, however, are the integration of multiple assays and functional components into a single chip to perform a total analysis of the sample.

5 PORTABLE DEVICES FOR DETECTION OF "TROPICAL" PATHOGENIC VIRUSES AND BACTERIA – EXAMPLES

5.1 Devices Already Operable in the Field

5.1.1 Tuberculosis

TB is a disease of increasing importance. Mortality is highest in the tropics, where over three-quarters of cases occur. Annually over three million people die from TB and one-third of the world population is infected with MTB. Patients spread the disease by producing aerosols containing the MTB bacteria. Persons that have close contact with these patients inhale these microdroplets and thus become infected. The mycobacteria multiply in the lung and can cause disease in 10% of infected people. It is estimated that the incidence of TB worldwide and the number of cases attributable to coexisting HIV infection will continue to increase substantially during the next decade. Most of this burden occurs among the low-income countries of the world, particularly those in southeast Asia and sub-Saharan Africa.

Few systems for the detection of the pathogenic bacteria MTB are disposable.

One qualitative diagnostic kit for detection of MTB DNA in clinical samples such as sputum, urine, and pleural aspiration contains TB lysis solution for DNA extraction and reagents for PCR. Part of the sputum samples that are prepared for culture may be used for PCR. Total processing time for 10 clinical samples is 5 h. Ready-to-use PCR mix, positive control, and other qualified reagents along with an easy to follow protocol are included. In this assay specific primers amplify conserved regions in the MTB genome. Using this kit, as few as 20 bacteria per ml can be detected (CinnaGen Inc.).

Another extensively used kit is provided by Gen-Probe (San Diego, CA) and is highly sensitive and specific for bacterial detection. The detection takes place in contaminated broth cultures in about 4 days, which is about 12 days shorter than when a non-amplified DNA probe alone[76] is employed. Other systems available for TB detection include the COBAS AMPLICOR (Roche, Switzerland) and the BD ProbeTec ET (Franklin Lakes, NJ). COBAS is a commercial amplification system less sensitive than mycobacteria growth indicator tubes (MGIT) but better than solid media (Ogawa).[77]

5.1.2 E. coli O157:H7

Portable detection kits available in the market have been developed for *E. coli O157:H7*. The Immunocard STAT system from Meridian Diagnostics (Cincinnati, OH) tests for the presence of *E. coli O157:H7* in just 10 min. The test can be performed either directly on stool or on overnight broth cultures. Stool material is diluted and added to the sample port of the device. The analyte is immobilized on gold particles coated with a monoclonal antibody specific for the *E. coli O157* lipopolysaccharide. This detection system apparently has a specificity of 95%.

A portable colorimetric immunoassay using antibody-directed liposomes encapsulating dye has also been developed. Antibodies (anti-*E. coli O157:H7*) thiolated by 2-iminothiolane were coupled to maleide-tagged liposomes encapsulating sulforhodamine (the marker dye). A sandwich format was used to allow for capillary migration of wicking agent. The color density of the measurements was directly proportional to the amount of the *E. coli* in the sample. The immunoassay does not need washing and incubation steps and can be performed in 8 min with a low detection limit of 10^4 cfu (colony-forming units) ml^{-1}.[78]

5.1.3 Gonorrhoeae

GonoGen™ is a monoclonal antibody-based coagglutination test intended for the confirmatory identification of *Neisseria gonorrhoeae*. The test does not require isolated, viable, or fresh cultures. After heating the specimen, a positive reaction will be indicated by the clumping with the detection reagent. GonoGen™ II is another monoclonal antibody–based colorimetric test intended for the confirmatory identification of *N. gonorrhoeae*. The test requires less than 7 min to run with no heating or pretreatment. The test does not require isolated, viable, or fresh cultures. GonoGen™ II eliminates agglutination, guesswork, by developing a red dot for a positive reaction (New Horizons Diagnostics, Columbia, MD). Finally, Roche developed a multiplex AMPLICOR PCR test for female urine samples and wet and dry endocervical swabs (ES) for *N. gonorrhoeae* detection.[79] The system makes use of an internal amplification control, being highly sensitive and specific in particular with ES.

5.1.4 Salmonella Typhi

Since typhoid fever annually afflicts about 17 million people (particular those from southeast Asia, Africa, and Latin America), scientists' efforts have been focused on early diagnostic techniques. Researchers have developed a dipstick assay for the detection of *Salmonella* Typhi–reactive immunoglobulin M (IgM) antibodies. The dipstick assay is a simplified version of the ELISA, providing either a positive or a negative result, and using stabilized components that can be stored for more than 2 years outside the refrigerator. ELISA for typhoid fever has been found superior to the Widal test.[80] The complete genetic code of *Salmonella* Typhi (responsible for typhoid fever)[81] has recently been deciphered, which raises hope for eventual complete eradication of the disease.

5.1.5 Visceral Leishmaniasis

Visceral leishmaniasis (VL) is the most severe form of leishmaniasis. The disease is fatal if left untreated. The diagnosis of human VL is difficult. The principal signs of VL are an enlarged spleen and a prolonged irregular fever. Other signs and symptoms include loss of weight, pallor, enlarged liver, enlarged lymph nodes, anemia, cough, and diarrhea. These signs and symptoms may mimic those of malaria, typhoid, TB, schistosomiasis, and a number of other diseases. Clinical suspicion may be confirmed directly by the detection of parasites in patient material or by culture. However, sample collection is invasive for the patient and parasite isolation by culture is time-consuming, expensive, and difficult. Owing to the limitations of the aforementioned direct diagnostic methods, a number of indirect immunological methods, such as indirect immunofluorescent-antibody tests, ELISAs, and a direct agglutination test (DAT), have been developed.

DAT remains the first-line diagnostic tool in many developing countries such as Ethiopia and

Sudan. It is a simple and economical test with a high sensitivity and specificity. A problem with DAT is that the often-used aqueous antigen is not heat-stable, thus requiring refrigeration. To obviate this limitation a freeze-dried antigen has been developed by KIT Biomedical Research that remains stable at ambient temperature for years. The sensitivity and specificity of this freeze-dried antigen is high. A drawback of the DAT is the relatively long incubation time (18 h; results available only the next day) and the fact that serial dilutions of blood or serum must be made, which makes the test less suitable for screening large numbers of samples. In order to circumvent this problem, a kit based on a fast agglutination screening test (FAST) for detection of anti-Leishmania antibodies in less that 4 h, in serum samples from human patients with VL, was developed.[82]

5.1.6 Leptospirosis

Leptospirosis is a zoonotic disease with a worldwide distribution. The disease, also known as *Weil's syndrome*, is easily confused with other febrile illnesses including viral hemorrhagic fevers. Laboratory confirmation is essential for a proper diagnosis. Culture, the microscopic agglutination test (MAT) and ELISA for the detection of specific immunoglobulin M (IgM) antibodies are important standard laboratory tests for the confirmation of leptospirosis.

KIT Biomedical Research has developed a number of rapid and robust tests including a dipstick assay, a lateral flow test, and a latex agglutination test that can be used outside the established laboratory as an alternative to the complicated standard tests.

The dipstick assay is based on the binding of specific IgM antibodies to a broadly reactive antigen prepared from a nonpathogenic strain. The use of the broadly reactive antigen ensures the detection of a wide range of commonly occurring serotypes. The assay utilizes a stabilized detection reagent that can be stored without the need for refrigeration for at least 2 years without losing reactivity. The result of the dipstick assay is obtained after 3 h and no special equipment is required to perform the assay.[82]

Like the dipstick assay, the lateral flow test is aimed at the detection of *Leptospira*-specific IgM antibodies in human serum or whole blood samples. The assay is similarly based on the binding of specific IgM antibodies to a broadly reactive antigen prepared from a nonpathogenic strain ensuring detection of a wide range of serotypes. The assay similarly utilizes stabilized components and the result of the lateral flow assay is obtained within 10 min. No special equipment is required to perform the assay.[82]

The LeptoTek Dri Dot assay consists of colored latex particles activated with a broadly reactive *Leptospira* antigen that is dried onto an agglutination card. The assay is based on the binding of *Leptospira*-specific antibodies, present in the serum sample, to the *Leptospira* antigen causing a fine granular agglutination that tends to settle at the edge of the droplet (the positive result). The assay is performed by suspending the dried detection reagent in 10 μl of serum using a plastic stick. The liquid is then mixed further by gently swirling the card. Agglutination occurs within 30 s.[82]

5.1.7 Mycobacterium Leprae

KIT Biomedical Research has also developed a simple dipstick assay for the detection of antibodies to a specific component of *Mycobacterium leprae* (ML). The dipstick assay is aimed at the rapid detection of ML-specific IgM antibodies in human serum or whole blood samples. The dipstick assay is based on the binding of specific IgM antibodies to a semisynthetic antigen consisting of the terminal sugar groups of the phenolic glycolipid-I (PGL-I) antigen of ML attached to bovine serum albumin. The use of this antigen ensures that the ML dipstick has a sensitivity and specificity that is comparable to ELISA. The assay utilizes a stabilized detection reagent that can be stored without the need for refrigeration for at least 2 years without losing its reactivity. The result of the dipstick assay is obtained after 3 h and no special equipment is required to perform or read the assay.[82]

5.1.8 Brucellosis

Brucellosis is an important but often neglected cause of morbidity in many regions of the world especially in developing areas of the Mediterranean region, the Middle East, western Asia, and parts of Africa and Latin America. The disease is most common in rural areas and among those involved in animal husbandry. Brucellosis also occurs in urban settings when animals are kept in compounds around houses, among meat packers, dairy workers, and veterinarians. *Brucella abortus*, *Brucella suis*, and *Brucella melitensis* are the causative agents. The treatment of chronic brucellosis is complicated and requires prolonged medication compared to acute brucellosis. The disease should be diagnosed early and the patient treated promptly. Typical severe acute brucellosis in its early stages cannot be diagnosed on clinical grounds alone.

KIT Biomedical Research has developed simple dipstick and flow assays for the detection of *Brucella*-specific IgM and IgG antibodies in human serum samples. The dipstick assay is aimed at the rapid detection of *Brucella*-specific IgM antibodies in human serum or whole blood samples from patients in the early stage of the disease. The dipstick assay is based on the binding of specific IgM antibodies to a lipolysaccharide fraction prepared from *B. abortus*. The assay utilizes stabilized nonenzymatic detection reagents that can be stored without the need for refrigeration for at least 2 years without losing reactivity. The result of the dipstick assay is obtained after 3 h and no special equipment is required to perform the assay.[82]

KIT Biomedical Research has also developed a new assay system for the serodiagnosis of human brucellosis. The assay consists of two tests: one for the detection of *Brucella*-specific IgM antibodies in the serodiagnosis of patients with an acute infection and one for the detection *Brucella*-specific IgG antibodies in the serodiagnosis of patients with persistent or recurrent infection. These *Brucella* IgM and IgG flow assays are very simple to apply, provide a quick result, and are most ideal for developing countries and rural settings.[82]

5.1.9 Malaria

KIT Biomedical Research has developed a qualitative nucleic acid sequence–based amplification (NASBA) procedure for the detection and semiquantification of *Plasmodium* parasites. The qualitative NASBA has recently been further developed into a quantitative NASBA that allows for the quantification of *Plasmodium falciparum* over a wide range of levels of parasitemia, 50–108 parasites per ml of blood. Since only 50 μl of blood is subjected to nucleic acid isolation, 50 parasites per ml was considered to be the lower threshold of detection, which is approximately 50× more sensitive than conventional microscopy.[82]

Given the distribution of *P. falciparum* in many third world countries and the relative paucity of well-equipped laboratories, the development of a rapid on-the-spot diagnostic for *P. falciparum* infection is very timely. Presently, two diagnostic test kits namely ParaSight™-F, a dipstick antigen-capture assay (Becton Dickinson, Sparks, MD), and immunochromatographic test (ICT Malaria Pf™, ICT Diagnostics, Brookvale, NSW, Australia) are available for rapid diagnosis of *P. falciparum* infection. Both these tests are based on detection of circulating antigen from *P. falciparum* histidine-rich protein-2 (Pf HRP-2) in whole blood. It is a water-soluble protein synthesized by the parasite and released from the *P. falciparum*–infected erythrocytes. Both these test kits are qualitative test methods based on the presence or absence of *P. falciparum*–specific antigen in the blood, and results may not always correlate well with the other methods that detect presence of parasite in the peripheral blood. However, these kits are simple to use for on-the-spot diagnosis of *P. falciparum* infection and do not require special equipment and can be performed with minimum skill. It takes less than 8 min to get a definite diagnosis so that treatment can begin without any delay; the latter being the prime cause of mortality due to malaria.[83]

5.2 Field-operable Devices Under Development

5.2.1 PCR Units

In late 1996, Lawrence Livermore laboratory delivered to the US Army the first fully portable, battery-powered, real-time DNA analysis system. To detect DNA in a sample, a synthesized DNA probe or primer tagged with a fluorescent dye is introduced into the sample before it is inserted into the heater chamber. Each probe or primer is designed for a specific organism that is the causative agent of diseases such as anthrax, plague, and so on. If that organism is present in the sample, the probe attaches to its DNA. By measuring the sample's fluorescence, the instrument reports the presence (or absence) of the targeted organism. In Livermore's portable unit, the thermal cycling process takes place in a micromachined, silicon chamber that has integrated heaters, cooling surfaces, and windows through which detection takes place. The PCR reaction and DNA analysis take place in a disposable polypropylene reaction tube inserted into the chamber. Because of the low thermal mass and integrated nature of Livermore's silicon chambers, they require very low power and can be heated and cooled much faster than conventional units. So the unit is not only portable but also much faster and more energy-efficient than benchtop models. A multiple-chamber unit that enables the analysis of many samples at the same time has been field-tested.

Multiplex PCR packages are still under development for the simultaneous detection of eight important viruses of swine: classical swine fever virus, African swine fever virus, porcine reproductive and respiratory syndrome virus, Aujeszky's disease virus, porcine parvovirus, swine vesicular disease virus, foot and mouth disease virus, and vesicular stomatitis virus. To simplify and accelerate the work, robotic nucleic acid extraction and pipetting are incorporated. This demonstrates that PCR diagnostic procedures can be rapid, robust, and automated.[84] In addition, the sensitivity of each nested PCR within these systems is very high; 1–10 genomic copies of the target viruses can be detected. As a result, the simultaneous nested PCR assays of the multiplex packages provide very high specificity.

5.2.2 Pen-side Test for Antibody Detection

Genesis Diagnostics designed a pen-side test for the reaction between antibody and antigen and based on a chromatographic technology that maybe replaced in the future by biosensors. Several other companies[85] are also working on this test. One prototype is based on a pan-serotype monoclonal antibody for capture of antigens. Pen-side tests do not require any specialized sample transport, have the advantage of a rapid time to result, and can easily be repeated.

5.2.3 Mini-flow Cytometer

Recently, Lawrence Livermore's laboratories reported a prototype for a miniature flow cytometer (*miniFlow*) that is portable and sensitive. Over the past two decades, flow cytometers have been used in laboratories to analyze cells and their characteristics, perform blood typing, test for diseases and viruses, and identify and separate out particular cells. These capabilities are enabled in the Livermore device through a technique that eases alignment and increases the accuracy of the standard flow-cytometric techniques. In a traditional flow cytometer, the cells flow in single file in solution while the experimenter directs one or more beams of laser light at them and observes the scattered light, which is caused by variations in the cells or DNA. Instead of using a microscope lens or an externally positioned optical fiber as a detector, the Livermore method uses the flow stream itself as a waveguide for the laser light, capturing the light and transmitting it to an optical detector. This approach not only eliminates the alignment problems that plague traditional flow cytometers but also collects 10 times more light than a microscope lens does. In the future, simplification of the alignment and more light collection will facilitate better and faster analysis. Bacteria are large enough for individual detection in the *miniFlow*, but viruses and proteins are not. So beads large enough to be detected are coated with an antibody and added to the sample. The virus or protein attaches itself to the bead and can then be detected. When different beads are coated with different antibodies, simultaneous detection of several biological agents is possible.

5.2.4 Virus Batteries

Belcher and her team are developing a way to actually "grow" rechargeable batteries with the help of viruses.[86] The team is forcing the viruses to interact with materials that they normally do not interact with. They use bacteriophage (M13) mixed together with a metal or other materials on which millions of them can align and stack themselves into orderly layers, creating a new material. The team also uses M13 viruses to synthesize and assemble nanowires of cobalt oxide at room temperature. By incorporating gold-binding peptides into the filament coat, hybrid gold–cobalt oxide wires were formed that improved battery capacity. Combining virus-templated synthesis at the peptide level with methods for controlling two-dimensional assembly of viruses on polyelectrolyte multilayers provides a systematic platform for integrating the nanomaterials to form thin, flexible lithium ion batteries.

5.2.5 Biochips

For rapid and accurate field identification of biological and chemical agents, biochips in concert with portable readers are being developed. One example is a system designed at Argonne National Laboratory for pathogen detection in less than 2 h. Argonne's biochips contain hundreds to thousands of test positions, each chip being a matrix of three-dimensional gel pads about $100 \times 100 \times 20\,\mu m$ in size. These pads are in defined locations such that reaction of a sample at a given position immediately identifies the sample. A segment of a DNA strand, protein, peptide, or antibody is inserted into each pad, tailoring it to recognize a specific biological agent or biochemical signature.[87,88] The biochip also makes use of PCR, enabling low-abundance bacteria and threat agents to be detected with relative ease, within hours instead of days.[89] With Argonne's miniaturized portable system one can simultaneously detect a multitude of bacteria and other agents from a single sample.[90]

5.2.6 BioPen

The BioPen is a penlike instrument that runs an automatic bioassay protocol at the end of a fiber-optic tip being developed in the group of Prof. R. Marks at Ben-Gurion University. The outcome of this assay, light radiation, is produced if the patient carries the pathogenic markers or if the target pathogen is detected. The light is translated ultimately into a number or a positive/negative answer on an LCD screen at the far end of the "cap" of the BioPen. The BioPen represents the first test that presents the full capabilities of bioassays in an easy-to-handle pen-shaped device. Bioassays on this device are based on specific modification of an optical fiber with a test-adapted biorecognition entity. The novelty lies in the development of original genetically engineered bioreceptors and the chemistry used to bind them covalently to the fiber surface. Necessary fluids are brought in through a unique shape-adapted microfluidic system (HydroLogic assembly) to the fiber-optic tip. A positive signal is reported by light emission transduced through the fiber to an innovative ultracompact photosensitive platform able to perform single-photon counting. The innovative technologies incorporated in the BioPen constitute major breakthroughs in several fields such as surface chemistry, microoptoelectronics, microfluidics, and biology.

In general, BioPen demonstrates several improvements over existing systems. First, the entire test device is expected to weigh no more than 300 g with a penlike size. It can be therefore used as a personal bioassay or serve for in-the-field testing without requiring any additional systems or laboratory installation. In addition, the disposable cone that contains the whole bioassay is completely closed and provides the user with a high level of biocontainment with no risk of contamination and can easily replace any desired bioassay. The bioassay itself will be performed in 20 min and demonstrates a better level of sensitivity than commonly used fluorescent tests due to the combined luminescence/fiber/photocounting assembly that results in a very low background. The multiplicity of bioreporters that can be coupled to the fiber (DNA/RNA, protein, whole cells, etc.) also makes the BioPen a remarkable tool for detecting not only biological entities but also a wide range of chemicals and toxins[91,92] that cannot be detected with standard antibody-based sensors.

5.2.7 Prototype of a Sensor to Smell Bacteria

KIT Biomedical Research, together with Cranfield University in the United Kingdom, have joined their expertise for the development and optimization of a portable prototype gas/volatile sensor array and associated computational analysis, for the rapid detection, identification, and analysis of MTB in culture, sputum, and ultimately breath samples. The aim is to develop a simple breath analysis device capable of distinguishing TB from other lung diseases.[82]

5.2.8 Lab-on-a-chip Portables

The development of a portable lab on a chip that could let doctors identify the strain of flu viruses in less than 2 h was reported in June, 2005.[93] The device, called the *Genotyper*, integrates fluidic and thermal components such as heaters, temperature sensors, and addressable valves to perform two independent serial biochemical reactions, followed by an electrophoretic separation. Integration of multiple steps of biological assays on a single device provides significant advantages in terms of sample/reagent consumption, process automation, analysis speed and efficiency, and contamination reduction. The key components (phase-change valves, thermally isolated reaction chambers, gel electrophoresis, and pulsed drop motion) of this device are electronically addressable and simple to operate—properties that could eventually lead to autonomous operation. The device's compact design and mass-production technology make it an attractive platform for a variety of genetic analyses. Ultimately, the *Genotyper* will be fully portable (about the size of a TV remote control), with wireless connectivity, and thus able to track the spread of existing or emerging flu strains around the world.

The *Genotyper* identifies flu type through a process resembling the genetic fingerprinting used in DNA identification. The process uses standard biochemical assays, only miniaturized. First, the influenza virus RNA is converted to DNA using reverse transcriptase. Then, a segment of the DNA is amplified manyfold through standard PCR. Finally, enzymes are released that "digest" or cut the amplified DNA in a sequence-specific manner. The DNA is then stained and pushed through a polymer gel matrix by an electric field. The DNA pieces move at rates controlled by their size. One type of flu, for example, would have DNA that is not cut by the enzyme, while another is cut. So, DNA bands are formed, whose locations within the gel determine if the DNA is cut or not. This provides a "fingerprint" that is diagnostic for the strain of flu. Different types of DNAs can be tested and identified using the *Genotyper* device by simply using different reagents. To demonstrate this versatility, DNA from a human and a mouse were "typed" with this device along with strains of influenza.

Currently, there are different versions of the *Genotyper* chip measuring between 1 and 2 in. in length and width. The chip does not yet include a purification unit, and there is still a need for external "readers" to see the DNA bands. Some versions require external air pressure while others generate pressure within the chip itself. The *Genotyper* is still in the development phase until its reliability can be established.

Recently, another research team from a Singapore-based medical firm Veredus Laboratories has designed a disposable laboratory on a chip that can detect various strains of avian influenza (H5N1) and other influenza viruses (influenza A or B), in a single test, rather than conducting a series of tests, as is now required. The chip is a plastic slide, about 75 by 25 mm, containing a microscopic laboratory with its own plumbing, pumps, and temperature controls. To use such a "lab", a droplet of a sample is placed on the slide and traces of RNA in the droplet are amplified using the reverse transcription PCR. The amplified DNA copies are pumped to other areas of the chip for comparison with particular DNA profiles, such as those of the flu virus strains. A match causes the appropriate part of the chip to fluoresce. Processing the sample takes about an hour, after which the results can be obtained in seconds with a portable optical reader. This lab-on-chip approach should also reduce the risk of cross-contamination inherent in conventional analytic methods.

6 CONCLUSIONS AND FUTURE TRENDS

The detection of some tropical pathogens and other dangerous microbial contaminants using conventional and improved identification technologies has been summarized in the present chapter. Also, the advantages and limitations of particular detection techniques for specific pathogens are briefly pointed out. With currently available technologies, medical staff faces a choice between faster, bedside tests that indicate the presence of a virus but do not specifically identify the lethal strain, or more precise testing that typically requires sending samples to a laboratory. To be safe, it is still better to perform routine laboratory methods in parallel with use of advanced technologies, even though classical technologies are often time consuming, to obtain a better track record of positive correlation. Although new and improved methods for detection will continually be developed, it must always be remembered that preventative measures are key for reducing the rate of disease appearance. In the same vein, scientists are making huge efforts to create vaccines to protect humans from developing harmful diseases. Thus, a new TB vaccine is set to undergo human trials, the first in 80 years. If successful, it will likely be used in tandem with the current bacillus Calmette–Guerin (BCG) vaccine.[94]

Nowadays, most of the existing immunological methods are coupled with electronic sensing modules.[95] An automated system based on solid-phase ELISA coupled with a multichannel optical flow cell with a sensor composed of a light emitting diode and a photodetector has been elaborated for the simultaneous detection of staphylococcal enterotoxin B, bacteriophage M13, and *E. coli*.[96] Another system that uses a light-addressable potentiometric sensor and a flow-through immunofiltration enzyme assay, which can simultaneously detect eight analytes in 15 min, has also been reported.[97] All these systems present multiple advantages such as automation, rapidity, and throughput. In the near future, advances in immunoassays coupled with biosensor technologies will provide simpler designs and higher sensitivity along with results being recorded in real time. Most importantly, however, it should be clear that the most crucial and pivotal component for any type of biosensor under development is the molecular recognition entity such as antibody, aptamer, enzyme, nucleic acid, receptor, and so on.

Sensitive and specific identification of pathogens such as bacteria and viruses has been dramatically increased with the introduction of immunoassays and nucleic acid technologies. These technologies have contributed in a major way to the advances made possible through the miniaturized devices such as microarrays or lab-on-chip systems. Already, several academic and nonacademic institutions such as Nanogen, Caliper Technologies, and Lawrence Livermore laboratories are deeply involved in the design of smart, miniaturized, and portable devices that will require only microliter quantities of reagents for specific microbial detection. Finally, continuous improvements in the affinity, specificity, and molecular recognition of bioreceptor capture moities will synergize with development of more accessible pathogen detection technologies that will easily be integrated in routine diagnostics.

REFERENCES

1. Y. Compton, *Military Chemical and Biological Agents*, Telford Press, Caldwell, NJ, 1987.
2. Y. H. Shangkuan and H. C. Lin, Application of random amplified polymorphic DNA analysis to differentiate strains of Salmonella typhi and other Salmonella species. *Journal of Applied Microbiology*, 1998, **85**, 693–702.
3. S. T. Nichol, J. Arikawa, and Y. Kawaoka, Emerging viral diseases. *Proceedings of the National Academy of Sciences of the United States of America*, 2000, **97**, 12411–12412.
4. L. H. Gould and E. Fikrig, West Nile virus: a growing concern. *Journal of Clinical Investigation*, 2004, **113**, 1102–1107.
5. O. Tomori, Impact of yellow fever on the developing world. *Advances in Virus Research*, 1999, **53**, 5–34.
6. S. B. Halstead, Observations related to pathogenesis of dengue hemorrhagic fever. VI. Hypotheses and discussion. *Yale Journal of Biology and Medicine*, 1970, **42**, 350–362.
7. WHO, *Dengue/Dengue Hemorrhagic Fever* [Online], 2004.
8. R. Cologna, P. M. Armstrong, and R. Rico-Hesse, Selection for virulent dengue viruses occurs in humans and mosquitoes. *Journal of Virology*, 2005, **79**, 853–859.
9. E. S. Lloyd, S. R., S. Zaki, P. E. Rollin, K. Tshioko, M. A. Bwaka, T. G. Ksiazek, P. Calain, W. J. Shieh, M. K. Konde, E. Verchueren, H. N. Perry, L. Manguindula, J. Kabwau, R. Ndambi, and C. J. Peters, Long-term disease surveillance in Bandundu region, Democratic Republic of the Congo: a model for early detection and prevention of Ebola hemorrhagic fever. *Journal of Infectious Diseases*, 1999, **179**, Suppl 1, S274–S280.
10. P. M. Sharp, D. L. Robertson, F. Gao, and B. H. Hahn, Origins and diversity of human immunodeficiency viruses. *Aids*, 1994, **8**, 27–42.
11. N. C. Rollins, M. Dedicoat, S. Danaviah, T. Page, K. Bishop, I. Kleinschmidt, H. M. Coovadia, and S. A. Cassol, Prevalence, incidence, and mother-to-child transmission of HIV-1 in rural South Africa. *Lancet*, 2002, **360**, 389.
12. S. S. Iqbal, M. W. Mayo, J. G. Bruno, B. V. Bronk, C. A. Batt, and J. P. Chambers, A review of molecular recognition technologies for detection of biological threat agents. *Biosensors and Bioelectronics*, 2000, **15**, 549–578.
13. P. Cattani, G. M. Rossolini, S. Cresti, R. Santangelo, D. R. Burton, R. A. Williamson, P. P. Sanna, and G. Fadda, Detection and typing of herpes simplex viruses by using recombinant immunoglobulin fragments produced in bacteria. *Journal of Clinical Microbiology*, 1997, **35**, 1504–1509.

14. C. Jehl-Pietri, B. Hugues, M. Andre, J. M. Diez, and A. Bosch, Comparison of immunological and molecular hybridization detection methods for the detection of hepatitis a virus in sewage. *Letters in Applied Microbiology*, 1993, **17**, 162–166.

15. N. Ahmed, A. K. Mohanty, U. Mukhopadhyay, V. K. Batish, and S. Grover, PCR-based rapid detection of Mycobacterium tuberculosis in blood from immunocompetent patients with pulmonary tuberculosis. *Journal of Clinical Microbiology*, 1998, **36**, 3094–3095.

16. X. Mao, L. Yang, X. L. Su, and Y. Li, A nanoparticle amplification based quartz crystal microbalance DNA sensor for detection of escherichia coli O157:H7. *Biosensors and Bioelectronics*, 2006, **21**, 1178–1185.

17. H. Yu and J. G. Bruno, Immunomagnetic-electrochemiluminescent detection of *Escherichia coli* O157 and Salmonella typhimurium in foods and environmental water samples. *Applied and Environment Microbiology*, 1996, **62**, 587–592.

18. Z. Bukhari, R. M. McCuin, C. R. Fricker, and J. L. Clancy, Immunomagnetic separation of Cryptosporidium parvum from source water samples of various turbidities. *Applied and Environment Microbiology*, 1998, **64**, 4495–4499.

19. Y. Y. Wong, S. P. Ng, M. H. Ng, S. H. Si, S. Z. Yao, and Y. S. Fung, Immunosensor for the differentiation and detection of Salmonella species based on a quartz crystal microbalance. *Biosensors and Bioelectronics*, 2002, **17**, 676–684.

20. S. Jindou, A. Soda, S. Karita, T. Kajino, P. Beguin, J. H. Wu, M. Inagaki, T. Kimura, K. Sakka, and K. Ohmiya, Cohesin-dockerin interactions within and between Clostridium josui and Clostridium thermocellum: binding selectivity between cognate dockerin and cohesin domains and species specificity. *Journal of Biological Chemistry*, 2004, **279**, 9867–9874.

21. A. Hiratsuka, H. Muguruma, K. H. Lee, and I. Karube, Organic plasma process for simple and substrate-independent surface modification of polymeric Biomems devices. *Biosensors and Bioelectronics*, 2004, **19**, 1667–1672.

22. C. A. Bell, J. R. Uhl, T. L. Hadfield, J. C., David, R. F. Meyer, T. F. Smith, and F. R. Cockerill III, Detection of bacillus anthracis DNA by lightcycler PCR. *Journal of Clinical Microbiology*, 2002, **40**, 2897–2902.

23. A. H. Peruski and L. F. Peruski Jr, Immunological methods for detection and identification of infectious disease and biological warfare agents. *Clinical and Diagnostic Laboratory Immunology*, 2003, **10**, 506–513.

24. H. R. Gelderblom and P. R. Hazelton, Specimen collection for electron microscopy. *Emerging Infectious Diseases*, 2000, **6**, 433–434.

25. W. Zhang, P. R. Chipman, J. Corver, P. R. Johnson, Y. Zhang, S. Mukhopadhyay, T. S. Baker, J. H. Strauss, M. G. Rossmann, and R. J. Kuhn, Visualization of membrane protein domains by cryo-electron microscopy of dengue virus. *Nature Structural Biology*, 2003, **10**, 907–912.

26. P. R. Hazelton and H. R. Gelderblom, Electron microscopy for rapid diagnosis of infectious agents in emergent situations. *Emerging Infectious Diseases*, 2003, **9**, 294–303.

27. A. D. Livingston, C. J. Campbell, E. K. Wagner, and P. Ghazal, Biochip sensors for the rapid and sensitive detection of viral disease. *Genome Biology*, 2005, **6**, 112.

28. N. V. Zaytseva, V. N. Goral, R. A. Montagna, and A. J. Baeumner, Development of a microfluidic biosensor module for pathogen detection. *Lab on a Chip*, 2005, **5**, 805–811.

29. W. Gopel and P. Heiduschka, Interface analysis in biosensor design. *Biosensors and Bioelectronics*, 1995, **10**, 853–883.

30. S. H. Jenkins, W. R. Heineman, and H. B. Halsall, Extending the detection limit of solid-phase electrochemical enzyme immunoassay to the attomole level. *Analytical Biochemistry*, 1988, **168**, 292–299.

31. W. G. Hu, H. G. Thompson, A. Z. Alvi, L. P. Nagata, M. R. Suresh, and R. E. Fulton, Development of immunofiltration assay by light addressable potentiometric sensor with genetically biotinylated recombinant antibody for rapid identification of Venezuelan equine encephalitis virus. *Journal of Immunological Methods*, 2004, **289**, 27–35.

32. W. E. Lee, H. G. Thompson, J. G. Hall, R. E. Fulton, and J. P. Wong, Rapid immunofiltration assay of Newcastle disease virus using a silicon sensor. *Journal of Immunological Methods*, 1993, **166**, 123–131.

33. I. Abdel-Hamid, D. Ivnitski, P. Atanasov, and E. Wilkins, Flow-through immunofiltration assay system for rapid detection of *E. coli* O157:H7. *Biosensors and Bioelectronics*, 1999, **14**, 309–316.

34. R. E. Ionescu, C. Gondran, L. A. Gheber, S. Cosnier, and R. S. Marks, Construction of amperometric immunosensors based on the electrogeneration of a permeable biotinylated polypyrrole film. *Analytical Chemistry*, 2004, **76**, 6808–6813.

35. R. E. Ionescu, G. Gondran, S. Cosnier, L. A. Gheber, and R. S. Marks, Comparison between the performances of amperometric immunosensors for cholera antitoxin based on three enzyme markers. *Talanta*, 2005, **66**, 15–20.

36. R.E. Ionescu, S. Cosnier, G. Herzog, K. Gorky, B. Leshem, S. Herrmann, and R.S. Marks, Amperometric immunosensor for the detection of anti-West Nile Virus IgG using a photoactive copolymer. *Enzyme and Microbial technology*, 2006, **40**, 403–408.

37. V. Kameswara Rao, G. P. Rai, G. S. Agarwal, and S. Suresh, Amperometric immunosensor for detection of antibodies of Salmonella typhi in patient serum. *Analytica Chimica Acta*, 2005, **531**, 173–177.

38. H. C. Berney, J. Alderman, W. A. Lane, and J. K. Collins, Development of a capacitive immunosensor: a comparison of monoclonal and polyclonal capture antibodies as the primary layer. *Journal of Molecular Recognition*, 1998, **11**, 175–177.

39. Z. Muhammad-Tahir and E. C. Alocilja, A conductometric biosensor for biosecurity. *Biosensors and Bioelectronics*, 2003, **18**, 813–819.

40. Y. M. Bae, B. K. Oh, W. Lee, W. H. Lee, and J. W. Choi, Immunosensor for detection of Yersinia enterocolitica based on imaging ellipsometry. *Analytical Chemistry*, 2004, **76**, 1799–1803.

41. Y. H. Chang, T. C. Chang, E. F. Kao, and C. Chou, Detection of protein a produced by Staphylococcus

aureus with a fiber-optic-based biosensor. *Bioscience Biotechnology and Biochemistry*, 1996, **60**, 1571–1574.

42. C. A. Rowe-Taitt, J. J. Cras, C. H. Patterson, J. P. Golden, and F. S. Ligler, A ganglioside-based assay for cholera toxin using an array biosensor. *Analytical Biochemistry*, 2000, **281**, 123–133.

43. T. Konry, A. Novoa, S. Cosnier, and R. S. Marks, Development of an "electroptode" immunosensor: indium tin oxide-coated optical fiber tips conjugated with an electropolymerized thin film with conjugated cholera toxin B subunit. *Analytical Chemistry*, 2003, **75**, 2633–2639.

44. S. Herrmann, B. Leshem, S. Landes, B. Rager-Zisman, and R. S. Marks, Chemiluminescent optical fiber immunosensor for the detection of anti-West Nile virus IgG. *Talanta*, 2005, **66**, 6–14.

45. A. Petrosova, T. Konry, S. Cosnier, I. Trakht, J. Lutwama, E. Rwaguma, A. Chepurnov, E. Muhlberger, L. Lobel, and R. S. Marks, Development of a highly sensitive, field operable biosensor for serological studies of Ebola virus in central Africa. *Sensors and Actuators B: Chemical*, 2007, **122**(2), 578–586.

46. B. A. Cavic, G. L. Hayward, and M. Thompson, Acoustic waves and the study of biochemical macromolecules and cells at the sensor-liquid interface. *Analyst*, 1999, **124**, 1405–1420.

47. A. Janshoff, H. J. Galla, and C. Steinem, Piezoelectric mass-sensing devices as biosensors-an alternative to optical biosensors. *Angewandte Chemie International Edition in English*, 2000, **39**, 4004–4032.

48. F. Aberl, H. Wolf, C. Kosslinger, S. Drost, P. Woias, and S. Koch, HIV serology using piezoelectric immuno-sensors. *Sensors and Actuators B*, 1994, **18**, 271–275.

49. J. S. Yu, H. X. Liao, A. E. Gerdon, B. Huffman, R. M. Scearce, M. McAdams, S. M. Alam, P. M. Popernack, N. J. Sullivan, D. Wright, D. E. Cliffel, G. J. Nabel, and B. F. Haynes, Detection of Ebola virus envelope using monoclonal and polyclonal antibodies in ELISA, surface plasmon resonance and a quartz crystal microbalance immunosensor. *Journal of Virological Methods*, 2006, **137**, 219–228.

50. R. Gomez, R. Bashir, A. Sarikay, M. R. Ladisch, J. Sturgis, J. P. Robinson, T. Geng, A. K. Bhunia, H. L. Apple, and S. Wereley, Microfluidic biochip for impedance spectroscopy of biological species. *Biomedical Microdevices*, 2001, **3**, 201–209.

51. J. Zhai, H. Cui, and R. Yang, DNA based biosensors. *Biotechnology Advances*, 1997, **15**, 43–58.

52. C. R. Graham, D. Leslie, and D. J. Squirrell, Gene probe assays on a fibre-optic evanescent wave biosensor. *Biosensors and Bioelectronics*, 1992, **7**, 487–493.

53. P. Feng, Commercial assays systems for detection foodborne Salmonella. *Journal of Food Protection*, 1992, **55**, 927–934.

54. K. B. Mullis, The unusual origin of the polymerase chain reaction. *Scientific American*, 1990, **262**, 56–61, 64–55.

55. B. Zhu and G. O. Wang, Current status and prospects for the nucleic-acid biosensors. *Progress in Biochemistry and Biophysics*, 1997, **24**, 510–513.

56. J. Wang, G. Rivas, X. Cai, N. Donatha, H. Shi-raishi, D. Luo, and F. S. Valera, Sequence-specific

electrochemical biosensing of M. tuberculosis DNA. *Analytica Chimica Acta*, 1997, **337**, 41–48.

57. J. Wang, J. R. Fernandes, and L. T. Kubota, Polishable and renewable DNA hybridization biosensors. *Analytical Chemistry*, 1998, **70**, 3699–3702.

58. B. G. Healey, R. S. Matson, and D. R. Walt, Fiberoptic DNA sensor array capable of detecting point mutations. *Analytical Biochemistry*, 1997, **251**, 270–279.

59. D. Kambhampati, P. E. Nielsen, and W. Knoll, Investigating the kinetics of DNA-DNA and PNA-DNA interactions using surface plasmon resonance-enhanced fluorescence spectroscopy. *Biosensors and Bioelectronics*, 2001, **16**, 1109–1118.

60. H. J. Watts, D. Yeung, and H. Parkes, Real-time detection and quantification of DNA hybridization by an optical biosensor. *Analytical Chemistry*, 1995, **67**, 4283–4289.

61. R. Gabl, H. D. Feucht, H. Zeiniger, G. Eckstein, M. Schreiter, R. Primig, D. Pitzer, and W. Wersing, First results on label-free detection of DNA and protein molecules using a novel integrated sensor technology based on gravimetric detection principles. *Biosensors and Bioelectronics*, 2004, **19**, 615–620.

62. B. M. Cullum and T. Vo-Dinh, The development of optical nanosensors for biological measurements. *Trends in Biotechnology*, 2000, **18**, 388–393.

63. R. A. Felder and G. J. Kost, Automation. Part 1. Modular stepwise automation and the future of diagnostic testing. *Medical Laboratory Observer*, 1998, **30**, 22–24, 26–27.

64. A. P. Turner, Immunosensors: the next generation. *Nature Biotechnology*, 1997, **15**, 421.

65. O. P. Lymans'kyi and O. Lymans'ka, Study of microorganism genome DNA by atomic force microscopy. *Tsitologiya I Genetika*, 2002, **36**, 30–36.

66. A. L. Weisenhorn, H. E. Gaub, H. G. Hansma, R. L. Sinsheimer, G. L. Kelderman, and P. K. Hansma, Imaging single-stranded DNA, antigen-antibody reaction and polymerized Langmuir-Blodgett films with an atomic force microscope. *Scanning Microscopy*, 1990, **4**, 511–516.

67. I. A. Budashov, I. N. Kurochkin, A. K. Denisov, V. V. Skripniuk, and T. A. Shabanov, The use of atomic-force microscopy for the analysis of microbiological objects. *Zhurnal Mikrobiologii Epidemiologii I Immunobiologii*, 1997, **6**, 11–15.

68. K. M. Radke, S. R. Nettikadan, J. C. Johnson, S. G. Vengasandra, and E. Henderson, ViriChip enhances reverse transcriptase polymerase chain reaction in biological fluids and environmental samples. *Analytical Biochemistry*, 2004, **330**, 350–352.

69. A. Manz, N. Graber, and H. M. Wildmer, Miniaturized total chemical analysis systems: a novel concept for chemical sensing. *Sensors and Actuators B*, 1990, **1**, 244–248.

70. L. Hood, J. R. Heath, M. E. Phelps, and B. Lin, Systems biology and new technologies enable predictive and preventative medicine. *Science*, 2004, **306** 640–643.

71. M. Hashimoto, P. C. Chen, M. W. Mitchell, D. E. Nikitopoulos, S. A. Soper, and M. C. Murphy, Rapid PCR in a continuous flow device. *Lab on a Chip*, 2004, **4**, 638–645.

72. N. Goedecke, B. McKenna, S. El-Difrawy, L. Carey, P. Matsudaira, and D. Ehrlich, A high-performance multi-lane microdevice system designed for the DNA forensics laboratory. *Electrophoresis*, 2004, **25**, 1678–1686.

73. D. C. Duffy, J. C. McDonald, O. J. A. Schueller, and G. M. Whitesides, Rapid prototyping of microfluidic systems in poly-(dimethylsiloxane). *Analytical Chemistry*, 1998, **70**, 4974–4984.

74. N. Chiem and D. J. Harrison, Microchip-based capillary electrophoresis for immunoassays: analysis of monoclonal antibodies and theophylline. *Analytical Chemistry*, 1997, **69**, 373–378.

75. M. A. McClain, C. T. Culbertson, S. C. Jacobson, N. L. Allbritton, C. E. Sims, and J. M. Ramsey, Microfluidic devices for the high-throughput chemical analysis of cells. *Analytical Chemistry*, 2003, **75**, 5646–5655.

76. X. Zheng, M. Pang, H. D. Engler, S. Tanaka, and T. Reppun, Rapid detection of Mycobacterium tuberculosis in contaminated BACTEC 12B broth cultures by testing with amplified Mycobacterium tuberculosis direct test. *Journal of Clinical Microbiology*, 2001, **39**, 3718–3720.

77. E. J. Oh, Y. J. Park, C. L. Chang, B. K. Kim, and S. M. Kim, Improved detection and differentiation of mycobacteria with combination of Mycobacterium growth indicator tube and Roche COBAS amplicor system in conjunction with duplex PCR. *Journal of Microbiological Methods*, 2001, **46**, 29–36.

78. S. Park and R. A. Durst, Immunoliposome sandwich assay for the detection of Escherichia coli O157:H7. *Analytical Biochemistry*, 2000, **280**, 151–158.

79. L. Mukenge-Tshibaka, M. Alary, F. Bernier, E. van Dyck, C. M. Lowndes, A. Guedou, S. Anagonou, and J. R. Joly, Diagnostic performance of the roche amplicor PCR in detecting Neisseria gonorrhoeae in genitourinary specimens from female sex workers in Cotonou, Benin. *Journal of Clinical Microbiology*, 2000, **38**, 4076–4079.

80. M. Hatta and H. L. Smits, Detection of Salmonella typhi by nested polymerase chain reaction in blood, urine, and stool samples. *American Journal of Tropical Medicine and Hygiene*, 2007, **76**, 139–143.

81. J. Parkhill, G. Dougan, K. D. James, N. R. Thomson, D. Pickard, J. Wain, C. Churcher, K. L. Mungall, S. D. Bentley, M. T. Holden, M. Sebaihia, S. Baker, D. Basham, K. Brooks, T. Chillingworth, P. Connerton, A. Cronin, P. Davis, R. M. Davies, L. Dowd, N. White, J. Farrar, T. Feltwell, N. Hamlin, A. Haque, T. T. Hien, S. Holroyd, K. Jagels, A. Krogh, T. S. Larsen, S. Leather, S. Moule, P. O'Gaora, C. Parry, M. Quail, K. Rutherford, M. Simmonds, J. Skelton, K. Stevens, S. Whitehead, and B. G. Barrell, Complete genome sequence of a multiple drug resistant Salmonella enterica serovar typhi CT18. *Nature*, 2001, **413**, 848–852.

82. *Royal Tropical Institute (KIT)*, Mauritskade 63, 1092 AD, Amsterdam, The Netherlands.

83. V. Dev, Relative utility of dipsticks for diagnosis of malaria in mesoendemic area for Plasmodium falciparum and P. vivax in northeastern India. *Vector Borne and Zoonotic Diseases*, 2004, **4**, 123–130.

84. S. Belák, P. Thorén, and M. Hakhverdyan, *Advances in Diagnostic Techniques*. In: *4th International Symposium on Emerging and Re-emerging Pig Diseases Rome*, 2003.

85. S. M. Reid, N. P. Ferris, A. Bruning, G. H. Hutchings, Z. Kowalska, and L. Akerblom, Development of a rapid chromatographic strip test for the pen-side detection of foot-and-mouth disease virus antigen. *Journal of Virological Methods*, 2001, **96**, 189–202.

86. K. T. Nam, D. W. Kim, P. J. Yoo, C. Y. Chiang, N. Meethong, P. T. Hammond, Y. M. Chiang, and A. M. Belcher, Virus-enabled synthesis and assembly of nanowires for lithium ion battery electrodes. *Science*, 2006, **312**, 885–888.

87. D. P. Chandler and A. E. Jarrell, Automated purification and suspension array detection of 16S rRNA from soil and sediment extracts by using tunable surface microparticles. *Applied and Environment Microbiology*, 2004, **70**, 2621–2631.

88. J. Zlatanova and A. Mirzabekov, Gel-immobilized microarrays of nucleic acids and proteins. Production and application for macromolecular research. *Methods in Molecular Biology*, 2001, **170**, 17–38.

89. S. V. Tillib, B. N. Strizhkov, and A. D. Mirzabekov, Integration of multiple PCR amplifications and DNA mutation analyses by using oligonucleotide microchip. *Analytical Biochemistry*, 2001, **292**, 155–160.

90. S. G. Bavykin, J. P. Akowski, V. M. Zakhariev, V. E. Barsky, A. N. Perov, and A. D. Mirzabekov, Portable system for microbial sample preparation and oligonucleotide microarray analysis. *Applied and Environment Microbiology*, 2001, **67**, 922–928.

91. B. Polyak, S. Geresh, and R. S. Marks, Synthesis and characterization of a biotin-alginate conjugate and its application in a biosensor construction. *Biomacromolecules*, 2004, **5**, 389–396.

92. B. Polyak, E. Bassis, A. Novodvorets, S. Belkin, and R. S. Marks, Bioluminescent whole cell optical fiber sensor to genotoxicants: system optimization. *Sensors and Actuators B-Chemical*, 2001, **74**, 18–26.

93. C. Allix, P. Supply, and M. Fauville-Dufaux, Utility of fast mycobacterial interspersed repetitive unit-variable number tandem repeat genotyping in clinical mycobacteriological analysis. *Clinical Infectious Diseases*, 2004, **39**, 783–789.

94. M. Tishler and Y. Shoenfeld, BCG immunotherapy–from pathophysiology to clinical practice. *Expert Opinion on Drug Safety*, 2006, **5**, 225–229.

95. A. Bossi, S. A. Piletsky, P. G. Righetti, and A. P. Turner, Capillary electrophoresis coupled to biosensor detection. *Journal of Chromatography A*, 2000, **892**, 143–153.

96. J. Kohn, An immunochromatographic technique. *Immunology*, 1968, **15**, 863–865.

97. K. A. Uithoven, J. C. Schmidt, and M. E. Ballman, Rapid identification of biological warfare agents using an instrument employing a light addressable potentiometric sensor and a flow-through immunofiltration-enzyme assay system. *Biosensors and Bioelectronics*, 2000, **14**, 761–770.

Lateral-Flow Immunochromatographic Assays

R. J. Davies, S. S. Eapen and S. J. Carlisle

Unipath Ltd., Bedford, UK

1 HISTORICAL PERSPECTIVE

In 1988 Unipath launched the first one-step rapid lateral-flow immunochromatographic assay (LFiA), a urine based pregnancy test simple enough for use by the untrained consumer. Prior to this, consumers were offered either mini-immunochemistry, or mini-laboratory kits. The immumochemistry kit involved the transfer of an antibody-coated peg through several solutions in the following sequence, urine sample, antibody-enzyme conjugate, wash buffer, and finally a substrate containing solution that left a colored precipitate on the peg if the urine donor was pregnant. The lab kit included sheep blood cells, pipette, purified water, a mirror, and test tube, and a two hour wait for a result (the first version of the e.p.t.™ by Warner–Chilcott). The e.p.t.™ test was affected by vibration and gave a high incidence of false nonpregnant results and could not detect pregnancy before the third day after a missed period.[1] Slower, and presumably less accurate tests involving rabbits and frogs were never used in the consumers hands. However, the earliest consumer pregnancy test was described in an ancient Egyptian papyrus of ca 1350–1200 BC.[2] Here a woman urinated on a mixture of wheat and barley seeds several times, if the wheat grew it indicated a female child, if the barley grew it indicated a male child, no growth indicated that she was not pregnant. In 1963 this test was replicated[3] and revealed that the urine of nonpregnant women did not promote growth, while that of pregnant women did in 70% of the tests, no indication of the sex of the infant could be discerned from this modern replication. The use of a nonenzymatic, color generating label greatly simplified pregnancy testing, and later other tests. The desire for home testing was the impetus for the first immunochromatographic assays. In its latest guise, Clearblue Digital™ (released 2003), utilizes an inexpensive optical reader to overcome any difficulties users may have in interpreting the test line.

To date many different LFiAs have been developed for use with different sample fluids, *viz*: sputum,[4] blood,[5] plasma and/or serum,[6] fecal,[7] food,[8] for the measurement of viruses,[9] drugs,[10] micro-organisms,[11] spores,[12] polymerized nucleic acid sequences.[13] Some of these tests are so easy to use that they have been CLIA waived United States.[14] However, the apparent simplicity of many of these tests belies the great pains the manufacturers have taken in their construction.

2 LFiAs: BASIC OPERATING PRINCIPLES

LFiAs utilize porous membranes, antibodies and usually a visible signal generating system to produce sensitive, disposable, and easy to use tests.

Handbook of Biosensors and Biochips. Edited by Robert S. Marks, David C. Cullen, Isao Karube, Christopher R. Lowe and Howard H. Weetall.
© 2007 John Wiley & Sons, Ltd. ISBN 978-0-470-01905-4.

Fluid flow is sustained by means of capillary forces and resisted by viscous drag generated by the materials used to construct the test strip. All components in the test strip must be water wettable and hydraulically connected—usually by means of small overlaps of the test strip components. These are either formed when the strip is laminated together on an adhesive tape, or by holding them together with pinch points designed into the plastic strip housing. In general the intrinsic capillary pull of the different functional zones of the strip increase with distance from the sample application zone, so as to produce a sustained capillary pull during the test; the overall hydraulic resistance to flow also increases as the sample migrates up the strip. For low volume LFiAs evaporative loss of liquid can stall flow, but this can be reduced by covering some of the components with a clear adhesive film.

The plastic casework has other functions apart from that mentioned above, it must direct the sample to the first capillary element (unless that is a protruding wick), usually via a sample well, and prevent the sample escaping to wet the interior of the housing. It can also act as a handle, and a means of carrying visual information for the user. In some devices the casework also holds desiccant. This maintains the sugar matrix and nitrocellulose (NC) glaze (see later sections in this chapter) in a glassy state in which molecular mobility and hence chemical reactivity, have almost ceased, extending product shelf life. Such is the sensitivity of LFiAs to water that they are packed in plastic heat-sealable films having a continuous metal layer, this is a far superior water barrier than a nonmettalized film. The casework may also hold the analytical membrane so that the test line(s) are precisely positioned, enabling a reader to make measurements of line intensity. Often with optically read strips, the casework will also contain a high level of filler, usually titania, to block light. Finally, the casework has to protect the test strip from mechanical damage.

There are two main LFiA formats; the sandwich assay used for the measurement of molecules large enough for two antibodies to bind simultaneously, and the competitive assay for small (hapten) molecules, which can only bind to one antibody at any particular time.

2.1 Sandwich Assays

Typical architecture is depicted in Figure 1.

With Clearblue™ the user urinates onto the wick, which will transport the urine by capillary forces to the conjugate pad. The wick contains solutes to condition the urine. The conjugate pad contains the signal generating system, in this case 400-nm diameter spheres of a blue-dyed polystyrene latex. Adsorbed on the latex surface is a monoclonal antibody (mAb), which binds to the α-subunit of the human chorionic gonadotrophin (hCG) molecule.[15] The latex is embedded in an amorphous sugar/protein matrix designed to prevent it adhering to the high surface area conjugate pad. This prevents latex aggregation upon drying and stabilizes the antibody over the *ca.* three year life of the product. Also, it facilitates rapid dispersion of the latex when wetted by urine. After liberation from the matrix, latex will bind a proportion of the hCG and the mixture (latex, sugar, free hCG, etc.) moves onto the analytical membrane. In Clearblue™ this is a custom manufactured, white porous NC membrane bonded to a clear Mylar backing during manufacture. This membrane allows passage of up to 700-nm diameter protein-coated latex. A second mAb, which binds specifically to the β-subunit of hCG, is adsorbed as a test line across the width of the membrane and a third is plotted downstream of the second to capture latex that has flowed through the test line. The NC is blocked with a proprietary treatment to prevent the nonspecific capture of the protein-coated latex. The mixture of latex with bound and free hCG travels through the test line whereupon the latex is bound, creating a blue line against a white background. The strength of the line color depends upon the number of latex particles captured and, to a lesser extent on the specific area of NC on which it is captured. The fraction of captured latex will, up to a point, increase with the number of hCG molecules on the latex. The assay is set up so that a visible test line develops above 25 mIU ml^{-1} hCG ($\sim$100 pg ml^{-1} based on 70 µg = 650 IU).[16] The control line develops regardless of hCG level, indicating that the test has run properly. As the hCG level increases a smaller fraction of latex passes onto the control line, and this would be increasingly feint unless corrected. To maintain a consistently strong control line the conjugate pad contains two lattices, one of which

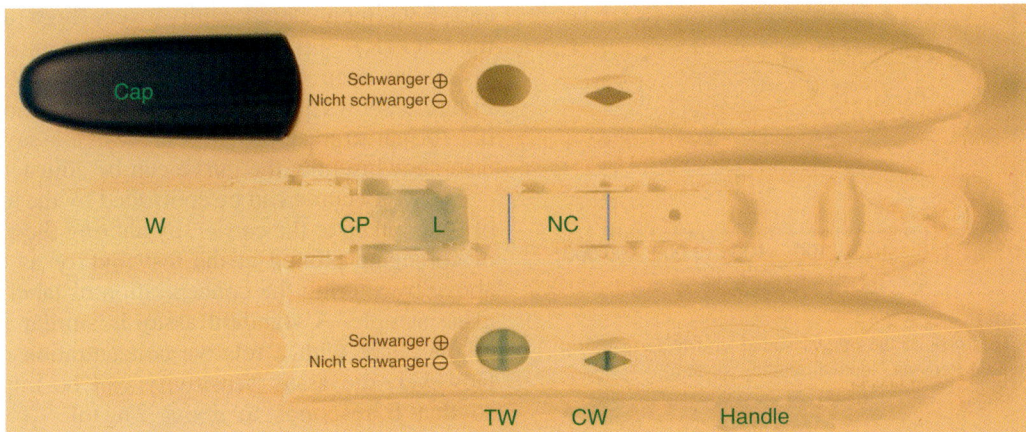

(a)

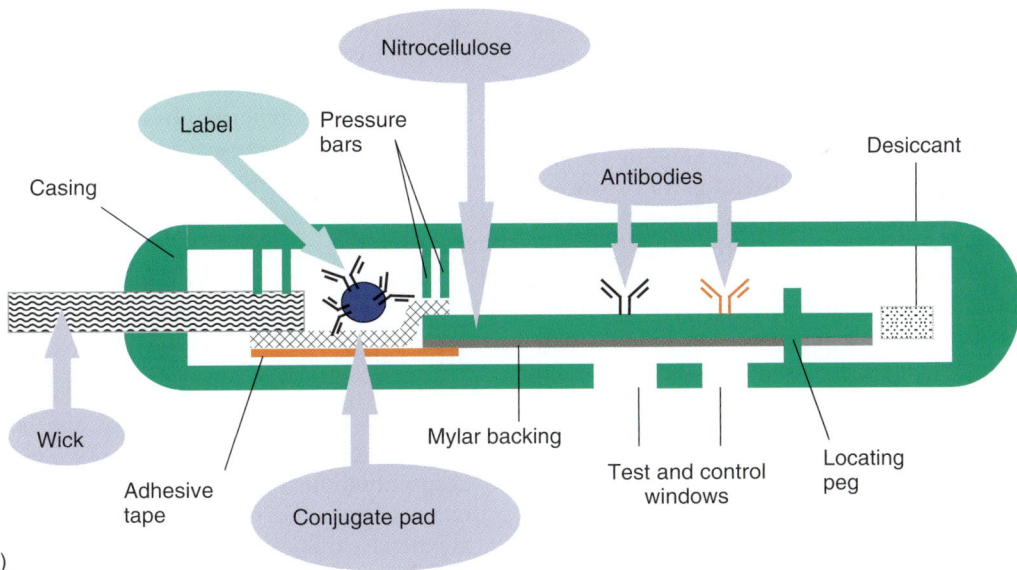

(b)

Figure 1. Photograph of the Clearblue™ pregnancy test. The center image shows the functional components, namely, W = wick, CP = conjugate pad, L = latex, NC = nitrocellulose. The two lines on the NC indicate the position of the test and control antibody lines plotted on the NC, these lie in the center of the test (TW) and control (CW) windows respectively. The lower image shows a developed test after running 100 mIU ml⁻¹ hCG in urine. The positive symbol indicates pregnancy and is made up of a latex line (vertical) and an ink line printed on the NC surface which only becomes visible when the NC becomes more transparent upon wetting by urine. (b) Schematic shows the main components of a Clearblue test strip.

does not have the anti-hCG antibody adsorbed and only captured by the control line. By careful design and choice of materials the test delivers a result in one minute and is over 99% accurate when used from the day menses are due.

Clearblue™ was originally designed as a visual read product, but it is possible to optically measure the signal and produce a semi-quantitative assay.

Figure 2 shows results from an hCG sandwich assay using the same monoclonal pairs used in Clearblue™ measured using a Persona™ optical reader. This reader measures the amount of light from a red light emitting diode (LED) (633 nm) transmitted through the wet NC, both at the test line and at regions either side of the line, which are used to set the baseline value. The

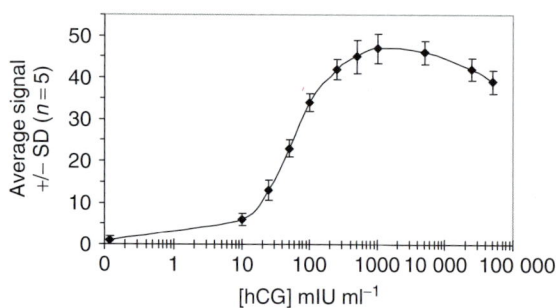

Figure 2. The form of the optical signal in a hCG test strip measured with a Persona™ optical reading system.

optical geometry is arranged so that when the latex at the test line blocks all light transmission the reading is 50%. In addition to showing the typical sigmoidal response of these assays to increasing analyte concentration, the curve also shows a high dose hook. This is because the concentration of hCG is so great that the latex immobilized mAb fails to bind enough of the free analyte to prevent substantial occupancies of the binding sites on the NC located mAbs. With such occupancy of binding sites on the latex and test line the probability of forming the immunological sandwich is reduced, at 100% occupancy of test line and latex no binding will occur. Some theoretical aspects of sandwich[17] and competitive[18] LFiAs have appeared recently, these begin to give a mathematical understanding to the importance of the various factors that can influence signal development at the test line.

2.2   Competitive LFiAs

These are used for the detection and semi-quantification of small molecules, typically drugs of abuse and their catabolites[19] or natural metabolites, such as estrone-3-glucuronide (E3G), an estorogen metabolite and one of the species measured in the LFiA contraceptive device, Persona™. In Persona™, E3G is conjugated to a carrier protein and plotted at the test line. Urine is sampled from a flowing stream by a wick, which transfers it to a conjugate pad containing two blue-dyed lattices. The first has adsorbed anti-leutenizing hormone for a sandwich leutenizing hormone (LH) assay conducted on the same strip, the second has an

anti–E3G monoclonal with a very high affinity for E3G ($K_D \leq 0.1$ nM), which is understandable given the deep binding cleft in this antibody.[20] Figure 3 shows a schematic for this assay, of similar format to many other competitive assays, and data showing how the curve can be adjusted.

The assay curve can be controlled by the amount of E3G linked to the carrier, by altering the amount of conjugate plotted at the test line, or as shown above by altering the concentration of label in the urine sample. A standard assay is similar to that of the curve with a relative concentration of 7.5. However, the assay sensitivity can be improved so that it responds to about 2 ng ml^{-1} (~5 nm) E3G. If we consider the sample volume, this test detects just 125 atto moles of E3G. In the presence of E3G concentrations above 20 ng ml^{-1} the line coloration is very weak. With visual read competitive LFiAs the lack of signal at high hapten concentrations presents some user difficulty as the result is counterintuitive. A control line or indicator must be present to reassure the user that the test has run properly.

3   MATERIALS USED IN LFiA SYSTEMS

3.1   Labels

The label generates a signal that can be seen by eye or an electronic reader. It can be a color due to absorption/reflection of visible light, fluorescence or even magnetic field set up by an assembly of magnetic particles.

Labels must stably hold upon their surface one of the antibody pairs used to ensure its capture at the line of immobilized capture species on the analytical membrane, and must be small enough to flow through the pores of the membrane. Consequently they tend to be colloidal particles and as such are susceptible to aggregation when the mechanisms promoting dispersion are overcome. Bare colloidal labels used in LFiAs have charged surfaces—due to adsorbed anions (e.g., citrate in the case of gold sol), or ionisable groups (e.g., sulphate and carboxylate based initiators used in the emulsion polymerization of lattices). This gives rise to sufficient electrostatic repulsion to prevent irreversible aggregation. Care has to be taken when coating labels with antibodies to keep the ionic

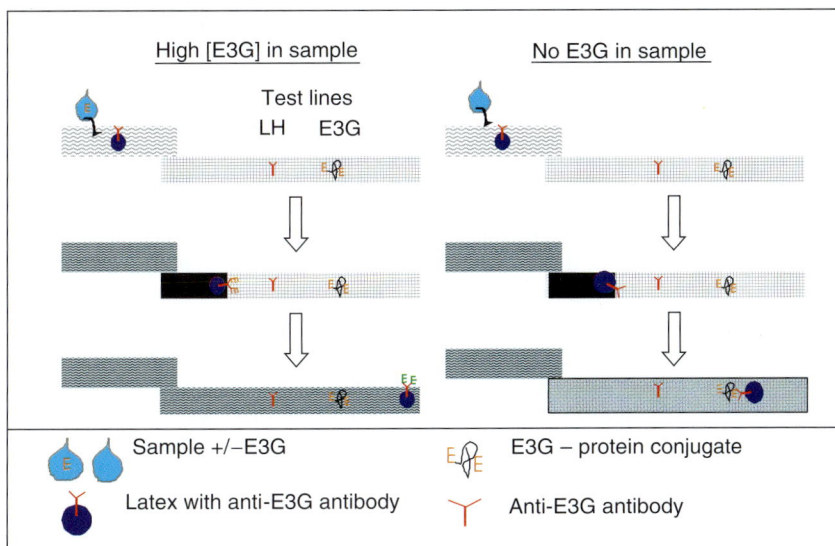

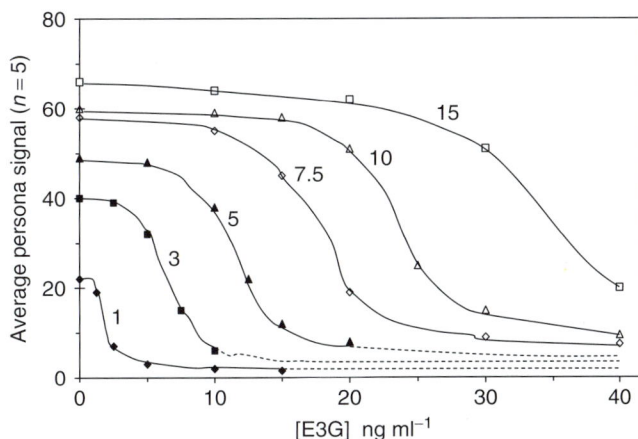

Figure 3. Schematic of a competitive assay for E3G and data showing displacement of the assay curve with increasing amounts of label. The numbers on the curves indicate the relative concentrations of blue latex-anti-E3G label incubated with urine spiked with E3G, a 25 μL volume of urine/label mix was passed up the 6-mm wide NC strip.

strength low and control pH to ensure electrostatic stability is not impaired. Once coated the particles usually tolerate environments of higher ionic strength found in standard samples, that is, *ca.* 0.15 mM in saline, saliva, plasma, sera, and urine. This is due to electrostatic and steric stabilization interactions from adsorbed antibody layer. An understanding of the interplay between the typically repulsive electrostatic and attractive van der Waals forces (DLVO theory) operating in aqueous suspensions of colloidal labels suggests useful guidelines to the process of antibody coating.[21] To produce a well controlled test the label should have

a low variation in size (low polydispersity), surface chemistry, and "signal content"—dye, fluor, or magnetic moment for a latex. It is also important to ensure that the label is uniformly coated with the antibody. For example, poor mixing of antibody and latex has been shown by flow cytometry to yield a mixture of particles, some fully coated with antibody and others with a sparse coverage (Inger Jonruup, personal communication). Good conjugation can be achieved with use of in in-line mixer.[22]

A pertinent question asked by those developing LFiAs is "what is the best label?". A trite answer is "the one that gives the most signal". Signal derives from a combination of the particle's signal generating properties, antibody binding activity when immobilized upon the surface, and its tendency toward nonspecific binding (NSB). With a machine read strip the result also depends upon the components used in the reader, which is governed by permitted expenditure. Further consideration must also be given as to how easy it is to load the label with antibodies (i.e., can it be washed between each coating stage in a centrifuge without aggregating, how susceptible is it to aggregation during the coating process, and how reproducibly can the base particle be made). Consequently there is no answer to this question. LFiAs have evolved around the traditional particles (latex, gold, selenium, and carbon sols), with new labels constantly being discovered and evaluated.

3.1.1   Traditional Labels

These yield visible lines and do not require an electronic reading system, although they are also used in conjunction with a reading system for semi-quantitative measurements.

Gold Sol
A preferred label in many tests is gold sol. Methods for preparing near monodisperse gold sol have been known from Zsigmondy's 1906 work onward,[23] but only recently has well-controlled material become commercially available. Gold sol has the advantage of not requiring dye, its color originates a strong, broad optical absorption band not present the bulk metal. The adsorbed photons set up collective oscillation of the conduction electrons (a plasmon), which extends to the surface as a surface plasmon wave, the frequency of this resonance (surface plasmon resonance (SPR)), is diameter dependant. Sols used in LFiAs have diameters from 20 to 60 nm and absorb green light (520–540 nm) making them appear red in white light.[24] Typically a 40-nm diameter sol, with a standard deviation on diameter of better than 5% is used, this produces a red test line under appropriate conditions. Another advantage of sol is that when coating with antibody or during the test, aggregation is easily observed. This is apparent when gold surface to surface separation is less

than the particle diameter, and is shown by a color change from red to blue/purple.[25]

Gold sol are typically 1/10th the diameter of the lattices used in LFiAs and have a high curvature, making it more likely that adsorbed antibody (of length ca 15 nm) protrudes from the surface of the sol. This increases mAb accessibility to binding analyte, particularly if it was inflexible and did not undergo structural rearrangement on the surface by the short range van der Waals dispersion forces. Measurements of the small decrease in the SPR wavelength when hCG binds to an antibody-coated 40-nm gold sol indicate that at least 25 hCG molecules bind per sol.[26] Measurements in our laboratory of the uptake of hCG by an anti-α-hCG coated 40 nm sol indicate *ca.* 100 molecules bound, with an expected 60 molecules of antibody adsorbed to the sol based on a coverage of 5 mg m^{-2}. This suggests that the binding activity of antibody on sol is very high. We found that about 400 hCG molecules bound to a 400-nm latex coated with the same antibody.

Polystyrene Latex
The quality of polystyrene and other polymer lattices has improved over the past 20 years. It is now possible to obtain from commercial sources almost monodisperse latex with controlled surface chemistries and reproducible dye or fluor levels that are surfactant free. (Surfactant free lattices are available from: Interfacial Dynamics Corp., Molecular Probes Inc., 29851 Willow Creek Rd., Eugene, OR 97402, USA. www.idclatex.com. Duke Scientific Inc, 2463 Faber Place, Palo Alto, USA. www.dukescientific.com. Polysciences Inc., 400 Valley Rd., Warrington, Pennsylvania, USA. www.polysciences.com.) The surfactant-free lattices remain stable in suspension because they have a greater surface charge density capable providing electrostatic stabilization. This simplifies protein adsorption as the protein no longer has to compete with surfactant for the surface, and the coating procedure does not require diafiltration, or the like, to gradually remove the surfactant while the protein adsorbs to maintain stability. Dyed lattices are good for visually read assays. Many dyed lattices are available with the dye permeating the bulk of the latex. The dye may be incorporated by solvent swelling of the latex in the presence of the dissolved dye, then washing with a nonsolvent to collapse the latex and remove the unincorporated dye.

Selenium Sol

Selenium is a metal, which in colloidal form like gold sol has a visible color (orange/red). Antibodies and other proteins will adsorb to it enabling it to be used in LFiAs. The sol is not commercially available and was used in Abbott Laboratories' Test Pak™ range of tests. It appears to have no advantages over gold sol.

Carbon Sol

Carbon sol, essentially soot from the controlled burning of oil, is typically used a filler in car tyres. It is the least expensive label and offers perfect contrast against the white NC membrane. Its major disadvantage is that it is usually quite hydrophobic and of an irregular size and shape. So while it will bind antibodies strongly, the preparation usually ends up highly aggregated even when detergents are used to disperse the starting material. It has been claimed that it is possible to select a carbon sol based upon measurement of its surface oxidation, highly oxidized sols are water dispersible and can be coated with antibody;[27] one commercial LFiA uses a carbon sol label. (The Contrast II hCG urine/serum LFiA, Genzyme Corp. 500 Kendall St., Cambridge, MA 02141, USA, www.Genzyme.com.)

3.1.2 Emerging Labels

Fluorescent Latices

Fluorescent lattices may improve assay sensitivity but need a light source and well defined (expensive) optical filters to discriminate the emission, which may only be 10–20 nm higher in wavelength (stokes shift), from the scattered incident light. Latices loaded with organic fluorophores have been available for many years, but have not been used in commercial LFiAs. Recently lanthanide chelates have been incorporated into lattices,[28] this circumvents one complication of the well known DELFIA (dissociation-enhanced lanthanide fluoroimmunoassay) microtitre plate based assay system. In this a lanthanide ion chelated to a secondary antibody is dissociated from the protein by an acid/detergent solution, which contains chelator and antenna molecules. The liberated ions become chelated in a micelle and thereby shielded from the massive quenching effect of water, allowing

them to emit light. The antenna molecules absorb UV (e.g., 340 nm) and pass this energy on to the chelated lanthanide: a europium chelate emits a sharp band at 615 nm (bandwidth *ca.* 10 nm) with a lifetime of about 1 μs. The long lifetime and large stokes shift allow the emission to be separated from the UV induced autofluorescence of the sample and materials used in the construction of the assay. This is achieved using crude optical filters and/or time-gating of the detection signal, and provides far better discrimination against background than is possible with more conventional organic fluors.

Quantum dots are monodisperse semiconductor materials, for example, CdSe, ranging in size from about 1 to 10 nm. They have very high extinction coefficients for two photon absorption, orders of magnitude higher than organic fluorophores,[29] absorb light over a wide range of wavelengths below their sharp emission peaks, and have high quantum efficiencies of typically 20–50%. Emission results from electron–hole recombination within a geometric space (the quantum dot core) smaller than the natural dimension of the hole-pair (exciton). Small changes in the size and shape of the core have profound effects on the emission wavelength. Sharp photoemission bands (in the 0.4–2 μm wavelength range) increase with Qdot diameter. Excitation of many different types (emission colors) of quantum dots (Qdots) is possible with one UV light source, making multiplexed assays possible. Although, the spatial separation of test zones or patches in LFiA makes this feature redundant. One problem with Qdots is that they are hydrophobic and colloidal stability is difficult to attain. They are often coated with a thin layer of ZnSe. In the case of CdSe this greatly increases the emission intensity of the core and allows for the attachment of other layers and biological materials.[30] Qdots have recently been used in bioassays[31] and their potential for use in LFiAs is underway.[32] Qdots can also be incorporated into latex particles (Qbeads) making them more tractable to those attempting antibody conjugation.[33]

Magnetic Latex

Superparamagnetic particles, which only become magnetic in a magnetic field have been produced using magnetite-loaded latex (Indicia Diagnostics, 33, Avenue de la Californie, 69600, Oullins,

France) and magnetite-silica aggregates coated with dextran.[34] Para- and ferro-magnetic particles are not suitable as they would aggregate due to their mutual magnetic attraction. The base magnetite is too small (*ca.* 10–20 nm) to maintain a permanent magnetic moment, and these particles can constitute up to 80% of the weight of a latex. The characteristics of these particles are not as well controlled as emulsion polymerized latex, they are available as stable dispersions in detergents, their mean magnetic moment and its variation requires tight control if they are to be used in assays. Although the particles are colored (brown to black, depending upon the loading and composition of the oxide), they are intended to be used with a reader that measures the magnetic properties of the captured material at the test line. Magnetic reading technology has been commercialized by Quantum Design—MAR[™] reader. (Quantum Design, 6325 Lusk Boulevard, San Diesgo, CA 92121. www.qdusa.com.) This moves the developed test strip line immediately beneath a thin sapphire window on whose distal surface is a series of flat, spiral conducting coils, and above a fixed magnet, which magnetizes the superparamangnetic latex. Movement of the magnetic latex line induces a voltage in the coils, the magnitude of which depends upon the magnetite content of the test line. This system appears to detect 30 pg ml^{-1} staphylococcal enterotoxin B.

Phosphorescent Particles
Inorganic phosphors, such as those coated on the interior of television screens are prepared by grinding a ceramic formed at high temperature. They are irregular in shape and too large (*ca.* 10 μm) to be of use in LFiAs. A phosphorescent particle has been claimed to be useful in NC based LFiA. The material has a particle size of between 5–200 nm, comprises a ceramic of aluminium oxide doped with europium, lanthanum, and neodymium oxides, and Sr_2CO_3: the latter compound appears to store energy from the UV incident light and also determines the color of the emitted light; strontium causes emission at 540 nm and calcium at 450 nm, the peak width at half maximum intensity as *ca.* 60 nm. The emission persists for several hundred seconds. These particles are coated with a silane coupling reagent, 3-aminopropyl-trimethoxysilane, and linked to antibody using carbodiimide. In LFiA format they are visible in subdued lighting and are claimed to be far more sensitive then gold sol.[35]

Up-converting phosphors are crystals of lanthanide elements, which absorb low energy infrared and emit at higher energy (higher frequency) visible wavelengths.[36] This is achieved by a multiphoton adsorption process and different colors are available, depending upon the lanthanide dopants. For example, ytterbium ions will adsorb two 980 nm photons and excite erbium ions to emit at 550 nm (green) close to the wavelength of highest visual acuity. Red and blue phosphors are also available. However, despite emitting visible radiation the intensity is too low to be detectable by eye, and an optoelectronic reader has to be employed. One benefit of these materials is that IR causes negligible fluorescence in biological samples and so the background signal is low. Wavelength-gated and time-gated detection and even phase-sensitive detection can be used to eliminate background, enabling extremely sensitive signal detection. These labels are eminently suited to highly colored and fluorescent samples. One technical issue is that the light output of the particle is exceptionally dependent on particle size; halving the particle diameter gives approximately a sixfold reduction in signal. It was not until the manufacture of spherical, low polydispersity particles[37] that they could be commercially exploited. To stabilize the crystals and make them water dispersible they are coated with a thin layer of silica. This can be chemically modified to permit covalent coupling of protein.[38] One company pioneering the use of these labels is OraSure Technologies, Inc, their products include saliva based drug tests and utilize a reader containing an LED and photodiode detector.

Liposomes
The aqueous lumen of bilayer lipid vesicles can be loaded with a variety of dyes/fluors and by means of the appropriate lipid and coupling chemistry it is possible to create a antibody-coated label. One problem is that large osmotic forces arising when the liposome is dried in the conjugate pad transiently break open the vesicle to release some of the entrapped signal molecules. However, by appropriate choice of the drying sugar most (>70%) of the liposomes and their contents can be recovered and mobilized up the NC upon rehydration.[39]

3.2 Analytic Membranes

NC is the most widely used polymer membrane for LFiAs. Initially the membrane was adhered to a carrier plastic using pressure sensitive adhesives. However, ingress of adhesive into the pores of the membrane gave rise to poor consistency and now leading manufacturers (Schleicher & Schuell, Millipore, Whatman, Sartorius, and MDI of India) cast the polymer directly onto a Mylar film to provide handling strength. Unbacked membranes find applications in flow-through immunoassays. As NC is a hydrophobic polymer manufacturers incorporate surfactants to enable wetting of the dry membrane. Properties such as the pore size and membrane thickness are carefully controlled during manufacture by manipulation of the evaporation rate and the film thickness of the polymer slurry on the Mylar. Figure 4 is an electron micrograph of the interior of a good NC membrane.

Substitute membranes for NC have been developed based on nylon (Cuno Corporation's Novylon™), and poly ethersulphone (Pall Corporation's Predator™), but these do not have all the attributes of a good NC membrane.

3.2.1 Characterization of NC

Manufacturers continually adjust their processes to maintain and improve consistency within and

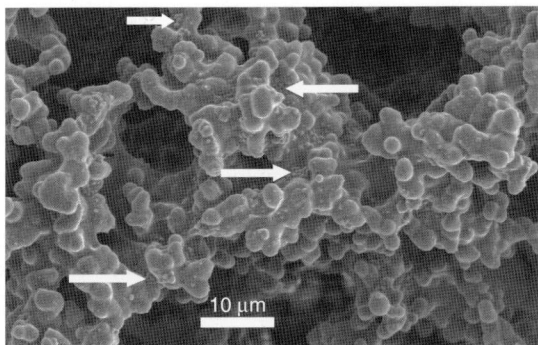

Figure 4. A scanning electron micrograph showing a cross-section through a NC membrane in the region of the anti-hCG test line. This test line was developed with 500-nm diameter polystyrene lattices in the absence of hCG and so has a sparse coverage of particles (some indicated by arrows). Note the typical nodular structure of the NC fiber and the variable "pore" dimension. A typical NC membrane is ~100 μm thick and about 80% air by volume.

between batches, and have devised standard tests to qualify their membranes.

Membrane Flow Rate
This common test is the time taken for water to wick 40 mm vertically up the membrane under defined conditions (RH, temp, etc.). It provides a measure of the pore size and aqueous wettability of the NC, which will depend on the type and amount of surfactant incorporated. Note that as the pore size decreases, the lateral wicking rate of the membrane also decreases, but the protein binding capacity of the membrane increases in line with its surface area. A slower wicking rate increases the effective sensitivity as it permits reagents to spend a longer time in the capture zone, allowing the label to make more collisions with the antibody on the test line.

Membrane thickness
This parameter is easy to measure and variations in membrane thickness will have a direct influence on flow rate and protein binding capacity, affecting signal. Membrane thickness is typically 50–200 μm.

Absorption Time and Capacity for Liquids
This test evaluates the time required for a membrane to take up a certain volume of test liquid under defined conditions, and the area of the resulting wet spot, which depends indirectly on void volume and thickness. These parameters are usually called *absorption time* and *absorption capacity*, and are respectively reported in seconds per μl and $cm^2 \; \mu l^{-1}$.

3.2.2 Deposition of Proteins onto NC

There are many systems with excellent volumetric precision available for dispensing reagents onto membranes. Generally they are of two types: contact and non-contact dispensers. Although contact dispensers are the most widely used, these can compress and damage the soft NC membrane leaving an indent at the application zone if they are not well set up. Non-contact dispensers, typically inkjet systems, are less likely to suffer from this problem and offer the added advantage of printing more complex patterns designed to give the user information on test performance (e.g., the tick mark that acts as a positive control in IMI's Test

Pak™ pregnancy test). Once proteins are deposited on the membrane they are often fixed by drying and then blocked using manufacturer's favored reagents.

3.3   Ancillary Membranes

At each end of the NC, either nearest the sample (proximal), or furthest from its point of application (distal), membranes of different functional requirement may be attached.

3.3.1   Conjugate Pad (Proximal End of NC)

This material is used to hold the dried label, typically $\sim 10^9$ particles, in an amorphous/crystalline sugar phase, containing other additives such as detergents and protein to reduce NSB, as well as buffer salts to control pH and other excipients. Ideally the label is uniformly held about the fibers of the pad so that dispersion is uniform and gradual into the flowing sample. Too slow release of label from the pad results in a high background of label on the NC. Very sharp release results in label only interrogating analyte from a very small fraction of the sample. Release characteristics are determined by the physical properties of the pad, formulation of the dried matrix holding the label, and sample viscosity.

3.3.2   Blood Separator (Proximal End of NC)

Red cells, flexible, biconcave discs of *ca.* $8 \times 2\,\mu m$, are typically larger than the "pores" of an analytic membrane. This would greatly reduce fluid flow, and may even cause haemolysis if they get into the membrane. With some labels, particularly gold and selenium sols, their contrast against the membrane would be reduced in the presence of released haemoglobin. Use of a cell separating membrane overcomes these issues and also provides a more accurate measurement of the analyte concentration in blood plasma, overcoming the need for haematocrit correction. Patents have been issued on several materials suitable for use as a plasma generator in a LFiAs, viz: glass fibers,[40] polymer fibers,[41] and asymmetric pore polymers[42]

amongst others. One of the most effective, in terms of plasma yield and rate of plasma delivery, is a mixed glass fiber matrix; it is also off patent and freely available for use. Red cell movement through this matrix is retarded leading to the development of a plasma front, which can be sucked up by the NC. This material may be used in a lateral or vertical flow mode if the material is thick and placed on top of the conjugate pad or analytic membrane. The label may be loaded into the blood separator in ways similar to those used to load the conjugate pad.

3.3.3   Sinks (Distal End of NC)

A sink is a piece of absorbent material placed in hydraulic contact with the membrane on the distal end of the device. Its main function is to ensure residual label is drawn away from the test and control lines, enabling a good contrast between these and the regions of the analytical membrane either side of them. In cases where a large volume of sample must pass through the conjugate pad and NC the sink must have enough capacity to imbibe this volume while maintaining a reasonable flow. As NC only has a small capacity for fluid this is used to increase the volume of sample interrogated to reduce the lower limit of detection of the test.

4   ATTACHMENT OF ANTIBODIES TO LABELS AND NC

4.1   Adsorption

Proteins irreversibly adsorb to nearly all surfaces by virtue of the many interactions shown in Figure 5.[43] Once the summed interactions exceed $\sim 10\,kT$ the protein is essentially irreversibly adsorbed. Strong adsorption is expected via hydrophobic interactions with latex and carbon sol due to strong van der Waals interactions, and for gold sol owing to its high Hamaker constant.[44] When coating a label by absorption the label must remain colloidally stable (dispersed as singletons), to achieve this it is important to keep the electrolytes concentration to a minimum while maintaining a pH at which the electrostatic charge on the label is sufficient to maintain stability. Fortunately antibody/protein adsorption is so

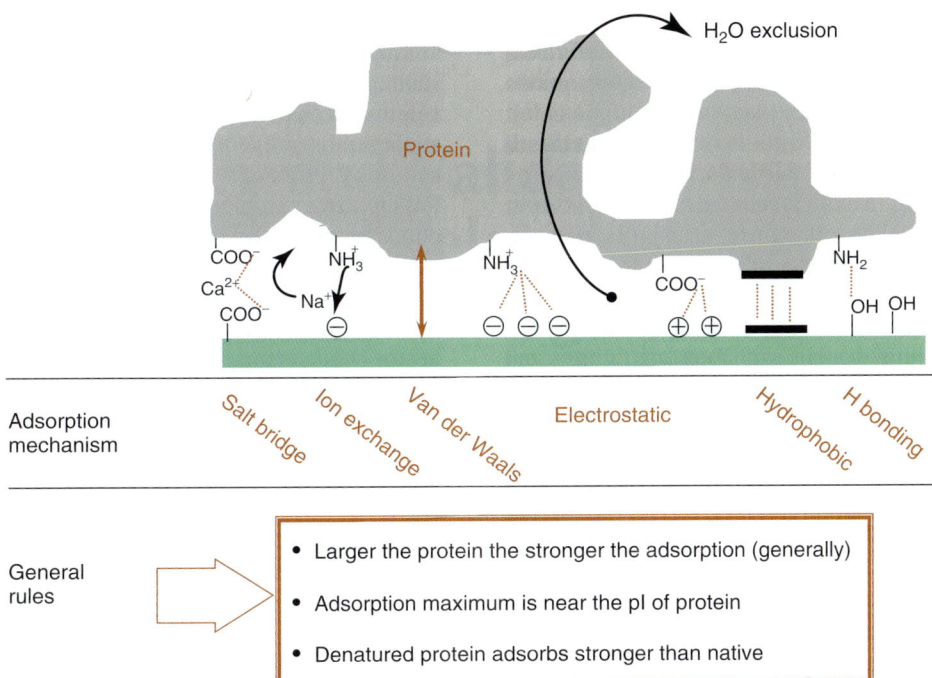

Figure 5. Protein adsorption mechanisms. Attractive interactions causing protein to adsorb to a surface will depend upon the properties of the surface (hydrophobicity, ionic nature, availability of H-bonding sites, roughness on a nanometer to sub-nanometer scale and the protein). Although the strength of the individual interactions may be less than kT, when added they can be sufficient to hold a protein at a surface irreversibly. The literature on protein adsorption is extensive, but three general rules appear to apply.

strong to most labels that only a low concentration of antibody is needed, typically $10–100\,\mu g\,ml^{-1}$,[6] this reduces the effect of the antibody's counterions on the bulk ionic strength. Gaps between the antibody adsorbed on the label are partially filled in a process called *blocking* with another smaller protein, such as bovine serum albumen to reduce the NSB of the particle to the test line. This is particularly important in sample matrices containing low levels of protein, such as urine. Complete blocking (i.e., preventing access to the latex surface by protein on the test line), is probably not achievable with one protein, but for most instances this suffices. As the adsorbed antibody is not orientated on the particle with its binding sites (paratopes) directed away from the surface, the analyte binding capacity of the label is reduced. While it is possible to measure the binding capacity of a label for analyte, what is really important is the fraction of these that project from the surface enabling sandwich binding to occur at the test line. This has not been measured, but if it could be it might offer a means

of evaluating different methods of attaching antibodies to labels. Immuno-reactivity changes with the combination of surface, protein characteristics, and adsorption conditions. With NC there is an increase in label binding capacity with antibody coverage ($mg\,m^{-2}$), but this tails off as the surface becomes crowded with antibody (authors unpublished observations), the same occurs for the binding of hCG to antibody adsorbed on planar surfaces.[45] The explanation is that adsorption of one antibody may obscure the paratope of another that has adsorbed earlier, and that this later adsorbing antibody does not provide a functional binding site.

Simplicity and wide applicability makes adsorption the most common route of attachment of antibodies to labels. It is also used to anchor antibodies to the NC membrane, where similar interactions to those depicted in Figure 5 occur. One specific interaction believed to predominate in the adsorption of protein to NC is the dipole–dipole interaction between the nitro group and the carbonyl of the peptide bond.[46] While it is clearly easier to apply protein to NC compared with colloids, as there are no concerns over stability, there are other factors to consider. The antibody needs to have a uniform binding activity throughout the volume of the test line, and the test line should be well defined spatially, particularly for an instrument read strip. Small levels (<10% v/v) of low alcohols (methanol, ethanol or propanol), may be added to facilitate wetting of the NC and to reduce the solubility of antibody so that it is strongly adsorbed. High levels of electrolyte may block pores in the dried strip and interfere with sample/label flow through the membrane; it may also reduce adsorption of the protein.

4.2 Covalent Immobilization

Covalent coupling of antibodies can be applied to analytical membranes, but is more commonly used with labels. While it is easy to give reasons for employing this more complex immobilization strategy, many of these are unreasonable given the paucity of publications that give experimental support to them, viz: to increase the loading of antibody to the surface; to prevent desorption of antibody by other materials (proteins) in the sample; to orientate the antibody and thereby increase the binding capacity of the surface. There are many publications where authors have gone through some chemical ritual designed to couple protein to latex, but they rarely demonstrate what fraction of the antibody has been covalently linked and the benefit over the control (adsorbed) system. Some authors appear to believe that an antibody having a reactive group at one end impacts upon the substrate to which it is to be coupled in such a way that a covalent bond is formed and the antibody has simply lost the ability to adsorb! Chemical coupling may appear useful, but its implementation in a manufacturing environment, which must produce a high yield of devices that function to certain analytical tolerances present added difficulties. Any coupling chemistries must be proven to have benefit and the reagents need a relevant, device independent quality control process.

Covalent coupling may, however, be of benefit in coupling antibody fragments and antibodies to protein adhesive surfaces and to systems exposed to harsh environments, for example, samples containing nondenaturing surfactants. To construct such a surface the label is first coated with a layer, a hydrogel, typically of a hydrophilic polymer such as dextran, polyethylene glycol, or ethoxylated celluloses either by adsorption or via chemical linkers, for example, thiols and silanes, which will bond to gold and silica based surfaces respectively. The bio-resistive polymer hides the label surface, and if suitably derivatized will couple protein without it encountering the denaturing effect of the underlying (hidden) surface. It also prevents nonspecific binding because it consists of a flexible, water soluble polymer holding a high level of water at the surface.[47] Binding of a protein to the coating, or binding of the coated latex to a surface, would require the entropically unfavourable restriction of mobility of the polymer.[48] When constructing a bio-resistive surface it is important not to make the hydrogel so thick that it contains antibodies within its thickness as these will simply deplete the sample of analyte and reduce the sensitivity of the assay.

Covalent immobilization of antibody to the analytical membrane does not appear to be of benefit judging by the absence of commercial tests using membranes that are available for such coupling, for example, Pall Corporation's Immunodyne ABC membrane.

5 SAMPLE MATRIX AND ISSUES

For a LFiA to meet its analytical performance interfering materials in the sample may have to be ameliorated to a tolerable level. The FDA has recommended that assay kits employing mouse mAbs include a warning that samples from donors who had received diagnostic or therapeutic antibodies may show a falsely elevated or depressed value when tested.[49] It is not possible to give a useful figure for the proportion of samples within a population that would give rise to false results, as this depends upon the materials used in the test. Recognizing the above caveat one can expect from 2–40% of samples to cause problems.[50] A powerful example of an interference problem is that of the blood hCG test, which indicated pregnancy in women who were not pregnant. As hCG

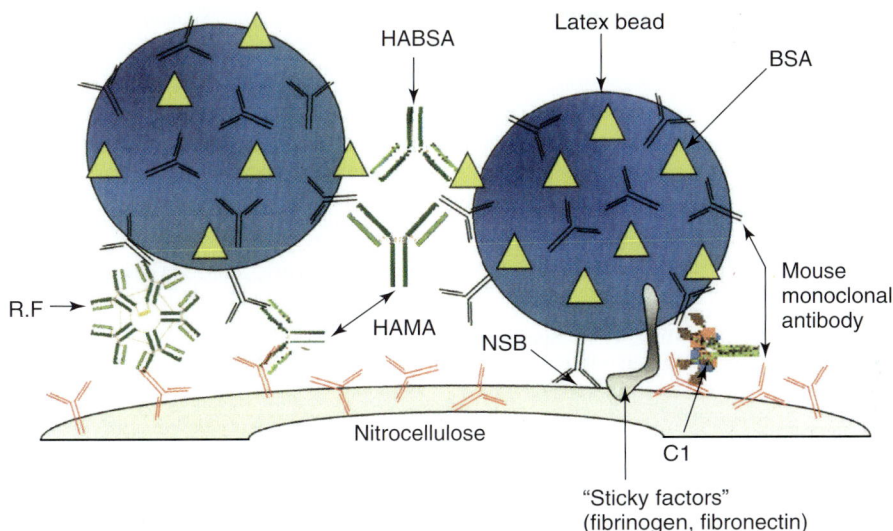

Figure 6.   Illustration of the types of protein—protein interactions that can cause problems in particle based LFiAs—note not all the possible interactions are shown. Particles may be aggregated by human anti-BSA (HABSA), human anti-mouse antibody (HAMA), Rheumatoid factor (RF) or the complement protein C1q. Small aggregates of particles may be linked to antibodies on the NC by similar interactions. Large proteins like fibrinogen and fibronectin may also link protein free surfaces together owing to their large size and tendency toward adsorption.[53] [Reprinted with permission Prime and Whitesides[53] copyright 1993, American Chemical Society.]

is also a marker for choriocarinoma these unfortunate women were subjected to unnecessary surgery because of the clinicians belief in the infallibility of the test.[51] The erroneous measurements ("false positives") were due to heterophile antibodies, a cause of concern in all blood assays. This type of interference can also cause "false negatives", if for example, the interfering protein binds and masks paratopes in a sandwich assay. Heterophilic antibodies are human antibodies, which can be of several classes (IgG, IgM, IgA), which will bind to the antibodies used in the LFiA.[50] Heterophiles usually arise as a normal immune response to the passage of dietary antigens through a compromised gut wall, as in celiac disease. Also through blood transfusions, vaccinations, antibody based drug targeting, and radio-imaging techniques.[49,52] Figure 6 shows most of the types of interference possible in blood and how they can act in an LFiA.

Other samples have their own problems, for example, mucus in sputum can aggregate latex. In some bacterial tests the antigen (e.g., Streptococcal protein A) has to be extracted from the cell wall, a procedure involving chemical extraction and its neutralization prior to applying to the test device as in the Clearview Strep A

test. (ClearviewTM StrepA details may be viewed at www.clearview.com/strepa.cfm.) In some urine based drug tests the sample can be easily adulterated in the privacy of the collection booth by the donor. Manufacturers strive to overcome these problems in order to meet the demands of the regulatory authorities, to better meet consumer needs and, of course, to secure the commercial advantage of a superior test.

6   CONCLUSION

LFiA offer an assay platform that can be as simple as unwrap, wet and measure, with a quick time to a visual result. With a reading system this can offer semi-quantitative measurements of large and small analytes in a variety of samples. New materials, reading systems and a better understanding of how to normalize samples will lead to quantitative assays and an even wider range of applicability than they currently enjoy. It is often overlooked that LFiAs are a massively parallel microfluidic assay platform, which has yet to be superseded as a mass market product, by the single, or multichannel, microfluidic systems that have been extensively researched over at least the past decade.

REFERENCES

1. T. A. Gossel and M. P. Mahalik, Ovulation and pregnancy testing products. *US Pharmacis*, 1993, **18**(7), 36–45.
2. http://history.nih.gov/exhibits/thinblueline, 2003.
3. P. Ghalioungui, *Magic and Medical Science in Ancient Egypt*, Hodder and Stoughton, London, 1963.
4. S. Jehanli, L. Brannan, L. Moore, and V. R. Spiehler. Blind trials of an onsite saliva drug test for marijuana and opiates. *Journal of Forensic Sciences*, 2001, **46**, 1214–1220.
5. Double CheckGold HIV 1&2. Orgenics, Yavne 70650, Israel. www.organic.com.
6. A. S. Valari, R. K. Hickman, J. R. Hackett, C. A. Brennan, V. A. Varitek, and S. G. Devare, Rapid assay for simultaneous detection and differentiation of immunoglobulin G antibodies to human immunodeficiency virys type 1 (HIV-1) group M, HIV-1 group O, and HIV-2. *Journal of Clinical Microbiology*, 1998, **36**(12), 3657–3661.
7. Y. Al-Yousif, J. Anderson, C. Chard-Bergstrom, and S. Kapil, Development, evaluation and application of lateral flow immunoassay(immunochromatography) for the detection of rotavirus in bovine fecal samples. *Clinical and Diagnostic Laboratory Immunology*, 2002, **9**, 925–926.
8. S. K. Sharma, B. S. Eblen, R. L. Bull, D. H. Burr, and R. C. Whiting, Evaluation of lateral—flow clostridium botulinum neurotoxin detection kits for food analysis. *Applied and Environmental Microbiology*, 2005, **71**(7), 3935–3941.
9. C. Quach, D. Newby, G. Daoust, E. Rubin, and J. McDonald, QuickVue influenza test for rapid detection of influenza A and B viruses in a pediatric population. *Clinical and Diagnostic Laboratory Immunology*, 2002, **9**, 723–724.
10. *Med Plus Drug Screen Test*, Applied Biotech, Incorporated, San Diego.
11. H. L. Smits, C. K. Eapen, S. Sugathan, M. Kuriakose, M. H. Gasem, C. Yersin, D. Sasaki, B. Pujianto, M. Vestering, T. H. Abdoel, and G. C. Gussenhoven, Lateral flow assay for rapid sero-diagnosis of human leptospirosis. *Clinical and Diagnostic Laboratory Immunology*, 2001, **8**, 166–169.
12. D. King, V. Luna, A. Cannons, J. Cattani, and P. Amuso, Performance assessment of three commercial assays for direct detection of bacillus anthracis spores. *Journal of Clinical Microbiology*, 2003, **41**, 3454–3455.
13. Y. Oku, K. Kamiya, H. Kamiya, Y. Shibahara, T. Li, and Y. Uesaka, Development of oligonucleotide lateral-flow immunoassay for multi-parameter detection. *Journal of Immunological Methods*, 2001, **258**, 73–84.
14. E. Lee-Lewandrowski and K. Lewandrowski, Regulatory compliance for point-of-care testing. A perspective from the United States (circa 2000). *Clinics in Laboratory Medicine*, 2001, **21**(2), 241–253.
15. M. Tegoni, S. Spinelli, M. Verhoeyen, P. Davis, and C. Cambillau, Crystal structure of a ternary complex between human chorionic gonadotrophin (hCG) and two Fv fragments specific for the a and b subunits. *Journal of Molecular Biology*, 1999, **289**(5), 1375–1385.
16. C. M. Sturgeon and E. J. McAllister, Analysis of hCG: clinical applications and assay requirements. *Annals of Clinical Biochemistry*, 1998, **35**(4), 460–491.
17. S. Qian and H. H. Bau, A mathematical model of lateral flow bioreactors applied to sandwich assays. *Analytical Biochemistry*, 2003, **322**, 89–98.
18. S. Qian and H. H. Bau, Analysis of lateral flow biodetectors: competitive format. *Analytical Biochemistry*, 2004, **326**, 211–224.
19. C. Barrett, C. Good, and C. Moore, Comparison of point of collection screening of drugs of abuse in oral fluid with a laboratory-based urine screen. *Forensic Science International*, 2001, **122**, 163–166.
20. C. H. Trinh, S. D. Hemmington, M. E. Verhoeyen, and S. E. Phillips, Antibody fragment Fv4155 bound to two closely related steroid hormones: the structural basis of fine specificity. *Structure*, 1997, **5**(7), 937–948.
21. J. Israelachvili. *Intermolecular and Surface Forces*, 2nd Edn, Academic Press, New York, 1991, pp. 246–254.
22. D. H. Ostrow, L. M. Cohen, K. S. Eble, P. J. Maney, E. W. Steckle, B. L. Rubin, C. A. Whittman, and P. A. Johnson. Procedure for Attaching Substances to Particles, International Patent Application. WO 98/08095, 1988.
23. D. J. Shaw. *Introduction to Colloid and Surface Chemistry*, 4th Edn, Butterworth-Heinemann, Oxford, 2003, p. 13.
24. W. R. Glomm, Functionalised gold nanoparticles for applications in bionanotechnology. *Journal of Dispersion Science and Technology*, 2005, **26**, 389–414.
25. R. Elghanian, J. J. Storhoff, R. C. Mucic, R. L. Letsinger, and C. A. Mirkin, Selective colorimetric detection of polynucleotides based on the distance-dependent optical properties of gold nanoparticles. *Science*, 1997, **277**, 1078–1081.
26. P. Englebienne, Use of colloidal gold surface plasmon resonance peak shift to infer affinity constants from the interactions between protein antigens and antibodies specific for single or multiple epitopes. *Analyst*, 1998, **123**, 1599–1603.
27. A. W. J. van Doorn, J. H. Wichers, W. Martinns, and J. van Gelden, Stable Aqueous Carbon Sol Composition for Determining Analyte, US patent 5,641,689, 1997.
28. J. R. Schaeffer, T. J. Chen, and M. A. Schen, Latex Particles Incorporating Stabilised Fluorescent Rare Earth Labels, US patent 4,784,912, 1988.
29. P. Alivisatos, The use of nanocrystals in biological detection. *Nature Biotechnology*, 2004, **22**(1), 47–52.
30. M. Danek, K. F. Jensen, C. B. Murray, and M. G. Bawendi, Synthesis of luminescent thin-film CdSe/ZnSe quantum dot composites using CdSe quantum dots passivated with an overlayer of ZnSe. *Chemistry of Materials*, 1996, **8**, 173–180.
31. W. C. W. Chan and S. Nie, Quantum dot bioconjugates for ultrasensitive nonisotopic detection. *Science*, 1998, **281**, 2016–2018.
32. A. M. Fisher, S. J. Sirk, J. L. Lambert, and M. P. Bruchez. Development of a Qdot Lateral Flow Assay. A Simple

Multiplexing Strip Test, 2003, Quantum Dot Vision, 8–9, October.

33. Y. Li, E. C. Y. Liu, N. Pickett, P. J. Skabara, S. S. Cummins, S. Ryley, A. J. Sutherland, and P. O'Brien, Synthesis and characterisation of CdS quantum dots in polystyrene microbeads. *Journal of Materials Chemistry*, 2005, **15**, 1238–1243.

34. *NASA Technical Reports Server, Nanomag Range of Particles from Micromod Partikeltechnologie GmbH*, Document ID 20060043413, Friedrich-Barnewitz Strasse, Rostock, Germany.

35. G. Li, H. Wu, and C. Liu, Europium Containing Fluorescent Nanoparticles and Methods of Manufacture Thereof, US patent 6,783,699. 2004.

36. D. A. Zarling, M. J. Rossi, N. A. Peppers, J. Kane, G. W. Faris, M. J. Dyer, S. Y. Ng, and L. V. Schneider. Up-Converting Reporters for Biological and Other Assays Using Laser Excitation Techniques, US patent 5,674,698 1997.

37. J. Kane, Method for Preparing Small Particle Size Fluoride Up-Converting Phosphors, US patent 5,891,361, 1999.

38. P. Corstjens, M. Zuiderwijk, A. Brink, S. Li, H. Feindt, R. S. Niedbala, and H. Tanke, Use of up-converting phosphor reporters in lateral—flow assays to detect specific nucleic acid sequences: a rapid, sensitive DNA test to identify human papillomavirus type 16 infection. *Clinical Chemistry*, 2001, **47**(10), 1885–1893.

39. D. Martorell, S. T. Siebert, and R. A. Durst, Liposome dehydration on nitrocellulose and its application in a biotin assay. *Analytical Biochemistry*, 1999, **271**(2), 177–185.

40. P. Vogel, H.-P. Braun, D. Berger, and W. Werner. Process and Composition for Separating Plasma or Serum from Whole Blood, US patent 5,055,195, 1984.

41. D. B. Pall and R. L. Manteuffel, Fibrous Web for Processing a Fluid, US patent 5,846,438, 1998.

42. D. M. Koenhen and J. J. Scharstuhl, Process and Device for the Separation of a Body Fluid from Particulate Materials, US patent 5,240,862, 1993.

43. W. Norde, *Driving Forces for Protein Adsorption at Solid Surfaces*, in *Biopolymers at Interfaces, Surfactant Science Series*, M. Malmsten, (ed), Marcel Dekker, New York, 1998, **75**: pp. 27–54.

44. J. Visser, On Hamaker constants: a comparison between Hammaker constants and Lifshitz-van der Waals constants. *Advances in Colloid and Interface Science*, 1972, **3**, 331–363.

45. H. Xu, J. R. Lu, and D. E. Williams, Effect of surface packing density of interfacially adsorbed monoclonal antibody on the binding of hormonal hCG. *Journal of Physical Chemistry B*, 2006, **110**, 1907–1914.

46. Rapid Lateral Flow Test Strips. Considerations for Product Development. Millipore Corp. Lit. No. TB500EN00, Revn. 06/02, 2002, 5.

47. R. G. Chapman, E. Ostuni, S. Takayama, R. E. Holmlin, L. Yang, and G. M. Whitesides. Surveying for surfaces that resist the adsorption of proteins. *Journal of the American Chemical Society*, 2000, **122**, 8303–8304.

48. E. P. K. Currie, J. Van der Gucht, O. V. Borisov, and M. A. Cohen-Stuart, Stuffed brushes: theory and experiment. *Pure and Applied Chemistry. Chimie Pure Et Appliquee*, 1999, **71**, 1227.

49. L. J. Kricka, Human anti-animal antibody interferences in immunological assays. *Clinical Chemistry*, 1999, **45**, 942–956.

50. C. Selby, Interference in immunoassay. *Annals of Clinical Biochemistry*, 1999, **36**, 704–721.

51. S. Rotmensch and L. A. Cole, False diagnosis and needless therapy of presumed malignant disease in women with false-positive human chorionic gonadotrophin concentrations. *Lancet*, 2000, **355**, 712–715.

52. T. H. Webber, K. I. Kapyaho, and P. Tanner. Endogenous interference in immunoassays in clinical chemistry. *Scandinavian Journal of Clinical and Laboratory Investigation*, 1990, **50**(Suppl. 201), 77–82.

53. K. L. Prime and G. M. Whitesides, Adsorption of proteins onto surfaces containing end-attached oligo (ethylene oxide): a model system using self assembled monolayers. *Journal of the American Chemical Society*, 1993, **115**, 10714–10721.

Chip-Based Biosensors for Environmental Monitoring

Kim R. Rogers

National Research Exposure Laboratory, Environmental Protection Agency, Las Vegas, NV, USA

1 INTRODUCTION

Monitoring the environment (air, water, and soil) for contaminants is a critical component for understanding and managing risks related to human health. Given the large and increasing number of environmental analyses as well as the future potential for monitoring biomarkers of human exposure to environmental pollutants, there exists a need for rapid, portable, and cost-effective analytical methods to support these measurements. Biosensors and bioanalytical methods appear to be well suited to complement standard analytical techniques for a number of environmental monitoring applications. In this regard, several recent reviews have addressed biosensor technology from perspectives that include agricultural monitoring,[1] groundwater screening,[2] ocean monitoring,[3] global environmental monitoring,[4] and recent advances.[5] The intention for this chapter is to discuss the topic of chip-based biosensors for potential environmental monitoring applications.

One of the areas of technological advancement that sensor and biosensor scientists have used to their advantage is miniaturization and adaptation of biochemical assays onto small planar surfaces (i.e., microchips) to form "biochip" arrays. A wide range of chip-based biosensors or biochips has been described in the literature.

Although there does not seem to be a specific definition for a biosensor chip, the consensus opinion would suggest a biological recognition element (i.e., enzyme, antibody, receptor, or microorganism) immobilized on a planar surface (e.g., glass, plastic, gold, silicon, etc.) that together respond in a concentration-dependent manner to a chemical species. One of the earliest immunoassays that could be considered to be chip-based was the microspot immunoassay introduced by Ekins, which used plastic substrates and confocal fluorescence microscopy.[6]

Although the most common substrate for immobilization of biorecognition elements is a glass microscope slide, a range of other formats has been reported. Other substrates include gold coated onto quartz or glass using a chromium adhesion layer that has been used for ellipsometry,[7] quartz crystal microbalance or surface plasmon resonance (SPR) techniques, carbon ink or nanoparticles adhered onto ceramic substrates that have been reported for electrochemical biosensors,[8] imaging fibers that have been etched to produce microwell arrays,[9] and microelectrode pipettes that have been used to electropolymerize polypyrrole microspots onto silicon chips.[10] One question, however, remains open to discussion concerning chip-based biosensors for environmental monitoring—Is smaller better?

The basic premise for adapting environmental biosensors into chip designs involves a range of potential technical, operational, and commercial advantages. Many of the potential advantages overlap in several of the following areas. The reduced use of construction materials (e.g., gold, platinum, etc.) can be advantageous in terms of fabrication cost and disposal of toxic materials (e.g., mercury, lead, etc.) both in fabrication and in use. The reduced volume required for each environmental sample can also be an advantage, especially for highly contaminated samples that must meet environmental disposal regulations or biomarkers in blood that must meet biohazard disposal requirements. The most compelling advantages for chip-based biosensor formats over larger biosensors, microwell plate assays, or standard test-tube methods are the potential for high sample throughput, rapid assay times, and the ability to determine multiple analytes in a single sample. Primarily because of considerations of scale, assay kinetics, and reagent delivery, the sample detection limits can also be significantly improved as compared to test-tube or, in some cases, microwell-type bioassays. Reduced instrument size and power requirement can result in mobile- and field-measurement capabilities. Manufacturing advantages for chip biosensors are numerous and include greater production throughput, lower cost per sales unit, greater versatility, and expanded application and market opportunities. Multiplexing a number of assays for different pollutants that can be detected using the same analysis platform could also provide a distinct advantage for chip-based biosensors. There are, however, some limitations with respect to potential environmental applications.

Although chip biosensors are typically formatted on a small planar surface, there are differences among the biochips that should be considered in any discussion of potential environmental applications for these devices. Environmental monitoring has significantly different requirements than that for clinical, diagnostic, or food industry applications. One way to discuss environmental applications for these devices would be to consider the requirements for several possible environmental monitoring scenarios. For the purpose of comparison to a well-established immunoassay format, one potential benchmark might be the 96-well ELISA. In the first example, the proposed task requires the assay of 96 samples for a single pesticide for which well-characterized antibodies are available.

This type of assay would require the chip to have a separate microfluidics flow channel to handle each sample. For example, a biochip with 4 parallel flow channels would require 24 separate runs (a single flow channel would require 96 sequential runs) not including calibration, performance references, and duplicate analyses. If the biochip cannot be regenerated, then a separate chip would be required for each run. An ELISA for this monitoring scenario could be performed on a single plate (excluding calibration, etc.).

In another scenario, consider analysis of multiple compounds (say 24) in each of the 96 samples. In this case, an array of 24 capture antibodies placed in separate locations on the chip could report 24 compounds in a single run with a single flow channel. For this format, the chip-based assay would show some advantages over a plate assay format that detects a single analyte in each well. Because the chip could be designed and fabricated to detect multiple compounds, it could also be subjected to the same sequence of assay solutions for all of the analytes of interest. Although reported biochip platforms are fairly versatile, they are still somewhat limited in their configurations, in that the chips must be specifically constructed for a particular group of compounds. By contrast to clinical applications, potential environmental application scenarios can be extremely variable ranging from analysis of hazardous waste, persistent organics, or biomarkers in matrices ranging from biological fluids to sewage sludge.

For certain environmental monitoring scenarios, sample preparation, and cleanup could limit the time, cost, and simplicity advantages gained by chip-based assay formats. For example, if each experimental sample requires a complex and time-consuming cleanup and preparation protocol, a simple and rapid assay format loses much of its advantage. Consequently, the answer to the question "is smaller better" depends to a large extent on the proposed environmental application. Unfortunately, a significant number of reports of novel chip-based biosensors for potential environmental applications have shown limited discussion concerning this topic. I am not suggesting that these reports have any less value; however, without solving these application challenges, it is unlikely that these techniques will be adopted for routine environmental monitoring.

2 BIOLOGICAL RECOGNITION ELEMENTS

The advantages and limitations for enzyme-, antibody-, receptor-, or microorganism-based chip biosensors for potential environmental applications are, to some extent, similar to those for biosensors in general. Advantages for the use of enzymes include a stable source of material primarily through biorenewable sources (e.g., bacteria, plants, or animal by-products), the ability to modify the catalytic properties or substrate specificity through genetic engineering, and the catalytic amplification of the biosensor response by modulation of the target analyte. Limitations of the enzyme biosensors with respect to environmental applications involve the relatively few environmental pollutants that interact specifically with enzymes either as substrates or inhibitors. Nevertheless, recent progress has been reported in areas such as genetic modification of enzymes to increase sensitivity, specificity, stability, and shelf life. Because assays for different enzymes used for environmental applications require different substrates, monitoring formats, and so on, it is unlikely that an array of enzymes on a single chip would be feasible. An inexpensive, disposable single-assay format that could be read with a small portable instrument, however, may be useful in a number of environmental monitoring scenarios.

Although antibodies are inherently more versatile than enzymes as biological recognition elements in biosensors for environmental applications, they show several inherent limitations. These limitations include the complexity of assay formats and the number of specialized reagents (e.g., antibodies, antigens, tracers, etc.) that must be developed and characterized for each analyte. Because antibodies can be selective for the detection of specific compounds, they are amenable to incorporation into sensor chip arrays to form multianalyte detection systems. In this case, the antibodies are different at each location in the array and a single sample, or in some cases multiple samples, would be exposed to each of these "sensor" sites. Small arrays have been demonstrated for the simultaneous detection of multiple analytes. For example, the detection of six hazardous bacteria and protein toxins has been demonstrated using a planar waveguide array biosensor.[11] This biosensor array was able to simultaneously measure six analytes for each run. In another microchip format, the six toxins ricin, viscumin, staphylococcal enterotoxin B, tetanus toxin, diphtheria toxin, and anthrax toxin were detected at nanogram per milliliter levels on a glass chip using a fluorescence microscope equipped to measure four wavelengths with a CD camera.[12] Although there are distinct advantages for chip-based biosensors that can analyze multiple analytes in single or multiple samples, this format requires each sample or small group of samples to use a separate chip. Because this process is sequential for each assay sample, such a scenario would limit many of the cost or throughput advantages that might potentially be gained over standard GC/MS or LC/MS techniques. An exception to this limitation would be in cases where a multichannel biosensor chip could be regenerated such that samples could be repeatedly analyzed using the same chip as in the case of the Automated Water Analyzer Computer-Supported System.[13] This chip-based biosensor system was demonstrated to simultaneously detect several compounds in a repeatable sequence.

Another possible array format might involve numerous environmental samples using the same antibody, with each sample analyzed at separate locations on the chip. Again, one of the challenges for this scenario would be the requirement for separate flow channels to each of the sensor locations. Although this biochip configuration is feasible and could be adapted to several of the previously cited systems,[11–13] it has not yet been reported for the simultaneous detection of large (i.e., >100) numbers of samples.

Although formatting issues for potential environmental applications continue to present challenges, advances in biosensor detection technology also continue to make chip-based biosensors an active research area and a promising solution for certain environmental applications. One area that forms the basis of advances in detection capabilities for biosensors involves the use of nanomaterials. For example, the use of metal nanoparticles in electrochemical detection has been shown to increase the surface area and enhance or modulate the conductivity properties of polymeric electrodes.[14] The use of gold nanoparticles has also been shown to increase the signal resolution for SPR-based biosensors.[15] This improvement could potentially allow the direct detection of small molecules—a problem that currently requires the use of more complicated assays that use indirect competitive formats.

In addition to electrochemical detection improvements, advances in optical detection capabilities have also been reported using multicolor luminescent semiconductor nanocrystals (quantum dots). Because of their well defined emission spectra, these labels have been used to simultaneously quantitate the binding of four separate immunolabels in a single assay.[16]

A wide range of cell-based biosensors has also been reported for potential environmental applications. The cell types used in these bioanalytical assays include bacteria, yeast, algae, and tissue culture cells. The most commonly reported microorganisms are genetically modified bacteria that respond to nonspecific stressors such as DNA damage, γ-radiation, heat shock, and oxidative stress; toxic metals such as lead, mercury, nickel, and zinc; organic environmental pollutants such as chlorinated aromatics, benzene derivatives, organic peroxides, trichloroethylene, and PCBs; and compounds of biological importance such as nitrate, ammonia, and antibiotics.[17]

Genetically modified microbial and cell-based biosensors show several advantages and limitations with respect to environmental biosensors, in general, and, more specifically, for chip-based biosensors. Bioreporter microorganisms show the potential to be interfaced to a wide range of transducers including those that operate using optical, electrochemical, and SPR techniques. Their limitations primarily involve the maintenance of a required cellular environment (e.g., nutrients, O_2, pH, ionic strength, etc.), and the time required for a response (e.g., hours for systems that require protein expression). One of the potential advantages for this type of biological recognition element involves the wide range of promoter genes that have been linked to reporter gene systems.

Unique chiplike platforms have also been reported for a number of cell-based biosensors for potential environmental applications. For example, a bioreporter strain of *Escherichia coli* that produces green fluorescent protein (GFP) in response to arsinite and antimonite was used to demonstrate a micofluidics platform.[18] Comparison of this system to a standard fluorescence cuvette platform showed that mixing and analysis using this centrifugal microflow system decreased the assay time and increased the assay reproducibility. Another

unique assay platform that has been demonstrated for an *E. coli* strain that is responsive to Hg (II), involved immobilization of individual bacteria into microwells on the face of an imaging fiber.[9] Recent advances in bacterial luminescence assays also include an automated and continuous toxicity monitoring system using a genetically engineered freshwater bacterium that is sensitive to heavy metals as well as a number of toxic organic compounds.[19] For this system, cell cultures were lyophilized in a series of 384-well plates, which were automatically reconstituted with wastewater, and the bacterial response was recorded allowing the determination of toxicity spikes in the sample stream. This type of system could potentially be adapted to a microchip system by immobilizing the microorganisms onto a planar array and sequentially directing water stream samples to microflow cavities above each microcell. The potential benefit of adapting these types of systems to chip-based assays would be the development of less expensive fluidics transfer systems. Another potential advantage for the incorporation of bioreporter microorganisms into biochips would be to form multiorganism–multianalyte arrays.

Bioreporter microorganisms that each respond to a different environmental pollutant and use the same biochemical reporter (e.g., luminescence, production of GFP, etc.) could be brought together in a single biochip array. In addition to reporting a range of compounds in a single assay, matrix interferences and cross-reactivity issues could be normalized through the integrated response of multiple organisms to the same environmental sample.

3   IMMOBILIZATION

A critical aspect of biosensor chip development involves immobilization of the biological recognition element onto the chip surface. These structures must be physically rugged as well as resistant to a range of chemical and thermal environments. Spatial resolution and sensor indexing are also important features in sensor design. Biological materials are typically imaged, micropipetted, or self-assembled onto meticulously cleaned planar surfaces. These materials can be chemically tethered or incorporated into solgel, hydrogel,[12] or polymer matrices.[20]

Another approach for spatial positioning of immunochemicals involves the use of an electrical field that attracts biotin-labeled capture antibodies to specific streptavidin-coated microelectrodes.[21] This electronic addressing technique was applied to the detection of fluorescently labeled staphylococcal enterotoxin B using a microelectrode array.[21] Each labeled toxin was concentrated from a mixture onto its specific antibody-coated electrode. In addition to the spatial positioning afforded by this technique, the time required for the antibody–antigen binding was shortened because of the imposed potential at the electrode surface.

Another approach to the spatial positioning of antibodies onto a sensor array involves the use of complementary oligonucleotides or peptide nucleic acids (PNAs) as a tether to attach analyte derivatives or capture antibodies.[22] There are two advantages to this immobilization scheme. First, specific sequences of DNA can direct immunochemicals to specific locations with attached complementary oligonucleotide sequences. Next, after the assay is complete, chemical disruption can be used to strip away the immunochemical, and another oligonucleotide-labeled conjugate can be bound to a specific location on the sensor surface.

Antibodies and enzymes can be covalently or noncovalently bound to chemically modified planar surfaces using a wide range of chemistries that have been well characterized over several decades. Although immobilization of enzymes typically reduces their initial activity, it also stabilizes their function over extended periods. Antibodies can also be immobilized with most of their binding function intact. Although most microorganism-based biosensors use microscale reaction chambers or flow cells, these bioreporters have also been tethered to planar surfaces using chemical linkers such as cysteine-terminated synthetic oligonucleotides.[23]

4 SIGNAL TRANSDUCERS AND COMMERCIAL PERSPECTIVES

The detection of biorecognition responses for chip-based biosensors is primarily based on optical, plasmon resonance, or electrochemical transducers. There are several inherent advantages and limitations for each technique. One advantage for optical transducers is that an entire array can be read using absorbance, fluorescence, or luminescence with a range of well-developed and relatively inexpensive optical imaging instruments. In addition, signal-processing software is widely available for a range of applications. Although multi-laser examination of individual locations in the array can reduce background signal noise and increase assay sensitivity, it also adds a degree of complexity and expense.

SPR has been well described as a detection strategy in a chip-based format. This technique is quite versatile in that it does not require the use of a probe to detect surface binding; however, small, environmentally important compounds typically require the use of indirect competition assays rather than measurement of direct surface binding. In addition, the chips mounted in flow cells are typically limited to sequentially reading one analyte at a time.

When discussing potential chip-based biosensors, there are a number of issues that should be addressed in the larger context of related bioanalytical assays. It is likely that biochips with the greatest potential for future development will become central components in commercial bioanalysis systems. Consequently, it is important to mention other bioanalytical systems and the general basis for their operation. Some of these systems include the following: The Luminex™ system is based on internally color-coded microspheres with surface linking chemistry to accommodate antibodies, receptors, or oligonucleotides. Beads containing from 1 of 100 specific dye sets can be differentiated using a flow-cytometry system. Binding of the surface label is indicative of analyte binding and can also be determined using a second, shorter-wavelength dye and a dual laser detector.[24] The use of this system has also been described for multiplexed sandwich immunoassays for an array of cytokines.[25]

The SearchLight system uses multiplexed sandwich ELISAs with up to 16 different capture antibodies prespotted onto 96-well plates or 4 capture antibodies spotted onto 384-well plates. Assays are run using standard ELISA methods with a secondary antibody tag of peroxidase activating a chemiluminescent substrate that is detected using a cooled CCD instrument. The SearchLight system has been recently compared to the Luminex and the FAST Quant for the detection of cytokines.[26]

The authors describe the advantages and limitations for each system.

The Meso Scale Diagnostics microplate system integrates microelectrodes into the bottom of each well of a 96-microwell plate. Various assays can be configured using electrogenerated chemiluminescent labels.[27] These labels, consisting primarily of $Ru(bpy)_3^{2+}$ analogs, luminesce only when they are in close proximity to the electrode surface so that bound and unbound tags can be differentiated without a plate washing step. The Multi-Spot plates use a patterned micoarray within each of the wells.[28]

Electrochemical transduction can be performed on a chip using electrodes that are patterned to form a disposable sensor or grouped together and incorporated into a microwell array similar to an ELISA microwell plate. An advantage of this type of system is that each of the electrodes in the array can be used for a separate assay and even subjected to different electrochemical conditions. One disadvantage of this configuration is that the microwell electrode plate limits many of the advantages of biochips (e.g., the assay becomes like an ELISA with electrochemical detection).[29]

5   SUMMARY

With the exception of the electrochemically based blood glucose monitor, few stand-alone biosensors have become commercially viable. Nevertheless, a wide range of biosensor concepts and technologies form the basis of or provide key components in clinical, diagnostic, and research instrumentation. In addition, with respect to environmental monitoring applications, several biosensor-based systems are becoming more widely used for a limited number of pollutant-screening applications.

Advances in microfabrication, biomolecule immobilization, and detection techniques have led to the conception and realization of chip-based biosensors. As a consequence of these advances, a wide range of chip-based biosensors, primarily using antibodies and some bacterial cells, has been reported. In addition, the use of nanomaterials to improve the operational characteristics of biosensors is becoming increasingly common.

Potential environmental applications present a range of challenges for chip-based biosensors. These challenges include the broad and shifting group of compounds of environmental interest as well as the wide range of contaminated matrices that might be involved. Chip-based biosensors, owing to their unique size and format characteristics, show the potential to provide rapid and cost-effective solutions to environmental monitoring problems.

NOTE

The US Environmental Protection Agency through its Office of Research and Development has funded and managed the research described here. It has been subjected to the Agency's administrative review and approved for publication as an EPA document. Mention of trade names or commercial products does not constitute endorsement or recommendation for use.

REFERENCES

1. S. Rodriguez-Mozaz, M. J. Lopez de Alda, M.-P. Marco, and D. Barcelo, Biosensors for environmental monitoring: a global perspective. *Talanta*, 2005, **65**, 291–297.
2. U. Bilitewski, A. P. F. Turner, *Biosensors for Environmental Monitoring*, Harwood Academic Publishers, New York, 2000.
3. M. N. Velasco-Garcia and T. Mottram, Biosensor technology addressing agricultural problems. *Biosystems Engineering*, 2003, **84**, 1–12.
4. S. Rodriguez-Mozaz, M. Farre, and D. Barcelo, Screening water for pollutants. *Trends in Analytical Chemistry*, 2005, **24**, 165–169.
5. K. R. Rogers, Recent advances in biosensor techniques for environmental monitoring. *Analytica Chimica Acta*, 2006, **568**, 222–231.
6. R. Ekins, F. Chu, and E. Biggart, Fluorescence spectroscopy and its application to a new generation of highly sensitive multi-microspot, multianalyte, immunoassay. *Clinica Chimica Acta*, 1990, **194**, 91–114.
7. Y. M. Bae, B.-K. Oh, W. Lee, W. H. Lee, and J.-W. Choi, Immunosensor for detection of *Yersinia enterocoitica* based on imaging ellipsometry. *Analytical Chemistry*, 2004, **76**, 1799–1803.
8. Y. Paitan, I. Biran, N. Shechter, D. Biran, J. Rishpon, and E. Ron, Monitoring aromatic hydrocarbons by whole cell electrochemical biosensors. *Analytical Biochemistry*, 2004, **335**, 175–183.
9. I. Biran, D. M. Rissin, E. Z. Ron, and D. R. Walt, Optical imaging fiber-based live bacterial cell array biosensor. *Analytical Biochemistry*, 2003, **315**, 106–113.
10. S. Szunerits, L. Bouffier, R. Calemczuk, B. Corso, M. Demeunynck, E. Descamps, Y. Defontaine, J.-B.

Fiche, E. Fortin, T. Livache, P. Mailley, A. Roget, and E. Vieil, Comparison of different strategies on DNA chip fabrication and DNA-sensing: optical and electrochemical approaches. *Electroanalysis*, 2005, **17**, 2001–2017.

11. C. A. Rowe-Taitt, J. W. Hazard, K. E. Hoffman, J. J. Cras, J. P. Golden, and F. S. Ligler, Simultaneous detection of six biohazardous agents using a planar waveguide array biosensor. *Biosensors and Bioelectronics*, 2000, **10**, 579–589.

12. A. Y. Rubina, V. I. Dyukova, E. I. Dementieva, A. A. Stomakhin, V. A. Nesmeyanov, E. V. Grishin, and A. S. Zasedatelev, Quantitative immunoassay of biotoxins on hydrogel-based protein microchips. *Analytical Biochemistry*, 2005, **340**, 317–329.

13. J. Tschmelak, G. Proll, J. Riedt, J. Kaiser, P. Kraemmer, L. Barzaga, J. S. Wilkinson, P. Hua, J. P. Hole, R. Nudd, M. Jackson, R. Abuknesha, D. Barcelo, S. Rodriguez-Mozaz, M. J. Lopez de Alda, F. Sacher, J. Stien, J. Slobodnik, P. Oswald, H. Kozmenko, E. Korenkova, L. Tothova, Z. Krascsenits, and G. Gauglitz, Automated water analyser computer supported system (AWACSS) part I: project objectives, basic technology, immunoassay development, software design and networking. *Biosensors and Bioelectronics*, 2005, **20**, 1509.

14. E. Katz, I. Willner, and J. Wang, Electroanalytical and bioelectroanalytical systems based on metal and semiconductor nanoparticles. *Electroanalysis*, 2004, **16**, 19–44.

15. W. P. Hu, S.-J. Chen, K.-T. Huang, J. H. Hsu, W. Y. Chen, G. L. Chang, and K.-A. Lai, A novel ultrahigh-resolution surface plasmon resonance biosensor with an Au nanocluster-embedded dielectric film. *Biosensors and Bioelectronics*, 2004, **19**, 1465–1471.

16. E. R. Goldman, A. R. Clapp, G. P. Anderson, H. T. Uyeda, J. M. Mauro, I. L. Medintz, and H. Mattoussi, Multiplexed toxin analysis using four colors of quantum dot fluororeagents. *Analytical Chemistry*, 2004, **76**, 684–688.

17. D. E. Nivens, T. E. McKnight, S. A. Moser, S. J. Osbourn, M. L. Simpson, and G. S. Sayler, Bioluminescent bioreporter integrated circuits: potentially small, rugged and inexpensive whole-cell biosensors for remote environmental monitoring. *Journal of Applied Microbiology*, 2004, **96**, 33–46.

18. A. Rothert, S. K. Deo, L. Millner, L. G. Puckett, M. J. Madou, and S. Daunert, Whole-cell-reporter-gene-based biosensing systems on a compact disk microfluidics platform. *Analytical Biochemistry*, 2005, **342**, 11–19.

19. J.-C. Cho, K.-J. Park, H.-S. Ihm, J.-E. Park, S.-Y. Kim, I. Kang, K.-H. Lee, D. Jahng, D.-H. Lee, and S.-J. Kim, A novel continuous toxicity test system using a luminously modified fresh water bacterium. *Biosensors and Bioelectronics*, 2004, **20**, 338–344.

20. I. Roy and M. N. Gupta, Smart polymeric materials: emerging biomedical applications. *Chemistry and Biology*, 2003, **10**, 1161–1171.

21. K. L. Ewalt, R. W. Haigis, R. Rooney, D. Ackley, and M. Krihak, Detection of biological toxins on an active electronic microchip. *Analytical Biochemistry*, 2001, **289**, 162–172.

22. M. G. Weller, A. J. Schuetz, M. Winklmair, and R. Niessner, Highly parallel affinity sensor for the detection of environmental contaminants in water. *Analytica Chimica Acta*, 1999, **393**, 29–41.

23. J.-W. Choi, K.-W. Park, D.-B. Lee, W. Lee, and W. H. Lee, Cell immobilization using self assembled synthetic oligopeptides and its application to biological toxicity detection using surface plasmon resonance. *Biosensors and Bioelectronics*, 2005, **20**, 2300–2305.

24. J. Dasso, J. Lee, H. Bach, and R. G. Mage, A comparison of ELISA and flow microsphere-based assays for quantification of immunoglobulins. *Journal of Immunological Methods*, 2002, **263**, 23–33.

25. D. A. A. Vignali, Multiplexed particle-based from cytometric assays. *Journal of Immunology Methods*, 2000, **243**, 243–255.

26. G. E. Lash, P. J. Scaife, B. A. Innes, H. A. Qtun, S. C. Robson, R. F. Searle, and J. N. Bulmer, Comparison of three multiplex cytokine analysis systems: Luminex, SearchLight™, and FAST Quant®. *Journal of Immunology Methods*, 2006, **309**, 205–208.

27. J. D. Debad, E. N. Glezer, J. N. Wohlstadter, and G. B. Sigal, *Clinical and Biological Applications of ECL*, in *Electrogenerated Chemiluminescence*, A. J. Bard (ed), Marcel Dekker, New York, 2003, pp. 43–78.

28. A. J. Bard, J. D. Debad, J. k. Leland, G. B. Sigal, J. L. Wilbur, and J. N. Wohlstadter, *Chemiluminescence, Electrogenerated*. In *Encyclopedia of Analytical Chemistry*, R. A. Meyers (ed), John Wiley & Sons, Chichester, 2000, pp. 9842–9849.

29. R. M. Pemberton, R. Pittson, N. Biddle, G. A. Drago, and J. P. Hart, Studies toward the development of a screen printed carbon electrochemical immunosensor array for mycotoxins: A sensor for aflatoxin B$_1$. *Analytical Letters*, 2006, **39**, 1573–1586.

Environmental Biochemical Oxygen Demand and Related Measurement

Yoko Nomura,[1] Mifumi Shimomura-Shimizu[2] and Isao Karube[2]

[1] *Department of Biomedical Engineering, University of California, Davis, CA, USA and* [2] *School of Bionics, Tokyo University of Technology, Tokyo, Japan*

1 HISTORY OF BOD MEASUREMENT USING BIOSENSORS

Biochemical oxygen demand (BOD) is one of the important pollutant indices of aquatic environments such as river or wastewater because it can measure organic pollutants. Excess contamination of organic substances in aquatic environment causes serious ecosystem damage.

BOD is conventionally measured by the BOD 5-day method, which is very commonly used around the world.[1,2] Although it is frequently used, the procedures include sample dilutions and 5-day incubation and require special skills and laboratory facilities. The BOD 5-day method also includes a titration procedure that requires a number of chemicals. The data obtained by this conventional method might be accurate, but it is not suitable for urgent and daily measurements such as monitoring of wastewater from factories.

To overcome these disadvantages, a biosensor for BOD estimations was first developed in 1977.[3,4] The biosensor was also the first whole-cell biosensor. Biosensors using whole cells of microorganisms are called *microbial sensors*, which exploit the metabolic functions of living cells. Since whole-cell biosensors or microbial sensors have advantages such as long life, low cost, and low environmental impact, they have frequently been applied to environmental monitoring.[5]

2 PRINCIPLE OF BOD SENSOR

Figure 1 shows the BOD sensor fabricated by Karube et al.[6] The operating principle of this biosensor is to measure the change in respiration activity of immobilized microorganisms. The biosensor consists of an oxygen electrode (Clark-type oxygen electrode) and an acetylcellulose membrane–immobilized omnivorous yeast *Trichosporon cutaneum*. The BOD 5-day method uses a consortium of microorganisms, but one of the microorganism species was used in this sensor to obtain high reproducibility of BOD estimation values. Immobilized *T. cutaneum* can oxidatively degrade most organic compounds (pollutants) in wastewater samples with high respiration activity. *T. cutaneum* consumes dissolved oxygen in samples when it oxidatively degrades organic compounds, and the oxygen electrode measures the reduction of the dissolved oxygen as current decrease (sensor response). The BOD sensor using *T. cutaneum* was able to estimate BOD in 20 min without special skills.[6] The correlation of the sensor response and the BOD values was linear between 0 and $60 \, \text{mg} \, \text{l}^{-1}$ when

Handbook of Biosensors and Biochips. Edited by Robert S. Marks, David C. Cullen, Isao Karube, Christopher R. Lowe and Howard H. Weetall.

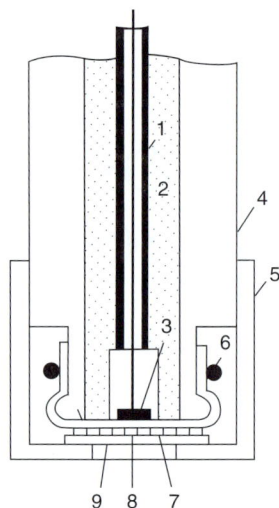

Figure 1. The operating principle of the first BOD sensor.[6] (1) Aluminum anode, (2) electrolyte, (3) platinum cathode, (4) insulator, (5) bored cap, (6) O-ring, (7) Teflon membrane, (8) immobilized microorganisms, and (9) acetylcellulose membrane. [Reprinted with permission Hikuma et al.[6] copyright 1979, Springer.]

glucose-glutamine (GGA) standard solution was measured. The BOD sensor was able to measure several kinds of untreated wastewater from fermentation plants.[6] The sensor was commercialized in 1983, and offered an environmentally friendly alternative to BOD measurement, only requiring phosphate buffer and GGA standard solution for the daily measurements.

3 EXAMPLES OF BOD SENSORS

To date, various improvements have been made on BOD biosensors, for example shorter response time (30 s), longer lifetime (16 months) and higher sensitivity (BOD $0.2 \, mg \, O \, l^{-1}$).[7,8] To compensate for the limitation of biosensor measurements caused by short response time or immobilization of only one or a few kinds of microorganisms, enzymes such as cellulase were also used for pretreatment when the BOD sensor was applied to some specific wastewater from pulp factories, and so on[8]. Several recent studies describe BOD sensors aimed for practical use and so flow injection analysis (FIA) BOD sensors were developed.[8–10] Other transducers such as optical fibers have been used for BOD sensors.[11]

Examples of BOD sensors were also summarized in current reviews,[8] and Liu et al. compared performance characteristics of BOD sensors in their review.[12] Several BOD sensors are described in the subsequent text.

3.1 Disposable BOD Sensor Using Micromachining Techniques

Conventional BOD sensors were bench top instruments because the transducers were not disposable. Yang et al. fabricated a BOD sensor using a disposable oxygen electrode as the transducer.[13] This oxygen electrode was a miniature electrode ($15 \times 2 \times 0.4$ mm, MOE05, Fujitsu, Japan) fabricated on silicon wafers using micromachining techniques. This BOD sensor measured $0.5 \, mg \, l^{-1}$ of BOD using GGA solution, which is a standard solution for the BOD 5-day method.

3.2 BOD Sensor for River Water Monitoring

BOD monitoring of river water (below BOD $3 \, mg \, l^{-1}$) was required in Japan but the conventional BOD sensor was not suitable for river water monitoring because of its insensitivity. The sensitivity of a BOD sensor depends on the biodegradation properties of immobilized microorganisms when enough dissolved oxygen is contained in a sample. Immobilized *T. cutaneum* on the oxygen electrodes cannot degrade the less-biodegradable organic substances, which are the main BOD pollutants in river water discharged from municipal secondary effluents after wastewater treatment.

Chee et al. fabricated a BOD sensor that combined an oxygen electrode and isolated bacteria strain of *Pseudomonas putida* from activated sludge used for municipal wastewater treatments.[14] This sensor estimated low BOD values of river water with the use of different dilutions of artificial wastewater for calibration. The correlation coefficient of the values by the sensor and the BOD 5-day method (r^2) was 96% when it measured river water, and the correlation improved by incorporating preozonation or photocatalytic oxidation (TiO_2) as a pretreatment.[15,16] This research was supported by the Japanese Ministry of Construction (Ministry of Land Infrastructure and Transport).

3.3 Mediator-type BOD Sensor

The sensitivity of a BOD sensor using an oxygen electrode also depends on the level of dissolved oxygen in the sample, but BOD sensors of mediator type are different.

The mediator type of BOD sensors is composed of microorganisms and redox-active substance (e.g., potassium ferricyanide or 2,6-dichlorophenolindophenol, DCIP).

Yoshida et al. fabricated a mediator-type biosensor as a new approach to BOD estimation. This amperometric biosensor integrated *Pseudomonas fluorescens biovar* V as an immobilized bacteria and potassium hexacyanoferrate(III) (HCF(III)) of potassium ferricyanide as a mediator. The operating principle is shown in Figure 2.[17]

The sensor does not directly measure oxygen uptake, but it measures the change in the current response due to the reduction of HCF(III) to HCF(II) during the metabolic oxidation of organic substances by immobilized *P. fluorescens biovar* V. HCF(III) was the mediator of the oxidized form, and HCF(II) was the mediator of the reduced form. When HCF(III) was present in the reaction medium, it acted as an electron acceptor and was preferentially reduced to HCF(II) during the metabolic oxidation of organic substances. The reduced HCF(III) was then reoxidized at a working electrode (+600 mV vs Ag/AgCl) which was held at a sufficiently high electric potential. As a result, a current was generated and detected using the electrode system. *P. fluorescens biovar* V was immobilized on a disposable sensor tip (working electrode) by photopolymerizing polymer poly(vinyl alcohol)-quaternized stilbazol (PVA-SbQ). The sensor was tested to measure BOD of the anaerobic synthetic wastewater in which the oxygen electrode could not detect dissolved oxygen. The estimated BOD values up to $150\,mg\,l^{-1}$ were almost the same as those when oxygen was supplied.[17] After optimization, the dynamic range of the sensor was $15–260\,mg\,l^{-1}$ and the reproducibility was also good (Relative Standard Deviation was 12.7%). This sensor was able to measure real wastewater samples.[17,18]

Several mediator-type BOD sensors such as MICREDOX® are also currently produced.[19] The mediator-type BOD sensors are summarized in the review by Moris et al.[20] Moreover, Kim et al. succeeded in the fabrication of a BOD sensor using a micro fuel cell without mediators.[21] BOD sensors based on bacterial luminescence were also developed.[22,23]

3.4 Commercial BOD Sensor

Several BOD sensors have been commercialized to date (Central Kagaku Co., Tokyo; Autoteam GmbH, Berlin, etc.) since the first commercial BOD sensor was produced by Nisshin Electric Co. Ltd in 1983.[24] Several companies in Japan and other countries tried to commercialize the BOD sensor. These sensors are based on the first BOD sensor, which detects the change in the respiration activity of the immobilized microorganisms by an oxygen electrode. These commercialized BOD sensors have been described in several reviews.[5,7,12,24] The market for BOD sensors was 400 million yen (approximately $3.6 million) in 2003,[25] and the interest in biosensors for environmental analysis is still high.

The BOD sensor method was established as Japanese Industrial Standard (JIS) in 1990 (JIS

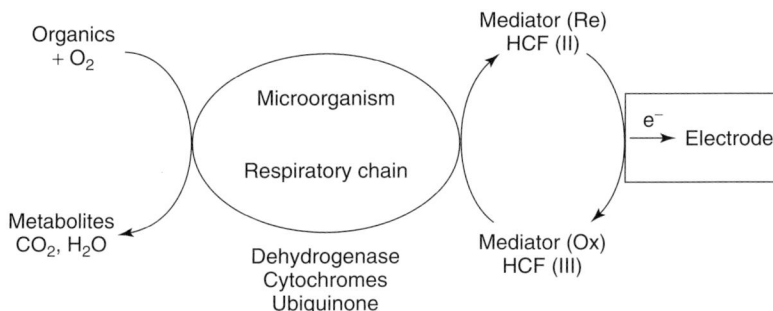

Figure 2. The principle of mediator-type BOD sensor.[17] [Reprinted with permission Yoshida et al.[17] copyright 2000, Royal Society of Chemistry.]

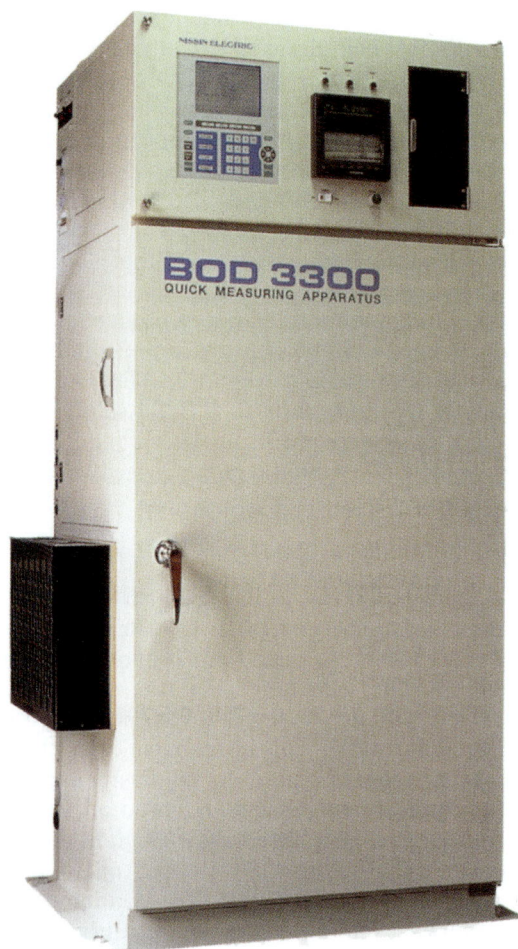

Figure 3. Commercial BOD sensor "BOD 3300" (http://www.aqua-ckc.jp/bod_2_Frame.html).

K3602).[26] Central Kagaku Co. offers several types of commercial BOD sensors according to JIS K3602 in cooperation with Nisshin Electric Co. Ltd, which manufactures the BOD sensors. Figure 3 shows a photograph of one of the commercial BOD sensors. This sensor "BOD 3300" is used for in situ, continuous monitoring of samples such as wastewater. The sensor can measure up to BOD 500 mg l^{-1} in 30–60 min. Their latest BOD sensor is a bench top instrument "QuickBOD α-1", whose detection limit is BOD 2 mg l^{-1} (http://www.aqua-ckc.jp/bod_Frame.html). The sandwich method using two porous membranes was used for immobilization of *T. cutaneum* to obtain high reproducibility and stability.

4 APPLICATIONS OF BOD SENSOR RESEARCH TO OTHER ENVIRONMENTAL MEASUREMENTS

Numerous microbial sensors have been fabricated since the development of the BOD sensor as the first microbial sensor. In environmental analysis, microbial sensors can detect and measure nitrogen compounds, sulfur compounds, cyanide, heavy metals (e.g., Hg or Cu, etc.), detergent, and toxic organic compounds such as phenols and polycyclic aromatic hydrocarbons (PAHs), and so on.[5,8] Some techniques developed for BOD sensors were useful for developing other microbial sensors. For example, genetically modified microorganisms have been used in whole-cell sensors to detect heavy metals.[8,27]

Cyanide compounds are highly toxic to fishes and animals, as well as humans. Since aquatic environments such as river water can be accidentally contaminated, rapid and continuous monitoring of cyanide is required. Cyanide has been detected by microbial sensors based on two different operating principles. Some of these sensors used a cyanide biodegrading strain and an oxygen electrode and measured the consumption of dissolved oxygen, similar to the first BOD sensor.[28] Other cyanide sensors detected the inhibition of the respiration activity of immobilized microorganisms by cyanide,[29] in contrast to the BOD sensors that measure increased respiration activity of the microorganisms.

A microbial sensor that measures linear alkylbenzene sulfonates (LAS), an anionic surfactant contained in detergents, was developed.[30] Detergent contained in domestic wastewater is not acutely toxic as is cyanide but excess detergent contamination in domestic wastewater can negatively affect the microorganism ecosystem in the activated sludge used at municipal sewage plants. The detergent biosensor developed was a bioreactor type sensor which consisted of LAS-degrading bacteria and a flow-cell-type oxygen electrode.[30] The oxygen electrode measured the consumption of dissolved oxygen caused by the biodegradation of LAS by the immobilized strain in the columns. This biosensor also responded to organic compounds in river water, but the selectivity of this sensor was improved by the use of another microbial sensor, immobilized *T. cutaneum.* The optimized LAS sensor could measure 0.2 mg l^{-1} of LAS in 15 min and the sensor was tested for continuous river water monitoring in situ.[31]

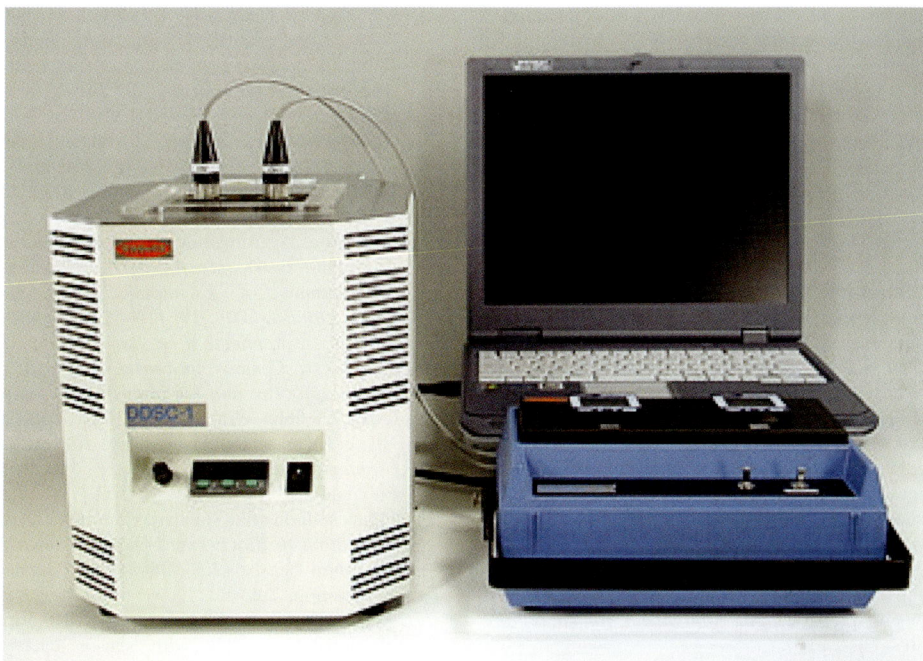

Figure 4. Soil quality sensor "Biosensor for soil diagnosis" (http://www.aist.go.jp/aist_e/latest_research/2004/20040402_1/ 20040402_1.html).

Currently, techniques developed for BOD sensors are being applied to soil quality measurement (http://www.aist.go.jp/aist_e/latest_research/2004/ 20040402_1/20040402_1.html). Our group and Sakata Seed Company fabricated a soil-diagnosis sensor that can rapidly detect soil diseases. Figure 4 shows the photograph of the "Biosensor for soil diagnosis", which is the first prototype instrument based on the measurement of respiration change in immobilized soil microorganisms by oxygen electrodes. The principle of the mediator-type microbial sensor is also applicable to such soil diagnosis.

Numerous microbial sensors were derived from the first BOD sensor, and some of these might be commercialized in the near future. Microbial sensors offer quick and convenient detection of environmental pollutants and toxicants.

REFERENCES

1. Japanese Industrial Standard Committee, *Testing Methods for Industrial Wastewater (JIS K0102)* 1993 Japanese Standards Association, Tokyo.

2. American Public Health Association, *Standard Methods for Examination of Water and Wastewater* 1986 American Public Health Association, Washington.

3. I. Karube, T. Matsunaga, S. Mitsuda, and S. Suzuki, Microbial electrode BOD sensors. *Biotechnology and Bioengineering*, 1997, **19**, 1535–1547.

4. I. Karube, T. Matsunaga, and S. Suzuki, A new microbial electrode for BOD estimation. *Journal of Solid-Phase Biochemistry*, 1977, **2**, 97–104.

5. I. Karube, Y. Nomura, and Y. Arikawa, Biosensors for environmental control. *Trends in Analytical Chemistry*, 1995, **14**, 295–299.

6. M. Hikuma, H. Suzuki, T. Yasuda, I. Karube, and S. Suzuki, Amperometric estimation of BOD by using living immobilized yeasts. *European Journal of Applied Microbiology and Biotechnology*, 1979, **8**, 289–297.

7. Y. Nomura, G. J. Chee, and I. Karube, Biosensor technology for determination of BOD. *Field Analytical Chemistry and Technology*, 1998, **2**, 333–340.

8. K. Riedel, G. Kunze, and A. Konig, Microbial sensors on a respiratory basis for wastewater monitoring. *Advances in biochemical engineering/biotechnology*, 2002, **75**, 81–118.

9. J. Liu, G. Olsson, and B. Mattiasson, Short-term BOD (BOD$_{st}$) as a parameter for on-line monitoring of biological treatment process Part I. A novel design of BOD biosensor for easy renewal of bio-receptor. *Biosensors and Bioelectronics*, 2004, **20**, 562–570.

10. J. Liu, G. Olsson, and B. Mattiasson, Short-term BOD (BOD$_{st}$) as a parameter for on-line monitoring

of biological treatment process Part II: instrumentation of integrated flow injection analysis (FIA) system for BOD$_{st}$ estimation. *Biosensors and Bioelectronics*, 2004, **20**, 571–578.

11. G. J. Chee, Y. Nomura, K. Ikebukuro, and I. Karube, Optical fiber biosensor for the determination of low biochemical oxygen demand. *Biosensors and Bioelectronics*, 2000, **15**, 371–376.

12. J. Liu and B. Mattiasson, Microbial BOD sensors for wastewater analysis. *Water Research*, 2002, **36**, 3786–3802.

13. Z. Yang, H. Suzuki, S. Sasaki, S. McNiven, and I. Karube, Comparison of the dynamic transient- and steady-state measuring methods in a batch type BOD sensing system. *Sensors and Actuators B*, 1997, **45**, 217–222.

14. G. J. Chee, Y. Nomura, and I. Karube, Biosensor for the estimation of low biochemical oxygen demand. *Analytica Chimica Acta*, 1999, **379**, 185–191.

15. G. J. Chee, Y. Nomura, K. Ikebukuro, and I. Karube, Development of photocatalytic biosensor for the evaluation of biochemical oxygen demand. *Biosensors and Bioelectronics*, 2005, **21**, 67–73.

16. G. J. Chee, Y. Nomura, K. Ikebukuro, and I. Karube, Development of highly sensitive BOD sensor and its evaluation using preozonation. *Analytica Chimica Acta*, 1999, **394**, 65–71.

17. N. Yoshida, K. Yano, T. Morita, S. J. McNiven, H. Nakamura, and I. Karube, A mediator-type biosensor as a new approach to biochemical oxygen demand estimation. *The Analyst*, 2000, **125**, 2280–2284.

18. N. Yoshida, J. Hoashi, T. Morita, S. J. McNiven, H. Nakamura, and I. Karube, Improvement of a mediator-type biochemical oxygen demand sensor for on-site measurement. *Journal of Biotechnology*, 2001, **88**, 269–275.

19. N. Pasco, K. Baronian, C. Jeffries, J. Webber, and J. Hay, MICREDOX®-development of a ferricyanide-mediated rapid biochemical oxygen demand method using an immobilised *Proteus vulgaris* biocomponent. *Biosensors and Bioelectronics*, 2004, **20**, 524–532.

20. K. Morris, H. Zhao, and R. John, Ferricyanide-mediated microbial reactions for environmental monitoring. *Australian Journal of Chemistry*, 2005, **58**, 237–245.

21. I. S. Chang, J. K. Jang, G. C. Gil, M. Kim, H. J. Kim, B. W. Cho, and B. H. Kim, Continuous determination of biochemical oxygen demand using microbial fuel cell type biosensor. *Biosensors and Bioelectronics*, 2004, **19**, 607–613.

22. C. K. Hyun, E. Tamiya, T. Takeuchi, and I. Karube, A novel BOD sensor based on bacterial luminescence. *Biotechnology and Bioengineering*, 1993, **41**, 1107–1111.

23. T. Sakaguchi, K. Kitagawa, T. Ando, Y. Murakami, Y. Morita, A. Yamamura, K. Yokoyama, and E. Tamiya, A rapid BOD sensing system using luminescent recombinants of *Escherichia coli*. *Biosensors and Bioelectronics*, 2003, **19**, 115–121.

24. K. Riedel, *Application of Biosensors to Environmental Samples*, in *Commercial Biosensors: Applications to Clinical, Bioprocess, and Environmental Samples*, G. Ramsay (ed) 1998 John Wiley & Sons, New York, pp. 267–294.

25. Nikkei Business Publications Inc., *Nikkei Bio Nenkan 2004 (Annual Report of Bio Market)* 2004 Nikkei Business Publications, Tokyo.

26. Japanese Industrial Standard Committee, *Apparatus for the Estimation of Biochemical Oxygen Demand (BODs) with Microbial Sensor (JIS K3602)* 1990 Japanese Standards Association, Tokyo.

27. A. Ivask, K. Hakkila, and M. Virta, Detection of organomercurials with sensor bacteria. *Analytical Chemistry*, 2001, **73**, 5168–5171.

28. J. I. Lee andI. Karube, A novel microbial sensor for the determination of cyanide. *Analytica Chimica Acta*, 1995, **313**, 69–74.

29. K. Ikebukuro, A. Miyata, S. J. Cho, Y. Nomura, S. M. Chang, Y. Yamauchi, Y. Hasebe, S. Uchiyama, and I. Karube, Microbial cyanide sensor for monitoring river water. *Journal of Biotechnology*, 1996, **48**, 73–80.

30. Y. Nomura, K. Ikebukuro, K. Yokoyama, T. Takeuchi, Y. Arikawa, S. Ohno, and I. Karube, A novel microbial sensor for anionic surfactant determination. *Analytical Letters*, 1994, **27**, 3095–3108.

31. Y. Nomura, K. Ikebukuro, K. Yokoyama, T. Takeuchi, Y. Arikawa, S. Ohno, and I. Karube, Application of a linear alkylbenzene sulfonate biosensor to river water monitoring. *Biosensors and Bioelectronics*, 1998, **13**, 1047–1053.

Optical Biosensor for the Determination of Trace Pollutants in the Environment

Güenter Gauglitz,[1] Güenther Proll[1] and Jens Tschmelak[2]

[1] *Institute of Physical and Theoretical Chemistry, Eberhard Karls University of Tuebingen, Tuebingen, Germany and* [2] *Bierlachweg, Erlangen, Germany*

1 INTRODUCTION

Clean water, its secure delivery to consumers, and protection of resources are among the most important problems for humans in the future. Thus, pollution of water sources, aquifers, and wetland systems caused by industry, agriculture, municipally created waste water, and recreational activities is identified as a Europe-wide problem, especially in the newly admitted European Union member states. Today, pollution of ground, surface, and riverwater by agriculture, industry and mining, traffic, boating, sports, and tourism presents a growing danger for drinking water. Therefore, the European Community has funded projects, beginning with the Framework IV programme, in order to implement strategies to measure and control pollution from various sources and to establish their practical use. These are demonstrated by the European Community Water Directives.[1,2] In parallel, projects were funded to establish fast, sensitive, cost-effective, and easy-to-use analytical systems capable of measuring a variety of small organic pollutants in aqueous systems, especially in the areas of herbicides, fungicides, insecticides, antibiotics, blue algae, toxins, endocrine disrupting compounds, and suspected carcinogens.

At present, technologies for water analyses are available, such as liquid chromatography (LC), high-performance liquid chromatography (HPLC), and gas chromatography (GC) in combination with detection principles such as mass spectrometry (MS) or diode array detection (DAD) and especially fluorescence detection (FLD). These methods have been reviewed in preceding years.[3–5] However, although these methods are well established and can look back on decades of developments and successful application, all of them require enrichment of water samples by several orders of magnitude prior to analysis to enable suitable trace analysis performance. Accordingly, a time-consuming solid phase extraction (SPE) is typically used as an enrichment procedure. This pretreatment step makes it difficult and rather expensive to develop suitable online monitoring systems and might be the reason why only a few such systems exist so far.[6,7]

2 IMMUNOASSAY-BASED SYSTEMS

Within the preceding context, immunoassays were introduced into water analysis since they offer very low detection limits. Thus, they are considered not to need any enrichment steps and require neither cleanup steps nor preconcentration. The common immunoassay formats of enzyme-linked immunosorbent assay (ELISA) are based

Handbook of Biosensors and Biochips. Edited by Robert S. Marks, David C. Cullen, Isao Karube, Christopher R. Lowe and Howard H. Weetall.
© 2007 John Wiley & Sons, Ltd. ISBN 978-0-470-01905-4.

on immunochemistry and are supposed to offer a high potential for new application in the future.[8]

However, all these new approaches lack the possibility of automation, are time consuming, and require personnel. A combination of these immunoassays and FLD techniques at least lowers the limit of detection and the limit of quantification. In combination with a suitable test format, for example, sandwich assay or noncompetitive, they can easily be carried out by flow-injection analysis (FIA).[9]

The new approaches in array technology, miniaturization and transduction in sensor signals, provide a new generation of immunosensor systems. Accordingly, the group of Ligler developed one of the first portable and fully automated total internal reflection fluorescence (TIRF)-based biosystems, suitable for multianalyte detection.[10,11] These newly developed strategies of biosensing can take advantage of arraying or layout array systems of recognition elements to allow simultaneous detection and quantification of multiple analytes.[12] In contrast to the above-mentioned well-established ELISA methods, biosensors[13] allow easy automation of immunoassays and therefore provide perfect features to meet current requirements in water analyzing systems. Some of the research groups currently using TIRF-based biosensors for a variety of applications are cited in Ref. 14.

3   AUTOMATIC MEASURING DEVICES

In our group, this approach of a fully automated system gaining low limits of detection and quantification by using the TIRF method and being based on either integrated optics or bulk-optic chips has been followed for many years. The first successful instrumentation was the river analyzer (RIANA) system (ENV4-CT95-0066).[15–17] Following the requirements of the EU water legislation and the needs of wastewater and drinking water facilities as well as according to the risk assessment/risk management approach,[18] target compounds are selected mainly from the groups of modern pesticides, endocrine disrupting compounds, pharmaceuticals, and toxins. In addition, a potential water-monitoring device needs to be robust, cost-effective, automated, and able to measure several tens of organic pollutants at low nanogram per liter levels in a very short time, preferably without any time-consuming sample concentrations and prior sample pretreatments. Another feature is considered to become more and more interesting for the future, namely the possibility of instruments with remote control, automated data processing, and the generation of alarm signals when the pollutant's concentration exceeds a preset threshold value.

The RIANA system mentioned above allows measurement of water samples without preconcentration and even in various matrices. In the case of many target samples, lower limits of detection than those defined by the European Commission at micrometer per liter levels could be achieved.[19,20] Data can be referenced by classical analytical methods.[21]

On the basis of this successful implementation of the RIANA system, a joint European project, Automated Water Analyzer Computer Supported System (AWACSS),[22] was launched 5 years ago, where 9 different groups from all over Europe started to set up a device meeting the above-mentioned prerequisites. Within the setup, flow injection was combined with a microfluidic cell, covering an integrated optical device in which laser radiation in the visible range is coupled in via optical fiber. The pigtailed IO chip provides waveguide arms with radiation totally internally reflected and supplying an evanescent field at the surface. At the surface, analytes are immobilized covalently using a biolayer interface, which supplies high loading of recognition elements and drastic reduction of nonspecific binding and fast biomolecular interaction processes. Modern spotting techniques using a Top Spot[23] offer the possibility of a spatially resolved surface chemistry to immobilize 32 recognition spots either for replica measurements or for multiple analyte analysis.

Pollutant-specific antibodies are added to the sample, the mixture is transferred by a flow-injection system, passing this optical transduction element. The added antibodies are all labeled with the same fluorophore, in this case Cy5.5. If the specific pollutant is present, the antibody is blocked in the sample solution, and the other nonblocked antibody can diffuse to the surface, which is monitored laterally resolved by an array of 32 fibers. This binding inhibition test format (see Figure 1) allows shifting the system via the antibody concentration to low limits of detection.

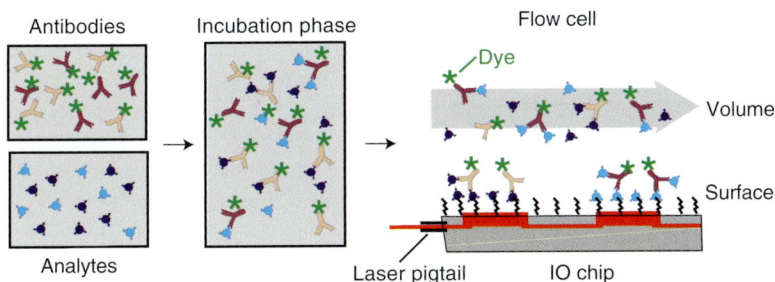

Figure 1. The AWACSS biosensor employs fluorescence-based detection of the binding of fluorophore-tagged biomolecules to the surface of an IO chip. The recognition is based on a binding inhibition assay.

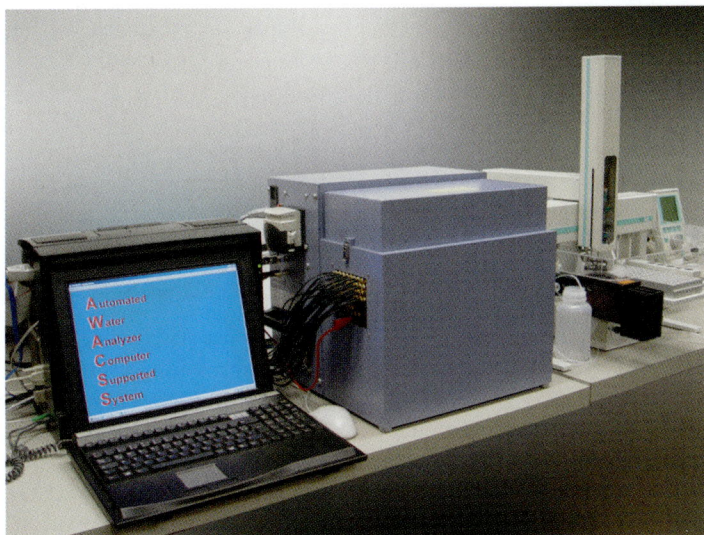

Figure 2. Picture of the setup with control unit, measuring system, and autosampler.

The setup is shown in Figure 2, including an autosampler for sample delivery, washing steps, and regeneration. The system includes a computer as a control unit for the measuring device and as a link to a web server, which contains a database as well as a threshold-based warning system.

A typical calibration curve is given in Figure 3. For accurate measurements with this heterogeneous noncompetitive assay format it is necessary to minimize effects that have an influence on the detected fluorescent light intensity. Beside a stabilized excitation light source, for example, a semiconductor laser with a monitoring diode, it is of great importance to reduce nonspecific binding of antibodies to the matrix or to the sensor surface. During assay design for the above-mentioned

systems we obtained best results by adding an excess of ovalbumin as a background protein to the antibody stock solution. Nonspecific binding to the sensor surface was suppressed by layers of bio-polymers like polyethylene glycol and/or aminodextrane.

In the following some details of a typical evaluation are discussed. Results are limit of detection, limit of quantification, and working range.

4   EVALUATION STRATEGIES

The standard experimental design for a calibration routine consists of nine independent blank (e.g., Milli-Q water) measurements and eight

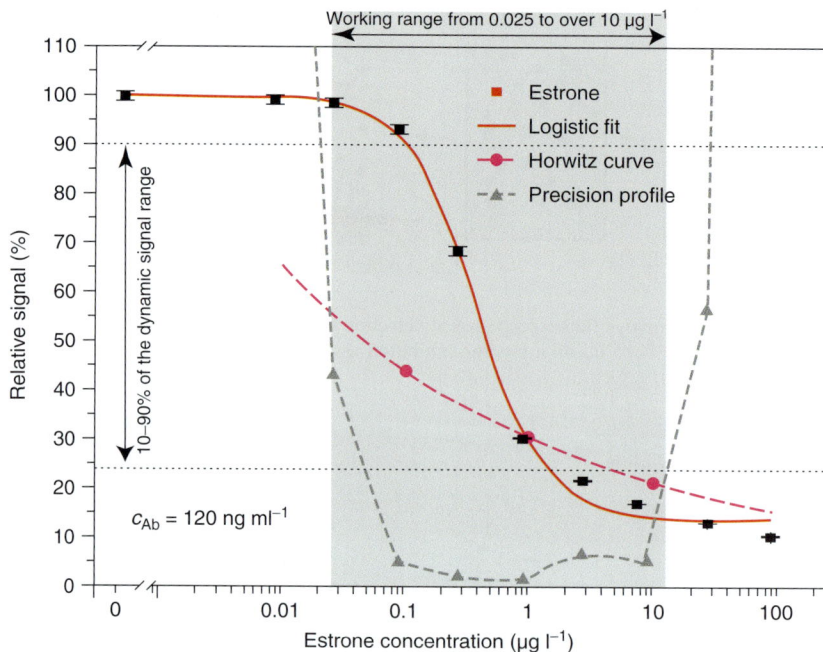

Figure 3. Calibration curve for estrone including logistic function, Horwitz curve, working range, and precision profile.

concentration steps (each measured as three replicas) of the analyte (e.g., spiked Milli-Q water). For all concentration steps and the blank measurements (nine replicas), the mean value and the standard deviation value (SDV) for the replica were calculated. The measured signal for the mean value of the blanks was set to 100%, and all spiked samples could be obtained as a relative signal below this blank value. To fit the data set a logistic fit function (parameters of a logistic function: A_1, A_2, x_0, and p)[24] was used:

$$y = \frac{A_1 - A_2}{1 + \left(\frac{x}{x_0}\right)^p} + A_2 \qquad (1)$$

A_1 is the upper asymptote and A_2 the lower one. The range between A_1 and A_2 is the dynamic signal range. The inflection point is given by the variable x_0 and represents the analyte concentration, which corresponds to a decrease of 50% of the dynamic signal range—the inhibitory concentration 50% (IC_{50}). The slope of the tangent in this point is given by the parameter p. Out of the logistic fit data, the 10–90% range of the dynamic signal can be calculated, which gives a

first impression of the possible utilization range of the received calibration curve. In compliance with the International Union of Pure and Applied Chemistry (IUPAC) rules "The Orange Book",[25] the limit of detection (LOD) is calculated as 3 times the standard deviation of the blanks (SDV_b) and the limit of quantification (LOQ) is calculated as 10 times the SDV_b.

The use of LOQ for logistic calibration curves is a contentious issue, because with its nonlinear behavior the results for immunoassays are often worse than they need to be. A real alternative is the use of the 95% confidence belt and the associated minimum detectable concentration (MDC) and reliable detection limit (RDL), which can easily be calculated for the sigmoidal calibration curves.[26] These authors reported on calibration and assay development using the four-parameter logistic model and assay quality control procedures.

To determine the working range, the precision profile ($x_{cv,i}$) and its intersections with the Horwitz curve[27,28] has to be calculated. On the basis of scores of Association of Analytical Communities (AOAC) intercomparison programs, Horwitz developed an empiric correlation between the comparative standard deviation (SD) and the concentration. For laboratory intercomparison programs Horwitz proposed an equation for the reproducibility $\sigma_R = f \cdot c^{0.8495}$ with a factor $f = 0.02$. The corresponding error is the relative standard deviation RSD $= 100 \, (\sigma_R c)$ that can be calculated with the reproducibility σ_R and the analyte concentration c. For intralaboratory reproducibility, Horwitz found a higher precision and consequently lower RSD values. In this case, the factor f can be reduced to two-thirds up to half of its former value. RSD values can be calculated for each concentration and they represent the Horwitz curve.

An applicable concentration determination is possible only if the precision profile is below the Horwitz curve. The SD values of the inverse function (SDVx_i) can be calculated using the SD values of the measured data (SDVy_i) and the associated values of the first derivative (y') of the logistic fit (y) for each concentration.

Then the variation coefficients ($x_{cv,i}$) can be calculated and plotted together with the values of the Horwitz curve and the calibration data including the logistic fit in the semilogarithmic graph. Finally, the range between the intersection points of the Horwitz curve and the precision profile represents the working range.

5 RECENT IMPROVEMENTS

Improvement of the optical transduction system and the quality of the evanescent field by the University of Southampton, the improvement of the electronic devices by Siemens, and the microfluidics by CRL resulted in a drastic improvement of the limits of detection obtained with this instrument in comparison with the first RIANA setup. At the final presentation, the consortium could demonstrate a fully automated system requiring water samples and sending the analytical results via the Internet to a host computer system in which the values and the database-based EU threshold values were compared. Accordingly, warning was sent via SMS to a mobile of a representative.

In the meantime, a large variety of analytes has been examined. Table 1 provides an overview of the samples, the limits of detection, and the limits of quantification.[29–35] All details are given in these publications. Since the apparatus is working

Table 1. Summarized results of trace measurements of pollutants in water

Analyte	LOD ($\mu g \, l^{-1}$)	LOQ ($\mu g \, l^{-1}$)	SDV (%)
Atrazine	0.0099	0.0710	2.70
Isoproturon	0.0008	0.0089	1.22
Propanil	0.0006	0.0045	1.29
Bisphenol A	0.0080	0.0710	0.90
Caffein	0.0008	0.0090	0.90
Sulphadimethoxine	0.0038	0.0447	1.56
Sulphatiazole	0.0027	0.0284	1.24
Sulphadiazine	0.0038	0.1167	1.26
Sulphadimidine	0.0032	0.0452	1.42
Sulphamethoxazole	0.0066	0.1781	1.06
Sulphamethizole	0.0042	0.0748	1.43
Sulphamethoxypyridazine	0.0042	0.0523	0.68
Tetrahydrocanabiol	0.0029	0.0424	1.52
Estrone	0.0002	0.0014	0.67
Progesterone	0.0002	0.0020	2.70
Testosterone	0.0002	0.0052	3.40
Before optimization			
Atrazine	0.03	0.30	3.33
Simazine	0.03	0.30	1.70
Isoproturon	0.11	1.10	2.08
Alachlor	0.07	0.70	3.98
2,4-D	0.07	0.70	2.15
Paraquat	0.01	0.09	10
PCP	4.23	45	4.59

highly reproducibly, the SDVs are small, and most recovery rates are in the ranges given by the AOAC International. Some of these values have been referenced to classical and standardized analytical methods in laboratories of the partners.[36]

6   RESULTS OF A FIELD TEST

Within the first field test of the instrument a small river in the southern part of Germany was selected, where the complete measurement station was installed together with a sampling setup in a watershed for raw tap water. The IO chip was modified and calibrated to measure atrazine, isoproturon, estrone, and bisphenol A. The simultaneous calibration was performed from 0 to 90 $\mu g\,l^{-1}$ in nine steps and resulted in four calibration curves (see Figure 4). For all compounds, the calculated LOD was below $7\,ng\,l^{-1}$.

The small river was successfully monitored over a period of approximately 24 h by using the AWACSS instrument in unattended and quasicontinuous mode. The fully automated system measured several times triplicates of riverwater

samples and Milli-Q blanks as a reference. On the basis of the calibration parameters, the individual concentrations of the four compounds were automatically calculated by the instrument's software package. The river has been sampled also manually for intensive comparison measurements with common analytical methods at an accredited laboratory.

For estrone, all biosensor measurements as well as the HPLC control measurements have been below the LOQ for both systems. Bisphenol A was not possible to measure because of an enormous leaching of this chemical substance out of the tubing of the sampling system. For the continuous monitoring of the two pesticides very good results that are confirmed by HPLC measurements could be obtained with the biosensor system. This field test demonstrated the ability of the AWACSS providing a tool for the continuous monitoring of organic pollutants in riverwater. Once a measurement station is installed only a little maintenance of the system is necessary. Nonexperts can carry out the exchange of buffer solutions and the replacement of the sensor-chip unit after 500 measurements.[37]

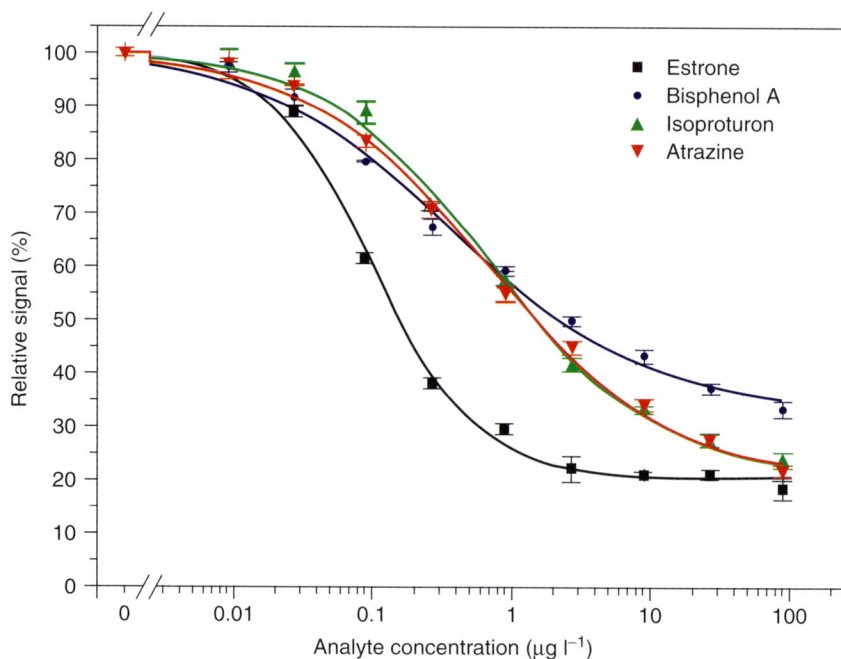

Figure 4. Simultaneous calibration for atrazine, bisphenol A, estrone, and isoproturon. For all compounds the calculated LOD is below $0.007\,\mu g\,l^{-1}$.

7   CONCLUSION

For the first time in water analysis an automated device facilitated the monitoring of a multianalyte system based on a fully automated immunoassay with limits of detection in the range of $1 \, \text{ng} \, \text{l}^{-1}$ for many analytes. The verification of the reproducibility, precision, and robustness of the optical sensor was also demonstrated by successful real sample measurements. Most recovery rates were between 70 and 100% for spiked tap water and different riverwaters. To avoid false-negative finds at very low analyte concentrations, a new strategy for matrix measurements was developed.[38] Very good recovery rates were achieved for ultra-sensitive measurements of various endocrine disrupting compounds,[39] algae toxins, and even measurements in nonaqueous matrices such as milk or sera.[40] All in all, time for one measurement includes incubation and a regeneration step below 15 min, where the real measurement takes just 2 min. The sophisticated surface chemistry allows regenerating the optical transduction devices more than 500 times without loss of quality.

ACKNOWLEDGMENTS

The work was done with some European-funded projects such as the River Analyzer project (RIANA ENV4-CT95-0066) by the European Commission under the Environment and Climate Programme (4th Framework Programme), the Automated Water Analyzer Computer Supported System project (AWACSS EVK1-CT-2000-00045), supported by the European Commission under the 5th Framework programme and contributing to the implementation of the key action "Sustainable Management and Quality of Water" within the Energy, Environment, and Sustainability Development Programme. Part of this work was also funded by the Bundesministerium für Bildung und Forschung (BMBF) within the PIWAS project (Parallisiertes Immunreaktionsbasiertes Wasseranalysatorsystem; 02WU0243). The antibodies were supplied by Ram Abuknesha, PhD, King's College London, London, UK. The integrated optics were optimized by James Wilkinson, Optoelectronics Research Centre, Southampton University, Southampton, UK; microfluidics were included in the system together with some electronics by the Central Research Laboratories Ltd., Hayes, UK; system integration and web-based process control was supplied by Mr. Kaiser, Siemens AG, Erlangen, Germany; reference measurements were carried out by Professor Barceló, Ministerio de Ciencia y Tecnología, Department of Environmental Chemistry, Barcelona, Spain; real water samples and reference methods were carried out by Frank Sacher, PhD, DVGW-Technologiezentrum Wasser, Karlsruhe, Germany, E. Korenková, Environmental Institute, Kos, Slovak Republic, and Jaroslav Slobodnik, PhD, Water Research Institute, Bratislava, Slovak Republic. Jens Tschmelak held a fellowship, and Günther Proll participated in the research training group "Quantitative Analysis and Characterization of Pharmaceuticals and Biochemically relevant Substances" funded by the DFG (Deutsche Forschungsgesellschaft) at the Eberhard Karls University of Tuebingen.

REFERENCES

1. Council directive (98/83/EC) of 3 November 1998 relating to the quality of water intended for human consumption. *Official Journal of the European Community*, 1998, **L330**, 32–54.
2. Directive 2000/60/EC of the European parliament and of the council of 23 October 2000 establishing a framework for community action in the field of water policy. *Official Journal of the European Community*, 2000, **L327**, 1–72.
3. D. Barceló and M. C. Hennion, *Trace Determination of Pesticides and their Degradation Products in Water*, Elsevier, Oxford, 1997.
4. D. Barceló and M. C. Hennion, Sampling of polar pesticides from water matrices. *Analytica Chimica Acta*, 1997, **338**, 3–18.
5. D. Barceló (ed), *Environmental Analysis: Sample Handling and Trace Analysis of Pollutants—Techniques, Applications and Quality Assurance*, Elsevier, Oxford, 2000.
6. B. Allner, G. Wegener, T. Knacker, and P. Stahlschmidt-Allner, Electrophoretic determination of estrogen-induced protein in fish exposed to synthetic and naturally occurring chemicals. *Science of the Total Environment*, 1999, **233**, 21–31.
7. P. Lopez-Roldan, M. J. Lopez de Alda, and D. Barceló, Simultaneous determination of selected endocrine disrupters (pesticides, phenols and phthalates) in water by in-field solid-phase extraction (SPE) using the prototype PROFEXS followed by on-line SPE (PROSPEKT)

and analysis by liquid chromatography-atmospheric pressure chemical ionisation-mass spectrometry. *Analytical and Bioanalytical Chemistry*, 2004, **378**, 599–609.

8. R. J. Schneider, Environmental immunoassays. *Analytical and Bioanalytical Chemistry*, 2003, **375**, 44–46.

9. P. M. Kraemer, A. Franke, and C. Standfuss-Gabisch, Flow injection immunoaffinity analysis (FIIAA)—a screening technology for atrazine and diuron in water samples. *Analytica Chimica Acta*, 1999, **399**, 89–97.

10. F. S. Ligler, J. P. Golden, C. A. Rowe-Taitt, and J. P. Dodson, Array biosensor for simultaneous detection of multiple analytes. *Proceedings of SPIE*, 2001, **4252**, 32–36.

11. K. E. Sapsford, Y. S. Shubin, J. B. Delehanty, J. P. Golden, C. R. Taitt, L. C. Shriver-Lake, and F. S. Ligler, Fluorescence-based array biosensors for detection of biohazards. *Journal of Applied Microbiology*, 2004, **96**, 47–58.

12. F. S. Ligler, K. E. Sapsford, Y. S. Shubin, J. Lemmond, C. A. Rowe-Taitt, J. B. Delehanty, and J. P. Golden, Array biosensor for environmental monitoring. *Proceedings of SPIE*, 2002, **4576**, 63–67.

13. H. H. Weetall, Chemical sensors and biosensors, update, what, where, when and how. *Biosensors and Bioelectronics*, 1999, **14**(2), 237–242.

14. O. S. Wolfbeis and R. Narayanaswamy (eds), *Optical Sensors*, Springer, 2004 (ISBN 3-540-40886-X).

15. RIANA ENV4-CT95-0066.

16. U. Schobel, C. Barzen, and G. Gauglitz, Immunoanalytical techniques for pesticide monitoring based on fluorescence detection. *Fresenius Journal of Analytical Chemistry*, 2000, **366**, 646–658.

17. A. Klotz, B. Brecht, C. Barzen, G. Gauglitz, R. D. Harris, G. R. Quigley, J. S. Wilkinson, and R. A. Abuknesha, Immunofluorescence sensor for water analysis. *Sensors and Actuators B-Chemical*, 1998, **51**(1–3), 181–187.

18. World Health Organization, *Guidelines for Drinking Water Quality*, 3rd Edn, World Health Organization, Geneva, 2003.

19. C. Barzen, A. Brecht, and G. Gauglitz, Optical multiple-analyte immunosensor for water pollution control. *Biosensors and Bioelectronics*, 2002, **17**, 289–295.

20. S. Rodriguez-Mozaz, S. Reder, M. Lopez de Alda, G. Gauglitz, and D. Barceló, Simultaneous multi-analyte determination of estrone, isoproturon and atrazine in natural waters by the river analyser (RIANA), an optical immunosensor. *Biosensors and Bioelectronics*, 2004, **19**, 633–640.

21. E. Mallat, C. Barzen, A. Klotz, A. Brecht, G. Gauglitz, and D. Barcelo, River analyzer for chlorotriazines with a direct optical immunosensor. *Environmental Science and Technology*, 1999, **33**, 965–971.

22. AWACSS EVK1-CT-2000-00045.

23. B. de Heij, M. Daub, O. Gutmann, R. Niekrawietz, H. Sandmaier, and R. Zengerle, Highly parallel dispensing of chemical and biological reagents. *Analytical and Bioanalytical Chemistry*, 2004, **378**, 119–122.

24. R. A. Dudley, P. Edwards, R. P. Ekins, D. J. Finney, I. G. M. McKenie, G. M. Raab, D. Rodbard, and R. P. C. Rodgers, Guidelines for immunoassay data processing. *Clinical Chemistry*, 1985, **31**, 1264–1271.

25. J. Inczedy, T. Lengyel, and A. M. Ure, *Compendium of Analytical Nomenclature. Definitive Rules 1997*, The Orange Book, 3rd Edn, Blackwell Science, Oxford, 1998.

26. M. A. O'Connell, B. A. Belanger, and P. D. Haaland, Calibration and assay development using the four-parameter logistic model. *Chemometrics and Intelligent Laborary Systems*, 1993, **20**, 97–114.

27. W. Horwitz, L. R. Kamps, and K. W. Boyer, Quality assurance in the analysis of foods and trace constituents. *Journal of The Association of Official Agricultural Chemists*, 1998, **63**, 1344–1353.

28. V. R. Meyer, Are duplicate determinations necessary? *Schweizerische Laboratoriums-Zeitschrift*, 2003, **60**, 63–65.

29. J. Tschmelak, G. Proll, and G. Gauglitz, Verification of performance with the automated direct optical TIRF immunosensor (River Analyser) in single and multi-analyte assays with real water samples. *Biosensors and Bioelectronics*, 2004, **20**, 743–752.

30. J. Tschmelak, G. Proll, and G. Gauglitz, Sub-nanogram per liter detection of the emerging contaminant progesterone with a fully automated immunosensor based on evanescent field technique. *Analytica Chimica Acta*, 2004, **519**, 143–146.

31. J. Tschmelak, G. Proll, and G. Gauglitz, Ultra-sensitive fully automated immunoassay for the detection of propanil in aqueous samples—steps of progress toward a sub-nanogram per liter detection. *Analytical and Bioanalytical Chemistry*, 2004, **379**, 1004–1012.

32. J. Tschmelak, G. Proll, J. Riedt, J. Kaiser, P. Kraemmer, L. Bárzaga, J. S. Wilkinson, P. Hua, J. P. Hole, R. Nudd, M. Jackson, R. Abuknesha, D. Barceló, S. Rodriguez-Mozaz, M. J. López de Alda, F. Sacher, J. Stien, J. Slobodník, P. Oswald, H. Kozmenko, E. Korenková, L. Tóthová, Z. Krascsenits, and G. Gauglitz, Automated water analyser computer supported system (AWACSS)—Part I: project objectives, basic technology, immunoassay development, software design and networking automated water analyser computer supported system (AWACSS)—Part II: Intelligent, remote–controlled, cost–effective, on–line, water-monitoring measurement system. *Biosensors and Bioelectronics*, 2005, **20**, 1499–1508; 1509–1519.

33. G. Proll, J. Tschmelak, and G. Gauglitz, Fully automated biosensors for water analysis. *Analytical and Bioanalytical Chemistry*, 2005, **381**, 61–63.

34. J. Tschmelak, G. Proll, J. Riedt, J. Kaiser, P. Kraemmer, L. Bárzaga, J. S. Wilkinson, P. Hua, J. P. Hole, R. Nudd, M. Jackson, R. Abuknesha, D. Barceló, S. Rodriguez-Mozaz, M. J. López de Alda, F. Sacher, J. Stien, J. Slobodník, P. Oswald, H. Kozmenko, E. Korenková, L. Tóthová, Z. Krascsenits, and G. Gauglitz, Biosensors for unattended, cost-effective and continuous monitoring of environmental pollution: automated water analyser computer supported system—AWACSS and river analyser—RIANA. *International Journal of Environmental Analytical Chemistry*, 2005, **85**, 837–852.

35. J. Tschmelak, G. Proll, and G. Gauglitz, Improved strategy for biosensor based monitoring of water bodies with diverse organic carbon levels. *Biosensors and Bioelectronics*, 2005, **21**, 979–983.

36. M. Petrovic, E. Eljarrat, M. J. Lopez de Alda, and D. Barceló, Recent advances in the mass spectrometric analysis related to endocrine disrupting compounds in aquiatic environmental samples. *Journal of Chromatography A*, 2002, **974**(1–2), 23–51.

37. G. Proll and G. Gauglitz, *Viable Methods of Soil and Water Pollution Monitoring, Protection and Remediation, NATO Science Series*, Irena. Twardowska, Herbert. E. Allen, M. H. Häggblom (eds), ISBN-10 1-4020-4728-2 (e-book), Springer 2006.

38. J. Tschmelak, G. Proll, and G. Gauglitz, Optical biosensor for pharmaceuticals, antibiotics, hormones, endocrine disrupting chemicals and pesticides in water: assay optimisation process for estrone as example. *Talanta*, 2005, **65**, 313–323.

39. J. Tschmelak, M. Kumpf, N. Käppel, G. Proll, and G. Gauglitz, Total internal reflectance fluorescence (TIRF) biosensor for environmental monitoring of testosterone with commercially available immunochemistry: antibody characterization, assay development and real sample measurements. *Talanta*, 2006, **69**, 343–350.

40. J. Tschmelak, N. Käppel, and G. Gauglitz, TIRF-based biosensor for sensitive detection of progesterone in milk based on ultra-sensitive progesterone detection in water. *Analytical and Bioanalytical Chemistry*, 2005, **382**, 1895–1903.

Food and Beverage Applications of Biosensor Technologies

Helge R. Schnerr

Department of Food Safety and Food Quality, Leatherhead Food International Ltd., Leatherhead, UK

1 INTRODUCTION

Food products are analyzed for a variety of reasons, for example, compliance with legal and labeling requirements, confirmation of quality and safety, hygienic aspects, nutritional adequacy, and authenticity of food products. To categorically confirm the quality and safety of final products, raw materials and products must be closely monitored for contaminants, disease-causing agents, and the presence of harmful compounds, such as allergens, as well as for nutritional evaluation purposes. The quality control of nutritional content and confirmation of food safety are major tasks for food and drinks manufacturers to ensure high-quality standards and minimize risk for the consumer. Since food products and their raw materials are complex mixtures of chemical compounds, highly specific, cost-effective, and reliable methods are increasingly needed.

For the past 25 years, we have witnessed remarkable progress in the development of affinity biosensors and their applications in areas such as medical diagnostics, drug screening, and food safety and security. Food analysis still involves extensive sample preparation, isolation, and concentration, and sometimes even derivatization of very complex matrices. Difficulties may be compounded by the fact that the components are present at very low concentrations.

Biosensors offer attractive alternatives to existing methods that can allow the creation of on line or on-site, sensitive, low-cost devices for routine use. The monitoring and control of manufacturing processes becomes possible owing to improvements in analysis speed and their application on line. Furthermore, portable biosensors can be used for monitoring products during manufacturing, distribution, and retail. The biomolecular interaction analysis (BIA) is not limited to proteins. Interactions between DNA–DNA, DNA–protein, lipid–protein, and hybrid systems of biomolecules and nonbiological surfaces can be investigated. It can be used to monitor whether two or more interactants bind to each other, how strong the interactions are or measure the actual association or dissociation rates. In addition, the binding of two interactants can be used to measure the concentration of one of the interactants using a calibration curve.

MindBranch, Inc. has estimated that the market size for worldwide biosensors at year-end 2003 was about \$7.3 billion. The market is projected to improve and grow to about \$10.8 billion by 2007 with a growth rate of about 10.4%.[1]

This chapter presents a review of the potential application of biosensors in the food and drinks industry, the commercially available systems, and some upcoming technology for future developments in this sector.

Handbook of Biosensors and Biochips. Edited by Robert S. Marks, David C. Cullen, Isao Karube, Christopher R. Lowe and Howard H. Weetall.
© 2007 John Wiley & Sons, Ltd. ISBN 978-0-470-01905-4.

2   NEEDS OF THE FOOD AND DRINKS INDUSTRY

Biosensors are analytical devices, which use biological interactions to provide either qualitative or quantitative results. They are extensively employed in many fields such as clinical diagnostics, biomedicine, process control, fermentation control and analysis, pharmaceutical analysis as well food and drink analysis.

It has been estimated that the food industry spends, on average, 1.5–2% of the value of its total sales on quality control and appraisal.[2] According to a new market report of Strategic Consulting Inc. entitled Food Micro 2005, the worldwide food microbiology market in 2005 represented over 625 million tests with a market value in excess of $1.65 billion.[3]

Characterizations of food products include determination of food components and nutritional composition (e.g., proteins, carbohydrates, and vitamins), food additives (aspartame, benzoate, and colorants), chemical contaminants (e.g., mycotoxins, pesticides, or veterinary drug residues), and contaminants of microbiological origin. Food produce may be microbiologically contaminated at source or at any stage during processing, packaging, and distribution. Contamination of food and beverages with bacteria (e.g., *Salmonella, Campylobacter*), viruses (e.g., noroviruses, hepatitis A), fungi (e.g., *Aspergillus* spp.), and parasites (e.g., *Giardia duodenalis, Cryptosporidium parvum*) can cause foodborne diseases.

Biosensors have a high potential for automation and allow the construction of simple and portable equipment for fast analysis. These properties will open up many new applications within quality and process control, including fermentation and quality and safety control of raw materials.

Because most food is highly sensitive to critical process parameters and can easy undergo rapid and destructive changes, process control is a key point in a modern industrial environment. Rapid feedback of information could help the food or drinks manufacturer to both reduce wastage from poorly controlled processes and increase productivity.

McMurdo and Whyard published a schematic diagram of feedback control in a theoretical process (Figure 1).[4] The data generated through external analysis of raw materials, intermediates, and

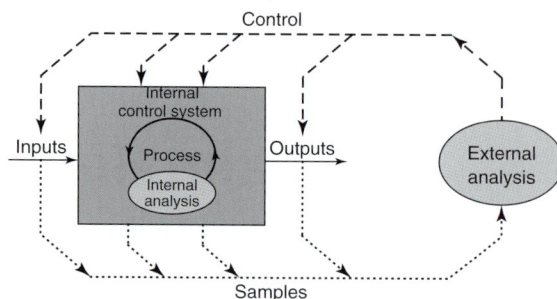

Figure 1. Desired feedback control of a process. [Extended adaptation from McMurdo and Whyard.[4]]

final products can be used to plan the production more safely and more efficiently. Nowadays, many operations in the food industry are continuous processes with a high level of automation. There is thus an increasing demand for analytical technology suitable for automatic quality control through the process and at the end of the line so that the real-time state of the process can be controlled. Biosensors do not offer just rapid results, on-line biosensor technology offers food industry monitoring in real time and an option of internal process control to fulfill the demands of a high standard of quality control.

However, biosensors have not been used extensively for on-line processing for several reasons. Firstly they are composed of biological material, which may be inactivated by sterilization or harsh processing conditions, like the extremes of pH and temperature, encountered in food processing. Furthermore, biosensors usually have a limited testing life owing to the stability of the biological recognition element.

3   GENERAL ASPECTS OF BIOSENSORS

A biosensor can be defined as a self-contained integrated device, which is capable of providing specific quantitative or semiquantitative analytical information incorporating a biological, biologically derived, or biomimetic sensing element either integrated within or intimately associated with an applicable transduction element.[5,6]

Accordingly, the basic principle of biosensor technology is to convert a biologically induced recognition event into either discrete or continuous

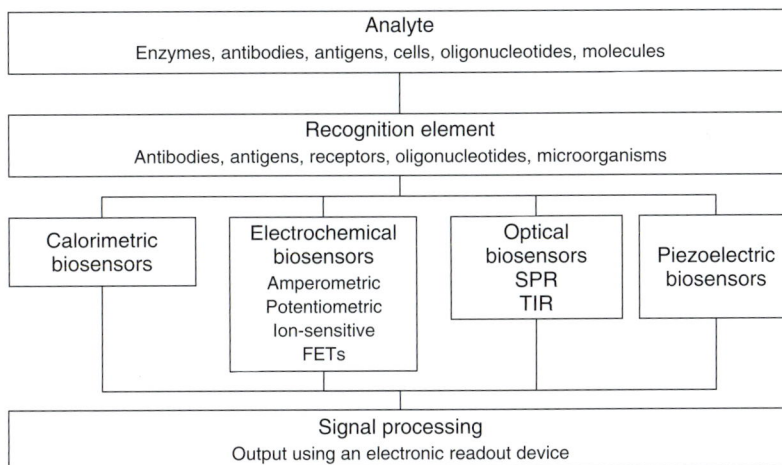

Figure 2. Key components of a biosensor showing examples of biocomponents, transducers, and signal display used in biosensor construction.

digital electronic signals that are proportional to a single analyte or a related group of analytes (Figure 2).

Biosensors may be classified according to the mechanism of biological selectivity (Table 1) or, alternatively, the mode of physicochemical signal transduction (see Table 2).

Biosensors offer several advantages such as small size, low cost, and selective and fast measurements even in complex environments, allowing direct applications in the field or in remote monitoring. While these advantages can be

significant, the potential disadvantages cannot be ignored. Owing to the instability of biomaterials biosensor reliability is reduced as measurements are repeated, thereby requiring frequent replacement of the sensing film. Additionally nonspecific binding to the recognition element can result in inactivation.

4 POTENTIAL APPLICATIONS

The food and drinks industry needs suitable methods for monitoring and control of key parameters

Table 1. Main biological recognition elements used for biosensors

	Biological component	Advantages	Disadvantages
Biocatalytic	Enzymes Catalytic transformation Inhibition	Simple in design and operation	Sensitive to ambient conditions
Bioaffinity	Antibody–Antigen Direct assay Indirect assay	Specificity Rapid	Cross reactions Regeneration
	Oligonucleotides Hybridization to species-specific sequence	High specificity Rapid	Preamplification
Microorganism based	Whole cells Increase in or inhibition of cellular respiration	Inexpensive Measures bioavailable fraction	Long response times Short lifetime
	Promoter regulation resulting in a gene expression		Variance

Table 2. Classification of biosensors according to the mode of the physicochemical transducers

Principle	Transducer	Detection	Example
Electrochemical	Amperometric	Detection of movement of electrons produced in a redox reaction	Pesticides[7] Phytase[8] Choline[9]
	Potentiometric	Detection of changes in the distribution of charges causing an electrical potential to be produced	Histamine[10]
	Conductometric	Detection of change of conductivity of the medium when microorganisms metabolize uncharged substrates	Pesticides in water[11]
Thermal	Calorimetric	Detection of the heat output (or absorbed) by the reaction	Process monitoring; enzyme activity[12] *E. coli* O157:H7[13]
Optical	SPR	Detection of changes in the plasmon resonance as a change in the angle of the incident light or shift in the wavelength of light absorbed	β-casein in milk and cheese[14] Protein in cell lysate[15]
	Chemi/bioluminescence	Detection of the change in the intensity of emitted light	Nitrate in water[16] Histamine in fish[17]
Acoustic	Piezoelectric	Detection of effects due to the mass of the reactants or products	GMOs[18] Toxic chemicals[19]

SPR: surface plasmon resonance; GMOs: genetically modified organisms.

not only for the characterization of the final product, but also for monitoring during the manufacturing process for process optimization and control. Additionally, the analytical monitoring should also deal with chemical and microbiological changes in food during storage and transport. Food analysis has to deal with both a wide range of analytes and within a variety of different matrices that consequently results in the use of a complex range of analytical methods. Since the pioneering work of the 1960s[20,21] there has been a phenomenal growth in the field of biosensors.

Table 3 shows a general survey of the great number of components to be determined. However, the difficulty of food analysis arises not only from the variety but also from the complexity of the food materials. Food products might be liquids, pastes, or solids, the last two forms being incompatible for direct analysis by most of the common existing analytical techniques. This requires the use of suitable extraction methods to bring the analyte of interest into a soluble form and secondly to remove some matrix effects which could mask the analyte or interfere with the detection method.

5 COMMERCIALLY AVAILABLE BIOSENSORS

Despite the great number of publications on biosensors applied in food analysis, only a few systems are commercially available. Key factors influencing the market are food safety requirements and legislation that may require due diligence testing. However it should be noted that quality assurance of raw materials is primarily supported by traceability systems owing to the sampling issues associated with large volume trading in bulk commodities.

The application and commercialization of biosensor technology has lagged behind the output of research laboratories. Although many biosensor-related patents are fielded each year, very few play a prominent role in the food and drinks industry. There have been many reasons for the slow technology transfer from the research laboratories to the marketplace: limited lifetime of the biological components, mass production, quality assurance, and instrumentation design. However, the important one is the lack of effective, systematic commercialization strategies.[37]

Studies of the market estimated that 96% of commercial biosensors are sold in the clinical

Table 3. Examples of biosensor applications in the food and beverage industry

Analyte	Function/property	Application area
Organic		
Alcohols	Indicator for quality	Process control[22,23]
Carbohydrates	Nutritional value	General analysis
Vitamins	Marker of nutritional damage during heat treatment	Quality control[24–26]
Proteins	Evidence of authenticity	Authenticity[27,28]
Inorganic		
Sulfite	Preservatives	Quality control[29]
Toxins		
Shellfish toxins	Contaminants/residues	Food safety[30–32]
Mycotoxins		
Pesticides		
Pathogens		
Salmonella	Food spoilage	Food safety[30]
Hepatitis A		
Toxic plankton	Contamination with toxins	
Pesticides		
Paraxon, carbaryl	Crop protection	Food safety[33,34]
Hormones		
Progesterone, clenbuterol	Growing promoters	Food safety[35]
Antibiotics		
Penicillin G, tetracyclines	Veterinary drugs	Food safety[36]

diagnostic sector, and only 4% are commercialized for use in the industrial biotechnology sector, in the food industrial sector, and in the environmental and wastewater sector.[37] This high difference might be at least be partially explained by the type of matrix in which the tests are carried out: medical biosensors are applied mainly on serum samples with a low variation of the matrix composition. On the other hand, biosensors in the food and drinks industry must be adapted to a wide range of diverse matrices, each of them associated with particular difficulties in terms of matrix effects and concentration range of interest.

After Clark and Lyon developed the glucose analyzer in 1962 on the basis of the amperometric detection of hydrogen peroxide, Yellow Springs Instruments Company put great efforts into biosensor commercialization, development, and production and entered successfully the market in 1979. This was the first of the many biosensor-based analyzers to be built by companies around the world. Since then, biosensors for determination of food components, pathogens, toxins, and pesticides have become commercially available, even if only in limited numbers. Some examples of commercially available biosensors and a summary of their characteristics are presented in Table 4.

Even though the commercially available biosensors for quantifying food components, pathogens, and toxins are in continuous development, the food and drinks industry is still not very receptive to biosensor technology.

Most of the biosensors available on the market are enzyme-based, but antibody, receptor, or DNA-based sensors have also been commercialized.

Bacteria can be detected by direct conductometry, which is achieved by monitoring changes in the growth medium or by indirect conductometry, which monitors changes due to evolution of CO_2 produced by the metabolism of substrates in the culture medium. Direct conductometry has been used extensively to detect both pathogens and nonpathogens, like *Salmonella*, *Campylobacter*, yeasts, or fungi, in foods.[50–54] Several instruments that monitor the change in electrical conductance

Table 4. Examples of commercially available biosensors for use in the food and beverage industry

Manufacturer	Instrument	Target compounds	Comments	Food sample	References
YSI Inc.	YSI 2700 SELECT™	Glucose	Real-time analysis of batches of up to 24 samples or on-line monitoring possible 1 min analysis time	Molasses	38
		(Dextrose) Sucrose Lactose L-Lactate L-Glutamate Ethanol Starch		Potatoes Cereals	
Biacore® AB	Biacore® Q	Folic acid	Real-time, automatic analysis of batches up to 40 samples Optimized, ready-to-use Qflex™ kits	Cereals	39
		Biotin Vitamin B_2 and B_{12} Pantothenic acid Veterinary drug residues Antibiotics Growth promoters		Milk powder Infant food Multivitamin Meat Honey Milk	
Texas Instruments Inc.	Spreeta™	Ingredients Contaminations	Hand-portable equipment Real time	Water Beverages	40
Research International Inc.	Analyte 2000™	E. coli O157:H7	Real time One sample per second	Hamburger	41
Distell	Distell Fish Fatmeter Distell Meat Fatmeter Distell Fish Freshness Meter	Water Water Appropriate dielectric properties	Hand-portable equipment Real time	Fish Meat Fish	42

Company	Product	Analytes	Description	Application	Reference
Sensolytics GmbH	OLGA	Glucose Lactate Sucrose Ethanol Glutamate Glutamine	On-line automated sterile filtration process monitoring and control	Beer	43
Chemel AB	SIRE®	Glucose Sucrose L-Lactate Methanol Ethanol H_2O_2 Ascorbic acid	Fast, easy, and accurate biochemical analyses	Soft drinks Beer Fermentation broths Yogurt	44
Universal Sensors Inc.	ABD 3000	Alcohols sugars Amino acids	Specific, fast, and inexpensive biosensor assay	Food Beverage	45
Innovative Biosensors Inc.	BioFlash™ system	E. coli O157:H7	Extremely rapid detection of pathogens at previously unseen levels of sensitivity and specificity	Lettuce	46
Malthus Instruments Ltd.	Malthus Systems	E. coli O157:H7 Fungi Yeasts	Conductometric assay using whole cells	Shellfish Evaluation of antifungal activity	47
Don Whitley Scientific Ltd.	RABIT	Food pathogens	Conductometric assay using whole cells	Vegetables	48
OptiSense BV	OptiSense Optical biosensor	Antibiotics Mycotoxins	Highly sensitive real-time test	Dairy industry	49

RABIT: rapid automated bacterial impedance technique.

of the media caused by the growth of microorganisms have been exploited by Malthus Instruments Ltd. or in RABIT by Don Whitley Scientific Ltd.

The SPR principle allows real-time detection of the specific interaction between an immobilized biorecognition element and a variety of analytes. Companies like Biacore AB or Texas Instruments Inc. have exploited SPR technology.[55,56] Biosensors are available in several forms, such as autoanalyzers, manual laboratory instruments, and portable devices.

6 UPCOMING TECHNIQUES, FUTURE DEVELOPMENTS IN BIOSENSOR TECHNOLOGY AND MARKETPLACE

Some promising developments in the biosensor field are listed in Table 5. These recent developments have already entered the market or will be available soon.

The most promising breakthroughs are to be expected in the area of sensor technology, that will allow the creation of on-line or on-site, sensitive, low-cost devices for routine use. Biosensors have a high potential for automation and allow the construction of simple and portable equipment for fast analysis. These properties will open up many new applications within quality and process control, including fermentation and quality and safety control of raw materials. The new application possibilities offered should be further explored and technologically evaluated.

Biosensor advancement in the commercial world could be accelerated by the use of intelligent instrumentation, electronics, and multivariate signal-processing methods. A biosensor array strategy, adaptable to multiple analytes detection, will allow the spread of development costs over several products. These future improvements will produce devices more competitive than the presently available instruments.

7 CONCLUSIONS

This chapter summarizes ongoing efforts, trends, and developments in the field of biosensors for food analysis.

Food has an important role in promoting and invigorating health and quality of life. To meet consumer requirements and provide healthy and high-quality food, the production- and processing-distribution chain has to be carefully monitored. Moreover, tracing possible sources of contaminants throughout the food chain and quantifying risk factors is also critical.

There is a great need for rapid and affordable methods to assure quality of products and process control in the food and beverage sector. The application of the biosensor techniques in the field of food processing and quality control offers advantages as alternatives to conventional methods due to speed, cost efficiency, high sensitivity, and specificity of measurements. Biosensors are a tool for determining and measuring a wide range of target analytes, occurring in a variety of food matrices, and within a diversity of process conditions.

Biosensors will complement or replace existing conventional analytical methods. However, as with most technologies, there is still room for improvement. The stability of the recognition element can be improved by using new immobilization technologies and developing novel protein stabilization, engineering, and automated manufacturing technologies, as well as by using more robust biological or nonbiological materials, like molecular imprinted polymers (MIPs).[63] But for a

Table 5. Examples for upcoming biosensors in the near future

Company	Development	Aim	References
AKUBIO Ltd.	RAP◆id™4	Resonant Acoustic Profiling (RAP™) technology for molecule interactions in complex matrices like serum or growth media	57
Axela Biosensors, Inc.	DOT™ sensor and DOT™ reader	Applications are potentially in agriculture, environmental, and food and beverage QC	58
Biophage Pharma Inc.	Phage Biosensor	*E. coli* O157:H7, *Campylobacter, Salmonella* in water	59
Universal Sensors, Ltd.	UTS™ technology	Aqueous-based samples	60
Sensortec, Ltd.			61
Stratophase, Ltd.	Refractive index sensor chips	Liquid samples	62

direct competition with established analytical procedures, biosensors must be designed to measure several different analytes in parallel using a single analyzer without having to change any component of the sensor system.[2]

Furthermore, the clearly defined and well-financed needs in medicine have driven some of the most successful biosensor developments to date, such as home blood glucose monitors. The food and beverage industry, however, has been characterized by low profit margins and a somewhat conservative approach to new technology. Public concern for food safety and increased demands for food labeling is likely to provide more impetus for innovative approaches to food analysis in the future.[64]

The promise shown by biosensor technology is real, but there are still technological problems to be solved. Additionally, the market penetration has to be improved for areas where biosensor technologies are ideal for upgrading food diagnostics. These potential opportunities include better quality control, on-line monitoring of processes and their control as well as seminal markets of authenticity and traceability of raw materials, intermediates, and food products.

In conclusion, as concerns regarding safe food and water supply increase, the demand for rapid detecting biosensors will only increase.

REFERENCES

1. U.S. & Worldwide Biosensor Market, *R&D and Commercial Implication*, Fuji-Keizai USA, Incorporated, Pub ID: FJ9744212004, 2004.
2. J. H. T. Luong, P. Bouvrette, and K. B. Male, Developments and applications of biosensors in food analysis. *Trends in Biotechnology*, 1997, **15**, 369–377.
3. *Food Micro-2005*, Strategic Consulting Incorporated, 2005.
4. I. H. McMurdo and S. Whyard, Suitability of rapid microbiological methods for the hygienic management of spray drier plant. *Journal of the Society of Dairy Technology*, 1984, **37**, 4–9.
5. D. R. Thevenot, K. Toth, R. A. Durst, and G. S. Wilson, Electrochemical biosensors: recommended definitions and classification. *Biosensors and Bioelectronics*, 2001, **16**, 121–131.
6. S. White, *Biosensors for Food Analysis*, in *Handbook of Food Analysis*, L. M. L. Nollet (ed), Marcel Dekker, Inc., New York, 2004, pp. 2133–2148.
7. M. Waibel, H. Schulze, N. Huber, and T. T. Bachmann, Screen-printed bienzymatic sensor based on sol-gel immobilized *Nippostrongylus brasiliensis*

8. acetylcholinesterase and a cytochrome P450 BM-3 (CYP102-A1) mutant. *Biosensors and Bioelectronics*, 2006, **21**, 1132–1140.
8. W. C. Mak, Y. M. Ng, C. Chan, W. K. Kwong, and R. Renneberg, Novel biosensors for quantitative phytic acid and phytase measurement. *Biosensors and Bioelectronics*, 2004, **19**, 1029–1035.
9. S. Pati, M. Quinto, F. Palmisano, and P. G. Zambonin, Determination of choline in milk, milk powder, and soy lecithin hydrolysates by flow injection analysis and amperometric detection with a choline oxidase based biosensor. *Journal of Agricultural and Food Chemistry*, 2004, **52**, 4638–4642.
10. K. M. May, Y. Wang, L. G. Bachas, and K. W. Anderson, Development of a whole-cell-based biosensor for detecting histamine as a model toxin. *Analytical Chemistry*, 2004, **76**, 4156–4161.
11. S. Suwansa-Ard, P. Kanatharana, P. Asawatreratanakul, C. Limsakul, B. Wongkittisuksa, and P. Thavarungkul, Semi disposable reactor biosensors for detecting carbamate pesticides in water. *Biosensors and Bioelectronics*, 2005, **21**, 445–454.
12. K. Ramanathan and B. Danielsson, Principles and applications of thermal biosensors. *Biosensors and Bioelectronics*, 2001, **16**, 417–423.
13. S. M. Radke and E. C. Alocilja, A high density microelectrode array biosensor for detection of *E. coli* O157:H7. *Biosensors and Bioelectronics*, 2005, **20**, 1662–1667.
14. S. Muller-Renaud, D. Dupont, and P. Dulieu, Quantification of beta casein in milk and cheese using an optical immunosensor. *Journal of Agricultural and Food Chemistry*, 2004, **52**, 659–664.
15. M. Kyo, K. Usui-Aoki, and H. Koga, Label-free detection of proteins in crude cell lysate with antibody arrays by a surface plasmon resonance imaging technique. *Analytical Chemistry*, 2005, **77**, 7115–7121.
16. A. G. Prest, M. K. Winson, J. R. Hammond, and G. S. Stewart, The construction and application of a lux-based nitrate biosensor. *Letters in Applied Microbiology*, 1997, **24**, 355–360.
17. Y. Sekiguchi, A. Nishikawa, H. Makita, A. Yamamura, K. Matsumoto, and N. Kiba, Flow-through chemiluminescence sensor using immobilized histamine oxidase from *Arthrobacter crystallopoietes* KAIT-B-007 and peroxidase for selective determination of histamine. *Analytical Sciences*, 2001, **17**, 1161–1164.
18. I. Mannelli, M. Minunni, S. Tombelli, and M. Mascini, Quartz crystal microbalance (QCM) affinity biosensor for genetically modified organisms (GMOs) detection. *Biosensors and Bioelectronics*, 2003, **18**, 129–140.
19. J. H. Sung, H. J. Ko, and T. H. Park, Piezoelectric biosensor using olfactory receptor protein expressed in Escherichia coli. *Biosensors and Bioelectronics*, 2006, **21**, 1981–1986.
20. L. Clark and C. Lyons, Electrode systems for continuous monitoring in cardiovascular surgery. *Annals of the New York Academy of Sciences*, 1962, **102**, 29–35.
21. S. Updike and G. Hicks, Enzyme electrode. *Nature*, 1967, **214**, 986.

22. M. Boujita, M. Chapleau, and N. El Murr, Biosensors for analysis of ethanol in food: effect of the pasting liquid. *Analytica Chimica Acta*, 1996, **319**, 91–96.

23. A. G.-V. De Prada, N. Pena, M. L. Mena, A. J. Reviejo, and J. M. Pingarron, Graphite-teflon composite bienzyme amperometric biosensors for monitoring of alcohols. *Biosensors and Bioelectronics*, 2003, **18**, 1279–1288.

24. H. E. Indyk, E. A. Evans, M. C. Bostrom Caselunghe, B. S. Persson, P. M. Finglas, D. C. Woollard, and E. L. Filonzi, Determination of biotin and folate in infant formula and milk by optical biosensor-based immunoassay. *Journal of AOAC International*, 2000, **83**, 1141–1148.

25. H. E. Indyk, B. S. Persson, M. C. Caselunghe, A. Moberg, E. L. Filonzi, and D. C. Woollard, Determination of vitamin B12 in milk products and selected foods by optical biosensor protein-binding assay: method comparison. *Journal of AOAC International*, 2002, **85**, 72–81.

26. D. Dupont, O. Rolet-Repecaud, and S. Muller-Renaud, Determination of the heat treatment undergone by milk by following the denaturation of alpha-lactalbumin with a biosensor. *Journal of Agricultural and Food Chemistry*, 2004, **52**, 677–681.

27. W. Haasnoot, K. Olieman, G. Cazemier, and R. Verheijen, Direct biosensor immunoassays for the detection of nonmilk proteins in milk powder. *Journal of Agricultural and Food Chemistry*, 2001, **49**, 5201–5206.

28. W. Haasnoot, N. G. Smits, A. E. Kemmers-Voncken, and M. G. Bremer, Fast biosensor immunoassays for the detection of cows' milk in the milk of ewes and goats. *Journal of Dairy Research*, 2004, **71**, 322–329.

29. D. Corbo and M. Bertotti, Use of a copper electrode in alkaline medium as an amperometric sensor for sulphite in a flow-through configuration. *Analytical and Bioanalytical Chemistry*, 2002, **374**, 416–420.

30. A. X. J. Tang, M. Pravda, G. G. Guilbault, S. Piletsky, and A. P. F. Turner, Immunosensor for okadaic acid using quartz crystal microbalance. *Analytica Chimica Acta*, 2002, **471**, 33–40.

31. J. S. Yoo, B. S. Cheun, I. S. Park, Y. C. Song, Y. Seo, N. G. Kim, H. W. Shin, and J. H. Lee, Use of sodium transfer tissue biosensor (STTB) for monitoring of marine toxic organism. *Journal of Environmental Biology*, 2004, **25**, 431–436.

32. L. Pogacnik and M. Franko, Detection of organophosphate and carbamate pesticides in vegetable samples by a photothermal biosensor. *Biosensors and Bioelectronics*, 2003, **18**, 1–9.

33. H. Schulze, E. Scherbaum, M. Anastassiades, S. Vorlova, R. D. Schmid, and T. T. Bachmann, Development, validation, and application of an acetylcholine-esterase-biosensor test for the direct detection of insecticide residues in infant food. *Biosensors and Bioelectronics*, 2002, **17**, 1095–1105.

34. Y. Zhang, S. B. Muench, H. Schulze, R. Perz, B. Yang, R. D. Schmid, and T. T. Bachmann, Disposable biosensor test for organophosphate and carbamate insecticides in milk. *Journal of Agricultural and Food Chemistry*, 2005, **53**, 5110–5115.

35. M. A. Johansson and K. E. Hellenas, Sensor chip preparation and assay construction for immunobiosensor

36. E. Gustavsson, J. Degelaen, P. Bjurling, and A. Sternesjo, Determination of beta-lactams in milk using a surface plasmon resonance-based biosensor. *Journal of Agricultural and Food Chemistry*, 2004, **52**, 2791–2796.

37. C.-T. Lin and S.-M. Wang, Biosensor commercialization strategy—a theoretical approach. *Frontiers in Bioscience*, 2005, **10**, 99–106.

38. http://www.ysi.com (2007).

39. http://www.biacore.com (2007).

40. http:// www.spreeta.com (2007).

41. http://www.resrchintl.com (2007).

42. http://www.distell.com (2007).

43. http://www.sensolytics.com (2007).

44. http://www.chemel.com (2007).

45. http://chemweb.ucc.ie/ (2007).

46. http://www.innovativebiosensors.com (2007).

47. http://www.labm.com (2007).

48. http://www.dwscientific.co.uk (2007).

49. http://www.optisense.nl/ (2007).

50. F. J. Bolton, *Conductance and Impedance Methods for Detecting Pathogens*, in *International Congress on Rapid Methods and Automation in Microbiology and Immunology*, A. Vaheri, R. C. Tilton, and A. Barlows (eds), 1991, Springer-Verlag, pp. 176–181.

51. J. Dupont, F. Dumont, C. Menanteau, and M. Pommepuy, Calibration of the impedance method for rapid quantitative estimation of Escherichia coli in live marine bivalve molluscs. *Journal of Applied Microbiology*, 2004, **96**, 894–902.

52. J. Sawai and T. Yoshikawa, Quantitative evaluation of antifungal activity of metallic oxide powders (MgO, CaO and ZnO) by an indirect conductimetric assay. *Journal of Applied Microbiology*, 2004, **96**, 803–809.

53. J. Sawai and T. Yoshikawa, Measurement of fungi by an indirect conductimetric assay. *Letters in Applied Microbiology*, 2003, **37**, 40–44.

54. J. E. Line and K. G. Pearson, Development of a selective broth medium for the detection of injured *Campylobacter jejuni* by capacitance monitoring. *Journal of Food Protection*, 2003, **66**, 1752–1755.

55. A. N. Naimushin, S. D. Soelberg, D. K. Nguyen, L. Dunlap, D. Bartholomew, J. Elkind, J. Melendez, and C. E. Furlong, Detection of *Staphylococcus aureus* enterotoxin B at femtomolar levels with a miniature integrated two-channel surface plasmon resonance (SPR) sensor. *Biosensors and Bioelectronics*, 2002, **17**, 573–584.

56. S. D. Soelberg, T. Chinowsky, G. Geiss, C. B. Spinelli, R. Stevens, S. Near, P. Kauffman, S. Yee, and C. E. Furlong, A portable surface plasmon resonance sensor system for real-time monitoring of small to large analytes. *Journal of Industrial Microbiology and Biotechnology*, 2005, **32**, 669–674.

57. http://www.akubio.com/ (2007).

58. http://www.axelabiosensors.com (2007).

59. http://www.biophage.com (2007).

60. http://www.universalsensors.co.uk (2007).

61. http://www.sensortec.uk.com (2007).

62. http://www.stratophase.com (2007).

determination of beta-agonists and hormones. *The Analyst*, 2001, **126**, 1721–1727.

63. S. A. Piletsky, S. Alcock, and A. P. Turner, Molecular imprinting: at the edge of the third millennium. *Trends in Biotechnology*, 2001, **19**, 9–12.

64. L. A. Terry, G. Volpe, G. Palleschi, and A. P. F. Turner, Biosensors for the microbiological analysis of food. *Food Science and Technology*, 2004, **18**, 22–24.

79

Agriculture, Horticulture, and Related Applications

Leon A. Terry

Plant Science Laboratory, Cranfield University, Silsoe, UK

1 INTRODUCTION

The application of biosensor technology is still fundamentally dominated by the blood glucose market. Transfer to the food sector has been slow and still represents a very small proportion of current sales. However, with the increasing trend from yield-driven to quality-driven provision of agricultural products, the concurrent demand for reliable and inexpensive methods for assessment of food quality, safety, traceability, and authenticity is set to expand; biosensors have the potential to fulfill this niche.[1]

Traditionally, agriculture and the food industry, in general, have taken a particularly conservative approach to the introduction of biosensors. Fundamentally, continued price deflation within the agricultural sector has discouraged investment despite rapid growth in food testing over recent years. Secondly, the problems posed by ensuring representative sampling and sample interference on sensor performance have often been overlooked during the development stage for sensors. This oversight or lack of recognition of the complexity of many foodstuffs has invariably inhibited the widespread introduction of biosensors within the agricultural sector. However, it is not necessarily the inherent specificity, selectivity, and adaptability of biosensors which make them ideal candidates for use throughout the food industry, but their relative low cost and ease of use, as compared with more conventional testing methodologies (e.g., HPLC, GC/MS), that really make them an attractive alternative quality assurance tool.

2 RECENT ADVANCES IN BIOSENSOR DEVELOPMENT FOR FOODSTUFF QUALITY ASSURANCE

A number of biosensor-based devices have been developed for agriculture and horticulture to quantify the concentration/presence of particular target analytes that may be indicators of food quality/acceptability or assist in selection (Tables 1 and 2). These devices have principally either relied on immunosensors or electrochemical biosensors.

2.1 Monitoring Livestock Condition

Reproductive management is a major financial concern of the livestock industry. For instance, predicting estrus onset in cattle, where artificial techniques are employed, can lead to considerable cost saving in herd management.[26] Despite visual observations still being used to detect estrus, it is recognized that considerable inaccuracies in predicting this time window are common since estrus does not always coincide with ovulation

Handbook of Biosensors and Biochips. Edited by Robert S. Marks, David C. Cullen, Isao Karube, Christopher R. Lowe and Howard H. Weetall.
© 2007 John Wiley & Sons, Ltd. ISBN 978-0-470-01905-4.

Table 1. Examples of the range of analytes monitored in food matrices other than fresh produce by amperometric biosensors

Analyte	Food matrix	Enzyme	Detection limit	References
Essential fatty acids	Fats and oils	Lipoxygenase, lipase, and esterase	0.04 mM in an FIA system	2
Lysine	Range of foods	Lysine oxidase	1×10^{-5} mol l^{-1}	3
Glucose and maltose	Beer	Glucose oxidase and amyloglucosidase	40 mM to glucose (only upper limit stated)	4
Glucose and glutamate	Beverages	Glucose oxidase and glutamate oxidase	10 μM for glucose and 3 μM for glutamate	5
Rancidification indicators	Olive oils	Tyrosinase	0.2–2.0 μM in different oils	6
Lactate	Wine and yogurt	Lactate oxidase	1.4×10^{-6} mol l^{-1}	7
D- and L-amino acids	—	Amino acid oxidase	0.1 or 0.2 mM for L- and D-amino acids	8
Choline	Dairy produce	Choline oxidase	5 μmol l^{-1}	9
Organophosphorous pesticide	Range of foods	Acetylcholinesterase	0.2–1.8 ppm	10
Insecticide residues	Infant food	Acetylcholinesterase	5 μg Kg^{-1}	11
Laminarin	Seaweed	β1,3-Glucanase and glucose oxidase	50 μg ml^{-1}	12
Alcohol	Beer and wine	Alcohol oxidase and horseradish peroxidase	5.3×10^{-6} mol l^{-1}	13

Adapted from Terry et al. 2005.

Table 2. Examples of the range of analytes monitored in fresh produce matrices by amperometric biosensors

Analyte	Food matrix	Enzyme	Detection limit	References
Fructose	Citrus fruits	Fructose dehydrogenase	10 μM	14
Amines	Apricots and cherries	Diamine oxidase and polyamine oxidase	2×10^{-6} mol l^{-1}	15
L-ascorbic acid	Fruit juices	Ascorbate oxidase	5.0×10^{-5} M	16
Sucrose	Fruit juices	Sucrose phosphorylase, phosphoglutaminase, glucose-6-phosphate 1-dehydrogenase	9.25 g l^{-1} in pineapple juice	17
Malic acid	Apples, Potatoes, and tomatoes	Malate dehydrogenase	0.028 mM	18
Polyphenols	Vegetables	Horseradish peroxidase	1 μmol l^{-1}	19
β-D-glucose; total D-glucose; sucrose; L-ascorbic acid	Tropical fruits (mango, pineapple, papaw)	Glucose oxidase; glucose oxidase, mutarotase; invertase, mutarotase, and glucose oxidase; ascorbate oxidase	7 mM	20,21
Cysteine sulfoxides	Alliums (e.g., onion and garlic)	Allinase	5.9×10^{-6} M	22
Pyruvic acid	Onion	Pyruvate oxidase	2 μmol g^{-1}	23–25

Adapted from Terry et al. 2005.

and can even occur in pregnant animals. As an alternative, sensors have been developed to detect bovine progesterone during milking.[27,28] Both approaches adopted a liquid handling system linked to a suitable immunoassay sensor. The approach adopted by Pemberton et al.[28] which relied upon a reduction in the binding to the sensor surface of alkaline phosphate–labeled progesterone in the presence of endogenous milk progesterone, was further improved by Mottram et al.[29] who demonstrated that an automated ovulation system was capable of detecting concentrations of progesterone between 3 and 30 ng ml^{-1} in whole fresh milk. Similarly, flow injection analysis (FIA) coupled to an electrochemical-based immunosensor system has been developed for the detection of gentamicin in milk.[30] Again the aim was to develop a field-based system that could be operated at-site. The system was able to distinguish between $100\,\mu g\,kg^{-1}$ gentamicin in milk in <10 min with no sample pretreatment required.

In addition to the use of antibody-based immunosensors, receptor-based devices have also been described. Setford et al.[31] employed an amperometric affinity sensor for the rapid quantification of β-lactam in milk. Biorecognition was achieved using an immobilized β-lactam antibiotic specific receptor binding protein to measure penicillin G levels in milk. This device was designed to be a one-shot disposable sensor, based on screen-printed sensors, making it an ideal field-based screening tool. Similarly, an enzyme immunoassay for β-lactam penicillins was reported by Delwiche et al.[32] but rather than using an ELISA affinity assay couple to amperometric determination of bound enzyme label activity,[31] the system used involved a photometric sensor as the transducer.

2.2 Improving Fresh Produce Quality Control

The fresh produce industry illustrates many of the problems encountered across the whole food industry in terms of still generally employing archaic quality control (QC) methodologies. For example, fresh produce quality in the intact or minimally processed/fresh-cut form is initially assessed by sight: other important quality attributes include taste, smell, and texture. Each of these four quality attributes can be assessed either subjectively or objectively. Typically, fruit processors reject approximately 10% of fruit intakes. Better selection through improved quantitative QC at low cost will inevitably result in improved overall quality for intact and minimally processed fruit products. Improved QC will probably, in the short-term, lead to increased rejections and reduce the number of "concessions" for fruit processors. A "concession" is where fruit material can be used but is expected to incur a cost penalty due to greater QC costs. Improved QC may result in approximately 25% saving on concessions for fruit processors.

In reality, current standard product-orientated QC operations are inadequate, and consider only appearance (e.g., colour, size, and shape), presence/absence of disease, and the concentration of total soluble solids (TSS). TSS is commonly expressed as degrees Brix (°Bx) and is typically still measured using a hand-held refractometer. There is often poor correlation between TSS and total sugar concentration.[33] Fruit sugars are one of the main soluble components in fresh produce that are important for flavor. In addition to sugar composition, fruit acid concentration can affect flavor directly and can regulate cellular pH, influencing the appearance of fruit pigments within the tissue during processing. The total titratable acidity (TTA) of fruit is not routinely measured as part of the standard QC procedures that are implemented by growers, suppliers, fresh produce distribution centres, and fruit processors. Titratable acidity is a measure of the buffering capacity of the fruit and is generally expressed as a percent of the predominant organic acid. Current standard QC operations do not use TTA because of the cumbersome and time-consuming nature of titrations. Fruit sugar/acid ratios can be used as an important index of consumer acceptability and act as one determinant of overall fruit quality. However, sugar/acid ratios are not frequently assessed for all fruit types due to primitive QC instrumentation and the requirement for skilled analytical scientists.

An initial step to improving routine QC assessment would constitute producing a simple and low-cost alternative to refractometry and titrations so that specific sugar and organic acid ratios can be standardized for fresh produce types. Biosensors may offer the opportunity to fulfill this niche and allow industry to adjudge fruit quality on the basis of taste (sugar/acid ratios) rather than just appearance alone. Introducing biosensor technology within the fresh produce industry (Table 2) may provide the ideal solution to providing improved QC, safety, and traceability methodologies.[20,25] It follows that biosensor applications could be extended across the whole food industry, for example, meat, dairy foods, and beverages (Table 1).

3 ADVANTAGES OF BIOSENSORS FOR FOOD ANALYSIS

Unquestionably, biosensors have made their greatest impact in the field of medical diagnostics. The use of screen-printed electrochemical biosensors

Table 3. Examples showing the use of biosensors to detect the presence of contaminating microorganisms in food

Target organism	Food matrix	Detection method	Detection limit	References
Escherichia coli	Dairy products	Acoustic	$3 \times 10^5 - 6.2 \times 10^7 \, ml^{-1}$	34
Staphylococcal enterotoxin A	Hot dogs, potato salad, milk, and mushrooms	Optical	$10-100 \, ng \, g^{-1}$	35
Salmonella typhimurium	Chicken carcass wash fluid	Optical	$1 \times 10^5 - 1 \times 10^7 \, ml^{-1}$	36
Staphylococcal enterotoxin B	Milk	Optical	$0.5 \, ng \, ml^{-1}$	37
Salmonella group B, D and E	Range of foods	Optical	$1 \times 10^7 \, CFU \, ml^{-1}$	38
E. coli	Range of foods	QCM (acoustic)	$1.7 \times 10^5 - 8.7 \times 10^7$ $CFU \, ml^{-1}$	39
E. coli O157:H7	Range of foods	Optical	$1 \times 10^3 \, CFU \, ml^{-1}$	40

QCM: quartz crystal microbalance.

by diabetics to regularly monitor their blood glucose levels has provided sufferers with a powerful method for controlling this pernicious condition. Using modern printing methods, these devices are manufactured on a scale of millions per month and sold globally. The accuracy, comparative low cost, and ease of use of biosensors have led to their widespread application. Adapting this technology for use not only for fresh produce (Table 2) but also in the wider food industry could lead to immense improvements in QC, food safety (*cf.* Table 3), and traceability. Abayomi et al.[25] demonstrated that pungency in bulb cvs. Renate and SupaSweet (SS1) onions (as measured by pyruvate concentration in macerated tissue) could be determined 20-fold more rapidly using a mediated biosensor format in comparison with the standard colorimetric assay used by the industry[41] (Figures 1 and 2) with no loss in resolution. Moreover, biosensors have been developed for determination of concentration of metabolites such as glucose, sucrose, lactate, alcohol, glutamate, and ascorbic acid, typically found in many food items (Figure 3)[42] and for rapidly detecting the presence of contaminating agents such as microorganisms, pesticide residues,[43] and antibiotics.[44,45]

Biosensor systems can be operated either as simple "one-shot" measurement tools or, when incorporated into a suitable fluid handling system, as multimeasurement devices. Both approaches can also be adapted to measure several different analytes, using the same sample solution. This versatility, coupled with a high degree of sensitivity and selectivity has prompted worldwide interest in both the fundamental research and commercial exploitation of biosensor technology. Biosensor systems can be designed such that they can be operated

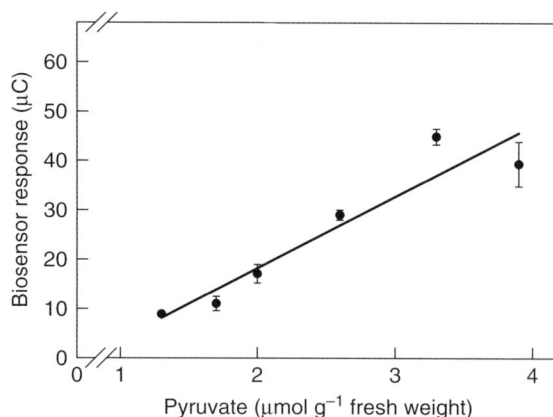

Figure 1. Mediated biosensor response to onion juices from six individual low pungency bulbs of increasing pyruvate concentration verified against conventional colorimetric analysis.[39] $R^2 = 0.83$; $y = 14x - 9$; $P < 0.001$. Standard error bars are from the mean of three experiments. $+200 \, mV$; phosphate buffer pH 5.7; cofactor mix B. [Reprinted with permission Abayomi et al.[25] copyright 2006, Elsevier.]

at-site on a real-time basis, removing the reliance on expensive centralized laboratory-based testing. Moreover, the process of miniaturization can be adapted to biosensors. Hence, an array of sensors can be integrated into a small portable device for multiple parameter determination for use by nonspecialized persons with a minimum of manual manipulation. This is one of the major advantages of using biosensors, as measurements can be made either during raw material preparation, food processing (e.g., as QC devices) or for checking the reliability of storage conditions. Hence, these devices can act as cost-effective tools for QC, process control, and the determination of food safety.

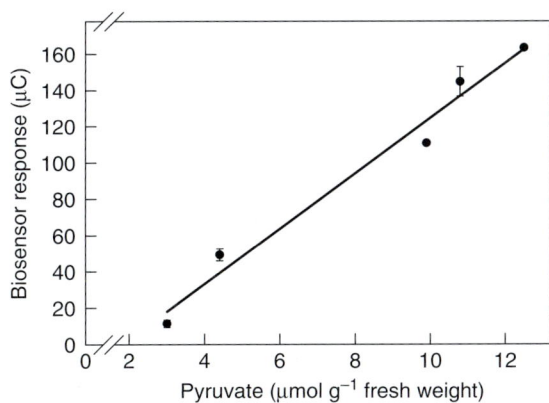

Figure 2. Mediated biosensor response to onion juices from two individual mild (cv. SS1) and three pungent (cv. Renate) bulbs of increasing pyruvate concentration verified against conventional colorimetric analysis.[39] $R^2 = 0.97$; $y = 15x - 27$; $P = 0.001$. Standard error bars are from the mean of three experiments. $+200\,$mV; phosphate buffer pH 5.7; cofactor mix B. [Reprinted with permission Abayomi et al.[25] copyright 2006, Elsevier.]

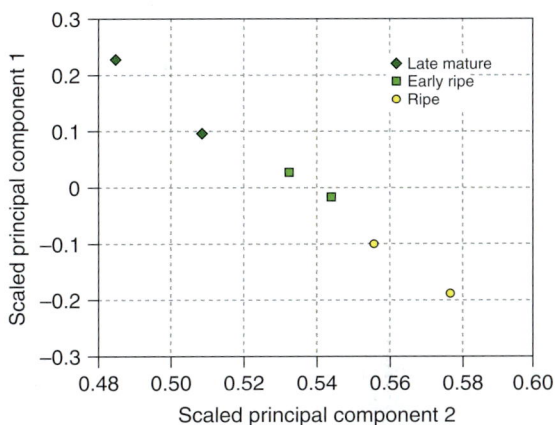

Figure 3. Chemometric (principle component analysis) output from biosensor array to score pineapple cv. Queen Victoria fruit ripeness. PCA scores plot for the scaled data matrix. PC1 and PC2 represent the first and second principal components respectively. [Reprinted with permission Jawaheer et al.[20,21] copyright 2002, 2003, Elsevier.]

Most biosensor research for the food industry has been done using enzyme-based amperometric electrochemical biosensors. However, to obtain functional biosensor devices, which can be manufactured within the necessary performance and cost constraints, other components and technologies must also be considered.

4 BIOSENSOR OPTIMIZATION FOR FOOD ANALYSIS

Almost all biosensors rely on membranes for improved functionality. Membranes can play different roles in the sensor format and are typically used to retain the biological component, while allowing the analyte to pass; one of the key features of a biosensor is the proximity of the biological recognition element to the transducer. Usually this is achieved using an immobilization process. Another useful function of membranes is their ability to extend the linear range of a biosensor by acting as a mass transport barrier. A limiting factor for a linear response from an enzyme-based biosensor may be the K_m of the catalyst. By imposing an analyte diffusion barrier (i.e., a membrane) over the enzyme a *pseudo* K_m may be produced, extending the linear range of the sensor. Membranes can provide a protective barrier for the sensor system, preventing fouling of the sensor by components in the sample matrix and, conversely contamination of the sample solution by the sensor. Membranes can also act to provide stability for the sensor, both for long-term storage and the operational capability of the sensor. By the use of a suitable membrane a high degree of selectivity can be achieved, either by allowing only the target analyte to reach the sensor surface or by eliminating other interfering compounds that may affect the sensor signal. Numerous materials have been used as membrane materials for biosensors including, cellulose acetate (CA), PVC, and Nafion (a polyfluorosulfonated hydrocarbon). Jawaheer et al.[20,21] demonstrated that interferences related to electrochemically active compounds present in tropical fruits could be significantly reduced by inclusion of a suitable CA membrane on a rhodinized carbon electrode. CA membranes improved the linear range of biosensors for β-D-glucose, total D-glucose, sucrose, and L-ascorbic acid by as much as 5-fold compared with sensors without an additional diffusion barrier.

Immobilization of the recognition element on or close to the transducer is a major factor in biosensor design and fabrication. Adopting this approach allows an efficient transfer of signal from the biological element to the transducer and hence to the biosensor user. In addition, with an immobilized biological element, the opportunities for reusing of the biosensor are greatly enhanced. The main methods of immobilization, particularly for enzymes, include physical adsorption, entrapment in a matrix (using gels, polymers, or printing inks), covalent binding, or electrochemical polymerization and photopolymerization. Physical adsorption is generally based on interactions such as van der Waals forces between the biological element and the transducer (e.g., a carbon electrode surface). Jawaheer et al.[21] demonstrated that pectin, a natural polysaccharide present in plant cell walls, could be used as a novel matrix to enhance enzyme entrapment on rhodinized carbon electrodes. Pectin also assisted in prolonging enzyme storage stability rather than improving operation performance and could be applied as a viscous paste of screen-printable consistency.

It has long been realised that advanced fabrication techniques are key to the successful development of commercially viable biosensors. Fortunately, many technologies have been developed for other applications, such as the microelectronics industry, that can be adapted to biosensor fabrication.

Screen-printing is a thick-film process, which has been used for many years in artistic applications and, more recently, for the production of miniature, robust, and cheap electronic circuits. This technique has been successfully exploited by electrochemical biosensor manufactures. The process has been one of the major reasons for the commercial success of many biosensors and is the process by which a number of medical diagnostic companies annually produce more than 1 billion (electrochemical) biosensor strips for home blood glucose monitoring. It follows, therefore, that the inexpensive nature of biosensor fabrication lends itself to an increasingly price-competitive industry, such as the fresh produce sector. Given this fact, it is perhaps surprising that biosensor fabrication techniques have not been transferred to the measurement of important target analytes in the food industry.

The ability to handle small volumes of liquids with high precision will be one of the key areas of development for some of the next generation of biosensors. In particular, where high-value reagents, such as particular enzymes or antibodies, are needed, screen-printing may not (because of cost implications) be the most appropriate method of production. For example, fructose, which normally increases during physiological fruit ripening, is a potentially desirable target for fresh produce biosensor development since it has been positively correlated to perceived sweetness.[46] However, the commercial availability of the fructose enzyme, fructose dehydrogenase, is relatively expensive and unstable and, therefore, not economically viable at present unless significant research is forthcoming. In addition, the same economic barrier exists for malate, which is the principle organic acid found in pome fruit (e.g., apples and pears) and an important parameter characterizing wine quality (e.g., malolactic fermentation). Malic acid measurement can be cost prohibitive when using either the malic acid enzyme (L-malate: $NADP^+$ oxidoreductase) associated with $NADP^+$ and pyruvate oxidase[47] or malate dehydrogenase and diaphorase immobilized on gold electrodes using glutaraldehyde.[48] Other printing methods can be used to overcome high-price enzyme usage, including ink-jet printing and other automated dispensing deposition systems. The deposition of biological agents, such as enzymes, can be carried out accurately and reproducibly using these print methods, which are suitable for depositing droplets of <1 nl in volume. Furthermore, non-contact technology, such as ink-jet printing, allows fluid to be placed on almost any surface, irrespective of texture and shape.

In addition to the variety of biosensor devices available, there are a number of methods that can be adapted to facilitate sampling. By their very nature, many food items are complex mixtures of many compounds; this can present a significant challenge to the efficient operation of a biosensor. Hence, whenever biosensor technologists design sensors for applications in the food industry, careful consideration must be given to sampling. For solid or semi-solid foods this, usually, involves an extraction process, possibly followed by a simple preparation step such as filtration.

Broadly, the main sampling methods that are used with biosensors can be categorized in several ways. At-site sampling involves taking a sample from the matrix and carrying out an extraction process followed by measurement by the biosensor. This approach to sampling can be detrimental in terms of efficiency. Manually removing and pretreating a sample prior to measurement may require some limited degree of technical skill. However, careful design of the sensor system should reduce this complexity to a level at which the procedure can be carried out rapidly and easily. At-site sampling is, probably, more suited to liquids (e.g., beer, wine, and fruit juices) where (generally) the target analyte of interest is more readily available. Similarly, sampling from fresh produce is usually not as challenging as for other more heterogeneous foodstuffs as many of the potential target analytes are in solution when tissue is disrupted/decompartmentalized and, thus, available for measurement in extracted fruit juice.[21,25] This said, significant variation in the spatial and temporal distribution of target analytes does exist in all fresh produce; a fact that reconfirms that sampling procedures must be optimized.

In situ sensors are placed directly in the matrix containing the target analyte. The use of in situ sensors has long been established in the bioprocess industry, where "dip in" devices are used to monitor a number of parameters such as pH and dissolved gas concentrations. A number of advantages are gained by operating sensors in this fashion including, real-time monitoring and a continuous output signal from the sensors (any rapid change in the analyte concentration can be readily observed). In addition, labour requirements are significantly reduced. This approach would be most applicable in the food processing field, where the careful (automatic) sampling and measurement of processed food (e.g., processes that incorporate a fermentation or distillation step) would be very useful. However, given the complex nature of many foods, in situ sampling and monitoring can be very difficult. Components in the food matrix can adhere to and foul the sensor surface, leading to erroneous signals. Calibration of the sensor (in situ) may be difficult. In addition, the sensor operation may be affected by varying conditions, during the process cycle, such as temperature, pH, and salinity.

Methods of on-line sampling involve the automatic removal and measurement of a sample, or sample stream, from the food matrix (e.g., FIA). FIA is a liquid handling technique that has proved flexible in adapting to most chemical and biochemical reaction procedures,[49] representing an effective compromise between the desirability of in situ monitoring and the technical ease of off-line measurements. The use of liquid handling systems can be used to present a sample in an appropriate format to the sensor. Flow operations are comparatively easy to automate, miniaturise,

and control as closed tubing avoids evaporation of fluids and provides exactly repeatable environment for highly reproducible mixing of compounds. Moreover, sensors can be protected from fouling during contact time and from interfering compounds that may be present in the food sample. This is especially relevant when considering target analytes within fresh produce matrices. For example, phenolic compounds (e.g., catechin and epicatechin) and ascorbic acid, which are commonplace in many fresh produce products, are electrochemically active and, thus, their influence must be greatly reduced or eliminated.[21]

5 FUTURE DEVELOPMENT IN BIOSENSOR TECHNOLOGY FOR FOOD

Overall, the use of biosensors for food analysis can provide a route to a specific, sensitive, rapid, and inexpensive method for monitoring a range of target analytes. This applies to monitoring carried out not only under laboratory conditions but also (e.g., with the use of screen-printed sensors) at on-site locations. These devices can be designed such that the non-specialist operator can use them effectively. However, as with most technologies, there is still room for improvement. Perhaps one of the main areas where biosensor technology can be improved is the actual recognition element itself. Developments in both enzyme (e.g., protein engineering) and antibody (e.g., antibody fragments) technology, and complementary DNA probes continue apace.[50] Advances in computational techniques now allow the modeling of both electron transfer reactions and receptor binding interactions with increasing accuracy. This not only enhances understanding of the receptor/transducer interface, but also allows consideration of the design of new receptors based on biological molecules.

In contrast to the development of purely biological recognition elements, the use of synthetic material (to perform the same task) has increased. Undoubtedly, these new techniques (e.g., molecularly imprinted polymers; MIPs) will impinge on the evolving biosensor field. Most importantly, biomimetics have the potential to overcome some of the shortfalls associated with biological components, primarily: poor stability and higher cost of production.[50] The success-ful introduction of such materials would enable biosensors to be used in many difficult environments. Along with these improvements in the biological recognition elements, other developments in areas such as further miniaturisation and advanced fabrication procedures should lead to more robust and inexpensive sensors. Linked to advances in sampling and extraction, the use of biosensors for monitoring target analytes in a range of foods that are at present difficult to access will become a real possibility.

The global food analysis market currently stands at €1.1 billion with rapid methods accounting for €115 million.[51] Therefore, although biosensors are likely to see growth in this area, it is probable that standard food analysis for microorganisms will remain a difficult market to penetrate. There are, however, several areas where biosensors are ideal candidates for improving food diagnostics. These potential opportunities include improved QC and assurance of food-derived raw materials,[25] testing for absence/presence of genetically modified constituents where feasible,[52] food authenticity/traceability, and incorporation into smart packaging. Fundamentally, however, the trend from yield-driven to quality-driven provision of agricultural products in response to consumer demands for improved food quality, safety, and traceability is set to grow in the developed world. The demand for reliable and inexpensive methods for assessment of quality is set to expand: biosensors offer the opportunity to fulfill this niche.

REFERENCES

1. L. A. Terry, S. F. White, and L. A. Tigwell, The application of biosensors to fresh produce and the wider food industry. *Journal of Agricultural and Food Chemistry*, 2005, **53**, 1309–1319.

2. M. Schoemaker, R. Feldbrugge, B. Grundig, and F. Spencer, The lipoxygenase sensor, a new approach in essential fatty acid determination in foods. *Biosensors and Bioelectronics*, 1997, **12**, 1089–1099.

3. A. Curulli, S. Kelly, C. A. O'Sullivan, G. G. Guilbault, and G. Palleschi, A new interference-free lysine biosensor using a non-conducting polymer film. *Biosensors and Bioelectronics*, 1998, **13**, 1245–1250.

4. F. Ge, X.-E. Zhang, Z.-P. Zhang, and X.-M. Zhang, Simultaneous determination of maltose and glucose

5. F. Mizutani, Y. Sato, Y. Hirata, and S. Yabuki, High throughput flow-injection analysis of glucose and glutamate in food and biological samples by using enzyme/polyion complex bilayer membrane based electrodes as the detectors. *Biosensors and Bioelectronics*, 1998, **13**, 809–815.

6. L. Campanella, G. Favero, M. Pastorino, and M. Tomassetti, Monitoring the rancidification process in olive oils using a biosensor operating in organic solvents. *Biosensors and Bioelectronics*, 1999, **14**, 179–186.

7. B. Serra, A. J. Reviejo, C. Parrado, and J. M. Pingarron, Graphite-Teflon composite bienzyme electrode for the determination of L-lactate: application to food samples. *Biosensors and Bioelectronics*, 1999, **14**, 505–513.

8. M. Varadi, N. Adanyi, E. E. Szabo, and N. Trummer, Determination of the ratio of D- and L- amino acids in brewing by an immobilised amino acid oxidase enzyme reactor coupled to amperometric detection. *Biosensors and Bioelectronics*, 1999, **14**, 335–340.

9. G. Panfili, P. Manzi, D. Compagnone, L. Scarciglia, and G. Palleschi, Rapid assay of choline in foods using microwave hydrolysis and a choline biosensor. *Journal of Agricultural and Food Chemistry*, 2000, **48**, 3403–3407.

10. K. C. Gulla, M. D. Gouda, M. S. Thakur, and N. G. Karanth, Reactivation of immobilised acetyl cholinesterase in an amperometric biosensor for organophosphorus pesticide. *Biochemica Et Biophysica Acta*, 2002, **1597**, 133–139.

11. H. Sculze, E. Scherbaum, M. Anastassiades, S. Vorlova, R. D. Schmid, and T. T. Bachmann, Development, validation and application of an acetylcholinesterase biosensor for the direct detection of insecticide residues in infant food. *Biosensors and Bioelectronics*, 2002, **17**, 1095–1105.

12. N. Miyanishi, Y. Inaba, H. Okuma, C. Imada, and E. Watanabe, Amperometric determination of laminarin using immobilised β-1,3-glucanase. *Biosensors and Bioelectronics*, 2004, **19**, 557–662.

13. A. G.-V. Prada, N. Pena, M. L. Mena, A. J. Reviejo, and J. M. Pingarron, Graphite-teflon composite bienzyme amperometric biosensors for monitoring of alcohols. *Biosensors and Bioelectronics*, 2003, **18**, 1279–1288.

14. K. T. Kinnear and H. G. Monobouquette, An amperometric fructose biosensor based on fructose dehydrogenase immobilized in a membrane mimetic layer on gold. *Analytical Chemistry*, 1997, **69**, 1771–1775.

15. M. Esti, G. Volpe, L. Massingham, D. Compagnone, E. La Notte, and G. Palleschi, Determination of amines in fresh and modified atmosphere packaged fruits using electrochemical biosensors. *Journal of Agricultural and Food Chemistry*, 1998, **46**, 4233–4237.

16. E. Akyilmaz and E. Dinckaya, A new enzyme electrode based on ascorbate oxidase immobilized in gelatin for specific determination of L-ascorbic acid. *Talanta*, 1999, **50**, 87–93.

17. E. Maestre, I. Katakis, and E. Dominguez, Amperometric flow-injection determination of sucrose with a mediated tri-enzyme electrode based on sucrose phosphorylase and electrocatalytic oxidation of NADH. *Biosensors and Bioelectronics*, 2001, **16**, 61–68.

using a screen-printed electrode system. *Biosensors and Bioelectronics*, 1998, **13**, 333–339.

18. M. Arif, S. J. Setford, K. S. Burton, and I. E. Tothill, L-Malic acid biosensor for field-based evaluation of apple, potato and tomato horticultural produce. *The Analyst*, 2002, **127**, 104–108.

19. L. D. Mello, M. D. P. T. Sotomayor, and L. T. Kubota, HPR-based amperometric biosensor for the polyphenols determination in vegetables extract. *Sensors and Actuators*, 2003, **B96**, 636–645.

20. S. Jawaheer, S. F. White, C. Bessant, S. D. D. V. Rughooputh, and D. C. Cullen, *Determination of Fruit Status Using a Disposable Multi-Analyte Biosensor Array and Principle Component Analysis*, In: *7th World Congress on Biosensors*, Kyoto, Japan, 2002 May 15–17.

21. S. Jawaheer, S. F. White, C. Bessant, S. D. D. V. Rughooputh, and D. C. Cullen, Development of a common biosensor format for an enzyme based biosensor array to monitor fruit quality. *Biosensors and Bioelectronics*, 2003, **18**, 1429–1437.

22. M. Keusgen, M. Junger, I. Krest, and M. J. Schoning, Development of biosensor specific for cysteine sulphoxides. *Biosensors and Bioelectronics*, 2003, **18**, 805–812.

23. L. A. Abayomi, L. A. Terry, S. F. White, and P. J. Warner, *Development of a Biosensor to Determine Pungency in Onions*, In: *The 8th World Congress on Biosensors*, Granada, Spain, 2004 May 24–26.

24. L. A. Abayomi, L. A. Terry, S. F. White, and P. J. Warner, *Defining Onion Pungency Using Amperometric Biosensors*, In: *International Workshop on Biosensors for Food Safety and Environmental Monitoring*, Agadir, Morocco, 2005 November 10–12.

25. L. A. Abayomi, L. A. Terry, S. F. White, and P. J. Warner, Development of a disposable pyruvate biosensor to determine pungency in onions (*Allium cepa* L.). *Biosensors and Bioelectronics*, 2006, **21**, 2176–2179.

26. M. N. Velasco-Garcia and T. Mottram, Biosensor technology addressing agricultural problems. *Bioprocess and Biosystems Engineering*, 2003, **84**, 1–12.

27. R. W. Claycomb and M. J. Delwiche, Biosensor for on-line measurement of bovine progesterone during milking. *Biosensors and Bioelectronics*, 1998, **13**, 1173–1180.

28. R. M. Pemberton, J. P. Hart, and J. A. Foulkes, Development of a sensitive, selective, elctrochemcial immunoassay for progesterone in cow's milk based on a disposable screen-printed amperiometric biosensor. *Electrochimica Acta*, 1998, **43**, 3567–3574.

29. T. Mottram, J. Hart and R. Pemberton, *A Sensor Based Automatic Ovulation Prediction System for Diary Cows*, In: *Proceedings of 5th AISEM Conference*, 2000, Lecce, Italy.

30. R. M. Van Es and S. J. Setford, Detection of gentamicin in milk by immunoassay and flow-injection analysis with electrochemical measurement. *Analytica Chimica Acta*, 2001, **429**, 37–47.

31. S. J. Setford, R. M. Van Es, and S. Kröger, Receptor binding protein affinity sensor for rapid β-lactam quantification in milk. *Analytica Chimica Acta*, 1999, **398**, 13–22.

32. M. Delwiche, E. Cox, B. Goddeeris, C. Van Dorpe, J. De Baerdemaeker, E. Decuypere, and W. Sansen, A biosensor to detect pencillin residues in food. *Transactions of the ASAE*, 2000, **43**, 153–159.

33. G. A. Chope, L. A. Terry, and P. J. White, Onion bulb storage is related to a temporal decline in abscisic acid concentration. *Postharvest Biology and Technology*, 2006, **39**, 233–242.

34. J. C. Pyun, H. Beutel, J.-U. Meyer, and H. H. Ruf, Development of a biosensor for *E. coli* based on a flexural plate wave (FPW) transducer. *Biosensors and Bioelectronics*, 1998, **13**, 839–845.

35. L. Rasooly and A. Rasooly, Real time biosensor analysis of staphylococcal enterotoxin A in food. *International Journal of Food Microbiology*, 1999, **49**, 119–127.

36. K. H. Seo, R. E. Brakett, N. F. Hartman, and D. P. Campbell, Development of a rapid response biosensor for detection of *Salmonella typhimurium*. *Journal of Food Protection*, 1999, **62**, 431–437.

37. J. Homola, J. Dostalek, S. Chen, A. Rasooly, S. Jiang, and S. S. Yee, Spectral surface plasmon resonance biosensor for detection of staphylococcal enterotoxin B in milk. *International Journal of Food Microbiology*, 2002, **75**, 61–69.

38. G. C. Bokken, R. J. Corbee, F. Van Knapen, and A. A. Bergwerff, Immunochemical detection of Salmonella group B, D and E using an optical surface plasmon biosensor. *FEMS Microbiology Letters*, 2003, **222**, 75–82.

39. N. Kim and I.-S. Park, Application of a flow-type antibody sensor to the detection of *Escherichia coli* in various foods. *Biosensors and Bioelectronics*, 2003, **18**, 1101–1107.

40. T. B. Tims and D. V. Lim, Confirmation of viable *E. coli* 0157:H7 by enrichment and PCR after rapid biosensor detection. *Journal of Microbiological Methods*, 2003, **55**, 141–147.

41. S. Schwimmer and W. J. Weston, Enzymatic development of pyruvic acid in onion as a measure of pungency. *Journal of Agricultural and Food Chemistry*, 1961, **9**, 301–305.

42. A. O. Scott, *Biosensors for Food Analysis*, The Royal Society of Chemistry, London, 1998.

43. L. Pogacnik and M. Franko, Detection of organophosphate and carbamate pesticides in vegetable samples by a photothermal biosensor. *Biosensors and Bioelectronics*, 2003, **18**, 1–9.

44. R. H. Hall, Biosensor technologies for detecting microbial foodborne hazards. *Microbes and Infection*, 2002, **4**, 425–432.

45. P. D. Patel, (Bio)sensors for measurement of analytes implicated in food safety: a review. *Trends in Analytical Chemistry*, 2002, **21**, 96–115.

46. L. A. Terry, K. A. Law, K. J. Hipwood, and P. H. Bellamy, *Non-structural Carbohydrate Profiles in Onion Bulbs Influence Taste Preference*, In: *Information and Technology for Sustainable Fruit and Vegetable Production. Frutic 05*, Montpellier, France, 2005 September 12–16.

47. N. Gajovic, A. Warsinkef, and W. Scheller, Comparison of two enzyme sequences for a novel L-malate biosensor. *Journal of Chemical Technology and Biotechnology*, 1997, **68**, 31–36.

48. M. Stredansky, B. Cini, C. Benetello, and S. Stredanska, *Biosensors for L-malate with Improved Sensitivity*, In: *8th World Congress on Biosensors*, Granada, Spain, 2004 May 24–26.

49. M. Kim and M.-J. Kim, Isocitrate analysis using a potentiometric biosensor with immobilised enzyme in a FIA system. *Food Research International*, 2003, **36**, 223–230.

50. J. D. Newman, L. J. Tigwell, A. P. F. Turner, and P. J. Warner, *Biosensors: A Clearer View*, In: *8th World Congress on Biosensors*, Granada, Spain, 2004 May 24–26.

51. L. Kahr, M. Brannback, and G. Von Blankenfeld-Enkvist, *Trends and Needs in Food Diagnostics*, In: *CE Food Congress*, Ljubljana, Slovenia, 2002 September 22–25.

52. I. Mannelli, M. Minunni, S. Tombelli, and M. Mascini, Quartz crystal microbalance (QCM) affinity biosensor for genetically modified organisms (GMOs) detection. *Biosensors and Bioelectronics*, 2003, **18**, 129–140.

80

From Earth to Space: Biosensing at the International Space Station

Christa Baumstark-Khan and Christine E. Hellweg

*Radiation Biology Department, Institute of Aerospace Medicine,
Köln, Germany*

1 HUMAN ACTIVITIES IN SPACE

The endeavor of men to escape the boundaries of Earth and to travel to the heavens is as old as the history of mankind itself. Religion, mythology, and literature reaching back thousands of years are sprinkled with references to magic carpets, flying horses, flaming aerial chariots, and winged gods. The ancient Greek legends tell the story of the first but unsuccessful space flight by Icarus and Jules Verne's classical fiction of a Moon voyage is the most prominent story known to nearly everyone.

Manned space flight started in the mid-twentieth century, when the development and refinement of aeronautics and the construction of rockets accomplished the necessary prelude to the exploration of near and outer space. The first milestone on the way to space was the launch of the Soviet Sputnik II, which carried its canine passenger "Laika" into orbit on November 3, 1957. Major achievements of the first era of space exploration were the launch of Yuri Gagarin on board a Russian Vostok rocket, thereby initiating the age of manned space flight, and the first spacewalk by the Soviet cosmonaut Alexei Leonov in 1965. Since that time, numerous manned missions to Earth orbit have been carried out routinely by the Americans and Russians and in 2003 by China, lasting from a few days in capsules to several hundred days in space stations.

A highlight in manned space flight was achieved in July 1969, when Neil Armstrong, commander of Apollo 11, set foot upon the Moon, declaring famously that it was "one small step for a man, one giant leap for mankind". The next notable achievement in space was the launch of the first space station, Salyut 1, from the USSR. After the first 20 years of exploration, focus began shifting from one-off flights to renewable hardware, such as the Space Shuttle program, and from competition to cooperation as on the International Space Station (ISS). With the beginning of not only a new century but also a new millennium, mankind extends its view to other planets in order to search for life out there. Recently, private interests have begun promoting space tourism, while larger government programs have been advocating a return to the Moon and possibly missions to Mars in the near future.

1.1 First Flight Programs

The years 1958–1961 were busy ones in both the United States and the Soviet Union for the development of manned space vehicles. Developmental works on the Soviet carrier rocket and the spacecrafts have commenced a series of five unmanned test flights. These Vostok precursor flights, Korabl

Handbook of Biosensors and Biochips. Edited by Robert S. Marks, David C. Cullen, Isao Karube, Christopher R. Lowe and Howard H. Weetall.
© 2007 John Wiley & Sons, Ltd. ISBN 978-0-470-01905-4.

Sputnik I through V, were designed to collect data on the effects of space environment (especially solar radiation) on biological specimens and to test the spacecraft systems. In late 1960 to early 1961 the Soviets were ready to begin manned space flight operations. The Vostok flights intended to collect data on weightlessness, a topic of considerable concern not only to Soviet flight physicians. For the second manned flight, the medical specialists insisted that the duration of the flight should not be longer than two or three orbits for judging the effects of zero gravity on physiology. But in accordance with political considerations the cosmonaut Titov and his advisors wanted to go for a day-long mission. During his flight Titov went through a phase of motion sickness,[1] a fact which held up the progress of the Soviet flight program for nearly one year.

The Project Mercury was the United States' first man-in-space program.[2] From 1958 till 1963 six manned space flights were accomplished as part of a 25-flight program showing that man can function as a pilot, an engineer and an experimenter without undesirable reactions or deteriorations of normal physiology for periods up to 34 h of weightless flight. Other facets of the Mercury experience were the development of manned space vehicles, accurate and detailed test procedures, and more responsive configuration control techniques. During the initial phase the Americans preferred ballistic shots resulting in only suborbital space flights. In February 1962, after a series of frustrating delays John Glenn became the first American to orbit the Earth with his space ship MA-6 Friendship 7.

Mercury and Vostok demonstrated the feasibility of placing a human being in orbit, observing his reactions to the space environment, and returning him safely to Earth at a known point. The second stage in human space flight was the development of multi-place spacecrafts for the conduction of more complicated missions—the era of Gemini and Voskhod. The program Gemini, which involved 12 flights, including two unmanned flights, provided techniques, equipment, and experience that helped bridge the difficult translation from experimental, Earth-orbiting Mercury to ambitious, lunar-landing Apollo.

The Gemini spacecrafts were equipped with a computer and radar to aid in solving the rendezvous problem. All of these systems went through troubled development and qualification periods and, in most cases, required extensive redesign. The same was true for design work on astronaut equipment and space suits to achieve a range of capabilities from extravehicular activity to cabin operations. The 10 manned flights of the Gemini program spanned 603 days (a flight every 60 days), and 1940 man hours in space during 970 mission-hours gave 16 different astronauts the chance for gaining experience. The length of the Gemini missions, for example, 330 h on Gemini VII in December 1965, reassured major medical concerns over man's ability to adapt to and function in space. It became more and more an accepted fact that man could withstand the space environmental conditions and would fly to the Moon, despite the rumors that run in some medical circles that the astronauts might die after being in long weightless flight. Gemini's most lasting contribution will probably be seen by future historians in the changing views toward our Earth and Earth's environmental problems, which was brought about by bringing back photographs from the blue planet from space.

For some time, the development phase of Gemini and Apollo proceeded along parallel lines. Experience from the Gemini program was passed on to Apollo, as 15 of the 20 Gemini astronauts subsequently flew in the lunar program.[3] The Apollo program was designed to land humans on the Moon and bring them safely back to Earth. It included a large number of unmanned test missions and crewed missions. The 11 crewed missions include two Earth-orbiting missions, two lunar orbiting missions, a lunar swing-by and six Moon landing missions. Lunar surface experiments included soil mechanics, meteoroids, seismic, heat flow, lunar ranging, magnetic fields, and solar wind experiments.

1.2   The Era of Space Stations

Designed for long-duration missions, space stations proved that humans could live and work in space for extended periods. Such missions expanded our knowledge of solar astronomy well beyond Earth-based observations. The Soviet Union launched the world's first space station, Salyut 1, in 1971, a decade after launching the first human into space. The United States sent its first space station, the larger Skylab, into orbit in 1973 and it hosted three crews before it was abandoned in 1974. Russia continued to focus on long-duration space missions and in 1986 launched the first modules of the Mir space station. In 1998, the first two modules of the ISS were launched and joined together in orbit. Other modules soon followed and the first crew arrived in 2000.

1.2.1 Salyut

The Soviet Salyut 1 was the first Salyut space station, and the first space station of any kind. It was launched on April 19, 1971. Its first crew was launched in Soyuz 10 but was unable to board it due to a failure in the docking mechanism; its second crew was launched in Soyuz 11 and remained on board for 362 orbits. During these productive 23 days different tasks were performed by the crew involving checking and testing the design and equipment of the orbital piloted station. Scientific topics were related to geological and geographical objects on Earth's surface and to physical characteristics and phenomena in the atmosphere and outer space (e.g., the spectrum of electromagnetic radiation). Beneath that, cosmonauts conducted medico-biological studies to determine the possibilities of performing various jobs by them in the station and to study the influence of space flight factors on the human organism. On June 29 the cosmonauts Dobrovolski, Patsayev, and Volkov returned to Earth, but unfortunately they were killed during reentry into the atmosphere due to a cabin pressure leak. The Salyut space station reentered Earth's atmosphere October 11, 1971 after 175 days in space.

During the years 1971 to 1991 the Soviets have operated seven Salyut stations in space for military and scientific purposes thereby collecting data on human physiology on long-duration space flight.[4] From 1977 until 1982 Salyut 6 was visited by five long-duration crews including cosmonauts from other Warsaw Pact countries. The longest flight onboard Salyut 6 lasted 185 days. Salyut 7 was inhabited for 4 years and 2 months, during which time it was visited by 10 crews constituting (including French and Indian cosmonauts). Salyut 7 de-orbited on February 7, 1991.

1.2.2 Skylab

The US space station Skylab was launched May 14, 1973, from the NASA's Kennedy Space Center by a huge Saturn V launch vehicle and reached its near-circular orbit at the altitude of 435 km. Shortly after liftoff, the meteoroid shield was torn off from the space station inadvertently by atmospheric drag. All other major functions and pressurization of the space station occurred as planned. The misfortune during launch resulted in considerable problems that had to be conquered before the space station would be habitable for the three manned periods of its planned 8-month mission. The manned Skylab missions were thus dedicated to repair works on the station in addition to their science program.

In the Skylab, both the man hours in space and the man hours spent in performance of extravehicular activities (EVA) under microgravity conditions exceeded the combined totals of all of the world's previous space flights up to that time. The effectiveness of the Skylab crews exceeded any expectations, especially in their ability to perform complex repair tasks within the station and during EVA, although all crew members had some initial problems with motion sickness. The capability to conduct longer manned missions was conclusively demonstrated in Skylab, first by the crew returning from the 28 day mission and, more forcefully, by the good health and physical condition of the second and third Skylab crews who stayed in weightless space for 59 and 84 days respectively.

Following the final manned phase of the Skylab mission, after return of crew 4 to Earth, Skylab was positioned into a stable altitude and systems were shut down. In the fall of 1977, it was determined that Skylab was no longer in a stable altitude and on July 11, 1979, Skylab impacted the Earth surface. The debris dispersion area stretched from the South-eastern Indian Ocean across a sparsely populated section of Western Australia.[5]

1.2.3 MIR

The Soviet Mir orbital station was the first consistently inhabited long-term research station in space.[6] Based on the previously launched Salyut series of space stations Mir was assembled in orbit by successively connecting several modules, each launched separately from 1986 to 1996. The major part of the entire orbital station was the core module combining its modules into a single complex. It accommodated control equipment for the station crew life support systems and science hardware, as well as crew rest locations. The Kvant astrophysics module consisted of a laboratory compartment with a transfer chamber and unpressurized compartment for science instruments. The Kvant 2 module consisted of three pressurized compartments (instrument–cargo, instrument–science compartments and airlock compartment) supplying the station with science hardware and equipment. It supported crew spacewalks as well as the performance of various scientific research and biotechnological experiments. The Kristall module consisted of two pressurized compartments (instrument–cargo and transfer–docking compartment) for the conduction of biological and materials processing research and experiments. The Kristall module was equipped with an active docking compartment for support of dockings of the US space shuttle orbiters to the MIR Station without modifying its configuration. The Spektr module (two compartments; pressurized instrument cargo and nonpressurized compartment with solar arrays and science instruments) was aimed to investigate atmospheric and geophysical processes. The Priroda remote sensing module, consisting of one pressurized instrument-cargo compartment, was the laboratory for studying geophysical processes in near Earth orbit, especially space radiation. By mid-1996, the completely configured station had a total mass of 11.5 tons and spanned an area of 31×33 m.

The station existed for 13 years until March 23, 2001, at which point it was deliberately de-orbited, and broke apart during atmospheric re-entry.[6] It was the first international collaboration bringing together cosmonauts and astronauts of many different countries to perform different scientific missions, such as the German–Russian MIR 92 mission,[7] various EUROMIR missions[8] and different missions from the US-Russian Shuttle–Mir program.[9] During that time, major experience was gained on international knowledge exchange, on space station operation and navigation, and on scientific experimentation in space. Prior to the Shuttle–Mir Program, NASA's longest duration experiences had been the Skylab missions of 1973–1974. Shuttle–Mir enabled astronauts to spend hundreds

of days on Mir, more time in orbit than had been accumulated in orbit since the Shuttle Program began in 1981. Shuttle–Mir's long-duration residences provided experience and data that will help in the ISS and in possible future missions to the Moon and Mars. Science conducted by astronauts aboard Mir included over 100 investigations in eight disciplines, including advanced technologies, earth sciences, fundamental biology, human life sciences, ISS risk mitigation, life support risk management, microgravity, and space sciences. The major lesson learned from the international cooperation on MIR was that space exploration is no longer a competition between nations. Working with Russia in the Shuttle–Mir Program provided the US with experience that is currently being applied to the ISS with its 16 participating nations. Implementation of long-duration missions on the orbital station and conducting of large amount of medical experiment have allowed to develop the national long-duration space flight support system. Achieved space flight duration (up to 438 days by the Russian cosmonaut Polyakov in 1995) has confirmed a principal opportunity for human interplanetary space flight.

1.2.4 The ISS

The most complex engineering and construction project in the world is currently taking place in space.[10] Sixteen countries and over 100 000 people are contributing to this monumental achievement. The ISS is a unique platform for conducting research in a variety of disciplines; to better understand the role gravity or its absence plays in biological and physical processes. As a long-duration laboratory, it enables research that was not possible on earlier platforms. In particular, ISS is ideally suited for studying the effects of long-duration space flight on humans, important for perfecting countermeasures to the deleterious effects to ensure crew safety and to enable exploration missions.

Configuration of the ISS
Construction of ISS began in late 1998, with the launch of the Russian module Zarya. Since that time, many flights of Russian and American vehicles have added the major elements to the station, enlarging the platform from its original 20,000 kg single module configuration to the

current 180,000 kg facility.[11] Up to May 2007 Russian Soyuz spacecrafts and American Space Shuttles have brought 15 long-duration crews to live aboard the station for 4–6 months each, resulting in a continuous human presence in space since October 2000. Concurrent with the launch of the first research rack in March 2001, the payload operations and integration center (POIC) at the Marshall Space Flight Center (MSFC) in Huntsville became operational. Additional payload-specific support beginning with Expedition 2 was provided by Telescience Support Centers at other NASA Centers, including Johnson Space Center for human life sciences investigations. Remote sites at investigators' institutes, domestic, and international, are also routinely tied in during times when those experiments are operating.

The ISS depends on the Russian Soyuz rockets and the US Space shuttles to deliver crews and payloads to the station. Prior to the space shuttle Columbia disaster in early 2003, when seven astronauts were tragically lost,[12] station planners were targeting 2004 for the completion of the US contribution to the project. But the accident delayed construction works considerably. During the two-half years, NASA spent recovering from the Columbia accident Russia's Federal Space Agency supported the ISS with a lifeline of unmanned Progress cargo ships and steady launches of crew-carrying Soyuz spacecraft. Unmanned Russian Progress vehicles have brought the required logistics to maintain the station and its crews. Among the payload items delivered so far have been seven large research racks and logistics totaling more than 6500 kg. NASA's three remaining orbiters (Atlantis, Discovery, and Endeavour) are vital to the space station's construction since they are the only vehicles capable of delivering major components to the orbital research platform, including heavy modules and supply containers.[13] Recently, NASA and its partner agencies have set a new plan to complete the ISS by 2010, delaying science utilization.[14] Assembly of ISS will continue in the near future with completion of the trusses, addition of solar arrays for power generation, and radiator panels for cooling, and the International Partner research modules, leading to a configuration represented in Figure 1.[15] The European Space Agency (ESA)'s Columbus module will increase the station's research capability with additional 10 research rack locations, with the Japanese Kibo module adding 11 more. The full set of research modules will also allow optimal location of life sciences research racks for maximal synergy.

Laboratory Modules

The US Laboratory Module DESTINY A significant milestone for ISS-based research was the addition of the US laboratory module Destiny,[16] launched on the STS 98/5A mission in February 2001. Weighing 14 000 kg at launch, Destiny is a cylindrical module, 8.5 m long with an external diameter of 4.3 m. Internally, it is configured with 24 racks lining the four surfaces. The racks contain various systems equipments such as life support, controls for the station's robotic arms, medical hardware, a crew sleep station, and up to 10 research facilities, of which 7 are already on orbit. The first research rack was the human research facility (HRF) Rack 1, installed in March 2001. The rack provides a core set of experiment hardware to support investigations, as well as power, data and commanding capability, and stowage. The second HRF rack, to complement the first with additional hardware and stowage capability, will be launched once Shuttle flights resume. Future years will see additional capability to conduct human research on ISS as International Partner modules and facility racks are added to ISS. Crew availability, both as a subject count and time, will remain a major challenge to maximizing the science return from the bioastronautics research program.

The Japanese Laboratory Module KIBO The Japanese Experiment Module (JEM) or Kibo (which means hope) will be the first Japanese manned experimental facility.[17] It is a combine module consisting of two different facilities composed of four components: the pressurized module (PM) and the exposed facility (EF). The core component PM is of cylindrical shape, 11.2 m long and 4.4 m in diameter. It contains 10 standard payload racks in which the astronauts perform scientific experiments on the effects of microgravity, space radiation, high vacuum, and abundant solar energy. The EF is a terrace located outside the port cone of the PM where experiments can be fully exposed to the space environment. The experiment logistic module (ELM) contains a pressurized section to serve the PM and a nonpressurized section to serve the EF. It is placed atop the port side of the PM and

(a)

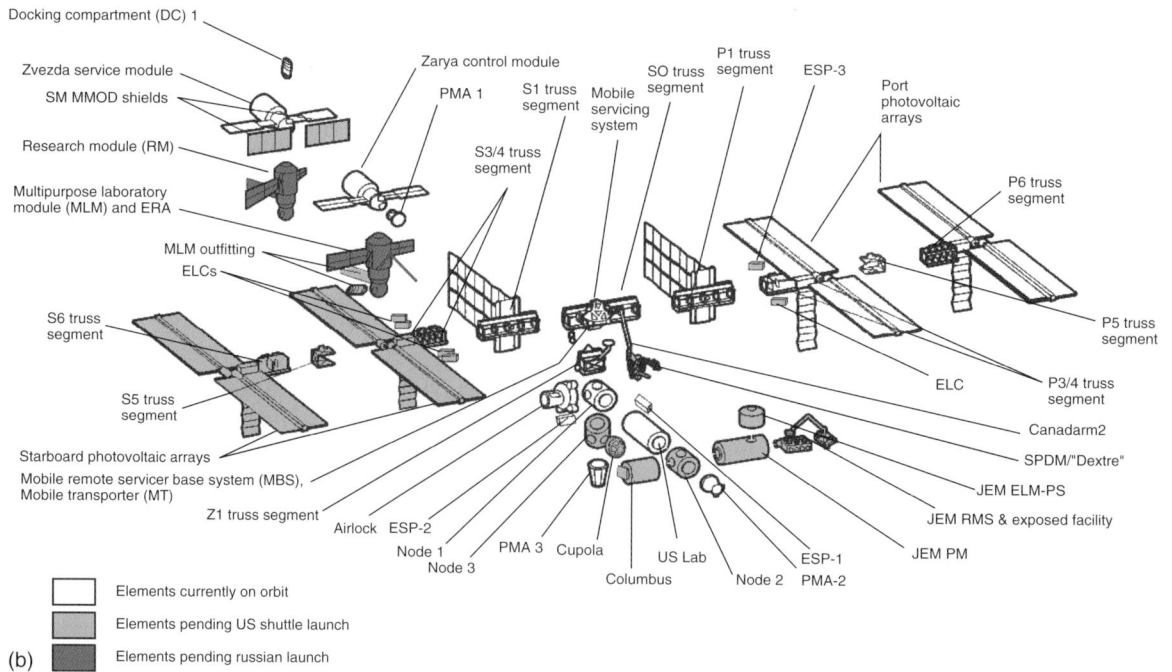

(b)

Figure 1.   (a) Artistic view of the complete configuration of ISS; (b) configuration of ISS including the International Partners' modules. [Adapted from Kitmacher et al.[15] with permission, copyright 2005, International Academy of Astronautics.]

is intended as a storage and transportation module. The Remote Manipulator System (JEM–RMS) is a robotic arm, mounted at the port cone of the PM, where it will serve the EF and move equipment from and to ELM. Kibo is intended to be docked to the ISS in 2008.

The European Module COLUMBUS   The Columbus laboratory is ESA's biggest single contribution to the ISS.[18] The 4.5-m-diameter cylindrical module (see Figure 2) is equipped with flexible research facilities that offer extensive science capabilities. During its 10-year projected life span, Earth-based researchers will be able to conduct experiments in life sciences, materials science, fluid physics, and a whole host of other disciplines, all in the weightlessness of orbit. Their efforts will be channeled through the Columbus Control Centre in Germany, which will interface with the module itself and also ESA's NASA partners in the United States.

The Columbus laboratory has room for 10 international standard payload racks (ISPRs), 8 situated in the sidewalls, and 2 in the ceiling area.[19] Each rack is the size of a telephone booth and able to host its own autonomous and independent laboratory, complete with power and cooling systems, and video and data links back to researchers on Earth. ESA has developed a range of payload racks,[20] all tailored to squeeze the maximum amount of research from the minimum

of space and to offer European scientists across a wide range of disciplines full access to a weightless environment that cannot possibly be duplicated on Earth.

The Biolab, for example, supports experiments on microorganisms, cells, and tissue cultures, and even small plants and small insects. Another rack contains the European physiology modules (EPM) facility, a set of experiments that will be used to investigate the effects of long-duration spaceflight on the human body. Experiment results will also contribute to an increased understanding of age-related bone loss, balance disorders, and other ailments back on Earth. The material science laboratory electromagnetic levitator (MSL-EML) is a facility for the melting and solidification of conductive metals, alloys, or semiconductors and a fluid science laboratory (FSL) will accommodate experiments on the strange behavior of weightless liquids. These too, could bring far-reaching benefits on Earth: better ways to clean up oil spills, for example, and even improved manufacture of optical lenses.

Outside its comfortable, pressurized body, Columbus has four mounting points for external payloads. Exposed to the vacuum of space, and with an unhindered view of the Earth and outer space, science packages can investigate anything from the ability of bacteria to survive on an artificial meteorite to volcanic activity 400 km below on the Earth.

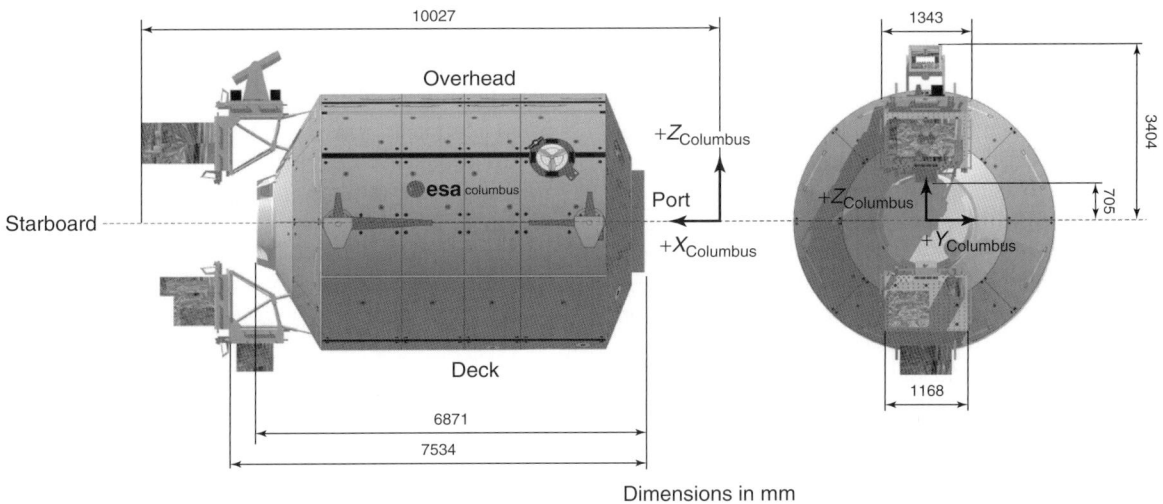

Dimensions in mm

Figure 2.   The European contribution to the ISS: the laboratory module Columbus with integrated external payloads.

The European laboratory module Columbus is scheduled to be delivered at the station in December 2007 with the carrier flight STS-122, the 24th mission to the ISS. The actual launch date of Columbus is not known at present, it is dependent on the success of the return to flight of the space shuttle.

1.3 Beyond the ISS

After the realization of the ISS, human exploratory missions beyond low Earth orbit (LEO) are widely considered as the next logical step in a worldwide peaceful cooperation in space. The NASA vision for space exploration calls for humans to return to the Moon by the end of the next decade, paving the way for eventual journeys to Mars and beyond. During the late 1960s and early 1970s, the Apollo program demonstrated US technical strength in a race against the Soviet Union to land humans on the Moon. Today, NASA's plans for a return to the Moon are not driven by Cold War competition, but by the need to test new exploration technologies and skills on the path to Mars and beyond.

NASA will begin its lunar test bed program with a series of robotic orbiter and landing missions. A human mission to the Moon will follow these robotic missions as early as 2015.[21] The purpose of a human mission to the Moon is to establish a base, perform surveys, establish infrastructure for permanent human base, collect and return material samples, and serve as an analogue environment for subsequent Mars exploration. This capability will be established as quickly as possible following the return of humans to the Moon. To best accomplish science and explorative goals, the outpost is expected to be located at the lunar south pole. The lunar outpost consisting of a habitat, power supply, communication, and navigation infrastructure, and surface mobility, and robotic systems shall already be emplaced when the first crew arrives at the Moon. Accordingly, construction work on the architecture of the habitat has to be done by robotic missions, and the crew will perform only some final assembly and verification tasks to make the outpost operational.

Starting in 2011, NASA will also launch the first in a new series of human precursor missions to Mars to demonstrate technologies and to obtain critical data for future human missions on chemical hazards, resource locations, and research sites. The first human mission beyond the Moon will be determined on the basis of available resources, accumulated experience, and technology readiness.[22] The timing of the first human research missions to Mars will depend on discoveries from robotic explorers, the development of techniques to mitigate Mars hazards, advances in capabilities for sustainable exploration, and available resources. The Mars exploration employs conjunction-class missions, often referred to as long-stay missions (1000 day mission), to minimize the exposure of the crew to the deep-space radiation and zero-gravity environment while, at the same time, maximizing the scientific return from the mission. This is accomplished by taking advantage of optimum alignment of Earth and Mars for both the outbound and return trajectories by varying the stay time on Mars, rather than forcing the mission through nonoptimal trajectories, as in the case of the short-stay missions (500 day mission). This approach allows the crew to transfer to and from Mars on relatively fast trajectories, on the order of 6 months, while allowing them to stay on the surface of Mars for a majority of the mission, on the order of 18 months. The habitat on Mars is implemented through a split mission concept in which cargo is transported in manageable units to the surface, or Mars orbit, and checked out in advance of committing the crews to their mission.

The establishment of such permanently inhabitable bases on Moon and Mars will add a new dimension to human space flight, concerning the distance of travel, the habitability of the base, and the durability of the astronauts. Above all radiation response, gravity related effects as well as psychological issues will become a possible limiting factor to human adaptability to the mission scenario.

2 FACTORS OF THE SPACE ENVIRONMENT

The story of Daedalus and Icarus in Greek mythology about the misfortune of Icarus when he ignored the risk limits set by Daedalus for flying toward the sun (not too high and not too low) might be one of the first written theses about the potential dangers from solar radiation in space.

Today, after decades of relatively frequent shuttle trips and orbital missions, and the realization that interplanetary missions "toward a human presence in space" are technically feasible, although not safe, space environmental factors are considered as major hazards. Humans in space, on board of the ISS as well as in interplanetary missions, are confronted to a complex matrix of a multitude of environmental factors of various kinds and intensities, with microgravity and cosmic radiation as the most dominant stressors. In the endeavor to assess the risks for humans in space—especially for long-duration missions—the concerted action of all stimuli has to be known and warning signals about changes of the "health" status of the environment are required.

2.1 Planetary Atmospheres and Space Vacuum

Stars, planets, and moons keep their atmospheres by gravitational attraction; accordingly atmospheres have no clearly defined boundary. The density of atmospheric gas simply decreases with distance from the object. In LEO the atmospheric density is about 1×10^{-7} Pa, still sufficient to produce significant drag on satellites. Beyond planetary atmospheres, space has been compared to a vacuum of about 1×10^{-16} Pa. On Earth, humans are subjected to an atmospheric surface pressure of 1.014×10^5 Pa, the surface pressure on Moon and Mars are about 3×10^{-10} and 636 Pa, respectively. There are limited data available from human accidents, describing the effects of exposure to near vacuum. Humans cannot survive the condition of vacuum. Exposed humans will lose consciousness after a few seconds and will die within minutes from ebullism and oxygen loss, before the body swells to about double size. The hazardous effects of vacuum can be omitted by containment in a space suit,[23] in which an artificial atmosphere of 2×10^3 Pa is maintained.

2.2 Gravity Levels

The most fascinating medical characteristic of space flight is undoubtedly weightlessness, an effect created by "zero g", the balance of centrifugal and centripetal forces. (The gravitational field at the Earth's surface, denoted g, is approximately $9.80665 \mathrm{m\,s^{-2}}$.) In the early days of space flight, speculation on organ failure in microgravity circled among aviation physicians, as well as worries on disturbed signal perception and cognitive functions. They feared that pressure on the nerves and organs, muscle tone, posture, and the labyrinth of the inner ear would give conflicting signals in the weightless state. The basic difficulty regarding long-term studies of weightlessness is the impossibility of simulating the exact condition on the Earth for extended time. The only possibility to achieve weightlessness is putting an object in a state of free fall. Microgravity experiments on Earth by using drop towers (about 5 s free fall), and aircrafts flying parabolic trajectories (up to 90 s weightlessness) have been conducted with biological samples. The best but most expensive device for zero microgravity experimentation is the sounding rocket; however, such sounding rocket flights are composed of an initial high g, an intermediate micro-g and a final high g phase. From 1951 on ballistic experiments were performed on mice, dogs, and monkeys, which revealed the survivability of these species exposed to varying levels of acceleration and deceleration.[24] Test pilots, who flew the first parabolic flights complained about spinning and described the feeling of "lost in space",[25] now known as *disorientation*. In 1955 the principal problems of weightless flight seemed solvable, as eating and drinking at zero microgravity were not troublesome when squeeze bottles and tubes were used, and urination presented no real difficulty. Some subjects suffered nausea, disorientation, loss of coordination, and other disturbances, but the majority reported that after they adjusted to the condition they found it "pleasant" and had a feeling of "well-being".

Another problem perplexing aeromedical experts at the beginning era of space flight was the effect on the human body of the heavy acceleration and deceleration forces, called *g loads*. Acceleration of a vehicle into space and the deceleration accompanying its return to Earth results in g loads several times the normal accelerative force of gravity, thereby imposing severe strain on astronauts' body organs. Accordingly tests were required before sending men into space. The ability to withstand immediate impact forces was shown by test pilots who survived rides on rocket-driven impact sledges at an impact force of 35–40 g. Experiments performed later on human centrifuges had more immediate relevance to space medicine than impact sledge tests, because of the relatively gradual buildup of g forces encountered during the launch and re-entry phases of ballistic, orbital, or interplanetary flight. Results from such centrifuge tests recommended acceleration forces from 3 to 8 g for takeoff from the physiological standpoint. As a consequence, human test subjects were exposed to peaks of 10 g for something over two minutes; chest pain, shortness of breath, and occasional loss of consciousness were the symptoms in response to the higher g loads.[26] Chimpanzees exposed to peak accelerations of 40 g for 1 min suffered internal injuries, including heart malfunctions. It appeared that prolonged subjection to high g forces might be severely injurious or perhaps even fatal to a man.[27] Further experiments with different materials led to the development of anti-g devices in the kind of contour seats, which allow short duration g loads of up to 20 g.

Under nominal launch and landing conditions impacts greater than 9 g are not to be expected. For the lunar roundtrip maximum g loads of 3 g during launch and ascend to the moon surface will not be exceeded; the gravity level during the aero capture manoeuvre at Earth arrival is calculated to lay in the range of about 6 g due to the atmospheric drag. Descending from LEO to Earth with a Space shuttle type spacecraft will result in 1–2 g. For the Mars mission, similar g loads are expected to occur during launch and landing; for aero capture and landing on Mars a maximum of 6 g will be reached. The interplanetary cruise is carried out a zero-gravity level. The two crew members orbiting Mars instead of landing on the planet's surface will cumulative zero-gravity periods of either 450 or 900 days (depending on the mission type), only interrupted by orbit insertion at Mars (6 g) and injection toward Earth (3 g). Especially the high gravity levels of 6 g at Mars arrival (aero capture) and landing on Earth (with a capsule) after the long period at microgravity seems critical for the health of the crew.

2.3 Radiation Environment

In addition to vacuum, weightlessness, and g loads, the problems of space flight included protecting the passenger from different kinds of electromagnetic radiation found above the atmosphere (Figure 3). The prime instrument for cosmic ray research from 1952 to 1958 was the oldest vehicle for human flight, the balloon. Most data on space radiation was obtained from satellites, probes and measurements performed during space flights, and on space stations.

In the interplanetary space, the radiation field is composed mainly of two groups: the solar cosmic radiation (SCR) and the galactic cosmic radiation (GCR). In the vicinity of the Earth, a third radiation component is present: the radiation trapped by the Earth's magnetosphere, the so-called van Allen belts.[28]

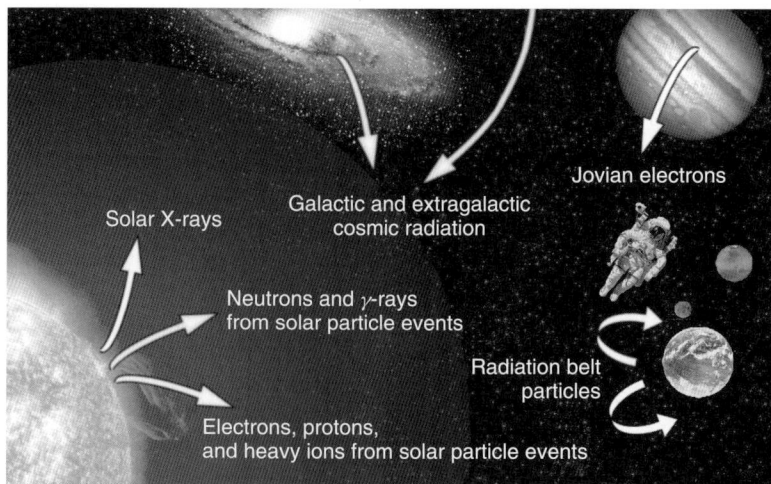

Figure 3. The radiation field in our solar system.

SCR consist of the low energy solar wind particles that flow constantly from the sun and the so-called solar particles events (SPEs) that originate from magnetically disturbed regions of the sun, which sporadically emit bursts of charged particles with high energies.[29] These events are composed primarily of protons with a minor component (5–10%) being helium nuclei (α particles) and an even smaller part (1%) heavy ions and electrons. SPEs develop rapidly and generally last for no more than some hours, however, some proton events observed near Earth may continue over several days. The emitted particles can reach energies up to several GeV. In a worst case scenario; doses as high as 10 Gy could be received within a short time. Such strong events are very rare, typically about one event during the 11-year solar cycle. Concerning the less energetic, though still quite intensive events, for example, in cycle 22 (1986–1996) there were at least eight events for proton energies greater than 30 MeV. For LEO, the Earth's magnetic field provides a latitude dependent shielding against SPE particles. Only in high inclination orbits and in interplanetary missions, SPEs create a hazard to humans in space, especially during EVA.

GCRs originate outside the solar system in cataclysmic astronomical events, such as supernova explosions. Detected particles consist of 98% baryons and 2% electrons.[30] The baryonic component is composed of 85% protons (hydrogen nuclei), with the remainder being α particles (helium nuclei) (14%) and heavier nuclei (about 1%). The latter component comprises the so-called HZE particles (particles of high charge Z and high Energy), which are defined as cosmic ray primaries of charges $Z > 2$ and of energies high enough to penetrate at least 1 mm of spacecraft or of spacesuit shielding. Though they only contribute to roughly 1% of the flux of GCR, they are considered as a potential major concern to living beings in space, especially for long-term missions at high altitudes or in high inclination orbits, or for missions beyond the Earth's magnetosphere. Reasons for this concern are based, on the one hand on the inefficiency of adequate shielding and, on the other hand on the special nature of HZE particle-produced lesions. If the particle flux is weighted according to the energy deposition, iron nuclei will become the most important component, although their relative abundance is comparatively small.

The fluence of GCR is isotropic and energies up to 10^{20} eV can be present. When GCR enter our solar system, they must overcome the magnetic fields carried along with the outward-flowing solar wind, the intensity of which varies according to the about 11-year cycle of solar activity. With increasing solar activity the interplanetary magnetic field increases, resulting in a decrease of the intensity of GCR of low energies. This modulation is effective for particles below some GeV per nucleon. Hence, the GCR fluxes vary with the solar cycle and differ by a factor of approximately five between solar minimum and solar maximum with a peak level during minimum solar activity and the lowest level during maximal solar activity. At peak energies of about 200–700 MeV per unit during solar minimum, particle fluxes (flow rates) reach 2×10^3 protons per 100 μm^2 (mean area of a human cell nucleus) per year and 0.6 Fe-ions per 100 μm^2 per year (100 μm^2 being the typical cross section of a mammalian cell nucleus). Although iron ions are 1/10th as abundant as carbon or oxygen, their contribution to the GCR dose is substantial, since dose is proportional to the square of the charge.

The fluxes of GCR are further modified by the geomagnetic field. Only particles of very high energy have access to low inclination orbits. Towards higher inclination particles of lower energies are allowed. At the pole, particles of all energies can impinge in the direction of the magnetic field axes. Due to this inclination-dependent shielding, the number of particles increases from lower to higher inclination.

The van Allen belts in the vicinity of the Earth are a result of the interaction of GCR and SCR with the Earth's magnetic field and the atmosphere. Two belts of radiation are formed comprising electrons and protons, and some heavier particles trapped in closed orbits by the Earth's magnetic field. The main production process for the inner belt particles is the decay of neutrons produced in cosmic particle interactions with the atmosphere. The outer belt consists mainly of trapped solar particles. In each zone, the charged particles spiral around the geomagnetic field lines and are reflected back between the magnetic poles, acting as mirrors. Electrons reach energies of up to 7 MeV and protons up to about 200 MeV. The energy of trapped heavy ions is less than 50 MeV, although their radiobiological impact is very small.

The trapped radiation is modulated by the solar cycle: proton intensity decreases with high solar activity, while electron intensity increases, and vice versa.

2.4 Resources and Waste

When humans embark on long-duration missions such as the establishment of permanent bases on the lunar surface or travel to Mars for exploration, they will continue to need food, water and air, and to produce metabolic and other waste products.

For these long-duration missions it may not be economical or practical to resupply basic life support needs from Earth. For long-duration missions it will therefore be essential to develop systems to produce food, purify and supply water, regenerate oxygen, and remove undesirable components of the air. A very important issue will be the waste treatment in general. It has to be considered that a variety of different waste products (liquid waste, gaseous waste, solid waste) from different sources (human, processes, power generation etc.) will be collected during long-term missions. Main aim should be to further process the waste products to get back all inherent consumables, so that a minimum of substances will really be lost. "Environmental control and life support systems" (ECLSS) are essential for all kinds of confined volumes or habitats within which people are supposed to live and to work over longer periods of time. The main, practically the only objective of any kind of life support system is to achieve and to control a physiologically acceptable environment within all kinds of confined habitats. In practice, ECLSS's must process human and other outputs, and must provide required input resources for humans and other biological species as required. The measurement and monitoring of relevant parameters from such ECLSS is absolutely necessary in order to know about and to control the actual conditions of the environment within which people are living and working, far away from a safe haven.

3 BIOSENSORS IN SPACE

3.1 Animals as Whole-organism Sensors

The first biosensors in space were whole organisms as substitutes for human beings, because

testing of survivability of spaceflight conditions was considered to be a prerequisite before manned space missions were attempted. In mid 1946 the first animals sent into space were fruit flies to explore the effects of radiation exposure at high altitudes. Numerous monkeys of several species were flown by the United States in the 1950s and 1960s. They were implanted with sensors to measure vital signs, and many were under anesthesia during launch. Able, a rhesus monkey, and Baker, a squirrel monkey, the first living beings, which successfully returned to Earth after traveling in space aboard an American Jupiter AM-18 rocket in 1959 experienced accelerations of $38 \times g$ and were weightless for about 9 min. Thirty-two monkeys flew in the American space program, each had only one mission. Numerous back-up monkeys also went through the programs but never flew. Monkeys from several species were used. In 1949 the rhesus monkey Albert II became the first monkey in space; however, he died during landing as a consequence of a parachute failure.

The first animal orbiting the Earth was launched on November 3, 1957. The Soviet dog Laika, had been intended to die, as the Sputnik 2 satellite was not designed to be retrievable. The Russian scientists had planned to euthanize Laika with a poisoned serving of food after 10 days of space flight. Russian authorities had previously circulated reports that Laika survived in orbit for four days and then died due to oxygen starvation when the cabin overheated from a battery failure. The physiological variables that were monitored and telemetered back to Earth included electrocardiogram (chest lead), blood pressure, respiration rate, and motor activity.[31] Prelaunch physiological parameters of Laika were normal, but pulse rate went up by a factor three at launch and at peak acceleration the respiration rate had increased three to four times above prelaunch values. At the start of weightlessness the pulse rate decreased to values near the prelaunch rate. However, it took three times longer than after a centrifuge ride on the ground to return to prelaunch values. Electrocardiogram traces also approached normal as the flight continued. However, telemetry showed that temperature and humidity in the dog cabin increased gradually. After 5–7 h into the flight no physiological parameters were transmitted and on the fourth orbit it was impossible to obtain any data on the condition of the dog. Postflight simulations showed that Laika probably died on the third or fourth orbit, due to overheating, fear and stress. But that was enough to prove that a living organism could tolerate a long time in Apollo, and later in Skylab.

Later animals were flown to investigate various biological processes and the effects microgravity and space flight might have on them. As of 2007, the national space programs of the United States, the Soviet Union, France, China and Japan have flown animals into space. Such missions, especially missions with "primate" astronauts, such as "Ham", demonstrated the ability to perform tasks during spaceflight.

3.2 Humans as Whole-organism Sensors

The first humans in space, the Russian cosmonauts as well as the US astronauts, can be considered as test objects for analyzing space environmental factors. As the test animals before, space travelers were connected to measurement devices and data were transmitted to Earth.[32] Nowadays, the conduction of these investigations has been facilitated by the HRF as part of the US laboratory Destiny.[33] From the short-term space missions in the early era of space flight, acute effects were observed and future mission planning was derived from such information. Up to now, more than 400 humans of both sexes and different age traveled into space and were confronted with the two main risk factors, namely microgravity and radiation, for various periods of time. The longest single flight (438 days) was performed by cosmonaut Polyakov in 1995; Expedition-11 Commander Krikalev became the human with the most cumulative time in space with a total of 803 days, 9 h and 39 min in space, including eight EVA's; and John Glenn became the oldest astronaut aged 77 years when flown the second time to space.

Cosmonauts' as well as astronauts' physiological data were monitored during performance inside the spaceship and outside during EVA's and analyzed nearly on-line after transmittance to Earth. Short-term effects induced by the state of weightlessness include "space sickness syndrome" which is very similar to seasickness or general travel sickness on Earth. The symptoms of normally disappear within a time span of a few hours up to five days. Changed hydrostatic pressure gradients along the body axes cause a fluid shift within the body, and the input to the body's many gravity receptors is altered significantly.[34] A long-term effect and a severe medical problem is the loss of bone and muscle mass during longer stays under microgravity conditions, which is based on an adaptation of the body to the lack of gravity (Table 1). So far, investigations have shown, that humans loose about 10% bone mass during a one year stay on a space station under microgravity. As a result, the physical and mental performance of the person in orbit drops and the ability to fast readapt again to gravity decreases. Furthermore, the decrease of performance is accompanied by cardiac arrhythmia, especially during extra vehicular activities. It is not clear so far, which level of gravity is sufficient to prevent the described problems.

The problem of potential hazard to astronauts from cosmic ray HZE particles became "visible", when the astronauts of the Apollo 11 mission returning from the Moon, reported of light flashes, that is, faint spots and flashes of light at a frequency of 1 or 2 min^{-1} after some period of dark adaptation. These events were observed during trans–lunar coast, in lunar orbit, on the lunar surface, and during trans-earth coast. Evidently, these light flashes that were predicted by Tobias in 1952, result from HZE particles of cosmic radiation penetrating the spacecraft structure and the astronaut's eyes and producing visual sensations through interaction with the retina. During the following six Apollo missions that carried the spacecraft outside the magnetic shielding of the Earth and on the Apollo–Soyuz Test Project (ASTP) in LEO as well as at ground-based accelerators, systematic observations demonstrated a variety of different types of flashes, such as thin short, or long streaks, double streaks, star like flashes or diffuse clouds, respectively, that were white in general. However, the pattern of types of flashes was different in LEO, in lunar missions, or at accelerators.

In order to record the passage of HZE particles through the astronaut's head and eyes and to correlate them with observed light flashes, a helmet-like device with nuclear emulsions was used by the crew of Apollo 16 and 17. This Apollo light flash moving emulsion detector (ALFMED) consisted of two sets of glass plates coated with nuclear emulsions, of which one set was fixed in position, whereas the second parallel located set was

Table 1. A comparison of some space-related health concerns with medical issues on Earth

Research area	Space	Earth
Bone	Bone loss and increased fracture risk Increased kidney-stone formation Injury to soft connective tissue	Osteoporosis and other bone disorders
Cardiovascular	Postflight orthostatic intolerance Cardiac atrophy Heart rhythm disturbances	Orthostatic intolerance Heart disorders, such as sudden cardiac death due to heart rhythm disturbances
Performance/Sleep	Errors due to sleep loss and disruption of the biological clock	Sleep problems due to jet lag, shift work and extended work schedules Accidents due to sleepiness
Immunology/Infection	Activation of dormant viruses in the body Increased infection risk Space flight-related anemia Interference with wound healing	Immune system disorders Viral outbreaks due to stress conditions (shingles, cold sores) Anaemia and other blood disorders
Muscle	Muscle loss and atrophy	Muscle wasting diseases Muscle weakening due to prolonged bed rest, immobilization, nerve crush injury and ageing
Neurovestibular	Space motion sickness and body orientation problems Re-entry vertigo Postflight dizziness, balance, posture and gaze stability	Vertigo and other balance disorders
Radiation effects	Cancer Damage to central nervous system Cataracts and other diseases	Risks from exposure to naturally occurring and work-related radiation

moved at a constant rate of $10\,\mu m\,s^{-1}$ for a total translation time of 60 min. Only in few cases the passage of an HZE particle through the astronaut's eyes coincided with a light flash event. However, the number of HZE particles traversing the eyes of the astronaut during the translation period agreed with the total number of flashes observed during this period.

Investigations on the frequency of visual light flashes in LEO and its dependence on orbital parameters were performed on Skylab 4, on ASTP and recently on MIR. The highest light flash rates were recorded when passing through the South Atlantic Anomaly (SAA). In this part of the orbit, the inner fringes of the inner radiation belts come down to the altitude of LEO which results in a 1000 times higher proton flux than in other parts of the orbit. These high light flash event rates during the SAA passages can be deduced either to the high proton fluxes or to the occurrence of some particles heavier than protons in the inner belts of trapped radiation. With the SILEY experiments on board of the space station MIR,[35] two separate mechanisms for the induction of light flashes

were identified: a direct interaction of heavy ions with the retina causing excitation or ionization, and proton-induced nuclear interactions in the eye (with a lower interaction probability) producing knockout particles. Possible mechanisms for stimulating the retina could be electronic excitation resulting in UV radiation in the vicinity of the retina, ionization in a confined region associated with δ rays around the track, or shock wave phenomena when HZE particles pass through the tissue matrix.

3.3 Sensing Radiation for Human Protection in Space

Knowledge of the radiation situation inside of a space vehicle is mandatory for each mission under consideration and shall be based on in-flight dosimetry data. Such measurements of radiation exposures (Figure 4) were performed during manned spaceflights at various altitudes, orbital inclinations, durations, periods during the solar cycle, and mass shielding, respectively.[36]

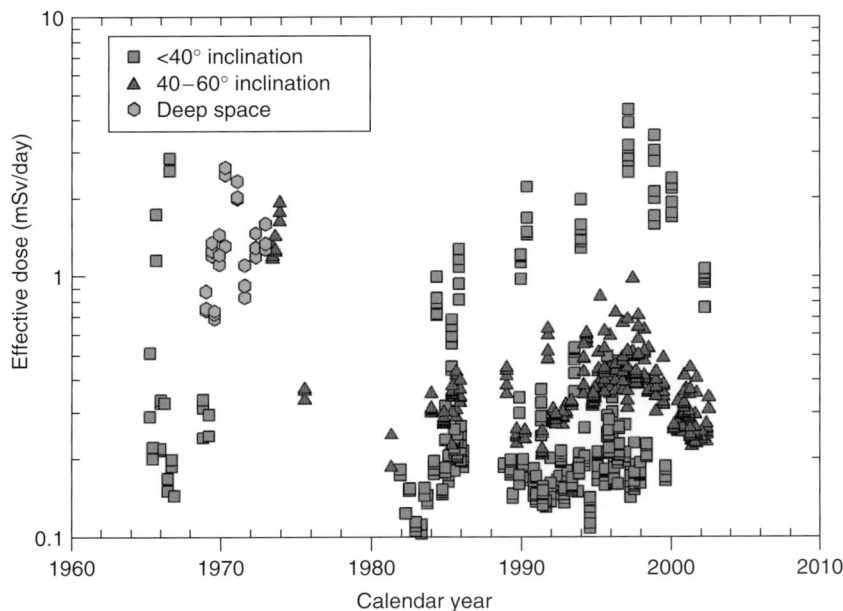

Figure 4. Effective doses, measured in low Earth orbit missions and missions to the Moon.

The deposition of energy by radiation strongly depends on the type of radiation under consideration, both macroscopically and microscopically. The contribution of the sparsely ionizing component of the radiation in space has been mostly determined by lithium fluoride thermoluminescence dosimeters (TLD), which "absorb" radiation dose by its valence electrons being excited to a higher energy state. The number of electrons at the higher energy state is directly proportional to the amount of ionizing radiation the crystal is exposed to. When the crystal is heated, these electrons fall back to their resting energy and emit photons, causing the crystal to glow. The emitted light intensity as a function of the temperature is called the *glow curve*. In a heating cycle the amount of emitted light, that is, the integral of the resulting glow curve, is proportional to the total dose received by the crystal since the last time it was heated ("annealed"). The sensitivity of TLDs is nearly constant in the energy range of interest. For densely ionizing radiation, the spatial pattern of energy deposition at the microscopical level is important. Their fluence has been mainly determined by use of plastic track detectors or nuclear emulsions. Plastic detectors are diallylglycol carbonate (CR39), cellulose

nitrate (CN) or polycarbonate (Lexan), covering different ranges of linear energy transfer (LET). The tracks of heavy ions are developed by etching; the track etching rate grows as a function of the LET. Plastic detectors allow determining the fluence, charge and LET spectrum of the heavy ions. These passive (in the sense that no power is needed during mission) dosimetry systems integrate over exposure time. Advantages are their independence of power supply, small size, high sensitivity, good stability, wide measuring range, and relatively low cost. However, long-duration space missions require time-resolved measurements, especially for protection purposes. This has been met by a small, portable, and space-qualified TLD reader suitable for repeated TLD readouts on board.[37]

In addition to passive dosimeters, active dosimeters have been developed to provide real-time dosimetry data. The measurement principle is based either on ionizations (e.g., ionization chamber, proportional counter, Geiger-Müller Counter, semiconductors, charged coupled devices (CCD)), or on scintillations (e.g., organic or inorganic crystals). A combination of two silicon detectors, the dosimetry telescope (DOSTEL), has been flown onboard of the Space Shuttle, the MIR station, and the ISS. Particle count rates, dose rates, and LET spectra were measured separately for GCR, the radiation belt particles in the SAA and SPEs.[38]

During human spaceflight a personnel dosimetry is required for each astronaut and separately for different activities, above all during EVA, when the astronauts are only shielded by the material of the space suit. A number of active devices such as small silicon detectors or small ionization chambers may be used but they need power and are difficult to design in sufficiently small dimensions. In most cases, passive integrating detector systems have been used, such as TLDs. However, these personal dosimetry systems provide only data on the "surface" or skin dose. In order to assess the depth dose distribution within the human body and especially at the most radiation sensitive organs, such as the brain, the blood forming organs, and the gonads, human phantoms are required equipped with different dosimetry systems at the sites of sensitive organs. The anthropomorphic phantom "MATROSHKA" was exposed for one year to the radiation in space outside of the ISS in order to determine the depth dose distribution of radiations within the human body during EVAs.[39]

Complementary to physical dosimetry, biological dosimetry systems have been developed and applied, which weigh the different components of environmental radiation according to their biological efficacy. They are especially important when interactions of radiation with other parameters of space flight, above all microgravity, may occur. Basically intrinsic biological dosimeters record the individual radiation exposure (humans, plants, animals) in measurable units; and extrinsic biological dosimeters/indicators that record the accumulated dose in biological model system. If applied in parallel with physical dosimetry, biological dosimetry systems can provide valuable complementary information because of reasons as follows: their ability to weigh the different components of environmental radiation according to their biological efficacy; their ability to give a record of the accumulated radiation exposure of individuals; their capacity to monitor the cumulative biological effects of all environmental stressors present; their high specificity; and their generally high sensitivity.

Intrinsic biological dosimeters are for example, G_0 lymphocytes, which respond to ionizing radiation with the expression of chromosome-type aberrations such as polycentric and ring chromosomes. The frequencies of these aberrations are correlated with radiation dose and can therefore be used as a biological dosimeter. After long-term space flights, for example, on board of the MIR station,[40] observed significantly elevated frequencies of chromosome-type aberrations with indications that the aberrations were radiation induced. The frequencies of dicentrics found in lymphocytes of MIR cosmonauts before and after their last space flight compared well with frequencies expected from doses of low and high-LET radiation to which they were exposed during flights. These data have also been used to predict the radiation risks of astronauts during interplanetary space missions. Another promising technique is premature chromosome condensation (PCC) that allows interphase chromosome painting and the detection of nonrejoining chromatin breaks without going though the first mitosis. This method is especially relevant for biological dosimetry for astronauts that are exposed to high doses of high-LET radiation in space, which may induce interphase death and cell cycle delay.

In addition to chromosomal aberrations, other intrinsic biomarkers for genetic or metabolic changes may be applicable as biological dosimeters, such as germ line minisatellite mutation rates or radiation induced apoptosis, metabolic changes in serum, plasma, or urine (e.g., serum lipids, lipoproteins, ratio of HDL/LDL cholesterol, lipoprotein lipase activity, lipid peroxides, melatonin, or antibody titers), hair follicle changes and decrease in hair thickness, triacylglycerol concentration in bone marrow and glycogen concentration in liver. Whereas the first three systems mentioned are noninvasive or require only blood samples for analysis, the latter systems are invasive and therefore appropriate for radiation monitoring in animals only. Dose response relationships have been described for most of the intrinsic dosimetry systems, yet their modification by microgravity remains to be established.

3.4 The ISS as Test Bed for Development and Validation of Biomonitors for Human Space Flight Activities and Earth-bound Applications

In the initial phase of human space flight, engineers, and scientists did not work closely together.

Engineers had concentrated on construction and operations work and most scientists had preferred to have their experiments ride aboard NASA's unmanned satellites. When it became clear in 1963, that the Gemini vehicle did not completely use its room capacity, a chance to set up an experiments program in orderly fashion was seized. Accordingly, more than 600 scientists were invited to send proposals for space experiments. With the establishment of the NASA Manned Space Flight Experiments Board in January 1964 science gained a permanent foothold in manned space flight operations. By the fourth Gemini flight—the second manned—experiments and principal investigators had been worked into mission operations with fair success; by the last flight, procedures had been sharpened sufficiently for the board to continue in Apollo, and later in Skylab. The ISS seems to be ideally suited as a technology test bed and an opportunity for verification and space qualification.[41] In the past, technology development suffered under the lack of flight opportunities. Initial experiments will consist of exposing materials, components, mechanisms, new sensors, spacecraft subsystems, etc. to the environment of space. In the preparatory phase of the Columbus multi-user facility for all four major microgravity research disciplines, namely materials science, fluid physics, biology, and human physiology became integrated in the design and developmental process of the ISPRs and the relevant user communities were actively involved in all reviews of these facilities.

3.4.1 The BIOLAB Facility aboard COLUMBUS

The Biolab is a biological research facility that will be a major contributor to the scientific capabilities of the European Columbus laboratory (Figure 5). It is designed to support biological experiments on micro-organisms, cells, tissue cultures, small plants, and small invertebrates. The effect of gravity on biological samples has been one of the most important areas of research on numerous space missions. The major objective of performing Life Sciences experiments in space is to identify the role that microgravity plays at all levels of an organism, from the effects on a single cell up to a complex organism including humans. The Biolab facility is integrated into an ISPR that will be launched inside Columbus on ISS flight 1E. Prepared standard experiment containers (ECs) and vials will be transported separately within the multi-purpose logistics module (MPLM), a cargo carrier that is located inside the Space Shuttle cargo bay, or other available transport means such as the European automated transfer vehicle (ATV), Russian Progress vehicles or the Shuttle's mid-deck lockers.[42]

After manual loading by the crew, all activities are automatic in the core unit on the left side of the Biolab. This drastically reduces the demand on crew time. The manual section (right side) is mainly for sample storage and specific crew activities. The main element of the core unit is the large incubator, a thermally controlled volume where the experiments take place. The incubator's thermal housing has been built by Ferrari, using mass-saving composite materials derived from Formula 1 racing technology. Inside the incubator are two centrifuges that can each hold up to six ECs and be independently spun to generate artificial gravity in the range from 10^{-3} to $2\,g$. Both are very precisely balanced, to avoid disturbing the space station's microgravity environment.

On top of the incubator, a drawer supports the handling mechanism, the automatic stowage (ambient and temperature-controlled) units and the two analysis instruments (a microscope and a spectrophotometer).[43] The handling mechanism is the key element of Biolab's automation, providing automatic operations such as insertion into the analysis instruments for immediate, in situ analysis. This handling mechanism can also place samples in the automatic stowage units for preservation or further analysis later. Biolab's manual section carries a laptop for crew control, two temperature-controlled units (TCUs) for sample storage and a BioGlovebox (BGB). The TCUs are cooler/freezers ($+10°$ to $-20\,°C$) for storing larger items and ECs. To minimize mass, the TCU skeleton is made of carbon fiber reinforced plastics on which layers of aluminized kapton as an infrared radiation barrier are glued. Taking advantage of the lack of convection in microgravity, the air between the layers also acts as thermal insulation. The BGB is a safe cabinet for handling toxic materials (such as the fixatives used to terminate biology experiments) and delicate biological samples that must

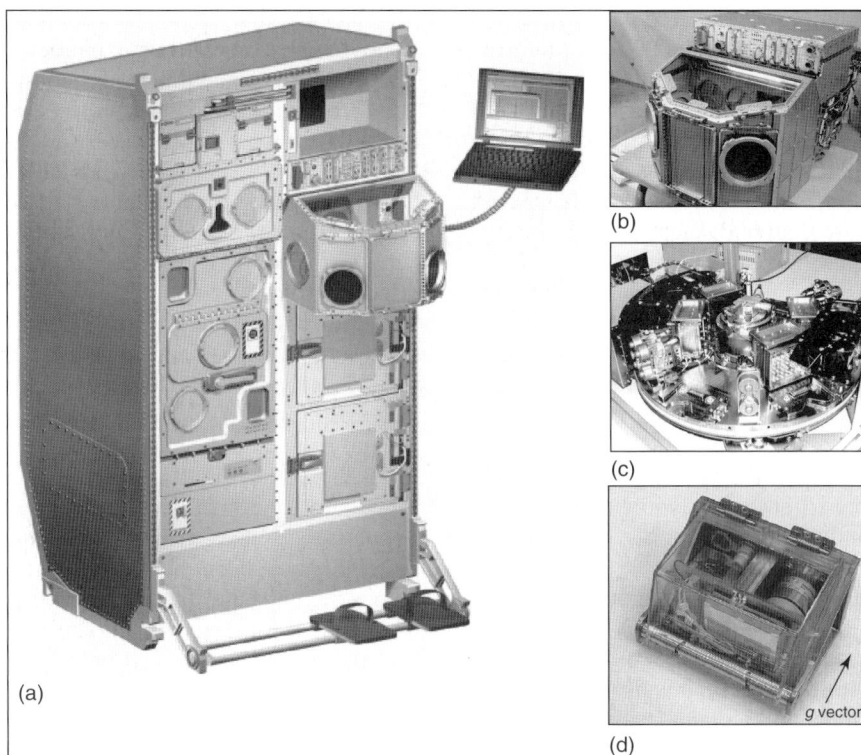

Figure 5. Biolab—one of the payloads racks that will fit inside Columbus. Biolab's right side is devoted to manual elements (a), such as the protruding BioGlovebox (b), while the left side houses the automated units, such as the incubator containing the centrifuge (c). A Biolab Experiment Container (d) with three sample vials. The large cylinder at right is part of the performance verification hardware in this test version.

be protected against contamination by the space station environment. An ozone generator ensures sterilization of the BGB working volume. This generator was developed specially for Biolab but it could also become a useful spin-off on Earth, where, for example, it could be used for sterilizing dental equipment.

The ECs are basically "empty boxes", with standard interfaces to Biolab that carry the biological samples.[44] Scientists need to concentrate only on the experiment hardware contained by the EC, with an available volume of $60 \times 60 \times 100\,mm^3$. The biological samples, together with their ancillary items are transported from the ground to Biolab, either within the ECs or in small vials. The latter case will apply if the samples require storage prior to use, in the minus eighty laboratory freezer for ISS (MELFI). In orbit, astronauts will manually insert the ECs into Biolab for processing. A frozen sample will

first be thawed out in the experiment preparation unit (EPU) and then installed inside the BGB.[45]

Once this manual loading is accomplished, the automatic processing of the experiment can be initiated by a crew member. The experiments are undertaken in parallel on a 0 and a $1\,g$ centrifuge respectively. The latter provides the flight reference experiment, while the ground reference experiment is performed at the facility responsible centre. During processing of the experiment, the facility handling mechanism will transport the samples to the facility's diagnostic instrumentation, where, through tele-operations, the scientist on the ground can actively participate in the preliminary in situ analyses. Typical experiment durations range from 1 day to 3 months. The facility responsible centre for the Biolab facility will have the overall responsibility of operating it according to the needs of individual EC providers.

The individual EC providers could monitor the processing of their experiments from their user home bases.

The selection, definition, and development phases of a Life Sciences flight research experiment have been consistent throughout the past decade. The implementation process is driven primarily by the shift from highly integrated, dedicated research missions on platforms with well-defined processes to self-contained experiments with stand-alone operations on platforms, which are being concurrently designed. For experiments manifested on the ISS or on short duration missions, the more modular, streamlined, and independent the individual experiment is, the more likely it is to be successfully implemented before the ISS assembly is completed. During the assembly phase of the ISS, science operations are lower in priority than the construction of the station. After the station has been completed, it is expected that more resources will be available to perform research. The complexity of implementing investigations increases with the logistics needed to perform the experiment. Examples of logistics issues include: hardware unique to the experiment; large up and down mass and volume needs; access to crew and hardware during the ascent or descent phases; maintenance of hardware and supplies with a limited shelf life; baseline data collection schedules with lengthy sessions or sessions close to the launch or landing; onboard stowage availability, particularly cold storage; and extensive training where highly proficient skills must be maintained. As the ISS processes become better defined, experiment implementation will meet new challenges due to distributed management, on-orbit resource sharing, and adjustments to crew availability pre- and post increment.[46]

3.4.2 The ESA Topical Team on Biomonitors

Research in Space Life Sciences has evolved rapidly through several stages. The first objective of this research was oriented toward the survival and safety of the crew. When the progress of technology made access to space easier, spaceflight, and especially microgravity became a real tool for research. Nowadays, well-controlled experiments are paving the way for consolidated research in Space Life Sciences as well as in other sciences.

In order to optimize the research outcome of these experiments, ESA's Life Sciences Working Group and the Microgravity Advisory Committee identified so-called focused science topics. These topics were chosen because they are the most promising research areas and in order to stimulate collaborative research initiatives.

ESA is encouraging and actively supporting the teaming-up of European scientists from life sciences, biotechnology, and physical sciences, who all share a common interest in performing experiments under microgravity conditions using the ISS, other carriers, or ground facilities. Such topical teams (TT) comprise researchers from academia and industry who are already actively involved in space experiments, as well as scientists who are presently not yet involved in space research. Topical Team meetings are an excellent forum for discussing international cooperation in space research with scientists within and beyond the European borders.

TT have been created in different fields of Life Sciences research using space environment. Their tasks are spread over the support of hardware concepts, their development, and adequate use, the promotion of coordination, collaboration, and cooperation, definition of control experiment, promotion of applications, ground research, and promotion of space research by education. TT will also define the objectives of the future research in space environment by identifying innovative high-priority research areas. The following TT in the fields of biotechnology are presently active: the Biomonitors TT, the Controlled Tissue Development TT, the Gas Liquid Transfer TT, the Microencapsulation TT, the Nutrition TT, the Tissue, and Cell Engineering TT, and the Preservation of Samples TT.

In 1999 the Biomonitors TT came together at first, aiming to identify the potential of the innovative biotechnology research area within a space application program. It assesses environmental factors of potential risks that can be addressed by biomonitoring systems, especially by systems that respond to the gravity stress, to confinement, and/or to an increased level of radiation. Such external biomarkers should complement the spectrum of physical instruments and devices, since biological systems are capable of weighting the effects, which space environments of differing complexities exert on living beings. From the planetary protection point of view biomarkers and biosensors would have the potential to add considerable value to the analysis of the bioburden of spacecrafts and spacecraft assembly facilities. In such artificial environments micro-organisms, if they withstand the controlled environmental conditions in the clean rooms, are expected to be the source of biological contaminations for spaceships and as a consequence for other planets. Accordingly, environmental monitoring for life-related compounds and factors plays an important role in defining and managing the risks to artificial ecosystems on other planets from biological contaminations. Complementary to currently available physical and chemical monitoring techniques, the use of biomarkers as well as of biosensors has the potential of recognizing the composition of microbial communities in closed systems.

3.4.3 Biotechnology for the ISS

Since the days of Mercury flights, astronauts have been connected to myriad electronic sensors that monitor their vital signs. Some astronauts are wired with electrodes, while some sensors might even be swallowed. However, nanotechnology inspired biosensors may change that, making spacecraft like the space shuttle simpler, safer, and more efficient operations. The creation of nanodevices such as nanobots capable of performing diagnostic or therapeutic functions *in vivo* is a destination within the emerging field of nanomedicine. Significant technological advances across multiple scientific disciplines, including miniaturization of analytical tools, led to the development of 3-D nanobot spheres. After being injected into the blood stream of an astronaut, the inexpensive sensor would be phagozytosed by macrophages or taken up into lymphocytes by receptor-controlled transport where it would monitor vital or death signals, for example, the presence of a virus or the expression of apoptosis related proteins. Such self-destructing cells express particular markers proteins, the CD-95 receptors, which can be tagged by specific antibodies delivered by the nanobots. By coupling with light-emitting reporter molecules, such tagged cells would be counted as they flow through capillaries beneath the eardrum by a tiny laser which is shaped like a hearing aide and is inserted in the astronaut's ear. A wireless link would relay the data to the spaceship's main computer for further processing. This scenario is at least 15 years away; however, many of the necessary pieces are already taking shape in the laboratory.[47]

DNA sequence biosensors for the detection of life signs are currently being developed as alternatives to conventional DNA microarrays and can be used to verify presence or absence of life signs on spaceship surfaces or in habitats. In the endeavor to develop a method significantly more rapid and sensitive than microarrays, most biosensors are at the proof-of-principle stage and only few are now in use as laboratory tools. The biosensor for DNA sequence detection relays on the molecular recognition event of a known probe with its counterpart directly on the surface of an active signal transducer. The most sensitive and functional technologies use fiber optics or electrochemical sensors in combination with DNA hybridization. Biosensors with single-molecule resolution represent an evolution of these systems, and may be the next generation of DNA sequence biosensors.[48]

Monitoring the environment for compounds and factors of concern plays an important role in defining and managing the risks to humans and ecosystems resulting from physical, chemical and biological contamination of the environment. Biological monitoring represents a relatively new branch in biotechnology with a worldwide rapidly growing market. Trends in biomonitoring are distinguished by miniaturization, implementation of new functions and cost-effective production. However, at present, only few devices have reached the state of standard environmental monitoring equipment. Therefore, it is expected, that the development of biomonitoring systems for applications in space will be performed concomitantly with the development of biomonitoring systems for routine applications on Earth. The unique monitoring challenges presented by the ISS and the demand for high reliability of space technology will be a driver in the development of multifunctional biological monitoring systems for a wide range of applications in space and on Earth. Complementary to currently available physical and chemical monitoring techniques, bioanalytical methods and the use of biosensors as well as internal and external biomarkers have the potential of providing solutions to a number of environmental problems.

Matters of concern are of global as well as of local nature and include closed habitats, such as given by the ISS as well as future extraterrestrial habitats accommodating artificial ecosystems for bioregenerative life support purposes.

Several of such biomonitoring systems suitable for applications in the ISS are based on gene expression, signal transduction, immune responses, and cell metabolisms. A step in this direction is the space experiment TRIPLE LUX (Gene, immune, and cellular responses to single and combined space flight conditions), which has been selected by ESA to be performed during the first increment of Biolab. The principal investigator and several co-investigators are members of the Biomonitors TT.

The SOS-Lux-toxicity test, as part of the TRIPLE-LUX project will provide an estimation of the health risk resulting from exposure of astronauts to the radiation environment of space in microgravity. The bacterial biosensor, developed at the German Aerospace Center (DLR), consists of recombinant bacteria are transformed with the pBR322-derived plasmid pPLS-1 carrying the promoterless lux operon of Photobacterium leiognathi as the reporter element controlled by a DNA damage-dependent SOS promoter as sensor element. The SOS-Lux-test indirectly monitors the kinetics of DNA damage-processing in the SOS system by measuring the luminescence emitted by the genetically altered bacteria. This system enables performing the test from radiation to data acquisition in microgravity, and therefore the estimation of the effect of the combination of space radiation and microgravity on the living cell, and particularly on the DNA damage repair system. Absorption is measured to provide data for the correction of a growth-related luminescence increase and for the detection of cytotoxicity. Automation and on-line data registration as early as 2 h after the test start satisfy minimal crew time requirements and make this test applicable for use on the ISS. During genotoxicity measurement campaigns, in comparison with other genotoxicity tests, the test proved its reliability, and low genotoxicity detection limit, as well as its short response time, in double-blind tests.[49]

A similarly designed biosensors for toxicity measurements uses brewer's yeast *Saccharomyces cerevisiae* modified with the jellyfish gene coding for green fluorescent protein (GFP) as a reporter

for the DNA damage-induced gene RAD54. A measure of the reduction in cell proliferation is used to characterize general toxicity. Biosensors like these have already found use in fields such as environmental monitoring, and even chemical weapons detection. This biosensor, codenamed FORRAY (fluorescence orbital radiation risk assessment using yeast), was send to space with expedition 9 in the spring of 2004 launching from the Russian Space Agency's launch site with a Soyuz rocket. For the experimental period of six days, the astronauts pressed a plunger every day, thereby mixing millions of dormant yeast cells with nutrients, forcing them into two compartments: one exposed to the damaging space radiation, and one shielded by aluminum. After return to Earth, the cellular fluorescence in the exposed and unexposed yeast samples was measured allowing them to link radiation levels to DNA damage. The same yeast, trademarked as "GreenScreen", is used by the University of Manchester Institute of Science and Technology (UMIST) spin out company Gentronix Ltd in products designed to detect potential DNA damaging agents in drug development and environmental samples.[50]

Because space research sheds light on the fundamental effects of gravity on tissue formation and development, continued cell culture research aboard the ISS will allow scientists to refine Earth-based biomedical techniques. By using space-based experimental platforms as a model, basic research will add valuable information to biomedicine and environmental sciences.

A biomonitoring system, based on the use of mammalian cells is the space experiment "Cellular Responses to Radiation in Space" (CERASP).[51] At the Radiation Biology Department of the German Aerospace Center (DLR), mammalian cell systems were developed as reporters for cellular signal transduction modulation by genotoxic environmental conditions. The space experiment CERASP to be performed at the ISS will make use of reporter cell lines thereby supplying basic information on the cellular response to radiation applied in microgravity. One of the biological endpoints will be survival reflected by radiation-dependent reduction of constitutive expression of the Enhanced variant of Green Fluorescent Protein (EGFP). A second endpoint will be gene activation by space flight conditions in mammalian cells, based on fluorescent promoter systems using the destabilized d2EGFP variant. The promoter element to be investigated reflects the activity of the nuclear factor kappa B (NF-κB) pathway. The NF-κB family of proteins plays a major role in the inflammatory and immune response, cell proliferation, and differentiation, apoptosis, and tumor genesis. Results obtained with X-rays and accelerated heavy ions produced at the French heavy ion accelerator GANIL (Grand Accelerateur National d' Ions Lourds) imply that densely ionizing radiation has a stronger potential to activate NF-κB dependent gene expression than sparsely ionizing radiation. The correlation of NF-κB activation to negative regulation of apoptosis could favor survival of cells with damaged DNA. A third endpoint to be examined will be DNA damage induced by combined exposure to radiation and microgravity and its repair.

The nematode *Caenorhabditis elegans* has been proposed as a whole organism experimental system for behavior and longevity studies in space due to its relative simplicity, short generation cycle, and life span. A miniaturized compact device for cultivating nematode worms in a microfluidic environment and monitoring them using shadow imaging was designed. It consists of a polycarbonate micro culture chamber with a volume of 4 μl, enabling gas exchange through a permeable membrane forming the top of the chamber. Images are acquired with a complementary metal-oxide semiconductor (CMOS) video camera chip attached to the bottom of the micro chamber. As an alternative to video acquisition, the filtered video output signal is used to determine worm activity, yielding a system that allows image acquisition in combination with a low-bandwidth activity measurement. Until now, five Space Shuttle missions have carried *C. elegans* experiments, but high cost and limited availability of space shuttles has led to the efforts to use satellites as a launch platform. Using the described system, the activity of *C. elegans* as a function of temperature was measured. Data obtained by manual counting of worm strokes and the video signal showed good correlation. The first demonstration of maintaining and monitoring *C. elegans* in a micro fluidic environment[52] makes the system attractive for research at the ISS.

3.5 Future Perspectives

NASA embodies the human spirit's desire to discover, to explore, and to understand. The Space Shuttle and ISS are not viewed as ends in themselves, but the means to achieving the broader goals of the nation's space program. Transportation and orbital facilities support and enable our efforts in science, exploration, and enterprise.

As NASA defines and expands its goals and objectives for long-duration exploration of space, interest in genetics, cell, and molecular biology have become key and critical topics. Increasingly, the capability to perform autonomous, in situ acquisition, preparation and analysis of biological samples and specimens to determine the presence and composition of biological components is required for both space biology and medical researchers. Technology developments and advances are needed to support applications across all of the relevant technology application areas, including bioastronautics, fundamental biology, and astrobiology.

Biological and biomolecular as well as genomic research is enabling unprecedented insight into the structure and function of cells, organisms, and subcellular components and elements, and a window into the inner workings and machinations of living things. Triggered by advances in microelectronics and related areas, we are now able to fabricate and construct devices and components such as sensors, actuators, machines, motors, valves, switches, pumps, and other items on the same scale as the biological targets of interest, even in some cases on the order of tens of nanometers in size. This directly scaled relationship allows for new strategies and interactions between physical devices and living systems.

These techniques and technologies have permitted the emergence of a new class of instruments and devices, generally described as mesoscale technologies. Many devices, techniques, and products are now available or emerging, which allow measurement, analysis, and interpretation of the biological composition at the molecular level, and which permit determination of DNA/RNA and other analytes of interest.

Finally, advances in information systems and technologies, and bioinformatics, provide the capability to understand, simulate, and interpret the large amounts of complex data being made available from these biological–physical hybrid systems. These synergistic relationships facilitate the development of revolutionary technologies in many areas, and bode well for the future of space biology research objectives.

It is evident that, biomonitors and biomedical sensors will keep improving. They will get a great, in many cases even vital, contribution to astronaut health. The biosensors, as being an integral part of biomedical devices, will continue to be a prosperous market segment. This suggests an ever-evolving trend coupled along with innovation. The biomedical sensors are also suitable and applicable for a broad spectrum of monitoring purposes in risky environments in space as well as on Earth.

REFERENCES

1. G. S. Titov, Preparation and results of a 24-hour orbital flight. *Life Science Space Research*, 1963, **1**, 128–140.
2. J. P. Loftus, Historical overview of NASA manned spacecraft and their crew stations. *Journal of the British Interplanetary Society*, 1985, **38**, 354–370.
3. C. A. Berry, Medical legacy of Apollo. *Aerospace Medicine*, 1974, **45**, 1046–1057.
4. N. L. Johnson, Military and civilian Salyut space programmes. *Spaceflight*, 1979, **21**, 364–370.
5. J. H. Disher, Manned space stations—a perspective. *Earth-Oriented Applications of Space Technology*, 1982, **2**, 167–177.
6. F. Sietzen, Mir: resting in peace. *Aerospace America*, 2001, **39**, 36–43.
7. H. Binnenbruck, H. W. Gronert, and C. Nitzschke, Mir 92 project. *Space Technology*, 1993, **13**, 205–208.
8. R. D. Andresen and R. Domesla, The euromir missions. *European Space Agency Bulletin*, 1996, **88**, 6–12.
9. J. Oberg, Shuttle-Mir's lessons for the international space station. *IEEE Spectrum*, 1998, **35**, 28–37.
10. L. David, International space station: becoming a reality. *Aerospace America*, 1999, **37**, 1–15.
11. J. D. F. Bartoe and L. Fortenberry, One year old and growing: a status report of the international space station and its partners. *Acta Astronautica*, 2000, **47**, 589–597.
12. J. Grey, Columbia—aftermath of a tragedy. *Aerospace America*, 2003, **41**, 26–29.
13. F. Culbertson, A tour of the ISS. *Acta Astronautica*, 2004, **54**, 793–797.
14. S. R. Morrissey, Completing the station. *Chemical and Engineering News*, 2006, **84**, 28.
15. G. H. Kitmacher, W. H. Gerstenmaier, J. D. F. Bartoe, and N. Mustachio, The international space station: a pathway to the future. *Acta Astronautica*, 2005, **57**, 594–603.
16. N. R. Pellis and R. M. North, Recent NASA research accomplishments aboard the ISS. *Acta Astronautica*, 2004, **55**, 589–598.
17. H. Kozawa, Japanese ISS program involvement. *Acta Astronautica*, 2004, **54**, 787–788.
18. A. Thirkettle, B. Patti, P. Mitschdoerfer, R. Kledzik, E. Gargioli, and D. Brondolo, ISS: Columbus. *European Space Agency Bulletin*, 2002, **109**, 27–33.
19. G. Reibaldi, R. Nasca, H. Mundorf, P. Manieri, G. Gianfiglio, S. Feltham, P. Galeone, and J. Dettmann, The ESA payloads for Columbus—a bridge between the ISS and exploration. *European Space Agency Bulletin*, 2005, **122**, 60–70.
20. B. Patti, R. Chesson, and M. Zell, A thirkettle, columbus: ready for the international space station. *European Space Agency Bulletin*, 2005, **121**, 46–51.
21. G. Horneck, R. Facius, M. Reichert, P. Rettberg, W. Seboldt, D. Manzey, B. Comet, A. Maillet, H. Preiss, L. Schauer, C. G. Dussap, L. Poughon, A. Belyavin, G. Reitz, C. Baumstark-Khan, and R. Gerzer, Humex, a study on the survivability and adaptation of humans to long-duration exploratory missions, Part I: Lunar missions. *Advances Space Research*, 2003, **31**, 2389–2401.
22. G. Horneck, R. Facius, M. Reichert, P. Rettberg, W. Seboldt, D. Manzey, B. Comet, A. Maillet, H. Preiss, L. Schauer, C. G. Dussap, L. Poughon, A. Belyavin, G. Reitz, C. Baumstark-Khan, and R. Gerzer, Humex, a study on the survivability and adaptation of humans to long-duration exploratory moissions, Part II: missions to mars. *Advances Space Research*, 2006, **38**, 752–759.
23. P. Webb, The space activity suit: an elastic leotard for extravehicular activity. *Aerospace Medicine*, 1968, **39**, 376–383.
24. A. G. Kousnetzov, Some results of biological experiments in rockets and Sputnik II. *Journal of Aviation Medicine*, 1958, **29**, 781–784.
25. H. J. von Beckh, Human Reactions during flight to acceleration preceded by or followed by weightlessness. *Aerospace Medicine*, 1959, **30**, 391.
26. A. C. Fisher, Aviation Medicine on the Threshold of Space, *National Geographic*, 1955, **108**, 241–278.
27. D. G. Simons, Review of biological effects of subgravity and weightlessness. *Jet Propulsion*, 1955, **25**, 209–211.
28. W. N. Spjeldvik, S. Bourdarie, and D. Boscher, Towards multi-dimensional space weather modeling for energetic oxygen ions in the earth's inner magnetosphere: equilibrium configuration. *Advances Space Research*, 2002, **30**, 2839–2842.
29. D. F. Smart and M. A. Shea, The local time dependence of the anisotropic solar cosmic ray flux. *Advances Space Research*, 2003, **32**, 109–114.
30. G. D. Badhwar and P. M. O'Neill, Long-term modulation of galactic cosmic radiation and its model for space exploration. *Advances Space Research*, 1994, **14**, 749–757.
31. J. B. West, Historical aspects of the early Soviet/Russian manned space program. *Journal Applied Physiology*, 2001, **91**, 1501–1511.
32. H. S. Fuchs, Hypertension and orthostatic hypotension in applicants for spaceflight training and spacecrews: a review of medical standards. *Advances in Space Research*, 1983, **3**, 199–204.

33. J. J. Uri and C. P. Haven, Accomplishments in bioastronautics research aboard international space station. *Acta Astronautica*, 2005, **56**, 883–889.

34. R. J. White and M. Averner, Humans in space. *Nature*, 2001, **409**, 1115–1118.

35. M. Casolino, V. Bidoli, A. Morselli, L. Narici, M. P. De Pascale, P. Picozza, E. Reali, R. Sparvoli, G. Mazzenga, M. Ricci, P. Spillantini, M. Boezio, V. Bonvicini, A. Vacchi, N. Zampa, G. Castellini, W. G. Sannita, P. Carlson, A. Galper, M. Korotkov, A. Popov, N. Vavilov, S. Avdeev, and C. Fuglesang, Dual origins of light flashes seen in space. *Nature*, 2003, **422**, 680.

36. G. Reitz, Space radiation dosimetry. *Acta Astronautica*, 1994, **32**, 715–722.

37. I. Apáthy, S. Deme, I. Fehér, Y. A. Akatov, G. Reitz, and V. V. Arkhanguelski, Dose measurements in space by the Hungarian Pille TLD system. *Radiation Measurements*, 2002, **135**, 381–391.

38. R. Beaujean, J. Kopp, S. Burmeister, F. Petersen, and G. Reitz, Dosimetry inside MIR station using a silicon detector telescope (DOSTEL). *Radiation Measurements*, 2002, **35**, 433–438.

39. J. Dettmann, G. Reitz, and G. Gianfiglio, MATROSHKA-The first ESA external payload on the international space station. *Acta Astronautica*, 2007, **60**, 17–23.

40. G. Obe, I. Johannes, C. Johannes, K. Hallmann, G. Reitz, and R. Facius, Chromosomal aberrations in blood lymphocytes of astronauts after long-term space flights. *International Journal of Radiation Biology*, 1997, **72**, 726–734.

41. S. A. Voels and D. B. Eppler, The international space station as a platform for space science. *Advances in Space Research*, 2004, **34**, 594–599.

42. A. Petrivelli, The ESA laboratory support equipment for the ISS. *European Space Agency Bulletin*, 2002, **109**, 35–54.

43. E. Brinckmann, New facilities and instruments for developmental biology research in space. *Advances in space biology and medicine*, 2003, **9**, 253–280.

44. E. Brinckmann, ESA hardware for plant research on the international space station. *Advances in Space Research*, 2005, **36**, 1162–1166.

45. E. Brinckmann, BIOLAB, EPU and EMCS for cell culture experiments on the ISS. *Journal of Gravitational Physiology*, 2004, **11**, 67–74.

46. L. J. Miller, C. P. Haven, S. G. McCollum, A. M. Lee, M. R. Kamman, D. K. Baumann, M. E. Anderson, and M. C. Buderer, The international space station human life sciences experiment implementation process. *Acta Astronautica*, 2001, **49**, 3–10.

47. E. S. Kawasaki and A. Player, Nanotechnology, nanomedicine, and the development of new, effective therapies for cancer. *Nanomedicine: Nanotechnology, Biology and Medicine*, 2005, **1**, 101–109.

48. W. Vercoutere and M. Akeson, Biosensors for DNA sequence detection. *Current Opinion in Chemical Biology*, 2002, **6**, 816–822.

49. E. Rabbow, N. Stojicic, D. Walrafen, C. Baumstark-Khan, P. Rettberg, D. Schulze-Varnholt, M. Franz, and G. Reitz, The SOS-LUX-TOXICITY-test on the international space station. *Research in Microbiology*, 2006, **157**, 30–36.

50. A. W. Knight, P. O. Keenan, N. J. Goddard, P. R. Fielden, and R. M. Walmsley, A yeast-based cytotoxicity and genotoxicity assay for environmental monitoring using novel portable instrumentation. *Journal of Environmental Monitoring*, 2004, **6**, 71–79.

51. C. E. Hellweg, M. Thelen, A. Arenz, and C. Baumstark-Khan, The German ISS-experiment cellular responses to radiation in space (CERASP): the effects of single and combined Space Flight Conditions on Mammalian Cells. *Advances in Space Research*, 2007, in press, doi:10.1016/j.asr.2006.11.015.

52. D. Lange, C. W. Storment, C. A. Conley, and G. T. A. Kovacs, A microfluidic shadow imaging system for the study of the nematode Caenorhabditis elegans in space. *Sensors and Actuators B*, 2005, **107**, 904–914.

Life Detection within Planetary Exploration: Context for Biosensor and Related Bioanalytical Technologies

David C. Cullen[1] and Mark R. Sims[2]

[1] *Cranfield Health, Cranfield University, Silsoe, UK and* [2] *Department of Physics and Astronomy, University of Leicester, Leicester, UK*

1 INTRODUCTION

Are we the unique example of life in the universe? If given appropriate conditions, is the emergence of life commonplace throughout the universe? These are fundamental questions for humanity and ones whose answers would be at the pinnacle of discoveries within human history to date. Biosensors, biochips, and other bioanalytical technologies have an emerging role in attempting to answer these questions due to their proposed inclusion as instruments on rover missions to the planet Mars. Mars is a focus as it is one of the best candidates within the solar system as a possible abode of life either now or in the past. Here the sensors would be used as part of a suite of complementary analytical instruments to search for molecular evidence of ancient and current life in the Martian environment. Finding evidence of life on Mars could indicate that life here on Earth is not necessarily unique and therefore potentially widespread within the universe.

This chapter's authors are the lead investigators for a project to develop an antibody microarray-based instrument called *SMILE* (*S*pecific *M*olecular *I*dentification of *L*ife *E*xperiment) and more generically termed the *life marker chip* (LMC) instrument. This instrument has been selected into the baseline instrument payload for the European Space Agency's (ESA) ExoMars rover mission[1]; a mission scheduled for launch in 2013. The development of the LMC is at an early stage and therefore this chapter will focus on describing (i) a brief scientific case for life on Mars, (ii) the analytical targets this implies, (iii) previous approaches to in situ life detection on Mars, (iv) the environmental, system-level, and regulatory constraints that all impact upon the design of biosensor and bioanalytical technology intended for such applications, and (v) an overview of the biosensor approaches being taken in the development of suitable instruments. The goal of the chapter is therefore to inform the biosensor and biochip community of the unique challenges that planetary exploration applications offer and encourage others within this community to consider these challenges. It is anticipated that addressing the challenges may produce solutions that will also benefit and advance biosensor technology for more routine Earth applications.

2 LIFE AND ORGANICS IN THE SOLAR SYSTEM

To place into context the requirement for in situ analytical techniques such as biosensors and

Handbook of Biosensors and Biochips. Edited by Robert S. Marks, David C. Cullen, Isao Karube, Christopher R. Lowe and Howard H. Weetall.
© 2007 John Wiley & Sons, Ltd. ISBN 978-0-470-01905-4.

related bioanalytical technologies for life detection in planetary exploration, a brief overview of the science case for life elsewhere in the solar system is given.

2.1 Formation of the Solar System

The current view of the formation of the solar system is summarized as follows.[2] The solar system initially formed approximately 4.6 Gyr ago over a period of approximately 100 Myr by the gravitational collapse of a dense molecular cloud. The gravitational collapse resulted in the formation of the early Sun, which heated the associated protoplanetary disk. The heating in the inner parts of the protoplanetary disk only allowed the condensation of refractory materials—metals and minerals—from the original materials of the molecular cloud. The rocky terrestrial planets, including Earth and Mars, eventually formed from the refractory-material-rich inner regions of the protoplanetary disk. Other gases were cleared from the inner early solar system by the Sun's solar wind and UV radiation effects to the outer and cooler parts of the solar system. Here the gas giant planets formed and beyond which the icy bodies (comets, dwarf planets, etc.) of the Oort cloud and Kuiper belt came to reside. The refractory-material-rich asteroid belt was formed due to the gravitational influence of Jupiter.

It is thought that the gravitational driven evolution of the solar system resulted in the migration of the orbits of the gas giants, which disturbed the asteroid belt and the Oort cloud about 4.1–3.8 Gyr ago.[3] This resulted in bodies from these regions, asteroids, and comets, being placed into eccentric orbits that crossed the inner solar system resulting in a large increase in the rate of impact events on the terrestrial planets. This has been termed the *Late Heavy Bombardment period* and would have delivered significant levels of volatiles such as water and organics to early Earth, Mars, and Venus as well as possibly providing planetary wide sterilization events for any early surface or near-surface emerging life.

2.2 Abiotic Organics in the Solar System

The initial input of organic material to the protosolar nebula comprised material synthesized in the interstellar medium and the molecular cloud from which the solar system formed. Subsequent processing of the material depended upon its position within the nebula and ensuing protoplanetary disc and may have included thermal and aqueous processing—the latter heated in water-rich planetesimals and asteroids.[4] The present evidence for such abiotically produced material is from meteorites. Some classes of meteorites—the carbonaceous chondrites—contain significant levels of solvent extractable organic material; up to 10% by weight and with polyaromatic hydrocarbons dominating this. The wide range of molecules detected in such classes of meteorites[5] includes amino acids at the tens of parts per million level and purines and pyrimidines at around the parts per million level. For amino acids, over 70 types have been identified, all without any appreciable chiral excess and with a structural population distribution different from the biotic amino acids on Earth.

As explained above, early Earth and other terrestrial planets would have most likely received a significant input of organics from extraterrestrial objects during the Late Heavy Bombardment period.[6]

Such abiotic organic material is still being delivered to all the planets via micrometeorites (derived from interplanetary dust particles) and for Mars at present an estimated 240 tonnes per annum is being delivered.[7] The possible detection of such abiotic organic material needs to be considered in any Martian life detection experiment.

2.3 Venus, Earth, and Mars

During the early evolution of the three outer terrestrial planets it is known for Earth, highly likely for Mars, and possible for Venus, that each had epochs or episodic periods with significant bodies of liquid water present. Additionally, the early organic molecular inventory of the three planets are expected to have been broadly similar—a combination of outgassed material from the planetary interiors and delivery of exogenous materials by impactors. This inventory will have been further developed by geochemical, atmospheric, and thermal processing. It is therefore possible that environments conducive to the emergence of life—i.e., liquid water and a complex inventory of organics—were present on all three planets.

However the subsequent evolutions of the three planets from such early periods differ markedly.

Earth has maintained a relatively stable environment suitable for the continuous presence of life since as early as 3.8 Gyr ago (see Section 2.4.1) and possibly much earlier. Active plate tectonics has resurfaced much of Earth so that it is extremely difficult to find unprocessed rocks from the early Earth that harbor evidence of very early life.

Venus has evolved a runaway greenhouse effect such that the current surface temperature ($> 400\,°C$) is not compatible with life. Additionally, volcanism and tectonics appear to have episodically resurfaced the planet and therefore destroyed any evidence of life that may have been present in an early and possibly clement Venusian environment.

Considering Mars, the current generation of Mars rovers and orbiters have recently allowed a rapid reappraisal of the history of Mars. A current view[8] is that an early warm, wet, and nonacidic chemistry phase was characterized by the formation of phyllosilicate ("clay type") minerals—the mostly likely epoch for the emergence of life (4.5–4.2 Gyr ago). A period of significant volcanic outgassing of volatiles including sulfur followed and resulted in an acidic wet phase with the formation of sulfate minerals (4.2–3.8 Gyr ago). This period may be correlated with the shutting down of the Mars geodynamo and gradual loss of a planetwide magnetic field with the resultant erosion and loss of a possibly dense atmosphere by the solar wind. The resulting final period, and extending to the present, is one of cold temperatures with a thin atmosphere and water-free mineral alteration with the formation of anhydrous ferric oxides (3.8–3.5 Gyr ago and continuing to the present).

There is also considerable speculation and research concerning recent and ongoing water cycling on Mars. Water volume, presence, and mobility due to obliquity changes,[9] seasonal variations, and even possible suggestions of liquid water or brines present near the Martian surface[10] are all under investigation.

An important point when comparing Mars to Earth and Venus is the lack of planetwide resurfacing on Mars since the earliest epochs such that mineral deposits from these periods remain accessible in the present. It is therefore expected that evidence of any early Martian life could be preserved in such deposits.

2.4 Life in the Solar System

There is evidence that life appeared rapidly on early Earth (see Section 2.4.1). It is therefore possible that life could have emerged independently—i.e., three or more independent geneses of life—on all three terrestrial planets under consideration given their likely similar environments. Alternatively, a single genesis of life could have evolved on any one of the three planets and seeded life on the other planets via the mechanism of lithopanspermia (see Section 2.4.3). Within this alternative scenario, life on all three planets would have a common single ancestor and hence a common fundamental biochemistry. This scenario has important implications for any life detection experiments as specific molecular indicators generic to Earth life are known.

2.4.1 Life on Earth

On Earth, evidence of life in the form of fossils of microorganisms can be found at 3.5 Gyr ago[11] and from isotopic evidence, possibly earlier at around 3.8 Gyr ago[12] although this interpretation is still hotly debated in the relevant communities. This latter date corresponds with the end of the Late Heavy Bombardment phase in the evolution of the solar system, which is likely to have resulted in major planetwide sterilizing impacts from asteroid and comets impactors. The implication is that life rapidly emerged on a geological timescale as soon as a stable environment was established. It is important to note that all present life on Earth appears to be the result of a single biochemistry, although other life chemistries may have existed in the distant past and may have been removed by evolution or impact events. This is the only example of life that we currently know.[13]

2.4.2 Life on Mars

At present, we can only speculate on the possibilities of life on Mars but it is important to consider the various possibilities as these not only have important scientific implications but are also crucial inputs into the design of analytical instruments to find evidence of life. Given the preceding discussion of a warm and wet early Mars, a number

Table 1. Various scenarios for the existence of life on Mars

	Scenario	Consequence
1.	Life never evolved on Mars	No evidence of life on Mars
2.	Life independently evolved on Mars in an early wet and warm epoch. Martian conditions changed and Martian life became extinct	Preserved evidence of extinct life present
		Biochemistry may be different to Earth life
3.	Life independently evolved on Mars in an early wet and warm epoch. Martian conditions changed and Martian life evolved to allow survival to the present day in protected environments—such as in a water-rich deep subsurface	Preserved evidence of extinct life present
		Evidence of extant life present
		Biochemistry may be different to Earth life
4.	Life evolved on a single unknown location in the early solar system. Via lithopanspermia, life on Mars and Earth had a common ancestry	Preserved evidence of extinct life present
	Then evolution as in 2	Common biochemistry with Earth life
5.	Life evolved on a single unknown location in the early solar system. Via lithopanspermia, life on Mars and Earth had a common ancestry	Preserved evidence of extinct life present
	Then evolution as in 3	Evidence of extant life present
		Common biochemistry with Earth life

of scenarios can be envisaged and are summarized in Table 1.

2.4.3 Lithopanspermia

Within the context of finding life on Mars, it is crucial to understand that it is possible that life on Mars and life on Earth could both be the same example of life. This understanding derives from a modern interpretation of the concept of interplanetary panspermia, specifically lithopanspermia.

The lithopanspermia[14] hypothesis states that viable microorganisms, protected within rocks, can be transported between planets as impact ejecta. The protection offered by a few centimeters of a mineral phase has been shown to protect organisms from the thermal, vacuum, and radiation environments expected during ejection, interplanetary transfer, and entry and impact phases of this process.[15] The observation that some meteorites on Earth have a Martian origin is further justification for the hypothesis. If this process occurred, then life could have emerged in principle on any suitable location in the solar system and seeded other bodies with suitable habitats. The consequence is that life on more than one body in the solar system could have had a common ancestor. For life detection, this means that generic biochemical indicators of life on Earth might be suitable for detecting life on Mars.

2.4.4 Other Locations

A number of other potential locations for life in the solar system are being considered.

The possibility of extant Venusian life has been considered.[16] At high altitudes in the Venusian atmosphere, temperature and pressure are similar to Earth and while water is present at low concentrations and the environment is highly acidic, these conditions are known to support life on Earth. Thus mission concepts are being developed that include the possibility of detecting life in the upper atmosphere of Venus (For example the possible inclusion of extant life detection instrumentation on a long-duration balloon-based aerobot within the European-led Venus Entry Probe concept; Cullen and Sims, personal communication).

The other class of possible locations for life are those that harbor liquid water. Of these, Europa—one of the Jovian moons—is the subject of great speculation[17] and possible future life detection missions (see Section 4.4.2). Europa is a world that comprises an ice crust covering a putative liquid-water ocean. The presence of a large and long-lived body of liquid water leads to the possibility of the emergence and support of life.

3 ANALYTICAL CONTEXT FOR LIFE DETECTION

The following section will limit discussion to the detection of life on Mars as this is the current community priority. Given the constraint that our understanding of life is limited to the single example of life on Earth, we assume that development of independent Martian life will, given similar environmental constraints, have followed broadly similar pathways to Earth, i.e., similar molecular evolution and solutions. Therefore assuming a water-based life with biomolecules providing membrane components and compartmentalization, energy production and storage mechanisms, storage of hereditary information, exploitation of chirality, and so on, does not appear to be an unreasonable assumption. Furthermore, as will be discussed later, the use of some approaches to biosensor detection of life requires a significant level of predetermination of the nature of the life to be detected.

The term *biomarker* is used within the astrobiology community to refer to any type of evidence of life. The major categories of biomarkers are listed in Table 2.

Only those directly relevant to biosensor and related techniques will be considered further, i.e. items 3–5 from Table 2. Furthermore, items 3 and 4 can be termed *molecular biomarkers*.

3.1 Molecular Targets

A summary of the classes or categories of molecular targets relevant to life detection and additional

Table 2. Classes of biomarkers for the detection of life

1. Structural evidence such as fossilized organisms (e.g., accessible by microscopic observations)
2. Isotopic fractionation between reservoirs (e.g., accessible by mass spectrometry)
3. Organic molecules of obvious biotic origin, i.e., no known natural abiotic synthetic route (possible biosensor targets)
4. Significant chiral excess within appropriate organic molecules (possible biosensor targets)
5. Metabolic activity (possible biosensor targets)

categories relevant to performing and interpreting a life detection experiment are given in Table 3. Some molecular targets can of course have a variety of origins, i.e., biotic and abiotic, see Section 3.1.2. Hence, in reality a number of different targets must be measured and detected to enable a case for positive indication of life to be made.

An international workshop on biomarkers for life detection on Mars was organized in May 2006 at the University of Leicester (UK) as part of the authors' ongoing SMILE LMC development program. The output of the workshop, including a detailed list of possible biomarkers, has been submitted for publication.[18] The following subsections describe the classes of molecular targets relevant to life detection and are partially based upon the publication resulting from the workshop.

3.1.1 Extant and Extinct Life Molecular Biomarkers

Within the context of biosensor and bioanalytical instruments, molecular biomarkers often fall into

Table 3. Categories of molecular biomarkers and other relevant targets for biosensor-based life detection

Category	Context	Typical properties
Molecular biomarkers of ancient extinct life	Stable breakdown/preservation products of life—typically membrane derived	Low-molecular-weight apolar
Molecular biomarkers of extant/recent life	Full range of molecules from metabolites to macromolecules	Low- to high-molecular-weight polar
Molecular biomarkers of microbial spacecraft contamination	Species-specific targets—antigens or nucleic acid sequence	High-molecular-weight macromolecules
Abiotic organics	Meteoritic in fall	Low-molecular-weight apolar and polar
Other targets	Manmade synthetic molecules as positive controls	Various
	Nonbiological contamination	Various—e.g., cleaning materials, rocket exhaust products

one of two main categories, those that are stable breakdown and preservation products of ancient extinct life and those that represent extant or recent life (recent in terms of significant degradation of the core molecular components to recalcitrant molecules). The bases for this discrimination are (i) the obvious scientific context, (ii) the difference in the typical molecular nature of such classes, and (iii) the consequential impact on instrument design and especially molecular biomarker extraction protocols and suitability of targets for biosensor detection.

Molecular biomarkers of ancient extinct life represent the result of degradation, alteration, and preservation and clearly will be influenced by environmental and geological factors including geochemistry, thermal processes, and radiation. On Earth we know the typical products of these processes for ancient Earth life. Nucleic acids and proteins are quickly degraded over geological timescales. Amino acids are modified with certain members of the biogenic amino acids degraded to simpler but longer-lived members. The key observation in Earth's biomarker record from ancient life on Earth is the dominance of degradation, alteration, and preservation products of membrane molecules. The assumption of a similar molecular evolution of function for Martian life leads to targets such as saturated hydrocarbon structures from classes such as hopanes, steranes, isoprenoids, porphyrins, and straight-chain hydrocarbons as well as chirally resolved generic amino acid detection.

Molecular biomarkers of extant and recent life again require some consideration of the nature of Martian life. Assuming similar molecular evolution of function, and even the possibility of a common ancestor with Earth life, leads to the identification of biomarkers generically associated with Earth life. Examples include adenosine triphosphate (ATP), generic pyrimidine and purine bases and their polymers—nucleic acids, amino acids, with chiral discrimination, and their polymers—polypeptides, carbohydrates such as trehalose, lipopolysaccharide, and possibly even conserved proteins such as the molecular chaperones (e.g., GroEL)—the latter examples are of course strongly associated with an expectation of a common ancestor between Earth and Mars life.

3.1.2 Other Molecular Targets

In addition to possible molecular biomarkers of Martian life, a number of other categories of organic molecular targets need to be considered when attempting an in situ Martian life detection experiment based upon molecular biomarkers.

Firstly, it is expected there will be abiotic organic molecules in the Martian environment. As previously mentioned (Section 2.2), one source of abiotic organic molecules is the continual delivery of micrometeorites to the Martian surface. The detection of such molecules is important due to (i) scientific questions concerning abiotic organic molecules in the solar system and on Mars, including questions concerning their degradation and preservation within the Martian environment[24] and (ii) contextual setting for the detection of molecular biomarkers as some abiotics, e.g., amino acids, could also have a biological origin. Examples of such abiotic molecular markers include polycyclic aromatic hydrocarbon (PAHs), isovaline, α-aminoisobutyric acid, and aromatic carboxylic acids.

To validate any in situ life detection measurement, it is critically important to confirm that a false positive signal has not occurred. An obvious source for false positive signals are Earth life molecular biomarkers that could be present as a result of contamination within the life detection instrumentation or elsewhere within a rover including any sample collection and preparation components. As will be described in Sections 5.1 and 5.2, minimizing this possibility is a major design input to any life detection mission and instrument. The detection of specific markers of contamination via the presence of species-specific antigens from microorganisms known to colonize/be present within spacecraft clean-room assembly areas is one approach. Hence in situ detection of such a biomarker would cast doubt upon validity of the co-detection of any generic molecular biomarker as the default interpretation would be that the generic signals came from the contamination.

Finally, to further validate in situ measurements, positive control molecular markers can be envisaged that would be dosed into an experiment. The choice of these is dictated by the criteria of avoiding molecules that have either natural abiotic or biological potential. Examples would be synthetic organic molecules for which specific receptors are readily available. (The authors are using a manmade herbicide—specifically atrazine due to its low molecular weight and relatively polar nature and ready availability of high affinity and specificity antibodies—as a positive control marker in development work for the SMILE LMC experiment.)

3.2   Environmental Context

When considering the detection of biomarkers, the environmental context of the sample in which biomarker detection will occur is crucial. Owing to the intense interest in the results of a life detection experiment on Mars, the limited number of opportunities for measurements and the difficult in revisiting a location at a future date to clarify any ambiguities, the choice of sampling site is crucial. Three major criteria contribute to the choice of a location and comprise (i) expectation for life to have been, or currently be, present, (ii) expectation for the preservation of evidence of life, and (iii) ability to physically access the site.

3.2.1   Suitable Locations

As our understanding of Mars continues to grow due to the variety of currently active surface rovers and orbiters, the possible number of locations that could be considered relevant for life detection also continues to grow. Obvious sites are those that have evidence of the presence of water, either historically or at present.

For extinct, ancient life, locations include weathered volcanic deposits exhibiting possible phyllosilicate signatures suggesting extended periods of water alteration[8] and bedded sediments again suggesting the long-lived presence of water.[19]

For extant or recent life, hydrothermal deposits, recent impact sites, numerous locations of ice, and areas that exhibit terrain features in common with terrestrial permafrosts are all possible locations. A very recent report has suggested the possibility of current outbursts of near-surface water or brines.[10]

Additionally the deep subsurface is an intriguing possible location and mirrors the similar interest directed toward Earth's deep subsurface.[20] Within a life detection context, it is not apparent how access to the Martian deep subsurface locations will be gained within the near future, without perhaps the presence of a manned expedition.

3.2.2   Preservation

For the detection of molecular evidence of life, not only must appropriate locations for life have existed in the past, but conditions suitable for the preservation of such evidence over intervening, potentially billions of, years must have been present. The apparent dominance of cold and dry conditions over a significant portion of Martian history is therefore favorable for preservation of such molecules. Radiation effects, a combination of solar UV radiation, solar energetic particles, galactic cosmic rays, and mineral radioactivity, will however have provided an energy source for the degradation of organic molecules. The depth profile of these effects is such that at depths below a few meters, only indigenous mineral radioactivity will have had a significant effect.[21] Additionally oxidation chemistry is expected to dominate the near-surface chemistry of Mars.[22] This results from the interaction of solar UV radiation with constituents of the atmosphere and the surface regolith producing putative superoxides and other oxidizing species. These are expected to oxidize organic material in the near-surface region. A conclusion therefore is that subsurface locations, of a few meters or more depth, offer the best situations for the long-term preservation of molecular biomarkers.

3.2.3   Accessibility of Locations

It must be remembered that in addition to science-driven choice of a sampling site, operation requirements restrict the choice of locations. This can occur at a global scale with landing site selection limited by factors such as (i) site altitude, (ii) latitude, e.g. due to reduced efficiency of photovoltaic power generation at high latitudes and restrictions imposed by the nature of launch and atmospheric insertion, i.e. from-orbit or by hyperbolic approach, and (iii) surface topography, i.e. topographical features that risk loss of the lander such as boulders and significant gradients. Additionally, local spatial restrictions apply due to limitations of rover mobility and access to areas of significant gradient, surface roughness, and unsuitable regolith consolidation.

4 MISSION HISTORY AND FUTURE OF LIFE DETECTION IN PLANETARY EXPLORATION

To date there have only been three attempts at direct, in situ life detection on another planetary body within the solar system. The first two attempts were the pair of NASA Viking landers that arrived on the Martian surface in July 1976 and September 1976. The third and most recent attempt was the UK's Beagle 2 probe that flew as part of the ESA Mars Express mission and is assumed to have been lost by an off-nominal landing on the surface of Mars in December 2003, which resulted in loss of contact with the lander and therefore no scientific return. All other Martian missions have concentrated on the geochemistry or geology of Mars.

Into the future, further attempts are being planned. The next life detection mission will be the ESA's ExoMars rover mission that is currently due for launch in 2013.[1] Additionally NASA has proposed an astrobiology-focused rover mission called the *Astrobiology Field Laboratory* (AFL)[23] with a possible Mars arrival date in the latter part of the next decade. Early-stage discussions are currently taking place concerning possible future Europa orbiter and possibly lander missions (for examples see Gershman et al.[24] and Atzei et al.[25]) as well as a long-duration upper atmosphere Venus probe with the potential for detecting life in the upper Venusian atmosphere.

4.1 Viking Missions

The NASA Viking missions were developed with detection of Martian life as the main objective. Moreover, there was premission expectation of extant life and this was reflected in the choice of life detection experiments flown. Developed in the late 1960s and early 1970s, four experiments are relevant to life detection[26]; three to detect metabolic activity within any regolith microorganisms and a GC/MS instrument for organic carbon detection.

4.1.1 GC/MS Experiment

A mass spectrometer together with a gas chromatograph and sample heaters was flown.[27,28] For organic detection, regolith samples were collected from within the top 10 cm of the Martian regolith (via a scoop and robot arm). The samples were heated in miniovens at various temperatures up to $500\,°C$ and evolved volatiles and volatile pyrolysis products transported into a GC and eluted material analyzed with a mass spectrometer. The detection limits for specific organic species expected to be eluted from the GC column varied but were typically in the parts per billion levels.

4.1.2 Carbon Assimilation Experiment

The carbon assimilation experiment was designed to detect life in Martian samples by supplying ^{14}CO and $^{14}CO_2$ to a Martian sample incubated in a closed chamber under Martian-like environmental conditions. The incorporation of ^{14}C into organic material within the sample would be taken as an indication of metabolic activity and hence life within the sample.[29] The detection of ^{14}C incorporation into organic material was achieved though the pyrolysis of the sample after incubation and selective trapping of the volatile pyrolysis products. The ^{14}C content of this fraction was determined with a simple radioactivity detector.

4.1.3 Labeled Release Experiment

The basis of the label release experiment was addition to Martian samples of ^{14}C-labeled organic substrates such as formate, lactate, glucose, and glycine in a closed chamber under varying environmental conditions. The release of ^{14}C into the gas phase would be been indicative of the presence of metabolically active life.[30]

4.1.4 Gas Exchange Experiment

The gas exchange experiment again incubated a number of Martian samples in the presence of experimentally added water, nutrients, and gases. A gas chromatograph periodically monitored the head space gases within the incubation container to observe changes in the gas composition due to chemical and possible biological processes.[31]

4.1.5 Results of the Viking Life Detection Experiments

The GC/MS experiment did not detect any organic carbon signal above an upper limit of a few parts per billion. This was surprising given the expected continuing delivery of abiotic organic carbon to the Martian surface by meteorites. Initial positive results from the life detection experiments were soon interpreted as resulting from abiotic chemistry within the Martian regolith, e.g. in some cases activity was still present after "sterilization" of the regolith samples by heating. The current widely held view is that an oxidizing chemistry within the regolith samples is the most likely explanation of the results.[32] Indeed, further consideration of this hypothesis could explain the lack of organic detection with the GC/MS as the metastable oxidation products of organic molecules, e.g. salts of carboxylic acids, while possibly present in the samples would not have been expected to generate volatile products amenable to GC/MS analysis.[22] The vast majority of the planetary community agree that the results of the three Viking life detection experiments and the GC/MS experiment indicated no evidence of extant life in the samples tested.

4.2 Beagle 2

The UK Beagle 2 lander was designed primarily to search for evidence of past and present life within the Martian environment.[33] The gas analysis package (GAP) was the primary scientific instrument on Beagle 2 and was central to the life detection aspect of the mission. Beagle 2 also included a percussive-driven "mole" sample acquisition system to enable regolith samples to be obtained from depths of up to 1.5 m[34] and therefore potentially below the proposed surface oxidant layer.

4.2.1 The Gas Analysis Package (GAP)

The GAP[35] was designed to include the ability to detect carbon, both organic and inorganic, within

the Martian environment. The GAP consisted of a number of sample ovens to receive samples and then to increasingly heat the samples in the presence of O_2 to perform stepped combustion of Martian samples. The emission profile of the resultant CO_2 combustion product (of any carbon) verses temperature would allow differentiation of the differing carbon reservoirs such as organic and carbonate reservoirs. The evolved CO_2 was to be detected using a mass spectrometer that would also enable analysis of the ^{12}C and ^{13}C isotopic constitution of the evolved carbon. The $^{12}C/^{13}C$ isotopic ratio is diagnostic of life as it is known from Earth that life favors the incorporation of ^{12}C compared to ^{13}C and thereby results in ^{12}C isotopic enrichment of biogenic carbon. The lower detection limit for carbon was stated as 0.02–0.01 ppm.

4.2.2 Mission Outcome

No surface experimentation on Mars occurred due to loss of contact with Beagle 2 after it was ejected from the Mars Express spacecraft within the vicinity of Mars as part of the Beagle 2 entry, descent, and landing phase of the mission.[36] The cause of the loss of contact is still unclear at the present time (early 2007).

4.3 ExoMars Mission

The ESA ExoMars mission[1] is design primarily as an astrobiology mission and therefore is expected to be the next Mars mission with the specific goal of detecting evidence of life. The mission is currently scheduled for launch in 2013 and with arrival at Mars approximately 2 years later.[1] The mission contains a solar powered surface rover (see Figure 1) with a mobility range of a few kilometers. A key novel feature of the rover will be a drill capable of obtaining samples from a depth of down to 2 m below the surface. This is crucial for evidence of life detection as the previously mentioned preservation issues of organic molecular biomarkers reduce the likelihood of detecting organics in the top few centimeters of regolith or consolidated material.

The rover will contain a range of possible instruments, which are currently divided into panoramic

2-m depth

Figure 1. Computer generated image (CGI) representation of the ESA ExoMars rover demonstrating the deployment of the drill. [Image courtesy of ESA.]

(panoramic camera, infrared spectrometer, ground penetrating radar), contact (Mössbauer spectrometer, Raman-LIBS), analytical (X-ray diffractometer, LMC, GC/MS, Urey), support (drill, sample handling, and distribution system), and environmental suites. Of these instruments, the three proposed organic detection instruments within the analytical suite are of most interest for evidence of life detection. The Urey instrument comprises an integrated system that includes a microfabricated capillary electrophoresis system for the detection of amine-containing organics after in situ derivatization with fluorescamine.[37] The LMC is an antibody microarray biosensor instrument (see Section 6.1.3). The third instrument is the GC/MS.

4.4 Other Future Missions

4.4.1 NASA Astrobiology Field Laboratory

While ESA's ExoMars mission is the most developed mission, the first, to propose the inclusion

of a biosensor-based experiment, and the next Mars astrobiology mission, NASA is developing an astrobiology Mars mission called the *AFL*. This has a primary scientific goal of life detection and within the baseline mission specification is consideration of antibody microarray devices.[38] A possible launch date is 2016.

4.4.2 Europa Lander

Apart from Mars, the other location in the solar system that is of significant interest to the astrobiology community is the Jovian moon Europa (see Section 2.4.4) due to the presence of a putative liquid-water ocean below an ice crust. The sampling of the ocean for evidence of life is a long-term objective. Two major issues need to be addressed for such a mission; they are (i) the intense radiation environment within the Jovian system due to the trapping of solar particle radiation by the Jovian magnetosphere and (ii) the penetration of the ice crust that may be many kilometers thick. For life detection instrumentation, and especially those that contain radiation-sensitive components such as bioassay and other chemical reagents, the radiation exposure dictates either significant levels of shielding and therefore all the associated detrimental issues of mass or the selection of compounds that are radiation tolerant. For molecular recognition sensors, it may be possible to exploit more robust receptor materials such as molecular imprinted polymers, in future applications. The interpretation of various Europa surface features as sites of contemporary ocean upwellings and melt-through may offer a route to the sampling of recently exposed ocean contents without resort to penetration of the ice crust.

4.4.3 Earth Missions

As an analog to possible Europa missions, the exploration of subglacial lakes in Antarctica[39] offers some similarity with liquid-water lakes of biological interest located beneath several kilometers of an ice overlayer. Thus Lake Vostok and more recently subglacial Lake Ellsworth[40] are being considered for exploration using robotic probes equipped with life detection instrumentation.

5 TYPICAL MISSION CONSTRAINTS FOR LIFE DETECTION INSTRUMENTATION

The detection of evidence of life in Martian samples would be relatively easy if samples could be returned to state-of-the-art Earth laboratories. The techniques, instrumentation, and protocols then used would be those that are already developed to analyze similar samples for biomarker detection in ancient Earth rocks and the full arsenal of techniques within the modern biotechnology sector. This scenario in the form of a Mars sample return mission[41] is being discussed but due to mission complexity is unlikely to happen until 2020 at the very earliest. Until such a mission occurs, samples need to be analyzed in situ on the Martian surface. This imposes a number of limitations and restrictions on analytical techniques due to engineering, environmental, and ethical considerations and requirements. It is therefore important that these limitations and restrictions are considered in any bioanalytical life detection instrumentation development and are described in the following subsections and summarized in Table 4.

5.1 Planetary Protection

A critical design constraint for any biosensor-type technology for use in life detection and planetary exploration concerns the topic and implications of planetary protection.[42] There is a risk that Earth organisms contaminating a spacecraft could be transferred to a planetary body—termed *forward contamination*. Such contamination of a planetary body raises ethical and scientific concerns. In essence, ethically one does not want to contaminate a pristine environment with Earth organisms with possible detrimental affect on any indigenous life and scientifically the contamination of an environment with Earth life makes scientific interpretation of that environment in terms of indigenous life more difficult. The level of concern is related to the nature of the planetary body and specific location being accessed and whether there is a possibility of there beginning or having been indigenous life and/or an environment conducive to maintenance and even colonization by contaminating Earth life.

Planetary protection for missions is overseen internationally by COSPAR (Committee on Space Research). COSPAR categorizes missions into a number of classes[43] depending on the objective of the mission, i.e., life detection or not, and the potential of the visited region to harbor life historically, at present or into the future. This categorization also includes setting upper limits for the degree of contamination or bioburden allowed. For example, Mars lander missions with extant life detection experiments will be either Category IVb or IVc. This requires the bioburden levels to be equivalent to, or less than, those of the NASA Viking landers after their sterilization step. The current default method for sterilizing spacecraft to reduce their bioburden was established during the Viking missions and comprises the dry thermal baking of a spacecraft and

Table 4. Extreme environments and other constraints typically encountered by instruments during development, build integration, and use within a planetary exploration mission

Sterilization	Default of dry heated treatment at $> 100\,°C$ although alternatives are possible
Contamination control	Need to reduce instrument contamination levels to below the detection limit of the analytical techniques deployed
Radiation	γ radiation, various particle radiations
Thermal environment	Both thermal cycling and extreme excursions during prelaunch and launch, cruise phase, and on-surface operations
Shock loadings	For example, during launch, pyrotechnic firings, entry, descent, and landing operations
Acoustic energy exposure	For example, during launch
Chemically aggressive sample matrices	For example, oxidants in Martian regolith
Extended storage times	Including prelaunch, cruise phase, and on-surface operations
Reduced gravity	On-surface gravity and microgravity conditions during cruise phase

any instrument payload at elevated temperatures ($>100\,°C$) for many hours. This obviously is problematic for most biosensor technology as the biologically derived components, e.g., antibodies, are unlikely to survive such treatment. Therefore alternative sterilization, or more specifically bioburden reduction approaches, are required and must be validated. Alternative approaches could include UV treatment, hydrogen peroxide plasma, γ irradiation, chemical treatments, and for some molecular reagents ultrafiltration. In reality, combinations of these approaches together with appropriate aseptic handling and assembly procedures are required to meet the stringent limits of contamination set. The most recent experience of preparation for a Category IVa (less stringent than Category IVb or IVc) mission that did not use terminal heat sterilization was gained by the Beagle 2 team.[44]

It is important to note that providing the validation of the bioburden reduction approaches and verifying the obtained levels of bioburden for the flight components are also major challenges. The current approaches for validation and monitoring involve various sampling techniques coupled with detection of viable organism by culture techniques as well as more modern approaches such as the limulus amebocyte lysate (LAL) assay, polymerase chain reaction, and ATP measurements.

5.2　Contamination Control

The normal approach to ensuring a limit on the number of viable organisms that are present on a spacecraft is sterilization by dry heat treatment. This and many other sterilization procedures render microorganism contamination nonviable but do not necessarily remove the physical presence of inactivated microorganisms. If, as is the case under consideration, the analytical targets of an instrument are the molecular components of life, then the presence of inactivated microorganism would provide a significant reservoir of molecular biomarkers and hence potential for false positive detection signals. Therefore the physical removal or prevention of contamination and understanding the resultant upper limits of contamination levels is a critical task. In essence, the level of contamination should be below the lower limit of detection of the most

relevant and sensitive analytical instrument within an instrument suite. The approaches to achieve this rely upon a combination of (i) cleaning techniques, (ii) choice of, or modification of, surfaces to optimize cleaning and minimize levels of contamination, (iii) handling and assembly of instruments in clean conditions, and (iv) appropriate instrument designs that limit the surface area of relevance and routes for the transfer of contamination.

5.3　Radiation Environments

A Mars mission has two major phases during which a biosensor instrument would be exposed to a variety of radiation environments and those are in addition to any radiation used to ensure instrument sterilization during assembly. These are the interplanetary cruise phase and the Martian surface operations phase. For the following discussion it will be assumed that the radiation environment at the Martian surface approximates to that during the interplanetary cruise phase—a reasonable assumption given the lack of a protective thick atmosphere and a planetwide magnetic field. During the interplanetary cruise phase the spacecraft will be subject to radiation primarily from the Sun and galactic cosmic rays. A wide range of radiation types are produced from these sources and comprise energetic particles such as electrons, protons, neutrons, and heavier ions and electromagnetic radiation including UV, X rays, and γ rays. The relevant energies of these radiations vary from a few megaelectron volts to thousands of gigaelectron volts. Additional sources of radiation can include radioactive sources within a spacecraft and rover, for example, radioisotope thermoelectric generators and radioisotope heaters units, and secondary radiation produced by the interaction of primary radiation with the structure of the spacecraft and rover. For surface operations, this can include secondary radiation products from the Martian surface.

The levels of radiation for a typical Mars mission can be calculated. Thus as an example for the ExoMars mission with a 2-year interplanetary cruise phase during solar maximum and 6 months surface operations, appropriate calculations produce a total radiation dose of 6.6 krad assuming a shielding level equivalent to 4 mm of aluminum from the spacecraft and rover structure (Alex Hands and David Rodgers, private communication).

For biosensor-type instruments, the components of concern, as far as radiation stability is considered, are biological materials and associated assay reagents. For the case of antibody microarray devices with fluorescent labels, there has been little study of the effect of the various radiations on these materials apart from limited studies of γ irradiation due to the interest in food sterilization[45] and sterilization of antibodies for biotechnological applications.[46] Therefore the current authors have initiated a test campaign to de-risk issues surrounding the stability of antibodies and fluorescent dyes to Mars mission radiation exposure levels. Preliminary data indicates that dried fluorescence dyes survive many times mission levels of proton and helium ion radiation[47] as well as γ radiation. Dried antibodies retain significant levels of activity when exposed to mission level radiation of proton and helium ions. It is also apparent that the storage of antibodies[46] or fluorescent dyes in a hydrated form is inappropriate as radiolysis products, such as free radicals, produced from water can inactivate both antibodies and fluorescent dyes. Thus for Mars missions there does not appear to be a fundamental issue concerning exposure to the expected radiation environment although for other missions, such as to Europa, the much higher levels of radiation exposure may provide a very significant challenge.

It should be noted that γ radiation doses reported to demonstrate sterilization in antibody preparations were 1.5–4.5 Mrad.[46] This illustrates that a γ radiation sterilization protocol may be applicable to address planetary protection requirements and if used would provide a significantly greater exposure to ionizing radiation than the mission itself.

5.4 Other Mission Constraints

In addition to the preceding three topics, a number of other constraints need to be considered when developing instrumentation for planetary exploration. While a number of these are typically considered during development of biosensor technologies for common Earth-based applications, they are often more severe for planetary exploration applications. For the sake of brevity, these will simply be listed in Table 4.

6 BIOSENSOR AND BIOANALYTICAL TECHNOLOGIES FOR LIFE DETECTION

The approach to the detection of evidence of life can take a number of directions, the specific details of which depend upon the exact questions beginning asked and also to a large extent of the mission opportunity. Historically, the Viking missions were searching for extant life and therefore looked for evidence of metabolism as this did not require a strong Earth-centric view of the nature of life. The current expectations of Mars are primarily focused on detecting preserved evidence of ancient extinct life while also maintaining as a secondary goal the detection of extant life. Therefore given that the most likely situation is that of detecting trace levels of recalcitrant preservation products of ancient life, antibody-based biosensor and bioanalytical systems are receiving the most attention. The drivers for this include (i) proven ability to detect a broad range of molecule types—from low-molecular-weight polar and apolar molecules through to macromolecules, (ii) proven ability to detect trace levels of molecules in environmental samples, (iii) a wide range of demonstrated instrument platforms such as lateral-flow immunodiagnostics devices, especially for simultaneous detection of multiple targets, antibody microarrays, and (iv) exploitation of antibody cross-reactivity to detect classes of related molecules. Two of these drivers warrant further discussion. The first listed driver is a powerful reason for the planetary exploration community's emerging interest in biosensor and bioanalytical technologies. The ability to detect a wide range of molecule types with a single instrument, i.e., by simply including different antibodies, enables a large scientific return for a single instrument. In spacecraft design where the need to minimize instrument mass and volume cannot be over estimated, this is a very appealing feature. The last listed driver is crucial as the obvious concern in the use of antibodies for life detection is the need to predetermine the biomarker targets to which to raise appropriate antibodies prior to the flight of the instrument. If a highly specific antibody was employed for a given target and that had little cross-reactivity to other related biomarkers that may belong to the same target class, a negative life detection signal could result indicating that the specific biomarker is absent even if there was a high level of other members of the class which would indicate life. Therefore selection of antibodies with appropriate levels of cross-reactivity to biomarker classes will be important. The types of targets for which antibodies will need to be raised have already been discussed (see Section 3.1.1) and have also been considered elsewhere.[48]

Although antibody-based systems are the present focus for development of biosensor technology for life detection on Mars, other approaches are being considered. These are briefly discussed at the end of this section.

6.1 Antibody-based Systems

Three groups are currently exploring antibody microarray-based systems and comprise (i) the Spanish Signs-of-Life Detector (SOLID) activity from the Centro de Astrobiologia in Madrid, (ii) the US-led Modular Assays for Solar System Exploration (MASSE) activity focused on Dr Andrew Steele's work in the Geophysical Laboratory at the Carnegie Institution of Washington, and (iii) the UK-led SMILE activity based on the work of the current authors.

6.1.1 SOLID

A Spanish group from the Centro De Astrobiologia (Madrid, Spain) have been developing an automated antibody microarray system for biomarker detection for robotic and remote field analysis. An application focus has been the Rio Tinto in southern Spain that offers a possible analog of a wet and acidic early Mars.[49] The initially reported SOLID prototype system[50] is shown in Figure 2. The system contains 12 independent sample-extraction and microarray reaction modules. It comprises:

1. A funnel input to receive up to 0.5 g of a particulate sample and a sample-distribution system that places the sample in 1 of 12 sample-processing modules.
2. A movable ultrasonic unit that acts as a reusable closure for each sample-processing module and enables application of ultrasonic energy to aid sample extraction. It also allows compression

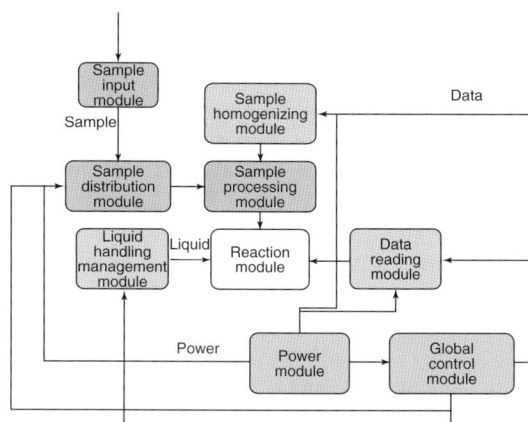

(a)

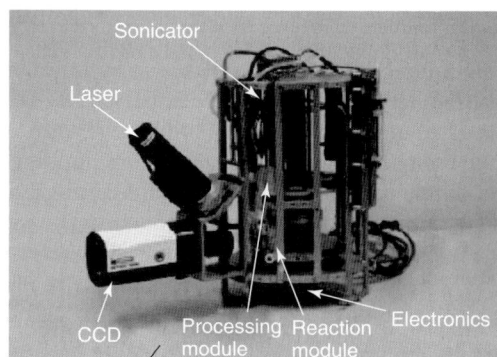

(b)

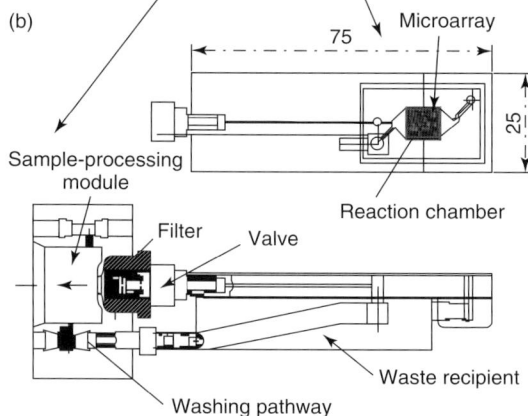

(c)

Figure 2. SOLID prototype. (a) SOLID module diagram, (b) SOLID picture showing the main components, and (c) details of sample-processing and reaction modules. [Reprinted with permission Parro et al.[50] copyright 2005, Elsevier.]

of the sample and expulsion of extraction fluid from the sample-processing module to a reaction module.

3. Each sample-processing module contains an exit valve protected by a filter to withhold particulate samples after extraction and during expulsion of the extracted material and solvent. It also provides additional chambers and fluidic conduits to enable the dosing of required solvents for extraction and assay reagents.
4. Expulsion of the extracted material, assay reagents, and solvent results in flow into a reaction module containing an antibody microarray within a thin-layer flow cell. Sandwich immunoassays occur on the microarray before washing with further solvent flow.
5. A laser excitation and CCD imaging system enables readout of the fluorescently labeled sandwich immunoassays.
6. The system has supporting control electronics and fluid pumps and valving and reservoirs for solvents and waste fluids.

The system has been demonstrated using real samples collected from Rio Tinto. After running the SOLID system with 0.25 g of sediment and an antibody microarray using a sandwich assay format and containing polyclonal antibodies against *Leptospirillum ferrooxidans* and *Acidithiobacillus ferrooxidans*, microarray spots with both these antibodies gave a positive signal. Both bacterial species were known to be present at the sampling location. The system was also demonstrated with a 0.25 g particulate sample that was flooded with 1 ml of an aqueous solution containing a number of protein antigens each at a concentration of $10\,ng\,ml^{-1}$. The relevant assay spots in the array gave a positive reading from the spiked particulate mineral sample.

More recently the group has demonstrated the ability to perform competitive multiplexed immunoassays in a microarray format.[51] This is an important demonstration as the vast majority of likely biomarker targets for an initial Mars mission will be low-molecular-weight molecules. Such molecules are normally detected using competitive immunoassay formats as their size and often lack of suitable reactive groups precludes the use of either sandwich assay formats or direct labeling of the targets as is commonplace in the mainstream proteomics applications of antibody microarrays. In the reported demonstration, various antigens were immobilized on the microarrays. An aqueous solvent with 10% methanol was used to extract antigens from mineral samples spiked with antigens. The extracted antigen contained within the aqueous solvent (with 10% methanol) was incubated with soluble antibodies and then exposed to the microarrays. Presence of extracted antigen occupied the appropriate antibody binding sites thereby inhibiting antibody binding to the appropriate microarray spot. Binding of the antibodies was detected after washing by a further incubation step with fluorescently labeled Protein A and reading in a standard microarray scanner. Depending on the specific antibody–antigen pair, detection levels ranged from sub–parts per billion to hundreds of parts per billion.

6.1.2 MASSE

The group of Dr Andrew Steele at the Geophysical Laboratory of the Carnegie Institution of Washington (USA) has long advocated the use of biotechnological approaches to life detection for planetary exploration. They proposed the MASSE concept that includes the use of antibody microarrays for the in situ detection of biomarkers.[52] Significant research and technology development is underway within this and associated groups, institutes, and companies funded by various NASA programs but full details of the outputs and current status are not currently available within the public domain due to governmental technology transfer policies. (The International Traffic in Arms Regulations (ITAR) is a set of US government regulations that limits sharing of information with non-US persons on defense-related information and technology. Space technology falls within its remit and inhibits the collaboration and interchange of information in international collaborators with NASA supported researchers.) Briefly, a handheld field-based fluorescent antibody microarray device has been demonstrated and field tested at the Arctic Mars Analog Svalbard Expedition (AMASE) site in 2006. Additionally a variant of this has been flown in 2007 on the space shuttle. A lipopolysaccharide bioassay system based upon LAL has also been field tested at the AMASE site.

6.1.3 SMILE

This chapter's authors are the proposers and lead investigators of the SMILE instrument (The current team developing SMILE for the ExoMars rover is UK led (by the authors) and with additional partners from the United Kingdom, The Netherlands, Germany, and the United States). Again this is an antibody, or more specifically an immunoassay, microarray device using fluorescent labels and therefore optical readout. It has been proposed and accepted into the outline and baseline instrument payload for the ESA ExoMars mission known as the *Pasteur package*. The decision as to which instruments will be included in the eventual mission is pending ongoing design and engineering studies at the time of writing. Within the context of ExoMars, the term *LMC* is used to generically denote such instruments. The primary focus for the SMILE instrument development is directed toward an implementation that is appropriate as a flight instrument for the ExoMars mission rather than Earth-based applications.

The SMILE instrument is in an early phase of development and therefore the following is a brief overview of the design and approaches being taken.

To avoid regeneration of microarrays, to minimize cross-contamination between samples and avoid the complexities of emptying, cleaning, and reuse of solid-sample containers and particulate filters, multiple single use components and subsystems are to be used. For the baseline instrument, up to 40 independent "assay channels" are included, therefore allowing up to 40 samples to be analyzed. An assay channel consists of:

1. Sample extraction and preconcentration

A sample container to receive approximately 1 g of Martian sample—either crushed regolith or crushed rock core—and that can be sealed and solvent added. Appropriate choice of solvent and application of heat and/or ultrasonic energy enables extraction of organic molecular biomarkers. A sample preconcentration step is included that offers the potential for (i) solvent volume reduction and hence preconcentration, (ii) exchange of solvent, and (iii) sample cleanup, e.g., removal of solvent soluble sample salts.

2. Immunoassay microarray with optical readout

The extracted components in a solvent suitable for performing an immunoassay are passed into a microfluidic portion of the assay channel. The channel contains predosed and dried immunoassay

reagents that dissolve into the solvent. The solvent, extracted components, and immunoassay reagents are then passed over an immobilized microarray of immunoassay components that are within the microfluidic channel allowing appropriate binding of fluorescently labeled assay components. Further flow of solvent flushes unbound assay components and washes the microarray. The baseline design includes the use of thin-film optical waveguides as the support for the immobilized microarray thereby allowing evanescent wave excitation of the fluorescent labels when at the surface of the waveguide. This enables (i) improved efficiency of fluorescent label excitation and (ii) the possibility of kinetic measurements that will aid in the interpretation of the assay, i.e., a definitive endpoint measurement will only be considered further if an appropriate kinetic binding profile is also seen. Optical imaging and detection of the fluorescent output of the array occurs with a single optical imaging subsystem.

3. Support components

Components include fluid pumps and valves to allow pressurization of the fluidic system and distribution of fluids through the sample channels, mechanisms to rotate a carousel configuration of up to 40 sample channels, heaters to raise the temperature to levels suitable for the immunoassays and handling of aqueous solvents, a box structure to contain the instrument, and support electronics and controllers.

The immunoassay microarray baseline design contains 100 assay spots and allowing for various control and calibration spots and triplication of spots, up to 25 independent immunoassays can be accommodated and therefore detection of 25 targets will be housed per assay channel. As previously described, the range of assay targets is wide; from low-molecular-weight molecules to macromolecules. The SMILE design is compatible with a variety of assay formats and that includes (i) sandwich assays for multiple-epitope-containing macromolecules, (ii) capture assays for target that are compatible with in situ fluorescent labeling, (iii) competition assays with prelabeled traces, and (iv) inhibition assays with immobilized antigens and soluble labeled antibodies.

An ongoing development program for the SM-ILE instrument includes (i) development of appropriate antibody libraries to populate the instrument, (ii) development of suitable sample-extraction and processing protocols, (iii) development of sterilization protocols for planetary protection, (iv) testing the radiation tolerance of assay reagents including antibodies, (v) breadboard de-risking studies of key components and subsystems, and (vi) production of detailed flight model designs.

6.2 Other Bioanalytical Systems and Components

6.2.1 Bioassays

Sensitive and generic life detection methods have already been developed for a number of terrestrial applications. ATP bioluminescence is commonplace in food hygiene and other applications. It has been considered for Martian use[53] but the most likely applications are in support of planetary protection issues during spacecraft, rover, and instrument assembly (see Section 5.1) and for Earth analogs of other solar system locations such as Antarctic subglacial lakes.[40]

The LAL assay again is well established as a sensitive bioassay for lipopolysaccharide from gram-negative bacteria. Similar to the ATP bioluminescence assay, it's current and near-future application is likely to be limited to study of Earth analogs of other solar system locations and support of planetary protection protocols.

6.2.2 Nucleic Acid Detection

At present, the lack of knowledge concerning life elsewhere in the solar system has resulted in little current consideration of nucleic acid amplification and detection techniques being employed as dedicated instruments for planetary exploration. The narrowness of the scientific return allied with the mass requirements for an instrument is an obvious driver for this situation. The application of life detection instrumentation to Earth analogs of planetary environments is a more likely application of these technologies and although development programs specific to these applications are not well

established, similar technologies are being rapidly developed for field-based applications within the security and defense sector.[54] Migration of such hardware to application within Earth analogs of planetary environments is an obvious step.

6.2.3 Artificial Receptors

Given the extreme environments that biological reagents will have to endure in some planetary exploration mission, e.g., to Europa, and the simplification to implementing planetary protection requirements if receptors could withstand heat sterilization, robust artificial sensors would be desirable. Molecular imprinted polymers are one such possible technology that is starting to receive interest for future applications. Both the current authors[55,56] and others are considering such technology.[57]

6.3 In situ Sample Acquisition, Processing, and Delivery

Although the preceding emphasis has been toward analytical techniques and the justification for their use, a major component of realizing in situ measurements is successful sample acquisition, processing into a suitable form, and delivery to an analytical instrument. The importance of this process cannot be over estimated. The difficulty in engineering such systems to operate within the constraints already described can consume equal, if not more, development and spacecraft/rover resources than the analytical instruments they serve.

It will suffice here to list a number of approaches that have been used or are currently being considered or developed. For sample acquisition these include (i) scoops on arms, e.g., Viking and the upcoming Phoenix mission,[58] (ii) the percussive-driven "mole"[34] on Beagle 2, and (iii) the drill proposed for the ExoMars mission. For sample processing, examples include (i) rock crushing, (ii) sample sieving, (iii) addition of reagents such as liquids and solutes, (iv) heating of samples in ovens, (v) solvent extraction, (vi) sublimation, and (vii) subcritical water extraction.

7   SUMMARY

We are at a unique and exhilarating point in time where two distinctive streams of endeavor are merging for the first time. Firstly humankind is actively planning and implementing missions to planetary bodies such as Mars to ask questions concerning the potential for, and existence of, life. Secondly biosensor technology has matured to a level where it can be seriously proposed as a tool to help answer such questions. This chapter has therefore given an overview of the context and issues surrounding the development and future use of biosensor, biochip, and other bioanalytical technologies for the detection of evidence of life in a planetary exploration context. Mars has been the focus of this chapter as it is the place thought most likely to harbor evidence of life and is the target of a number of upcoming missions although other planetary bodies such as Europa are of interest.

For the biosensor community, the unique combination of challenges and drivers coming from the planetary exploration community results in a development environment that may stimulate novel design solutions which could also benefit more traditional Earth-based biosensor applications. We hope that this chapter will stimulate others within the biosensor community to look toward some of the challenges of detecting life elsewhere in the solar system using biosensor and related technology.

ACKNOWLEDGMENT

We would like to thank Daniel Thompson and Paul Wilson for their critical reading of the manuscript and suggestions.

REFERENCES

1. J. Vago, B. Gardini, G. Kminek, P. Baglioni, G. Gianfiglio, A. Santovincenzo, S. Bayón, and M. van Winnendael, ExoMars Project Team, ExoMars—searching for life on the red planet. *ESA Bulletin-European Space Agency*, 2006, **126**, 16–23, (ISSN 0376–4265) (Note the launch date information within this publication has been superseded and the launch date is currently 2013 with arrival at Mars in 2015—correct January 2007).

2. I. de Pater and J. J. Lissauer, *Planetary Sciences*, Cambridge University Press, 2001.

3. R. Gomes, H. F. Levison, K. Tsiganis, and A. Morbidelli, Origin of the cataclysmic late heavy bombardment period of the terrestrial planets. *Nature*, 2005, **435**, 466–469.

4. J. F. Kerridge, Formation and processing of organics in the early solar system. *Space Science Reviews*, 1999, **90**, 275–288.

5. O. Botta and J. L. Bada, Extraterrestrial organic compounds in meteorites. *Surveys in Geophysics*, 2002, **23**, 411–467.

6. C. Chyba and C. Sagan, Endogenous production, exogenous delivery and impact-shock synthesis of organic molecules: an inventory for the origins of life. *Nature*, 1992, **355**, 125–132.

7. G. J. Flynn, The delivery of organic matter from asteroids and comets to the early surface of Mars. *Earth Moon and Planets*, 1996, **72**, 469–474.

8. J.-P. Bibring, Y. Langevin, J. F. Mustard, F. Poulet, R. Arvidson, A. Gendrin, B. Gondet, N. Mangold, P. Pinet, and F. Forget, Global mineralogical and aqueous mars history derived from omega/Mars express data. *Science*, 2006, **312**, 400–404.

9. J. W. Head, J. F. Mustard, M. A. Kreslavsky, R. E. Millikien, and D. R. Marchant, Recent ice ages on Mars. *Nature*, 2003, **426**, 797–802.

10. M. C. Malin, K. S. Edgett, L. V. Posiolova, S. M. McColley, and E. Z. N. Dobrea, Present-day impact cratering rate and contemporary gully activity on Mars. *Science*, 2006, **314**, 1573–1577.

11. J. W. Schopf, Microfossils of the early Archean apex chert—new evidence of the antiquity of life. *Science*, 1993, **260**, 640–646.

12. S. J. Mojzsis, G. Arrhenius, K. D. McKeegan, T. M. Harrison, A. P. Nutman, and C. R. L. Friend, Evidence for life on earth before 3,800 million years ago. *Nature*, 1996, **384**, 55–59.

13. P. C. W. Davies and C. H. Lineweaver, Finding a second sample of life on earth. *Astrobiology*, 2005, **5**, 154–163.

14. C. Mileikowsky, F. A. Cucinotta, J. W. Wilson, B. Gladman, G. Horneck, L. Lindegren, J. Melosh, H. Rickman, M. Valtonen, and J. Q. Zheng, Natural transfer of viable microbes in space–1. From Mars to earth and earth to Mars. *Icarus*, 2000, **145**, 391–427.

15. G. Horneck, C. Mileikowsky, H. J. Melosh, J. W. Wilson, F. A. Cucinotta, and B. Gladman, *Viable Transfer of Micro-Organisms in the Solar System and Beyond*, in *Astrobiology: the Quest for the Conditions of Life*, G. Horneck and C. Baumstark-Khan (eds). Springer, Berlin, 2002, pp. 57–76.

16. D. Schulze-Makuch, J. M. Dohm, A. G. Fairén, V. R. Baker, W. Fink, and R. G. Strom, Venus, Mars, and the ices on Mercury and the Moon: astrobiological implications and proposed mission designs. *Astrobiology*, 2005, **5**, 778–795.

17. C. F. Chyba and C. B. Phillips, Possible ecosystems and the search for life on Europa. *Proceedings of the National Academy of Sciences*, 2001, **98**, 801–804.

18. J. Parnell, D. C. Cullen, M. R. Sims, S. Bowden, C. Cockell, R. Court, P. Ehrenfreund, F. Gaubert, B. Grant, V. Parro, M. Rohmer, M. Sephton, H. Stan-Lotter, A. Steele, J. Toporski, and J. Vago, Searching for life on Mars: selection of molecular targets for ESA's aurora ExoMars mission. *Astrobiology*, submitted.

19. A. H. Squyres and A. H. Knoll, Sedimentary rocks at Meridiani Planum: origins, diagenesis and implications for life on Mars. *Earth and Planetary Science Letters*, 2005, **240**, 1–10.

20. K. Pedersen, Exploration of deep intraterrestrial microbial life: current perspectives. *FEMS Microbiology Letters*, 2000, **185**, 9–16.

21. L. R. Dartnell, L. Desorgher, J. M. Ward, and A. J. Coates, Modelling the surface and subsurface martian radiation environment: implications for astrobiology. *Geophysical Research Letters*, 2007, **34**, L02207.

22. S. A. Benner, K. G. Devine, L. N. Matveeva, and D. H. Powell, The missing organic molecules on Mars. *Proceedings of the National Academy of Sciences*, 2000, **97**, 2425–2430.

23. MEPAG Astrobiology Field Laboratory Science Steering Group, *The Astrobiology Field Laboratory*, JPL Document Ref. # CL#06-3307 2006.

24. R. Gershman, E. Nilsen, and R. Oberto, Europa lander. *Acta Astronautica*, 2003, **52**, 253–258.

25. A. C. Atzei, P. Falkner, M. L. van den Berg, and A. Peacock, The Jupiter minisat explorer, a technology reference study. *Acta Astronautica*, 2006, **59**, 644–650.

26. H. P. Klein, J. Lederberg, and A. Rich, Biological experiments: the Viking Mars lander. *Icarus*, 1972, **16**, 139–146.

27. D. M. Anderson, K. Biemann, L. E. Orgel, J. Oró, T. Owen, G. P. Shulman, P. Toulmin, and H. C. Urey, Mass spectrometric analysis of organic compounds, water and volatile constituents in the atmosphere and surface of Mars: the Viking Mars lander. *Icarus*, 1972, **16**, 111–138.

28. K. Biemann, J. Oro, P. Toulmin, L. E. Orgel, A. O. Nier, D. M. Anderson, P. G. Simmonds, D. Flory, A. V. Diaz, D. R. Rushneck, and J. A. Biller, Search for organic and volatile inorganic compounds in two surface samples from the chryse planitia region of Mars. *Science*, 1976, **194**, 72–76.

29. N. H. Horowitz, J. S. Hubbard, and G. L. Hobby, The carbon-assimilation experiment: the Viking Mars lander. *Icarus*, 1972, **16**, 147–152.

30. G. V. Levin, Detection of metabolically produced labeled gas: the Viking Mars lander. *Icarus*, 1972, **16**, 153–166.

31. V. I. Oyama, The gas exchange experiment for life detection: the Viking Mars lander. *Icarus*, 1972, **16**, 167–184.

32. H. P. Klein, The Viking biological experiments on Mars. *Icarus*, 1978, **34**, 666–674.

33. I. P. Wright, M. R. Sims, and C. T. Pillinger, Scientific objectives of the beagle 2 lander. *Acta Astronautica*, 2003, **52**, 219–225.

34. L. Richter, P. Coste, V. V. Gromov, H. Kochan, R. Nadalini, T. C. Ng, S. Pinna, H.-E. Richter, and K. L. Yung, Development and testing of subsurface sampling devices for the beagle 2 lander. *Planetary and Space Science*, 2002, **50**, 903–913.

35. D. Pullan, M. R. Sims, I. P. Wright, C. T. Pillinger, and R. Trautner, *Beagle 2: the Exobiological Lander of Mars Express*, in *ESA Special Publication*, SP-1240, The European Space Agency, 2004, pp. 165–204.

36. M. R. Sims (ed), *Beagle 2—Mission Report*, University of Leicester, 2004.

37. A. M. Skelley, H. J. Cleaves, C. N. Jayarajah, J. L. Bada, and R. A. Mathies, Application of the Mars organic analyzer to nucleobase and amine biomarker detection. *Astrobiology*, 2006, **6**, 824–837.

38. A. Steele, D. W. Beaty, J. Amend, R. Anderson, L. Beegle, L. Benning, J. Bhattacharya, D. Blake, W. Brinckerhoff, J. Biddle, S. Cady, P. Conrad, J. Lindsay, R. Mancinelli, G. Mungas, J. Mustard, K. Oxnevad, J. Toporski, and H. Waite, *The Astrobiology Field Laboratory: Final report of the MEPAG Astrobiology Field Laboratory Science Steering Group (AFL-SSG)*, Jet Propulsion Laboratory document Ref. CL#06-3307 2006.

39. M. J. Siegert, S. Carter, I. Tabacco, S. Popov, and D. D. Blankenship, A revised inventory of Antarctic subglacial lakes. *Antarctic Science*, 2005, **17**, 453–460.

40. M. J. Siegert, A. Behar, M. Bentley, D. Blake, S. Bowden P. Chritoffersen, C. Cockell, H. Corr, D. C. Cullen, H. Edwards, A. Ellery, C. Ellis-Evans, G. Griffiths, R. Hindmarsh, D. A. Hodgson, E. King, H. Lamb, L. Lane, K. Makinson, M. Mowlem, J. Parnell, D. A. Pearce, J. Priscu, A. Rivera, M. A. Sephton, M. R. Sims, A. R. Smith, M. Tranter, J. L. Wadham, G. Wilson, and J. Wodward, Lake Ellsworth Consortium, Exploration of Ellsworth subglacial lake: a concept paper on the development, organisation and execution of an experiment to explore, measure and sample the environment of a West Antarctic subglacial lake. *Reviews in Environmental Science and Biotechnology*, 2007, **6**, 161–179.

41. B. Sherwood, Mars sample return: architeture and mission design. *Acta Astronautica*, 2003, **53**, 353–364.

42. J. D. Rummel and L. Billings, Issues in planetary protection: policy, protocol and implementation. *Space Policy*, 2004, **20**, 49–54.

43. J. D. Rummel, P. D. Stabekis, D. L. Devincenzi, and J. B. Barengoltz, Cospar's planetary protection policy: A consolidated draft. *Advances in Space Research*, 2002, **30**, 1567–1571; and updates for the introduction of category IVc missions to Mars at http://www.cosparhq.org/Scistr/Pppolicy.htm—published 25 March 2005.

44. J. M. Pillinger, C. T. Pillinger, S. Sancisi-Frey, and J. A. Spry, The microbiology of spacecraft hardware: lessons learned from the planetary protection activities on the beagle 2 spacecraft. *Research in Microbiology*, 2006, **157**, 19–24.

45. S. Lee, S. Lee, and K. B. Song, Effect of gamma-irradiation on the physicochemical properties of porcine and bovine blood plasma proteins. *Food Chemistry*, 2003, **82**, 521–526.

46. T. Grieb, R.-Y. Forng, R. Brown, T. Owolabi, E. Maddox, A. McBrain, W. N. Drohan, D. M. Mann, and W. H. Burgess, Effective use of gamma irradiation for pathogen inactivation of monoclonal antibody preparations. *Biologicals*, 2002, **30**, 207–216.

47. D. P. Thompson, P. K. Wilson, M. R. Sims, D. C. Cullen, J. M. C. Holt, D. J. Parker, and M. D. Smith, Preliminary investigation of proton and helium ion radiation effects on fluorescent dyes for use in astrobiology applications. *Analytical Chemistry*, 2006, **78**, 2738–2743.

48. B. L. Tang, A case for immunological approaches in detection and investigation of alien life. *International Journal of Astrobiology*, 2007, **6**(1), 11–17.

49. D. Fernandez-Remolar, J. Gomez-Elvira, F. Gomez, E. Sebastian, J. Martin, J. A. Manfredi, J. Torres, C.G. Kesler, and R. Amils, The Tinto river, an extreme acidic environment under control of iron, as an analog of the Terra Meridiani hematite site of Mars. *Planetary and Space Science*, 2004, **52**, 239–248.

50. V. Parro, J. A. Rodriguez-Manfredi, C. Briones, C. Compostizo, P. L. Herrero, E. Vez, E. Sebastián, M. Moreno-Paz, M. García-Villadangos, P. Fernández-Calvo, E. Gonzalez-Toril, J. Perez-Mercader, D. Fernandez-Remolar, and J. Gomez-Elvira, Instrument development to search for biomarkers on Mars: terrestrial acidophile, iron-powered chemolithoautotrophic communities as model systems. *Planetary and Space Science*, 2005, **53**, 729–737.

51. P. Fernandez-Calvo, C. Nake, L. A. Rivas, M. García-Villadangos, J. Gómez-Elvira, and V. Parro, A multi-array competitive immunoassay for the detection of broad-range molecular size organic compounds relevant for astrobiology. *Planetary and Space Science*, 2006, **54**, 1612–1621.

52. J. G. Maule and A. Steele, *A Prototype Life Detection Chip*, In: *35th Lunar and Planetary Science Conference*, League City, Texas, 2004 March 15–19, Abstract no. 2091.

53. R. K. Obousy, A. C. Tziollas, K. Kaltsas, M. R. Sims, and W. D. Grant, Searching for extant life on Mars: the ATP-firefly luciferin/luciferase technique. *Journal of the British Interplanetary Society*, 2000, **53**, 121–130.

54. D. V. Lim, J. M. Simpson, E. A. Kearns, and M. F. Kramer, Current and developing technologies for monitoring agents of bioterrorism and biowarfare. *Clinical Microbiology Reviews*, 2005, **18**, 583–607.

55. O. Y. F. Henry, S. Piletsky, W. D. Grant, M. R. Sims, and D. C. Cullen, *Robust Molecular Imprinted Polymer Thin-Films for An Astrobiology Biomimetic Sensor Array*, in *Proceedings of the First European Workshop on Exo-Astrobiology*, Huguette Lacoste, Graz, Austria, ESA SP-518, 2002 September 16–19, 513–514.

56. M. R. Sims, D. C. Cullen, N. P. Bannister, W. D. Grant, O. Henry, R. Jones, D. McKnight, D. P. Thompson, and P. K. Wilson, The specific molecular identification of life experiment (SMILE). *Planetary and Space Science*, 2005, **53**, 781–791.

57. N. R. Izenberg and O. M. Uy, *Molecularly Imprinted Polymer Geochemical Detectors for Outer Solar System Exploration*, in *Forum on Innovative Approaches to Outer Planetary Exploration 2001–2020*, Lunar and Planetary Institute, Houston, TX, 2001, p. 42.

58. R. Shotwell, Phoenix—the first Mars Scout mission. *Acta Astronautica*, 2005, **57**, 121–134.

Part Nine

Commercialization, Business and Regulatory Issues

Commercialization, Business and Regulatory Issues

Christopher R. Lowe

Institute of Biotechnology, University of Cambridge, Cambridge, UK

As noted earlier in this handbook, there is a substantial demand for chemical and biological intelligence in almost all aspects of human lifestyle and welfare. Such demand stems from a better appreciation of the major issues that our planet will face in the future, notably, the impact of understanding the human genome, proteome, and metabolome, the spread of infectious diseases via mass movement of people, bioterrorism, new forms of energy, global warming, and climate change, and the consequent requirement for making measurements of key analytes in a variety of samples and circumstances. In principle, technologies for these measurements have been in existence for many decades. Thus a key question to pose, is why have biosensor, bioarray, and biochip technologies not had a greater impact on these unmet needs? As a proportion of the total analytical market for instruments, services, and consumables, at present, the biosensor market is negligible and still dominated by a few key analytes. Why is this?

First, the old adage is that to be successful in the marketplace, a new technology has to be ten-times better or ten-times cheaper than an existing product to be acceptable and to displace sunk investments in hardware, software, and humanware. Sadly, most biosensor technologies do not meet these criteria and merely offer the advantage of convenience of use. This is particularly true of the glucose monitoring market, where the price of the sensor replacing the traditional colored paper test was not very different and the advantage was user friendliness and accuracy. Consequently, these glucose sensors have displaced the majority of other glucose tests and now account for the majority of biosensor sales worldwide.

A second issue is whether current biosensors are fit for the purpose, i.e. are they capable of measuring analytes quantitatively with the accuracy and speed required at a price the customer is willing to pay? For example, in the heady days when biosensors were expected to replace all other enzymes and immunoassays, newer techniques such as surface plasmon resonance (SPR) were considered to address the deficiencies of the then current assays. In reality, SPR could not cope with the throughput, sensitivity, and harsher environment that most of these assays were conducted in. Unsurprisingly, therefore, the technique has found a natural niche in the well-controlled laboratory, being used by skilled operators for interaction analysis. The chapter on the Biacore by Löfås describes the rocky path from biosensor concept to the creation of the business of label-free protein interaction analysis (*see* **Biacore – Creating the Business of Label-Free Protein-Interaction Analysis**). A similar story could be told for other sensors-turned interaction analysis equipment based on the resonant mirror and the quartz crystal microbalance. In a similar vein, Abramowitz describes a case study on the development of the Affymetrix DNA

Handbook of Biosensors and Biochips. Edited by Robert S. Marks, David C. Cullen, Isao Karube, Christopher R. Lowe and Howard H. Weetall.
© 2007 John Wiley & Sons, Ltd. ISBN 978-0-470-01905-4.

chip (*see* **Commercialization of DNA Arrays – Affymetrix a Case Study**) and Anderson the development hurdles facing a fiber optic biosensor (*see* **RAPTOR: Development of a Fiber-Optic Biosensor**).

The final issue relates to how the technologies are validated and regulated for analytes where these issues have important consequences, for example, in the clinical, environmental, security, and health and safety sectors. The article by Sergeev describes a number of these regulatory and validation issues for putative biosensors; without these, no biosensor will be commercially acceptable (*see* **Regulatory and Validation Issues for Biosensors and Related Bioanalytical Technologies**).

82

Biacore – Creating the Business of Label-Free Protein-Interaction Analysis

Stefan Löfås

Biacore AB, Uppsala, Sweden

1 INTRODUCTION

The impact of biosensors on life science research has gone through a tremendous development since the early 1980s. Significant advances have been made: new detection principles for different types of applications, improved commercialization of biosensor instruments, and a much better general understanding of the potential and limitations of biosensors. Despite these major advances, there are not many good examples of successful business cases directly related to biosensor technologies. One clear exception, however, is the company Biacore AB in Sweden, and its development of commercial optical biosensor instruments based on surface plasmon resonance (SPR) detection. The company started research and technology development in the mid 1980s and, because of major breakthroughs in surface chemistry, detector development, and microfluidics-based sample handling successfully launched the first in a subsequent series of bioanalytical instrumentation in 1990.

The label-free, real-time interaction analysis made possible by these technological developments has proven to be a versatile, general tool for studies of different types of protein and biomolecular interactions within academic life science research, drug discovery and development, manufacturing and quality control, and, more recently,

within a completely new market of food analysis. The instruments are used for a wide range of applications with typical data including specificity assays, binding affinities and kinetics, concentration measurements, as well as thermodynamics. Over the years, new systems have been launched to meet the needs within the widening range of user environments.

This chapter describes the Biacore story from a business case perspective, with special emphasis on the reasons for the successful outcome, both from the technology and corporate perspectives. Recent developments in the technology, applications, and new instruments are also included.

2 BACKGROUND AND RATIONALE BEHIND BIACORE

Sweden has a long tradition in the development of instrumentation and other tools for the life science industry. This tradition began with Professor Theodor Svedberg at Uppsala University, who was instrumental in developing the ultracentrifuge in the 1930s, and Professor Arne Tiselius, inventor of the electrophoresis technique, both later rewarded with the Nobel prize. Commercialization of these technologies was successfully achieved via the Swedish companies Pharmacia Fine Chemicals AB and LKB Produkter AB (now both part

Handbook of Biosensors and Biochips. Edited by Robert S. Marks, David C. Cullen, Isao Karube, Christopher R. Lowe and Howard H. Weetall.
© 2007 John Wiley & Sons, Ltd. ISBN 978-0-470-01905-4.

of GE Healthcare), which subsequently have been very active in providing tools for biomolecule purification and analysis. Realizing the potential in new biosensor technologies, Pharmacia, in 1984, decided to establish the new company, Pharmacia Biosensor AB (later renamed to Biacore AB), to develop bioanalytical systems based on new basic research developments.

Close contacts and collaborations with the academic community in Sweden had laid the foundations for this new company. In particular, the team formed by Professor Ingemar Lundström at Linköping University in the late 1970s was very important in this context. Lundström initiated the research around various physical detection technologies for use in new areas such as chemical, biochemical, and clinical analysis. At the same time, the National Defense Research Institute in Sweden had started a program in close collaboration with Professor Lundström, focused on research into new technologies for the detection of chemical and biological warfare agents. Pharmacia Biosensor was formed with the belief that the new technologies would result in a shift in the way analysis of biomolecules would be performed. Other ongoing biosensor-related activities within Pharmacia were also consolidated into this new entity, into which staff were initially recruited both from research groups and Pharmacia.

During a three-year period from 1985 to 1988, basic technology development was carried out for a totally new instrument concept based on SPR detection. This was made possible by creating a truly cross-disciplinary team working in all areas, from mechanics, optics, electronics, and software development to chemistry and biochemistry. As described in more detail, novel technology solutions were achieved within surface chemistry, microfluidics, and in the configuration of the optics. The company included former Pharmacia staff with experience, and mind-set, from product development. Following these research activities, a seamless commercialization phase then resulted in the first product launch in 1990. This phase was based on the evaluation of various business concepts for the technology, which resulted in a primary focus on products for biomolecular interaction analysis studies aimed toward the life science research community in academia and the

pharmaceutical industry. Other potential applications, such as concentration analysis of analytes in the clinical diagnostics area were also evaluated, but it was realized that both the technical and commercial competition was significant within this area of use.

The availability of SPR-based instruments for real-time, label-free biomolecular interaction analysis was well timed to meet the growing needs for better understanding of the relationships between structure and function of biomolecules, particular proteins, and nucleic acids. With the molecular biology revolution in the 1980s, entirely new collections of tools were created for efficient construction of antibodies, mutant variants of proteins with important biological functions, as well as oligonucleotide collections. Consequently, quantitative determination of function in relation to structural variations in biomolecules had become a real need and with the launch of Biacore's first systems, the company had created a user-friendly, high-quality, and versatile tool that provided an entirely new type of approach for obtaining important data. In particular, reliable information could be obtained about relevant parameters like specificity, activity, affinity, and binding kinetics of the biomolecules.

From a funding perspective, this timing was also important. Significant investments were made initially by Pharmacia and later by Swedish risk capital during years of positive, forward-looking attitudes. The situation was more difficult during the early 1990s, when the funding climate became much more restricted in conjunction with a more narrow focus on the genomics area. The areas of protein analysis and antibody-based therapeutics became less favored, a situation that did not change until the proteomics revolution came into being in the late 1990s.

3 THE TECHNOLOGY BEHIND BIACORE SYSTEMS

In Biacore systems, the phenomenon of SPR is used to measure mass concentration-dependent changes in refractive index close to a sensor surface. The potential of SPR-based optical detection for biomolecular interaction studies was first described in 1983 in a publication from Professor Lundström's group.[1] They used an experimental setup with reflectivity measurement at a fixed angle of incidence halfway down the reflectance minimum. There are also several other possible configurations of SPR detectors, which are described elsewhere in this book, and have been reviewed elsewhere, see, for example, Garland.[2] However, the application focus for the commercialization of Biacore instruments required a new detector concept. Limitations in prior art solutions for parameters like detection rate, sensitivity, and dynamic range, as well as the dependence on specialists for the acquisition of data were strong driving forces for the creation of the system. This involved the integration of exchangeable sensor surface with stationary optics, miniaturized sample handling, and real-time data readout and interpretation.[3] In particular the development of three cornerstones was instrumental for the successful launch of an easy-to-use and reliable new technology: (i) integrated optical detection, (ii) sensor surfaces with biospecific coatings, and (iii) the integrated microfluidic cartridge for controlled sample delivery (Figure 1).

3.1 Detection

The optical detectors in the original Biacore systems where constructed using a Kretschmann configuration, with back-illumination of the chip via a prism and a light emitting diode as light source. This creates a wedge-shaped light beam with a range of incident angles, which employs a photodetector array as an imaging system (Figure 2). This enables the monitoring of changes in the reflection angle over time without the need for any movable parts, with high dynamic range, and high linearity over the whole range.[4] The configuration also allows simultaneous monitoring of multiple sensing spots. In the first commercialized Biacore systems, four spots were defined by docking the flow cells in the integrated microfluidics, toward the sensor chip.

To enable a user-friendly handling of the sensor interface, innovative solutions were developed. A separate sensor surface consisting of a glass substrate with a thin gold coating for the resonance interaction is used for quick exchange after usage. Efficient optical contact between the prism

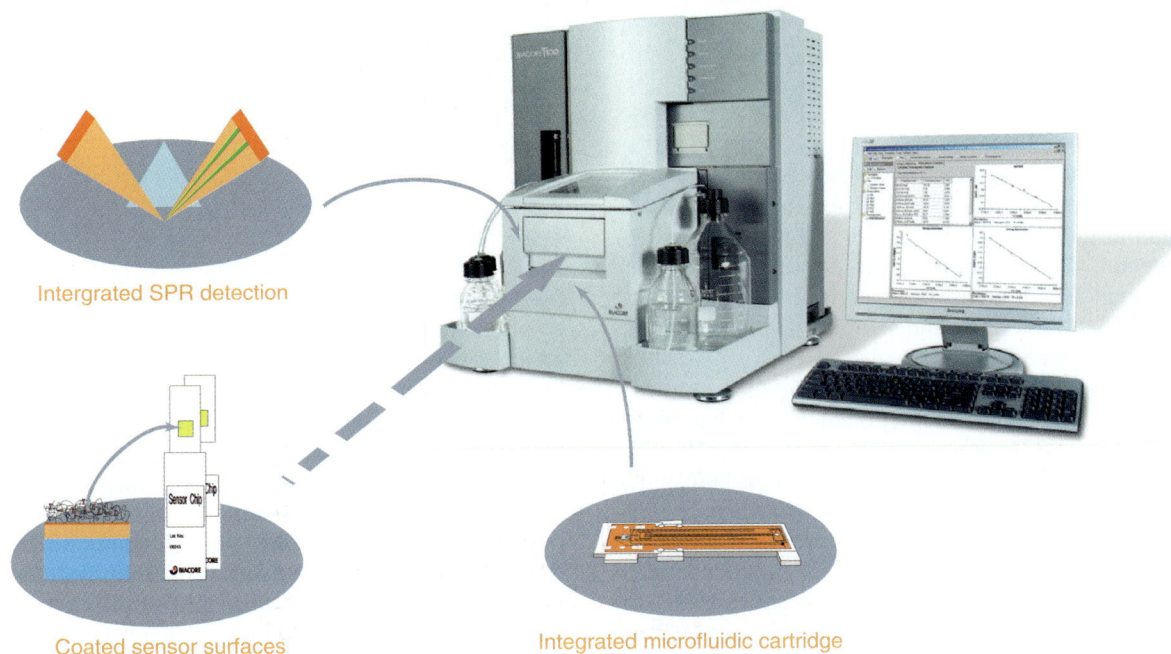

Intergrated SPR detection

Coated sensor surfaces

Integrated microfluidic cartridge

Figure 1. Picture of a Biacore instrument with the cornerstones outlined: optical detection system; integrated microfluidic system; sensor chips with dextran chemistry coating.

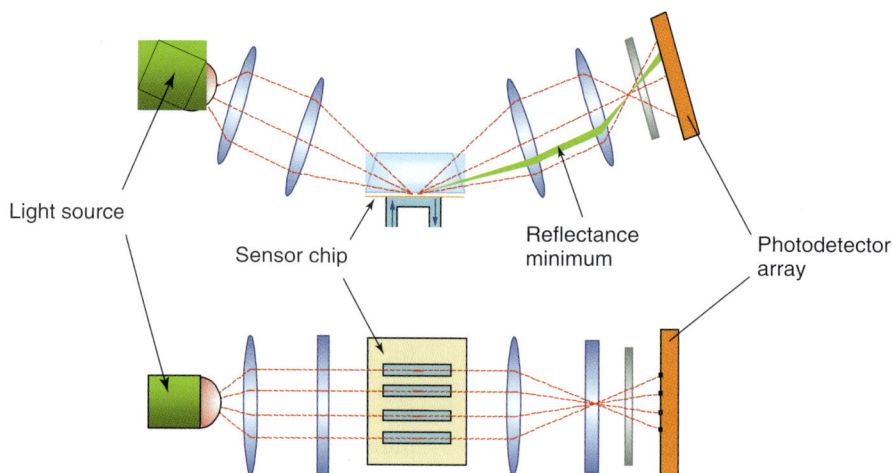

Figure 2. Schematics of the optical detection system in Kretschmann configuration with light source, sensor chip, flow cells, and array detector.

and the sensor surface is achieved by a novel thin interfacing unit (optointerface) consisting of a glass substrate coated with an elastomer layer with matching refractive index. This eliminated the inconvenient use of matching oil. A docking mechanism built into the instrument facilitated a defined and repeatable assembly of the detector unit, the optointerface, sensor chip, and microfluidics.

3.2 Sample Handling

The intended applications also placed special demands on the quality, precision, and amount of sample delivery to the sensor areas in the SPR detector. The possibility for multiplex detection in different sensor areas also has to be matched with flexible addressing of the samples, simultaneously or in sequence.

These requirements led to the development of a totally new type of integrated fluidic handling cartridge (IFC), connected to pumps for liquid delivery over the sensor surface, and an autosampler module for sample transfer from microtiter plates or other types of vials to the IFC.[5]

The fluidics cartridge consists of microchannels and diaphragm valves that are pneumatically operated to facilitate precise switching between buffer flow and delivery of a well-defined sample plug (Figure 3). The valves are placed near the sensor areas, thereby reducing dispersion to a minimum.

The flow channels are formed by precision casting in a hard silicon polymer plate. The thin-layer flow cells in the sensor area are formed by pressing the sensor chip against a part of the IFC with multiple open rectangular grooves at the outlet side of the microchannels. Typical dimensions for the flow cell are 1.6-mm long, 500-μm wide, and 50-μm high, with a total cell volume of 40 nl. The thin-layer flow cell enables optimal mass transfer conditions with linear flow rates from 1 to 100 μl min^{-1}.

3.3 Surface Chemistry

In the first publications describing the use of SPR for biochemical detection, the surface-associated component (a potential interaction partner) was attached by simple adsorption to the metal surface. However, owing to the high tendency for the fragile proteins to denature on adsorption to solid surfaces resulting in a loss or alteration of binding properties, a more elaborate strategy was needed to meet the demands for a general-purpose sensor surface with reliable immobilization methods for different kinds of biomolecules. Prior art techniques for immobilization of proteins to various types of surfaces were developed for label-based detection, but because of the relative limitation in sensitivity of label-free techniques, novel surface-attachment approaches that

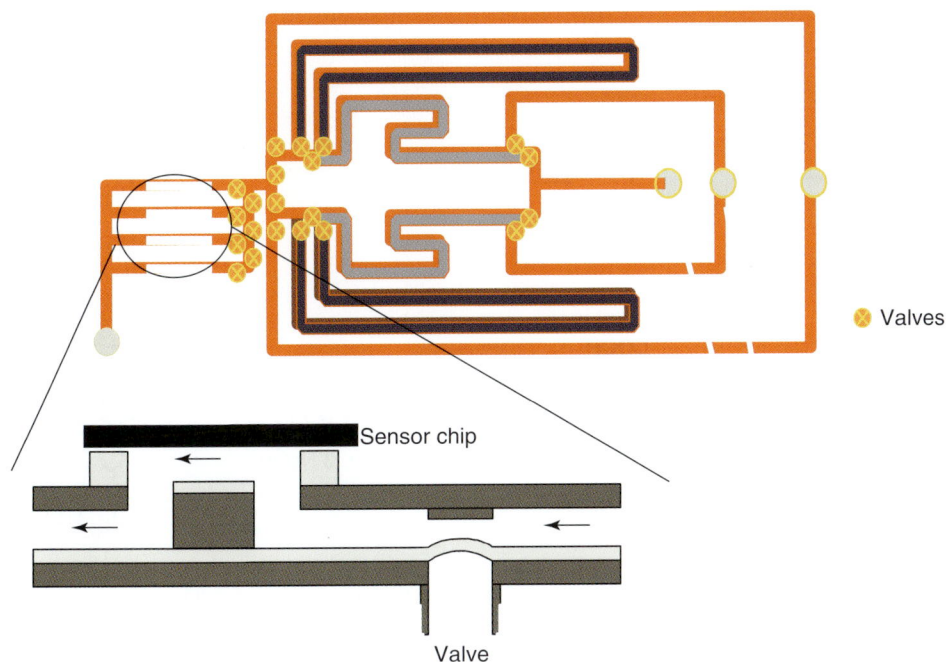

Figure 3.   Schematics of the integrated microfluidic cartridge of original Biacore instrument with flow cell dimensions and membrane valves outlined.

preserved binding activity were required. Development of such techniques focused on (i) minimizing nonspecific binding and denaturation, (ii) maximizing immobilization capacity, and (iii) introducing functional groups for the covalent coupling of biomolecules.

On the basis of the concept of self-assembled molecular monolayers (SAM) of thiol or disulfide molecules on metal surfaces, originally developed by Allara and Nuzzo as models for interface studies,[6,7] hydroxyl-terminated long-chain thiol alkanes were designed for the formation of the SAM on gold.[8] This layer forms the basis for a low adsorbing interface that can be further derivatized in a stepwise manner with carboxymethyl dextran (CM-dextran). This thin hydrogel-like polymer layer, based on dextran polymer, is composed of mainly unbranched glucose units, providing high flexibility, and water solubility. The introduction of carboxymethyl groups facilitates further activation for subsequent immobilization of different biomolecules (Figure 4).

This type of surface modification successfully meets the multiple objectives outlined previously:

- The hydrogel-like layer provides a highly hydrophilic environment with an intrinsically low nonspecific protein adsorption.
- The extended polymer structure increases the binding capacity severalfold compared to flat surfaces.
- Dextran is also highly suited for well-defined covalent immobilization of proteins that rely on a wide variety of chemistries.
- Finally, this thin-layer extension is well matched to the penetration depth of the SPR evanescent wave.[9]

The carboxylic acid residue in the CM-dextran coating acts as a functional group that can either be used for direct coupling or be switched to other functionalities. Figure 5 shows how the carboxylic groups can either be directly reacted with amine groups or be converted for use in coupling chemistries based on thiol reactions, aldehyde and carboxylic acid condensations, and biotin capture techniques. The amine coupling method in combination with the CM-dextran surface is by far the most widely used immobilization strategy in Biacore instruments.[10]

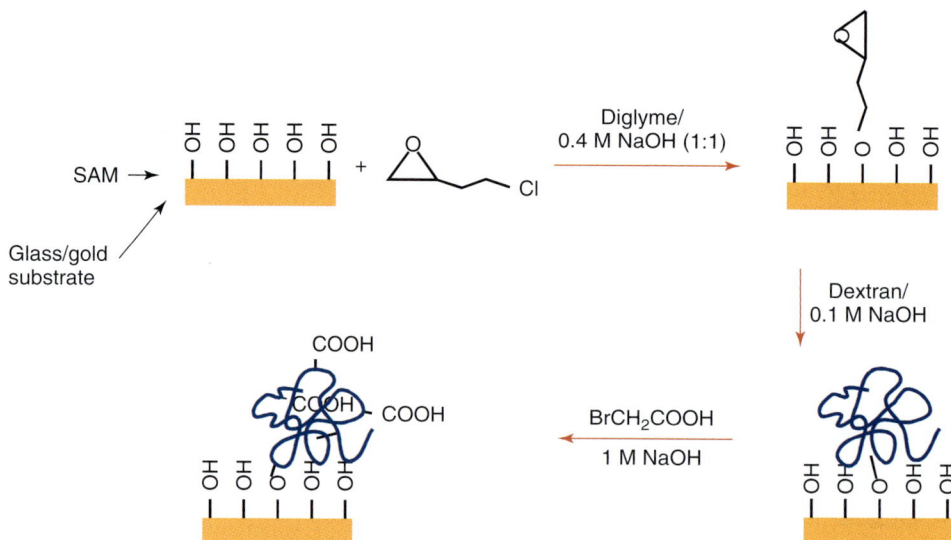

Figure 4. Chemistry modification sequence for the carboxymethylated dextran–coated sensor chip.

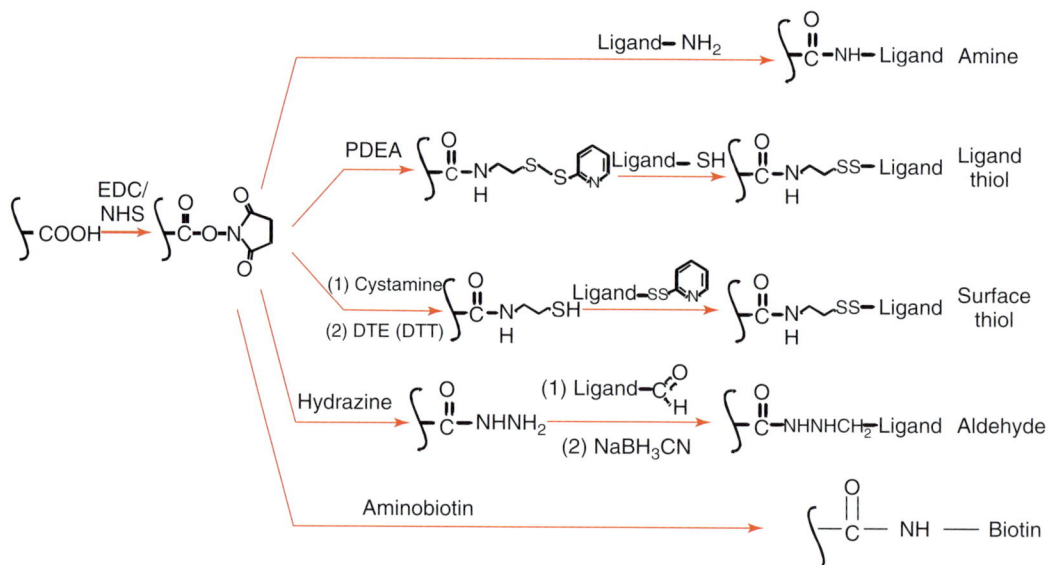

Figure 5. Alternative attachment chemistries starting from the carboxylic functions of the sensor chip, including direct amine coupling, surface thiol, ligand thiol, aldehyde, and biotin coupling.

Achieving high-capacity immobilization levels relies on a novel approach, with electrostatic attraction of the proteins to an N-hydroxysuccinimide (NHS) ester-activated, carboxylated surface, on which a fraction of the carboxylic groups remains unreacted.[11] Much lower protein concentrations than those normally used in coupling to solid phases can consequently be employed. The reaction times are also considerably shorter, in the range of 1–10 min. Coupling occurs under very mild conditions, where only a small fraction of the nucleophilic groups on the protein are reactive (e.g., the amino groups on the lysine residues are unprotonated), which results in very few immobilization points, little or no cross-linking, and a high likelihood of preserving activity.

4 ASSAYS AND METHODOLOGY

The data output from assays performed by Biacore systems required a new set of terminology.[12] The real-time monitoring of the responses generated by association or dissociation of molecules to the sensor chip was termed *sensorgram*, by analogy with *chromatograms*, for example (Figure 6). Responses resulting from mass concentration-dependent changes in the plasmon resonance angle were given an arbitrary unit called *resonance unit* (*RU*). Verification experiments have shown that 1 RU is approximately $1 \, \mathrm{pg \, mm^{-2}}$ for a typical protein.[13] Responses can also be recorded at specific times and are called *report points*. The inclusion of a step called *regeneration*, where the bound analyte is removed from the surface by the injection of a suitable solution, is enabled by the continuous monitoring of the *absolute response* of the surface coverage, showing that the step is quantitative and that the surface is ready to be used in a new analysis cycle.

The most-used assay format is a *direct binding assay* where the interaction between an immobilized interactant and an injected analyte is measured. Affinity and kinetics measurements are normally performed using this format. Subsequent injections of binding molecules can also easily be performed in a so-called *sandwich assay*, using an antibody for verification or enhancement of the first binding for example. Alternative assay formats include *competition-based assays*, where an analyte in solution competes for binding to the interacting partner with a derivative of the analyte attached to the surface. This assay format is most commonly used for concentration determinations, in food analysis applications for example.[14]

5 APPLICATIONS

There is no doubt that the main reason for the wide acceptance of Biacore's technology has

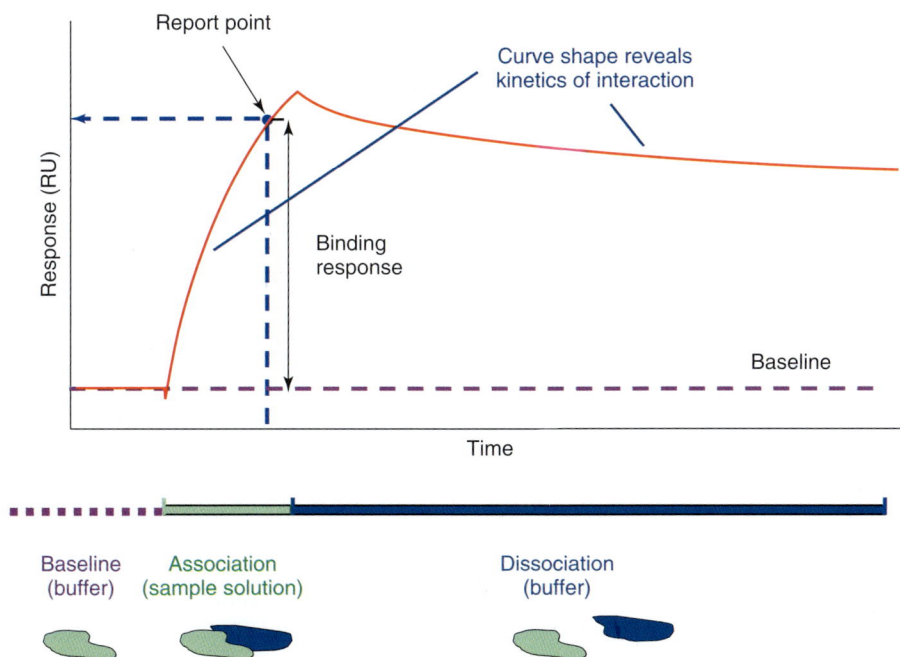

Figure 6. The basics of a sensorgram: a real-time monitoring of the change in the reflectance minimum angle (response) as a function of time. The response is expressed in the arbitrary unit RU. Response levels at specific time intervals are called *report points*.

been driven by its use in quantitative kinetics and affinity analysis of biomolecular interactions within basic and applied research.[12] This is evident from the steady stream of peer-reviewed papers published since 1990. Until 2005, more than 4500 papers based on data obtained with Biacore instruments can be tracked. Thorough reviews of the content of these articles have been published on a regular basis from Myszka and collaborators.[10,15–18] The wide spread of journals where these papers appear underline the versatility of this technology and demonstrate that it now has emerged as a well-established, common biophysical tool, alongside the likes of analytical chromatography, mass spectrometry, and light spectroscopy.

Among the plethora of different applications, those related to studies involving antibodies or fragments thereof are most abundant. Interactions studies involving antibodies were amongst the earliest important applications after the launch of the first Biacore systems as a result of the development of monoclonal antibody technology in the 1980s. Antibodies or similar binding molecules have become an indispensable tool for use in various types of diagnostic assays and more recently as reagents in biomarker research. But perhaps more importantly, antibody development has, over the last couple of years, finally reached a level where the therapeutic applications have reached maturity, largely owing to the introduction of humanized antibodies and variants thereof. Today some 20 antibodies are approved as therapeutic drugs and several hundred are in various stages of clinical trials. Biacore systems have become standard tools in the selection, characterization, and optimization stages of the antibodies, where data of both the affinity and the binding kinetics to the target molecule are central.

The use of Biacore technologies in monitoring the adverse effects of unwanted immune responses has also developed as a natural consequence of the increasing success of biotherapeutic drugs, where immunogenicity is a common side effect of this drug class.[14] Here the real-time monitoring of weak-affinity antibodies with rapid kinetics in human plasma, which have developed as a result of the administration of the therapeutic drug, can be performed with Biacore systems. These antibodies can be clinically significant and crucially, such data are hard or impossible to achieve with more common techniques like enzyme-linked immunosorbent assay (ELISA).

Within basic research, there are a large number of publications related to a wide variety of protein–protein interactions in the context of various disease areas such as cancer, neuroscience, and infection. Most prominently, Biacore systems are used within cancer research, aiming at a deeper understanding of the cellular signaling cascades involved, which present potential therapeutic targets for the development of better cancer drugs. Examples include studies performed to evaluate binding properties of substances that potentially intervene in growth factor interactions with cell-surface receptors in order to regulate the downstream signaling.[12]

The performance of Biacore systems has steadily improved, and studies of direct, small-molecule binding to immobilized proteins or oligonucleotide structures are now routinely performed. This has been made possible by improved detector sensitivity, efficient signal referencing, and improved data handling. Kinetic analysis of small-molecule binding can also be achieved, which has opened up entirely new applications within the drug discovery area. In particular, the detailed characterization of the kinetic binding properties of lead compounds to their target proteins has recently attracted great interest as a means for obtaining additional information in the selection and optimization process to the final drug candidate.[19–21] Traditionally, the potency of a drug candidate is measured by the equilibrium binding affinity measurements. However, information about the on-rate and off-rate binding gives an additional dimension for understanding the drug interaction with the target, which can be directly relevant to their therapeutic efficacy.

With the advent of fragment-library screening approaches as an alternative to traditional compound libraries, drug discovery requires a novel analytical solution. These small compounds have weak-affinity binding to the target proteins, but have the potential to be better in the optimization to the final candidate. Biacore systems provide an attractive alternative in this screening process, as weak, transient binding down in the millimolar range can be detected.

6 BUSINESS ASPECTS

At the launch of the first commercial system in 1990, Biacore had already established its own sales and marketing organization in the United States and in Europe. This was achieved with key people partly recruited from the sister company, Pharmacia Biotechnology, who already had knowledge about a potential customer base and an understanding of the needs for a new technology for protein-related analysis. In Japan and other parts of Asia too, an early introduction was made possible by the use of the Pharmacia organization as a distributor. The market introduction was initially limited to few applications and mostly related to antibody-based studies, but close relationship with the early users successively built up examples for wider applications. One important vehicle here was the introduction of a user-group meeting forum called *Biasymposium*, the first in London in 1991 followed by annual events in the United States, Europe, and later in Japan. Here the users are given possibilities to present data, and participants can interact with other users, potential buyers, and experts in the use of Biacore systems.

Biacore sales have steadily increased over the years to reach an annual revenue level around 500–600 million SEK ($50–70 million, depending on currency exchange rates). More than 2500 instruments have been sold and besides the instrument revenues, consumables like sensor chips and coupling reagents generate a significant recurrent business stream. The early establishment of a complete organization with all functions from R&D and manufacturing to sales and marketing has also been important in the development of a profitable company, which happened as early as 1995. The sales organization has grown accordingly, with multiple sales and support teams in each of the different regions; the latest addition being the establishment of an organization in China in May 2006.

The wide range of applications in various fields, as witnessed by the reference list covering more then 4500 peer-reviewed publications (2006), has become of great importance for successful marketing. As already described, the technology has gradually expanded from initial use for protein–protein interactions in academia and pharmaceutical research into further downstream applications in the drug discovery and development process. During recent years, interesting applications have also increased within the food analysis area.[14] Concentration determinations of additives of vitamins in food and testing of drug residues have traditionally been made with standard techniques like ELISA and microbiological methods. By applying immunobased competitive methods on the Biacore platform, reliable, automated analysis using ready-to-go kits is now possible. To date, kits have been developed for a range of B-vitamins such as folic acid, vitamin B12, and biotin, as well as for drug residues such as the antibiotics sulfonamides, beta lactams, and hormones such as clenbuterol.

7 RECENT DEVELOPMENTS AND FUTURE OUTLOOK

It is evident that the technology underlying Biacore systems has improved greatly since the introduction in 1990. The original instrument was designed to measure interactions based on mass changes from proteins and peptides larger than 5000 Da. More recent instruments such as Biacore S51 (2001) and Biacore T100 (2005) now readily detect binding of small molecules with molecular weight down to 100 Da. Likewise, the measurable intervals for the association rate constant have improved from 10^5 to $\sim 10^8 \, M^{-1} \, s^{-1}$ and dissociation rate constants from 10^{-2} to $\sim 1 \, s^{-1}$. Equilibrium affinity measurements can now be made within the millimolar to picomolar range. This has been accomplished by improvements in the detector hardware, reference subtraction, and numeric analysis of the sensorgrams. The user friendliness is another part that has developed tremendously and guided methods for the control and evaluation software (e.g., software wizards) speed up experiments and secure data quality. For example, in Biacore T100 a thermodynamics wizard is implemented that runs an experiment with automatic changes in temperatures and subsequent data evaluation for generation of the energy parameters.

Another example of how Biacore technology has developed and adapted is related to fulfilling the requirements placed by regulatory authorities such as the Food and Drug Administration (FDA) in the United States on instruments providers. These have increased significantly during recent years. In particular, instruments for use in drug development and quality control have to follow rules to meet the demands related to good practice (GxP) including traceability of electronic records. Therefore, all recently developed Biacore instruments that are intended for use within these environments are fully compliant with these regulations.

The proteomics era has increased the demands for higher throughput in the generation of protein-interaction data. The increased use of Biacore systems in the pharmaceutical and biotech drug discovery has also generated a demand for increased throughput for applications in drug candidate screening. A first major step to meet these demands was made with the launch of Biacore A100 in 2005. On the basis of the same core technology as its predecessors, this system is configured with a fluidic system with four parallel flow cells, each including five detection spots (Figure 7). Individual addressing of each spot using hydrodynamic flow steering enables precise in situ immobilization of different interaction partners. Flexible assay configurations facilitate generation of several hundred up to thousands of data points per day with the same quality of data as the previous systems. Built-in software for quality control of the sensorgrams enables quick processing of and selection from the large volume of data generated.

In keeping with its position as global leader in the protein-interaction-analysis market, Biacore has also begun to acquire externally developed technology that complements its business. Flexchip (acquired in 2005) has an SPR-based, interaction-array format that enables analysis of a single sample against up to 400 interactants simultaneously, thus facilitating rapid interaction profiling and selection in upstream research applications.

We have come a long way from the first data on the original Biacore to today's advanced instruments. The technology is now totally accepted as a central biophysical tool for characterizations of interactions with proteins and other biomolecules. It is likely that the future will see wide-ranging further improvements, such as in small-molecule studies, higher throughput, assay configurations, and data handling. The proteomics area will be central for life science research and drug discovery for several decades to come and the continuing development of high-performance systems from Biacore should contribute significantly to this next exciting phase in biomedical research and development.

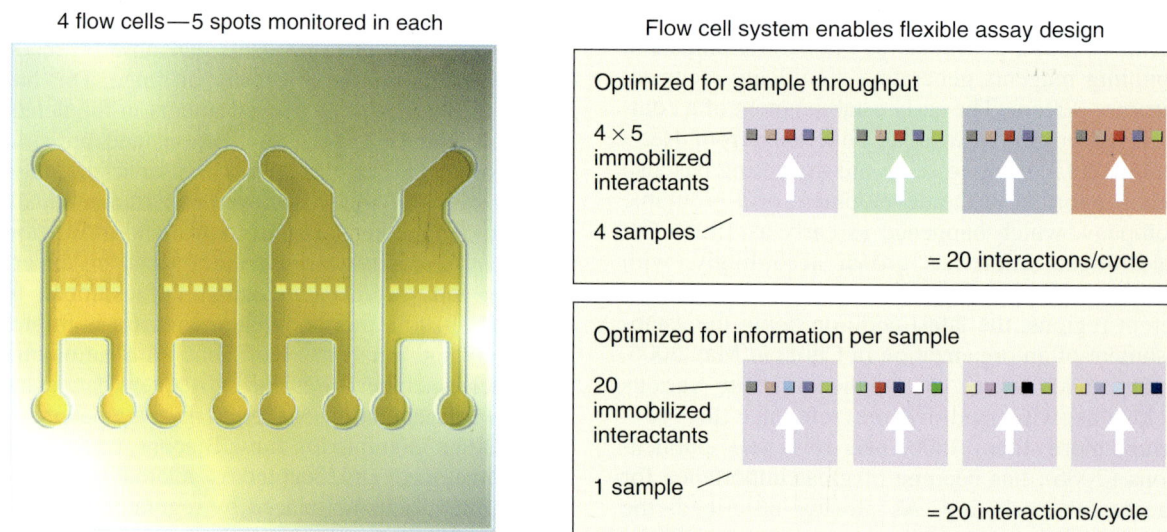

Figure 7. Schematics of the flow cell system in Biacore A100, comprising four parallel flow cells, designed for hydrodynamic addressing. Five detection spots can be simultaneously monitored in each flow cell and up to 20 interactions can be characterized during each analysis cycle. Examples of different assay designs are outlined.

REFERENCES

1. B. Liedberg, C. Nylander, and I. Lundström, Surface plasmon resonance for gas detection and biosensing. *Sensors and Actuators*, 1983, **4**, 299–304.
2. P. Garland, Optical evanescent wave methods for the study of biomolecular interactions. *Quarterly Reviews of Biophysics*, 1996, **29**, 91–117.
3. U. Jönsson, L. Fägerstam, B. Ivarsson, B. Johnsson, R. Karlsson, K. Lundh, S. Löfås, B. Persson, H. Roos, I. Rönnberg, S. Sjölander, E. Stenberg, R. Ståhlberg, C. Urbaniczky, H. Östlin, and M. Malmqvist, Real-time biospecific interaction analysis using surface plasmon resonance and a sensor chip technology. *Biotechniques*, 1991, **11**, 620–627.
4. B. Ivarsson and M. Malmqvist, *Surface Plasmon Resonance-Developments and Use of BIACORE Instruments for Biomolecular Interaction Analysis*, in *Biomolecular Sensors*, E. Gizeli and C. R. Lowe (eds), Taylor & Francis, London, 2002, pp. 241–268.
5. S. Sjölander and C. Urbaniczky, Integrated fluid handling system for biomolecular interaction analysis. *Analytical Chemistry*, 1991, **63**, 2338–2345.
6. R. G. Nuzzo and D. J. Allara, Adsorption of bifunctional organic disulfides on gold surfaces. *Journal of the American Chemical Society*, **105**, 1983, 4481–4483.
7. A. Ulman (ed), *Thin Films: Self-Assembled Monolayers of Thiols*, Academic Press, San Diego, 1998.
8. S. Löfås and B. Johnsson, A novel hydrogel matrix on gold surfaces in surface plasmon resonance for fast and efficient covalent immobilization of ligands. *Journal of the Chemical Society, Chemical Communications*, 1990, **21**, 1526–1528.
9. S. Löfås, M. Malmqvist, I. Rönnberg, E. Stenberg, B. Liedberg, and I. Lundström, A novel hydrogel matrix on gold surfaces in surface plasmon resonance sensors for fast and efficient covalent immobilization of ligands. *Sensors and Actuators B*, **5**, 1991, 79–84.
10. D. G. Myszka, Survey of the 1998 optical biosensor literature. *Journal of Molecular Recognition*, 1999, **12**, 390–408.
11. B. Johnsson, S. Löfås, and G. Lindquist, Immobilization of proteins to a carboxymethyldextran modified gold surface for biospecific interaction analysis in surface plasmon resonance. *Analytical Biochemistry*, 1991, **198**, 268–277.
12. G. Franklin and A. McWhirter, *Seeing Beneath the Surface of Biomolecular Interactions: Real-Time Characterization of Label-Free Binding Interactions Using Biacore's Optical Biosensors*, in *Protein Microarray Technology*, D. Kambhampati (ed), Wiley-VCH Verlag GmbH & Co. KGaA, Weinheim, 2004, pp. 57–106.
13. E. Stenberg, B. Persson, H. Roos, and C. Urbaniczky, Quantitative determination of surface concentration of protein with surface plasmon resonance by using radiolabeled proteins. *Journal of Colloid and Interface Science*, 1991, **143**, 513–526.
14. R. Karlsson, SPR for molecular interaction analysis: a review of emerging application areas. *Journal of Molecular Recognition*, 2004, **17**, 151–161.
15. R. L. Rich and D. G. Myszka, Survey of the 1999 optical biosensor literature. *Journal of Molecular Recognition*, 2000, **13**, 388–407.
16. R. L. Rich and D. G. Myszka, Survey of the year 2000 commercial optical biosensor literature. *Journal of Molecular Recognition*, 2001, **14**, 273–294.
17. R. L. Rich and D. G. Myszka, Survey of the year 2001 commercial optical biosensor literature. *Journal of Molecular Recognition*, 2002, **15**, 352–376.
18. R. L. Rich and D. G. Myszka, Survey of the year 2003 commercial optical biosensor literature. *Journal of Molecular Recognition*, 2005, **18**, 1–39.
19. R. L. Rich, L. R. Hoth, K. F. Geoghegan, T. A. Brown, P. K. LeMotte, S. P. Simons, P. Hensley, and D. G. Myszka, Kinetic analysis of estrogen receptor/ligand interactions. *Proceedings of the National Academy of Sciences of the United States of America*, 2002, **99**, 8562–8567.
20. S. Löfås, Optimizing the hit-to-lead process using SPR analysis. *Assay and Drug Development Technologies*, 2004, **2**, 407–416.
21. W. Huber, A new strategy for improved secondary screening and lead optimization using high-resolution SPR characterization of compound-target interactions. *Journal of Molecular Recognition*, 2005, **18**, 273–281.

83

Commercialization of DNA Arrays – Affymetrix a Case Study

Stanley Abramowitz

Advanced Technology Group, Silver Spring, MD, USA

1 INTRODUCTION

Medicine will be practiced in the twenty-first century with the benefit of genotype phenotype relationships. The knowledge of genetics of the human and model organisms has increased exponentially during the last decades along with the knowledge of biological pathways. This knowledge base coupled with the constantly developing technologies of sequencing and resequencing has enabled this coming transformation in the practice of medicine. The Human Genome Program, a cooperative international effort to sequence the human genome and model organisms and to develop advanced methodologies to sequence and resequence genomes has fueled and continues to advance these changes. The advancement of technology continues and the goals of the Human Genome Project in the United States are to develop methodologies for sequencing the human genome for $1000.[1,2] The Human Genome Program in the United States is a cooperative venture of the DOE (Department of Energy) and the NIH (National Institutes of Health). More information concerning these programs can be found on the NIH, NHGRI (National Human Genome Research Institute) website: www.genome.gov. These technologies hold the promise for everyone to have their own DNA sequence available in a credit card–sized document with them at all times. This coupled with the ever-expanding knowledge of genotype phenotype relationships will lead to an era of personalized medicine. Medicine will treat diseases before they are clinically apparent. NCI (National Cancer Institute) and NHGRI have recently announced a $100 million cooperative effort to accelerate the understanding of the molecular basis of cancer through genome analysis technologies including large-scale genome sequencing.[3] This three-year project will determine the feasibility of a full-scale effort to systematically study the genomic changes involved in all human cancers. Medicine will treat diseases before they are clinically apparent. Appropriate drugs for the treatment of disease will be chosen, taking into account the patient's genetic makeup. The number of bad interactions of drugs with the patient or the use of drugs that do not have therapeutic value will be drastically reduced. Much of this development has been and will be realized through the use of DNA arrays. We are already seeing the benefits of this personalized medicine. Herceptin is being used to treat those persons whose breast cancer has the genotype that will be responsive to this medication. Similarly Gleevac is being prescribed for those having a particular type of leukemia that is identified by a DNA diagnostic. Studies have indicated that Iressa is useful for those 10% of lung cancer patients with a particular mutation but is hardly useful for others. The stratification of patients by genotype has the promise

Handbook of Biosensors and Biochips. Edited by Robert S. Marks, David C. Cullen, Isao Karube, Christopher R. Lowe and Howard H. Weetall.
© 2007 John Wiley & Sons, Ltd. ISBN 978-0-470-01905-4.

of affording more effective medicine while not using precious time to treat those who will not be responsive to the medication, or worse, damaging the patient. It also will be utilized in drug trials so that those people with genotypes that allow them to be potentially receptive to a particular drug will enter the trials. It will also allow the introduction of drugs for people with particular genotypes. Drugs that will only be useful for small percentages of the total population, which would not be easily approvable without specific knowledge of the patient's or the cancer's genotype, will be acceptable for use. This holds the promise of aiding the drug discovery process and possibly leading to fewer failures in phase III clinical studies.

2   DNA ARRAYS – DESCRIPTION

In the late 1980s and early 1990s three research groups working independently of each other were actively doing research that would result in the technology we now know as DNA microarrays. Edward Southern, developer of the Southern blot, developed a methodology to synthesize DNA probes, in situ, using ink-jet technology and in this manner build an array. This development ultimately led to the company Oxford Gene Technology. Other companies, as discussed later, have adopted this technology to construct DNA arrays. Stephen Fodor, working at Affymax, was working on using photolithographic masks to provide the template for the in situ synthesis of DNA probes one oligonucleotide at a time. This technology, which was later developed in Affymetrix, a spin-off of Affymax, afforded the advantage of making very dense DNA microarrays. At about the same time, Patrick Brown at Stanford University was developing a distinct technique for fabricating what he described as DNA microarrays. Brown's group used previously constructed oligonucleotides and deposited these on glass substrates. Brown popularized this technology by actually publishing detailed methods and plans for the apparatus necessary for the synthesis of DNA microarrays on the web and in the open literature. The technologies he developed were adapted by many companies including Incyte and Agilent. All three of these generic technologies and other

variants that followed afford the researcher the possibility of doing a multitude of experiments in parallel with one of several DNA microarrays.

One can use DNA arrays in a multitude of technology areas such as health and health care, clinical diagnostics, basic biology research to assess pathways for biological processes, genome diversity, toxicology, biomass diversity, infectious agent monitoring, agriculture including assessing and maintaining diversity, improving yield, monitoring and improving disease resistance, and assessment of nutritional value, biotechnology, environmental monitoring, measurements of pathogens in food and water, animal husbandry; personal identification, and forensics.[4,5] Arrays are used for gene expression, genotype discovery and detection, resequencing, whole-genome analyses, forensics, personal identification, and disease detection and monitoring.

In DNA arrays, target single-stranded DNA binds to immobilized DNA probes generating double-stranded DNA. Many different formats can be used in these array technologies, varying from the use of short probes (typically about 25–60 oligonucleotides), to the use of large sequences up to gene sizes, such as those used in FISH (fluorescent in situ hybridization) technologies. There are several methodologies utilized for the fabrication of DNA arrays depending on the length and the density of probes desired. If one is dealing with large lengths of DNA consisting of gene fragments or cDNAs one can use a spotting technique that can deliver DNA via piezoelectric techniques, similar to those used in ink-jet printers or quills. The matrix can be glass, silicon, or a polymer substrate. The ink-jet spotters are commercially available or can be fabricated with available instrumentation. Companies including Agilent,[6] Motorola,[7] Clinical Microsensors (now Motorola),[8] and GE (General Electric)[9] utilize these technologies for the fabrication of DNA arrays.

DNA arrays based on small polynucleotides with lengths of about 25 oligonucleotides can be made by photolithographic techniques similar to those used in the semiconductor industry. Many probes can be made on a wafer by the use of suitable masks and light-directed synthesis chemistries to construct the desired oligonucleotides on the substrate. The power of this method is that four steps are required for each nucleotide of the probe. So that 100 steps or masks are required for a polynucleotide containing 25 oligonucleotides. Using this technology, more than 10^6 probes/cm^2 can be fabricated. Many arrays can be fabricated on a single wafer and then diced up and packaged. Present technology is producing 10 million probes per DNA array. Each wafer can be diced into tens or hundreds of identical DNA arrays. A "foundry" has been built by Affymetrix, the principle purveyor of this technology, in order to manufacture arrays in bulk and to provide quality control.

There has been considerable effort to replace the photolithographic masks with multimirror technologies to direct the light for synthesis. Several companies are using this technology including Nimblegen[10] and Febit.[11] These companies are not synthesizing the massively dense arrays described in the preceding text. Nimblegen fabricates arrays that are "Affymetrix like" and can be used with the Affymetrix instrumentation. They currently collaborate with Affymetrix, fabricating arrays for fast turnaround, in a program called *NimbleExpress*. With the maskless technology that is utilized, as many as 12 000 transcripts can be determined with 282 000 unique oligonucleotide probes. Febit is fabricating custom arrays for researchers and others and has introduced a "desktop" array synthesizer, which designs the array and fabricates it from the sequence given to the system. Using this system, a researcher can fabricate several DNA microarrays containing 7600 individual probes.

Nanogen[12] utilizes a unique system to make short polynucleotide arrays. They utilize the electronic charge inherent in DNA to perform directed synthesis onto an array having as many as 400 probes, each of which can be electronically addressed, thus, allowing the fabrication of the DNA array. Clinical Microsensors (now Motorola) makes use of the fact that doubly stranded DNA is a conductor and allows electron transfer, but is not a "molecular wire". Therefore, one can determine whether hybridization has taken place by measuring the conductivity of each site in the array.

Other technologies including those of Illumina,[13] Luminex,[14] and Pharmaseq[15] utilize "non-geometric arrays". Illumina uses randomly ordered, addressable high-density optical sensor arrays. DNA is immobilized in microspheres on an optical substrate containing thousands of micrometer-scale wells. The sensors occupy different locations from array to array. A two-color technique is used to separately identify the probe and the hybridized DNA. Luminex and Pharmaseq utilize "homogeneous assays" in which DNA probes are attached to microspheres or transponders and the targeted DNA is hybridized on these probes. The analysis is then done via a flow cell cytometer.

In most of the applications fluorescent dye readouts are being used for the assay. Lasers excite the fluorescent dyes, which are chemically attached to the DNA targets, and the emission is analyzed by confocal microscopy or CCD arrays. Other methodologies have also been used including ^{32}P radioactive labels including mass spectroscopy. Some exceptions to this are explained in the preceding text. There are also a large number or industrial, academic, and government laboratories that utilize other technologies for DNA sequence analysis including electrophoresis and mass spectroscopy.

3 AFFYMETRIX – INTRODUCTION

Affymetrix, a spin-off of Affymax, was founded in 1993 by Stephen Fodor with funding and inspiration from Alejandro Zaffaroni. Zaffaroni, a noted biochemist and experienced and famous venture capitalist, is the founder of many biotechnology companies including Alza (now Johnson & Johnson), Affymax, Dynax, Surromed, Maxygen, Symyx, and Alexza. Before this he did pioneering research at Syntex and served as the president of the company and its research institute. Affymax was founded before 1989. Affymetrix was founded to exploit the developments in the synthesis of DNA microarrays using photolithographic technologies and to provide these DNA microarrays to researchers and ultimately to the Clinical Diagnostic market. Affymetrix received its first patent in 1992 and its first round of funding in 1993. Their Initial Public Offering (IPO) was in 1996. In a similar manner to other biotechnology companies and those in the DNA analysis area they parlayed federal grants to demonstrate their technology. NIH awarded a RO1 grant in 1992 and they received a DOE, Small Business Innovation Research (SBIR) grant in 1993. This was followed by an Applied Technology Program (ATP)[16] grant in 1994. This award, a joint award with Molecular Dynamics (now GE), the largest award given by the ATP, was designed to demonstrate the ability to analyze DNA completely under computer control from sample preparation through to data analysis. Affymetrix utilized its DNA array technology, while Molecular Dynamics used electrophoresis microchannel plates to analyze the DNA. Both companies were successful in demonstrating their technologies. During this 5-year grant period, they developed other products including the DNA sequencers that were ultimately used together with those from Applied Biosystems to sequence the human genome and other model organisms and the development of DNA microarrays for human immunodeficiency virus (HIV), P53 gene (a gene that normally inhibits the growth of tumors, thereby, preventing or slowing the growth of cancers), and the first gene chip for assaying 1500 single nuclear polymorphisms (SNPs). Moreover the ATP grant provided both companies with exposure and a verification of their technologies, which made subsequent funding and scientific collaborations easier and also provided additional visibility to the scientific community.

Affymetrix supplies a complete system for DNA analysis consisting of sample preparation modules with fluidics station, read-hybridization apparatus, data analysis software, and associated workstations. There are over 2000 installations of these systems. The software and workstations provide the researcher with the genes and/or sequences that are found in the target samples. One can also obtain the "raw" data and do the analysis with alternative software. They also built a foundry for the fabrication of chips in 1999 in order to ramp up production of DNA arrays and obtain a greater degree of quality control and reproducibility from batch to batch. By the end of 2004 there were over 3000 archival publications from government, academic, and industrial entities utilizing the Affymetrix technology. Most of these studies focused on humans, but these arrays have also been used to study biology, gene expression, and sequence variation of any organism with a sequenced genome. Affymetrix dominates the market for DNA microarrays controlling about 70% of the market. They have many DNA microarrays in their catalog and also fabricate custom arrays for those requiring small numbers of arrays for research purposes. Many government, academic, and industrial research laboratories, which had fabricated their own DNA microarrays using

technologies based on ink-jet printing or quill printing for their research needs, have switched to Affymetrix DNA microarrays. Presumably, they have switched because of the reproducibility and quality control available in Affymetrix DNA arrays. Some have also utilized DNA microarrays of other companies including Agilent, Motorola, Nanogen, and those nongeometric arrays such as those of Illumina presumably for the same reason.

Throughout its corporate existence Affymetrix has been very aggressive in seeking patent protection for its technology in the Americas, Europe, and Asia. They zealously protect these assets and have brought suit with other companies when deemed necessary. They have engaged in many collaborations allowing the cross-licensing of technologies when it is in their interest. They have been successful in settling their most serious patent litigations namely those with Oxford Gene Technology, HySeq, and Incyte. The Incyte litigation also involved Stanford University. The successful settlement of these patent litigation cases has allowed Affymetrix and the others to cross-license their technologies and practice their art.

Affymetrix and the other companies exploiting DNA array technologies have had to overcome several hurdles in order to successfully commercialize their technologies. The technology was a new one and a bit complicated for many of the potential users. It marked a distinct change in the way biology research is done. Automated systems coupled with the use of powerful computer technologies replaced the traditional "wet chemistry" that biological researchers were used to. This coupled with the many different arrays that are available from different suppliers and the "homemade" arrays based on Brown's published technologies for fabricating apparatus for the synthesis of arrays and the analysis of the results made the choice of these expensive systems difficult. This was exacerbated by the lack of compatibility between the various technologies. The patent litigations also added to the problems that these technologies were involved in and did not make these choices easier. Happily for Affymetrix and the other companies involved in DNA array technologies, most of these problems have been resolved.

4   AFFYMETRIX PRODUCTS

Affymetrix provides DNA arrays for a multitude of organisms including human, mouse, Drosophila, *Caenorhabditis elegans*, rat, zebra fish, agricultural species, Arabidopsis, barley, maize, sugarcane, tomato, grape, wheat, and bacteria including *Escherichia coli*, Plasmodium, and *Bacillus subtilis*. With these microarrays one can study gene expression, species differences, and so on. By comparing diseased and nondiseased tissue one can find which genes are upregulated and which are downregulated as a result of the disease state. These arrays can also be utilized for the discovery and determination of genotype via SNPs. This allows the researcher to correlate disease states with the existence of SNPs within the diseased genome.

High-density DNA microarray technology affords the researcher the ability to analyze nucleotide sequence as well as to measure gene expression, gene diversity, and so on, for essentially the whole human and other genomes in a single experiment, often on a single array.[17] Whole-genome analysis can be performed and hence elucidates the molecular basis of disease as well as genetic pathways that are not properly operating in a wide variety of disease states including cancers, infectious diseases, multiple sclerosis, sudden infant death syndrome, and so on. The high-density DNA arrays have been designed to differentiate between splice variants of a single gene each of which yields a different protein. They have also been designed to differentiate between polyadenylation variants. A more complete description of the technology, the available apparatus and microarrays, and the data acquisition systems can be found at the Affymetrix website: www.affymetrix.com.

5   AFFYMETRIX – ACQUISITIONS AND COLLABORATIONS

Early in their corporate life, Affymetrix developed liaisons with academic, industrial, and government laboratories. These collaborations afforded Affymetrix a window into current biological research and how these technologies could be used for the research developments that later could lead to products. Affymetrix supplied these entities with DNA microarrays for their particular needs and the associated instrumentation.

5.1   Acquisitions

Affymetrix has also been quite active in purchasing other companies with core competencies that would be useful to their technology. They acquired Genomic Microsystems, a company that has developed a rapid technology for scanning arrays using confocal microscopy and have also developed methodologies for fabricating low-density microarrays. This methodology could be useful for rapid turnaround times for custom arrays that would not necessarily warrant the expense of making the large number of photolithographic masks necessary for the Affymetrix high-density fabrication technology discussed in the preceding text. Affymetrix acquired other companies for the computer algorithms they developed including Neomorphic, a computer genomics company, which has developed successful algorithms for the discovery of unique genes for the identification of potential drug targets. ParAllele a company that has developed unique genotyping reagents for use with the Affymetrix DNA SNP microarrays was acquired in 2004.

5.2   Collaborations

Affymetrix has been very active in fostering collaborations with other companies, academia, and government agencies both in the United States and in other countries. A shortlist of their industrial collaborators listed at their website for 2004–2005 includes Invitrogen,[18] Caliper,[19] Aventis,[20] Curagen,[21] GlaxoSmithKline,[22] bioMerieux,[23] Serono,[24] Stratagene,[25] and Veridex.[26] Of particular interest is their collaboration with PreAnalytix,[27] a coventure of Qiagen[28] and Becton Dickinson,[29] to make kits for blood RNA for use in Affymetrix DNA microarrays. These kits are very useful in improving gene expression assays using the Affymetrix technology. Their academic collaborators include Fred Hutchinson Cancer Center,[30] Broad Institute[31] (a coventure of Harvard and

Massachusetts Institute of Technology), Cambridge University,[32] Institut Curie,[33] Karolinska Institute,[34] University of California,[35] and Rockefeller University.[36] They are collaborating or doing research and development for government laboratories including NHGRI,[37] NHLBI[38] (National Heart Lung Blood Institute), EPA[38] (Environmental Protection Agency), NIAID,[39] (National Institute of Allergy and Infectious Diseases), NIMH[40] (National Institute of Mental Health), DOE[41] and agencies in other countries including Wellcome Trust Case Control Consortium[42] and Wellcome Trust Sanger Institute[43] and Max Delbruck Center for Molecular Medicine.[44] Of particular interest is their work for NIAID, for whom they are fabricating a DNA microarray for the detection of pathogens of interest to biodefense. They will produce DNA microarray(s) for 26 different bacterial and 10 viral species and many of their subspecies. A smallpox DNA microarray is being fabricated with a grant from CDC[45] (Center for Disease Control). They are also collaborating with nonprofit entities including National Alliance for Autism Research[46] and the Michael J. Fox Foundation.[47]

5.3 Clinical Diagnostics

Affymetrix has been very active in translating research results into clinical diagnostics. The development of standards for the introduction of DNA microarrays into the drug approval process and for clinical diagnostics has been a complicated issue for the FDA.[48,49] They have, with the inputs of academia, industry, and government, issued several guidelines. Affymetrix has a long-standing relationship with Roche[50] for the design and fabrication of DNA microarrays for clinical diagnostics. They signed an Easy Access agreement in 1997, in which they agreed to share information for use in the development of clinical diagnostics. In 2003, Roche licensed Affymetrix GeneChip technologies for the development of clinical diagnostics for a host of human ailments including cancer, osteoporosis, and cardiovascular, metabolic, infectious, and inflammatory diseases. This collaboration with Roche has resulted in the introduction of a complete system for clinical diagnostics (similar to those developed by Affymetrix for research uses, but tailor-made for the clinical diagnostic market) that has received FDA

and European approval. The first DNA microarray for use in this system has also recently received approval from FDA[51] and European regulators.[52] The DNA microarray measures a person's genotype for the P450 enzyme, CYP4502D6, which is implicated in the metabolism of prescription drugs. It is estimated that the metabolism of 25% of prescribed drugs could be assayed with this DNA microarray. A patient's genotype for these enzymes is determined with the DNA microarray, and the results are interpreted with Roche software, which indicates the patient's ability to metabolize and the prescribed drugs and its toxicology based on the patient's genotype. This DNA microarray is expected to find wide use in the choice of appropriate prescription drugs for a great variety of diseases. This is only the first of what one can expect to be a steady stream of DNA microarrays for a variety of clinical diagnostic uses.

There are home brews developed on the basis of scientific papers for a host of applications that utilize Affymetrix DNA microarrays. These DNA microarrays mostly in the gene expression area have been developed and marketed for research use and have not received FDA approval for clinical diagnostic uses. Nevertheless there are laboratories that offer these analyses for a variety of diseases including breast and other cancers, inflammatory, cardiovascular, and so on. These laboratories run these assays and deliver the results to clinical physicians for interpretation. It is still an open question whether gene expression profiles as determined from DNA microarrays or other DNA analysis technologies will receive FDA approval for clinical diagnostic uses.

6 PERLEGEN

Perlegen,[53] a spin-off of Affymetrix, was founded in 2000. Its mission is to develop the pharmacogenomics information necessary to predict drug response. Affymetrix retained a majority interest in Perlegen after a $1 000 000 private placement, in 2001. The two facets of drug response, efficacy and safety, are being addressed. This is accomplished by measuring genomic diversity at single-base resolution. They utilize Affymetrix Gene Chip technology for this. Currently, among other DNA microarrays, they are utilizing a two-microarray set capable of assaying 500 000 SNPs. These microarrays have more than 10 million probes. They have sequenced 50 human genomes to assess human diversity. They have also identified SNPs in 71 individuals of European American, African-American, and Han Chinese American ancestry. Perlegen also does contract research for others and is involved in assaying diversity in other species including rice. They will be collaborating with the International Rice Research Institute to identify SNPs in 15 rice strains.

7 AFFYMETRIX – COMMERCIAL STRATEGY

Affymetrix regards itself as a foundry similar to the foundries that are used for electronic chip fabrication. They have the basic know-how in the design of DNA microarrays and the instrumentation to prepare samples, amplify DNA, hybridize the target DNA on arrays, detect the hybridized DNA using confocal microscopy, and analyze the results using computer algorithms using workstations. They supply all the instrumentation for these tasks as well as data handling systems. They have also developed instrumentation, as discussed earlier, for the analysis of DNA samples in the clinical diagnostic arena. Others including many of their collaborators in industry, government, and academic laboratories have the capability to develop the detailed genetic information necessary for the determination of the genetics and genetic pathways of particular diseases, toxicology, and the choice of drugs for particular disease states and ultimately the development of clinical diagnostics. Affymetrix also collaborates with groups interested in other genomes, including those interested in agricultural, bacterial, viral, and other model organisms. Their website indicates the breadth of their DNA microarray catalog offerings and their research and development interests. They also have a program for the fabrication of custom DNA arrays for research uses. The researcher provides the sequence information and Affymetrix will design and fabricate the DNA microarray for their use.

Their industrial collaborators including Roche, bioMerieux, and PreAnalytix take the developed clinical diagnostic through FDA and European approval and market them. Affymetrix receives up-front payment and royalties on the DNA microarrays sold. The companies receive DNA microarrays for their research and development of the eventual product, presumably, at a lower price. Affymetrix will also entertain proposals for collaboration with biotechnology companies including small ones who have genetic expertise, who can supply the necessary genomic information. They would then work collaboratively toward the development of clinical diagnostics. The developed clinical diagnostics could then be brought through the FDA and/or European approval processes and marketed by Affymetrix's larger collaborators as discussed in the preceding text.

REFERENCES

1. NHGRI Seeks Next Generation of Sequencing Technologies, 2004, www.genome.gov/12513210.
2. NHGRI Expands Effort to Revolutionize Sequencing Technologies, 2005, www.genome.gov/1501528.
3. NIH Launches Comprehensive Effort to Explore Cancer Genomics, The Cancer Genome Atlas Begins with Three-Year, $100 Million Pilot, 2005, www.cancer.gov/newscenter/pressreleases/CancerGenomeLaunchPressRelease.
4. S. Abramowitz, Towards inexpensive DNA diagnostics. *Trends in Biotechnology*, 1996, **14**, 397–401.
5. S. Abramowitz, DNA analysis in microfabricated formats. *Journal of Biomedical Microdevices*, 1999, **1**, 107–112.
6. www.chem.agilent.com.
7. www.motorola.com/lifesciences.
8. www.amershambiosciences.com.
9. A. C. Pease, D. Solas, E. J. Sullivan, M. T. Cronin, C. P. Holmes, and S. P. Fodor, Light generated oligonucleotide arrays for rapid DNA sequence analysis. *Proceedings of the National Academy of Sciences of the United States of America*, 1994, **91**, 5022–5026.
10. www.nimblegen.com.
11. www.febit.de.
12. www.nanogen.com.
13. www.illumina.com.
14. www.luminex.com.
15. www.pharmaseq.com.
16. www.atp.nist.gov.
17. J. A. Warrington and T. B. Broudy, *Microarrays: Human Disease Detection and Monitoring in Nucleic Acid*, in *Testing and Human Disease*, Ch. 3, A. Lorincz (ed), CRC Press, 2006.
18. www.invitrogen.com.
19. www.caliperls.com.
20. www.aventis.com.
21. www.curagen.com.
22. www.gsk-us.com.
23. www.biomerieux.com.

24. www.serono.com.
25. www.stratagene.com.
26. www.veridex.com.
27. www.preanalytix.com.
28. www.qiagen.com.
29. www.bd.com.
30. www.fhcrc.org.
31. www.broad.mit.edu.
32. www.cambridge.ac.uk.
33. www.curie.fr.
34. www.ki.se.
35. www.ucdavis.edu.
36. www.rockefeller.edu.
37. www.genome.gov.
38. www.epa.gov.
39. www.idri.gov.
40. www.nimh.nih.gov.
41. www.energy.gov.
42. www.ccc.sanger.ac.uk.
43. www.sanger.ac.uk.

44. www.mdc-berlin.de.
45. www.cdc.gov.
46. www.naar.org.
47. www.michaeljfox.org.
48. Multiple Tests for Heritable DNA Markers, Mutations, and Expression Patterns; Draft Guidelines for Industry and FDA Reviewers, 2003, www.fda.gov/cdrh/guidance/1210. html.
49. E. F. Pretricoin III, J. L. Hackett, L. J. Lesko, R. K. Puri, S. I. Guttman, K. Chumakov, J. Woodcock, D. W. Feigal Jr, K. C. Zoon, F. D. Sistare, Medical applications of microarray technologies: a regulatory perspective. *Nature Genetics Supplement*, 2002, **32**, 474–479.
50. www.roche-diagnostics.com.
51. New Device Clearance: P450 Genotyping Test and Affymetrix Microarray Instrumentation System— K042259, 2004, www.fda.gov/cdrh/mda/KO42259. html.
52. www.devicelink.com/ivdt/archive/04/11/007.html.
53. www.perlegen.com.

84

RAPTOR: Development of a Fiber-Optic Biosensor

George P. Anderson[1] and David A. McCrae[2]

[1] Center for Bio/Molecular Science and Engineering, US Naval Research Laboratory, Washington, DC, USA and [2] Research International, Monroe, WA, USA

1 INTRODUCTION

Since the discovery of the evanescent wave, scientists have taken advantage of its unique properties to develop biosensor systems.[1,2] The maturation of fiber-optic technologies provided new optical platforms for the continued advancement of evanescent wave-based systems.[3,4] One particular implementation, fluoroimmunoassays utilizing evanescent wave excitation, had many perceived advantages: high sensitivity, remote near-real-time detection, and surface-sensitive detection for analyses in complex matrices. Among those attracted to exploit this promising avenue were scientists at the US Naval Research Laboratory (NRL). Anticipating the growing threat of biowarfare and bioterrorism, they decided to pursue development of a fiber-optic immunoassay system and evaluate its potential to meet current and future biothreat-detection needs. Anticipation of future military needs is a central theme of NRL, which traces its beginnings back to 1915, when worry over the great European war prompted Thomas Edison to reflect that, "The Government should maintain a great research laboratory. ... In this could be developed ... all the technique of military and naval progression without any vast expense." [5] While over time expenses have risen, the value placed in protecting our sons and daughters from hostile elements is, if anything, more paramount.

2 NRL FIBER-OPTIC BIOSENSOR

The first NRL fiber-optic biosensor (Figure 1) was assembled on an optical table using off-the-shelf components and utilized step-index plastic-clad silica optical fiber (200-μm core) to form the sensing surface.[6] While crude by today's standards, it did allow testing and validation of the fundamental principles inherent in the development of an efficient evanescent wave sensor system. The keys to creating a highly sensitive immunosensor were efficient evanescent wave stimulation of surface-bound fluorophores with subsequent recovery of that fluorescence and the successful attachment of proteins, primarily antibodies, to the silica surface of the optical waveguides necessary to specifically capture the targeted molecules. The first breakthrough came when Thompson and Villarruel deduced the fiber geometry required to efficiently excite fluorophores at the sensing surface.[7] Soon after, Golden and Anderson described in great detail, both theoretically and experimentally, the mismatch in the mode carrying capacity created when an optical fiber's cladding is removed and its core is placed in a fluid having a lower

Handbook of Biosensors and Biochips. Edited by Robert S. Marks, David C. Cullen, Isao Karube, Christopher R. Lowe and Howard H. Weetall.
© 2007 John Wiley & Sons, Ltd. ISBN 978-0-470-01905-4.

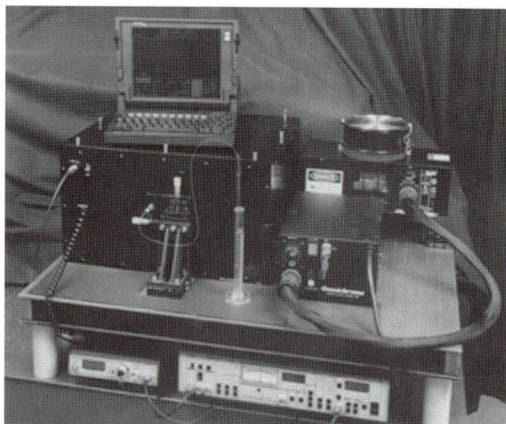

Figure 1. Photograph of the NRL fiber-optic biosensor.

index of refraction.[8–10] This understanding permitted the design of optimally efficient tapered optical probes. This advance, when coupled with highly effective immobilization chemistry developed earlier at NRL, facilitated the creation of very sensitive sensing elements.[11] Once a simple means of encasing the fiber-optical probe in a flow chamber was developed, numerous sensitive immunoassays were readily demonstrated, albeit with considerable difficulty owing to the fragile nature of the fiber-optic probe and the limitations of a single-channel instrument.[8,12,13] Most assays developed were sandwich fluoroimmunoassays (Figure 2), where capture antibodies immobilized on the surface of the optical probe bind the antigen. Fluorescently labeled antibody then binds to the captured antigen, and the instrument

detects the fluorescent complex. Assays for a variety of targets such as ricin, botulinum toxoid A, or *Yersinia pestis* F1 antigen were tested.

3　ANALYTE 2000

While the laboratory breadboard fiber-optic sensor assembled at NRL proved to be a highly effective instrument, it had little else to recommend it. Primarily because of the large size, immense weight (150 lb), and high power requirements, any potential for field use was quite limited. To meet the need for a small, portable, rugged, multichannel instrument capable of performing in the field, NRL requested proposals to construct such an instrument in 1994. Among the many responses was one from Research International (RI), who proposed the Analyte 2000, an instrument that easily met those requirements (Figure 3). Described in detail elsewhere, the instrument used four pairs of diode lasers and photodiodes, which provide the excitation and monitor fluorescence, respectively. Each pair was placed on one of four printed-circuit cards. The cards were mounted on a controller board enclosed in an $8 \times 11 \times 17$-cm box.[14] Each card was connected via a fiber bundle jumper to a separate 600-μm core diameter plastic-clad silica optical probe similar in design to the 200-μm core optical probes used on the NRL device. The fiber bundle consisted of a central quartz fiber that carried the excitation light to the probe and a circular array of plastic fibers surrounding the

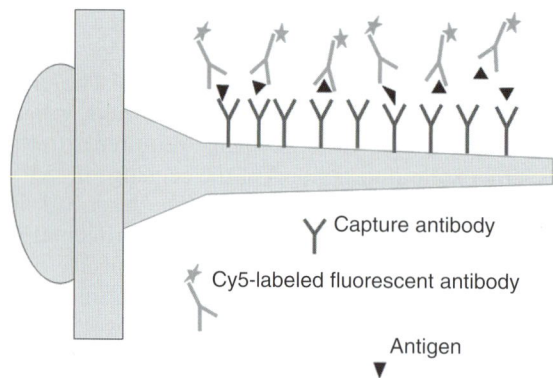

Figure 2. Schematic of a sandwich fluoroimmunoassay on an optical probe.

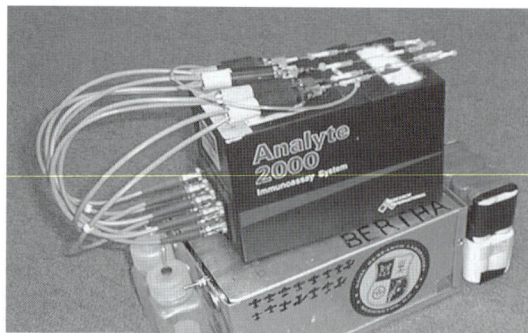

Figure 3. Photograph of the Analyte 2000 paired with the NRL automated fluidics unit developed in 1997. Note airplane icons represent number of flights on which this unit was tested, broken wing airplane icons represent crashes.

excitation fiber, which carried the returning fluo-rescence signal. The tapered probe surface readily converted the lower-order modes launched into the fiber into higher-order modes useful to excite fluorescence, and although the tapered probe func-tioned to convert the higher-order modes of the returning fluorescence into lower-order modes, the majority of the signal was still captured by the cir-cular array of collection fibers. Thus, the arrange-ment of collection fibers surrounding the excitation fiber proved highly efficient. A laptop computer controlled the instrument through an RS-232 con-nection. Numerous immunoassays were developed using the Analyte 2000, and its utility for field testing was demonstrated on a number of occa-sions. The first field test occurred in Korea (1995) immediately following delivery of the first Analyte 2000s (see Figure 4).

The Analyte 2000 saw extensive use with many immunoassays being demonstrated on this platform.[15–18] In addition, it became the first auto-mated biosensor to fly on an unmanned aerial vehicle (UAV) during testing at Dugway Prov-ing Ground, UT in 1997 and again in 1998 (see Figure 5).[19,20] For this test the Analyte 2000 was paired with an NRL custom-built fluidic unit that utilized a microcontroller to manage commercial peristaltic pumps and solenoid valves, and a ram-air powered cyclone aerosol collector, built by RI. In a sense, this unit presaged the future devel-opment of the BioHawk biosensor described at the conclusion of this chapter. Although NRL moved beyond the Analyte at this point to the

Figure 5. Photograph of the Analyte 2000 paired with the RI ram-air cyclone and NRL automated fluidics unit developed in 1998. [Reprinted with permission Ligler et al.[19] copyright 1998, American Chemical Society.]

testing of follow-on systems, it remains a useful, well-built instrument; other researchers and orga-nizations have continued to utilize this versatile assay platform, aided by the fact that it has been upgraded to allow the use of the same polystyrene fibers utilized by the RAPTOR.[21–23]

4 RAPTOR

While the Analyte 2000 was a huge improvement over the laboratory breadboard device, it still had operational limitations. Perhaps the most critical were the fragile optical probes, each of which had both fluidics and optical connections that had to be made individually. This limited its operational use to laboratory assay development and the occa-sional field demonstration. Thus, to meet the desire to have a truly man-portable biosensor, the Navy Technology Insertion Program provided funding in 1997 to develop just such an instrument. This pro-totype instrument, a precursor to the RAPTOR, was also developed by RI (Figure 6); called the *MANTIS*, it had many of the desirable features that are still found in the current RAPTOR design. It was a semiautomated self-contained sensor sys-tem, with integrated optics, fluidics, electronics, and computer control to perform four parallel immunoassays on a single sample. All the compo-nents were enclosed in a hardened $10 \times 10 \times 7$-in. package weighing less than 13 lb, and included a battery to provide power sufficient to operate for over 8 h. The fragile quartz waveguides of the Analyte were redesigned to be manufactured

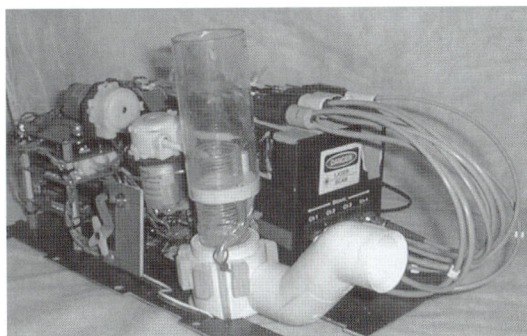

Figure 4. Photograph of the Analyte 2000 being demon-strated at the Demilitarized Zone, Korea, 1995.

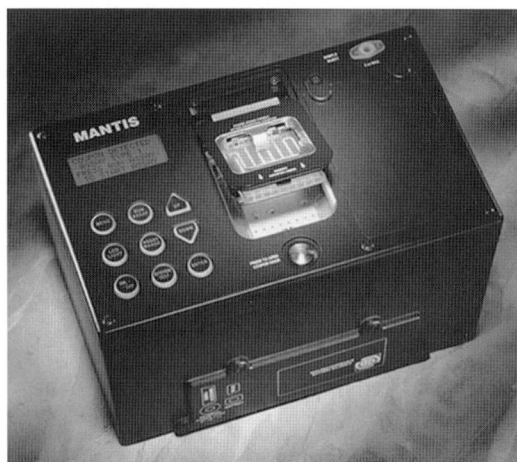

Figure 6.   Photograph of the Mantis.

Figure 7.   Photograph of the RAPTOR. [Reprinted with permission Jung et al.[25] copyright 2003, IEEE.]

from injection-molded polystyrene and included a patented "concentrator" to maximize evanescent excitation and an integral lens to focus the return signal on the photodiode detector. This change not only permitted mass production of the optical probes, but simplified the antibody attachment process as well, thereby enhancing manufacturability. Simple adsorption of the antibody or adsorption of NeutrAvidin followed by a biotinylated antibody replaced the relatively more complex silane/chemical cross-linker chemistry utilized for the silica probes. Once the wave-guide were coated, they were mounted four at a time into assay cartridges. The assay cartridges could then be easily inserted into the instrument and, upon door closure, the optics were aligned and fluidic connections made. The MANTIS was used effectively in several tests, and immunoassays were successfully demonstrated.[24] The MANTIS used a pneumatically powered fluidic system both to move fluids and to actuate valves for power conservation and noise control reasons. Unfortunately, because materials with requisite properties to construct reliable, long-endurance pneumatic valves could not be found, the device had an unacceptably high failure rate.

To overcome these limitations, the instrument was redesigned to use peristaltic pumps for fluid movement, which had the added benefit of supplanting the need for additional fluid-control valves. Numerous design improvements were made over the course of the development

of the new instrument, and it was also renamed RAPTOR. The design of the RAPTOR has been described in detail, so only the highlights will be mentioned here (Figure 7).[25] The RAPTOR has three reversible peristaltic pumps: one single-channel pump and two four-channel pumps. The single-channel pump provides the wash buffer. One of the four-channel pumps moves the sample or wash buffer over the four optical probes, and the other four-channel pump moves the reagent from separate tracer reagent vials over each respective probe and then back again. The use of peristaltic pumps solved all reliability issues, exemplified by the fact that we have had a RAPTOR operating now for years without a failure. Additionally, the pumps allowed the use of separate tracer reagents (the MANTIS had a single reservoir in which all four tracers were mixed), thereby obviating the issue of reagent compatibility. Interestingly, separate tracer reagents allowed us to demonstrate the detection of eight different analytes using only four waveguides. To accomplish this, we coated the optical probes with multiple capture antibodies and used mixed tracer reagents to generate a pattern of responses, which was then used to identify which target was present. While the overall sensitivity decreased as expected, since total reagent concentration was held constant, it was a nice example of what the future held in store.[26]

The RAPTOR has been tested in numerous trials and found to be an ideal instrument for the automated detection of threat agents. It can be paired with and control the function of the Smart

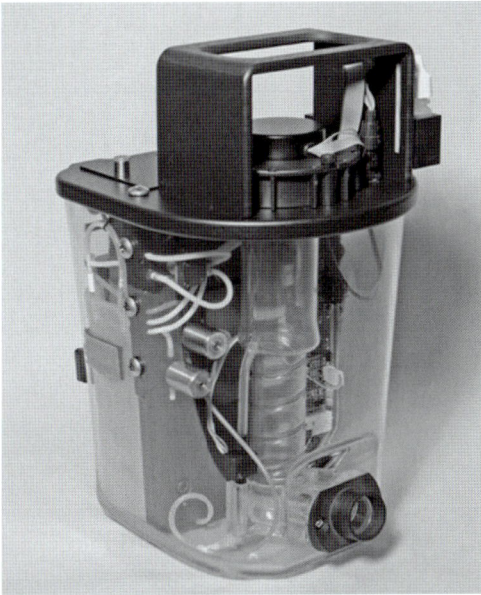

Figure 8. Photograph of the SASS 2000.

Air Sampler System (SASS 2000) (Figure 8), a portable wetted-wall cyclone, which RI developed independently following the initial test with the ram-air-driven prototype used for collection of BW simulants using UAVs. Since an ideal application of the RAPTOR is as an unattended sentry for detection of aerosolized threat agents, the next generation RAPTOR, the BioHawk, has an integrated SASS, whose features are described in the subsequent text.

The SASS series of portable wetted-wall cyclones has two key features: high energy efficiency, permitting 24 h operation on a single battery, and water inventory control. The water is maintained at a constant volume, replenishing that lost to evaporation, allowing very high concentration ratios or extended sampling periods.

The cyclone has four main sections: (i) a cyclonic cup, (ii) a mixing column, (iii) a cistern, and (iv) a water feedback loop. A high-efficiency, battery-operated centrifugal blower pulls air into the cyclonic cup, which has a rapidly swirling film of water on exposed walls. This water film also passes across the inlet nozzle region, in essence forming a water curtain through which the incoming air passes. In parallel with this process, additional water continuously enters through a nozzle mounted in the cup base. This nozzle distributes

a fine spray of droplets into the cup and, along with the wetted walls and water curtain, provides effective air-to-water mixing. Next, air flows from the cup into the stripping tube that is connected to the cup's upper surface. Air entering the smaller-diameter stripping tube increases in rotational velocity, enhancing particulate collection by the wet stripping tube ID. The airflow rate and tube diameter are selected so that adequate shear force is produced to create a concurrent flow of water on the tube wall: the stripping tube is operated beyond the so-called flooding limit, meaning that liquid introduced at the base of the tube cannot flow counter to the upwelling air.

The upward-moving film of particulate-laden water is captured in a cistern at the top of the unit. From that point, a small-diameter tube allows the water to flow back into the cyclone cup via the spray nozzle previously mentioned. The circulating water provides a pumpless, three-tiered capture strategy through the action of the water curtain, cyclone cup, and stripping column.

5 BIOHAWK AND THE FUTURE

The latest incarnation of the RAPTOR technology is a completely redesigned system, designated the BioHawk, which incorporates both air sample collection and analysis into a single man-portable package (Figure 9). The BioHawk is built of two main modules; the instrument unit and a separate liquids package. The liquids package has containers for water for the air sampler, buffer for the analyzer, and waste. Within the instrument unit are several components: the air sampler (which is the cyclone from the SASS), the analyzer (which is a highly modified version of the RAPTOR fluoroimmunoassay optics), the fluidics, as well as the electronics for control and communication, and a battery.

One key requirement for the redesigned system is the ability to conduct eight simultaneous assays rather than four. To meet this requirement and still maintain portability and relative simplicity of manufacture, the basic optical interrogation mode was changed. The optical waveguide is now of noncylindrical symmetry. Excitation light is still introduced to the waveguide via a "concentrator" on axis. However, emitted fluorescence is now

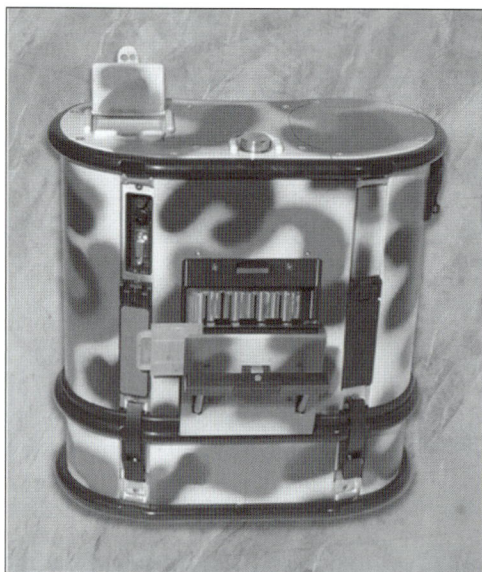

Figure 9.   Photograph of the BioHawk.

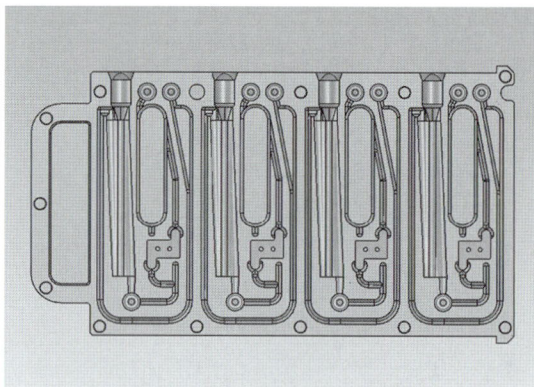

Figure 10.   Drawing of the BioHawk's assay cartridge.

detected perpendicular to the waveguide axis with a linear photodiode array. This approach allows the waveguide to be partitioned spatially into two portions, with a different capture antibody spray coated on each portion, permitting two independent assays to be conducted on each waveguide. The decrease in surface area of the waveguide available for the immunoassay is counterbalanced by the decrease in noise afforded by perpendicular detection and the S/N ratios remain about the same for the two configurations.

As with the RAPTOR, the waveguides are coated with capture antibody. Unlike the RAPTOR, the secondary antibodies are also stored on board the disposable cartridge (Figure 10). Two different labeled antibodies are lyophilized in each of the four reservoirs; mixing two reagents is a fair compromise to increase capability while adding minimally to the fluidics complexity. During instrument start-up, buffer is introduced to the reservoirs and agitated gently to redissolve the antibodies. This critical design change allows the user to switch assay panels simply by replacing the cartridge, whereas the RAPTOR had required that both the cartridge and the accompanying reagent vials be switched out. Combining both functions also avoids the inevitable human error such a process portends, avoiding potentially disastrous consequences.

The resulting biosensor unit can collect particulates and toxins from the air or can accept fluid samples manually, and then perform simultaneous immunoassays for up to eight target analytes. Like the RAPTOR, assays take about 10–20 min each and the device can perform 20 or more sequential assays on a single disposable cartridge. Results are displayed on an LCD touch screen or remotely via RS-232/RF link. It is a knapsack-size, man-portable, MILSPEC product that offers integrated performance for both sampling and analysis of aerosolized BW threats.

ACKNOWLEDGMENTS

We would like to thank Dr Frances Ligler, for it was her vision and determined efforts that sustained this project over the years. We also thank Joel Golden, Lisa Shriver-Lake, Keeley King, Dave Cuttino, Dr James Whelan, Dr Richard Foch, and Dr Chris Taitt to name but a few of the scientists at NRL, for their indispensable efforts on this project; and acknowledge the key contributions of Elric Saaski, Chuck Jung, and our many colleagues at RI. We thank the Department of Defense, the Office of Naval Research, and the US Marine Corps for their support.

REFERENCES

1. M. N. Kronick and W. A. Little, A new immunoassay based on fluorescence excitation by internal reflection

spectroscopy. *Journal of Immunological Methods*, 1975, **8**, 235–240.

2. C. R. Taitt and F. S. Ligler, *Evanescent Wave Fiber Optic Biosensors*, in *Optical Biosensors: Present and Future*, F. S. Ligler and C. A. Taitt (eds), Elsevier Science B.V., Amsterdam, 2002, pp. 57–94.

3. T. B. Hirschfeld and M. J. Block, Fluorescent immunoassay employing optical fiber in capillary tube. 1984, US Patent 4,447,546.

4. J. D. Andrade, R. A. Vanwagenen, D. E. Gregonis, K. Newby, and J. N. Lin, Remote fiber optic biosensors based on evanescent-excited fluoroimmunoassays: concept and progress. *IEEE Transactions on Electron Devices*, 1985, **ED-32**, 1175–1179.

5. The Naval Research Laboratory's web page. About NRL's history and Edison's role; 2002, http://www.nrl.navy.mil/content.php?P=HISTVISION.

6. J. P. Golden, L. C. Shriver-Lake, G. P. Anderson, R. B. Thompson, and F. S. Ligler, Fluorometer and tapered fiber optic probes for sensing in the evanescent wave. *Optical Engineering*, 1992, **31**, 1458–1462.

7. R. B. Thompson and C. Villarruel, Waveguide-binding sensor for use with assays, U.S. Patent 5061857, 1991.

8. G. P. Anderson, J. P. Golden, and F. S. Ligler, Fiber optic biosensor: combination tapered optical fibers designed for improved signal acquisition. *Biosensors and Bioelectronics*, 1993, **8**, 249–256.

9. G. P. Anderson, J. P. Golden, and F. S. Ligler, An evanescent wave biosensor-Part I: fluorescent signal acquisition from step-etched fiber optic probes. *IEEE Transactions on Biomedical Engineering*, 1994, **41**, 578–584.

10. J. P. Golden, G. P. Anderson, S. Y. Rabbany, and F. S. Ligler, An evanescent wave biosensor-Part II: fluorescent signal acquisition using tapered fiber optic probes. *IEEE Transactions on Biomedical Engineering*, 1994, **41**, 585–591.

11. S. K. Bhatia, L. C. Shriver-Lake, K. J. Prior, J. H. Georger, J. M. Calvert, R. Bredehorst, and F. S. Ligler, Use of thiol-terminal silanes and heterobifunctional crosslinkers for immobilization of antibodies on silica surfaces. *Analytical Biochemistry*, 1989, **178**, 408–413.

12. F. S. Ligler, J. P. Golden, L. C. Shriver-Lake, R. A. Ogert, D. Wijesuria, and G. P. Anderson, Fiber-optic biosensor for the detection of hazardous materials. *Immunomethods*, 1993, **3**, 122–127.

13. L. K. Cao, G. P. Anderson, F. S. Ligler, and J. Ezzell, Detection of *Yersinia pestis* F1 antigen by a fiber optic biosensor. *Journal of Clinical Microbiology*, 1995, **33**, 336–341.

14. J. P. Golden, E. W. Saaski, L. C. Shriver-Lake, G. P. Anderson, and F. S. Ligler, Portable multichannel fiber optic biosensor for field detection. *Optical Engineering*, 1997, **36**, 1008–1013.

15. L. A. Tempelman, K. D. King, G. P. Anderson, and F. S. Ligler, Quantitating staphylococcal enterotoxin B in diverse media using a portable fiber-optic biosensor. *Analytical Biochemistry*, 1996, **233**, 50–57.

16. G. P. Anderson, K. D. King, L. K. Cao, M. Jacoby, F. S. Ligler, and J. Ezzell, Quantifying serum anti-plague antibody using a fiber optic biosensor. *Clinical and Diagnostic Laboratory Immunology*, 1998, **5**, 609–612.

17. C. A. Rowe, J. S. Bolitho, A. Jane, P. G. Bundesen, D. B. Rylatt, P. R. Eisenberg, and F. S. Ligler, Rapid detection of D-dimer using a fiber optic biosensor. *Thrombosis and Haemostasis*, 1998, **79**, 94–98.

18. I. B. Bakaltcheva, F. S. Ligler, C. H. Patterson, and L. C. Shriver-Lake, Multi-analyte explosive detection using a fiber optic biosensor. *Analytica Chimica Acta*, 1999, **399**, 13–20.

19. F. S. Ligler, G. P. Anderson, P. T. Davidson, R. J. Foch, J. T. Ives, K. D. King, G. Page, D. A. Stenger, and J. P. Whelan, Remote sensing using an airborne biosensor. *Environmental Science and Technology*, 1998, **32**, 2461–2466.

20. G. P. Anderson, K. D. King, D. Cuttino, J. Whelan, F. Ligler, J. MacKrell, C. Bovais, D. Indyke, and R. Foch, Biological agent detection with the use of an airborne biosensor. *Field Analytical Chemistry and Technology*, 1999, **3**, 307–314.

21. D. R. DeMarco, E. W. Saaski, D. A. McCrae, and D. V. Lim, Rapid detection of *Escherichia coli* O157:H7 in ground beef using a fiber-optic biosensor. *Journal of Food Protection*, 1999, **62**, 711–716.

22. N. Nath, M. Eldefrawi, J. Wright, D. Darwin, and M. Huestis, A rapid reusable fiber optic biosensor for detecting cocaine metabolites in urine. *Journal of Analytical Toxicology*, 1999, **23**, 460–467.

23. H. J. Kwon, S. C. Peiper, and K. A. Kang, Fiber optic immunosensors for cardiovascular disease diagnosis: Quantification of protein C, factor V leiden, and cardiac troponin T in plasma. *Advances in Experimental Medicine and Biology*, 2003, **510**, 115–119.

24. K. D. King, G. P. Anderson, K. E. Bullock, M. J. Regina, E. W. Saaski, and F. S. Ligler, Detecting staphylococcal enterotoxin B using an automated fiber optic biosensor. *Biosensors and Bioelectronics*, 1999, **14**, 163–170.

25. C. Jung, E. W. Saaski, D. A. McCrae, B. M. Lingerfelt, and G. P. Anderson, RAPTOR: a fluoroimmunoassays-based fiber optic sensor. *IEEE Sensors Journal*, 2003, **3**, 352–360.

26. G. P. Anderson, B. M. Lingerfelt, and C. R. Taitt, Eight analyte detection using a four-channel optical biosensor. *Sensor Letters*, 2004, **2**, 18–24.

85

Regulatory and Validation Issues for Biosensors and Related Bioanalytical Technologies

Nikolay V. Sergeev,[1] Keith E. Herold[2] and Avraham Rasooly[1,3]

[1] Center for Devices and Radiological Health, FDA, Silver Spring, MD, USA, [2] Department of Bioengineering, University of Maryland, College Park, MD, USA and [3] Diagnostic Biomarkers and Technology Branch, NIH-National Cancer Institute, Rockville, MD, USA

1 FOREWORD

The usage of the term *biosensor* in this chapter is consistent with the definition of *biosensor* by the International Union of Pure and Applied Chemistry (IUPAC) as a "device that uses specific biochemical reactions mediated by isolated enzymes, immunosystems, tissues, organelles, or whole cells to detect chemical compounds usually by electrical, thermal, or optical signals". Thus, clinical assays that utilize ligands such as antibodies or nucleic acids as recognition elements in combination with a signal transducer meet this definition. The term *biosensor* is used along with *assay*, *bioanalytical technique*, *device*, and *test* interchangeably in the chapter to signify clinical tests carried out with biosensors. There are many types of diagnostic tests and each type has its own capabilities, limitations, standardizing procedures, and acceptance criteria. This chapter is a review of the most common measurements and metrics that characterize performance of biosensors intended for use as clinical tests. Statements in this chapter do not necessarily represent the official opinion of any of the organizations with which the authors are associated.

2 INTRODUCTION

Biosensor-based clinical diagnostics is a dynamic and rapidly evolving area. Biosensors and bioanalytical technologies have the potential to decrease detection time, lower the detection limit, increase specificity and sensitivity, and allow high-throughput analysis of biologically important molecules. These biosensor characteristics position them as extremely powerful tools for in vitro diagnostics (IVDs). IVDs are considered to be medical devices and are defined as "reagents, instruments, and systems intended for use in the diagnosis of disease or in the determination of the state of health in order to cure, mitigate, treat, or prevent disease" 21 CFR § 809.3(a). According to the Food, Drug, and Cosmetic Act, 21 U.S.C. 321(h) enacted by the US Congress in 1976, IVD tests that are to be commercialized for the diagnosis and management of patients are subject to FDA (United States Food and Drug Administration) regulation. The main safety concerns associated with biosensors for IVD arise from the risk of misdiagnosis due to a false-positive or a false-negative result, leading to improper patient

Handbook of Biosensors and Biochips. Edited by Robert S. Marks, David C. Cullen, Isao Karube, Christopher R. Lowe and Howard H. Weetall.
© 2007 John Wiley & Sons, Ltd. ISBN 978-0-470-01905-4.

management and, in some cases, inappropriate public health response. To reduce risk to the public health associated with such devices, International Standardization Organization (ISO) and regulatory agencies have developed regulations, guidelines, and policies applicable to major aspects of biosensors manufacturing, validation, and use. From a medical device regulatory perspective, it is critical to define key quality system requirements and ensure that they are met or exceeded enabling the development of a safe and effective diagnostics device. In all cases the work performed before routine implementation of IVD tests is divided into two major components: analytical verification and clinical validation. This chapter is a review of the key regulatory and validation issues affecting overall (both analytical and clinical) performance of biosensors for IVD, such as intended use, accuracy, precision, analytical sensitivity, analytical specificity, cross-reactivity, interference, sample matrix effects, clinical accuracy, and predictive positive/negative values with prevalence.

3   INTENDED USE

Clear definition of the *intended use* of an IVD test or device is one of the main steps in the development of IVD tests. Claimed performance characteristics of a clinical test are always evaluated by regulatory agencies (such as FDA) in the context of the intended use. The intended use specifications should include the analyte the device is intended to measure (biomarker and its type), proof of concept under intended conditions of use (showing that the biomarker has a diagnostic value), the clinical purpose of measuring the analyte (disease or infection), the populations for which the device is indicated, and the intended setting where the test or device will be used (e.g., home use or clinical/point-of-care settings) by the intended user (clinical personnel or an individual). The analytical and clinical data required greatly depends on both the detection principle and the claims made regarding intended use. For new IVD products, it is recommended that manufacturers consult with an appropriate regulatory agency for premarket device approval or clearance. This practice was introduced by FDA to help manufacturers avoid wasting resources on studies that would not support FDA approval for their intended use. This is especially important for devices in emerging fields, such as microarray-based diagnostics, where technology, review policy, and regulatory science are continually evolving.

4   ANALYTICAL PERFORMANCE

The aim of *analytical verification*[1] is to determine that the biosensor can measure its intended analyte reproducibly and accurately when the measurement is performed by the intended user. The analytical performance of a biosensor or a bioanalytical technique is most properly characterized by a number of interrelated parameters such as accuracy, precision, sensitivity, specificity, detection/quantitation limits, and linear/measuring range. Very often, different types of bioanalytical tests, for example, immunoassays or detection of nucleic acids, utilize different metrics and have specific performance evaluation criteria. Additionally, it should be mentioned that in different regulatory documents one or even several of the mentioned parameters may have different meanings, implying different evaluation procedures. On the other hand, analytical performance parameters with identical meaning may be defined differently in international, regional, and local regulatory documents. To avoid confusion, users and developers of IVD tests are encouraged to contact an appropriate regulatory agency for updated guidelines that regulate specific types of assays in particular areas.

In general, the combination of performance parameters allows verification of the analytical performance of a biosensor, specifically whether or not it meets the requirements of the intended application. Any analytical or operational limitations of the bioanalytical test can be revealed and worked out before conducting diagnostic evaluations on clinical samples. Sources of analytical variability are inherent in any analytical procedure and can be divided into three main groups: preanalytical variables—including sampling procedure, sample preparation and handling, sample loading, storage, and transport; analytical variables—including precision and accuracy of the test method and factors which may interfere with an assay performance; and postanalytical variables—including result validation, interpretation of the result, data entry, data transfer, and the method used to report the results. In this chapter only the major parameters related to analytical and clinical performance of bioanalytical tests are described. As a rule, more technologically complex devices involving multiple steps (e.g., sample handling and processing) present a more challenging analytical validation requirement. Also, different analytical performance parameters are usually evaluated for quantitative and qualitative tests as listed in Table 1. It is important to point out that since biosensors combine biological recognition elements with physical transducers, two very different elements, each of the components uniquely affects the device's performance. In general, the variability of the biological ligands will be far greater than that of the electronic transducer, but from the regulatory viewpoint, the analytical performance of the device is a measure of combined performance of both elements, that is, the whole system. The validation of the individual biological and physical elements separately is highly recommended but may not be required.

4.1 Analytical Accuracy (Trueness and Precision)

The biosensor's *accuracy* (analytical) is defined as "a closeness of agreement between a test value and the accepted reference (established) value".[2] This term sometimes is used interchangeably with *trueness*[3] and is usually understood to be the total error of the measurement. Trueness in turn is defined by ISO as "the closeness of agreement between the average value obtained from a

large series of test results and an accepted reference value"[4] and along with the "precision"[4] is the measure of analytical accuracy. Accuracy of a biosensor is usually verified using protocols specific to the type of instrument, detection principle, and the assay output. Typically, for biosensors that measure the concentration, amount, or activity of analyte present, the accuracy can be verified by analyzing a large number of samples and comparing the measured values with the true (known) or accepted reference values obtained with a "gold standard method". The systematic signed deviation (systemic error) of the test results from the "true value"[5] is called *bias* and is usually specified along with the trueness of the bioanalytical test. Bias can arise if the sample preparation step systemically fails to fully extract the analyte of interest or if other substances present in the sample interfere with the detection of intended analyte.

The term *precision* is used as a measure of "closeness of agreement between independent test/measurement results obtained under stipulated conditions".[4,6] While trueness and bias are the measures of a systemic deviation of the tests results, the precision is an estimate of random errors inherent in every measurement procedure, that is, the factors that cannot be completely controlled. In other words the analytical precision accounts for unidentified variables coming from various sources such as reagent lots, instruments, and operators. The precision is calculated as a standard deviation (SD) or a coefficient of variation (CV%), which expresses the SD as a percentage of the mean value of the replicate measurements. The precision of a bioanalytical procedure is usually verified in a series of individual measurements of an analyte "when the analytical procedure is applied repeatedly to multiple aliquots of a single homogeneous volume of biological matrix".[3] Determination of the precision requires knowledge of the true levels of the analyte of interest in the samples being tested and is usually specified in terms of three separate parameters: repeatability, "intermediate precision", and reproducibility.[4,6] *Repeatability* is determined by measurements performed in one laboratory by the same operator using the same equipment on the same day (also called *within-day* or *intraday* precision) and represents the irreducible inherent precision of the method. Repeatability should be tested by a minimum of five determinations at

Table 1. Performance parameters used for the analytical verification of quantitative and qualitative tests

Parameter	Qualitative tests	Quantitative tests
Trueness	−	+
Precision	−	+
Specificity	+	+
LoD	+	+
LoQ	−	+
Linearity	−	+
Measuring range	−	+

+ Signifies that this parameter is normally evaluated. − Signifies that this parameter is not normally evaluated.
LoD: limit of detection; LoQ: limit of quantitation.

three different concentrations (low, medium, and high) in the range of expected concentrations.[3] The *intermediate precision* shows the variations of day-to-day analysis, by different operators and different instruments. *Reproducibility* (also called the *interlaboratory precision*) is defined as the measured precision in multiple measurements obtained by different laboratories (different operators, equipment, and laboratories). However, in some regulatory documents[7] interlaboratory reproducibility is referred to as a clinical performance parameter because such performance is recommended to be assessed with clinically relevant specimens in laboratory settings typical of where the device test system will likely be used. As a rule the reproducibility needs to be established by conducting repeatability experiments in different locations (e.g., multicenter trials) and isolating between-laboratory precision from the within-laboratory components. Variability between laboratories (i.e., reproducibility) is often the largest single component of precision while repeatability is often the smallest component of precision. Intermediate precision (between runs, between days, or between operators) is also usually small since it is controlled by an individual laboratory. The detailed calculations for multilaboratory or multiinstrument study designs can be found in references on the analysis of variance and are usually performed according to recognized protocols.[4,6]

4.2 Sensitivity/Specificity

The purpose of many IVD biosensors is to detect the presence of small amounts of analytes in various matrixes and often in the presence of interfering substances. This implies that the device must not only be capable of detection of low levels of the intended analyte (sensitivity) but also should be able to selectively differentiate between target and nontarget analytes (specificity). In general, sensitivity and specificity characterize the ability of the biosensor to selectively recognize the intended analyte and produce specific interpretable signal under the specified assay conditions. There are several alternative common uses of the term *sensitivity*. The IUPAC defines the *analytical sensitivity* as the "the slope of the calibration curve." The calibration curve relates the mean of the measured signal to the actual (known) concentrations; the steeper the slope of the calibration curve, the more sensitive the assay is to slight changes in amount of analyte. However, the term sensitivity is also used in several other contexts either alone or with modifiers. An example is functional sensitivity, which is used to describe the interassay precision at very low analyte concentrations, for certain diagnostic assays with high precision requirements at low concentrations.

In many clinical laboratories and diagnostic applications, "sensitivity" or "analytical sensitivity" are used interchangeably with "limit of detection (LoD)," "lower LoD," or "detection limit."[8] In these cases the sensitivity of a biosensor system generally corresponds to the LoD and is understood to be the minimal detectable analyte concentration typically given in units of particles or mass per unit volume (particles/liter or mass/liter). The preferred use of the term *sensitivity* in relation to qualitative IVD tests is the ability of the test system to detect the presence of the target analyte in samples where it is actually present at concentrations equal to or exceeding the LoD.

Specificity refers to the ability of a biosensor to correctly identify (and) quantify an analyte in the presence of interfering variables. In qualitative testing, specificity is "the ability of a measurement procedure to determine only the component it is intended to measure,"[9] that is, to produce negative results in concordance with negative results obtained by a reference method. In other words, the specificity is the ability of a biosensor to respond only to the specified analyte and not to other substances present in the sample, and very often is a function of experimental conditions. Often, the specificity of an assay can be expressed in a unique way accepted in the area of application (e.g., length of a PCR amplicon in a PCR-based test). The specificity is inversely related to the cross-reactivity of the test, that is, a lower cross-reactivity corresponds to a higher specificity. To determine the specificity of a bioanalytical method, a panel of similar analytes and nonrelated interfering substances inherent in the sample matrix, should be tested.[10] Frequently, nonidentifiable interferents come from the sample matrix requiring that the specificity evaluation studies be performed in each specimen matrix indicated for use. Interference[11] can also occur owing to competitive inhibition when other structurally related analytes are present in the sample or, for example, for PCR-based assays when amplification kinetics between an internal control and the target are unbalanced. From the regulatory viewpoint, the specificity testing should quantify the interference and demonstrate that potentially interfering substances or conditions encountered in actual specimens do not confound assay results.

4.3 Limits of Detection/Quantitation

The LoD is an important performance criterion for qualitative tests when results are reported as "positive" or "negative". When the sample is diluted below the minimal detectable concentration of the analyte, the test will not produce a positive output even though the target is present. Sometimes the term *LoD* is used interchangeably with sensitivity and "minimum detectable concentration" and generally corresponds to the lowest amount of analyte in a sample that can be detected with stated probability, although perhaps not quantified as an exact value.[8] Other "limits" that are widely used in clinical diagnostics include, but are not limited to, "limit of blank" (LoB), LoQ, "lower end of the measuring range" (LMR), and "lower limit of linear range" (LLR). The use of these parameters is closely associated with the particular assay design and its intended use. For example, the LoD is usually reported both for qualitative and quantitative assays, while LoQ is relevant only to quantitative assays. Depending on the assay type, there is often an overlap in the definition of these parameters. In general, these quantities are related such that LoB < LoD ≤ LoQ. The placement of LMR and LLR in relation to these depends on the particular biosensors system.

As a rule, the lowest "limit" is the *LoB*, which is the highest value of the test output signal obtained in a series of experiments on clinical samples that do not contain the intended analyte. In other words LoB is a measure of the background signal "noise" produced by the device in response to sample matrix only and is sometimes called the *critical value*.[12] It is important to note that the LoB refers to an observed test result only, while all of the other mentioned limits refer to actual concentrations of the analyte.

The *LoD* is the analyte concentration at which a biosensor's response exceeds the LoB and may therefore be declared as "positive". The LoD is typically determined by a standard technique consisting of testing the biosensor against samples of the target analyte prepared as a serial dilution. While LoD is a necessary performance parameter mainly attributed to quantitative tests, the knowledge and regular confirmation of the LoD for a qualitative test is a valuable tool for monitoring the consistency of the test performance. It is common to see changes in LoD of bioanalytical tests due to, for example, minor variations in reagent lots. An unknown increase of LoD for a clinical test can increase the rate of false-negative results, because samples with theoretically detectable levels of an analyte may fall below the actual LoD, while samples at higher concentrations may still give correctly positive responses. Consequently, frequent experimental confirmation of the LoD is important for interpretation and for monitoring consistency of the performance of many qualitative tests.

The *LoQ* is the lowest actual concentration at which the analyte is reliably detected *and* at which the analyte can be quantitatively determined with stated acceptable precision and trueness, under stated experimental conditions.[8] The LoQ is also called "lower limit of determination", "lower limit of quantitation" (LLoQ), and *LMR* and has meaning only when specified along with the coupled parameters (precision and trueness). Depending on the intended use and objectives of the test, the LoQ can be equal to or significantly higher than the LoD.

The *LLR* is the lowest concentration at which the method response has a linear relationship with the true concentration of the intended analyte.[13] The LLR is widely used in tests where accurate analyte quantitation is required and may be equal to or higher than the LoQ.

Currently, many devices report LoD and LoQ that come from measurements of blank samples.[3] In this approach, the mean value of LoB is taken as a basis which then is adjusted by a factor of 10 times the SD to estimate the LoQ and a factor of 2 or 3 to estimate the LoD. However, for clinical tests it is recommended that one follow the approach described by the ISO defining LoD in relation to stated levels of Type I and II errors[14] (also called α and β *errors* in CLSI guidelines[8]). α error (Type I error) is a probability of falsely claiming that a substance is present when it is true that substance is absent. LoB is established in such a way that the test is positive for the samples without analyte only $\alpha\%$ of the time. β error (Type II error) is a probability of falsely claiming that a substance is absent when it is true that substance is present. LoD is then defined as an amount of analyte in a sample for which the probability of falsely claiming the absence is β, given a probability α of falsely claiming its presence. As a rule, LoD and LoQ values should be determined and specified along with a confidence interval (e.g., 95%).

4.4 Linearity and Range

The *linearity* of a bioanalytical test is defined as "the ability (within a given range) to provide results that are directly proportional to the concentration (amount) of the analyte in the test sample".[13] Linearity is typically measured by testing serially diluted samples containing levels (concentrations or activities) of the target analyte, which are known by formulation or known relative to each other (not necessarily known absolutely). When the overall system responses (measurement results) are plotted against known values of the analyte levels, the degree to which the plotted curve conforms to a straight line is a measure of system linearity. It should be mentioned that some bioanalytical tests (i.e., competitive immunoassays) are nonlinear by nature, so the test results may not be linear even after data transformation such as logarithmic conversion. The linear relationship between the observed system response values and the true levels of the analyte is valuable for interpretation of quantitative test results through simplified interpolation. Furthermore, for a quantitative bioanalytical method (i.e., real-time qPCR), it is impossible to accurately interpolate between points unless the functional form of the results is known and the simplest functional form is a linear relationship. The evaluation of the test linearity should include the following important concentrations: minimum analytical concentration or LLR, various medical decision limits, and maximum analytical concentration or the "upper limit of the linear range" (ULR). The test linearity can be verified with a calibration (standard) curve generated with six to eight nonzero standard samples prepared in the same matrix as the samples in the

intended study.[3] A serial dilution often gives a better correlation coefficient (r), but could give a false regression coefficient if an error has occurred in the preparation of the initial sample.

The *linear range* of an analytical method is defined as the span of analyte concentrations for which the system output results are directly proportional (with stated trueness and precision) to the input analyte levels. The *measuring range* (also referred to as the *reportable range* or *working range*[15]) is similar to the linear range but without the requirement for the direct proportionality. In other words, within the measuring range the test system response may have acceptable accuracy and precision, but have a nonlinear relationship to the analyte levels.

In some cases the linear range and the measuring range coincide but more typically the linear range is contained within the measurement range. The boundaries of the measuring range are LoQ and "upper limit of quantitation" (ULQ), while the linear range is characterized by LLR and ULR that are usually higher and lower than LoQ and ULQ, respectively.

5 CLINICAL PERFORMANCE (DIAGNOSTIC UTILITY)

While analytical validation presents its own challenges, the process of clinical validation[1] of bioanalytical tests is an even larger issue, because in addition to all of the analytical variables discussed previously it accounts for biological variations between individuals in the target population. One of the major issues in validation of clinical performance of IVD tests is the validation of biomarkers the IVD test is intended to measure. This is a challenging stand-alone problem that is discussed in detail elsewhere.[16] Clinical performance of bioanalytical tests has to be evaluated in clinical studies in order to establish safety and effectiveness. In cases such as microbial identification assays, the goal is to define "positives" and "negatives" with a high degree of certainty because mistaken diagnosis can lead to adverse health effects due to inappropriate treatment. In other cases, such as the measurement of a blood-borne marker that is elevated in certain disease states the diagnosis typically depends on the relative concentration of the marker compared to a disease-free population. In both types of cases, the diagnosis depends on the accuracy of the data from the tests. From a clinical perspective there is no "linear range" for the disease quantitation, instead the quantitative determination of a biomarker in a clinical sample can be used for giving the categorical "answer" to the question about the patient status. In other words, during the clinical validation it is necessary to access the ability of the test to correctly classify individual persons into two subgroups, for example, a subgroup of persons affected by some disease (and therefore needing treatment) and a second subgroup of unaffected (healthy) persons.

Because of this requirement for the highest possible predictive power of the assay, one may say that from the clinical validation point of view all IVD tests are considered to be purely qualitative, where the "disease status" of a patient is the intended "analyte". We believe that this simple and not entirely correct perspective will be helpful in understanding the meaning of the process of clinical validation of IVD tests.

5.1 Diagnostic Accuracy

While many sources on the verification of analytical performance are available, the validation of clinical performance is much less extensively described in the literature. The cornerstone of the establishment of the clinical performance of a diagnostic device or IVD test is to prove that it provides information about a disease or a clinical condition that is useful and/or meaningful to the health care provider and the patient. This characteristic is called the *diagnostic accuracy* and may be expressed in a number of ways (clinical sensitivity and specificity, positive and negative predictive values with prevalence, test efficiency) depending upon its intended use.[17] As a rule diagnostic accuracy refers to the comparison of the results between the test system under evaluation and the true diagnosis obtained with the best available technique (e.g., a combination of methods and procedures) and should be evaluated during extensive trials using clinical samples from the intended-use population.

The exact types of clinical data that are needed depend greatly on the claims that are made in the intended use but, in any case, should include

examination of both normal and affected patient samples. In addition, the test population should be clearly defined (e.g., infants, pregnant women, or men over age 50). The number of samples used for clinical validation should allow calculation of confidence intervals (as a rule 95% confidence interval) for each of the clinical performance parameters. Several interrelated parameters are used to validate the diagnostic accuracy of bioanalytical tests and are described in the subsequent text.

5.2 Diagnostic Sensitivity/Specificity

The terms *diagnostic* or *clinical sensitivity*[9] are used to characterize or compare the performance of diagnostic tests in relation to prespecified clinical information (e.g., presence or absence of a target analyte related to a disease). Diagnostic sensitivity and specificity are related to the performance of the test in terms of the detection of the disease and may not directly relate to the analytical sensitivity and specificity of the biosensor itself. In other words, analytical performance of the test provides a basis (i.e., test "score") for the interpretation of test results (e.g., "positive" or "negative"), which are then used to classify patients into two clinically relevant groups (e.g., "diseased" or "normal"). Comparison of the distribution of the test results within the groups with true values determined by a reference method is then used to calculate the clinical sensitivity and specificity of the test in relation to the particular disease. For a qualitative diagnosis, diagnostic sensitivity for a patient's "diseased" status (or other relevant target condition) is the proportion of patients with a positive output value in the test device system determined to be infected by the reference method. Diagnostic specificity is usually calculated as the proportion of "normal" patients with a "negative" output value in the test results.

5.3 Predictive Values with Prevalence/Test Efficiency

Although diagnostic sensitivity and specificity are commonly used to describe the clinical accuracy of a test, in practice, predictive value of the test provides a different perspective that includes disease prevalence. The risk to the patient of false-positive or false-negative results is often tied to the clinical decisions made on the basis of the result. It is therefore important to establish valid, accurate predictive values of the test allowing definition of a patient's status with high confidence. Predictive values of the test are a function of the diagnostic sensitivity and specificity. The proportion of total positive test results that are truly positive is called the *predictive value of a positive test result* (*PVP*).[9] In other words, the PVP is the measure of the probability that an individual with a positive test result actually has a particular disease or condition the test is designed to detect, so the PVP is a function of both sensitivity and prevalence. Alternatively, the proportion of total negative test results that are truly negative determines the *negative predictive value of the test* (PVN)[9] and is a function of specificity. The "prevalence of disease" is not one of the biosensor performance parameters, but directly affects the predictive value of a test result and is usually considered during the test validation. *Prevalence* represents the total number of cases per unit of the targeted population (which differs from incidence representing the number of new cases per year per unit of population). *Estimated efficiency of the test* (EET) is the total agreement of the test results with the true diagnosis and is represented by the percent of all results that are true results, whether positive or negative. By the definition, the EET is a function of both clinical sensitivity and specificity and can reach its maximal value (100%) only if both sensitivity and specificity are 100%. All clinical performance parameters are affected by occurrence of cases of mistaken diagnosis based on the test results. They are described by the *false-positive fraction* (FPF), which is a fraction of "positive" test results declared "negative" by a reference method, and

the *false negative fraction* (FNF), which is a measure of "negative" test results obtained on truly "positive" samples. A useful way of organizing the clinical performance parameters of a test is a 2×2 contingency table[2] as shown in Table 2. Using the values given in Table 2 we get the following:

$$\text{Diagnostics sensitivity} = 100\% \times A/(A + C),$$

$$\text{Diagnostics specificity} = 100\% \times D/(B + D),$$

$$\text{PVP} = 100\% \times A/(A + B),$$

$$\text{PVN} = 100\% \times D/(C + D),$$

$$\text{EET} = 100\% \times (A + D)/N,$$

$$\text{FPF} = 100\% \times B/(B + D),$$

(can also be defined as 1-specificity)

$$\text{FNF} = 100\% \times C/(A + C).$$

5.4   Interrelation between Clinical Performance Parameters

For any hypothetical clinical test results, the distribution of test responses to "diseased" and "normal" samples from the intended population can be plotted as two corresponding histograms, giving the "test-performance curve" as in Figure 1. The two histograms show distributions of "scores" of test results plotted from biosensor responses to samples identified as "normal" (left histogram) and "diseased" (right histogram). Both histograms have a range of values because of biological variations between individuals. In an ideal case, when the distribution of test results does not exhibit any overlap, as in Figure 1(a), the test can correctly identify all positive and negative samples by defining a cutoff score at the base of the trough

Table 2. Comparison of test results with the known true diagnosis

Test results	Known true diagnosis		Total number of test results
	Positive (D+)	Negative (D−)	in each category
Test positive (T+)	A	B	A + B
Test negative (T−)	C	D	C + D
Total number of samples with known diagnosis	A + C	B + D	N

Key: test/disease (T/D).
A: T+/D+ (true positive); B: T+/D− (false positive); C: T−/D+ (false negative); D: T−/D− (true negative); N: total number of samples used in the study.

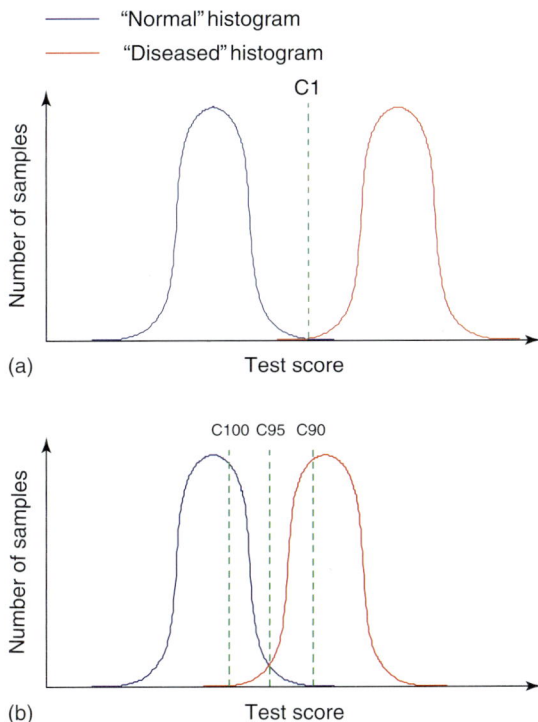

Figure 1. Test-performance curves. Simulated data produced by tests with nonoverlapping (a) and overlapping (b) distributions of "negative" and "positive" test results.

between the two peaks. The cutoff score (decision level, decision threshold) is used as a criterion to classify the subjects as "positive" or "negative" on the basis of test results. At the selected cutoff (C1 in Figure 1a) all samples are correctly identified (both FPF and FNF are 0) giving "zero" B and D values in Table 2. In this case, the clinical sensitivity of the test is 100% (all true-positive samples are correctly identified as "positives" by the test), specificity is 100% (all true negative samples are identified), the EET is 100% (all samples both "positive" and "negative" are correctly identified). Furthermore, the ability of the test to correctly predict either "positive" or "negative" patient status is 100% (both PVP and PVN are 100%).

For any test in which the distributions of results from the two categories of subjects overlap (Figure 1b), there are inevitable "trade-offs" between clinical performance parameters. The overlap between tests results can occur if some fractions of the test results were assigned to wrong categories (as determined by a true diagnosis), that

is, the test results contain false-positive and false-negative fractions. The amount of the overlap may be strongly dependent on how closely the used biomarker is correlated with the disease state. The sensitivity and specificity (and other related parameters discussed in the preceding text) of this test will depend on the choice of the cutoff value. Outside the overlap region, either clinical sensitivity or clinical specificity is 100% and both are unvarying. Within the overlap region, neither specificity nor sensitivity is 100% and both are varying along with other clinical performance parameters as the cutoff varies.

By plotting and analyzing "diseased" and "normal" histograms, it is easy to appreciate the interrelationship between clinical performance parameters and to determine cutoffs, above which lie biosensor responses with preferred (depending on the intended use) performance characteristics. As a rule, high cutoff value minimizes false-positives while a low cutoff value minimizes false negatives. For example, for screening tests used to test entire populations for the presence of an analyte (biomarker or infectious agent), it is usually desired to have a high sensitivity to ensure that true-positive results are detected at the cost of producing more false-positive results, as long as the negative consequences of a false-positive diagnosis are not severe. The highest theoretical test sensitivity (100%) can be achieved only if 100% of the test responses to "diseased" samples lie above the test cutoff. At the cutoff value C100 (Figure 1a), truly "positive" samples in the tested population are correctly identified by the test with no "false-negatives", giving the clinical sensitivity of the test and PVN of 100%. However, a fraction of negative test results (i.e., 10%) lies above the C100 thus giving a nonzero FPF value of 10%, reducing the clinical specificity (not all "negative" samples are identified) and PVP (not all "positive" results are actually "positive").

This lower specificity can be tolerated if a good confirmatory test exists and if the social/economic consequences of the false-positive results are not too severe. Confirmatory tests are designed to be specific (at the expense of sensitivity, if necessary) and have a high positive predictive value. These performance characteristics of the test can be achieved at the assay cutoff value labeled as C90 in Figure 1(b). With the cutoff set at C90, 90% of the biosensor responses to positive samples lie above the cutoff, with no "false-positives" ($FPF = 0$). However, 10% of test responses to "positive" samples are mistakenly declared as "negative", giving FNF of 10%. At the C90 cutoff value the clinical performance of the test is characterized by the following values: sensitivity is 90%, specificity is 100%, PVP is 100%, PVN is ≈91%, and EET is 95%, which may be acceptable for a confirmatory test.

For diagnostic tests, which are used to diagnose a particular disease or condition on the basis of a clinical suspicion that it might be present, both high sensitivity and specificity are usually desired. By choosing a cutoff value between C90 and C100, different unique combinations of sensitivity and specificity can be obtained. In other words, within the overlap region there is a continuum of sensitivities and specificities in which each sensitivity value has a corresponding specificity value. The "trade-off" between sensitivity and specificity in the overlap region in Figure 1(b) is quantitatively captured by the receiver's operating characteristics (ROC)[18] in Figure 2. Typically, on the Y axis, sensitivity, or the true-positive fraction, is plotted. On the X axis, false-positive fraction ("1-specificity") is plotted. In the ROC plot, the various combinations of sensitivity and specificity possible for the test in a given setting are readily apparent. As the decision level changes, sensitivity improves at the expense of specificity or vice versa. At the cutoff value C95 (Figure 1b)

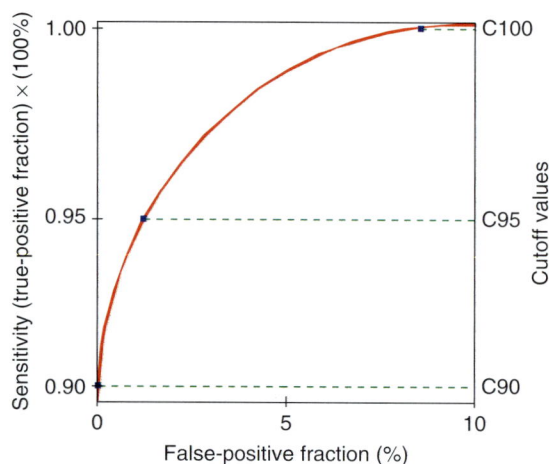

Figure 2. Receiver operating characteristic (ROC) of the test in the overlap area illustrated in Figure 1(b).

both clinical sensitivity and specificity are 95%, which may be an acceptable clinical accuracy for diagnostics tests.

6 CONCLUSION

In the last 10 years the methodology of method validation has been extensively developed. Analytical verification and clinical validation of IVD tests are required to demonstrate the performance of the clinical test and the reliability of results obtained with the test. Numerous guidelines have been published by recognized authorities establishing performance parameters and acceptance criteria for different types of IVD biosensors. These guidelines provide a standardized and systematic approach for a bioanalytical method validation, helping manufacturers to establish a validation plan at an early stage of assay development. As the technologies becomes more complex, regulatory and validation policies are evolving to assure efficacy and safety of biosensors, but in many cases it is still left to the manufacturers to establish quality requirements of novel IVD tests. Appropriate validation policies for novel clinical tests can be very dependent on the nature of the sample, the type of analytical methodology, and the intended use of the assay. For newly developed devices analogous to ones already existing on the market (approved for IVD), the validation process usually consists of systematically comparing the performance of the new test with the accepted performance characteristics of the existing device. For novel clinical assays, the validation process requires rigorous demonstration of analytical performance and clinical validity through extensive laboratory and clinical testing.

In recent years biosensor technologies have developed rapidly, combining advances in biological recognition elements and physical transducers. One clear trend is the development of multiple sensing integrated instruments enabling automated multianalyte analysis in one chip. The use of such integrated biosensor chips may allow a complete analysis of a complex sample, reporting the presence of a wide range of chemical and biological analytes. The recent development of nanotechnology offers promise in several new areas including transducers, recognition ligands, and labeling technology. The new nanosensors may offer new capabilities such as miniaturization to nanoscale size, higher sensitivity down to detection of one or a few molecules, reduced cost of production, and high throughput. Novel advanced technologies implemented in biosensors are expected to bring new, not yet identified regulatory and validation issues, which are to be resolved through the combined efforts of analytical scientists, test developers, clinicians, and regulatory agencies.

ACKNOWLEDGMENTS

The authors thank Larry Bockstahler (NIST), Brandon Gallas (CDRH/OSEL/DIAM), and Marina Kondratovich (CDRH/OSB/DBS) for valuable suggestions and critical review of the manuscript.

REFERENCES

1. ISO, *Quality Management Systems–Fundamentals and Vocabulary. ISO 9000*, International Organization for Standardization, Geneva, 2000.
2. ISO, *Statistics–Vocabulary and symbols–Part 1: Probability and General Statistical Terms. ISO 3534-1*, International Organization for Standardization, Geneva, 1993.
3. Food and Drug Administration, Center for Drug Evaluation and Research (CDER). Center for Veterinary Medicine (CVM), Guidance for Industry: *Bioanalytical Method Validation*, US Department of Health and Human Services, Rockville, 2001.
4. ISO, *Accuracy (Trueness and Precision) of Measurement Methods and Results–Part 1: General Principles and Definitions. ISO 5725-1*, International Organization for Standardization, Geneva, 1994.
5. *Method Comparison and Bias Estimation Using Patient Samples*, Approved Guideline- 2nd Edn, CLSI document EP9-A2, Clinical and Laboratory Standards Institute, Pennsylvania, 2002.
6. *Evaluation of Precision Performance of Clinical Chemistry Device*, Approved Guideline- 2nd Edn, CLSI document EP5-A2, Clinical and Laboratory Standards Institute, Pennsylvania, 2004.
7. Food and Drug Administration, Center for Devices and Radiological Health, Office of In Vitro Diagnostics Device Evaluation and Safety, Division of Microbiology Devices, Draft Guidance for Industry and FDA Staff, *Nucleic Acid Based In Vitro Diagnostic Devices for Detection of Microbial Pathogens*, Rockville, 2005.
8. *Protocols for Determination of Limits of Detection and Limits of Quantitation*, Approved Guideline. CLSI document EP17-A, Clinical and Laboratory Standards Institute, Pennsylvania 19087-1898 USA, 2004.
9. *Molecular Diagnostics Methods for Infectious Diseases*, Proposed Guideline. CLSI document MM3-P2, Clinical and Laboratory Standards Institute, Pennsylvania, 2005.
10. *Evaluation of Matrix Effects*, Approved Guideline- 2nd Edn, CLSI document EP14-A2, Clinical and Laboratory Standards Institute, Pennsylvania, 2005.
11. *Interference Testing in Clinical Chemistry*, Approved Guideline- 2nd Edn, CLSI document EP7-A2, Clinical and Laboratory Standards Institute, Pennsylvania, 2005.
12. ISO, *Capability of Detection–Part 3: Methodology of Determination of the Critical Value for the Response Variable When no Calibration Data are Used*, ISO/DIS 11843-3, International Organization for Standardization, Geneva, 2000.
13. *Evaluation of the Linearity of Quantitative Measurement Procedures: A Statistical Approach*, Approved Guideline. CLSI document EP6-A, Clinical and Laboratory Standards Institute, Pennsylvania, 2003.
14. ISO, *Capability of Detection—Part 1. Terms and Definitions*, ISO 11843-1, International Organization for Standardization, Geneva, 1997.
15. *Clinical Evaluation of Immunoassays*, Approved Guideline. CLSI document I/LA21-A, Clinical and Laboratory Standards Institute, Pennsylvania, 2002.
16. W. A. Colburn and J. W. Lee, Biomarkers, validation and pharmacokinetic-pharmacodynamic modelling. *Clinical Pharmacokinetics*, 2003, **42**(12), 997–1022.
17. P. M. Bossuyt, J. B. Reitsma, D. E. Bruns, C. A. Gatsonis, P. P. Glasziou, L. M. Irwig, D. Moher, D. Rennie, H. C. de Vet, and J. G. Lijmer, Standards for reporting of diagnostic accuracy. The STARD statement for reporting studies of diagnostic accuracy: explanation and elaboration. *Clinical Chemistry*, 2003, **49**(1), 7–18.
18. *Assessment of the Clinical Accuracy of Laboratory Tests Using Receiver Operating Characteristic (ROC) Plots*, Approved Guideline. CLSI Document GP10-A, Clinical and Laboratory Standards Institute, Pennsylvania, 1995.

Part Ten

The Future

The Future

Christopher R. Lowe

Institute of Biotechnology, University of Cambridge, Cambridge, UK

The current substantial demand for chemical and biological analysis in almost all aspects of human lifestyle and welfare is likely to increase as the major issues facing our planet become more pressing in the future and we are likely to require more convincing metrics to monitor and subsequently control seriously disruptive events. Major sociological disturbances involving climate change, energy usage, carbon footprints, environmental pollution, exposure to toxic substances, population demographics, and infectious diseases all require careful monitoring of key biomarkers. Thus, future developments in biosensors and biochips will have to address several key issues, including the ability to address point-of-sampling measurement, real-time monitoring, systems integration, and remote interrogation. Each of these scenarios will require the development of specific technologies and approaches suited to that environment.

Perhaps the fastest growing sector of analytical science is that of point-of-sampling analysis, i.e. the ability to perform analyses of complex samples at the point at which the sample is collected and without returning the sample to a central laboratory. Such techniques are required in point-of-care analysis of clinical samples, where the analytical procedure is performed in the home, workplace, physician's office, day-care centre, ward, or operating theatre and where time is of the essence. Similar methodologies are required for on-the-spot detection of alcohol, illicit drugs, or other behavior-altering compounds at the roadside, workplace, or transport hub, or in security,

military, or first responder situations where conventional explosives, chemical or biological agents are suspected of being deployed. The ability to analyse samples on-the-spot would permit early detection and identification and allow delineation of the extent and geographical area of the affected location. Forensic analysis of scenes of crime would benefit immeasurably from point-of-sampling analysis of body fluids to identify the nature, source, and potential characteristics of individuals suspected of committing crimes. However, all these applications, and those in food, water, and raw material quality monitoring, require analytical measurements to be made without the benefit of clean, well organized laboratories with skilled operators. Thus, the requirements to perform precise, and in most cases, quantitative measurements in an uncontrolled environment in an inexpensive, relatively small "handleable" sensor which is capable of sample pre-processing and has low logistics burden is a real challenge to the sensor community. Field samples will often be cluttered with pollen, dust, and other natural biological and chemical constituents that will require separation, extraction, or filtration to allow measurement of the target analytes. Biochip technology based on silicon or glass devices with on-chip microfluidics, valves, and separation systems are potentially capable of resolving some of these issues, although the conversion of micro devices into instruments with macroscopic human interfaces still poses a significant hurdle. Devices constructed of inexpensive materials, particularly plastics, present a less

Handbook of Biosensors and Biochips. Edited by Robert S. Marks, David C. Cullen, Isao Karube, Christopher R. Lowe and Howard H. Weetall.
© 2007 John Wiley & Sons, Ltd. ISBN 978-0-470-01905-4.

expensive alternative, although the techniques for fabricating plastic fluidics, valves, pumps, channels, filters, and separation systems are less well developed than those constructed in more conventional inorganic materials. Similarly, the fabrication of plastic electronic devices based on amorphous silicon and polymer-based optical waveguide, grating and organic light emitting diode systems is still very much in its infancy. Nevertheless, inexpensive polymer sensors offer substantial benefits of manufacturability, lightness, low power burden, and disposability. Not surprisingly, such polymer-based sensors have been employed in the home measurement of glucose for at least two decades since they offer inexpensiveness combined with adequate accuracy.

Real-time measurements of key analytes using biosensor and biochip technologies poses several quite difficult issues to resolve, not least ensuring that the sensor response times are within the rise and fall times of the analyte levels itself. Sadly, for most analytes, particularly those that require sample preparation or are at a low concentration, the sensor response is often too slow for real-time monitoring and discrete sampling and analysis has to be performed at discrete intervals. In most cases, where the analyte concentration changes rapidly, physical or spectroscopic techniques involving UV-VIS, near-IR, or Raman spectroscopy are often the only option. However, for analytes at higher concentrations, such as glucose, lactate, and certain other key metabolites, enzyme and chemical recognition systems, allow reversible measurements with fast kinetics. For analytes at low concentrations, such as hormones, high affinity recognition systems based on antibodies or receptor proteins are required, and this, combined with avidity affects created by immobilisation on the transducer surface, creates effectively an irreversible sensor with exceptionally slow off-constants, which requires a regeneration process to present a new binding surface. Thus, real-time monitoring of low concentration analytes is a challenge that needs to be addressed, possibly by using recombinant engineered proteins where the affinity for the analyte is modulated by some characteristic, such as heat, light, sound, or magnetism, of the transducer, i.e. the transducer becomes a transceiver.

A third key challenge is to develop integrated networks of sensors that could report events over a substantial area. This is particularly apposite for military, civilian, and environmental analyses where spatial concentration profiles are required in real-time. Such systems would be ideal for plotting the occurrence, source strengths, distribution, and fate of toxic substances through food or beverage intake, inhalation or contact in order to protect human health and the environment. Thus, miniaturized postage stamp-sized biosensors containing biochips for monitoring battlefields, water courses or civilian events may be distributed widely at fixed locations, on vehicles, or worn on protective clothing or about the person as arm bands or wrist watches. In sufficient quantity, these inexpensive miniature sensors could provide hundreds or thousands of monitoring points for target molecules in the air, water, or on personnel. *In vivo* biosensors could monitor physical reactions to adverse stimuli injurious or detrimental to the health of exposed person. Again, like the point-of-sampling sensors, the requirement is for inexpensive, lightweight, low-power sensors with appropriate sensitivity and selectivity. In addition, the sensors should ideally be biodegradable, to pose no ongoing threat to persons or the environment, and remotely interrogateable.

Remote sensors integrated into an extensive network will present additional challenges to the sensor developer; ideally, they should be fabricated from inexpensive glass or plastic materials and be interrogateable via optical, electromagnetic, or radio-frequency radiation. Such technologies exist in other sectors such as radio frequency identification (RFID) tags and there are signs that this know-how will be applied to sensors. Remote sensing is particularly important for monitoring toxic industrial chemicals, noxious and polluting gases and in circumstances, such as space exploration, where human contact is impossible.

A further significant challenge is to integrate sensor technology into actuator systems that allow an in-built response to the measured analyte. Such systems will find application *in vivo* for delivering therapeutic or modulating drugs in response to changes in the concentration of clinically significant analytes, the so-called pharmacy- or pill-on-a-chip. The best example of such a putative approach would be the release of human insulin in response to rising levels of blood glucose; however, key considerations include the size of the implanted device which would have to include a reservoir of the insulin, the stability of the protein, its release kinetics, and the algorithm relating insulin dose to glucose concentration, power delivery and longevity, and the potential fouling and de-calibration of the sensor. Other systems involving the sensor-controlled release of analgesics and anti-hypertensives offer more tractable alternatives, particularly in the former case, if there is a manual override via a remote activation mechanism. Such systems could be fabricated in more durable substrates such as silicon or hybrid technologies since any implanted device would require *in situ* use for a minimum of six months.

Another challenge would be to exploit sensors or sensor arrays into autonomous or robotic systems in order to collect real-time data on local concentrations of analytes and pollutants in aerial and aqueous environments. Such devices would require a radio feedback loop and report on the presence, source, and identity of suspicious components in the air or water supplies. Data handling, with or without sophisticated artificial intelligence algorithms, and the fusion of data from several different autonomous devices would be an essential component for the interpretation of events monitored by these integrated systems.

Many of these future sensor devices and systems will require highly selective, reversible and durable recognition systems to discriminate the target analyte from the morass of non-specific chaff. Consequently, future sensor technologies will require the development of novel recognition elements based on engineered biosystems, stabilized cell lines, antibody and receptor fragments, catalytic antibodies, and various biomimetic systems centred on molecularly imprinted polymers. For any *in vivo* applications, it will be important to ensure that such systems are durable and do not leach toxic components into body fluids. An important effect of creating biomimetic recognition elements is that such systems could be integrated into the fabric of the transducer to create a combined sensor/transducer in the same way that the holographic sensors described in this handbook are fabricated.

Author Index

Subject Index